LOST LAND OF THE DODO

DEDICATION

We would like to dedicate this book to the memory of our friend the late France Staub (1920–2005[1]), doughty Mauritian naturalist, artist, dentist and *bon viveur*, who did so much, in his eclectic way, to keep wildlife issues before the Mauritian public in the later 20th century.

In addition we would like to recall some of the great characters in the natural history of the Mascarenes: Peter Mundy for suspecting evolution 200 years ahead of his time; François Leguat for the 17th century's best natural history travel book and pioneering observations on bird behaviour and territory; Sieur Dubois for his precise enumeration of wildlife in near-pristine Réunion; Joris Laerle for providing the inspirational and unique images of Mauritian birds drawn from life; Gui Pingré, Jean-François Charpentier de Cossigny and Jean-Baptiste de Lanux for sticking to facts in the 18th century; Philibert Commerson for initiating *in situ* illustration; Pierre Poivre for the foundations of conservation legislation; Hugh Strickland for pioneering and perceptive bio-historical research; François Pollen and Edward Newton for the first properly documented collecting in the 19th century, and also the pioneers of old bones, Alfred and Edward Newton (again!), whose interest in Mascarene fossil history formed the basis for all subsequent palaeontological research; also Théodore Sauzier, George Clark and Etienne Thirioux, three 19th-century amateur fossil collectors and natural historians who have never been fully appreciated, and without whose efforts our knowledge of the Mascarene fauna would have been significantly diminished; in the 20th century Reginald Vaughan, Jean Vinson, and Harry Gruchet for conservationist thinking and action against the odds; and finally Alfred North-Coombes, champion of Leguat and historian of Rodrigues and the discovery of the islands, and Carl Jones, author of Chapter 10, who has done so much to establish in Mauritius one of the most successful hands-on conservationist organisations in the world.

LOST LAND OF THE DODO

AN ECOLOGICAL HISTORY OF MAURITIUS, RÉUNION & RODRIGUES

ANTHONY CHEKE AND JULIAN HUME

Yale University Press
New Haven and London

Published 2008 in the United Kingdom by T & AD Poyser, an imprint of A&C Black Publishers Ltd.
Published 2008 in the United States by Yale University Press.

Text copyright © 2008 by Anthony Cheke and Julian Hume.
Artwork copyright © 2008 by Julian Hume, except for the following:
François Leguat Reserve, Rodrigues: colour section page 12 tl, bl; Owen Griffiths/François Leguat Reserve, Rodrigues: 13 t, b; 14 t; 15 b; 16 t; Sonia Ribes/Museum d'Histoire Naturelle de La Réunion: 7 tl, tr, b; 9 br; 10 b; 11 bl.

All paintings in the colour section are by Julian Hume.

All rights reserved.
This book may not be reproduced, in whole or in part, including illustrations, in any form (beyond that copying permitted by Sections 107 and 108 of the U.S. Copyright Law and except by reviewers for the public press), without written permission from the publishers.

Commissioning Editor: Nigel Redman
Project Editor: Jim Martin

Design by Alliance Interactive Technology, Pondicherry, India

Printed and bound in China through Lion Production Ltd

Library of Congress Control Number: 2008920613
ISBN 978-0-300-14186-3 (hardcover : alk. paper)

A catalogue record for this book is available from the British Library.

The paper in this book meets the guidelines for permanence and durability of the Committee on Production Guidelines for Book Longevity of the Council on Library Resources.

10 9 8 7 6 5 4 3 2 1

Front cover: Dodo watching as a Dutch ship arrives by Le Morne
Spine: Newton's Day-gecko
Back cover: Calling male Rodrigues Solitaire

Title page: Male Rodrigues Solitaires in territorial dispute

CONTENTS

Preface	6
Acknowledgements	8
Introduction	10
1 Geography of the Mascarenes	13
2 First contact	21
3 The pristine islands	32
4 Where did the Dodo come from?	53
5 Early settlement	75
6 United under France	90
7 A century of sugar	116
8 The limits to growth	155
9 A miraculous survival	203
10 Practical conservation on Mauritius and Rodrigues	226
11 Reflections	260
Tailpiece: *La Ravine Saint-Gilles*	274
Notes on Chapters 1–11	276
Appendix 1. Land vertebrates introduced to the Mascarenes	368
Appendix 2. Native fauna in Mauritius	371
Appendix 3. Native fauna in Réunion	374
Appendix 4. Native fauna in Rodrigues	376
Appendix 5. Introduced fauna in Mauritius	378
Appendix 6. Introduced fauna in Réunion	381
Appendix 7. Introduced fauna in Rodrigues	383
Appendix 8. Faunal observations on the northern islets of Mauritius	384
Appendix 9. Sources for appendices 2–8.	388
Notes for Appendices 2–9	390
Appendix 10. Contributions of travellers, collectors and fossils to knowledge of the Mauritius land vertebrate fauna	395
Appendix 11. Contributions of travellers, collectors and fossils to knowledge of the Réunion land vertebrate fauna	396
Appendix 12. Contributions of travellers, collectors and fossils to knowledge of the Rodrigues land vertebrate fauna	397
Appendix 13. Excavations in 2005 and 2006	398
Appendix 14. Species lists for artwork in the colour section	400
Appendix 15. Local names of Mascarene vertebrates	402
Bibliography	405
Index	453

PREFACE

So rapid and complete was their extinction, that the vague descriptions given of them by early travellers were long regarded as fabulous or exaggerated, and these birds, almost contemporaries of our great-grandfathers, became associated in the minds of many persons with the Griffin and the Phoenix of mythological antiquity.

Hugh Strickland, 1848, on the Dodo and the Rodrigues Solitaire[1]

[The Dodo] *has remained famous as the bird whose deadness so many people and things are in danger of emulating...*

Australian travellers Leslie and Coralie Rees visiting its lost land in 1952[2]

Since its its 'rediscovery'[3] in the mid-19th century, the Dodo has been synonymous with extinction – indeed the catchphrase 'as dead as a dodo'[4] has become so ingrained in the English language that we use it in almost any context. The bird itself has become a byword for stupidity – for being clumsy, flightless and allowing itself to be caught and eaten without running away[5]. What is less often realised is that until the question of the past existence of the Dodo was raised and resolved, the entire concept of 'extinction' did not exist. God had created all the plants and animals in, according to a famous calculation, 4004 BC, and there was no question of mere Man being capable of destroying what He had created. The fossils in the rocks, considered to be animals that had died in Noah's Flood (or alternatively put there by the deity to fool inquisitive humans), were the only creatures allowed to be extinct, and that because God himself made them so[6]. From this theological viewpoint, adhered to by society's leaders if not so fully by explorers and naturalists, extinction could not, and did not, happen. Hence when it was discovered in Europe that the Dodos reported by early Dutch visitors to Mauritius no longer existed on that island, and had never been found anywhere else, it became fashionable to believe they had never existed at all![7] The eventual understanding that they had existed (once remains were rediscovered in museums) was part of the jigsaw of ideas that prepared the ground for Darwin and Wallace's revolutionary (if not entirely novel) theory of evolution published in 1858. Both of these very different biologists got their ideas through studying islands – Wallace in the 'East Indies' (now Indonesia), Darwin (especially) in the Galapagos.

Darwin also visited Mauritius (in 1836), but was by then tired, ill and on his way home, and had little to say about his visit, though he did collect some marine life and a frog[8].

So Mauritius became the island where extinction not only had occurred, but where it was, so to speak, discovered. Out of this discovery arose the concept of 'preservation' of species, the first step to current ideas of environmental conservation. When Edward Newton arrived in Mauritius from Britain to take up the post of Assistant Colonial Secretary in 1859, he rapidly noticed that some birds only recently still common had apparently vanished. With the first finding of Dodo bones in 1865, and then those of other extinct Mauritian species, the issue of preventing further extinctions took on a new urgency, and in 1878 Newton initiated the first laws anywhere specifically designed to protect indigenous land birds from persecution[9] – until then the only animals that attracted any legal protection were either game, quarry or food, or deemed useful in some other way; pest controllers, perhaps, or providers of commodities such as feathers or fur.

There is a sense in which the old philosophy of the sanctity of creation has indirectly infused the modern concept of conservation – the idea that the earth and its wildlife are not 'ours' to wilfully damage, that we have a responsibility to look after what we also have the power to destroy. In a materialistic world where everything has its price, where the rule of the market is no longer seriously questioned by alternative political philosophies, it is hard to make a case for anything having an intrinsic value outside the confines of human economics. Witness the relentless destruction

'How the beasts got into the Ark – Noah kicked out the Dodo'. An illegibly signed unpublished cartoon bound with other material collected by Alfred Newton in a set titled 'Indian Ocean. Madagascar – Mascarene Islands (MSS)' held in the Newton Library, Cambridge University, Zoology Department.

their despoilers took good notes. We have a lot to learn from what they recorded.

Postscript

Much of our understanding of the Mauritian extinct fauna has been derived from subfossil bones collected in the Mare aux Songes swamp. After more than a century of neglect, excavations recommenced there in 2005, and were significantly expanded in 2006[11]; further work is intended, with JPH intimately involved. This book went to press before the palaeontological results had been studied or published, but exciting new discoveries are expected (see Appendix 13).

A note on English names

Throughout this book, we have generally used the most widely accepted English names for all taxa, with the exception of some of the cagebirds that have been released onto the islands; in these cases, we decided that it was preferable to use the long-established avicultural names as used in the Mascarenes, which better reflect how these species reached the islands.

Two of the bird names are more problematic. The Réunion Solitaire (formerly thought to be a type of dodo) has been shown relatively recently to be an ibis, hence we refer to this species as Réunion Ibis. However, this bird was always called a 'solitaire' by the islanders, and most of the literature refers to it by that name, so we have followed this where appropriate. We have given both names in the tables and appendices, and in Box 8.

The native bulbuls of the islands have always been locally known as 'merles' (from the French for the superficially similar Blackbird), and this name persists to this day. By contrast, the introduced Red-whiskered Bulbul is frequently referred to on the islands as simply 'bulbul'. The native species belong to the genus *Hypsipetes* and are closely related to the Black Bulbul *H. leucocephalus* of mainland Asia. They should properly be called Mauritius Bulbul and Réunion Bulbul. However, they are frequently referred to as 'merles' in the literature and we have chosen to retain the name 'merle' in the main text. Where appropriate (such as in the tables and in Box 25) we have given both names.

of irreplaceable forests with their immensely complex and subtle ecology for the sake of a few million dollars worth of dead wood. Fortunately many humans retain an inherent appreciation of beauty, a curiosity about their world, a fascination with the grandeur and absurdity of nature that allows them to see that all this does have real, if intangible, value. The Dodo was both grand and absurd, but above all, like all lost species, a precious example of the process of evolution through aeons of time in its own little world of Mauritius.

We are used to seeing on our television screens fictional cultures that have evolved on other planets throughout the cosmos, and enjoy seeing the weird aliens that life has thrown up in far galaxies. Our own world is in fact remarkably like the fictional universe – think of islands and continents as planets where each has its own special circumstances and generates its own peculiar life-forms. There are huge ones and tiny ones, each with its irreplaceable specialities. They have all been invaded, but do we want them all to be subject to the uniform bleak conformity of the 'Empire' or the 'Dominion'[10] – the local inhabitants exterminated by the rats, bulldozers, pesticides, concrete and blind incomprehension of a 'western culture' (and its imitators) out of control?

This book tells the story of three such 'planets' – islands far out in the Indian Ocean that escaped the heavy hand of man until fewer than 500 years ago. Their history is more complete than most because

ACKNOWLEDGEMENTS

AUTHORS' NOTE
The style we have chosen for this book, a clear text annotated with numbered endnotes, is more usual in historical or literary than in scientific works. However, we felt that text unencumbered by endless 'Harvard references' and explanatory byways was easier to read, while including all the source material in the notes still makes it possible for scholars and specialists to follow up any lines of particular interest. The endnotes also contain anecdotes, biographical details and other snippets that would distract from the flow of the main text, but are nevertheless part of the story. For details of abbreviations and acronyms used in the text and endnotes, see the preamble to the bibliography. ASC is responsible for the main text, JPH for the box features and Appendix 13. All translations from French are by ASC unless otherwise stated.

Acknowledgments for Mascarene work done in the 1970s and 1980s will be found in earlier publications[12]. For this book I would like to thank Carl Jones and the Mauritian Wildlife Foundation who have provided accommodation and sometimes transport on visits to Mauritius in 1996, 1999 and 2003, and Auguste and Christel de Villèle who have kindly done the same in Réunion and helped with local literature research; Aleks Majlkovic played host in the MWF house on Rodrigues in 1999. I have had long and fruitful discussions with many people, notably (for Mauritius and Rodrigues) Carl Jones, the late France Staub, Vikash Tatayah, John Maureemootoo, Roger Safford, Nik Cole, Philip LaHausse, Alan Grihault, Owen Griffiths, Jean-Michel Vinson, Rachel Atkinson and Aleks Majlkovic, and (for Réunion) Jean-Michel Probst, Pierre Brial, Sonia Ribes, the de Villèles, Alain Vauthier, Bruno Navez, and Jacques Trouvilliez, plus much correspondence with Mathieu Le Corre, Marc Salamolard, Thomas Ghestemme, and Christophe Lavergne. Roger Safford and Dennis Hansen contributed helpful feedback on drafts – David Bullock and Steve North kindly commented on Chapter 9 and inspired much useful discussion, and Yousoof Mungroo of the Mauritian National Parks and Conservation Service facilitated my visit to Round Island in 2003. For correspondence or discussion on more general issues I would like to thank Jeremy Austin, Aaron Bauer, Jim Groombridge, Dennis Hansen, David Hershey, Ivan Ineich, Christian Jouanin, Zoltan Kórsós, Jannie Linnjeberg, Christophe Lavergne, François Moutou, John O'Brien, Wayne Page, Dave Roberts, Gordon Rodda, Herbert Rösler, Guy Rouillard, Samuel Turvey, Carlo Violani, Miguel Vences, John Williams and Ralfe Whistler. From the museum world I am indebted to Robert Prys-Jones (NHM Tring), Christian Jouanin & Roger Bour (MNHN Paris), Mike Brooke (Cambridge), Malgosia Nowak-Kemp (Oxford), Rainer Günther (Berlin), Franz Tiedemann (Vienna), Marie-Dominique Wandhammer (Strasbourg) and Bob McGowan (Edinburgh). Evelyne Couteau, Gabrielle Baglione and Jeanine Monnier most helpfully sent me, respectively, copies of manuscript material from Jean-Baptiste de Lanux, Peron & Lesueur and Philibert Commerson, Ran Meinertzhagen allowed me access to Richard Meinertzhagen's diaries, and Pascale Heurtel showed us Commerson's manuscripts in the MNHN archives in Paris. Henk Beentje kindly translated some key texts from 17th century Dutch. My library research has largely been in the various parts of the Oxford University Library System, notably the Bodleian, Rhodes House, Indian Institute, Sackler and Radcliffe Science Libraries – plus the Alexander Library (Edward Grey Institute of Field Ornithology), and the Plant Sciences and Geography Departments, all of whose libraries are now part of the OULS, and also the Taylorian Institution. I am most grateful to the staff of all these libraries for unflagging work ferreting out endless requests for old books and journals from stacks and locked cupboards, and particularly to Linda Birch, who retired in 2005 after many years as Alexander Librarian. I have also used the libraries at the Natural History Museum in London and Tring, the British Library, Cambridge University Library (Alfred Newton papers), the Zoological Society of London, the Mauritian Wildlife Foundation and the Muséum d'Histoire Naturelle in Saint-Denis (Réunion). I would like to thank Nigel Redman and particularly Jim Martin at A&C Black, and Julie Dando for producing the original maps in this volume. My son Oli kindly scanned most of the text illustrations. Finally I must thank my friend and co-author Julian Hume for collaborating so cheerfully through often difficult times, and my wife Ruth for putting up with too-long absences from our business at the Inner Bookshop while the book project stretched from months into years to everyone's frustration, not least our patient publisher.

Anthony Cheke

So many people over the years have given their encouragement and support – it is almost impossible to put my gratitude into words. I will start by thanking Storrs Olson and Helen James, who launched my interest in palaeontology; I have never looked back. In the UK, I would particularly like to thank Robert Prys-Jones and Carl Jones, who have helped make Mascarene research so rewarding. My deep gratitude goes to Mike Barker, Dave Martill, Bob Loveridge, Darren Naish, Dave Hughes, Errol Fuller, Cyril Walker, the late Colin Harrison, Glyn Young, Nick Arnold, Graham Cowles, Paula Jenkins, Sandra Chapman, Andy Currant, Ann Datta, Ray Symonds, Mike Brooke, Andrew Kitchener, Michael Walters, Jo Cooper, Mark Adams, Martin Staniforth, Ralfe Whistler, Rungwe and Claude Kingdon, Owen Griffiths, Aurele Andre, Daniel Sleigh, S. Abdhoolrahaman, Alan Grihault, Clem Fisher, Justin Gerlach, and in particular Effie Warr, former librarian at Tring.

In the Netherlands, I thank Perry Moree and Pieter and Else Floore, whose knowledge of early Dutch history has been an inspiration. On Mauritius I thank the late France Staub, the late Claude Michel, Jean-Michel Vinson, the late Alfred North-Coombes and all of the team past and present at MWF. On Rodrigues I thank Richard Payendee, Arnaud Meunier, Emmanuella Biram, Alfred Begué, Mary Jane Raboude and Sweety Sham Yu. I thank Cécile Mourer-Chauviré, Christophe Thébaud, Christian Jouanin, Sonia Ribes, Roger Bour, Jean-Michel Probst, Pierre Brial, Dominic Strasberg and Auguste and Christel de Villèle for their support on Réunion. I am indebted to the Palaeo Research Group, School of Earth & Environmental Sciences, University of Portsmouth, Department of Palaeontology and Bird Group, NHM London and Tring, the Muséum d'Histoire Naturelle in Saint-Denis (Réunion) and Troyes, Muséum National d'Histoire Naturelle, Paris, University Museum of Zoology, Cambridge, Mauritius Institute, Royal Museum of Scotland, NHM Liverpool, La Vanille Crocodile Park, Algemeen Rijksarchief, The Hague, State Archives, Cape Town, Arsib Nasional, Jakarta and the Artis Bibliotheek, Amsterdam, whose provision of facilities and expertise made my endless requests for documents and specimen material attainable. I thank Anthony Cheke, co-author, colleague and friend, who has provided so much interest, support and enthusiasm, despite the sometimes turbulent times during the book's preparation.

I am forever in the debt of my wife Jenny and my children, Jade, Jasmin and Jeradine, and my parents, brothers and sister, whose patience and understanding have made this project possible.

Julian Hume

An 1874 engraving by Butterworth and Heath of a Dodo preening its foot, derived from Roelant Savery's painting 'Landscape with exotic birds' *in the collection of the Zoological Society of London.*

INTRODUCTION

These singular birds ... furnish the first clearly attested instances of the extinction of organic species through human agency
                                    Strickland (1848) on the Dodo and Rodrigues Solitaire[1]

To simplify a little, one could say that the ecosystem is like our homes. It's what allows us to live. Like all houses, it's made up of components which, like bricks, are assembled side by side to create the totality of the unit. If you take a brick out, the building doesn't fall down – but you risk having draughts, leaks, and if you keep on taking out bricks the system ends up unable to perform its function. It is the same with nature – there comes a point when it falls apart.
                                    Vincent Florens, explaining ecosystem degradation to a journalist in 2005[2]

Imagine you are a Dutch sailor in 1598. You have been on a ship in bad weather for weeks since your last landfall at the Cape, eating dry biscuits and suffering from scurvy – and suddenly your prayers have been answered. Land has been sighted! Up ahead is one of the mythical islands on the dodgy Portuguese maps that the captain boasted of. Cloud-topped mountains and seductive forests promise fresh water and food – and as you land after negotiating the reefs, dinner walks up and asks to be eaten! Giant juicy tortoises ten times the size of anything seen in Europe, and weird fat flightless birds that don't know how to run away. Welcome to Mauritius, a land untouched by man, and about to be changed forever.

Oceanic islands, created by undersea volcanoes and isolated by hundreds or thousands of miles of open sea, are very special. They are natural experiments in evolution, each one unique depending on its age and size, on which land is nearest, on whether it is tropical or temperate, on the direction of prevailing winds and ocean currents, and on which animals and plants were the first to arrive. On islands several millions of years old far from sources of immigration, evolution can proceed so far that almost all the biota can be endemic, i.e. specific to that place. Less obvious are 'islands' in the middle of great continents – islands of habitat, equivalent to sea-girt islands in their history and evolution. They are typically isolated lakes or high mountains, separated by wide stretches of unsuitable terrain, where endemic species have similarly evolved[3].

Many far-flung islands in the temperate zones were colonised by man in early or prehistoric times, but a number of tropical islands were not reached until much more recently. Island ecosystems do not mix well with humans – one of the common characteristics of isolated islands is that there are few (or no) predators, and only limited numbers of herbivores to influence the vegetation. Under these circumstances species evolve which have no defences against organisms from the large continents[4]. Humans not only kill directly, but introduce all kinds of animals, plants and diseases that can overwhelm these fragile ecosystems. Islands colonised by people a long time ago have always lost many, sometimes all, of the endemic species that evolved there, and their past history can only be reconstructed from material such as old bones and subfossil seeds and pollen, dug out of swamps and caves.

Easter Island[5] is a small (166km²) and immensely isolated subtropical volcanic island in the south-east Pacific Ocean, 2,250km from the nearest land (the Pitcairn Islands) and 3,747km from the nearest point of South America. Today it is a treeless savanna with a mysterious vanished culture that carved massive monolithic statues. Pollen deposits show it was originally a forested island, but it was subsequently clear-felled by its Polynesian population. When first described by Western visitors in 1722, little trace remained of the original flora, and none of any fauna there might have been, and even the folklore gave only fragmentary glimpses of the former ecosystem – for palaeoecologists, everything has had to be reconstructed from subfossil material. Although often cited as a directly human-induced eco-catastrophe, it now appears that exploding numbers of Polynesian Rats arriving with the immigrants caused a forest decline through seed predation, before the people finished it off. A sobering fact is that the first people did not get there until about 1200; although deforestation was complete by about 1650, final cultural and economic collapse only followed after the introduction of Western diseases and removal of people as slaves after 1722.

Hawaii[6], a volcanic archipelago, was colonised by Polynesians (around 300-400 AD); New Zealand[7], an old continental island group, a little later (*c.* 900 AD). These are much larger islands, and by the time Europeans arrived (1700s AD) were only partly deforested and still had a great deal of their endemic fauna. Nonetheless, hundreds of years of occupation by colonists who habitually introduced rats, pigs and dogs wherever they went had taken their toll on the fauna. Some of the losses were hinted at in folklore, and in New Zealand it didn't take long for European settlers to find bones of enormous flightless birds (moas) that confirmed the stories. Again there were no written records, and it has taken a further two centuries to rediscover all the extraordinary lost birds of the Hawaiian Islands[8]. Meanwhile European colonisation has proved almost as destructive as Polynesian, and a host more species, especially in Hawaii, have become extinct.

The history of the great continental island of Madagascar is even more tantalising. Although only 300km from the African coast and nearly 1,600km long, it remained almost completely free of human influence until colonised from across the Indian Ocean by settlers and traders as recently as *c.* 800 AD[9]. When Flacourt explored parts of it in 1648–52 he was told that flightless elephant birds and giant lemurs still existed in remote areas[10], though he did not see them himself. These now-extinct creatures were part of normal life to the Malagasy – but they too had no written language, and no records of their discovery and settlement of the island. Being a continental island, with an ancestral fauna (some of which had walked onto it before its isolation by continental drift around 80–88 million years ago[11]), there developed a range of herbivores and predators; much of the fauna, especially the mammals (lemurs, tenrecs etc.) and flying birds, has been able to withstand human impact better than on truly oceanic (and much smaller) islands.

The shipwreck of the Arnhem *off Mauritius in 1662, from Stokram (1663), illustrating the dangers of tropical storms and reefs. The* Arnhem *was actually lost in the open sea about 192km (120 miles) east of Mauritius, but over the years many Dutch, French and English ships were wrecked on the reefs.*

> **GENERAL COMMENTS ON THE BOX FEATURES ON SPECIES AND SPECIES GROUPS**
>
> All of the most important contemporary accounts are provided for extinct species, as this provides useful morphological and ecological information about these species that are often unrepresented by skin remains or detailed descriptions or illustrations. Examples of contemporary accounts describing species that are still extant are also included, as their present day distribution is often an artefact of human intervention.
>
> Probst and Brial (2002) have collected all the old wildlife accounts for Réunion (in French), but there has been no such complete compilation for Mauritius and Rodrigues. Ours here (much of it translated into English for the first time) does not claim to be complete, but is a selection of the most informative descriptions from early visitors to all three islands.
>
> **Specimen collections**
> Specimen collections are biased toward extinct species represented by fossil material and museum skin specimens. Extant species are mentioned only if they are represented in the fossil record. For a full listing of museum skin material see Cheke and Jones (1987).
>
> **Abbreviations**
> The following abbreviations are used:
>
> Institutions: BMNH, The Natural History Museum, London, England (formerly the British Museum (Natural History)); UMZC, University Museum of Zoology, Cambridge, England; UMO, University Museum, Oxford; RMS, Royal Museum of Scotland, Edinburgh; MHNH, Muséum National d'Histoire Naturelle, Paris, France; UCB, Université Claude Bernard-Lyon 1, France; NHMV, Natural History Museum of Vienna, Austria; RMNH, Rijksmuseum van Natuurlijke Histoire, Leiden; MI, Mauritius Institute, Port Louis, Mauritius.

The Mascarene islands of Mauritius, Réunion and Rodrigues, the subjects of this book, together with Christmas Island[12] and the Atlantic's answer to Easter Island, St. Helena[13], are the most substantial tropical oceanic islands where the first visitors were Westerners, some of whom arrived complete with pens (or quills) at the ready. They described the islands as they found them, and also the consequences of their arrival. Brief early visits by the Portuguese excepted, we have a more or less continuous record, beginning with the early travellers' encounters with the unusual creatures they met, supported by a wealth of subfossil material that corroborates the animals they described, such that the islands' ecological history is almost fully discoverable[14]. Having this much more precise handle on exactly what happened and what went wrong, we are in a position to learn the lessons more fully, and to use this information to work out how to reverse ecosystem damage where possible. Parallels can also be drawn with islands and ecosystems in other places where we have less complete histories. It is a sad but salutary story – with a bit of a happy ending (so far) for the animals and plants that have managed to survive the last four centuries.

Dodos, parrots and tortoises imagined in primeval Mauritius. From Ramdoyal (1981).

CHAPTER 1

GEOGRAPHY OF THE MASCARENES

AND HOW ANIMALS AND PLANTS COLONISE ISLANDS

Mauritius is an Ile scituate within the burning Zone close by the Tropick of Capricorne, but in what part of the World is questionable, participating as well in part with America, in respect of the immense South Ocean, as bending towards the Asiatique Seas from India and Iava. But most properly adioyning the great Ile Madagascar, from which it is distant two hundred leagues or sixe hundred English miles, whereby I judge it placed in the Afrique seas, and thereby imcorporated into Afrique. But how ever doubtfull of what part of the three it is, of this I nothing doubt, that for varietie of Gods temporall blessings, no part of the Universe obscures it

Thomas Herbert having trouble with his geography in 1634[1]

Réunion, Mauritius and Rodrigues are three strikingly different islands united only by their relative proximity and volcanic origin, lying at the southern edge of tropics in the Indian Ocean east of Madagascar. Réunion, the largest and most southerly, is nearest to Madagascar, but is nonetheless 665km offshore at 21°S 55.5°E. Mauritius (20.25°S 57.5°E), next in size, is 164km east north-east of Réunion. The smallest, Rodrigues (19.75°S 63.5°E) is by far the most isolated, being 574km east of Mauritius and some 4,800km west of Australia, the next landfall[2].

Réunion has both the highest peak in the Indian Ocean, and its only active volcano outside the subantarctic. It sits just to the east of a volcanic hotspot that during the past 66 million years is believed to have given rise to the Deccan Traps lava fields in India, the Maldives, the Chagos, the Saya de Malha and Nazareth banks (large undersea plateaux), the St. Brandon (Cargados Carajos) atolls and Mauritius[3]. On our mobile planet these hotspots are relatively stable in position relative to the poles, but the tectonic plates slowly move over them, producing a series of islands, or lava fields if the plate is land rather than sea-floor[4]. As the great Mesozoic continent of Gondwanaland broke up, the final split around 84 million years ago (mya) saw a large section (now India) break off and drift north; around 64–66 mya it passed over the current site of Réunion, where the flood basalts erupted, and eventually crunched into Central Asia. The result of India's impact with Asia is the massive crumple zone of the Himalayas[5]. Another part, now Madagascar, initially (*c.* 120mya) broke off with India, but got stuck off the African shore, while a small fragment, now the Seychelles, ended up stranded part-way between Madagascar and India. The rest of Gondwanaland is now split into South America, Africa, Australia, New Guinea, New Caledonia, New Zealand and Antarctica.

According to potassium–argon and stratigraphic dating, the hotspot began to generate Mauritius some 10 mya. The island remained volcanically active as it drifted off the main magma source, often with long periods of quiescence, until *c.* 25,000 years ago[6]. About 3 mya the first eruptions forming Réunion broke the surface, and the island has grown (and eroded) actively since then[7]. Off to the east, on a separate volcanic upwelling on a fracture zone that intersects the track of the plates over the hotspot, lies Rodrigues. According to a much-cited but very limited study of Rodriguan rocks, it is commonly supposed to be the youngest of the islands at around 1.5 my[8], but the geomorphology and the degree of plant and animal endemism argue for a much older origin, confirmed by more recent geological work[9]. The hotspot island chain between Mauritius and India is considered further in Chapter 4.

RÉUNION

Réunion is the largest and most spectacular of the islands, 70km long by 50km wide, covering 2,512km². The great bulk of the central massif, centred on the Piton des Neiges (3,069m) and the Piton de la Fournaise (or 'Le Volcan', 2,631m), creates a rain shadow. The east of the island, exposed to the prevailing easterly winds, is wet and cloudy every day of the year, an annual average of 6,000–7,000mm of rain falling in the wettest part of the coast, and 10,000–12,000mm a few miles inland (e.g. Hauts de St Rose, 860m)[10]. In the west the coastal fringe is very dry (540mm/yr at St-Gilles), though it is wetter and cloudier higher up. This cloud, with a base at around 1,500m often shades the coast in the afternoon, even though the sky directly overhead is clear. The centre of the island is dissected into huge 'cirques' and very deep gorges. Cirques are roughly circular caldera-like valleys caused by erosion under very high rainfall, but with very narrow outlets to the sea through dramatic gorges. In the south-east of the island, around the active volcano, the valleys are as deep (cliffs up to 1,000m), but narrower, apart from the fairly smooth slope of the *pays brulé* (the 'burnt land') where regular lava flows reach the sea. All round the island the land slopes steeply upwards towards the centre, 61% of the land surface being above 1,000m[11]. Frosts are regular above about 1,500m, and it occasionally snows on the summits[12].

Around Réunion the land shelves steeply into deep water, in places under massive sea cliffs. Apart from stretches of the sheltered west coast there is no fringing reef; even where some coral has developed it is close inshore enclosing only a very narrow lagoon. There are no natural harbours and only one area with a good anchorage (Saint-Paul Bay) – this was important in the human history of the island[13].

MAURITIUS

Mauritius is smaller, roughly 60km long by 40km broad, covering 1,865km². It is also much lower and flatter, though even so it is markedly wetter in the windward east than the leeward west. In the north there is a broad undulating lava plain rising to around 150m above sea level, though in the middle of the island there are steep sharp mountains rising out of the plain, remains of a huge ancient volcano that was once much taller. The land rises in the south-central part to a plateau around 550–600m, dissected in the area of the oldest lavas in the south-east by one large and several smaller valleys, the principal one being the Black River Gorges. The highest point is Black River Peak at 828m, but several others reach up to 760m[14]. The gentle slopes of the late and intermediate lavas contain a large number of caves (lava tubes), up to 6m high and 9m wide, some of which extend for hundreds of metres underground, often just a few metres beneath the surface[15]. Annual rainfall ranges from 1,400–1,800mm in the coastal south-east to 3,600mm on the high plateau and down to as little as 800mm in a narrow dry western fringe[16].

The coast is fringed by a broad reef-fringed lagoon mostly about half a mile (0.8km) wide, except for two short stretches of coast, in the west and the south, where the sea breaks directly onto low cliffs. There are numerous islets in the lagoon where it is particularly broad, mostly along the east coast where the lagoon is 5km wide in places. Off the north shore, on an undersea platform 45–63m deep, there are four biologically significant islands: Round Island, Serpent Island, the Flat Island group, and Gunner's Quoin[17].

RODRIGUES

Rodrigues is much the smallest of the islands (104km²), but is surrounded by a very wide submarine platform, part of which is occupied by an

Figure 1.1. The *enclos* (main lava flow zone) of the Volcan on Réunion, showing lava flows, coastal cliffs, islands of vegetation and the abundance of palm-trees. From Bory de St.Vincent's *Voyage* (1804).

extensive shallow reef-fringed lagoon about twice the area of the land surface. The island, 17.7km long by 8.45km wide, is in the form of a hog's back, with steep valleys extending off the central ridge; the highest point, Mont Limon, is only 398m. Like the other islands the main relief is composed of basaltic lavas, but Rodrigues also has an area of limestone plateau, Plaine Corail, consisting of consolidated coral sand (calcarenite) accumulated through wind action. Plaine Corail is studded with caves created by water dissolving and eroding the limestone[18]. The climate is relatively dry, with rainfall ranging from under 800mm on the southern edge of Plaine Corail to more than 1,600mm in the centre-west uplands. Most of the island gets 1,000–1,400mm annually, about the same as the northern plains of Mauritius. The lagoon is studded with islets to the south and west, mostly flat limestone (calcarenite) or sandbanks, though some have small basaltic hills.

HOW REMOTE ISLANDS ARE COLONISED

Plants and animals can only colonise a volcanic island once eruptions have ceased in enough of the island to allow life to establish. Even if we know the maximum age of the island from dated rocks, it is difficult to estimate how long the land has actually been habitable. Inferences can be drawn from the plant and animal species that have evolved – the more different they are from their nearest relatives, the longer they are likely to have been there. Thus an overgrown and flightless Dodo, only distantly related to other pigeons, has been there a long time, whereas a Common Moorhen differing little from those in Madagascar (or Europe for that matter) has probably arrived relatively recently. This is only a guide; actual rates of evolution are not known, and some groups evolve quicker than others. The fact that a species is relatively unchanged does not prove it has not been on an island a long time, it only makes it less likely.

For this reason the relative age of island biotas is often studied using the sum of endemic species, genera and families – the higher the proportion of endemics, and the more there are in higher taxonomic categories, the longer the island has been there. Isolation is also an important factor; the further away from sources of immigration the island is, the fewer colonisers are likely to arrive. There will be less turnover, and a greater chance for those that do get there and survive to evolve into forms adapted to that particular island environment without being threatened by new competitive arrivals[19].

Isolated islands of continental origin, such as New Zealand, Madagascar and the Seychelles 'microcontinent', started off with a pre-established fauna and flora which then evolved in isolation for 60–80 million years; any immigrants from outside would have to compete with the established biota. Volcanic islands are all younger, and often even more isolated, although there are also groups near continents (e.g. the Comoros between Madagascar and Africa), where the influence of their large neighbours is dominant. However, the age and distance of an individual island may be less significant than the history of the hotspot over which the island formed. The Hawaiian chain, the clearest demonstration of the hotspot phenomenon, consists of a long line of islands lying north-west from the large, high, recent and still volcanically active island of Hawaii, similar to Réunion (though much larger)[20]. It extends through older and progressively more eroded high islands (Maui, Oahu, Kauai; like Mauritius), through low islands with broad lagoons and residual basaltic hills (Niihau, Necker; resembling Rodrigues),

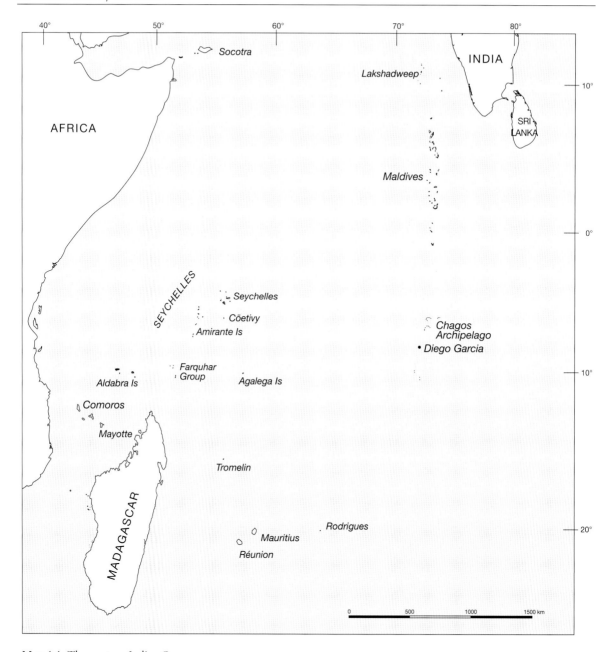

Map 1.1. The western Indian Ocean.

atolls with no volcanic material above sea-level (Laysan, Midway, like Aldabra) to submarine seamounts or guyots that no longer reach the surface. The age of the oldest Hawaiian atolls is around 28 my[21]. As new islands appear in a hotspot chain they can be colonised from their older neighbours.

Newly formed volcanic islands are, of course, entirely without life. Plants and animals will start arriving from the beginning, though which ones and how many will depend on the island's isolation, size and topography, and the prevailing winds and currents[22]. Some plant seeds are salt-resistant and waterborne, and others, together with fern and moss spores and many small invertebrates, are carried widely by air currents – these are the first to colonise new islands. Wind-drifted birds and bats, and irruptive species (some rails), will arrive fairly early on, though if the islands are extremely remote (Easter Island,

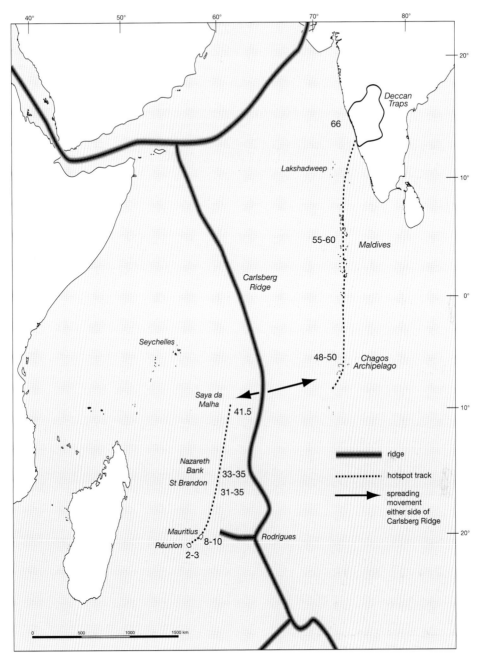

Map 1.2. The volcanic origin of western Indian Ocean islands. The dates and current positions of the islands along the Deccan–Réunion hotspot track are shown, along with their relation to the Seychelles mini-continent. Adapted from Kearey & Vine (1996), Courtillot (1999) and O'Neill *et al.* (2004).

Hawaii) the number of founding species may be very small. These birds and bats will bring further plant seeds in their guts or stuck to feathers or fur. Larger animals cannot establish until there is vegetation for them to live in and on, and chance plays a much larger part in colonisation – for instance, if a single bird arrives it will be unable to breed; bats might do better as most females carry their young with them while they are small. It is surprising what can travel across oceans, as there is an alternative to being able to fly or swim. If a tree, or better still a raft of vegetation, is swept out to sea by a storm, any animals on it, and

BOX 1   PETRELS AND SHEARWATERS (PROCELLARIIDAE)

It is now difficult to ascertain the former diversity and abundance of seabirds. Early observers report large breeding colonies of various species, particularly on the islets around Mauritius and Rodrigues, but today many species are locally extinct or comprise non-breeding migrants. Furthermore, some very distinct but enigmatic species are known from just a few sightings and even fewer specimens. The Réunion Black Petrel *Pseudobulweria aterrima* has only been seen on a few occasions, and it was not until the late 1990s that an exhausted but live specimen was photographed; their breeding site has only been located very recently. Another rare endemic, Barau's Petrel *Pterodroma baraui*, is a nocturnal species breeding on Réunion and occasionally Rodrigues in the past, and a rare vagrant to Mauritius. The northern islets off Mauritius and Round Island in particular have remained a sanctuary for petrels, though the 'Round Island' Petrel, currently considered (pending DNA studies) to be a population of Trindade Petrel *Pterodroma arminjoniana*, may be a recent colonist. Two widespread shearwaters nested on Mauritius and Réunion, Tropical Shearwater, now *Puffinus bailloni* (following recent DNA work), which is no longer resident in Mauritius, and the Wedge-tailed Shearwater *P. pacificus*, which is still common.

Petrels have always been considered fair game for mariners, and their habit of nesting in burrows has made them extremely vulnerable to egg and chick predation by humans and introduced predators, particularly rats and cats.

Accounts (a selection)
Jolinck (in Keuning 1938–51) on Wedge-tailed Shearwaters in 1598 (translation by Henk Beentje):

> Also there are birds that shriek like humans, they are in holes under the earth and if one had not found them, one would have said that it was a rabble, as they shriek all night long especially in the morning, because there were sailors that went towards the shrieking and took the birds out of the holes, and they were good grey birds good to eat. I have been on a river at night with our sloop and there was such shrieking as if 50 people complaining at each other, but it was all birds.

Bory (1804) on Barau's Petrels in 1801, at the Caverne à Cotte, above the Rivière des Remparts:

> All the cave's vicinity was strewn with seabirds' heads, of the petrel kind. The not inconsiderable heaps of bodiless debris having attracted my attention, I learnt that these were heads of *fouquet*. The *fouquet* appeared to me to be the same bird of which Labat had spoken so much, and which in his time in the Antilles were called diablotins*. Not finding myself in the season when *fouquets* were taken, I shall confine myself to reporting what I was told. In spring, a brown coastal bird, strongly reminiscent of a gull ['goëland'], and called taille-vent, abandons the shoreline and comes to lay its eggs in these areas of cliffs, which nature appears to have wished to render inaccessible. It is at the time of the solstice that the nestling taille-vents have acquired a certain size and excessive fat deposits that make them sought after. Then the creoles go seeking them, and all those they find are preserved in salt that they carry with them. These birds, thus salted, keep for some time, and take on a taste more or less like old hareng-saurs [red herrings]. Their fat is unctuous, and smells somewhat of fish oil, as with all seabirds. The Morne de Langevin, the Volcan, the heights of the rivière de l'Est and the Salazes are the places where the *fouquets* are most often found.
>
> [Later, near the Piton Dolomieu of Le Volcan] The wind carried the [volcanic] vapours to the opposite side from where we were. I saw some of those seabirds called in this country *fouquets*, and which retire for the night in the high mountains, passing to leeward of the crater without appearing inconvenienced. One of them crossed the crater itself without altering course and without appearing dazzled.

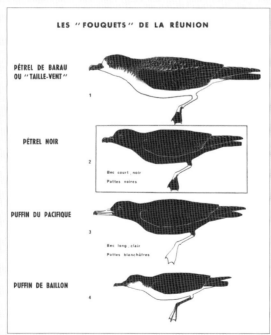

Petrels in Réunion. Publicity poster illustrated by Paul Barruel, c. 1970.

Lesueur (1803, unpublished MSS in Le Havre), on the Tropical Shearwater in Mauritius in 1803:

> We often saw tropic-birds in the mountains where these birds come to make their nests in rock cavities, and a bird known in the island by the name of coupe vent ['wind-cutter'] and which hunters regularly shoot by lying in wait of an evening towards Montagne Longue. It is a petrel which approaches [in appearance] the Pètrel obscur (Lin.) [Little Shearwater, now *Puffinus lherminieri*] of northern seas, [but] I do not think it is the same species. It appears that its habits are to go out a considerable distance from the shore, as one does not see them, or at least I never saw one, flying about on the island's coasts. In the evening these petrels regain the coast and head for the mountains to spend the night there. It is in the moment they pass [over] that the hunters shoot them.

Tafforet (1726), in Rodrigues. Note that *fouquet* is a general name in the Mascarenes for shearwaters and petrels; the first bird described here is the Wedge-tailed Shearwater:

> The *fouquet* is the colour of *fols* [noddies] but a little larger, and its beak is longer and hooked like a frigate-bird. They do not go far [out] to fish, and normally [do so] at night. There are those who say they cannot fly by day as the light is too bright for them, but I have however seen them flying about in daylight not sticking to a known route. They are [found] in holes under rocks, and cry like little children. At night when they go out to fish, I have made many fall to earth in this way: as they leave their holes and you hear them call, you need a dry latan leaf to set fire to suddenly. As soon as they see the light they fall to the ground. In contrast, if they do not see it, they continue on their way . . . There are also mountain *fouquets* [probably gadfly-petrels *Pterodroma* spp.] but very few; I have only seen them in flight which is why I cannot speak [more] pertinently; they nest in holes in the ground high in the hills.

---

* Bory was spot on here; the *diablotin* is the very similar-looking Black-capped Petrel *Pterodroma hasitata*, now rare, but formerly widespread in the West Indies (Brooke 2004).

even some of the plants (or at least their seeds), may be able to survive for days or weeks drifting across the sea. This is how lizards, snakes and many invertebrates (e.g. snails) are believed to colonise islands, both as living adults and as eggs[23]. Many reptiles (and/or their eggs) are resistant to sea-water, and can survive long journeys on floating logs or vegetation; some, notably tortoises, float very well by themselves[24]. Some snails can seal themselves into their shells, which enables their otherwise vulnerable soft bodies to survive on floating vegetation. Other vertebrates, notably amphibians and mammals (other than bats) are normally unable to cross sea-water barriers. The presence of endemic rats on isolated islands such as Christmas and the Galápagos[25] is exceptional, and until recently it was thought that amphibians and freshwater fish never reach truly oceanic islands[26], but island age and DNA studies have shown that some frogs must have crossed the sea to Madagascar, Mayotte and the Seychelles[27].

Figure 1.2. "This lizard is hopeful of arrival on an island by means of the log raft on which it has inadvertently become a traveller"; illustration from Carlquist's classic *Island Life* (1965).

Movement through the islands

While life starts arriving as soon as the first island becomes habitable, some of the biota can 'hop' to newer islands as they emerge over the hotspot[28] while the original point of colonisation erodes over millions of years. Thus the age of the chain may be more important than the age of individual islands, at least for those animals or plants that can still disperse. Only part of an island's biota will be able to hop to a newer island; those that have become flightless, or have developed heavy short-lived seeds, become extinct as the island becomes an atoll or submerges altogether. In the Indian Ocean colonisation was also facilitated during the extended Pleistocene periods of lowered sea-level[29] by the presence of many more islands, some very large, providing 'stepping stones' for good fliers from southern India. Many of the more geologically recent arrivals amongst birds and flying-foxes, most of which have Asian affinities, probably reached the islands at this time (pp. 63–65)[30]. Winds and ocean currents nowadays favour dispersal from Rodrigues to Mauritius and not the other way, though this may have been different in the past, and in any case the mass of the Nazareth and St Brandon banks creates current eddies that can back towards Rodrigues[31].

There are good biological reasons to suppose that Réunion was colonised well after the other two Mascarene islands. It had no flightless birds, only one tortoise species, fewer endemic forms in relation to its area, and many species shared with Mauritius, the likeliest source. This is to be expected as it is clearly a geologically young island; steep, high, with an active volcano and little reef development. Although the island is believed to have emerged about 3 mya, and the oldest dated rocks are from 2.1 mya, there was a cataclysmic series of eruptions from the dying Piton des Neiges volcano over the period 223,000 to 188,000 years ago, during which much of the island would have been smothered in pyroclastic flows or subject to fires they initiated[32], as happened at Krakatau off Java in 1883[33]. Many species must have been lost in this holocaust, hence much of what was recorded by the first human visitors will have colonised since these eruptions. By contrast, Mauritius and Rodrigues shared similar flightless birds (rails and the oversized pigeons, the Dodo and Rodrigues Solitaire), large-headed parrots, big endemic day-geckos, two tortoise species each, and an extra fruitbat. Together with a much more eroded topography[34], these factors suggest a much longer timescale for species to evolve. Most of this is also mirrored in the flowering plants and snails[35], though not in very mobile groups like ferns, orchids or butterflies[36]. Flightless birds cannot cross the sea, so the Dodo and Rodrigues Solitaire's ancestors, and those of the flightless rails, must have flown to Mauritius and Rodrigues – their descendants becoming flightless on these islands before Réunion emerged (and well before the pyroclastic eruptions), which explains why that island had no dodo equivalent. Other animals and plants did hop successfully from Mauritius to Réunion, only 164km away, its cloud bank (and rarely the mountains themselves) visible from high up on a clear day, which is why they share so many species (or sibling species) today (x-refs Chapters 3 and 4)[37]. Only one bird, two fruitbats and a handful of snails (but no reptiles) were shared between Rodrigues and Mauritius, the bird and one bat being mobile enough to colonise Réunion also[38]. Mauritius also suffered a devastating eruption well after its origin; the phase of 'intermediate lavas' in Mauritius ended with an explosive episode around 500,000 years ago[39], which must have been almost as destructive to the biota as the later events in Réunion, though we know that many species did in fact survive; as some reptiles (and no doubt plants) also did in Réunion[40].

The generally accepted dates, originating in the 1960s, suggest that Mauritius is the oldest island (8+ my) and Rodrigues the youngest at 1.5 my, but the composition of the flora and flora is more consistent with these two islands being of similar age, and the potassium–argon dates for Rodrigues are based on just two rocks currently above sea-level[41]. Rodrigues would have been nearly ten times larger when the sea-level was 100m or more lower during the Ice Ages[42], but that only takes us back two million years or so. The key appears to lie in the age of the Rodrigues Fracture Zone which began its activity 8–10 mya[43]. The lack of raised beaches indicates that Rodrigues is sinking[44] (as volcanic islands always do with time[45]), so it is probable that 8–10mya Rodrigues emerged as a high island, a view supported by more recent work which has identified a series of older lavas[46]. Either island could also have been seeded from an earlier island spawned by the hotspot – e.g. St Brandon, the nearest. This archipelago of atolls and banks[47] some 385km northwest of Mauritius may still have been a 'proper' island when Mauritius and Rodrigues first emerged; its submarine basalts are dated to 31 mya[48], similar in age to Midway atoll in the Hawaiian chain. There is a small guyot (Soudan Bank) much nearer, rising to within 13m of the surface only 175km northwest of Mauritius, and another at 46m depth that lies 147km east of Rodrigues, both of which may have at one time been high islands, as well as re-emerging during the Ice Ages[49]. The discrepancy between the traditional dating of rocks in Rodrigues and the apparent age of the biota remains to be fully resolved[50]. The origins of animals that reached the Mascarenes will be discussed further in Chapter 4.

CHAPTER 2

FIRST CONTACT

FINDING AND DESCRIBING THE MASCARENES

The five ships ... being severed beyonde the Cape of buona speranza from the other three of their company, and having quite lost them, came all of them shortly after under an island called (as it is thought) by the Portugals Isola de Don Galopes: but they named it the island of Mauritius. Here they entered an haven, calling the same Warwicke, after the name of their vice-admirall, wherein they found very good harborow in twenty degrees of southerly latitude. This island ... is a very high, goodly and pleasant land, full of green and fruitful vallies, and replenished with palmito-trees, from the which droppeth wholesome wine ...

The first Dutch account of Mauritius (only the English version has survived)[1]

The twentie seventh, latitude twentie one degrees, then we saw an Iland West South-west, and South-west by West some five leagues from us being very high land ... The Ile is like a Forrest, and therefore I called it Englands Forrest; but others call it Pearle Iland, by the name of our ship [marginal note: *A new iland discovered in 21. degrees*]

J. Tatton, on 'discovering' Réunion, 27 March '1612'[2]

More than two thousand years ago, traders from the Mediterranean and Arabian regions were already making the sea crossing to India, and penetrating south along the African coast to Zanzibar. Mariners from the east, probably from trading posts in northern Sumatra, were independently exploring, and they reached Madagascar first, around 800 AD. They knew Aldabra and other islands north of Madagascar, and the Maldives nearer India, but appear to have largely ignored them. Well to the south of regular trade routes, the Mascarenes evaded early discovery. There is no evidence that the proto-Malagasy people (from Sumatra or Java) ever encountered them, and the Arabs apparently only did so around 1300 AD. Arab mariners apparently did little except note them on their charts, which were later acquired by the Portuguese[3]. To avoid their discoveries being leaked to rival European powers, the Portuguese in the 16th century wrote no reports on the lands they explored, so details of their itineraries are often obscure. Through careful study of surviving records of their voyages, Mauritian historian Alfred North-Coombes concluded that the islands were first seen by the Portuguese in 1510 (Réunion), 1516 (Mauritius) and 1528 (Rodrigues)[4]. They appear to have landed only rarely – to reduce risk of detection they were under instruction to avoid intermediate landfalls and to hurry inconspicuously to and fro with their spices (and booty stolen from Arab traders). North-Coombes traced only one record of a Portuguese landing; in 1528 *Santa Apelonia* (Réunion) was said to have "plenty of fresh water, trees, birds and fish" – indicating a previous landing. When they thought an island could prove useful, the Portuguese planted bananas and left livestock to multiply and provide fresh meat on future visits; they left nothing on Mauritius or Rodrigues, but may have left goats on Réunion[5].

The arrival of the Dutch

The first properly documented landing was in 1598, when Admiral van Warwyck from the newly independent Netherlands led a fleet into what is now Mahébourg Bay in Mauritius. Unlike the Portuguese, the Dutch recorded everything they found, including material from a shipwreck, though there was no indication of human survivors[6]. Rodrigues was visited briefly by the Dutch in 1601[7], and frequently sighted thereafter. The lack of a natural harbour and the formidable defence of encircling reefs persuaded most

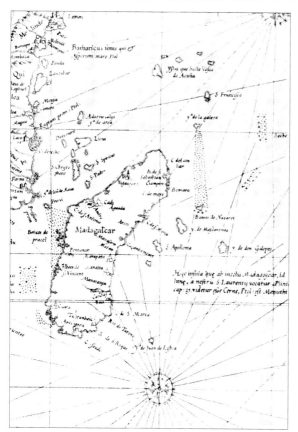

Figure 2.1. Mercator's map of Madagascar and islands nearby, 1569. From Visdelou-Guimbeau (1948).

passing ships to stay well clear, and no useful descriptions exist before Leguat's account of his two year exile there in 1691–3[8]. Réunion was the last to have its first descriptive visit, until recently thought to be by the English East-Indiaman the *Pearl* in 1613, under Captain Samuel Castleton. However, Bontekoe reported that in 1619 he had found a plaque recording an earlier Dutch visit by 'Commander Ariean Maertsz Block' with 13 ships. Adriaan Martensz Blok had indeed left Holland in December 1611 with 13 ships and stopped at Réunion during 6–23 August 1612; his brief account has only recently been discovered[9]. However this does not explain how the chronicler of the remnant of Verhoeff's flotilla, sailing past without stopping in late December 1611, knew that the island abounded in tortoises, birds and fish – perhaps someone had good Portuguese contacts[10].

To voyagers in those days it was a surprise and a wonder to find these lush islands uninhabited, 'desert' islands in the original sense of 'deserted'. It was not long before reports of shipwrecked sailors living for years on wits and wildlife (before being eventually rescued) led to romantic novels on the same theme. There were several such incidents in the Mas-carenes[11], which was perhaps why Henry Neville included a Dodo-like bird in the first of these novels, *Island of Pines*, published in 1668[12]. The proliferation of such novels in the early 1700s, typified by *Robinson Crusoe* in 1725, led to real accounts becoming suspect, with unfortunate results for the credibility of Rodrigues's first inhabitant, as we shall see.

Interpreting early descriptions

One of the problems confronting any historian is the use of names. If an object, plant or animal mentioned in an account is given a name but not actually described, how do we work out what it is? If a whole culture, fauna and flora is unfamiliar to an explorer, how does he (mariners of the 16th and 17th centuries were all men) name the novelties, and if he does describe them, is the description too entrenched in the writer's own background to be easily interpretable? These problems, particularly the second, are well illustrated in early accounts of the Mascarenes. There is a further complication; many of the travels were rapidly translated into other European languages by people who knew nothing of the islands being described, often compounding confusions inherent in the original.

Let us start with the famous engraving of the first Dutch camp on Mauritius, made to accompany the account of the voyages undertaken under Admirals Cornelisz van Neck and Wybrant van Warwijck, and published in 1601 (Figure 2.3)[13]. The wildlife in the picture consists of six birds, a bat, tortoises and fish, plus a number of trees including two palms. Three kinds of bird, the bat and the two palms are given explanatory text.

Three birds are named specifically: *walckvogel*, made up on the spot by the sailors, *rabos forcados*, a Portuguese name presumably borrowed from other mariners, and *Indische raven*, a term (together with its French equivalent *corbeau indien*) that came to mean 'hornbill' in the East Indies. Two of these birds are also described, so they can be identified directly with a high degree of confidence, although the pictures are rudimentary. The other biota, given names for their perceived equivalents in Europe, are not described and can only be identified inferentially from other evidence: 'tortoise', 'bat', 'turtle dove', 'parrot', 'palm-tree', 'date-palm'[14]. It is also instructive to compare our understanding today with what Hugh Strickland made of the picture in 1848, before any subfossil bones had been found.

The walckvogel

The *walckvogel* was completely new to everyone on the ships, so they described it fairly fully, and, though the picture is sketchy, it would have been enough to identify the bird once specimens had been brought to

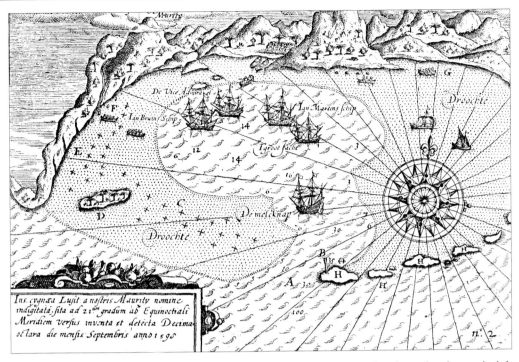

Figure 2.2. Van Warwijk's fleet in Grand Port harbour, Mauritius in 1598. The islet with palms to the left ('D') is Ile aux Aigrettes, the series on the right ('H') the group where Leguat was exiled a century later. From a facsimile in Bonaparte (1890).

Europe. One of the sailors on the same expedition used a quite different alternative, *doederssen*[15]. The next visitors (Harmensz's voyage) called the same birds *griff-eendt* or *kermisgans*, and *dronte* (the *Gelderland* crew)[16], though the captain of one ship in the fleet (the *Zeeland*) called them *dronten* or *dodersen*[17]. By 1602 we have *dod-aars*, *dronte* and *walghvogel*, and in 1606 they have become (in German) *totersten* and (again) *walckvogel*[18]. Only in 1628 does 'dodo' first appear in English for the mystery flightless bird[19]; the alternative form 'dodar', current from 1638, survived well into the 18th century[20]. The bird was so unique that only two of these appellations borrow from another bird-name; *dodaars* is used in Holland for the Little Grebe (*Tachybaptus ruficollis*), a small, round-bodied 'tail-less' bird, while *griff-eendt* and *kermisgans* were used almost as a joke; van Warwijck's fleet anchored in Grand Port Bay on 20 September 1598, the day before Amsterdam fair or *griff*, better known as a *kermis*[21]. *Griff-eendt* ('fair-duck') and *kermisgans* ('festival-goose') refer to fowl fattened for the fair. Dodos were as big and fat as '*kermis*-geese', so acquired the nick-name, revived (and first recorded) when the *Gelderland*, van Warwijck's flagship, returned to Mauritius on almost the same date in 1601; there is no suggestion that anyone thought Dodos were related to real geese or grebes. In 1601, though not rediscovered till the 1860s, there was an official artist on the *Gelderland*, who made accurate drawings of fish, a turtle and four species of bird, including a freshly killed Dodo. Although the bird in the drawing is not named, a map in the same folio indicates where the '*kermis*-geese or *dronten*' were caught, thus clinching the identification[22].

Rabos Forcados

The plumage and behaviour of the *rabos forcados* are described well enough to enable anyone familiar with tropical seabirds to recognise frigatebirds (*Fregata* spp.), as Strickland correctly noted. In the 1840s the taxonomy of this group had not been worked out, and the name Strickland used for the Mauritian bird, *F. aquila*, is now restricted to the Ascension Island Frigatebird in the Atlantic. Both the species likely to have occurred in Mauritius, the Lesser and Greater Frigatebirds (*F. ariel* and *F. minor*), have blackish females and young with white underparts, so the bird in the engraving must remain indeterminate[23].

The Indian Raven

The 'Indian raven', of which only the colour is confusingly described, was pictured as a large bird with a substantial beak and a kind of projection on its forehead, perched up a tree. The birds were often referred to in subsequent reports, the Dutch always calling them *ind(ian)ische/indiaensche ravens*, translated

Figure 2.3. Dutch life on Mauritius in 1598. From a facsimile in Bonaparte (1890). Numbers have been overwritten with modern versions to improve clarity. We have used Strickland's translation into English (1848, Appendix A) from the French of an early version of the voyage; names in the original Dutch and French are italicised in brackets. Moree's (1998) English version, direct from the Dutch, is less faithful to the original (i.e. it contains more interpretation); for the original Dutch see Keuning (1938–51, vol. 3).

1. Are tortoises [*schildtpadden*] which frequent the [high] land, deprived of paddles for swimming, of such size that they load a man [and can still walk very upright]; [they also] catch crawfish [*escriuisses/ecrevisses*] a foot in length which they eat.
2. Is a bird called by us *walckvogel* [*oiseau de nausée*] the size of a swan. The rump is round, covered with two or three curled feathers; they have no wings, but in place of them three or four black feathers. We took a number of these birds, together with turtle doves [*turtelduyven*] and other birds, which were captured by our companions when they first visited the country, in quest of a deep and potable river where the ships could lie in safety. They returned in great joy, distributing their game to each ship, and we sailed the next day for this harbour, supplying each ship with a pilot from among those who had been there before. We cooked this bird, which was so tough that we could not boil it sufficiently, but eat it half raw. As soon as we reached the harbour, the Vice-Admiral sent several men ashore to seek for inhabitants but we found none, only turtle doves and other birds in great abundance, which we took and killed, for as there was no one to scare them, they had no fear of us, but kept their places and allowed us to kill them. In short, it is a country abounding in fish and birds, insomuch that it exceeded all the others visited during the voyage.
3. A date tree, the leaves of which are so large that a man may shelter himself from the rain under one of them, and when one bores a hole in them and puts in a pipe, there issues wine like [sack]; a mild and sweet flavour; but when one keeps it three or four days it becomes sour. It is called palm-wine.
4. Is a bird which we called *rabos forcados*, on account of their tails which are shaped like [tailors'] shears. They are very tame, and when their wings are stretched they are [easily] a fathom in length. The beak is long, and the birds are nearly black with white breasts. They catch and eat flying fish, also the intestines of fish and birds, as we proved with those we captured, for when we were dressing them, and threw away the entrails, they seized and devoured the entrails and bowels of their comrades. They were very tough when cooked.
5. Is a bird which we called Indian Crow [*Indische raven/corbeau indien*], [about the size of a] parroquet [*papagaien*; macaw implied], of two [and] three colours.
6. Is a wild tree, on which we placed (as a memorial in case ships should arrive) a tablet adorned with the arms of Holland, Zealand and Amsterdam, so that others arriving might see that the Dutch had been here.
7. This is a palm tree [*palmite*]. [A good] many of these trees were felled by our companions, and they cut out the bud marked A, a good cure for pains in the limbs. It is two or three feet long, white within and sweet; some ate as many as seven or eight of them.
8. Is a bat with a head like a meerkat*. They fly here in great numbers, and hang in flocks [on] the trees; they sometimes fight and bite each other.
9. Here the smith set up a forge, and wrought his iron; he also repaired some of the ironwork on the ships.
10. Are huts which we built of trees and leaves, for those who aided the smith and cooper [in smithing and coopering in order to depart] at the first opportunity.
11. Here our chaplain [Philips Pietersen of Delft], a sincere and plain-spoken man, preached a very severe sermon, without sparing anyone, twice during our stay on the island. [Half went on land to attend it in the morning, the other half in the afternoon]. Here was [a man named Laurens, born on the island of Madagascar] baptised, along with one or two of our own men [who had not been baptised].
12. Here we applied ourselves to fishing, and took an incredible quantity, to wit, two barrels and a half at one haul, [of all] different colours.

---

* 'meerkat' (literally 'lake-cat') *must* have been a colloquial term for a familar Dutch animal, probably an otter; this was before the South African animal now known as Meerkat *Suricata suricatta* had been discovered. The French translation used 'marmolot', also unidentified, but translated by Strickland as marmot – but that alpine animal would have been unfamiliar to Dutch sailors.

Figure 2.4. Head and foot of a freshly killed Dodo, drawn by Joris Laerle in 1601. From Moree (2001).

into French as *corbeau indien*. Although also applied in France to macaws in the 1650s[24], the term was widely used in Dutch, French and English in the East Indies to denote hornbills[25]. Confusion persisted for well over a century – in the 1770s Buffon was still mistakenly assuming that some sort of crow (as in the everyday use of '*corbeau*') was being discussed[26]. Strickland, reflecting the most widespread usage, and influenced by the projection over the bill of the bird in the engraving, assigned the '*Indische raven*' rather definitively to 'a species of *Buceros*' i.e. a hornbill. Although quoting Thomas Herbert at length, Strickland uncharacteristically failed to notice that Herbert's 'cacato' was curiously similar to the bird in the Dutch engraving (though mirror-reversed), and represented the same species. A 'cacato' (cockatoo) is likely to be a kind of parrot, and indeed Herbert described them succinctly as "birds like parrats, fierce and indomitable". Nonetheless, later writers were confused by Dutch voyagers always referring quite separately to '*Indianische ravens*' and '*papagaien*' as two separate classes of birds, while failing to give useful descriptions, though Pastor Hoffman in 1675 did his best: "red crows with recurved beaks and blue heads, which fly with difficulty and have received from the Dutch the name of '*Indianishe ravens*'"[27]. Thus until subfossil material of a large parrot was found in the 1860s, the Indian Raven was misidentified or ignored; mentions of parrots, even from English visitors who never used the term 'Indian Raven', were taken as referring to something quite different. Even finding the bones did not immediately clinch the argument; Emile Oustalet was still arguing in 1897 that '*corbeaux indiens*' were hornbills whose bones were yet to be found, formerly present in addition to the extinct parrots by then named *Lophopsittacus mauritianus* (and hereafter called Raven Parrots)[28]. In 1983 Pierre Verin, compiling a book of old voyages to Mauritius, called the bird a hornbill, and in 1993 France Staub revived the idea again[29]. No hornbill bones have been found. In fact, their presence would be most improbable on zoogeographical grounds; hornbills are poor at sea crossings and unknown on oceanic islands[30]. The artist on the *Gelderland* also drew this bird – unequivocally a large crested parrot. An analogous saga is still being played out in relation to the supposed Réunion dodo (of which more later).

The Red Hen

The *Gelderland* artist also drew another flightless bird, with a slender, curved bill. The contemporary Dutch voyagers never described such a bird, but their lists included the 'hen' words *feldhüner* or *veldthoenders*, used in Germany and the Netherlands for grouse and partridges. Thomas Herbert mentioned 'hens' in 1629, providing a sketchy drawing but no description, and Cauche referred to "red hens [*poules rouges*] with woodcocks' beaks"[31]. Strickland, whose intuition failed on the big parrot, did better here. He assimilated these birds, a drawing from van den Broecke (visiting in 1617), together with birds Leguat later called 'gelinottes' (another 'grouse' name), as representing the same species, but was unable to identify it, not having access to the *Gelderland* journal or subfossil bones. Nearly 20 years later bones of a large flightless rail, a 17th-century picture of a flightless bird with chestnut plumage and a long decurved bill, and Peter Mundy's manuscript with another description and sketch were discovered. The mystery was solved; these birds were all *Aphanapteryx bonasia*, the extinct Red Hen or Red Rail[32] – which leads us on to an awkward case of mistaken identity.

In the 1620s and 1630s visitors to Mauritius expected to see Dodos, having read the popular Dutch accounts widely published in friendly European languages – French, English and German (the Dutch were at war with Spain and Portugal[33]). However by

Figure 2.5. Two Raven Parrots, drawn by Joris Laerle in 1601. From Moree (2001).

Figure 2.6. A freshly killed Red Hen, drawn by Joris Laerle in 1601. From Moree (2001).

the late 1630s visitors were failing to find them, Peter Mundy in 1638 having to be content with two captive ones seen a decade earlier at the Indian port of Surat, on his previous trip[34]. After 1640 there was a period in which no-one saw or mentioned the birds[35], until in 1666 John Marshall saw "dodos or red hens which are larger a little than our English hens, have long beakes and their wings so little it is not able to support their bodies"[36]. His 'dodos' were clearly not true Dodos *Raphus cucullatus*, but explicitly equated with Red Hens. In 1675 Pastor Hoffman, whose description of Indian Ravens we met earlier, reported a "particular sort of red bird known as *toddarschen* which is the size of an ordinary hen"[37]. As Alfred Newton pointed out in 1868 when the Hoffman account first surfaced: "it would appear from this [usage] that in Hoffman's time one common name of the Dodo had been transferred to another species of bird, in accordance with that odd process of substitution which has obtained in so many countries, where the rightful owner expiring bequeaths (as it were) its titles to a survivor"[38]. All late references to 'Dodos' or '*Dodaersen*' must therefore be examined very critically; they are as likely to be referring to Red Rails – the oft-cited last observation of the Dodo, by Benjamin Harry in 1681, is very much at issue here[39]; we will return to it in Chapter 5.

Ben Van Wissen pointed out in his Dodo book that in the 17th century "copyright or author's rights on works of art did not exist. People simply borrowed, pirated, or stuck in bits at will, with or without acknowledging their sources, just so long as the result was tasteful and saleable"[40]. Such plagiarism was rampant, and often makes it difficult to assess whether a published account is genuinely that of the voyager, or has been amplified or amended later by the author or an editor. Van Wissen illustrated this with an extreme example where a Dodo picture was 'borrowed' to illustrate a voyage through the Straits of Magellan, where the mariners were said to have caught and salted large numbers! In the Mauritian context the reverse occurred; penguins were used to illustrate Dutchmen catching Dodos in the published version of Harmensz's voyage, and cassowaries do duty for Dodos in another engraving, from De Bry's collection of voyages[41]. Thomas Herbert's engravings of a 'palmeto tree' (*Latania*), and 'a tropique bird' (Frigatebird or *rabos-forcados*) are clearly copied from the classic 1598 Dutch engraving discussed earlier (Figure 2.3), and the earlier picture appears also to have influenced his Dodo and 'cacato', though unlike the Dutch illustration, his rather odd 'batt' is shown correctly hanging head down. Mauritius Dodos were twice borrowed by publishers to illustrate different versions of Bontekoe's account of his visit to Réunion in 1619. A Dutch edition of 1646 used a Savery sketch, and Thevenot used van den Broecke's drawing to illustrate a French edition in 1663 – contributing to the misidentification of the Réunion 'Solitaire' (see p. 30) that was not cleared up for over 300 years. Strickland spotted both the transposition and the identity problem, but died in 1853 before the real nonsense started. Bontekoe (or his Dutch editor) compounded the problem by claiming to have seen *dod-eersen* in Réunion, but as he was expecting to go to Mauritius (his ship having missed the target), he seems to have added in the iconic Mauritian bird as an afterthought from earlier descriptions of that island, as Strickland surmised in 1848[42].

Due to this epidemic of copying, it is difficult to assess reports that closely mirror previous accounts – the authors (or editors) could be copying, or the accounts could be similar simply because the same animals and plants were seen. The faunal lists given by early Dutch visitors to Mauritius subsequent to 1598 are suspiciously similar and may not be independent, and this problem gets worse in the early 1700s when compilers, both French and Dutch, recycled old reports and presented them as up-to-date accounts of the islands[43]. One fascinating story is that of the use of red cloths to catch Red Hens (see Box 11, p. 127). This tale developed as the 17th century wore on, but it is difficult to tell whether through copying or observation; Leguat independently said the same of the flightless rail in Rodrigues[44]. In general, voyagers were most interested in wildlife they could kill and eat, so anything ridiculously easy to catch warranted a comment. This may partly explain the decline in Dodo and Red Hen reports as the 17th century progressed – even if the birds still existed in small numbers, there were too few to be worth hunting or writing about.

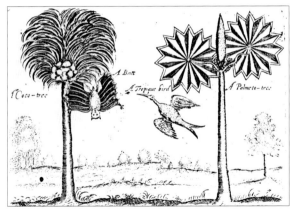 

Figure 2.7. Animals and trees in Mauritius, from Thomas Herbert's travels (1634). Note how the 'Cacato' (Raven Parrot), the Dodo, the 'Tropique bird' (Frigatebird) and 'Palmeto-tree' are copied from Figure 2.3. The bat also features in Figure 2.3; only the 'Coco tree' (Coconut) and 'Hen' (Red Hen) are new.

Leguat's giant

Perhaps the most bizarre example of mistaken identity gone rampant is the story of Leguat's *géan(t)* or 'giant' bird (Figure 2.8). François Leguat, in his otherwise meticulous account of his time in Rodrigues and Mauritius in 1691–95, combined this name, a picture borrowed from Adriaan Collaert's century old *Avium vivae icones*, and a fairly detailed description to create an entity that has generated argument and a pile of literature ever since[45]. Take away the distractions of the name and the picture, and it is easy enough to recognise the birds seen in Mauritius from Leguat's description:

> One sees many of those birds known as 'géans' because their head stands about six feet [high]. They are extremely high-mounted and have a very long neck. They are completely white, except for an area under the wing which is a bit red. They have a goose's beak, but a little more pointed, and the toes of the feet are separate and very long. They graze in marshy areas, and dogs often surprise them because it takes them a long time to get into the air . . .[46]

As Buffon first noted in 1781[47], this description comes very close to a flamingo, particularly in the diagnostic red patch under the wing, though it fails on beak shape and the fact that flamingos have webbed feet. However Leguat got his name from Marquis Duquesne's prospectus for establishing an island paradise, for which Leguat and his companions were supposed to be the advance party[48]. Duquesne was aiming for Réunion (not Rodrigues), and copied, almost word for word, *his* details from Dubois's voyages published in 1674[49]. However he made one curious alteration in his bird list, substituting *géants* for Dubois's *flamants* (flamingos, supported by an adequate description)[50]. Nonetheless Leguat's 'wrong' (unwebbed) feet and the illustration, albeit a hundred years old and of an unidentified 'Avis indica', led Henri Schlegel to combine them in 1858 to create *Leguatia gigantea*, allegedly a giant extinct rail six-feet tall[51]. This bird had immediate detractors, but other ornithologists, usually citing Leguat's legendary veracity in other respects, believed in its existence. This belief was in the face of a complete lack of any other eye-witness support for the *géan(t)*[52], and an absence of subfossil material. Flamingos, on the other hand, were frequently mentioned by other early travellers and their bones have been found in the Mare aux Songes, the swamp near Mahébourg where Leguat probably actually saw the supposed *géants*[53]. Despite everything the *géant* won't lie down[54]; it is impossible to *disprove* its erstwhile existence, and one can only rely on the weight of evidence against it.

The *géant* was neither the first, nor by any means the last, Mascarene bird to be given a scientific existence based on traveller's tales alone. Buffon devoted five pages to a '*oiseau de Nazareth*' and its relationship to the Dodo and the Rodrigues Solitaire, this entity acquiring the latin name of *Didus nazarenus*[55]. In fact the name was just a tag added by Cauche to his description of the Mauritian Dodo; he said "we called them '*oiseaux de Nazareth*' perhaps for having been found in the Island of Nazareth, which is above that of Prince Maurice, in 17 degrees of south latitude". Strickland recognised in 1848 that Cauche had probably simply modified '*oiseau de nausée*' (= *walckvogel*) into '*oiseau de Nazareth*'[56]. Although Strickland was persuaded that it did not exist, Nazareth Bank remains the name of a barely submerged plateau north of St Brandon, although Cauche's Ile de Nazaret, which appears on old Portuguese maps, has been identified as Tromelin. Oudemans argued in 1917 that this island might conceal another dodo[57], but, as Renaud Paulian pointed out in 1961, ten minutes on Tromelin would put paid to that idea[58]; it is a flat sandy islet of less than 1km^2, with neither forest or land birds[59].

Figure 2.8. Leguat's 'géant' and its prototype, Collaert's 'auis indica'. From Leguat (1707) and Collaert (1580–1600).

Speculation on the identity of birds reported by early voyagers is a valid and necessary exercise in understanding the pre-human ecology of oceanic islands, but it can be overdone. Schlegel, Rothschild and Hachisuka used nuances between different accounts and simple errors to erect no less than nine imaginary species[60]. We have discussed Schlegel's giant rail (i.e. Leguat's *géant*), but he also accorded species status to Herbert's 'hen', a move Hachisuka endorsed on the grounds that it had (in Herbert's illustration) a straight bill; thus '*Pezocrex herberti*'[61]. Hachisuka invented a third Mauritian flightless rail by arguing that Peter Mundy's 'hen' was different again, because its colour was 'wheaten' rather than chestnut, hence '*Kuina mundyi*'. As we have seen, all these 'hens' behaved identically when presented with red cloths; also each observer saw only one kind – where were the other two hiding?[62] Hachisuka also felt inspired to create a species for a pied bird seen in 1726 by Tafforet on an islet off Rodrigues, although everyone else had linked it to subfossil starling bones. As this bird ate dead tortoises, he argued that its "carnivorous habit" made it "impossible to place . . . among the starlings" – so it had to be a corvid, a sort of chough, '*Testuphaga bicolor*'. In fact most starlings are omnivores, and on a remote oceanic island without competition might well expand their niche to include carrion[63].

Cauche – a cautionary tale

A few travellers were economical with the truth. François Cauche, who commented on Red Hens, and was the only person to describe the Dodo's nest, egg and call, probably never went to Mauritius at all[64]. He was on a 22-gun French ship, the *Saint-Alexis*, under captain Alonse Gouverte. Cauche's account mentions only this one ship, claiming that in 1638 he went in it to Rodrigues, Mauritius, Réunion and Madagascar. However, Dutch records in Mauritius record two visits of a French ship, in 1638 and 1640[65] – an unnamed 14-gunner, captained by Salomon Gouverte. Cauche mentioned Salomon as Alonse's son, and said it was he who went ashore at Rodrigues. The Dutch also reported that the French said on their 1640 visit that they had left Dieppe with another ship, a 22-gunner (i.e. the *Saint-Alexis*), with which they expected to rendezvous at Mauritius. It appears that for some reason Cauche, in his published account, wanted to conceal the existence of the second smaller ship, possibly because it was illegally harvesting ebony in Mauritius (which the small Dutch outpost was powerless to stop). The smaller vessel spent five months in Mauritius in 1638, leaving in December for Dieppe but returning for a brief 11 days in 1640. The date Cauche said he arrived in 1638 coincides with that given by the Dutch for the 14-gunner's visit in that year (early July), but the shortness of the visit ('a fortnight') echoes the 1640 visit, as does Cauche's mention of an English ship, confirmed by the Dutch as the *William*[66]. Furthermore, Cauche made much of claiming the three Mascarene Islands for the French crown in 1638, whereas the Dutch confirmed that was the smaller ship's mission for Mauritius in 1640, but made no mention of such claims in 1638[67]. Cauche and the *Saint-Alexis* perhaps only joined the smaller ship on its second journey, in 1640. Cauche's account of birds and other animals appears to draw heavily on previous works, which he acknowledges here and there by referring his readers to other named accounts. He commented on only six species for Mauritius: fruitbat, Dodo, frigatebird, Red Hen, Pigeon Hollandais and an alleged tiny parakeet the size of a lark; he claimed the rail and pigeon were also found in Madagascar. His description of the first three appears to be lifted straight from the van Warwijk accounts (and indeed he, or his editor, cited them for the frigate). Cauche's Red Hen was given a straight "woodcock's" bill like Herbert's 'hen', but no one had previously reported its colour, so his report appears valid. 'White red and black turtle doves' fits the colour scheme of the Pigeon Hollandais very well; the Dutch had not published descriptions of this pigeon, so again, this looks original, and there *are* blue pigeons, albeit less showy, in Madagascar. His '*paroquets*' "yet smaller [than a thrush] in Prince Maurice's island, with yellow necks and the rest green, no bigger than a lark" cannot be identified[68] and may relate only to Madagascar[69]. Overall his story in relation to the Mascarenes seems fairly unreliable, though it may contain genuine material gained second-hand from the mariners on the smaller ship that spent so long ashore in Mauritius in 1638. His report of the Dodo's egg, nest and its call "like a goose" are unsubstantiated from any other source and are best treated with caution, although they are entirely plausible – his description of the nest and egg matches Leguat's for the Dodo's closest relative, the Rodrigues Solitaire (Box 17)[70]. According to modern historians, Cauche had commercial and political reasons for pretending he was in Madagascar in 1638 rather than 1640[71].

Leguat's Voyage et avantures
While many accounts conceal minor plagiarisms and exaggerations, and Cauche dissimulated to further his business, one famous book acquired the reputation among literary historians of being not the true story of experiences in the Mascarenes, but rather what the French call a *robinsonade* – a desert island novel[72]. The book is François Leguat's *Voyage et avantures*, in which he described two years spent stranded on Rodrigues before sailing to Mauritius in a home-made boat, only to be imprisoned by the Dutch on a tiny islet in Mahébourg Bay[73]. Leguat's story had detractors from the start, apparently arising from petty disputes in the expatriate Huguenot community in The Netherlands and England[74]. These antagonists established a tradition in France that the book was a novel, that 'Leguat' had never existed and that the book was written entirely by others. In the islands no one doubted the essence of his story, as they were familiar with the environments Leguat discussed and recognised them. Abbe Gui Pingré, visiting Rodrigues in 1761, wrote that "this work is dismissed as a tissue of fables; I have found it a great deal less so than I expected". In France some naturalists were influenced by the literary critics; while Buffon praised Leguat's detailed observations on the Rodrigues Solitaire, Cuvier rejected everything he wrote as fiction. In the mid-19th century, subfossil bones were found that matched Leguat's account of animals in Rodrigues, and another early account of Rodrigues turned up which confirmed Leguat's observations[75]. Even so, Pasfield Oliver, a Mascarene expert chosen by the Hakluyt Society to edit their edition of the *Voyage*, retained some doubts about the book's authenticity[76]. In the 1920s an American literary historian, Geoffrey Atkinson, set out to 'prove' that Leguat's book was fiction, that it was written by François-Maximilien Misson and that it was part of a tradition of fabulous voyages. Atkinson argued that Leguat's genuine sources of background material were in actuality his *only* ones, the apparently original material being pure invention; some eminent biologists fell for it[77]. During the 1920s and 1930s more documentary material came to light in Europe that confirmed Leguat's account, and in any case contemporary Dutch documents confirming Leguat's arrival at, imprisonment in and banishment from Mauritius to Java had been published in the 1890s in Cape Town and soon after in Mauritius[78]. This did not prevent Percy Adams repeating Atkinson's 'proof' in his 1962 book *Travellers and travel liars*[79], leading Rodrigues's historian, retired Mauritian agriculturalist Alfred North-Coombes, to set about rehabilitating Leguat once and for all[80]. Even after North-Coombes's detailed study was published in 1980, Adams (who ignored or had not seen it) was still claiming in 1983 that Misson had "published his amazing invention – not by any means out of whole cloth – the *Voyages et avantures de François Leguat*"[81]; doubts still persist in some quarters[82]. Leguat's champions recognise that Misson, always known to be Leguat's editor, added homiletic commentary designed to boost the story's value in promoting the Protestant cause, and to demonstrate God's actions in the everyday affairs of men[83]. From a biologist's perspective, the remarkable thing is that almost everything that Leguat said about wildlife in Rodrigues *can* be confirmed, both by other accounts and by the wealth of subfossil material found in the caves on Plaine Corail[84]. The name '*solitaire*' that Leguat used for the large flightless bird on Rodrigues was another borrowing from Duquesne/Dubois, but

the birds he described in such detail were unrelated to the Réunion 'Solitaire', and agree closely with the subfossil remains of the Dodo-relative *Pezophaps solitarius*. The prejudices of literary 'experts' apart, he has proved to have been a reliable witness to Rodrigues. Leguat's case is not unique in the Indian Ocean; recently another 'novel', Robert Drury's account of his long captivity in Madagascar soon after Leguat's adventures, has been shown to be a true story, not, as was generally supposed, a novel by Daniel Defoe[85].

The saga of Leguat's veracity is a warning to the unwary on the real problems surrounding the interpretation of voyagers' tales, and also a testimony to the various people who, over the years, have devoted meticulous attention to setting the matter straight (the contributions of travellers, subfossils and collectors to our knowledge of Mascarene vertebrates are summarised in Appendices 10–12). Chapter 3 discusses the current state of knowledge of the primeval fauna of the islands prior to human arrival, based on a massive literature of interpretation and identification, of which we have here cited only a few cases and pitfalls. However, errors and misinterpretations are not confined to those interpreting books – bones and even museum skins can mislead too. For the Mascarenes, the most notorious "banquet of codswallop" concerns a whitish passerine in Liverpool museum, labelled 'Madagascar', hyped as a second Rodrigues starling by H. O. Forbes in 1898, then transferred arbitrarily to Mauritius by Walter Rothschild. It was widely accepted as such, despite serious doubts, until being finally debunked over a century later as an albino specimen of the Martinique Trembler from the West Indies![86] A number of species erected in the 19th century from bone material have been re-assessed and, in taxonomic jargon, 'sunk'. These include a grebe, a darter and a moorhen from Mauritius, and a second owl on both Rodrigues and Mauritius[87]. In the other direction, subfossil bones originally assigned to the large Raven Parrot are now known to include two species[88]. Finally, as recently as 1987, a bone found in Réunion in 1974 and originally identified as from a mysterious 'stork' turned received history upside-down; there never was a dodo there, and the Réunion 'Solitaire' was in fact an endemic semi-flightless ibis![89]

The Réunion dodo

Over the years this non-existent Réunion dodo has generated a huge literature, based on travellers' descriptions of the 'solitaire', the borrowed picture and description in Bontekoe's account, and a set of 17th-century paintings of white dodos. Whereas Mauritius Dodos had been sent alive to Europe and were in consequence well-illustrated, and Leguat had left a detailed description and passable drawing of the Rodrigues Solitaire, the accounts from Réunion were fragmentary and somewhat contradictory, and the only published pictures were in editions of Bontekoe, borrowed from Mauritian originals. Bontekoe was also the only visitor to use a dodo name ('*dodaarsen*') for the Réunion bird; the French always called the bird a *solitaire*. Although the best account, from Dubois[90], did not evoke a Dodo, the size and turkey-like feet apparently overruled in the minds of European naturalists the slight problem that it had a long beak 'like a woodcock' and could fly (though it rarely did so). By the mid-18th century the French encyclopaedists accepted that Dodos inhabited Mauritius and Réunion[91], tending to lump the islands together without much discrimination. As the birds had disappeared, no new information was forthcoming until 19th-century writers started looking into the history. First out was Auguste Billiard, who wrote in 1820 that in the time of Governor Labourdonnais (1735–46) the '*dronte*' (i.e. Dodo) or solitaire was still around, and that Labourdonnais had sent one as a curiosity to a director of the French East India Company[92]. Hugh Strickland re-published Dubois's account in 1844, and in 1848 compiled what was known of the Réunion bird, being careful to note that the accounts differed from those of the true Dodo, a point also made by Charles Coquerel in Réunion itself[93]. Then in 1856 William Coker discovered the first 'white dodo' picture (Figure 2.9b), stimulating a rash of publications in which the identity of these birds with the Réunion Solitaire was promoted. This was cautiously endorsed by Alfred Newton, zoology professor at Cambridge, whose imprimatur, repeated in his *Dictionary of Birds* in 1896, confirmed in the minds of most ornithologists the existence of the 'white dodo of Réunion' – a belief enhanced by star billing in Walter Rothschild's stunning, if flawed, 1907 tome, *Extinct Birds*[94]. By 1938 the 'white dodo' was so entrenched that Graham Renshaw not only assigned to it the manifestly *grey* Dodo painted by Jan Savery in Oxford's zoology museum, but even claimed for it the Prague skull, and "two skeletons at Cambridge"[95]. Around the same time Masauji Hachisuka became interested and 'resolved' the various inconsistencies in the pictures and accounts by dividing the Réunion species in two: the 'white dodo' and a 'solitaire' like the one in Rodrigues, amplifying his interpretation in his Dodo book of 1953[96]. Although Hachisuka's 'two species' scenario had few takers, the dodo was definitely 'fact'; the Réunion birds volume of the prestigious series *Faune de l'Empire Français* had one on its cover[97], and the local natural history magazine *Info-Nature* used a dodo standing on a tortoise as its logo until 1989. These castles in the air were sustained more by wishful thinking than by facts, and again it is the cautious and careful Strickland who comes out of it best. He presented the facts he had, and simply

Figure 2.9. Salomon Savery's engraving of a Mauritian Dodo (**a**), used to illustrate Bontekoe's account of Réunion (from Strickland 1848), and the *Illustrated London News* of 20 September 1856, the first public presentation of a white dodo painting (**b**) (with a Red-breasted Goose *Branta ruficollis*); this was billed as 'Persian' by Coker, but later found to be by the Dutch artist Pieter Withoos. Note that the beak of the bird was cropped in the original publication.

pointed out that there had been a large, easily caught and *possibly* flightless bird in Réunion.

The case of the 'white dodo' slowly began to unravel in 1958, when James Greenway published an important book on extinct birds in which he expressed mild doubts as to the origin and authenticity of the 'white dodo' paintings, and emphatically rejected the 'two species' hypothesis. Some subsequent compilers then began to show caution in dealing with the Réunion species[98]. In 1987, before the ibis was described, one of us showed that the bird Labourdonnais had supposedly sent to France was probably a *Rodrigues* Solitaire, and that the paintings of 'white dodos' derived from an original series by Pieter Holsteyn in Holland. There was no evidence connecting the 'white dodo' to Réunion – and it was unlikely that a bird from there would have reached Holland in the mid-17th century. Hence the 'white dodo' paintings were probably derived from an albinistic Mauritian Dodo (Figure 2.9)[99], and so it has proved. In a previously undocumented painting by Roelant Savery, completed in Prague in 1611, a Dodo gleams white with yellow wings in the sun, facing left – exactly as depicted, only slightly modified, by Holsteyn and Withoos decades later. The model was a specimen from Mauritius in the Imperial collection in Prague, described in the contemporary manuscript catalogue as 'dirty white'[100]. Until the late 1980s, whatever one may have thought of the paintings, the accounts of a 'solitaire' in Réunion could not easily be disentangled from the presumption that there was a dodo there. There was, however, one perceptive exception; Robert Storer correctly predicted in 1970 that "if and when remains of such birds are found on Réunion, they will prove to be unrelated either to the Dodo or the [Rodrigues] Solitaire, and I would not be surprised if they proved to be derived from rails or some group other than pigeons"[101]. Then the ibis was discovered . . .

Mythical birds are, however, quite hard to kill, and the 'Réunion Solitaire', masquerading as a Dodo, still appears in recent checklists and books[102]. Even a definitive book on pigeons published in 2001 still promoted it as fact, rejecting the ibis as irrelevant and accepting the 'white dodo' paintings as authentic pictures of Réunion birds[103]. Perhaps in keeping with its imaginary nature, the Réunion dodo also survives as the logo and nickname of the local brand of lager beer, each bottle sporting its jaunty image (Figure 2.10).

Figure 2.10. Label current in 2003 from Bourbon Beer, from Réunion.

CHAPTER 3

THE PRISTINE ISLANDS

The Mascarene biota at the time of discovery

We found all the view [before us] admirable. We never tired of looking at the little mountains of which it is almost entirely composed, so richly were they covered in great and beautiful trees. The streams that we could see flowing fell into valleys whose fertility we could not doubt . . .

François Leguat in 1691 on seeing Rodrigues for the first time, from just offshore[1]

As the Ile is prodigall in her water and wood, so shee corresponds in what else a fruitfull mother labours to be excellent in, not only boasting in that varietie of feathered creatures, but in the rarenesse of that varietie . . .

Thomas Herbert, extolling Mauritius in 1634[2]

The hand of humanity, or at least the plants and animals imported by colonists, is all too evident almost everywhere in the Mascarenes today – only parts of upland Réunion remain largely untouched[3]. What remains is nonetheless recognised as one of the world centres of plant diversity[4]. While we have a good idea of the primeval fauna, it has proved quite difficult to reconstruct a clear idea of the original vegetation in the most devastated parts of the islands, the coastal and lowland areas. Early travellers were faced with hundreds of tree species they had never seen before, so they picked only a few to mention (ebony, palms etc.) and ignored the rest in their writings; they might discuss individual tree species, but rarely described their setting. Over time, after settlement, names were given to the different plants and their properties discovered[5], but by then the lowland forest had been effectively destroyed, except in the high-rainfall danger zone where lava flows reach the sea in Réunion. Only in Rodrigues do we have a fairly adequate description of the appearance of lowland vegetation[6].

From early reports and ecological inference from what is left of the original vegetation, all three islands were completely forested when discovered, apart from the highest elevations in Réunion and on fresh lavas near its volcano[7]. As discussed in Chapter 1, the two larger islands have a wetter zone on the windward (eastern) side and a drier one in the lee (western). The coastal vegetation differed in the two zones; the tall, dense, mixed rainforest reached the coast on the windward side, but on the dry side there is some dispute as to the composition of the vegetation.

The coastal dry zone of Réunion and Mauritius, vulnerable to fires and probably forming the habitat of several of the most spectacular endemic animals, has been characterised by botanists as being a 'palm savanna'[8]. We do not believe any typical savanna vegetation existed in Mauritius, and consider it was much more restricted than supposed in Réunion[9]. While there is no doubt that this zone on all three islands was characterised by the endemic fan-palms, latans *Latania* spp. and an abundance of Hurricane Palms *Dictyosperma album* with their edible cabbage, there is nothing in the early Mauritian literature to suggest a 'savanna' – open grassland with occasional trees and shrubs[10]. The supposed extent of this 'palm savanna' was the area of less than 1,000mm rainfall in Mauritius and Réunion, but in fact on both islands early visitors described forest reaching to the shore almost everywhere. In Rodrigues (coastal rainfall 800–1,200mm) we have Leguat's rather fuller account, describing fairly open mixed woodland rich in palms and screw-pines (*Pandanus*) in the lowlands; we believe the dry lowland forest in both Mauritius and Réunion was similar[11]. The concept of the palm savanna arose by extrapolation from the surviving vegetation of Round Island, 21km off the north coast of Mauritius[12]. On this steeply sloping islet with little soil, no water table, and rainfall of only *c*. 850mm, there is a thicket heavily dominated by palms and screw-pine[13]. There is a short narrow strip on the west coast of Mauritius where the rainfall is under 800mm, and even the area within the 1,000mm isohyet is only a band about 24km long and up to 5km wide from Port Louis to Tamarin – all the northern plains being a bit wetter[14], and always referred to by visitors as 'wooded'. In Réunion parts of the north-west coast have only 500mm of rainfall each year, but no trace of the original vegetation remains. Of all the numerous early visitors, only Guillaume Houssaye in 1689 actually gave a useful description of the vegetation in the dry zone – wooded throughout except for the stretch of coast from the Cap la Houssaye to Etang Salé, which was "burnt [i.e. parched] country of almost nothing but rocks where nothing grows except *benjoin* trees and *lataniers* on which the goats feed". We think there was, in Réunion only, palm/

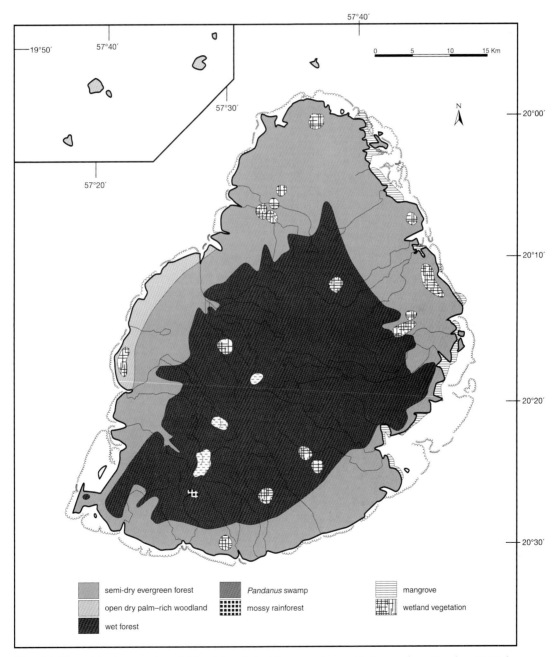

Map 3.1. The original vegetation of Mauritius, derived by extension and inference from Vaughan & Wiehe (1937), together with readings of the early accounts.

benjoin savanna along the west coast where the rainfall is under 750mm[15], except in the region of the Pointe de Galets, where Houssaye specifically stated it was wooded[16]. Many early visitors to Réunion commented on the abundance of 'aloes' along the dry coast – the endemic *Lomatophyllum macrum*[17]. On all three islands the palm-rich woodland was closely associated with tortoises that fed on the fruits, dispersing the seeds of the latans and screw-pines[18].

It is likely that on stable dunes and sandy upper beaches there would have been some open grassland, grazed by the numerous giant tortoises reported on each island. Only Heyndrick Jolinck in 1598 and Cornelisz Matelief in 1606 specifically mentioned grassy areas[19], though no one described anything recognisable as the ultra-short 'tortoise turf' that occurs on Aldabra[20]. Dunes exist in several coastal areas of Mauritius and along parts of the dry western coast of Réunion[21]. In Rodrigues, where the reported tortoise density was much higher than on the larger islands[22], there are dunes only in the south near Graviers and Mourouk[23]; the sandy spit at Port Mathurin dividing the sea from the tidal mudflat behind (now reclaimed) was free of trees but swept by the sea in heavy weather[24]. Tafforet reported a shortage of grass everywhere, with the tortoises having to eat leaves and seeds fallen from the trees. Mauritius also had some mangrove forest, roots encrusted with oysters, extending in some places "100 fathoms" (600ft/185m) into the lagoon[25], though its original extent is not known.

While travel in Réunion was generally impeded by the very rugged terrain, in Mauritius numerous early visitors complained that it was difficult to penetrate the forest because the trees grew so close together[26]. This might sound like hyperbole, but in 1638 the commander of the first Dutch settlement, Cornelis Gooyer, reported that six men were unable to force a passage for a simple footpath through the forest from the northwest port to the southeast port (Port Louis–Grand Port) because it was too thick[27]. In 1880, forester Richard Thompson was impressed by the dense upland wet forest:

> It is only in these forests that an idea can be gained of the grandeur and composition of what are essentially known as Evergreen Tropical Forests, and which at one point must have covered with the densest tree vegetation it is possible to imagine four fifths of the area of the island. The Tree-ferns rising to heights of 25 to 30 feet, the countless other Ferns, the Peppers, Creepers and Turners, the mass of tall clean stemmed under growth packed so closely together as not to give passage to a man through them, and above all the dense almost black shade of these forests, are something to see and admire.[28]

Reginald Vaughan and Octave Wiehé, pioneering the study of Mauritian vegetation in the 1930s, commented that tree density in the native Mauritian upland forest was astonishingly high, 4–5 times that of comparable forests elsewhere[29] – and that was after three centuries of degradation by invasive animals and plants. Very high tree density appears to be an adaptation to withstand cyclones, and is now known to be matched on other islands subject to intense tropical storms[30].

The ease with which hunters were able to chase and catch feral cattle, goats and pigs in northern Mauritius during the 1600s indicates a more open forest in the drier areas, confirmed by Leguat in 1693[31]. A drawing of agricultural clearance by the Dutch near Poste de Flacq in 1670 shows thick but not impenetrable lowland forest, with latans and Hurricane Palms even in this fairly wet zone (Figure 3.1)[32]. This more open forest provided tall, straight trees for ship's masts and for construction, and also the best ebony; judging by reports from visiting British ships, the tallest wood was in the lowland plain under Black River Peak[33].

Although the forest under 300m in Réunion appears not to have been as dense as that in central Mauritius at 125–150m, higher up it was just as impenetrable, and it remains so to this day[34]. The lowland semi-dry forest above the savanna zone in the

Figure 3.1. *The farm at Vuijle Bocht.* Part of a drawing from 1670 in the Dutch Rijksarchief (State Archives), showing settlement activity, forest clearance and wildlife at 'Foul Bay', now Post de Flacq, Mauritius. Hurricane Palms and latans are seen, with a variety of broad-leaved trees, plus wildfowl and an eel in the river, a probable sheldgoose on the land between the streams, a deer, a goat, some pigs in a pen, and an unidentified bird on a cut stump.

west was similar to that in Mauritius, fairly open with tall trees, but the wet forest higher up was very dense, thickly populated with emergent Upland Hurricane Palms, although these have now largely vanished from the otherwise fairly intact forests[35]. In the mid-altitude mixed forest, and especially in the ultra-wet screw-pine thickets in the east, travel is virtually impossible where paths have not been cut, and even where paths exist progress is often prevented by deep ravines. Tree-ferns, three species of *Cyathea* reaching 10m in height, are also characteristic of the mixed forests and the screw-pine thicket. Above about 1,500m and occasionally up to nearly 2,400m there are tracts dominated by the 'tamarin des hauts' *Acacia heterophylla*, with a light canopy but huge girth. This forest, associated with the endemic bamboo *Nastus borbonicus*, is a fire climax, regenerating only if the tree cover is burnt off (or artificially cleared), so its past extent will have been variable, depending on the frequency and size of volcanic eruptions[36]. Higher still, starting at around 1,850m but extending lower on exposed ridges, there is a dwarf forest dominated by giant heather *Philippia montana*, in cloud most of the day and heavily festooned with grey wispy *Usnea* lichen. At its most exuberant the heather reaches 6–7m tall, under which develops a deep layer of organic matter consisting of rotting trunks covered with a thick layer of moss and epiphytic ferns, and the large sedge *Machaerina iridifolia*. The vegetation becomes shorter with altitude, and several other shrubs become co-dominant, one of the more prominent being the endemic yellow-flowered St John's Wort *Hypericum lanceolatum*, which provides nectar for the Réunion Olive White-eye *Zosterops olivaceus*. Over about 2,800m on the Piton des Neiges (and rather lower around the Volcano) above the daily cloud-banks, the heath gives way to sparse grass with occasional shrubs. August is the coldest month; frosts strike irregularly in winter above about 1,500m, burning the leaves of the upper reaches of the mixed forest, its frequency clearly defining the upper limit of true trees (*Acacia heterophylla, Sophora denudata*).

The forests of Rodrigues

As already discussed, the forest in Rodrigues was rather open – Leguat said it was easy to walk around everywhere as there was little or no undergrowth[37], possibly a result of the high densities of giant tortoises. Leguat and Tafforet found tortoises in all parts of the island, though the large aggregations of 2,000–3,000 together seem to have been coastal. Hurricane Palms and latans grew mostly in the valleys[38], and the tallest trees grew in the deeper soil where the valleys opened out toward the coast. Leguat clearly stated that these tall lowland forests had a closed canopy; he was particularly impressed with large spreading strangling figs, *Ficus reflexa* and *F. rubra*[39]. The Huguenots had to fell trees near the shore to make space for their huts and gardens – i.e. there was no natural open 'savanna' by the coast. However, Pingré in the 1760s said that in some bays, especially at Port Mathurin, beach sand reached some 25–30 *toises* (48–58m[40]) inland, on which grew nothing but "a few weeds"; the present Port Mathurin is on a sand spit which may have been open and grassy, while Leguat's settlement was around the stream (Rivière Cascade Pigeon) to the east of the sand spit on basaltic soil. Large trees grew even where the soil depth was negligible, giving the impression from a distance, as Leguat put it, "of an island more advantageous than it merits, because one thinks it consists throughout of excellent soil". Even before deforestation the streams were seasonal, shrinking to a trickle in the drier months[41].

Surviving cyclones

All the islands are subject to violent tropical storms, or cyclones, during the summer, principally between December and April[42]. Although cyclonic depressions occur every year and contribute substantially to the annual rainfall, big storms strike the islands on average every 4–5 years, with a severe one every 15 years or so. The worst damage occurs if two or more violent cyclones strike in succession in a single season, as happened in Réunion in 1806–7, Mauritius in 1824 and 1960, and Rodrigues five times since 1850, most exceptionally in 1875–76 when four struck in two months[43]. Cyclones have affected the evolution of the forest and may also have similarly influenced the animals. Severe storms partially defoliate the forest and destroy flowers and fruit, though the trees themselves are rarely uprooted unless isolated or ancient[44]. The native forest, in good condition, is rarely over 20m high; it has a canopy of interlaced branches, very wide spreading roots and many trees have buttressed stems[45]. In addition there is a very high trunk density. Although some species are stimulated to flower and fruit by cyclones[46], the immediate result of a storm is a sudden shortage of foliage, fruit, flowers and cover for the native fauna. Several of the animals, in the primeval state, had a marked fat-cycle[47], which may have evolved to enable them to survive these shortages, unpredictable and infrequent though they are. Unfortunately most animals with seasonal fattening are now extinct; those that survive (merles *Hypsipetes* spp. and flying-foxes) appear to have lost the habit, perhaps because their densities are now so low compared with pre-human times that the effect is no longer triggered or functional. A behavioural adaptation survives, though: Golden Bats readily forage on the ground. This extraordinary behaviour in a bat must be an adaptation to finding food when none

36 Lost Land of the Dodo

remains on trees[48]. The fruitbats may even need to descend to the ground during cyclones to avoid being torn from their perches by the gales – they are surprisingly poor at negotiating even moderate winds[49], and once swept into the air and out to sea in a cyclone would almost certainly be doomed. Visiting the ground would have been perfectly safe in Rodrigues before cats and dogs were introduced, as there were no native predators large enough to catch them.

The early visitors were not ecologists, and we have little evidence of how these habitats were used by the original fauna. Inferences can be made from the flowering and fruiting habits of Mascarene plants, and from similar animals in other parts of the world, to partly reconstruct the ecosystem before it was disturbed. We know most of the parts but not really how the jigsaw fitted together. All known land vertebrates native to the islands are listed in Table 3.1 – to which

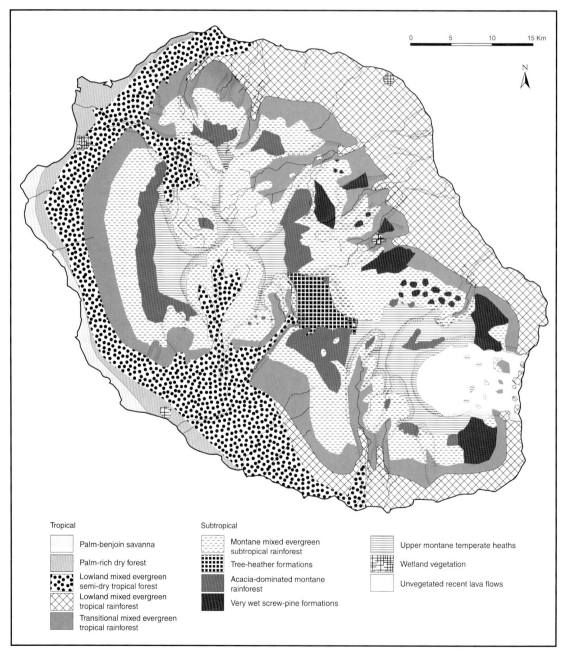

Map 3.2. The original vegetation of Réunion. From Cadet (1977) and Girard & Notter (2003), amended by re-interpretation of early accounts.

we have added turtles, seabirds and dugongs, whose fate was closely tied to that of the land fauna. A number of these species, known only from subfossil remains, apparently disappeared before the first humans who recorded wildlife arrived – probably victims of the rats that reached Mauritius and Rodrigues before any reports were made on the fauna[50]. Although it is likely that small lizards and the blind-snake would have been overlooked, it was mostly the big reptiles that disappeared first[51]. In Mauritius the fortuitous failure of rats to reach Round and Serpent Islands enabled both boas and three lizard species (and a large tarantula) to survive after they vanished on the mainland, allowing their ecology (at least in its restricted residual habitat) to be studied; two further small lizards survive only on offshore islets (Chapter 9)[52]. Various birds and bats were never (or not clearly) mentioned by early travellers, mostly no doubt through oversight, though for Mauritius the harrier and the Reed Cormorant were both conspicuous species unlikely to be overlooked – their counterparts in Réunion were frequently mentioned[53]. No one ever reported the Réunion Lizard-owl, nor the merle and 'babbler' in Rodrigues, and there are no unequivocal references to the night heron or Abbott's Booby in Mauritius. Travellers and settlers also failed to record a third fruitbat in Mauritius, or more than one in Rodrigues, but as most did not notice there were two (of hugely different size) in Mauritius and Réunion, their failure to recognise a third (intermediate in size) is not surprising[54].

THE FAUNA OF PRISTINE MAURITIUS

In pristine Mauritius there were no terrestrial mammals, and the equivalent ground-dwelling niches were taken by reptiles, birds and land crabs[55]. Judging by Round Island today and rat-free islands in the Seychelles and elsewhere, it is likely that ground-living lizards and land crabs were extremely abundant in Mauritius and Réunion[56]. In Mauritius there were four species of forest skink ranging from the huge 680mm (27") Didosaurus, through the large Telfair's Skink, the medium sized Bojer's to the small Macabé Skink; a fifth species, Bouton's Skink, lives only on coastal rocks[57]. Large skinks tend to be omnivorous, and this is certainly true of Telfair's Skink on Round Island today – it is a predator of other reptiles and insects, eats fruit, and will scavenge anything edible. The smaller skinks are principally insectivorous[58]. Didosaurus was big enough to take other lizards and also hatchling tortoises. Some trees have flowers and fruit that sprout from the trunk near the ground[59], suggesting pollination and perhaps dispersal by reptiles or ground-living birds. The surviving *Nactus* long-fingered night-geckos[60] are largely terrestrial on Round Island and Gunners Quoin, but unlike the skinks are nocturnal and wholly insectivorous.

The only terrestrial herbivores were two species of giant tortoise, one high-backed, allowing browsing up to 1m up, the other domed, presumably principally a grazer[61]. One of the characteristics of the Mascarene flora is the occurrence of heterophylly in many species of tree – saplings carry curious long thin foliage, sometimes bizarrely coloured, while adult trees have more familiar leaves. Heterophyllous species are mostly lowland dry-forest species, and they are particularly prominent in Rodrigues[62]. Aldabran tortoises kept on Ile aux Aigrettes largely ignore plants showing juvenile leaves, suggesting that heterophylly arose as a defence against browsing by tortoises, though authors of a study on the chemical defences of Mascarene heterophyllous plants preferred leaf-eating birds as the evolutionary agents[63]. Another probable adaptation to tortoise browsing is the spiny trunk of some palms when young[64], though this does not explain why other species, notably latans and lowland Hurricane Palms *Dictyosperma album* are not spiny and clearly thrived in the presence of abundant tortoises; the *Hyophorbe* species (Bottle Palms) are toxic, at least to humans[65]. While adult tortoises had no predators, the eggs and young would have been eaten, as on Aldabra today, by land crabs and rails[66], and no doubt also by the endemic night herons and Didosaurus.

The absence of large predators allowed birds to evolve flightlessness in safety – the Dodo was the size of a large goose, and the Red Hen equivalent to a domestic fowl. The large Raven Parrot *Lophopsittacus mauritianus* is often cited as flightless in the literature; this is an error[67], but its anatomy does suggest that both it, and Thirioux's Parrot *Psittacula bensoni*, which we believe to be the 'grey parrot' of early accounts, were largely terrestrial in habits[68]. The endemic *Dryolimnas* wood-rail also had reduced wings, as did the wood-rail in Réunion[69]. The ecology of these birds has been open to a great deal of speculation, the Dodo in particular. Only one visitor mentioned its food – 'raw fruit' according to anonymous sailor writing in 1631, though he and others also described its powerful bite[70]. France Staub believed Dodos mainly ate palm fruits, and attempted to relate their fat-cycle to the fruiting regime of these trees[71]. A whole scientific myth has grown up around Stanley Temple's attractive but flawed hypothesis that the Dodo's extinction caused regeneration failure in a large forest tree, the Tambalacoque *Sideroxylon grandiflorum*, with its seeds supposedly unable to germinate without passing through a Dodo's gut (this is discussed fully in Chapter 7)[72]. Hachisuka speculated that Dodos would have eaten 'crabs and shellfish', pointing out that the large terrestrial crowned pigeons *Goura* from New Guinea wander about on river banks eating small crabs[73]. Hachisuka was given to flights of fancy, but in this case he may be right – *Goura* is quite closely allied to dodos and solitaires[74]. Dodos swallowed a large stone which was held in the gizzard, and presumably had an important digestive and grinding function[75]. Whatever Dodos ate in nature, they must have been able to cope with a wide range of food, as they were easy to transport on ships and keep in captivity, some apparently surviving for years[76].

Red Hens, large flightless rails with long decurved bills, look well-adapted to taking invertebrates or reptiles found in or on the soil[77], though we have no direct reports of their diet; like Dodos, they were reported to use their beaks aggressively in defence[78]. The hen's bill closely resembled that of the endemic ibis (or Solitaire) in Réunion[79], which was said to feed on soil invertebrates[80], and the Limpkin *Aramus guarauna*, a North American snail specialist which also resembles the Red Hen in size and shape[81]. Mauritius was well supplied with large endemic land snails, many now extinct[82], which could well have been the Red Hen's principal food, and many subfossil snail shells show damage consistent with attacks from a Red Hen's beak[83]. There was also Sauzier's Woodrail, a smaller ground feeder whose habitat on Mauritius is unclear, though bones are known from both swamp (Mare aux Songes) and dry rocky areas (Le Pouce range). The closely related White-throated Rail in Madagascar frequents marshes and watercourses, but in Aldabra, a very dry atoll, it occurs in all habitats from mangroves through scrub to open beaches[84]. On Aldabra this rail has a relationship with giant tortoises, feeding on bloodsucking insects; tortoises respond to the birds by standing up 'on tiptoe', allowing the birds access to soft areas of skin[85]. There is no report of such behaviour in the Mascarenes, but given the abundance of tortoises, similar associations may well have developed.

In South America *Anodorhynchus* macaws are palm-seed specialists that once depended on the now-extinct large-mammal 'megafauna' to eat the fruit, digest the flesh and excrete the seeds, which the parrots then ate. They have more recently latched on to domestic cattle to perform the same function[86]. Carlos Yamashita, who discovered this, suggested to us that the Raven Parrot, with its macaw-like bill, may have specialised similarly in Mauritius, either feeding directly on the palm seeds, or waiting for them to be 'cleaned' by tortoises or Dodos[87]. If true, this would help explain two things – the early disappearance of this parrot, and why none reached Europe. Palms were cut down in enormous numbers from the early days, and the large-seeded Latans had the most restricted, coastal, distribution. It had puzzled us why these large birds never reached Europe alive as did Dodos (and a Red Hen). Parrots were notoriously popular with mariners[88], so one would have expected

Figure 3.2. Profile of native upland forest in Mauritius, showing the structure of the best tract remaining in the 1930s; the large tree is a Makak *Mimusops maxima*, one of the dominant Sapotaceae. From Vaughan & Wiehe (1941), scales redrawn.

Raven Parrots to have been brought home to Holland or England – but if they refused to eat anything but the seeds of palms or large trees they would not have survived the journey. Raven Parrots were strongly sexually dimorphic in size, more than any other parrot[89], so the sexes possibly exploited different-sized foods.

Also putatively terrestrial was the enigmatic grey parrot of the early accounts, of which we know little except that it occurred in large numbers and was easy to catch[90]. We assume that tarsi and bill elements Holyoak described as *Lophopsittacus bensoni* belong with parakeet breastbones also found in subfossil deposits, and that this was the grey parrot of the eyewitness accounts[91]. Given the abundance of these birds, they may have been exploiting the small fruit of Hurricane and Bottle Palms[92]. Like Raven Parrots, no grey parrots reached Europe; it may have been too unimpressive, or again it may have been too specialist a feeder. The still surviving Echo Parakeet was never successfully kept in captivity until modern dietary techniques made this possible in the 1990s[93].

The last of Mauritius's ground-living animals is the Malagasy Turtle Dove. It has usually been considered an introduction, but subfossil remains have been identified from all three islands, showing that it is a native species[94]. This bird is largely a seed-eater, also taking small snails and feeding almost exclusively on the ground[95]; it would have been the only granivore on the island. It is fairly scarce in deep forest, and may have originally been a bird of the more open lowland woodland.

Where these terrestrial animals were distributed in Mauritius is not known. It seems likely that Dodos, Raven Parrots, tortoises and Bojer's Skinks were mostly found in coastal or lowland areas[96], while the other skinks, Malagasy Turtle Dove, grey parrots, Red Hens, snakes and night-geckos were widespread, but there is no unequivocal information in the old accounts. In addition to fallen fruit, trees fruiting at ground level and the eggs and young of other creatures, there would have been some input into the diet of ground-feeders from seabirds, both from discarded food and their guano. These led to increases in invertebrate levels and thus densities of their predators, in this case including land crabs as well as lizards[97]. There was at least one colony of Abbott's Booby (probably small), frigatebirds nested and roosted in the vicinity of the early Dutch settlements, and there were originally shearwater colonies too (Chapter 6).

Predators

Mauritius had several dedicated vertebrate predators, three birds (an owl, a harrier and a kestrel) and two snakes. Nothing was ever reported about the habits of the largish long-legged owl, known only from fossils, an 18th-century drawing and a good plumage description, but its anatomy suggests it specialised in terrestrial lizards[98]. The harrier was never unequivocally reported by visitors, but its subfossil bones are indistinguishable from those of the Réunion Harrier[99], unusual in the genus *Circus* in having relatively short wings and being adapted to hunting in forest. In Réunion the harrier feeds nowadays on small mammals, and to a lesser extent on birds, introduced lizards and grasshoppers[100], but originally skinks, especially the large Telfair's, would no doubt have replaced the mammals. The Mauritius Kestrel also has short wings and behaves more like a sparrowhawk than a kestrel; it specialises on the endemic arboreal *Phelsuma* day-geckos, though some individuals also take small birds[101]. All three birds of prey are likely to have occurred in all forest types; the owl was last reported in forested areas, and the kestrel is known to have been common throughout the island[102].

The Keel-scaled Boa, which reaches about 1.42m in length and is now confined to Round Island (Chapter 9), is known from subfossil deposits on the mainland. It is another specialised lizard-predator, feeding on geckos and skinks mostly at night, on the ground and in low vegetation. Females are much larger than males, and they may exploit different foods[103]. The Burrowing Boa, also presumably once present on the mainland, became extinct on Round Island before its ecology was studied. It was a burrowing species with a typical blunt snout, but its principal prey is a mystery, as only Carié's Blind-snake lived underground. Perhaps it specialised in buried eggs (of skinks, tortoises or turtles) or the chicks of burrowing seabirds, or, more probably, it used its fossorial morphology to sneak up on lizards hiding under leaf litter. It had the same curious lizard-trapping jaws as the Keel-scaled Boa[104].

Birds and lizards on Mauritius would also have been subject to predation from spiders: there were large tarantulas apparently similar to the lizard-eating species on Serpent Island (Chapter 9), and orb-web spiders with 3m webs that can catch small birds and arboreal geckos[105]. Ground-nesting birds may also have suffered from eggs or young taken by land crabs, originally abundant in coastal areas.

Animals of the canopy

The forest canopy sheltered a varied selection of animals. The colourful *Phelsuma* day-geckos live mostly in the tree-tops throughout the island[106], but the large, dull brown Günther's Gecko probably favoured the lowland palm-rich forests[107]. There was another parrot, the Echo Parakeet, and two pigeons, six woodland passerines, and three flying-foxes. The smaller day-geckos and three small birds, the Mauritius Fody and the White-eyes, while taking a lot of insects, also like flower nectar, and so act as pollinators. The Olive White-eye's long bill is particularly adapted for probing flowers, but the more generalist Grey White-eyes can also be active pollinators[108]. Some Mauritian flowers have red nectar – particularly attractive to the day-geckos which pollinate them. At least one plant is not only pollinated but also dispersed by day-geckos, which (accidentally?) ingest the seeds[109]. The parrots, pigeons, fruitbats, the Mauritius Merle, both white-eyes and the fody would all have taken fruits, mostly smaller than those that could be tackled by Dodos, Raven Parrots and tortoises, though the larger fruitbats can take quite large fruits[110]. There are no specific observations on what the bats ate in pristine Mauritius, but even now in Mauritius a high proportion of the Black-spined Flying-fox's diet is formed by the fruit of native trees, especially ebonies and the large sapotaceous canopy trees – even the Natte, which has a thick sticky latex like chewing gum. The Natte shows signs of coevolution with bats as dispersers, and there is some evidence of enhanced germination of seeds from fruit eaten by these bats; the bats' role in seed dispersal was clearly very important. Flying-foxes also visit flowers, damaging some by eating them whole, but probably pollinating others[111]. When undisturbed the fruitbats fly by day as well as by night, though the smallest, the extinct Rougette *Pteropus subniger*, was said to be strictly nocturnal[112].

The Echo Parakeet survives; it takes unripe fruit, and both it and Pink Pigeons also eat foliage, particularly when fruit is out of season[113]. Pink Pigeons are also partial to flowers, particularly of a common forest tree *Nuxia verticillata*. Merles can take the smaller palm fruits[114], but both Merles and Pink Pigeons generally concentrate on smaller items. The diet of the extinct Pigeon Hollandais can be inferred from surviving blue pigeons in Madagascar and the Seychelles, but there is one record of stomach contents. Charpentier de Cossigny dissected one in 1755 and reported four 'nuts' in its gut, which he was told were those of either *takamaka* or the *natte à petites feuilles*, both rainforest trees with seeds up to an inch long and two-thirds as broad (*c.* 25 × *c.* 15mm)[115] – this bird clearly took much larger fruit than the Pink Pigeon does. In 1801 the Pigeon Hollandais was reported to live on fruit and freshwater snails[116].

Mauritius Cuckoo-shrikes feed in the canopy on large insects and geckos (and their tails!), while Mascarene Flycatchers feed on small insects in the under-storey. Grey White-eyes and Mauritius Fodies are generalist feeders, taking insects, nectar and small fruit; the white-eye feeds by gleaning, while the fody spends more time probing rotten wood and moss for grubs. The fody's adaptability was shown in the 1670s, when Governor Hubert Hugo reported 'sparrows' had become an agricultural pest when provided with a new and abundant food supply of cultivated grains[117]. The Mauritius Merle, primarily a frugivore, also regularly takes insects and geckos, and rarely also flowers; Mauritius Cuckoo-shrikes and Merles have been observed feeding geckos to their young[118].

All these birds of the canopy probably originally occurred throughout the island, though to judge by Réunion today, the flycatcher may have been more common in the lowlands.

A cave-nesting swiftlet and a swallow by day, and two microbats by night were the only aerial insect-feeders on pristine Mauritius; all four are shared with Réunion, one of the bats not differing from those found in Madagascar and Africa, the other apparently endemic to the Mascarenes[119]. The Mascarene Swiftlet and the smaller microbat, the Mascarene Free-tailed Bat[120] use the same lava tunnels, the swiftlets returning at dusk as the bats emerge – both are now much reduced from population levels in the 19th century when the first detailed reports were made, so one must presume they were originally abundant, although they went unmentioned by the first visitors and settlers. The bat roosts in thousands packed tightly together on the cave roofs, while the swiftlet, related to species exploited in Asia for birds'-nest soup, glues its little nest-cups of lichen and saliva to vertical fractures in the highest parts of the caves. The Mascarene Swallow, a cliff-nester, may never have been very common. The larger bat, although called the Grey Tomb Bat, roosts by day in cliffs and palm trees. It is largely confined to the drier side of the island where it is still found today[121].

Aquatic communities

The numerous ponds and rivers on Mauritius were home to a small waterbird community, but there were no amphibians, aquatic reptiles or true freshwater fish. There were three endemic waterbirds – the Mascarene Teal and the Mascarene Coot shared with Réunion, and a sheld-goose related to counterparts in Réunion and Madagascar; these shared the waters with two widespread species, the Reed Cormorant and the Greater Flamingo[122]. Little suitable nesting habitat for flamingos exists (or existed) on any of the Mascarenes, but nevertheless the condition of subfossil bones from Réunion indicates birds in breeding condition, so it appears that some nesting took place, although it is likely that the majority were non-breeding visitors from Madagascar[123]. The teal, related to Bernier's Teal of Madagascar and the Indo-Australasian 'grey teals' (Chapter 4), may, like them, have nested in tree-holes[124]. Two herons common when the island was discovered, the endemic Mauritius Night Heron and the widespread Dimorphic Egret, seem to have been coastal rather marshland birds, but details are sparse. Judging by its wing structure, the night heron had poor powers of flight[125], and may, like its counterpart in Rodrigues, have fed largely on lizards and invertebrates on land rather than in wetlands or on the shore, as some night herons do in Cuba today[126]; it probably also enjoyed hatchling tortoises. Two other waterbirds now present, Common Moorhen and Striated Heron, lack early records or subfossil bones, so are possibly recent colonists[127]; they were not introduced by human agency.

Few visitors commented on shorebirds, which no doubt reached Mauritius then as they do now. 'Curlews' were noticed as good game-birds; the common visitors are Eurasian Whimbrels, but Common Curlews occur also. Around 25 species of migrant shorebird, mostly wintering Palaearctic breeders, have been recorded in the Mascarenes[128]. These birds apparently find the islands with ease, whereas equally far-travelled passerine migrants, which turn up regularly in the Seychelles, are virtually unknown in the Mascarenes[129].

Seabirds

The abundance of seabirds caught the imagination of early visitors to many remote islands, but not in Mauritius. Abundant landbirds for the pot, and a lack of conspicuous seabird colonies on the mainland, seem to have caused visiting mariners, alert (at sea) for seabirds as a welcome sign of nearby land[130], to largely ignore them once on shore. One account from 1598 discussed what appear to have been Wedge-tailed Shearwaters, found near their anchorage at Grand Port[131]. Apart from this the early Dutch accounts emphasised only one seabird, the *rabos forcados* (frigatebirds) that apparently attended their camps to steal fish-guts and other refuse[132]. The only evidence that they bred is a passing reference on how easily they could be caught, as "feeling safe on their nests", they allowed themselves to be taken by hand[133]. There may also have been non-breeders using the island as a temporary base, as happens commonly elsewhere; the next nearest known colony was on islets off Rodrigues. The only other breeding seabird reported early on was John Marshall's account in 1668 of a pair of goose-sized whitish birds by a nest in a "very high tree" – probably Abbott's Booby[134]. Although the Dutch visited the islands in Mahébourg Bay, nothing was said about the seabirds until Governor Deodati incarcerated Leguat on Ile aux Vacoas in 1694 – he reported abundant nesting terns and shearwaters on neighbouring Ile aux Fouquets[135]. However, the Dutch name for Serpent island, still home to innumerable seabirds (Chapter 9) was Meeuwe Klip[136] – 'Gull Rock', even if the birds there were actually Sooty Terns, Brown and Lesser Noddies and Masked Boobies[137]. Wedge-tailed Shearwaters presumably bred then as now in large numbers on Round Island, where both Red- and White-tailed Tropicbirds also nest. The most interesting seabird found there today, the 'Round Island Petrel' (currently considered to be Trindade Petrel *Pterodroma arminjoniana*) was not reliably recorded before 1932; although probably present in the mid-1800s, it may be a relatively recent colonist (Chapter 9). Gunners Quoin and Flat Island (with its satellites) also had colonies of both tropicbirds and the shearwater, possibly with boobies and terns also. On the mainland, only three seabirds besides Abbott's Boobies are known to have nested: White-tailed Tropicbirds in cliffs and tree-holes in the interior (where they are still found), Wedge-tailed Shearwaters and Tropical Shearwaters[138]. One Dutch report in 1598 reported tropicbirds, frigatebirds and probable boobies offshore with what were apparently pelicans – perhaps Pink-backed Pelicans, which formerly bred in the Amirantes atolls north of Madagascar[139]. There are no other records, so the birds were probably just visitors.

Marine life

The lagoon and beaches around Mauritius were home to Dugongs *Dugong dugon*[140] and turtles. Most visitors merely mentioned Dugongs as good eating, but Hoffman in 1673–75 considered them "more abundant in the vicinity of Mauritius than anywhere else"; he described them well, and noted that they grazed on sea-grass in shallow water. Wreeden, then Dutch governor, found them "in large quantity" at Flat Island in 1672[141]. Ships calling at Mauritius in the 17th century regularly sent out parties to 'turn turtle' – i.e. to catch turtles hauling out to lay eggs and turn them over to prevent them escaping. The favourite place soon earned the name of Turtle Bay, which it has kept to this day[142]. No indication of numbers was ever given, but as they were found commonly throughout the year, the breeding population must originally have run into thousands. The majority were no doubt Green Turtles *Chelonia mydas*, but Hoffman remarked on combs etc being worked from carapaces of (presumably) Hawksbills *Eretmochelys imbricata*, and the *Gelderland* artist drew Loggerhead Turtles *Caretta caretta* caught at Mauritius in 1601[143].

THE FAUNA OF PRISTINE RÉUNION

The original fauna of Réunion was similar to that of Mauritius, but much poorer in reptile species, lacking snakes, the largest skinks and geckos, and also flightless birds[144], reflecting the more recent emergence of the island. Many species are or were shared with Mauritius, or closely allied to Mauritian forms, indicating that much of the vertebrate fauna colonised Réunion from the neighbouring island, by far the nearest land. The reptile fauna is more impoverished than the birds, illustrating reptiles' poorer dispersal abilities (Chapter 11). Two of the three Mascarene fruitbats occurred in Réunion – one of these, the Black-spined Flying-fox, was, with the Malagasy Turtle Dove, the only native land vertebrate shared by all three islands. It suggests their propensity for inter-island flight exceeds all but the highly migratory visiting shorebirds.

The ecology of Réunion's fauna would have differed little from that of their relatives in Mauritius, although the greater altitude was reflected in a seasonal vertical migration of many birds, noted by earlier visitors. It is difficult to assess the effect of the absence of Dodos and Raven Parrots, though the ibis (or 'solitaire') seems to have more or less taken the niche the Red Hen occupied in Mauritius. One might have expected the lack of snakes to have resulted in higher lizard numbers, or for other lizard predators, such as the owl, with less competition, to have been more common, but there is no evidence from early visitors to suggest either; reports of lizards are even fewer than for Mauritius and no one ever saw the owl! The Réunion Forest Day-gecko would have originally been widespread throughout the island, spread through niches occupied by three species in Mauritius. The only other day-gecko on Réunion, the Manapany Day-gecko, is known only from the dry west coast, both now and in the past[145].

The extinct Hoopoe Starling, found only on Réunion, may have taken lizards in its diet, but was basically a generalist (as are most starlings)– Desjardins said captive ones would 'eat anything', and Levaillant was told that flocks of them damaged the berries in coffee plantations[146]. The starling probably nested in

> **BOX 2**   **TROPICBIRDS (PHAETHONTIDAE)**
>
> Two species are known from the Mascarenes, the Red-tailed Tropicbird *Phaethon rubricauda* and the White-tailed Tropicbird *P. lepturus*, locally called *Paille-en-queue*. They remain fairly common on the islands but suitable nesting habitat has been reduced. Red-tailed Tropicbirds are abundant on Round Island off Mauritius, and although once rare, White-tailed Tropicbirds are beginning to return in numbers to Rodrigues and have begun to nest in the cave and gorge areas of the Plaine Corail.
>
> *Accounts*
> Leguat (1707) in 1691–3:
>
>> There's another sort of bird as big as a pigeon all over white, its beak is short and strong, it has a feather in its tail a foot and a half long, from whence it takes its name, being call'd straw-tail. These birds made a pleasant war upon us, or rather upon our bonnets; they often came behind us, and caught 'em off our heads before we were aware of it: This they did so frequently, that we were forc'd to carry sticks in our hands to defend ourselves. We prevented them sometimes, when we discover'd them by their shadow before us; we then struck them in the moment they were about to strike us: We cou'd never find out what use the bonnets were to them, nor what they did with those they took from us.
>
>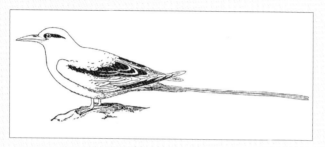
>
> Tafforet (1726) in Rodrigues in 1726:
>
>> There are many Boatswain-birds (Paille-en-queue) which are all white, and others of white red. The Boatswain-birds nest ordinarily in the holes of the cliff or in the hollow trees which abound, especially the Benjoin.
>
> White-tailed Tropicbird *Phaethon lepturus* from Berlioz (1946).

tree-holes, as would have all the parrots and possibly the Mascarene Teal. In other respects, with the exceptions noted below, Réunion shared its terrestrial vertebrate fauna with Mauritius, having the same sibling or replacement species of flying-fox (Black-spined and Rougette), tomb bat, free-tailed bat, harrier, kestrel, wood-rail, owl, parakeets (Echo and Thirioux's), pink pigeon, blue pigeon, swiftlet, swallow, cuckoo-shrike, merle, paradise flycatcher, grey white-eye, olive white-eye, fody, day-geckos, night-gecko, Slit-eared Skink, large skink (like Telfair's), and Bouton's Skink. While birds mostly matched one-to-one, there were fewer than half as many lizard species. The few wetlands also had a similar fauna: cormorant, egret, night heron, teal, sheldgoose, flamingo, and coot. Dubois's Kestrel and the Réunion Night Heron were apparently not derived from their Mauritian counterparts, but from separate colonisations from beyond the Mascarenes, as neither was as specialised as the forms on Mauritius.[147]

Forest birds

In addition to the Hoopoe Starling, Réunion had two forest parrots not shared with Mauritius – the Mascarin Parrot and Dubois's Parrot, of whose ecology nothing was recorded. The former, a middle-sized bird, was not a specialist feeder, as several were kept alive in Paris in the late 1700s[148]. Dubois's Parrot, about the size of an Echo Parakeet but with red head, wings and tail, was reported only by Dubois himself, who said nothing of its habits[149].

Occurring in forest glades and in open areas is the insectivorous Réunion Stonechat, the only avian insectivore on the islands that feeds largely on the ground. In more open areas Réunion also has Madagascar Buttonquails, which did not occur naturally on Mauritius. Although no bones have been found to confirm they are native, the 'small grey partridges the size of quails' described by Dubois fit this species, as does Père Vachet's description of the repeated very short flights made when pursued[150].

Both islands had a fody, the Réunion bird resembling the Mauritian species closely. Dubois described the male as red on the head and breast, but its habits, however, appeared to be quite unlike anything reported in Mauritius, until Hugo's similar account was published in 2002 (Box 31, p. 228). The Réunion Fody occurred in considerable flocks, was a pest of grain in the fields and even a nuisance in people's kitchens – behaviour reminiscent of the typically granivorous Madagascar Fody, and unlike the habits seen in the predominantly insectivorous island fodies today. However in the mid-19th century, before Cardinal Fodies were introduced to Rodrigues, Rodrigues Fodies also formed flocks in open grassland, probably feeding on seeds[151]. As it has never been formally named, we propose to call the Réunion Fody:

Foudia delloni sp. nov. Size as other fodies (roughly that of a House Sparrow *Passer domesticus*); breeding male bright red on head, neck, throat, and upperparts of wings, brown on back, paler on belly; tail brown. Female and eclipse male brown on head, neck and wings where male red, paler brown on throat, otherwise similar to male. Differed from *F. rubra* in having the wing coverts red not grey-brown (male in full plumage), and from *F. madagascariensis* in the restricted area of red in the male. Known only from Réunion Island, Indian Ocean, where extinct soon after 1672. It is named after Gabriel Dellon, the first traveller to describe it; the description is based on two contemporary accounts, Dellon's in 1668 and Dubois's in 1671–72[152].

Above the tree-line

Réunion is unique in the Mascarenes in having large areas above the tree-line, ranging from giant heather forest around 1,700–1,800m through to scant grass at the highest levels. Only four birds are known to have exploited this zone, the stonechat, the two white-eyes and the '*oiseau bleu*' – a large ground-dwelling bird, reluctant to fly. The Réunion Olive White-eye, like its Mauritian counterpart a nectar specialist and important pollinator, is particularly attracted to the seasonally abundant yellow flowers of an arborescent St John's Wort and a small laburnum-like tree in the heath zone, and still makes altitudinal migrations to exploit these food sources, also used by the Mascarene Grey White-eye[153].

The *oiseau bleu* lived, at least latterly, in the open woodland with temporary marshy pools of the Plaine des Cafres at around 1,600–1,800m[154]. Now that the mystery of the 'solitaire' has been solved (see p. 30), the *oiseau bleu* is the most enigmatic bird from the old accounts. It was a quasi-flightless bird the size of a 'large capon', blue with red bill and legs[155]. The description could fit the Purple Swamphen *Porphyrio porphyrio*, but the *oiseau bleu* lived, not in lowland swamps as befits the gallinule, but in subalpine 'forest-steppe'. Furthermore the birds escaped hunters by running, neither flying (though they could) nor hiding, which is the Purple Swamphen's tendency[156]. Although many writers have considered that these birds were Purple Swamphens[157], a species common in Madagascar, it seems more probable that the *oiseau bleu* was an endemic derivative of this bird, rather than identical to it. Purple Swamphens would surely also have occupied typical habitat such as the Étang de Saint-Paul, where *oiseaux bleus* were never reported. Dubois classed them as terrestrial birds (i.e. not waterbirds like ducks), and everyone else said they only lived on the Plaine des Cafres. This situation seems analogous to the 'solitaire', which in Réunion did not inhabit typical ibis habitat (wetlands again), but lived in the forest, and was also classed by Dubois as a land-bird. A possible explanation is that these two colonised Réunion before any wetlands developed, and by the time geomorphological processes had created swamps they had become irretrievably adapted to other types of habitat. It is rather surprising

Figure 3.3. Impressions of the different habitats in Réunion, drawn by Nicolas Barré for his paper on bird ecology (Barré 1983). The letters code for: S: dry savannah (now entirely of introduced species), LF: lowland mixed wet forest, UF: upland mixed wet forest, T: tamarin forest, SF: very wet screw-pine forest, H: high-altitude heath. The tree heights are to scale.

that neither colonised Mauritius, but perhaps the pre-existence of Red Hens prevented their establishment. Wild-breeding Purple Swamphens were established in Mauritius from the mid-1800s until the 1960s (Chapter 7), and possibly in Réunion in the 1800s, but the evidence suggests they were introduced[158], though additional vagrants from Madagascar cannot be excluded. No subfossil material of the *oiseau bleu* has yet been found; the bird's diet and habits are unknown.

The Réunion Tortoise

Unlike the other two islands, Réunion had only one species of tortoise, found most abundantly on the dry west coast, but also high up in the interior[159]. It has been claimed that these animals were absent from the wetter windward side, but the discovery of Melet's manuscript confirms Tatton's observation in 1613 that they were also numerous in the northeast, at least around Saint-Denis[160]. Remains of Réunion Tortoises, notably bigger than those in Rodrigues and especially Mauritius, vary from domed to high-backed, but DNA analysis shows they were all from a single population.

This variability is similar to that found in the Galápagos, whereas on the other Mascarenes the pairs of species present were distinguished by being either high-backed or domed[161]. Only one visitor mentioned the tortoises' food – leaves fallen from trees[162] – but we assume they also grazed, browsed and ate fallen fruit. In the dry west tortoises may have had to make seasonal migrations to higher and wetter country to find sufficient food, as they do in the Galápagos[163], returning to the coast to nest on the sandy beaches; others appear to have lived and bred in the mountains (p. 88).

Native mammals

It is from Réunion that we have good early accounts of the habits of the Black-spined Flying-fox and Rougette. Administrator and keen naturalist Jean-Baptiste de Lanux, disturbed that Buffon had followed tradition in including fruitbats amongst the carnivores in his classification, wrote him a long and informative letter in 1772 that Buffon printed in 1776[164]. He emphasized first and foremost that these bats ate nothing but fruit and flowers, not specifying the fruit (apart from cultivated ones) but reporting that in January and February (mid-summer) the flowers of the '*bois puant*' *Foetidia mauritiana* attracted them to coastal areas in huge numbers to feast on the nectar, scattering the stamens beneath the trees. For the Black-spined Flying-fox he noted feeding by day as well as by night, but said only odd ones flew around in the daytime unless disturbed. Mating generally took place in May (early winter), with young born in late October, becoming free-flying by the following mid-winter solstice, i.e. mid-June[165]. The Black-spined bats roosted as expected on tree branches, but the smaller Rougette, which was strictly nocturnal, roosted in groups of up to 400 inside hollow trees – he was told (but could not verify) that these groups were of females accompanied by a single male[166]. Both species were said to be at times very fat, but Lanux only specified the dates for the Black-spined Flying-fox: summer and early autumn. He noted that Black-spined Flying-foxes sometimes flew very high, and suggested they might easily make the crossing to and from Mauritius.

Bory de St Vincent reported seeing in 1801 a tiny all-white bat roosting in latan palms in Réunion, this remains unidentified and cannot be associated with any known species. Close associations of bats with specific trees are known elsewhere; there is a small all-white bat in central America that roosts only in *Heliconia* leaves, while the Malagasy endemic *Myzopoda* species apparently require Travellers' Palms *Ravenala* to roost in[167]. Réunion had three other microbats, two (Grey Tomb and Mascarene Free-tailed) the same as in Mauritius, but there were also Pale House Bats *Scotophilus borbonicus*, now vanished, said to have been prevalent at higher altitudes[168]. Mascarene Swallows and Mascarene Swiftlets still occupy the daytime aerial niche, as in Mauritius.

Seabirds

The highest parts of Réunion are home to two endemic seabirds, Barau's Petrel and the Réunion Black Petrel. So difficult is the terrain on the highest crumbling cliffs of the Piton des Neiges that Barau's Petrel nests were not actually found until 1995, although the population can be counted in the thousands (Chapter 8)[169]. The Réunion Black Petrel is known from only a few specimens spanning the last 150 years, and may always have been rare, as it was only once reported before the 19th century[170], while Barau's Petrel was abundant enough to have been systematically trapped using flares, as reported by Bory in 1801 (Chapter 6). The birds are attracted to lights at night, crash-land, and are easily caught (Chapter 8)[171]. Large numbers of Tropical Shearwaters and some Wedge-tailed Shearwaters also nest on cliffs and steep slopes on the coast and inland. As Réunion has only one small offshore stack (Petite Ile), and no shallow marine shelf, there are only limited opportunities for other seabirds. A few Brown Noddies and Wedge-tailed Shearwaters nest on Petite Ile, and the noddies also possibly breed on mainland cliffs at Cap Méchant, though in the 19th century there was a huge noddy roost on the cliffs near Saint-Denis. The only other seabird is the ubiquitous White-tailed Tropicbird, which, as on the other two islands, nests scattered on cliffs throughout the island[172].

The rapid drop into deep water around the island and the very small area of reef lagoon made Réunion unsuitable habitat for Dugongs, whose absence was noted by travellers. Green Turtles nested commonly on the beaches, especially at Saint-Paul; Hawksbills were only mentioned by one early visitor, Bellanger in 1691, but probably bred in small numbers[173].

THE FAUNA OF PRISTINE RODRIGUES

Apart from the absence of snakes, skinks and hawks, the reptile and large-bird fauna of Rodrigues closely matched that of Mauritius, while small birds were poorly represented. The aerial feeders, swifts and swallows were apparently absent[174], and there were only four known species of arboreal passerines, two of which became extinct before being recorded in life. While the palm fruits supported Solitaires, two species of parrot, and a large, frugivorous day-gecko, the same abundant geckos plus snails fed the endemic night heron, owl and Leguat's Rail; the rail also ate tortoise eggs, dug up from the ground[175]. The large, nocturnal and carnivorous Liénard's Giant Gecko probably preyed on smaller lizards and on birds and their nests, and is known to have raided seabird eggs and young on islets[176]. The two species of giant tortoise, a high-backed browser and a domed grazer (as in Mauritius), were extraordinarily abundant, keeping the understorey open and penetrable, and competing with Solitaires for fallen latan palm fruit. The seabird colonies on the offshore islets supported an endemic pied starling that was apparently an egg specialist and opportunist scavenger of carrion[177]. A further four geckos are known as subfossils but were not seen by visitors.

The Solitaire

Leguat was so impressed by the Solitaire that he wrote a three-page essay on its habits; this is one of the first coherent observational accounts of animal behaviour in the wild ever published[178]. He described the differences in the sexes, growing a little nostalgic about claimed breast-like tufts of feathers on the hens' thorax; males were greyish and brown, weighing up to 20kg (45lb), females were paler brown or 'blond', and much given to preening[179]. They had black eyes, and a band above the tan-coloured bill like a 'widow's headband', clearly shown in Leguat's rather crude woodcut (Box 17). They were monogamous, made nests 30–45cm high out of fallen palm leaves, and laid a single egg, larger than a goose's, which took seven weeks to hatch, during which time both sexes incubated in turn[180]. They then looked after their offspring for several months before it joined a flock of other young, the adults then returning to their territories. The males displayed vigorously, twirling on the spot and making a rattling sound with their wings that was audible for 200 paces. They defended their territories, especially when they had small nestlings, fighting off intruders (including humans) using the round, musket ball-like mass on their wings and their sharp beaks as weapons. According to Leguat only males would chase off males, and females chased females. They were fat from March through September, and every bird had a large stone in its gizzard (which Leguat and his friends used to sharpen their knives!). They could run swiftly, outrunning men in dense forest though not in more open areas[181]. Dodos likewise swallowed single gizzard stones, which may imply similarities between these two related birds' diets – though the Solitaire had a much smaller, more 'normal' bill. Leguat was amazed to find that even 'nestling' Solitaires had gizzard stones; he assumed they hatched with them *in situ*, but presumably the adults in fact fed them the stones early on to help them digest their food. Despite nesting on the ground, Solitaires were clearly nidicolous as are their pigeon relatives, though Leguat says nothing about how the young were fed; other pigeons provide hatchlings with pre-digested 'pigeon milk'[182].

Other endemic land birds

Leguat's Rail, related to the larger Red Hen of Mauritius, was a flightless grey bird with a red bill and legs, and a red wattle around the eye[183]. Leguat bemoaned his inability to find their well-hidden nests, so was unable to sample their eggs. He described their reaction to the colour red in similar terms as travellers to Mauritius did for the Red Hen (Box 11, p. 127), though Tafforet did not remark on it. They had decurved bills of very variable length, the longer ones more decurved than the shorter. Leguat stated that the bill was 2" (5cm) long and straight, whereas Tafforet called it "more or less like a curlew but a bit thicker and not quite as long". This disparity is reflected in the bones, but it is odd that each observer 'saw' only the one sort. Although Leguat said the sexes were similar, the variation in subfossil bones led Günther and Newton to suggest possible dimorphism in bill size[184], which may also have been reflected in their diet. Both Leguat and Tafforet said they were very fat, Tafforet saying that at times it prevented them from running, which they were usually good at. Since they fed on rich chelonian eggs, a good deal of fat is to be expected; this may have been seasonal.

Rather less was said about the Rodrigues Night Heron, but that is is still more than we know about the Mauritian or Réunion species. Leguat noted that they were particularly fond of lizards (Box 6, p. 83; Box 33, p. 243), and Tafforet commented that they rarely flew, but were able to run very well when chased. Neither traveller described their plumage.

Leguat enjoyed the company of the then-abundant slate-coloured pigeons – they flocked around his table eating melon seeds he rejected, though he said nothing of their natural food. He noted, as did Tafforet, that they only nested on offshore islands. Leguat perceptively assumed this was to avoid rats, which had not then reached the lagoon islets. Subfossil bones indicate there were originally two pigeon species present, the Malagasy Turtle Dove and another smaller but related species, the Rodrigues Dove. Whether both species survived (and were not distinguished) or one was already extinct when Leguat arrived we may never know[185]. Tafforet, the last to see them, did not describe his *tourterelles*.

There were at least two species of parrot on Rodrigues, though Tafforet described three (Box 20, p. 181). As only two kinds have been found as subfossils, we provisionally accept, despite a size discrepancy, the interpretation of Tafforet's bird with red wing patches as a colour morph (or simply a full adult male) of the Rodrigues Parakeet, which is known to have been turquoise[186]. We have no other source for the red-winged parrot, as Leguat gave a rather unclear description of "blue and green parrots

... especially of mediocre and equal size", while Pingré in 1761 mentioned *perroquets* but only described entirely green *perruches*[187]. Leguat's Parrot was a heavy-headed bird somewhat resembling the Raven Parrot, but smaller and less of a specialist; Leguat kept several captive without difficulty, even training them to talk. Tafforet reported that they ate the seeds of an (unidentified) lemon-scented shrub on the southern islets, as well as the berries of *bois de buis* (*Fernelia buxifolia*) on the main island, while Leguat noted that they liked the nut of a common tree with "a fruit like enough to an olive" (*Cassine* [= *Elaeodendron*] *orientale*). Neither observer mentioned palm fruit in connection with parrots, though they probably took at least the smaller ones on the tree before fruitfall, the resident fruitbats likewise. The Rodrigues Parakeet, the size of a Ring-necked Parakeet *Psittacula krameri*, may have fed extensively on leaves as does the Echo Parakeet in Mauritius; it outlived Leguat's Parrot, surviving well after the forests had been devastated, implying a less vulnerable ecology[188].

The lizard-owl was said by Tafforet to live in trees and to eat small lizards and small birds; it is mentioned only in passing by Leguat, as his only ally against rats. Tafforet reported that it sang only on fine nights, but not in bad weather.

The absence of rival small passerines can be seen in the adaptations of the noisy and conspicuous Rodrigues Fody[189], which, in addition to occupying niches similar to the Mauritius Fody on that island, has developed an elaborate brush-tongue to exploit nectar as efficiently as the olive white-eyes on Mauritius and Réunion[190]. By contrast, the secretive insectivorous Rodrigues Warbler, the only member of its subfamily in the Mascarenes, is little differentiated from its close relative in the Seychelles and indeed other members of the genus *Acrocephalus*[191]. Apart from the Rodrigues Starling mentioned above, the only other passerines were a black bulbul (or merle) and an unidentified form, both known only from subfossil bones and apparently extinct before the first reports[192]; the merle's ecology was no doubt similar to those of its congeners on the other islands.

The first visitors to land on Rodrigues reported 'geese' as well as doves, parrots, 'dodos' (i.e. Solitaires) and 'other birds'[193]. However, given that the report was second-hand, 'geese' was probably a term sailors had used for some other large bird, most likely boobies, which are goose-sized with webbed feet. No bones of any duck or goose have been found in Rodrigues, and there was no suitable wetland habitat[194].

Bats

Although no visitor noticed more than one species of fruitbat, subfossil evidence shows that both Golden Bats and Black-spined Flying-foxes were present at one time[195]. We cannot be sure whether the now vanished Black-spined Flying-fox, better known from Mauritius and Réunion, was still around in 17th century Rodrigues and overlooked, or had already disappeared. At its present size Rodrigues is very small for supporting two species of flying-fox, so they must have colonised when the island was larger during a period of lower sea level. However, undisturbed, and perhaps with somewhat divergent habits (they are different in size), both might have persisted until human interference disturbed the balance. Only Abbé Pingré, in 1761, gave a description of the bats' coloration[196] – his account rules out the Black-spined Flying-fox, and is consistent with the Golden Bat. However Leguat said he saw bats whose wings were (each) two feet long (French measure: 64cm = span of 128cm), whereas Pingré said his were 1'–1'6" (32–48cm, spans of 64–96cm)[197]. These perceived differences may refer to the different species, the actual wing-spans being 90–102cm for the Black-spined Flying-fox, 66–76cm for the Golden Bat[198]. Pingré's size estimate, like his pelage description, matches the Golden Bat, while Leguat's is too large for either, but does rather imply he saw Black-spined Flying-foxes. Leguat, the only early visitor to comment on diet, reported that the bats were very fond of wild figs [199]. Another native tree is known locally as *bois chauve-souris*: "bats feed greatly on its fruits"[200]. Contrary to the accepted modern view of flying-fox reproduction, Leguat reported that the bats had two young at a time, an observation also supported for Golden Bats by a 20th-century visitor – in fact twins have recently been recorded both in the wild and in captive populations, so it is possible that under the right conditions this may be regular[201].

Land reptiles

Large nocturnal lizards as long as a man's arm (Leguat) and as thick as a man's wrist (Tafforet) must have been the giant gecko collected by Captain Descreux on Frégate Island in 1842, now known to have been not just the largest *Phelsuma* day-gecko, but one of the largest geckos of any kind[202]. The smaller, brighter diurnal lizards about a foot long confused early reporters by seeming to be all sorts of colours and possibly species, but in 1761 Abbé Pingré finally discovered the truth – they could change colour with startling speed[203]. The bright-to-black lizards were Newton's Day-gecko, which Leguat said normally ate palm fruit (and the melons at his table). Day-geckos are normally insectivorous with a strong interest in nectar[204], but Newton's Day-gecko's teeth suggest frugivory, consistent with Leguat's report, though Marragon reported it eating a lot of insects[205]. Was there a shortage of insects when the ancestor of these geckos first colonised Rodrigues?

BOX 3 — BOOBIES (SULIDAE)

Boobies breed on many island archipelagos in the Indian and Pacific Oceans; at least three species once inhabited the Mascarenes. The largest species, the tree-nesting Abbott's Booby *Papasula abbotti*, once occurred on Mauritius and Rodrigues; old accounts have been confirmed by rare fossil remains. This species now breeds only on Christmas Island. The Masked Booby, *Sula dactylatra*, still breeds on Serpent Island in the North. The islets off Rodrigues, particularly Ile Frégate, once harboured large colonies of boobies; at least two species, Abbott's Booby and Red-Footed Boobies *Sula sula* bred there. As boobies were hunted for meat and eggs, this may have been the reason why they were quickly eliminated from mainland Rodrigues and most of the offshore islands. Abbott's Booby probably died out in the 1830s, but Red-footed Boobies survived until at least the mid-1870s.

Accounts
John Marshall (Khan 1927, Cheke 1987a), in Mauritius in 1668:

> *I see upon the Island 2 birds by a nest upon a very high tree. They were much bigger than geese as seemed to mee, had long beakes and nests [necks?] and were of a whitish colour. [probably Abbott's Boobies]*

Tafforet (1726) in Rodrigues in 1726:

> *The Boeufs [Abbott's Boobies] are the size of a good capon; their plumage is all white with the exception of the wing and tail feathers which are black; its beak is about 5 inches long, coming to a point at the tip, and inside it is like a saw; it is called boeuf because it calls like an ox; it often makes a noise with its wings when flying, that one might think was a gust of wind if it continued after the bird had passed; they normally lay on the branches of trees where they make their nests, and the male and female take turns to incubate the egg, for they lay only one, while one or the other goes fishing. The [Red-footed Booby] is so called because it always calls like that; it is a bird which is not as big as a boeuf, has a beak approaching that of the boeuf, and is grey, a bit white on the belly; they perch and make their nests in trees and incubate by turns, but are in larger numbers than the boeufs. When they are little they are all white and the beak all black, and when they are big they are grey with the beak greenish; the frigates dare not approach them when they are landed on the trees, or in the water where they defend themselves, and once left alone they fly off to go to where they have their nests, and never make a mistake although it is often nearly dark when they arrive; one sees them coming in prodigious quantities from 4 in the afternoon until nightfall.*

Pingré (1763) in Rodrigues in 1761:

> *I can find no traveller who has spoken of the boeuf, at least by this name. It is a bird bigger than a [domestic] duck, which it resembles in the general form of the body; its beak is very strong, about as long as a duck's, pyramidal in shape, or rather conical, ashy coloured, with a hint of red; the point, which is a bit hooked, is black; the eyes, which are precisely where the beak ends, are fine, large and black; the neck and all the body is covered with dazzlingly white down; the feathers of the wings and tail are black; the colour of the feet is blackish grey; the toes are joined by a membrane or web. The animal does not stand tall, its cry is very raucous, somewhat resembling the lowing of an ox. The flesh is almost black; its taste, approaching that of our [= French] seabirds, is not unacceptable. we found the flesh tough, it is true, but that is because we did not give it time to tenderise . . . Tratras are so called because of their cry. I only saw very young ones; they were covered with down, extremely white. I would not be sure that this whiteness was unchangeable."*

Specimen collections
Fossil evidence, albeit scarce, confirms that boobies were once resident on the Mascarenes. Only two fossil wing bones of Abbott's Booby exist, which are kept in the Mascarene fossil collections at UMZC (collected from the Mare aux Songes) and UCB (collected on Rodrigues). Fossil material of *Sula* sp. are housed at BMNH and UMZC.

Abbott's Booby *Papasula abbotti* displaying. From Nelson (1971).

The early accounts did not report the Giant Gecko's diet, but Liénard was told that on Ile Frégate (by then its only location) it ate both figs and seabird's eggs, and even sucked the blood of the nestlings.

Four more species of lizard have been identified amongst the subfossil bones from the caves on Plaine Corail. Two are night-geckos in the genus *Nactus*, both larger than any of the others in the genus[206], and the other two remain undescribed, though one appears to be a very unusual form[207]. There is no unequivocal mention of night-geckos in the early accounts, but they may have been the lizards mentioned by Tafforet as owl food, as a collection of their bones, possibly from owl pellets, has been found in a niche in a cave on Plaine Corail[208].

Both Leguat and Tafforet stated that there were three species of tortoise, but failed to describe their differences, Tafforet confining himself to saying that the largest he saw measured 3' to 3'8" in carapace length (French measure: = 0.97–1.19m), while Leguat estimated maximum size by weight: "around 100 [French] lbs" (48.5kg)[209]. Pingré recorded two sorts, large ones, relatively scarce, called *carosses*, and smaller, unnamed ones[210]. Subfossil remains and surviving museum specimens indicate only two species, a large high-backed one weighing up to 60kg, and a smaller domed sort of around 12kg[210]. Tafforet said they ate fallen leaves and tree seeds, Leguat and Pingré naming specifically the fallen fruit of latan palms. Tafforet observed that they were not very fat, attributing this to their huge numbers[211] and shortage of grass, and related the higher density in valleys to a shortage of water in the dry season.

Seabirds

Unlike Mauritius, where seabirds were largely ignored both by settlers and visiting ships, we are very well informed about the rich variety inhabiting the many islets in the Rodrigues lagoon. Leguat and his companions had no boat, but Tafforet did, and he used it to explore and describe each islet in turn. Rodrigues boasted more seabirds, both in species and numbers, than either Mauritius or Réunion: two boobies, frigatebirds (possibly two species), at least three petrels, the two tropicbirds and five species of tern. Of these the tropicbirds and two gadfly-petrels nested on the main island, while the others nested on islets. Ile Frégate, rocky with tree cover, was where both Abbott's and Red-footed Boobies nested, together with large numbers of frigatebirds and Wedge-tailed Shearwaters, while Sooty Terns and the two noddies favoured the large flat calcarenite islands in the south, Gombrani and Pierrot (= Chat), where there were also a few Fairy Terns. Noddies and shearwaters also nested on several of the smaller islets[212]. Pingré's description of the *boeuf* (Box 3) is so precise that it confirmed the former presence of Abbott's Booby, now confined to Christmas Island off Java, and from Tafforet and Leguat we know that the commoner Red-footed Boobies were largely of the white-tailed brown morph[213]. The only frigatebird collected in Rodrigues is a Lesser Frigatebird *Fregata ariel*[214], but it is likely that Great Frigatebirds *F. minor* were also originally present[215]. Frigatebirds and boobies are generally held to have nested only on Ile Frégate, but Pingré reported both on Ile Coco in 1761, and in March 1846 large numbers of boobies were roosting on Ile Crabe[216]. Tafforet recognised both tropicbirds, nesting in trees and cliffs on the main island, as they still do[217]. According to Tafforet, Sooty terns, his *equerets*, nested only on Pierrot, though Pingré reported large numbers on Gombrani and also a colony on Coco – no doubt they moved sites at times. Tafforet included as a 'seabird' the *sentinelle* which "fished on the banks of streams or pools", were "blackish mixed with greyish white" and "flew up calling incessantly" when disturbed. The Striated Heron, still common, best fits this description; it appears not to have colonised the other islands at this date[218].

Marine animals

The wide expanse of lagoon around Rodrigues was originally a haven for large numbers of Dugongs, and the beaches provided ample space for both Green and Hawksbill turtles to lay their eggs. Leguat reported Dugongs up to 20 (French) feet long (6.5m), and herds of three or four hundred "grazing like sheep" in 3–4 feet (1–1.3m) of water. Tafforet's figures were more modest: herds of 30–40, the largest individuals being 15–18 feet (5–6m) long[219]. Dugongs graze on marine grasses, and the seagrass beds are still there, albeit now degraded, in the Rodrigues lagoon[220]. Both Leguat and Tafforet described Dugongs suckling a single young, Leguat being very scathing about a standard encyclopaedia of his time that asserted they had twins[221]. Leguat's picture of the Dugong correctly indicated the bilobed horizontal tail fluke and the presence of tusks[222].

Green Turtles were also abundant, these large reptiles laying up to 200 eggs at a time on sandy beaches, always at night. Leguat asserted they laid 1,000–1,200 eggs each year from several landings, the eggs taking six weeks to hatch in the warm sand; these figures are in line with current knowledge[223]. The heavy toll taken by "frigate-birds, boobies and other seabirds" was estimated at 90% by Leguat, while Tafforet added that few escaped sharks and other fish. Pingré was told that the main laying season was in October and November; neither Leguat not Tafforet, while mentioning seasonality, gave dates. Pingré was the only early naturalist to mention '*carrets*'[224], the standard local name for Hawksbills, the source of commercial tortoiseshell. Hawksbills must surely have been frequent nesters on Rodriguan shores, as they still turn up in reasonable numbers even today[225].

Crabs

Although outside the main scope of this book, we should mention land crabs, crustaceans that were of major concern to Leguat, and clearly enormously important in the lowland ecology of pristine Rodrigues[226]. Apart from rats, crabs were the principal seedling predators in the Huguenots' vegetable gardens. Their numbers were prodigious, living in extensive underground galleries that Pingré said made walking dangerous in areas near the shore, though they were absent on higher land. At the July and August full moons they flocked in their thousands to the sea to lay their eggs. Leguat said his group could kill three thousand in a night and see no diminution; 70 years later Pingré reported thousands killed by every ship's crew that arrived, again without any visible reduction. This crab-dominated ground fauna can still be seen today on Aldabra and Christmas Island, where the crabs are the most important scavengers and detritus feeders, recycling everything organic that falls on the

Table 3.1 The native tetrapod fauna of the Mascarenes.

Symbols after each taxon:

§ = family or subfamily endemic to the Mascarenes ▲ = genus endemic to Mascarenes ▲ = genus endemic to island
● = species endemic to Mascarenes ● = species endemic to island ○ = subspecies endemic to Mascarenes
X = species totally extinct

The following sections are separated by a / and appear under each island:

Species status
S = survives   E = extinct   [ ] = indicates some uncertainty   ES = extinct on main island, but survives on offshore islets
= = never present, but related form existed or exists; see adjacent species

Abundance code
c = common   f = fairly common   r = rare   v = rare and vulnerable   e = endangered   i = offshore islets only

Records (bracketed where there is doubt)
o = observed by visitors   [o] = probable sighting but no description   s = specimens collected and still preserved
[s] = specimen(s) of uncertain provenance   d = good illustration or description of a specimen
b = bones found in subfossil deposits   exists, but no specimen is known to survive

The principal references for this table are as follows. Bats: Hume 2005, Cécile Mourer pers. comm.; birds: Mourer *et al.* (1999, Réunion), Newton & Gadow (1893, Mauritius), Günther & Newton (1979, Rodrigues), supplemented by Cowles (1987, 1994), Hume (2005a,b) and Bourne (1968, 1976); reptiles: Arnold (1980, 2000), Austin & Arnold (2001, 2006), Austin *et al.* (2004a).

	Species	Réunion	Mauritius	Rodrigues
Mammals — fruitbats	Black-spined Flying-fox *Pteropus niger* ●	E/odb	S/v/osb	E/[o]b
	Golden Bat *Pteropus rodricensis* ●		E/b	Sv/osb
	Rougette *Pteropus subniger* ●X	E/odb[s]	E/osb	
Mammals — microbats	Mascarene Free-tailed Bat *Mormopterus acetabulosus* ●	S/c/os	S/c/os	
	Grey Tomb Bat *Taphozous mauritianus*	S/c/os	S/c/os	
	Pale House Bat *Scotophilus borbonicus*	E/s		
	Bory's White Bat '*Boryptera alba*' [▲]	E/o		
	Dugong *Dugong dugon*		E/o	E/o
Birds — petrels and shearwaters	Wedge-tailed Shearwater *Puffinus pacificus*	S/f/osb	S/ci/os	S/ei/osb
	Tropical Shearwater *Puffinus bailloni*[1]	S/c/osb	E/osb	
	'Round Island' (Trindade) Petrel *Pterodroma arminjoniana*[2]	=	S/ri/os	=
	Kermadec Petrel *Pterodroma neglecta*[2]	=	S/ri/os	=
	Barau's Petrel *Pterodroma baraui* ●[3]	S/f/osb	=	[E]
	Bourne's Petrel *Pterodroma* sp. (undescribed) ●X[4]	=	=	E/ob
	Réunion Black Petrel *Pseudobulweria aterrima* ●	S/e/os		E/ob
	Bulwer's Petrel *Bulweria bulwerii*		S/ri/o	
Birds — cormorants and allies	White-tailed Tropicbird *Phaethon lepturus*	S/f/osb	S/c/osb	S/r/osb
	Red-tailed Tropicbird *Phaethon rubricauda*		S/ci/os	S/r/o
	Abbott's Booby *Papasula abbotti*		E/ob	Ei/ob
	Red-footed Booby *Sula sula*			Ei/osb
	Masked Booby *Sula dactylatra*		S/vi/os	
	Reed Cormorant *Phalacrocorax africanus*[5]	E/o	E/b	
	Great Frigatebird *Fregata minor*[6]		[E]/o	[E]/i/o
	Lesser Frigatebird *Fregata ariel*[6]		[E]/o	E/i/os
Birds — herons and allies	Réunion Night Heron *Nycticorax duboisi* ●X	E/ob	=	=
	Mauritius Night Heron *Nycticorax mauritianus* ●X	=	E/ob	=
	Rodrigues Night Heron *Nycticorax megacephalus* ●X	=	=	E/ob
	Dimorphic Egret *Egretta dimorpha*	E/os	E/ob	
	Striated Heron *Butorides striata*[7]	S/r/os	S/c/os	S/c/os
	Réunion Ibis (Réunion Solitaire) *Threskiornis solitarius* ●X	E/ob		
	Greater Flamingo *Phoenicopterus roseus*	E/ob	E/ob	
Birds — ducks & geese	Réunion Sheldgoose *Alopochen kervazoi* ●X	E/ob	=	
	Mauritius Sheldgoose *Alopochen mauritianus* ●X	=	E/ob	
	Mascarene Teal *Anas theodori* ●X	E/ob	E/ob	
	pochard *Aythya* sp. (probably referrable to *A. innotata*)	E/b		

Table 3.1 (cont.):

	Species	Réunion	Mauritius	Rodrigues
Birds / raptors	Réunion Harrier *Circus maillardi* • [8]	S/f/os	E/[o]b	
	Dubois's Kestrel *Falco duboisi* •x	E/ob	=	
	Mauritius Kestrel *Falco punctatus* •	=	S/f/os	
gulls and terns	Roseate Tern *Sterna dougallii*			S/ei/os
	Sooty Tern *Sterna fuscata*[7]		S/ci/os	S/fi/os
	Fairy Tern *Gygis alba*			S/vi/os
	Great Crested Tern *Thalasseus bergii*			Ei/os
	Brown Noddy *Anous stolidus*	S/ri/os	S/ci/os	S/ci/os
	Lesser Noddy *Anous tenuirostris*		S/ci/os	S/ci/os
rails and allies	Dubois's Wood-rail *Dryolimnas augusti* •x	E/ob	=	
	Sauzier's Wood-rail *Dryolimnas* sp. (not described) •x [9]	=	E/b	
	Red Hen *Aphanapteryx bonasia* ▲•x		E/odb	=
	Leguat's Rail *Erythromachus leguati* ▲•x		=	E/ob
	Mascarene Coot *Fulica newtoni* •x	E/ob	E/ob	
	Common Moorhen *Gallinula chloropus*[7]	S/f/o	S/c/os	
	'oiseau bleu' ?*Porphyrio* sp. [•]x	E/o		
	Madagascar Buttonquail *Turnix nigricollis*[10]	S/c/os		
pigeons[11]	Dodo *Raphus cucullatus* ▲•x [11]		E/osb	=
	Rodrigues Solitaire *Pezophaps solitarius* ▲•x		=	E/ob
	slaty pigeon ?*Alectroenas* sp. •x	E/o	=	
	Pigeon Hollandais *Alectroenas nitidissima* •x	=	E/os	
	Réunion Pink Pigeon *Nesoenas duboisi* •x	E/ob	=	
	Mauritius Pink Pigeon *Nesoenas mayeri* •	=	S/e/osb	=
	Rodrigues Dove (?*Nesoenas*) *rodericana* •x	=	=	E/ob
	Malagasy Turtle Dove *Nesoenas picturata*[7]	S/c/osb	S/c/osb	E/b
parrots	Raven Parrot *Lophopsittacus mauritianus* ▲•x		E/ob	=
	Leguat's Parrot *Necropsittacus rodricanus* ▲•x		=	E/ob
	Echo Parakeet *Psittacula eques* •	E/od	S/e/osb	=
	Rodrigues Parakeet *Psittacula exsul* •x	=	=	E/osb
	Thirioux's Grey Parrot *Psittacula bensoni* •x [12,13]	E/o	E/ob	=
	Dubois's Parrot '*Psittacula*' (?)*borbonicus* •x[13]	E/o	=	=
	Mascarin *Mascarinus mascarinus* ▲•x	E/osb		
owls	Réunion Lizard-owl *Mascarenotus grucheti* ▲•x	E/b	=	=
	Commerson's Lizard-owl *Mascarenotus sauzieri* ▲•x	=	E/odb	=
	Tafforet's Lizard-owl *Mascarenotus murivorus* ▲•x	=	=	E/ob
	Mascarene Swiftlet *Aerodramus francicus* •	S/c/os	S/v/os	
passerines	Mascarene Swallow *Phedina borbonica* ○	S/f/os	S/f/os	
	Réunion Cuckoo-shrike *Coracina newtoni* •	S/v/os	=	
	Mauritius Cuckoo-shrike *Coracina typica* •	=	S/v/os	
	Réunion Bulbul (Réunion Merle) *Hypsipetes borbonicus* •	S/c/o	=	=
	Mauritius Bulbul (Mauritius Merle) *Hypsipetes olivaceus* •	=	S/v/os	=
	Rodrigues Bulbul (Rodrigues Merle) *Hypsipetes* sp. (undescribed) •x	=	=	E/b
	Mascarene Paradise Flycatcher *Terpsiphone bourbonnensis* •	S/c/os	S/e/os	
	Réunion Stonechat *Saxicola tectes* •	S/c/os		
	Rodrigues 'Babbler' (undescribed) ▲•x			E/b
	Rodrigues Warbler *Acrocephalus rodericanus* •			S/e/os
	Mascarene Grey White-eye *Zosterops borbonicus* •	S/c/os	S/c/os	
	Réunion Olive White-eye *Zosterops olivaceus* •	S/c/os	=	
	Mauritius Olive White-eye *Zosterops chloronothos* •	=	S/e/os	
	Réunion Fody *Foudia delloni* •x [14]	E/o	=	=
	Mauritius Fody *Foudia rubra* •	=	S/e/os	=

Table 3.1 (cont.):

	Species	Réunion	Mauritius	Rodrigues
Birds	Rodrigues Fody *Foudia flavicans* ●	=	=	S/v/os
	Hoopoe Starling *Fregilupus varius* ▲●X	E/osb		=
	Rodrigues Starling *Necropsar rodericanus* ▲●X	=		E/ob
chelonians	Réunion Tortoise *Cylindraspis indica* ▲●X	E/osb	=	=
	Mauritius High-backed Tortoise *Cylindraspis triserrata* ▲●X	=	E/ob	=
	Carosse Tortoise *Cylindraspis vosmaeri* ▲●X	=	=	E/osb
	Mauritius Domed Tortoise *Cylindraspis inepta* ▲●X		E/ob	=
	Rodrigues Domed Tortoise *Cylindraspis peltastes* ▲●X		=	E/osb
	Green Turtle *Chelonia mydas*[15]	E/o	E/os	E/o
	Hawksbill Turtle *Eretmochelys imbricata*[15]	E/o	E/os	E/os
	Loggerhead Turtle *Caretta caretta*[15]		E/od	
Reptiles — lizards	Telfair's Skink *Leiolopisma telfairii* ▲●X [16]		ES/ri/osb	
	Arnold's Skink *Leiolopisma* sp. (undescribed) ▲● [16]	E/b	=	
	Didosaurus *Leiolopisma mauritiana* ▲●X	=	E/b	
	Bojer's Skink *Gongylomorphus bojerii* ▲● [16]	=	ES/ci/osb	
	Réunion Slit-eared Skink *Gongylomorphus borbonicus* ▲●X [16]	E/os	=	
	Macabé Skink *Gongylomorphus fontenayi* ▲● [16]		S/r/os	
	Bouton's Skink *Cryptoblepharus boutonii* ○ [17]	S/e/os	S/f/os	
	Mauritius Night-gecko *Nactus serpensinsula* ● [18]	=	ES/ri/osb	=
	Réunion Night-gecko *Nactus* sp. (undescribed) ●X [18]	E/b	=	=
	Small Rodrigues Night-gecko *Nactus* sp. (undescribed) ●X	=	=	E/b
	Lesser Night-gecko *Nactus coindemirensis* ●		S/ri/os	
	Giant Night-gecko *Nactus* sp. (undescribed) ●X			E/b
	Mourning Gecko *Lepidodactylus lugubris*			S/c/os
	Réunion Forest Day-gecko *Phelsuma borbonica* ○ [19]	S/f/os	=	
	Blue-tailed Day-gecko *Phelsuma cepediana* ● [20]	=	S/c/os	
	Upland Forest Day-gecko *Phelsuma rosagularis* ● [20]		S/f/os	
	Manapany Day-gecko *Phelsuma inexpectata* ●	S/v/os	=	
	Vinson's Day-gecko *Phelsuma ornata* ●	=	S/c/os	
	Guimbeau's Day-gecko *Phelsuma guimbeaui* ● [20]		S/f/os	
	Günther's Day-gecko *Phelsuma guentheri* ●		ES/ei/os	
	Newton's Day-gecko *Phelsuma edwardnewtoni* ●X		=	E/osb
	Liénard's Giant Gecko *Phelsuma gigas* ●X			E/odb
	undescribed gecko 1 [▲]●X [16,21]			E/b
	undescribed gecko 2 [▲]●X [16,21]			E/b
snakes	Keel-scaled Boa *Casarea dussumieri* §▲● [22]		ES/vi/osb	
	Burrowing Boa *Bolyeria multicarinata* §▲●?X [22]		E[Si]/os	
	Carié's Blind-snake '*Typhlops*' *cariei* ●X		E/b	

Notes on Table 3.1

1. Specimens of small shearwaters from Durban and Réunion were described as a new species, *Puffinus atrodorsalis* by Shirihai et al. (1995); however Bretagnolle & Attié (1996, 2000) have shown that the Réunion specimen is simply a juvenile *P. bailloni*, as conceded by Shirihai himself (2001). The Durban specimen remains enigmatic, but probably has no connection to the Mascarenes. Austin et al. (2004) have revised the small *Puffinus* shearwaters using DNA analysis, with the Indian Ocean forms, formerly *P. lherminieri*, becoming *P. bailloni*.
2. Brooke et al. (2000) have established that two species of gadfly petrels nest on Round Island. However, the identity of the commoner one, usually considered to be *Pterodroma arminjoniana*, is unresolved.
3. Barau's Petrel has been recorded breeding in Rodrigues; once, in 1974 (Jouanin 1987).
4. Cowles (1987) reported bones of an unknown large *Pterodroma* from Rodrigues, but did not name it.
5. The Little Cormorant *Phalacrocorax niger* is not ruled out, as subfossil bones were only compared to the African *P. africanus* (Olson 1975b).
6. The only frigatebird specimen dating from near the time when they still bred in the islands is a *F. ariel* collected by Bewsher, labelled 'Mauritius' but probably collected in Rodrigues in 1874 (Cheke 2001a). Bourne (1968) and Staub (1973) have argued that both species bred in Rodrigues, based on early visitors reports of

two breeding seasons, and on their feeding habits. There is no indication which species actually bred in Mauritius: frigatebirds of some sort ('*rabos forcados*') nested near Mahébourg Bay in the early 1600s (Chapter 2).

7. Some species appear to suffer extinctions followed by recolonisation: e.g. Striated Heron in Réunion (19th/20th centuries). The Sooty Tern has recently re-established itself in Rodrigues after nearly a century's absence (following massive persecution for its eggs). Common Moorhens may not have been present in Mauritius or Réunion before the late 18th century. It is unclear whether the Malagasy Turtle Dove, known subfossil from both Mauritius and Réunion, survived throughout in small numbers, or died out and was reintroduced from Madagascar.
8. The extinct harrier in Mauritius is now known to have been conspecific with the Réunion Harrier (Mourer *et al.* 2004).
9. Mauritian bones formerly referred to *Dryolimnas cuvieri* (Cowles 1987) are larger and are referable to a possibly flightless form allied to *D. augusti*. The status of the single record of *D. cuvieri* (the type specimen) is unclear – it may have been a vagrant (Chapter 7).
10. In the 1660s and 1670s several visitors to Réunion reported small '*perdrix*', which Dubois described as 'little grey partridges the size of quails'. These are likely to have been button-quails, but *Turnix* are not formally recorded until 1831 (Sganzin 1840).
11. The Dodo and the Rodrigues Solitaire are usually classed in the endemic family Raphidae, though they are sufficiently different for some to believe they should be split into a family each (Raphidae and Pezophapidae: see Cheke 1985). These differences based on morphology have been challenged by DNA studies which strongly suggest that both birds should be included within the Columbidae (Shapiro *et al.* 2002), a position we have followed.
12. The grey parrot of the early accounts was tentatively matched by Cheke (1987a) with the bones named *Lophopsittacus bensoni* by Holyoak (1973), but the type specimen (a sternum) is from a large *Psittacula* allied to *P. eupatria*, and so the name *bensoni* must now be reallocated to this genus (Hume 2005, Chapters 3 & 4).
13. Dubois's Parrot, which he described as similar in size to the Echo Parakeet, but 'having the head, upper-side of the wings and tail the colour of fire' was called *Necropsittacus borbonicus* by Rothschild (1907) – however, it is unlikely to be related to the much larger *Necropsittacus rodricanus* of Rodrigues, but sounds more like a *Psittacula* similar to *P. eupatria*, which would connect it to the recently identified bones of a similar sort from Mauritius (now called *P. bensoni*), and to the Rodrigues Parakeet *P. exsul*, of which some individuals were described by Tafforet (1726) as green with red in the wings (Hume 2005). Also, we follow Jones's recommendation (1987) and recent specialist literature (e.g. Thorsen & Jones 1998, Jones *et al.* 1999, Stattersfield & Capper 2000) in treating the entirely green *Psittacula* parakeets on Reunion and Mauritius as conspecific under the oldest name *eques* (given originally to a Réunion specimen), although the surviving Mauritian bird is commonly known as *P. echo*.
14. The previously unnamed species is given the scientific name *Foudia delloni* in Chapter 3.
15. Sea turtles are still seen around the islands and occasional breeding attempts by Green Turtles are reported (Cheke 1987a, Chapter 8), but the large colonies were wiped out in the 18th century. It is not known if Loggerheads bred, but the fact that the *Gelderland* crew caught at least two of different ages in 1601 (Moree 2001) suggests that they did, at least on Mauritius. We have not included the recent Green Turtle nesting in Réunion in the table, as it is not yet a fully re-established population.
16. The *Leiolopisma* and *Gongylomorphus* skinks in Réunion had diverged enough to be treated as full species (Arnold 2000, Austin & Arnold 2006, *contra* Arnold 1980). *G. borbonicus* is closer to *G. bojerii* than to *G. fontenayi* (Jeremy Austin, pers. comm.), and *G. fontenayi* in Mauritius includes the so-called 'Orange-tailed Skink' from Flat Island, which appears to be a race of it (Freeman 2003); it has not been formally described.
17. Bouton's Skink has been recently rediscovered in Réunion after not being seen for over a century (Probst 1999); it may have recolonised rather than been present throughout, but it is very small and notoriously easy to overlook.
18. The night-gecko on Réunion is considered to be specifically distinct from the larger Mauritian form *Nactus serpensinsulae* (Arnold 2000). This latter species has also developed separate races on Serpent Island and Round Island (Arnold & Jones 1994). *Nactus* geckos are generally called 'slender-toed geckos', but it is clumsy for individual species names, and 'night-gecko', though inadequately specific across the family as a whole, has been used for the Mascarene *Nactus* for many years.
19. There are two subspecies of *P. borbonica* on Réunion (Meier 1995, Probst 2002), with another race on Agalega Island (Cheke 1982b, Cheke & Lawley 1983, McKeown 1993), 619 miles (990km) north of Mauritius.
20. DNA studies indicate that the taxon *rosagularis*, originally described as a race of *P. guimbeaui*, is better considered a full species more closely related to *P. cepediana* (Austin *et al.* 2004a).
21. Two more lizards have been found in subfossil deposits in Rodrigues, one very unusual (Arnold 2000 & work in progress).
22. The two boas are now classed in their own family, Bolyeridae (e.g. Vidal & Hedges 2002). The Burrowing Boa has not been seen since 1975 despite much searching (North *et al.* 1994, Korsós & Trocsányi 2001b), but it was always very hard to find, so may just possibly still survive on Round Island.

ground[227]. On Christmas Island the annual shoreward breeding migration of millions of red crabs is something of a tourist attraction – something similar must once have occurred on Rodrigues.

Hermit crabs were also common, but the massive Coconut Crab seems to have been absent from the Mascarenes, though formerly common on islands further north, as it still is on Aldabra[228]. Oddly, Leguat did not mention the large orb-web spiders, which were probably the only small-bird predators on the island, bar the owl; the endemic Rodrigues species, *Nephila ardentipes*, is very abundant today[229].

CHAPTER 4

WHERE DID THE DODO COME FROM?

THE ORIGINS OF THE MASCARENE VERTEBRATE FAUNA

Of these two sorts of fowl, for ought we yet know, not any are to be found out of this island, which lieth about a hundred leagues from St.Lawrence. A question may be demanded how they should be there and not elsewhere, being so far from other land and can neither fly nor swim; whether by mixture of kinds producing strange and monstrous forms, or the nature of the climate, air and earth in altering the first shapes in time or how . . .
          Peter Mundy, commenting on the Dodo and Red Hen in 1638[1]

Rodrigues, of more ancient volcanic origin than Réunion, and thus emerging from the waters at a more distant era, had long before her supported plants and animals. Further from her infancy, nature must necessarily, on this untamed land, present species that carry fewer traces of imperfection, than those which populate a barely consolidated soil, on which the vegetation and the animals had not and perhaps still have not [attained] stable forms . . .
          Bory de St Vincent (1804) on the issue of differing island biotas[2]

To consider where the fauna and flora of the Mascarenes might have come from, the obvious first step is to look at a map of the Indian Ocean (Map 1.1). The nearest land is Madagascar – so much nearer than any other that one might consider any other source too improbable to take seriously. But history is rarely so simple, and as one looks deeper into the history of the Mascarene biota, it turns out that Madagascar's contribution, while important, is less than overwhelming, while the rise and fall of islands over time, along with ocean currents from the east, have been very significant.

The curious case of the tamarin des hauts

We'll start with a puzzling endemic plant, the massive straggly tree that dominates upland forest below the tree-line on Réunion – the *tamarin des hauts*, *Acacia heterophylla*. The oddest thing about this tree is not its phyllodes (leaf-mimicking stems), nor its ecology geared to regeneration after volcanic fires, but the fact that its nearest relative occurs over 17,000km away, in Hawaii, the most isolated archipelago on Earth. The Koa *A. koa* is so close in appearance and ecology that it has sometimes been considered the same species[3]. Hawaii, being the older island group, is presumably the source of the Réunion trees, but if so how could they have got there? These acacias have smallish hard seeds, long-lived but not particularly thick-skinned. They are too large to be taken up on long-distance air currents, and too small and thin-walled to survive long periods in sea water. In any case neither air- nor sea-currents normally cross the equator, and neither in Hawaii nor in Réunion do the trees grow anywhere near the sea. In fact, neither grows below about 450m (on Réunion it grows still higher, at 1,200–2,400m)[4]. This makes little sense until one considers that an endemic Réunion seabird, Barau's Petrel, is possibly derived from the Hawaiian Petrel[5] – and both species nest high in the mountains of their respective islands. Although on both islands petrels now nest above the upper limit of acacia forest, this was not the case in the past. In Hawaii, until decimated by introduced animals, the petrels nested under tree-roots and rocks mostly between 450 and 1,500m, precisely the altitude range of the Koa; likewise in Réunion, where at the turn of the 19th century Barau's Petrel still nested at around 1,800–2,000m on forested cliffs close to acacia stands[6]. Réunion may have been colonised by lost petrels from Hawaii that strayed into the wrong ocean, finding there a new mountainous island at the right latitude (albeit south instead of north). The birds could have had seeds attached to their feathers that they had picked up in their burrows – and thus the tree reached Réunion. Oddly Hawaii itself is well outside the normal distribution of phyllodic acacias in Australasia; apart from Réunion's *tamarin* and three further allied species in Hawaii itself, the most similar trees to the Koa are the Tasmanian Blackwood *A. melanoxylon* and *A. simplicifolia* from Fiji and Samoa[7].

Biogeography is not an exact science – there is always a degree of speculation. In the case discussed above there are two unproven links: that Barau's Petrel's ancestor came from Hawaii, and that, if it did, it brought viable acacia seeds on its plumage. The first link is very plausible, through similarities in voice, plumage, nest-site and breeding latitude (DNA studies are awaited). The second link is more problematic, though the chances of little 5mm discoid seeds travelling on a bird that nests under the parent tree is inherently more probable than sea- or wind-transport over such a huge distance. Furthermore, gadfly petrels rarely land on the sea, and only come on to land to breed, so a long-lived[8] seed concealed in the plumage would have a good chance of surviving intact until the bird landed by its chosen nest site.

In earlier chapters we have seen which species were present on the Mascarenes before humans arrived, and, in a general sense, how they would have got there, but we can also explore to a surprising degree where they came from, and what geological history, weather systems, ocean currents and the insights of taxonomy and phylogenetics can tell us about their geographical origins and the routes they used to reach the islands[9]. The more an island species resembles its relatives elsewhere, the easier it is to see where it originated, unless the species is so widespread that it could have come from anywhere. The more it has evolved its own particular island characteristics, the harder it becomes to trace its past. Some species in the Mascarenes, largely seabirds and waterbirds but including some bats and lizards, are members of widespread species that have reached the islands recently, or whose members in different areas regularly mix so that no population separates out. A few species are shared with Madagascar but are unrelated to forms outside the region. A substantial group are distinct species, but not greatly changed from related ones in Asia, Australasia, Madagascar or, for seabirds, other oceans. A final group consists of animals whose immediate relatives are far from obvious, or which have no close relatives at all – the stars of the show, most of them, sadly, now extinct.

Each of these groups in turn gives us a kind of snapshot of conditions prevailing for colonising species at different times in the past. Some of the widespread species of any given era, and those from the nearest land mass, are always going to be present on oceanic islands, and for these species the existing Mascarene fauna reflects the current world situation for mobile animals. The second batch, distinct species with close relatives elsewhere, can have more than one kind of origin. They may represent more recent arrivals in a freak or one-off colonisation that speciate rapidly. Others illustrate geologically recent periods of easier access to the Mascarenes, now cut off, while a few may represent slowly evolving forms that have been around since the islands first formed. Animals without close relatives are likely to represent relicts from the very first colonisation of the islands, possibly from earlier islands where they had already diverged from their own ancestors. Mascarene species are listed with their relatives and details of their source areas in Table 4.1.

WIDESPREAD SPECIES

Looking at the widespread species, particularly those that are basically land animals, gives us some idea of what is moving around between islands now, and which colonists might arrive in a period of considerable isolation. While birds, bats and flying insects are affected by wind direction and distance, non-flying creatures are dependent on flotation, and thus the direction of ocean currents. The prevailing winds in the Mascarenes blow from the southeast, familiar to mariners as the 'Southeast Trades' or trade-winds[10]. Since there is no land in that direction (even Australia is too far north), these winds inhibit land-animal colonisation by flying, though it could bring in southern or subantarctic seabirds. However, translocation from Rodrigues to Mauritius would be relatively easy, but not the reverse, a problem well-known to Mauritian seafarers in the days of sail[11]. The ocean currents, likewise, sweep from east to west, though they can originate, depending on the time of year, either from western Australia or from the Lesser Sunda Islands. Drift across the Indian Ocean from east to west was graphically demonstrated following the Krakatau eruption of 26–27 August 1883. The volcano lies between Java and Sumatra, some 5000km ENE of Mauritius; floating pumice reached Mauritius by February 1884, then Madagascar, and, a year after the eruption, South Africa[12]. Although they vary in speed, the direction of currents is constant, carrying flotsam from Rodrigues to Mauritius, but not the other way – as late as 1846 messages were sent from Rodrigues to Mauritius in bottles if no ships were available[13]. However the current can eddy south of St Brandon back towards Rodrigues[14], so occasional Mauritius-to-Rodrigues transport cannot be ruled out. Between Mauritius and Réunion the winds and currents are affected by the presence of the islands,

Table 4.1 The origins of the Mascarene fauna

Af = Africa	**In** = continental India plus Sri Lanka	**PT** = pantropical
Aa = Australasia	**IO** = widespread in Indian Ocean	**PT+** = pantropical extending to subtropical
AO = Atlantic Ocean	**IP** = Indo-Pacific	**Sey** = Seychelles
EI = 'East Indies' (Indonesia, Malaysia, Philippines)	**Md** = Madagascar	**W** = widespread
G = Gondwana	**Pm** = Palearctic migrant	**wIP** = widespread Indian/ Pacific Oceans
H = hotspot island chain	**PO** = Pacific Ocean	
x = extinct in the Mascarenes	xx = totally extinct	2xx = two species extinct
[Ré x] = extinct in Réunion	[M x] = extinct in Mauritius	[Ro x] = extinct in Rodrigues
[2xx] = two species extinct but at least one survives		

Notes:
1. New Guinea counts as EI or Aa depending on the remaining range of the taxon.
2. Landbirds from EI, In, Pm and Sey will mostly have arrived via re-emerged islands in the Pleistocene.
3. A '/' is used to separate proximate origins from ultimate ones (e.g. Md/Af indicates an African taxon that arrived via Madagascar).
4. For areas other than the Mascarenes, [E] indicates currently extinct at that location.

b) Widespread species that have colonised the Mascarenes

Group	Species	Code	Nearest site/probable source	World distribution
Bats	Taphozous mauritianus	Md/Af	Madagascar/Madagascar	Africa [a1]
	Scotophilus borbonicus ?x	Md/Af	Madagascar/Madagascar	Africa [a1]
	Dugong dugon x	Md/wIP	Madagascar/Madagascar	Indo-Pacific [a2]
Seabirds [a3]	Puffinus pacificus	wIP	Seychelles, Chagos/?	Tropical Indo-Pacific
	Puffinus bailloni	wIP	Seychelles, Chagos/?	Tropical Indo-Pacific
	Phaethon lepturus	PT	Seychelles, Chagos/?	Tropical oceans worldwide
	Phaethon rubricauda	wIP	Agalega [E], Seychelles/?	Tropical Indo-Pacific
	Sula sula x	PT	St Brandon [E], Tromelin/?	Tropical oceans worldwide
	Sula dactylatra	PT	St Brandon, Tromelin/?	Tropical oceans worldwide
	Fregata ariel x [a4]	wIP	St Brandon, Tromelin/?	Tropical Indo-Pacific
	Sterna dougallii ?x	W	St Brandon, Seychelles/?	India/Atlantic/Australasia
	Sterna fuscata	PT	St Brandon, Agalega [E]/?	Tropical oceans worldwide
	Gygis alba	PT	St Brandon, Agalega/?	Tropical oceans worldwide
	Anous stolidus	PT	St Brandon, Agalega/?	Tropical oceans worldwide
	Anous tenuirostris [a5]	IO	St Brandon, Agalega/PO	Tropical Indian Ocean
Land birds	Phalacrocorax africanus x [a6]	Md/Af	Madagascar/Madagascar	Africa
	Egretta (gularis) dimorpha x [a7]	Md/Af	Madagascar/Madagascar	Africa, India
	Butorides striata [a8]	EI	Madagascar/East Indies	Worldwide tropics
	Phoenicopterus roseus x	Md	Madagascar/Madagascar	Africa, Asia, Caribbean
	Gallinula chloropus	Md	Madagascar/Madagascar	Almost worldwide
Turtles [a9]	Chelonia mydas	PT+	St Brandon, Agalega/?	Tropical seas worldwide
	Eretmochelys imbricata	PT+	St Brandon, Agalega/?	Tropical seas worldwide
	Caretta caretta x	PT+	Madagascar/?	Tropical seas worldwide
Land reptiles	Lepidodactylus lugubris [a10]	wIP	Coëtivy (Seychelles), Chagos/?Aa	Indo-Pacific, widespread
	Cryptoblepharus boutonii [a10]	Md/wIP	Madagascar/?Aa	East Indies/Australasia/East Africa

b) Species shared with Madagascar only

Group	Species	Code	Comments
Land birds	Aythya ?innotata x	Md	Malagasy endemic species of widespread genus
	Turnix nigricollis	Md	Malagasy endemic species of widespread Old World genus
	Nesoenas picturata [Ro x] [b1]	Md	Malagasy region endemic genus
	Phedina borbonica [b2]	Md	Malagasy endemic; another species in central Africa

Table 4.1 (cont.)

c) *Species with congeners elsewhere (or same species with restricted range)*

	Species	Code	Nearest relative (location)	Original source
Bats	*Pteropus niger* [Ré x, Ro x] c1	Md/Ind	*Pteropus seychellensis* (Seychelles)	*Pteropus giganteus*, India/Maldives
	Pteropus subniger xx c2	Ind/EI	?*Pteropus hypomelanus* (Maldives)	Indo-Malesian islands
	Pteropus rodricensis [M x] c1	?EI	?*Pteropus vampyrus* (Greater Sunda Is.)	Indo-Malesian islands
	Tadarida acetabulosus c2	Md/Aa	*Mormopterus jugularis* (Madagascar)	?Australasia
Seabirds	*Pterodroma 'arminjoniana'* c3	AO/PO	*Pterodroma arminjoniana* (Trindade I.)	*P. heraldica*, Pacific
	Pterodroma neglecta c3	Aa/PO	same species (south Pacific)	Pacific Ocean
	Pterodroma baraui c4	?PO	?*Pterodroma sandwichensis* (Hawaii)	Pacific Ocean
	Pseudobulweria aterrima c5	PO	*Pseudobulweria rostrata* (Pacific)	Pacific Ocean
	Papasula abbotti x c6	IO	same species (Christmas I., Assumption I. [E])	formerly distributed across the Pacific
Land birds	*Nycticorax* spp. 3xx c7	Md/?	*Nycticorax nycticorax* (Madagascar)	?Africa or Asia
	Threskiornis solitarius xx c8	Md/Af	*Threskiornis bernieri* (Madagascar)	*Threskiornis aethiopica*, Africa
	Alopochen spp. 2xx c8	Md/Af	*Alopochen sirabensis* [E] (Madagascar)	*Alopochen aegyptiacus*, Africa
	Anas theodori xx c9	EI	*Anas gracilis* complex (Indo-Australasia)	?East Indies
	Circus maillardi [M x] c10	Md/Pm?	*Circus macrosceles* (Madagascar)	Eurasia
	Falco endemic spp. [Ré xx] c11	Md/Af	*Falco rupicoloides* (Africa)	Africa
	Dryolimnas spp. 2xx c12	Md/?	*Dryolimnas cuvieri* (Madagascar)	?
	Fulica newtoni xx c13	Md/Af	*Fulica cristata* (Madagascar)	Africa
	Porphyrio sp. x[x] c14	Md/?	*Porphyrio porphyrio* (Madagascar)	Africa, but Asia possible
	Nesoenas spp. [Ré xx] c15	Md	*Nesoenas picturata* (Madagascar)	Malagasy endemic genus; Asian source
	Rodrigues Pigeon xx c16	?H/Md	*Nesoenas mayeri* (Mauritius)	?Asia via Madagascar
	Alectroenas nitidissima xx c17	Md/EI	*Alectroenas madagascariensis* (Madagascar)	Malagasy endemic genus; Asian source?
	Psittacula eques [Ré x] c18	In	*Psittacula krameri* (India/Sri Lanka)	genus centred on southeast Asia
	Psittacula exsul xx c18	?In	*Psittacula (e.) wardi* (Seychelles)	*Psittacula eupatria*, India
	Psittacula bensoni xx c18	In	*Psittacula (e.) wardi* (Seychelles)	*Psittacula eupatria*, India
	Aerodramus francicus c19	In	*Aerodramus elaphra* (Seychelles)	?*Aerodramus unicolor*, Sri Lanka
	Coracina spp. c20	?Aa	?*Coracina tenuirostris/schisticeps*	?Australasia
	Hypsipetes endemic spp. c21	In(?Md)	?*Hypsipetes crassirostris* (Seychelles)	*Hypsipetes poliocephalus*, India
	Terpsiphone bourbonnensis c22	Md/?	*Terpsiphone mutata* (Madagascar)	?; genus widespread in Africa, Asia
	Saxicola tectes c23	Md/Af	*Saxicola (torquata) axillaris* (Madagascar)	widespread in Africa
	Acrocephalus rodericanus c24	Md/Af	*Acrocephalus sechellensis* (Seychelles)	?Africa via Madagascar
	Zosterops borbonicus c25	In	Comores, Seychelles	?India
	Zosterops olivaceus/chloronothos c25	In	Comores, Seychelles	?India
	Foudia spp. [Ré xx] c26	Sey/Af	*Foudia sechellarum* (Seychelles)	Africa
Reptiles	*Nactus* spp. [Ré xx, Ro 2xx] c27	Aa	*Nactus arnouxii*	Australasia
	Phelsuma spp. [Ro 2 xx] c28	Md/?G	? (Madagascar)	?
	Typhlops cariei xx c29	?Md	? (Madagascar)	?; widespread genus

Table 4.1 (*cont.*):

d) Species in endemic genera with no obvious affinities

	Species	Code	Putative affinities
Land birds	*Aphanapteryx* xx; *Erythromachus* xx (Rallidae) [d1]	?H/Aa/G	affinities unknown
	Raphus xx; *Pezophaps* xx (Columbidae) [d2]	?H/EI	closest to *Caloenas nicobarica*
	Mascarinus mascarinus xx (Psittacidae) [d3]	?H/IP	affinities with psittaculines
	Lophopsittacus xx; *Necrospittacus* xx (Psittacidae) [d3]	?H/IP	affinities with psittaculines
	Mascarenotus 3xx (Strigidae) [d4]	?H/??	affinity with Asian *Otus*?
	Fregilupus xx; *Necropsar* xx (Sturnidae) [d5]	?IP/?Pm	affinities with Asian starlings
Reptiles	*Cylindraspis* 5xx (Testudinidae) [d6]	H/Md	distant affinity with Malagasy forms
	Leiolopisma [Ré xx, 1 in M xx] (Scincidae) [d7]	Aa	affinity with Australasian *Emoia*
	Gongylomorphus [Ré xx] (Scincidae) [d8]	H/Md	affinity with Malagasy skink radiation
	Rodrigues geckos (undescribed) 2xx (Gekkonidae) [d9]	?	?
	Bolyeria ?xx, *Casarea* (Bolyeridae) [d10]	?H/?EI	distant affinity with Asian *Xenophidion*

Notes on Table 4.1

a1. While there is no confusion about the tomb bat, the house bat *Scotophilus borbonicus* was thought to be confined to Réunion and possibly Madagascar, but the only surviving Réunion specimen proved to be indistinguishable from the *S. leucogaster/viridis* complex of Africa (Hill 1980, Cheke & Dahl 1981) and some Malagasy specimens are apparently the same (Eger & Mitchell 2003, Goodman *et al.* 2005). If Mascarene, Malagasy and African examples are indeed the same species they should be united under the oldest name, *S. borbonicus* (Cheke & Dahl 1981).

a2. The Dugong's original range was the tropical and subtropical Indian and western Pacific Oceans, but it is now much fragmented (Reynolds & Odell 1991).

a3. For seabird distribution see Tuck & Heinzel (1978) or Harrison (1985); for widespread land birds see Walters (1980) or Dickinson (2003), subject to taxonomic revisions.

a4. The Lesser Frigatebird also has one station in the Atlantic – Trindade Island off Brazil. It may not be coincidental that this is also the only colony outside the Indo-Pacific for the Herald Petrel complex *Pterodroma arminjoniana/heraldica*. The Greater Frigatebird *Fregata minor* is omitted from this list as there is no proof of breeding in the Mascarenes. However its distribution is similar, again with a lone outstation on Trindade.

a5. The Lesser Noddy is replaced by the very similar White-capped Noddy *Anous minutus* in the tropical Atlantic and Pacific Oceans (Tuck & Heinzel 1978, Harrison 1985).

a6. Although bones were identified as *Phalacrocorax africanus* (Olson 1975b), the Indian *P. niger* is not excluded.

a7. The taxonomy of the *Egretta* superspecies including the reef herons and the Little Egret is in contention, with the Malagasy *dimorpha* being batted about between *garzetta* (Little Egret), the two reef herons *gularis* and *sacra*, and being treated as a full species in its own. Assuming one accepts the larger-billed coastal reef herons, with grey and white forms, as separate from the usually freshwater Little Egret (almost always white), then it probably makes most sense to treat *dimorpha* as a race of the Western Reef Heron *E. gularis*, which encloses its range (Hancock & Elliott 1978, *contra* Hancock & Kushlan 1984 who treated it as *garzetta*). Hancock (1999) has reported substantial recent hybridisation between reef herons and Little Egrets in several areas, so the situation is clearly unstable.

a8. See text for the curious distributional history of the Striated Heron in the Indian Ocean.

a9. For a summary of turtle distribution in the Indian Ocean see Frazier (1984). The African crocodile *Crocodilus niloticus*, widespread in Madagascar (Kuchling *et al.* 2003), and adapted to sea water as well as fresh, colonised Aldabra and the Seychelles (Stoddart 1984c) but not the Mascarenes.

a10. The Mourning Gecko *Lepidodactylus lugubris* is as likely to have arrived directly from Australasia or the East Indies as from Chagos or the Seychelles, as it has only reached Rodrigues and not the western Mascarenes. In contrast, Bouton's Skink is absent from Rodrigues and, although originally also Indo-Pacific, it could also have reached the Mascarenes from Madagascar, although the distinct DNA profile of Mauritian animals suggests they arrived there first and later colonised Madagascar (Rocha *et al.* 2006). The Indo-Pacific blind-snake *Ramphotyphlops braminus* is generally thought to have been introduced to the Mascarenes; however, this tiny and easily overlooked snake is parthenogenic (see Nussbaum 1984) and could have arrived on its own.

b1. This pigeon has been traditionally been placed in *Columba*, but recently into *Streptopelia* (Sibley & Monroe 1990, Gibbs *et al.* 2001). DNA analysis explains this ambiguity, as this species, together with the Mauritius Pink Pigeon and '*S.*' *chinensis* and *senegalensis*, form a separate grouping that split off near the base of the *Columba/Streptopelia* divergence. To clarify the relationships, we are using *Nesoenas* for both *picturata* and the Pink Pigeon *mayeri*, and *Stigmatopelia* for the other two (Cheke 2005b). *Nesoenas* is thus a Malagasy region endemic genus, possibly with Asian affinities; *Stigmatopelia* is also probably Asian in origin, though *S. senegalensis* is widespread in Africa.

b2. *Phedina* is apparently a relict genus of swallow with no near relatives, and there are doubts (Brooke 1972) about its affinity with its only congener, *P. brazzae* of central Africa (Hall & Moreau 1970).

c1. See Anderson (1912) and O'Brien (2005) for inter-generic relationships in *Pteropus*, Mickleburgh *et al.* (1992) for current distribution, and Moutou (1989) for a biogeographical discussion.

c2. Apart from *Mormopterus acetabulosus*, apparently a Mascarene endemic, and the Malagasy *M. jugularis*, the small molossids in the genus *Mormopterus* are confined to South America and Australasia (Eger & Mitchell 1996).

c3. See text and Chapter 9 for discussion of the complexities surrounding Herald Petrel group *Pterodroma arminjoniana/heraldica* and the Kermadec Petrel *P. neglecta* on Round Island.

c4. The relationships of Barau's Petrel are disputed. The birds closely resemble in plumage the Caribbean *Pterodroma hasitata* and Bermudan *P. cahow*, which Imber (1985) put in a different sub-genus, while Bretagnolle & Attié (1991) related Barau's, on the basis of call notes, to *P. inexpectata* and *P. phaeopygia*. See text for the particular link to the latter's Hawaiian race (now treated as a full species, *P. sandwichensis*).

c5. The Réunion Black Petrel is part of a relict group, the others all from the Pacific apart from an extinct species on St Helena.

c6. In historical times Abbott's Booby had at least four colonies, all in the Indian Ocean, but now survives only on Christmas Island (Feare 1978, Stoddart 1984, Cheke 2000a). It has no close relatives (see Nelson 1978, 2006; Olson & Warheit 1988).

c7. The extinct endemic night herons were close to the widespread Black-crowned Night Heron (Mourer *et al.* 1999).

c8. Mourer *et al.* (1999) are followed for the relationships of extinct Réunion birds known from subfossil bones. Both the ibis and the sheldgoose link via Malagasy species to African ancestors.

c9. The Mascarene Teal is closer osteologically to the Indo-Australian Grey Teal *Anas gibberifrons* than to Bernier's Teal *A. bernieri* from Madagascar (Mourer *et al.* 1999).

c10. We follow Simmons (2000) and Bretagnolle et al. (2000b) in treating the Réunion Harrier as a full species, closely allied to the Malagasy Harrier *Circus macrosceles*. However, Simmons's DNA phylogeny of the genus matches neither morphological similarities nor biogeographical likelihood, and we consider it remains an open question as to which member of the marsh harrier complex provided the ancestor to the Malagasy-Réunion species pair. Simmons proposed the Australian *C. approximans* whereas previous workers have all favoured the east Asian *C. spilonotus* which has very similar plumage (Berlioz 1946, Clouet 1978, Cheke 1987b). The extinct Mauritian *Circus 'alphonsi'* is now known to have been a Réunion Harrier (Mourer et al. 2004).

c11. The Mauritius Kestrel *Falco punctatus* is closer to the African Greater Kestrel *F. rupicoloides* than to the Malagasy *F. newtoni* or the Seychelles *F. araea*, which are later derivatives of the same source (DNA studies by Groombridge et al. 2002). The extinct *F. duboisi* on Réunion (DNA not studied) was osteologically close to the Common Kestrel *F. tinnunculus* (Mourer et al. 1999) and may have come from a later separate colonisation.

c12. *Contra* Cowles (1987), Mauritian subfossil *Dryolimnas* proves to be from a large undescribed flight-reduced form (Hume & Prys-Jones 2005), as in Réunion (Mourer et al. 1999). Flightlessness develops rapidly in this genus: the Aldabran race (or species) *D. (c.) aldabranus* has become flightless with considerable osteological alteration (Mourer et al. 1999) since the atoll last emerged from oceanic flooding some 100,000 years ago (Taylor et al. 1979, Rohling et al. 1998). According to Livezey (1998), *Dryolimnas* is related both to the widespread genus *Rallus* and to Australasian and Pacific *Gallirallus*; it is not clear from where it reached Madagascar or the Mascarenes.

c13. While not ruling out a derivation from the widespread Common Coot *Fulica atra*, Mourer et al. (1999) considered Mascarene bone material closer to the Malagasy and African Crested Coot *F. cristata*.

c14. This is the as-yet unidentified '*oiseau bleu*' of the early accounts in Réunion.

c15. The Mauritius Pink Pigeon, and its extinct Réunion counterpart *Nesoenas duboisi*, are often placed in *Columba*, but there are enough anomalies for the genus to be justified (Goodwin 1983); DNA work (Johnson et al. 2001) additionally associates these birds with the Malagasy Turtle Dove. They appear to represent a colonisation, presumably from Madagascar, of stock also ancestral to *N. picturata*, which itself re-invaded later.

c16. Although Mourer et al. (1999) re-evaluated the Rodrigues Dove as closer to the Indo-Pacific ground dove genus *Gallicolumba* than to other Mascarene forms, further study with more material (JPH, work in progress) reveals that it is in fact another early *Columba/Streptopelia* offshoot close to or possibly within *Nesoenas*.

c17. Goodwin (1983) considered the blue pigeons related to *Ducula* and *Ptilinopus* (Indo-Pacific fruit pigeons), and unrelated to anything in Africa; DNA confirms this relationship, adding *Drepanoptila* of New Caledonia (Shapiro et al. 2001). We have omitted from the table the ill-defined 'slaty pigeon' of Réunion, for which no bones are known, usually presumed to have been an *Alectroenas*.

c18. While the Echo Parakeet *Psittacula eques* is close to the Ring-necked Parakeet *P. krameri* (Groombridge 2000), the extinct turquoise Rodrigues Parakeet *P. exsul* is osteologically closer to *P. eupatria* (Hume 2005), and presumably arose from an earlier invasion (Keith 1980). *P. bensoni*, another *eupatria*-type parakeet (Hume 2005), from Mauritius and Réunion may represent another offshoot of the line that led to *P. exsul* in Rodrigues.

c19. There are several candidates for the ancestral swiftlet that reached the Seychelles and the Mascarenes (see text); the nearest geographically is the Indian Swiftlet *Aerodramus unicolor*.

c20. The two Mascarene cuckoo-shrikes are quite unlike the Malagasy species *Coracina cinerea* (Benson 1971, Cheke & Diamond 1986) but resemble two Australasian species, *C. tenuirostris* and *C. schistacea*. However, pending DNA analysis, cuckoo-shrike relationships, and hence the source of Mascarene forms, are unresolved.

c21. Mascarene merles, ultimately from Asia, are bracketed by their immediate ancestors in the Seychelles to the north and descendants in Madagascar to the west – our interpretation of Ben Warren's DNA studies (Warren 2003, Warren et al. 2005). The birds in Madagascar spread out again, so that Keith (1980), since partly supported by Warren's DNA work, argued that *madagascariensis*-type bulbuls originating in Asia have spread around the Indian Ocean twice, an earlier wave giving rise to the olive-tinged birds in the Seychelles (*Hypsipetes crassirostris*), Mauritius (*H. olivaceus*) and Réunion (*H. borbonicus*), whereas a later wave is the origin of *H. madagascariensis* itself, a greyish bird virtually indistinguishable in plumage (though not voice) from *H. (m.) poliocephalus* in Sri Lanka and India. In the Comoros descendants of two waves (olive *parvirostris* and grey *madagascariensis*) coexist on Moheli and Grande Comore (Louette & Herremans 1985, Louette 1988, Warren et al. 2005). The combination *H. poliocephalus* used by Keith (1980) from the oldest name of the forms in southeast Asia has not been widely adopted; many authors use *H. leucocephalus* (e.g. Grimmett et al. 1999) or still consider the Asian and the grey Malagasy forms all to belong to one wide-ranging species '*H. madagascariensis*'.

c22. Mascarene Paradise Flycatchers have been associated in the past with the Asian *Terpsiphone affinis* and *T. paradisi* (which breeds in Sri Lanka) (see Benson 1971), though in our view its size, 'jizz' and song place it very close to Malagasy *T. mutata* (see Cheke & Diamond 1986), of uncertain origin. In the Seychelles there is a larger species, *T. corvina*, of different origin, possibly related to the African *T. viridis* (Hall & Moreau 1970, Benson 1971).

c23. The Réunion Stonechat, while distinct (Cheke 1987b, Wink et al. 2002a,b), is not long differentiated from the very widely distributed Common Stonechat complex *Saxicola 'torquata'* (now split into three species), and even in the past often treated as a race of it (Cheke 1987b, Urquhart 2002). A race of the African Stonechat *S. (t.) axillaris* occurs in Madagascar (Sinclair & Langrand 1998, Urquhart 2002).

c24. The closely related Rodrigues and Seychelles Warblers were formerly paired in the bitypic genus *Bebrornis*, but their ancestor over the water appears to be the visually rather different *A. newtoni* from Madagascar, itself related to the African *Acrocephalus gracilirostris* (Dowsett-Lemaire 1994, Leisler et al. 1997), the ancestor originally suggested by Hall & Moreau (1970) and Diamond (1980). *Acrocephalus* warblers have colonised many Pacific islands, but these are considered derivatives of the migratory Great Reed Warbler *A. arundinaceus* (Milder & Schreiber 1989), another branch of the same section of the genus (Leisler et al. 1997). The song of the Rodrigues and Seychelles species pair is much closer to *A. gracilirostris* and *A. newtoni* than to likely Palearctic migrants (Diamond 1980, Catchpole & Komdeur 1993, Dowsett-Lemaire 1994), so they and also Malagasy birds appear to be part of a past radiation of *gracilirostris*-type birds from Africa.

c25. Despite ample study of the ecology, calls and morphology of the Mascarene white-eyes, no one had come up with a convincing reason, apart from proximity to Madagascar, to relate these birds to any particular ancestral source, until DNA analysis allied them to a colonisation from Asia (Warren 2003, Warren et al. 2006). The plumage types seen in the three Mascarene species are not matched elsewhere in the genus (Moreau 1957, Gill 1971). *Zosterops modestus* from the Seychelles has become grey, but has not developed the spectacular white rump or axillary tufts of *Z. borbonicus*, and nowhere else have white-eyes become 'functional sunbirds' like *Z. olivaceus* and *Z. chloronothos* (Moreau et al. 1969) with their long bills and habit of defending flowering trees from other birds (Gill 1971, Cheke 1987a,b). The DNA work supports an early invasion of Indian Ocean islands from India, followed by a later wave from Africa. The Seychelles also had two species, the greyish *Z. modestus* (ex-African wave) and the bright yellow-green *Z. semiflavus* (extinct, from the earlier Asian wave) almost indistinguishable morphologically from *Z. mayottensis* on Mayotte (Comoros) (Moreau 1957). Louette (1988) claimed that this resemblance was just fortuitous, a view that has been confirmed by DNA, with the Mayotte birds being from the African wave (Warren 2003). There is a single generalised greenish-yellow species *Z. maderaspatanus* (African wave) on Madagascar (Moreau 1957, Morris & Hawkins 1998, Warren et al. 2006).

c26. The fody genus *Foudia* is endemic to the Malagasy region, considered by Moreau (1960a,b) to be near to the basal stock that gave rise to *Euplectes* bishop birds in continental Africa. This has been confirmed by DNA analysis (Warren 2003), but it now appears that Madagascar may have been colonised via the islands and not vice-versa. The departure from Africa must have been long before the present generic pattern there had evolved.

c27. *Nactus* night-geckos (five species in all) were present on all three islands, and now appear from DNA studies to have descended from a single colonisation of a form close to the Australian *N. arnouxii* (Arnold et al. in prep. a, *contra* Bullock et al. 1985, Arnold & Jones 1994). There are only three species outside the Mascarenes, all in Australasia and the Pacific (Kluge 1983).

c28. Like *Nactus*, the day-geckos are descended, according to DNA analysis, from a single colonisation. Austin et al. (2004a) now recognise nine Mascarene species, with the two extinct Rodrigues forms and the Round Island species *Phelsuma guentheri* closest to the ancestral form; Mascarene day-geckos are not particularly close to any of the existing species on Madagascar. The day-geckos are a distinct group whose centre of radiation lies in Madagascar, but their original affinities are obscure.

c29. Carié's Blind-snake is known only from subfossil vertebrae (Hoffstetter 1946) from which it was not possible to decide on its affinities, but there are several species in Madagascar (Guibé 1958, Preston-Mafham 1991, Raxworthy 2003a), the nearest likely source.

d1. The affinities of the two large flightless rails remain undecided, but Livezey (1997) tentatively related them to *Cabalus* and *Capellirallus* from New Zealand and a northwest Pacific cluster of large rails he placed in *Habropteryx* (split from *Gallirallus*, also a Pacific group). Rails date back to the Cretaceous (Keith 1980), so the ancestral stock for all these can presumably be sought in Gondwana.

d2. Dodos and solitaires have long been generally accepted as derived from pigeons, but DNA studies (Shapiro *et al.* 2001) show them firmly embedded in the Columbidae, derived from the same stock as the Nicobar Pigeon *Caloenas nicobarica*.

d3. These extinct parrots are osteologically close to *Psittacula* (Holyoak 1973, Hume 2005) but clearly derive from much earlier colonisation(s) than gave rise to *P. bensoni*, *P. exsul* and *P. eques*. It is likely that *Mascarinus* is from a more recent source than *Lophopsittacus* and *Necropsittacus*.

d4. The lizard-owls are very distinct derivatives of the widespread scops owl genus *Otus* (Mourer *et al.* 1994, 1999), which speciates readily on islands (e.g. Rasmussen *et al.* 2000); no attempt has been made to discover to which living owls they are closest, so the source area remains unknown.

d5. The two Mascarene starlings are derived from members of the genus *Sturnus* (JPH work in progress, Zucca 2006) whose present range is centred in continental east and southeast Asia, some species migrating into India and Malaysia in winter. They are unrelated to the Madagascar Starling *Hartlaubius auratus*, whose affinities are African (Zucca 2006).

d6. The Mascarene tortoises, in a recently recognised endemic genus allied to Malagasy *Dipsochelys* and *Asterochelys* (all segregates from the old inclusive *Geochelone*), separated from their relatives before the Mascarenes emerged and so must have used the hotspot chain. Their differences from other similar genera are well-marked and long-established (Arnold 1979, Austin & Arnold 2001). Other Indian Ocean islands (Seychelles, Aldabra) were colonised by *Dipsochelys dussumieri* (= *Aldabrachelys gigantea* auct.), closely related to the extinct Malagasy form *A. grandidieri* (Arnold 1979, Austin *et al.* 2003). Austin & Arnold (2001) have established by DNA analysis that all five Mascarene species derive from a single radiation.

d7. *Leiolopisma*, formerly a large genus including many Australasian species, was restricted in serum albumin studies to the Mascarenes (Hutchinson *et al.* 1990), and more recently found by DNA work to be allied to the Australasian genus *Emoia* (Austin & Arnold 2006). There are now considered to have been three species, two on Mauritius and one on Réunion (Arnold 1980, 2000, Austin & Arnold 2006). Largish skinks on other Indian Ocean island groups (Seychelles, Comoros, Madagascar) are all members of the unrelated genus *Mabuya* (Cheke 1984), widely distributed in the lands surrounding the ocean, but absent in the Mascarenes. No particular existing Australasian skink is considered especially close or ancestral to *L. telfairii* (Hutchinson & others 1990, Greer 1974, Austin & Arnold 2006); the large extinct *L. mauritianus* (formerly *Didosaurus*) was only recognised as a *Leiolopisma* in 1980 (Arnold 1980).

d8. DNA studies place Bojer's and Macabé Skinks nested within a large radiation of Malagasy skinks, closest to the *Amphiglossus/Androngo* clade (Carranza *et al.* 2001, Carranza & Arnold 2003, Austin & Arnold 2006).

d9. The affinities of the newly discovered subfossil endemic gecko genus from Rodrigues are as yet unknown (Nick Arnold, work in progress).

d10. There has been much debate about the origin and relationships of the endemic boa family Bolyeridae, but recent work puts them closest to the recently discovered genus *Xenophidion* in Malaysia and Borneo (Wallach & Günther 1998, Lawson *et al.* 2004), and not close to South American tropidophiids as once thought. How they reached the Mascarenes is an open question.

and may at times blow or flow one way or the other, though westerlies are rare and short-lived[15]. In practice Réunion has received fauna from Mauritius but there is no evidence of colonisation in the reverse direction, though some flying species (tomb bats, swifts, swallows, Malagasy Turtle Doves, Common Moorhens and Striated Herons) may move back and forth to some extent, and in the past fruitbats, flamingos, cormorants and the endemic ducks and coots may also have done so.

Most of the seabirds, the Reed Cormorant, Common Moorhen, Striated Heron, Dimorphic Egret, two microbats, two lizards, the turtles and the Dugong are widespread tropical species found in suitable places throughout the western Indian Ocean and African coast, or, in the moorhen's case, almost the whole world[16]. While seabirds and turtles range widely across the oceans and can be expected to colonise remote islands free from continental predators, it is less easy to see why the cormorant and moorhen, freshwater species, and the largely coastal herons and Dugong should be such avid explorers across hundreds of miles of inhospitable sea – but over time they have been. We will not comment further on widespread seabirds unless there is a particular point of interest.

The moorhen, egret and probably the cormorant reached our islands from Madagascar[17] – the moorhen apparently within historical times, as no subfossil bones have been found, and the first mention (from Mauritius) dates from the late 1700s[18]. Striated Herons may also be a recent arrival; the only putative early record is Tafforet's from Rodrigues in 1725 (p. 48). From there they presumably spread to Mauritius, where, again, there are no subfossils and no one reported them until the early 1800s; the first mention in Réunion is not until the 1830s[19]. The subspecific affinities of these birds show that they have colonised Indian Ocean islands more than once. Common Moorhens in the Seychelles appear to be of African or Asian rather than Malagasy origin[20], while Striated Herons in the Mascarenes are indistinguishable from those in Java, and distinct from their relatives in Madagascar and Africa. These Javanese-type herons are also found in the Chagos archipelago (a possible stepping stone), the Aldabra group and the Comoros, while the central Seychelles (between Chagos and Aldabra) have birds of an Afro-Malagasy type[21]. It would appear, on a time-scale of a few thousand years, that Moorhens and Striated Herons have been criss-crossing the Indian Ocean in all directions! This kind of activity can be seen among other widely distributed herons and rails. Cattle Egrets *Bubulcus ibis* briefly colonised Mauritius in the late 1960s (soon dying out), and are also seen occasionally in Réunion[22]. Black-crowned Night Herons *Nycticorax nycticorax* (of which more below) have recently established themselves in the Seychelles, where they were virtually unknown before 1995[23]. Also in the Seychelles there is a not only a population of Cattle Egrets but also Yellow Bitterns *Ixobrychus sinensis*, unchanged from their relatives in mainland Asia where they breed south to Sri Lanka[24]. Finally, White-breasted Waterhens *Amaurornis phoenicurus*, widespread in Asia, colonised Diego Garcia (Chagos) sometime in the mid-20th century, apparently spreading down from the Maldives[25].

BOX 4   CORMORANTS (PHALACROCORACIDAE)

Cormorants were reported on Réunion during the early years of settlement; Dubois in 1672–73 and Feuilley (1705) gave good descriptions, but no fossils have been found. They were encountered on the lakes and, when young, made reasonably good eating. They were not recorded again and presumably died out in the early 18th century. They also occurred on Mauritius but were never mentioned by early mariners. The Mauritius skeletal material has been identified as the Reed Cormorant *Phalacrocorax africanus*, a wide ranging African species that is represented on Madagascar by the race *pictilis;* presumably the same subspecies occurred in Réunion and Mauritius.

Account
Feuilley (1705):

The cormorans are the size of a duck and the same form, except their neck is a bit longer; [they] *live in the meres like flamingos and live in the same way. They are not eaten unless very young. They smell strongly of marsh and the wild.*

Specimen collections
The Mauritian fossil material can be seen in the UMZC; it was originally assigned in the 19th century to a darter *Anhinga,* but was reassessed by Olson (1975) and found to be the Reed Cormorant.

Reed Cormorant *Phalacrocorax africanus*. From Bannerman (1953).

Seabird populations

Even amongst oceanic seabirds, where one might imagine there were no boundaries, there are populations which rarely mix and so begin to differentiate. Although most species are the same throughout the Indian Ocean, several species inhabiting the Mascarenes show signs of this kind of separation. Indian Ocean Red-footed Boobies exhibit two colour phases, all-white and 'white-tailed brown'. The proportions vary from colony to colony, suggesting minimal inter-colony exchange. The extinct Rodrigues colony consisted mostly of brown birds; in its nearest neighbour (St Brandon, also now extinct) they were almost all white, while in the colony one along (Tromelin) a third are brown, two-thirds are white; and so on[26]. The various colonies of Tropical Shearwater across the Indian Ocean have recognisable plumage differences: Réunion birds resemble those on Aldabra more than those from the granitic Seychelles, although the Seychelles are between Réunion and Aldabra[27]. This phenotypic pattern is not reflected in the DNA, which shows that the Arabian, Comoros and Réunion populations form one lineage, while the Aldabra and Seychelles birds hardly differ from birds in the western Pacific. It appears that Melanesian/Polynesian islands were colonised from an old western Indian Ocean population, and later the Seychelles and Aldabra were recolonised from the Pacific, both lineages later developing minor differences between sub-populations[28]. White-tailed Tropicbirds have developed distinctively golden-coloured populations on opposite sides of the Indian Ocean: on Europa Island (Mozambique Channel) and Christmas Island (off Java), although central Indian Ocean birds, the Mascarenes included, are uniformly white. Masked Boobies show no visible differences, but their DNA indicates minimal mixing between Indian Ocean colonies[29].

Widespread bats

Only Réunion and Mauritius have insectivorous bats, sharing two species, with Réunion, nearer to Madagascar, adding two more[30]. Two species, the Grey Tomb Bat and Pale House Bat, are not distinguishable from those in Madagascar and/or mainland Africa[31], so they have either arrived recently, or are there has been enough gene exchange with Madagascar through inter-island flights to prevent speciation; indeed, the periodic appearance and disappearance of Tomb Bats in Réunion may indicate migration between islands[32]. The third species, the small free-tailed bat *Mormopterus acetabulosus*, remains a bit of a mystery. It is usually cited as occurring in Madagascar and Africa, but there are only two isolated (and doubtful) African records, and the Malagasy reports are now thought all to refer to *M. jugularis*. In addition, the genus *Mormopterus* is otherwise confined to Sumatra, Australasia and South America, with recent workers discounting the African occurrences[33]. *M. acetabulosus* thus seems to be a Mascarene endemic of eastern origin, which we are calling the Mascarene Free-tailed Bat. The fourth species, the enigmatic small white bat reported roosting in latan palms by Bory de St Vincent in 1801 (p. 44) is without physical remains and remains unidentified[34].

Small African or Malagasy bats have also reached other remote Indian Ocean islands: the Aldabra group (three species) and the granitic Seychelles (one species), while a third Asian genus *Emballonura* (in addition to *Mormopterus* and flying-foxes) has reached Madagascar. Only the Seychelles form *Coleura seychellensis* is an indisputably distinct endemic species, though the Aldabran Free-tailed bat *Tadarida 'pusilla'* is a well-marked variety of an African and Malagasy species[35].

Widespread lizards

The two widespread lizards are Bouton's Skink *Cryptoblepharus boutonii* and the Mourning Gecko *Lepidodactylus lugubris*[36]. Bouton's Skink was first discovered in Mauritius, but it was later found to inhabit rocky shores across the Indian Ocean and the Pacific. Like the seabirds discussed earlier, the skink, of Pacific-Australasian origin, has developed distinguishable races on different Indian Ocean islands, including Mauritius. DNA work suggests that the Mauritian population is rather distinct from other Indian Ocean races, and from the pattern of haplotypes we suggest that Mauritius was colonised first, Madagascar and the other islands being reached from there[37]. The Mourning Gecko, which in the Mascarenes occurs only on Rodrigues, is an odd lizard. Most populations consist entirely of females that reproduce by parthenogenesis. Although some widespread Pacific clones have been traced to a hybrid origin about 8,000 years ago, the Indian Ocean populations appear to differ from these, and both sexual and parthenogenic lineages occur on Rodrigues, the latter possibly having arisen *in situ* fairly recently[38]. Both these lizard species are well-adapted to sea-drifting. Bouton's Skink is tolerant of salt water, feeding on rocky shores in the spray zone[39], while parthenogenic forms of the Mourning Gecko can colonise an island from just a single individual or egg – and the hard-shelled eggs of geckos are quite resistant to sea water[40]. Even so, Mourning Geckos, of largely Indo-Pacific distribution, have only reached the easternmost islands of the western Indian Ocean: Chagos, Coëtivy (Seychelles) and Rodrigues[41]. The South Asian Cheechak *Hemidactylus frenatus* may be native to some western Indian Ocean islands, but the Mascarene populations are certainly introduced[42].

Dugongs

Dugongs, formerly abundant in the Mascarenes, are coastal feeders, but they can and do explore new areas. Although 390km across open ocean from their range in northern Madagascar, Dugongs were recorded at Aldabra in 1970, 1976 and 2001; one is even said to have turned up back at Rodrigues in 1970[43]. Wandering to the Mascarenes may not have been unusual in the pre-human past; Dugongs perhaps drifted from Australasia[44]; found the lagoon to their liking and then stayed on to breed. Dugongs have declined enormously throughout the Indian Ocean, so a recolonisation of the Mascarenes is unlikely[45].

Before leaving the widespread species we should mention the migratory birds that regularly reach the Mascarenes – temporary visitors, not breeders. Waders nesting in arctic regions migrate south in large numbers in autumn to spend the winter on the shores of lakes and seas far to the south. Several species are common in the Mascarenes (and all Indian Ocean islands) during the northern winter, with a few lingering all year[46]. These birds, in large and regular numbers, are not strays, and their ability to return northwards has been shown by recoveries of ringed birds[47]. While the birds presumably find Mauritius and Réunion directly from the north, they appear to reach Rodrigues eastwards from Mauritius[48]. Why waders should be so adept at commuting to our islands, when other migrants do not, is a mystery[49]. A few seabirds are the only other regular migrants, most consistently Common Terns from Eurasia and the subantarctic southern skuas and Wilson's Stormpetrel. None of these waders or seabirds stop to breed, presumably because the habitat and latitude does not match their normal nesting criteria. There is however a solitary record of breeding Whimbrels (on Flat Island in 1910), and one of us saw a territorial display (which other Whimbrels simply ignored) in 1974[50].

SPECIES SHARED WITH MADAGASCAR

Only four Mascarene vertebrates, all birds, are otherwise found only on Madagascar. All but one are non-passerines. The exception is the Mascarene Swallow *Phedina borbonica*, which is partly migratory in Madagascar, and hence more inclined to inter-island flight than most passerines and so less likely to speciate. Nonetheless Mascarene birds form a distinct subspecies, larger and darker than Malagasy birds[51]. Malagasy Turtle Doves have successfully colonised most western Indian Ocean islands[52], although it is a mystery why this generally sedentary species should fly off over open ocean. Buttonquails are also normally considered poor fliers, yet they have colonised islands across the East Indies and the western Pacific[53], so reaching Réunion from Madagascar is not unexpected. The Madagascar Pochard *Aythya innotata* is known from two subfossil bones found on Réunion – the attribution to species is thus somewhat tentative, and its erstwhile status on the island uncertain[54].

Both the swallow and the turtle dove are anomalous

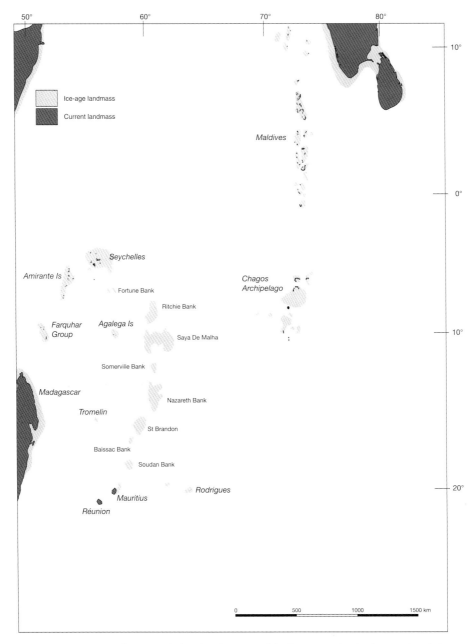

Map 4.1. Position and extent of the islands exposed by lower sea levels during the Pleistocene ice ages. Derived from Admiralty Charts and bathymetric map in Burnett *et al.* (2001).

amongst their allies. There are only two species in the swallow genus *Phedina*, the other occurring in the Congo basin in central Africa, while the dove is hard to place and has been shuffled between *Streptopelia* (standard turtle doves) and *Columba* (the genus including ordinary grey pigeons) by recent authors. It has no very close relatives apart from the Mauritius Pink Pigeon, and its ancestral origin is open to speculation[55].

ENDEMIC SPECIES WITH CLOSE RELATIVES ELSEWHERE

If recent colonists such as Common Moorhens and Striated Herons are so biogeographically complex, how can one sort out species that have been in the islands for much longer and have separated into endemic species and genera? Perhaps surprisingly, some cases are straightforward, though in others the origins are too far back to offer any certainty; worse, some apparently simple cases have been thrown into the melting pot by the advent of DNA analysis, which sometimes produces results that conflict with conventional taxonomy and biogeographical logic.

Animals with endemic species in the Mascarenes but congeners elsewhere can be roughly grouped into four classes: those with affinities in Madagascar, those with Asian connections, those with allies in Australasia, and finally seabirds with diverse or uncertain origins. There are no vertebrates with affinities to Africa (though this has been suggested for the kestrels; see below).

Animals with affinities directly with Asia

The granitic Seychelles, distant as they are from the nearest continent, are visited every year by numerous migrants from northern continents, many of them small passerines. These travellers do not reach the Mascarenes, where the only regular landbird migrants are numerous Palaearctic shorebirds and, much less consistently, a very few Eurasian falcons. Even in the Seychelles none of the smaller migrants are from eastern Asia – they are all European or western Eurasian species that appear to have overshot from Africa, which is much nearer. Larger east-Asian landbirds do reach the Seychelles: Cinnamon Bittern *Ixobrychus cinnamomeus* (as we have seen, the related Yellow Bittern has colonised), Amur Falcon *Falco amurensis*, White-breasted Waterhen (which has colonised the Chagos), Little Cuckoo (which has colonised Madagascar), Brown Fish Owl *Ketupa zeylonensis*, and Pacific Swift *Apus pacificus*[56]. Looking at the Mascarene land birds and bats that are shared with other areas (i.e. those which have arrived relatively recently), we find that all are either from Madagascar or are very widespread and mobile, with none solely from Asia. The inference from both non-arrival of migrants and the connections of undifferentiated species is that our islands are simply too isolated at present for most Asian birds, especially passerines, to reach them – yet Indo-Malesian forest birds and fruitbats are well represented in the fauna.

The answer to this conundrum lies in geological history. Glacial conditions lock up enormous amounts of water in ice-sheets; leading to a significant lowering of sea-level and exposing old islands that had sunk beneath the sea (Map 4.1)[57]. Recent work extends the onset of the most recent bout of glaciations rather further back than previously thought; there is some evidence of a lowering of sea-level around 5 million years ago (mya), and clear indications of a low-stand generated by a glaciation at 2.4mya[58]. However until the mid-Pleistocene, glacial development appears to have been somewhat uneven in different parts of the world, and the effects on sea-level unclear. Given the severity of several ice advances from 1.6 to 0.5mya, the islands would have been exposed more than once[59]. During the last 500,000 years sea-level fell by over 60m on at least five occasions, each time re-creating large islands across the Indian Ocean between the Mascarenes and India[60]. There was thus a series of time windows, from at least 2.4mya until the end of the last glaciation, when lowered sea-levels exposed islands that provided 'stepping stones' along which Asian species could penetrate southward from India across the Indian Ocean (Map 4.1). Since these 'temporary' islands were present for up to 50,000 years at a time (Figure 4.1) there was plenty of time for forests to become established[61] and for them to be gradually colonised by wind-drifted birds and bats, some of which reached the Mascarenes. Sea-levels rose rapidly after *c*. 14,000 years ago, drowning most of the Indian Ocean low islands by or before 7,500 years ago – thus cutting off this avenue of colonisation until the next ice age. Early colonists by this route will have had nearly 2.5 million years to evolve, the latest arrivals only a little over 10,000 years.

These low islands, though flat and rather featureless, would have provided forested habitat for good numbers of birds and fruitbats, and any reptiles and invertebrates that could reach them, much as Aldabra does today. No less than four different lineages of flying-fox reached the southwestern Indian Ocean (three to the Mascarenes), and some bird genera (*Psittacula* parakeets, *Coracina* cuckoo-shrikes, *Terpsiphone* flycatchers, *Zosterops* white-eyes, *Hypsipetes* bulbuls) spread around at least twice, though from which direction is not always entirely clear. Cave-nesting swiftlets are an Asian-Pacific group unrepresented in Africa or even Madagascar, and the parakeet genus *Psittacula* is largely Asian, with only one species extending to northern Africa (but nowhere near Madagascar). Likewise the flying-foxes so characteristic of Indian Ocean islands are part of a large Indo-Australasian and Pacific genus, completely absent in Africa.

Flying-fox colonisation from Asia

Research on Indian Ocean flying-fox DNA indicates that Knud Andersen was right, nearly a century ago, to claim that there had been four independent colonisations from Asia – however, the detail of Andersen's supposed relationships with Indo-Austronesian species does not stand up to DNA scrutiny[62]. Unfortunately recent DNA studies have excluded the extinct Rougette, which Andersen associated with the Island Flying-fox *Pteropus hypomelanus*, found until recently in the Maldives south of India. The Rougette evolved the habit of roosting in tree-cavities or caves, which may indicate an early establishment, but such a change, while very unusual in the genus, might develop

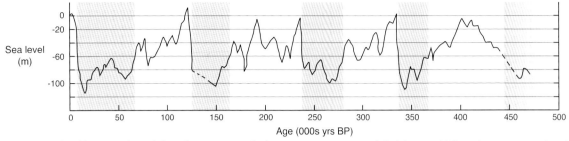

Fig. 4.1. Sea-level low-stands and their duration over the last 500,000 years. Simplified from Siddal *et al.* (2003). Grey bands indicate periods where sea-level was 60m or more below current levels.

quite quickly as an adaptation to cyclones[63]. The DNA of living bats revealed an early invasion with two surviving species, in the Comoros and Pemba (off the Africa coast), while the next oldest appears to be the Golden Bat, whose ancestor split off before the widespread large Asian species *P. vampyrus* and *P. giganteus* evolved. From the latter, still found as near as the Maldives, have evolved the rest of the Indian Ocean forms, now distributed in Madagascar, Aldabra, the Comoros, Seychelles and the Mascarenes; our Black-spined Flying-fox is genetically very close to the Seychelles species *P. seychellensis*. Much of this evolution is apparently recent, though an earlier study that included some of the same species made date estimates that suggested that the Rougette and the Comoros/Pemba pair may have reached the region via the old hot-spot islands (see Chapter 1 and below). Ancestors of the Golden Bat and Black-spined Flying-fox arrived during the Ice-Age low sea-levels[64], during which time there may have been regular movement along the chain to and from the Seychelles. Fruitbats seen on their way to feeding grounds may not look like long-distance fliers, but a Little Red Flying-fox *P. scapulatus* is recorded as having reached New Zealand alive after flying 3,200km from Australia[65], so inter-island hops of a couple of hundred miles do not look unreasonable[66]. The Black-spined Flying-fox demonstrated its mobility by being almost the only vertebrate to spread to all three of our islands; we would not be surprised if, in the future, subfossil material of the Golden Bat turns up in Réunion[67].

Birds of Asian origin

When Con Benson studied Indian Ocean cuckooshrikes, he noticed that the striking sexual dimorphism in colour and pattern seen in the Mascarene pair was mirrored in two Indo-Australasian species, the Common Cicadabird *Coracina tenuirostris* and Black-tipped Cicadabird *C. schisticeps*, occurring widely in the East Indies and in New Guinea, respectively, while the male Madagascar Cuckoo-shrike *C. cinerea* recalled the Black-headed Cuckoo-shrike *C. melanoptera* from India and Sri Lanka[68]. The inference was that a cicadabird-like species colonised the Mascarenes, while Madagascar received its cuckooshrikes from India. The direct flight from the East Indies looks rather far for a cuckoo-shrike, so perhaps a cicadabird relative such as the Indochinese Cuckooshrike *C. polioptera* extended its range down through India at some time in the past. The Mascarene pair are clearly closely related to each other, whereas the Malagasy species, differing sharply in both plumage and voice, looks like an independent colonist from Asia[69].

Another clear-cut case of Asian origin is the presence in the Seychelles and the Mascarenes of cave swiftlets. This group has an Indo-Pacific distribution similar to that of flying-foxes, with one species, the Indian Swiftlet *Aerodramus unicolor*, reaching Sri Lanka[70]. The Seychelles Swiftlet *A. elaphrus* and Mascarene Swiftlet *A. francicus* share a common ancestry according to DNA studies, but the research did not cover the whole genus, so we cannot be certain that the Indian bird was the source; there are other possible candidates[71]. The Pleistocene stepping-stone islands would have had steep coral sides with former under-sea caves and solution caverns for swiftlets to nest in, much as Glossy Swiftlets *Collocalia esculenta* do on Christmas Island[72].

Recent DNA work on *Psittacula* parakeets supports the generally accepted view that the Echo Parakeet *Psittacula eques* is close to the Ring-necked Parakeet *P. krameri*, and furthermore establishes that it is closer to Indian birds than is the African form, normally considered a subspecies[73]. The Mauritian bird differentiated from Asian stock about 1.7mya on molecular clock dating, around the time of the second glaciation severe enough to substantially lower sea-levels[74]. However some doubt must attach to this DNA dating, as the Indian Ring-necked Parakeets introduced to Mauritius are suggested by the molecular clock to have separated from Indian stock 400,000 years ago, when in fact we know they were introduced from India in the 1880s![75] This is possibly explained

▲ A Mauritian lowland dry forest; a Dodo *Raphus cucullatus* preens in a woodland of Black Ebony *Diospyros tessellaria*, Tambalacoque *Sideroxylon grandiflorum*, screw pine *Pandanus* sp. and palm *Hyophorbe amauricalis*. For a full species list see Appendix 14c.

▲ One of the world's largest parrots, the now-extinct Raven Parrot *Lophopsittacus mauritianus* (male on right, female left) extracts seeds from a *Sideroxylon* fruit, alongside the only surviving Mascarene parrot, the Echo Parakeet *Psittacula eques*.

▲ Mauritius Sheldgeese *Alopochen mauritianus* (left) feed in a lakeside forest with Mascarene Teal *Anas theodori* (bottom right) and Mauritius Night Heron *Nycticorax mauritianus* (centre right); all are now extinct.

▲ Commerson's Lizard-owl *Mascarenotus sauzieri* with Telfair's Skink *Leiolopisma telfairi* prey. This reconstruction is based on a drawing and description of the extinct owl; the skink is now restricted to Round Island.

▲ A Mauritius Bulbul (Merle) *Hyspipetes olivaceus* (bottom) and male Mauritius Cuckoo-shrike *Coracina typica* (top right) mob a Réunion Harrier *Circus maillardi.* (top left). The harrier is only known from fossils on Mauritius but it survives on Réunion.

▲ A Pigeon Hollandais *Alectroenas nitidissima* on the critically endangered ebony *Diospyros egrettarum*.

▶ The snakes and lizards of Round Island, with Serpent Island in the background. For a full species list see Appendix 14e.

▲ A day in the life of the Mare aux Songes. See Appendix 13 for details on recent excavations at the site, and Appendix 14d for a list of the species illustrated here.

▲ Beneath the Trois Mammelle and Rempart Mountains. Mauritius Pink Pigeon *Nesoenas mayeri* (centre); Mauritius Fody *Foudia rubra* (right, on tree); Mauritius Olive White-eye *Zosterops chloronothos* (bottom) feeding on the critically endangered *Trochetia blackburniana*; Mauritius Kestrel *Falco punctatus* soaring above (left) with Mascarene Swiftlets *Aerodramus francicus* (right); Upland Forest Day-gecko *Phelsuma rosagularis* (far right, on tree).

▼ A Red Hen *Aphanapteryx bonasia* locates food in the form of the large, extinct Mascarene snail *Tropidophora carinata* (bottom left). One of the world's largest skinks, Didosaurus *Leiolopisma mauritiana*, bask and feed on the fruit of *Pandanus* sp.; they dwarf a Blue-tailed Day-gecko *Phelsuma cepediana* (right, on trunk).

▲ The Dutch on Mauritius in 1602. The woodcut on which this reconstruction is based constitutes the only known image of Thirioux's Grey Parrot *Psittacula bensoni*. A captured bird was made to squawk, attracting other members of the group, so that all of the flock could be taken. Dodos are being clubbed in the background.

▲ The large, flightless Dubois's Wood-rail *Dryolimnas augusti* feeds among Réunion Tortoises *Cylindraspis indica* in a grove of Screw Pines *Pandanus utilis* and Hurricane Palm *Dictyosperma album*.

▲ Mascarin (Mascarene Parrot) *Mascarinus mascarinus*, depicted in its natural coloration of greys rather than the browns seen in the two surviving museum skins.

▲ Réunion Echo Parakeet *Psittacula eques*. This parrot disappeared from Réunion in the 18th century, not long after settlement of the island began.

▲ Dubois's Parrot. This enigmatic bird was described from life only once, and never seen again. Its relationships are uncertain.

◀ Réunion Ibises (Réunion Solitaires) *Threskiornis solitarius* on the side of the Piton des Neiges volcano.

▲ Breeding pair of 'oiseaux bleus' on the Plaine Caffres. This species is known from only a few reliable accounts, and appears to have been a large gallinule *Porphyrio* sp. In the tree above, Réunion Olive White-eyes *Zosterops olivaceus* feed on *Hypericum lanceolatum*.

▼ Etang de Saint Paul, Réunion. This large lake once harboured a number of waterbird species. For a full species list see Appendix 14a.

▲ Dubois's Kestrel *Falco duboisi* hunts a Manapany Day-gecko *Phelsuma inexpectata* (left, on trunk), while Grey Tomb Bats *Taphozous mauritianus* (left, below) roost in a tree cavity; White-tailed Tropicbirds *Phaethon lepturus* (centre left) and a Réunion Harrier *Circus maillardi* (centre right) fly over the forest canopy. The spectacular Palm Orchid *Angraecum palmatum* (right) has not been seen since the 1980s and may well be extinct.

▼ Réunion Fody *Foudia delloni* nov. sp. (male top left, female top right) with male Réunion Stonechat *Saxicola tectes* (bottom left) and male Réunion Cuckoo-shrike *Coracina newtoni* (bottom right).

▼ Mascarene Teal *Anas theodori*. This small duck occurred on both Réunion and Mauritius.

▲ Seabirds above the cliffs at La Possession. For a full species list see Appendix 14f.

▲ Black-spined Flying-foxes *Pteropus niger* flying over Red Latan Palm *Latania commersoni*.

▼ Arnold's Skink *Leiolopisma* sp., with a damselfly *Enallagma glaucum*. This undescribed skink is only known from the fossil record.

▲ Rougette *Pteropus subniger* (roosting), compared with Golden Bat *P. rodricensis* (centre) and Black-spined Flying-fox *P. niger* (right). The Rougette is one of only three *Pteropus* bats known to roost in hollow trees.

▼ A Hoopoe Starling *Fregilupus varius*. This starling was abundant on Réunion and a popular cagebird, but it disappeared suddenly and mysteriously in the mid-19th century.

▼ Réunion's pigeons in a gorge above St Louis. Malagasy Turtle Dove *Nesoenas picturata* (top left); Réunion Pink Pigeon *Nesoenas duboisi* (centre left); and slaty pigeon *Alectroenas* sp. (centre right).

▲ Rodrigues Night Herons *Nycticorax megacephalus* stalk Newton's Day-geckos *Phelsuma edwardnewtoni* on a Rodrigues Spindle Palm *Hyophorbe verschaffeltii*.

▼ Carosse Tortoises *Cylindraspis vosmaeri* feeding on the fruit of Rodrigues Spindle Palms *Hyophorbe verschaffeltii*.

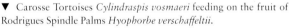

▲ Blue morphs of the Rodrigues Parakeet *Psittacula exsul* feed on *Hibiscus lilliflorus* (above) and *Syzygium balfouri* (below).

▼ Liénard's Giant Gecko *Phelsuma gigas* raids a Rodrigues Fody *Foudia flavicans* nest, which hangs from a bough of the endemic *Scyphochlamys revoluta*. A Rodrigues Warbler *Acrocephalus rodericanus* (centre top) and Rodrigues Bulbul (Rodrigues Merle) *Hypsipetes* sp. (bottom) are attracted to the commotion.

▲ Rodrigues Starlings *Necropsar rodericanus* attack hatchling Green Turtles *Chelonia midas* on Ile Pierot, situated in the lagoon; a Lesser Frigatebird *Fregata ariel* soars overhead.

▲ Rodrigues Doves ?*Nesoenas rodericana* on the critically endangered *Zanthoxylon paniculatum*, above the Anse Quitor Valley.

▲ Ile Frégate, Rodrigues. Dugongs *Dugong dugon* abound in the surrounding lagoon, while seabirds breed in huge numbers on the islet. For a full species list see Appendix 14b.

▼ Carosse Tortoises *Cylindraspis vosmaeri* (left) and Rodrigues Domed Tortoises *Cylindraspis peltastes* (centre) graze on the Plaine Corail, while Leguat's Rails *Erythromachus leguati* hunt for disturbed insects.

▲ Tafforet's Lizard-owl *Mascarenotus murivorus* pounces on an undescribed giant night-gecko *Nactus* sp. (centre bottom), the largest of the Mascarene species. Golden Bats *Pteropus rodricensis* are one of the few *Pteropus* species to forage on the ground. Land crabs (left) and hermit crabs (centre) were once an important part of the terrestrial fauna on Rodrigues.

▼ The large-headed, long-tailed Leguat's Parrot *Necropsittacus rodericanus*, feeding on the nuts of Bois d'Olive *Cassine orientale*.

▲ Rodrigues above present-day Port Mathurin as described by Francois Leguat, where he claimed he was able to walk for a hundred paces solely on the backs of giant tortoises.

▼ A pair of Rodrigues Solitaires *Pezophaps solitaria* chase an intruding male from their territory. Leguat's Rails *Erythromachus leguati* (bottom left) skulk under the cover of *Zanthoxylum paniculatum*, a tree that, like *Scyphochlamys revoluta* (right), is now critically endangered.

by Mauritian birds being of the northern race *borealis*[76], whereas the Indian ones sampled were of the southern race *manillensis*. While 400,000 years of divergence between the two races is possible, some Mauritian birds actually show characteristics of *manillensis* (females with the lower mandible dark not red), so this explanation is not very convincing[77]. Another parakeet, closely related to the Indo-Malaysian Alexandrine Parakeet *P. eupatria*, survived into the 19th century in the Seychelles, and subfossil bones reveal similar birds in Mauritius, apparently the 'grey parrot' of the old accounts[78]. Furthermore, the Rodrigues Parakeet, also extinct, was also derived from the 'alexandrine' type but had a much modified breast-bone, and was apparently an earlier invader than the 'ring-neck'-type birds that led to the Echo Parakeet[79]. *Psittacula* parakeets colonised twice or possibly three times during the Plio-Pleistocene, while their ancestors also invaded earlier (p. 71).

It was not only the Mascarenes that received animals from Asia during the Pleistocene sea-level lowstands. Genera shared with Madagascar and some other relevant species are discussed below, but in addition both the Seychelles and Madagascar have a magpie-robin species each, the Seychelles an oriental scops owl, Madagascar also a hawk owl, a cuckoo and a nightjar, and the Comoros a drongo, all of Asian affinities[80].

Conventional taxonomy never made much progress in elucidating the origins of the Mauritius Kestrel, differing sharply from its nearest neighbours in Madagascar and the Seychelles. The sexes are alike in both the Mauritius Kestrel and the Madagascar/Seychelles forms, but in Mauritius they have typical kestrel 'female' plumage, whereas in the Seychelles they are 'male'-type and in Madagascar there are two colour phases, one male-type, the other female-type. The Mauritius bird also has unusually short wings and long legs, like a sparrowhawk, an adaptation to hunting under the forest canopy[81]. DNA study suggests that the Mauritius Kestrel derives from ancestral kestrel stock allied to the Greater Kestrel *Falco rupicoloides*, that reached Mauritius directly from Africa, or via a brief stay on Madagascar, around 1.7mya – well before the now widespread Common Kestrel *F. tinnunculus* evolved[82]. The molecular clock, if correct, suggests that ancestral kestrels reached Mauritius, as with the parakeet, around the time of the glaciation on the Pliocene-Pleistocene boundary[83]. Whatever the source and date for the Mauritius Kestrel, the extinct Dubois's Kestrel in Réunion is apparently a later independent arrival. It was larger and lacked the extreme wing-shortening adaptations shown by the Mauritian bird, but there is too little bone material to explore its relationships further. Dubois called it an *émerillon*, or merlin, so was it predominantly grey – i.e. 'male'-type, like the Madagascar and Seychelles species?[84]

Animals whose ancestors have affinities with or close relatives in Madagascar

In connection with Madagascar, we find a further three sub-groups – animals with an Asian affinity, those from an originally African source, and Malagasy endemic genera with diverse or unknown origins. The first group includes the Black-spined Flying-fox, the teal, the harrier, the merles, the white-eyes and the free-tailed bat. The second group comprises mostly waterbirds (herons, ibis, geese, coots and moorhens), and several small passerines (stonechat, flycatcher, warbler). The third group consists of *Dryolimnas* rails, blue pigeons, fodies and day-geckos.

Madagascar: Asian affinities

Of species with close relatives on Madagascar but ultimately with Asian affinities we cannot tell whether the Mascarene populations reached the islands from Madagascar via the Seychelles, or colonised the Mascarenes first. The Black-spined Flying-fox is part of the *rufus* group represented in Madagascar, but its DNA places it closest to the Seychelles Flying-fox, not the Malagasy *rufus*[85]. Likewise, free-tailed bats may have reached Madagascar first then the Mascarenes, or the other way round, or, given that the Mascarene species is "without close relatives in *Mormopterus*"[86], there may have been two separate colonisations from the east.

The merles (*Hypsipetes* bulbuls) in Mauritius, Réunion and the Seychelles look at first sight like the survivors of an earlier colonisation than the birds now found on Madagascar – these last are so like their Asian cousins that they are often put in the same species. It has been suggested in the past that Madagascar and the Seychelles/Mascarenes may have been colonised independently[87], but DNA studies support a single invasion from Asia, the birds initially spreading from the Seychelles west to the Comores and south to the Mascarenes. Madagascar was then colonised from the Mascarene population[88], reverting to its Asian appearance on the continental island. From Madagascar a second wave then fanned out to Aldabra and the Comoros, where on Grande Comore and Mohéli there are now two species, an upland greenish species from the first batch, and a lowland grey one from the second[89]. The merles' initial spread from Asia is placed (with caveats) at more recent than 2.6mya – i.e. towards the beginning of the Ice-Age lowering of sea-levels.

Two waves of white-eyes have colonised the Mascarenes, but from where? The Mascarene species don't look or sound anything like the ordinary-looking *Zosterops maderaspatana* on Madagascar, and are equally unlike anything in the other likely source, Asia, though Reg Moreau thought the bills of olive white-eyes recalled the Sri Lankan species *Z. ceylonensis*[90]. The olive white-eyes, with extra-long, thin nectar-feeding bills, brush tongues, habits mimicking sunbirds, and a distinct species each on Mauritius and Réunion[91], are clearly the earlier invaders, while the Mascarene Grey White-eye, a generalist shared between the islands, came later. Ben Warren's recent DNA work suggests the Indian Ocean has indeed been invaded by two waves of white-eyes, the first from Asia, the second from Africa, followed by a later surge from the latter's descendants in Madagascar. But *both* Mascarene types belong to the earlier Asian wave, which must have invaded our islands twice, perhaps in different Ice-Age periods when the stepping stone islands were above water. According to the DNA study, the oldest Indian Ocean species is the recently extinct Seychelles Chestnut-flanked White-eye *Z. semiflava*, suggesting that for white-eyes, as with fodies, merles, swifts and warblers, the Seychelles were a key focal point for birds moving along the island chain – no doubt because, unlike the coral islands, they did not regularly disappear under water.

The cases just cited represent a reversal in the previously assumed pattern of avian (and reptile) dispersal. It has generally been considered that "insular endemics are evolutionary dead-ends" resulting from an "apparent unidirectional movement from continents to islands over evolutionary time". However, it is now clear that merles, white-eyes and fodies, and possibly flying-foxes differentiated on the islands and then colonised Madagascar, a pattern now known to have also happened in the Pacific and, for lizards, the Caribbean[92].

Subfossil bones show the Mascarene Teal was closely related to Bernier's Teal from Madagascar, and closer still to the Sunda Teal, both members of a group known as 'grey teals' – another case of dispersal from the Indo-Australasian east. The Madagascar bird has brown plumage, but Mauritian ducks were described as 'grey teal' in the only surviving description, also suggesting a greater similarity to the ancestral group than to the Malagasy species[93].

What used to be thought a simple Asia–Madagascar–Réunion sequence for harriers has been upset by recent DNA work. The Réunion and Malagasy Marsh Harriers, considered either races of one species or sister species, have generally been associated with the Eastern Marsh Harrier *Circus spilonotus* from China[94], but recent work implies that the Indian Ocean forms and the migratory Western Marsh Harrier *C. aeruginosus* derive from the Australian Swamp Harrier *C. approximans*[95]. This is despite Réunion and Madagascar birds looking neither like Swamp Harriers nor Western Marsh Harriers, but resembling Eastern Marsh Harriers (and even Pied Harriers *C. melanoleucos* of India/Sri Lanka); harrier plumage may be very plastic. Very few Palaearctic migrants have established breeding populations south of the equator: Booted Eagles and Black Storks breed in South Africa[96], and Little Cuckoos colonised Madagascar, with their descendants recently elevated to full species status[97].

Madagascar: African affinities

The extinct Réunion Ibis, the sheldgeese and the Mascarene Coot are all clearly descended from Malagasy forms with affinities in Africa. The sheldgoose genus *Alopochen* is purely African, and although there are sacred ibises *Threskiornis* in Asia and Australia, and coots all over the world, the Mascarene and Malagasy species relate best to African forms[98]. The same may apply to the '*oiseau bleu*', presumed to be a *Porphyrio*-type swamphen (though swamphens are also common in South Asia) and the night herons, but we have no bones of the *oiseau bleu*, and the Black-crowned Night Heron, source of the Mascarene forms, is so widespread that the original colonists could as easily have come from Asia as from Madagascar. Night herons must in any case have colonised twice, as the Mauritius and Rodrigues species were nearly flightless, so could not have spawned the longer-winged Réunion birds[99].

The affinity of Mascarene Paradise Flycatcher *Terpsiphone bourbonnensis* with the Malagasy species *T. mutata* is evident to those who know both in the field, but it is less clear what the source for *T. mutata* might have been, although Africa is likely. Seychelles Paradise Flycatchers *Terpsiphone corvina* are larger with a very different feel, and must have had a different ancestor[100]. The Seychelles Warbler *Acrocephalus sechellensis* appears from song-type and DNA analysis to be very close to the Madagascar Swamp Warbler *A. newtoni*, in turn derived from African reed/swamp warblers of the *gracilirostris* group[101]. Rodrigues Warblers *A. rodericana* are so similar to Seychelles birds that one must presume the same origin, probably from the Seychelles down the island chain when the sea-level was lowered. An African ultimate origin is clear enough for the Réunion Stonechat, via the Malagasy race *sybilla* of the widespread African Stonechat *Saxicola axillaris* (now taxonomically split from European and Asian Stonechats)[102].

Madagascar: Groups of indeterminate origins

Turning to genera endemic to Madagascar, *Dryolimnas* rails have spread to Aldabra, Assumption, Réunion and Mauritius. An early colonisation of the Mascarenes led to near-flightless wood-rails on both islands, both now extinct[103]. That *Dryolimnas* rails, despite being normally sedentary, can and do cross the ocean is proved by the collection in 1809 of a Malagasy White-throated Rail *D. cuvieri* in Mauritius. These birds are not particularly close to other rail genera, though morphological study associates them with the widespread *Rallus* and the Indo-Pacific *Gallirallus*[104]. The Buff-banded Rail *G. philippensis* breeds in Indonesia and Christmas Island, and has also straggled across to Mauritius[105], but *Dryolimnas*'s ancestors must have reached Madagascar in the Pliocene or earlier.

The Mauritius Pink Pigeon *Nesoenas mayeri*, and its extinct Réunion counterpart *N. duboisi*, are fairly ordinary pigeons, generally put in their own genus by characters intermediate between typical woodpigeons *Columba* and turtle doves *Streptopelia*. DNA studies show that their common ancestor diverged long ago into four clades: two large ones (woodpigeons, turtle-doves), and two small ones, one of which consists of the Malagasy Turtle Dove and the pink pigeons[106]. This does not resolve their ultimate ancestral origin, though the two pink pigeons presumably came to the Mascarenes from Madagascar, followed later by the Malagasy Turtle Dove as we now know it. These three, in the Indian Ocean endemic genus *Nesoenas*, are relics of an early radiation of proto-woodpigeon/turtle dove stock, which also includes *Stigmatopelia* (Spotted and Laughing Doves, forming the other small lineage). The big woodpigeon/turtle dove divide began about 8–9mya, while the two small sub-groups split into different lines around 7–8mya[107]. While Spotted and Laughing Doves evolved and radiated through Asia (both) and Africa (Laughing Dove), the pink pigeon/Madagascar Turtle Dove line developed separately on Madagascar, splitting around 1.5mya, when a proto-pink pigeon colonised Mauritius and Réunion. A second invasion by what had become the Malagasy Turtle Dove happened much more recently. While two of the three Mascarene columbiform groups appear to have connections to New Guinea (dodos and *Alectroenas* blue pigeons), there are no turtle doves and only three woodpigeons in Australasia, so we need to seek the origins of the *Nesoenas* pigeons elsewhere. Since both *Stigmatopelia* doves occur in India, that area may be the original source of *Nesoenas*. 7mya is too late for birds to spread along the original hotspot islands, and too early for the Ice-Age re-emergence islands. However the Malagasy Turtle Dove is one of the few land birds to colonise low coral islands, so its ancestor may have spread from Asia to Madagascar using atolls.

The Rodrigues Dove ?*Nesoenas rodericana* is known only from brief accounts and a few subfossil bones. The single sternum then known led Cécile Mourer and her colleagues to associate it with *Gallicolumba* ground doves, centred in New Guinea and the Pacific[108]. However more material has come to light and the bird now appears to be another early offshoot of the same proto-*Columba*/*Streptopelia* stock as the pink pigeons[109].

Blue pigeons and fodies are two other bird groups endemic to the Malagasy sub-region with no close relatives elsewhere. Both have spread to most western Indian Ocean islands, forming new species where they have settled. Madagascar has always been presumed to be the centre of these radiations, but with minimal diversification within the big island itself (one blue pigeon and two fodies only), the evidence for this is not good. However the fodies' affinity is, distantly, with African bishop-birds *Euplectes*[110], so Madagascar or the Comoros are the obvious obvious stepping stones. But, as so often is the case, the obvious is not supported when looked at in detail; DNA evidence suggests the oldest fodies are the most far-flung: Seychelles then Rodrigues, with the Mauritius Fody derived from them, and the Comoros species descended from the Mauritius bird, not the other way round[111]. The two Malagasy species arose independently from the Mauritius Fody's precursor. It therefore appears that the original colonisation was from Africa to the Seychelles, birds then spreading down the Ice-Age islands to the Mascarenes, whence they colonised *westwards* (with the wind) to the Comoros and Madagascar. There a species adapted to grasslands developed, becoming in its turn, with humanity's help, a very successful coloniser. Once suitable (agricultural) habitat had been created on the smaller islands, these Cardinal Fodies (*Foudia madagascariensis*, usually referred to as Madagascar or Red Fodies in the ornithological literature) established themselves almost wherever they were released[112].

The blue pigeons, *Alectroenas*, on the other hand, have distant affinities with Indo-Pacific fruit pigeons, so could have colonised the Mascarenes or the Seychelles (or a hotspot island now drowned) and evolved into a separate genus before arriving in Madagascar. At present we do not know which of the 4–5 species is nearest to the ancestral form, though DNA analysis indicates that *Alectroenas*'s nearest relative is the Cloven-feathered Pigeon *Drepanoptila holosericea* from New Caledonia, from which, on a rough calculation, it separated about 8–9mya; *Ptilinopus* fruit doves are ancestral to both [113].

The day-geckos *Phelsuma*, very diverse in Madagascar and with species on most western Indian Ocean islands (and an outlier on the Andamans), may have spread around and speciated while the extra islands allowed shorter-distance rafting, but the Mascarenes were apparently colonised only once, by an ancestor that evolved consistent differences with existing species elsewhere[114]. Günther's Gecko, the surviving form nearest to the ancestral Mascarene colonist, has a vertical pupil in the eye, whereas in the genus as a whole the pupil is round; this looks on the face of it like a link to the bronze geckos *Ailuronyx* of the Seychelles, very *Phelsuma*-like lizards that still retain claws, nocturnal behaviour and a strongly vertical pupil[115]. Mascarene day-geckos (and bronze geckos) all show another putatively primitive character – they glue their eggs to a surface, whereas most other *Phelsumas* lay loose eggs in crevices. Egg-glueing is characteristic of rock-living geckos rather than tree-living ones, suggesting that Mascarene forms are closer to a rocky country ancestor than Malagasy species, which have mostly lost the glueing habit[116]. DNA evidence, however, does not support any link to bronze geckos, but a radiation spreading from Africa via Madagascar to the Mascarenes around 4–5mya. The two extinct Rodrigues species show up, together with Günther's, as basal to the Mascarene radiation, their nearest sampled relatives being a cluster of Seychelles, Aldabra and northern Madagascar species, the bronze geckos being a separate, much older lineage. At the date suggested there would have been no large islands available to facilitate dispersal, though forested atolls would have sufficed. West of Réunion the currents divide north and south round Madagascar, the northern component no doubt responsible for carrying Réunion day-geckos to colonise Agalega (990km north of Mauritius), the only low island with a Mascarene, not Malagasy, *Phelsuma* lineage[117].

The extinct Carié's Blind-snake is probably best placed in this group of species of indeterminate origins. The genus *Typhlops* is almost world-wide and very ancient, so it is difficult to know where a blind-snake might have come from. There are several species in Madagascar, but a drift from India/East Indies cannot be ruled out; the Australian blind-snakes are in a different genus[118]. There is too little subfossil material to assess the affinities of this Mauritian endemic.

Animals with Australasian affinities
Apart from the debatable case of the cuckoo-shrikes, the only Mascarene vertebrates originating in Australasia were lizards (geckos and skinks). If the currents in the southern Indian Ocean during the Pleistocene were similar to the present, then drift from Australia, distant though it is, would be more likely than from Madagascar – the South Equatorial Current flows steadily from both western Australia and along its northern coast directly towards the Mascarenes[119]. Rodrigues would be the obvious first landfall. A population established there could then raft on to Mauritius (and thence to Réunion). However, there is no trace of skinks in Rodrigues, so the ancestors of *Leiolopisma* skinks, and more recently the tiny *Cryptoblepharus*, may have rafted straight to Mauritius, or (in the case of *Leiolopisma*) to the older island of St Brandon (see below). As *Leiolopisma* is now considered an endemic Mascarene genus, these skinks are considered below (p. 72).

The genus *Nactus*, night-geckos with claws (not pads), comprises three species in Australia and the Pacific, from one of which sprang the five Mascarene species, of which only two survive, both on islets off Mauritius. The two extinct Rodriguan species were the largest in the genus. These two probably represent the earliest colonisation, though Mauritius was reached from there before they split into two species, the first Mauritian line, *N. coindemirensis*, colonising Réunion once it became habitable. After a further period, but still before its two species had diverged, more Rodriguan geckos rafted to Mauritius giving rise to *N. serpensinsula*, now divided into two well-marked subspecies[120]. Earlier morphological work had suggested the two living Mauritian species derived from different Australasian ancestors, but DNA studies have since shown this was not the case[121].

Seabirds of unclear origin and affinities
The birds discussed here are petrels, all traditionally placed in the gadfly petrel genus *Pterodroma*. However, a morphological reassessment supported by DNA work has established that the former subgenus *Pseudobulweria* consists of a rather ancient group related to shearwaters with a quite different history to the true gadfly petrels, whose affinities lie with the fulmars[122]. The rare Réunion Black Petrel *Pseudobulweria atterima* joins in *Pseudobulweria* a group of petrels that are almost all enigmatic, being either extinct or known from only a handful of specimens. Only the Tahiti Petrel *P. rostrata*, widespread across the Pacific, is reasonably well-known. Most of the known forms are from the Pacific, but the extinct St Helena Petrel *P. rupinarum* lived in the Atlantic, and Réunion's *P. aterrima* is known from subfossils to have also occurred on Rodrigues. The known breeding sites of *Pseudobulweria* petrels are all high up in mountainous islands in a band between 20°S and 5°S, spread across the three oceans, birds coming on land only at night; given the apparent age of the genus it is not possible to say in which ocean it originated[123].

Like *Pseudobulweria* petrels, the herald petrel group breed widely across the Pacific, with one station each in the Indian and Atlantic Oceans, all at roughly the same latitude; Mauritian birds on Round Island are generally considered conspecific with *Pterodroma arminjoniana* from Trindade Island, off Brazil (though this is under review). The herald group differs from *Pseudobulweria*, however, in nesting in the open or under rocks on low islands, and in being active on land during daylight. They are closely allied to another low-island surface-nester, the Kermadec Petrel *Pterodroma neglecta*, normally based in the South Pacific near New Zealand. A Pacific origin is likely for this complex, but there is conflicting evidence as to whether the 'Round Island Petrels' originate from Herald, Trindade or Kermadec Petrel stock – and the recent presence of actual Kermadec Petrels, identified by calls and white primary-shafts, only confuses the issue[124]. 'Round Island Petrels', very variable in plumage, are genetically more diverse than Pacific Herald Petrels or Kermadec Petrels, so a hybrid orgin, or introgression between wings of a widespread species is not ruled out. The recent arrival of Kermadecs suggests that these and other petrels range widely at sea and sporadically colonise new sites – as seen also by the appearance in the past few years of typical Herald Petrels *P. heraldica*, Black-winged Petrels *P. nigripennis* and Bulwer's Petrels *Bulweria bulwerii*, the latter now regularly breeding on Round Island[125]. Black-winged Petrels are also from Kermadec Island (north of New Zealand), and like Kermadec Petrels are subtropical feeders, whereas the herald group are basically tropical. At 20°S, Round Island is near enough the southern edge of the tropics to be potentially attractive to both tropical and subtropical species, which could forage respectively north and south from the island. Unfortunately little is known of the range of 'Round Island Petrels' at sea.

The third group of petrels on the islands, represented by Barau's Petrel *Pterodroma baraui* on Réunion, are high mountain nesters that come ashore at night. We discussed these birds in the opening paragraphs of this chapter to illustrate the conundrum of the endemic acacia tree in Réunion. An origin in Hawaii, or at least the Pacific, remains the best hypothesis, but until more extensive work has been done on the DNA, calls, feather lice and other critical taxonomic characters in *Pterodroma*, it may not be possible to fully unravel this very tangled genus – particularly where there is an ongoing flux of intermixing and hybridisation as suggested for the Herald/Kermadec complex. Barau's Petrel is, however, a well-defined species with no plumage variation and a regular breeding season during the southern summer, so its relationships may be readily resolved once fuller studies are done.

Although not a Mascarene endemic, we include Abbott's Booby here to emphasise its ancient relict nature and the fact that two of its four recent breeding stations were in the Mascarenes. This large, long-winged, tree-nesting booby is not closely related to other tropical boobies or to cold-water gannets. Although only known historically from the tropical Indian Ocean, it formerly occurred also in the Pacific; the last surviving colony is on Christmas Island south of Java[126].

MASCARENE ENDEMIC GENERA WITH OBSCURE OR DISTANT AFFINITIES

Animals in this section range from genera quite close to ancestral forms, such as the *Leiolopisma* skinks, to creatures so far removed from anything else on the planet that they are put in a family of their own, known only from the Mascarenes (the Mauritian snakes). For these animals, the oldest elements in the islands' fauna, the time-series of islands appearing at intervals as the Indian and later the Afro-Malagasy plate moved over the Réunion hotspot becomes essential to our understanding of their biogeography. As India moved north during the late Cretaceous and early Tertiary, new eruptions every few million years produced a series of islands that have since dwindled to coral atolls or submerged undersea plateaux (Maps 1.2 and 4.1)[127]. The earliest series, now the Lakshadweep Islands, has apparently not been dated, but must be between 65 and 57mya, the latter date being recorded for the northern Maldives, the next batch. The Lakshadweeps and the large Padua Bank are only about 220km off the (modern) coast of southern India, within fairly easy range of flight and rafting from the mainland[128]. The Lakshadweep/Padua group would still have been large, high islands 57mya when the Maldives started emerging, another 240km to the south, with Maliku (or Minicoy) between, and still within range of India (330km). At its peak the Maldive group would have been a large island chain stretching some 850km from north to south, with the largest island at least 600km long. Only 9 million years later, at 49mya, the Chagos emerged, a further 400km south including an island of 180 × 120km over what is now the Great Chagos bank. The next step in the chain is complicated by the spreading sea floor of the Mid-Indian Ocean (or Carlsberg) Ridge, splitting off the next eruption of the hotspot from its original proximity to the Chagos. This new eruption, at *c.* 45–38mya, formed the Saya de Malha Bank. We do not know how close it was originally to the Chagos, but looking at their current positions relative to the mid-ocean ridge, it may have been as little as 160km. Saya de Malha was one of the biggest Indian Ocean islands (c. 280 × 210km), and over geological time it has drifted south and west as the Chagos drifted north and east, so their separation is now 875km.

The opening of the Carlsberg Ridge may have been critical in the moulding of the fauna to be passed on to the later islands in the chain, as it would have been harder and harder for animals to reach Saya de Malha. The ancestors of dodos, flightless rails and snakes may have reached Saya de Malha from India via the Maldives and the Chagos[129]; after about 30mya Saya de Malha was progressively cut off from further immigration from that direction, by the expanding sea floor and by the erosion and subsidence of the Chagos Bank into a series of small atolls[130]. Saya de Malha is quite close to the granitic Seychelles, a fragment of Gondwanaland that split off from India and become isolated in the central Indian Ocean. There ancient lineages of plants, insects, amphibians and lizards indicate the continuous presence of land throughout its separate history[131]. There may therefore have been some interchange with the Seychelles (plants and some insects, perhaps lizards) before Saya de Malha passed some of its fauna on south. Nazareth Bank (35mya) is only 210km south of Saya de Malha and more or less contiguous with St Brandon, dated at 31mya, together making an island pair similar in size to Saya de Malha. The distance between Saya de Malha and Nazareth is similar to that between Mauritius and Réunion today, and it is likely that these islands would have shared many species evolving over their period of existence. The interval from St Brandon's emergence to the appearance of Mauritius and Rodrigues is the longest in the hotspot's history, over 20 million years; by that time, around 10mya, St Brandon would have been an ageing island with a well-developed endemic fauna and flora. Mauritius is 385km southwest of the nearest part of the St Brandon bank, with two guyots in between that may also have been high islands for a few million years – the larger, Soudan Bank, only 13m below the surface even now, has an area half the size of Mauritius above the 200m undersea contour. At around 15mya volcanic activity began around the Rodrigues triple point, where three mid-ocean ridges meet, with the island itself appearing around 10 million years ago.

We think Rodrigues and Mauritius were independently colonised from Nazareth/St Brandon by proto-dodos, big rails and parrots, owls, and tortoises, while Mauritius alone was reached by snakes and skinks. Blue pigeons may have had a history on these hotspot islands before reaching the Mascarenes; other candidates, such as fodies and day-geckos, now look from DNA studies too recent to have evolved on the old, high islands. As in the Mascarenes today, St Brandon 10 million years ago would have had a mixture of old and new faunal elements. Rodrigues is 450km southeast of St Brandon, 130km closer than it is to Mauritius. We have no information about late Miocene ocean currents, but as the equator and continents were not far from their present positions, the currents may have been similar. This would have made it relatively easy for drifted species to come from Australasia, and for seeds and rafts to go from Rodrigues or St Brandon to Mauritius (Réunion did not yet exist), but harder to go from St Brandon to Rodrigues, and very hard for anything to reach Rodrigues directly from Mauritius.

Birds
The big flightless rails on Mauritius and Rodrigues, the Red Hen and Leguat's Rail respectively, are not easy to relate to any existing rails elsewhere, though they are clearly allied to each other[132]. The best that can be said is that they show some apparent affinity to equally flightless extinct rails from New Zealand, one of which, the Giant Chatham Island (or Hawkins's) Rail was as large as the Mascarene pair[133]. As rails can become flightless and speciate very fast, these birds may not be as ancient as they appear, but the loss of feather structure in the Red Hen is more marked than in any surviving flightless rail, suggesting a long history. The best guess is that a lineage of volant Asian rails made its way down the hotspot islands to Mauritius and Rodrigues[134].

Dodos and solitaires diverged so far in shape and size from their pigeon ancestors that they have generally been accorded their own family, the Raphidae[135]. So odd are they that early taxonomists (without, at that time, specimens) aligned them with ratites, swans and even vultures[136], and they have more recently been associated with rails[137]. However since Professor Reinhardt of Copenhagen suggested to Strickland in 1845 that their real allies were pigeons, this has been the consensus view, amply confirmed by detailed statistical analysis of skeletons and also by DNA from subfossil bones[138]. But which pigeons are they closest to? The front-runner, from Strickland onwards, has generally been the Tooth-billed Pigeon *Didunculus strigirostris* from Samoa in the Pacific, mainly because it is the only living pigeon with a large, vaguely dodo-ish beak. Hachisuka, author of the other classic Dodo book, championed the *Goura* crowned pigeons from New Guinea – they are the biggest living pigeons, and, like Dodos, semi-omnivorous ground feeders. René Verheyen, in a pioneering osteological study published in 1957, allied the dodos with crowned pigeons and the Nicobar Pigeon *Caloenas nicobarica* in a new family, Caloenididae. Recent cladistic studies of bone characters and DNA analysis are somewhat conflicting, but they tend to support Verheyen's ideas; all place crowned pigeons close to dodos. While both of Anwar Janoo's cladistic studies make *Goura* the dodos' 'sister group', they differed on other affinities, with one version bringing in the Nicobar Pigeon and ground doves, the other ordinary *Columba* woodpigeons. The DNA study singled out the Nicobar Pigeon as closest, followed by *Goura* then *Didunculus* (which comes nowhere near in Janoo's or Verheyen's schemes). As both osteological and DNA analysis show dodos and solitaires evolving within the pigeon family rather than being an early or tangential offshoot, we consider them members of the Columbidae, although subfamily status (Rhaphinae) should perhaps be retained to emphasize their morphological distinctiveness.

Figure 4.2. The Dodo's nearest relative, the Nicobar Pigeon, with its distribution. From Goodwin (1983).

Until recently, dodos and solitaires were considered unique among pigeons, but a substantial subfossil extinct fauna found in Fiji includes a flightless pigeon almost as large as a Dodo, but more closely related to *Goura* than to the Nicobar Pigeon[139].

The Nicobar Pigeon, small though it is, is actually a very good candidate for the dodos' ancestor. We do not know past its distribution, but its current range in the East Indies and New Guinea is compatible with colonising the Mascarenes, and it is clearly an ancient line. Furthermore it is nomadic and migrates around seeking new feeding grounds on islands. In his monograph on pigeons Derek Goodwin stated that "few . . . pigeons are more distinct from all others than is this species"[140], and noted that it differs from fruit pigeons in being able to digest hard seeds in addition to fruit pulp. This ability provides the most fascinating link to dodos and solitaires; each bird has a gizzard stone used to help grind and crush seeds[141]. Many birds eat small bits of grit for minerals and to help digestion, but single stones appear to be rare, though we are unaware of a full study of the issue[142]. Nicobar Pigeons are ground feeders that use their bills to dig around in the soil, as surmised for Dodos, but this is not normal pigeon practice[143]. The other near-relatives, crowned pigeons, are found only in New Guinea, within the Nicobar Pigeon's range. The authors of the DNA study suggested dodos and solitaires split from the Nicobar Pigeon line around 43mya, and dodos from solitaires around 27mya. This, as they pointed out, is long before the Mascarenes emerged, so the birds' early evolution would have been on the hotspot islands. If correct, the ancestral line would, like the tortoises discussed below, have split into a species each on Saya de Malha and Nazareth Bank/St Brandon (emerged 30–35mya), colonising Mauritius and Rodrigues separately from there. However using the dating accepted by the other pigeon DNA study, the divergence of the dodo line from the Nicobar Pigeon would have been c. 10mya, and between Dodo and Solitaire c. 6mya[144] – a scenario that situates the primary split at about the right age for the emergence of Mauritius and Rodrigues. Either way, this lineage probably remained fairly normal smallish flighted pigeons until they reached the Mascarenes. There are no related pigeons on Madagascar.

All the extinct Mascarene parrots, and indeed the only surviving one, are referable to the tribe Psittaculini, the most widespread group in the Old World. They are all osteologically close to the mainly Asian parakeet genus *Psittacula*[145], which itself includes three of the Mascarene species. The large Raven Parrots, with their massive oversize bills, were found only on Mauritius, while their counterpart in Rodrigues was the less extreme (but still large-billed) Leguat's Parrot. These forms were very specialised, and their ancestors may have begun to diverge on the hotspot islands. Less differentiated from typical parrots was the Mascarin, known only from Réunion, but here again, while apparently allied to *Psittacula*, it differs clearly from any living genus. Finally we have the three *Psittaculas* discussed in an earlier section (Echo, Rodrigues and Thirioux's Parakeets)[146]. This group has thus invaded southwards (including the Seychelles) as many as six or seven times altogether, *Lophopsittacus*, *Necropsittacus* and possibly *Mascarinus* before the original islands sank beneath the sea, *Psittacula* later when they re-emerged in the Pleistocene. The Psittaculini are absent in Madagascar[147], which does however have two interesting parrots in the peculiar genus *Coracopsis*; this genus lacks close relatives, but is generally associated with the two ancient African genera *Psittacus* and *Poiocephalus*[148]. The Mascarin has sometimes been associated with the black parrots, but they are structurally rather different[149].

The endemic owls are now all placed in the endemic genus *Mascarenotus*, which we are calling lizard-owls – largish, long-legged, eared owls allied to the wide-ranging scops owl genus *Otus*[150]. The scops owls occurring in the Seychelles, Madagascar and the Comoros are all normal *Otus*, and indeed the Seychelles Scops Owl *Otus insularis* is vocally so similar to the Indo-Pacific *O. magicus* that it has sometimes been regarded as a subspecies[151]. The Mascarene owls were much larger, with extra-long tarsi typical of island forms on which there are (or were) no mammals[152]. The Mascarene group's nearest relatives among scops owls have yet to be identified, but they are so distinct that they may have started diverging on old islands of the hotspot chain, initially colonising Mauritius and Rodrigues from St Brandon.

Being called a *huppe* (i.e. a hoopoe) by the locals confused European taxonomists for a while, but *Fregilupus varius* of Réunion has been allied with starlings since the mid-19th century, and when starling bones were found in Rodrigues it was clear the two species were closely related[153]. Andrew Berger dissected a pickled Hoopoe Starling in 1957 and

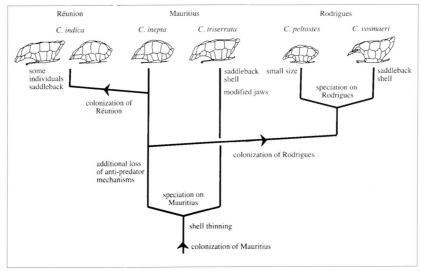

Figure 4.3. The evolution of Mascarene tortoises. From Austin & Arnold (2001).

found enough differences from the two more ordinary starlings to raise doubts as its to status, but more recent work with a bigger selection of starlings confirmed the earlier diagnoses for both this and the Rodrigues bird. Although the Mascarene species have osteological idiosyncrasies that indicate a fairly long separation, it appears from both DNA and morphology they are close to Asian species of *Sturnus*, some of which have similar patterns to the *huppe*, though without the crest[154]. However the DNA suggests *Fregilupus* diverged from Asian *Sturnus* some 8–9mya, which, if true, presents problems in terms of ocean crossing with few or no intermediate islands. The Mascarene pair are unrelated to the only starling in Madagascar or to any of the numerous species in Africa[155].

Reptiles
Until recently, most large tortoises were placed in one large worldwide genus *Geochelone*, which included the giant tortoises of the Galápagos, Aldabra and the Seychelles, the Mascarenes, Madagascar and India (the last three all extinct). Recent work has shown that the genus is heterogeneous, and that the Mascarene species, with their lightweight shells, are best placed in their own genus *Cylindraspis*[156]. While *Cylindraspis* evidently originated from Madagascar[157], unlike the Aldabran and Seychelles forms there is no congener there – i.e. the genus evolved after leaving the big island. The dates derived from different DNA studies are inconsistent, but all indicate a split from the Malagasy ancestor before Mauritius emerged[158]. As *Cylindraspis* itself divided rapidly into two lineages, it looks as if tortoises from Madagascar simultaneously colonised different hotspot islands (probably Nazareth Bank and St Brandon)[159],

floating on to Rodrigues (*inepta* line) and Mauritius (*triserrata*) when they emerged. The *inepta* lineage then colonised Mauritius from Rodrigues and Réunion from Mauritius (forming *indica*), while dividing into two species in Rodrigues itself (*vosmaeri*, *peltastes*). The principal DNA study of these tortoises did not consider the hotspot possibility, and proposed that Rodrigues was colonised (against the currents) from Mauritius[160], but given these currents, the St Brandon>Rodrigues>Mauritius route is much easier to envisage than St Brandon>Mauritius>Rodrigues. There is a similar 'hotspot option' in relation to *Leiolopisma* and *Gongylomorphus* skinks (below), and an analogous difficulty with ocean currents in relation to the large day-geckos (p. 67–68).

Telfair's Skink, the extinct Didosaurus and the large Arnold's Skink in Réunion, all in the endemic genus *Leiolopisma*, are, according to DNA studies, descended from a single initial colonist from Australasia[161]. A plausible molecular-clock calculation suggests that *Leiolopisma* split from *Emoia*-like ancestors some 17mya, i.e. well before Mauritius existed. This suggests to us that the ancestral forms drifted to St Brandon, and from there colonised Mauritius after the island's emergence, failing to reach Rodrigues against the currents. Between about 2.3 and 3.4mya. the line radiated into two lineages in Mauritius, one of which drifted to Réunion following its emergence. Although when first described from a few bones the Réunion species was thought to be identical to Telfair's Skink, recent molecular work suggests it was probably closer to the larger Didosaurus[162].

The slit-eared skinks of the genus *Gongylomorphus* have ear- and nose- closing mechanisms unique in their family, though we do not know their purpose. Nick Arnold has described these defences, possibly

BOX 5 FRIGATEBIRDS (FREGATIDAE)

Frigatebirds have a wide Indo-Pacific distribution and have occurred on oceanic islands throughout the Indian Ocean. Two species are thought to have once bred on the Mascarenes, the Great Frigatebird *Fregata minor* and the Lesser Frigatebird *F. ariel*, though only the latter has been confirmed by specimens. They are generally associated with seabird colonies and nest among other species. On Mauritius they were resident breeders in the early 17th century, but all later records appear to refer to birds visiting from elsewhere. On Rodrigues, frigatebirds occurred on many of the islets within the lagoon, and early observers reported them as common. They had probably ceased to breed on Rodrigues by c.1850. They still turn up as rare vagrants in the Mascarenes.

Accounts
Jolinck (in Keuning 1938–51) at sea off the north coast of Mauritius in 1598 (translation by Henk Beentje):

> . . . *and here flew many rabos juncos [tropicbirds] and raby forcados [frigatebirds] and white herons almost as large as white storks [boobies!] and another kind of bird, ash-grey with flat beaks and long necks [possibly Pink-backed Pelicans* Pelicanus rufescens*], and more and different birds.*

Tafforet (1726) in Rodrigues in 1726:

> *These frigates are all so lazy, that they perch by day on the trees, at the edge of the sea, awaiting the other birds who go to fish. They make them disgorge; after an ineffectual resistance, they are constrained to vomit the fish which are in their gullets when the frigate devours it before it reaches the water. But when the largest of these birds are four or five together, the frigates, however strong and agile they may be, do not attack them, and thus they can feed their little ones who wait for them above. The males of the frigates have beneath the throat a red skin, which, when they are breeding, is swollen, and becomes round and as large as a chopine [bottle], and red as scarlet, and at other times this skin is quite flat.*

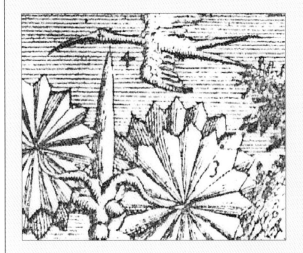

From de Bry's (1601) version of the first Dutch account of Mauritius in 1598:

> *Furthermore there are also other birds that breed there [that are] well suited for eating, they are called Rabos forcados for the simple reason that they have their tails split like a pair of shears. These birds were so tame that, feeling safe in their nests, they could easily be caught by hand or slaughtered with clubs – to the extent that in the space of half an hour we could have completely filled a ship's boat with their carcasses. This being the nature of the said birds, one was entitled to infer that no humans had ever lived there, since on our arrival they were so lacking in fear that they all but settled on our heads.*

Specimen collections
The skeleton of a frigatebird is extremely light and delicate; therefore, skeletal elements are extremely scarce in the fossil record. No Mascarene frigatebird fossil remains have yet been discovered.

Frigatebird *Fregata* sp. From *Het Tweede Boeck* (Anon. 1601).

---

against an insect parasite, as "almost paranoid"[163]. These skinks have generally been associated with, and sometimes included in, the widely distributed burrowing-skink genus *Scelotes*, and they have been considered particularly close to Seychelles forms now placed in *Janetaescincus* and *Pamelaescincus*. However, old-style '*Scelotes*' is now treated as a polyphyletic collection of only distantly related forms[164]. Apart from the specialised nose and ear characters, *Gongylomorphus* appears morphologically primitive in the broad *Scelotes*-related assemblage, and has been considered a relict of an early radiation in the Indian Ocean area[165]. While DNA work incidental to another study placed slit-eared skinks and *Pamelaescincus* close to the Malagasy skink genera *Androngo* and *Amphiglossus*[166], more comprehensive studies show the Seychelles forms diverging before the entire assemblage of Malagasy skinks separated from their African relatives. The Mascarene slit-eared skinks show up as an early divergence from the Malagasy radiation[167]; *Gongylomorphus* appears to have diverged about 18mya, and hence probably evolved on a hotspot island before reaching Mauritius[168]. Unlike *Gongylomorphus*, the related group of Malagasy skinks mostly have very reduced limbs[169].

Last, and far from least, are the snakes. The Dodo may have pride of place in the popular imagination, but in terms of biological interest the endemic Mauritian snake family Bolyeridae is a clear winner. These snakes have some quite unique characters, the oddest of which is an upper jaw hinged in the *middle*[170], and their only (distant) relatives are in a recently discovered genus, *Xenophidion* from Southeast Asia; they derive from stock basal to all other pythons and boas[171]. There are snakes in Madagascar and the Seychelles, but the Malagasy boas are a quite different group from the Bolyerids[172], and the single endemic genus in the Seychelles is a colubrid[173] – a more recently evolved lineage than the boas. Unlike Dodos, at least one of these snakes is still alive and thriving, if only on the arid rock of Round Island (Chapter 9).

Ordinary snakes (the blind-snakes Typhlopidae are different) rarely succeed in colonising truly oceanic islands, the only other cases being the Galápagos, Vanuatu, Fiji and the Seychelles; only in Fiji, and possibly the Seychelles (islands much older than the Mascarenes) have there been multiple invasions[174]. Nowhere apart from Mauritius are oceanic islands inhabited by an endemic snake family. Ron Nussbaum, discussing the Seychelles species, argued that snakes have trouble establishing themselves because they are all carnivorous and need a good supply of lizards, small mammals, birds etc. to survive[175]. While true for a freshly emerged island, this would not apply to mature islands, and indeed some snakes, introduced to islands by people, establish themselves all too easily. The Wolf Snake *Lycodon aulicus* in Mauritius and Réunion is an example, and the devastation of Guam's native fauna by the Brown Tree Snake *Boiga irregularis* is a classic ecological disaster story[176]. We think the infrequency of snake colonisations has more to do with their being good swimmers, and thus inclined to escape off floating rafts long before they reach distant islands; they can certainly live for long enough without eating, so starvation is not the problem.

So how did Mauritius acquire these unique snakes? Although recent discoveries have shifted the immediate relationships of the bolyerids from South America to Asia, an originally Gondwanan origin for the whole python clade is still postulated, though with a different history from the Malagasy boas, iguanas and terrapins that have South American relatives[177].

However the bolyerids' reassignment to an Indo-Australasian lineage does not alter the problem of how they reached the Mascarenes. A rough molecular-clock calculation on when the Keel-scaled Boa *Casarea dussumieri* separated from *Xenophidion* gives *c*. 22mya[178], consistent with arrival on a hot-spot island before arriving in Mauritius, perhaps through rafting directly from the East Indies rather than island-hopping down the whole chain[179].

Why these particular snakes rather than the numerous other types managed to drift successfully across the sea is not something herpetologists have yet addressed, though it is generally accepted that the founding ancestor was a generalist like the Keel-scaled Boa, with the more specialised Burrowing Boa *Bolyeria multicarinata* evolving later. The Keel-scaled Boa is a good climber[180] but so are many other snakes; it would be interesting to see if it has a dislike of water or of swimming, which would have made it less inclined to slope off a floating raft or tree trunk than other snakes.

Whatever the snakes' claim to greater biological uniqueness, they still just look like, well, snakes – we have to admit that the Dodo is more impressive, being large, incongruous, a bit cuddly-looking, readily cartoonable and thoroughly dead. We need potent symbols on which to focus our ideas – where would the World Wildlife Fund (as was) be without its Giant Panda? A vanishing snail or endangered rodent would not have the same impact or appeal. Overgrown pigeon the Dodo may have been, but it makes a very fitting symbol of extinction; it is entirely appropriate that the Dodo was an island species, as it is on islands that most species have become extinct over the last few hundred years. In the next chapters we shall see why and how.

CHAPTER 5

EARLY SETTLEMENT

The 17th Century: 1598–1715

Tortoises were ... caught so extensively by Lamotius that scarcely a single one is found now. ... Lamotius allowed the ebony sawyers in the forest to go out in gangs to catch them, one gang going out as soon as the other had returned. This is why the ebony sawmill took so long to make. Often they caught 70 or 80, but the wickedest thing was that they did not use the meat. This, the best part of the creature, was left rotting in the field...

Governor Deodati of Mauritius in 1698[1]

... the rats, brought here by ships thrown onto the shore, abound and are very troublesome; one should perhaps have brought some weasels which are their sworn enemies and which would very quickly have exterminated them. But who knows whether these beasts would not have subsequently become even more of a nuisance than the rats themselves in destroying birds and their eggs!

Gerrit van Span, describing Mauritius in 1694[2]

In the 16th and 17th centuries, Europeans had no concept of environment or ecology in the modern sense, and little thought for the consequences of spreading their eurocentric cultures and creatures around the planet. Apart from recording their activities in writing, their attitude to new lands was little different to that of cultures such as the Polynesians, who spread across pristine Pacific islands, or the people from the Greater Sundas who invaded the island continent of Madagascar. On island stopovers, the main concern of the East India companies, the syndicates that were developing trade with the Orient, was to ensure that sources of food, water and wood were available to ships, and to extract any commercial value from these places if possible[3].

While the Portuguese had a policy of minimum landfall on the route to India[4], the emerging Dutch, English and French fleets operated differently. Unfavourable conditions could mean months at sea without landfall, with the crew sick or dying of scurvy and the ship perhaps damaged by storms, so the availability of safe havens was an important part of their routeing strategy. A trading journey to the Indies commonly took two to four years to complete, so a network of way-stations was the only way to avoid excessive loss of ships, cargo and personnel. At first, uninhabited islands such as St Helena, Mauritius and Réunion were used more or less communally by the rival fleets, but as time went by strategic considerations resulted in their being claimed and settled by the different companies – Mauritius by the Dutch in 1638, St Helena by the English in 1659 and Réunion by the French in 1663. In parallel, trading posts were established at inhabited ports, the most important of which for the Western Indian Ocean in the 17th century were the Cape of Good Hope (Dutch), Mozambique (Portuguese), Fort Dauphin, Madagascar (French) and 'Mohilla' and 'Johanna' (Mohéli and Anjouan) in the Comoros (English). In the early 1600s, the Dutch, when needing to stop, favoured Mauritius in both directions, but English ships going east generally used the Mozambique channel, stopping at the Comoros, touching at Mauritius only on the way back. Later in the century, after the Dutch East India Company (the VOC) established a major base at the Cape (1652), the Dutch rarely visited Mauritius *en route*, but kept their settlement there to extract ebony and collect ambergris[5]. Réunion and Rodrigues, lacking natural harbours, were generally avoided, though the latter served as a marker on the way to Mauritius. Once the French had taken possession of Réunion, their ships called there in both directions, with frequent trips to and from Madagascar[6].

While indigenous animals and plants were exploited to the full by passing ships, it was also standard practice to release livestock (pigs, goats, cattle and chickens) to multiply in the woods, to provide food on future visits[7]. Rats arrived off ships and shipwrecks by accident, and once the islands were settled, cats were released to control the rats, and hunting dogs imported to help catch wild goats and pigs. Fruit trees (especially citrus and bananas) and coconuts were planted prior to settlement, but only after colonisation was a wide range of plant species introduced. The detailed history of subsequent ecological degradation differs from island to island. The late arrival of rats in Réunion is significant, helping us to make inferences about their effects on the other islands which they reached before anyone wrote detailed accounts.

MAURITIUS

This island ... is a very high goodly and pleasant land, full of green & fruitfull vallies, and replenished with palmito-trees, from the which droppeth holesome wine. Likewise here are very many trees of right ebenwood as black as iet [jet], *and as smooth & hard as the very ivory: and the quantity of the wood is so exceeding, that many ships may be laden herewith ... Here they tarried 12 daies to refresh themselves, finding in this place great quantity of foules twise as bigge as swans, which they called walghstocks or wallowbirds being very good meate* [Dodos]. *But finding also abundance of pidgeons & popiniaes* [parrots], *they disdained any more to eat of those great foules, calling them (as before) wallowbirds, that is to say, lothsome or fulsome birdes. Of the saide pidgeons and popiniaies they found great plenty being very fat and good meate, which they could easily take and kil even with little stickes; so tame they are by reason y the Isle is not inhabited, neither be the living creatures therein accustomed to the sight of men. Here they found ravens* [Raven Parrots] *also, and such abundance of fish, that two men were able to catch enough for all five ships. Tortoises they found so huge, that tenne men might sit and dine in one of their shelles, and one of them would creepe away, while two men stood upon the back thereof ... We sought all the island over for men, but could find none, for that it was wholly destitute of inhabitants.*[8]

Thus did the Dutch mariners describe Mauritius on first landing in September 1598, the first and last visitors to see the island in its pristine state. As was the practice at the time, they soon started 'improving' it.

Ever since Thomas Herbert asserted in 1634 that the Portuguese had introduced livestock and monkeys[9] (Crab-eating Macaques), this myth has held sway in the literature[10]. Although scholarly studies in the 1980s showed this was improbable, and there is no mention of such animals in the eight contemporary manuscript accounts of the first Dutch visit[11], the story keeps re-appearing even in serious history books. Even those studying the monkeys are not immune[12]. As Alfred North-Coombes pointed out in 1980, if the Portuguese had left animals on Mauritius, the Dutch would have said so, as they did for St Helena[13] – whereas on Mauritius, finding there was no livestock, they promptly set about introducing some, albeit in 1598 this included only chickens[14]. The 1598 accounts mention no mammals at all except bats, but by 1602 cats are supposed to be the only four-legged beasts, while in 1606 it is rats and monkeys[15]. We believe that the Dutch '*Katten*' (in 1602) was a misprint or transcription error for '*Ratten*', as there is no further mention of cats for over 100 years; feral cats are unlikely in the absence of a rat 'problem' to be dealt with, and cats rarely survive shipwrecks[16]. Pelage and DNA data point to the macaques having originated in Sumatra, visited regularly by the Dutch, or in Java, where they established their East Indies headquarters, so despite the old rumour that Portuguese were fond of monkey meat, the animals were almost certainly pets that escaped or were released from the first Dutch ships returning from the east via Mauritius, in 1602. DNA studies indicate a very small founding population, consistent with an unintended introduction, though a larger deliberate one (e.g. 20–30 by the Portuguese for meat) would still be 'small' from a population genetics perspective[17].

The first Dutch visitors in 1598 found evidence of a recent shipwreck, which they presumed was of a Portuguese vessel[18]. It may not have been the first (and plenty more would follow), so rats will have become established on the island some time before 1598. It is odd the Dutch saw no rats in 1598, but perhaps the rodents took a while to home in on the new visitors[19]. Meanwhile these rats had already started the gamut of extinctions that the Dutch would shortly cause to accelerate.

Several species known from subfossils were never reported by early visitors to Mauritius, with some probably extinct before the Dutch arrived. Among the lost species are three large lizards and all three snakes[20], animals notoriously susceptible to rat predation if they have evolved in the absence of such predators. Early travellers occasionally remarked on the absence of snakes[21]; it was some time before some were discovered on Flat Island. While the Dutch were clearly only interested in reptiles they could eat (i.e. tortoises and turtles), chunky (and no doubt inquisitive) lizards like the Didosaurus, if still extant, could hardly have avoided mention as curiosities.

Given their prodigious breeding rate, it might take only a few years for rats to spread across the island and exterminate vulnerable species. Two of the lizards, and both bolyerid snakes, survived on offshore islets

Figure 5.1. Dutch sailors imagined in Mauritius in 1602, showing palms, tortoises and an anachronistic wild goat. From Soeteboom (1648).

which rodents did not reach until the 19th century, and still survive today on Round Island[22] (though the Burrowing Boa, last seen in 1975, is probably now extinct) – this effectively confirms rats as the culprits. Also unrecorded, or not explicitly noted by early travellers, were several birds and a flying-fox[23], but most were probably overlooked or lumped in with other species rather than disappearing before the Dutch arrived; none are likely candidates for being wiped out by rats. Failure to distinguish two species of flightless rail or three fruitbats can scarcely be blamed on sailors and travellers, so the absence of early records of Sauzier's Wood-rail or Golden Bat does not imply that they had already vanished

Although van West-Zanen, on Mauritius in 1602, mentioned the subsequent introduction of goats and pigs, his account was not published until 1648, and there is no reason to suppose that any livestock, beyond Warwijk's short-lived chickens, was released before Matelieff's visit in early 1606. His ships left 24 goats and some pigs[24], and the same year the *Concord* was sent via Madagascar with specific orders to collect "cattle and animals" for Mauritius. This was presumably done, as in November 1607 Steven van der Hagen saw five healthy cows hanging around near the landing place, and goats had increased to 17; Verschoor's account of the same visit reported feral pigs, goats and cows[25]. By 1611 the island was "full of tortoises, goats and birds". Indeed, all the livestock bred very fast – by the 1620s, pigs and cattle had joined goats in being seen as 'abundant' by visiting ships [26]. In 1598, Warwijck's men cleared ground and put in unspecified plants, and in 1606, Warwijck's and the *Concord*'s crew planted coconuts, oranges, bananas and cotton. More oranges were spread about by Matelieff's crew later in the year, who reported the earlier sowings to be thriving[27].

During their early visits, before the livestock had become numerous, the Dutch slaughtered large numbers of birds, fruitbats, tortoises, turtles, dugongs and fish. Although some crews found Dodos unpalatable, preferring parrots and pigeons[28], others gorged themselves on them, salting more away for the journey. West-Zanen wrote of at least 45 Dodos killed during his visit in 1602[29]. Some were killed just for the gizzard which was "so large that it could provide two men with a tasty meal and was actually the most delicious part of the bird"[30]; similarly, tortoises would be killed solely for their livers[31]. The greatest slaughter was reserved for the tasty Thirioux's Grey Parrots, killed for fun as well as food: "Besides their usefulness to us [as food], there was also much amusement to be got from them. Sometimes when we had caught a grey parrot, we made it call out, and at once hundreds more came flying around, and we were able to kill them with sticks", wrote Steven van der Hagen in 1607[32]. Pigeons were not spared: "the sailors catching as many as one hundred and fifty of them in a single afternoon, and had they been able to carry more, they would have caught with their hands, or killed with sticks, as many as they desired"[33]. The flying-foxes were "as fat and even fatter than pigeons . . . they are cooked and then taste very good" according to Matelieff in 1606[34]. Van Hagen's crew caught four dugongs which "weighed probably quite five or six hundredweight, and though their exteriors were fishlike, yet when they had been boiled their flesh resembled exactly the flesh of cows. They were very fat, tasted very good, and had fully an inch and a half of blubber on them"[35]. The same men assembled "ten tons of dried fish"[36] to take on the onward voyage! And so it goes on.

This slaughter, while heavy, may not have been insupportable for many of the species involved. The visitors were few, infrequent, and confined their activities to restricted coastal areas. Furthermore, once the cattle, pigs and goats had become numerous in the 1620s, meat-starved crews turned their attention to this much bigger game, and the native birds were henceforth largely ignored, though tortoises, turtles and dugongs were still harvested.

Once the Dutch had got over their excitement at finding Dodos and other naïve animals they could walk up to and kill, they stopped mentioning them in their accounts. After the 'Verhoeff' voyage in 1611, when the sailors had to be "careful not to let their legs or arms be seized by the great curved beaks with which these animals could wound them rather gravely", the

Dutch were curiously silent about Dodos until a survivor of the shipwrecked *Arnhem* recorded them in 1662; there is just one mention in the interval, in 1631[37]. Similarly, parrots and monkeys escape comment, and likewise most other wild birds until well into the 1670s. Over this period visitors from Portugal, England, France and Germany help fill the gaps.

Apparently discouraged by the disastrous loss of three major ships on the reefs in 1615, together with the Governor of the Javan settlements Pieter Both, no Dutch ships visited Mauritius from 1617 to 1625[38]. From then until the establishment of their first settlement in 1638, VOC ships called only irregularly[39]. English ships, however, started using the island frequently from the late 1620s, and most good faunal accounts for the rest of the century come from English visitors or ships' logs[40]. Occasional Portuguese, Danish and, more often, French ships also made stops to re-victual and cut ebony[41].

The island's ecology experienced two phases in the 17th century. In the early period, when ships stopped only for food, water and wood, direct human impact was quite low except in the vicinity of the usual landing places (Grand Port and Port Louis), but monkeys, pigs, cattle and goats were introduced, adding to the rats already present. After settlement began in 1638, human impact was more severe; forest cutting for ebony was a major industry, clearings were maintained for agriculture, and, most significantly, there were up to a few hundred people permanently on the island, feeding themselves and passing vessels through livestock and arable farming, supplemented by extensive hunting and fishing. In addition to the official colonists, a variable number of escaped slaves lived in the remote forests, raiding the Dutch farms in addition to harvesting forest produce[42].

These two phases allow us to compare the impact of introduced animals with the effects of human settlement. Only one new animal was introduced by the first Dutch settlers – Rusa Deer *Cervus timorensis* from Java, in 1639[43]. As large numbers of ungulates were already wild in the forests, the deer will not have had much impact on the fauna[44]. Peter Mundy's account of the island just days before the settlers landed provides a baseline for the comparison, supplemented by Cauche's story, the meat of which, if it is to be believed, comes from his associates' long visit later in 1638 (pp. 28–29)[45].

In fact the faunal lists (Appendix 2) before and after 1640, up to Leguat's visit in 1693, hardly differ in species composition, though some details are unclear, as few visitors troubled to describe the parrots, pigeons or waterbirds beyond a simple mention. One difference stands out – all early reports mention Dodos, but after 1638 they do not[46]. However there are three kinds of exception to this: two cases where the word 'dodo' (or equivalent) is used for well-described Red Hens (p. 26), one case where real Dodos are found isolated on an islet in the lagoon, and the rest where the word '*dodaarse(n)*' or '*dodaersen*' is used but the birds are not described.

The last unequivocal eye-witness account of a living wild Dodo was in 1662, when some survivors from the shipwrecked *Arnhem* were exploring the coast. They waded at low tide over to an islet, where they found *doddaersen* "larger than geese but not able to fly; instead of wings they had small flaps, but they could run very fast". Significantly they found Dodos only on an island, relatively inaccessible to introduced animals. The writer, Volkert Evertsz (or Iversen), found goats there, but mentioned no other feral livestock – goats, deer, monkeys and rats will sometimes swim, pigs do not. He continued: "one of us would chase them so that they ran towards the other party who then grabbed them; when we had one tightly held by the leg it would cry out, then the others would come and could be caught as well"; his group may well have killed all the Dodos on the islet! Evertsz was the last to report *both* Dodos and Red Hens[47]. Since the island had a stream, it must have been Ile d'Ambre off the north-west coast, one of the few that can be waded to at low tide (Ile aux Cerfs and Ile de l'Est are lower without fresh water)[48]. On the mainland some Dodos must have lingered on after 1638, as birds taken alive to Batavia (see p. 83) were probably caught after settlement, possibly as late as 1647. After this all 'dodo' records are open to question.

In 1674 Governor Hugo interviewed a slave, Simon, recaptured after 11 years in the forest, and asked him about Dodos – Simon said he had only seen a Dodo twice in all that time[49], though when in the interval we do not know. Since 'dodo/*dodaerse*' was being used for Red Hens from at least 1668, and was the term used by Hoffman, pastor to the settlement during Hugo's governorship (p. 27), we cannot be sure that Hugo used *dodaerse* in its original sense, though taking the trouble to ask about Dodos implies that he had seen none and was wondering what had happened to them. However, Simon may have understood the term differently from Hugo, or lied to give an acceptable answer (his life was at stake), so the 'record' has to remain in doubt[50]. Hugo also reported *dodaarsen* killed by the settlement's hunters in 1673[51]. Benjamin Harry, visiting Mauritius in 1681 on the *Unicorn* mentioned eating "Dodos whose fflesh is very hard"[52], recalling John Marshall's comment in 1668 that "Dodos or red hens" had "skin like pig skin when roasted, being hard"[53] – suggesting Harry's tough meat was that of Red Hen. Finally there is a series of *dodaersen* collected by hunters in the 1680s and recorded by Governor Isaac Lamotius in his diaries.

Between mid-1685 and late 1688 "official hunters and fugitive slaves took at least fifty dodos" on 12 separate occasions, but the birds so named were not described, neither was any other type of flightless species reported – the other native game listed were ducks, geese and flamingos[54]. Opinions vary as to what these (and Hugo's) *dodaersen* really were. One view is that Lamotius, an educated man interested in natural history, would not confuse a Dodo with a Red Hen (so the birds were Dodos), the other is that he was simply using the name then current, and the birds were Red Hens[55]. Lamotius was certainly interested in trees and timber, and in fish – an impressive set of his fish paintings survives in the Cape Archives – but his writings as published show little interest in terrestrial wildlife[56]. Nonetheless he should, in principle, have known about Dodos from the Dutch travel literature, though the books had no good pictures. Against the 'genuine Dodo' hypothesis is the problem that no late reports mentioned two kinds of flightless bird, yet we know that Red Hens were still around in 1673–75 (Hoffman), and Leguat's *gelinottes* of 1693 were probably this species. As Red Hens were apparently still extant in the 1680s, but there had been no confirmed mainland Dodos since 1638, this line of reasoning suggests the 'dodos' were actually Red Hens, Lamotius (not having two kinds to compare) simply calling the surviving species by the then-current name. The only other good contemporary faunal account is the log of the *President*, coinciding at Mauritius with Harry's ship in 1681. It mentioned neither Dodos nor Red Hens, only ducks, geese and flamingos[57]. The question, and hence the true date of the Dodo's extinction, is unlikely to be resolved unless actual descriptions of '*dodaersen*' seen or caught in the 1670s and 1680s come to light in the archives[58].

Causes of the Dodo's demise

Being flightless, Dodos must have nested on the ground, Cauche claiming they laid a single egg on heaps of grass in the woods[59]. Given that humans were too few, and their activity too concentrated around the two landing places, to have been the cause of the Dodo's disappearance, we need to look at the introduced animals. Ground nesting birds on tropical islands where land crabs are common are usually able to withstand rats, being pre-adapted to this functionally similar predator[60]. Dodos had survived with rats for some decades at least, so that leaves monkeys and pigs as the likely suspects. Either is possible, but pigs, abundant by the 1620s, have a formidable reputation for seeking out and finding the eggs and hatchlings of birds and reptiles. In the later 1600s the Dutch frequently reported pigs digging up and eating both turtle and tortoise eggs[61], as they also do in the Galápagos Islands. They dig up and eat eggs and young of burrowing seabirds, and are thought to have eliminated the tuataras *Sphenodon* spp. from the New Zealand mainland, not to mention their damage to endemic snails and plants on islands all over the world[62]. Although we think pigs prevented Dodos from breeding on the mainland, the isolated survivors on pig-free Ile d'Ambre were probably all killed by humans (i.e. Evertsz and his group); they were not reported subsequently. Hugo had goatherds stationed on Ile d'Ambre in 1673, but only goats and tortoises were reported there[63].

The Dutch established a small fort and settlement in 1638, and until they abandoned it in 1658 they used the island as a base to cut ebony and, initially, to re-victual ships. Apart from the diminishing supplies around Grand Port, the cut ebony had to be dragged to the sea then ferried to the stores. This was laborious and time-consuming, so exploitation was confined to accessible areas near the coast, principally Turtle Bay/Port Louis, Baie du Jacotet, and Grand-Baie/Cap Malheureux[64]. During the five years of van der Stel's governorship (1639–44) 5,000–6,000 'pieces' of ebony were exported. Presumably only ebonies large enough to yield the saleable black heartwood were cut. Re-victualling VOC ships became sporadic after the Dutch base was established at the Cape of Good Hope in 1652, though English ships continued to stop by fairly frequently[65]. The VOC, enjoying a monopoly of supply, fearing a glut and having full stores, restricted ebony harvesting from 1000 to 400 'pieces' a year in 1650. The Dutch re-occupied the island in 1664 and resumed ebony cutting, this time without restrictions; much was wasted through fires, cyclones and the VOC sending too few ships to pick up the timber. Hugo (1673–77) had a road "18–19 thousand paces" long built from the Flacq plains to the mouth of the Groote Rivier (Grande Rivière Sud-est), and introduced ox-carts to haul ebony, but the amount exported does not appear to have much increased, since the company's ships called so infrequently. Right up until the Dutch left in 1710, most transport around the island was still by coasting boats, so inland districts were spared major interference[66].

So little lowland forest remains that we cannot now be sure how densely large ebonies originally occurred. Judging by forest fragments remaining in the Magenta reserve under Trois Mamelles mountain, there could have been four to five ebonies over a foot (30cm) in diameter at breast height per acre (10–12/hectare), mixed with the co-dominant *bois d'olive* and other trees[67]. Ebony harvesting varied in intensity through Dutch rule, but at a rough estimate of 500 trees per year, this works out at around 40–50ha effectively clear-felled of large trees – the other species were used for construction, boat-building and repair, fuel etc.[68]. Over the 66 years of Dutch settlement some 27–33km² of ebony forest would have been cut over, perhaps 6–7% of the primeval lowland forest[69]. The limited amount of farming was largely practiced in areas already cut for ebony, so need not be added to this total[70]. Invasive forest weeds had not arrived, so cut-over lands would have regenerated to some extent, though no doubt feral ungulates would have slowed this down significantly.

Habitat destruction was thus probably not a major source of extinctions in the 17th century. The greatest toll on vegetation was not on ebony but on palms, which were cut for their 'cabbage' (Hurricane Palms) and for thatch (latans). By as early as 1638 Mundy noticed that the palms around the northwest harbour (Port Louis) had mostly been cut down[71], and in the 1670s Hugo banned the cutting of 'palmettos' (Hurricane Palms) to preserve the stock – to no avail[72]. His successor, Lamotius, complained in 1683 that the crew of the *Berkeley Castle* had "destroyed 40 or 50 palmetto trees", and visitors late in the century no longer mentioned palm-cabbage or palm-wine[73]. There were still enough palms in the Flacq area for Lamotius to produce and export up to 30 barrels a year of arrack (distilled palm wine), but it is not clear how many survived his tenure. Unlike the heavy ebony logs, palm cabbage can be quite easily cut in remote areas and carried along footpaths by men working alone (as still happens in Réunion, see Chapters 7 and 8). Hurricane Palms, and possibly also latans, may have been sufficiently reduced to have affected animals that depended on their fruits for food. Regeneration, at least of the very palatable Hurricane Palms, would have been restricted by goats and deer. Throughout the Dutch period escaped slaves lived in the forest[74]. Palm cabbage would have been their only easy source of vegetable food[75], and they may, over several decades, have made significant inroads.

Between the decline of the Dodo in the 1630s and the turn-of-the-century extinctions, two tree-nesting seabirds and probably the Raven Parrot disappeared. Frigatebirds, nesting in large numbers near Grand Port when the Dutch first arrived, are not recorded on land after Cauche's account of 1638; the sole observation of Abbott's Booby was in 1668, and Raven Parrots are not confirmed after Hoffman's sojourn during 1673–75. Direct human predation following settlement may have eliminated the frigatebirds. They often nest on easily accessible vegetation, and are notoriously sensitive to disturbance – the colony was dangerously close to the Dutch HQ. Abbott's Boobies on Christmas Island nest at the top of tall trees[76], and those on Mauritius presumably did the same, making them less vulnerable to humans. However, this would have offered no defence against monkeys, which are known to have wiped out Red-footed Boobies (also tree-nesters) on an island off Puerto Rico[77]. The two boobies that Marshall saw by a nest in a tall tree in 1668 may have been the last old birds of a species that had been unable to breed successfully on the island for decades[78]. Finally, a large parrot like the Raven Parrot must have nested in big tree-holes[79], which monkeys could have easily entered and robbed, particularly if (like other Mascarene endemics) they had no notion of defending themselves. Like seabirds, parrots are long-lived, and old birds would have survived for decades after the last young bird was successfully reared. Raven Parrots, possibly palm-fruit specialists[80], may also have suffered after settlement from the massive palm-harvest discussed above. Monkeys have not stopped White-tailed Tropicbirds nesting successfully in small numbers in Mauritian forests – but this is a widespread species with well-developed defensive tactics against such predators[81].

Rats, cats and pigs

Throughout their period of settlement the Dutch had trouble growing annual crops because of rats, which appeared to grow more and more numerous during the century. Almost every despatch during the second Dutch settlement (1664–1710) mentions their depredations, and the consequent failure of beans, rice, indigo, wheat and other crops, with only sweet potatoes and sugar cane being relatively immune[82]. Intensively managed garden crops could be grown, but field crops were a disaster[83]. Oddly these despatches never mention introducing predators – they used arsenic poison and direct slaughter, but where were the cats? Gerrit van Span's speculation about weasels (quote on p. 75) suggests that as late as *c*.1690 the VOC still thought no predators had been imported, and Leguat, whose visit spanned 1693–96, said that "several regiments of cats *would* [our italics] infallibly deliver upon them [rats] a vigorous war". However, Leguat also reported a recent decline in ducks, geese, 'waterhens' (probably coots) and '*gelinottes*' (probably Red Hens), all formerly common but by now rare[84]. All these are ground-nesters, except perhaps the Mascarene Teal, which may have nested in tree-holes (p. 40), but the ducklings would still need to walk to water from their nesting tree. In 1681 there were still "great numbers of . . . grey teal and geese" hunted by the sailors of the *President* and the *Berkeley Castle*, and Harry on the latter also ate 'dodos' (probably Red Hens as already discussed), so it seems there was a subsequent rapid decline[85]. Leguat was the last to report these birds as still extant; after that there are only equivocal mentions of 'wild ducks' in 1696, and 'ducks' in 1706 (both probably referring to farmyard ducks), with Governor Roelof Deodati's positive statement in 1698 that "not a single wild goose is seen here any more"[86].

This rapid decline of ground-nesting birds (and birds with terrestrial young) that had been unaffected by pigs is the typical result of introducing cats onto islands where birds are naïve to predators[87]. There is no evidence that the Dutch authorities introduced them, so they may have come from a visiting (English?) ship in the mid-1680s, probably at the northwest harbour, whence they would have eventually spread across the island[88]. Although the Dutch never mentioned cats at all, La Merveille, visiting in 1709, saw "cats, grown wild, that had been put on the island to try to destroy the rats, of which there are a very large quantity"[89]. Merveille's crew were given wild chickens to eat by Dutch hunters, feral descendants of birds turned loose when the bulk of settlers left. They were not reported again, probably succumbing to the cats after easier prey became scarce.

Meanwhile, the Dutch settlers discovered that pigs were more of a pest than a boon. In 1666 pigs were reported to "destroy the young cattle [livestock in general?] running wild everywhere", and 27 years later, Leguat noted that they "were very injurious to the inhabitants because they eat all the young animals that they are able to catch"[90]. Governor Wreede reported in around 1670 finding tortoises common only on east coast islets because pigs ate their eggs on the mainland[91]. By 1673 Hugo's administration was trying to kill wild pigs with screw-pine 'hearts' boiled in arsenic, to prevent them killing suckling kids and eating tortoise and turtle eggs, and also ambergris washed up on the beaches[92]. By 1709 Merveille reported that pigs were so destructive that "recently the order was given for a general beat to destroy them, and, the inhabitants having assembled, more than 1,500 were killed in a single day"[93].

Effective as pigs were as predators of tortoise and turtle eggs, they do not seem to have entirely prevented them from breeding, possibly the sheer numbers of tortoises ensured some successful reproduction. The Dutch settlers, however, also took a heavy toll. In the mid-1670s Hugo complained of tortoises being massacred purely for their fat: 400–500 tortoises killed for a barrel of grease, 30–40 for half a pint, sometimes 50 being opened and mortally wounded before a fat one was found. The fat served to make candles and a much esteemed butter-substitute[94]. Hugo explored the northern islets and found good numbers of tortoises on Flat Island in 1674 (removing 320)[95]. Nonetheless tortoises remained reasonably common on the mainland into the early 1680s, but then seem to have crashed, being only infrequently reported later in the decade[96]. Leguat reported them rare in 1693, and Deodati, unfairly blaming their destruction on wasteful slaughter encouraged by Lamotius, said in 1698 that in five years he "had not had them twice on his table"[97]. After that there is silence until one last mention of mainland tortoises by a French ship in 1721[98]. It seems unlikely that all the old tortoises would die out simultaneously, and in any case they should have lived for a century or more beyond the last successful breeding[99] – something significant must have changed.

Tortoises had remained common until the early 1680s despite heavy harvesting by men and predation by pigs. Assuming that they took about 20 years to mature[100], it would be 5 or 10 years before a total reproductive failure would have a noticeable effect on the stock of harvestable animals. This timing fits with the bird declines mentioned earlier, and the new factor again appears to be cats. There were also escaped hunting dogs running wild in the 1670s, but it is not clear for how long they survived[101]; in any case they hunted larger game, and the timing is too early. Studies in the Galápagos suggest that while pigs dig up tortoise nests and eat both eggs and hatchlings (and can take 5 year-old young), cats are particularly severe predators on juveniles up to two years old. Pigs concentrate on tortoise nesting areas on the beaches, but cats hunt out vulnerable baby tortoises anywhere. It appears that giant tortoises can survive pigs *or* cats, but not both together[102]. Mascarene tortoises were very thin-shelled[103], and would have been vulnerable to cats for longer than Aldabran or Galápagos animals. In pre-human times they would have had to face land crabs, herons and rails as predators, but none of these would have been able to tackle them after the first year or so. Once cats were added to the damage from pigs, all recruitment ceased, and the surviving large animals were removed over the next decade by the human population.

By the time the Dutch left in 1710 all the ground-nesting birds and the Raven Parrots were extinct, and the tortoises were all but gone. Dimorphic Egrets were not reported again, but they appear to have lingered on as the French named an islet off Mahébourg 'Ile des Egrettes' (now Ile aux Aigrettes)[104]. The forests, while cut back in the lowlands, and the palms seriously reduced, were otherwise largely intact, their ecology still unaffected by the invasive plants that would become so damaging later on.

Dodos abroad – menageries and curiosities

Before we move on to the other islands, there is a byway in the history of Dutch Mauritius that merits discussion – the story of Dodos and Red Hens carried alive to Europe and Asia[105]. Although keeping and feeding living wild birds on long sea journeys must have taken forethought and dedication, there is nothing in the ships' logs that have come to light referring to these unexpected passengers, though they routinely carried livestock to eat or release. Yet we know that several Dodos made these long journeys, though we have *direct* evidence of only one. In 1628 Emmanuel Altham called at Mauritius with an outward fleet of East India Company (EIC) ships, and wrote to his brother Edward (via a home-bound ship): "This iland haveing many goates hogs and cowes upon it and very strange fowles called by ye portingals DoDo, which for the rareness of the same the like being not in ye world but here I have sent you one by Mr Perce, who did arrive with ye ship *William* at this island ye 10th of June", with a postscript adding tellingly "if it live"[106].

We can infer from various sources how many Dodos may have reached Europe and elsewhere alive, but there are only three actual eye-witness accounts of living birds seen outside Mauritius. The first is from Peter Mundy, passing Mauritius without landing in March 1634, who wrote in his journal that Dodos were found only there, and that he had seen "two of them in Suratt howse that were brought from thence"[107]. Mundy was based at the EIC's Surat factory between September 1628 and February 1634, and the presence of Dodos in India is confirmed by an exquisite painting by Ustad Mansur, Emperor Jahangir's court artist, probably done between 1624 and the emperor's death in 1627. Jahangir was an enthusiastic naturalist and wrote commentaries on many animals and birds from his own observations, often illustrated by Mansur. He stopped writing three years before he died, having said nothing about Dodos. Knowing his love of exotic creatures, visitors and petitioners would present him with animals from distant lands, and no doubt English or Dutch traders gave him a Dodo to further a commercial facility[108]. This suggests that at least three live Dodos reached India, as Jahangir's bird (1624–27) would pre-date the two Mundy saw in *c*.1628.

Another eye-witness account is from Sir Hamon Lestrange, who in 'about' 1638 came across a live Dodo exhibited as a curiosity in London, being fed pebbles to amuse the public[109]. Earlier, in 1634, a Mr Gosling had "bestowed the Dodar (a blacke Indian bird) upon ye Anatomy School" in Oxford; this may also have reached the country alive[110]. By 1676 the Anatomy School had two stuffed Dodos, and John Tradescant, sometime before 1656, had acquired the famous specimen whose head and foot still grace Oxford University's zoology museum[111]. At that time most dead specimens from the tropics were preserved as dried heads or feet, only very rarely as skins, and then only of small animals (e.g. birds of paradise, prepared for sale by 'natives')[112], so the existence of whole stuffed birds is significant. It is unlikely that EIC ships carried taxidermists, and preservation in spirit was not yet in use, so the best explanation is that these Dodos reached England alive and were preserved after they died. Hence probably at least three live Dodos reached England – it is generally assumed (with no direct evidence) that Lestrange's bird ended up as a museum specimen; if it did not, the total would be four[113]. Another Dodo's foot, first recorded in 1665, made its way via the Royal Society's collections to the British Museum, but this bird may not have arrived in England complete[114].

It is less clear how many birds reached The Netherlands. Although the majority of known pictures of the Dodo were done in Holland, almost all can be traced back to a sketch and a clutch of paintings by Roelant Savery. Savery's earliest Dodo painting dates from around 1611, painted while working for Emperor Rudolf II in Prague. The basic template was recycled in most of his other paintings up to and including 1626, when he probably drew his famous sketch of three Dodos (Figure 5.2)[115]; the sketch must pre-date the derivative Dodo by Savery's associate Gillis de Hondecoeter, painted in 1627[116]. Figure 5.2 shows three lively Dodos, but one would have sufficed to produce it (the two in the foreground look suspiciously similar). Only four of Savery's paintings, post-dating the sketch, show Dodos in different poses, mostly derived from it.

Ostensibly unconnected with Savery's work is a pen sketch by Adriaen van der Venne, also from 1626, and an aquarelle by Cornelius Saftleven from 1638. The artist wrote on the former that it is "a faithful portrait of a *walchvogel* . . . which was transported alive from the island of Mauritius, as it was seen in Amsterdam in the year 1626" (possibly the same bird reportedly seen and sketched in Leiden in, supposedly, 1624) – the third eye-witness report. However, Venne's bird, fat and bulgy, looks like an early Savery image; to record the presence of the live bird, Venne apparently did a pen-and-ink copy of a Savery or Hondecoeter painting![117] The Saftleven picture is very different and may be from a bird imported later[118]. An earlier Dutch painting, Jan Brueghel the Elder's *The Element of Air* dated to 1611, half conceals a Dodo's head and shoulders, however its cadaverous air suggests that Brueghel probably only saw a preserved specimen[119]. From these indications possibly only two Dodos reached Holland alive (in *c*.1626 and *c*.1638)[120]; another, sent to Holland in 1599, may have been diverted to Prague. One famous Savery painting shows a bird often regarded as a Red Hen half hidden behind a large Dodo, catching a frog in its beak[121]. However this 'Red Hen', and its companion behind, is dark streaky brown with a straight beak, looking to us like Savery's attempts at Bitterns in other pictures; we do not consider it evidence of any Red Hens reaching Holland alive, as is often implied[122].

Apart from a crude sketch in a book by Clusius[123], the earliest Dodo picture produced in Europe from a specimen dates from around 1610. Compiled between 1607 and 1611, a catalogue of the Emperor Rudolf II's

Figure 5.2. Savery's sketch of three Dodos, preserved in the Crocker Gallery, Sacramento, California. From Eeckhout 1954.

collections in Prague included "1 Indian stuffed bird called a *walghvogel* by the Dutch according to Carolus Clusius". Illustrations of animals from the collection were commissioned in 1610; in oil on parchment by unknown artists these include from Mauritius a Dodo, a Red Hen and a Black-spined Flying-fox. Since the collection's specimens came from animals that had died in the royal menagerie, the pictured Dodo had probably lived in the zoo for a while, alongside the Red Hen and the bat[124]. The illustration shows a rather shrunken specimen (perhaps embalmed or dried rather than stuffed), very dark in colour, with an unhooked lower mandible (possibly young), whereas the written catalogue described the Dodos's colour as dirty off-white[125] – exactly as depicted in Savery's picture from 1611, painted while in Prague[126]. So probably two Dodos (with the Red Hen and the bat) reached Prague[127]. *How* these animals reached Prague is not documented, but to create his menagerie, Rudolf must have had agents in major European ports, ready to snap up exciting new creatures as ships came back from their travels. Rudolf had several Dutch artists working for him, so he must have had good relations with the Dutch; some think the Dodo supposedly sent back on the *Vriesland*, returning via Mauritius in 1599, was acquired for Rudolf's menagerie[128].

Aside from Europe, the other destination of Dodos leaving Mauritius was the VOC's eastern headquarters at Batavia in Java. In 1647 the Governor of Batavia sent a live Dodo to the VOC's outpost at Nagasaki in Japan[129]. John Nieuhoff described animals found in the 'Indies' in a book of travels published in 1682. He never visited Mauritius, but on the strength of his description is often thought to have seen a Dodo in Batavia in 1653 or 1658. However his book was published long after his visits, and his Dodo description is lifted from a book published in 1658[130]; he did not claim to have seen all he described, and no doubt included the bird for completeness. More than one Dodo may have reached Batavia; there was plenty of traffic between Mauritius and Java during the first Dutch settlement from 1638 onwards, diminishing after their new base at the Cape was founded in 1652[131]. The 1647 report is well after most Dodos seem to have disappeared on the Mauritian mainland. Did the bird survive for a decade or so in captivity, or was it a rare survivor, caught and exported for its scarcity value?

Although only five Dodos were directly reported as seen alive outside Mauritius[132], probably at least 11, and one Red Hen, reached foreign destinations alive. There is little to tell us how long individuals survived in captivity; the European ones appear to have been short-lived, but the bird(s) taken to Java may have survived for several years.

RÉUNION

Réunion was little visited until it was settled in 1663; there is no natural harbour and in most places the land shelves so steeply into deep water that there is no safe holding ground for anchors. Samuel Castleton saw no introduced animals during his three-day visit to 'England's Forrest' in 1613[133], but Adriaen Blok had seen flocks of goats in 1612, and Willem Bontekoe also encountered them in 1619[134]. Given that there were several flocks, the animals must have been released some years prior to 1612, probably by the Portuguese (p. 21)[135]. The next known visitor, Thomas Herbert on the *Hart* in 1629, wrote that it was "without hogs or goats till our captain bestowed some there now as we passed"; five goats and three pigs were released, but the visit was so short that they recorded nothing else except "an infinite multitude of land-turtles"; Herbert himself did not go ashore[136].

After this there are no recorded visits until the French East India Company's commander at Fort Dauphin (Madagascar), Jacques Pronis, exiled a dozen dissidents to 'Mascareigne' in 1646. After Etienne de Flacourt replaced Pronis in 1649, he recalled the men, who gave a glowing account of the island's qualities. In his pioneering book *Histoire de la grande isle Madagascar* Flacourt gave a good description of the island, noting a complete absence of rats, mice, fever, biting insects and 'snakes harmful to man'[137]. Amongst the edible creatures – tortoises, turtles, doves and parrots – were plenty of pigs; the *Hart*'s release was clearly successful. The pigs were very tasty; attributed to feeding almost exclusively on tortoises. Flacourt claimed the island for France and renamed it Bourbon after the French royal dynasty[138]. He introduced cattle, sending four heifers and a bull from Madagascar in 1649, and further animals in 1654, by which time the first batch had increased to around 30. Another 13 Frenchmen were voluntarily exiled (as opposed to being banished) there from 1654 to 1658, one of whom, Antoine Thoreau, made the first tour around the island. This group were tricked into leaving for India on an English ship that needed more crew in 1658; the island remained deserted again until 1663, when two Frenchmen from Fort Dauphin and some slaves (including women) were landed by the *Saint-Charles* – meeting the Dutch ship *Lansmeer* searching for survivors of the *Arnhem* shipwreck.

These temporary French settlements will have made little impact on the island's native wildlife, the inhabitants doing no more than cultivate small plots on the edge of the Étang de Saint-Paul. Meanwhile, the introduced livestock had time to spread and multiply; in 1665 the abundant feral pigs were again noted as living on tortoises[139]. The newly established *Compagnie Orientale des Indes* (COI) took over the previous company's rights in 1665 and made a major move on Madagascar, establishing at the same time a re-victualling station on Réunion, leaving a further 20 men. The numbers of people, now with more women and some children, had risen to 76 by 1671 and to about 200 in 1674. The initial settlement was by the best anchorage at Saint-Paul, but in 1669 the new governor, Etienne Regnault, moved his 'capital' to Saint-Denis. By 1671 there were houses at Sainte-Marie and Sainte-Suzanne, east of Saint-Denis, the latter becoming a recognised settlement by 1686. This pattern remained, with steadily increasing numbers until the early 1700s. In 1713 the population was 1,171, nearly four times the population of Mauritius when the Dutch abandoned it in 1710[140].

In contrast to Mauritius, where the main aim of the settlement was ebony extraction, the colony in Réunion was explicitly agricultural from the start. Land clearance was more organised and more concentrated than in Mauritius – initially there were no equivalents of the Dutch free burghers who set up tiny scattered holdings. Since agricultural land was limited at Saint-Paul, confined to lake margins surrounded by mountains and the arid western coastal zone, the settlers soon spread to the wider and better-watered lowlands in the north and northwest. Contact was maintained largely by boat, as Saint-Paul and Saint-Denis are separated by a 10km stretch of high sea-cliffs interrupted by precipitous gorges, making the land journey, by foot or pack animal, slow and difficult. In 1704 the *Compagnie* sent Sieur Feuilley from France to survey the colony and its potential; his report, including his itinerary around the entire coast, gives a vivid picture of the state of the island[141]. The VOC never managed to extract a similar document out of their officials in Mauritius, despite endless requests in the 1690s and early 1700s[142].

Feuilley's account, accompanied by a rough map, shows how limited the colonists' penetration of the island had been by 1704. Saint-Paul was quite densely settled and farmed, but south to Saint-Leu there were just a few scattered livestock herders, and in the northeast the arable settlements were scattered along the coast as far as what is now Bras-Panon, a few penetrating inland behind Saint-Denis. Hunting, periodically banned and un-banned by the *Compagnie*, was widespread, but ostensibly directed largely at feral livestock. However Feuilley said very little about tortoises, covering up for a ghastly massacre. Antoine Boucher, the company's storekeeper/treasurer in Réunion during Feuilley's visit, released in 1710 a withering broadside against the company's management of the island, in which the profligate destruction of tortoises featured prominently:

> *You say, sirs, that you have governed this colony, but that is the greatest misfortune that has befallen it, as it is during this unhappy government . . . that you have cut the root and entirely destroyed this celestial manna that the Lord so liberally spread over all the island by prodigalities so outrageous that some amongst you that I know very well have killed up to forty thousand land tortoises to fatten up just one of your pigs, making use solely of the livers of these tortoises, not thinking that the flesh of these animals was delicate enough as food for pigs, whereas among humans it passes for an exquisite dish. This one item, sirs, cries to heaven for vengeance, and should make you subject to a just punishment from God.*
>
> *What has become, sirs, from your glorious reign, of the quantity prodigious and innumerable of game such as flamingos, pigeons, doves, teal, water-hens, geese and wild ducks, whimbrels, waders, snipe, merles, Hoopoe-starlings and others whose details would be too long [to enumerate], not mentioning an infinity of other game from this country whose names are unknown in France. The abundance of those I have mentioned as well as the others was such that they ate one another and darkened the air. All this has much changed, one now finds just the feeble remnants which have managed to escape your insatiable gluttony. But, I say, you did not eat them: these viands fit to feed a king were too gross for delicate mouths like yours. Had you just been eating them you would not have all but destroyed them, it required what you actually did, massacring by the thousand these innocent animals which should have been reserved for better men than you, filling your stomachs with just their juices, their grease supplying you abundantly with oil when melted, which only proves the bounty that these game animals ought to have been . . .*[143].

Boucher, later to govern the island from 1723–25[144], was notorious for being intemperate and rude about everyone[145], but something fairly dramatic must have fired him up, and the fact that other contemporary accounts said nothing about it, suggests a conspiracy of silence within the colony about the slaughter of wildlife, though, as we shall see, humans were possibly not the worst culprits (except for the tortoises). Hunting pressure was just as severe on the introduced 'big' game – goats and pigs. Settlers had to go progressively further afield to find them, and by 1716 the only region where they were still abundant was in the far south, between the Rivière Saint-Etienne and the Grand Brûlé, the lava flows of Le Volcan[146].

To clarify the picture we have to go back to the first decade of settlement. At the beginning we are treated in numerous accounts to an island paradise in which the air was so pure that all nasty things, even rats, were supposedly killed simply by breathing it![147] Game – feral livestock, tortoises and birds – was abundant and so tame it could be caught by hand, and once bees had been introduced (in 1666), honey could be "found in the woods whenever one wants"[148]. On top of that there were so many fish in the Étang de Saint-Paul that seine-nets broke with the weight of the catch[149]. Melet's account, from 1671, is typical:

> There are birds in such great confusion [sic] and so tame that it is not necessary to go hunting with firearms, they can so easily be killed with a little stick or rod. During the five or six days that we were allowed to go into the woods, so many were killed that our General [de La Haye] was constrained to forbid anyone going beyond a hundred paces from the camp for fear the whole quarter would be destroyed, for one needed only to catch one bird alive and make it cry out, to have in a moment whole flocks coming to perch on people, so that often without moving from one spot one could kill hundreds. But, seeing that it would have been impossible to wipe out such a huge quantity, permission was again given to kill, which gave great joy to everyone, because very good fare was had at no expense. The commonest species were wood-pigeons and turtle-doves, thrushes, parrots, pepeux [?] and little partridges, large excellent buzzards and other birds called solitaires which are very good and the beauty of their plumage is most curious by way of the diversity of brilliant colours that shine on their wings and around their neck . . .[150].

However, the inevitable soon happened: some time in the mid-1670s a long-boat with rats as unintended passengers was thrown onto the shore[151]. The rats enjoyed a dramatic population explosion after their arrival as they exploited the untapped resources of this new territory[152]. By 1676 they were already causing havoc to crops – governor Germain de Fleurimond lamented to the French minister Colbert in November 1678 that: "since 3 years [ago] such large quantities of rats have thrown themselves onto the land, that there is nothing we can do that these miserable animals do not ruin and lose all [sic], even out to the most un-inhabited places"[153]. The first extinction soon followed. The endemic fodies were so abundant that they destroyed grain harvests and even flocked into peoples' kitchens, burning themselves on the cooking fires. Yet after Dubois's account of 1672, they are never heard of again[154]. Island fodies are notoriously vulnerable to rat predation on their nests – Seychelles Fodies *Foudia sechellarum* survive only on rat-free islands, the nesting success of Mauritius Fody *F. rubra* is near-zero where rats and monkeys can get at their nests, and the Aldabra Fody *F. aldabrana* only nests successfully up coconut trees, out of rats' reach[155].

Unlike the Dutch, the French introduced cats fairly soon after rats became established[156]. Although François Martin made no mention of cats when compiling (*c.* 1685) his account of the arrival of rats, by 1703 Borghesi reported that they had gone wild, multiplied, and already decimated the endemic slaty pigeon, details confirmed by Feuilley in 1705[157]. Like the fody in the 1660s, the pigeons used to come into houses after food, and were so numerous that a cook could catch enough for dinner by killing birds flying around the kitchen! Ducks and geese, still abundant in 1687, were not reliably reported thereafter[158]; Feuilley's faunal list in 1705 mentioned no wildfowl (or coots), and Boucher emphasised their extinction in 1710. The inference is that cats arrived in the mid-1680s, the much hunted ducks and geese succumbing sooner than inedible cormorants and unapproachable flamingos, which had gone by 1710. In 1711 Governor Parat was ordered to have feral cats hunted and shot. Earlier, in 1708, *Compagnie* troubleshooter Hébert, governor of Pondicherry in India, gave instructions to governor Villers to ensure no monkeys were allowed into Réunion, given the disastrous effects of their introduction to Mauritius[159].

Another bird which soon disappeared was the ibis, known to the settlers as the *solitaire*. The first visitors found them near their landing sites, but by 1667 they were said to be found only in the 'most remote places'. In the 1660s and 1670s Carré, Melet, Dubois and Bellanger said they were good eating, but by 1705 Feuilley said their flesh was hard and 'tasted really bad', and that the birds, although numerous, were found only 'on the mountain tops'. As this bird was nearly flightless, it probably nested on the ground, suffering devastating nest predation from pigs. The only report of 'solitaires' being seen near the coast in the west, and that unreliable (Bontekoe in 1619; see pp. 26, 30–31), pre-dates the release of pigs in 1629, but the birds appear to have persisted in the eastern lowlands until the early 1670s[160]. By the time humans settled at Saint-Paul any western birds had retreated to high ground, much of which in Réunion is rugged enough to exclude pigs, so the birds remained common. They may have changed their diet as their habitat became restricted, rendering their flesh less palatable. After 1705 there seems to have been a dramatic crash; there is one further report (in 1708, laconic and uninformative[161]) – then nothing. A 'very old slave', supposedly questioned about *drontes* by governor

BOX 6   HERONS AND EGRETS (ARDEIDAE)

Early travellers to the Mascarenes mentioned the presence of a number of herons, and bone remains confirm that several species, including three endemics, once inhabited the islands. The endemics were all night herons of the genus *Nycticorax*. Although they could still fly when hard pressed, the Mauritian and Rodrigues birds had become 'behaviorally' flightless, with correspondingly reduced wing elements. The leg bones of the Mauritian and Rodrigues forms are particularly robust, a probable adaptation for a terrestrial mode of life.

Mauritius Night Heron *Nycticorax mauritianus*
This species was not clearly mentioned in any early account and it is therefore difficult to determine when they became extinct. Leguat's 'Bitterns' in 1693 may have been night herons and it is generally assumed that the birds disappeared shortly after this date.

Réunion Night Heron *Nycticorax duboisi*
Dubois in Réunion mentioned a heron in 1672–3, a species now called *N. duboisi* and confirmed by skeletal remains, but he appears to have described a juvenile. This was the largest Mascarene species, but it showed no reduction in its wing elements, unlike the other two. Night herons were not mentioned again after Dubois's account.

Account
Dubois 1672–3 (in Mourer-Chauviré *et al.* 1999):

> *Bitterns or great gullets, large as big capons* [chickens] *but fat and good* [to eat]. *They have grey plumage, each feather tipped with white, the neck and beak like a heron and the feet green, like the feet of the* 'poullets d'Inde' [turkey]. *They live on fish.*

Rodrigues Night Heron *Nycticorax megalocephala* by Julian Hume, 2006.

Rodrigues Night Heron *Nycticorax megacephalus*
This was a robust species, with a stout, straight bill and short strong legs. On Rodrigues, the French Huguenot refugee François Leguat described the difficulty in preventing the night herons from snatching and devouring the large day geckos *Phelsuma edwardnewtoni* once the lizards had accidentally been knocked out of the palm trees. This protective action on Leguat's part was taken because the geckos had become regular and favourite visitors to Leguat's table during meal times. During the Transit of Venus expedition in 1761–63, the astronomer Alexandre-Gui Pingré made a detailed survey of the Rodrigues fauna, but he failed to mention the night herons. They may well have become extinct by the time of his visit. Why the Mascarene night herons died out so quickly is now very difficult to determine. A combination of habitat loss, introduced animals (particularly cats) and perhaps hunting were probably responsible.

Account
Tafforet (1726) (from Oliver 1891):

> *There are not a few Butors* [night herons], *which are birds which only fly a very little, and run uncommonly well when they are chased. They are the size of an egret, and something like them."*

Other species
Dimorphic Egrets *Egretta dimorpha*, a species common on Madagascar and a vagrant to other Indian Ocean Islands, once occurred on Mauritius and Réunion. Willem van West-Zanen mentioned them in 1602 on Mauritius, whilst Dubois in 1672–3 and Feuilley in 1704 described them on Réunion. They seem to have disappeared from Mauritius about the same time as the night herons, presumably for the same reasons, but they persisted in Réunion until the mid 19th century.

Accounts
West-Zanen in Mauritius in 1602 (Soeteboom 1648), translated by Henk Beentje and amended by ASC:

> *. . . white and black herons . . . the herons were less tame than the other birds found here, flying* [up] *amongst dense tree-branches.*

Feuilley (1705) in Réunion in 1704:
"The aigrettes are ashy grey in colour; there are [also] *completely white ones. Their size is that of a small hen, with very long neck and legs. They feed along rivers, streams and meres, and rubbish heaps. They are good to eat but rarely fat.*

The Striated Heron *Butorides striata* is resident on all of the Mascarene Islands, and a number of races are described from other Indian Ocean island groups. They appear to be of the southeast Asian race *javanica* and may be a recent arrival.

Specimen collections
Fossil material of the endemic Mascarene night herons can be seen at the BMNH, UMZC and the UCB.

Louis-Henri de Freycinet in the 1820s, claimed the birds still existed around Saint-Joseph in his father's infancy. If the man was 80 and his father 30 at his birth, then we are already back around 1710–15, but the story may be apocryphal[162]. We suspect that once feral cats had cleaned up the lowland wildlife they penetrated inland, and climbed their way into areas pigs had not reached, wiping out the ibises by 1710–15. The mysterious *'oiseau bleu'*, presumably a large gallinule (p. 43), also retreated to high ground between the 1670s and 1705, but it may have outlived the ibis (p. 106).

Dubois recorded three further birds in 1671–72 which were never reported again, but, knowing so little about them, we can only speculate about their fate[163], though all have counterparts in Mauritius to give us clues. The birds were the endemic Kestrel *Falco duboisi*, Pink Pigeon *Nesoenas duboisi* and Wood-rail *Dryolimnas augusti*, all confirmed by subfossil bones found in recent years (Chapter 3). The Mauritian equivalents of the first two are still with us, but both 'disappeared' for a century. While we know Mauritius Pink Pigeons are vulnerable to monkey and rat predation at the nest and to cats when feeding on the ground, it is not known why Mauritius Kestrels apparently became so rare that they were not reported from 1666 to 1753 – though nest predation will have played a part[164]. While monkeys were never introduced to Réunion, the arrival of rats and cats may have been enough to eliminate the kestrel and pink pigeon there, while some difference in ecology allowed their counterparts to survive on Mauritius. Sauzier's Wood-rail *Dryolimnas* sp. in Mauritius is as enigmatic as Dubois's Wood-rail in Réunion – apart from bones, there are no records. The flightless race of the related White-throated Rail *D. cuvieri* on Aldabra can survive rats but not cats[165], so cats probably wiped out the equally flightless Dubois's Rail in Réunion.

Tortoises seem to have fared better in Réunion than in Mauritius. As late as 1703, Borghesi's account suggests that they still bred successfully, laying their eggs in sand to be hatched by the sun. Despite heavy predation from the abundant feral pigs, Réunion tortoises were apparently less vulnerable than their Mauritian counterparts, though why is not clear. We suspect (see p. 81) that it was cats and pigs combined that finally finished tortoises off in Mauritius, so the arrival of cats must also have increased pressure on them in Réunion. However, human harvesting was prodigious, and if, as it appears, they required sand to lay in[166], expanding penetration and settlement down the west coast would eventually have made breeding difficult, even if many adults were safely up in the hills for most of the year. To get an idea of the quantities taken for food, Borghesi told of six canoe-loads presented to his superior (the papal legate Maillard de Tournon) by Governor de Villers in August 1703. A canoe-load was 50–60 animals, so Maillard was given over 300; this was overkill, so they released 3 boat-loads back into the forest. Luillier's ship, visiting in March the same year, was given 200 tortoises. Back in 1671, Melet reported that the crew of La Haye's squadron got through about 200 a day during their six-week stay, some 8,400 in all![167] No doubt such quantities were regularly given to ships, in addition to those squandered locally – one can see why Boucher got so vehement![168] By 1709, however, the *Compagnie* flagship *St Louis* which stayed 6 months, and on which Villers and Boucher then travelled home, got three bullocks weekly but only one boat-load of tortoises per month; a longboat could no doubt carry more than a dug-out canoe, but it was probably only single-figures per day[169].

In Réunion as in Mauritius, tortoises triggered the first attempts at conservation, with orders banning hunting dating from as early as 1671[170]. In 1674 the COI's viceroy in India, Jacob de la Haye, found the colony in disarray, and laid down a draconian set of rules for the inhabitants. These included banning all hunting (birds, game, tortoises) completely, on the grounds that it made settlers lazy and distracted them from cultivation and livestock rearing; the penalty was death for a second offence! All this was hardly popular in a society that had always lived off wildlife, and the governor La Haye appointed in 1674, Henri d'Orgeret, rapidly relented and allowed a moderate amount of hunting[171]. By 1687 it was actually made compulsory for settlers to hunt goats twice a week and tortoises once, the logic apparently being that, since hunting could not be stopped, some constraint (i.e. not doing it daily) was better than a free-for-all. To preserve stocks, no one was supposed to go south across the river at Saint-Gilles; the package was, as usual unenforceable[172]. Law and day-to-day realities were never very close in late 17th-century Réunion; it was almost *de rigeur* to break or ignore the company's rules, and officious governors were jailed, killed or exiled by the colonists with depressing regularity[173]. Enforcement and laxity alternated, bad habits having started early on; Hébert said in 1708 that "formerly there were so many [tortoises] that they were a nuisance in people's land-holdings. They were obliged to kill them and feed them to their pigs"[174]. Indeed tortoises on the wet windward side had effectively vanished by 1694[175]. De Villers, governor from 1701–11, frequently altered the rules. Hunting (of goats and pigs) was allowed once a week in 1702, banned in 1704, and permitted again the same year, although tortoise collecting was not constrained until 1708, when inhabitants were restricted to one per month each. These were fetched by canoe from Boucan Laleu (now Saint-Leu), halfway down the west coast[176]. Feuilley commented in 1705 that:

> . . . *in the woods there are also feral cattle, goats and pigs, the hunting of which is banned absolutely, as much for males as for females. It has been made clear to the inhabitants that that if, other than tortoises taken for their own food, they carried on hunting pigs and goats (I exclude cattle the hunting of which has always been banned), all would be destroyed, and of the two things [= options] one [should be chosen]: that it was necessary either to keep to hunting and save the tortoises, or to take tortoises and stop hunting [game]. It was judged appropriate to be content with tortoises, being something that in thirty years one cannot perceive any increase. In that way the inhabitant is not forced to destroy his own animals for food, so that they can be reserved for ships which have need of them*[177].

In a footnote he added:

> *The inhabitants have been given permission to hunt [game] again to prevent them eating their personal livestock, as a result of appeals that the inhabitants have made to sieur Villers* [signed by Feuilley, de Villers and Boucher].

Later events (pp. 104, 140) suggest that there were effectively two groups of tortoises on the island: a coastal/lowland population that nested on beaches and dry lowland river-beds, and a montane population that did not. The former were very vulnerable once all coastal areas were settled or frequented, whereas the others could cling on in areas inaccessible to pigs and unattractive to cats – as François Martin remarked: "what is surprising is that one finds these tortoises on mountains where men can get to only with much difficulty and great risk"[178]. In 1708 Hébert complained that people from Saint-Paul had to go seven leagues (28km), roughly to Étang Salé, to get tortoises. By the end of the first phase of colonisation (1715) lowland tortoises were (like feral pigs and goats) largely confined to the remote southwestern coast, and subject to increasingly desperate official attempts at conservation – further restrictions in 1710–11 and 1715–16 were eased in 1718 (officially to supply ships), then reinforced again in 1728[179]. Tortoises were so important to ships' crews to prevent scurvy that the *Compagnie* considered their increasing rarity a major disaster, but their best efforts were in vain. Tortoise poaching was never fully controlled, and there was no way of stopping feral pigs eating them[180].

Rules governing turtles were equally fruitless. In 1701 the *Compagnie* ordered De Villers to restrict settlers to (a very generous) two per week[181], which was clearly unsustainable. Feuilley mentioned landings in March and April in his 1705 report, but by then they were generally rarely mentioned, and disappear from accounts completely after 1710.

As in Mauritius, some animals were never reported by early visitors, and are known only from subfossil remains. The most surprising perhaps was the endemic owl *Mascarenotus grucheti*. Mascarene lizard-owls were large birds, but the only references to nocturnal 'birds' are to '*fouquets*' (petrels) and to fruitbats, bats being thought of as birds. No one mentioned Arnold's Skink *Leiolopisma* sp., the Réunion equivalent of Telfair's Skink *Leiolopisma telfairii*, a large, bold and very inquisitive lizard. There was no specific report of the pochard *Aythya* sp., but it could have been subsumed amongst 'ducks and teals'. Given the late arrival of rats, it seems unlikely that Arnold's Skink had become extinct before humans arrived, and indeed the (then) common, though smaller, Réunion Slit-eared Skink *Gongolymorphus borbonicus* also went unmentioned until the 19th century, as did the day-geckos. Neither Dutch nor French designed to describe lizards when there was other more edible fare to report, although one or two visitors to Réunion did notice unspecified *lézards*. The lizard-owl on Mauritius *Mascarenotus sauzieri* was very rarely seen by visitors, so the Réunion one could have died out early and escaped notice altogether. It may have had a restricted distribution, perhaps to the dry western coast whose vegetation suffered first from clearance and overgrazing.

Part of the lawlessness of late 17th-century Réunion was due to chronic neglect by the *Compagnie*. Between 1676 and 1702 the island sometimes went for years without a ship calling[182], and it was in one of those intervals, 1690–95, that Duquesne thought it had been abandoned and plotted to settle French protestants there – but Leguat's advance party had to divert to Rodrigues when they found Réunion still inhabited (p. 29). Only with the arrival of Villers and Boucher in 1701–02 did the island begin to revert to some sort of order, but by then the islanders were set in their profligate ways with wildlife.

RODRIGUES

In the period covered by this chapter very little happened in Rodrigues. After the first brief encounters in 1601 and 1611 in which only sketchy wildlife observations were made (pp. 21 and 46), a few landings and shipwrecks were recorded at long intervals during the 17th century[183], from one of which rats must have landed. Aside from the rats, there is no indication that anything had happened to upset the natural conditions on the island before Leguat and his party arrived in 1691[184]. Leguat's group of eight men cleared a small area around their settlement to grow crops, which they struggled to preserve from rats and land crabs. They ate tortoises, dugongs and all manner of birds, but were too few to have any significant ecological impact. After building a boat, they sailed to Mauritius in 1693, leaving the island much as they

Figure 5.3. The frontispiece to Leguat's *Voyage et avantures* (1707), showing his settlement in Rodrigues, with tortoises, rats, crabs, a Solitaire and a lizard on a palm trunk. The tortoises are evidently the small species with the domed carapace, *Cylindraspis peltastes*.

found it, perhaps supplementing the flora with a few weeds.

Two English pirates were stranded on Rodrigues for two years around 1707–09[185], and a British naval survey team visited between 1710 and 1714, reporting to the Réunion governor Antoine de Parat that there were plenty of tortoises, but leaving no other record of their landing[186]. When Julien Tafforet landed in October 1725 and was accidentally stranded for nine months[187], nothing had changed in the 37 years since Leguat left, as his detailed report showed (p. 45). It was only when the French imported settlers to raid the island's tortoises that the rot set in – as we shall see in Chapter 6.

As on the other islands, some creatures disappeared without having been seen alive. There is no indication that either Leguat or Tafforet saw the *Nactus* night-geckos, of which there were two species, one unusually large. The fossil record has also yielded two further endemic lizards, and two birds that no one ever saw – a merle and an unidentified passerine (p. 46). Neither early visitor distinguished two sorts of pigeon, whereas we know both Malagasy Turtle Dove *Nesoenas picturata* and a related endemic species were present at one time. While rats may well be to blame for the loss of geckos and small birds, they are less likely to have wiped out the pigeons, particularly the turtle dove, which co-exists with rats elsewhere[188]. Unlike the other two islands, Rodrigues shrunk dramatically when sea-levels rose as the last Ice Age ended, between 14,500 and 10,000 years ago – a large coastal plateau was flooded by the encroaching sea, reducing the island's size by 90% in a few thousand years, though the shallow lagoon flats, twice the size of the current island, may have been dry until about 7,000 years ago[189]. Such a dramatic reduction in habitat would probably have caused a number of extinctions. As no subfossil bones have been carbon-dated, we do not know for certain which of these 'pre-historic' species survived the island's spectacularly sudden shrinking.

CHAPTER 6

UNITED UNDER FRANCE
The Mascarenes from 1715 to 1810

All around one, at a great distance, are discovered the great coastal [mountain] chains that mark the horizon of the entire space within one's sight. This space is covered with large trees whose ancient crowns shade the areas that Man has not yet enslaved to his industry. The love-calls of the monkeys, parrots and other solitary birds, which seek peace to allow themselves pleasure, announce to the traveller that he is in one of those places on the globe where nature seems to have preserved for itself an asylum for those creatures at whose expense we augment our domains...

Bory de St Vincent on Mauritius in 1801[1]

The Compagnie des Indes had set aside for these forges an extent of woodland known as the Reserves, of ten thousand arpens [= acres] if I recall [correctly]. They thought that in making regulated fellings in these high canopy woods, they would regrow the following year, and that the saplings would serve to recreate the canopy – but how many generations will elapse before this fine forest reproduces itself? Whatever they may say, the observations I made, with great care, have proved to me that woods at the Ile de France [Mauritius], once cut, do not come back ... thus the forest that currently fuels the Mon Desir foundry will soon become an immense desert.

Guy le Gentil on Mauritian forest policy in the 1760s[2]

Whereas the 17th century was one of pioneering exploration and ineffectual attempts at settlements, the politics of imperial expansion and increasing world trade in the 18th century brought new vigour and a much clearer sense of purpose to the French East India Company, and this was soon reflected in the Mascarenes. In 1717 a new governor, Beauvollier de Courchant, was sent out to Réunion, with Antoine Boucher (now styling himself Desforges-Boucher) as his deputy. Their mission was to set up the colony as a viable supply base for ships, and to establish a plantation economy based on coffee. An upsurge in piracy in the Indian Ocean, partly as a result of European pirates being driven out of the West Indies, soon made them realise that Mauritius had to be secured as well, or its untenanted status would leave Réunion perpetually open to raids, or to another power moving in to replace the Dutch. A formal act of possession was made in 1715 but not followed up until 1721; after a French captain found a plaque claiming the island for the Holy Roman Emperor, the Réunion authorities sent over a small detachment to establish a base. This was followed in April 1722 by an official governor and group of settlers organised from France[3].

Mauritius had a good natural harbour at Port Louis, and Mahé de Labourdonnais, appointed by the Company in 1735 to sort the islands out, realised that it was a far better place to use as a staging post and refurbishing centre for shipping. He moved his administration there, and so Mauritius became the centre of operations with Réunion taking a back seat; this is reflected in the lack of information we have for that island in the latter half of the 18th century. While Réunion remained agricultural, providing grain and other supplies for Mauritius, the latter became (by the standards of the time) somewhat industrialised, with shipbuilding and iron-smelting works putting pressure on the forests, in addition to agricultural clearance. Mauritius was also the focus of important expeditions to the Far East and Madagascar first led by and later (in the 1770s) sponsored by Pierre Poivre[4], who by then was civil governor (*intendant*). These were to collect 'useful' plants for cultivation in the islands, several of which were planted in the forests and went wild. The upheavals in France following the 1789 revolution largely bypassed the Mascarenes, which were effectively independent for several years. Slavery was banned by the revolutionary government in France, but the island elites were unwilling to give up the main engine of their economy. The islands' strategic importance eventually led to them being captured in 1809–10 by the British during the Napoleonic Wars[5].

MAURITIUS

Only a few escaped slaves and a handful of sick Dutchmen and deserting sailors remained in Mauritius after the Dutch withdrew from the island in 1710[6], and for the next 12 years the ecosystem had a breathing space while the French dithered about settling the island. A few ships, mostly pirates, are known to have visited during the interval, but we have no descriptive accounts. Even when the French did finally move in, in 1722, the settlement was initially small and tentative, and the settlers were almost immediately beset by disaster. Attempts at agriculture were overwhelmed by monkeys and rats, and the settlers themselves were harassed by *marrons*, escaped slaves whose numbers increased with each contingent the French imported to work the land[7]. From the beginning the settlers had to live off wildlife – by then there were no tortoises to speak of, but there were plenty of deer and wild pigs in the forests (hence, we suspect, the importance of deer-hunting in Franco-Mauritian culture that has endured to the present). As had occurred earlier in Réunion, such easy pickings led to abuse: 50 or 60 deer or pigs were killed to get a pound of fat, the rest of the carcass being left to rot in the woods. Pierre Lenoir, overall commander of the French establishments in the Indies, ordered a complete hunting ban in 1726, unless organised by the Company specifically for food[8]. However, by 1731 food supplies were so short after a cyclone that "everyone, white and black, are in the woods living off game for want of rice", though two residents were nonetheless fined for a 'massacre' of game in the Company's *Grandes Reserves*. The settlers resorted to wild game again after cyclones followed by a drought in 1734–35[9]. The presence of the *marrons* also influenced the make-up of the colony, as there had to be an unusually high proportion of soldiers, and special detachments of men from Réunion with experience in pursuing fugitive slaves. Indeed, such was the attraction of this 'sport'[10] that it was successfully used by governor La Bourdonnais as a lure to entice settlers to Mauritius from Réunion in the late 1730s[11]. In 1732 a murderous raid by *marrons* drove all the settlers out of Flacq to take refuge in Grand Port[12], and pillage by "escaped slaves, rats and monkeys" induced Jean-François Charpentier de Cossigny to hand his concession at Moka back to the Company[13].

Until the Company sent its new troubleshooter Bertrand Mahé de Labourdonnais to run the islands in 1735, the new colony probably made little impact on the environment. Cultivation simply re-used the old Dutch clearances around Grand Port and Port Louis, though the presence of slaves and their pursuers penetrating throughout the forests must have caused disturbance to native wildlife as well as the deer, pigs and goats they largely subsisted on. The total number of people in 1735 was still small, only about a thousand (there were by then over 8,000 in Réunion), including only 61 settled land-holding families, concentrated around Grand Port[14]. Our faunal lists from the first half of the century are not very full, so it is difficult to pinpoint the few changes that appear to have taken place between the last good list (Leguat's account from 1693–96) and 1735. No French account mentions either the endemic night heron *Nycticorax mauritianus* nor the Dimorphic Egret *Egretta dimorpha*, though the name 'Ile des Egrettes' for an islet in Mahébourg Bay suggests that the latter survived for some time after the French takeover. The night heron may have been vulnerable to cats, at least at its nest, while the egret is likely to have been a victim of human persecution if it only nested in a few crowded colonies on inshore islets. The captain of the *Courrier de Bourbon*, who brought the first settlers in 1721, left a journal in which we find the last reliable mention of tortoises on the mainland; which were said to be rare compared with the large numbers on Flat Island and Gunners' Quoin; he also reported large numbers on "a small islet two leagues from Grand Port covered in nothing but coconut trees"[15]. Two French leagues was about 8km, so the likely island is Ile des Deux Cocos along the coast southwest of Mahébourg Bay[16]. This was, for tortoises, analogous to the island where Evertz found the last Dodos in 1662 – a predator-free refuge, until the French arrived. Lenoir, who had tried to ban hunting on the mainland, wrote to the Company in 1727 that "vessels bound for India or on the homeward journey must not be allowed to remove tortoises at their discretion from the neighbouring islands, and the captains must be forbidden to send out their boats to collect tortoises without informing the island commandants and stating the number they required[17]", but it was probably too late. Gandon commented ambiguously in 1732 that "tortoises begin to become rare, though they are pretty numerous", possibly reflecting the contrast between the mainland and the outer islets, or a recent memory of it[18], before the Company began bringing supplies from Rodrigues. Since the only game mentioned by official reports and other visitors (e.g. Père Ducros in 1725) was deer, pigs and goats, it seems unlikely that tortoises survived on the mainland beyond the mid-1720s. At this time in

Réunion there was a rash of legislation aimed at protecting tortoises. There was nothing similar in Mauritius – only Lenoir's rules to preserve deer. Famines notwithstanding, these had some success, Gandon commenting in 1732 that "there is no lack of deer, goat and wild-boar hunting, and this may long continue, with hunting managed economically, private persons and soldiers being forbidden to hunt". In the early years birds were still alarmingly tame; Ducros in 1725 observed that "there are infinitely many turtle-doves, which may be caught by hand, and parrots, some green and others grey. When one parrot is made to cry out, all the others come to the cry, and may easily be captured"[19].

The dynamic new approach to the islands inaugurated by Mahé de Labourdonnais in 1735 resulted in a rapid expansion of agriculture. The first clearances on high ground, around Moka, had been made a little earlier for planting coffee (hence the name), but the introduction of manioc and the development of sugar-cane and indigo cultivation, together with increased immigration, soon resulted in widespread forest clearance. By 1739 the population had tripled[20], and by 1740 about an eighth of the island had been cleared[21]. Already the hills around Port Louis were barren grasslands, fired by fugitive slaves in the dry season; Baron Grant complained in 1741 that the "destruction of all the wood in the environs of Port Louis was a fatal error" that exposed the hills and town to winds and scorching sun, and dried up the streams[22]. Faced with famine conditions on his arrival, La Bourdonnais immediately sent boats to Rodrigues to fetch tortoises to supply passing ships and preserve the settlers' meagre meat resources, leaving a small detachment there to gather them for the shuttle boats to collect[23]. As late as 1740 "Negroes and labourers have necessarily been sent to live by hunting in the woods" because of a poor harvest, although by then Mauritian land-owners were growing beans and 'corn' (cereals) as well as the cash crops[24].

Agriculture was no sinecure – we'll let Baron Grant take up the story, writing the following after deforesting his concession by slash and burn in 1742:

> I sowed the ground which I had cleared, with rice and maize; but the rainy season was not yet commenced, it being the early part of January; and the latter grain was exposed to the devastation of the rats; so that I sowed another crop when the rains came on, which may be considered the manure of this soil: indeed no other is necessary, as each year yields a two-fold harvest. The corn, which is excellent, though its grain is small, remains but four months in the earth; and French-beans are equally rapid in their progress to maturity: but our harvests are infested by those very destructive enemies, the locusts, which fly like birds, and come no one knows from whence, in such clouds as to darken the sky. They eat the plants down to the very earth; then they lay their eggs, which are speedily hatched, and the ground is covered with them: they soon hop about, and would shortly rise upon the wing if they were not destroyed. The mode of attaining this very important object is as follows: small holes are made in the ground about the size and depth of an hat, into which the Negroes, with small brooms, sweep the young tribes, and having covered them with earth, they press it down with their feet. As the practice is universal over the island, this mischievous insect is at last destroyed.
>
> The rats are very large, and issue from the woods at night. In order to destroy them, the Negroes set traps along the border of the woods: they consist of wooden balls cut in half, the flat side set towards the ground; these are supported by three small pieces of wood, and some grains of maize are up them; so that when the rats seize on the bait, the semi-ball falls and crushes them.
>
> The monkies, very fortunately for us, never quit their retreats but during the day. Negro boys are placed to make a noise about the woods, in order to frighten them; but these cunning animals will, if possible, discover some avenue which is not guarded, and carry off what they can find. . . . They have also a discipline, which I know not how to attribute to mere instinct; as, on their marauding parties, they have sentinels, who are placed in every necessary point, to give signal of approaching danger. To these mischievous animals may be added the hurricane and the dry seasons: the lightning is also very violent, and the thunder louder than is heard in other parts, from the position of the mountains. Torrents of rain and whirlwinds precede the hurricane, which nothing can resist; but they are necessary evils, like our winters in France.[25]

Monkeys and rats we have met before, but the rats have grown 'large' and locusts have arrived. These rats, described by another eyewitness as "at least as big as rabbits, with white bellies"[26] and coming out at night in countless numbers, were in fact a new invasion. Brown or Norway Rats *Rattus norvegicus* had reached European ports between 1700 and 1716, and spread very rapidly[27]. They arrived in Réunion around 1735 (pp. 105–106), and soon or simultaneously made the crossing to Mauritius. Cossigny reported them first, stating that one planter had killed 7,000 'gros rats' in the year from August 1737; by 1741–42 their population had already escalated to plague proportions[28]. The locusts likewise appeared first in Réunion (p. 105), causing havoc from 1727 onwards, and spreading to Mauritius sometime after 1735, either by flying or (more likely) as eggs in soil[29]. Around the same time, game species were deliberately introduced: hares, 'partridges' (= francolins) and guineafowl. Guineafowl (probably brought from Réunion) were first noted in 1732, when hares and *perdrix* were still absent[30]. By early 1741 Baron Grant

Map 6.1. Mauritius agriculture and forest distribution in 1763. Map by Jacques N. Bellin (from Lenoir 1979). Note that the diagonal bottom left to top right represents north to south on this map.

was able to console himself for agricultural disasters by shooting, in addition to deer, parrots and fruitbats, three sorts of *perdrix*, though he complained that "there were plenty of white hares, whose flesh is indifferent, but they are now become very rare". These may not have been the Black-naped Hare *Lepus nigricollis* present later in the century[31], and it has been argued that 'white [-fleshed] hares' were actually rabbits[32]. La Caille also reported three *perdrix* in 1753; we cannot be sure which species they were, although by the 1770s Chinese and Grey Francolins *Francolinus pintadeanus* and *F. pondicerianus*, and the Madagascar Partridge *Margaroperdix madagascariensis* were present. Gazelles from Senegal were released by Governor David around 1747, but soon hunted out[33].

In the mid-1750s the island was accurately mapped for the first time by Abbé de la Caille, who said that in 1753 a tenth of the island was under cultivation; the rest, apart from the dry hills around the port, being "almost entirely covered in with woods, which are of handsome appearance". Like the early visitors he observed how difficult they were to penetrate: "a passage through is rendered very difficult and troublesome, from the quantity of fern and creeping plants. These plants, whose branches, like those of our ivy, wind about and interlace themselves with the shrubs and dead wood, render the forests in a great measure impassable"[34]. La Caille listed 'useful' plants, the list rapidly expanding as the century advanced, Fusée-Aublet's and Bernardin de St Pierre's accounts from the 1760s reading like a general checklist of cultivated tropical plants. From 1748 to 1755 Pierre Poivre, later to administer the islands, was engaged by the Company in bids to import living spice plants from the East Indies to Mauritius, in an effort to break the Dutch monopoly. Labourdonnais had set up a trial garden at his mansion *Mon Plaisir* at Pamplemousses where he experimented with new crops. Poivre's early importations were planted there, while in the 1750s Fusée-Aublet ran a rival garden at the governor's new residence at Réduit[35]. Poivre acquired the Pamplemousses estate in the late 1760s and expanded the garden, selling it to the King of France for a royal botanic garden on leaving in 1772. In the 1780s, under Poivre's protégé Nicolas Céré, Pamplemousses became the world's pre-eminent tropical garden, the director corresponding and exchanging with botanists and planters all over the world[36]. Abbé Rochon, before accompanying Poivre back to France in 1772, explored Madagascar and brought to Mauritius a huge consignment of plants, which were planted at Pamplemousses in 1771[37].

These botanical transfers led to exotic species escaping cultivation and going wild. Several invasive plants that now damage the native forest and inhibit its regeneration arrived in the mid-1700s; Strawberry Guava *Psidium cattleianum*, Rose-apple *Syzygium* (= *Eugenia*) *jambos*, Traveller's Palm *Ravenala madagascariensis*, White Popinac (*Acacia*) *Leucena leucocephala* and Mauritius Hemp *Furcraea foetida* are among the most pernicious[38]. Until the mid-19th century there is little information on how far these invaders penetrated the native vegetation, but by the early 1800s Matthew Flinders found upland forests around the Mare aux Vacoas full of weeds such as 'wild tobacco' and the tropical raspberry *Rubus rosifolius*, the latter introduced by Commerson in 1768[39].

As the islanders became more settled and affluent, ornamental plants and cage-birds were imported; it was soon realised that bringing in some of the latter was a big mistake. The easiest cage-birds to keep are seed-eaters, which, when they escape, do what comes naturally – which can be disastrous in grain-fields. Cage-birds were carried on ships as pets; Bernardin noted sparrows and canaries on the ship taking him to Mauritius in 1768[40]. First out in Mauritius was the little Grey-headed Lovebird *Agapornis canus* from Madagascar, released around 1739, and already in "astonishing multitudes" by 1755. During the Seven Years War (1756–62) Yellow-fronted and/or Cape Canaries[41] were brought from the Cape as "presents for the ladies" – but, as Le Gentil put it, "it was one of the most deadly presents ever made to the island" – to which were soon added Java Sparrows *Lonchura oryzivora* and *moineaux de Chine* from Asia – the latter were Scaly-breasted Munias *Lonchura punctulata*, known in aviculture and on the islands as Spice Finches. "These birds had multiplied, by 1765, to an incredible level; they fell in flocks of two or three hundred on a field of oats or wheat, and in short order ruined it irretrievably"[42]. The species became such serious pests that in 1766, following an appeal from planters, governor Antoine-Marie Desforges put a price on their heads, quite literally. Each slave-owning inhabitant had to produce annually, per slave that they owned, at least 15 bird's heads, 15 rat's tails and 4 pounds of locusts, plus one monkey's tail per three slaves[43]. Poivre's more detailed regulations in 1770 counted Java Sparrows and *moineaux de Chine* as double the others, and the person submitting the most birds' heads each year got a prize – an Indian slave, to be presented "in the name of his Majesty"[44]. In 1792, during another food production crisis, these rules were revived, now requiring birds' heads by area (1 per *arpent* = 0.4ha), rather than per slave[45].

Locusts and mynas

The bird pests settled in just as the locusts, so disastrous since the late 1730s, were to meet their match. They had been partly controlled over the years by driving hoppers into trenches and burying or burning them. However, this was extremely labour-intensive, and some land-owners were less than assiduous, allowing new generations to develop[46]; laws requiring the locusts' destruction were decreed in 1762[47]. In 1759 governor Desforges obtained from India some Common Mynas *Acridotheres tristis*, birds with a reputation for eating grasshoppers and locusts[48]. They were sent to Réunion, where after some initial success, they were supposedly killed off by the planters, who mistook their feeding technique for robbery of sown grain. Desforges then asked the Comte de Maudave, French representative at the Danish trading post at Tranquebar, to fetch some more. Maudave asked Foucher d'Obsonville to help:

> As locusts had sometimes done considerable damage in the Isles of France [Mauritius] and Bourbon [Réunion], M. Boucher des Forges, Governor-general of these establishments, wrote to the Coromandel coast to have some pairs of these birds; they were sent to him, but they were soon destroyed, the colonists having presumed that they ate their seeds. However, the imputation having been recognised as rash, and the locusts having [again] caused serious damage, this banning was soon repented. As a result, towards the end of the last war, M. Boucher des Forges wrote to the Count of Maudave, colonel, then Resident for the nation at Tranquebar the Danish establishment on the Indian coast, [to ask him] to try and send him a not inconsiderable number. M. Maudave asked an officer named M. Beylier and myself if we could help him to that effect. The task was not very difficult. The second shipment then took place, and appears to have succeeded all too well. In fact, the Governor-general, and thereafter M Poivre, who was subsequently appointed intendant [civil governor] of these colonies, having thought best, as did their successors, to take the greatest precautions to assure the existence of these birds, they have multiplied so much that they are [now] a drain on the inhabitants.[49]

The war Foucher mentioned was the Seven Years War, and Céré, writing in 1777, gave the date as 1762; Maudave was employed in India from 1761–63 to stir up Indian princes against the British[50]. As hinted by Foucher, one of the first things Poivre did on his return to Mauritius in 1767 was to issue an edict protecting the mynas, with swingeing fines of 500 *livres* for killing one or disturbing a nest[51]. After a brief resurgence in 1770, the locusts were not seen again, their disappearance generally attributed to the mynas, though a cyclone in February 1771 seems to have been the proximate factor[52]. Mynas were already abundant by 1668–70, Bernardin de St Pierre describing their "prodigious consumption of locusts", roosts of thousands of birds in trees, and their loud evening twittering[53]. Poivre encouraged agriculture in both islands in 1767 by fixing the price government paid for foodstuff, setting a higher figure for Mauritius "in consideration of the risks run by the inhabitants for their crops from the innumerable locusts... The Isle of Bourbon having been delivered of this scourge..."[54]. Although there was a last upsurge of locusts there the following year[55], this suggests that the Réunionnais had not in fact wiped out their mynas in 1759 – the 9-year interval from plague to control (1759–68) being exactly the same as later achieved in Mauritius (1762–71). Poivre made sure that the farmers did not rely on mynas alone – they were legally required to carry on with existing control methods[56].

Assuming the combination of physical destruction plus the mynahs achieved results and the birds thereafter kept locusts in check, this ranks as the first known successful application of biological control[57]. The insect involved has been identified as the Red Locust *Nomadacris semifasciata*, but this may not have been the only species; they were always simply called *sauterelles* (grasshoppers) by the islanders[58]. As often happens with biological control, the mynas turned to other sources of food once the job was done – taking fruit, raiding dovecotes in pursuit of the squabs, and generally becoming a nuisance themselves. Buffon commented that they would be "even harder to extirpate [than the locusts], unless by propagating bigger birds of prey; but such a remedy would without doubt bring its own inconveniences. The great secret would be to maintain at all times enough mynas to do the necessary against the noxious insects, but to maintain control over their multiplication beyond a certain point"[59]. In fact there was little subsequent complaint about mynas; after an initial population explosion, numbers apparently stabilised at a level that caused few problems. Naturalist Bory de St Vincent, however, saw it differently in 1801: "the [mynas] have now ruined the entomology of the island, which nonetheless provides some fine insects"[60].

Biological control was popular in mid-late 18th-century Mauritius. Bernardin reported that several pairs of (unidentified) crows had been released to control rats, but that islanders had killed all but three because they raided chicks in their gardens; a shrike called *ami du jardinier* was also released, but likewise did not persist[61]. Commerson and the younger Cossigny corresponded on the subject in 1770, Commerson wanting to introduce "shrikes, *dominicains* [?*Paroaria* spp.], tyrant-flycatchers, flycatchers, woodpeckers, *arniers* [?]" to eat insects, "little falcons, *bouchers* [more shrikes?], night birds [owls?]" to pursue seed-eating birds, and harmless snakes to eat rats. "It isn't just about the frogs that it was thought useful to import to purge stagnant water of the prodigious quantity of gnat larvae that swarm there. It would please me if you would take these ideas into consideration... and tell me what species of animal you think it would be advantageous to acquire"[62]. Cossigny himself wanted to control rats by introducing bigger ones; he enthused over "Siamese bamboo rats the size of cats which make war on ordinary rats and pass in that country for an exquisite dish"[63] – it evidently did not occur to him that rats fond of bamboo shoots might go for sugar-cane or maize. A few years later he recommended flooding the woods of both islands with more cats to kill rats, mice and small birds, asserting that losing a few chickens was a small price to pay given the potential gains[64]. Sonnerat suggested releasing "large birds of prey" to eat bird pests[65], and more shrikes were imported in 1774 to "destroy little birds", but again, despite being given the same protection as mynas, they failed to establish[66]. The frogs Commerson mentioned, released sometime before 1768, failed to establish; in 1792 another attempt successfully installed the Malagasy Grass Frog *Ptychadena mascariensis*, which of course did not eliminate mosquitos[67]. Probably a little earlier, Common Tenrecs *Tenrec ecaudatus* were introduced from Madagascar, whether this was for food or as some form of insect control is not recorded. They were already widespread and abundant in the forests when first reported in 1801, introduced "in the latter part of the 18th century by a Mr Mayeur, commander of a slaver in the Madagascar trade, who let loose two or three individuals at Trois Islots" (in the west, north of the Bamboo Mountains). The ever-keen younger Cossigny recommended their introduction in 1803; they must have escaped his notice on his final visit to the island in 1800[68]. Also imported, probably for food, was the giant snail *Achatina fulica*. The mistake was soon realised. Céré complained in 1781 to Poivre (by then back in France) that the "large snails from Madagascar", already common and widespread, were worse than chickens, ducks and pigs for ruining gardens, and were not even good eating![69] The arrival of House Shrews *Suncus murinus* was no doubt unintentional, as these unwelcome immigrants travel on ships like rats. They had reached Réunion in the 1730s, and are first mentioned for Mauritius around 1773, though they had arrived "a while back", probably in the 1760s[70].

Conservation measures
In the mid-1700s, political changes took place that had profound long-term effects on the Mauritian environment, and led to a framework of game and forest legislation that was to endure, in essence, for nearly 200 years. In 1763 the *Compagnie des Indes* was bankrupt. In an attempt to stay afloat, the Company sold the islands of Mauritius and Réunion to the French crown. The transfer of administration took effect in 1767, when a system of double governorship was installed – a military supremo and a civil one. The newly appointed leaders were Daniel Dumas

BOX 7 FLAMINGOS (PHOENICOPTERIDAE)

During the early years of occupation, Greater Flamingos *Phoenicopterus roseus* were present and apparently breeding on Mauritius and Réunion, with stragglers occasionally reaching Rodrigues. Flamingos originally numbered many thousands and were still considered common in the early 18th century, but like all waterfowl they were systematically hunted for food. They had completely disappeared (or dispersed) from Mauritius by c.1770 and from Réunion by c.1730.

Greater Flamingo *Phoenicopterus roseus*. From Herbert (1638).

Accounts
Dellon (1685) in Réunion in 1668:

The only [bird] for which a gun is necessary is called the flamant; they are as large as turkeys; their legs are five or six feet long, and the neck the same. They taste exquisite, but are very canny and are not taken without difficulty.

Dubois (1674) in Réunion in 1671–2:

These are big birds the height of a man due to their neck and legs being are very long. They have bodies a large as a goose, and white plumage, black at the end of their wings. These birds have red flesh; they are very good and tender.

Feuilley (1705) in Réunion in 1704:

The flamingos normally stay in the meres or lakes, especially at the Etang du Gaulle [= Gol] and the Rivière d'Abord. They live only on mud and little fish. Their size is that of a goose. They have a very long neck and taste good. The season when they are best is during the months of June, July and August, because these times are the driest of the year, and the meres being low they find food easily.

Feuilley (1705, with additional comment by Antoine (Desforges-) Boucher:

There are sometimes up to three or four thousand on the aforesaid Etang du Gaulle. Hunting them is altogether difficult as these birds always have their heads in the air and see [the hunter] from afar, which means one cannot approach them. Thus to kill them one has to go [out] at night and crawl on one's stomach as near as possible to the edge of the mere, and there wait without making the least noise for the day to begin dawning, in order to shoot them. Sometimes one even shoots them at night if they can be discovered, and it can be seen that they are in a group. If the first shot misses it is pointless firing a second as they fly off and will not return to the lake for two or three days.

and Pierre Poivre, who never saw eye to eye. Poivre, with his long years of travel through Asia and the East Indies, was not only experienced and learned, but, unusually for colonial administrators, had vision. La Bourdonnais had wanted to turn Mauritius into a military powerhouse, but Poivre intended to create a stable self-sufficient agricultural base, ensuring the long-term future of the island's land quality and water supplies by erosion control and forest management. By contrast, Dumas was by all accounts an unimaginative and self-interested soldier, largely unable to grasp the importance of Poivre's proposals, and was obstructive or at best unhelpful in their implementation.

Poivre and Dumas inherited in Mauritius an island with a population of about 20,000, permanently on the verge of famine. Deregulated wood-cutting and clearances under Magon's governorship (1756–59)[71] had led to large areas of grassland that burnt in the dry season, often deliberately fired by fugitive slaves. The fires caused erosion[72], preventing regeneration of forest on uncultivated land, and the loss of wild game; in the rains the grasslands were also a breeding ground for locusts[73]. Poivre issued an inter-related set of laws to remedy the situation, including social legislation for improving conditions for slaves. He had genuine humanitarian motives, but in addition fewer slaves would run away, lowering the threat to landowners and the consequent drain on resources[74]; Cossigny the elder noted that the *marrons* were preventing expansion into certain parts of the island because of the danger to life[75]. Poivre estimated that 10% of the island was under cultivation in 1767, but that 70% was suitable for agriculture or plantation[76]. His travels had alerted him that deforestation led to erosion, drying of streams and reduction in rainfall, a view he had successfully transmitted to Navy Minister, the Duke of Praslin, who included forest conservation measures in his instructions to Poivre on his appointment[77]. He therefore wanted to reserve as forest cover the 30% of Mauritius deemed uncultivatable, and prevent it being clear-cut for timber and left bare – a regular local practice of absentee landlords speculating in land[78]. Various regulations were issued creating wooded reserves along streams and on mountain ridges and summits, keeping a quarter of every concession wooded, leaving wind-breaks around clearings, requiring any clearance made solely for timber to be replanted and browsing animals fenced out, even that boundaries be of living hedges or stone walls to avoid the waste of wood in making palisades[79]. Concessions already clear-felled had to be replanted up to the required quarter, with preference given to native species, the six best timber trees being listed in the regulations[80]. To conserve timber for construction and shipbuilding, buildings in Port Louis had to be made of stone, and the use of domestic firewood and industrial fuel was restricted to dead and storm-felled trees, and less 'desirable' species[81]. Landowners along public roads and paths had to plant the preferred timber trees at 15-foot (4.6m) intervals along the roadsides. Finally, the narrow fringe of forest around the coast, protected in theory since 1723 and reinforced in 1736, was re-confirmed. The function of this fringe, 50 paces wide and known as the *pas géométriques*, was not conservationist but defensive – to stop hostile ships from seeing what was happening on-shore, and to allow free passage of troops along the coast without fences or buildings to obstruct them[82]. The impetus outlived Poivre himself, for the extensive woodlands formerly reserved to supply coppiced charcoal for the ironworks near Pamplemousses were acquired by the Government in 1774 when the forges closed down[83].

Poivre arrived in the middle of yet another food shortage, during which the populace had again been decimating deer in the woods to get meat[84]. He saw it as a strategic necessity to maintain a large pool of wild game in the forests in case of siege or other difficulties of overseas supply, but in the absence of any such crisis he decided to once again ban hunting to allow stocks to multiply, importing meat from Réunion and Madagascar in the meantime. There had been an earlier hunting ban under governor David around 1750; La Motte was worried about overhunting in the mid-1750s, though the elder Cossigny was by 1764 optimistic that protection would soon allow game to recover its former abundance[85]. Nonetheless, the preamble to Poivre's regulations asserted that deer were almost wiped out[86]. The new rules banned all hunting of deer and goats, but allowed landowners to shoot "hares, partridges and other small game" on their own land; wild pigs, oddly, are not mentioned[87]. Gamebirds were not to be hunted or their nests robbed during the breeding season (actual dates were not specified), on pain of a 300 *livre* fine. 'Blacks' (i.e. slaves) were forbidden to hunt at all, apart from "birds and destructive animals"; 'birds' were the small seed-eaters, which caused such trouble to farmers. Selling game was banned, apart from 'small game' displaying the landowner's permission. To prevent attacks on fawns, dogs were not to be released into the forests, and no one passing through the forest could be accompanied by more than one dog. It was this legislation that first gave special protection to the myna. A further comprehensive set of rules was drawn up to control fishing, including a total ban on catching turtles, on pain of confiscation of boats and nets and three months in prison[88]. In the same spirit, the 1770 regulations on killing seed-eaters specifically excluded birds "such as partridges, *mesanges* [?], merles and other insect-destroying birds", which, rather than be killed, the government "recommends, on the contrary, to manage as much as possible these birds that are useful to growers"[89]. Legal protection of birds other than for food or sport was an innovation that took nearly a century to catch on elsewhere. Although the Common Myna was the only bird whose killing was made a crime, the official encouragement of other species considered useful for agriculture showed the beginnings of a broader perspective. The more explicitly utilitarian law of 1792 (p. 94) listed "mynas, *mésanges*, merles and turtle doves" as unacceptable for the bird-head quota, but there was no longer any suggestion that insectivores be encouraged. The anomalous addition of doves (*tourterelles*), largely seed-eating, is not easily explained[90], unless it was to preserve them as game; though not a danger to crops at harvest, they can take sown seed.

Well-intentioned as all this legislation was, enforcing it was another matter. Already in 1769 Bernardin de St Pierre was party to the netting of three turtles at Black River while failing to catch dugongs, and on another occasion he watched hunters with dogs chasing a deer in the same area[91]. However he said deer were very common at Black River, in contrast to the dire shortage alleged in the legislation. Nicolas Céré, whom Poivre had installed as director of the King's Garden at Pamplemousses, wrote to his mentor in 1781 that no-one was enforcing the forest laws, fires were used to drive game, and all the *nattes*, *canelle* and most of the ebony had gone from the Pamplemousses area[92] – from the 1760s until at least 1806 ebony was a major export to China[93]. On the plus side, 100,000 *bois noir* [*Albizzia lebbek*] were planted as shade trees and to reduce erosion in the valleys around and behind Port Louis, but the scheme, run by the younger Cossigny, faltered due to inadequate financial control[94]. Details of forest clearance from 1770 to 1810 are sketchy, though a survey in 1785 claimed that 4/5 of the Grand Port district was still under forest, much of it of good quality[95]. Matthew Flinders observed around 1805 that between the Moka Range and the south coast (as viewed from Grand Bassin peak) "not one half, probably not one third part of the primitive woods are cut down"[96]. There was increasing pressure on upland forests for shipbuilding and housing while the British blockade

Figure 6.1. Deforestation on Le Pouce mountain, Mauritius, *c.* 1801, as seen by Milbert (1812).

prevented wood being shipped around the coast from Grand Port, extraction being facilitated by new roads through forested areas after 1795[97]. Houses in Port Louis were supposed to have been stone-built since Poivre's time, but this was rarely if ever enforced[98]. Decaen's administration (1803–10) was alarmed that existing conservation regulations, especially on steep slopes and river banks, were being ignored and leading to a serious long-term risk to the water supply. He revived and extended Poivre's forest legislation, and appointed a conservator of water and forests, supported by forest guards (the basis of forest governance under British administration), but it was apparently no more effective than earlier attempts in reducing the destruction of accessible forests. Flinders described the clandestine arrangements land-owners made with contractors to cut ebony for export during Decaen's rule[99].

When Poivre was not trying to save crops and forests, and import spices to undermine the Dutch monopoly, his natural history interests turned to introducing 'useful animals'. As early as 1750 he had bought peacocks and other birds in China to import to Mauritius, but whether these arrived alive is uncertain[100]. In 1767 he sent Abbé Galloys off to China in search of a quite staggering list of plants and animals – he had clearly learnt an enormous amount about the Asian biota on his trips to the Far East. In addition to a long catalogue of plants, Galloys was to look out for several species of pheasant, bustards, Purple Swamphens *Porphyrio porphyrio*, wild ducks, peacocks, plus an "aviary of green turtle doves and one of common merles that destroy locusts". Further, he was to keep an eye open for pairs of various birds "eaters of those ravaging insects that are a plague at the Ile de France, in particular the *hoa-mi* (or *hou-me*)"[101]. This was a tall order; however Galloys sent five cases of plants in 1768, and returned in 1769 with another six – but no birds; none, at least, were recorded in the documents or subsequent correspondence[102]. However, someone did import Purple Swamphens, though not from the Far East; de Querhoënt reported in 1773 that "*poules bleus* from Madagascar have had young at the Ile de France [Mauritius]", presumably in captivity[103] as they were not reported as feral until well into the 19th century. The younger Cossigny was also keen on bringing in gamebirds and domestic fowl. He released *tourterelles* (doves) from Bengal in 1767; these were probably Spotted Doves *Streptopelia chinensis*, as Zebra Doves *Geopelia striata* do not occur in this region[104]. Bernardin reported two species of introduced '*tourterelle*' in 1768–70, equally without description or detail[105]. The Zebra Doves were first confirmed in a book by Pierre Sonnerat published in 1782, but as he made many visits to Mauritius between 1768 and 1780, this says little about its date of introduction[106]. Milbert also added peacocks, European pheasants, tortoises and 'mountain ducks' from the Cape to the list of Cossigny's importations, saying that the ducks were excellent game – evidently hearsay, as Cossigny himself said that the ship's crew had eaten them all before reaching the island![107] Nonetheless there may have been feral ducks for a while in the late 18th century, as Bernardin had reported around 1770 that "on certain meres geese and wild ducks have been released", and James Prior mentioned them in 1810; against this, Matthew Flinders, held on the island during 1803–10 and a frequent visitor to the Mare aux Vacoas, said there were none[108]. Bernardin also mentioned pheasants as recently released in the woods, but a "fine pheasant from China" rather than the European variety; no pheasants persisted. By the early 1800s various European cage-birds had been imported and escaped, Milbert reporting canaries, bullfinches and goldfinches – they were not heard of again[109].

Native species under pressure

While most of the attention was on the fortunes of deer and other classic 'game', the native wildlife was being shot and eaten in some quantity. In the 1750s La Motte reported that in a day a hunter could kill three or four dozen "green parrots with long tails", i.e. Echo Parakeets. From July–September 12–15 dozen Mauritius Merles, "a ball of fat worthy of an Ortolan", could be killed in a morning! The rest of the year they were too thin to be worth eating. He wondered how, after this annual slaughter, they were still as abundant as ever[110]. La Motte praised the parakeets as very good to eat, though the elder Cossigny thought them "always skinny and very tough, whatever sauce you put them in"[111]. In 1759 both green and grey parrots were still apparently common, but Cossigny's is the last positive report of Thirioux's Grey Parrot, the only Mascarene bird still alive in the mid-18th century of which no details reached Europe and the encyclopaedists[112]. It may have survived a little longer; Bernardin de St Pierre wrote unhelpfully of seeing in 1768–70 "several species of parrot, though of mediocre beauty"[113]. Finally dugongs, seen by Bernardin in the 1760s, were still considered occasional in 1799[114] but they are not mentioned thereafter.

Other native birds that were good eating were also becoming scarce. Already in 1741 Baron Grant noted that "birds very much diminish in the woods, as the monkies, which are in great numbers, devour their eggs." The elder Cossigny wrote in 1755 of the Pigeon Hollandais that, common 23 years before, was now rare because of forest clearance and hunting by escaped slaves where forest persisted[115]. Another juicy endemic, the Mauritius Pink Pigeon, had acquired notoriety for being toxic, at least at certain times of year. If eaten it was said to cause painful convulsions and partial paralysis, with the victim recovering after a few days[116]. It was generally accepted that some seasonal fruit was responsible, and Cossigny tried unsuccessfully to discover what it was[117]. This mystery has in fact never been solved, though the tradition of seasonal toxicity persisted on the island right up until the species became too rare for anyone to eat it[118]. La Motte also claimed the problem affected the Echo Parakeet[119], plausible enough if both were feeding on a seasonally abundant fruit toxic to humans but not to the birds.

Although, as usual, edible or pest species attracted most attention, some of the others got noticed as the island's improving economy allowed some writers the leisure to explore and observe from interest rather than necessity. While the endemic owl remained in obscurity, the Mauritius Kestrel bounced back in the mid-1700s, reported by La Caille in 1753 and then by Bernardin 15 years later, who encountered one on a latan palm in then-untouched forests near Souillac[120]. Bernardin's enjoyable account of his walk around the island is all too brief and lacks any overview on the general state of the forests, but it does give a picture of how little frequented the southern woods were between Black River and Mahébourg Bay in 1769; his account can usefully be compared to Milbert's wanderings in 1801–04 and similar excursions in the mid-19th century[121]. Bernardin recorded the demise of the island's flamingos – only three were left, which he did not see himself[122]. The native passerines were rarely mentioned, although they did not go entirely unnoticed (p. 100). Someone even managed to bring a Mauritius Fody to England alive in the 1770s, where it lived for "several years" in Marmaduke Tunstall's aviaries. For a small insectivorous bird this was some achievement. Its stuffed remains featured in Peter Brown's *New Illustrations of Zoology* in 1776, before ending up in the Newcastle (now Hancock) Museum[123]. A living Pigeon Hollandais was sent to Holland in 1790, where it lived for several months, allowing Vosmaer to record its display and calls[124]. Only the elder Cossigny reported endemic snakes outside the northern islets; in 1764 he wrote that "I should say that there are snakes, quite long and big, marked with brown and green, on the islets marked on my map. I have seen several shot on the Gunner's Quoin. There were also some on the islet by the Grand Port pass [i.e. Ile de la Passe] . The labourers working on the battery killed several"[125].

In more than a century since some of the last Dodos were shipped out as curiosities, virtually no animals from Mauritius had reached Europe for study, and European natural history compilations depended on travellers' accounts. From the 1730s a few people on Réunion and Mauritius started corresponding with members of the *Académie des Sciences* in Paris, notably with Réne-Antoine Réaumur. Lanux's letters will be discussed under Réunion, but the elder Cossigny, first from Réunion then from Mauritius, kept up a correspondence for more than 23 years[126]. Cossigny also occasionally sent specimens, but there was no systematic collecting until Poivre diverted Philibert Commerson from Bougainville's major South Seas expedition in 1768. Commerson was primarily a botanist, whom Poivre wanted around to identify spice and other exotic crop plants that he was again importing from the East Indies. Here again Poivre's vision was wide, and Commerson was given a commission to explore and collect the natural history of the Mascarenes and also Madagascar, and he was even assigned two talented artists to draw the specimens, Paul Jossigny and Pierre Sonnerat. The tragedy of this enterprise was that Commerson died in Mauritius before he was able to write up his discoveries. His collections of plants, fish and birds were shipped to Paris, together with many incomplete manuscripts, only to disappear into the depths of the Paris Natural History Museum, where they were inadequately sampled by the encyclopaedists of the day. As early as 1791, Poivre's friend, astronomer and navigator Abbé Alexis Rochon, was complaining about the scandalous neglect of Commerson's work. Although Commerson's qualities as a naturalist and collector are now fully recognised, his manuscripts have still not been properly studied[127]. The manuscripts include Jossigny's animal drawings (birds, bats, tortoises, insects etc.), some of which are extremely important to Mascarene wildlife history. One, for example, is a full-size drawing of the lizard-owl from Mauritius, not unearthed until 60 years after its extinction[128]. Sonnerat included some material worked on with Commerson in his books, notably a painting and description of the Pigeon Hollandais, but his works contain errors of date and locality. Some birds he claimed to have found in China and Madagascar may in fact have been collected in Mauritius[129]. Another important source of information was the Viscomte de Querhoënt; Buffon included material from his letters in the *Histoire Naturelle*[130]. It was through these collectors and correspondents that the smaller birds first became known, though Buffon also assigned to the Mascarenes a number of birds belonging elsewhere. By contrast, Buffon's description of the 'Grand Traquet', collected by or for Commerson but given no land of origin, was not recognised as a Mauritius Cuckoo-shrike *Coracina typica* until the 1980s![131]

The Dodo remembered

It took a long time for anyone to notice, but by the late 18th century a few people finally asked 'where are the Dodos?'. In 1760, when George Edwards published *Gleanings of Natural History*, featuring an article on the Dodo and a new picture based on an old Savery painting he owned, he clearly did not know the bird had not been seen for a century[132]. Buffon, writing around 1770, was puzzled by the confusion in names and supposedly recorded localities of *drontes* (= dodos: Mauritius and Réunion), *solitaires* (Réunion and Rodrigues) and Cauche's *oiseau de Nazareth* (Mauritius and 'Ile Nazare'). He knew the birds were confined to the Mascarenes, but had no idea they had disappeared; indeed he strongly recommended visiting naturalists to collect specimens to help sort out the confusion[133]. Responding to Buffon's appeal, Morel, chief scribe of the Port Louis hospitals, pointed out in 1778 that big, flightless birds were extinct on all three Mascarene Islands – the first time an extinction had been formally documented in the scientific literature. Morel stretched the facts a little, claiming that no such birds had been seen in any of the Mascarenes or the Seychelles in "over 60 years that these areas have been visited and inhabited by French colonies" – his informants apparently unaware that the Rodrigues Solitaire had in fact survived until around 1760. Morel added:

> As to the Solitaire of Rodrigues . . . although there were scarcely ever more than 5 or 6 persons white and black in residence on this little island, which used to furnish us many tortoises, this small number was enough to wipe these tortoises out, and send away the turtles that no longer come to lay in large numbers, as were found 20 years ago; even more reason to no longer find there any large bird that could not fly and was good to eat, heavy and defenceless animals being soon destroyed in an inhabited area[134].

Mauduyt, in an encyclopaedia that otherwise slavishly followed Buffon, acknowledged in 1784 that "it thus appears, either because the species has been entirely destroyed, or because the species consists now only of a very small number of individuals pushed back into little frequented parts, that one no longer finds the Dodo in the same places where those who first landed discovered them"[135]. Milbert confirmed in 1801 that "it is absolutely gone, and no trace is to be found of it"[136]. Within a few years the very existence of Dodos and Solitaires came to be doubted. By 1828 even the rediscovery of the Oxford specimen was not enough to dissuade leading French ornithologist René-Primevère Lesson, who had visited Mauritius in 1824, from repeatedly asserting that the 'Dodo' was really a cassowary![137]

Deserters from the Baudin expedition

At the turn of the century governmental interest in science in France revived after the rigours of revolution, and in 1801 a mission to the South Seas under Nicolas Baudin on the corvettes *Géographe* and *Naturaliste* reached the islands. Once at Mauritius several scientists and artists deserted, among them Jacques Milbert, Jean-Baptiste Dumont, Jacques Delisse and Jean-Baptiste Bory de St Vincent. Milbert and Bory later claimed they were too ill to continue, but François Peron, who did proceed with Baudin, explained in the official account of the project that several desertions were due to Baudin's poor leadership and insensitive treatment of the men[138]. Pierre Bernard Milius, then second in command on the *Naturaliste*, wrote in April 1801 that

> Some officers and naturalists disembarked due to disagreements. Our best sailors deserted to go off with privateers, and those who remained only continued the cruise with regret. After the impediments we experienced at Port N.O. [Nord Ouest = Port Louis] *one should consider the Isle de France* [Mauritius] *a dangerous port of call for discovery expeditions. It offers too many advantages to young artists for them not to decide to disembark. The hope of making a fortune and the fear of exposing themselves to greater dangers makes them break all manner of agreements.*[139]

However bad for the expedition, these desertions triggered a new era in our knowledge of the islands' natural history. Bory, a young man of only 23, soon went on to do a remarkable study of Réunion (see below), while Milbert, an artist and keen naturalist, stayed on in Mauritius for three years, his *Voyage pittoresque...* giving us an invaluable picture (both verbal and visual) of Mauritius at the time. Delisse and Dumont settled in the island and became founder members with Milbert, Bory, Lislet-Geoffroy and others of the *Société des Sciences et Arts*, the island's first scientific society, and then in 1805 of its successor the *Société Littérraire de l'Ile de Maurice* (soon to become the *Société libre d'Émulation*). Dumont also collected birds for the Paris Museum[140]. Peron and Lesueur, who stayed with Baudin's expedition, made the first scientific collections of snakes and lizards from the island on the return journey in 1803, and left useful unpublished notes on natural history, although their excursions were mostly in the area between Pamplemousses and Reduit[141]. Around the same time the first bat specimens reached Paris[142], a doctor, Charles Chapotin, who was on the island in 1805 and was another founder member of the *Société littérraire*, published natural history observations. A fourth founder member was naval artillery captain Mathieu, who made a good collection of animals for Louis Dufresne and the Paris Museum. Clearly both competent and well-informed, he collected three species new to science on a relatively well-worked island. In addition to type specimens of the Malagasy Turtle Dove, White-throated Rail and Grey Tomb Bat, he got one of the only surviving specimens of the Pigeon Hollandais, and the first Echo Parakeet and Bojer's Skink from Mauritius[143]. Mathieu's White-throated Rail remains the only record for the island; it was gravid, and may have been breeding, possibly a recent colonist[144]. A few years earlier Common Moorhens *Gallinula chloropus* appear for the first time – Edward Newton traced memories back to around 1790, though the first contemporary mention was from the younger Cossigny in 1799[145].

Milbert's eye for detail and appreciation of wildlife and scenery makes his book a landmark in descriptive accounts of Mauritius. He travelled extensively around the island, vividly describing the landscape, plantations and people. There are no statistics, but plenty of observations on forests and deforestation – sometimes rather contradictory. He was clearly schooled in the Poivre-type view of the value of forest cover, yet as he railed against deforestation and how the woods were almost gone, he also painted a picture of the still extensive tangled forests of the interior, in which it was still easy to get lost. He joined a hunting party one day, but refused to take an active part, drifting away from the group as he and his slave admired plants and wildlife. They had no food with them, but when the slave suggested finding a cabbage-palm to eat, Milbert unexpectedly refused:

> *Unfortunately it* [the palm] *can only give this delicious cabbage once: the head cut off, the tree does not regrow. From this has resulted the extreme rarity of the* palmiste, *formerly so common in these countries; they are hardly still to be found at the Ile de France. This consideration did not permit me to consent to the destruction of so fine a tree; I expressly forbade my negro to use the hatchet he had brought for cutting our way through the woods.*

They compromised by eating a cyclone-felled palm[146].

Milbert also discussed the increasing rarity of ebony and latan palms (though his slave found one to make a fan from at Petite-Rivière), and he commented on the magnificence of the forests around Tamarin Falls, where the trees would be commercially very valuable, but luckily could not be extracted. Around Grand Bassin he extolled the abundance of huge lianas and tree-ferns. He saw day-geckos[147], and "among the birds which abound in these dense trees, I noted the *cardinal*, the brightness of whose plumage vies with rubies. The *coq des bois*, the large and small parakeets, and the turtle doves are in good numbers there, but these last are almost the only birds whose songs interrupt the silence of this desert"[148]. At one point he found "on various clumps of bamboo, spiders of an immoderate size; a little *bengali* had been caught in their perfidious webs"[149]. We can only give

Figure 6.2. Wayside scene in forest, Mauritius, c. 1801, as seen by Milbert (1812). Note old native trees, palms, regenerating scrub, and the palanquin used as transport.

a glimpse of Milbert's observations here, but of all old books on the island, this is the one, together with its unique set of engraved plates, that should be reprinted, read and appreciated[150].

While Milbert's writing is very personal and evocative, his ex-colleague François Peron wrote a more penetrating analysis of the question of forest cover versus stream-flow. He noted that all the most educated and long-standing planters complained that cutting the forests had reduced rainfall, and hence the water level in the rivers. However, he preferred an alternative hypothesis; that the rainfall had not changed, but that greater evaporation from cleared land reduced water reaching the streams. He added that "whatever the value of this latter suggestion, it remains nonetheless undeniable that deforestation in almost all parts of the island has been pursued in a most culpable fashion"[151]. These observations were published after Decaen had already revitalised forest protection legislation (p. 98), but Peron, who had the ear of the governor, may well have been involved in formulating policy while spending seven months on the island in 1803–04[152].

Peron's friend and collaborator Charles-Alexandre Lesueur reported, as did others, on the hunting of deer, hares and tenrecs, but he was alone in recording an organised shearwater hunt in Mauritius. Echoing Bory's discovery of Barau's Petrel *Pterodroma baraui* in Réunion (p. 109), it was Baudin expedition scientists who first noted hill-nesting shearwaters in Mauritius. At that time a petrel called a *coupe-vent*, "similar enough to the *Petrel obscur* of northern seas" nested commonly on Montagne Longue (near Port Louis), and waiting hunters shot them as they flew inland at dusk. Lesueur did not describe the bird, but the '*petrel obscur*' is Audubon's Shearwater *Puffinus lherminieri*, which looks very like the the Tropical Shearwater *P. bailloni* found in the Mascarenes[153]. No doubt Tropical Shearwaters then nested all over Mauritius, as they still do in Réunion; they were not to see the century out. Lesueur was also the first to refer to "partridges which do not exceed the size of a sparrow" (i.e. Painted Quails *Coturnix chinensis* from Asia), also reporting another (unidentified) quail the same size as those in Europe[154]. There is no documentation, but as Painted Quails from Mauritius match specimens from Sumatra and Java[155], they were probably brought back from spice-seeking journeys in the 1770s (one carried by the ornithologically inclined Pierre Sonnerat[156]), or possibly in the 1760s by the younger

BOX 8   IBISES (THRESKIORNITHIDAE)

Réunion Ibis (Réunion Solitaire) *Threskiornis solitarius*

More speculation, assumption and misinterpretation has been made about a bird known as the Solitaire of Réunion than any other, save the Dodo of Mauritius; indeed for a long time it was thought to be a kind of dodo. However, the discovery of skeletal remains revealed that the Solitaire was in fact an ibis, a bird that better corresponds with the early accounts. Morphological details suggest that it was closely related to the Sacred Ibis *T. aethiopicus* of Africa and the Straw-necked Ibis *T. spinicollis* of Australia.

After its discovery, the Réunion Ibis, now *Threskiornis solitarius*, gradually retreated into the remote mountaintops as Réunion became more populated. The last account to refer to them dates from 1708; they must have died out shortly afterwards.

Accounts

Melet (1672) in Réunion in 1671:

The commonest species there were pigeons ramiers and touterelles [miscellaneous pigeons], grives ['thrushes' = bulbuls], parrots, pepeux [?] and little partridges, excellent large buzzards [harriers] and [an]other sort of bird called solitaires [ibises] which are very good [to eat] and the beauty of their plumage is most fascinating for the diversity of bright colours that shine on their wings and around their neck.

Dubois (1674) in Réunion in 1671–2:

These birds are so-called [solitaires] because they always go alone. They are as big as a large goose and have white plumage, black at the tip of the wings and tail. At the tail there are feathers approaching the tail of the ostrich. They have a long neck and the beak made like that of the woodcock, but bigger, and legs and feet like the turkeys. This bird is caught by running after it, since it flies only very little.

Réunion Ibis *Threskiornis solitarius* by Jean-Michel Probst. From Probst and Brial, 2002.

Feuilley (1705) in Réunion in 1704:

The solitaires are the size of an average turkey cock, grey and white in colour. They inhabit the tops of mountains. Their food is only worms and filth, taken on or in the soil.

Specimen collections

Fossil material is housed at UCB and MNHN. This includes the supposed fossil stork *Ciconia* sp., which was described by Cowles (1987) and is now known (Cowles 1994) to be referable to the ibis.

Cossigny. Visiting Toussaint de Chazal's land near Grand Bassin, Lesueur reported a snake found 8 inches (20cm) underground as forest was cleared. It was "a very small species, at most 4–5 inches" – clearly a blind-snake, and presumably a Flowerpot Snake *Rhamphotyphlops braminus*, which must have already been long-established to have reached so remote a location[157].

While Poivre and Céré promoted spices and indigo, the younger Cossigny, another scientist-administrator, encouraged sugar-cane, introducing new varieties from the East Indies in 1782[158]. An influential islander, he was delegated by the island's colonial assembly to the revolutionary government in Paris in the late 1790s[159]. In 1795, events on the other side of the world triggered a trend that was to revolutionise agriculture in Mauritius and would eventually spell the end for most of the native forests. The slave revolt in Sainte-Domingue (now Haïti) from 1790 onwards completely destroyed the thriving sugar plantations that supplied the entire French market[160]. When news of this reached Mauritius in 1795, planters were quick to respond to a market opportunity. They increasingly abandoned indigo as their principal cash crop in favour of sugar, and refused to implement revolutionary laws freeing slaves[161]. Although Napoleon's governor, General Decaen, attempted from 1803–10 to reinstate coffee and food crops[162], the die was cast. On top of the political events, a series of severe cyclones in 1806–07 persuaded planters that the only good crop was one that withstood hurricanes, and only one did – sugar cane[163].

Whatever they were growing, the planters used little or no manure or fertiliser, preferring simply to cut more forest when land became unproductive; exhausted fields were often turned over to slaves to cultivate[164]. Charles Grant wisely remarked in 1801 that

> *... this felling of the woods ought to be observed and controuled by the administration; as the inhabitants, to accelerate their fortunes, will soon have laid waste the whole island; so that it will then become uninhabitable. It is high time to remedy it. There is still more than half the island covered with most beautiful woods, and it*

would be very impolitic to let one of the finest and most productive spots of the globe be destroyed, blest as it is with an healthy climate and magnificent harbours[165].

By the end of the French rule only 18–25% of the island was being actively cultivated, though a third had been cleared of forest at some time[166]. In 1809 the human population stood at just under 69,000, having more than tripled in the 43 years since 1766[167].

RÉUNION

In Réunion the revitalisation of the Company's interests, and the introduction of coffee as a cash crop from 1717 onwards, accelerated the coastal clearances that had been going on for decades. New concessions were granted more or less sequentially southwards along the west coast. Until 1717 land south of Saint-Gilles had been little used for anything but grazing, hunting, woodcutting and tortoise collection, but systematic clearances for coffee under deputy-governor Desforges-Boucher from 1718 pushed rapidly south. By 1736 the area around Saint-Pierre was well settled; the church built there in 1737 was roofed in hollowed-out Hurricane Palm trunks![168] Desforges granted himself a concession at Le Gol in 1719, where later in the century his son created an important centre for introducing 'useful' plants and animals. The northeast coast was already settled sparsely as far as Saint-André, but being wetter it was less suitable for coffee, so by 1740 serious settlement only extended as far as Saint-Anne. Given the low population density, forest clearance was only partial where land was settled. A detailed map from 1718 shows that even around the oldest settlement at Saint-Paul, open farmland was not extensive; most of the area was still wooded, though no doubt hunting pressure was severe[169]. Nonetheless, by 1731 the administration was complaining that there was no timber big enough to make *pirogues* (dug-out canoes) within 20 leagues (80km) of Saint-Denis – this was the distance by sea, i.e. roughly as far as Saint-Pierre[170]. The human population also rapidly increased as the Company brought in thousands of slaves to service the new coffee enterprise. In 1735 the white population was still under 600 (from 507 in 1708) but slaves had increased from 387 to 7,573[171].

The first casualty of the expansion along the dry west coast was the tortoise. In 1717 and 1721, to judge by Le Gentil and Père Gaubil's reports, they were still ordinary fare – tortoise soup rapidly cured the crew of Le Gentil's ship of scurvy[172]. However, by this time they had "practically disappeared from the zone between Boucan de Laleu [now Saint-Leu] and the Rivière Saint-Étienne", and in 1722 were common only in the deep southwest, both wood and tortoises being fetched from Grand Bois to supply ships[173]. Despite numerous attempts at conservation measures (pp. 87–88), nothing could stop pigs "eating the precious beasts", and although occasional poachers were fined, taking tortoises was clearly socially acceptable[174]. Restrictions were re-imposed in 1716 and 1718, but relaxed later that year by Desforges on becoming deputy-governor – a little surprising, given his tirade against slaughter less than a decade earlier! He had special responsibility for the leeward side of the island, and given the fraught politics of tortoise protection, took care not to report his hunting dispensation to his boss in Saint-Denis[175].

Also under Desforges's aegis, emigrations towards the south began immediately. Yesterday's truth is no longer today's. Yesterday tortoises were loved, today they are burnt. All those whom Parat had prosecuted [c. 1715] for their time in the Pays des Vivres [Land of Supply], returned to take possession. Desforges invited them to legitimise their titles, and to help them along, leading by example. He solicits for himself and his family, next to the mere [Etang du Gol] the first concession that is accorded (5 March 1719).[176]

The last official mention of tortoises was in two further decrees in 1728, which banned all tortoise hunting under any circumstances[177]. Four cyclones and a catastrophic drought in 1734 had the settlers out hunting anything that they could get[178], which probably put paid to the last accessible animals. By 1732 tortoises had already been declared extinct by Abbé Gandon and the normally careful elder Cossigny, and Père Caulier stated baldly in 1764 that they "had been extinct down to the last *poulet* [hatchling] for over 30 years"[179]. Nonetheless, a specimen reached France in 1729 and two more, alive, in 1736, the latter illustrated by Petit in 1741[180]. In fact it seems that tortoises along the western coastal zone were wiped out (the eastern ones had gone by 1694), while upland populations inland survived. In 1743 they were said to be, with wild goats, the main food of fugitive slaves in the interior[181]. Lanux sent tortoise lice to Réaumur in 1754, but the tortoise may not have been native[182]. Bory tried and failed to find any alive in 1801, but was shown the carapace of a juvenile found (alive?) 24 years earlier near Saint-Philippe; however he did not penetrate the cirques[183]. After that nothing was heard of them until the mid-19th century (pp. 141–142).

Visitors and residents in the 1720s and 1730s enjoyed hunting parrots, Whimbrels, fruitbats and, especially, merles[184]. Père Gaubil in 1721 noted that there were "several species of parrot which are most savory and fat", but only Cossigny in 1732 gave any indication of appearance; grey and green ones were still common at this date[185].

Ships heading for Réunion in the 1710s had orders to collect and introduce animals. In November 1714 the Company told ships collecting coffee plants in Moka (Arabia) that "if anywhere the vessel makes

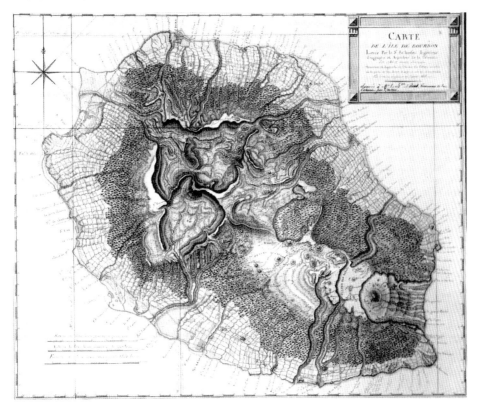

Map 6.2. Map of Réunion in 1818, by Selhausen. Despite the apparently forensic accuracy of this map, we believe from descriptive material that Selhausen exaggerated the degree of forest loss in the heights of Saint-Pierre and in the coastal region by Saint-Philippe, and in general set the lower margin of the forest too high up the slopes.

landfall they find chickens, guinea-fowl or any other transportable winged fowl like partridges, pheasants and other birds, we ask that males and females are embarked with a view to populating Bourbon [Réunion]"[186]. Similarly in 1717 the *Courrier de Bourbon*, due to travel to Réunion, was told to stop in Tenerife and load "some stock of animals as diverse as partridges and camels, a male and two females, seeds of useful plants, silkworm eggs, mulberry plants and vines as well as wine"[187]. It seems the 1714 instructions were followed; Le Gentil reported in 1717 that "from the Indies have also been brought rabbits, quails, partridges and guinea-fowl. The rabbits were unable to dig burrows, the quails, migratory, have barely stayed, and the partridges have disappeared. Thus it is only the guinea-fowls that have multiplied". Cossigny confirmed the establishment of guineafowl in 1732[188]. Vines (and new deputy governor Desforges) were brought on the boat from Tenerife, but no partridges or camels[189]. The guineafowl were included in a 1728 law banning hunting of pigs, goats, tortoises and the taking of honey and gum (benzoin)[190]. By around 1740 a species of perching partridge was said to be common[191] – this cannot have been the indigenous buttonquail *Turnix nigricollis*, but was probably the Grey Francolin *Francolinus pondicerianus*, which frequently takes cover up trees rather than on the ground[192].

Introduced pests

As in Mauritius, agricultural progress had its setbacks. Coffee was immune, but from 1727 locusts ravaged other crops[193], though d'Heguerty claimed that by around 1740 they had been brought under some control by pitfall trenches and fire. This may have applied only to the windward (northwest) coast; Frère Lecoq reported farmers only just holding their own against the ravages on the leeward side in 1740, and there was a disastrous plague on both the large islands in 1744[194]. According to the elder Cossigny in 1732, the locusts had come from Madagascar "in bundles of soil enclosing [the roots of] certain plants or shrubs that were brought from there"[195]. Rats were also troublesome. In 1731 rats and locusts together devoured 95% of the harvest[196], and in 1735 a new sort of rat "as big as little cats" with "a white belly and a short tail" (i.e. Norway Rats) came ashore from the good ship *Venus*, wrecked at Saint-Denis. The rats set about attacking chickens and even suckling pigs and goats; they were said to be immune to cats, and only killed by good dogs[197]. By 1741 they appear to have decimated domestic poultry (chickens, geese and turkeys) in the Saint-Denis/Sainte-Suzanne area, but they had not spread across the cliffs and mountains to Saint-Paul and the west coast[198]. D'Heguerty's disquisition on rats gave them almost supernatural powers:

> One sees there [in Réunion] *different species of rats, which ordinarily inhabit tunnels and cellars, and from which the odour leaves an impression so strong that if they pass over a stack of bottles, all the wine becomes musky to the point that it is no longer possible to drink it. Others get into farmyards where they cause terrible damage, not only to large and small poultry, which they kill and drag into their burrows, where they make a kind of store-room, but they equally attack sheep in their pens, goats, kids and pigs. They attach themselves to the crown of their heads and suck out the brain, and destroy many. Others again, equally destructive under water, go into rivers diving for shellfish of all sorts, which they have the instinct to carry onto rocks which emerge above water level and with the aid of a stone which serves as their hammer, break and eat them; one can see overnight piles of these shells on the rocks*[199].

'Rats' that sour wine can only be House Shrews, and d'Heguerty's report was the first from the Mascarenes[200]. However, House Shrews appear to have died out, as they were absent in the 1820s and not recorded again until the 1860s (p. 141). During their initial phase of superabundance, Norway Rats, living in underground tunnels[201], clearly exceeded their usual level of predatory activity. Some of the exploits sound exaggerated, but these rats are known to swim, dive and hunt underwater[202]. Mice are first mentioned by Lanux as abundant in 1754[203], but they may have been around for some time by then.

As with locusts and Ship Rats, bird pests reached Réunion before Mauritius, no doubt because it was the older French colony, and until 1735 the principal port of call for shipping. Small birds called *bengalis* that ravaged crops of rice and wheat had become abundant by 1740. The name used is too vague to allow identification. Though *bengali* was later used for Red Avadavats, these birds were probably the same as the *jacobins* of the 1750s[204]. A *jacobin* was sent by Lanux to Réaumur in 1754 and described by Brisson in 1760 – this bird was a known cereal pest now called the White-rumped Munia *Lonchura striata*[205]. It was no doubt brought from India as a cage-bird; it remains one of the most popular worldwide, still called 'Bengalese Finch' by aviculturalists.

Although bird pests turn up in accounts of Réunion[206] and in legislation throughout the century, nowhere are they adequately identified, though it seems likely that at least Java Sparrows, Yellow-fronted Canaries, Cardinal Fodies and Red Avadavats *Amandava amandava* were present by the 1770s. Bird pests were already subject to compulsory slaughter by 1761[207], but a pest-control decree in 1769 mentioned only *oiseaux de Malgache* by name; these are not Cardinal Fodies as one might expect (since they come from Madagascar), but *jacobins* by another name[208]. There were no further reports of *jacobins*, although they were probably established for three or four decades[209], apparently dying out once other seed-eaters were introduced. An enigmatic finch depicted by Martinet in the *Planches Enluminées* as the '*bruant de l'Ile Bourbon*' (Buffon's '*mordoré*') appears to be a colour variant of the Cardinal Fody, evidence that it had reached Réunion before 1770[210]. Buffon stated rather vaguely that *bengalis* and *senegalis* (i.e. small estrildids) were present in both Mauritius and Réunion[211]. The Java Sparrow, Yellow-fronted Canary, Cardinal Fody and Red Avadavat featured under local names in a pest control law of 1820. Local hunter, trapper and diarist Jean-Baptiste Renoyal de Lescouble also reported Spice Finch and Common Waxbill *Estrilda astrild* around the same time, the latter in 1825 being "new to the island". The first ornithological confirmation of the species involved came in 1860s[212]. In addition to the seed-eaters, the Zebra Dove was probably, as in Mauritius, introduced in the late 1700s, though the first report is not until 1820[213]. This uncertainty reflects the dearth of information about Réunion wildlife between 1740 and 1860; we have no clear picture of the pattern of introductions.

As already discussed, Common Mynas were released in 1759 to control locusts, and after some initial success were persecuted for appearing to take sown seed. However, we suspect (pp. 94–95) that they were not completely wiped out as then believed, although they were probably boosted by further releases. Locusts were still reported to be a pest in 1761 and 1763, though Poivre clearly thought they had been eliminated by 1767[214]. In 1768 there was a final outbreak, in the east only; islanders were again required to turn out and fight them, this time needing to provide 5 *livres* (weight) of locusts per slave to the authorities on pain of a 500 *livre* (monetary) fine[215]. Mynas were protected by law in 1767 (as in Mauritius), with this renewed in 1774 and 1786[216].

The native fauna

In 1717 the enigmatic '*oiseau bleu*' was still said to occur on the Plaine des Cafres "a little plain high in the mountains"[217]. In 1763 an English prisoner-of-war described the Plaine des Cafres, adding that "there are also some curious birds that never frequent the shoar, who are so little accustomed to the sight of men, that apprehending no danger, they come so near as to be knocked down with sticks"[218]. Strickland and others[219] have supposed that these 'curious birds' were the same as the '*oiseau bleu*' last reported nearly 50 years earlier, but given the general tameness of the avifauna when the island was first colonised, and that the (then) inaccessible Plaine de Cafres was so little visited, almost any upland bird that rarely encountered humans might be meant. No indication of the bird's size or colouring was given, and the writer had evidently not seen the Plaine des Cafres himself – not surprisingly, since he was in detention.

The last person to describe the Mascarin Parrot *Mascarinus mascarinus* on its home ground in Réunion was Feuilley in 1704 (Box 19). Dubois noted in the 1670s that it was not edible, so perhaps this is why it was ignored. However, it survived, and specimens reached Paris dead and alive during the middle and latter part of the century. Brisson described a living bird around 1760, and Buffon a specimen in the Cabinet du Roi, illustrated in the *Planches enluminées*. Neither were of known origin, though Buffon was aware from De Querhoënt that the species was found in Réunion[220]. Mauduyt reported that there were 'several' alive in Paris around 1784[221], but De Querhoënt's note to Buffon in the 1770s was the last report from the wild. Bory made no mention of it in 1801, and the captive birds in Paris had also died by then, Levaillant reporting only stuffed birds (three) that year. Only one survives today, in the Paris Museum[222]. At least one specimen reached England, acquired at an unknown date before 1780 by Sir Ashton Lever; when his collection was sold, the bird was bought by the Vienna museum, where it is still preserved[223]. Following Hahn's claim, published in 1834, to have seen one alive in the Bavarian royal menagerie, the species is generally thought to have lingered in captivity until that date[224]. However a printed inventory dated 1826 listed no such bird[225], and Hahn's picture, although claimed as 'from life', was clearly copied from the *Planches enluminées*. Hahn's clear statement, for the Mascarin and other species, that he had seen living birds in Bavaria, suggests there had been one there, but as his visit is undated, it could have been long before his book was published. Of the other two parrots no more is heard; a stuffed Echo Parakeet from Réunion was described by Brisson, and later Buffon (illustrated in the *Planches enluminées*)[226], but the grey parrot disappeared without trace (as in Mauritius), not being unequivocally reported after 1704[227].

Jean-Baptiste de Lanux spent most of his life in Réunion, arriving in 1722 and active till his death in 1772[228]. He was in charge of slave trading in Madagascar, then Secretary to the Provincial Council and chief registrar in Saint-Denis[229], but his interest to us is that he was the first resident naturalist on the island. He corresponded with Réaumur in Paris in the 1750s, sending many bird and bat specimens to enrich Réaumur's cabinet[230]; in 1771 he played host to Commerson, ill after a trip to Madagascar[231]. Brisson studied the specimens, describing most of the endemic species, including the passerines on which travellers' accounts were usually silent. Lanux recorded local names, several of which are still in use today[232], including *tuituit*, still the name for the Réunion Cuckoo-shrike – a bird then completely forgotten for the next 110 years![233] While the large birds were already named by the early 1670s[234], one might imagine from their non-appearance in accounts that small birds were ignored. Not so. The Réunionnais clearly knew and named all their birds, although, aside from the Paris encyclopaedists and their correspondents[235], only Bory, visiting in 1801[236], wrote about them before the 1860s. Lanux's correspondence with Réaumur, containing snippets of ecological interest, remains unpublished[237].

Bats

In 1776 Buffon published an important letter from Lanux on flying-foxes[238]. His ecological observations were discussed in Chapter 3 (p. 44), but he also commented on their decline. "When I arrived [in the island] these animals [flying-foxes] were as common, even in the settled areas, as they are rare today", he wrote, going on to attribute this scarcity to deforestation, low reproductive rate (one young per year) and hunting:

> They are hunted for their meat, for their fat, for young individuals, throughout all the summer, all the autumn and part of the winter, by whites with a gun, by negros with nets. The species must continue to decline, and in a short time. In abandoning populated areas to retreat to those that are yet to be so, and into the interior of the island, [they encounter] *fugitive negros* [= slaves] [who] do not spare them when they can [get them]. . .

He described the ease of trapping Rougettes roosting in tree-holes by blocking their exit and suffocating them with smoke; up to 400 might roost together in one hollow tree. Regularly fatter than the larger Black-spined Flying-fox, they were a principal source of cooking fat after the tortoises gave out[239]; the East India Company even considered harvesting bat-oil for export in the 1730s![240] The tone of Lanux's account suggests that finding bat-trees, "for that is what we used to call the roosts of our Rougettes", was no longer commonplace in the 1770s. By 1801, when Bory wanted to find "bats of the large species", he failed: "the number of these animals is much diminished, and the species itself will soon disappear, because its flesh being so delicious, the hunters and *marrons* seek it out to eat". Yet in 1805 Charles Tombe was able to report 'considerable quantities', presumably of Rougettes[241]. Bory was unaware that there were two species; and indeed the larger species was probably extinct in

Réunion by then, only the Rougette surviving, despite its greater apparent vulnerability. We do not know why the Black-spined died out first, but perhaps the Rougette was better able to exploit high altitude forest where it was less subject to hunting pressure[242].

Lizards continued to be ignored by almost all 18th century writers, so we have no idea how they responded to changes in the environment – new rats, shrews, human activity. Lanux mentioned 'lizards' in passing among other predators on his precious silk-worms[243]; Père Brown's enigmatic species is discussed in Chapter 3.

Environmental change in the latter half of the century

Unlike Mauritius, the environmental changes that took place during the last two-thirds of the 18th century are poorly documented, at least in print. From the mid-1730s onwards slave imports continued at high levels to provide labour for the expanding coffee, cereal and livestock estates, and so, it appears, did the spread of settlement, though details are lacking. The total population grew very rapidly from 1735 to 1779; most of the increase was due to imported slaves (8,153 to 37,138), less to Europeans (580 to 6,464)[244]. Livestock increased similarly, Le Gentil hyperbolically describing mountains between Saint-Denis and La Possession in the 1760s as "appearing from a distance as if white, from the number of herds of goats which are always there"; there were also good numbers of cattle – neither animal was plentiful in Mauritius. Around the same time Pingré reported a scarcity of wild game, so addicted were the inhabitants to hunting, and noted the use of sap from the *bois de natte* for making bird-lime[245].

In a bid to ease the lot of a growing population of *petits blancs* (poor whites), the far south around Saint-Joseph, previously untouched, was conceded in 1785. Families from the relatively populous northeast were resettled there, a small-holding 'poor-white' and 'free-black' culture surviving in that area to this day. Expansion of cultivation in most areas was by infilling, and by the end of the century settlements still occupied only a relatively narrow strip around the coast. Only the extreme southeast was left alone, for that was dangerous country, where eruptions from the volcano flowed to the sea[246]. Some attempts were made to make use of the cooler climate of the heights – Bouvet de Lozier unsuccessfully promoted cattle grazing on the Plaine des Cafres in the 1750s (hence, perhaps, the disappearance of the 'oiseau bleu'). He also organised systematic hunts of fugitive slaves, successfully reducing the risk to planters and plantations. Nonetheless there was little reason for whites to colonise remote areas, though some groups of ex-*marrons* settled in the rugged interior of Mafatte in the later 1700s[247]. The forbidding surroundings of the volcano were occasionally explored from 1761 onwards, most famously by a high-powered team including Joseph Hubert, *ordonnateur* Crémont, governor Bellecombe and others in 1768, and again in 1771 with Crémont, Commerson and Jossigny guided by the young Lislet Geoffroy[248]. Around 1800 this area was also frequented by inhabitants of Saint-Joseph collecting nestling petrels for food[249].

Sometime in the mid-1700s deer were introduced from Mauritius. By 1758 interim governor Antoine-Marie Desforges-Boucher had "herds of deer" on his estate at Le Gol, and by 1770 they had "dispersed into the mountains and forests"[250]. Desforges is our old friend the champion of the mynas, who retired to his family estate in Réunion after the royal régime replaced the French East India Company in 1767; he maintained extensive gardens of tropical plants and a menagerie of 'rare animals' on his estate[251]. Feral deer were protected by game laws in 1767 and 1786[252], but this did not prevent them being wiped out: Bory reported that by 1801 "there used to be deer in Réunion; these deer were, so it is said, the same as those in Mauritius. One was killed eight years ago near the Ravine Blanche, in the heights of the Saint-Pierre district. I do not think any have been seen since"[253]. Desforges's estate was near the coast north of Saint-Pierre, so the deer had not spread very far before succumbing to hunting pressure; their absence was confirmed by 19th century writers[254]. Hares and quails crept in between 1767 and 1786 – there is no mention of them in the 1767 game law, but both feature in the later one, though no more is heard of them until the 1820s (p. 138). In 1827 the hares were said to have been released some 50 years before, i.e. in *c.* 1775[255]; these Black-naped Hares were probably imported from Mauritius, where they were well-established by the 1770s. While quite possibly Painted Quails (first confirmed in the 1820s, p. 138), the *cailles* reported in the 1786 game law cannot be identified further as there is no late 18th-century description[256]. Although not mentioned until the 19th century, giant snails, imported as a cure for a chest complaint, were released in the 1780s, followed by Malagasy Grass Frogs around 1790[257]. Despite letting in all the bird pests, the authorities were alert enough to maintain the ban on importing monkeys[258].

While on his travels around the island Bory came across *perroquets noirs* (black parrots). He left no description, but identified them as '*Psittacus niger*', citing Gmelin's edition of Linnaeus's *Systema Naturae*[259]. He shot (and ate) two, and given his close view, we can be reasonably sure that they were black parrots from Madagascar (now *Coracopsis* spp.), the two species not then distinguished; the date of introduction is unknown. Although reported up to the 1860s, no specimens reached museums, and they were never positively identified (p. 142).

Plant introductions were rife, being actively promoted in the 1760s and 1770s by Desforges at Le Gol in the west and Joseph Hubert at Saint-Benoit in the east, both liaising with Nicolas Céré at Pamplemousses Gardens in Mauritius. The principal (and successful) intent was to develop the spice industry and to make tropical fruit-trees available in the island. Hubert's careful culture of cloves saw Réunion become a major producer from the 1790s until the 1850s, before competition from elsewhere made planters turn to sugar[260]. However, one of his introductions, the Rose-apple, went wild, becoming the major invader of native forest in the wet east of the island. The equally invasive Chinese Guava may have arrived earlier, as it was already cited as the "chosen food of our large [fruit] bats" in 1764 – both of these plants had reached Mauritius 15 or 20 years earlier[261]. In terms of ecological impact tree crops are generally less devastating than grasses such as sugar-cane and cereals, because they retain more of the three-dimensional structure of forest. Thus coffee, a small tree shaded by tall *bois noirs* (*Albizzia lebbeck*), thriving in the dry west and the taller clove in the humid east – both provided some sort of evergreen forest-substitute for birds and geckos. Indeed the endemic Hoopoe Starling was unpopular around 1800 for feeding on coffee berries[262].

Bory de St Vincent's visit to Réunion in 1801 mirrors his ex-colleague Milbert's to Mauritius at the same time. Whereas Milbert was a gregarious and somewhat romantic artist, albeit with a passion for natural history, Bory was more of a loner, and a dedicated geologist and botanist much in tune with Hubert, who he visited. His interest in wild nature kept Bory's wanderings away from cleared and cultivated areas; he wrote little about them, but had plenty to say on the untouched vegetation. Cabbage Palms were still found in the forests above Saint-Denis on the way to the Plaine d'Affouches, and latans were still common on the west coast between Saint-Paul and the Rivière des Remparts. He described animals he and his porters killed and ate (merles, black parrots), but unlike most travellers he also commented on other wildlife: stonechats in the upper limit of heath vegetation, 'swallows' over Grand Étang and the mere at Saint-Paul, feral pigeons, Brown Noddies on seacliffs and tropicbirds on inland ones. Merles were in 'prodigious quantity' on the Plaine des Chicots, and of such 'incredible stupidity' that they could, in less frequented places, be killed with sticks; there was no need to carry meat when hiking! He saw wild goats in several places, and observed that, unlike in Mauritius, no tenrecs had been introduced. In fact, at almost the same time (1801 or 1802), Jean-Baptist Renoyal de Lescouble released some, shipped from Madagascar, at La Montagne, across the river west of Saint-Denis[263] Bory noted Tomb Bats roosting in tamarind trees at Saint-Paul, and also a mystery tiny all-white bat in latans further along the coast. He enquired about Dodos on both islands, but could find "no hunter, even amongst the oldest, who could tell me a single thing on the subject". It was much the same for tortoises, though they were still remembered, and he was shown a carapace being used as a lamp. He found a big pile of Barau's Petrel corpses in a cave above the deep gorge of the Rivière des Remparts, and was told that the locals collected fledglings for food in season. Being there in November he did not see this harvest (the birds fledge in March–April), though he saw a few petrels overflying the active crater of the volcano. He remarked here and there on introduced plants – the wild raspberry already common in Mauritius, *Rubus rosifolius*, was only just beginning to take hold in Réunion, on the path up to Salazie. Wild strawberries *Fragaria vesca* were well-established in extensive carpets on the Plaine des Cafres[264]. A few years earlier surgeon/botanist Jean Macé had collected the first specimens of the ill-fated Pale House Bat *Scotophilus borbonicus*, which quietly vanished a few decades later (p. 142)[265].

As in Mauritius, Decaen's administration of 1803–10 attempted to rationalise and improve forest legislation and promote self-sufficiency in food crops, while promoting coffee, cloves, cotton, and pepper for export, and it finally dawned that Réunion could not continue to also feed Mauritius[266]. However coffee was beset by insect pests in the dry west, and in the wet east by giant snails that had "multiplied to an astonishing degree"[267]. Worse still, the agricultural plans were upset, even more dramatically than in Mauritius, by the weather. Two severe cyclones in early 1806 were followed by unprecedently long and heavy rainfall in December–January and then a severe drought. The island's ecological and agricultural future was disastrously transformed:

> *From the 12th to 23rd December 1806, there was an extraordinary downpour of rain; on the 26th the rain began again and lasted until the 6th of January 1807. For 12 days the water fell in torrents without interruption. It was what was called, in an absolute sense, an* avalasse[268], *about which eye-witnesses, for a long time afterwards, spoke only with a kind of horror. Due to their quantity and the general steepness of the island, the waters acquired a devastating impulse. The soil was washed [off], scoured to the bedrock; all crops were uprooted and dragged off. The soil was removed to such a degree that the sea was yellow*[269]. *To ensure that everything died off, an obstinate drought followed the* avalasse *which ended on March 14th in a hurricane lasting several days, which would have completed the devastation had there been anything left to destroy. ... The grain harvest failed completely: there was not just a shortage, there was famine.*[270]

Figure 6.3. View of the Morne du Bras Panon, northeast Réunion, 1801, as seen by Bory de St Vincent (1804); note the abundance of Hurricane Palms.

Not only was there an immediate famine (eventually resolved by imports from Madagascar), but long-established coffee plantations and their shade trees were either directly destroyed or died from the after-effects of storms and drought[271]. Attempts at replanting failed because the soil had been so denatured that coffee would no longer grow properly. Although the clove plantations survived, coffee cultivation never fully recovered, and planters turned instead back to food crops, and, increasingly, to sugar[272]. As in Mauritius at the same time, the scene was set for a dramatic decline in the fortunes of the native biota, which the British capture of the islands in 1809–10 then catalysed.

RODRIGUES

During the 18th century Rodrigues regressed from desert-island paradise to ecological ruin. The administrators in the two large islands bled it dry of its only 'useful' resource (tortoises and turtles) then abandoned it, burnt-over and desolate, to a handful of farming (and feuding) entrepreneurs.

Rodrigues was an unimportant outlier, little more than a signpost to ships that Mauritius was nearby, until governor Desforges-Boucher got wind in 1724 that a trading company based in Ostend, then under Austrian rule, intended to occupy it. Suddenly it became important to defend the integrity of French territory, however useless, and Desforges decided to take possession. The expedition he sent from Réunion, which left Julien Tafforet and his shipmates stranded for 9 months in 1725–26, was intended to set up a small defensive and agricultural settlement there, but a storm drove the ship back to Réunion after the advance party alone had disembarked, and the project was abandoned. Desforges's death and the Paris Company's disapproval of the plan then prevented its revival; the Ostenders never showed up[273]. Although there were no major changes between Leguat's time and Tafforet's visit in 1725–26, there were some more subtle shifts that indicate that all was not well. Tafforet only rarely saw doves on the mainland, noting that they were common only on the southern islets, and that, to a lesser extent, the same applied to parrots. He also saw a bird that Leguat missed; the Rodrigues Starling, but only on the lagoon islet Ile au Mat, which Leguat did not visit. These birds may all have been victims to rats raiding their nests on the main island[274].

Rodrigues was not to be left in peace for long. While there were still tortoises to be had in southern Réunion and on the northern islets of Mauritius, the islands' administrators were prepared, in public at least, to abide by Company orders not to collect them from Rodrigues. They chafed at the bit somewhat, since passing ships of all nations, including those of their own Company, stopped off and took all they wanted, often showing off or trying to sell their haul in Port Louis or at Saint-Paul[275]. Furthermore the Rodrigues tortoises were bigger and better than the remnants still available on the other islands in the 1720s. As Gandon remarked in 1732: "vessels coming from India don't miss, when they can, the opportunity of dropping anchor there. They take lots of tortoises . . ."[276]. One of these ships must have left goats, as there were none in 1726, but Gennes recorded them in 1733 (see below).

Knowing full well how the inhabitants had annihilated the animals in Reúnion and Mauritius, the Company's ban on the islanders going to Rodrigues was to ensure that a supply of tortoises remained available for their own ships[277]; they wanted to avoid a repeat of this calamity in the last island then known to harbour these precious beasts[278]. When Mahé de Labourdonnais, already no stranger to the islands, arrived as governor on 4 June 1735, he brought authority from the Company over Rodrigues tortoises, and permission to collect them for distribution to passing ships and to make "soup for the sick in the hospital"[279]. Mauritius was desperately short of meat, with no surplus to supply ships or to give to scurvy sufferers, so Labourdonnais immediately organised a tortoise run to Rodrigues. He sent "le Sr Oriol" on the *Hirondelle* on July 10 to establish a small post there, manned by "two whites and six blacks", to collect and pen the animals between ships' visits. Already by 15 December he could report to the Company that he was giving two Rodrigues tortoises per sailor to every ship returning to France, and in March 1736, that "there is no ship to which I don't give 200 chickens and 400 tortoises"[280].

Gennes de la Chancelière gave us a last glimpse of Rodrigues before the tortoise outpost was installed, landing en route to Mauritius in 1733:

At the head of this little port is found the Enfoncement [bay] of François le Gac, where there are many tortoises. It was there where we sent our boats which had a lot of difficulty landing, but which however returned after 24 hours loaded with 500 of different sizes. . . . These tortoises feed on latan fruit. This tree is something like a palm-tree, that is that the latter has leaves cut and entirely separated, whereas those of the latan are only divided, and are held together like a fan or parasol. On top of the mountains tall trees are seen, and in general the island is completely covered in woods. Our men told of seeing some goats and a large quantity of birds of different species. They brought, amongst others, two which were a third bigger than the largest turkey [i.e. Rodrigues Solitaires]. They nevertheless appeared still young, still having down on the neck and the head; their wings were but sparsely feathered, and they were without a proper tail. Three sailors told me of having seen two others, of the same species, as big as the biggest ostrich[281]. *The young ones that were brought had the head made more or less like the latter animals, but their feet were similar to those of turkeys, unlike the ostrich [foot] which is forked and cloven in the shape of a hind's foot. These two birds, when skinned. had an inch of fat on the body. One was made into a pie that turned out to be so tough as to be uneatable. It is not the case for tortoises, whose flesh is delicate, having the taste of heifer's meat. In a word, it is a grand refreshment for a crew. If the Company were to throw onto the island a few pairs of cows and bulls, one would notice their usefulness to Mauritius and Réunion even after twenty years.*

Returning in 1735, Gennes de la Chancelière wrote:

The only birds we saw on the sea crossing were a few cordeliers [noddies] and some flèches-en-queue [tropicbirds]. We only saw fonds [error for fou(d)s, boobies] and frégattes [frigatebirds] within 40 leagues of Rodrigues . . . As we approached the coast, we saw two ships under the land, beating about with trimmed sails, waiting for their boats which they had sent to take tortoises . . . Our original intention had been also to collect tortoises on Rodrigues, but the wind appearing a little too strong, we decided it was better to not to stop, and we set a course for Mauritius.[282]

From the beginning, the tortoise trade employed two small ships, with each making two trips per year. However, since the handful of men at the outpost were neither naturalists nor, in all probability, literate, we have no more accounts of the island until Pingré went there in 1761 to observe the transit of Venus[283]. In the meantime there are only secondhand glimpses of Rodrigues Solitaires and their fate, and the ever mounting head-count of transported tortoises, logged in ships' records and Company documents. Coming in batches, but used up steadily, tortoises were kept in a holding pound in Port Louis harbour; there was also a sea-water enclosure for turtles[284]. Before returning to the tortoises, we will look at the fragmentary reports of Solitaires.

The end of the Solitaire

Gennes's account of 1733 makes it clear that Solitaires were then still common. Sometime in the 1730s one or more birds were taken to Réunion, where d'Heguerty saw one in captivity. His little-read report describes all three islands, starting with Rodrigues, encountered first by mariners coming from Europe round the Cape of Good Hope[285]:

It [Rodrigues] is about 100 leagues east of Mauritius and abounds in tortoises and turtles. No one is unaware that tortoise soup is the best anti-scorbutic; crews need it after a long voyage. One also finds various kinds of birds there that are often taken by coursing, amongst others Solitaires that have almost no feathers on their wings. This bird is bigger than a swan, with a sad countenance. Tamed [= in captivity] one sees them always walking along the same line as far as the space allows, then returning the same way without deviation. When they are cut open, one normally finds bézoards [stomach stones] which are valued and which are useful medicinally. The Dugong fishery is considerable and a great

BOX 9 DUCKS AND GEESE (ANATIDAE)

Ducks and geese were almost exclusively mentioned as being good to eat and easy to catch, but with few descriptive details. It is hardly surprising that early settlers rapidly hunted out 'tame' ducks and geese, and they became extinct on Mauritius and Réunion very early in the islands' history, sometime during the 1690s. An account by Bouwer, sailing with the fleet under Wolfert Harmenszoon in 1601, mentions that he landed on Rodrigues and collected some 'wild geese'. These may have been boobies (also web-footed) rather than wildfowl, as Rodrigues has no wetland habitat, and there is no other indication of wildfowl there.

Mascarene Teal *Anas theodori* by Jean-Michel Probst. From Probst and Brial, 2002.

Mascarene Teal *Anas theodori*
Fossil material has revealed the presence of a small teal *Anas theodori*, most similar to the Sunda Teal *Anas gibberifrons* complex, inhabiting both Mauritius and Réunion. *A. theodori* was probably quite capable of flying between the islands.

Mauritius Sheldgoose *Alopochen mauritianus* and Réunion Sheldgoose *A. kervazoi*
The Egyptian Goose *Alopochen aegyptiacus* colonised Mauritius and Réunion and evolved into two species, *A. mauritianus* on Mauritius and *A. kervazoi* on Réunion; the last-named has only been recently confirmed in the fossil record. Fossil remains of *A. mauritianus* are extremely rare, and only one account gives any indication of what this species looked like in life. On Mauritius, *A. mauritianus* was mentioned in 1681 as inhabiting the 'woods and dry ponds'; like many geese they probably grazed on dry land. Fossil material from Réunion indicates that *A. kervazoi* differed from *A. aegyptiacus* by more robust leg bones and a short, deep bill.

Accounts
Marshall in 1668 on the Mauritius Sheldgoose (in Khan 1927):

> Here are many geese, the half of their wings towards the end are black, and the other halfe white. They are not large but fat and good [to eat].

Log of the *President* in 1681 (in Barnwell 1950–54):

> On shore at these bays are fine plains and great store of goats, the best on the island, also deer and hogs. Up a little within the woods are several ponds and lakes of water with great numbers of flamingoes and gray teal and geese; but for the geese these are most in the woods or dry ponds. No ducks on the island*.

Dubois (1674) in Réunion in 1671–2 on the Réunion Sheldgoose:

> . . . wild geese, slightly smaller than the European geese. They have the same feathering, but with the bill and feet red. They are very good [to eat].

Pochard *Aythya* sp.
Fossil remains of a pochard *Aythya* sp., a genus present on Madagascar, Southeast Asia and Australia, were recently discovered on Réunion. The Madagascar Pochard *Aythya innotata* is most similar to the Réunion *Aythya* bones so it may well have been this species or a closely related one that once occurred on Réunion. The only historical indication is an ambiguous mention of 'ducks and teal' in 1710, which may indicate this species.

---

* presumably meaning no large meaty ducks, as opposed to the small teal.

---

resource for Mauritius, where many are taken salted in addition to tortoises.[286]

Jonchée de Goleterie had included a *solitaire* in a list of Mauritian birds of 1729, so perhaps he too saw captive birds from Rodrigues, imported with illicitly collected tortoises[287]. Auguste Billiard reported in 1822 that governor Labourdonnais had sent a "*dronte ou solitaire*" from Réunion to the Company in Paris during his term of office (1735–47). As we have seen (p. 85), the Réunion 'solitaire' (the ibis) was almost certainly extinct well before the 1730s, so d'Heguerty's account provides an alternative explanation – that Labourdonnais sent a *Rodrigues* Solitaire, if indeed this belated account is to be believed at all[288].

Silence then falls until 1755, when the elder Cossigny wrote to Réaumur that:

> For 18 months I have been trying without success to procure a Solitaire from Rodrigues Island where we have a little post for collecting tortoises which are fetched by our corvettes. Those who captain the corvettes are well disposed to oblige me, as is the sergeant who is posted there. I have promised all one could want, in spirits or piastres [money] to whoever brings me at least one alive. It is claimed that cats, which have gone wild on this little island, have destroyed this species of bird that has only

stumps for wings, but I am strongly inclined to believe that these cats are the men of the post who have eaten all those they have found, as they are very good. At last, I have been given hope of obtaining one, which, so it is said, has been spotted.[289]

The problem was probably both men *and* cats. The birds had become scarce and hard to find, and by 1761 survival was even more tenuous. Astronomer Guy Pingré spent four months in Rodrigues, yet never saw a Solitaire, though:

> Mr de Puvigné assured me that the race was not entirely destroyed, but that they had retreated into the most inaccessible areas. I heard tell of neither gelinottes [rails] nor butors [night herons] nor alouettes [waders] nor snipes; it is possible there were some in Leguat's time, but either they have retreated into inaccessible places or, more likely, the species no longer survive, since the island has been populated with cats.[290]

Pingré's wildlife observations

Pingré's visit was most fruitful in terms of wildlife, because, despite the catalogue of extinctions, there was a lot he did see; he was an astute observer, always careful to distinguish hearsay from the evidence of his own eyes. He described the trees, but neglected the woods, though he did note that there appeared to be no regeneration of Hurricane Palms and latans, whereas *Pandanus* screw-pines germinated freely[291]. Various fruit trees had been introduced, but as yet none that would invade the native forest[292]. Cats had been imported to deal with the ever-troublesome rats, but, as usual, "they retreated into the woods and went wild. They seem to have made an alliance with the rats; the farmyard [i.e. chickens and ducks] was the theatre of their depredations". Pingré not only described the edible creatures (tortoises, turtles, fruitbats, parrots and Whimbrels) but also many without culinary interest. Of game, he regretted most the scarcity of parakeets and the great rarity of green (i.e. Leguat's) parrots, because they were such good eating. He was particularly taken with the abundant seabirds, his precise description of beak and eye clinching the identification of Abbott's Booby[293]. Like Tafforet, he visited lagoon islets, listing the birds on each; the huge numbers of Sooty Terns on Ile 'Monbrani' (= Gombrani) were particularly impressive. The absence of pigeons led him to suppose that Leguat had used 'doves' for Fairy and Sooty Terns – wrong, if understandable, since the native pigeons, confined to offshore islets by 1726[294], were by then extinct. The large nocturnal geckos described by Leguat had apparently gone, but Pingré saw Newton's Day-gecko and was astonished by its ability to change colour from bright greens and blues to 'hideous black' in just seconds. Land crabs were enormously abundant and their holes were dangerously easy to fall into while walking. As Pingré's account was not published in full until 2004 (though snippets leaked out in the 19th century), none of his observations reached the French encyclopaedists. Buffon had given Pingré's assistant Thuillier an allowance to collect specimens, but there is no sign in Buffon's writings that he consulted either man on their return; his Rodrigues Solitaire account relied entirely on Leguat with no mention of the more recent visitors[295]. The astronomers' precious collections were lost *en route* home, due, once again, to interference by the British (p. 114); was Buffon too disgusted by the loss to talk to them, or had he simply forgotten their existence?

Whereas in the 32 years between Leguat's and Tafforet's visits nothing that Leguat recorded had become extinct, the 35 years from Tafforet to Pingré witnessed a catastrophic collapse of the native fauna. There were no longer rails, night herons, owls, red-winged parrots, pigeons, starlings or giant geckos, and both the Solitaire and the big green parrot were almost gone[296]. Given the abundance of tortoises, seabird eggs and lagoon fish to feed the small settlement, it seems unlikely that any of these birds, except perhaps the Rodrigues Solitaire, would have been hunted hard enough to cause extinction. Rats had been there for a long time, and anything susceptible to them would have been long gone by 1726. There were apparently still no pigs – Pingré mentioned cattle, goats and a few sheep as domestic animals, but not pigs, and emphasised that tortoises were the only wild quadruped. Once again this suggests cats as culprits; abundant in 1755, they must already have been there a decade or more. And it is cats that contemporary commentators blamed for destroying tortoises and Solitaires, though none left any direct evidence. Once they had finished off the nests and even adults of the most vulnerable birds[297], the cats would have fallen back on tortoise and turtle hatchlings. The absence of pigs will have allowed some successful breeding, and turtles could also use the numerous islets, but the intense exploitation of these animals for human use would have eventually overwhelmed the few young tortoises that escaped the cats. By 1795 feral pigs had become established[298]; as we have seen on the other islands, cats and pigs together were apparently able to prevent any tortoise recruitment[299]. Pingré also saw a wide area west of Baie aux Huitres devastated by a fire in February 1761[300] – fires were not yet widespread, but by the early 1800s large areas of the island had been burnt over, and the impact on wildlife must have been severe.

Like Leguat and Tafforet, Pingré was stuck on Rodrigues for far longer than he expected or intended, this time because British sailors stole his ship and burnt a tortoise-shuttle that had brought rice from Mauritius. Hence no one could leave until the next ship arrived from Mauritius in September[301]. The British left us mere snippets of information; Nichelsen, with a larger fleet later that year, noted approvingly that "here is great plenty of sea and land turtle, particularly the latter; we fed our sick people entirely upon them, and served the ship's company several days in a week with them. Here are manatees [= Dugongs], which are good and wholesome food."[302]. By the time he left the island, Pingré had tired of eating tortoise, delicious though it was: "In the three and a half months I spent on the island we ate practically nothing else: tortoise soup, tortoise fricassée, stewed tortoise, tortoise *en godiveau* [?], tortoise eggs, tortoise liver. . ."[303].

Tortoises traded and extinguished

Pingré's visit coincided with the end of the main exploitation of Rodrigues tortoises; the British threat halted operations for some years, and after that the numbers no longer justified a full resumption of shipments. Because the tortoise trade was an official Company operation, records were kept of all the ships involved and their loads, providing an unusually forensic record of the extinction of a species. Using surviving statistics, Alfred North-Coombes calculated how many tortoises were taken. With some assumptions to cover lost data, he reckoned that between 1735 and 1771, a total of 280,000 animals were taken[304]. Rodrigues is a small island, and one might think it unable to support so many tortoises, but in fact the number is not only reasonable but probably about right. On Aldabra giant tortoise density in 'woodland' areas is around 20 animals per hectare (2,000/km²), rising to much higher numbers in areas of 'tortoise turf' with occasional shade trees[305]. There is no evidence of extensive open areas on Rodrigues equivalent to the closely grazed turf on Aldabra (p. 34), so we have taken Aldabra woodland density as a reasonable estimate for Rodrigues. The island covers 109km², and tortoises were found all over it. This gives a total population of 109 × 2,000 = 218,000 animals. The total number removed will have been higher than the standing population at any one time, as in the early years, the removal of a few thousand animals will have made space for more young to grow to adulthood, topping up the population[306]. However, cats will then have curtailed further recruitment, and there would have been increasing failure to replace removals.

After the debacle with the British in 1761, a resurgence in the tortoise population was claimed by Yves Julienne, commander on the island. He reported to Poivre in 1767 that there were "still lots of tortoises, that they are multiplying well and that the wild cats, and the dogs left by the English, have been destroyed, and if there were good boats an abundant turtle fishery could be had"[307]. This optimism was undoubtedly exaggerated, especially the alleged destruction of feral cats! Poivre intended, after finding new sources in the Seychelles and adjacent islands, to remove the settlement (then still containing only 32 people) and "allow the tortoises to multiply at leisure for several years to establish an abundant resource in case of great demand"; this too was wishful thinking. In 1767 Julienne managed to assemble a shipment of 1,215 big *carrosses* (the large saddle-backed *Cylindraspis vosmaeri*), normally too large to collect and carry to the holding park. However after that it turned out to be difficult to collect even a few hundred animals for each of two trips in 1768[308]. In 1769 the settlement was withdrawn, leaving only a token presence (a soldier and a few slaves) to indicate the island was French; they were supposed, according to instructions in 1771, "to husband the small number of tortoises still remaining". The French naval cartographer Après de Mannevillette published his mammoth sea-router in 1775, including a comment on Rodrigues that "only a guardhouse and a few blacks is maintained on this island, to collect tortoises the quantity of which diminishes daily; it is to be feared that the rats and wild cats, which multiply abundantly there, will soon have destroyed the species". His book included a map of the island, showing a turtle pound by Port Sud-Est[309]. In 1778 Morel, working for the Port Louis hospital, the main recipient of Rodrigues tortoises, wrote that they were 'destroyed'[310]. For the rest of the century the rare despatches from Rodrigues never mention tortoises (only fish and turtles), though a visiting captain found a single animal on Plaine Corail in 1786. So forgotten had the island become that, without supplies or relief, the last commander died in office in 1791; two surviving slaves signalled a passing ship, which alerted Mauritius to send a rescue boat[311].

Sometime around 1770 Commerson had examples of wildlife from Rodrigues sent to him. He intended to visit the island, but there is no evidence that either he or his artist Jossigny ever did so. Jossigny's excellent drawings of the Rodrigues Parakeet and the Rodrigues Domed Tortoise *Cylindraspis peltastes* are the only ones of either species from life[312]. In 1786 a visitor picked up some large bird bones in a cave – all that was left of the Solitaire. These bones, eventually transmitted to Georges Cuvier in 1830, finally convinced the great naturalist that Dodos and Solitaires had actually existed[313].

After only a few months respite from human interference, colonists began to return to Rodrigues in 1792, this time to farm and fish. Among them, in 1794, was Philibert Marragon, appointed civil agent by the Mauritian authorities. Liking being the big cheese on a small island, he discouraged too much immigration, and by 1804 there were still only 82 settlers. In 1795 he wrote a report showing all too clearly how much had changed. Although there were still upland forests and abundant latan palms, the exposed east coast was now just scrub. Wild cats were rampant and native land 'birds' reduced to "flying-foxes the size of pigeons, parrots likewise and sparrows [= fodies]; all good to eat, especially the bats; as for the sparrows, though not timid, they are not worth bothering to hunt". Seabirds, *crabiers* (Striated Herons), Whimbrels and migrant waders were much as before. Lizards 8–10 inches (20–25cm) long "of the prettiest dress" (i.e. Newton's Day-geckos) were appreciated for eating insects in and around the houses. Tortoises, however, were almost gone:

> The tortoises formerly so common seem to be completely destroyed. In more than a year since I have been here I have only seen two, and those in almost inaccessible gorges. This is a great shame for, in addition to their goodness, they would have been a great succour in this land totally without butcher's meat. Luckily turtles supply it during the laying season that lasts 6–7 months. One can provide for the whole year by putting them in enclosures; they could even be exported. The Hawksbill shows up sometimes; its meat is not as good as [Green] turtle's, but the shell is pretty.[314]

Marragon, like Pingré before him, blamed cats:

> I have little difficulty in believing that they have contributed to the destruction of tortoises, even turtles, by digging up their eggs or eating the hatchlings, especially on their way to the sea or when they swim along the shore for a while before diving. I have seen cats watching for and catching fish; it would not be astonishing if they took hatching turtles.

He complained of "grasshoppers, crickets and other small insects", pests that would be controlled if mynas were introduced. Caterpillars were also a nuisance, but domestic ducks ate them avidly. Land crabs were still abundant, and Dugongs were still found in the lagoon – though this was the last report.

In 1802, in a further report, Marragon expanded on some themes of his earlier one, notably that slaves were inclined to roam the island on the day of rest wantonly setting fire to "the bushes and young plants that are growing, and making perish any tortoises that might be found, which is one of the causes rendering them extremely rare"[315]. A last few may thus still have lingered until 1802, but this is the last positive indication of survival[316]. Marragon's description of the bleak aspect of the windward slopes suggests that fire-raising had been going on for a long time, perhaps right back to the outbreak Pingré had seen in 1761. In 1804 a 'square league' (*c.* 15km^2) burnt for a month, threatening to engulf the whole island until put out by cyclonic rains[317]. Heaps of leaf litter collected under latans and burnt vigorously in dry periods[318] – Leguat had reported no signs of fire, and it appears wildfire was rare or unknown in Rodrigues, and the wildlife had no way of coping with it.

Marragon proposed to introduce not only mynas but some sport: "deprived at Rodrigues of many species of game, I am doing my best to introduce hares, rabbits, partridges, guineafowl and even deer. There are several suitable islets to put them on, and two issues which indicate them – the first is to avoid causing damage to crops, the second is to keep from the feral cats those animals that cannot defend themselves from such a dangerous enemy." Feral cats were not the only potential enemy; he thought slaves raiding islets for Sooty Tern eggs would steal his game, so he planned to put the birds on different islands. Before he was able to do much of this, Decaen's administration took over in Mauritius, and threatened to evacuate everyone from Rodrigues, ostensibly to avoid giving assistance to the British. This was never done, but uncertainty hung over the settlement for the rest of the decade[319]. Marragon may have released guineafowl, reported as common in 1833, and also rabbits, bones of which were found as subfossils in the caves in 1874[320]; the other game animals had to wait several decades (p. 150). His suggestion to export Green Turtles was taken up while the British blockade was preventing imports of cattle and sheep from Madagascar[321]. In 1795 Marragon said he had seen only one centipede. To anyone familiar with Rodrigues before the House-Shrew decimated the centipede population that may seem extraordinary, but the two large species formerly so common are both introduced, no doubt accidentally with plants or agricultural produce[322].

In 1809, exactly as originally planned in 1761, the British took Rodrigues and assembled a fleet there, this time carrying it through to their capture of Mauritius. We have endless military accounts of these actions, but next to nothing on the island itself: "The island consists of a range of hills of no very great elevation, and mostly covered with palmiste or mountain cabbage, citron trees, manioc and other shrubbery, there is little cleared land, the hills are bestrewn with loose stones, and the rocks appear frequently through the soil..."[323]

CHAPTER 7

A CENTURY OF SUGAR

The Mascarenes 1810–1914

Fine crops, vigorous plantations intercut with deep ravines, agreeably situated houses, circle the island in a belt rising unequally to a third of the height of the highest peaks, forests form a second belt, dominated by the ramparts of one and the other volcano.

Auguste Billiard on Réunion in 1817[1]

Droughts occurred periodically; and when cattle and crops suffered from a scarcity of water, those who attributed such visitations to the cutting down of forests appealed to the Government, which published an Arrete, or a proclamation, or passed an Ordinance. But when the drought was over, and as long as abundant rains and favourable weather prevailed, the new law was forgotten, or at all events became a dead letter. The destruction of forests carried on as before, until another drought came, which was followed by another prohibitive law. The consequence was that the more numerous the forest laws became, the more the forests disappeared.

Dr Charles Meldrum on Mauritian forests in 1881[2]

The British take-over in Mauritius did little there that was not echoed by the continuing French presence in Réunion. Both régimes oversaw a rapid march to a sugar economy in the early-middle 19th century, with acreages increasing dramatically until stabilising in the 1860s. The overall environment in Réunion suffered less than Mauritius, because a much smaller proportion of the island is suitable for sugar cultivation; however other cash crops stimulated massive deforestation at higher elevations in the late 19th century onwards. Until the mid-1870s in Mauritius, and throughout this period in Réunion, despite much forest legislation largely aimed at watershed protection, there was little to stop vested interests, both private and governmental, from cutting native forest if the economics looked right. Rodrigues, meanwhile, pursued a different path to ecological degradation – slash and burn peasant agriculture and a policy of maintaining a high population of free-range livestock for export to Mauritius. As the island is so much smaller and less mountainous than the other two, this resulted rapidly in loss of almost all the forest cover; only a few vestiges of the native vegetation survived.

MAURITIUS

In 1810 Decaen negotiated a surrender to the British in which he persuaded the victors to allow the inhabitants to retain their property, laws and language. The influences impacting on the environment were thus largely external at first, most notably the removal of tax discrimination favouring West Indian sugar in 1825, allowing the Mauritian crop full access to the British market. The planters were already moving towards sugar as the most reliable crop in a cyclone-prone island, and more storms in 1818, 1819 and especially 1824 confirmed them in their views; coffee, cloves and cotton suffered severe losses[3]. One of the leaders in developing the sugar industry was Charles Telfair, navy doctor in the campaign to capture the island turned administrator and then planter, but also a passionate naturalist who coordinated the foundation of the *Société d'histoire naturelle de l'Ile Maurice* in 1829[4].

In the early years the British administration under Sir Robert Farquhar governed with a light touch, making little visible attempt to enforce the forest laws re-introduced by Decaen. Indeed, Farquhar quickly sold off for agriculture the "impenetrable" woods of the '*grandes réserves*' at Bois Rouge in 1816[5], though he did appoint (briefly, 1811–13) a Waters and Forests Department, soon absorbed into a minor branch of the Surveyor-General's Department. Internal transport was still rudimentary; beyond Moka the route south through the upland forests was still by foot or palanquin (a carriage carried on poles) in 1824[6]. British capital was invested in local estates, and after Port Louis was reopened to international shipping in 1816, the stimulus to trade resulted in a rapid increase in acreage under sugar. Initially this was at the expense of other crops, but after the British market was opened in 1825, sugar started spreading to lands then still forested in Flacq and the northern plains[7]. A visitor in 1827 vividly described the forest clearance:

> ... the clearing of land seems to proceed with some alacrity.... The land which I saw clearing was covered with trees, whose roots appeared to be in possession of the sub-surface, and whose stems (or stumps after the trees had been cut down) occupied (two or three feet from each other) the surface above ground. The brush-wood had been burnt, but the larger trees were still strewed indiscriminately about the field that was clearing.[8]

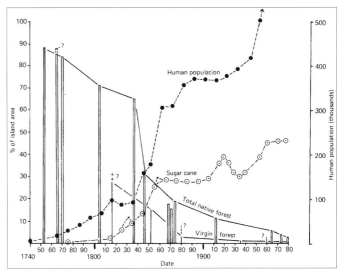

Figure 7.1. Deforestation in Mauritius, compared with the growth of sugarcane cultivation and human population. From Cheke (1987a).

Admiral Cyrille Laplace, visiting in 1830 on a French round-the-world expedition, commented unfavourably on the loss of shade trees and food crops to sugar. He had been in Mauritius as a naval cadet in 1810, and enjoyed scrumping oranges in the then abundant citrus groves, and the sight of coffee plantations shaded by *bois noirs*. By 1830 about 10% of the island's area was under cane, while forests still covered around 65%, the rest being land under other crops or formerly cleared but abandoned (Figure 7.1); most of the forest was in the uplands over 200m[9]. Forest clearance ran ahead of canefield expansion due to other demands – domestic fuel-wood and charcoal, construction, and fuel for the new steam-powered sugar mills, which rapidly replaced water and animal power between 1822 and 1840[10]. In the early years ebony was still exported; 750 tons of heartwood per year during 1812–17[11]. Slaves who previously worked in the docks, on arable land or in coffee cultivation were diverted to the sugar estates, where work was harder and more unpleasant; this period became known to them as *letam margoz*, the bitter times[12].

A revival of natural history in the islands in the late 1820s and early 1830s gives us an insight, albeit incomplete, into the state of wildlife before the chopping expanded dramatically around 1835[13]. Julien Desjardins in particular was recording and collecting animals, and after the *Société d'histoire naturelle* was founded, also started preparing reports for a wider audience. The early reports were circulated in manuscript, eventually collected and published in 1972; Desjardins's own notes are lost, but before they disappeared Emile Oustalet extracted much of what he had written about birds[14]. Desjardins recorded the shooting of the last Pigeon Hollandais in 1826 and the last endemic lizard-owl in 1836[15]. He accidentally released Hoopoe Starlings from Réunion, then a popular cagebird, from his aviary in Flacq in 1835, one of which was subsequently shot in 1836 in Savanne[16]. He was the first to identify the introduced Painted Quail, and to record migrant waders other than Whimbrels. Bojer's Skinks *Gongylomorphus bojerii*, which Desjardins described, were common in coastal areas, and the Rougette was included without comment in his list of mammals[17], suggesting it was still common. Laplace, returning on another world tour in 1837, visited Desjardins at home in Flacq, and admired his personal museum. Discussing the forests:

> He [Desjardins] explained to me the anomalous way in which the cultivation of sugar cane had developed over the previous 20 years, and how far it had extended at the expense of forests and of other kinds of plantation. These developments had been undertaken so thoughtlessly that fields of food-crops for feeding the blacks [slaves] and pastures have almost disappeared, and the colony can no longer subsist without rice from India and livestock from Madagascar, and is even obliged to import the coffee necessary for its own consumption.[18]

The lowlands around Le Morne in the south-west were clearly still well-wooded, and Laplace seemed to see no contradiction in suggesting they be cleared for cultivation! Desjardins himself left for Paris in 1839 intending to write a natural history of Mauritius, but he died in 1840 before he could do so[19].

With fingers in many pies, Charles Telfair was also on the supervisory commission of the Royal College (then in Port Louis), and in 1826 persuaded the Governor to fund a chair of natural history, to which he appointed Wenceslas Bojer, a Bohemian botanist settled in the island. Bojer taught all the sciences, but a dispute with the Catholic bishop (who oversaw religious teaching) persuaded the next governor to withdraw the post in 1832, much to Telfair's and Desjardins's fury. Bojer was eventually reinstated, 23 years later, in 1855. Formal science teaching was then unusual anywhere in the world, and had it not been suppressed Mauritius might have become an important scientific centre. As it was, the *Société d'histoire naturelle* remained an active focus, and Bojer saw students for private tuition[20]. Telfair corresponded with the Zoological Society in London, sending them sent numerous specimens (now mostly lost). He was the first to describe the annual hibernation and diet of the Common Tenrec – they ate "worms, insects, lizards and eggs of snails"[21].

The early 1830s also saw visits from Algernon Strickland, Victor Sganzin and Charles Darwin[22]. Strickland was serving in the navy, but his more academic brother 'Hughie' had persuaded him to collect birds; Algernon's diary shows little personal interest in natural history, though he was befriended by the Telfairs. His specimens are preserved in Cambridge, and Hugh's interest in Mauritius led later to his famous monograph on the Dodo. Sganzin was a French military naturalist who wrote an important paper on birds and mammals in Madagascar, in which he secreted snippets on the Mascarenes. Charles Darwin's visit in 1836, homeward bound and exhausted from his travels, produced nothing of ecological interest, though he was impressed by the bountiful fields of sugarcane![23] James Holman, another friend of Telfair's, toured the island in 1829–30, but his account is short on natural history, which is not surprising since he was blind. He visited Bassin Blanc (where his guides saw tropicbirds) and reported the hilly hinterland of Baie du Cap as "covered with excellent timber too expensive to extract". He also noted imports for the cage-bird market of Grey-headed Lovebirds, Java Sparrows, black parrots, cockatoos and lories[24].

The end of slavery

After much argument and resistance from the 'plantocracy', and pressure from Britain, the Governor finally announced the abolition of slavery in 1835. There were nearly 77,000 slaves, 76% of the population, who were supposed to serve a period of 'apprenticeship' before being fully freed; 34,000 were sugar estate workers. In practice many abandoned their old masters at once; others worked so poorly that they were let go. It proved cheaper to import Indian 'coolies' under contract ('indentured labour') to replace the ex-slaves, than to pay wages sufficient to keep them in the harsh field-work they so detested. The ex-slaves settled on unclaimed land to practice peasant agriculture (mostly slash and burn). In 1850, long before white Mauritians colonised Curepipe, the outer slopes of the Trou aux Cerfs crater were "covered ... with vegetation, among which there is a large native [*sic*] population; many inhabitants there are Malagasy"[25]. Ironically an important centre for *affranchis* (freed slaves) was the southwest corner singled out by Laplace – Case Noyale is still today the focus of authentic creole ex-slave culture. While in the years up to 1843 the Indians merely replaced the slaves, thereafter not only was immigration massively stepped up[26], but so was forest clearance for cane. This renewed expansion was driven by the use from 1843 onwards of Peruvian guano as fertiliser, enabling sugar to be planted continuously without rotation or fallow. While the ex-slaves carved out tiny smallholdings in out-of-the-way parts of the island[27], the sugar barons invaded the central highlands of Moka and the gentle slopes of Savanne, while the remaining forests of Flacq were decimated; sugar acreage increased by 270% in those districts from 1852 to 1864[28]. Completing a proper road from Port Louis via Curepipe through to Souillac (1830) and Mahébourg (1832) greatly facilitated the exploitation of the Savanne district[29]. The total acreage under cane stabilised after 1858 until the end of the century[30], but pressure on the forest did not let up. Published figures suggest that 53% of the surviving forest was cut between 1836 and 1846, but only a further 19% by 1874. The sources on which the 1846 figure is based are of doubtful accuracy, and as the maximum forest destruction is quoted as a decade ahead of the fastest

Figure 7.2. Tamarin Falls, Mauritius, and surrounding native forest, the area explored by Matthew Flinders in the early 1800s and later by Edward Newton in the 1860s, as seen by Bradshaw (1832).

extension of cane (much better documented), the 1846 figure is probably exaggerated[31].

As early as 1838, local botanist Louis Bouton prepared a memorandum against forest destruction, but to no avail. He tried again in 1853, this time inspiring a new ordinance legally protecting mountain and riverine forest, but the reserves were not demarcated and not enough enforcement staff were provided[32]. In 1857 Bouton complained that the "guardian of woods and forests" would be better off planting trees to conserve water flow than worrying about Guinea Grass clogging up the dry water-courses around Port Louis[33].

From today's perspective it is instructive to note that Bouton's reasons for forest conservation centred on protecting the water supply through rainfall, which he believed would dry up if the forest were cut down. Although a keen naturalist, there is no hint of ecological considerations. He was essentially advancing an 18th-century argument (basically Poivre's) in an unreceptive 19th-century capitalist milieu. In 1860 British engineer James Mann emphasised how little forest cover contributed to annual rainfall, while strongly supporting the value of forests in "preserving and husbanding the rain water after it has fallen". However he felt that Mauritian forests were doomed, and that the solution to water supply lay in reservoirs[34]. The various forest laws that evolved during the rest of the century in essence accepted the protection of headwater and catchment forests as 'sponges', but preserving forests was only seriously considered in areas unsuitable for sugar production. While forestry consultant Richard Thompson recommended in 1880 that one tract of native upland forests be retained "intact for their glorious beauty which surpasses that of all other natural phenomena met with in the island", it was not until Vaughan and Wiehe's pioneer ecological studies in the 1930s that anyone seriously allowed them an intrinsic value outside prevailing economic imperatives[35]. Sir Joseph Hooker of Kew Gardens, leading a project to catalogue the flora of the Empire, wrote to Bouton in 1859 that "ere long, by the increased and increasing cultivation, the native vegetation of the Island of Mauritius, as in St Helena, will be destroyed"[36]. His interest, stimulating collections in Mauritius and expeditions to the island's dependencies, did result in Baker's 1877 *Flora*, but no conservation initiatives[37].

The dying trees

Another factor, overlooked in recent ecological studies, was also eroding the forests. In 1845 Bouton's attention was drawn to extensive areas of dead and dying trees around Piton du Milieu:

> *On an excursion I have just made to the Piton du milieu de l'Ile, I was surprised, despite having been forewarned by local inhabitants, by the large quantity of trees, dying or already dead that one meets in the fine forests in this part of the island. From the summit of the mountain where I was, the view to the limit of my vision passed in all directions over these great skeletons which stood out in the middle of the vegetal world surrounding them as if predicting an end both soon and inevitable. . . . Examining the facts from closer to, it was easy to see that the old trees were not the only ones to succumb, that the disease had equally struck tender saplings and that it seemed, given the frightening progress it had already made, to threaten to take down the entire forest.*[38]

According to Bouton's guide, the soil had turned swampy and the trees had slowly drowned at the roots. Whatever the explanation, the phenomenon of mass dieback was clearly fairly recent, and was also very widespread. Skeletal trunks of large trees overtopping 'small wood' had already been noted by Thomas Wise in 1838 near Dr Moon's estate at Henrietta[39]. The area known as Bois Sec (near Grand Bassin) was so named because, as Nicholas Pike described it in the late 1860s, "thousands of dried-up skeletons of trees blanched to a ghastly whiteness meet the eye on every side, and but for a tangle of lianes and plants at their feet showing life, it might be a forest of primeval days over which some blighting plague had passed"[40]. Arthur Gordon contrasted the green trees in Kanaka to the dead ones at nearby Grand Bassin in 1874[41]. A similar dead forest was seen by Vincent Ryan in 1855[42] and George Clark in the late 1850s just south of Curepipe (see quote, p. 122), and a little to the southwest Edward Newton wrote in 1863 that "in a few years the forest around Doody . . . [illegible] much changed, the forest trees are dying out & ravenals and palmistes spring up everywhere, these two latter will soon take the place of the former". Nearby, a few years later, Pike also noted "many dead and dried trunks of large trees" by the Diamamou Falls on the Grande Rivière Sud-est[43].

The distribution of these dead trees over a large part of the upland plateau suggests a major disaster. One theory was wind; a similar tree-kill happened after cyclone Carol in 1960 which "destroyed a very large percentage of the large trees standing along the edge of the eastern escarpment of the Black River Gorges", attributed to a freak wind-tunnelling effect[44]. Its extent was, however, tiny compared to the 1840s event, yet Carol was one of the severest storms on record. There were no cyclones as violent as Carol in the years before 1845, though one in 1836 was fairly severe[45]. The boggy ground theory does not look likely – much of the forest land was naturally very wet, and there was no obvious reason for a change in areas then largely untouched by human activity. Also, the affected areas are topographically different: the land is relatively flat around the Piton du Milieu, but sloping and better drained at Forest Side (Curepipe)

BOX 10   HAWKS AND FALCONS (ACCIPITRIDAE)

Birds of prey have populated many island groups throughout the Indian Ocean, and far-ranging migrants regularly visit the Mascarenes. Two genera with endemic species are certainly known from Réunion and Mauritius, i.e. *Circus* and *Falco*, while vagrant species occasionally arrive.

Mauritius Kestrel *Falco punctatus*
The Mauritius Kestrel, *Falco punctatus*, was once considered the most endangered bird in the world, with numbers down to six birds in 1974. However, it has now increased to 2,000+ individuals, probably as many as the island's current habitat can support, and is one of the great success stories of conservation. The Mauritius Kestrel has sparrowhawk-like adaptations (short broad wings; long legs) indicating an ability to hunt within forest, and it feeds primarily on lizards and small birds.

Accounts
Bernardin (1773) in 1768 (translated by Jason Wilson):

> There is a kind of hawk called chicken-eater; they say it also lives on grasshoppers. It remains near the shore. The sight of a man scares it.

Mauritius Kestrel *Falco punctatus*. Illustrated by Jean-Michel Vinson, 1974.

Later in the same year:

> On the way I saw one of those hawks called chicken-eaters. It was perched on a branch of a latan tree. I aimed at it with my gun, lit both my primes, but no shot sounded. The bird remained calmly on the branch and I left it there.

Dubois's Kestrel *Falco duboisi*
A morphologically distinct species, *Falco duboisi*, inhabited Réunion until the 1670s. Unlike the Mauritius Kestrel, which has rounded short wings; Dubois's Kestrel lacked these and was morphologically similar to the Eurasian Kestrel *F. tinnunculus* complex. Dubois called it '*emerillon*', and since it was capable of taking the inhabitants' chicks it was probably persecuted accordingly. Despite some suggestions to the contrary, it appears that Dubois was describing the kestrel and not a falcon. The *emerillon* was small (like a kestrel), whereas the '*pieds jaune*' was 'the size and form of falcons' – and at the time that would have implied Peregrine *F. peregrinus* (kestrels may be *Falco*, but were not 'falcons' in the sense used by falconers). Dubois's description of *pieds jaunes*, which he contrasted with the *papangues* (harriers), were probably the migratory falcons that still regularly visit the islands, i.e. Peregrine, Eleonora's Falcon *F. eleonorae* and Sooty Falcon *F. concolor*. More recently *pieds jaune* has been used for female Réunion Harriers.

Réunion Harrier *Circus maillardi*
The Réunion Harrier *Circus maillardi* survives but remains vulnerable on Réunion, and also formerly occurred on Mauritius. *Circus* '*alphonsi*', which was described from fossil remains collected on Mauritius, is extinct, but is now considered conspecific with *C. maillardi* (Mourer 2004).

Accounts
Dubois (1674) in 1671–72:

> There are three sorts of rapacious bird which cause much damage to the island's game and to the inhabitants' domestic fowls. The first are called papangues [harriers]. They are the size of capons, but made like eagles. They have no difficulty in unhatting people, for they fly by at head height, and with their hands or talons carry off the hat or bonnet. If it is a woman with white in her hair, they don't take it well. These birds destroy many pigs and goats. These papangues are as good as a chicken in the pot, but they are eaten but little, there being many better game animals. The second [sort] are called pieds jaunes ['yellow feet'], of the size and form of falcons. They cause much damage to settlers' fowl and to the island's game. The third [sort] are emerillons [literally 'merlins', but presumably the kestrel] which, although small, can nevertheless carry off baby chicks and eat them."

Specimen collections
Fossil material of *C. maillardi* (described as *C. alphonsi*) from Mauritius is kept at the UMZC; MNHN and UCB; *F. duboisi* is at UCB.

Réunion Harrier *Circus maillardi* pursuing white-eyes. From Barré and Barau (1982).

and Bois Sec[46]. There was an unprecedented series of droughts at the right time – 1832–33, 1839 and 1842–43, with the most severe in 1839, but there were worse droughts in 1869–70 and 1896–97, after which no additional forest dieback occurred, although the original decline continued[47]. An epidemic thus seems the most likely explanation, and indeed 35 years after Bouton's melancholy discovery, Indian forest expert Richard Thompson observed trees being killed by infestations of bark beetles, though he believed the attacks were initiated by clearance of adjacent forest and consequent exposure to wind. By 1895 Paul Koenig, another believer in 'wind initiation', reported about the beetle damage that:

> *The intensity of this evil is now frightful . . . in certain badly injured localities, where favourable breeding places are offered to all sorts of insects, they spread out in increased numbers and set to attack healthy trees. . . . we thus see the insects otherwise slightly injurious, because they feed on dead and dying trees, become highly injurious when attacks take place to a calamitous extent* [i.e. on healthy trees].[48]

The discovery of beetle attack did not gel with the general Mauritian public, and Gleadow's report in 1904 emphasised the contrast between observational science and popular myth:

> *It is very commonly alleged that all the best indigenous* [tree] *species are fast becoming extinct on account of some occult reason of climate or soil . . . The skeletons of dead trees at Bois Sec, Kanaka, and other places are a prominent feature of the landscape, and the public mind has sought in vain for an explanation. Most persons put the death of the trees down to wind, although the skeletons were standing, and the place* [Bois Sec] *named, before the great cyclone of 1896. Moreover that cyclone, with all the smashing that it did, has not produced any similar effect in general. Finally the places in question are not particularly exposed to winds, whereas the finest forest in that neighbourhood covers a crater, the Trou de Kanaka or Grand Trou, that rises up exposed to all the winds that blow. Some of the best Mauritian species, once firmly established on solid ground, can defy, as they have often defied, the most violent gales. The theory of wind had to be abandoned, but investigation soon revealed the real enemy. This report will be in print before its proper name is settled, but it is a small brown pimple-backed beetle, with a long beek* [sic] *apparently tipped with steel. It belongs to the great tribe* Curculio [weevils], *bad characters all in their degree . . . The particular vice about this beetle is that its grub eats the cambium layer, which is the seat of life and growth beneath the bark. The cambium layer is to a tree very much in function what the veins are to a man. When the galleries of the grub encircle the tree completely the tree dies.*[49]

Gleadow illustrated the beetle he thought was the culprit, calling it a *Ceutorhynchus*, though it is probably a *Palaeocorynus* (both are curculionid weevils)[50]. As mentioned above, the whole episode has slipped under the radar of modern researchers; no such beetle is listed in recent accounts of forests pests[51], and no papers on beetles highlight any insect that kills native trees[52]. Given the pattern of attack (Gleadow also illustrated the under-bark galleries), the culprit is more likely to have been a bark beetle (Scolytidae) than a weevil[53], like the insect and its associated fungus that devastated Britain's elm trees in the 1970s[54]. Indeed Thompson reported fungal attack possibly associated with the beetles: "very few of the larger mature trees now met with in the forests are free from dry rot", and local entomologist Donald d'Emmerez de Charmoy examined about a hundred trees in 1901 without finding an obvious causative insect, but found ample evidence of fungal attack in the sapwood[55]. It looks as if a bark beetle was introduced to Mauritius in the 1830s that went through an epidemic phase, spreading a fungus that killed large areas of native forest with no resistance to this type of attack, before settling down around 1910 to an endemic and less damaging level[56]. Such epidemics, once the beetle is present, can also be triggered by storms[57] (or by droughts?), so periodic cyclones may have kept it active. A trifle too late, the importation of unbarked timber was banned in 1910 "to prevent the spread of bark beetles"[58].

By the 1850s we begin to see the extent to which exotic species were penetrating native forest and taking over in secondary growth. Strawberry Guava was well established, spread by pigs eating the fruit, Travellers' Palms (*Ravenala*) and Rose-apple (*Syzygium jambos*) were invasive in the Mare aux Vacoas, Midlands and Quartier Militaire areas, *bois d'oiseaux* (*Litsea glutinosa*) was abundant enougth to be considered the monkeys' favourite fruit, and the drier slopes in the southwest were covered in *jamblong* (*Syzygium cumini*)[59]. The bramble *Rubus rosifolius* had by this time got everywhere; its relative *R. alceifolius*, later to become one of the worst invaders on both Mauritius and Réunion, was already invasive after only 30 years on the island, and the sub-shrub *Wikstroemia indica* was rapidly expanding its range[60]. In the later 1860s between Fressanges and 'Dhoodie', Pike "crossed a plain covered with the Ravenala or Travellers' trees as far as we could see . . . seeming to rejoice in the swampy land. . . . just beyond this grove was another entirely of dark jamrose [Rose-apple] in full flower". In 1871 Bouton listed some of the most successfully invasive trees: Strawberry Guava, two *Litsea* species, *Ehretia serrata* (now *E. acuminata*), *bois de campêche* (*Haematoxylon campechianum*) and White Popinac *Leucena leucocephala*, the last two in dry places (though he did not say so)[61]. Mauritius Hemp *Furcraea foetida* was established along the dry road from the port to Moka in 1853[62]. However, no studies were done, and while it was known *which* species were naturalised [63], their extent and penetration was still a matter of incidental observation.

Further retreat of forest and wildlife
Population pressure was rocketing. By 1846 58,000 Indians had arrived, and by 1851 over 80,000, doubling the 1830 population. Ten years later a further 110,000 had arrived – human numbers had tripled in 30 years. Regulations now required better housing, and these people all cooked on wood fires, so once the wood cleared for cane was used up there was a continued steady drain on the forests[64]. This intensified when the first railway opened in 1864, using wood-burning steam engines to ferry loads of sugar-cane from the factories around the island to Port Louis. Then in 1865 malaria-transmitting mosquitos slipped in. The disease became a raging plague, killing 50,000 people by 1868. Malaria was initially restricted to the lowlands, so the richer inhabitants of Port Louis fled uphill, founding new settlements; in reverse order of social and financial status at Beau-Bassin, Rose-Hill, Vacoas and Curepipe, they moved along the line of the new railway. The wealthiest settled in the damp and relatively cool forests of Curepipe on the plateau at 550m, commuting in daylight to the capital by train[65]. Above the ex-slave settlement of Phoenix, the area was still mostly jungle, so every new dwelling cut into untouched forest. Where people went so followed sugar, as it could grow up to 550m on the drier (west) side of the plateau, though yields were smaller. Edward Newton's hunting companion, planter James Currie, was "credited with having turned the wilds of Midlands and Fressanges into beautiful fields of sugar cane" in the mid-late 1800s[66]. Some low-altitude estates, especially in the dry west, were abandoned for many years, becoming a scrubby forest of exotics[67]. In the 1860s no one knew how malaria was transmitted, so various theories arose to explain the epidemic. One was that forest clearance encouraged the spread of disease, and this helped in the 1870s to support the introduction of tougher forest legislation. Malaria control measures, begun around 1900 (see pp. 125, 164–165), were to affect the environment for the next century.

After Desjardins's untimely death in 1840, the *Société d'histoire naturelle*'s activities centred largely on botany. The island's animal life was apparently largely ignored, though Desjardins's widow gave his collections to the Society's new museum, opened in 1842 in a wing of the Royal College in Port Louis[68]. Edward Newton wrote in 1862 that "it is quite disheartening [having] anything to do with the museum, there is not one soul who cares or knows about ornithology in the island, though perhaps some of them would be much offended at my saying so"[69]. However, British amateur naturalists began to arrive: teachers, administrators, engineers. The first of these was teacher George Clark, who arrived in about 1836, becoming head of the government school in Mahébourg in 1851[70]. He did not join the Royal Society of Arts and Sciences, as the society had grandly become, but worked quietly on his own, keeping a menagerie of local wildlife. In 1859 he published (anonymously) an excellent natural history of Mauritius, accompanied by a detailed topographical tour of the island[71]. Clark's *Ramble round Mauritius*, based on his own observations and largely uninfluenced by the older literature, provides a vivid snapshot of the island as the headlong extension of sugar at the expense of forest peaked. Since Clark had lived through the period of maximum forest clearance, he often commented, as he toured the island, on what an area was like before. The Plaine des Roches, largely covered in lava and thus uncultivatable, had nevertheless been deforested. The country around 'Mont Piton' (Piton du Milieu), had been until "a few years ago covered with fine timber which was considered of so little value that it was given to anyone who would be at the trouble of clearing the land of it, the ebony trees alone being reserved for the proprietor." At Trois Islots, in the island's centre, indigo had given way to cane; along the dry west coast (Petite Rivière – Bambous) grazing land had disappeared under sugar, or, in the absence of livestock, had become thorny acacia scrub[72]. Cossigny's pride and joy, *bois noir*, had suffered from being too palatable as cattle fodder – the slopes between Black River and Le Morne were covered with jamblong, which "had outlived the Bois Noir and other trees which formerly existed here"[73], suggesting some deforestation here preceded the freeing of slaves (pp. 117–118). The lowlands of Savanne, forested in the 1830s, were now covered in sugarcane. In the uplands on the road from Mahébourg back to Port Louis forest was vanishing fast, and in the part of Curepipe still known as Forest Side:

> Many huts are seen by the road side, chiefly occupied by woodcutters and charcoal burners, whose donkey carts are to be met with on the road at all hours and in all weathers. Vast numbers of dead trunks of trees were formerly standing near the road side, and a few still remain, picturesque though melancholy objects. In wet weather there are many pools of water in this part of the forest, generally covered with the pretty Nymphoea stellata [a water lily]. Previous to the destruction of the forests these pools always contained water, but they are now frequently dry.

These woodcutters were ex-slaves, living on tenrecs and monkeys, and *inter alia* supplying a nearby sawmill making roofing shingles for the housing boom – there were no longer latans to supply leaves for roofs, though grass thatch was widely used. The

Chapter 7: A century of sugar   123

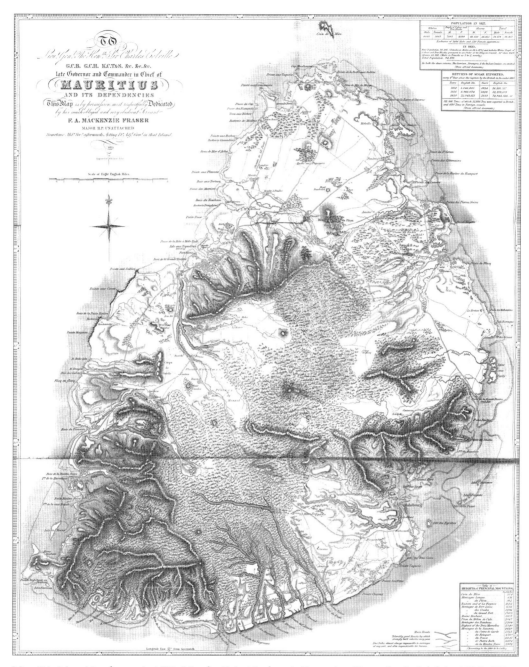

Map 7.1. Mauritius forests in 1835. Map by F. A. Mackenzie Fraser (see Toussaint & Adolphe 1956: 675).

uplands then vaguely known as 'Vacoas', stretching from Quatre Bornes to Tamarin Falls and the Mare aux Vacoas, had long been sparsely settled, but by 1859 had acquired the island's highest sugar estate and were being rapidly deforested, though "some valuable forests still remain". The Mare aux Vacoas itself was a shallow mere and swamp characterised by endemic screw-pines (*vacoas*), full of goldfish; in the dry season many were stranded and eaten by rats![74]

The lizard-owl had gone, and the Pigeon Hollandais is not even mentioned. Clark's school in Mahébourg was not far from where the last owl was collected (Grand Port), and there was local knowledge:

A species of horned owl existed here as lately as the beginning of this century, and was tolerably plentiful in the woods, but I believe there are no more remaining. . . . Mr Dalais, who was an experienced hunter at the beginning of the present century, and who is still able to enjoy field sports, tells me that he shot several in his youth; and from his description of them they must have been very much like the greater horned owl of England [= Long-eared Owl, Asio otus]. The destruction of these birds is much to be regretted, as they would have done a good service in helping to keep down the rats.[75]

If, as we suspect, the owl lived largely on forest skinks, it would have been having a thin time since rats eliminated Telfair's Skink, exacerbated by the arrival of House Shrews and tenrecs, which would have decimated the small Macabé Skink (Bojer's being largely coastal). The lizard-owl's survival may thus already have been tenuous by the time the massive forest clearances started. Perhaps it could not adapt to new prey (a failure apparently shared by the owls in Réunion and Rodrigues), and/or forest destruction removed too many nest-sites (probably hollow trees)[76]. The Pigeon Hollandais was last reported from Savanne, the next district to the west, whose lowland forests, still intact in the 1830s, had largely gone by 1859; if this was its preferred habitat then little was left across the island by 1859. Also frugivorous species often need large home ranges, as trees fruit irregularly and optimal fruit availability will 'move' between forest types. If, like the lowland parrots, the Pigeon Hollandais fed on palm-fruit, an important food source (Hurricane Palms) was in free-fall long before the Savanne forests were cut. Finally, blue pigeons often perch conspicuously on bare branches, thus exposing themselves to hunters[77]. Clark was the last to describe living Rougettes, albeit only captive; he was told that they were rarer than the larger Black-spined Flying-fox.

"Taking a walk in the woods with an Irish gentleman, he remarked to me that he found an *abundant scarcity of birds*; and such must be the impression of most persons accustomed to England and France", a state of affairs Clark blamed on "the numbers destroyed by monkeys". He knew of three pigeons – Malagasy Turtle Dove and the introduced Spotted and Zebra Doves; no mention of either Mauritius Pink Pigeon or Pigeon Hollandais, though the former may have been confused with the turtle dove, both then being called *ramier*[78]. Common Moorhens and Striated Herons he noted in passing, together with 'the curlew' (= Whimbrel) and 'the sandlark', an amusing anglicization of the local *allouette* [*de mer*] (small migrant shorebirds). Echo Parakeets were "still to be found on the wooded mountain tops, but are rarely seen in the plains"[79]. Mauritius Merles had experienced "a vast diminution in their numbers" from the days (remembered by his informants) when they were "so numerous they could be knocked down with a stick" when attracted to a stuffed cat or hare. Clark described the native small passerines and their habits: while Mascarene Grey White-eyes were common everywhere, as they still are, Mauritius Olive White-eyes were "not very common", frequenting banana plantations for the nectar. Bojer's Skinks were still common in coastal areas.

Introduced animals

While native animals declined, introduced ones prospered. The Common Tenrec, principal source of meat to rural islanders, "has only existed here for about sixty years, and from a short period after its introduction has been unsparingly destroyed, yet it is swarming in every part of the island". House Shrews, rats, mice, hares, feral cats, mynas, Cardinal Fodies, Spice Finches, waxbills, two species of canary and lovebirds all thrived. British soldiers had recently introduced House Sparrows *Passer domesticus* to Port Louis[80], but Java Sparrows had declined and were only seen around Black River. Clark was the first to report Feral Pigeons *Columba livia*, though domestic pigeons were an old import, so had probably gone wild long previously[81]. Two 'partridges', three quail species and guineafowl provided 'sportsmen' with game[82], though the last had become rare. The only wild duck was the White-faced Whistling Duck (or Teal); there was no mention of Purple Swamphens. A persistent pest in gardens were giant snails, a second species of which had been introduced in 1847[83].

Clark did not mention chameleons, though there is evidence suggesting that two species were feral for some time in the 19th century. Although Desjardins stated in 1837 that there were no chameleons, he had collected a Panther Chameleon *Furcifer pardalis*, labelled Mauritius and studied by the encyclopaedists Dumeril and Bibron in Paris in 1836; Lesson and Garnot, in Mauritius in 1824, brought back another. A Warty Chameleon *F. verrucosus* in Paris collected by Quoy and Gaimard was labelled 'Madagascar', but the collectors never went there, so it probably came from Mauritius. An unidentified chameleon was seen in the 1840s, and two more Warty Chameleons were caught in the wild in 1882 and 1885[84].

Clark sometimes picked up information that escaped others. On the cliffs carved out by the Grande Rivière Sud-Est "sea fowl breed in the numerous cavities afforded by these banks, and their nests often induce the young creoles to risk life and limb to obtain them". 'Sea fowl' is pretty imprecise, but they were probably Tropical Shearwaters[85]. Some were collected around 1860 in the gorge of the other Grande Rivière, Nord-Ouest[86], but any surviving the attentions of the 'young creoles' probably succumbed to the mongoose in the early 20th century (pp. 136–37).

Clark's account, and those of other mid-19th century writers, reveals how many wetlands there once were in Mauritius. Clark saw an abundance of moorhens in "many marshy spots covered with sedge" along the dry coast near Petite Rivière, in the upland wet forest pools, etc. Belle Mare, "one of the few spots on the island where teal breed", was also full of Common Moorhens. Edward Newton found abundant moorhens in a small swamp near Henrietta and the Tamarin Gorge. The Belle Mare marshes are still there, much reduced, the others mostly gone, with the biggest recorded on a map published in 1918[87]. Once it was recognised around 1900 that malaria was spread by mosquitos, systematic draining of the island's marshes and pools and canalisation of streams was undertaken, which continued with more or less vigour until 1958[88]. Newton's moorhens would not have appreciated the eminent malaria expert Ronald Ross, who on 20 January 1908

> Gave a picnic tea to the governor, Miss Lane and most of the notable ladies and men of Mauritius in the middle of the deadly Phoenix Marsh near Vacoas . . . the marsh was drained a little later and the malaria vanished.[89]

The effect of this wetland drainage on wildlife has never been examined, but it no doubt contributed to the disappearance of the *sarcelle* (White-faced Whistling Duck) and the Purple Swamphen in the 1950s, and to the rarity of Meller's Duck – all introduced species which the sporting fraternity would have been happy to keep[90]. In 1910 Richard Meinertzhagen watched swamphens at the Mare St Amand, large marshy lakes on the Rivière Sèche on La Lucie estate near Bel Air (Flacq district), pictured in *Mauritius Illustrated* around the same time. Although recommended for draining in 1921, the La Lucie marshes apparently survived until the river was canalised in 1945[91]. Moorhens, and the introduced frogs, nonetheless survived this decimation of their habitat.

Edward Newton and Nicholas Pike

In the same year Clark's *Ramble* was published, another British public servant arrived who was to have a more superficial but ultimately more enduring impact on the island's natural history. Edward Newton was in the island's senior administrative stream. Arriving at 27 as an Assistant Colonial Secretary, he remained on the island for 19 years, acting as governor by the 1870s when the governor was on leave and presiding over the Royal Society of Arts and Sciences. In addition to being able to influence the island's government and elite, his brother Alfred was an ambitious academic zoologist, elected Professor of Zoology and Comparative Anatomy in Cambridge in 1866[92]. The contrast with the unassuming Clark is striking. Unlike Clark, who was interested in everything from worms to monkeys, Newton was almost exclusively concerned with birds, particularly their nests and eggs. In his private writing he comes over as somewhat arrogant and racist[93], but his passion for ornithology enormously advanced the knowledge of birds in the southern Indian Ocean, as he, and his circle of amateur collectors, explored the Seychelles, the Comoros and Madagascar as well as Mauritius and Rodrigues. Although he published little on Mauritius, he was very active in the field in his early years on the island, his letters and notebooks full of his birding, until he left for England via the Seychelles in 1867; his large specimen collection is preserved in Cambridge. After returning in 1868 as Colonial Secretary his duties and poor health largely kept him from further fieldwork[94], though he actively promoted that of others, up to his final departure in 1878.

Newton's notebooks give us our first reasonable idea of the distribution of both native and introduced birds in Mauritius, when much of the main upland areas were still forested, although rapidly becoming fragmented and degraded. He soon noticed the effects of forest destruction on the birds. Writing to his brother Alfred in 1863 he summed up the situation clearly:

> I do not think you are right in your . . . [illegible] *as to the rarification* [sic] *of species here, the 'struggle for existence' is between them and man and not acclimatized species of birds, I take it. Probably with the exception of the sparrow (& these are confined to the town) the other introduced species have decreased in the last twenty years, as I fancy the great introduction of them took place in the French time, and I should very much doubt that any species has been established since then – the cause, I believe, is simply the destruction of forest for cultivation of sugar cane and the enormous increase of population. When you know that in 1859 alone we had an Immigration of more than 49,000 coolies you can fancy the effect it must have produced on a small place like this. All the species decreasing so fast here (with the exception of* Phedina madag. [Mascarene Swallow]) *are explicitly forest ones, & I have but little doubt that before many years if sugar continues to be a profitable speculation, the kestrel will become extinct & the* Oxynotus [Mauritius Cuckoo-shrike] *and Thrush* [Mauritius Merle] *confined to one or two patches of forest in the uncultivatable land, while* F. erythrocephalus [Mauritius Fody] *&* Z. chloronothus [Mauritius Olive White-eye] *will only be found much more sparingly than now in the higher parts of the island . . . Of the pigeons I know very little, but it is clear they are much less common now than they were said to be years ago, the acclimatization of exotic species can have nothing to do with this, as none of them live in the same forest as the fruit eating pigeons, whereas the destruction of forests must necessarily* [have] *considerably narrowed their feeding grounds.*[95]

He rarely described the forests themselves, but mentioned specific nest-trees and birds' habits enough to be fairly sure what kind of habitat he was operating in. What is clear from his notes and snippets in other writings of the period is that all the surviving native birds, except the Pink Pigeon, were widespread and locally common throughout the upland forest that still extended from Baie du Cap north-west to Nicolière and west through the Bamboo Mountains, a broad swathe across the middle of the island. The forest was far from pristine; much had been selectively cut over, it was patchy and invaded by Travellers' Palms among others, but it was good enough to support the native birds. Newton even found a Mauritius Fody's nest in a lemon tree on the St Aubin estate near Souillac[96]. Unlike the others, the Mauritius Pink Pigeon was already confined to the southwestern uplands, seen only around Grand Bassin and nearby. Newton apparently never visited the Bamboo Mountains, Montagnes Blanche or Fayence, nor Nouvelle Decouverte-Nicolière in the centre-north, where it is likely that most native birds still occurred; another writer recorded Echo Parakeets, merles and flycatchers on Mt Camisard in the Bamboo range[97]. The parakeet was common enough to provide a meal for Charles Boyle's party in the Black River Gorges in the mid-1860s[98]. Aerial feeders are rarely mentioned, but Newton and Pike visited various swift caves, and Newton reported swallows so reduced by a cyclone in 1861 that they were still rare in 1878. They recovered; by 1904 Paul Carié found them abundant around Port Louis and Corps de Garde mountain, though absent elsewhere[99].

Newton was keen to find the Pigeon Hollandais, and early on believed it still existed, but he was confused by local use of the same name also for the Mauritius Pink Pigeon. His notes record two conversations he had with people who remembered the real *Alectroenas nitidissima*. Much of his fieldwork was done on the Mondrain estate (now Henrietta) near Tamarin Falls, staying at weekends with 'old Moon' and his wife, living on her late father's estate[100]. Malcy Moon, a gifted botanical artist born in 1803, was the daughter of Toussaint-Antoine de Chazal, a friendly neighbour to Matthew Flinders when billeted next door from 1806–1810[101]. In November 1863 she told Newton that

> When she was a girl and used to go into the forest with her father de Chazal, she has seen quantities of Pigeon Hollandais and Merles, both species were so tame they might be knocked down with sticks, & her father used to kill more that way than by shooting them, as she was a nervous child, her father always warned her before he fired, but she would entreat him to knock the bird down with his stick & not to shoot it – she said the last Pigeon Hollandais she saw was about 27 years ago just after she married poor old Moon, it was brought out of the forest by a marron. She said it was larger than a tame pigeon & was all the colours of the rainbow, particularly about the head, red green & blue.[102]

Newton noted later that the she mostly saw them around the swampy area known as Petrin, not far from Grand Bassin. Still common around 1815, the species had all but gone by '27 years ago', i.e. 1826. However, Malcy de Chazal actually married William Moon in July 1831[103], so one can presume a date around then. Earlier that year, Newton had met a Mr Ducasse:

> A coloured man, who told me that only once in his life had he killed a pair of 'Pigeon Hollandais' (?Alectroenas nitidissima) & that was when Col. Simpson (now Gen.; old Simpson) was here (this must have been at least 20 years ago – but I do not know precisely) & had not seen a single one since. The Pigeon Ramier (?Trocaza meyeri) he tells me is not uncommon near Grand Bassin, & is still caught by snares with the end of a long stick being placed round their necks as they sit on the boughs, the birds are so stupid they do not fly away – the skin I got two years ago was from that neighbourhood.[104]

Ducasse was clearly talking of Mauritius Pink Pigeons being snared, as the other *ramier*, the Malagasy Turtle Dove, would never allow such liberties – so it appears he was genuinely referring to *Alectroenas*. Col. James Simpson commanded the 29th Worcestershire Regiment on the island, not '20 years ago', but from 1826 till 1837 or 1838[105], so again around 1830 seems likely for Ducasse's birds. These dates match the last specimen acquired by Desjardins in 1826, and his remarks in 1832 that they were "still found towards the centre of the island in the middle of those fine forests which by their remoteness have escaped the devastating axe"[106].

The third significant expatriate naturalist of the period was Nicholas Pike, American consul 1867–72, a gregarious enthusiast "determined to note everything" he saw in an island whose literature was old, scattered and hard to find[107]. His main interest was inshore marine life, but on land he liked ferns, and he had a keen eye for all aspects of natural history. He spent so much time exploring the island, both socially and physically, that his official duties must have been rather light. This activity resulted in his entertaining book *Sub-tropical rambles*, a general account of the island, to be followed, so he professed, by "a second volume, nearly completed . . . treating more fully on the Fauna and the Flora of Mauritius", which never appeared[108]. Rediscovering the manuscript for this second volume, and also a paper on the island's birds offered to the RSAS in 1871 (but apparently never sent)[109], would be of major importance to Mauritian natural history. In the 1920s E.W. Gudger tried hard to find the text of the second book, to no avail[110]. Despite their manifest common interests and active membership of the RSAS, Pike and Newton were not on good terms. In 1869 Newton said Pike was "*supposed* to be a great naturalist, but is a most awful liar & humbug you ever came across". Pike repaid the compliment by writing Newton entirely out of his book[111].

BOX 11   RAILS (RALLIDAE)

Rails have managed to reach almost every oceanic archipelago throughout the world and can, within a few generations, evolve flightlessness, which leads easily to oceanic island speciation and endemism.

Red Hen *Aphanapteryx bonasia*

The monotypic genus *Aphanapteryx* is characterised by hair-like plumage, a long decurved bill, vestigial wings and strong robust legs. *A. bonasia* appears to have been an opportunist omnivore and able to survive for a long time alongside people and rats. The introduction of cats in the late 17th century, however, proved disastrous, and the rails were not seen again after 1693.

Red Hen *Aphanapteryx bonasia*. Drawn from life by Peter Mundy, c. 1638.

Accounts

An anonymous Dutchman (in Servaas 1887, tr. by Henk Beentje) in 1631:

The soldiers [red hens] were very small in stature and slow of foot, so they could be caught easily by hand, their armour or gun was their mouth [beak], which was sharp and pointed, and which they used instead of a dagger, were very naked and [unrecognisable word], not hewing about like soldiers, run about in great disorder, now here, now there, not being true to each other at all.

Herbert (1634) in 1629:

The hens in eating taste like parched [roast] pigs, if you see a flocke of twelve or twentie, shew them a red cloth, and with their utmost silly fury they will altogether flie upon it, and if you strike downe one, the rest are as good as caught, not budging an iot till they be all destroyed.

Mundy (1608–67) in 1634, from hearsay:

. . . a fowl called Mauritius hens: if one is captured, the rest will come around you so you may catch them alive with your hands.

Mundy in 1638 after landing on Mauritius for the first time (from Barnwell's modernised 1948 version):

A Mauritius hen is a fowl as big as our English hens, of a yellowish wheaten colour, of which we only got one. It hath a big long crooked sharp pointed bill, feathered all over, but on their wings they are so few and small that they cannot with them raise themselves from the ground. There is a pretty way of taking them with a red cap, but this of ours was taken with a stick.

Marshall on Mauritius in 1668 (in Khan, 1927):

Here are also great plenty of Dodos or red hens which are larger a little than our English henns, have long beakes and no, or very little Tayles. Their fethers are like down, and their wings so little that it is not able to support their bodies; but they have long leggs and will runn very fast, and that a man shall not catch them, they will turn so about in the trees. They are good meate when roasted, tasting something like a pig, and their skin like pig skin when roosted [roasted], being hard.

Hoffman (1680) on Mauritius 1673–5:

. . . [there is also] a particular sort of bird known as toddaerschen which is the size of an ordinary hen. [To catch them] you take a small stick in the right hand and wrap the left hand in a red rag, showing this to the birds, which are generally in big flocks; these stupid animals precipitate themselves almost without hesitation on the rag. I cannot truly say whether it is through hate or love of this colour. Once they are close enough, you can hit them with the stick, and then have only to pick them up. Once you have taken one and are holding it in your hand, all the others come running up as if to its aid and can be offered the same fate.

Leguat's Rail *Erythromachus leguati*

Although both were flightless and highly derived, the Rodrigues species *E. leguati* is distinct from *A. bonasia* in a number of characteristics. They were reported by Leguat in 1691–3 as common, fat and unable to run after gorging themselves on tortoise eggs. They were also mentioned by Tafforet in 1626; however, Pingré in 1761 stated that they were by then extinct. Again, cats seem to have been the culprits.

Accounts

Leguat (1707) in 1691–3 (from the original English translation of 1708):

Our Wood-hens are fat all the year round, and of a most delicate taste. Their Colour is always of a bright Gray, and there's very little difference in the plumage between the two sexes. They hide their Nests so well, that we cou'd not find 'em out, and consequently did not taste their Eggs. They have a Red List about their Eyes, their Beaks are straight and pointed, near two Inches long, and red also. They cannot fly, their fat makes 'em too heavy for it. If you offer them any thing that's red, they will fly at you to catch it out of your Hand, and in the heat of the Combat, we had an opportunity to take them with ease.

Tafforet (1626) (from A. Newton's 1875 translation):

There is a sort of bird, of the size of a young hen, which has the beak and feet red. Its beak is a little like that of a curlew, excepting that it is slightly thicker and not quite so long. Its plumage is spotted with white and grey. They generally feed on the eggs of the land tortoises, which they find in the ground, which makes them so fat that they often have difficulty in running. They are very good to eat, and their fat is of a yellowish red, which is excellent for pains. They have small pinions [wings], without feathers, on which account they cannot fly; but, on the other hand, they run very well. Their cry is a continual whistling. When they see any one who pursues them they produce another sort of noise, like that of a person who has the hiccups.

> Rails related to *Aphanapteryx* appear never to have reached Réunion. The Réunion Ibis (or Réunion Solitaire) *Threskiornis solitaria* probably occupied the same ecological niche (Chapter 3).
>
> **Wood-rails *Dryolimnas* spp.**
> The White-throated Rail *Dryolimnas cuvieri* is widely distributed in Madagascar, where the volant nominate form is reasonably common. The type specimen was collected on Mauritius, but it is not clear whether there was a population there or the bird was just a vagrant. There are two other subspecies, the flightless Aldabra Rail *D. c. aldabranus* from Aldabra, which holds the distinction of being the last surviving Indian Ocean flightless rail, and the recently extinct Assumption Island Rail *D. c. abbotti*, which was also in the process of becoming flightless. A second species of *Dryolimnas* has recently been described from fossil remains collected on Réunion. Dubois's Wood-rail *D. augusti* was morphologically similar to the Aldabra species, although larger, and was probably also flightless. Recent work (JPH in progress), however, has confirmed that Mauritian bones formerly attributed to *D. cuvieri* are in fact from a distinct large endemic *Dryolimnas*.
>
> **Oiseau bleu**
> The most enigmatic of all rails once occurring on the Mascarenes is the 'oiseau bleu,' a mysterious bird that lived on the Plaine des Cafres, Réunion. They were considered good game and, although able to fly, could easily be caught and killed with sticks. As their colour was described as blue with red beak and legs, the oiseau bleu is generally considered to represent a large *Porphyrio* gallinule. However, no skeletal evidence of any kind has been found to resolve the taxonomy. Whatever its generic placement turns out to be, the 'blue bird' was last reported in the early 18th century.
>
> *Accounts*
> Dubois (1674) in 1671–72
>> *Oiseaux bleus, as large as solitaires. Their plumage is entirely blue, the beak and feet red, made like hen's feet. The do not fly, but they run extremely fast, such that a dog has difficulty catching them in a chase. They are very good* [to eat].
>
> Feuilley (1705) in 1704:
>> *The Oiseaux bleuff live in the plaines on top of the mountains, and especially on the Plaine des Cafres. They are the size of a large capon, blue in colour. Those that are old are worth nothing to eat because they are so tough, but when they are young they are excellent. Hunting them is not difficult because one kills them with sticks or with stones.*
>
> De Villers (1708):
>> *One sees there* [the Plaine des Cafres] *a great numbers of* oiseau bleus *which nest amongst grasses and aquatic ferns.*
>
> Le Gentil (1727) in 1717, extended by 'Père Brown' (1773):
>> *Towards the east of the island there is a little plateau up a high mountain called the Plaine des Cafres where one finds a large blue bird whose colour is very striking. It resembles a wood-pigeon. It flies but rarely and always barely above the ground, but it walks with surprising speed. The inhabitants have never called it anything other than* oiseau bleu; *its flesh is quite good and keeps well.*
>
> *Specimen collections*
> Fossil material of *A. bonasia* can be seen at UCB; UMZC; BMNH; and the only almost-complete associated and articulated specimen is on display at the MI. *E. leguati* fossil material is found at the UMZC; BMNH; MNHN; and UCB. All *Dryolimnas augusti* fossil material, which probably represents one individual, is deposited in the UCB. Fossil Mauritian *Dryolimnas* is found at the UMZC.

Though not in the same league as Clark, Newton and Pike, Charles Boyle deserves a mention. He was chief commissioner of railways, and keen on the great outdoors. His book *Far away* is appallingly racist, but contains numerous natural history notes, though his main list of animals is partly borrowed from Clark[112].

Forest legislation – progress at last

During Newton's period as colonial secretary, forest protection was again thrust to the fore, though, notwithstanding his earlier observations, there is no evidence that he took part in the debate. In 1859 the Chamber of Agriculture had produced a report noting that "yearly as land is cleared the amount of water flowing in our rivers diminishes" and also complaining of a shortage of firewood. They also claimed that "the climate . . . has generally become too dry" and, probably exaggerating, that "nearly all our building timber is imported". Another ineffective law to protect streams was made in 1863; as Pike noted in describing the forests' retreat: "strict laws have long been in existence for the preservation of the forests, but they do not seem to have been enforced. As wood and charcoal are the only things used as fuel, the destruction is still going on"[113]. In 1867 a commission under Dr Charles Regnaud was appointed to look into the origins of the malaria epidemic, unanimously agreeing that "the diminution . . . of forests had occasioned increased radiation of heat from the soil, the drying up of numerous springs, the reduction of rivers to a lower level, the pollution of their waters, the formation of torrents, and the deposit of alluvial matter in marshy places and at the mouths of rivers", all this being responsible for the outbreak. At the same time Regnaud also chaired a Water Supply Commission to "consider the best means of keeping the mountains wooded and of improving the island's water supplies"[114], which, with Louis Bouton,

Charles Meller and James Caldwell as the other members [115], unsurprisingly made the same points as the malaria inquiry, but focused more specifically on the drop in river flow and soil erosion from cleared areas. They too stressed a direct effect of deforestation on climate[116] – the Poivre-type beliefs of the locals overriding the science of the colonial power, with Mann's observations of a decade earlier forgotten or more probably ignored[117]. No doubt the direct observations of reduced water flow were correct, though no account was apparently taken of the effect of increased *demand* that the burgeoning population and use of irrigation (via culverts) in cane-fields must have had on river volume.

The attempt to legislate in 1870 for forest protection on the recommendations of the Water Supply commission was stymied by the Chamber of Agriculture who had done a *volte-face* since 1859 on the question of keeping 10% of their estates wooded[118]. There was a particularly severe bout of deforestation in 1871–72, mostly in the central southern uplands[119], so the new Governor Arthur Gordon, seeing an urgent need to do something[120], oversaw revised regulations in 1872. After further reports from a reconstituted commission (again chaired by Regnaud), the regulations were extended in 1875, after bolder proposals were mangled by the governing council[121]. While all this politicking was going on the forests were still being ravaged – the annual figures for ebony exports telling the tale. Some 250 tonnes were traded between 1859 and 1862, declining to a few tonnes per year until 1876 when 51 tonnes left the island. In a final orgy of cutting before forest legislation took real effect, a staggering 686 tonnes were exported in 1877, followed by *c.* 260 tonnes in 1878, after which exports collapsed, never again reaching a tonne[122]. As Gleadow put it in 1904: "The Colony used to export ebony to a considerable amount. It was nearly all stolen, and now the trade is extinct for want of material"[123].

The Regnaud commissions not only recommended montane and riverine forests be protected, but that whole tracts of the uplands be positively replanted, the government to acquire land for the purpose. They recognised for the first time the importance of cloud forests in capturing moisture (though misunderstanding the mechanism), and emphasised the moderating influence of forest on runoff – deforestation causing 'torrents', i.e. flash floods, resulting in erosion and damage[124]. They listed trees to be planted in different parts of the island, emphasising that native trees were too slow-growing for restoring watershed protection, so recommending a tranche of foreign ones. No nature conservation here, then – and, as Noel Brouard has pointed out, many species they recommended were wholly unsuitable![125] Still in a Poivrean mode they reported that "the legislature would confer an immense boon upon cultivation of all kinds, by preserving every kind of bird at present found in the island, and more particularly the insectivorous birds, and in introducing new species of them into the island", with the caveat that if grain were again to be cultivated, granivorous birds would have to be controlled.

The 1872 legislation, extended and superceded in 1875, set up a Woods and Forests Board (with forestry staff), a land purchase fund (fed by a special sugar tax), and properly defined mountain and riverine reserves to be protected from all cutting, or replanted if already cut over. However, although the infrastructure was set up, little actually happened after Gordon left in 1874 until Frederick Napier-Broome took over as acting Governor in December 1878. He invited Richard Thompson from the Indian Forest Service to study and prepare a plan for Mauritian forests – the first systematic survey of the situation. Thompson's report was very thorough and he saw much that Mauritians had missed through familiarity and lack of experience of other countries[126].

Richard Thompson, regeneration and the 'Dodo tree'
As a forester, Thompson's brief was not ecology or nature conservation, but as he was keen to preserve the native forests in areas where watershed protection had highest priority, his remarks have ecological force. He was the first to collate the gamut of threats to the native forests – deer, monkeys, hares (in the lowlands), giant snails, and beetles were all seen as problems (he failed to spot that rats are major seed predators). He considered snails the worst menace, and, noticing their scarcity in native forest where tenrecs were abundant, he wanted tenrecs protected from hunting in reserved forests, though this was not done until 1900[127]. Monkeys "devour and throw down the unripe seeds of all the principal and important forest trees, so that it is scarcely possible to secure ripe seed that will germinate". They also ate eggs and young of birds – hardly news, since Baron Grant had reported the same in 1741 (p. 99). To the elite of Mauritius, deer had long acquired a status akin to cows in India. Not that Mauritians avoided killing them, quite the reverse, but they protected herds at all cost to indulge in *la chasse*, the island's principal upper-class male-bonding ritual[128]. It therefore took an outsider to notice that

> Deer do considerable damage to young growth... whole acres of young transplants of which they are fond are found cut back. Unfortunately the deer seem to prefer eating such species as Makak, Natte and Ebony more so than the others; the first two yield decidedly the most durable and valuable timber among the indigenous trees.

He wanted upland plantations, including degraded native forest areas being restocked, fenced off against

Figure 7.3. Native upland forest along the Rivière du Poste, Mauritius, with deer, by Edouard Pitot, 1830s. From Unienville (1991).

deer. Although the Woods and Forests Board re-emphasised the deer damage problem in 1882, and wanted a law to permit deer control on government land[129], by 1895 the local code of silence had reasserted itself. "Very few details could be collected on the influence of deer on forest vegetation ... the rangers were unanimous in saying that the damage done by deer in our forests is practically nil, on account of the great quantity of grass at their disposal"[130].

Despite the activities of monkeys, Thompson noted good regeneration of the native upland forests on the plateau, though not on slopes or valley bottoms, where exotics predominated. However, some trees were not performing: "it is a remarkable fact that no natural seedlings of the 'Tambalacoque' are now to be found in the forests, though the tree is common. Colophane likewise is peculiar in this respect; with these two exceptions all the other species reproduce largely and freely from seed". The Colophane *Canarium mauritianum* is the largest native Mauritian tree, and as late as 1877 featured in the standard list of local trees as "often attaining a diameter of 6 feet. Pirogues, canoes, are often hollowed out of its trunk"[131]. Both it and the Tambalacoque *Sideroxylon grandiflorum* have extremely hard seeds. The continued failure of these trees to regenerate led to speculation by Reginald Vaughan and Octave Wiehe in the 1930s that "the germination and distribution of these [the Tambalacoque's] remarkable woody seeds were probably assisted by their passage through the alimentary canal of the Dodo, and young seeds of this species have been unearthed with Dodo remains"[132]. Vaughan would live to regret this speculation.

In 1977 Stanley Temple made headlines in the scientific press by claiming that only 13 Tambalacoque survived and that, because the Dodo's gut was required to prepare the seed for germination, no seedling had appeared for over 300 years. He supported his argument by feeding a turkey 17 seeds and getting three to germinate of the 10 that survived its gut. This 'obligate mutualism' seemed such a good story that it was rapidly incorporated into textbooks, despite failing to stand up to closer examination. The seed did not need to pass though a bird's gut. Mauritian foresters (and even an English botanist at Kew) were germinating Tambalacoques in the 1930s and 1940s without the help of turkeys or artificial abrasion. Furthermore, in pre-human Mauritius there were tortoises, parrots and giant skinks around, any or all of which might have been involved in cleaning flesh off the tough seeds. Germination was indeed poor, but it was later thought to be due to a fungus rotting most seeds as soon they fell; in any case, few seeds survived the long maturation period (18 months) without being destroyed by monkeys. As Kew botanist A. W. Hill had shown in 1941, the seed has a ready-made zone of weakness along which it always splits when germinating. There were far more trees than 13 – stated as 'hundreds' in the early 1990s, a 2001–03 survey estimated 900–1000, with 296 found and measured. In the 1970s the smallest tree found was around 10cm dbh; it was thought to be under 100 and possibly as young as 30 years old. The recent study revealed more small trees, albeit too few to ensure survival of the species. The seeds found with Dodo remains were not 'young' Tambalacoque nuts, but a related species with smaller seeds, *S. sessiliflorum*, and their presence with Dodos in alluvial deposits did not imply a dietary connection. In his final years Vaughan was a vigorous opponent of this 'tambalacoque and bull' story, and the last thing he published was a letter to *Animal Kingdom* refuting an article by Temple![133]. As Stephen Jay Gould put it in 1980 "most 'good' stories turn out to be false . . . but debunking doesn't match the fascination of a clever hypothesis. Most of the 'classic' stories of natural history are wrong, but nothing is so resistant to expurgation as textbook dogma"[134]. Natural seedlings have now been found in the weeded and protected fenced-off 'conservation management areas' in the Brise Fer forest. Trees reinvigorated by the removal of weeds fruited better, and reduced monkey damage, the primary cause of loss once fruit had set, had finally allowed seeds to mature, fall, and germinate naturally on the forest floor, with no problem from fungus and no help from Dodos, parrots, tortoises, or humans[135]. The problem of forest regeneration in Mauritius has broader and more insidious causes than a simple absence of Dodos – we will come back to this later.

Thompson's recommendations on forest management formed the basis of forestry policy, and thus by default conservation policy, for the next 70 years. He essentially endorsed the Regnaud/Bouton reports, but backed them up with better science and more specific proposals. The headwater catchment of the Black River and the Rivière du Poste, the Les Mares/Petrin area of stunted forest and screw-pine swamp was to be retained for watershed protection, together with adjacent areas such as Grand Bassin and Kanaka. Native forests should in general be preserved (including on private land), with only dead and diseased trees cut for timber, to be replaced with natives or a mixture with camphor and pine. The best native stands should be left "intact for their glorious beauty". He also recommended restocking the degraded forests around Piton du Milieu with native species and Toon *Cedrela odorata*, an Indian evergreen, and the purchase of further areas for watershed protection. He wanted to strengthen the river reserves, and create a proper Forest Department to administer all state ('Crown') lands. The Woods and Forests Board (created in 1872) approved the bulk of this in November 1880, and Napier-Broome then pushed ahead with yet another set of forest laws in 1881, greatly strengthening the administrative base for forestry operations. Tree-cutting was banned on specified private lands, and many old concessions bought back for the government forest estate[136]. His farewell speech in 1883 announced the re-purchase of 15–16,000 *arpents* (*c.* 6,500ha) of land, though due to unprecedented pressure from landowners he resisted the board's attempt to extend protection to all scarps and river banks, whether formally reserved or not[137]. By 1886 some 29,000 acres (11,736ha) had been bought, but the programme was arbitrarily halted by acting Governor William Hawley, and the whole process stalled until Sir Charles Bruce took over in 1897[138]. Although wholesale forest destruction was stopped, enforcement was patchy and clandestine felling continued. On becoming Governor in 1883, Sir John Pope-Hennessy complained that "the little that was left in Mauritius of primeval forest is being steadily destroyed by Indians who have leased the land as speculators of timber. Everyone remarks that the daily process of destruction exceeds the comparatively feeble attempt at tree planting"[139]. It was not just 'Indians': by the turn of the century a new report and programme had to be initiated.

Even the apparently straightforward operation of exploiting the numerous dead trees in the forests caused much controversy. An operation in 1877 had to be stopped because the contractors were illegally removing live trees and exceeding the boundaries of their concession, and the opposition of estate owners around the newly acquired government forests, frightened of unsanctioned clear-felling and drying up of their water supplies, prevented further action for some time. Meanwhile the dead timber deteriorated, as beetles, termites and fungi neglected to put their activities on hold while the politics was sorted out[140]. Delimiting mountain and river reserves was problematic – landowners resented the Government telling them not to cut trees on their land, and were inclined to bribe surveyors and forestry staff to alter boundaries in their favour, with government commissions legalising the corrupt surveys in some cases[141]. By the time Bruce took over in 1897, much of the land scheduled for government acquisition had been broken up and sold on, and money was no longer available for land-purchase. In 1903, attempting to make a new start, and to persuade reluctant colonial authorities in London that forestry was important, Bruce invited Frank Gleadow from the Indian Forest Service to spend a year developing another plan for the forests and their future[142]. Gleadow's 1904 report was very thorough, but was notorious for the blunt way in which he detailed bribery, corruption, nepotism and other illegal practices, and a "general want of moral tone" permeating Mauritian society. It is surprising that someone with oriental experience should react with such shock and horror to the alleged misdeeds of the Mauritian forestry staff, magistrates, surveyors and estate owners, Gleadow exempting only the recently appointed Director of Forests himself (Paul Koenig)[143]. True or not, Gleadow's report was certainly unpopular, and a censored version without the accusations was published in 1906[144]. Unfortunately the net result was that his plans were not implemented and his very pertinent observations on the forests themselves largely ignored. As with Thompson, conservation in the modern sense was not part of his brief, and although he recommended preserving the remaining untouched forests, this was mainly for reasons of water conservation, as in the past; however, he did propose research into growth rates of native trees. Invasion by Strawberry Guava, the two *Litsea* species and Rose-apple seemed to have increased in the 23 years since Thompson. Nonetheless, the best stands of native forest, though invaded by *yatis* (*Litsea monopetala*), were still not penetrated by Strawberry Guava; he said little about natural regeneration, though Koenig had considered it still 'satisfactory' in 1895.

In his own writings Paul Koenig never acknowledged responsibility for spreading what was to become one of the forests's worst weeds. In 1902 he started planting Ceylon Privet *Ligustrum robustum* as "an understorey and useful fuel crop", also later widely used as a 'nurse' for forestry saplings. Koenig was so keen on this plant that he is reported to have carried seeds when travelling and scattered them out of train windows into the forest edge. In the 1930s Ceylon Privet was widespread in secondary scrub-forest, and praised for its ability to rapidly colonise 'blank ground'. By 1939 it was hard to find any native forest free from its penetration, but it was only in 1980

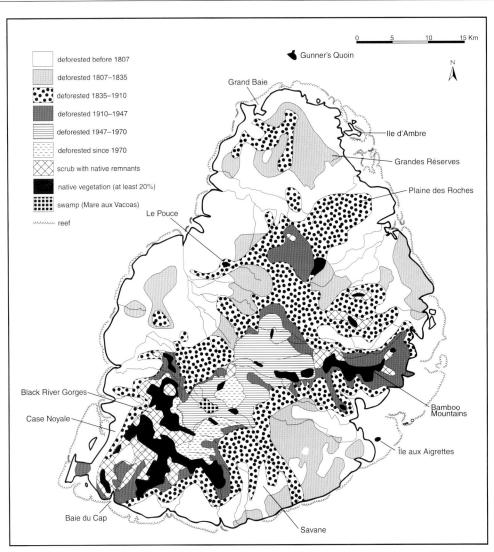

Map 7.2. The phases of forest clearance in Mauritius, 1800–2000. Adapted from maps by Lislet-Geoffroy (1807), Fraser (1835), Maisonneuve (1851), UK Colonial Survey (1902–03), Gleadow (1904), Walter (1908), Vaughan & Wiehe (1937), Tillbrook (1968), Directorate of Military Survey (UK) (1971), Cheke (1987c), Safford (1997b).

was it was finally recognised by the Conservator of Forests as no beneficial 'nurse', but an important factor hindering regeneration of native forest[145].

A glut of exotic gamebirds

While Bouton fought his long battle for the forests, dying in 1879 before seeing its eventual partial success, those with 'sporting' interests had other things in mind. By the mid-19th century introducing gamebirds was again fashionable, led this time by British expatriates. No doubt Franco-Mauritians also indulged, though it was not until 1877 that they emulated their colleagues in Réunion by forming an acclimatisation society, run for some time by the ubiquitous Albert Daruty de Grandpré[146]. From Madagascar in the 1860s came shiploads of ducks, partridges, Purple Swamphens and Grey-headed Lovebirds[147], and, from India, quails. We have no detail as to who organised all this; indeed it appears to have been somewhat haphazard. We know from George Clark that in 1859 only the White-faced Whistling Duck was established, and an additional (unidentified) quail from India possibly so[148]. He noted that the Madagascar Partridge (locally called a 'quail') was released in large numbers but rarely seen again; Edward Newton regretted having shot one in 1865 at Henrietta, where Caldwell had only recently released a dozen. They were again being imported in 1876 and 1906, but apparently never persisted for long[149] – the original 18th-century introduction (p. 93) seems to have petered out some time in the early 19th century. Newton noted in 1863 that his friend Currie had released 'several kinds of Madagascar ducks' on the Mare aux Vacoas, and found one species, Meller's Duck, breeding nearby at Mare Longue[150]. From Newton's specimens we know Clark's Indian quail was the Jungle Bush Quail *Perdicula asiatica*. Common Quail *Coturnix coturnix* were also released, probably from the 1880s, as Newton neither recorded nor collected them; they may never have been fully self-sustaining in the wild[151]. The arrival of Madagascar Buttonquail is a mystery; Sganzin claimed they were present in 1831, but there are no other records before 1910, when Meinertzhagen referred to their introduction "at a fairly recent date". Given their abundance in Réunion, they could well have been introduced early in the century, died out, then been imported again around 1900[152]. As noted earlier, Purple Swamphens became widely established in meres and marshes on the island[153].

The Mare aux Songes

One of these marshes proved to be a lot more interesting than a home to introduced waterfowl. Our friend George Clark, having written a natural history of the island not matched for 130 years[154], turned his attention to the ex-fauna. There had been a flurry of interest in Dodos in the RSAS when Hugh Strickland was seeking more bones around the time his *Dodo and its kindred* was published (1847–51), but the only bones found had been in Rodrigues (of Solitaires) and the topic was again forgotten[155]. Around 1860 Dr Philip Ayres, who had himself found a Dodo bone in a cave on Mauritius[156], visited Clark in Mahébourg, the subject of Dodos' remains coming up as they explored the Dutch ruins at Grand Port. Ayres favoured digging in the ruins, but Clark considered alluvial deposits a better prospect. In September 1865 Clark was alerted by his pupils to tortoise bones turned up during peat-digging at the Mare aux Songes near Mahébourg. He got permission from the marsh's owner, Gaston de Bissy, to explore, and after some frustrating failures in the peat horizon, Clark eventually sent men out into the middle of the marsh "to enter the dark coloured water about three feet deep and feel in the soft mud with their feet". Here they started to find bird remains including Dodo bones, which he promptly sent to the museum in Port Louis. There they were received without much enthusiasm, so when Clark had accumulated a more representative collection he sent them off to the British Museum, publishing an account of his finds in a local newspaper, the *Commercial Gazette*, to bypass the supercilious savants[157].

In Europe the interest was explosive, setting off an orgy of Dodo studies, both osteological and historical, that caught the public imagination; the Dodo was forever enshrined as the archetypal symbol of extinction caused by man, and also as a bumbling ineffective oddity that almost deserved its fate[158]. Although the history of subfossil discoveries in Mauritius has no bearing *per se* on the ecological history, our knowledge of the appearance and anatomy of many early extinctions comes almost entirely from these bones, so the excavations are of paramount importance in understanding the wildlife of pre-human Mauritius. The Mare aux Songes has produced the greatest quantity of animal remains, and was for 40 years the only site that yielded bird and lizard bones (tortoise bones were more widespread). It was re-worked, after Clark, by Theodore Sauzier in the 1890s and Paul Carié (who had inherited the Mon Desir estate) in the early 1900s. In 1899 when Étienne Thirioux began to search caves and screes in the valleys behind Port Louis, a second, and very important, source of bones was discovered, including the only skeletons of the Didosaurus and the Red Hen, and the sole Dodo skeleton where all the bones belong to a single bird – these precious specimens are still preserved in the Mauritius Institute in Port Louis[159]. Although Clark's and Sauzier's bird bones were thoroughly worked up, Carié's and Thirioux's were not, although the reptile material was studied, much later, in the 1940s and 1970s[160]. The bat bones, ignored at the time, are unexpectedly interesting, confirming that all three Mascarene flying-foxes inhabited Mauritius sympatrically (pp. 37, 39)[161]. Thirioux's collections from the valleys behind Port Louis also included bones of the Radiated Tortoise *Asterochelys radiata* from Madagascar, presumably escapes from captivity; Lesueur's drawings include a young Radiated Tortoise, probably drawn in Mauritius in 1803[162].

Toads, terrapins and snakes

As with game birds, accidental introductions of other vertebrates underwent a revival in the later 19th century, after a long spell in which no new species had become established. Desjardins had reported the presence of toads in the 1830s, but these, if briefly established, soon died out; Clark was emphatic that there were none around in 1859[163]. Bouton sent an 1830s specimen to London in the 1870s, which Albert Günther identified as the large Black-spined Toad *Bufo melanostictus* from India[164]. Toads may have failed, but snakes had no problems establishing themselves in the 1860s and 1870s. Odd snakes reported before appear to have been one-off escapes[165], but by the late 1860s the tiny Flowerpot Snake, first reported in 1803 (p. 103), was turning up regularly, and a decade later the Wolf Snake *Lycodon aulicus* was well-established as well[166]. Neither was deliberately introduced, the Flowerpot Snake probably arriving in soil with plants on several occasions, the Wolf Snake in cargo from India; the subterranean Flowerpot Snake is so small it had been overlooked from 1803 until the 1860s[167]. In 1871 East African Box Terrapins *Pelusios subniger* were discovered in a pond at Beau Plan (Pamplemousses), and one was found in a Port Louis sewer. They were still thriving in 1878 when Albert Daruty received some live specimens from Diego Garcia, probably the original source of the Beau Plan animals; there seem to be no records after 1882, though there was no follow-up until the 1980s[168]. In 1860 Evenor Dupont proposed introducing crows to control rats; this was not pursued, though his other suggestion, releasing House Sparrows (to control the cane borer), had been anticipated for quite other reasons by British soldiers (p. 124) – with no effect on the insect pest[169].

Conservation beginnings

The mid-19th century saw the disappearance of the Rougette, and of Bojer's Skink from the mainland. After Clark's discussion in 1859, the only confirmed record of the Rougette is two specimens in the Natural History Museum (London) collected by Henry Whiteley around 1864–65. There is a specimen in Strasbourg dated 1876, but it came from a dealer and '1876' appears to be the acquisition date[170]. Although the species appeared in faunal lists as late as 1972, and rumours persist of a small fruitbat surviving to the 1920s, it is more likely that it had died out by around 1870. Fruitbats were largely ignored by local naturalists after Desjardins, but the Black-spined Flying-fox was evidently the only species known to Pike. Increasing deforestation, with the selective removal of the large trees in which the bats roosted, together with the ease with which these hole-roosters could be caught *en masse*, were probably the main causes of the Rougette's extinction[171].

The disappearance of Bojer's Skink coincides with the establishment of the Wolf Snake; there are some reports of lizards that post-date Clark's account, but none refer unambiguously to Bojer's Skink. Edward Newton saw a "lizard which is new to me, but I could not catch the beast" at 'Black River' in February 1862, and Karl Möbius supposedly collected a specimen in the same vague location in 1874[172]. However, doubtful localities dog Möbius's Indian Ocean reptile collection, and Newton failed to describe his 'beast'. Pike watched what appears to have been Bouton's Skink on the shore at Tamarin Bay (Black River district), so perhaps that was what Newton saw; in 1870 Pike reported Bouton's "common on all the shores in Mauritius" but Bojer's absent[173]. D'Emmerez de Charmoy referred in 1914 to "the small species of the Grande Terre [mainland] are skinks, which frequent the bare rocks of the coast", probably also Bouton's not Bojer's[174]. From being common in the 1850s Bojer's Skink apparently rapidly declined and disappeared, like the Rougette, without anyone noticing. The Wolf Snake is a known lizard specialist, and is the best candidate for eliminating the skink; rats, shrews and tenrecs had been around far too long to be responsible for its mainland extinction, and the Bloodsucker and mongoose were introduced 30–40 years too late. It is nonetheless a little odd, as Mauritian skinks should have been pre-adapted to predatory snakes – both endemic bolyerids are lizard specialists, yet Bojer's Skink thrives amongst them on Round Island (Chapter 9)[175]. In addition to the northern islets, Bojer's Skink survives on Ilot Vacoas off Mahébourg, having disappeared from nearby Ile aux Fouquets sometime between 1972 and 1987 after House Shrews arrived (probably in someone's picnic basket)[176].

While lizards and bats were vanishing in silence, the surviving endemic birds did get some attention. The fruitless quest for the Pigeon Hollandais and the focus on extinction centred around the Dodo bones slowly concentrated Edward Newton's mind. Just before he left for good he proposed that five endemic species, plus the Malagasy Turtle Dove, be given legal protection under the 1869 game regulations[177] –

Figure 7.4. Trois Mamelles and Rempart Mountains from the direction of Vacoas, Mauritius, c. 1870, with forest on the plateau. From Pike's *Subtropical rambles* (1873).

> **BOX 12**   **COOTS AND MOORHENS (RALLIDAE)**
>
> Coots and moorhens are adaptable aquatic rails. The Common Moorhen *Gallinula chloropus* is found on most western Indian Ocean islands and appears to be a recent colonist to Mauritius, whilst the only surviving coot in the region is the Red-knobbed Coot *Fulica cristata* of Madagascar. A coot of the genus *Fulica* was once found on Mauritius and Réunion. From fossil skeletal material collected in the Mare aux Songes, a Mauritian species *Fulica newtoni* was described, which appeared to be a large, flightless derivative of Common Coot *F. atra* or of *F. cristata*. More recently *F. newtoni* has also been identified from fossil material in Réunion.
>
> *Accounts*
> François Martin in Réunion in 1665 & 1667 (in Lougnon 1970):
>
> *[In 1665] The river basin [at Saint-Gilles] was covered in geese and water-hens, and the depths full of fish . . . the water-hens allowed one to approach almost [close enough] to catch them by hand; we sent them all on board. [But by 1667] We saw neither geese nor water-hens on the Etang de St Paul which was formerly covered in them.*
>
> Dubois (1674) in Réunion in 1671–2:
>
> *Water-hens, which are as big as chickens. They are completely black and have a big white crest on the head.*
>
> *Specimen collections*
> A series of fossil elements are housed at the BMNH, UMZC and UCB (Réunion specimens).
>
>
> Mascarene Coot *Fulica newtoni*. From Milne-Edwards, 1872

framed (possibly by Newton) to allow species to be added on the Governor's say-so. His suggested list of birds was queried by members of the *Société d'Acclimatation* whose members wanted to protect introduced species and popular birds such as tropicbirds. They completely misunderstood his reasons for selecting endemic species, which were more or less the same criteria we use today:

> *The great object of all naturalists is to protect as far as possible all existing fauna and flora, and the question of protecting animals and plants which has [sic] been introduced is quite another affair. I need not now go into the very serious errors which have been made by ardent acclimatiers in introducing what they believed to be of great benefit, but which afterwards proved to be of the most serious inconvenience and loss to the human inhabitants, beyond mentioning that in New Zealand the English rabbit has been so destructive as to starve out the sheep, and every effort has been made to get rid of it, even to the extent of introducing its natural enemies of the genus* Mustela *[stoats and weasels].*[178]

Newton's opinion on tropicbirds was that, since they were widely distributed elsewhere, their disappearance from Mauritius would not mean they were "lost as a species to the world" – true, but hardly helpful to Mauritians wanting them protected! Reading his justification for the species he wanted to protect, it appears that Newton thought the principal threat was shooting. Oddly, to us, he ignored the greater threat; despite knowing the extent of deforestation during his 19 years on the island (as his diaries tell us), there is no hint that he concerned himself with habitat conservation[179]. In the event, it was not until July 1880, two years after Newton's departure, that Lt-Governor Napier-Broome promulgated a bird-protection ordinance. Presumably to keep everyone happy, it included not only *all* the endemics, but also many popular introduced species, including several that, not long before, were cereal pests with a price on their heads![180] Although the law was aimed at land-birds, the contentious Red-tailed Tropicbird *Phaethon rubricauda* was included. Seabirds on Flat and Gabriel Islands were covered in 1887, with Red-tailed Tropicbirds on Round Island getting additional special protection in 1897[181]. The game laws were similarly used to protect the Common Tenrec on Crown Lands in 1900 – protection that must always have been entirely nominal[182]. The original legislation was for five years only, being regularly renewed and amended until replaced by new regulations in 1939[183]. Once again Mauritius was ahead of the game in wildlife legislation – the first general legislation to protect ordinary land birds elsewhere, passed in Britain in the same year (1880), was much more selective and inadequate compared to the Mauritian law[184].

Three additional birds became established in the late 19th century – Ring-necked Parakeet *Psittacula krameri* (c. 1886), Village Weaver *Ploceus cucullatus* (c. 1886) and Red-whiskered Bulbul *Pycnonotus jocosus* (1892). All were apparently released accidentally from aviaries, the parakeets around Grand Port, the weavers at Cap Malheureux in the north, and the bulbuls from an aviary in Moka broken open by the severe cyclone of 1892 – though some suspected their owner Gabriel Regnard released them deliberately. The parakeet and the weaver spread rather slowly, but bulbul numbers exploded, the population covering nearly 200km² in 8 years, reaching the south of the island in 10, and colonising all suitable habitat (most of Mauritius) in 18. While the parakeet and weaver had no immediate effect on the native fauna, preferring man-made habitats, the bulbul was soon accused of killing adults and robbing nests of the two white-eyes and of eliminating the large orb-web spiders *Nephila inaurata*, which were previously abundant. Paul Carié wrote: "one of my employees brought me on several occasions the birds [white-eyes] found dead beside paths. The popular rumour accused the bulbuls, and after a while one had to face up to the evidence. They attacked these little birds, and stole their nests, after having devoured their eggs"; adding that Donald d'Emmerez de Charmoy had confirmed this through additional observations[185]. The bulbuls' predation on the spiders was later repeated in Réunion; their webs were still abundant in the drier parts of Réunion in the 1970s, but by the late 1990s, by which time the bulbuls were abundant, they were becoming decidedly scarce (p. 182)[186]. Mauritius Fodies and Mauritius Merles, never again reported as common as they had clearly been in the 1860s–1880s, may also have been affected by the bulbuls[187]. A fourth immigrant showed up around 1900 – Indian House Crows *Corvus splendens* introduced themselves, apparently from free-flying birds travelling on ships from India. They established a small population around Port Louis harbour, and then at Roche Bois slaughterhouse nearby[188].

While these newcomers were settling in, some long-established exotics were fading away. Java Sparrows and Red Avadavats had been declining for decades; the former was last reported in 1892, the latter in 1904. Cape Canaries, so common in the 1860s, crashed dramatically; they were confined to the Trou aux Cerfs at Curepipe by 1913, and died out in the 1920s[189]. The decline of Java Sparrows is clearly connected to the abandonment of cereal crops, probably accelerated by the arrival of House Sparrows[190]. Red Avadavats were apparently unable to co-exist easily with Common Waxbills *Estrilda astrild*, a later arrival. Cape Canaries declined after the arrival of Village Weavers – even ceding their local name *serin du Cap* to the incomers. All three were popular cagebirds, Java Sparrows being supplemented by imports in the 1830s once they began to decline, and Red Avadavats fetching high prices just before they died out[191]. All three also declined in similar circumstances in Réunion (p. 140; Chapters 8 and 11), though the canary survived there by retreating to high altitudes (a choice absent in Mauritius); Red Avadavat has clung on in tiny numbers in the dry west.

Biological control disasters

Although Edward Newton had wanted in 1864 to introduce coucals from Madagascar to control borers, caterpillars attacking sugar cane stems, it was not until the turn of the century that biological control involving vertebrates began again. The Bloodsucker, an agamid lizard originating in Asia, was already established in Réunion (p. 141); around 1900 d'Emmerez released some from there as part of a general campaign against the Pink Stem Borer *Sesamia calamistis*. The lizards spread rapidly, with no recorded effect on the moth[192]. Around the same time a much bigger mistake was made in the name of biological control – the introduction of the Small Indian Mongoose *Herpestes auropunctatus*. In 1899 there was an outbreak of plague in the port, and in a somewhat panicked response, Governor Charles Bruce decided to release mongooses, owls and snakes to control the rats transmitting the disease, despite admitting to being aware from his earlier postings in the West Indies of the "disastrous consequences that have followed the multiplication of this little creature [the mongoose]"[193]. The same mongoose, released clandestinely in Jamaica in 1872, had by 1883 caused such damage to poultry, game and wildlife that it was banned – too late[194]. In the 1850s some Mauritians had kept tame mongooses to keep rats down, but they destroyed them after finding they preferred chickens[195]. After a vigorous press campaign against mongooses led by the Chamber of Agriculture (citing disasters elsewhere), the government nevertheless "after full consideration" allowed in a batch from India. Most discussions suggest the experimental group was supposed to be single sex, though the Governor was more equivocal, writing of "every precaution being taken to prevent their over-multiplication", suggesting he expected some breeding! In the event, either through incompetence or bad faith, 16 males and three females were released in March 1900. The detractors' predictions were soon fulfilled: "these *few males* bred with prodigious rapidity; their descendants now [1914] overrun the whole island, and, having naturally discovered a more toothsome prey than the rat, turn by preference to partridges, quails, hares, pigeons and chickens"[196]. There was, in fact, some control of rats, but also an almost complete elimination of gamebirds over the next 15 years or so. As early as 1901, complaints from poultry farmers and hunters poured in, but by the time a price was put on the mongooses' heads in 1905 it was too late to wipe them out. By 1910, when d'Emmerez studied their diet, their main food was tenrecs, shrews and rats, but hares, birds (mostly chickens and partridges), lizards, frogs, snails and insects also featured strongly[197]. While Meinertzhagen found five species of gamebird

Date	Sugar		Coffee		Cloves		Wheat		Rice		Maize	
	area	weight	area	weight	area	weight	area	weight	area	weight	area	weight
1820	-	4,500	-	1,948	-	466	-	515	-	456	-	-
c. 1825	8,241	-	8,909	-	4,993	-	-	-*	-	-*	28,840	-*
1827	11,805	15,000	8,845	-	3,401	-	-	-	-	-	-	-
1836–37	14,839	24,900	4,180	988	2,980	193	1,253	345	2,099	650	23,587	-
1840	16,000	28,000	-	-	-	-	-	-	-	-	-	-
1847	23,442	-	-	362	2,346	-	-	-	-	-	-	-
1851	25,800	-	2,715	-	1,246	64	826	-	518	-	19,280	-
1855	55,200	56,000	2,342	-	794	-	25	-	174	-	19,280	-
1861	62,000	81,600	2,156	-	321	-	10	-	92	-	18,700	-

Table 7.1 Area (ha) and production (tonnes) of selected crops in Réunion, 1820–61. In 1851 there were also 3,385 ha in beans, 2,782 in manioc, 600 in tobacco and 1,500 in sweet potato. * the 'maize' figure for 1825 is for all cereals (including wheat and rice, not given separately)[206].

still widespread in 1910 (he missed Common Quails), by 1916 Chinese Francolins *Francolinus pintadeanus*, bush quails, Common Quails, Painted Quails and buttonquails were all but extinct. As Henri Antelme wrote in 1914:

> At Henrietta, during the lifetime of the Hon. George Robinson ... it was rare that at [his] gatherings more than sixty of the birds [Chinese Francolin] were not shot. I was for a long time a hunter of this partridge, and more than once I bagged thirty of them in a single morning, with the co-operation of one or two friends. It is sad indeed that such interesting game is doomed to destruction by the depredations of the mongoose, which devour all the eggs and young ones.[198]

Hares, guineafowl, Grey Francolins and the formerly abundant tenrecs all survived in reduced numbers[199]. The status of Meller's Duck was never clear; Henry Slater found it common on Mare aux Vacoas in 1875 and one writer said it was plentiful in marshes before the mongoose, but others suggested it was never abundant – nonetheless its restricted 20th-century distribution may have been be due to the mongoose. This predator, combined with marsh drainage for malaria control, probably led to the eventual disappearance of Purple Swamphens and White-faced Whistling Ducks (p. 169)[200]. Joseph Huron commented in 1923 that "all the poachers of the island acting together could never have done as much damage as this quadruped"[201].

Between Newton's notes in the 1860s and Richard Meinertzhagen's visit in 1910–11 (before the bulbul was abundant in forest), the status of endemic birds appears unchanged, though Meinertzhagen made relatively few visits to forest areas. He saw 'lots' of Mauritius Kestrels and flying-foxes at Chamarel Falls (then still in native forest), and exploring the area between Grand Bassin and Alexandra Falls in February 1911, saw plenty more kestrels and good numbers of all endemic species except the cuckoo-shrike (one only) and Pink Pigeon (a pair), though he heard more of the latter "cooing in the forest"[202].

RÉUNION

After five years of British rule, Réunion was returned to the French in April 1815[203], but despite becoming an important source of sugar for metropolitan France, it remained a colonial backwater little visited by outsiders. The conversion to a sugar economy was slower than in Mauritius, with little extension of cultivated land into native forest before 1848 – at first sugar replaced coffee and spice plantations. This spared the forest, but the tree crops (e.g. coffee grown in the shade of large *bois noirs*), were an environment suited to many endemic birds and geckos, while cane fields offer nothing to the native fauna.

The new post-Napoleonic royalist regime abolished or declared void most of the legislation established by Decaen[204], including that promoting food crops over sugar. At the same time, the loss of Mauritius as a replacement for Saint-Domingue (Haïti) meant that Réunion became the obvious new source. The enterprising planter Charles Desbassyns made good use of the British occupation and his friendship with Matthew Flinders to import from England a revolutionary (for Réunion) new steam-powered sugar mill in 1815, launching industrial sugar manufacture on the island. At first many planters simply added cane to the mix on their concessions, but over time the larger landowners bought out most of the smaller holdings, and concentrated on sugar, which paid 4–5 times more per hectare than grain crops or maize. Nonetheless, in 1851 there were still 19,000 hectares of maize against 25,000 in cane. As late as 1837 there were still 701 coffee plantations to 159 in cane, though the latter were much larger[205].

The French traveller Auguste Billiard has left us a detailed account of the island during 1817–1820, which while giving us an idea of the degree of agricultural settlement and forest cover contains virtually nothing on wildlife. Betting de Lancastel's useful geography from 1827 is better on fauna and forests[207]. Billiard noted that coffee, ruined by the 1806 cyclone, had largely given way to maize on the western slopes cultivated to 400 *toises* (780m) above sea-level, that cabbage-palms still towered over the forests from 400 to 700 *toises* (1360m) but were absent from the lowlands, and latans were widespread up to 400 *toises* (780m). Around the Ravine à Marquet (above La Possession) and above St-Leu there were nonetheless still extensive coffee plantations. The Étang de Saint-Paul, haunt of waterbirds a century earlier, had been largely converted to rice-paddies, cane-fields and vegetable gardens, intercut with canals. The road up to the Plaine des Cafres was still covered in virgin forests. On the windward side, sugar plantations were already well-established around Saint-Denis and Desbassyns's factory at Sainte-Marie. Further south Billiard visited elderly botanist Joseph Hubert, whose estate at Bras Mussard was a shaded plantation of clove and nutmeg trees, hedged with rose-apple. Cloves were badly hit in the cyclone Billiard witnessed in February 1818 – even native forests were left with patches of uprooted trees 'acres in extent'. Billiard criticised the way in which the original concessions, extending in narrowing bands from the coast to the heights, precluded any sensible forest management and indeed promoted their destruction, resulting in reduced rainfall and an increasing shortage of construction timber[208].

We have very few observations on the wildlife in Réunion in this period of agricultural transition – just passing mentions in the diaries of Renoyal de Lescouble during 1812 to 1838, a few notes from Victor Sganzin's visit in 1831, and a partial list in Betting's *Statistique*[209]. Renoyal, who had introduced Common Tenrecs in 1801–02 (p. 109), was a teacher and musician. He was keen on aviculture and hunting, but said little on changes in status, and lacked insight as to why. In 1834, hunting in the heights of Sainte-Suzanne, he saw only two Réunion Merles where ten years earlier friends had shot 'several dozen' in a day; feral goats had similarly declined. Betting's book and Renoyal's diaries provide the first reliable record of several introduced animals in addition to the tenrecs: Malagasy Grass Frog (1827), Painted Quail (1827), Common Waxbill (1825), Zebra Dove (1827), Malagasy Turtle Dove (1824), Black-naped Hare *Lepus nigricollis* (1827) and a probable Spice Finch (1822). Renoyal had mentioned quails earlier (1812–13), but lacking description they are unidentifiable, as Betting reported a small 'caille de Madagascar' in 1827 and Sganzin mentioned Madagascar Buttonquails in 1831–32. Unidentified doves ('*tourterelles*') had been exempted from persecution in bird-pest legislation in 1820, but Renoyal named both *tourterelle de Sumatra* (Zebra Dove) and *touterelle malgache* (Malagasy Turtle Dove). Buttonquails and Malagasy Turtle Doves were originally native (pp. 42–43), but may have died out and been re-introduced. Hares and frogs had been around unreported for some time, and quails, Zebra Doves, and the pests specified in the 1820 law (Java Sparrows, Cardinal Fodies, Yellow-fronted Canaries and Red Avadavats) are likely to have arrived well before being noted in legislation and by Betting and Renoyal (p. 106). Renoyal called the *sénégali* (Common Waxbill) "new in the country and very pretty" – despite the 1820 law banning both import and possession of seed-eating cage-birds of any kind![210] Hoopoe Starlings feature only in 1822, Renoyal reporting the capture of six, four being sent to Mauritius as cage-birds[211]; Betting noted black parrots as already rare by 1827. Tenrecs had become common by 1827[212]; Renoyal did not mention their being eaten, but Sganzin remarked on this in 1831–32. Betting gave a good description of the Manapany Day-gecko *Phelsuma inexpectata*, and clearly considered the Réunion Slit-eared Skink *Gongylomorphus borbonicus* common; he emphasised that House Shrews, so common in Mauritius, were absent from Réunion[213].

Sganzin and collecting expeditions

Victor Sganzin, who considered himself a serious scientific naturalist schooled in Buffon's tradition, visited both islands *en route* to becoming military administrator of the French outpost at Sainte Marie island, Madagascar. He was much more interested in Madagascar than the Mascarenes, but his passing remarks frequently compare Malagasy and Mascarene species. However, he only collected in Madagascar, so it is difficult to disentangle his own observations (or those directly reported to him) from his gleanings from Buffon's encyclopaedias. Worse, he confused many species; in short, he was pretentious and not entirely reliable[214]. He discussed lemurs at length without any hint of feral ones in Réunion, and was vague on flying-foxes, implying that large ones were still to be seen there as well as in Mauritius[215]; Auguste Vinson also reported flying-foxes, the smaller Rougettes, in 1831 (p. 139)[216]. Sganzin asserted that Madagascar Buttonquails were present on both Réunion and Mauritius – if correct, the only record for a long time on either island[217]. Captive Purple Swamphens were breeding on a lake in the *Jardin du Roi* (now the museum gardens) in Saint-Denis. His observations on common species (Striated Heron, Cardinal Fody etc.) are probably sound, as is his record of Moorhen, the first for Réunion[218]. However he clearly never entered native vegetation in Réunion, nor spent much time on Mascarene wildlife.

Various collectors, who wrote nothing themselves, passed through Réunion – the Paris museum holds reptiles collected *inter alia* by Joseph-Fortuné Eydoux and Emmanuel Rousseau. In 1830, Eydoux, travelling with Laplace on the *Favorite*, collected the first specimens of the Manapany Day-gecko and several Réunion Slit-eared Skinks[219]. Rousseau's visit in 1839 provided the last known specimens of the skink, and the first of the Wolf Snake from India, already reportedly common[220]. The decline of the skink (still common in 1830) and the arrival of this lizard-eating snake are unlikely to be coincidental – as we have already seen, there was similar pattern of extinction in Mauritius. According to Louis Maillard the skink survived, though rare, until around 1860; there are no later reports[221]. De Nivoy, who more famously collected a Hoopoe Starling in 1833, also brought back a Panther Chameleon and the type specimen of the Malagasy Grass Frog. The frog had arrived around 1790 (p. 108), but we have no date for the chameleon, though they had been brought from Madagascar as curiosities since the 1750s at least[222]. The old restrictions must have lapsed, as monkeys and lemurs were being imported in 'large numbers': given the understandable paranoia about monkeys becoming feral (due to the Mauritian experience), the import of all primates was banned (again!) in 1822, anyone with a pet being given two months grace to get rid of it, either by export or death[223]. An attempt to obtain deer and Chinese Francolins from Mauritius in 1844 to release at Aurère (Cirque of Mafatte)[224] appears to have come to nothing.

Exploiting the *cirques* and uplands

1848 was a key year for the environment in Réunion. The French authorities belatedly abolished slavery, with similar consequences as seen in Mauritius a decade earlier. Most slaves left the plantations to set up independent smallholdings in marginal land in the lowlands, and planters started importing contract labour to replace them. However, the scale of immigration in Réunion was never as overwhelming as in Mauritius, though the population did jump 65% in the 20 years following 1848[225]. The principal difference between the islands was the emergence in Réunion of a poor white peasantry (the '*petits blancs*'). Prior to emancipation, the disinherited younger children of white planter-farmers lived on small-holdings with a few slaves. These holdings became uneconomic without the slaves, so they were forced to become wage-slaves themselves or to sell up (to large land-owners) and go somewhere cheaper. Many chose the latter course, migrating to the remotest parts of the island – the 'cirques', deep gorge bottoms, and the cold misty uplands, previously inhabited only by bands of fugitive slaves, who for decades had deterred settlers by their menacing presence. Although a few *marrons* lingered in remote areas, this danger had in practice evaporated by the 1820s, and some planters ruined by the cyclones of 1825 and 1829 had already established themselves in Salazie, discovering the hot springs in 1830. This settlement expanded slowly during the 1830s and 1840s (the population was around 2,000 in 1848), but there was no major movement into the cirques, the large cliff-enclosed mid-altitude zones, until after 1848. However such was the post-emancipation upheaval and the rush to cultivate that the relatively accessible Salazie, still largely forested in 1846[226], had been largely trashed (despite protective forest regulations) by 1850, and Cilaos and Mafatte by 1868, the latter in just 6 years (1862–68). Cilaos had been practically inaccessible until a vertiginous single-track footpath was completed in 1845. The *petits blancs* practised shifting cultivation in the nutrient poor uplands, thus removing much more forest per farmer than settled agriculture would have done[227].

Only amongst the explorers of Salazie have we any record of what the cirques were like before the massive deforestation. In 1831, Auguste Vinson, then 11, accompanied his father to Salazie to study the medicinal properties of the hot springs. Vinson junior (also a doctor) became one of a group of enthusiastic naturalists who documented the fauna in the 1860s. His glowing account of the forests of Salazie, written up 57 years later, is coloured by the extinctions that took place in his youth, but is powerfully evocative of this lost world:

> We passed through the shadiest coverts, next to the most magnificent trees, through dense refreshing thickets where the soil disappeared under a carpet of green moss as if an emerald coat spangled with strawberries as red as rubies. . . . On the plateau where today exists the true village of Salazie – the village of Petit Sable – there was a thick forest, frequented by clouds of black and white huppes [Hoopoe Starling] *the size of a pigeon. This powerful insectivore whose unique habitat was Réunion, has today completely disappeared from creation. Around the Mare à Poules d'Eau it was still virgin forest and its thousand streams, the forest always full throughout of black parrots of which one can no longer find a single pair in the whole island. From the mere, it required a full day, by a path cut with great effort, to reach the hot springs. There in the hollows of dead trees there were still those giant bats, called by the name of collet rouge* [Rougette], *which numbered 30 in a single roost. These mammals with rusty throat and breast made a most delicious meal.*[228]

When Vinson wrote this in 1888, the Hoopoe Starling, the introduced black parrot and the Rougette had all vanished. Vinson's father had been so impressed by the quality of the forest sheltering the headwaters of the Rivière du Mat, that from 1839 onwards he attempted in vain to prevent its clearance. The best he got was a law requiring landowners to keep a fifth of their land wooded, but, as the younger Vinson complained in 1888: "How many of these concessionaires or their successors have observed this clause?"[229].

Although plans for developing the island's heights had been mooted for some time[230], it was only after 1848 that changed circumstances brought it about. As sugar progressively claimed the slopes from below, the small cultivators were forced upwards. Between 1848 and 1880 forest was cleared up to about 500m on the windward side and to 700–800m in the west. Freed slaves supplemented their small-holdings by making charcoal from the forest[231]. After forest laws enacted in 1827 and 1833 proved ineffective, in 1853 governor Henri Hubert-Delisle attempted a remedy by creating a forest department, but it was "quickly paralysed by ignorance and the timidity of the majority of its officers and the untamed hostility of the population"[232]. A link road begun in 1857 between the 500 and 700m contours, while never completed, proved an attractive line of settlement in the drier west. This road (now the D3) is still the principal mid-altitude link in the west and south. The two rather extensive *plaines* (Palmistes and Cafres) that link east and west in the saddle between the volcanic massifs, were also settled. Here there were official development schemes, though, given the cussedness of the *petits blancs* and the difficult conditions, nothing went according to plan. The idea was for the Plaine des Palmistes to grow crops to feed the sugar-dominated coast, but the poor soils and ever-wet climate soon put paid to that, and the poor white settlers opted, as usual, for subsistence agriculture; by 1859 there were already 1,400 inhabitants. The Plaine des Cafres was gazetted for pasture, to supply the island's meat and dairy products, but by 1872 there were barely 100 inhabitants. At an altitude of around 1,600m, most afternoons saw a cold mist, and this, combined with thin soil, required the settlers to be even hardier than usual. Few were up to the task. The Plaine des Palmistes settlement remained very restricted in area, with much of the surrounding forest surviving; the Plaine des Cafres vegetation also largely survived this stage, but was to succumb later[233].

Naturalists and collectors

Only in 1855 did an equivalent of the Mauritian *Société d'histoire naturelle* emerge in Réunion, with returning Paris-educated intellectuals forming both a *Société des Sciences et Arts*[234] and an acclimatisation society, both of which published journals. The previous year, Hubert-Delisle decreed that the old island council building in the *Jardin du Roi* (the Saint-Denis botanic garden) be converted into a natural history museum; mayor Gustave Manès organised the conversion and opened the museum in August 1855[235]. The first book to attempt a full faunal list was published in 1862 by Louis Maillard, its inaccuracies soon rectified by papers from Charles Coquerel and Auguste Vinson. Beginning in 1860, publisher and lithographer Albert Roussin published a five-volume encyclopaedia on the island's history, culture and natural history, with individual articles by various authors on many of the birds, surviving and extinct[236]. These compilations were nonetheless deficient in that the various authors, apart from outsider François Pollen, appear to have done little or no fieldwork outside the lowlands, apparently gleaning information on endemics from hearsay and specimens brought to the museum or Vinson's aviary. While easily-seen exotic species were quite well-covered, they failed to document the disappearance of the Hoopoe Starling and Rougette. Pollen, working with D. C. van Dam collecting in Madagascar and Réunion for the Dutch natural history museum in Leiden, was the first to properly research the Réunion Cuckoo-shrike – and indeed made the only detailed observations of the species before the 1970s![237] Of the contemporary Réunionnais, only botanist Eugène Jacob de Cordemoy, and possibly Auguste Lantz who took over the museum in 1865, seem to have been inclined to penetrate the forests; neither contributed much to the 1860s publications. Coquerel noted in the mid-1850s that Norway Rats had confined Ship Rats to the hill forests, where they were still common[238]. The discovery of Dodo bones in Mauritius produced a flurry of interest in Réunion, although Coquerel had earlier pointed out that "the island contains many caves which have never been visited from the palaeontological point of view, and which may perhaps contain the remains of lost birds"[239]. However, it was more than a century before anyone investigated (pp. 191–192).

The local authors established adequate faunal lists in the 1860s, but Pollen provided the only good notes on abundance and distribution[240]. By this time Common Waxbills had overtaken Red Avadavats in abundance, Java Sparrows had become rare, various quails had been introduced, and Grey-headed Lovebirds had become established. House Sparrows, escaped from an aviary in 1845 and then encouraged by local enthusiasts[241], had grown common and widespread by the 1860s. Auguste Vinson, reminding his readers what pests small granivores had been in the past, noted wryly in 1868 that "today, given over to the exclusive culture of cane, Réunion has watched with indifference the introduction of the House Sparrow, which propagates itself with a disastrous fecundity"[242]. Malagasy Turtle Doves were "pretty rare", inhabiting forests and mountains, and moorhens fairly common, especially at the Étang de Saint-Paul[243]. Edward Newton stopped briefly in Saint-Denis in 1867, coming away with a specimen of a Dimorphic Egret

Egretta dimorpha (given him by Lantz), which supports the hints in the local literature that egrets still bred in Réunion at this late date; they were not reported after 1868[244]. Pollen saw only Striated Herons[245], not listed by the locals, who appear to have conflated them with dark-phase Dimorphic Egrets. Finally, Pollen was the only commentator in the 1860s to take much notice of seabirds. He described large roosts of Brown Noddies on the northen sea-cliffs, down which intrepid bird-catchers would abseil to snare birds for sale in the Saint-Denis market, though noddies were reported only to nest, as now, on Petite Ile, a stack off the Manapany coast. He also commented on the use of flares to catch petrels, and on the nesting habits of White-tailed Tropicbirds *Phaethon lepturus*, noted as fairly common[246].

Introduced gamebirds and other animals

The 19th century fashion for introducing gamebirds hit Réunion before Mauritius. Judging by Vinson's remarks in the 1860s, Grey Francolins and Jungle Bush Quail had been introduced some 20 years before, and efforts (allegedly unsuccessful) had been made to establish Madagascar Partridges (first released around 1845) and *cailles nattées* – Rain Quails *Coturnix coromandelica* or more probably Common Quails[247]. In fact the Madagascar Partridges had simply abandoned the coast and settled in the uplands[248], beyond the reach of coast-bound hunter-naturalists! The Chinese Francolin, so popular with Mauritian hunters, was still lacking in Réunion, so Vinson wrote in 1876 to the RSAS in Mauritius asking for some. If any were sent, no one noticed them for a long time; they were not formally recorded before 1948[249]. Although Grey-headed Lovebirds first appear in the ornithological literature in the 1860s, Réunionnais poet Charles Leconte de Lisle mentioned green *perruches* ('small parrots') and other wildlife in a poem published in 1857, whose inspiration dated from the 1840s[250]. Crows (species unknown) were released and apparently bred briefly before disappearing, and White-faced Whistling Duck, dabbling ducks and Egyptian Geese *Alopochen aegyptiaca*, regularly imported from Madagascar, were said to fly back there[251]. Vinson and Coquerel were indefatigable, even obsessive, 'acclimatisers', proposing to import and release all manner of 'interesting' birds. Coquerel wanted to bring in tits to control the cane borer, and nightingales to liven up the forests, while Vinson imported doves, pheasants, California Quails *Callipepla californica*, Australian parakeets and several finches and weavers, though it is unclear how many of these he released[252]. Palm Squirrels *Funambulus palmarum* were released around 1855, Maillard noting their establishment in Saint-Denis around 1860, but they appear to have soon died out[253]. Feral lemurs, apparently from escaped pets, survived rather longer: although not reported in print until 1859, they appear to have been established for some time, possibly since the 1820s, given the large numbers brought in before their import was banned in 1822. They apparently survived until 1878, but were not reported again. The Ruffed Lemurs *Varecia variegata* (and possibly Brown Lemurs *Eulemur fulvus*; confirmatory specimens are lacking) lived at the Plaine des Makes in the west ('*make*' = maki, lemur in Malagasy) and in the the Rivière des Marsouins gorge in the east[254]. In the 1880s, Lantz reported that the island administration wanted to introduce a gamut of insectivorous birds from Madagascar for biological control: two owls, a nightjar, a roller, a bee-eater, sunbirds, warblers, white-eyes, a drongo, vangas, a crow, coucals, couas and the Cuckoo-roller *Leptosomus discolor*![255]. This plan, fortunately, was never carried out.

While prone to introduce animals deliberately, Vinson also recorded accidental arrivals – the Bloodsucker came in with sugar-cane plants on a ship from Java around 1865[256]. In 1868 Vinson noted a severe drop in Forest Day-geckos *Phelsuma borbonica*, which he blamed on an unidentified snake supposedly imported from Madagascar some 15 years earlier (c. 1853) that had "multiplied in a remarkable manner". However, no Malagasy snake has ever been confirmed for Réunion, and the animal is more likely to have been the Wolf Snake, present since before 1839. It may have gone for day-geckos after wiping out the ground-living skinks. By 1891 the apparent total disappearance of the other day-gecko, the Manapany Day-gecko, which was common in the early 1860s, was being blamed on the Bloodsucker[257]. However, the lizard's Mauritian counterpart, Vinson's Day-gecko *P. ornata*, does not appear affected by Bloodsuckers, so Wolf Snakes spreading to the west look a more likely cause, though competition from exotic house-geckos is also possible[258]. House Shrews also reappeared in the mid-1860s. The Saint-Denis museum acquired a specimen in 1864, and there are others (from 1865 and 1874) in Strasbourg and Leiden, although it escapes mention in local literature until 1916[259]. They probably came from Mauritius, though they may have come direct from Asia like the Bloodsucker, or from Madagascar.

More extinctions

While the acclimatisers were busy importing exotics, endemic species were disappearing, regretted a little but neither studied nor documented. Surprisingly, tortoises reappear in the story here: Vinson wrote in 1868 that, following reports of animals around the turn of century in the mountains between Saint-Denis and Saint-Paul, "the last of all were seen at Cilaos only a few years ago". This was echoed, also for Cilaos, by Jules Hermann in 1898; he reported having spoken (when we do not know) to an old man, Ambroise Rochefeuille, "perhaps the last of our contemporaries to have seen them", who told of seeing them climbing cliffs and recumbent trees[260]. Beginning earlier in the century, enormous numbers of Radiated Tortoises were imported from Madagascar to satisfy the Réunionnais' appetite for tortoise meat[261]; some no doubt escaped and were reported as wild. However, the fact that the last reports were from Cilaos does suggest the native species. Cilaos was effectively isolated from the rest of the island through inaccessibility until the 1840s. It could thus have been a refuge for the native tortoise, although it was here, on the Ilet à Cordes, that a whole village of *marrons*, escaped slaves, existed in the 1740s and 1750s. In the 1740s *marrons*, according to a contemporary account, lived largely on tortoises and wild goats, but the report was from slaves living in the heights of Saint-Paul not Cilaos[262]. Vinson and his contemporaries seem to have made no attempt to check the reports they cited, nor to find living tortoises. Reviewing the evidence, Paris tortoise expert Roger Bour concluded that they probably died out around 1840 – too early for the 1860s naturalists to have found them, even if they had tried.

The Rougette's fate is almost as obscure as the tortoise's. From alleged abundance at Salazie in 1831, by 1862 they were said by Maillard to very rare and confined to "the old forests in the *cirques* of the interior". There is no reliable 19th-century specimen from Réunion, and in the mid-1860s Pollen failed to find or even hear of them[263]. Eugène Jacob de Cordemoy, who had seen Hoopoe Starlings in his youth, wrote in 1908 as if he had never seen the bat, and noted that there were none in the local collections. The Saint-Denis museum accession book, started in 1855, contains no Rougettes, and no specimens are preserved there[264]. It thus seems likely that Maillard's remarks in 1862 were over-optimistic, and possibly memories of how things were when he first arrived in the late 1830s. The last Rougettes probably died out around 1850, victims of the invasion of the cirques by white colonists, who while clear-felling for agricultural projects, removed the large hollow trees the Rougettes roosted in, and ate any bats and tortoises they came across in the process. The introduced black parrot vanished at the same time, probably for the same reasons; with no specimens we cannot tell which species of *Coracopsis* it was[265]. As well as these large species, the small insectivorous Pale House Bat was last seen in 1867. Maillard called it *chauve-souris des hauts* (upland bat), saying it "lives generally in the forests and is also found in the coastal zone"; the cause of its apparent extinction on the island remains obscure[266].

The Hoopoe Starling excited a little more interest, but was nonetheless also allowed to disappear without trace. As with the Rougette and the black parrot, the naturalists of the 1860s considered it very rare, the clearest statement being from A. Legras, who wrote in 1861 that "the *huppe* has become so rare that we have hardly seen a dozen in our wanderings in search of birds"[267]. Pollen tried quite hard in the mid-1860s:

> This species has become so rare in Réunion that one has not heard tell of it for a decade. It has been destroyed in all coastal areas, the same in upland areas not far from the coast. Believable persons have however assured me that it should still exist in the forests of the interior near St-Joseph. Old créoles [those born in the island] I asked on this subject told me that, in their youth, these birds were still common and that they were so stupid that they could be killed by hitting them with a stick.[268]

In the 1870s Vinson and Lantz were still holding out hope. Lantz wrote to Professor Alphonse Milne-Edwards saying that he had reason to believe the bird was not extinct and that he would be able to procure a specimen. Vinson wrote with evident uncertainty that there "may not be a single one left"[269]. With hindsight it is clear that all 1860s and later records are hearsay – there are no eyewitness accounts by this time, and we have to go back a little to establish a more realistic date of disappearance. We know from several accounts and a plethora of specimens collected that the Hoopoe Starling was still common in the 1830s[270]. In 1844 a Father Lombardi, a Corsican priest formerly working in Réunion, presented six or seven fresh specimens to Italian museums[271] – we know neither the collection date(s) nor where in the island he worked, but clearly around 1840 the birds were still easy to collect. The last useful witness was Eugène Jacob de Cordemoy, who wrote in 1910 to Neville Manders:

> I have known the bird you ask me about since childhood, namely the *Fregilupus varius*, which has in fact entirely disappeared ... When I was a boy this bird lived in the forests of the interior of the island and never set foot nor wing in towns or inhabited places. It remained faithful to the forests where it was bred, which it enlivened with its clear notes. I used to hunt it at an age when one is pitiless ... By no means shy, it was not frightened even by the sound of firearms, and after a regular slaughter, one went off with dozens of these poor victims in one's game-bag. After ten years spent in Paris I did not find a single one in the forests where formerly they flew about in flocks ...[272]

Jacob de Cordemoy, born in 1835, completed his medical training in Paris in 1859, and was certainly back in Réunion in 1861. He is unlikely to have left for Paris before 1850, and can hardly have been hunting as a toddler in the 1830s, so his letter indicates that the bird was still common in forested areas into the late 1840s or early 1850s, but it had vanished before 1860[273].

BOX 13   PIGEONS AND DOVES OF MAURITIUS (COLUMBIDAE)

Pigeons or doves were mentioned in most early accounts, but unfortunately, very few descriptive details were given. During the 17th and 18th centuries, indeterminate pigeons were found and slaughtered in vast numbers. By the 1830s all but two Mascarene species had become extinct. Only the Malagasy Turtle Dove *Nesoenas picturata* (possibly re-introduced) and Mauritius Pink Pigeon *Nesoenas mayeri*, which was reduced to 10-15 birds by the late 1970s, survive. The Pink Pigeon has now recovered to approximately 350 individuals thanks to intensive conservation efforts, but numbers fluctuate greatly.

Pigeon Hollandais *Alectroenas nitidissima*

This species was so named because its red, white and blue plumage represents the colours of the Dutch flag (unlike the pre-Revolutionary French flag). It is only known from three specimens, the most recent taken in 1826. It died out shortly after this date. The extinction of this spectacular species appears to be associated with deforestation, direct hunting and nest predation by monkeys and rats. A live bird in captivity c. 1790 was illustrated in colour.

Pigeon Hollandais *Alectroenas nitidissima*. Depiction of a live bird in 1790 from Tuijn, 1969.

Accounts
Cossigny (1732-55) in 1755, writing from Mauritius to Réaumur in Paris

What you will find best in the little barrel is a pigeon hollandois, at least that is the name given here to this species of bird which has aspects of pigeons and parrots. This one is female, I have not been able to get a male, infinitely more handsome. These birds, which I saw commonly enough in the island 23 years ago, are nowadays very rare, for as plantations are cleared, forests cut, they have retreated and keep to the regions of the island where the escaped slaves take and eat them ... [he opened up the bird and found 4 large seeds] I'm told [by some] that these are seeds of tacamacha [Calophyllum tacamahaca], by others that they are bois de natte à petite feuilles [Labourdonnaisia calophylloides], but whatever they are, this bird must have a pretty large gullet to be able to swallow such large nuts ... These must be very good food for the birds, as it was extremely fat, and its gizzard contained nothing but this nut that filled it [the other three were in the crop].

Vosmaer (c. 1790, in Tuijn 1969):

"These [are] dark-blue with lead coloured head-feathers, which they can turn upwards just as a collar, were sent to me from the Cape, but originated from an Isle Mauritius. Presented to the court by the baron I. N. E. van Lynden 1790, and were called Pavillons Hollandais. One only I have received alive but died later after a few months from dropsy. In the evening 11 till 12 o'clock and many times during the night it made nice sounds 10 to 12 times quickly after another like Baf Baf [prounounced Barf Barf], and during the day a kind of cooing sound.

Milbert (1812) in 1801–03

The inhabitants of the Ile de France [Mauritius] call it pigeon hollandais; the head, neck and chest are adorned with long pointed white feathers which it can raise at will; the rest of the body, and the wings, are a fine deep violet; the end of the tail is a purplish red. It is one of the finest species of its kind [i.e. pigeons] ... it lives solitarily in river valleys, where I have often seen it without being able to secure one. It eats fruit and fresh water molluscs.

Mauritius Pink Pigeon *Nesoenas mayeri*

Surprisingly, the pink pigeon managed to survive, along with Echo Parakeet *Psittacula echo*, in the Black River Mountain range located in southwest Mauritius. From just a handful of individuals, the population has recovered to a few hundred, but the population continues to fluctuate. It is the only surviving endemic Mascarene pigeon.

Accounts
Cossigny (1732–55) in 1732:

The pigeon-ramiers [Pink Pigeons] in Mauritius are of great beauty, but there is a season of the year in which one must respect them. Many having eaten them at this time have found themselves very sick. They provoke in you a kind of paralysis with contortions of all the limbs and intolerable pains, which last for 24 hours following; I don't however know of anyone [afflicted] who actually died. These pigeons must no doubt feed themselves on various seeds that produce these effects. They are never so fat as at this time, and it is true are very tempting as they are so appetising. As for me, I don't want to taste them at any season.

Malagasy Turtle Dove *Nesoenas picturata*. From Stephens, 1819–26.

Malagasy Turtle Dove *Nesoenas picturata*

Some early accounts mention 'turtle doves' on Mauritius and a good candidate for this species is *N. picturata*, a pigeon widely distributed in the Indian Ocean. The Malagasy Turtle Dove occurs on Mauritius today but may be a re-introduction; fossil material has also been collected.

It remains a mystery why so common a bird, that had withstood man, rats, cats and environmental change for 200 years, should have so suddenly disappeared. The decade 1850–60 was of course a period of intense forest clearance in the cirques, and we know the birds were common in Salazie before it was settled (p. 139). However, there is no suggestion that Jacob needed to make the arduous trek into the cirques to find his birds, and he mentioned returning to the *same* forests where he had hunted them, evidently not cut down. He was born, and presumably brought up, in St-André, not far from the entrance to Salazie, but the nearest forests would have been the eastern slopes of the Plaine des Fougères, near where Renoyal's Hoopoe Starlings were caught in 1822. The number of specimens collected in the 1830s suggests they were widespread, or at least easily obtained near Saint-Denis; there are more Hoopoe Starling specimens from the 1830s and 1840s than all other Réunion birds together for these dates, and it was not a sought-after rarity. Furthermore the bird was not generally eaten; it is not listed among gamebirds after Dubois's list of 1672 until 1839, and Vinson specifically noted that it was locally regarded as 'impure' as an insectivore[274]. Jacob de Cordemoy did not explain why he and his friends killed so many – probably just for 'fun'. Louis Brasil, reviewing the history of its extinction in 1912, suggested a parasite or disease. This remains a very likely cause[275]. Large Mauritian passerines (Mauritius Merle, Mauritius Cuckoo-shrike) feed their young where possible on day-geckos, so perhaps the Hoopoe Starling also fed its young on lizards. Its strong legs suggest it was primarily a terrestrial feeder, so it may have gone for skinks, and as we have seen, the Réunion Slit-eared Skink crashed after 1840. Nothing was recorded of the Hoopoe Starling's breeding habits. Starlings normally nest in holes, and forest clearance may have removed suitable overmature nest-trees. A flocking species, it may have nested colonially and required a social group to breed succesfully, though we are not aware of any starlings that do this. It was omnivorous and easily kept in cages[276], though there is no record of any having been captive-bred. If only those naturalists had gone out and caught the last few. . .

Forest clearance for new crops

The period 1850–60 was prosperous, fuelled by good sugar prices and cheap imported labour, but in 1863 the Réunion economy began a long slide. The sugar-cane borer arrived in 1857[277], sugar prices in the island's only market (France) fell with competition from Cuban cane and European beet-sugar, malaria struck in 1868, and the opening of the Suez canal in 1869 took Réunion off regular trade routes[278]. With other crops forced aside for 'king cane' there was little to fall back on; coffee, such as was still grown, was devastated by a fungal parasite which struck in 1882[279]. As the economy staggered along, new cash crops were frantically sought, but few showed much promise. Vanilla was the first to do so, and spread along the wet east coast and parts of the southwest in the 1860s and 1870s. These areas were already long cleared of native forest, so the environmental impact was small – indeed, possibly beneficial. Vanilla is a climber, so many trees were planted to support the crop, though the total area was never very large[280].

The next successful crop was a different matter entirely. In 1882 oil was successfully distilled for the first time on the island from rose geranium, a high-value, low-volume crop, ideally suited to being grown in areas with poor road communications. It grew well in the drier heights of western Réunion, and created what Defos du Rau called a kind of 'gold rush'[281]. By 1889 it was being exported in tonnage quantities and during the next two decades a swathe of deforestation took place at altitudes between 600 and 1,500m, starting at Le Tampon and progressing northwards. Because its ideal growing conditions were found at an altitude previously untouched, all expansion cut into fresh forest. On top of that the stills to extract the oil, built by the fields, required an enormous amount of wood fuel. Then it was found that the crop exhausted the fields in five to six years, so more forest had to be cleared to allow the soil to recover; erosion and the loss of fragile forest top-soil was rife[282]. In some areas, at least, the old fields were planted with a quick-growing fuel crop, an Australian acacia, but this then became an invasive weed[283]. The enthusiastic growers were mostly *petits blancs* – well-versed in shifting cultivation.

Judging by the map in Defos du Rau's monumental study, at least 200km² of virgin forest, mostly montane *bois de couleurs*, was lost to geranium between 1880 and 1925. This all happened despite new forest protection legislation enacted in 1874, designed to protect water catchments. However, the legislation only applied on state land, and although the highest parts of the old sea-to-summit concessions had been re-nationalised in 1871, little land below the 1,600m contour on the western Grand Bénare massif was under control of the *Eaux et Forêts*[284]. Nonetheless some forest down to 1,100m was retained in the northern sector, though very little on the Plaine des Cafres or in the heights of Saint-Leu and Trois Bassins. In the cirques of Mafatte and Cilaos much of the land remained in state hands or was reclaimed, and special permission had to be sought from Saint-Denis even to have timber to repair houses – the law, once enacted, was initially enforced much more rigorously than in Mauritius, drawing opprobrium from

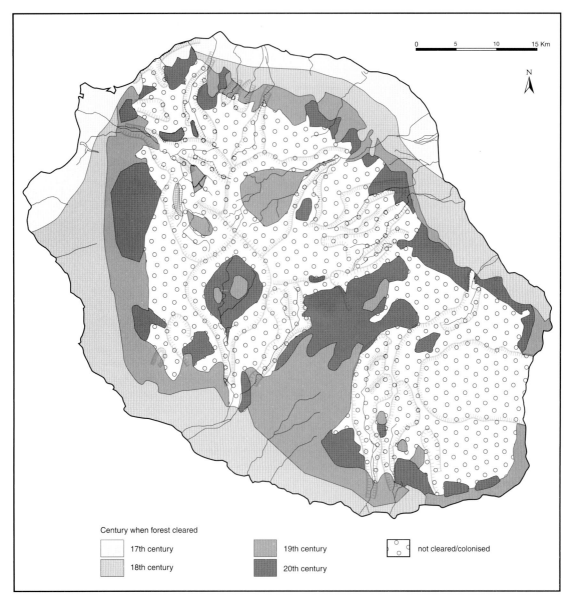

Map 7.3. Phases of land settlement and forest clearance in Réunion, 1663–2000. Adapted from Defos du Rau (1960), Bertile (1987) and Girard & Notter (2003).

Mauritian visitors, and creating a sort of low-level guerilla warfare between the *petits blancs* and the forest service[285]. Some re-afforestation was already well under way in Mafatte in 1876, presumably with exotic species[286]. In 1900 the director of the forest service, G. Kerourio, wrote:

> Until recent years almost all our houses and other buildings were constructed entirely in wood; framework, boarding, roof, all were made entirely of wood. The majority of our bridges are similarly built in wood. The development of the railway led to a big demand for sleepers from our forests. The consumption of firewood is no less considerable. Oil and its derivatives are very little used; and rare are the factories that use it. All our sugar factories burn fuel-wood, and the entire population uses, for domestic purposes, only fuel-wood or charcoal. The industry extracting essential oils installed in the colony in recent decades has consumed a considerable quantity of wood fuel. It is principally to this industry that we owe the large-scale clearances of recent years in privately-owned forests [287].

With "rare exceptions", privately-owned forests, by 1900, consisted of "thickets of inferior species of little value", from which all the best timber had been ruthlessly exploited, especially during the recent economic difficulties[288]. Visiting in 1895 William Oliver described *petit-blanc* agriculture in lurid, even racist, terms:

> The rainfall has not become less but has become more uncertain, and one cause of this is the reckless way in which so much of the forest has been cleared. The great idea of the Creole is to clear the forest and plant beans or maize. By doing so he not only insures [sic] a supply of food, but is enabled to indulge his proclivity for destruction. He gets one or two crops, and then the earth, deprived of its natural covering and under the influence of a tropical rainfall, washes away, leaving a wilderness of rocks, among which the tabac marron [Solanum mauritianum], that most accursed of weeds, springs up and flourishes. Then landslips, more or less serious, occur, and outraged nature takes her revenge.[289]

Kerourio, too, was concerned about erosion, but there is no evidence any action was taken. He died soon after, and the *Eaux et Forêts* was left without competent direction for over 20 years[290]. The book in which Kerourio wrote, a survey of the island's products and activities for an exhibition in Paris, carefully avoids mentioning geranium by name at all; there is astonishingly little even about sugar![291]

Kerourio also disingenuously claimed that all fuel wood used in the lowlands (apart from some charcoal) now came from new she-oak plantations around the coast. A massive government land-purchase of sand dunes around Étang Salé in 1874 was the main focus of these plantations, which had been started by the previous owner; by the 1920s there were said to be over a million she-oak trees there[292]. In addition to promoting three species of she-oak, Kerourio favoured plantations (on derelict land) of Silver Oak *Grevillea robusta*, *bois noir* (*Albizia lebbek*), miscellaneous acacias, gums and Neem *Melia azederach*[293]. In 1889 the forestry service also started managing a plot in Bélouve to regenerate the endemic *tamarin des hauts*, *Acacia heterophylla*.

The initiative in Bélouve arose out of an economic survey of that area in 1867, followed by a pioneering, if flawed, ecological survey by foresters and botanists in 1888[294]. The government commissioned the 1867 survey to see if the area was suitable for cultivation and concessions, but the report concluded it was not, principally to preserve the headwaters of several important rivers, adding that the location was too inaccessible for transporting any crops that might be grown. In 1888, Goizet, the first fully trained forester to take charge of the *Eaux et Forêts* in Réunion[295], accompanied by botanist Jacob de Cordemoy, visited the area and came up with a plan to regenerate the forest. Goizet described the forests in the following apocalyptic terms:

> The commission regrets to report that the forests that cover the high plateaus of the island – the only ones that the axe has so far respected – are everywhere afflicted by decrepitude and their existence is gravely threatened. Parasites [sic] (heathers, heath shrubs, ferns etc.) have invaded them in every area and tend to displace more and more the valuable species. In certain areas the wild vine [Rubus alceifolius], a veritable destructive plague, has made its appearance.
>
> The robust canopy, formed principally of tamarins des hauts still holds out and resists this invasion. But already many individuals amongst these giant trees of imposing stature, have perished miserably and lie helpless on the ground. Those whom life has not completely abandoned extend their twisted branches in all directions, convulsed in agony, as if to reclaim the air and light which escapes them. One can forsee the moment when, defeated and brought to earth, they will succumb in their turn without leaving behind any further generations. Of these superb plants, formerly the pride of our woods, there will then remain only formless skeletons, sad debris of a forest richness wiped out. . . . In consequence, the commission unanimously proposes adopting the project to manage the forest of Bélouve.[296]

The seeds of the distorted understanding of the forest ecosystem afflicting Réunion foresters for most of the 20th century are already visible: the lack of spontaneous regeneration under *tamarins* was seen as a failing of nature, and the 'right' response was to clear-fell and open the ground to stimulate germination of seedlings. It would take 90 years and several proper ecologists to persuade the Office National des Forêts (as it became) that the *tamarin* was a fire climax whose very nature was not to regenerate without a burn (p. 178).

The invasion of the upland mixed evergreen forest by geranium cultivators was probably also responsible for the devastation, well beyond the clearances, of cabbage palms – an important food source for frugivorous wildlife. While the Hurricane Palm had long disappeared in the wild from coastal regions, its upland relative, the *palmiste rouge*, was still abundant mid-century; Pasfield Oliver published a photograph of palms covering the eponymous Plaine des Palmistes, taken in 1864 from the Grande Montée[297]. The palm-cabbage was (and is) a delicacy for which the more affluent were willing to pay good money, so once they had access, it was open season for palm poachers. This process appears not to have been documented, but can be followed in photographs; no wild palms can be seen in photographs of forest and scenery in books published around 1900, though in less-frequented areas, as noted by William Oliver in 1895, palms were still to be found[298].

Meanwhile the forests were being invaded by exotics. As in Mauritius, roadside raspberries were welcomed by travellers, already "growing in thousands the length of the path" to Salazie in 1837[299]. Rose-apple and Strawberry Guava have been mentioned already, but the menace *par excellence* in Réunion's wet forests was (and remains) the bramble

Rubus alceifolius, called '*vigne marrone*' (wild vine) after its leaf-shape. Apparently introduced around 1840, by 1888 it was already the island's worst forest pest, invading clearings and clambering into the canopy, suffocating the understorey[300]. Ironically, this plant is pollinated and its fruits dispersed on Réunion by endemic birds[301]. The other most pernicious pests of wet forests were species of ginger, which came to dominate the ground flora; neither guava nor Rose-apple became (relative to Mauritius) seriously pathological in intact forest, probably due to the absence for 150 years of wild pigs and fruitbats. There are many rampant exotics in the dry lowlands, but even in the 19th century there was virtually no native lowland forest left to invade. *Litsea glutinosa*, penetrating dryish habitat as in Mauritius, has been invasive since at least the 1890s, e.g. in the best remaining dry forest of the Grande Chaloupe[302]. On the Plaine des Cafres, at high altitude, a prickly European invader was already widespread in 1895; William Oliver commented that "a large quantity of gorse grows about here, and I could easily have fancied that I was walking across some English common"[303]. The *tabac marron*, already present in 1825 and decried by Oliver (p. 146) as a pest of abandoned fields, was also an invader of any opening in low-mid altitude forest[304]. There seems to be no record of when the species of fuchsia that are now naturalised were introduced, though it is likely to have been in the mid-late 19th century[305].

The long spell of economic depression coincides with another period when we have little direct information on the fauna. Réunion was visited by Mauritian journalist Charles Leal in 1877; he made incidental natural history notes on his perambulations, but his attempt at a formal bird list was muddled and inaccurate. His main contribution was to notice a temporary revival in feral pigs (he saw "two or three groups of five or six passing through the woods" above Salazie) and providing a rare report of swift caves (in the Cilaos gorge)[306]. Auguste Vinson continued writing into the 1880s, but his paper on Mascarene pigeons was thoroughly confused, even inspiring a corrective discussion from Albert Daruty in Mauritius[307]. Auguste Lantz, the long-standing and internationally admired museum curator, travelled to Madagascar and the Seychelles, but appears mostly to have used the native Réunion fauna for exchange with foreign museums, without actually studying it; visitors regularly lavished praise on him and 'his' museum. He only wrote one short paper, an incomplete annotated list of birds and mammals, clearly dashed off to keep an editor happy[308]. However, he knew how to find rarities if he needed to. In 1888 Alfred Newton in Cambridge was seeking specimens of the little known Réunion Black Petrel, and approached Lantz. Lantz got agriculturalist Auguste de Villèle onto the case, and he turned up eight specimens in 1889–90 – this rare endemic seabird was not seen again until 1970![309] Endemic passerines were marketed commercially as delicacies: Paul Carié noted "in 1898 . . . at the thermal springs at Salazie, bird hunters sold them [Réunion Merles] by the dozen to bathers, together with *Zosterops* [white-eyes]; this trade was happening on a large scale, and the birds thus destroyed were sold even in Saint-Denis"[310]. For the same period we have retrospective (but not contemporary) information about the introduction of Village Weavers. Some escaped around 1880 from a cage on a boat loading sugar at the short-lived jetty at Bois Rouge, then property of Adrien Bellier – hence the local name *Oiseau Bellier*[311]. If they weren't bothered by snakes, Bloodsuckers, shrews and weavers, at least the authorities were still keen on keeping monkeys out. William Oliver enjoyed the story of one that was solemnly escorted by four policemen from a shipwreck near Sainte-Marie to a cage in the Jardin Colonial in Saint-Denis (by the museum) to ensure it did not escape[312]. Deer were another matter. Land-owning Réunionnais were always jealous of the pleasures of *la chasse* enjoyed by their counterparts in Mauritius, and in *c*. 1900 deer were once again introduced, this time to a very restricted *îlet* (small plateau), Terre-Plate, in Salazie, where by 1909 there were said to be 300, already reaching the limit of their habitat[313]. Another target for enthusiastic hunters was also released around the turn of the century at Bras Panon (on the northeast coast) – wild Red Junglefowl *Gallus gallus*[314].

In the first years of the 20th century the economy was still in the doldrums, the only bright spark being essential oils. Vanilla, vetiver and ylang-ylang were largely planted on land already cultivated, while upland forest clearance continued for geranium, whose production was still rapidly rising in 1914. Sugar production had meanwhile fallen to an average well under half of its peak in the early 1860s, and much land was left fallow, too low or wet for geranium, too high for vanilla or ylang-ylang. There appears to be no contemporary account of what happened to this land, but in retrospect it looks like the origin of the large areas now under Rose-apple and Strawberry Guava that are so prominent in the mid-altitude zone on the windward slopes. Natural history activity was minimal. The only known naturalist to be active was Paul Carié from Mauritius, who apparently visited several times during 1898– 1925; he (and his assistant Majastre) collected specimens, but never wrote up any proper account of these trips[315].

RODRIGUES
Once the British troops left Rodrigues to invade Mauritius in 1810, the island reverted to its previous state

– a few settlers, with their unruly slaves, bickering amongst themselves. The new British rulers in Mauritius did not even send a representative to the island until 1821, and then not for the purpose of administering the island, but because new abolitionist laws in Britain required all slaves to be formally registered. Marragon continued as *de facto* island chief until 1826, but unfortunately stopped writing letters, so we have very little information until a Mauritian surveyor was sent in 1825 to map the island and sort out concession boundaries. C. T. Hoart made only sparse comments on the vegetation, noting that "there is no timber in the west and only a cover of *squine* grass, but on the limestone outcrop [Plaine Corail] trees of *bois puant* and *benjoin*, though small, are hard and very plentiful". He reported abundant fish in the lagoon, but did not mention Dugongs – by then presumably extinct. Ile Frégate was covered in fig trees and sheltered "lots of excellent birds called *fouquets* [shearwaters]" – their excellence presumably culinary. The seabird islands of Sable and Cocos were "low islands near the edge of the reef and good for nothing"! Rats and caterpillars were abundant, damaging crops, Hoart reporting several unsuccessful attempts to introduce mynas to control insect pests. The human population was still only 123 in 1826, consisting of three landowning white families and 100 slaves[316].

As noted (p. 114), the Rodrigues Solitaire bones found in 1786 reached Europe in 1830 and caused a bit of a stir. 'Mr Duncan', curator of the Ashmolean Museum in Oxford, where the Dodo's head and foot resided, inspired Charles Telfair to try to seek Solitaires in Rodrigues[317]. Telfair did not go himself, but persuaded a Colonel Francis Dawkins to make enquiries on an official visit in 1832. Dawkins found a few bone fragments in a cave, and established to his satisfaction that no such birds still existed[318]. Honoré Eudes, an island resident, subsequently found better bones which he forwarded to Telfair, telling him also that "a bird of so large a size as indicated by the bones has never been seen by M. Gory [sic] who has resided forty years on the island" – Gorry had been among the new colonists arriving in 1792. Dawkins and Eudes also commented on the *boeuf*, which Eudes reported, in a letter accompanying specimens, as so named "from its cry which is absolutely that of a calf. I recall that a . . . [illegible] having heard it cry, I and some blacks [slaves] we went searching for a calf and we found this bird in a tree ... Mr Gorry not being in the same bay was luckier, his blacks got two of these animals"[319]. The skin that reached London was identified as a Red-footed Booby *Sula sula*, but Eudes's description of the call is so vivid that he must have encountered Abbott's Booby, apparently in a tree on the mainland. Telfair reported that one specimen

Figure 7.5. View of Port Mathurin and the north coast of Rodrigues, looking west, in 1846, as seen by Edward Higgin. Note the bare hillsides with only occasional trees. From Strickland (1848).

of a *boeuf* was "so eaten up by insects" that he refrained from sending it to the Zoological Society – an ignominious fate for the last Abbott's Booby from Rodrigues![320] Dawkins was also the first to report feral guineafowl.

Another relict of the original fauna, unseen since Tafforet's day, made a last appearance in 1842. Mauritian sea-captain and keen naturalist François Liénard received from a Captain Descreux not only a Newton's Day-gecko but also, from Ile Frégate, several giant geckos *Phelsuma gigas*, including one that lived for two months in Mauritius. Descreux said the geckos were abundant on the islet, living on figs and seabird eggs[321]. No one ever saw the species again. On the mainland they apparently survived Ship Rats, but disappeared after cats were released between 1732 and 1755. However, no lizards were found on Ile Frégate in 1874, and by 1914 there were large rats, rabbits, and ample shearwaters (thus probably no cats) on the islet[322], but no sign of the lizards, so it looks as if the Norway Rat helped the Giant Gecko into oblivion.

Rodriguan slaves, semi-freed as paid 'apprentices' in 1835, were fully liberated in 1839, Eudes bringing a special proclamation from Mauritius[323]. Unlike in Mauritius and Réunion, this important socio-cultural change had little environmental effect. Many slaves already cultivated patches on unconceded or abandoned land. Soon deciding to stop working as apprentices, they simply moved to their gardens, where they practiced shifting cultivation and built huts out of latan leaves.

Corby, Gardyne and Higgin

The first clear idea of the environment in 19th-century Rodrigues does not emerge until Thomas Corby, another surveyor, reported on the island in 1845. The details are in a long letter to the Surveyor-General,

J. Augustus Lloyd[324], first explorer of Round Island (p. 210). Corby and Lloyd were both interested in natural history, and Corby's report is full of useful observations. Sent to look into the possibility of Rodrigues replacing Madagascar as a source of beef, Corby wandered over much of the island, but kept mostly to the coast and the spine of the island between Mont Limon and the east coast at St. François, apparently not penetrating the south-facing valleys where the best vegetation survives today. The coastal zone was by then largely grassland, fired in the dry season to stimulate forage for free-ranging livestock. In the uplands there were two miles of abandoned cultivation east of Mont Limon "covered with wild raspberries and other rank thorny weeds. The half-calcined trunk of a gigantic Bois Puant or Palm Tree here and there gave evidence of the fine forests which had once existed in this spot". Eastwards the land was "thickly studded with wild Vacouas [screw-pines] and Lataniers, whose trunks all bear witraces [sic] of fire", but descending towards the coast this gave way to scorched grass and barren rocks, except in the St. François valley itself. In the low interior around La Fouche and Montagne Topaz in the west he found "trees of small growth, palms and Vacoas are [more] plentiful here than in any other part of this island", commenting that the soil was too thin and rocky for agriculture. Plaine Corail was already reduced to "a few dwarf trees in the interstices of the rocks and some scanty herbage". Corby's list of wildlife differs little from Marragon's 50 years earlier, apart from the now-abundant guineafowl and centipedes. He commented on Rodrigues Fodies *Foudia flavicans*, and indeed probably collected the first museum specimen[325]. Like Marragon and Hoart, he wanted mynas introduced to keep down insect pests, and also, for the same purpose, House Shrews ('musk rats'). Corby did not visit the lagoon islets and mentioned seabirds only in passing, but some were collected on the trip, possibly by W. Kelly, captain of the *Conway*, the ship Corby travelled on. Kelly had been asked by George Cunninghame, Collector of Customs in Port Louis, to look for Solitaire bones, but found none[326]. The human population had almost doubled since 1825, but still numbered only 323.

It is clear from Corby's account that most of the island had lost any reasonable forest, with, in some areas, only the relatively fire-resistant latans and screw-pines still thriving. Given that we know that fire-raising had already begun in 1761, and had been rampant since the 1790s, it is hardly surprising that most forest trees had succumbed. It certainly appears that fire, rather than deliberate clearance, was the most significant agent of environmental change on Rodrigues, where rainfall is everywhere low, especially all round the coast[327]. Mauritius suffered the same pattern of fires in the dry area around Port Louis in the mid-18th century (p. 92). There was also significant agricultural clearance but the population was still small, and it is clear from Corby's observations that even on abandoned concessions (as many were in 1845), fires prevented any regeneration that escaped the feral pigs, goats and cattle[328]. Corby recommended that "with a few improvements" the island could support 10–12,000 cattle and render Mauritius self-sufficient in beef. His report was shelved by the Colonial Office[329], though the principle of Rodrigues as a grazing economy remained *de facto* policy, to the long-term detriment of the ecosystem (p. 153). Among his suggested improvements was the planting of *bois noir* and *filao* [she-oak] in the uplands, Rose-apple along streams and coconuts around the coast.

The year 1846 brought the shipwreck of the *Trio* on its way to Mauritius from India, and two interesting accounts of the island to complement Corby's from the previous year[330]. While neither Alexander Gardyne nor Edward Higgin were as informed observers as Corby, they had more to say on seabirds and on day-to-day life. Higgin's sketches of the island, showing the pitifully few trees remaining on the hills above Port Mathurin, were almost the first pictures of the island from life ever published, and have pride of place in Strickland's Dodo book[331]. Gardyne's diary reported uncontrolled raiding of Sooty Tern eggs by the *Trio*'s crew on Ile Gombrani, wanton killing of boobies on Ile Crabe, and an abundance of centipedes[332]. The most lasting outcome was Higgin's throwaway remark that the island was composed of granite. This made Rodrigues a potentially exciting geological rarity, as the only mid-oceanic islands known to be granitic were the Seychelles, which harboured a strange fauna and flora. This error from a geologically ill-informed traveller was the principal inspiration for the Royal Society to attach biologists to the 1874 Transit of Venus expedition[333], which generated the first proper inventory of the biota.

In 1856, following two severe cholera epidemics, and despite new facilities being built on Flat Island (p. 208) and Cannonier's Point, the Mauritian government had the bright idea of turning Rodrigues into a quarantine station, and sent a commission to report. The idea, driven by the panicked Mauritian public wanting diseases kept as far away as possible, was both impractical and expensive, and quietly dropped. The report reprinted much useful earlier material, and was the first since Marragon's to mention the importance of tern eggs from lagoon islets as a regular local food source. Latans were both praised as a free source of food for pigs, and blamed (with screw-pines) for producing so many dead leaves that "when they happen to catch fire the whole island is soon in a blaze"[334]. Around the same time, Edward Messiter,

> **BOX 14   PIGEONS AND DOVES OF RÉUNION (COLUMBIDAE)**
>
> On Réunion at least three pigeons and dove species were mentioned. Some were extremely abundant, tame and confiding. They succumbed rapidly to cats around 1700 and were also much hunted.
>
> **Réunion Pink Pigeon *Nesoenas duboisi***
> A species akin to the Mauritius Pink Pigeon *Nesoenas mayeri* once occurred on Réunion. It was described by Dubois in 1672–3.
>
> **Malagasy Turtle Dove *Nesoenas picturata***
> Although not clearly reported in the early literature, subfossil material has confirmed that the Malagasy Turtle Dove *Nesoenas picturata* was also native to Réunion.
>
> **Slaty pigeon ?*Alectroenas* sp.**
> A second pigeon, probably an *Alectroenas*, was described by Bontekoe in 1619 and Dubois in 1672–3. Based on these descriptions, this pigeon lacked the blue body coloration of the Pigeon Hollandais *A. nitidissima* of Mauritius, being instead slate-coloured. Like the Malagasy *A. madagascariensis*, the Réunion bird may have also lacked the white head, neck and shoulders of *A. nitidissima*.
>
> *Accounts*
> Bontekoe (1650) in Réunion in 1619:
>
>> We found large numbers of ramiers of the species which has blue wings. They let themselves be taken by hand, or we knocked them down with sticks and canes, without their making any effort to fly away. In one day we killed a good 200.
>
> Dubois (1674) in Réunion in 1672–3:
>
>> . . .wild pigeons, everywhere full with them, some with slaty-coloured feathering [*Alectroenas* sp.?], the others russet-red [*Nesoenas duboisi*]. They are a little larger than the European pigeons, and have larger bills, red at the end close to the head, the eyes ringed with the colour of fire, like pheasants. There is a season when they are so fat that one can no longer see their croupion*. They are very good tasting. Wood-pigeons and turtle-doves, as one sees in Europe and as good.
>
> Borghesi in 1703 (in Lougnon 1970):
>
>> "*In* [but, see below, probably should read 'before'] *the days we passed in the island, there were also found those wild pigeons we call detour*† *in such large numbers that the women, at the very moment they were preparing the meal, killed them by dozens with sticks, even right inside their kitchen. But since in recent years one began to see rats . . . it was necessary in Ile Bourbon [Réunion] to introduce cats to ward off the great damage these rats caused. But these* [the cats] *confronted with such an abundance of food in the countryside, did not stay in the houses or plantations; dispersed in the woods they multiplied in great numbers. Befriending the rats, and in league with them, they destroyed entirely the above-mentioned pigeons.*
>
> Feuilley (1705) in 1704:
>
>> Since a while back one no longer sees ramiers, either they have abandoned the island or the cats have destroyed them.
>
> * *croupion* normally means either a bird's rump or the 'parson's nose' of a cooked chicken; here we suspect it must mean 'cloaca'.
> † 'of towers' – the writer clearly mistook them for rock doves/feral pigeons.

the new administrator (now styled 'magistrate') got very exercised about rats, and wanted legislation to compel each cultivator to produce 15 tails per acre per month (shades of Poivre!). The Mauritian government turned him down, suggesting he introduced "weasels, ferrets or black snakes" to do the job – advice fortunately not taken up[335].

Solitaire bones

In 1864 Edward Newton, by then with five years of Mauritian birdwatching under his belt, managed to get a trip to Rodrigues on government business, though his main intent was clearly to go birding. However, he had no control over the ship's movements, and only got a couple of days free – time enough to 'discover' the Rodrigues Fody and Rodrigues Warbler *Acrocephalus rodericanus*, record some seabirds and waders, and pick up a few bones in caves on Plaine Corail[336]. Fired by finding the bones, and by more sent by Jenner the next year, Newton got funding from England to send Jenner a posse of labourers from Mauritius to dig for more. This resulted in a huge haul of bones that Newton and his brother Alfred presented in London in 1868[337] – the extinct Rodrigues Solitaire had become osteologically almost the best known animal on the planet! On his brief visit Newton also recorded several 'new' introduced birds – mostly ('dove' = Zebra Dove, 'bengali' = Common Waxbill) through hearsay, though he saw Grey-headed Lovebirds[338]. Newton was told that the lovebirds came from an American whaler which had picked them up in Madagascar. Also during Jenner's time, Grey Francolins were brought from South India on the *Teemayma*, captained by Guinol – the francolins and some quails (which failed to establish) were exchanged for chickens to feed the ship's crew[339]. By 1916 these importations, including also Zebra Doves, had been assimilated in local lore to have arrived together on the '*Gemima*', captained by Genaud, in 1862 (though the waxbills remained unaccounted for). In 1923 Bertuchi wrote:

> *Red-leg Partridges* [= Grey Francolins] *are to be met everywhere on the island. They were introduced about*

1862, when the French sailing ship La Gemima visited the island for repairs. She carried an aviary which, amongst other birds, included parakeets, zebra-doves and partridges, which were set free on the island.[340]

No doubt Bertuchi was told this as historical fact, but 'Gemima' is not a French name, and the alleged captain, 'Genaud', sounds very like 'Jenner'. Whatever the exact events, increased contact with Mauritius and American whalers from the 1850s on resulted in birds being imported. Rusa Deer *Cervus timorensis* were also introduced in 1862–63, allegedly at Jenner's instigation[341]. Numbers reached *c*. 180 in 1882, and, with new game regulations instituted in 1883, to 1,500– 2,000 in 1892[342]. In the 1880s and 1890s a stop at Rodrigues to hunt deer, guineafowl and 'partridges' (i.e. francolins) was popular in the Royal Navy and with VIPs from Mauritius, accounts appearing in the memoirs of various admirals and government officials[343]. Mary Broome, wife of the first Governor of Mauritius to visit Rodrigues, expressed the seriousness of this pursuit in 1881 from a cynical angle:

> Our last day at Rodrigues held, indeed, hard work, for we spent it from an early hour en chasse, *the paraphenalia of which might have served for at least a small punitive expedition. Such munitions of war, in the shape of guns and cartridges! And the commissariat* [food supply] *was on an equally liberal scale*.[344]

All for nothing, since the first lieutenant of the *Euryalus* had scared away all the game the previous day! By 1893 the next visiting governor, Sir Hubert Jerningham, found the whole island virtually "given over to the deer in the forests . . . and to herds of cattle which were practically wild" and were in urgent need of control. Although there seems to have been no formal instruction, after this intervention deer declined sharply, recovering only after being protected again in the early 1900s[345].

The 1874 expedition

In 1874 came the first Transit of Venus since 1769 and the return of scientists to Rodrigues for the occasion [346]. Bayley Balfour led the biological team, and made a thorough inventory of the plants, with general observations on vegetation; his colleagues George Gulliver and Henry Slater were collectors whose specimens were worked up by others, although Slater published a report on the caves and wrote a 'bird report' for Edward Newton. Balfour rapidly discovered there was no granite, and was not impressed by the state of Rodrigues, "fully one half of the island" being "entirely bare of vegetation"[347]:

> The great and tall trees [described by Leguat] *have now almost entirely disappeared, the eternally verdant canopy formed by their boughs no longer exists, and the "little Eden" is now a dry and comparatively barren spot, clothed with a vegetation mainly of social weeds, and destitute of any forest growth save in unfrequented and more inaccessible parts in the recesses of the valleys . . . we now have a bare parched volcanic pile, with deep stream-courses for the most part dry, in place of the verdant well-watered island of 200 years ago*.[348]

He went on to pronounce on the processes he thought had caused this massive loss of native forest: goats, fires and invasive plants, notably White Popinac (locally *acacia*). Introduced some 30 years earlier near Port Mathurin, it was by 1874 "filling up completely many of the valleys and destroying the native vegetation", being spread about by livestock. Balfour noted that cultivation had declined since slaves were freed, the plantation economy reverting to self-sufficiency; wheat was no longer grown "mainly because of the parroquets [Grey-headed Lovebirds] and Java sparrows which abound". This was the first and only report of Java Sparrows – Slater also saw them, but they were not recorded again; doubtless the loss of wheat crops contributed, as perhaps did competition with House Sparrows, which arrived soon afterwards. Large trees survived here and there from abandoned coffee plantations. Balfour also noted that indigo plantations had formerly occupied "some of the central portions of the island", explaining some upland deforestation. In coastal areas there were some dense thickets of the pan-tropical Cuban Bast *Hibiscus tiliaceus* and associated strandline trees, while inland the streams were bordered with popinac and higher up with Rose-apple. Overall, east of a line from Rivière Saumatre in the North to the mouth of Rivière Coco in the south, where the topography is steepest, there were valleys "in their upper parts filled with a tolerably dense growth of trees and shrubs", whereas the lower basaltic west was largely "barren of any trees or shrubs, save perhaps a stray stunted Vacoa [screwpine], Palmiste [Hurricane Palm], Latanier [latan] or Citron [citrus]". On Plaine Corail there were still abundant, if stunted trees, especially *benjoin*. Balfour argued that the striking difference between east and west was due to fires spreading west in the prevailing wind, uninterrupted in the western parts, but obstructed by topography in the east.

Many trees now endangered were still 'abundant' or 'very common', while some now invasive had hardly arrived at all: Strawberry Guava and Traveller's Palm, for example, were scarce[349]. Two invasive 'century plants', Mauritius Hemp and a sisal, were already very common in some areas, and citrus trees gone wild formed 'impenetrable thickets'[350]. Latan palms, though still widespread, were becoming scarcer; felling them for timber had been banned, though leaves were still used for thatching, room partitions and baskets. Henry Slater found some large native trees on Plaine Corail in "a sort of ravine terminated at each end by a cavern" where, for want of domestic livestock in those parts, he shot roosting guineafowl for his meals while digging for bones in the caves[351]. The heads of valleys had no doubt been spared from clearance as too steep for cultivation, though any large timber had been cut out. The island's population was still small (c. 1,200) in 1874, but increasing fast[352].

Between them Slater and Gulliver made a thorough inventory of the island's animal life, though Gulliver, apparently misled by the granite hypothesis, started off wasting a lot of time searching for frogs[353]. Despite their efforts they failed to locate three endemic animals thought by locals to survive – the two large day-geckos and the Rodrigues Parakeet. A specimen of the last had been obtained in 1871 by Jenner before he finally left the island[354], but the geckos were known to science only from Liénard's descriptions of 30 years earlier. Gulliver reported that he had been "told of a much larger lizard [than house geckos] which inhabits a certain part of the island, and have myself searched the spot, but have been unable to find it. I have also offered a reward for a specimen, but have not yet procured one. The same has been the case with regard to another lizard which lives on Frigate Island"[355]. Slater saw a parakeet on Plaine Corail when he had no gun, but a second (and last) specimen was shot by local pilot William Vandorous in 1875, who passed it to James Caldwell (who also saw 'several' himself). Newton's Day-gecko turned up again in 1876, magistrate Henri Desmarais sending a specimen to Edward Newton, no doubt in response to Caldwell who had, in 1875, also offered a 'good reward' for a specimen "of the long lizard called *coulevec* also supposed to be extinct"[356]. However no trace was found then or later of Liénard's Giant Gecko, but two more Newton's Day-geckos, almost the last, were caught in 1884[357]. Significantly the expedition also failed to collect any Mourning Geckos *Lepidodactylus lugubris*, suggesting that this Pacific drifter may have arrived subsequently[358]. Ile Frégate was visited by Slater and Balfour and, less happily, by the crew of HMS *Shearwater*, their ship. Balfour commented on the presence of *Pisonia* trees which were "the favourite nesting place of the Fou [Red-footed Booby]", with Slater adding that:

> There is a large colony of these birds in a grove on Frigate Island, that being the only islet sufficiently wooded, and they never come to the main island. . . . the skins of the young form a good substitute for swans down; when HMS Shearwater *left Rodrigues, her rigging was festooned with skins belonging to the officers, hung there to dry.*[359]

Slater noted that frigatebirds, although present, no longer bred[360], and 1874 was also the last year the boobies were recorded breeding – the scientific expedition apparently doing them a lot more harm than the Rodriguans, echoing Gardyne's account of 28 years earlier. However, Gardyne and co. had slaughtered boobies roosting in trees on Ile Crabe – Slater's remarks on Frigate suggest that Ile Crabe had since been deforested. Slater also collected Wedge-tailed Shearwaters on Frigate – he described their calls after dark – but he saw no large geckos. The other seabirds were much as before, including Great Crested Tern *Sterna bergii*, though Slater commented on seeing no Red-tailed Tropicbirds – indeed none had been reported since Pingré in 1761[361].

The surviving landbirds, the Rodrigues Warbler and Fody, were, as a decade earlier, common and widespread, with no particular indication of habitat preference, apart from needing woody or shrubby vegetation; the fody's nest was then "usually placed in young trees of the Bois d'Olive"[362]. Golden Bats had been familiar to residents since Leguat's time, but Gulliver's specimens were new to science. Neither Slater nor Gulliver commented on their ecology, though Balfour noted that *bois chauve-souris* ('bat-tree') was so named because they liked its fruit[363].

Slater was the first to record mynas (still scarce) and (with Balfour) Java Sparrows, though he found the latter "not at all common", unlike their crop-pest comrades the Grey-headed Lovebirds that had "multiplied to a vast extent", going about in 'huge flocks'. However the love-birds were feeding on wild grass seeds, whereas Java Sparrows require the larger fare of cereal crops, which, as Balfour noted, had been abandoned. Zebra Doves were still scarce enough to have eluded Slater (who doubted their existence), but he did not visit the 'north end of the island' where more Rodrigues Parakeets had been reported. Gulliver's Stump-toed Geckos *Gehyra mutilata*[364] were also new, and amongst Slater's bones were those of rabbits – another unnoticed introduction, possibly dating back to Marragon. The first time they were explicitly noted alive was not until c. 1914, when they were common on Ile Coco and Ile Frégate but did not occur on the mainland[365].

Cyclones and the parakeet

Rodrigues was struck by four bad cyclones in 1875–76, followed by further severe storms in 1878 and 1886. It is probably no coincidence that the endemic parakeets disappeared at this time. By all accounts their numbers were down to a handful by 1875 – even Caldwell, though he saw several and was keen to provide for his friend Newton, could not get one until helped out by Vandorous (above) in August 1875[366]. The run of cyclones started in December, the one on 27 February 1876 one of the worst on record[367]. With so little tree-cover left, the birds may have simply been caught in the wrong place – the parakeet was never seen again. Caldwell did better underground, getting not only two complete Solitaire skeletons, but even finding the gizzard stones Slater had sought in vain, and a skull of the Black-spined Flying-fox[368].

Last-ditch forest conservation

Apart from his comment about latans, Balfour said nothing about forest protection, and indeed until the 1880s there was little but *ad hoc* rules from the magistrates to discourage excessive tree-cutting, which were invariably ineffective. Although Messiter had made some attempts to control wood-cutting in the 1850s, it fell to his successor George Jenner to make serious efforts to limit the damage. In general inhabitants had had indiscriminate permission to cut trees for their own needs and to supply wood to the now regular numbers of American whalers visiting the island. Jenner appointed the first forest ranger, though his job seems to have been more concerned with keeping livestock under control than preventing tree-felling. Jenner also regularised land-ownership and attempted unsuccessfully to control livestock roaming on state land by means of pasturage rents. His plans were in part stymied by the collapse of the livestock export market to Mauritius following the loss of both life and confidence in the great malaria epidemic of 1866–67[369]. Desmarais, also keen to curb the rampant abuse, introduced she-oaks in 1876[370].

Only in 1881 was the new and initially dynamic magistrate Joseph O'Halloran finally permitted to make regulations with the force of law; in 1882 he established mountain and river reserves, but there was no funding to back up the regulations[371]. O'Halloran's successor, Barthèlemy Colin, had less interest in forests and was on bad terms with the island doctor, L. E. Roussel. Roussel recommended general re-afforestation, especially planting up Cascade Pigeon and keeping the stream feeding Port Mathurin free of livestock, but this was ignored by Colin[372]. At some point one persistent source of wood-loss was curbed – the practice of bartering for water, livestock and wood-fuel with American whalers, which Jenner had attempted unsuccessfully to control in the 1860s, was finally banned, and by 1902 the whalers had deserted Rodrigues for the Seychelles[373]. Much of the gain in forest cover between 1880 and 1914, when the next report was made, was apparently due to invasive expansion of Rose-apple into old clearances rather than deliberate planting. Rampant cutting appears also to have been at least locally curtailed, as Paul Koenig was able in 1914 to report passable stands of native trees, where apparently there were none in 1874, notably on Grande Montagne. A. J. Bertuchi, on the island from 1914–17, noted "a rapid reforestation taking place" since tree-cutting had been restricted, though Koenig noted a widespread lack of natural regeneration due to pigs and goats, and heavy cutting of native trees for timber[374]. Whereas only the upper valleys could be described as wooded in Balfour's time, Koenig found that 'where unleased' (i.e. without grazing animals) lee slopes were clothed in Rose-apple, with occasional native trees (mostly *gandine*, latan and screw-pine), Strawberry Guava and *bois noir*. Cascade Pigeon valley, supplying Port Mathurin's water, "calls for urgent attention" as the upper parts were bare. Other parts of the central table land owned by absentee landlords were "under almost pure jamrosa [Rose-apple], which it would be useful for the government to acquire", although the "tops of mountain slopes, when crown land, are bare" (Monts Lubin, Limon, Malartic and Le Piton). Bertuchi's photographs confirm that Mont Malartic was bare to the south (windward), with thickets apparently of Rose-apple and popinac around the church at St Gabriel, whereas Mont Limon was a patchwork of thickets and open fields[375].

The native vegetation might have fared better against exotics without the epidemic of scale insects during 1884–88 that attacked both native trees and introduced citruses. Many trees, especially *bois puant*, one of the island's larger trees, were weakened by the insect and succumbed during the fierce cyclone of 1886[376].

While there had clearly been some environmental improvement in the upper parts of the island, the same could not be said for the lowlands. Ever since Marragon's time the lower slopes, burnt free of any tree cover, had been used largely for grazing, only the fertile valley bottoms being cultivated. Almost all this land remained in state hands, the majority of concessions being in the central uplands. This area became known as the 'cattle-walk' and was from 1871 onwards semi-formalised, and for a time (until 1881) leased to a single company for grazing. The Cattle Walk became a legal entity in 1893 with regulations enabling the magistrate to define its limits[377]. Apart from copses of shade trees noted by Corby, it had been virtually treeless for a century – in 1913 Koenig commented that all the useful trees were gone and that "the cattle-walk is now practically bare apart from scattered latanier trees in the south-east". Even the supposed river reserves were denuded of vegetation and "Rivière Papayes and its tributary Rivulet Citrons ... is gradually becoming a torrent carrying away all the fertile soil of the uplands"[378]. This probably explains why he wanted in many areas to see no cutting of popinac stems or even harvesting of seeds, fearing the same fate, although outside valleys and hilltops harvesting was acceptable. Over the years seed from this invasive exotic had become an important export to Mauritius for cattle feed[379]. Plaine Corail was entirely within the cattle-walk, but Koenig had little to say about it; Bertuchi's only photo shows more or less bare ground, with a fringe of gnarled native trees (?*bois d'olive*) facing Ile Crabe, though large trees still survived in the collapsed caves[380].

Clearly the lush, fire-controlled grass reported by Corby in 1845 and Gulliver in 1874 had deteriorated to heavily over-grazed pasture controlled by livestock pressure. Neither régime did trees any favours.

Our period ends with a very clear picture of the island – two detailed reports and the first book on Rodrigues since Leguat. The visit by the directors of forestry and agriculture from Mauritius in 1914 was the first by senior technical staff; they each produced detailed and very useful analyses. Bertuchi's book complements these with a general picture of island life at the same time, copiously illustrated with the first published photographs of Rodrigues[381]. Agricultural director Frank Stockdale made a number of suggestions in relation to agriculture in already cultivated areas, but also suggested forestry should include more planting of Rose-apple and also Jack-fruit (both for food). He expressly warned against introducing mongooses, Red-whiskered Bulbul, prickly pear, lantana and *herbe condé* – the last "would ruin pasturage and would kill out *acacia* [popinac] as it has done in parts of the Pamplemousses district" [in Mauritius][382]. O'Halloran had requested mongooses in 1887 to help control rats, but had second thoughts when two pairs arrived from Calcutta; he killed them when warned how destructive they were to poultry and game[383].

Koenig recommended the re-afforestation of denuded summits, the re-purchase of 207ha of abandoned upland property, and the planting of she-oak on light soils in the cattle-walk to give shade and improve grass cover. Another recommendation had a direct bearing on the whole style of Rodriguan agriculture. As already mentioned, much of the uplands was under shifting cultivation; by 1914 this was done legally, the magistrate giving leases on patches of land to be cleared and cultivated. After a few years (or less) of intensive cropping the soil was exhausted, and another patch would be leased, the first reverting to scrub:

> There still remains a source of [forest] *destruction, in that the inhabitants are in the habit of acquiring year by year fresh tracts of woodland, the undergrowth of which they cut down and burn, and here they plant their haricot beans. They utilize a tract of land for one season, and abandon it the next. Thus the work of destruction continues.*[384]

Many Rodriguans who owned their own land preferred to keep it wooded, and lease crown land to cultivate, thus gradually degrading the 'government's land' and destroying any remaining native vegetation. As North-Coombes pointed out, Rodriguans traditionally respected their own and neighbours' land, but cared little what they did to land belonging to the state. Koenig thus proposed banning leasing of land to those already owning property, but increasing access to lands already bare around La Ferme, which were in any case more suitable for cultivation[385].

Neither Koenig nor Stockdale mentioned wildlife, though Bertuchi made a noble attempt to enumerate it. However, while noting the Golden Bat, commonly seen in the deepest ravines according to H. J. Snell, Bertuchi saw neither Rodrigues Warbler or Rodrigues Fody, though he remarked on Cardinal Fodies and all the introduced passerines. He could have missed the unobtrusive warbler, but the Rodrigues Fody is noisy and conspicuous. This suggests the native fody had declined dramatically from its abundance in the 1870s, possibly through being confined to the fragmentary tall forest by Cardinal Fodies and House Sparrows, which had arrived since 1874. Since both were clearly abundant by 1914, they must already have been there some time. Judging by Bertuchi's photographs, there was ample Rose-apple thicket for warblers to thrive in, so they were possibly overlooked.

Bertuchi also discussed seabirds, commenting that Sooty Terns no longer bred because "their eggs were collected in boat-loads and eaten by the natives". Noddies, Fairy Terns, White-tailed Tropicbirds and *fouquets* (shearwaters) still nested[386]. There were rabbits on Iles Coco and Frégate, and also rats, "some of great size" (i.e. Norway Rats), on the latter. Great Crested Terns, reported as common by Slater in 1874 and with an established local name *goilon* (in other words they were not vagrants), are no longer mentioned[387]. The former abundance of seabirds attested by Tafforet and Pingré was echoed in the exploitable quantities of guano found on Iles Chat, Gombrani and Cocos, and exported to Mauritius from *c*. 1900 until exhausted[388]. Cocos and Sables had been untouched until she-oak was planted on them in 1884[389]; it is not clear how much native vegetation (including *Pisonia* trees) was cut to clear space, and whether this affected the nesting terns; Bertuchi's photos show that Fairy Terns, at least, had adapted by 1914 to nesting in she-oak trees.

CHAPTER 8

THE LIMITS TO GROWTH
The Mascarenes in the 20th century

Perhaps the most regrettable thing is the fact that the lessons of the past seem to have come to nothing. Destruction of the environment continues, introductions still take place. The genetic potential present is ignored, the biology of surviving species still too little known. Réunion, like too many other lands unfortunately in the same situation, is losing little by little its natural heritage amidst general indifference.
François Moutou on island degradation, writing in 1983[1]

The relative absence of native birds from the lowlands, where most people live, seems common to several islands on which the native habitats have been directly or indirectly destroyed (e.g. the Hawaiian and Mascarene Islands). For city dwellers, intentional releases can be a means to construct a pleasant environment, and for planners, a means to restore a kind of biodiversity. The poorer the species richness of native birds, the more common are bird introductions.
Philippe Clergeau and Isabelle Mandon-Dalger on wildlife psychology on islands[2]

Up until at least the 1880s, the majority of inhabitants of the Mascarenes had at least some contact with the original flora and wildlife of the islands. However, by 1914 this link to the islands' biotic heritage was fading fast, especially in Mauritius. As society grew more urbanised, and rural settlements became surrounded by canefields, people increasingly saw the widespread introduced plants and animals as the 'real' wildlife of their island. This perception spread to the intelligentsia, and even otherwise reputable biologists. For instance, Donald d'Emmerez de Charmoy and Paul Koenig were respected in their fields (entomology, forestry), but wrote widely in the early to mid-1900s on other wildlife topics – much of it deplorable nonsense. This lack of educational and intellectual continuity with the past resulted in the new generation of proactive conservationists in the 1970s having an uphill struggle to persuade Mauritians and Réunionnais of the importance of their remaining endemic biota. Nonetheless, there was a quiet revival of scientific forest ecology on both islands in the 1930s and 1940s, which laid the foundations for later conservation work.

Meanwhile, under pressure from increasing demand for wood, economic criteria (as opposed to water catchment priorities) were applied for the first time to upland forests, and plantations of pines (Mauritius) and Japanese Cedar *Cryptomeria* (Réunion) began to replace native vegetation. Large swathes of forest were also devastated by temporary excursions into cash-crops: tea in Mauritius and scented geranium in Réunion. All three islands had become crowded, with endemic under-employment, and, in addition to the cash-crop projects, other job-creation schemes impacted on the natural environment, most notably the ill-considered *Travail Pour Tous* forestry scheme in Mauritius in the 1970s.

As environmental concern grew in the later years of the century, the interests of evolutionary scientists and conservationists converged on the vulnerable biota of oceanic islands, and Mauritius became the centre of a major international conservation effort. As a result of another economic initiative, however, Mauritius became better known to most people in the West as a luxury tourist destination, in which wildlife scarcely featured as an attraction at all.

MAURITIUS

There were few credible champions of Mauritian wildlife in the first half of the 20th century. The most competent was Paul Carié, who had successfully re-excavated the Mare aux Songes, and researched and collected birds on both Mauritius and Réunion. However, he failed in his political ambition to return Mauritius to French rule, and left Mauritius for good in 1913[3]. Bouton's indefatigable successor, Albert

> **BOX 15**   **PIGEONS AND DOVES OF RODRIGUES (COLUMBIDAE)**
>
> Rodrigues had at least two native pigeons, both confirmed by skeletal material and a few descriptions. The Rodrigues Dove ?*Nesoenas rodericana* has skeletal characters unlike any other pigeon and it is clearly highly derived, apparently from a *Nesoenas*-type ancestor (JPH, work in progress). Some fossil elements referable to Malagasy Turtle Dove *Nesoenas picturata* have also been found, but it does not occur on Rodrigues today.
>
> The pigeons were already avoiding mainland Rodrigues and nesting on the islets in 1691–92, due to persecution by rats. Tafforet made a brief mention of them in 1726, stating that they were rare on the mainland and common only on the islets. The astronomer Pingré declared them totally extinct by 1761; the early accounts did not distinguish the species.
>
> *Accounts*
> Leguat (1707) in 1691–93 (from the English translation of 1708):
>
> > *The pigeons here are somewhat less than ours and all of a slate colour, fat and good. They perch and build their nests upon trees; they are easily taken, being so tame, that we have had fifty about our table to pick up the melon seeds which we threw them, and they lik'd mightily. We took them when we pleas'd, and ty'd little rags to their thighs of several colours, that we might know them again if we let them loose. They never miss'd attending us at our meals, and we call'd them our chickens. They never built their nests in the Isle, but in the little islets that are near it. We suppos'd 'twas to avoid the persecution of rats, of which there are vast numbers in this Island...*
>
> Tafforet (1726):
>
> > *The doves there are in great numbers, but on the mainland very few are seen, because they go to feed on the islets to the south, as well as the parrots, and come to drink likewise on the mainland.*
>
> *Specimen collections of Mascarene pigeons and doves*
> Skins of *Alectroenas nitidissima* are held at the RMS; MNHN: and MI. Rare subfossil material is held at UMZC. Subfossil *Nesoenas mayeri* and *N. picturata* from Mauritius is held at UMZC and Rodrigues fossil material is held at the MNHN. Fossil *N. duboisi* can be seen at the UCB.

Daruty de Grandpré, was still president of the Royal Society of Arts and Sciences, but the Society had gone into decline when the government withdrew its subsidy in 1901, its function in Mauritian polity usurped by its offshoot the Chamber of Agriculture, the newly professional Forest Service, and new government-funded scientific laboratories[4]. The Museum, long at the society's core, had been absorbed into the government-run Mauritius Institute in 1885, and when d'Emmerez was promoted to Government Entomologist in the new Department of Agriculture established in 1913, he was replaced by a poet, William Edward Hart![5]

The result of this decline in scientific endeavour was that the large landowners, reviled as forest-abusers by Gleadow, became *de facto* the principal agents of nature conservation, while the Forest Department's position became increasingly ambivalent. Although not generally naturalists, the landowners needed to preserve tracts of forest because of the paramount importance of *la chasse* – deer require ample habitat in which to thrive, of course. A few landowners, like members of the Antelme family, notably Henri and Georges from the turn of the century to the 1930s, were interested in other wildlife as well. Georges Antelme was keen on birds and kept (and tried to breed) Mauritius Pink Pigeons in captivity, although he was best-known for a classic book on deer hunting, which unfortunately contains nothing of note on other wildlife[6]. Another hunter, Joseph Huron, published books with wildlife content in the 1920s; he was concerned about game lost to the mongoose, but also noted a decline in Malagasy Turtle Doves and an increase of deer in peri-urban areas, both of which he related in 1923 to massive forest clearances for sugar cultivation "in the past thirty years". He reported Mascarene Grey White-eyes recovering after nearly disappearing due to the explosive expansion of Red-Whiskered Bulbuls, and was the first to draw attention to Mer Rouge in Port Louis harbour as an important site for migrant waders[7]. Some of these hunters were good observers, and their memories have provided distributional information for birds and bats in the 1920s to 1940s when very little was published[8]. Forest on private deer hunting estates has generally become more degraded than government forests, in part through being opened up to create open glades for grazing. Nevertheless, without the hunting fraternity it would all (apart from 'mountain reserves' on steep slopes) have long ago disappeared. Also, with little state land at low elevations, most remaining lowland forest patches are on private *chasse* land - e.g. at Magenta, Baie du Cap and Yemen[9].

The development of forest policy

Although little of Gleadow's forestry report was put into effect (p. 131), the Forestry Service, under its first trained forester Paul Koenig, largely continued the conservative policies based on Thompson's 1880 report, though he also expanded timber plantations on degraded land (mostly in Chinese Pine *Pinus tabuliformis*). Standing dead trees were finally exploited, even using a tramway to assist extraction in the

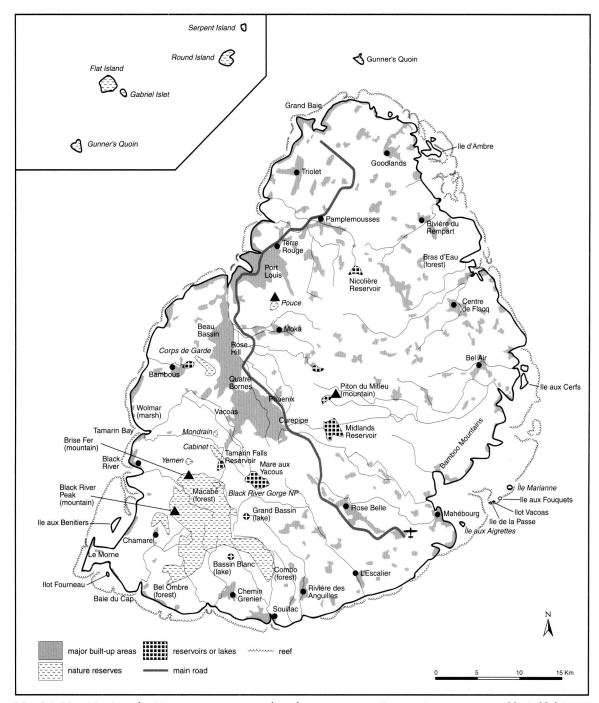

Map 8.1. Mauritius into the 21st century: towns, roads and nature reserves. From various sources, notably Saddul (2002) and recent IGN 1:100,000 maps.

Kanaka block, the blanks being infilled with mixed exotics, more for continued cover than for timber production[10]. Between 1910 and 1926 Koenig experimented fairly successfully with growing selected native trees under pine nurses, but the pine had to be cut in the 1939–45 war, leaving the natives, which had not formed a full canopy, exposed to cyclones; the severe storm in 1945 destroyed the plantation[11]. The 1914–18 war disrupted timber imports, and native trees in reserved areas (especially around Grand Bassin and Piton du Milieu) were felled to meet requirements, including what were in Thompson's time "the finest private forests on the island" (on H. Pitot's land at Bois Sec and Combo). The old Woods and Forests Board, abolished in 1914, was reconstituted in 1923, and after public protests over massive felling of native forest, re-established in 1926 the primacy of water-catchment conservation over timber production in the uplands, against Koenig's protests. At the same time the Director of Agriculture, Harold Tempany, strongly advised that the upland marshy forest at Les Mares should be considered a natural reservoir and preserved. Commercial felling of native trees supposedly stopped after 1923 (apart from clearance for the Midlands reservoir), but, in the context of the times, this probably meant 'felling of *large* native trees *for timber*' rather than any general protection. The restriction may also have applied only on Crown Land, as 699 ebony trees were "utilised" at Yemen (a private lowland estate) in 1926–27[12]. When Koenig retired in 1929, new director Gilbert Sale re-organised the Department and put forward a massive rolling plantation plan, largely for pine and Japanese Red Cedar *Cryptomeria japonica*, to eventually cover 40,000 acres (16,187ha), nearly 60% of Crown Forest land[13]. His economic analysis was considered faulty, and the scheme, in any case unaffordable in the 1930s recession, abandoned. A financial commission in 1932 attempted to abolish the Department and constrain forestry activity again to watershed protection; Sale managed to prevent this, though his budget and staff were cut. Since the 1880s various species of pine had been tried[14], but Sale established that the Slash Pine *Pinus elliotti*, introduced in 1929 from North America, was the most resistant to cyclones, and also the fastest growing; since the 1940s foresters have favoured this tree for upland timber production. Sale was succeeded in 1936 by J. E. Carver, who expanded the thrust towards economic forestry, though his assistant H. C. King also experimented with planting native species, probably encouraged by Koenig's pilot plots, then still growing well. Although Carver and King still designated certain areas of native forest as 'non-productive', it seems the 'watershed only' lobby had ceased to be effective by the later 1930s[15].

Although the broad view of forest changes is clear enough over the first half of the century, the detail is harder to follow, as the published reports give little detail of the exact location of new plantations, and less on what exactly they replaced. As far as we can tell, using Thompson's report and map from 1880 as a starting point, extensive pine plantations initially replaced degraded native forest in what foresters saw as the best land for commercial forestry – Nouvelle Decouverte/Nicolière and Quartier Militaire/Vuillemin in the centre-north of the island (c.1900–1940). However Koenig logged Parc-aux-Cerfs and Montagne Perruche south of Curepipe in the early 1920s, and there were plantations of pine and juniper east of Grand Bassin and 'improvement felling' in Combo. A plan for a large reservoir at Midlands where the growing stock was "a high forest of ruling species of good size" (i.e. the large Sapotaceae *etc.*) led to felling during 1923–1930, although the reservoir was not built until 2003! By the 1940s pines were being planted around Piton du Milieu. This was presumably following felling, "in [unspecified] degraded areas areas selected for afforestation with pine". Timber cut during the 1939–45 war included "200 baulks ... supplied for the Stevenson dry dock in Port Louis; species were *Canarium mauritiana* and *Calvaria major*", i.e. Colophane and Tambalacoque, both now endangered. Clear-felling was avoided in better quality native forests[16]. While large tracts suitable for birds were lost, enough good and degraded forest remained in 1945 to maintain adequate if small populations of all the native birds and bats; indeed, the only loss since 1836 was the Rougette, in the 1870s (p. 134). The Echo Parakeet's distribution in the 1940s, as compiled by Carl Jones from people's memories, almost exactly matches the main block of then-surviving (if partially logged) native forest, extending across Midlands to the Piton du Milieu[17]. In the 1930s small relict forest patches, such as around Candos Hill in Quatre Bornes, and between Vacoas and Trois Mamelles, held native birds such as Mauritius Fodies, but these areas near conurbations were cut over during the privations of the 1939–45 war, and planted to cane thereafter[18]. "An area of relatively homogeneous indigenous forest had to be cleared for the construction of Mare Longue reservoir" in 1946–47 – some 100ha in one of the best forests, albeit "invaded with guava and other exotics". Also in 1946, Japanese Red Cedar was being planted in "indigenous pole forests"; one of these plots would later become the vital nesting site of 'Pigeon Wood' (p. 164)[19].

Since most native forest felled was for plantations, the big increase in land under cane between 1910 and 1923, driven by rapidly rising sugar prices, must have come from secondary scrub/forest on long abandoned cultivation, though sugar gains around Baie du Cap were probably from native forest. Land shown wooded on the Colonial Survey map in 1905 in western and northern coastal areas and parts of Flacq, and subsequently cultivated, was certainly not native forest, as these areas had been cleared in the 18th century, and are shown unforested on the 1854 Maisonneuve map. However they were no doubt home to wildlife, and their clearance would explain Huron's observations on deer and doves mentioned above (p. 156). Sugar prices collapsed in the mid-1920s, and land under cane fell back, not rising again until the late 1940s[20].

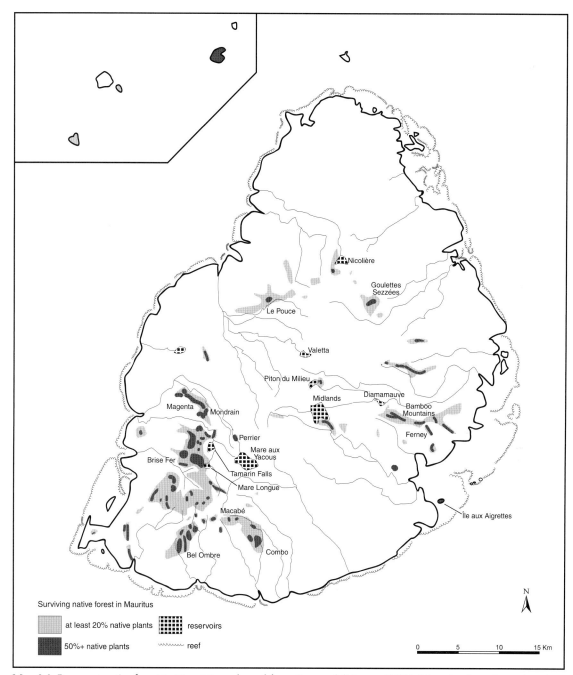

Map 8.2. Remnant native forest in Mauritius; adapted from Page and d'Argent (1997). Note that in order to show it at all this scale, the amount of passable/good forest (i.e. with more than 50% native woody plants) has to be exaggerated, thus there is even less in fact than the tiny amount shown.

Biological control: vertebrate failures, insect successes

Meanwhile, in 1911 a new sugar-cane pest erupted, a scarabeid beetle *Phyllophaga* (= *Phytalus*) *smithi*[22], provoking a new round of attempted biological control. The agents were mostly fungi and parasitic insects (which were ultimately successful), but two toads were also released for this purpose. The small Guttural Toad *Bufo gutturalis*, successfully introduced from South Africa in 1922 by Gabriel Regnard (the Red-whiskered Bulbul man again)[23], failed to feed on the beetle, so Cane Toads *B. marinus* were brought in from Puerto Rico. Although considered effective in controlling some pests, this animal was well-known to be predatory and poisonous, and the first batch in 1933 was destroyed on arrival. This caution was overridden in 1936 by the Government's Phytalus Investigation Officer, W. F. Jepson, who enthused about apparently successful control by Cane Toads of related beetle pests of sugar in Puerto Rico. Following Jepson's strong recommendation, 80 imported toads, followed by locally bred tadpoles and immatures, were released in 1938; when these failed to establish, a further 164 were set loose in 1950–51, equally unsuccessfully. As Cane Toads have become all too easily established in tropical countries and islands elsewhere, most famously (and disastrously) in Australia, it is odd (though fortunate) that they failed in Mauritius[24].

The successful toad was generally thought to be relatively harmless until it was shown in the 1990s to feed extensively on native snails, and may have caused the extinction of many endemic species[25]. Another cryptic menace is the ubiquitous Cheechak *Hemidactylus frenatus*, which appears to have eliminated native night-geckos from some offshore islets – at least, where there are Cheechaks, the night-geckos are, with one exception, absent[26]. On Ile aux Aigrettes there is competition between Cheechaks and Vinson's Day-gecko, the former breeding more quickly and able to occupy habitat faster. Although termed a 'house gecko', the Cheechak is versatile enough to occupy almost any habitat from houses through forest to bare rocks on islets[27].

In addition to toads, the introduction of Lesser Hedgehog Tenrecs *Echinops telfairi* as beetle predators was tried in 1932, but the animals all died before release. The long-established Common Tenrec was dismissed at the time as of no benefit to agriculture, though more recently was said to be "a great destroyer of *vers blancs* [beetle larvae], but is kept down to small numbers by the mongoose introduced in 1899 [*sic*] to destroy rats"[28]. The release of Wattle-necked Softshell Turtles *Palea steindachneri* at Moka around 1920 was apparently accidental, with captives escaping during a flood. These large, aggressive terrapins have since spread throughout the lowlands, and could well have helped finish off the whistling ducks and Purple Swamphens already under pressure from mongooses and drainage. In addition to fish, frogs and crustaceans, softshell turtles are known to attack young waterbirds by approaching from underneath and dragging them down[29]. East African Box Terrapins (p. 134), on the other hand, have apparently died out[30].

While some early 20th century biocontrol attempts (mongoose, Lesser Hedgehog Tenrec, Cane Toad) were official acts of Government agencies, others (Bloodsucker, Guttural Toad) were effectively 'freelance' jobs. There is no evidence that d'Emmerez consulted anyone before importing Bloodsuckers from Réunion, and the almost total silence on the Guttural Toad's arrival suggests it was known to be clandestine. All countries, and most particularly ecologically vulnerable oceanic islands, need controls to prevent anyone with a madcap idea from importing whatever animals take their fancy. Indeed Mauritius introduced regulations to that end in 1905, and some attempt (entirely unsuccessful) was made to eradicate Regnard's toad; such laws have always been hard to enforce[31]. Also most important is the issue that if one sectional interest, in this case agriculture, is allowed a free hand to import animals to solve its particular problems, such animals may create devastating knock-on effects elsewhere[32]. The official introductions in these Mauritian cases were actually (mongoose) and potentially (Cane Toad) worse in their side-effects than the unofficial ones.

A classic example of the interaction of introduced plants and animals is the case of the invasive shrubby weed *Cordia curassavica* and the Red-whiskered Bulbul. Both arrived around 1890, the plant probably with sugar-cane varieties from Guyana, the bird escaping from an aviary (pp. 135–136). *Cordia* was originally encouraged as some parasites introduced to control the cane borer fed on its sugary secretions. The bulbul soon revealed a particular liking for its fruit; the connection so obvious that the plant even acquired the bird's name: *oiseau condé* to *herbe condé*. Already an agricultural nuisance by 1910, *Cordia* seeds were so efficiently disseminated by bulbuls that, free of the limiting factors in its homeland, it invaded everywhere, reaching densities of 10,000 plants per hectare on rough ground in the lowlands. By the 1940s manual control in cane-fields and other crops became impossible. After studies by Octave Wiehe in its native tropical America, and tests that proposed biological control agents would not have unwanted side-effects (at last!), insects were imported that succeeded in controlling the weed by 1952–53[33]. The near-disappearance of the plant unfortunately had no effect on bulbul numbers – they simply shifted to other food sources.

Beginnings of ecological study

The 1930s saw the beginning of ecological studies on the island, from which developed the first nature reserves. Reginald Vaughan, who had come to Mauritius in 1923 to teach chemistry at the Royal College, developed an interest in the vegetation. In the 1930s, with his student Octave Wiehe, he made the first adequate survey of the remaining native vegetation, also investigating the exotic communities that had become established. They made a very detailed analysis of a plot in Macabé forest, chosen as the best surviving example of upland rainforest[34]. Although in their papers they attempted to separate the structure of native forest from the secondary vegetation dominated by introduced species, they clearly saw how the forest was being degraded and were concerned for its survival. H. C. King, the forester working from the late 1930s onwards on regenerating native trees, implicitly acknowledged Vaughan as motivating his studies[35]. When native forest was under threat during the 1939–45 war, Vaughan lobbied for its protection, being rewarded by an ordinance in 1944 that provided, *inter alia*, for the setting up of a heritage board and reserves. In 1946 he was appointed the first Director of a reinvigorated Mauritius Institute, chairing the Ancient Monuments and National Reserves Board set up under the 1944 Ordinance[36]. The Board created the first nature reserves in Mauritius (proclaimed in 1951), selected to preserve different types of native vegetation. Wildlife was not formally taken into account, though in 1950 Vaughan did express the hope that "if properly managed [the reserves] would serve as important sanctuaries for what remained of our native bird-life"[37]. In practice only two of the reserves (Macabé and Bel Ombre) were big enough to help birds or bats, though by the time the birds were more thoroughly studied in the 1970s neither proved to be the best areas for most of the surviving endemics[38].

What Vaughan and Wiehe did not do was look at forest dynamics. They were well aware of the damaging effects of invasive plants, but paid little attention to how introduced animals might degrade the ecosystem. Although monkeys were known to damage fruit of some trees, and even to defoliate Colophanes, Thompson's strictures on deer, snails, beetles etc. in the 1880s were long forgotten by the Forest Service of the 1950s. As recently as the 1940s, King had noted severe deer attack on native seedlings in plantations, though he stated that "in natural forest attack is inconspicuous"[39]; this 'inconspicuousness' is especially true for those unfamiliar with deer-free forests, as many Mauritian foresters have been[40]. Indeed in the 1960s the main complaint against deer was that the hunting leases in Crown forest prevented foresters from freely shooting monkeys damaging pine saplings![41] In 1969, when the government considered exploiting deer as an additional source of meat, field studies by a New Zealand deer ecologist, the then Conservator of Forests and a German deer-farming consultant looked at diet, but noticed no adverse effect on the forest. John Procter's otherwise excellent conservation report in 1975 concluded (echoing the foresters) that deer had little impact[42]. However Mauritian deer density hopelessly exceeded levels considered acceptable in managed European forests or native forest in New Zealand[43], and much patient citing of data from around the world eventually brought home the fact that deer (and pig) damage, though inconspicuous, was likely to be insidious and all-pervasive in Mauritian native forest[44]. By the early 1980s the negative impact of deer and pigs as well as monkeys was beginning to be accepted, though rats and giant snails were still underrated[45]. Finally addressing the issue of browsers, it was proposed in the 1980s that deer be reduced to under 1 per km^2 in the (then) proposed Black River National Park, and current policy is to bring deer and pigs in the park down to "near-zero numbers"[46]. Definitive work, comparing the effects of different browsers, seed predators, fungal parasites etc. on forest dynamics, is still yet to be done[47], but the conservation exclosures first established in the 1980s, and extended since (pp. 239–40), show that keeping out pigs and deer, combined with weeding invasive plants, allows spectacularly improved forest regeneration[48]. Exotic animals also interact among themselves; we saw in Chapter 7 how mongooses reduced tenrec numbers and eliminated several gamebirds, and more recently rats have been found to keep shrews down; indeed, on Ile aux Aigrettes, conservationists were unaware of shrews until Ship Rats were eliminated in 1991, after which they were suddenly everywhere, reaching a density of 30 per hectare[49].

After Meinertzhagen's visit (1910–11) and Paul Carié's final departure around the same time, Mauritian birds were little documented until the late 1940s. Although Georges Antelme was interested, and Réné Guérin wrote an unwieldly three-volume compilation on Indian Ocean birds, little of substance was recorded[50]. Memories from elderly nature-lovers and hunters have been gleaned subsequently[51], but the status of birds was based on rumour, hearsay, and even gross error[52] until Frank Rountree did some limited investigative bird-watching during 1949–52[53]. His observations fleshed out a checklist he co-authored in 1952, which formed the basis for ornithological work thereafter[54]. However even Rountree's distributions are generally rather vague, reflecting the absence of serious surveys. In 1950 he proposed long-term studies on rare birds, setting up an ornithological body in Mauritius and publication of a popular bird identification guide[55]. Although the RSAS did establish an ornithological sub-committee, it was 25 years before the other two aims were fulfilled.

The ornithological sub-committee, which became in 1955 the Mauritius branch of the ICBP, concentrated on improving legal protection of Mauritian birds, but appeared resigned to losing the forest outside the early nature reserves[56]. As the then Colonial Secretary, Robert Newton, was a keen bird-watcher, the committee were able to instigate new legislation, introduced in 1957 – though Newton himself saw little direct threat to birds from people, except on

BOX 16   DODOS (COLUMBIDAE) I

Dodo *Raphus cucullatus*

The Dodo has been the subject of more debatable literature than almost any other bird, yet little is known about its ecology. It was first mentioned in the literature in 1599 and first illustrated in 1601, but died out by the second half of the 16th century. A few specimens (dead and alive) were exported from Mauritius, one of which ended up as the unique stuffed head and foot that has survived to the present day in Oxford. As many accounts exist and they have been reproduced on numerous occasions in other literature (e.g. Wissen 1995, Fuller 2002), only three accounts are mentioned here. The first is by the Dutchman Jolinck, who was sent to survey the Mauritius coastline a few days after the island was discovered in September 1598. The second, albeit short, has rarely been reprinted and constitutes the only account that mentions a (possibly) juvenile Dodo. The third was written by an anonymous sailor, which beautifully epitomises the vulnerability of Dodos to humans, yet also illustrates their ability to defend themselves if approached too closely.

Accounts
Jolinck in Mauritius in 1598 (in Keuning 1938–51, translated by Henk Beentje).

> . . . also we found large birds, with wings as large as those of a pigeon, so they cannot fly, and by the Portuguese called penguins. These birds have a stomach so large that 2 men can do a wonderful meal with it, and it is also the best-tasting part of the bird. Also we found here many kinds of birds such as pigeons, herons, water-snipes [= waders], grey parrots; Indian river woodcocks [= Red Hens] as large as a hen; raboforcadoes [= frigatebirds] were many here, but not very good to eat and this bird's wings were so long that I could not measure them. We found here also handsome geese, ducks, cranes [= flamingos] and more, other birds so tame that most of the ones we caught were hit with a stick.

Manoel d'Almeida in 1617, on the Mauritius Dodo (from Grandidier *et al.* 1903–20; also Verin 1983):

> We took in the island an ostrich still very young, although already larger than an ordinary turkey.

Anonymous account from 1631 published by Servaas (1887). Translated by Henk Beentje:

> These mayors [Dodos] are superb and proud they displayed themselves to us with a stiff and stern face and wide open mouth, very jaunty and audacious of gait and would scarcely move a foot before us, their war weapon was the mouth, with which they could bite fiercely, their food was raw fruit, they were also well-adorned, but were abundantly covered with fat, and so many of them were brought aboard, to the delight of us all.

Specimen collections
Most museums around the world hold at least some fossil Dodo material, almost all of which was obtained from the Mare aux Songes marsh on Mauritius. Etienne Thirioux collected the only known collection of bones from a single individual, from caves on Le Pouce, Mauritius, and this specimen can be seen on display at the MI. The remains of the unique stuffed Dodo (head and foot) are held at the UMO.

Dodo *Raphus cucullatus*. A painting based on a live Dodo by Mansur c. 1625.

Dodo *Raphus cucullatus* head study. From Strickland and Melville (1848).

Round Island (Chapter 9). Like the local naturalists, Newton felt habitat loss was unavoidable: "these birds also suffer from the changes to their habitat due to the inescapable need for economic development to provide a living for an over-large population: for instance by the expansion of tea production and by the continued spread of sugar cultivation"[57]. Nonetheless, in the 1950s and 1960s the degraded upland forests still held fair numbers of endemic passerines, though there was more concern about the larger birds[58]. Legal protection was extended to endemic reptiles in 1973, but the Black-spined Flying-fox was not protected until the inclusive Wildlife Act in 1983[59].

Tea as an additional cash crop

The 'wind of change' that swept the globe in the aftermath of the 1939–45 world war did not pass Mauritius by. The island emerged well aware of its vulnerability to a one-crop economy, inadequate wood reserves, and an excessively high human population. During the war, to keep everyone fed, 27% of sugar land was appropriated for food crops. Rice, coconut oil and meat imports from, respectively,

Burma/India, Agalega and Madagascar ceased, and people were forced to eat home-grown maize and potatoes as staples[60]; local wood, both native and planted, had to be harvested (p. 158). The population, which had more or less stabilised in the 1880s, began to rise again in the 1920s, and by the mid-1940s was beginning to rocket, largely due to a declining death rate as malaria was overcome and sanitation improved. Unemployment was high, and it had become a priority to find labour-intensive, but also profitable, work. And what could be more labour-intensive than tea?

Tea was brought to Mauritius from China around 1770, but was grown only as a minor crop until a research plot was established in 1886 (extended in 1893), using seed from Ceylon. During the 1939–45 war, when imports became unreliable, acreage was rapidly expanded to satisfy local demand. In 1944 Alfred North-Coombes, then Senior Agricultural Officer, proposed that 'scrub-lands' on higher ground be developed for tea as an export crop, to diversify the economy and provide employment. In 1950 an expert from Ceylon identified some 40,500ha, including some under sugar, as suitable for tea[61]. The Forest Department naturally regarded this with some alarm, but combined with the Agriculture Department to develop a policy combining forestry requirements (both for protection and timber) with the new thrust for tea. Allan and Edgerley's report suggested releasing only 6,186 acres (2,503ha) of forest land for tea, pointing out that some 40,000 acres of sugar land, and another 7,000 of 'unused or undeveloped' land, all in private ownership, was also available[62]. However by 1955 economic pressures persuaded the legislature to seek up to 15,000 acres (6,070ha) for tea, and in 1957 a further report aimed to divert 9,000 acres (3,642ha) from forestry. By 1966 the Forest Department had been relieved of 6,273ha intended for tea plantations, and with it any idea of watershed protection. However, in 1968 only some 2,023ha of Crown forest land had actually been planted to tea (with about 1,200ha from other land)[63]. The total under tea reached a peak of some 10,000 acres (4,221ha) around 1971[64], so much allocated forest was never taken up. Nonetheless, a swathe of degraded native forest, bird and bat habitat, was lost in a triangle from Curepipe in the west, Quartier Militaire in the north and Montagne Table in the east[65] – an area equivalent to all the surviving native forest in the current Black River Gorges National Park[66]. This was all ultimately for nothing; in the 1990s the tea industry collapsed under a manpower shortage and an inability to compete on quality in world markets. In July 1996 bulldozers were active in the Piton du Milieu area, destroying native forest fragments as well as tea. From 3000ha in 1994, by 1999 some 2,300ha had been ripped out: converted to sugar or abandoned[67]. One large area originally earmarked for tea, the Parc-aux-Cerfs block south of Curepipe, retained its degraded forest (and native birds) until 1974, when it was lost to the next major attack on the forests[68]. Scrubland at Midlands cleared for a projected reservoir in the 1920s (p. 158), by the 1980s surrounded by tea plantations, finally fulfilled its destiny in 2003 with the inauguration of a huge new dam and storage lake[69].

Travail pour tous

The prosperity of the 1990s seemed unattainable in the early 1970s, when the newly independent government was struggling with overpopulation and chronic unemployment. Industrial development was in its tentative early stages, so planners, running out of steam on tea, returned to the forests for a new make-work scheme. The tea lands had been largely prepared by 'relief workers' in an earlier workfare scheme, reworked in the 1971–75 Four-year Plan into a programme called *Travail pour Tous* (Work for All)[70]. The programme included further continued land clearance and planting, using the same techniques as before, but replacing 'scrub' with pines instead of tea[71]. Like the tea scheme it was funded by the World Bank or its offshoots[72], at a time before environmental impact statements were required. A new body, the Development Works Corporation (DWC), created to run the workfare projects, was given control of the forestry scheme, despite the existence (and experience) of the long-established Forest Department[73]. Administration with no local knowledge was brought in from outside (India), and land that the Department had preserved untouched as protection forest was removed from its control, to be returned only after the new plantations were in place. The native forest areas given over to the DWC included Les Mares, the swampy headwaters of the Black River and Rivière du Poste at the head of the Gorges, the Kanaka-Bois Sec area and the Parc aux Cerfs block referred to above, amounting in all to over 2,000ha, or about two-thirds of the remaining native (good and degraded) plateau forest that had escaped the tea clearances[74]. The loss of Les Mares, linking the Savanne range with the Macabé-Brise Fer forests was a disaster for the birds, though the full impact was not immediately apparent, as the destruction took place just before critical ecological studies began. This was the very area that in 1929 Tempany had been so keen to preserve for its water-retaining capacity.

In the wake of international attention arising from the studies described below, and especially Gerald Durrell's high profile involvement from 1976, the government moved from policies that effectively ignored the ecosystem to ones which embraced a responsibility to preserve it. In the post-independence optimism, the 1971–75 Four-Year Plan did call for a national park and efforts to preserve the native flora and fauna, but equally made it clear there was no money forthcoming. By the mid-1970s recession (after the Scott and Procter reports had been issued) the park plan had dwindled in the 1975–80 plan (dated 1976) to a two-line comment that "part of the area under forest reserves will be converted into a National Park". However, there was movement, and although the 1984–86 plan did not mention conservation at all, the government had in 1983 approved a formal and costed *Wildlife research and conservation programme* drawn up by ICBP and the Jersey Wildlife Preservation Trust, and passed a new and comprehensive Wildlife Act[75]. Despite this progress a new road was driven in the mid-1980s from Les Mares past Bassin Blanc towards the south coast, allowing woodcutters with lorries to ruin the formerly near-pristine surroundings of the crater lake[76].

Status of native land vertebrates

In 1973 two teams from the USA and UK started programmes of ecological research on Mauritian birds[77]. Their census work during 1973–75 revealed, despite the ongoing loss of forest, apparently viable populations of endemic passerines, despite small numbers, though the position of the three larger species was dire. However, it later became clear that all the birds displaced from recently cut forest had made the passerine situation look much better than it really was[78]. The densities in the remaining forest in 1973–75 were not sustainable, and when next censused in 1992–93, with little change in forest area, numbers of Mauritius Fodies, Olive White-eyes and Mascarene Paradise Flycatchers had plummeted. Cuckoo-shrikes and Merles, on the other hand, were stable or slightly increasing – the Mauritius Merle is a relatively adaptable species not confined to native forest[79]. The picture was grim in the mid-1970s for the Mauritius Kestrel, Pink Pigeon and Echo Parakeet; kestrels were down to single figures (6 individuals), pigeons around 25–30 and parakeets 30–40[80]. The kestrel had already hit rock-bottom for reasons unconnected with forest loss (see below), but the other two, especially the pigeon, were badly affected. Cyclone *Gervaise* in February 1975 then halved pigeon numbers, but unlike the aftermath of *Carol* in 1960, they did not recover. Later research indicated that losing Les Mares was critical for the pigeon, fody and Mauritius Olive White-eye; indeed, the Pink Pigeon's close association with that area had earned it the local name *Pigeon des Mares*[81]. Detailed studies on pigeon and fody nesting habits and success then showed that only a tiny proportion of nests survived predation by rats and monkeys, and that success depended on nesting in areas unattractive to the predators. Since the 1970s the only such site has been a tiny plantation of Japanese Red Cedars, known as 'Pigeon Wood'; neither monkeys nor rats like climbing about in these trees, and their thick foliage hides the nests[82]. Les Mares was a very wet swampy forest, where much of the vegetation was on rocky islets interspersed with marsh, or was screw-pines growing directly out of water – unpleasant habitat for the predators, and now thought to have offered nesting birds similar protection to that now found in Pigeon Wood[83]. An apparent invasion of upland forest by Norway Rats, previously scarce in native vegetation, may be related to the increased human activity, as may an upsurge in feral cats, lethal to fledgling Pink Pigeons, in both plantations and adjacent native forest[84]. Another Pink Pigeon predator, the mongoose, prefers scrubby habitat to tall forest[85], so also probably increased following the clearances. Les Mares was also important for feeding: although parakeets did not nest, they regularly foraged there, and it was rich in nectariferous plants favoured by the olive white-eyes[86]. Its 500ha or so were also strategically placed in the centre of the then remaining native forest area.

The Echo Parakeet's continued decline after 1975 was exacerbated by the loss of Les Mares, but it was more directly related to the long-term decline in overmature trees with suitable nesting holes. By the 1970s the few remaining sites were subject to intense competition from Ship Rats, mynas, White-tailed Tropicbirds, Ring-necked Parakeets, Mauritius Kestrels and bees, rats also being the worst predator; nests also failed through infestation by tropical nest-flies, probably introduced[87]. Like Pink Pigeons, but unlike the small passerines apart from the Olive White-eyes, Echo Parakeets range over large areas of forest seeking seasonally abundant fruits and foliage, so any reduction in total forest area will reduce carrying capacity; food shortage is a frequent cause of nest failure[88]. Like all frugivores (and nectar-feeders) needing a large minimum home-range, there is clearly some threshold (not yet known) below which the forest area becomes too small to maintain *any* Echo Parakeets.

To understand the Mauritius Kestrel's decline we have to go back to the late 1940s. At that time the bird was widespread throughout the mountains and hills of Mauritius. It was rare but not threatened, although sometimes shot because of an undeserved reputation for killing chickens, its local name being *mangeur de poule*[89]. In the 1950s it disappeared throughout its range except for the Black River Gorges, a change that was a mystery at the time[90]. However this contraction coincided with the revamped anti-malaria campaign from 1949 onwards, when spraying with DDT against mosquitos replaced the extensive drainage works previously practised. Although spraying was largely in buildings, breeding places such as marshes and ponds were also treated, and some DDT was used in agriculture. The area furthest from human habitation (and thus insecticide treatments) was the forest block in the southwest around the Gorges. By the 1970s it was well-known that organochlorine insecticides were particularly hazardous to birds of prey. Birds were not necessarily killed by DDT, but breeding was severely impaired through reduced fertility and thinned eggshells, which were easily broken. Mauritius Kestrels mostly eat day-geckos, themselves high up the food chain, so even if the geckos' insect food contained only small amounts of DDT, it would have accumulated in the lizards before the kestrels ate them. It is now thought that this poisoning, rather than predation or habitat loss, reduced the kestrels to tiny numbers in the 1970s. The Mauritius Cuckoo-shrike, an insectivore that also takes day-geckos, showed the same pattern and timing of decline, and both species have been able to re-establish in long-abandoned territory in the 1980s and 1990s, areas without habitat restoration or other discernible environmental improvement. Widespread insecticidal treatment was stopped in 1973 when the island was declared malaria-free, so the birds' subsequent recovery fits with DDT being to blame for the decline. The kestrel's resurgence was much assisted by captive breeding and release (p. 251), but their re-establishment in the Moka range and the Bamboo Mountains would have been impossible if toxic insecticide was still in use. The cuckoo-shrike's return to the lower Gorges, lower Bel Ombre and Combo has been entirely unaided, but they have not recolonised the Bamboo Mountains, probably due to a reluctance to disperse across unforested land[91].

The recovery of the three non-passerines under intensive conservation management is told in Chapter 10, but while they were seeing better days, some of the passerines, left to their own devices, did less well. The two rarest, the Mauritius Fody and Olive White-eye, were surveyed again in 1998–2001. Their range has continued to contract as isolated groups in scattered native copses fail to maintain themselves, a continuing legacy of the 1970s clearances. Since almost the only fodies successfully nesting in recent years have been in the Japanese Cedars at Pigeon Wood, almost all birds in the core population will have been raised there, and, perhaps in consequence, are increasingly using conifers for breeding – pines where there are no cedars. By 2001 Les Mares, by then covered in 25-year-old pines, had been recolonised by 21 pairs. Although the Bel Ombre population had shrunk by 75% since 1992–93 (from 20+ pairs to 5), expanding into Les Mares has kept the overall total stable for a decade, an optimistic sign[92]. A captive breeding programme begun in 2002 has enabled fodies to be released on Ile aux Aigrettes, where breeding has been so successful in the absence of rats that by February 2006 there were 132 birds[93]. In contrast, the Mauritius Olive White-eye has continued to decline, by at least 40% per decade, a clearly unsustainable fall; numbers have been maintained only in the Combo area. The precise reason remains unknown, though Les Mares appears to have been a productive ('source') area supplying birds to other areas ('sinks'), but unlike fodies, Olive White-eyes have been unable to find another good breeding place. Currently in their best area, Combo, the birds largely occupy dense patches of introduced Rose-apple; this is good for feeding (nectar and insects) but provides no protection against nest-predators. Red-whiskered Bulbuls, blamed for a decline in white-eyes in the early 1900s, may prove to be more important nest predators than rats. Olive white-eyes and merles, unlike grey white-eyes, never recovered their former abundance, still seen in their analogues in Réunion where rats are equally common (p. 189), and where bulbuls have yet to have an effect; detailed studies there are awaited. A captive breeding programme for olive white-eyes began in 2005, and releases on Ile aux Aigrettes took place in December 2006[94]. Common though it is, only 8% of grey white-eye nests produce fledged young; most fail through predation, with Red-whiskered Bulbuls the prime suspects[95]. Olive white-eyes even suffer competition for nectar from introduced honey-bees![96]

Of the native birds, Mascarene Grey White-eyes have adapted best to man-modified habitats, occurring wherever there is tree cover[97], though even in native forest this ubiquitous species has very low nesting success. The Mascarene Paradise Flycatcher, apparently in free-fall in native habitat since the 1970s, was discovered around 1970 in good numbers in forestry plantations at Bras d'Eau, some 16km north of the nearest good native forest in the Bamboo Mountains. This population has remained vigorous, nesting successfully in the spindly coppiced understorey under the shade of evergreen *Araucaria*, mango and other exotic trees in a plantation formerly free of monkeys; their recent appearance seems not to have affected the flycatchers[98]. Tiny groups of flycatchers are also found at intervals in other exotic wooded patches around the island, but these are often short-lived. Some groups may have hung on since before the native forest was totally destroyed, but the flycatchers seem to be better at dispersal than other native passerines, and can sometimes colonise new areas, though they are absent from much apparently suitable woodland[99]. In the native forest areas, only the Combo and Pigeon Wood populations (*c.* 35 pairs) appear truly healthy[100]. While elsewhere in the native forest flycatchers have declined or vanished, numbers recorded at Combo have increased steadily since the 1970s, and those at Pigeon Wood have

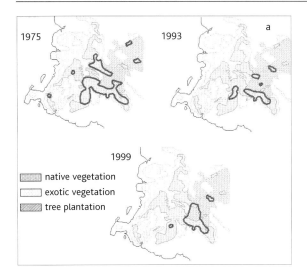
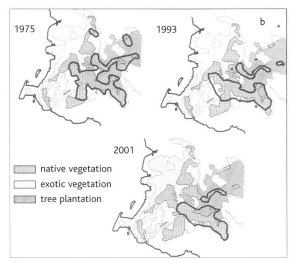

Figure 8.1. Maps showing the shrinking distribution of Mauritius Fodies (a) and Mauritius Olive White-eyes (b), 1975–1999/2001. From Woolaver (2000) and Nichols et al. (2004).

recovered after a crash in the early 1993; the disparities are not so far explained, though the Combo 'increase' may be due to intensive fieldwork revealing birds previously overlooked[101]. Small groups still turn up elsewhere, e.g. at Baie du Cap in 2003 and nearby Chamarel in 2004[102]. The flycatcher's nest is so easily found that it is remarkable that the species survives at all.

The two aerial feeders, the Mascarene Swiftlet and the Mascarene Swallow, have been less studied than the forest endemics, because they are shared with Réunion, and are not generally seen as endangered[103]. However, swiftlet numbers crashed between 1975 and 1985[104], and while they appear to have stabilised at the lower level (c. 750 pairs in 1996), the species faces serious problems. Together with the Mascarene Free-tailed Bat, it nests on the roof of underground lava tunnels, too many of which are easily accessible to people who use them for purposes incompatible with wildlife. Organised refuse collection is mostly urban, and it has long been a rural habit to dump rubbish down caves – out of sight, out of mind. At first, especially in larger caves, this is not a problem, but eventually the cave entrance fills up and is blocked, preventing birds and bats from reaching their nest and roost sites. Partly collapsed caves just north of Surinam were chosen as the official rubbish tip, blocking the entrances to intact tunnels in the same system. Use of caves for ritual purposes ('magic') is widespread, with practitioners lighting fires, often using rubber tyres, which produce toxic smoke. Ritualists and ordinary vandals also directly destroy the nests with poles[105]. Finally, the Chinese community likes birds'-nest soup – though demand has apparently diminished since the gourmet variety from Asia became readily obtainable, Mauritian nests being of a poorer quality and the supply unreliable[106]. Attempts to put bars across cave entrances have not been a success, as, undefended, they are soon broken into[107]. A more satisfactory approach has been to gain help and understanding of local residents. One of the caves near Vacoas where ASC did research in 1973–75 was still being looked after by local families in 1996, and the number of nests had nearly tripled[108]. An educational project in the 1980s involved local school children in cave conservation in the Roches Noires area – some at least of these caves were still in relatively good shape in 1996 and 2005, though bats were doing better than swiftlets. The bats come and go; a cave at Roches Noires with good numbers in 1996 had none in 2003, but huge numbers in 2005. This mobility makes it all the more important to investigate and protect the entire cave network[109]. Mascarene Free-tailed Bats are still very numerous (80,000–100,000 estimated in lava tubes in 1996, c. 250,000 in 2005), but they are as vulnerable as the swiflets to disturbance or blocking up of their roosting caves. As in Réunion, some also use much smaller cavities, such as on cliff faces, which are relatively safe[110]. An important report on cave conservation in 1998 has yet to make much impact; caves are not mentioned in the government's biodiversity conservation report of 2000, which did not even acknowledge the existence of microbats![111]

The Mascarene Swallow, nesting in scattered small colonies on cliffs and bridges throughout the south and west of the island, seems to be maintaining its numbers. Indeed, in 1996 ASC recorded the largest aggregation ever for the species in Mauritius (about 40 birds), ironically feeding over the same official rubbish tip that plugged the entrances to the swiftlet

BOX 17 DODOS (COLUMBIDAE) II

Rodrigues Solitaire *Pezophaps solitaria*

The Solitaire was closely related to the Dodo of Mauritius, but was quite distinct in shape and form. The Solitaire survived for longer than the Dodo, as Rodrigues's isolation prevented occupation by people until the arrival of Francois Leguat and his Huguenot followers in 1691, when the Solitaires were still common. The Solitaire was next mentioned by second mate Tafforet, while he and four other men were marooned on Rodrigues in 1725–26 for eight months. Tafforet said that they were still common, but Pingré, in 1761, was told that they still existed only in secluded places, although he never saw one. The speed with which the Solitaire disappeared can be correlated with the increased tortoise trade between Rodrigues and Mauritius/Réunion from 1730–1750. The tortoise hunters burnt off the vegetation and surely would have captured Solitaires at every opportunity for food. Cats and pigs had also been introduced by this time and may have been serious predators of eggs and chicks. Inevitably, probably by the late 1760s, the Solitaires had become extinct, as the discovery of the first bone material in 1786 did not evoke any recollection of a living bird from residents.

Accounts
Leguat (1707) in 1691, from the original English translation of 1708:

"*Of all the Birds in the Island the most remarkable is that which goes by the name of the solitary, because it is very seldom seen in company, tho' there are abundance of them. The Feathers of the Males are of a brown grey Colour: the Feet and Beak are like a Turkey's, but a little more crooked. They have scarce any Tail, but their Hind-part covered with Feathers is roundish, like the Crupper [rump] of a horse; they are taller than Turkeys. Their Neck is straight, and a little longer in proportion than a Turkey's when it lifts up his Head. Its Eye is Black and lively, and its Head without Comb or Cop. They never fly, their Wings are too little to support the Weight of their Bodies; they serve only to beat themselves, and flutter when they call one another. They will whirl about for twenty or thirty times together on the same side, during the space of four or five minutes. The motion of their Wings makes then a noise very like that of a Rattle; and one may hear it two hundred Paces off. The Bone of their Wing grows greater towards the Extremity, and forms a little round Mass under the Feathers, as big as a Musket Ball. That and its Beak are the chief Defence of this Bird. 'Tis very hard to catch it in the Woods, but easie in open Places, because we run faster than they, and sometimes we approach them without much Trouble. From March to September they are extremely fat, and taste admirably well, especially while they are young, some of the Males weigh forty-five Pounds.*

The Females are wonderfully beautiful, some fair, some brown; I call them fair, because they are the colour of fair Hair. They have a sort of Peak, like a Widow's upon their Breasts [Beaks], which is of a dun colour. No one Feather is straggling from the other all over their Bodies, they being very careful to adjust themselves, and make them all even with their Beaks. The Feathers on their Thighs are round like Shells at the end, and being there very thick, have an agreeable effect. They have two Risings on their Craws [crop] and the Feathers are whiter than the rest, which livelily represents the fine neck of a Beautiful Woman. They walk with so much Stateliness and good Grace, that one cannot help admiring them and loving them; by which means their fine Mein often saves their Lives.

Rodrigues Solitaire *Pezophaps solitaria* depicted by Francois Leguat in 1708. This is the only known illustration of the Solitaire drawn from life.

Tho' these Birds will sometimes very familiarly come up near enough to one, when we do not run after them, yet they will never grow Tame. As soon as they are caught they shed Tears without Crying, and refuse all sustenance till they die.

We find in the Gizzards of both Male and Female, a brown Stone, of the bigness of a Hen's Egg, 'tis somewhat rough, flat on one side and round on the other, heavy and hard. We believe this Stone was there when they were hatched, for let them be never so young, you meet with it always. They never have but one of 'em, and besides, the Passage from the Craw to the Gizard is so narrow, that a like Mass of half Bigness cou'd not pass. It serv'd to whet our Knives better than any other Stone Whatsoever. When these Birds build their Nests, they choose a clean Place, gather together some Palm-Leaves for that purpose, and heap them up a foot and a half high from the Ground, on which they sit. They never lay but one Egg, which is much bigger than that of a Goose. The Male and Female both cover it in their turns, and the young is not hatch'd till at seven Weeks' end : All the while they are sitting upon it, or are bringing up their young one, which is not able to provide itself in several Months, they will not suffer any other Bird of their Species to come within two hundred Yards round of the Place; But what is very singular, is, the Males will never drive away the Females, only when he perceives one he makes a noise with his Wings to call to the Female, and she drives the unwelcome Stranger away, not leaving it 'tis without her Bounds. The Female do's the same as to the Males, whom she leaves to the Male, and he drives them away. We have observ'd this several Times, and I affirm it to be true.

The Combats between them on this occasion last sometimes pretty long, because the Stranger only turns about, and do's not fly directly from the Nest. However, the others do not forsake it till they have quite driven it out of their Limits. After these Birds have rais'd their young One, and left it to itself, they are always together, which the other Birds are not, and tho' they happen to mingle with other Birds of the same Species, these two Companions never disunite. We have often remark'd, that some Days after the young leaves the Nest, a Company of thirty or forty brings another young one to it, and the now fledg'd Bird, with its Father and Mother joyning with the Band, march to some bye Place. We frequently follow'd them, and found that afterwards the old ones went each their way alone, or in Couples, and left the two young ones together, which we call'd a Marriage.

This Particularity has something in it which looks a little Fabulous, nevertheless, what I say is sincere Truth, and what I have more than once observ'd with Care and Pleasure.

Tafforet (1726), from A. Newton's 1875 translation:

The solitaire is a large bird, which weighs about forty or fifty pounds. They have a very big head, with a sort of frontlet, as if of black velvet. Their feathers are

> neither feathers nor fur; they are of a light grey colour, with a little black on their backs. Strutting proudly about, either alone or in pairs, they preen their plumage or fur with their beak, and keep themselves very clean. They have their toes furnished with hard scales, and run with quickness, mostly among the rocks, where a man, however agile, can hardly catch them. They have a very short beak, of about an inch in length, which is sharp. They, nevertheless, do not attempt to hurt anyone, except when they find someone before them, and, when hardly pressed, try to bite him. They have a small stump of a wing, which has a sort of bullet at its extremity, and serves as a defence. They do not fly at all, having no feathers to their wings, but they flap them, and make a great noise with their wings when angry, and the noise is something like thunder in the distance. They only lay, as I am led to suppose, but once in the year, and only one egg. Not that I have seen their eggs, for I have not been able to discover where they lay. But I have never seen but one little one alone with them, and, if any one tried to approach it, they would bite him very severely. These birds live on seeds and leaves of trees, which they pick up on the ground. They have a gizzard larger than the fist, and what is surprising is that there is found in it a stone of the size of a hen's egg, of oval shape, a little flattened, although this animal cannot swallow anything larger than a small cherry-stone. I have eaten them: they are tolerably well tasted.
>
> Gennes (1735):
>
> *Our men told of having seen goats and a large quantity of birds of different kinds: they brought, amongst others, two of which were bigger by a third than the largest turkey; they appeared, nevertheless to be still quite young, still having down on the neck and head; their wingtips were but sparsely feathered, without any proper tail. Three sailors told me of having seen two others, of the same species, as big as the biggest ostrich. The young ones that were brought had the head made more or less like the latter animal, but their feet were similar to those of turkeys, instead of that of the ostrich which is forked and cloven in the shape of a hind's foot. These two birds, when skinned, had an inch of fat on the body. One was made into a pie, which turned out to be so tough that it was uneatable.*
>
> Cossigny (1732–55) in 1755:
>
> *. . . for 18 months I have been trying without success to procure a solitaire from Rodrigues Island. . . . I have promised all one could want, in spirits or piastres, to whoever brings me at least one alive. It is claimed that cats, which have gone wild on this little island, have destroyed this species of bird that only has stumps for wings, but I am strongly inclined to believe that these cats are the men of the post who have eaten all those they have found, as they are good to eat. At last, I have been given hope of obtaining one, which, so it is said, has been spotted.*
>
> d'Heguerty (1754), around 1735:
>
> *One finds there* [in Rodrigues] *birds of various species that are commonly taken by giving chase, and amongst others Solitaires which have almost no wing feathers. It is a bird bigger than a swan, with a melancholy appearance. Tamed* [probably meaning simply 'captive'], *one sees them always walking the same line till they run out of space, then retracing their steps without deviating. Cut open, one normally finds* Bézoards [internal stones] *which are valued and useful in medicine.*
>
> *Specimen collections*
> Mounted skeletons can be seen at a number of museums, including UMZC, BMNH, MNHN, and the MI.

caves. However, the total population is small, in the low hundreds at best, and colonies under bridges are vulnerable to vandals[112].

After centuries of being shot as game, the Black-spined Flying-fox was finally given legal protection in 1983. Within the national park at least, this seems effective, with good numbers roosting in the lower Black River Gorges, Combo and in Bel Ombre but pressure from fruit-growers led the government to permit limited culling in 2006. The Grey Tomb Bat has always been elusive, with the only known roosting sites holding insignificant numbers, though in August 1996 30+ could be seen hunting at dusk from the pier at Black River. Almost all records are from Black River – Tamarin area, but the bats can be seen inland on the slope up to Vacoas[113]. Apart from rather random collecting of museum specimens, nothing was done on the lizards until Jean Vinson started investigating, first on offshore islets and then on the mainland, from the 1940s onwards. This turned up several new species and culminated in a monograph published in 1969[114]. On the mainland the common day-gecko was revealed in the 1960s to actually consist of three species (now extended to four), and a new small skink turned up in 1973 in Macabé Forest[115]. Since then distributional studies have shown that the day-geckos are all widespread and common within their habitats, and that Macabé Skinks, originally thought rare, are widely distributed in upland forest[116]. On the coast, sporadic nesting attempts by Green and Hawksbill Turtles, formerly so abundant, were reported in the mid-1970s until at least 1987, but little seems to have been formally recorded; turtles have been protected by law since 1980, but are still caught and killed by fishermen when found[117].

Further introductions and status of earlier arrivals

After the disasters earlier in the century, one might have expected more caution on introductions. No such luck. Although legal constraints are in place, they have not proved effective in practice. Great Green Day-geckos *Phelsuma madagascariensis*, Laughing Doves *Stigmatopelia senegalensis* and three species of wildfowl escaped or were deliberately released in the 1980s and 1990s. The focus for most of these has been a commercial zoo, Casela Bird Park, off the main orbital road near Flic en Flac. This was designed in 1976 by the then leader of the captive-breeding project, moonlighting for the sugar estate that financed it[118]. The risk of birds escaping in cyclones due to cages being built around trees was soon noted[119], but it was not anticipated that they would deliberately *release* species they bred in excess, some of which, Laughing Doves, the day-gecko, White-faced Whistling Ducks and Egyptian Geese, have become established in the zoo's vicinity[120]. Over 100 Diamond Doves *Geopelia cuneata* were also released from a private aviary at Black River in the 1976, but failed to take hold[121]. Great Green Day-geckos have also been released by others; there is a colony in Floreal, a suburb of Curepipe a long way from Casela, and other groups elsewhere[122]. Mallard *Anas platyrhynchos* were released in 1979 on several upland waters for hunting purposes. They appeared well-established at Tamarin Falls reservoir in 1985, and by the early-1990s outnumbered the long-established Meller's Duck[123]. None of these releases had any official sanction, yet there has been no attempt to eradicate any of them. Doves and wildfowl will probably not affect the native fauna, and indeed the Egyptian Goose could be seen as a valid (if unintentional) ecological replacement for the extinct endemic sheldgoose[124]. The day-gecko is another story; *Phelsuma madagascariensis* is a large, aggressive lizard, likely to prove a major problem for the endemic day-geckos if it spreads. Popular pet terrapins, notably Red-eared Sliders, are imported in some numbers, and the sliders appear to be established in ponds attached to west-coast tourist hotels and a lake at Palmar (east coast). A second Indian blind-snake, discovered in recent years, may have been present undetected for a long time[125]. It remains to be seen whether the recently established Invasive Alien Species Committee[126] will have more success than past initiatives in reining in feral citizens and over-enthusiastic would-be biological controllers. The current distribution of introduced reptiles on Mauritius and associated islets is given in Table 8.1

In addition to the DIY introductions, there was an official biological control initiative aiming to control giant snails. In 1959 the Department of Agriculture imported Rosy Wolf Snails *Euglandina rosea* from Hawaii and the Caribbean, and in 1961 another predatory species *Gonaxis quadrilateralis* from East Africa. The former were bred up artificially and released in batches from 1960 until 1969, and also distributed to Réunion and Rodrigues in 1966[127]. It is now well established that Rosy Wolf Snails do not control *Achatina*, but instead often devastate native and endemic species [128]. In Mauritius they do not appear even to eat giant snails, let alone control them, though they arrived too late to be blamed for the large number of endemic snail extinctions which preceded their introduction, for which rats, shrews, toads, and tenrecs were probably responsible. Nonetheless, Rosy Wolf Snails eat native snails preferentially, increasing the pressure on their survival, but are relatively scarce in areas of good native forest, apparently discouraged by the presence of native carnivorous snails[129]. Recent investigations have shown that the two giant snails occupy different habitats; *Achatina fulica* is found in primarily agricultural and urban areas, while *A. immaculata* occurs in forests, both native and secondary[130].

Introductions from previous centuries have had mixed fortunes. Ground-nesting gamebirds and waterfowl were already retreating fast in the face of the mongoose by 1914 (pp. 136–137). Madagascar Partridge and Jungle Bush Quail probably did not survive into the 1920s, but Painted Quails lingered into the 1940s, while Chinese Francolins were sighted as late as 1956[131]. Only Grey Francolins and Helmeted Guineafowl survive; the former remains common, but the latter persists only in small groups in the Yemen/Black River Gorges area; Common Quail are maintained only by artificial releases[132]. The waterbirds lasted longer, perhaps because their habitat offered some protection from mongooses, though not from drainage for malaria control. White-faced Whistling Duck (a tree-nester) survived in small numbers until around 1956, later re-installed in the 1990s with new releases. Purple Swamphens were last recorded in 1951[133]. Meller's Ducks, unlike the others living on upland streams and reservoirs, survived better, but have apparently died out since Mallards arrived; the last confirmed record was in 1992. In 1977, Mauritian birds were taken for captive breeding in Jersey, since the parent population in Madagascar is considered endangered[134].

Java Sparrows and Red Avadavats had already died out around 1900 (p. 136); post-1904 records of Red Avadavats appear to have been escapes or re-releases[135]. Cape Canaries, although reduced to a rump by 1913, apparently lingered into the early 1920s at the Trou aux Cerfs, Curepipe, eventually succumbing to intense trapping for the cage-bird trade; new releases were reported as unsuccessful in 1953[136]. Grey-headed Lovebirds, still common in 1916, declined rapidly, surviving in small numbers until about 1950[137] – the common factor in both these last two may have been the spread of Village Weavers. In a throwback to 18th century concerns, 'birds' (unspecified) were regarded in 1948 as a problem in the small area devoted to rice cultivation; hares and deer had to be fenced out[138]. The newer arrivals, Red-Whiskered Bulbul, Village Weaver and Ring-necked Parakeet, continued their expansion, the bulbul becoming and remaining the island's commonest bird in almost all habitats, though one of its early victims, the large orb-web spider, was by 2005 showing some recovery in the forests[139]. Village Weavers mostly keep to inhabited lowland areas, while Ring-necked Parakeet remained restricted to the Grand Port/Mahébourg and Pamplemousses areas until the 1940s, then expanded first across the lowlands then into the cultivated uplands by 1956[140]. Only then did it start penetrating native forest seeking tree-holes to nest in, for which they compete with Echo Parakeets. Ring-necked Parakeets

Table 8.1. Past and present distribution of introduced reptiles on Mauritian islands. Data derived from Bell *et al.*(1993), Cole (2005), Cole *et al.* (2005) and Carl Jones & members of MWF 1993–2005. Symbols: ✻ = feral population present; X = feral population extinct; C = captive only.

Species	Mauritius mainland	Flat Island	Ile Gabriel	Pigeon Rock	Gunner's Quoin	Round Island	Serpent Island	Ile aux Aigrettes	Ile aux Singe	Ile de la Passe	Ilot Vacoas	Ile aux Fouquets	Ile d'Ambre	Ile Bernache	Ile aux Cerfs	Ile Mangénie (= de l'Est)	Ile aux Benitiers	Ilot Fourneau
Aldabra Giant Tortoise *Dipsochelys dussumieri*	C	X						✻										
Wattle-necked Softshell Turtle *Palea steindachneri*	✻																	
Red-eared Slider *Trachemys scripta*	✻																	
African Box Terrapin *Pelusios subniger*	?X																	
Panther Chameleon *Furcifer pardalis*	✻																	
Cheechak *Hemidactylus frenatus*	✻	✻	✻					✻	✻	✻		✻	✻	✻	✻	✻	✻	✻
Indian House Gecko *Hemidactylus brooki*	✻																	
Common Worm Gecko *Hemiphyllodactylus typus*	✻							✻									✻	✻
Stump-toed Gecko *Gehyra mutilata*	✻	✻						✻				✻					?✻	
Square-toed Gecko *Ebenavia inunguis*	✻																	
Great Green Day-gecko *Phelsuma madagascariensis*	✻																	
Bloodsucker *Calotes versicolor*	✻							✻						✻	✻	✻	✻	✻
Wolf Snake *Lycodon aulicus*	✻							✻									✻	
Flowerpot Snake *Ramphotyphlops braminus*	✻	✻	✻					✻				✻			✻			✻
Slender Worm Snake *Typhlops porrectus*	✻								✻							✻		

in the secondary forest periphery may inhibit Echo Parakeets from using those areas, but there is little or no direct food competition. Ring-necked Parakeets were formerly a serious pest of maize plantations in the Case Noyale/La Gaulette area near Le Morne, but these have been abandoned for sugar for many years; the effect of losing this maize is unknown, but the birds have very catholic feeding habits[141]. House Crows expanded less vigorously, with only two foci (Roche Bois and Case Noyale) by 1939. The 1945 cyclone almost wiped them out, but more birds arrived off a ship in 1950, numbers gradually expanding to cover much of the urban and suburban north of the island, with a large roost in the popular Company's Gardens in central Port Louis. The population, around 100 birds in 1976, grew to c. 850–1,000 by 1988, and c. 6,000 by 2002; an eradication programme due to start in 2002 made no visible impact on the Port Louis roost by mid-2003[142].

The mongoose remains as abundant as it so quickly became after its introduction. Although its density in native forest is relatively low, and it feeds largely on mammals and invertebrates, it is nevertheless (with cats) a persistent predator of ground-feeding (and congenitally naïve) Pink Pigeons, so is selectively controlled (Chapter 10)[143]. The Wolf Snake remains widespread if inconspicuous in lowland areas, and its presence on Ile aux Aigrettes is a bar to reintroducing endemic lizards there – attempts to eliminate it have failed[144].

Scapegoated for centuries of wildlife abuse, the monkeys finally got their own study in the 1980s – the results disconcerting conservationists by apparently absolving the monkeys from guilt[145]. However the researchers were more interested in how the immigrants had adapted over time to their new environment than in conservation issues. Finding monkeys most abundant in secondary scrub, they concentrated on this habitat. In native forest monkeys were relatively scarce[146], and while some damage to endemic trees was recognised, the study, spread discontinuously over several years, failed to reveal the way in which the animals move around seasonally, targeting fruiting guava, favourite trees and other likely food sources, including active 'birdwatching', presumably to find nests[147]. Numerous subsequent (and some earlier) conservation-oriented investigations have pinned the blame back where it always belonged – monkeys seek and consume eggs and nestlings, and are devastating even at quite low densities, rivalling Ship Rats[148]. Mauritius's monkey population was estimated at 23,000–35,000 in the mid-1980s, revised upwards to 40,000 using different assumptions in the mid-1990s, and estimated at about 60,000 in 2005[149]. These rising figures are probably not a population increase, but simply better estimates of a roughly stable population. Mauritian Crab-eating Macaques are disease-free, a very desirable characteristic for medical research, so with a view to controlling numbers, two companies were licensed in the 1980s to catch and export monkeys, with an levy on each animal exported to go towards conservation-related projects[150]. In practice most monkeys exported are captive-bred by the companies (easier to work with), and as trapping hardly reduces overall numbers unless it becomes intensive, the main wildlife benefit is the levy, which provides significant funding to the National Parks and Conservation Service[151]. As Owen Griffiths, Mascarene snail expert and founder of one of the exporting companies pointed out, this money

Goes straight to the national parks conservation fund that runs conservation projects in Mauritius and includes weeding forests, getting rid of exotics and building predator fences to keep out (so far) pigs and deer. It is a fundamental part of the conservation program in Mauritius. It is always frustrating for us that although animal rights people, specifically from the UK, claim to be conservationists and say "ban the use of monkeys, ban the export of monkeys from Mauritius"; yet this quarter of a million dollars a year is simply fundamental to conservation in Mauritius. Clearly there is no other equivalent source of funds available.[152]

Conservationists wanting to control rabbits on Round Island in the 1970s encountered similar problems (pp. 220–221). Use of monkeys in medical research is genuinely contentious, but the issue here is about a lot more than captive animal welfare alone[153].

The island's natural visitors, migrant waders and terns from the northern hemisphere, lost their best feeding ground in the late 1970s, a 100ha tidal mudflat known as Mer Rouge on the northern side of Port Louis harbour. In the mid-1970s this was the best place to see Palaearctic waders, with 20 species recorded, plus Common Tern *Sterna hirundo*, in addition to less distant migrants: Crab-plover *Dromas ardeola*, and Great and Lesser Crested Terns (*Sterna bergii* and *S. bengalensis*)[154]. A town and harbour planning report in 1953 scheduled this area for reclamation, but nothing happened until the early 1970s. Most of the area was reclaimed by about 1980, but infill was not complete until 1991[155]. Some good mudflats remaining on the northern side of the harbour in the Terre Rouge estuary were designated a 26ha bird sanctuary in 1999, and a wetland of international importance ('Ramsar site') in 2001, although subject to considerable pollution from effluent[156]. The commoner migrants can also be seen at various points around the shore[157].

In 1994 the national park envisaged in the 1971 development plan and recommended by Sir Peter Scott and the Procter and Salm Report in 1973–75 was finally created, under a new department within the Ministry of Agriculture. Its first director, Yousoof Mungroo, was the first graduate of a conservation training programme for overseas students initiated at Jersey Zoo in the mid-1970s[158]. The Black River Gorges National Park initially consisted of 6,574ha, encompassing the bulk of surviving native upland forest, but unfortunately excluded the important forest around Bassin Blanc (in private hands) that links the western and eastern blocks of the park. Plans to include Bassin Blanc have been announced, and a comprehensive biodiversity conservation plan published[159].

Shortly before the park was created, a little noticed social revolution had a major impact on the forests: about 1985 the government removed a tax on bottled gas, and people formerly cooking with wood fuel largely converted to gas. In the 1970s the Forestry Service still permitted hundreds of woodcutters into the forests, supposedly cutting just guava and privet for fuel-wood, but in practice taking anything in the permitted 1–3" (25–75mm) size range, endemics included. It was an extraordinary sight each evening to see these people streaming out of the forest towards Vacoas with large bundles of poles tied precariously

BOX 18   PARROTS (PSITTACIDAE) OF MAURITIUS

It is very difficult to determine how many species of parrot once inhabited the Mascarenes, and all species are now extinct bar the Echo Parakeet *Psittacula eques* of Mauritius. Almost every contemporary account mentions parrots and they were depicted by a number of artists and travellers, but most have been described from museum skins and fossil remains only. The Mascarene parrots appear to be derived from the parakeets of the genus *Psittacula* (Hume 2005), which are small- to medium-sized parrots with long tails.

Raven Parrot *Lophopsittacus mauritianus*

The Raven Parrot *Lophopsittacus mauritianus*, also known as the Broad-billed Parrot, Indian Crow or Raven, survived until at least 1673–74 and had evolved a huge bill for cracking palm and other forest-tree nuts. Despite suggestions to the contrary, there is no evidence that this species was flightless. The Raven Parrot also exhibited the greatest size sexual dimorphism known in any parrot. Being large and presumably easy to catch, the Raven Parrot probably disappeared as a result of hunting, deforestation and nest predation by introduced monkeys and rats.

Accounts
Het Tweede Boeck (1601):

. . . Is a bird which we called the Indian Crow, more than twice as big as the parroquets, of two or three colours.

Reyer Cornelisz in 1602 (in Strickland and Melville 1848):

In this country occur Tortoises, Wallichvogels [Dodos], Flamingos, Geese, Ducks, Field-hens, large [male] and small [female] Indian Crows [= Raven Parrot], Doves, some of which have red tails (by eating which many of the crew were made sick), grey and green Parrots with long tails, some of which were caught . . .

Jacob Granaet in 1666 (in Barnwell 1948):

Within the forests dwell parrots, turtle and other wild doves, mischievous and unusually large ravens [= Raven Parrot], falcons, bats and other birds whose names I do not know, never having seen before.

Hoffman in 1673–75 (in Cassel 1934):

There are also geese, flamingos, three species of pigeon of varied colours, mottled and green perroquets [parakeets], red crows with recurved beaks and with blue heads, which fly with difficulty and have received from the Dutch the name of 'Indian crow'.

Raven Parrot *Lophopsittacus mauritianus*. From Newton & Gadow (1893–96), based on a drawing of live birds by Joris Laerle in 1601.

Thirioux's Grey Parrot *Psittacula bensoni*

This grey, long-tailed species was particularly sought after as game. Despite this persecution, grey parrots remained reasonably common until the 1750s, but the population must have crashed shortly afterwards, as it was last mentioned in 1759. It was during the 1730s that the French instigated large-scale slash-and-burn forest clearance, and this no doubt had a serious effect on tree cavity nesting species, including parrots.

Accounts
West-Zanen in 1602 (in SoeteBoom 1646):

. . . some of the people went bird hunting. There were so many birds around they could pick them out of the sky or pull them off the ground. It was an entertaining sight to see. There are many grey parrots and once caught they were tamed and made a particular cry and hundreds of the birds then fly around your ears, which were then knocked down with little sticks out of the sky. Also just as tame are the pigeons and turtle doves, that let themselves be caught easily . . .

Cossigny (1764) in 1759:

The woods are full of parrots, either completely grey or completely green. One used to eat them a lot formerly, the grey especially, but the one and the other are always thin and very tough whatever sauce one puts on them.

Catching and killing parrots on Mauritius in 1602. This is probably the only depiction of Thirioux's Grey Parrot *Psittacula bensoni* in existence. From Soeteboom, 1646.

Echo Parakeet *Psittacula eques*

The Echo Parakeet, once described as the world's rarest parrot, was reduced to as few as six individuals in the 1980s. It has now recovered to some 340 birds and is the only surviving parrot in the Mascarenes.

Accounts
La Motte (1754–57), on Mauritius in 1754 and 1756:

One eats here [in Mauritius] a good number of long-tailed green parrots called perruches *whose flesh is black and very good. A hunter can kill three or four dozen in a day. There is a time of year when these birds eat a seed that makes their flesh bitter and even dangerous.*

Echo Parakeet *Psittacula eques*. From Michel (1981)

onto their bicycles[160]. Until word got around, firewood use continued to rise (50,000m^3 in 1985, 60,000m^3 in 1990), then crashed to 10,000m^3 in 1995, dropping further to c. 4000m^3 in 2005 – the number of households involved fell from 58,000 (25%) in 1990 to 12,000 (4%) in 2000. In 2001 official plans included a proposal to cut invasive guava and privet in native forests for use in wood-burning power stations, to reduce imports and carbon dioxide output, but did not address how to avoid the same problem of indiscriminate cutting[161]. Although pressure on forests for fuel has declined, agricultural and housing demands, and also road-building, are claiming forested land, and the tourist industry has gobbled up much of the coastal *pas géometriques* formerly kept planted up and supposedly inalienable[162]. Meanwhile, the Mauritius Institute museum, previously an important player in wildlife knowledge and conservation through its collections, director, curators and publications, went into decline in the 1980s when an unsuitable curator was kept on for nearly 20 years as acting director, until his death in office in 1999 allowed a new start[163].

RÉUNION

The European disaster of the Great War (1914–18) proved something of a boon for the Réunion economy, as growing and trading of sugar beet was seriously disrupted, so colonial cane sugar was suddenly once again in demand in France[164]. Most of the accrued profits went into re-organising and upgrading the industry, so there was little direct effect on the environment. However a post-war surge in demand for geranium oil[165] caused a renewed bout of deforestation: "the woodcutters, as always from Le Tampon, hired out their labour ever northwards. They arrived at Le Guillaume around 1916–18, then crossed the Rivière des Galets and clear-felled the Dos d'Ane, which, since the cane crisis, had lost half its inhabitants."[166]. This frenzy stopped around 1924–25, the peak years of essential oil production, by which time little untouched forest survived below 1,400m in the western heights. Only the Plaine des Makes (state-owned) and Tevelave were spared:

> As to privately-owned forests, they have been, for [the last] sixty years, completely devastated, either by abusive exploitation, without foresight, taking no account of rules which should govern the utilisation and management of forest resources, or because they have been quite simply taken out to make room for new crops. It is only, I think, the Crédit Foncier Colonial that not just preserves forests on its lands, but also looks after them and tries to regenerate them.[167]

By the mid-1920s mid-altitude native forest had mostly gone apart from in the slopes of the volcano, the *pays brûlé*, where no-one lived, and even there little was mature forest due to the regular lava flows and resulting fires. However, in addition to the patches in the west, some good tracts survived between the *enclos* of the volcano and Basse Vallée, and also in the very wet zone north of the *enclos* towards Sainte-Rose[168]. Hunting, legal and illegal was however rife; from 1900 to the 1920s shooting 150 Réunion Merles a day in forests around the Volcan was a normal day's hunting, and one professional bird-catcher at the Plaine des Cafres was said to have lived on the *gizzards* of his catch, after selling the gutted bodies. Emile Hugot reported a decline in the merles after 1915 (reason unknown) after which they never recovered their previous abundance[169].

Whereas in Mauritius there was support and legislation for some degree of nature preservation from the 1870s onwards, there is no sign of conservationist thinking in Réunion until the 1930s, and then only for plants. Making a case in 1934 for selective regeneration of mixed evergreen forests to promote the commercially valuable native species, Marcel Rigotard also proposed the 'excellent measure' of setting up botanical reserves protected from both felling and hunting[170]. He failed to back up his suggestion with any details, and nothing seems to have resulted from this initiative. The post-war boom fizzled out around 1925, and the island reverted to its languid state of international isolation, quietly growing sugar and geranium with little further impact on forests and wildlife. Geranium production diminished after the 1925 peak, secondary scrub forest developing on fallow land; the native forest did not re-establish[171].

One significant environmental event punctuated this quiet time – toads arrived. Malaria was still a major problem, so thinking to control mosquitos, Auguste de Villèle, whom we last met catching Réunion Black Petrels (p. 147), brought some toads over from Mauritius around 1927 and released them[172]. The toads did not noticeably reduce mosquitos, but spread rapidly during the 1930s, and their raucous chorus at breeding ponds was already a noise nuisance in St-Denis by 1937; they did not reach Cilaos till after 1939[173]. An attempt to establish ducks failed again – '*sarcelles*' (probably whistling duck) released at Grand Étang (eastern uplands) probably in the 1920s or 1930s, all 'flew away to Saint-Paul'[174].

The forests in wartime – shortages and poaching

Even more than in Mauritius, the exigencies of the 1939–45 war resulted in shortages and recourse to forests for construction wood and fuel. The island authorities accepted the Vichy government in France, which resulted in a blockade by British forces which reduced shipping movements by 90%, until the island capitulated to the Free French in November 1942[175]. Réunion was forced to become self-sufficient, though the government was alert to the risks of rampant deforestation, replacing the 1874 legislation with new forest restoration laws in 1941, which were still effectively in force in 1980[176]. A thousand hectares were cut at Étang Salé and 500 in Cilaos, though the former was all plantation she-oaks largely used to fuel the railway[177]. This was on government land; what was felled on private estates is not recorded. Even after the war, a system of concessions in state forests of *tamarin des hauts*, the large endemic acacia, had to be revoked in 1949 after excessive felling, though its replacement with managed *tamarin* plantations soon led to much more extensive clearances[178]. Geranium cultivation, at a standstill during the blockade, picked up to pre-war levels from 1944 onwards, with further forest destruction. Hildebert Isnard commented in 1951 that:

> *These forest soils have a precarious fertility that is exhausted in a few harvests. It is necessary to ceaselessly proceed to new clearances to obtain new land, and also for the wood necessary to fire the stills. The soil abandoned after cultivation supports meagre grasses that do not protect against erosion: the great "avalassess"* [downpours with massive runoff] *of the rainy season cause gullying and degradation.*[179]

Pierre Rivals's photos show fresh forest clearance for geranium above Moka (lower Plaine des Fougères) in about 1945, and this area was still subject to some clear-felling as late as 1974, during a prolonged (if capricious) post-war geranium boom that only began to fade in 1977. Like Mauritian tea, geranium, though lasting longer, eventually fell foul of world markets (and synthetic substitutes); by the mid-1990s production had collapsed to negligeable levels[180].

While the histories all refer to general shortages and near-famine during 1941–46, extolling inventive uses of wild plants for clothing, cleaning and food, none mention the war's biggest impact on wildlife – poaching. Food from the forest has always been, as in Mauritius, an important component of life in Réunion, but whereas in Mauritius it became formalised into the upper-class deer hunt[181], in Réunion it largely went the other way, becoming a way of life for the poor and marginalised. The *braconnier* became almost an institution, hunting tenrecs with terriers and small birds with lime-sticks charged with glue made from the sap of local trees.

> *Poaching has become a tradition in Réunion, the techniques of bird-catching transmitted from father to son, from friend to friend. The majority of Réunionnais have set lime-sticks. The majority have equally no idea that this activity could have negative results.*[182]

In the 1970s older inhabitants still recalled the cry of the bird vendor, and described how one famous poacher would pack his catch into a hollowed-out pumpkin, with the stalk carefully replaced, to elude police attention[183]. Noddies caught on Petite Ile were sold at 3 for 125 (old) francs during the 1940s, and harvesting wild palm-cabbage was an ordinary part of life for inhabitants of the cirques in the same period[184].

While the tradition was of long-standing[185], the privations of the war years must have made poaching a necessity for many, hence, we suspect, the long shadow haunting conservationists into the 21st century. As Daniel Vaxelaire has pointed out, memories of the famine influenced island politics well beyond the return to normality: "the long penury of the war would remain, for several decades, the most visceral of the arguments of the anti-autonomists"[186]. So also did the habit of catching wild protein die hard. Visiting the island in 1948, Philippe Milon was struck by the ubiquity of bird-catching:

> *If the few native native species that survive have shown that they can, up to a certain point, resist human destruction, one should not conclude that they can persist indefinitely in this island where the very dense human population do not apply even the most elementary rules of nature protection. During the breeding season, nestlings are destroyed indiscriminately, not only by children but by a fair numbers of grown-ups. At all seasons, a whole variety of methods are used. Firstly the majority of children (and some men) have in their pockets catapults made with old inner tubes; these are called* flèche [=, literally, 'arrow'] *and, with stones as ammunition, they kill whatever* [they fancy].
> *After the* flèches, *which are principally a children's weapon, which does not prevent them being dangerous, are the* lacs, *that is to say snares. These are mostly used for systematic trapping. Quails, partridges* [= francolins], *and doves are also caught* à la gobe (*it is a sort of deadfall trap usually set up* [with sticks] *in a figure-four*[187]). *But the most murderous of these hunts is done* à la colle, *i.e. with bird-lime, and with call-birds. Poachers, known as such, make most of their income this way, and, as they usually hunt in the forest, their victims are mostly Merles and the two species of white-eye. In many of the island's restaurants, one is offered, on the* [day's] menu *or à la carte 'little birds'! I watched with horror and stupefaction Olive White-eyes being plucked for the table.*[188]

Bird poaching was still widespread in the 1970s, though somewhat less intense than in Milon's time. Due to a recent crash in bird numbers, the Plaine des Cafres in 1974 could only support two or three professional bird poachers, whereas there had been 'many more' only six or seven years earlier. Le Brûlé above Saint-Denis was a notorious poacher's nest, providing beaters for the deer-hunt at the Plaine des Chicots who set lime-sticks before they started work, collecting their catch before going home in the evening[189]. By 1989 poaching was again panicking conservationists trying to establish a nature reserve there for the endangered Réunion Cuckoo-shrike[190]. In the 1970s and 1980s the island was still penetrated by a "dense network of paths in remote and apparently inaccessible areas, frequented by poachers of cabbage-palms, crayfish, trout, tenrecs, deer and birds"[191]. However by 2000 poaching appeared to be gradually diminishing, as more affluent younger generations felt less need to make such a hard living.

> *Poaching was at one time a dietary necessity. Indeed until the 1960s–70s it was common for people to go hunting or fishing to augment their meals, especially in the uplands of the island . . . In our time it is no longer a necessity to hunt these birds to feed oneself. The practice has* [nevertheless] *remained for some a habit, more or less occasional. It is a sort of pastime that deserves criticism . . . There are apparently* [no longer] *professional poachers devoted solely to birds.*[192]

While professional bird-catchers have dwindled, amateurs after cagebirds and others catching birds to eat as a hobby are still numerous, and possibly still doing significant damage to bird populations[193]. Seabirds are still at risk (pp. 187–188), harriers are wantonly shot, and more valuable plunder still has *afficionados*; once the deer fence at the Plaine des Chicots deteriorated in the early 1990s, the animals dwindled very rapidly, and no cabbage-palm is safe if it can be reached and retrieved[194].

Wildlife status 1945–1970s

Although his visit was very short (12 days), Milon was the first person in 85 years to attempt a systematic inventory of the birds *in situ*, and also the first to flag up the rarity of the Réunion Cuckoo-shrike[195]. Although quite right to single out this species, he set a bad precedent by second-guessing its status ("on its way to extinction") on the basis of very little data. A similar report from Christian Jouanin in 1965, "unlikely that more than 10 pairs of the birds survive", led to its inclusion in the first *Red Data Book*. Frank Gill's fieldwork in 1967 on Mascarene Grey White-eyes did not add further clarity, although one of his study areas included the cuckoo-shrike's range at the Plaine des Chicots[196]. By chance, the first systematic survey coincided with forestry plans to expand plantations of Japanese Red Cedar into the cuckoo-shrike's only known habitat. The 1974 study showed that the birds, although very restricted in distribution, had a probably stable population of some 120 pairs, but emphasised the urgent need to preserve their habitat from clearance and forestry plantation[197].

Before proceeding to the controversial question of forest management that was to dominate conservation priorities in Réunion from 1970 for the rest of the century, we need to review the status of animals in the post-war years. Milon's productive visit established much needed first-hand data for birds[198]. The Chinese Francolin was 'not rare' in 1948, and Grey-headed Lovebirds were widespread, but both were to die out by the mid-1970s. The Madagascar Partridge, formerly common, had begun a decline in the 1920s and was scarce; the Grey Francolin was said to be common[199]. Village Weavers had spread all over the lowlands outside the forest, while Spice Finches were declining and Red Avadavats were now rare. Heavily trapped as cagebirds, Red Avadavats have hung on on in the extreme south[200]. Among the natives, Réunion Merles appeared scarce after the 1948 cyclone, and were subject to heavy poaching[201]. The House Shrew was widespread and common in the 1950s, but rodents and reptiles were not surveyed until the 1970s[202], although the day-geckos had provided a surprise (see below) in the 1960s. François Moutou noted feral cats during his rodent studies: "not very abundant", but nonetheless a threat to native birds (and introduced gamebirds) in the uplands, as scats contained mostly bird bones and feathers[203]; as in Mauritius, they appear since to have increased.

While compiling a monograph on Mascarene lizards, Jean Vinson sent day-gecko specimens from Réunion to Professor Robert Mertens in Germany, who realised that they not only differed from Mauritian Blue-tailed Day-geckos with which they were traditionally combined, but that there were two species. One, the Manapany Day-gecko, was no sooner discovered than on the endangered list, inhabiting in the 1970s only a tiny area on the south coast; when resurveyed in the mid-1990s it had spread (or been spread?) along 10km of coastline centred on Manapany[204]. Once the news got out, this rare lizard was targeted by unscrupulous dealers servicing the growing trade in pretty reptiles for terraria: "the main danger to the gecko is collectors, who fly from Europe to catch them . . . in order to sell them"[205]. The other species, the Réunion Forest Day-gecko, is widespread, mostly in the east, often in low densities. Although introduced, Panther Chameleons are appreciated in Réunion as attractive animals, their scarcity and limited distribution around Saint-Paul leading to ecological studies in the 1960s, and a special protection order in 1969, even featuring on a unique (for Réunion) nature protection postage stamp in 1971. They have since increased both in numbers and range, becoming generally common in the west, and spreading all round the coast[206].

Controversial forest policy

The presence of the cuckoo-shrike as a 'flagship' species eventually focused a debate already raging on the island between conservationists and the Office National des Forêts (ONF). Forestry was re-organised

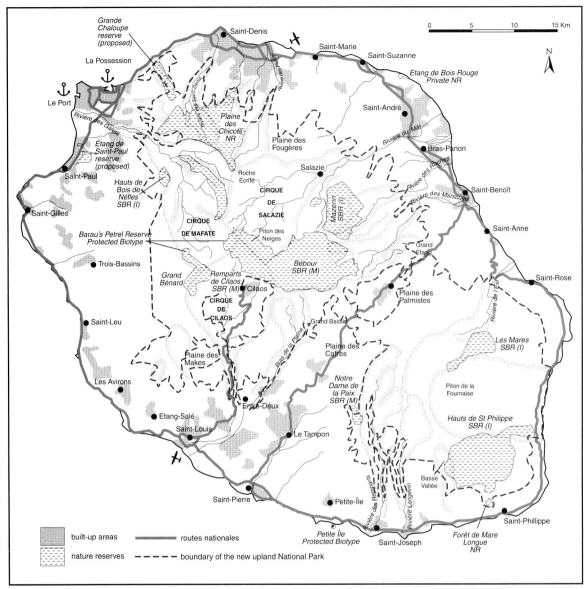

Map 8.3. Contemporary Réunion: towns, roads and nature reserves and the boundary of the new uplands National Park. From Bertile (1987), Girard & Notter (2003), www.para-national-reunion.prd.fr and recent IGN 1:100,000 maps.

and revitalised on Réunion in 1948 after the island became an overseas *département* of France[207]. Since useful native timber had been more or less used up, the intention was to seek new and productive species for reafforestation. One such was the Japanese Red Cedar, a cyclone-resistant tree that grows very well in the cooler uplands of Réunion. There were also plans to create plantations of native species, especially the big upland acacia, the *tamarin des hauts*, and the former lowland timber trees (*nattes* etc.)[208]. The foresters saw no intrinsic value in the native forest ecosystems to be cleared for their schemes, and had a distorted idea of the local forest ecology. They did set aside one small 21ha patch of lowland wet forest in 1958, but then cleared the irreplaceable forest all around it for managed regeneration of their chosen species. Higher up they created two tiny (3–4ha) reserves in 1963[209], but otherwise progressively clear-felled native forest in Cilaos, lower Plaine des Chicots, Basse Vallée and part of Bébour for Japanese Cedar, and Bélouve and the slopes of the Grand Benard for *tamarin*. In 1970 a local nature protection society, the SREPN[210], was established, and soon began clashing with the foresters:

> Réunion is the only one of the three Mascarenes to still preserve extensive remains of native vegetation. It is

BOX 19   PARROTS (PSITTACIDAE) OF RÉUNION

Echo Parakeet *Psittacula eques*
The parakeet on Réunion appears to have been conspecific with the Echo Parakeet of Mauritius. It is known only from a number of paintings and descriptions of specimens, now lost. This species died out very swiftly on Réunion, with the last mention of the parakeets dating from 1732.

Accounts
Brisson (1760) from a stuffed specimen:

> Two small rings, one rose-coloured and the other blue, entirely encircle the neck of this parakeet, which is of the size of a turtle dove; the rest of the plumage is green, which is darker on the back, yellowing under the body; and in several parts with a dusky streak on the middle of each feather; below the tail and laid out on each quill is a yellowish fringe bordering the brown-grey; the superior half of the beak is a fine red; the inferior is brown; it is probable that this parakeet, coming from the island of Bourbon, is also found on the corresponding continent, either Africa or India.

?Thirioux's Grey Parrot *Psittacula ?bensoni*
A grey parrot was mentioned on Réunion, and appears to be either closely related or identical to Thirioux's Grey Parrot from Mauritius. It became extinct sometime after 1732.

Accounts
Dubois (1674) in 1671–72

> . . . grey parrots, as good as the pigeons . . . The sparrows [Réunion Fodies], the grey parrots, pigeons and other birds, the bats [fruitbats], do plenty of damage, as much to seeds as to fruit.

Feuilley (1705) in 1704:

> There are several sorts of parrot, of different sizes and colours. Some are the size of a hen, grey, the beak red [Mascarin]; others the same colour the size of a pigeon [Thirioux's Grey Parrot], and yet others, smaller, are green [Echo Parakeet]. There are great quantities, especially in the Sainte-Suzanne area and on the mountainsides. They are very good to eat, especially when they are fat, which is from the month of June until the month of September, because at that time the trees produce a certain wild seed that these birds eat.

Mascarin (Mascarene Parrot) *Mascarinus mascarinus*
Two preserved skin specimens of Mascarin *Mascarinus mascarinus* survive, both collected during the 18th century, and a few fossil elements have also been collected. Dubois described this species from life in 1674 and a few live birds made it to France. Hahn stated that he saw the last living bird in the King of Bavaria's menagerie around 1834, but there is doubt concerning the date he saw it, and it is more likely that this species died out at the turn of the 19th century.

Accounts
Dubois (1674) in 1671–72:

> . . . parrots a little bigger than pigeons, with plumage the colour of [Red] squirrel fur, a black hood on the head, the beak very large and the colour of fire.

Brisson (1760), from a captive bird in Paris:

> Upperparts of head and neck clear (ash) grey. Back, rump, underparts of neck, breast, belly, sides, legs, scapular feathers, uppercoverts of tail very-dark (ash) grey. Wing feathers of the same colour. The tail is composed of 12 feathers: the two median ones are also very-dark (ash) grey. All the lateral ones are of the same colour, except that they have a little white at their base. The eyes are surrounded by a naked skin, bright red. Prunelle black, iris red. The base of the superior half of the beak is also surrounded by a red naked skin in which the nostrils are placed. Beak similarly red. Legs pale flesh. Claws grey-brown. I am unaware from which country it is found. I have seen it living in Paris.

Mauduyt (1784):

> The Mascarin is found at Ile Bourbon [Réunion]; I have seen several alive in Paris, they were rather gentle birds; they had in their favour only that the red beak contrasted agreeably with the dark background of their plumage; they had not learnt to talk.

Head and foot study of the Mascarin *Mascarinus mascarinus*. From Forbes, 1879.

---

absolutely indispensable to safeguard them. Biological reserves should cover adequately large areas, with as far as possible natural geographic boundaries, such that ecological factors such as temperature, rainfall and evaporation remain constant over time. It is obvious that a hectare of native forest isolated in the middle of Japanese Cedars, crops or secondary vegetation, will rapidly lose its original characteristics as a result of such major modifications to its environment.[211]

The forestry plans were encapsulated in a map published in 1968 when only a fairly small proportion of their proposed surface had been cleared and planted[212]. This map showed that if fully implemented, plantations would annihilate the entire cuckoo-shrike habitat, the whole of the near-pristine Bébour-Bélouve plateau, all the surviving lowland forest at Saint-Philippe (bar the tiny reserve), and convert most of the existing natural *tamarin* forest into managed plantations – a major ecological disaster in the making. The plan's mastermind, Jean-Marc Miguet, with the ONF in Réunion since 1949, justified the scheme to environmentalists with an ecological rather than economic argument, based on the conclusions of Goizet and Jacob de Cordemoy in the 1880s (p. 146). Miguet's view was that the natural upland climax tree was the *tamarin*, whose 'natural' dominance over the 'feeble mixed evergreens in the uplands' was failing through lack of recruitment, while in the lowlands the tall mixed evergreens, here the climax, were regenerating inadequately. In general the native forest was seen as '*figée*' ('congealed'), no longer self-sustaining[213]. He cited Rivals's vegetation studies in support of his views, though Rivals had commented only that due to slow growth the forests *gave the impression* of being in a time-warp. Thus to Miguet the necessary action was to promote the regeneration of *tamarins* by clear-felling, and 'assist' the mixed evergreens by more felling and then encouraging selected saplings (basically Goizet's system from the 1880s for *tamarins* and Rigotard's 1930s proposals for the mixed-evergreens); Miguet called this 'ecological silviculture'[214].

Whereas in Mauritius Vaughan and Wiehe's ecological studies in the 1930s directly informed policy for native forest and led to the creation of botanically reasonable reserves in the 1940s, in Réunion it was totally different. Notwithstanding having read Rivals's detailed vegetation studies, there is in Miguet's early writings[215] not the slightest inkling that he understood the concepts of endemism or natural ecosystems. Thus the entire thrust of his plans was to replace native forest with managed woodland, using a very limited range of species selected entirely for their usefulness as timber; no value was placed on the native forest *per se*. For the *tamarin* Rivals had already pointed out that it was dependent for natural regeneration on fires from volcanic activity, and would in quiet times be gradually superceded by the much smaller mixed evergreens, a finding amply confirmed by Thérésien Cadet's study in the 1970s[216]. We suspect it was anathema to foresters to imagine the grand giant of the forest being usurped by spindly good-for-nothing-but-charcoal trees, so they refused to contemplate the reality. As to the lowland mixed evergreens, they are simply slow-growing: there was no tangible evidence that regeneration had stopped, and the crippling invasion by exotics seen in Mauritius had not by 1950 seriously affected the undisturbed wet forests around Saint-Philippe, nor had there been any fungal epidemic. Neither Rivals not Cadet said much directly about regeneration, but implied it was adequate. Recent lava flows are still colonised primarily by native species, although trees with bird- and bat-dispersed fruit are slow to do so in the near-absence these days of suitable dispersers[217].

Miguet's 'ecological' argument coincided with the ONF's declared aim to make Réunion more self-sufficient in timber and boost rural unemployment[218]. However environmentalists soon pointed out that these aims could easily be achieved without cutting a single native tree – there were thousands of hectares of degraded or secondary forest where no one would object to plantations of Japanese Cedar or whatever they chose[219]. The snag, from the ONF's point of view, although they were rarely explicit, was that this was largely private land and thus not under state control; until the mid-1980s the ONF resisted the idea of recycling this secondary forest, apart from the small amount already in the *domaine*[220].

Conservationists challenge forest policy

The forestry plans formulated in 1950 began with a massive construction of new roads in forest areas previously inaccessible to vehicles[221]. In addition to the damage the foresters would shortly do, this let poachers, particularly of palm-cabbage and tree-ferns, reach new territory, and facilitated the spread of exotic plants into previously 'clean' vegetation: "As soon as a route penetrating the native forest is opened all harvestable cabbage palms rapidly disappear, and they persist now only in areas sufficiently remote and inaccessible to make collecting them on a large scale difficult"[222]. The clearances for *tamarin* plantations began at Bébour in 1950, and for mixed evergreens at Saint-Philippe in 1953. By the mid-1960s forest had been cleared and planted with Japanese Cedar at the lower Plaine des Chicots and Cilaos, and grandiose plans made to create 3000ha of cabbage-palm plantations. As the plantations advanced in the early 1970s, the local objectors wanted all cutting of native forest stopped (as unnecessary for timber production), but made no specific suggestions for reserves. The 1973–75 cuckoo-shrike survey enabled one of us (ASC) to formulate a clear case for a large reserve (*c*. 25km^2) to protect this species, and thus the forests within that space[223]. As Japanese Cedars were already encroaching on its boundaries (and had already swept away the bird's 19th-century territories), the matter became urgent.

Until the mid-1970s the ONF, charged by default with overseeing almost all the surviving native forest in Réunion, was a law unto itself, with little or no contact with bodies outside the regional government. The local environmentalists were treated with disdain as ignorant objectors to 'progress'. In 1973 Miguet petulantly withdrew his contribution to the SREPN's special 'forests' issue of their journal, and threatened legal steps to ban its distribution once he discovered that it also contained articles attacking ONF local and national policy[224]. He was, superficially, more accommodating with foreigners, facilitating the BOU researchers' work by allowing free use of ONF *gîtes*, forest cabins dotted about in remote areas. A deaf ear, however, was turned to data that did not suit Miguet's views. In 1974 a forestry document blithely trumpeted that "birds nest willingly in the canopy of these trees [Japanese Cedars]"[225], whereas in fact birds avoided them almost completely, not only for nesting but for everything else, the occasional individuals seen only passing through to better habitat[226]. Miguet even claimed in the withdrawn article that "it is hardly an exaggeration to say that with the spread, in any case limited, of 3–4,000ha of Japanese Red Cedars, man is, for once, correcting in a certain measure a gap in nature: to sum up, the Japanese Cedar 'belongs', through a deep calling, to Réunion, at least as much as the sugar cane, or the potato to Europe"[227]. In François Moutou's diplomatic understatement: "The sad thing is that all this [forestry activity] is conducted by the Office National des Forêts (ONF), which does not seem to understand the tropical rain forest in Réunion"[228].

The cuckoo-shrike question brought international conservation bodies into play, confronting the ONF with messages it had previously refused to hear. In addition, Cadet's masterly vegetation thesis published in 1977 completely undermined Miguet's ecological theories; unlike Rivals in the 1940s, who was a *zoreille* (incomer from France), Cadet was a Réunionnais and a highly respected member of the local university. By 1980, though still a powerful figure in Réunion society, Miguet was fighting a rearguard action for his views, attempting to justify his policies by writing in the French ecological journal *Terre et Vie*[229]. He cited Professor Renaud Paulian in support of his ideas, eliciting a response from the veteran tropical entomologist and biogeographer[230]. Paulian criticised Miguet for not conducting ecological studies on fauna and soils to see whether 'ecological silviculture' preserved biodiversity, and for failing to create adequate reserves for trees he deemed of no commercial value. He further implied that Miguet's methods would be fine for *expanding* rather than *replacing* areas under native trees. However, from experience in Madagascar, he thought controlling fire and grazing would improve regeneration, though neither problem was an issue in the wet and herbivore-free Réunion lowland forests.

The spread of nature reserves

By the late 1970s the stand-off between the SREPN and the ONF had eased enough for serious contacts and discussions to begin, leading to a joint statement on nature reserves in 1978, and an official request from the ONF in Réunion to the French research institute ORSTOM to send an expert to suggest areas to be gazetted[231]. This project in early 1982 coincided with Miguet's retirement. After a lifetime of assaulting the native forest, Miguet may in his last days in office have changed his perspective, as he then spent much of his retirement promoting the Conservatoire Botanique de Mascarin, an educational and research body dedicated to preserving Réunion's flora[232]. Jean Bosser's mission marked the beginning of a profound change of attitude and policy that developed over the next two decades. In 1985 a list of sites of ecological significance (ZNIEFF) had been drawn up[233], and by the late 1980s the ONF had appointed a forest botanist, Pierre Sigala, who developed a new conservation strategy, formally entrenched from 1993 in a policy document informed by ecosystem conservation – no doubt helped by the creation in 1992 of a new French environment agency (DIREN) with regional offices including Réunion[234]. Bosser's approach had been rather strictly botanical (echoing Vaughan's in Mauritius 30 years earlier), and while he acknowledged the cuckoo-shrike, he proposed a reserve area (2km^2) that was far too small to safeguard the bird, and ignored several other proposals made by the SREPN[235]. Several of his other proposals were already judged insufficient by 1988–89[236]. However things were now moving and several reserves, including two large ones of 19km^2 and 41km^2 in the uplands, were created in 1985 and 1987, with several others under evaluation[237]. By 1998 these studies had resulted in adding a 51km^2 upland reserve at Bébour, and a further 800ha in Cilaos, with, in December 1999 and 25 years late, a 36.4km^2 reserve at the Plaine des Chicots/Roche Écrite, home of the cuckoo-shrike[238]. The later is innovative in Réunion for not being run purely by the ONF, but by a steering committee drawn from three organisations – the ONF, SREPEN (ex-SREPN) and SEOR, the local ornithological society. Although gazetted in 1999, it was transferred to the tripartite administration only in January 2003, and formally inaugurated in July[239]; by then the man in charge of the ONF in Réunion, Jacques Trouvilliez, boasted a Ph.D in avian ecology![240]

Despite the recent progress in protecting native ecosystems in Réunion, there are still important areas not covered; four of the eight 'Important Bird Areas' designated by Birdife International in 2001[241] remain outside reserves, essentially because (apart from the cuckoo-shrike) decisions have been driven by botanical considerations, though several ZNIEFF sites (mostly botanical) also remain unprotected. A major project to designate the entire uplands as a National Park finally became reality in March 2007. In France, as in the UK, this implies preservation of landscape and traditional land-use rather than creating a huge nature reserve, so human activities and habitation would remain. However it encompasses the four unprotected bird areas and gives them some passive protection and possibly more, as park documents look forward to increasing protected areas. At last, also in the context of the park, serious consideration is being given to using degraded land for production forestry, 13km² being mentioned, apparently mostly formerly under geranium[242].

Miguet's 'ecological sylviculture' at Saint-Philippe, instead of using degraded land, unnecessarily destroyed most of the only surviving area of fully developed lowland wet forest, leaving only the absurdly small Mare Longue reserve, isolated in an area of mixed secondary forest and vanilla plantation. The reserve was increased from 21ha to 68ha in 1981[243], but by then there was little else at this altitude to save. Some of the managed plantation nearby has begun to take on the aspect of a mixed evergreen forest, but is of course very poor in species[244]. Some 3.1km² of primary forest was converted to managed regeneration, with another 2.9km² to camphor a bit higher up – as late as 1989 the plan had been for 30km². Fortunately only about half of the *tamarin* forest was subjected to management, so good tracts still survive at the Plaines des Chicots, d'Affouches and des Fougères, and Bébour. In the late 1950s some 200km² were scheduled for *tamarin* plantations, but by 1966 this had been scaled down to 45–47km². The final figure is about half that, 22.4km², of which 19.7km² is for sustained timber production, the rest essentially re-afforestation for erosion control in degraded areas[245]. The original plan for Japanese Cedar was for 30–50km² of plantation, almost all replacing upland mixed evergreen forest. In the event a smaller but still substantial area of some 18.8km² was clear-felled and planted[246]. Severe cyclone damage in 2002 to some stands and persistent objections to the species by ecologists appear to have dampened the ONF's enthusiasm for cedars. There are no current plans to extend these plantations, and, in some environmentally sensitive areas (e.g. Bébour), the intention is to replace cedars progressively with indigenous species as the conifers mature and are cut[247]. While the unnecessary replacement of native forest by plantations in Réunion is regrettable, the total of 47–48km² converted post-1950 compares favourably with the 127km² of forestry plantations in Mauritius, mostly created in the 20th century from native forest, albeit often partly degraded, on top of which there was also the large area of forest lost to tea[248].

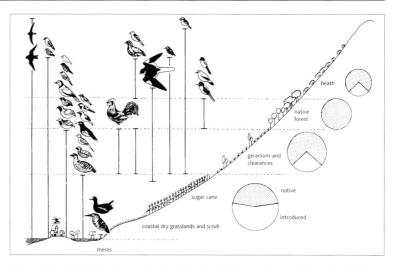

Figure 8.2. Altitudinal zonation in Réunion birds, native and introduced, from Barré & Barau (1982). Pie-charts show the proportion of introduced (white) and native (shaded) species in different habitats – this illustration dates from before the Red-whiskered Bulbul's explosive expansion.

The spread of invasives and the role of deer

As the ONF themselves admit, their silvicultural techniques encouraged the very invasive pest species that they elsewhere decried, especially bramble in *tamarin* plantations[249]. Already widespread and invasive in the late 19th century, dispersed by birds bramble now permeates all levels up to 1,700m, even though it cannot set seed above 1,100m[250]. It is particularly invasive, not only in plantations, but under the rather open canopy of primary *tamarin* forest, and ironically another unwelcome alien, deer, are rather good at controlling it. In the 1970s, when deer numbers peaked at the Plaine des Chicots[251], there was very little bramble there, whereas across the Rivière des Pluies gorge at the Plaine des Fougères, superficially similar habitat was heavily invaded, with big old *tamarins* strangled by the rampant vine[252]. Deer numbers (outside a special 300ha enclosure) are now much lower, and bramble and ginger have re-invaded throughout[253]. Before exploring the other invasive plants, we need to catch up on the history of deer and other introduced animals in Réunion.

The herd introduced around 1900 was still thriving in Salazie in 1937, but appears to have largely disappeared during the war years, probably eaten out of necessity. A few, perhaps having somehow escaped and climbed the rampart, inhabited the Plaine des Chicots in the late 1940s and 1950s; by around 1960 the does were said to be old and infertile. Mauritian Rusa Deer were released in five batches at the Plaine des Chicots from 1954 onwards, and also seven Red

> **BOX 20** **PARROTS (PSITTACIDAE) OF RODRIGUES**
>
> **Rodrigues Parakeet** *Psittacula exsul*
> The Rodrigues Parakeet was considered common and good to eat during the late 17th and early 18th centuries, but became extremely scarce by the mid-19th century. It is known from two skin specimens, a female taken in 1871, and a male, probably the very last of the species, taken on August 14, 1874, and a few fossil bones. Unfounded rumours of its continued existence persisted until the 20th century but this species is now long gone.
>
>
>
> Rodrigues Parakeet *Psittacula exsul*. Drawing by Jossigny (c. 1770) based on a live specimen.
>
> **Leguat's Parrot** *Necropsittacus rodericanus*
> The largest Rodrigues psittaciform, the all-green Leguat's Parrot was first clearly mentioned by ship's mate Tafforet in 1726. This species was characterised by a large head and bill, and a long tail. Pingré listed *N. rodericanus* as very rare in 1761, and this was the last time it was mentioned. It presumably died out due to forest clearance and direct hunting by tortoise-hunters and rat predation of eggs and chicks.
>
> *Accounts*
> Francois Leguat (1707) in 1691–93 (from the original English translation of 1708):
>
> *There are abundance of green [Leguat's Parrot] and blew [Rodrigues Parakeet] Parrets, they are of a midling and equal bigness; when they are young, their Flesh is as good as that of young Pigeons.*
>
> In reference to his house:
>
> *you see' twas between two Parterres, and upheld by a great Tree, which also cover'd it on the side of the Sea. This tree bore a Fruit something like an Olive [= Bois d'Olive* Cassine orientale*]; and the Parrots lov'd the Nuts of it mightily.*
>
> *We often delighted ourselves in teaching the Parrots to speak, there being vast numbers of them. We carried one to Maurice Isle, which talk'd French and Flemish.*
>
> Tafforet (1726), from A. Newton's 1875 translation:
>
> *The parrots are of three kinds, and in numbers. The largest are larger than a pigeon, and have a very long tail, the head large as well as the beak. They mostly come on the islet (Ile aux Mat) which is to the south of the island, where they eat a small black seed, which produces a shrub whose leaves have the smell of the orange-tree, and come to the mainland to drink water. The second species is slightly smaller and more beautiful, because they have green plumage like the preceeding [Leguat's Parrot], a little more blue, and above the wings a little red as well as their beak [Rodrigues Parakeet, ?full male]. The third species is small and altogether green, and the beak black [Rodrigues Parakeet].*
>
> Pingré (1763) in 1761:
>
> *The perruche [parakeet] seemed to me much more delicate [in flavour, when compared to fruitbat]. I would not have missed any game from France if this one had been commoner in Rodrigues; but it begins to become rare. There are even fewer perroquets [parrots], although there were once a big enough quantity according to François Leguat; indeed a little islet south of Rodrigues still retains the name Isle of Parrots.*
>
> *Specimen collections of Mascarene parrots*
> Fossil material of *L. mauritianus* and *P. bensoni* can be seen at BMNH; UMZC; MNHN; and UCB. Two skins of *P. exsul*, the holotype 18/Psi/67/h/I female, Rodriguez, 1871 and paratype 18/Psi/67/h/2 male, Rodriguez, 1874 are housed at UMZC. Fossil material held at UMZC and BMNH. Two *M. mascarinus* skins exist. The holotype MNHN211 is found in the Muséum National d'Histoire Naturelle, Paris, France and a second partially albinistic specimen is in the NHMV. Fossil material can be found at UCB.

Deer *Cervus elephas* from France in 1964. A hunting reserve was set up in 1961, and the Plaine keepered actively to discourage poachers[254]. Théophane Bègue was a wildlife-friendly gamekeeper, and his general constraint on poaching had a generally beneficial knock-on effect on the birds[255]. Réunion Cuckooshrike numbers in the mid-1970s were well above the estimates made in the 1960s, although those were not very reliable. The downside was that the beaters working on *chasse* days enjoyed catching small birds by setting lime-sticks before the hunt began.

Not content with one *chasse*, the ONF released deer in the *tamarin* plantations at Bélouve in 1974, having to cull them soon after to contain damage to young trees; a few persist, reduced by poaching to around 15 animals in 2002. Further groups were released in the late 1970s at the Rivière des Roches, and more recently at the Rivière des Remparts and Marla (Cirque de Mafate), in addition to herds reared commercially for meat in large, allegedly escape-proof enclosures[256]. There appears to be a rising interest in hunting (as opposed to poaching) in Réunion, coming as fewer and fewer animals are permitted as game; most wild birds are now protected. The Black-naped Hare bears the brunt of the hunters' guns, while tenrecs are hunted seasonally with terriers[257]. One hopes that the foresters have absorbed the impact of overstocked deer on ecosystems unadapted to large herbivores, though deer are not mentioned in most recent papers on the Bélouve *tamarin* plantations. In tropical New Caledonia the same introduced Rusa Deer, at densities ranging from 1 to 17 head/km², ruin enrichment plantings of endemic Kauri *Agathis moreli*, and the ONF themselves commissioned a report on the impact of deer at the Plaine des Chicots published in 1994[258]. However, to the general public the island lacks 'animals' and 'game', and there is always pressure to 'remedy' this. Echoing Miguet's claims for the Japanese Cedar, huntsman Charles Cazal in 1974 claimed that "nature was wild, a disorder reigned in the vegetation, life with deer re-established the balance"[259]. A special issue of *Marchés Tropicaux* in 1971, on the French overseas *départements* ('DOM'), claimed that "populating the wild massifs of Réunion in wild animals [is] self-evidently indispensable"; admitting forestry caution about deer, they proposed chamois, mouflons, ibex and feral goats![260] In fact feral goats still survived in small numbers in the 1970s, but have since disappeared, presumably through vigorous poaching[261]. "Hunting is written in the genes of the Réunionnais" said the hunting association's Secretary-general in 2003, which probably explains why there are still estimated to be four or five times as many poachers as holders of hunting permits[262].

Unsolicited introductions

The lull in new introductions after 1900 came to an end after the 1939–45 war. Before anything actually arrived, ornithologist Philippe Milon was advocating 'enhancing' the island with a few extra birds, choosing species he hoped would avoid negative ecological impact: Dimorphic Egret, Madagascar Snipe *Gallinago macrodactyla*, Hottentot Teal *Anas hottentota* and Hoopoe *Upupa epops*[263]. The first would have been a reintroduction, but the snipe and duck were more to please hunters. There had been an endemic duck, but the proper ecological replacement would be Bernier's Teal or its relatives, not the Hottentot Teal[264]. The Hoopoe was just to 'make pretty', though no doubt Milon had the extinct Hoopoe Starling in mind. In the event different and less desirable birds arrived.

The first were the Ring-necked Parakeet and the notorious Red-Whiskered Bulbul. Both were declared pests, and the parakeet was shot out by Armand Barau in 1976[265]. The bulbul, first reported in 1972 as groups of birds in canefields but probably released around 1970 by tourists returning from Mauritius, was not so easily dismissed[266]. The bird spread slowly at first from a focus in the southeast, but by the early-mid 1980s was well-established and extending up the east coast. By 1990 it was widespread and abundant in the wet east, though still absent from the upland mixed forests and the dry north-west. The dry coast was colonised in the mid-1990s, and by 1999 they could be seen everywhere but the lava deserts of the Volcan, though still scarce in upland native forest[267]. Rapid micro-evolutions has already been reported; and knock-on effects are apparent – most noticeably a big decline in large orb-web spiders, exactly as in Mauritius nearly 90 years earlier[268]. The loss of flycatchers from Étang Salé forests may be related to bulbuls. Bulbuls are the main disperser of a recently introduced weed, *Clidemia hirta*, that has become abundant in eastern Réunion with their help, and the birds also enhance germination of giant bramble, Lantana and Brazilian Pepper over uneaten fruit[269]. Another insidious change is the conversion of introduced fuchsias from 'non-pathogenic' in Bébour and Plaine des Palmistes in the mid-1970s to 'radically altering the physiognomy of the vegetation' by 1989[270]. Bulbuls reached the Plaine des Palmistes and Bébour–Bélouve from the east coast in the 1980s[271], and could distribute fuchsia fruit that previously went uneaten, though a subtle species swap in fuchsias is also involved. The bird has become a serious agricultural nuisance, being transferred from 'game' (killing restricted) to 'pest' (no constraints) status in 2000. Bulbuls attack fruit, bees, and even vegetables, and subsidised cage-traps are available to growers. 3,000 birds were caught in two months (Nov–Dec 2000) in the Saint-André/Sainte-Suzanne area[272], but bulbuls are so numerous and so mobile that any trapped in populous districts will be rapidly replaced by a surplus from wooded areas. Many are caught more traditionally using bird-lime, which also indiscriminately snares native species[273]. The bulbul story is a textbook case of enthusiastic but ignorant bird-fanciers releasing a known pest on a thoughtless whim – as so often before in these islands and elsewhere. Destabilising an ecosystem can cost nothing and take no time, re-establishing it is expensive and can take forever, if it can be done at all.

Nonetheless, hot on the heels of one cage-bird came another, this time the Peking Robin *Leiothrix lutea* (usually referred to in the ornithological literature as Red-billed Leiothrix). Like the bulbul, this species is widely traded and frequently released into the environment, though less widely established[274]. Although not noticed by ornithologists until the late 1990s, enquiries showed it had been known to local nature wardens and bird trappers since the late 1980s, with breeding first recorded in 1988[275]. It had probably established in the mid-1980s, passing unnoticed by birders in the little visited secondary thickets of eastern Réunion. A frugivore like the bulbul, experiments have shown it not only eats the fruits of invasive privet, ginger and guava, but enhances the

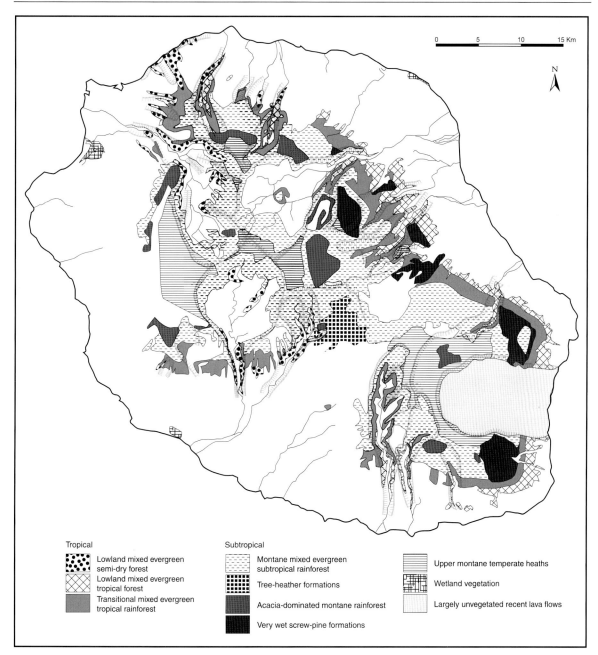

Map 8.4. Remnant native forest in Réunion. Adapted from Cadet (1977) and Girard & Notter (2003).

germination rate of the first two[276]. It thus has the potential, like the bulbul, to increase the rate at which exotic weeds penetrate native forests, into which the bird has also since spread.

A whole catalogue of other escaped cage-birds have been recorded since the early 1990s, only two of which have become definitely established. These are Pin-tailed Whydah *Vidua macroura* and Barbary Dove *Streptopelia risoria*. The whydah is an African weaver-finch which is a brood parasite of related species. In Réunion only Common Waxbills and possibly Spice Finches are suitable hosts, so the whydah is unlikely to impact on the native vegetation. The species was released in several western lowland localities in 1990, but by 2003 it had reached Sainte-Suzanne (northeast) and was penetrating inland in the west. The Barbary Dove is a long-domesticated form which has appeared recently in Saint-Denis and in the Possession – Dos D'Ane area; it seemed well-established in 2003. Two other species that had died out, Grey-headed Lovebird and White-rumped Munia, have been seen again sporadically in numbers that might suggest re-establishment. Other weaver-finches and Ring-necked Parakeets have been recorded, and there have even been colonies of up to 60 nests of Red-billed Queleas *Quelea quelea* at Saint-Leu in 2001 and 2005, and Saint-Gilles in 2006, source unknown[277]. Attempts made since the late 1970s to establish Common Pheasants *Phasianus colchicus* as game have apparently not created a viable feral population, but birds bred and released can be met with over most of the island[278]. In 2004 came the first reports of Indian House Crows, including two together in Saint-Denis. This bird establishes very readily. There is no suggestion of deliberate release; it is more likely they hitched a ride on ships from Mauritius[279].

After a long gap since the establishment of the Bloodsucker in the 1860s, lizards have been prominent among new arrivals. The first in was the small Malagasy day-gecko *Phelsuma lineata* released in the 1940s by Vincent de la Giroday in his garden at La Revolution, Sainte-Marie. These animals have been well-behaved and remained on site, possibly contained by the surrounding canefields; there are no native day-geckos present. Three other day-geckos introduced more recently are less restrained. These colourful lizards caught the public imagination around the world in the 1970s, especially in Germany and Holland, but also in Réunion and Mauritius. Many Réunionnais visit Madagascar where there are lots of common and conspicuous species, much more obvious than the scarce Réunion endemics. Many had never seen the native geckos, so, as with birds, felt a need to brighten their environment – bringing in day-geckos from Madagascar, Mauritius and the Seychelles, to release in suburban gardens. An early colony of Blue-tailed Day-geckos *Phelsuma cepediana* around 1960 at Sainte-Marie was apparently wiped out by spray-drift from nearby fields, but in the 1970s and again in the 1990s more were released at La Montagne near Saint-Denis, where they have established. Great Green Day-geckos *P. madagascariensis* were released in the same area in 1995, and earlier (1994) at Saint-André, and have since occupied some 75km² around Saint-André and Sainte-Suzanne. Gold-dust Day-geckos *P. laticauda*, released near Saint-Gilles by Christian de Villèle in 1975, are now common in Saint-Gilles, La Saline on the north-west coast, inland to Saint-Gilles les Hauts & l'Hermitage, and have spread (perhaps with human help) north to Saint-Paul and (again) La Montagne. Lastly, a thriving population of Seychelles Small Day-geckos *P. astriata* was found at La Possession in 2004[280-281]. Human agency has also boosted the endemic Manapany Day-gecko: a population established at 600m altitude at Le Tampon arose from eggs accidentally carried on palms and banana plants from the coast[282]. Given that the species was previously known only from the hot dry coast, its ability to live and breed in the cooler temperatures of Le Tampon suggests a much wider former distribution. Two small dull-coloured anthropophilous 'house geckos' were formally recorded in Réunion for the first time only in the 1960s, although one had been present for a century – arriving unnoticed in cargo from other tropical areas, house geckos get little attention unless specifically studied[283].

Bigger lizards have also been on the agenda. Rainbow Lizards *Agama agama* were first noted in 1996 by dockers at the new port on the northern side of the Pointe des Galets, probably self-introduced from Africa in cargo. They have spread around the dry shingle delta of Pointe des Galets/La Possession, with some individuals captured and released behind the Étang de Saint-Paul[284]. Naturalists were first alerted in October 1998, when the population was still restricted to a relatively small area and could have been eradicated, but unfortunately no-one tried[285]. Perhaps more insidious are the increasing reports of Green Iguanas *Iguana iguana* seen roaming the forests in various places, including the young lava flows near the south-east coast. No breeding has been recorded, but the sighting frequency suggests there may be a feral population of this large South American lizard, now a popular pet worldwide. It established rapidly in Hawaii, in habitat very similar to southern Réunion[286]. More dangerous for wildlife is the possibility that ferrets or mongooses could become naturalised. There have been several sightings, captures and road-kills of ferrets and one of a mongoose[287]. Finally (or more probably not!) Red-eared Sliders are, as in Mauritius, imported as pets, and may have established in the Étang de Saint-Paul[288]. None of these new arrivals have yet impacted on the Bloodsucker or the Wolf Snake, which remain common in the lowlands[289].

Giant snails had received almost no attention for generations until a misguided attempt to control them with Rosy Wolf Snails from Mauritius was made in 1966[290]. As elsewhere this failed, and the still-abundant giant snails were found in 2001 to be gravely damaging attempts to re-establish the endemic aloe *Lomatophyllum macrum*, once so abundant along the dry west coast (p. 34). The snails' demolition job is so thorough that they are now thought to be the main agent of the plant's near-disappearance (and possibly also several other species), although habitat destruction and loss of seed-dispersing frugivores is also important[291]. Only *Achatina immaculata* is common in Réunion, the earlier introduction, *A. fulica*, is now scarce and local[292].

Fewer new invasive plants have become naturalised recently than animals, although one of the worst

BOX 21   OWLS (STRIGIDAE)

True owls have colonised all of the main island groups in the Indian Ocean. The surviving Indian oceanic island species are all scops owls in the genus *Otus*, and it is from this genus that the extinct endemic Mascarene genus *Mascarenotus* was derived. The Mascarene species were characterised by large size and long legs, presumably adaptations for catching reptiles and possibly small birds.

Commerson's Lizard-owl *Mascarenotus sauzieri*. Drawing based on a recently killed specimen by Jossigny c. 1770.

Commerson's Lizard-owl (Mauritius Scops Owl) *Mascarenotus sauzieri*

The Mauritius species was a large, white-faced, eared owl with bare tarsi. The drawing by Jossigny in the 18th century is the only known illustration of the genus. One specimen was described in detail by Desjardins, while a few others were reported around the same time. In 1859, the amateur naturalist George Clark stated that they were by then extinct after being formerly plentiful.

Accounts
Desjardins in 1837 (from Oustalet 1897):

> I shall give no name to the Scops I describe, although I have every reason to think it new. The toes and even the tarsi are without feathers, only on the front part of the latter does one see some short dense feathers descending to a point to about its middle. The toes are very strong, armed with hooked claws. The beak is very strong, arcuate from its base; the upper mandible much longer than the other and covering it, and as if frankly cut off at its tip. The nostrils are pierced fairly high in the horny part of the beak. The eyes, of which I have not been able to see the colour are round, situated like all the family, to the front. The are encircled with a disk of stiff narrow feathers, interrupted at the sides. A kind of ruff can be seen under the throat. Two plumes ['ears'], analogous to those of Eagle and Long-eared Owls, and very noticeable, are seen behind the eyes and near the top of the occiput. The wings are a little longer than the tail, the 4th and 5th primaries are the longest, the 3rd and 6th are next, then the 2nd, equal to the 8th; the 1st is the shortest of all. The tail is as long as the tips of the toes; it is rounded and of little extent; all the retrices are the same length. The ears are brown, with fawn nuances, the feathers of the ocular disks are white with fawn overtones. All the upperparts are dark brown, the feathers of the head, neck and back are edged with russet, but not strikingly; on the scapulars it is more marked and some even have on the outer edge one or two white spots surrounded with brown. The large feathers of the tail are less brown and more russet with some marbling of a clearer russet mixed with brown. The tertials have a brown band centrally and their outer edge is pleasantly marked with squarish spots or ocelli or irregular bars in white, fawn and brown. The greater remiges or flight feathers have the same pattern but more developed, with spots of fawnish white on the inner border, which produces outwardly a regular speckling on a brown ground; the tip of these big feathers is finely mottled with brown on a fairly pale ground; in addition there is a fairly large white area under the wings. The throat and underparts are very agreeably decorated with rather bright fawn feathers, the middles of which are blackish brown, and which in addition have 2 to 4 surprisingly large roundish white spots. The large feathers of the thighs are whitish with a little fawn on the edges, and a browner centre line. In general all the parts that are well feathered, like the back, belly, flanks, legs show a very dense blackish down which is covered by the patterned feathers already described. The colour of the beak and feet is a rusty brown. The length of the beak to the claws is one foot one and a half inches. The wingspan [from beak to claws – error by Oustalet] is one foot ten inches.

> In September 1837 several inhabitants of the Savanne area told me they had seen owls in their forests; Dr Dobson, of the 99th Regiment, assured me he had killed one in the woods of Curepipe. It could well happen that in a few years, the species, if it is one, will have disappeared completely because of the destruction of our forests, and the large number of poachers who roam the woods that remain.

Réunion Lizard-owl (Réunion Scops Owl) *Mascarenotus grucheti*
Owls were never reported historically on Réunion, but bone remains confirm that *M. grucheti*, which was similar if slightly smaller than the owl on Mauritius, once existed there. It is not known when this species became extinct.

Tafforet's Lizard-owl (Rodrigues Scops Owl) *Mascarenotus murivorus*
This species was briefly mentioned by Leguat (1707) in 1691–93 and described by Tafforet in 1726, the latter stating that the small brown owls fed on lizards and birds and lived almost exclusively in trees. The astronomer Pingré in 1761 never reported them, so they presumably disappeared sometime between 1726 and 1761. All the Mascarene lizard-owls probably nested in tree-holes, so deforestation would have been particularly devastating.

Account
Tafforet (1726):

> A bird is seen which is very like the brown owl, and which eats the little birds and small lizards. They live almost always in the trees; and when they think the weather fine, they utter at night always the same cry. On the other hand, when they find the weather bad they are not heard.

Specimen collections
Mascarene owls are particularly scarce as fossils, but they are also known from descriptions and one drawing. No skins have survived; Desjardins's specimen disappeared from the museum in Mauritius in the mid-1800s, apparently destroyed during a cyclone. Fossil material of *M. sauzieri* can be seen at UMZC and UCB, *M. grucheti* at UCB, and *M. murivora* at UMZC and BMNH.

possible has got a grip in some areas – the Ceylon Privet. It was introduced from Mauritius to Cilaos for hedging in the 1960s, spreading into local waste ground in the 1970s and into native forest in the 1980s, although it was some time before botanists noticed[293]. Spectacularly invasive in Mauritius, it has penetrated native forest equally effectively in the Cirque of Cilaos, despite a vigorous attempt since 1989 to eradicate it. Although most of the prolific fruit falls uneaten beneath the parent bushes, some birds, notably introduced bulbuls and mynas, like it and spread it around; adding Peking Robins to the mix will only make things worse, though Réunion Merles also eat privet fruit once it has penetrated native forest[294]. By 1998 some 30km^2 in Cilaos and Salazie were infested, while only 'a few hundred hectares' were subject to active weeding by the ONF, though biological control measures were being studied. In addition to spreading itself, gardeners have disseminated it widely as an easily propagated fast-growing hedge plant[295]. Research in Réunion (which should have been done 80 years ago in Mauritius) shows that the privet is self-pollinated, germinates freely in low light, grows up to twice as fast as native species, lacks insects or diseases to attack it, and is probably toxic to immediate neighbours – i.e. nothing can stop it invading all the island's forests unless some control method is found[296].

We have highlighted the two worst plant invaders of native forest, giant bramble and Ceylon Privet, and mentioned the fuchsia, but there are others that cause concern. Gingers, usually cited as *Hedychium gardnerianum* but in fact a cluster of several similar species[297], are extremely invasive, spreading both by rhizomes and through birds eating the fruit. Like privet, gingers take over the forest floor, producing an extremely dense shade and effectively preventing any regeneration of native trees. As in Mauritius, Strawberry Guava is very widespread and seriously invasive, but its density is lower in Réunion's native forests, perhaps because there are no feral pigs and monkeys[298]. Secondary forest on the lower slopes of the wet west, where not under Rose-apple, can be dominated by guava, and in 1991 it was ranked above even the bramble as the invasive species most requiring study and control[299]. The top 10 also included the potato tree (*tabac marron*), two tree-nettles and a daisy invasive of upland forest, and Lantana which favours lower drier areas; most lowland exotics came lower in the list, probably because there is so little native vegetation left to invade[300]. A more recent popular account rightly included the Helicopter Vine, Black Wattle, Brazilian Pepper, Gorse and even the tropical cut-flower staple White Calla, which carpets mid-altitude open spaces with white monocultures[301]. A common cause of invasion into good forest is the opening of new trails for forest management and hiking, though the problem can be avoided by keeping the the trails narrow enough to avoid making breaks in the canopy[302]. Research into possible biological control of Strawberry Guava, privet and the giant bramble, in conjunction with Mauritius, is under way, with several promising leads. However, great care is needed to ensure that agents chosen do not impact adversely on native species; also in both islands guava is an important fruit crop (and in Mauritius also for construction/fuel-wood), which makes it harder to devise a control strategy[303].

'Lost' species no more

While forests were being cut and exotics invading there was also some good news in the second half of the century. The first concerned two petrels. In April 1963 a petrel caught alive on the Saint-Denis seafront was brought to the island's only known bird-watcher, Armand Barau. Not recognising the bird, he sent it by air (still alive) to Paris, where Christian Jouanin described it as a new species, naming it after its discoverer. Further investigation unearthed misidentified old specimens in the Saint-Denis museum, and also Bory's account of the harvesting of nestlings at the Caverne à Cotte on the Rivière des Remparts cliffs in 1801 (p. 109). Barau's Petrel may have been unknown to science, but had been known locally for two centuries, and still carried the name, *taille-vent*, that Bory recorded in 1801[304]. However no-one had any idea how many there were, nor where they nested, though Jouanin located the general area on cliffs in the Cirque de Cilaos. Information on numbers accumulated during the 1970s (with the population in the low thousands), but the daunting cliffs of the Piton des Neiges defied those seeking nests[305]. Observations on the island and at sea suggested a population of some 3,000 pairs in the 1980s, but it was not until 1995 that the first nest was found, at 2,500m on the slopes of the Piton des Neiges massif[306]. By 2001 the population was being estimated at *c.* 6,000 pairs[307] – this apparent rise probably reflects better site coverage, as ornithologists with mountaineering skills, notably Jean-Michel Probst, explored areas previously thought inaccessible. However, since the species went unnoticed from 1801 to 1963, yet is now so obvious offshore and streaming inland in the breeding season, a long period with low numbers looks likely, from which it recovered in mid-century. A continued increase through the 1970s and 1980s has, however, been questioned. Although numbers seen from land rose, the sighting frequency out at sea was similar in 1967 and 1978–88[308]. Nonetheless, in 2003 more than a thousand birds per evening could be seen flying into Mafate, and much larger numbers over Saint-Philippe/Rivière Saint-Etienne[309].

All is not, however, well with Barau's Petrel. The birds are easily attracted to light, and most of the examples collected or handled by scientists in the 1970s and 1980s were birds grounded in built-up areas, often juveniles. Since they cannot easily take off from a flat surface, grounded birds, if not rescued, are killed by cats and dogs or run over[310]. Increased prosperity since the mid-1970s has resulted in an enormous increase in new housing, especially in the north and west, with associated road-building and streetlighting[311]. By the mid-1990s birds were being grounded in their hundreds, and something had to be done. In four breeding seasons 1996–99, a rescue network with widespread publicity collected 2,348 seabirds, 70% of them Barau's Petrels, most of which were released alive on the coast. Numbers rose to 797 in 2001, then dropped to 225–409 in 2002–04, with only 205 in 2005, a decline reflected in maxima counted offshore. In 2006 a communication failure led to Cilaos church being spotlighted during fledging, and more than 100 petrels were killed by flying into it[312]. Light pollution is not the petrels' only problem. Feral cats kill adults and young at the nest, while rats attack eggs and nestlings. Some large colonies produce few or no young because of cats; all colonies show evidence of rats, though there are some that cats have not reached[313]. These mammals appear to have spread into petrel nesting areas since the opening up of new hiking trails and 'canyoning' sites – rats following the litter trail and cats following the rats, as we have already seen in Mauritius (p. 164). Réunion's feral cat population was estimated at 100,000 in 2001![314] Human predation has been a feature for centuries; some nest sites show archaeological signs of past occupation[315], and *èn cari fuké* (petrel curry) is still a popular rural dish, if more often of Tropical Shearwater than Barau's Petrel. In 1992–93 poachers shot dozens, possibly hundreds, of Barau's Petrels as they flew inland at dusk near Saint-Pierre. Although this stopped after offenders were arrested in 1993 and 1994, the total bird-kill over four years of shooting may have run into four figures, prompting a claim that the population had been halved[316]. The nesting areas on the Piton des Neiges and Grand Bénard massifs were declared nature reserves in January 2001[317].

In the late 19th century Auguste Lantz made museum curators happy supplying Réunion Black Petrels (p. 147), but when Christian Jouanin, fired by the discovery of Barau's Petrel, started investigating in the 1960s, he soon discovered that none had been seen since. As his paper went to press bearing what appeared to be a sad tale of extinction, he got news of a fresh specimen, grounded at Entre-Deux between the ravines of Grand Bassin and Cilaos. Another turned up in 1973, in, bizarrely, Armand Barau's garden[318]. At least the species survived, but there was no hint of numbers or nesting area. It was 20 years before another was found, but meanwhile identification at sea had been sorted out, and oceanic observations during 1978–95 suggested a population of at least 45 and probably some 250 pairs. A grounded specimen was run over in 1995, and the campaigns to rescue petrels brought down by lights had turned up another 13 by December 2004[319]. Finally, in a zone pinpointed in 1999, SEOR members recorded unfamilar petrel calls in 2002 from cliffs in the isolated Grand Bassin ravine (Plaine des Cafres). They did not see the birds, but the locals called them *timize* and said they had been much more numerous 20–30 years before[320].

In March 1999 another 'lost' species turned up: Jean-Michel Probst and Paul Colas discovered a small population of Bouton's Skink when searching for endemic snails on a sea cliff near Petite Ile in southern Réunion[321]. One caught for examination was later released, and although in 2003 Probst failed to rediscover them, some turned up on Petite Ile itself in 2006[322]. The lizard is very small, shy, well camouflaged and thus easily overlooked, so may still survive on the mainland. Its inclusion in Maillard's faunal list of 1862 is the only previous record, and there are no Réunion specimens in any museum[323].

After several attempts since the mid-1980s, some possibly successful, a Green Turtle nest was watched from laying in August to successful hatching in October 2004[324]. The laying female was a wild adult, not one of the 140 released in Réunion since 1998 from a former commercial turtle farm. The farm was set up in 1977, following preliminary studies begun in 1972, to provide meat for the local market and export, but international rules (CITES) restricting trading in turtles undermined the project, and after various relaunches it eventually metamorphosed into a centre for turtle study and conservation in 1998. In recent years it has conducted publicity campaigns, surveyed the increasing numbers of turtle movements around the island, and enhanced various beaches to encourage landings, finally crowned with success in 2004[325]. Moves are also afoot to establish a tortoise park in western Réunion as "a microcosm of habitat where tortoises could live in their original conditions and be observed by scientists". The species proposed, unlike in Mauritius and Rodrigues, is the Radiated Tortoise from Madagascar, which is widely kept as a pet on Réunion. The species was imported in huge numbers in the 19th and early 20th centuries, and even used as ballast on ships! Although most were promptly eaten, enough were kept for a widespread culture of tortoise breeding to develop, with numbers estimated at 40,000–200,000 in 2003. Unlike in Mauritius where Aldabra Giant Tortoises were (and are) kept largely by large landowners and successor sugar estate companies, in Réunion most Radiated Tortoises are kept in back gardens by private individuals. Technically they are in breach of CITES regulations for this species, endangered in its native Madagascar. However, as no-one wants to discourage this thriving DIY (and largely inadvertent) conservation effort, a light touch has so far been applied, with public education programmes and encouragement for breeders to get permits and come within the rules[326].

Despite increasing human pressure around much of the coast, Red-tailed Tropicbirds are trying to colonise Réunion. Although not formally recorded before 1998, a bird ringed as a nestling off Australia was found in a garden at Saint-Gilles-les-Hauts in September 2001, and more significantly, a pair was found displaying above Saint-Leu in December 2002, and further pairs nearby in May 2003 and 2004 showed signs of nesting activity[327]. The recent rapid increase on Round Island off Mauritius may be stimulating some birds to seek less crowded nesting sites elsewhere, though numbers are still well short of the totals estimated in the 1860s (Chapter 9). Over 200 pairs of White-tailed Tropicbirds still breed on cliffs in and around the island; regularly taken for food in 1948, they were still subject to sporadic poaching for the table and as stuffed ornaments 50 years later[328]. Since the lure technique generally used attracts mainly young birds seeking a mate, a single poacher working on a fixed site can catch a succession of birds, sucking in all the local unpaired individuals.

Owl rumours had been circulating for years before tourist Eric Renman reported seeing a scops-owl-like bird at the Plaine d'Affouches in 1993. At the same time there was an old and persistent legend that Réunion was inhabited by a night-creature of ill omen, the *bébet tou't* ("spirit that goes toot")[329]. These strands merged into an suspicion that the mystery *bébet* was an owl. However a little investigation in the mid-1990s confirmed what one unbeliever had discovered back in 1972 – that the *bébet tou't* was in fact the Malagasy Buttonquail, which has a 'toot toot' mating call uttered especially at dusk and dawn[330]. However, this did not eliminate the putative owl sightings, and some recordings sounding like scops-owls. The birds seen, about the size of a typical *Otus*, are too small to be the extinct Réunion Lizard-owl. The question remains open, but Jean-Michel Probst, who has both searched for the elusive birds and followed this story for many years, is fairly convinced there is no resident owl, though there may be occasional vagrants, as in the Seychelles[331].

An unexpected re-emergence was the discovery in 2002 of a substantial thicket of Red Latan palms in a remote rocky part of the south coast. This grove, although known to locals who exploit the seeds, came as a surprise to botanists, who thought the species effectively extinct in the wild[332]. The grove, within which there is some regeneration, appears to consist of wild trees with associated native coastal flora, though an ancient plantation for thatching is not ruled out. There is severe rat predation on seeds and seedlings, as formerly happened on Gunner's Quoin, Mauritius (p. 218), and invasive plants appear to inhibit regeneration.

Status of native land birds and bats

Until relatively recently, birds in Réunion had no status unless they were vermin (*nuisible*) or game (*gibier*). Even the Réunion Harrier was classed as vermin until 1966, and not fully protected until 1974. Although still fairly common in 1948, anxiety about its status led to the change in designation in 1966, though there is little objective evidence of rarefaction – Frank Gill described it as "widespread and regularly encountered" in 1964[323]. Harriers had always been and still were regularly persecuted for alleged crimes against poultry. The first study, in 1974–76, combined with data from the BOU expedition, suggested a population of 130–200 pairs, with casual observations suggesting a gradual increase thereafter[334]. It was thus somewhat surprising to find that a census in 1997–98 indicated that there might be only 50–100 pairs. The authors argued that with no evidence of a decline, the earlier estimates must have been over-optimistic, but re-examining their figures suggested less of a difference than at first appeared, and another published analysis of the same data yielded an estimate of 120–180 pairs. Intensive surveys in three sites in 1999 found 22 pairs where the earlier census had found 17 (29% more), and even that was considered an underestimate[335]. Apparently stable over 30 years, the local estimate in 2007 was around 200 pairs. Whatever figures are accepted, the harrier is one of the world's rarest raptors and is now treated as endangered in the international Red Data Book[336]. Both in the 1970s and in the 1990s the birds were notably scarce or absent in the three large *cirques* of the interior, where persecution has traditionally been most severe. During 1997–2001 the SEOR recovered 26 wounded harriers, of which half were definitely shot or illegally captured, and another 38% probably so. These were just the birds found alive and reported – more had no doubt been killed. Shooting was recorded in several areas, notably, still, the cirques of Cilaos and Mafate, but also in the west (Tampon, Saint-Louis) and the central east (Sainte-Suzanne, Saint-Benoit). Birds are killed for fun, for stuffing and for supposed risk to chickens, though SEOR surveys recorded no poultry captured or brought to the nest in 1,500 hours of watching[337]. Although most recovered birds are subadults wandering out of safer territories, this level of persecution probably keeps the population below the island's carrying capacity[338], and may

BOX 22 SWIFTS (APODIDAE)

The Mascarene Swiftlet *Aerodramus francicus* occurs on Mauritius and Réunion and breeds in caves. It has declined on Mauritius but it is still relatively abundant on Réunion.

Account
Buffon (1770–83):

M Commerson was the first to bring this new species from the Ile de France [Mauritius]; it is not very numerous, although there are plenty of insects there. They have very little flesh, and are not good to eat; they occur equally in town and in the country, but always in the vicinity of fresh water. One never sees them land; their flight is very brisk. Their size is that of a tit and their weight two gros and a half [5.7g]. The Viscount de Querhoënt has often found them in the evenings along the wood-edge, so he presumes it is in the woods that they pass the night.

Mascarene Swiftlets *Aerodramus francicus* (upper) and Mascarene Swallow *Phedina borbonica* (lower); from Barré & Barau, 1982; for swallows see Box 23, p. 193.

---

explain some of the anomalies in estimating numbers; 40 years of formal protection is clearly not enough to remove the entrenched antipathy of rural people to birds of prey.

As the only landbird then considered endangered, the Réunion Cuckoo-shrike had the lion's share of ornithological attention in the late 20th century. In spite of all this, the reasons for its restricted distribution and decline from locally common to seriously at risk remain unclear. Pollen's pioneering study in the 1860s revealed the birds at lower elevations than today and their stomachs full of endemic beetle larvae of a genus, *Oryctes*, that he said was found on cabbage palms – so the loss of palms and the supposed extinction of the beetles was the first obvious inference. The early work in the 1970s, since confirmed, showed that territories are smaller in mixed forest than in *tamarins* higher up, but that even in the better habitat it takes far more visits per hour to feed young than in the Mauritian sister species – i.e. the food items found are smaller than ideal. These findings inform the current consensus that the present habitat is suboptimal, and that originally the cuckoo-shrike, like most of its genus, would have inhabited lowland and mid-altitude forest[339]. A large proportion of the diet is beetles, but the insects in its present range are mostly relatively small, whereas Pollen's *Oryctes* were larger, and are in fact neither extinct nor apparently connected to palms[340]. Hence the obvious solution of planting palms may thus be irrelevant. There is a cabbage-palm plantation in forest at the old (19th century) altitude at the Ilet à Guillaume in the Rivière de Saint-Denis below the Plaine des Chicots, but there were no cuckoo-shrikes there in 1974, nor in 2005[341]. Animals in suboptimal habitat are more susceptible to adverse pressures than in good areas, and the Plaine des Chicots is unfortunately not free of these. Poachers still active in the east of the reserve, furthest from the tourist accommodation and where the gamekeeper used to live, may have been responsible for the loss of several territories between 1990 and 2000. Fires, suspected of being set deliberately, destroyed nearly 100ha of forest at the Plaine d'Affouches in 1991–94 and 60ha of the Plaine des Chicots in 2006. Ship Rats are abundant, and feral cats common in the reserve; dummy nests are raided more often where rats are abundant than where they are fewer. Cuckoo-shrike nesting success reflects the same pattern, and rat control in 2005–2006 increased nesting success[342]. Disturbance from deer hunting has reduced, but conversely there are far more tourists using the reserve's paths. Most are not naturalists, but hikers more interested in striking views and personal fitness than in wildlife. They are noisy and leave litter, which attracts Norway Rats and thence cats. Low-flying tourist helicopters can seriously disturb breeding birds[343]. The new reserve is a huge step in the right direction, but much more needs to be done to understand the ecological dynamics of this secretive bird – although numbers were still estimated at 120+ pairs in 1998, they had apparently slumped to only 50 pairs by 2005. Recent studies show a massive but unexplained preponderance of unpaired males, which may have biased earlier song-territory based surveys[344].

Although the Réunion Cuckoo-shrike has had the most investigative attention, other native birds have not been ignored. In fact the first detailed study was on the commonest species, the Mascarene Grey White-Eye. Visiting the Mascarenes in 1964 on the International Indian Ocean expedition, Frank Gill was struck by the unexpected plumage variation in the Réunion population, and returned in 1967 to study it, picking up ecological information on the side; he also investigated Réunion Olive White-eyes and the interaction between the two. He was the first to attempt to enumerate any of the common species, suggesting some 556,000 for the Grey and 154,000 for the Olive. These figures have stood the test of time quite well, fairly similar estimates (by different methods) being made in 1974 and in 1983, and still quoted in 2002[345]. The other endemics received some attention in the 1970s, with all the avifauna included by Nicolas Barré the early 1980s; at that time, in the absence of Red-whiskered Bulbuls and Peking Robins, the only exotic penetrating the forest in any numbers was the Cardinal Fody[346]. Despite massive ecological changes, traces remained of the seasonal migrations up and down the mountains noted by the early visitors (p. 41), Réunion Merles and both white-eyes doing this. Baseline population estimates made for the other forest endemics contrasted in their thousands with the relatively tiny numbers of their relatives in Mauritius: Réunion Merle (20,000+), Mascarene Paradise Flycatcher (c.50,000), Réunion Stonechat (180,000–200,000)[347]; these estimates should be repeated, to monitor the effect of bulbuls and other environmental changes. So far the bulbul's spread has not had any documented impact on numbers of other birds, but paradise flycatchers had disappeared by the mid-1990s from the mixed exotic and she-oak plantations at Étang Salé (west coast) where they were not uncommon in the 1960s and 1970s – possibly an advance warning of worse to come[348].

Of small birds, only the aerial feeders appeared less prosperous in Réunion than Mauritius in the 1970s. The Mascarene Swallow, given as 200 pairs in 1974–75, had improved to 300–500 pairs in 1979–81 (possibly through better coverage)[349], while the Mascarene Swiftlet was noted as "very much less" common than in Mauritius in 1973–75, with up to 5,000 pairs around 1980. Both have increased since, swallows being seen almost anywhere in the late 1990s, and swiftlets now far more common than in Mauritius, with several colonies into four figures, including in the old railway tunnel between Saint-Denis and La Grande Chaloupe, a rare case of technology favouring wildlife! One colony, at La Chapelle (Cilaos), estimated at 10,000–15,000 pairs, had boasted a mere 10 nests in 1974 (locals devastated it when a new road made access easy), but it has become by far the island's biggest colony. It had been large before; hundreds of nests were recorded there in the 1940s[350]. A massacre of hundreds of birds at a large colony in 2003 was followed by persistent taking of nests, which have apparently acquired a reputation for enhancing the effects of cannabis[351]. This kind of 'rhinoceros horn' effect can easily completely undermine conservation efforts if the myth takes hold. The swiftlet's habitual associate, the Mascarene Free-tailed Bat, received some attention in the 1980s, but its distribution and numbers, while apparently satisfactory, are unknown in detail. The largest known roost, at a cave in Trois Bassins, suffered 90% destruction during the long wet cyclone *Hyacinthe* in January 1980, but plenty of bats survived elsewhere, as the numbers in the cave had substantially recovered by the following summer. The Grey Tomb Bat, as in Mauritius found in the drier lowlands, remains tolerably common, though was noted in the 1980s to disappear (where to?) during the southern summer. This interesting observation has yet to be followed up[352].

Seabird status

The first serious attempt to estimate the distribution and numbers of shearwaters nesting on sea-cliffs and inland ravines had to wait until the 1990s. Until 1990 only about seven sites were known for Tropical Shearwaters, increasing to 34 colonies by 1995, 235 by 1997 and 300 by 2002[353]. Colonies turned out to be small, the total estimated at 3,000–5,000 pairs in 1997, 4,000–6,000 pairs in 2002. Although birds had been heard in various places, the only confirmed colony of Wedge-tailed Shearwaters before 1990 was a few pairs on the little stack of Petite Ile off Manapany, but by 2002 13 colonies on coastal and inland cliffs had been found, one with 300+ pairs, for a total of some 1,000 pairs[354]. There is no reason to suppose that these recent discoveries represent an increase in these birds, just better data; in fact Tropical Shearwaters are very possibly far fewer than formerly.

Although there is no mention of shearwaters in the 2001 special issue of the SEOR's journal devoted to poaching, Jouanin pointed out in 1987 that "it [Tropical Shearwater] is also the *fouquet* best known to the islanders, the only one whose nest they can find with certainty . . . it is difficult to explain how the birds have been able to remain abundant in Réunion when nests are so easy to find and the chicks are considered good eating". The answer may lie in Milon's observation that while "some créoles take, for eating, adults and young that they can catch at the nest, the majority are unwilling to eat them, considering them birds of bad omen; when they are heard calling at night, it is a sign of bad luck; thus in a land which lacks owls, other nocturnal birds are the heralds of ill-fortune which humans require"[355]. It is possible that poaching mortality has declined, only to be replaced by the birds being attracted to lights at night, numbers of Tropical Shearwaters recovered being second only to Barau's Petrels[356]. Both shearwaters have suffered from the nets placed along the the motorway cut along under the cliffs between Saint-Denis and La Possession. The cliffs are up to 250m high, unstable, and subject to rockfalls especially during cyclonic rains. Vast steel-mesh nets have been hung down the cliffs to prevent rocks falling on the road, and birds trying to nest are often trapped behind them and die. It is not recorded whether the huge Brown Noddy roosts reported in the 1860s by Pollen persisted into the 20th century, but when the road was first constructed in 1959–63 no one was aware of the shearwater colonies on these cliffs, nor yet during the extensive reconstruction in 1973–76, when there was large-scale dynamiting and many nest-sites were surely destroyed. Although further from the cliff, the extended road was still subject to rock-falls (around 10,000 tons per year), and in the early 1990s the first nets were installed, added to over the years. By 2004 245,000m² of cliff-face were covered, with a further 440,000m² projected for the end of 2006, subject this time to a public enquiry and environmental impact assessment. Fortunately neither shearwater, nor the White-tailed Tropicbird, is restricted to this area, but more nets will certainly mean more dead birds, in addition to the loss or damage of many rare plants, some now confined to these cliffs[357].

The Brown Noddy is the only other breeding seabird; in company with Wedge-tailed Shearwaters and White-tailed Tropicbirds there is a colony of c.250 pairs on Petite Ile. Others regularly sit on the cliffs in the far south (Cap Méchant and elsewhere), but there is some doubt as to whether they nest there[358]. Petite Ile (c.2ha) is Réunion's only offshore islet, fortunately rat-free. Although a nature reserve since 1986, it was connected to the mainland from 1939 to 1962 by a simple 'ski-lift', and used by its then owner as a weekend retreat. He populated it with rabbits and chickens (now long gone); most of the natural vegetation was destroyed[359].

Status of introduced species

During his studies around 1980, Nicolas Barré estimated numbers of introduced doves and passerines in addition to natives, with Cardinal Fodies and Common Waxbills the most abundant at 144,000 and 116,000 respectively. Zebra Doves rated 14,000, but Malagasy Turtle Doves only 1,300; Red Avadavats were rarest, at under 300[360]. Surveys in 2001–02 on game-birds and pigeons gave the first real picture of their distributions, though no attempt was made to estimate numbers. Red Junglefowl were still largely restricted to the same mid-western secondary forests inland of Bras Panon where they were first released over a century ago. Grey Francolins, already declining in the 1970s, were confined to two patches in the north (Sainte-Marie) and mid-west (Étang Salé), many informants reporting a drastic range reduction (hunters had even voluntarily stopped shooting them). Jungle Bush Quails were concentrated in the driest zone, Saint-Paul to Saint-Leu, while Painted Quails were found in three disjunct areas: the northwest, Saint-Paul and the Saint-Pierre area; to judge by earlier (if vaguer) accounts, both continue to decline[361]. Madagascar Partridges were distributed throughout the uplands, while Common Quails were widespread in the west and the Plaine des Cafres, scarce in the east. Similarly distributed, Madagascar Buttonquails, the only native 'game-bird' and also the most abundant, were commonest in the lowlands under 200m. Of the two pigeons, Zebra Doves (still legally game) were, as ever, widespread in the lowlands outside forest, while Malagasy Turtle Doves could be found almost anywhere below the tree-line, but more frequently in the lowlands[362]. This large dove has increased markedly in recent years, helped no doubt by formal protection since 1989; quite sparsely distributed in the 1970s, it is now very common[363].

The revised hunting regulations of 1974, using powers in French legislation, formally protected all birds not specifically scheduled as vermin or game, though no list was included. No distinction was made between native and introduced species, and Réunion Merles were still officially game, though the open season was reduced from 3 months in 1974 to one in 1981[364]. In 1989 a completely new law included a specific list of all protected species, including for the first time mammals, reptiles and some butterflies[365]. The downside of this approach is that newly discovered or rediscovered animals are not covered until added – thus Bouton's Skink is not covered, though Réunion Merles and Malagasy Turtle Doves, formerly game, are now both officially fully protected. Old habits, however, die hard, and both are subject to considerable poaching, the former as a cage-bird as well as being good eating – in 2004 a poacher was surprised with 33 caged merles[366]. The vermin category was inexplicably abolished, and the Red-whiskered Bulbul became game with a limited open season, though it was re-designated a pest in 2000. Newly arrived pests such as Peking Robin, Rainbow Lizard and iguanas, like rediscovered natives, do not feature. Non-game introductions are also altogether omitted, whether benign or a danger to native fauna and flora, though there is a blanket prohibition on using birdlime.

Belated discovery of subfossil remains

It was more than a century after Dodo bones were found in Mauritius that extinct Réunion species started to be exhumed. Apart from tortoise bones found by Émile Hugot in the 1950s which vanished somewhere in the Paris museum[367], the first successful digs were not until 1974, when first Bertrand Kervazo then Graham Cowles dug in caves under the cliffs south of Saint-Paul, where the first French settlers supposedly camped in the 17th century. Cowles, ASC and Auguste de Villèle also investigated Caverne Vergoz near La Saline. In 1980 two small caves at Saint-Gilles yielded more bones, and in 1989 a major discovery was made south of Saint-Gilles in a small swamp, Marais de l'Hermitage. This was subsequently worked systematically for several years by Sonia Ribes of the Saint-Denis museum, Roger Bour and Cécile Mourer, yielding thousands of tortoise bones and a steady if limited number of bat, bird and lizard bones; further material emerged from another cave near La Saline, La Caverne de la Tortue. These digs revealed that the Réunion Solitaire was an ibis, confirmed many of Dubois's birds from the 1670s, and added the previously unknown lizard-owl and several new species of its presumed prey. The picture of Réunion's past at last caught up with information from the other islands – regrettably a bypass around l'Hermitage started in 2003 has destroyed half the swamp, and compromised further investigation of this apparently unique Réunion bone deposit[368].

Future prospects

The creation, mentioned earlier, of a huge national park embracing all the Réunion uplands, with sustainable land-use outside the core areas, has the potential to be an example to islands everywhere. Whereas in Mauritius the Black River Gorges national park is underfunded and undermanned (especially in qualified personnel), the Réunion proposals look lavish. If the final result is anything like as multidisciplinary, co-ordinated and well-funded as the preparatory groundwork has been, then it will be a triumph for vision and conservation, and a vindication of the long and lonely battles fought by the SREPN in the 1970s. The 'state of the nation' document produced by the park project authority with input from all local agencies and departments, is itself the best ever atlas of the island. The maps include wildlife data never before published (e.g. bat and swiftlet caves), the present extent of forestry plantations and vegetation, as well as standard geographical information[369]. Given the rugged terrain, accurate vegetation mapping has always been a problem in Réunion, but under DIREN auspices this is being tackled by sophisticated aerial photography and use of GPS for ground-truthing[370].

As in Mauritius, the overall prosperity of the island has increased enormously since the 1970s; thus human pressure on the environment has outstripped the also substantial population increase, 774,600 in January 2005. However, in general new housing, industry and road-building has been on former agricultural land in lowland areas, much of it in the dry east, abandoned as sugar acreage declined, or on dry savanna areas cleared but never cultivated[371]. In fact the greater prosperity and educational level has so far benefitted what is left of the natural environment by increasing the number of those concerned to protect it[372]. However this prosperity is mostly primed from France and fuelled by cheap imported energy. Should another crisis arise that isolated the island from outside supplies (as in the 1939–45 war), the situation, desperate then, would now be far more difficult – with many times more mouths to feed, and very little experience of self-sufficiency to fall back on.

RODRIGUES

In 1914 Rodrigues was still a backwater, with all transport still on foot or by pack-animal. What Bertuchi grandly termed 'main roads' were simply earth tracks assisted by a little engineering on the steep hill out of Port Mathurin. Their condition horrified governor Sir Hesketh Bell in 1916. In 1923 there were still no wheeled vehicles[373]. Due to the way the island was settled, there were, outside the capital, no real villages, but houses dispersed across the landscape, still largely constructed from latan leaves. Bertuchi's photographs show good stands of latans in one or two places, and a picture published in 1922 shows 80+, with most of their leaves cut off, in one view[374]. Hurricane Palms were still amongst the 'principal trees', and Travellers' Palms were clearly invading the Cascade Pigeon valley[375]. The population had risen by 1911 to 4,667 and to 6,573 in 1921. Trading with Mauritius during the 1914–18 war reverted to sailing ships, but was otherwise normal[376].

Etienne Thirioux, the barber responsible for some of the best Mauritian subfossil material (p. 133), went to Rodrigues in the early 1900s, and there collected two specimens of Newton's Day-gecko, the last ever seen. We do not know where on the island the lizards were collected or when, but he wrote to Mauritius from Rodrigues in 1915–16, and died there aged 71 in June 1917[377]. Although little was done to follow up Stockdale's agricultural recommendations, Paul Koenig made sure his ideas on re-afforesting Rodrigues were put into effect "with commendable enthusiasm and good planning"[378]. The assistant director of forests was sent to establish nurseries, and a forest ranger posted to Rodrigues on a permanent basis. Lessees were moved off 30ha of high ground, and an attempt was made to remove goats from the cattle-walk. He extended planting of she-oak in coastal areas, and planted up 174ha of mixed trees on the central tableland, and 49ha in the dry west at La Ferme. However over the same period ship's captain William Brebner had acquired the old concessions at l'Union and Solitude, and exploited most of the timber "consisting of fine old *bois noirs*, *bois puants* and *bois d'olive*" and cleared 81ha of Rose-apple and Strawberry Guava to plant maize and beans. The

> **BOX 23**   **SWALLOWS AND MARTINS (HIRUNDINIDAE)**
>
> The Mascarene Swallow *Phedina borbonica* is found on Madagascar, Mauritius and Réunion, but it is scarce everywhere other than Madagascar; it is a vagrant to the Seychelles. For an illustration see Box 22, p. 189.
>
> *Accounts*
> Buffon (1770–83):
>
> > This last name ['Hirondelle des blés'] *is that by which this species is known at the Ile de France* [Mauritius]. *It inhabits areas sown with barley, clearings in woods and for preference higher ground. They perch frequently on trees and stones, and follow herds* [of livestock] *or rather the insects that torment them. From time to time one sees them flying in large numbers behind ships that are in the island's roadstead, always in pursuit of insects. Their call is much like that of our* hirondelle de cheminée [Barn Swallow *Hirundo rustica*]. *The Vicount de Querhoënt has observed that the* hirondelles des blés *frequently flew about in the evening around a cutting that had been made in a mountainside, from which he judged that they pass the night in holes in the earth or cracks in the rock, like our* hirondelles de rivage [Sand Martin *Riparia riparia*] *and our swifts* [Common Swift *Apus apus*] *... M. de Querhoënt has only heard about the nesting of this bird from an old Creole from Bourbon* [Réunion], *who told him that it happens in the months of September and October, that he had taken these nests several times in caves, rock holes etc., and that they are composed of straw and a few feathers, and that he had only ever seen two eggs, grey speckled with brown.*
>
> Buffon's information on swifts and swallows is in general very accurate (see Cheke 1987c), though the swift normally weighs nearly double Commerson's figure, and of course nests and roosts deep in caves not in the woods (see Box 22 on p. 189) – it is rather surprising this was not known to the locals in the islands.

island was gaining plantations but still losing native forest. Magistrate L. Ulcoq, on the island 1919–20, wanted the government to acquire l'Union and Solitude, but the new director of agriculture in Mauritius, Harold Tempany, sent to report in 1920, was against this, largely because he saw these sites only in terms of a proposed agricultural station, not as valuable for watershed protection or forestry [379]. Sometime before 1940 Solitude was, however, acquired and plantations begun that were to become key to the endemic fody's survival [380]. Wallace Hanning, magistrate from 1922–25, finally acted on Stockdale's warning about introducing invasive species, banning "the introduction into Rodrigues of any mongoose or any bulbul *Pynonotus* [sic] *jocosus*", any other "mammal, reptile, insect or bird" without a permit and Prickly Pear, Lantana or *herbe condé*. At the same time he promulgated curious bird-protection rules, which ignored the endemics in favour of waxbills, mynas and Cardinal Fodies, together with the two tropicbirds[381]. Although Rodrigues remains free of the other pests mentioned, Lantana is now everywhere, and dominant on Plain Corail. It appears to have arrived in the late 1920s; first found by Wiehe in 1938, by 1949 it was "spreading considerably on the southern coast", and in 1970 was widespread in the lowlands[382].

Tempany disagreed with Stockdale and Koenig's ideas, and indeed throughout most of the 20th century Rodrigues was plagued by an ever-changing succession of conflicting notions on how best to manage the island's land. Tempany preferred coconuts to she-oak for coastal planting – a good idea had the Rodriguans not decided to dig up and eat the germinating nuts, rather literally nipping the idea in the bud. Limits of the cattle-walk and how much land within it could be cultivated were frequently changed. Natural disasters also played their part. After cyclones there were food shortages and sudden demand for wood to rebuild shattered dwellings. A severe drought combined with insect attacks in 1929 caused crops to fail, and many of the by-then 8,000 inhabitants, starving, cut down thousands of latan palms to eat the cabbage. The latans never recovered[383]; despite the enormous utility of these trees for thatching and pig food, no one initiated a replanting programme until the 1940s. There had been little or no regeneration for decades, with rats and pigs eating all the seeds. Nonetheless Octave Wiehe reported latans abundant in 1937–41[384] – this must have been relative to the state of other native species. It is not clear what happened to the "fairly large" plantations initiated by the agriculture department in the 1940s. There was no trace of them the mid-1970s, so they may have fallen foul of Hotchin's 'agricultural revolution' of 1955–68 (p. 195).

Octave Wiehe visited Rodrigues in 1937 and 1941 to investigate an outbreak of citrus canker. On his recommendation large areas of citrus scrub, as well as individual householders' trees, were destroyed in a failed bid to stop the disease spreading, a serious economic blow to the locals. At the same time he surveyed the island's vegetation for the first time since the Transit expedition of 1874[385]. His map scale is too small to get a detailed picture of wooded areas (and forestry plantations are excluded), but there was evidently significant forest cover (mostly Rose-apple) from Mont Malartic to Mon Limon, at the heads of all the southern valleys from Rivière Coco westwards to Mourouk, and in the two valleys leading into Baie aux Huitres, together with the ridge between them. Wiehe considered the upper Cascade Victoire valley to retain the best native vegetation, but he appears not to have visited Cascade Saint-Louis[386]. Some 809ha was under Rose-apple, but the valleys of Cascade Pigeon and Solitude are shown as 'cultivated land',

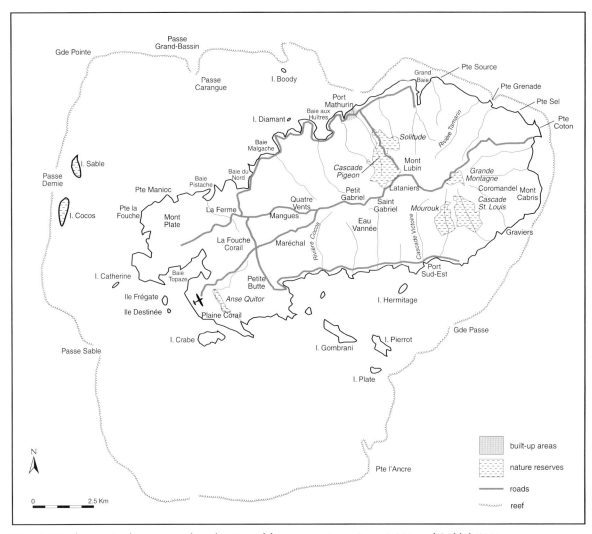

Map 8.5. Rodrigues: Settlements, roads and protected forest areas. From Jauze (1999) and Saddul (2002).

possibly including plantation. The former, at least, as the principal water catchment, must have been at least partly wooded[387]. His photographs show the ridge road from Mont Lubin to Montagne Malartic arched over with large Rose-apples, where he, and others in the 1930s and 1940s, saw endemic warblers and fodies[388]. Wiehe noted that some plants pathologically invasive in Mauritius (Strawberry Guava, *bois d'oiseaux*), were much less pernicious in Rodrigues, attributing this to the absence of Red-whiskered Bulbuls to spread the seeds around[389].

The 1929 drought revealed the seriously overgrazed condition of the cattle-walk, and damage caused by grazing outside it, provoking a commission from Mauritius to recommend the complete removal of goats from the island. Although environmentally sound, this would have been a disaster for rural Rodriguans, dependent on these animals for exchange, meat and meagre supplies of milk. A change of governor, and a new magistrate's determination to confine goats to the cattle-walk, avoided the economic disaster, but only served to perpetuate the system of institutionalised overgrazing[390]. Always seeking to raise revenue from the impoverished island, an export tax had been imposed on popinac seeds. This was revoked in 1930, but demand in Mauritius was dropping; by the end of the decade, while some seed export continued, the extensive popinac thickets were also being let for making charcoal (supposedly leaving native trees intact), followed by a couple of years of cropping[391]. Although never quantified, the increasing population was having an escalating effect on forest from the concomitant requirement for fuelwood, the only source of energy for cooking, washing etc. The primary source, inevitably, was the 'government's' forest.

Unlike the straitened conditions in Mauritius, the 1939–45 war boosted the Rodriguan economy, as troops stationed there brought money into the island and ate local produce. However, the rising population was causing concern, and a report by the Natural Resources Board in 1948, chaired by botanist Reginald Vaughan, insisted that a "reduction of the present numbers living on the island [by then 13,326] should be regarded as a matter of the utmost urgency", emphasising that Rodriguans had outstripped Malthus's expectations of doubling in 25 years, by doing it every 20 over the previous century[392]. Given the island's meagre medical, hygiene and water facilities this was fairly remarkable, though an absence of major tropical diseases (notably malaria) certainly helped. In 1948 North-Coombes was struck by "the gradual encroachment of cultivation in wooded areas which should have remained untouched. Land is being cleared for crops around centenarian mango trees . . . The next stage will be the cutting down of the mango trees and further desolation"[393].

Hotchin's axe

Various proposals to promote emigration from Rodrigues came to nothing, so 'plan B' was to improve the agricultural base to support a larger population. To this end Philip Hotchin, veteran of schemes in West Africa, was drafted in by Director of Agriculture Lucie-Smith in 1955, staying until 1968[394]. Initially hailed by officials and observers as Rodrigues's saviour, he turned previous policy upside-down, particularly in relation to forest cover. In North-Coombes's polite understatement he "pursued an unwise policy in regard to the conservation of adequate cover of trees in the most vulnerable parts" of the island – i.e. he chopped them all down wherever he could. He was basically given *carte blanche* by the governor to do whatever he thought appropriate, and funded to do it. Hence he could overrule forestry service objections, and convert into arable fields most of the protection forests in the uplands that had been carefully built up since the 1880s, and subsequently enhanced by Koenig. While some of the summits, too rocky to cultivate, retained their cover, wooded land further down was completely cleared, notably the large tracts in the Baie aux Huitres hinterland shown as wooded on Wiehe's 1949 map[395]. Only some privately owned patches in Camp Baptiste and Vainqueur escaped Hotchin's axe. Octave Wiehe's 'best' native forest in Cascade Victoire had completely disappeared by 1964; it was too steep to terrace and cultivate, so was probably simply cut down once locals saw that wholesale butchering of trees was now official policy[396]. Hotchin's plan was to create terraced agriculture over as much of the island as possible, apparently irrespective of local conditions. He even eradicated most of the popinac, although it provided fuel-wood, forage for goats and seeds for export; from 130 tons sent to Mauritius in 1963, production fell to 10 tons in 1967, and none thereafter[397]. Exports of garlic, onions and pigs rose spectacularly, new dams were built and water pumped uphill to new reservoirs, and crop rotation and fertilisers were introduced. Short-term visitors were taken in; France Staub was effusive in 1967:

> *Hotchins* [sic] *has perfectly understood the Rodriguan spirit; he has succeeded in awakening their pride and given them the taste for a national effort. Under his aegis, the Rodriguan countryside, formerly desert-like and showing from all sides the poverty of its carpet of vegetation, has returned to a picture of prospering hillsides. The contoured steps have controlled erosion, manure has enriched the soil, and the forests of tecomas and she-oaks protect the tops of the hills and soften the trade-winds loaded with salt . . .*[398]

There were, however, downsides to all this frenetic and well-intentioned activity. Most obviously, it did not even work. By 1974, only six years after Hotchin left, the system had largely broken down. The terraces, possibly built too hurriedly, were seriously damaged in the severe cyclones of 1968 and 1972, and also constantly degraded by wandering cattle. Without Hotchin's whip hand the Rodriguans reverted to type – being as laid-back as possible consistent with survival. Maintaining terraces was not part of this world-view[399]. As Daniel Gade saw it in 1983: "The island culture, African in its farming and herding methods, yet disconnected [from African norms and within itself] in its social organisation, developed no conservation ethic of its own"[400]. Already in 1971 the government's four-year plan talked as if Hotchin had never been:

> *With 2500 to 3000 farm families seeking to earn a livelihood from subsistence crop and livestock, the land has become seriously eroded, overgrazed and denuded of forest cover. The most urgent and important task in Rodrigues is to restore physically the capacity of the land for crop and livestock production. Because most of the land is on steep slopes, this will require terracing . . . On the basis of a land capability survey to be carried out in the latter part of 1971, the island will be divided into arable areas, pastures and areas to be planted and maintained under forest.*[401]

Nevertheless, by 1981 a quarter of the terraced land was fallow due to soil infertility. The terracing was all too late, the good soil having been eroded off years before[402].

Insidiously, all the foresters' work over decades to get the people to respect trees and forest cover was undone. Seeing the government cutting its own forests was a green light for everyone else to do so, and this legacy severely affected attempts at re-greening the

BOX 24   CUCKOO-SHRIKES (CAMPEPHAGIDAE)

Cuckoo-shrikes are small to medium-sized birds that skulk in the forest canopy. Two endemic species occur on the Mascarenes, *Coracina typica* of Mauritius and *Coracina newtoni* of Réunion. No direct written evidence from early accounts refers to them but they would certainly been considered fair game along with the *Hypsipetes* bulbuls. The Réunion species is endangered and restricted in range. The Mauritius species is more widespread, but it is rare and also considered endangered.

Account
Quoy & Gaimard (1824), in Mauritius in 1818, give the earliest eyewitness account:

In that colony [Mauritius] *this bird is confused with Merles, despite the difference in appearance and habits. It does not go in flocks like the latter,* [but] *lives alone, isolated, abruptly traversing the great woods; it is in those of least elevation that it seems to be best pleased. This shrike is quite rare; we only saw this one, which one of us killed in an excursion around the island.*

Mauritius Cuckoo-shrike *Coracina typica* by Jean-Michel Vinson, 1974.

---

island right through the 1980s. Hotchin's insensitivity to island ecosystems was matched only by Miguet's in Réunion at the same time, but it was more immediately damaging – the rapid and extensive loss of tree cover, combined with severe cyclones, very nearly caused the extinction of the Golden Bat and the two surviving endemic birds. The already tenuous patches of native vegetation came under intense pressure. Cascade Victoire has been mentioned, but in the 1970s and 1980s several endemic plant species were found to be down to their last one or two specimens, most famously the *café marron*, an attractive white-flowered coffee relative, by 1980 reduced to a single individual by a much-used road. This bush became a sort of pilgrimage site for botanists, and suddenly a plant that had never had any medicinal or other particular use became a cure-all for venereal disease and a potion for liver cleansing after alcohol abuse[403]. "Ordinary people could not understand the fuss being made [over endangered species] if the plant had no practical uses"[404]. After being nearly killed several times by persistent 'medicinal' pruning, the plant survives only by being surrounded by three fences and attended by a watchman. Fortunately, apart from the endemic hibiscus, the other equally rare species have received less unwelcome attention from the public[405]. Hotchin's arboricidal tendencies even extended to such iconic Rodriguan trees as latans. By 1974 so few remained that they were in danger of being killed by excessive leaf harvesting, and even collecting seeds for replanting was difficult as they were 'eaten by small children'[406].

As Staub's quote indicates, some trees were planted under Hotchin's regime, principally *tecoma* and she-oak. These plantations, in only very limited areas, were intended to combine timber production and watershed protection; while they produced wood, more often firewood coppice than timber, they were inferior both in protection and wildlife habitat to the Rose-apple they replaced. Although Rodrigues Fodies like the flowers when the trees are in leaf, *tecoma* is deciduous in the dry season, becoming unsuitable for birds, and allowing soil to dry out and erode. She-oak is no use to the endemic birds or bat for food or shelter, and gives such light shade that it too is poor cover on dry slopes[407].

Hotchin's programme also recklessly ignored the endemic wildlife. While no one but local foresters and a few trigger-happy Mauritian visitors noticed the bat between 1917 and 1963, the birds had received enough attention for Jean Vinson, visiting in 1930 and again in 1963, to see a serious decline. More sanguine than Staub, Vinson blamed the "intensive clearance of wooded areas for agricultural reasons" for a big decline in fody numbers and the perceived impending extinction of the warbler[408]. Vinson and his companions found no Rodrigues Warblers at all in 1963, and the following year Frank Gill saw only one, promptly shooting it for the Smithsonian collections in Washington[409]. Gill was part of the US contribution to the International Indian Ocean Expedition, whose ornithologists were guided by a *Preliminary Field Guide to the Birds of the Indian Ocean* which actively encouraged collection of specimens, with not a word of restraint in relation to rare or endangered species. This was before the first *Red Data Book* appeared, but Greenway's seminal *Extinct and Vanishing Birds of the World* had been published in 1958 – which, while not mentioning the Rodrigues species, treated Mauritius at some length. As Bill Bourne pointed out when reviewing the *Preliminary Field Guide*, its attitude to collection was cavalier, not to say irresponsible[410]. However, the warbler held on, filmed at the nest by a French dentist and amateur bird-watcher in 1971–72, and Staub claimed that "for the last seven years [1963–70] there had been no noticeable decrease in the *Foudia flavicans* population". This was before the worst cyclone ever recorded in Rodrigues, *Fabienne*, in February 1972[411].

Introduced species also succumbed to Hotchin's operations. Wild guineafowl, heavily persecuted for eating planted seed, died out around 1963. Grey-headed Lovebirds, blamed for attacking ripening maize, crashed suddenly in 1956, were very scarce in 1964, but persisted in tiny numbers until late 1974[412]. Grey Francolins, also persecuted, managed to survive. Deer, which had already declined to a handful during 1937–54, had disappeared by 1956[413]. In the other direction, a new exotic was first noted in 1963, the Yellow-fronted Canary *Serinus mozambicus*; uncommon, it could have been there unnoticed for decades[414].

Fodies, warblers and bats

In early 1974, Rodrigues Warbler and Fody numbers were about 23 and 26 pairs respectively, with a maximum of around 70 Golden Bats – dangerously small populations for any species. Not only that, 75% of the fodies were clustered in about 16ha, a tiny area for nearly a bird's whole population[415]. Foresters who had been in Rodrigues in the 1960s revealed the detail of Hotchin's clearances – the loss of Rose-apple on the hilltops removing around 75% of the birds' habitat and a good food source for the bat, and the removal of large fruiting trees, Mango, Indian Almond and especially Tamarinds, devastating the bats' favourite roosts and most reliable food supply[416]. There had been a large Tamarind copse at the eponymous locality Tamarins, where bats had regularly roosted and fed, and in general the rest of the upper Baie aux Huitres Valley (Jardin Mamzelle/ Camp du Roi) had been their stronghold, but had been almost completely denuded of trees by Hotchin. The summit ridge around Les Choux and Malartic, formerly "dark with *jamroses* which overlooked roads and made them slippery with fruit", was now bare and windswept, and most large trees had been felled in St Gabriel nearby. Bats had also favoured Anse Mourouk valley, but there too the large old trees had gone. In the mid-1950s bat numbers had been around 1000, crashing precipitously in the late 1960s as felling intensified before Hotchin left; fody numbers, already declining, were still around 100 pairs in 1964[417].

Once the habitat had been so reduced and fragmented, all three species became extremely vulnerable to cyclones, and (Staub notwithstanding) numbers of the easily counted fody, had in fact dropped to as few as five or six pairs in December 1969, following cyclone *Monica*. There were still only 10 pairs at most in December 1972 after *Fabienne*, from about 25 pairs 14 months earlier – about as near to extinction a bird can get and still pull through. The warbler's minimum numbers may not have been much higher – 15 pairs or so in 1972, but probably fewer post-*Monica* in 1969, when three naturalists "searched Cascade Pigeon and Solitude for a week without finding any warblers"[418]. In 1971–72 a maximum of only 10 bats were seen, all at Jardin Mamzelle, with none in Cascade Pigeon[419]; the observer may have missed some, but projecting back from the rate of increase during 1974–78, the numbers after *Fabienne* can hardly have exceeded 30. Bats were also sporadically hunted, largely by Mauritians posted to the island; they were used to hunting fruitbats at home and were probably unaware that numbers in Rodrigues were dangerously low. Even when bats were still relatively common in 1963, Jean Vinson thought hunting was excessive and recommended restrictions, though nothing was done[420].

Frank Gill's visit in 1964 was the first by a non-Mauritian scientist since the 1874 Transit expedition, but did not inspire any conservation initiative, notwithstanding Jean Vinson's worries published the same year. Other locally published reports described the low numbers of birds, but no action resulted, though a land use report, apparently shelved before publication, recommended nature reserves for plants and birds. No such proposals appeared in the government four/five year plans for 1971–74 or 1975–80[421]. It was coincidental that on the Transit's centenary both the BOU expedition and conservation-minded French botanists arrived, bringing the plight of Rodrigues's devastated biota once again to the attention of the international scientific community[422]. Appeals to protect both birds and bat, the latter most urgent, led to the retiring magistrate in 1974 attempting to protect Golden Bats under the 1923 bird protection regulations. This breach of logic inspired the Forestry Service, on behalf of government, to draft new rules, though these never became law[423]. The endemic vertebrates were finally legally protected by the *Wildlife Act* of 1983, which extended Mauritian protective legislation to Rodrigues for the first time[424].

In 1974 the rotating magistrates from Mauritius were for a few years replaced by a new style of administrator, brought in, like Hotchin, from outside. Nigel Heseltine, again like Hotchin, had more or less a free hand to rule as he chose[425], but differed in being much more sensitive to environmental issues. He was sympathetic to wildlife, and although the new regulations never materialised, he apparently prevented any bat-hunting during his watch. Mauritians again began shooting bats in 1977 just after Heseltine left, though in the absence of cyclones numbers had continued to increase[426]. The 1974 BOU report had reached Jersey Zoo, and Gerald Durrell, already interested in Round Island and Mauritian endemic birds, decided in discussion with Wahab Owadally, Conservator of Forests, to add Rodrigues to his itinerary, catching the first bats for captive breeding in May 1976[427]. Further bats, and also some fodies, were caught for the breeding programme in October 1978; bat numbers were up to at least 151 (despite shooting) and fodies to at least 100 pairs[428]. Golden Bats bred easily both in Jersey and at the Black River facility in Mauritius, and are now a feature in zoos all over the world. Plans to introduce captive-bred bats to other Indian Ocean islands to provide a reserve population came to nothing, the urgency in any case waning as the species recovered in Rodrigues itself[429]. Fodies were less amenable, and although a number were bred in captivity, the on-going increase in the wild population led to a run-down of this programme[430]; however, the lessons learnt were put to good use in breeding Mauritius Fodies from 2003 onwards. The October 1978 captures were timely, as Rodrigues was hit by another severe cyclone in February 1979, halving the wild bird and bat populations[431].

Continued pressure on forests

Severe drought conditions during 1973–78 forced the authorities to allow livestock, their normal pastures bare, to graze in the protection forests, particularly Cascade Pigeon – a decision which, however necessary, led to increased disturbance and unauthorised wood-cutting. It was very evident in 1974 that foresters were fighting a losing battle trying to grow trees when the local population considered forested land to be a source of free fuel as of right. There was an obvious solution: given that both fuel and forestry were necessary, the government should subsidise paraffin, paying for it (in the long-term) by diverting foresters' time from the fruitless struggle against woodcutters into productive forestry[432]. The suggestion met with no response. By 1978 it had got worse:

> In 1974, to relieve the plight of livestock dying from prolonged drought, permission was given to graze cattle and goats in the forest land; this facility, once granted, has proved impossible to withdraw, or at least it has been impossible to enforce its withdrawal. Cattle, goats and pigs and attendant herders now wander through forest lands where before they were not, and the suitability of habitat for the warbler [is] drastically impaired. This degradation is worst at St Gabriel and Sygangue, but also evident at Mont Limon and Cascade Pigeon, where it may eventually affect the Fody and the bat, especially as illegal woodcutting has also increased sharply since 1974.[433]

Ten years later, Wendy Strahm wrote that "the destruction of trees for firewood for cooking" was still a major problem; plantations for firewood were being established, but, echoing the earlier suggestion "a programme to decrease the need for wood should be implemented"[434]. In respect of firewood she was over-optimistic about mangroves being propagated in northern bays, though Réunion botanist Thérésien Cadet had pointed out years earlier that similar plantations started in 1959 were "doomed to failure". The substrate was not sufficiently muddy, so germinating plantlets from fruiting trees could not penetrate sufficiently and did not establish[435] – no doubt why no mangroves were native to Rodrigues (unlike Mauritius). Indeed, the original attempt had failed completely; Strahm was referring to a new initiative started in 1980, whose trees barely reached headheight 20 years later, though they were beginning to accumulate their own silt[436]. They may eventually produce spawning grounds for fish, but their growth rate is far too slow for firewood production! By the end of the century the mangroves were seen primarily in terms of filtering silt eroding off the land to prevent it spreading through the lagoon[437]. Although there had belatedly been a special temporary import of paraffin in 1978 to reduce pressure on forests, no proper fuel policy was initiated, and in 1983 90% of Rodriguans still cooked with wood[438]. However, due to other socio-economic factors, the percentage using wood-fuel declined to 55% by 1990, but against a rapidly rising population (by then 34,000) the impact was blunted[439]. Nonetheless, one of us, returning in 1999 for the first time in 21 years, noted a vast improvement in the condition of the forests (as well as a larger area wooded), and MWF botanist Alecks Maljkovic confirmed that cooking with bottled gas had by then largely replaced the use of wood[440].

Rodrigues, estimated to be able to support 3,000 cattle and 7,000 sheep and goats, had 4,000 and 12,000 respectively in 1981. In 1995 there were no less than 9,800 cattle, and measures were yet again being proposed to prevent free-ranging and confine them to the cattle-walk, which despite good intentions from Hanning in 1923 onwards, had never been fenced[441]. However, there was progress for some areas of native vegetation. Various proposals for nature reserves had been made since the 1970s, most of them focusing on Cascade Pigeon for the vertebrates and Grande Montagne, Anse Quitor, and Cascades St. Louis and Mourouk for the vegetation[442]. Following EEC-funded field-work from 1982 onwards, Grande Montagne and Anse Quitor valley were declared nature reserves and fenced in 1986[443]. A decade later Wendy Strahm told an interviewer the irony of how the reserve fencing was achieved:

On Rodrigues, she says, her project ran on a ten-thousand dollar budget. The FAO had a big project on the island to revitalise agriculture. It was valued at thirty-three million ECUs. "They called the manager Monsieur Trente-trois Millions – every Rodriguan knew he had thirty-three million ECUs; he had a bigger budget than the entire budget given to Rodrigues from Mauritius." Wendy observed the FAO workers putting up fencing to keep the goats out of agricultural areas, and mentioned in a friendly way to Monsieur trente-trois Millions that she could certainly use some of that fencing for a nature reserve. With never a question, he gave it to her. "It was a drop in the bucket to them," she comments, adding that the head of the program justified the gift as protecting watersheds. The important point is that we did things through that agricultural programme that we could never have achieved through our small conservation grants. I used to dream wouldn't it be nice to get 10 kilometers of fencing to put around this little reserve? And then one day it happened.[444]

Cascade Pigeon, most of Cascade St Louis and the twin valleys of Mourouk were also fenced in 1986 to keep livestock out, principally for watershed and erosion control, but with conservation input. As with rabbit removal on Round Island (pp. 223–224), this has had the unwanted effect of allowing invasive weeds, previously controlled by grazing, to flourish, greatly diminishing the benefit to native plants of fencing herbivores out[445]. This difficult issue remains unsolved. The flora clearly needs some mild herbivorous input, preferably from tortoises. Since 1996 the UNDP's Global Environment Facility have funded intensive weeding and restoration operations in Grande Montagne and Anse Quitor, aimed at recreating the original habitat[446], but manpower and funds have not extended to doing the same in St Louis and Mourouk. The progress at Anse Quitor was nearly derailed in 1996 by a plan to extend the nearby airstrip into the reserve, but an international campaign in 1999–2000 led by MWF persuaded the government in 2001 to extend in the other direction, and also to acknowledge the permanent importance of the reserve[447]. Recent conservation policy has actively involved Rodriguans in protecting and rehabilitating their natural heritage. By 2000 the MWF project leader Richard Payendee was a Rodriguan, and aid money was also funding Mary Jane Raboude to give talks in villages, schools and youth groups, not only helping Rodriguans to understand and appreciate their unique biota, but also inspiring volunteers to come forward and help with replanting and other work[448].

Current status
The Rodrigues Fody, Rodrigues Warbler and Golden Bat have benefitted significantly from increased forest cover since the mid-1980s, and indeed fody and bat numbers have risen to levels that are difficult to count.

From 1981 to 1999 warblers and fodies first spread out more evenly throughout the wooded areas, then recolonised Grande Montagne, a kilometre across cultivated land from the nearest 1980s habitat (for warblers) at Mont Limon, and 1.5km from earlier fody territory[449]. Such distances may look trivial, but these birds are so extraordinarily reluctant to disperse that spreading to new areas is a major achievement[450]. Rodrigues Warblers were slow to respond to better conditions, censused numbers remaining between 10 and 30 pairs from 1972 to 1991, after which they increased by 1999 to an estimated 150 birds[451]. The fluctuations during 1972–91 can only be partly accounted for by cyclones, *Fabienne* (1972) and *Celine II* (1979) having far more effect than *Damia* (1982) and *Bella* (1991) despite being of similar intensity[452]. Perhaps by 1982 forest regrowth had reached a stage where warblers were better protected from storms, but disturbance from woodcutting restricted breeding success until wood use for fuel declined in the 1990s. For Rodrigues Fodies the pattern is clearer – a steady increase, with stepped setbacks after severe cyclones. From 60–75 pairs in 1981–83, the numbers rose to *c*.100 in 1989 and *c*.150 in 1991, reaching at least 334 pairs in 1999 (plus many unpaired birds), spread over an area five or six times larger than in 1978[453]. The bat's progress has been even more spectacular: from about 70 after cyclone *Celine II* in 1979, numbers rose steadily to 800+ in 1991. Cyclone *Bella* then halved numbers to 400+, after which they rose to 1,200 in 1997 and 1,500+ in 1998, reaching *c*.3,500 in 2001–02[454], by which time the former 'world's rarest bat' was beginning to be seen by some Rodriguans as a menace to their fruit crops![455] New roosts have become established outside the 1970s/1980s range, the animals returning under now improved conditions to Jardin Mamzelle, Acacia (= Camp du Roi) and Mourouk, and spreading out to Vainqueur, Anse Baleine and Rivière Coco. Unlike the birds, Golden Bats are very mobile, and, while needing tall trees for roosting, will go anywhere on the island to find fruit[456]. If this successful conservation story is to have a happy ending, some solution has to be found to the conflicting interests of wildlife preservation and the needs of fruit growers, whether commercial or self-sufficient. It goes against the grain for an endangered species, but Rodrigues is a very small island, and limited official culling may have to be sanctioned, as was being openly discussed on the island in 2003[457]. Programmes increasing native plant numbers may reduce the pressure, but the bats have long depended on cultivated fruit trees, as well as the ubiquitous Roseapple. The issue was put into temporary abeyance by cyclone *Kalunde*, which struck in March 2003, followed by rains heavy enough to cause the normal post-cyclone flowering and fruiting to abort. Bat

> **BOX 25** **BULBULS (PYCNONOTIDAE)**
>
> Bulbuls are medium-sized passerines that frequent forests and scrub, feeding on fruit and insects. They are noisy birds and can be approached quite closely. Both Mauritius *Hypsipetes olivaceus* and Réunion *H. borbonicus* Bulbuls (known as merles on the islands) were considered delicacies and because of this, they are almost the only passerines that can be positively identified from the early literature. In Réunion as late as the 1920s, up to 50 merles could be shot in one outing, and they were often used to make paté (Cheke 1987b). While the Réunion bird remains fairly common, the Mauritian species has become very rare. A *Hypsipetes* bulbul once inhabited Rodrigues, as confirmed by fossil remains collected in 1974, but was never recorded in the early literature.
>
>
>
> Réunion Bulbul *Hypsipetes borbonicus*. From Berlioz, 1946.
>
> *Accounts (selected from many)*
> Feuilley (1705) in Réunion in 1704:
>
> *The huppes and merles are the same size as those in France, and are marvellously flavoured. They are fat at the same time as the parrots, living on the same food. Hunting [them] is done solely with sticks or rods six or seven feet long with which they are caught, though the [this type of] hunt is little practised. Feral cats destroy a lot of them. These birds allow a very close approach; the cats can take them without them [even] moving from their perches.*
>
> Le Gentil (1729), in Réunion in 1717:
>
> *In the months of July and August, the months when winter reigns, one sees descending from the mountains a species of thrush, a very fat bird of exquisite flavour. It lives on rice and wild coffee. One catches them using a running noose attached to a rod. They are so unfearful that they often perch on the hunter's arm. The least blow knocks them down, and they are so fat that they have trouble flying.*
>
> Bory (1804) on merles in Réunion in 1801:
>
> *We stopped at Trois-Jours [uphill from Le Chaudron, St Denis] to dine on the merles we had killed en route. These merles are not the same as European ones: their plumage tends to the slaty and dark brown. They make a sort of tremulous creaking call, which appeared to me their sole song. They taste very good, and are incredibly stupid. In certain little-frequented areas one can kill them with sticks. They hardly go away when a gun is fired, and I have seen them killed after a first shot missed without their having shifted at all.*
>
> La Motte (1754–7), in Mauritius 1754 and 1756:
>
> *In the months of July, August and September our forests are full of small birds the size of a quail, grey in colour with yellow beak and feet, which are called quite improperly* merle. *This small bird is a ball of fat worthy of an Ortolan [bunting]. A mediocre shot can kill twelve or fifteen dozen in a morning. I'm surprised that there are any left after the prodigious destruction that happens every year. During the rest of the year they are thin, and are not eaten.*

numbers fell to under 2,000, heavy juvenile mortality adding to deaths caused directly by the storm, but had recovered to over 4,000 by late 2006[458]. Warbler and fody numbers appear to have been less affected. In 2005 the native forest restoration on Grande Montagne was supporting increasing numbers of the two endemics[459].

New introductions

Rodrigues did not escape the epidemic of new introductions affecting Réunion and Mauritius in the 1980s and 1990s, acquiring several animals already long-established in Mauritius. Around 1986 an anonymous islander imported Bloodsuckers to liven up his garden at Quatre-Vents, and in about 1997 Blue-tailed Day-geckos arrived, apparently as eggs in crates of Coca-Cola![460]. The Bloodsuckers rapidly spread over the island, becoming a pest to honey producers, an important new island enterprise – the lizards sit on hives and pick off the bees[461]. The day-gecko, still restricted in 2004 to a small area around Port Mathurin, could be considered an analogue for the lost Newton's Day-gecko, but ecologists would have chosen a different species[462]. In 1997 House Shrews were first seen in Port Mathurin, and by 1999 had spread widely, rapidly decimating the large centipedes formerly so characteristic of Rodrigues: in 1972 "each large stone lifted sheltered a centipede", and they were still abundant in 1995[463]. The shrews are thought to have introduced themselves via a ship from Mauritius[464]. Other presumed 20th century immigrant vertebrates are two house geckos: the Worm Gecko *Hemiphyllodactylus typus* (first recorded in 1930), and the Indian House Gecko *H. brookii* first noted in 1983; even the Cheechak was not formally recorded before 1930, but is likely to have arrived in the 19th century, although still absent in 1874[465]. The MWF/forestry house at Solitude must now be one of the world's best places to watch house geckos – five species can be seen together clustered round the light over the porch[466]. There is nothing to say when the Flowerpot Snake *Ramphotyphlops braminus* arrived – it could have been on the island since Marragon's time[467]. In 1999 a toad was reported, but apparently there was just one, and they have not established[468]. Finally, the giant snails: absent in 1874 and not mentioned by the reports around 1914–17, they probably arrived sometime in the early-mid 20th century[469]. By the mid-1960s their abundance stimulated a control attempt through the introduction of Rosy Wolf Snails[470].

Keen as he was on nature conservation, Nigel Heseltine is blamed for one most unfortunate plant introduction, *piquant lulu*, a very spiny African acacia[471]. This small tree thrives in semi-dry areas such as Rodrigues and western Mauritius. Intended as quick-growing protective ground-cover, it was officially promoted in the early 1980s, but rapidly became invasive and is now as much of a pest in Rodrigues as in Mauritius[472].

Seabird status

Following the devastation in the 19th century, life for Rodrigues seabirds was relatively quiet in the 20th. Little had changed on the islets between 1917 and the next report in 1942 – some hundreds of both noddies and Fairy Terns nested on Iles Coco and Sable, with unknown numbers of Wedge-tailed Shearwaters on Frégate. The latter were still numerous enough to produce "the most extraordinary uproar" at night[473]. In 1952, possibly having been left alone during the relative prosperity of the war years, there were "innumerable" noddies and Fairy Terns, though in 1958 (July) and 1964 (September) only some 300 Brown and 600 Lesser Noddies were counted, with about 500 Fairy Terns (both islets combined)[474]. The birds breed on and off throughout the year in varying numbers, so a single-date count may be misleading, and certainly underestimates the overall total using the island[475]. By 1974, although noddies had apparently increased (1,500–2,000 Lesser Noddy nests), Fairy Terns numbers had collapsed, with only 15 pairs or so surviving[476]. The islands, though not then officially nature reserves, had a resident warden who was supposed to protect the birds, but was in fact doing the opposite. Large numbers of eggs were collected annually, and adult Fairy Terns were being slaughtered – more than 400 corpses were found in a pit on Ile Coco in 1969[477]. Administrator Nigel Heseltine sacked the offending warden (causing a confrontation with his trade union), but the problem was not solved. By 1979 most birds had abandoned Ile Coco for the less disturbed Sable – there were about 800 individual Lesser and 200 Brown Noddies, and still only some 30 Fairy Terns[478]. By the mid-1980s the numbers were similar, but now Sable was too disturbed, with cats at large and eggs being collected, so the birds had moved back to Coco. In 1985 there were also about 30 Sooty Terns, including four pairs with eggs, the first confirmed breeding since the 19th century[479]. In the interim the islands had finally in 1981 been declared nature reserves[480]. Over the same time the shearwaters on Frégate had not been fairing too well – at least, they were considered from 1963 onward to be in low numbers, though no nocturnal census was attempted. In 1974 there was ample evidence of human predation, with bodiless wings scattered around, as with tropicbirds on Round Island (see Chapter 9); as all informants denied that anyone ate shearwaters, they may have been used as fish bait[481].

The lagoon islets were all thoroughly surveyed by a conservation team in July 1993, who proposed the eradication of mice (no rats present) from Coco and Sable, and restore Ile Crabe (the biggest) to native vegetation. Oddly, no firm proposals were made for Frégate (rich in rats!), despite the shearwater population and an unusually good surviving collection of native trees[482]. The mice on Coco and Sable were poisoned in November 1995. The 1993 visits were out of season for shearwaters, the team accepting a 'given' figure of "less than 50 pairs", while numbers of terns had gone up to *c.*5000 and *c.*500 pairs of Lesser and Brown Noddies respectively on Coco (with none nesting on Sable), rising further by 2005 (8,000+ Lesser, 3,000+ Brown). In 1993 there was no sign of Fairy Terns returning to their former numbers: there were only about 25 pairs, divided between the two islets[483]. Numbers, still low in 1996 and 1999, had risen to *c.*150 pairs by late 2003, although only 30–70 pairs were seen in 2005; however the total population numbers several hundred, since up to 500 birds together have been seen on Sable. Sooty Terns were not breeding in July 1993, but a few were nesting in January 1996. There were 50+ pairs in May 1999, rising to *c.*200 pairs in October 2003 (with a few also on Sable), and *c.*1,500–2,000 in 2005[484]. In 1993 rotating wardens were looking after Ile Coco, but failing to prevent tourists entering the supposedly off-limits core seabird nesting area, and they were not preventing disturbance on Sable either. More recently, the higher profile of conservation in the island and the increasing importance of tourism (with the seabirds a major attraction) has sharpened up the wardening, leading to the encouraging results seen in recent reports[485].

The danger of untimely visits and reliance on secondhand information is amply illustrated by the outcome of Jean-Michel Probst's visit in January 1996[486]. He and his associates concentrated on shearwaters, and in contrast to the low numbers previously estimated, reckoned there were *c.*750 pairs on Ile Frégate. They stayed overnight in full breeding season, and censused calling birds in a 10 × 10m quadrat, then estimated the nesting area (1800m²), so getting their figure – no one had done this before. In addition, they found small numbers nesting on mainland cliffs between Graviers and Pointe Coton. They warned against the impending construction of a house, but (unlike Coco and Sable) Frégate is privately owned, and the house was subsequently built[487]. In May 1999 (outside the breeding season), Dave Showler could only report an old nest and rumours from fishermen of birds coming there at night. However a feral cat was seen in 1996, and by 1999 there were clearly several. It appears that attempted eradication in 1999 failed, as in October 2003, when the birds should have been breeding, there were signs of cats but no evidence of shearwaters at all[488]. Both in Mauritius and Rodrigues, unwanted dogs and cats are often dumped on islets instead of being 're-homed' or killed[489]; this particular 'gift' to Frégate appears to have wiped out the ancient colony of shearwaters, so recently showing signs of recovery. Rabbits, still abundant in 1970, declined through the 1970s and had died out by 1983. Ship Rats were still common on the islet in 1993, but by 1999 the cats appeared to have eliminated them and were subsisting on centipedes[490].

From the days of Leguat onwards tropicbirds have been a characteristic sight in Rodrigues, and Jean Vinson bemoaned their loss in 1963. However his pessimism was premature, as shortly afterwards France Staub found both species breeding on the cliffs of Cascade Victoire, an area Vinson had not visited. White-tailed had probably been present throughout, but the absence of Red-tailed in 1874 and in Bertuchi's account (p. 154) suggest they went extinct and recolonised more recently. Numbers of Red-tailed Tropicbirds remained at around 10 pairs from 1974 to 1999, while White-tailed Tropicbirds increased from around six to 20 pairs, with a few elsewhere, including Ile Coco; both species were estimated at between 50 and 75 pairs in 2005[491]. In the 1970s there was a minor industry selling stuffed tropicbirds to the (then) rare tourists and to officials returning to Mauritius – in 1974 a policeman, ending his tour of duty, proudly carried a mounted Red-tailed Tropicbird home on the ship[492]; so much for the Bird Protection Regulations 1923, then still in force, which *did* cover tropicbirds! Taxidermists also favoured Hawksbill Turtles; immatures were regularly on sale in 1974, and still in 2001 three years after turtle hunting was banned. In 1974 one of us felt reluctantly obliged through politeness to eat Green Turtle offered as a special treat, though by 2004 one part of the reef was being touted to tourists as a place to see them. Both turtles would undoubtedly recolonise Rodrigues if hunting was successfully controlled[493].

A surprise in March 1974 was the discovery of a Barau's Petrel nest under a hilltop boulder at Quatre-Vents. "The bird made the mistake of calling in its burrow in the daytime when there happened to be people sitting on the rock outside". They killed and stuffed the well-grown chick, thus providing concrete if regrettable evidence of the first nest of the species ever found. It seems to have been a one-off; a bird was flying around the area that December, but there have been no records since. Although some locals said a few pairs bred in the cliffs, there was no lore to suggest the bird was an established breeder[494]. This Mascarene endemic is otherwise found only on Réunion.

The future

Rodrigues has never had so much international attention as in the early 2000s. The high-profile Shoals Rodrigues lagoon conservation programme, a visit by UK royalty with Raleigh International, a prestigious conservation prize for John Mauremootoo's community-based ecosystem rehabilitation work, and the island appearing on widely disseminated tour brochures, has transformed the island from an obscure backwater to a flagship site for ecological restoration of devastated islands[495]. With a human population in 2005 of over 37,000[496], it remains to be seen if the island's fragile and intermittent water supply can also reliably supply the new international hotels, whose consumption is typically many times that of local residents. If that hurdle can be overcome (perhaps by solar-powered distillation plants), then tourism (including eco-tourism) may be able to provide this previously subsistence economy with enough surplus to continue the recent progress in bringing a nearly defunct ecosystem back to some semblance of a future[497]. With the seabird islets already attracting foreign visitors, and endemic birds, bat and plants for connoisseurs, interest should be enhanced by a new combined tourism/conservation project on Plaine Corail to recreate a 20ha Rodriguan landscape with 75,000 endemic plants and 1,000 Aldabra tortoises (opened in July 2007), which should also prove educational[498] and a welcome change from the nadir of 1955–68 and its legacy.

CHAPTER 9

A MIRACULOUS SURVIVAL

The curious story of Round Island and the other islets off northern Mauritius

Expedition to Snake Island to kill land turtle . . . the island is about three miles in circumference, not inhabited . . . [there are] large bats and doves and a prodigious number of large snakes that we were fearful of sleeping at night. Likewise abundance of cabish trees, some now thirty feet high, very sweet and good to eat.

From the log of the *Scarborough* in October 1703[1]

A few years ago, when a ship called the Sumatra *was lost on a little Island, or l'Isle Ronde, the Crew, who saved themselves thereon, remained many days, before they were discovered, or taken off, by a smoke and fire, that they at last made, and on their return, reported that Island was full of large Serpents and Snakes.*

Charles Noble writing in 1755[2]

On a shelf of shallow sea north of Mauritius lie four small islands, very different in topography, which have served over the centuries as refugia for species long exterminated on the main island. They are largely composed of a fine-grained volcanic tuff with basaltic lava flows on some of them, and were erupted originally through coral reefs – chunks of coral are embedded in the tuff right up to the summit of Round Island[3]. Separated from the mainland by a shelf less than 35 fathoms (63m) deep, they would have been hills in an extension of the northern plains during the low sea-level of the last Ice Age. While joined to the rest of Mauritius, they shared its fauna, but this was stranded when sea level rose around 12,000 years ago. Their history since humanity's arrival is so different from the mainland, and so important for the understanding and conservation of the Mauritian fauna, that they warrant a chapter of their own.

From the nearest to the furthest from the coast these islets are Gunner's Quoin, the Flat Island group, Round Island and Serpent Island (Map 9.1). Flat Island, the biggest at 292ha, is broken into three parts by arms of the sea – Flat Island proper (253ha), Gabriel Island (42ha) and Pigeon House Rock. Pigeon Rock is a small flat-topped stack with vertical sides, whereas the other parts are low-lying and fairly flat, with a 102m hill at one end of Flat Island. The other three are much more rugged and surrounded by cliffs. Gunner's Quoin covers 65ha rising to 162m over high sea-cliffs on the western end, Round Island is larger at 219ha, 267m at the summit, and Serpent Island, the smallest at 31ha, is a curious dome rising to 162m, with vertical or undercut cliffs almost all the way round[4]. As the local joke goes, Serpent Island is round with no snakes, while Round Island, which does have snakes, is not round. Gunner's Quoin is only 4km from the mainland, whereas Serpent Island is 24km offshore.

Although occasionally visited, the islets were not described in any detail until the mid-19th century, so their original vegetation has to be inferred from snippets of earlier information and from relicts surviving into the mid-1800s. We know from the weather station on Flat Island that rainfall is lower (950mm) than anywhere in Mauritius apart from the dry west-coast fringe (p. 33)[5]. Flat Island was originally forested with enough diversity to attract fruitbats and pigeons from the mainland. It was described as "covered in latans", another common palm, in 1714[6]. By the time the islands were botanically described, there were few trees left on Gunner's Quoin or Flat Island bar a few latan palms and screw-pines[7], though Round Island in the 1840s still retained a cap of stunted mixed evergreen

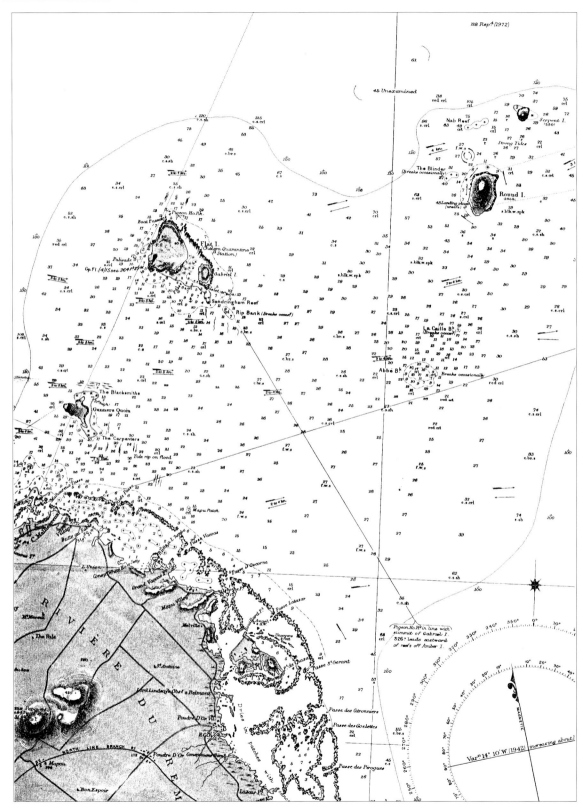

Map 9.1. The northern islets of Mauritius as seen by 19th century seafarers. From the Admiralty Chart of 1879, very slightly updated to 1974.

> ### BOX 26 THRUSHES AND CHATS (TURDIDAE)
>
> The Réunion Stonechat *Saxicola tectes* is a common endemic on Réunion, occurring at higher elevations. As it was rarely mentioned in the early literature, it may always have been confined to the montane regions of Réunion.
>
>
>
> Account
> Bory (1804) in 1801:
>
> [around the Piton de Villers, Plaine des Cafres] *Flies were common in this high-up region, as was a small bird which jumped from branch to branch, and is called in this country* tec-tec . . . *The insects we have described in the heights of the Rivière des Remparts, also live in the last* [= uppermost] *ambavilles* [heath vegetation] *of the Gros Morne, where the tec-tec comes to make war on them.*
>
> Variations in the male plumage of the Réunion Stonechat *Saxicola tectes*. From Barré and Barau, 1982.

forest near the summit, and palm-rich woodland lower down. Originally the three larger islets would all have been largely covered in palm-rich woodland, probably most varied on Flat Island, which has a water table, swampy areas and soils enriched with calcareous beach-sand. Serpent Island, called Meeuwe Klip ['gull rock'] by the Dutch[8], was presumably then, as now, a barren rock covered in thousands of nesting seabirds.

17th- and 18th-century visitors

The first recorded visits were in the 1670s. Flat Island, already called Snake Island by the Dutch in 1672, was rumoured to harbour good numbers of tortoises and Dugongs. Governor Hugo organised an exploration of the northern islets in May 1674, but the sailors were unable to land on Round or Serpent Islands; they reported a forest of palms and screw-pines on Round Island, but Serpent Island was bare and white like salt. Landing on Flat Island, they got their legs bitten by a multitude of hungry and cannibalistic tortoises![9] The sailors removed 320 tortoises, but the population of Flat and Gabriel islets must then have been around 6,000 animals [10]. They found only one snake, but Flat Island remained 'Snake Island' to both Dutch and English for several decades thereafter[11]. The Dutch apparently rarely exploited the islets, but visiting English ships certainly did so. By 1700 when tortoises had become very scarce around the normal anchorage at North-west Harbour (Port Louis), visitors sent boats to Flat Island to collect them. The *Rising Sun* reported in August 1702 that on "Snake Island about 14 league distance hence is great plenty of large land-turtle, from whence we used to fetch them for change of diet for our people". The *Scarborough* sent an "expedition to Snake Island to kill land-turtle" in October 1703, and the *Westmoreland* collected 68 'turtle' there in August 1706. As late as April 1722 the *Lyon* "sent the long-boat and cutter to Long Island [= Flat Island] for turtle", getting 120. The last two may have been catching sea-turtles, but the quantities collected make this unlikely[12].

The French also took to raiding Flat Island and Gunner's Quoin. A passenger on the *Reine d'Espagne* reported Flat Island "covered in tortoises" in January 1714, and in 1721 Garnier du Fougeray sent boats to "the little islands several leagues from the large, from which they brought a large quantity of *tortues*"; very welcome, the hunting being otherwise indifferent[13]. Adding to the *Lyon*'s haul, the *Diane* took 80 tortoises from Flat Island in April 1722, and the *Courrier de Bourbon* reported "large quantities" there and on Gunner's Quoin in February 1723[14]. By 1726 the French East India Company's supremo in Asia, Pierre Lenoir, was trying to ration harvesting from the northern islets by getting ships to register their requirements first (p. 91). In 1732 the Company still wanted the Mauritian authorities to get tortoises from Round and Flat islands, while admitting they were small and inferior to those from Rodrigues – indeed, in 1731 La Fontaine and Tafforet had been ordered to Rodrigues to "take as many tortoises as they can to populate the islets adjacent to our island". The islanders themselves stopped mentioning the islets in relation to tortoises; Gunner's Quoin became instead a source of latan leaves for roofing[15]. Once Labourdonnais arrived to take charge of the Mascarenes in 1735, there was no longer any serious question of local sourcing, and he got the go-ahead to get supplies from Rodrigues (p. 111).

As the century progressed landings on Gunner's Quoin and Flat Island were occasionally reported. Round Island, hard to land on, normally featured only as a landmark to mariners, as did 'Ile Paras' or 'Parasol', later mis-named Serpent Island. The sole recorded landing on Round Island was by accident, when sailors from the shipwrecked *Sumatra* found refuge there around 1750 – finding lots of snakes but not mentioning tortoises. The anomaly of snakes and large lizards occurring on these islets but not on the mainland caught the interest of the more enquiring travellers. Nicolas de La Caille, visiting in 1753, commented that "there are no snakes in Mauritius; it is claimed they cannot live there, but that in the neighbouring islets, called *Isle Ronde* [Round Island], *Isle Longue* [Flat Island] and *Coin de Mire* [Gunner's Quoin] [there are] lots of snakes and serpents. I cannot confirm this fact; what I do know is that on the islet called *Coin de Mire*, I saw lizards a foot long and a good inch in thickness [Telfair's Skinks], whereas in Mauritius [mainland] I only saw very small ones running on walls and on rocks, much as one sees in France"[16]. The elder Cossigny, in Mauritius on and off between 1732 and 1759, reported seeing "several [snakes] shot on Gunner's Quoin", which was also reported in the 1770s to harbour Red-tailed Tropicbirds and other (unnamed) seabirds[17].

Sometime in the mid-late 18th century a Mauritian tortoise was acquired by Sir Ashton Lever for his famous museum[18]. When or how it reached England is a mystery, since, due to persistent hostilities in Europe, British ships were rare visitors to Mauritius after 1722, and in any case mainland tortoises were extinct so it must have come from an offshore island[19]. Possible occasions would have been during the British attempt to invade Mauritius in 1748, or the rescue of the 1750 shipwreck survivors on Round Island. The acquisition was presumably before Lever died in 1788, and in any case the museum was dispersed in 1806[20]. A live Mauritius Fody reached England in equally unexplained circumstances around 1770 (p. 99); Lever also acquired at least one parrot from Réunion, so British collectors may have had friendly French contacts. Although tortoises survived on Round Island until at least 1844 (p. 211), the last mention before this date was in 1731.

Early 19th-century visitors

After the 1750s there is a gap until Lislet Geoffroy visited Flat and Gabriel Islands in 1790. The islands were then mostly covered in "a reed resembling *Calamus aromaticus*" and 'pourpier' (*Portulaca oleracea*), with large numbers of latans and some *veloutiers* [beachside shrubs][21]. He clearly had prior knowledge, as he reported that recent fires had much depleted the snakes, though he did find one "28 to 30 inches long which I am preserving in spirits of wine"[22]. A likely reason for fires was the island's use for grazing; in 1803 (and no doubt earlier) goats were kept there, and burning to encourage forage grasses (at the expense of woody vegetation) was standard practice on Mauritius. Fishermen also had huts there for regular, if temporary, accommodation, and the islet had served as a temporary quarantine station during a smallpox epidemic in 1782[23]. During the British blockade prior to the capture of Mauritius, Flat Island was taken in 1809 and used as a hospital for injured sailors, run by Charles Telfair, who cultivated vegetables to provide fresh food for the sick[24]. Although a keen naturalist after whom Desjardins later named the large endemic skink, he unfortunately wrote nothing about Flat Island or his later collecting activities in Mauritius (p. 118). James Prior landed on Flat Island just before the British invasion in 1810. In the 20 years since Lislet's visit the palm forest had been destroyed: "Flat Island is about four miles long and two broad, covered by strong grass, among which is that tufted species which, when split and dried, is made into hats by the seamen. The cotton-shrub also grows here". He mentioned no livestock, but described Telfair's Skinks and noted that the island "abounds with small but not ill-flavoured hares"[25]. He saw 'curlews' (i.e. Whimbrels), and described Pigeon Rock as "a great resort for sea-fowl". Charles Chapotin, on Mauritius in and around 1805, also recorded Telfair's Skinks on Flat Island[26], and a Captain Lumley, whose ship was quarantined on Flat Island in 1820, had "only rabbits for company"[27]. No early visitor reported anything corresponding to Günther's Gecko, but its former presence is proved by subfossil egg remains found in a cave[28].

Figure 9.1. Gunner's Quoin, looking south towards Pieter Both Mountain on the mainland. Note the already sparse palms. From Bory (1804).

Milbert visited Gunner's Quoin around 1801, observing that "scrawny Latans stagnate on this arid rock, and stunted shrubs emerge bleakly amid coarse herbage". Here also the original vegetation had already been badly damaged and mammals introduced: "The timid hare finds among the plants an insubstantial nourishment; this game is not found in large groups, isolated animals finding their shelter in the interstices of the volcanic rocks"[29]. Unfortunately he recorded nothing about seabirds or reptiles. Prior landed here too, noticing the hares plus "an abundance of sea-fowl, particularly the tropic bird". He found tomatoes growing wild, but, like Milbert, mentioned no reptiles.

The identity of the lagomorphs released on the islets is a puzzle and also biologically important – rabbits do more damage than hares. While there is no dispute about the rabbits that later devastated Round Island, there is doubt about which species were present on Flat Island and Gunner's Quoin. Both Milbert and Prior called them hares, and in his faunal discussion for Mauritius Milbert gave a passable description of the Black-naped Hare, adding that "I was assured that there is also a smaller species [?=rabbit], but I did not see it"[30]. 'Lièvre' was apparently used in Mauritius for both hares and rabbits in the 18th century (p. 93), and as late as the 1820s Julien Desjardins, an experienced naturalist, described the animals on Ile d'Ambre as '*lièvres*' when all later observers insisted they were rabbits[31]. Prior, however, was English, so the argument does not apply to him. In 1828 visiting French naturalist Jean-Réné Quoy was told by Desjardins that on Gunner's Quoin there was "such a large quantity of a species of *lièvre* that they can be killed with sticks" – hares never reach such densities, so presumably the animals were rabbits. Quoy and his colleague Paul Gaimard tried to land there and enjoy this 'sport', but were prevented by heavy surf[32]. Gunner's Quoin apparently hosted both species at various times; hares were present until recently, but a rabbit skull has also been collected there. Whatever animals were already there, further hares were released in 1934[33]. There were rabbits on Gabriel Island in the 1940s and 1950s, but the animals on Flat Island proper were generally reported as hares, though botanist John Horne implied the presence of rabbits in 1885[34].

Serpent Island

Another puzzle is how Serpent Island, in the mid-late 18th century, acquired the reputation for harbouring snakes; as we have seen, the 'Snake Island' of the early 1700s was Flat Island. Serpent Island escaped mention by the French until 1753, when La Caille called it 'Ile Paras' on his definitive map; although a text table gives Isle aux Serpents and Isle Parasol as alternatives, he did not include Serpent Island in his list of islands with snakes[35]. By the mid-1770s the alternative name had vanished, and clearly the Serpent Island myth was thoroughly entrenched; it appears so named on Après de Mannevillette's map of 1775, Bonne's map of 1782 and Grant's map of 1801[36]. Bory de St Vincent wrote that "Serpent Island, much smaller [than Round Island] is a rock nearly 5 leagues from the mainland; it is claimed that small snakes are found on it, whereas none exist either on the neighbouring rocks nor on Mauritius". Charles Tombe even used it in speculative biogeographical remarks: "To all appearances it seems that these little islands and those near them are parts of Mauritius detached by some earthquake. Serpent Island, however, appears to have been always separate because of these animals [i.e. snakes] of which there has never been any trace on the main island". This makes little sense since Lislet-Geoffroy, by then a prominent surveyor and engineer, had caught a snake on Flat Island in 1790, and had even had a conversation about the islet with Bory, who published his remarks about its geology[37].

As late as 1818 the zoologists from Freycinet's *Uranie* expedition were claiming that there were no snakes on Réunion or Mauritius, except possibly on Serpent Island[38]. Peron and Lesueur, who had arrived with Bory, brought back a Burrowing Boa collected on an unidentified islet in 1801 or 1803[39]. Lesueur's notes suggest he thought it came from Serpent Island: "snakes do not exist on this island [mainland

> **BOX 27** **OLD WORLD WARBLERS (SYLVIIDAE)**
>
>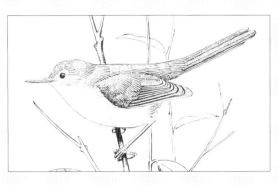
>
> **Rodrigues Warbler** *Acrocephalus rodericanus*
> Only one species of warbler occurs within the Mascarenes, the Rodrigues Warbler *Acrocephalus rodericanus*, endemic to Rodrigues. Although now restricted in range, it once occurred all over the island. This species has now been identified from deposits collected in the Plaine Corail. For the only early mention, see under Rodrigues Fody (Box 31).
>
> Rodrigues Warbler *Acrocephalus rodericanus* by Jean-Michel Vinson, 1974.

Mauritius], but they are found on an island close by which carries the name of the Ile of Serpents – it is a very remarkable thing that in two places so close together . . . there are these reptiles on one and that they are totally lacking on the other"[40]. However in 1805 the short-lived *Société Libre d'Émulation de l'Isle de France*, whose membership included Lislet, was proposing *inter alia* to investigate on Gunner's Quoin and Flat Island the "plants, the animals, the soils and the species of snakes that are found on them, and the differences in natural history between our island [Mauritius] and these islets; and what is on those whose access is impossible?"[41]. The snakes may have become too scarce on Flat Island and Gunners Quoin to be seen by casual visitors, and folk memory then banished them to another island. Why Serpent Island rather than Round Island should have been chosen is obscure, especially as no-one had landed there – so how could anyone know? Since snakes were not mentioned by Milbert, Chapotin or Prior, livestock grazing and the use of fire may have rendered much of both Flat Island and Gunner's Quoin unsuitable, although the snakes were still there. The *Uranie* naturalists, Quoy and Gaimard, returning in 1828, were this time shown a snake from Gunner's Quoin in Desjardins's collection. Lesson was told of their presence on Flat Island in 1824, and Magon de Saint-Elier reported them to still occur on both Gunner's Quoin and Flat Island in 1839[42]; Thomas Corby supposedly collected one on the former at an unknown date[43]. As mentioned above, the first scientific specimen was acquired by the Baudin expedition in 1803 – but Peron and Lesueur muddled their sources and labelled the Burrowing Boa as from 'Nouvelle Hollande' (Australia)[44]. Their surviving notes suggest they never visited the northern islets themselves[45], so they must have got their specimen from others. The first Keel-scaled Boa was probably collected in 1825–26, Dussumier getting one on Round Island[46]. Corby's specimen from Gunner's Quoin, still in the Desjardins Museum in 1870, did not "appear to agree with any of our Round Island specimens" according to Sir Henry Barkly[47]. Unfortunately it subsequently vanished, as did Desjardins's specimen, leaving us uncertain which snake(s) inhabited Gunner's Quoin and Flat Island.

Although there are specimens of Telfair's Skink in Paris that may have come from Peron and Lesueur[48], it was not until 1830 that the various skinks were formally described, when Julien Desjardins realised that, however common locally, they were unknown to science. He described three skinks, Telfair's, Bojer's and Bouton's, as endemic to Mauritius[49], although Bouton's was later found widely distributed on coasts around the Indian and Pacific Oceans. Telfair's was still common on Gunner's Quoin, Flat Island and Round Island, and Desjardins assumed "probably also on Serpent Island". Bojer's and Bouton's he mentioned only for the mainland, though later on the islets became important refuges for both.

Gunner's Quoin and Flat Island in the mid-19th century

After the flurry of activity early in the the century, Gunner's Quoin and Flat Island were almost ignored until Dr Philip Ayres was appointed superintendent of the quarantine station on the latter, hastily built in 1856 after Mauritius was struck by an epidemic of cholera brought by incoming ships. The previous year a lighthouse had been built there, and it is likely that rats reached the island during this construction activity. Already by 1860 there were feral cats, to which Ayres attributed an accumulation of hare bones found in a small cave. Ayres's paper, largely on geology, said nothing about reptiles, but by 1872 Louis Bouton reported that the snakes (and, wrongly, latans) had gone from Flat Island; he did not mention lizards, but there are also no further records of Telfair's Skinks[50]. Edward Newton visited Flat Island in 1860 and 1862, but recorded little; he saw Whimbrels and a Red-tailed

Chapter 9: A miraculous survival

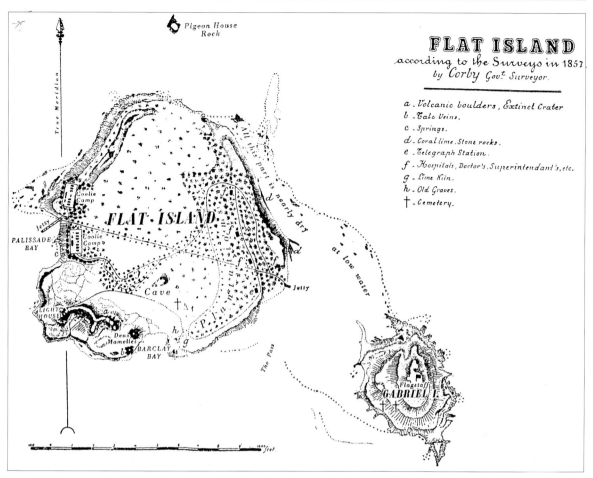

Map 9.2. Flat Island in 1857. Map by Thomas Corby, from P. J. Barnwell's collection (courtesy of Richard Barnwell).

Tropicbird landing "as if it had a nest". Pike's visit in 1869 adds nothing on land fauna or plants[51]. In addition to buildings and permanent staff, the quarantine station maintained herds of cattle and goats, though by 1885 the cattle had been recently taken off[52]. A fuel-wood plantation of she-oak started in 1874 was still actively maintained in 1903[53]. John Horne, then in charge of Pamplemousses botanic garden, made the first systematic plant survey when quarantined there in 1885, by which time native woody vegetation was reduced to latans and screw-pines (still common), the coastal fringe of *veloutiers*, and the shrubby *baume de l'Ile Platte*[54]. Since the removal of cattle, screw-pines had begun to regenerate, but Mauritius Hemp and Prickly Pear were threatening to overrun the island, and the invasive exotic grass *Heteropogon contortus* was common; the forage plant White Popinac had been recently introduced[55]. Horne was called away before he could explore the cave where Ayres had found bones in 1860: "It would be interesting to know what sort of beasts, if any, lived on Flat Island before the advent of man on it. If any record of such exists, it most likely lies hidden in the alluvium which forms the floor of this cave". This investigation still remains to be done – indeed the cave's location had been forgotten! However it features on Corby's 1857 map (Map 9.2), so perhaps Horne's ambition can belatedly be fulfilled.

Walter Besant, then teaching at the Royal College, spent a week with Ayres on Flat Island. He admitted to being no naturalist, but his account provides the only 19th-century impression of the seabird colony on Pigeon House Rock:

In one place where the coral reef stopped there was a curious pillar of rock about forty feet above the water and twenty or thirty feet in diameter. It stood a few yards from the shore, and was covered with innumerable wild birds. My friend would never shoot them. We would sit by the shore and watch this multitude flying, screaming, fishing and fighting all day long. I know nothing about birds, but of their numbers I have a vivid recollection.[56]

Gunner's Quoin remained almost totally neglected in the literature until the 1970s, apart from a brief botanical survey in 1935 (p. 215). We know only that

Corby and Lloyd were aware of Bojer's Skinks and both tropicbirds there in 1844, good numbers of Red-tailed Tropicbirds and some Wedge-tailed Shearwaters were nesting in the 1860s, and that by 1881, snakes were reported as confined to Round Island[57]. Rats had presumably arrived at Gunner's Quoin some time earlier, wiping out both snakes and Telfair's Skinks. The rats on Gunner's Quoin and Flat Island were respectively Norway Rats and Ship Rats, suggesting they reached the islets at different times and by different means[58].

Round and Serpent Islands in 1844

The smaller endemic lizards, extinct or about to be so on the mainland, survived the rats on Gunner's Quoin and Flat Island, the night-geckos remaining undiscovered there until the 1980s; for snakes and the larger lizards, Round Island was the last refuge after the 1860s. The first detailed account arose from an expedition in December 1844, when the Surveyor-General, Col. John Lloyd, took a team to Round and Serpent looking for guano to use as fertiliser. They succeeded in making the first known landing on Serpent Island, describing its seabirds for the first time, but they were disappointed in their quest for guano – the island was too steep-sided for the droppings to accumulate, being washed off every time it rained.

Serpent Island is the only islet apparently completely unchanged from its natural state. In 1844 it was, as now, "covered with sea birds with their young", Corby reporting the same species still present today: Masked Boobies, noddies (the two species not distinguished) and Sooty Terns, though without hazarding a guess as to their numbers. Lloyd noted that "the air was darkened with sea-birds near the shore". Given that there are, and presumably have been for centuries, several hundred thousand pairs of seabirds nesting there, it is remarkable that in the entire history of Mauritius to 1844 they were never mentioned. Boobies, probably Masked, were recorded at sea near Mauritius in 1818, but the only mention of Sooty Terns, the most numerous species, was in the *Westmoreland*'s log in 1704[59] yet they were well-known to islanders. They all had local names (which Corby used), the same names Pingré had recorded in Rodrigues 80 years earlier[60]. Corby also found a centipede and "a species of lizard (*Scincus*) like that of the Gunners Quoin", confirmed in 1949 as a race of Bojer's Skink[61].

The guano team's visit to Round Island was a landmark in the history of the northern islets. The four companions were caught in an unexpected cyclone and marooned for a week "without fire or the means of cooking our provisions, without any shelter but the crevices in the rocks, deluged with torrents of rain and all our stock of fresh water washed away by the furious surf". But they survived, and Lloyd's description of the island's vegetation and fauna has served ever since as the baseline against which to compare later observations. Much of the islet was thickly wooded with palms and screw-pines, especially in the ravines where soil was plentiful, supplemented by "some belts of forest wood on the upper part, but much stunted, such as 'bois de ronde', ebony and benjoin". Lloyd's terminology for palms was eccentric, but his descriptions are recognisable, and clearly all three species were then almost equally abundant: Latans, Hurricane Palms (of which they gladly ate the cabbage) and Bottle Palms. Most importantly, Lloyd recorded 3 feet (0.9m) of soil on the upper slopes of the north of the island, and that soil was "by no means wanting" over all but the exposed southwest side. He noted lizards "growing to a size of 14 to 18 inches in length and a very few of the serpent tribe, specimens of which are already in the Museum"[62]. He described the nests, flying abilities and fishing methods of the two tropicbirds, emphasising that "myriads of these birds exist on this Island". White-tailed Tropicbirds, he said, were the commoner on Gunner's Quoin, while Red-tailed Tropicbirds were more abundant on Round Island. He also noted "the 'fouquet' known from its black bill and its dismal cry of several notes when about to roost" [Wedge-tailed Shearwater] and "the stately Frigate bird". Lloyd's writing suggests a

Figure 9.2. Flat Island (**a**) and Gunner's Quoin (**b**) as seen by Pike (1873).

non-breeding roost of frigates on the island; there is no indication of nesting around Mauritius at this late date. Layard shot one there 12 years later, and Edward Newton saw one in 1860; there have been occasional records, usually after cyclones, over the years since[63]. The most sinister observation, in the light of later developments, was the "immense numbers" of rabbits: "swarms" sheltering in the stunted woodland high up and "swarms" near the landing place – in short, everywhere[64]. Finally, and most importantly, Lloyd saw "a few very large species [sic] of the land tortoise of the Angola description". It is unlikely that anyone would have released imported tortoises on Round Island, so here on this inaccessible islet, endemic Mauritian tortoises had apparently survived unseen since the 1730s. Henry Barkly later reported that "the Hon'ble Mr Kerr" (William Kerr, one of the 1844 party) had told him that

Figure 9.3. Red-tailed Tropicbirds by the Landing Rock pillar, Round Island. From a drawing by Edward Mortelmans in Durrell (1977b).

> *the late Mr Corby captured a female land tortoise in one of the caves on Round Island and brought it to Mauritius, where it produced a numerous progeny, which were distributed among his acquaintance. I am not aware whether any of the animals are now alive, or whether anything is known about them, but it is a great pity should the opportunity have been lost of examining to what species they pertained, and of ascertaining if it was the same as that which existed in Mauritius on its discovery, and of which the Carapaces and other remains are still found in such quantities in Mares [swamps] of this land.*[65]

A great pity indeed – with a little perseverence and imagination this species, and indeed the endemic Mascarene genus *Cylindraspis*, could have been saved from extinction! It is possible, indeed likely, that some of those 'numerous progeny' *were* still alive in 1870. . .

Visitors of the mid-19th century

In late 1856 Edgar Layard, curator of the South African Museum in Cape Town, cruised the western Indian Ocean on the British warship HMS *Castor*, making several visits to Round Island for bird skins and eggs. He must also have visited Serpent Island as he collected Masked Boobies and a Sooty Tern (plus eggs), though he conflated the two islets in his records. His brief account of a three-day visit refers only to Red-tailed Tropicbirds and Wedge-tailed Shearwaters (plus Sooty Terns and a 'gannet' i.e. a Masked Booby, which were really on Serpent Island). Layard collected an egg from a Tropical Shearwater, a species about to disappear from the Mauritian mainland, and only sporadically found since on Round Island[66].

Soon afterwards, in November 1860, Edward Newton visited Round Island, accompanied by a Captain Stokes and the ever-present Thomas Corby, by then the islet's lessee. Newton's published account refers only to birds, but his diary gives more detail, and suggests considerable degradation of the ecosystem since 1844:

> *There is very little vegetation on the island, there being hardly any soil. A species of* vacoa *(Pandanus) [screwpine] is by far the commonest plant, there is also a species of* palmiste *which is common, whether the same as the true one or not I do not know, but it looks very different, this may however be caused by the —* [illegible] *which they have sustained by the wind, and some of them are most curious growing in this shape, the stem presenting a most —* [illegible] *carroty appearance*[67]. *Besides these two I noticed a stunted forest tree or two of one species, a sort of juniper [sic], a few grasses, lichens, mosses [sic] and ferns. No vegetation is found within a hundred yards of the sea, probably on account of the heavy swell causing so much spray; there are besides large tracts of perfectly barren rock, inclining at an angle of perhaps 45° without a vestige of vegetation (these are certainly slipping and uncomfortable to walk over) with their edges clearly defined as if they had been marked out by hand. On other places there are vast tracts of the same sort of rock though flat and covered with low pieces of basalt, this latter is principally on the south side. On landing Stokes and myself proceeded to the south side of the island for the purpose of shooting a goat; we each shot one. He had a rifle and made a very fair shot at an old billy. I killed a 'nanny' unfortunately, but as they swarm it can not signify much.*[68]

There was still a huge population of Red-Tailed Tropicbirds, overwhelming even this keen egg-collector: "I was rather short of baskets for carrying eggs, and consequently I did not get as many as I might have done. Certainly I had been told that the eggs might be picked up by the thousand, but I had not believed the statement"[69]. In his paper he simply mentioned 'very large numbers', but the diary is more specific: "The numbers of the birds on the island might be numbered

> BOX 28   MONARCHS (MONARCHIDAE)
>
> **Mascarene Paradise Flycatcher** *Terpsiphone bourbonnensis*
> Although the small *Tersiphone* paradise flycatchers in the Seychelles and Madagascar have distinct sexual dimorphism in plumage colour and tail length, the Mascarene species *Terpsiphone bourbonnensis* exhibits only slight differences, with the males resembling females of the Seychelles and Madagascar species. They are rare on Mauritius but less so on Réunion, and are easy to approach quite closely. There is only one early account. They have not been identified from the Mascarene fossil record and are not known from Rodrigues.
>
> *Account*
> De Querhoënt (1773, unpublished MSS in the Bibliothèque du Muséum, Paris) in Mauritius:
>
>> The coq maron *or* colin *is native to the Isle de France* [Mauritius]; *it is the size of a* [European] *Robin* [Erithacus rubecula] . . . *It is given the name of* coq maron *because it is the first bird to make its song heard as dawn breaks, this song is a monotonous little whistle. Although it always lives in the woods, it is not wild; one can attract them easily by imitating their call. They are, from what I have been told, always paired,* [the male] *never separated from the female. They nest in September on trees in the woods . . . they lay four greenish eggs. They live on insects.*
>
>
>
> Mascarene Paradise Flycatcher *Terpsiphone bourbonnensis*. From Berlioz, 1946.

by 10,000s and not 1,000s". Newton also found Wedge-tailed Shearwaters nesting, saw a few White-tailed Tropicbirds, and was the first to note land-birds: a pair of Zebra Doves. Newton's is also the first record of goats. Barkly wrote later that Corby himself had introduced them[70] – clearly several years before Newton's visit, given their numbers. Newton's diary is silent on snakes or lizards[71].

The indefatigable naturalist and raconteur Nicholas Pike decided to pay a visit in 1868. Since Corby's death in 1865, the island's lease had passed to his ex-deputy A. Vandermeersch[72], who accompanied Pike. Pike found abundant screw-pines and palms (of the three usual species), but missed the hardwood trees: "they must all have been nearly all cut down or destroyed, as we found few traces of them". He noted "a few patches of grass" fed on by "numbers of rabbits and goats". An avid collector of everything that moved, he accumulated not only birds and reptiles but all manner of invertebrates. However, he was not the only collector; one of the boatmen was, when surprised by a large snake, "carrying a long pole on his shoulder, at each end of which were suspended several *pailles-en-queues* or tropic birds". Neither Lloyd nor Newton mentioned seabird harvesting, but here was a man going equipped with a pole to carry tropicbirds. It appears that that a new and unfortunate tradition had arisen, perhaps since Vandermeersch had taken over as lessee.

Pike returned the next year on the first formal scientific investigation of the island, led by the Governor of Mauritius himself, Sir Henry Barkly. After receiving Pike's report from 1868, Sir Joseph Hooker, director of Kew Gardens, had asked Barkly to investigate the islet's unusual biota[73]. As was the fashion among the colonial hierarchy and Franco-Mauritian society, the visit was also the excuse for a splendid picnic, nearly aborted by an impending storm, "the excitement of the weather giving a double zest to our viands". Most of the party left in a hurry after the meal, but Pike, Vandermeersch and their servants decided to stay on, though after a bruising night fighting the water gushing over the island's bare rocks (shades of Lloyd), they were in no fit state to do much work before being rescued, but they luckily managed to salvage their collections[74].

Barkly had invited John Horne along to collect plants for Hooker, and in the 'forest' near the summit he identified three species of ebony, the *benjoin*, and a couple of other trees, all stunted by the wind and some spreading much wider than they were high. One tree, *Fernelia buxifolia*, was common throughout the island, oddly overlooked by previous visitors. However, the overall impression of the island was of "sloping sides with bare rock scarred and furrowed with water-courses and studded here and there with groups of stunted palms or twisted screw-pines, crowded near the summit into a forest". The rabbits and goats had become "in one sense a nuisance, being very injurious to the trees on which, from the scarcity of grass, they are often forced to browse; the curious shape of the Bottle Palm enabling the goats to climb up even to the cabbage, whilst the seeds no sooner fall to the ground than they are nibbled, or, if they succeed in germinating, the young shoots are immediately eaten down"[75]. Thus did Barkly succinctly forecast the fate of the island's flora over the next century.

Barkly and Pike's writings on the animals they saw in November 1869 are confused and contradictory, though their collections were most important. Barkly sent the reptiles to the British Museum. They included a new species, the large Günther's Day-gecko, only alluded to vaguely in his own report. Both he and Pike thought they had "four very distinct varieties of serpent", all unknown to science, but they proved to be the two already known[76]. On lizards Barkly and Pike were at cross purposes, muddling Bojer's and Bouton's skinks and even (in Pike's case) apparently confusing them with Vinson's Day-gecko[77]. Another puzzle in Pike's reports concerns finding some petrels nesting in the same place as the tropicbirds, in little rock hollows at the seaward end of a "great gorge" (Big Gully). "I saw a good many petrels (*Puffinus chlororhyncus*) sitting in the same locality. These birds also lay a single egg, quite white, as large as a hen's. There were no young about, and the eggs were all fresh that I took". As the Latin name he used suggests, the only petrel then confirmed from the islet was the Wedge-tailed Shearwater, but the location, nest site and even the description of the eggs are typical of 'Round Island' (Trindade) Petrels, not formally discovered there until 1948 (though there are eggs from 1932)[78]. In his copy of the RSAS's *Transactions*, Edward Newton wrote in the margin of Pike's report that: "the birds breed in holes of the rocks quite in out of sight; he could not *see* them sitting". This is true of shearwaters but not of 'Round Island' Petrels, which nest out in the open, often in cavities in the tuff as described by Pike, and in roughly the same part of the island[79]. However Pike noticed no conspicuous display flights, and found only fresh eggs and no young, both of which point towards shearwaters. 'Round Island' Petrels nest throughout the year, but by December usually have young but few eggs, whereas the shearwaters mostly lay in November, their young hatching in late December onwards. The petrels fly about noisily in the daytime, whereas shearwaters are either at sea or in their holes during the day and only fly around the island at night[80]. However, Newton's letters reveal that he too was on the island with Pike and Barkly in 1869, writing that "the only new bird I got was a true petrel, a large one larger than Bulwer's, I think brown, it was amongst the skins the Lloyds gave us". In 1864, Newton had written that his associate S. Roch had "a large petrel (not shearwater) which he got on Round Island a short time ago & which I did not see when I was there; he has also got some eggs of it – he had others but he stupidly broke them, and the one he has is badly smashed." Newton referred to it later as a 'dark petrel', but Roch never took it to Newton's brother Alfred in Cambridge for identification as Edward had hoped[81]. The whereabouts of Lloyd's and Roch's specimens are unknown; unless these specimens or Pike's eggs turn up somewhere, we cannot be certain that these were 'Round Island' Petrels, but on balance it seems likely that a small population was present in the 1860s.

The most bizarre item concerns not a vertebrate, but a "brown beetle about 1½ inch in length, tubercled all over . . . of which I can find neither figure nor description"[82]. This giant tenebrionid beetle, *Pulsopipes herculeanus*, is now found only on Frigate Island in the Seychelles, an island also visited by Pike, but not until 1871. Furthermore he made no mention of it in his Seychelles trip report[83], and in any case the specimen was in Pike's "prepared case" of Round Island animals put together for Barkly to send to London in December 1869, and reached the British Museum before Pike had been anywhere near the Seychelles. So did this strange large beetle once inhabit Round Island? No-one has seen it there since, and the record has generally been considered a mistake. Jean Vinson's contention that the case also contained an endemic Seychelles snake sent to Pike by a contact appears to have been another error, the animal in fact coming from Round Island. Hence it now seems probable that this odd and rare beetle, as well as other invertebrates (other beetles and also scorpions), was found both on Round Island and the Seychelles; if also formerly present on mainland Mauritius, they have long since disappeared due to predators or other environmental changes[84].

Despite the scant published information, the islands were not ignored by Mauritians. The lighthouse and quarantine station meant that there was continuous human activity on Flat Island. Barkly alluded to a plantation of the cultivated screw-pine *Pandanus utilis* (*vacoa*) there[85] – perhaps staff were kept busy when no ships needed quarantine cutting *vacoa* leaves to make the bags in which sugar was exported in the 19th century. Newton revisited Flat Island in July 1874 to inspect the quarantine facilities in his capacity as acting Governor, but left no record of wildlife[86]. Round Island was visited, without accounts being available, by Evenor Dupont in 1872, a collector for Karl Möbius in 1874, Albert Daruty in 1873 and 1879, and German reptile collectors in the 1880s[87]. The last recorded 19th-century visit to Round Island was by keen amateur botanist Dr Henry H. Johnston in November 1889. By then it had "a very barren aspect, its steep brown rocky sides having only small clusters of and scattered plants of palms and screw-pines to enliven the desert appearance of the island". Most of the palms were growing out of bare rock, the previously reported soil "having been washed away by the heavy rains into the sea". Johnston was the last to see the "small wood on the south side of the hill 450–550 feet above sea-level", and noted that many of the more widespread *Fernelia* bushes were dying or dead "from the effects of the drought"[88]. The only animals he mentioned were rabbits. In 1906 a geological exploration company from South Africa (again after guano) landed there and collected subfossil bird, tortoise and flying-fox bones in a rock fissure; the tortoise bones were unfortunately never identified and are lost, but the bat skull turned out to be probably from a Golden Bat, now confined to Rodrigues. Its former presence in Mauritius has since been confirmed from bones found in mainland deposits (p. 394, Table 3.1)[89]; it may well have been the bat seen on Flat Island in 1703.

The persecution of tropicbirds first noted by Pike was causing concern in the 1870s. In 1878 Edward Newton drew up a list of six birds to be protected in Mauritius, five of them endemic landbirds. Local naturalists were upset by his exclusion of Red-tailed Tropicbirds, which he justified on the grounds that while "aware [it] is occasionally shot by thoughtless persons . . . this species has a very wide range . . . so that if it was exterminated here, it would not be lost as a species to the world as would be the case if five, out of six species I propose should be protected, were exterminated"[90]. The objectors did not want it 'exterminated here', and when the protection order was eventually issued in 1880 (p. 135), '*Phoeton phoenicurus, Paille en Queue – Ile Ronde*' was included (along with other additional species). Round Island still held 'immense numbers' of Red-tailed Tropicbirds around 1880, but another proclamation was issued in 1886 "to protect certain seabirds on Flat Island and Gabriel Island". By 1897 Round Island was the target, and a law was passed specifically protecting Red-tailed Tropicbirds there[91]. Harvesting these birds was clearly out of control.

Meinertzhagen's and Wiehe's visits

In 1910 Flat Island was visited twice (September and November) by the buccaneering British soldier and birdwatcher Richard Meinertzhagen. He found a thriving population of giant tortoises, and a Whimbrel's nest! Meinertzhagen later acquired a reputation for creative exaggeration and economy with the truth[92], but there is little evidence of these tendencies when he was in Mauritius[93]. There *were* feral tortoises: he photographed them (preserved in his diaries), stating on his first visit that "there are several giant tortoises which roam about freely, I have not been able to get any information about their origin . . . they obviously breed on the island, as there are many of all sizes". In fact five young animals had been released in 1883 to establish a breeding population outside Aldabra, in response to a conservation campaign organised by Albert Günther in England. Pressure on successive Mauritius governors from Günther and the (London) Royal Society finally led to Sir John Pope-Hennessy seeking animals for breeding. Only the local acclimatisation society offered tortoises, but specified that they be released to freedom on Flat Island, not enclosed as Günther had requested – a wise move, unexpected for the period, since tortoises in traditional Mauritian enclosures rarely breed successfully[94]. On his second visit Meinertzhagen "found the giant tortoises breeding, which is satisfactory, and the Whimbrel nesting, which is remarkable". In his paper he recorded "a nest with three eggs was found on Flat Island on Nov. 18 1910; it was among some rough grass and close to the sea". Whimbrels are common visitors to the Mascarenes in the northern winter, but return to the Arctic to breed. A few birds remain during the northern summer, and perhaps, with the seasons switched, some might be stimulated to nest in November – but no one else has recorded breeding in Mauritius, and Meinertzhagen left no photographic evidence[95]. He photographed free-range donkeys, recorded feral guineafowl, and saw (and shot) a feral cat. His photographs show old *veloutiers, Tournefortia argentea*, near the shore, with no sign of regeneration.

In October 1910 Meinertzhagen also visited Round Island. His photographs show thickets of all three palms and screw-pines and he found good numbers of Red-tailed Tropicbirds. "Not wishing to fire at any more goat [he had just shot 4], I walked to the summit of the rock. Here I found the red-tailed Bo'-sun birds nesting everywhere. Their nest is a mere scratching under a ledge of rock or among vegetation. I found fresh eggs, hard-set eggs, newly hatched young and young just able to leave the nest"[96]. Despite their numbers, he noted that "this bird, although protected by law, is terribly persecuted by the fishermen, who slay it indiscriminately on Round Island when opportunity offers, and sell it as food to the Chinese at Port Louis, who consider it a great delicacy"[97]. He saw adults but found no nests of White-tailed Tropicbirds, and "flushed and shot a brown petrel which proves to be the *Puffinus obscurus*, which flew out from under some dead palm leaves, but though I made a long search I found no sign of its breeding". In both diary and paper he used the wrong Latin name *obscurus* [= *bailloni*, Tropical Shearwater] for Wedge-tailed Shearwaters, but clearly saw no sign of 'Round Island' Petrels. He was there in mid-season, and the summit where he found so many tropicbirds is now one of their regular nesting sites, so they must still have been scarce. In addition to goats, he noted that "the rabbits . . . are all black and white, very fat and large". He saw centipedes and lizards (not specified), but no snakes.

Over twenty years passed before another scientist visited the islets – botanist Octave Wiehe explored all but Serpent Island in 1935, writing it up with Reginald Vaughan in the classic ecological study of Mauritian vegetation[98]. The published account is very brief, but for Gunner's Quoin and Flat Island it is important as the first vegetational account for many decades. On Gunner's Quoin the palm forest was reduced to a 'scrub' of latan and screw-pine around a sheltered south-facing bay, the rest, apart from cliffs and the spray zone, being mostly grassland dominated by a native grass *Dicanthium nodosum*. This 'savanna' was being invaded by the introduced grass *Heteropogon contortus*, and woody exotics such as lantana and *bois noir*[99]. Hares were present; whatever their earlier status, more had been released in 1934, and deer were added in 1938[100]. Flat Island had lost its indigenous vegetation almost entirely, being largely covered in she-oak, with a ground layer of *Stenotaphrum dimidiatum*, a common tropical grass. "A few native species [unspecified] still maintain a precarious existence on the rocky cliffs to the south of the island"; no mention of swampy areas or bullrushes. Gabriel islet had fared better, with the native shrub *Psiadia* still dominating, though only a few scattered latan and screw-pines were left. Round Island was now largely "bare of vegetation, but remnants of a palm thicket formerly of much larger extent may still be found". Only dead stumps persisted of the former broadleaved 'forest'. "Goats and rabbits have been introduced on Round Island . . . natural regeneration has thus been arrested by the destruction of young seedlings and rapid erosion of the soil is taking place. It is evident that unless drastic steps are taken the indigenous flora on the islets will become extinct in a comparatively short time". The botanists made no mention of birds, lizards or snakes, nor of any mammals other than rabbits, hares and goats.

The last sixty years

Flat Island

Photographs taken on Flat Island by Patrick Barnwell in 1944 show the island as open grassland and she-oak plantation, with dead and dying overmature screw-pines here and there. There were still some large old tortoises[101] – the reason for their presence long forgotten, they were apparently taken off in the 1950s and pensioned off to Pamplemousses gardens[102]. No doubt breeding success had declined due to feral cats. We have been unable to trace when the island stopped being used for quarantine (probably soon after Meinertzhagen's visit)[103], but the lighthouse was manned until the 1980s, and since at least the 1940s the island has been, with Gabriel Islet, a popular picnicking place for Mauritians with access to sea-going boats. Unlike Round Island, and since the 1970s Gunner's Quoin, which have had considerable scientific attention, Flat Island and Gabriel Islet have been largely neglected. Robert Newton saw a few Red-tailed Tropicbirds nesting on Flat Island in November 1955, but found the island 'disappointing':

"The greater part of the island consists of a flat plain with coarse grass concealing the rocks and volcanic tuff; it is practically treeless except for groves of casuarinas [she-oak] at the eastern end"[104]. Introduced landbirds no doubt colonised Flat Island from Mauritius when it became inhabited in the 1850s, and Meinertzhagen had seen Zebra Doves and Common Waxbills in 1910, but it was not until 1971 that Alec Forbes-Watson's party provided the first landbird list: Zebra Dove, Red-whiskered Bulbul, Yellow-crowned Canary, Cardinal Fody, House Sparrow, Common Waxbill and Common Myna – plus the ubiquitous native Striated Heron[105]. By 1993[106] the grassland had become largely overrun with White Popinac, brought in, as Horne noted, in the 1880s. Lantana and *Heteropogon* grass were still invasive; a last few screw-pines and latans were still holding out on the hill, though the coastal vegetation appeared to be maintaining itself better than in the 1940s[107]. Fires accidentally or deliberately set by visitors were regular. Hares appeared to have died out, but feral cats were still present, Ship Rats and mice widespread, and there were two introduced house geckos, Stump-toed Gecko and Cheechak. Grey Francolins were added to the bird list, but canaries had dropped out. Of native lizards Bojer's and Bouton's Skinks were recorded, as was Vinson's Day-gecko. Later in the 1990s the exciting discovery was made that the Macabé Skink, formerly thought to be a race of Bojer's and confined to the mainland, was living on Flat Island sympatrically with *bojerii*, and thus had to be considered a full species[108]. In 2003 a patch of tuff on the hill was found to harbour Lesser Night-geckos, the first time a *Nactus* had been found sympatric with Cheechaks. Since Cheechaks were generally considered responsible for extirpating night-geckos, this was rather surprising. Further investigation showed that the night-geckos survived only because their tuff substrate was so crumbly that the Cheechaks' toe-pads, adapted to smooth solid surfaces, could not get a grip, whereas the night-geckos' clawed toes were ideal. On equal terms Cheechaks excluded night-geckos from refugia, and also directly attacked, injured and even ate them[109]. Rats, mice and cats were eradicated in 1998[110]. The recent discovery of the two extra lizards on Flat Island enhances its conservation value, and makes it a prime candidate for serious ecological restoration, being the only one of the northern islets large enough and with enough of a water table to support a rich forest.

Gabriel Islet

Gabriel Islet in 1993 still retained much of the character reported for Flat Island by Horne in 1885. Inland of a coastal shrubby fringe like that on Flat Island, the native *Psiadia* scrub survives, covering about half the

BOX 29   STARLINGS (STURNIDAE)

Starlings are gregarious birds of forests and open woodlands. The Mascarenes were once inhabited by two very distinct species, both accurately described in early accounts, but only the Réunion species was preserved as skins. An endemic starling has yet to be discovered on Mauritius.

Hoopoe Starling *Fregilupus varius*

The Hoopoe Starling or *Huppe Fregilupus varius* was a largish crested bird that was reported as common until the late 1840s. It disappeared with rapidity, with the final records being made in the 1850s; reports in the 1860s are but hearsay. It had lived alongside a number of introduced, potentially dangerous predators and competitors, e.g. the Ship Rat *Rattus rattus* and the Common Myna *Acridotheres tristis*, for decades and it was more likely that disease was to blame for its extinction. The last specimens were taken in the early 1840s.

Hoopoe Starling *Fregilupus varius*. Drawing based on a live specimen by Jossigny, c. 1770. From Oustalet 1897.

Accounts

Dubois (1674) in 1671–72:

> Huppes* or Callendres* with a white tuft on the head, the rest of the plumage white and grey, a long beak and feet like a raptor; they are a little smaller than baby pigeons. It's another good game [gamebird] when fat.

Levaillant (1807, quoted by Probst & Brial 2002), citing information from an anonymous Réunion resident:

> [this species] *lives in large flocks in* Bourbon [Réunion] *where it frequents humid places and marshes, and causes big damage to coffee trees.*

Desjardins in *c.* 1836 (in Milne-Edwards and Oustalet, 1893):

> *My friend Marcelin Sauzier brought me four of them* [Hoopoe Starling] *alive from Bourbon in May 1835. They eat anything. Two escaped a few months later, and it may be that they populate our forests.*

Desjardins in 1837 (manuscript quoted by Oustalet 1897):

> We do not hesitate today to report that the Huppe, which lives in the forests of Bourbon and which is kept as a cage-bird, may have become naturalised in Mauritius. Mr Autard has presented us one which had recently been killed in the District of Savane, and we ourselves had 4 which our colleague the elder Mr Sauzier, long term correspondent from Bourbon, had brought us from that neighbouring island, and which, after several months in captivity, took flight, and made off towards the woods, now so rare, in the District of Flacq.
>
> At the session of 5 January 1837, an example was presented stuffed by Mr Liénard senior [François Liénard], who had got it from Mr Autard who had killed it in the Savane [district]. The latter had assured him [Liénard] that he had often seen considerable flocks of them.

Eugène Jacob de Cordemy in 1910 (in Manders 1912):

> I have known the bird you ask me about since childhood, namely the Fregilupus varius *(old writers called it F. capensis), which has in fact entirely disappeared . . . When I was a boy this bird lived in the forests of the interior of the island and never set foot nor wing in towns or inhabited places. It remained faithful to the forests where it bred, which it enlivened with its clear notes. I used to hunt it then at an age when one is pitiless. I can see it now, a little larger than the white blackbird [merle blanc =* Coracina newtoni*], with a white crest on the head in the case of the male, the wings a blackish grey on the upper surface, the beak and feet yellowish. By no means shy, it was not frightened even by the sound of firearms, and after a regular slaughter one went off with dozens of these poor victims in one's game-bag.*
>
> *After ten years spent in Paris I did not find a single one in the forests where formerly they flew about in flocks. All ruthlessly destroyed. I shall never forgive myself for the part, slight though it was, which I took in the matter. I lost my taste for sport and the best bag would not tempt me . . . We will now consider the feeding habits of this bird. Having raised several in the aviary, I can risk talking about it though I never saw one feeding in the wild state. In my aviary its food consisted of bananas, potatoes, and choux-choux,* [Sechium edule, *boiled]. But when left to its own instincts, it must, like the other winged denizens of the forest, have eaten insects as is done by its companion in the forests, the Bourbon Blackbird (*Hypsipetes olivaceous*) [now* H. borbonicus*], and as is the habits of most fruit-eating birds.*

Rodrigues Starling *Necropsar rodericanus*

The Rodrigues Starling *Necropsar rodericanus* is known from skeletal remains and one detailed account. Its leg bones are particularly robust and the bill musculature strong. Tafforet's account describes the starling's ability to tear dead (juvenile?) turtles or tortoises out of their shells – a strong bite would have been useful for this purpose – and the birds could be easily reared using meat for food. They, like the pigeons and parrots, were scarce or absent on mainland Rodrigues, living only on the islets by 1726. In 1761, Pingré did not report them, so it is likely that they had died out in the intervening years.

Account

Tafforet (1726), from A. Newton's 1875 translation:

> A little bird is found which is not very common, for it is not found on the mainland. One sees it on the islet au Mât [= Ile Gombrani], which is to the south of the main island, and I believe it keeps to that islet on account of the birds of prey which are on the mainland, as also to feed with more facility on the eggs of the fishing birds which feed there, for they feed on nothing else but eggs or turtles dead of hunger, which they well know how to tear out of their shells. These birds are a little larger than a blackbird [Réunion Merle], and have white plumage, part of the wings and tail black, the beak yellow as well as the feet, and make a wonderful warbling. I say a warbling, since they have many and altogether different notes. We brought up some with cooked meat, cut up very small, which they eat in preference to seeds.

Specimen collections

Fossil material can be located at UMZC and BMNH.

---

* Huppe in French normally refers to the unrelated Hoopoe *Upupa epops*, but was always used in Réunion for the endemic starling; *callendre* does not appear in standard dictionaries, but is clearly cognate with the English 'Calandra Lark' for *Melanocorypha calandra*, called by similar names in all European languages.

island, though now invaded by Lantana. There is grassland with scattered screw-pines and latans; 60% of the species were still indigenous, despite evidence of recurring fires. The rabbits had died out but Ship Rats were common. Wedge-tailed Shearwaters (*c*. 250 pairs) bred, but suffered some poaching, and a handful (*c*. 5 pairs) of White-tailed Tropicbird held on. Robert Newton had reported 'large numbers' of tropicbirds there in November 1955, though Beau Rowlands did not see many in October 1970[111], no doubt decimated by poaching. The same three native lizards as then seen on Flat Island were present, plus Cheechaks. Rats were eradicated in 1995 but a latan seed showing signs of rat-gnawing was found in 2006[112].

Pigeon House Rock

Pigeon House Rock was not scaled until 1995, so earlier information depended on views from Flat Island or a boat. The century between Besant's visit and Newton's in 1955 is a complete blank. Newton noticed only Red-tailed Tropicbirds on the stack, though in October 1971 both tropicbirds were present[113]. In the early 1970s small numbers of Masked Boobies, Sooty Terns and both noddies were sometimes seen on the top; nesting could not be confirmed[114]. The 1993 surveys omitted Pigeon Rock, but an award-winning video made in 1995 records the stack's exploration by Réunion ornithologist Jean-Michel Probst and some rock-climbers: they found both tropicbirds and Wedge-tailed Shearwaters nesting and two resident lizards, Bojer's Skink and Lesser Night-gecko[115].

Serpent Island

Serpent Island was at long last revisited in 1948, by a team led by Barnwell and including zoologist Jean Vinson. This fruitful trip yielded not only confirmation that the skink was Bojer's, but also the unexpected discovery of a completely new species of gecko[116]. The Serpent Island Night-gecko *Nactus serpensinsula* was the first to be discovered of a whole radiation of the genus *Nactus* in the Mascarenes[117]. The seabirds were as before, plus a few Red-tailed Tropicbirds. Robert Newton landed, with difficulty, in November 1954 and January 1957, after which there was no detailed report until 1992, although there had been a couple of undocumented landings and birds noted from offshore. In October 1973 "hardly a nestable surface on the islet [was] not covered with birds, which got up like a huge cloud when we passed"[118]. Newton reported only on seabirds (no change)[119], but visits in the 1990s revealed a fascinating micro-ecosystem on the islet. The remains of fish etc. brought by seabirds to their young, dead birds, their excreta and damaged eggs support a thriving population of insects on which the larger fauna largely lives. In 1992 Roger Safford discovered a large tarantula, whose main prey appears to be night-geckos, while Bojer's Skinks ate broken eggs as well as insects. Most night-geckos have regenerated tails, ascribed to predation by skinks and/or spiders[120]. Another predator, an 80mm-long centipede, was discovered in 1995[121], also possibly capable of taking hatchling lizards. In the 10,000 or so years since Serpent Island has been isolated from the Mauritian mainland, there appears to have been intense selection on the lizards stranded in this seabird-dominated, waterless and almost plant-free habitat. Both are markedly larger than their counterparts on Round and Flat Islands – possibly a response to a long absence of larger lizards, still present (or around until recently) on the other islets and the mainland. The Serpent Island Night-gecko retains an invertebrate diet, while the island's population of Bojer's skink has become more omnivorous, echoing Telfair's Skink on

Table 9.1. Numbers of pairs of breeding seabirds on Serpent Island as estimated by various observers[123].
● = present   * = abundant   – = not recorded   ○ = positively absent

Year and month	Masked Booby	Red-tailed Tropicbird	All terns	Sooty Tern	Common Noddy	Lesser Noddy	Source (see note 123)
1844/Dec	●	–	*		*		Lloyd & Corby (1)
1856/Nov	●	–	●				Layard (2)
1948/Nov	●	few	300,000	*	*	*	Vinson (3)
1954–55/Oct	25/c.60	–		4,000	1000	scores	Newton (4)
1957/Jan	120	few		20 x thousands	thousands	*	Newton (5)
1970/Nov	●	●	30,000	*	*	*	Rowlands (6)
1973/Oct	–	–		c.50,000	5,000–10,000	1,000+	Cheke (7)
1973/Oct	10–20	–		500,000	100,000	250,000	Cheke (7)
1973–75/?	50	20–30		thousands	thousands	hundreds	Temple (8)
1992/Nov	c.50	○		10 x thousands	thousands	thousands	Safford (9)
1993/Sep	c.125	○		250,000	1,000	10,000	Bell *et al.* (10)
2003/Nov	40–50	○		200,000	40,000	20,000	Tatayah (11)

Round Island. Both lizards are distinct enough to be considered separate subspecies[122].

There has been no apparent overall change in the last 150 years in seabird composition or numbers on Serpent Island. The number of breeding boobies appears stable at up to 120 pairs (varying year to year), while the numbers of terns, although always clearly in the thousands, have been given such different estimates over the years, that it is difficult to be sure what they mean.

The absurd variation in published figures serves little more than to show that when faced with huge numbers of birds, the average observer has no realistic way of assessing them, one of us (ASC) being as guilty as the rest. The variation is clearly not due to seasonality[124] as all the visits were around the same time of year. However, numbers nesting may differ between years and indeed breeding may be aseasonal. Given the importance of this colony for terns and boobies in the Mascarenes and indeed the western Indian Ocean as a whole, a proper survey is long overdue. A mosaic of photographs taken from offshore all round the island at different times of year, plus aerial shots of the summit, would be necessary for any real hope of accuracy. Seabird numbers estimated in 2005 for all Mauritian islands are given in Table 9.2.

Gunner's Quoin

Robert Newton reported briefly on Gunner's Quoin in November 1955 – still dominated by Lantana, *bois noir* and the endemic aloe, with some screw-pine and latans. He noted some 95 pairs of Red-tailed Tropicbirds, with a few White-tailed Tropicbirds and Wedge-tailed Shearwaters. Apart from this visit, the islet was neglected until the 1970s, when biologists studying Round Island reptiles began looking for places to establish 'reserve' populations of endangered species. Gunner's Quoin was the obvious choice – uninhabited, nearby and with a similar ecology – so it was visited to see what survived and what needed doing to make it safe for Round Island's unique creatures. In 1973 the situation was not encouraging. The few surviving latans and screw-pines, almost all elderly, fruited abundantly, but no regeneration resulted, while rat and hare droppings were plentiful. Only Vinson's Geckos and Bouton's Skinks were seen, but no Bojer's Skink, which had been collected in the 1960s[125]. Small numbers of both tropicbirds were nesting, and Wedge-tailed Shearwaters were still reported to do so as well[126]. Visits in 1978 and 1982 confirmed the parlous state of the island's ecology[127], though Bojer's Skinks were seen and a new nightgecko was discovered, *Nactus coindemirensis,* in 1982. It was at first thought to be a small version of the ones on Serpent and Round Islands, but it later proved to be different species (subsequently found on Flat Island, Pigeon House Rock and Ilot Vacoas in Mahébourg Bay)[128]. The vegetation was thoroughly surveyed in 1982, much of the island dominated by introduced scrub consisting largely of *Cordia macrostachya*, Indian Plum *Flacourtia indica* and Sandalwood *Santalum album*. Apart from some stunted *Eugenia lucida,* the only surviving native trees were screw-pines (including some young plants), latans with inadequate regeneration (seeds mostly eaten by rats), and a healthier population of Mauritian Dragon Trees *Dracaena concinna*[129]. The deer introduced in 1938 were allegedly shot out in 1980, though as none were seen or heard of during visits in 1973–75 and 1978 the numbers must always have remained very low, if any persisted at all after the 1940s[130].

Following an assessment visit and a new botanical survey in 1993, in late 1995 rats and hares were eradicated from Gunner's Quoin – a prerequisite to any release of Round Island reptiles. In August 1996 the island was re-visited by David Bullock and Steve North, authors of the 1982 survey, assisted this time by a team from the National Parks service and the Mauritian Wildlife Foundation; in 2003 a further NCPS-led project visited to establish guidelines for restoration and reintroductions, and a first batch of Telfair's Skinks was released in December 2006[131]. No sign of surviving rats or hares was found in 1996, but in 1998 there was evidence of grazing which turned out to be from rabbits, surreptitiously released by hunters deprived of their former quarry; they were eliminated the same year[132]. During 1982–1996 the population of mature latans remained stable at around 20 individuals, though some seedlings were establishing following removal of rats and hares. By 2003 regeneration was clearly healthy. Screw-pines also showed little change, with "large numbers of *Pandanus* seedlings having recently germinated beneath 'mother' trees". In 1993 there were fears that the Lesser Night-gecko discovered in 1982 might already be extinct, but plenty were found in 1996, more widespread than in 1982; the rats only recently removed, the other lizards showed little change. Seabirds (not fully surveyed) seemed stable, with perhaps some increase in Wedge-tailed Shearwater activity; 3,000–5,000 pairs were estimated in 2003. Human predation has continued on Gunner's Quoin seabirds: 20–30 corpses of Red-tailed Tropicbirds in 1978, "two large piles of seabird remains" (mostly adult shearwaters) in 1993, further remains in 2003. No doubt this is why numbers are still relatively low. In 1982 there were 50–70 active nests of Red-tailed Tropicbirds and 15–20 of White-tailed, rising to some 200–300 and 50–100 respectively in 2003[133]. Like Flat Island, Gunners Quoin supports introduced land birds – Grey Francolins, Spotted and Zebra Doves, Red-whiskered Bulbuls, Common Mynas and Common Waxbills;

	Mauritius mainland	Round Island	Serpent Island	Flat Island	Gabriel Islet	Pigeon House Rock	Gunner's Quoin	Ile de la Passe	Ilot Vacoas	Ile aux Fouquets	Rodrigues mainland	Ile Cocos and Ile aux Sables
Island size (ha)		219	31.6	253	42	0.63	65	2.19	1.06	2.49		14 & 8
'Round Island' (Trindade) Petrel *Pteroderoma arminjoniana*		150-200										
Kermadec Petrel *Pterodroma neglecta*		10-15										
Bulwer's Petrel *Bulweria bulweria*		3										
Wedge-tailed Shearwater *Puffinus pacificus*		40,000-80,000	5-10	10-20	250-400	30-50	3,000-5,000	1-5	40-70	10-30	50-75	
Red-tailed Tropicbird *Phaeton rubricauda*		1,000-2,000		15-30		75-100	200-300			50-75		
White-tailed Tropicbird *Phaeton lepturus*	500-1,000	750-1,500		5-10	15-30	30-50	50-100			1-5	50-75	
Masked Booby *Sula dactylatra*			40-60									
Sooty Tern *Sterna fuscata*			200,000-300,000									1,500-2,000
Fairy Tern *Gygis alba*												30-70
Brown Noddy *Anous stolidus*			20,000-30,000									3,000-5,000
Lesser Noddy *Anous tenuirostris*			20,000-30,000									8,000-12,000

Table 9.2. Numbers of annual breeding pairs of seabirds on Mauritius, Rodrigues and satellite islands. Estimates gathered 1993–2005 by Carl Jones, Vikash Tatayah and MWF staff; all islands visited several times.

the bulbuls and mynas have no doubt contributed to the introduction and spread of exotic plants, and bulbuls have been seen commuting from the mainland; Striated Herons nest in thickets of Indian Plum[134].

Round Island

The jewel in the islands crowning the north of Mauritius is undoubtedly Round Island. Because of its unique reptiles, the colony of 'Round Island' Petrels, the large numbers of Red-tailed Tropicbirds and the only surviving wild location for two palm species, Round Island has been well covered since Jean Vinson revived interest in it in 1948. The decline (and partial recovery) of the palms, the devastating human toll on tropicbirds, the loss of the Burrowing Boa, the ups and downs of the other fauna, and the eradication of goats and rabbits (and the extraordinary politicking this engendered) have all been documented in a now enormous literature.

The group that visited Serpent Island in November 1948 also explored Round Island, Vinson making the first serious report on its fauna since 1869[135]. The thousands of Red-tailed Tropicbirds had dwindled to barely half a dozen pairs, such was the "inordinate destruction by fishermen" despite the law protecting them; indeed White-tailed Tropicbirds, though still few, were more numerous; harvesting may have intensified during wartime shortages. Wedge-tailed Shearwaters proved still abundant, and the team discovered a previously unrecorded petrel nesting amongst boulders at the summit. Specimens sent to Dr Cushman Murphy in America were identified as Trindade Petrels, whose only other known nesting site is the eponymous island in the western Atlantic off Brazil[136]. Although palms and screw-pines were all still fairly common, a photograph of the western slopes looking downhill (with Flat Island beyond) shows large areas bare of vegetation and only old tall palms. Vinson recorded abundant rabbits and around 100 goats, and, like Barkly in 1869, was "unable to find a single young palm or screw-pine as they are all destroyed as soon as they germinate". They found no snakes, but had more luck the following December, catching two Keel-scaled Boas. One kept for months afterwards ate day-geckos, but refused house-geckos. When Vinson gave a talk on the trip, a member of the audience offered a recently caught Burrowing Boa specimen; it was clear neither snake was abundant. They found reasonable numbers of Bojer's Skinks and Günther's Geckos, but Telfair's Skinks were few and oddly shy; Vinson's Day-geckos were common in the latans. The scorpions, centipedes and land snails recorded by Pike and Barkly had apparently largely disappeared.

In the period up to 1963 not much happened. Vinson's collections stimulated study of the boas' affinities and the discovery of their bizarre hinged jaw mechanism; they had already (in 1946) been elevated into a new subfamily (now a full family, Bolyeridae; pp. 73–74)[137]. In October 1954 Robert Newton commented that "the piles of dismembered corpses [of Red-tailed Tropicbirds] are a distressing sight for any visitor to Round Island in the breeding season between October and January which, unfortunately, is the most favourable period of the year for landing", but nevertheless found 62 active nests. Although Red-tailed numbers had in fact risen a little since 1948, Newton thought fishermen might wipe them out, together with the petrels, and, as Colonial Secretary, got the island declared a nature reserve in 1957. He hoped (over-optimistically) that "with the co-operation of the Police and District Administration, and by suitable propaganda, we will be able to stop this destruction"[138]. Visits to the islet in 1952, 1954 and 1957 revealed the vegetation apparently stable, but Vinson was shocked to find in 1963 that the severe cyclones of 1960 and 1962 had thrown down most of the palms and screw-pines; latans had been decimated and the other species "practically wiped out". He made it clear that eradicating rabbits and goats was the only way to preserve the island's unique species[139]. Realising that action was urgent and not likely to happen with local energy alone, Vinson toured international conservation bodies in 1964, and submitted a special report to IUCN in 1965, also probably prompting the first article on the island's predicament to appear in an international magazine[140]. Also in 1965 a local action committee was convened[141], but apart from sporadic hunting parties to shoot goats and rabbits not much actually happened; in any case the impetus was largely lost when Vinson died unexpectedly in May 1966[142]. The committee was reconstituted in 1969, but concentrated more on preventing the slaughter of tropicbirds than on habitat loss[143]. From 1966 to 1974 some regeneration of latans and screw-pines was noted, coinciding with a temporary reduction in rabbits caused by the 1960–62 cyclones, and ongoing shooting of goats[144].

The wheels of international interest had in fact slowly begun to turn, and in January 1973, at the government's invitation, Mauritius was visited by Sir Peter Scott, then chairman of the World Wildlife Fund. His recommendations for eliminating goats and rabbits were soon attempted by Dr Stanley Temple, in Mauritius to save the dwindling kestrel population. In November, assisted by a team of helpers (including ASC), he tried poisoning rabbits with Lucerne (alfalfa) soaked in strychnine. This bait was unattractive to reptiles, but like manna to rabbits on the parched waterless islet – it worked, but using strychnine provoked an international controversy that set the eradication programme back a decade. A more detailed conservation report in 1975[145] endorsed the "practical techniques" already established (i.e. strychnine poisoning), but conservationists did not anticipate that poisoning furry bunnies to preserve creepy snakes looked like criminal lunacy to the animal welfare movement, particularly as strychnine leads to a painful death. Behind the scenes lobbying locally by the MSPCA, and the banner being taken up by UFAW in UK[146], put a stop to further poisoning until an alternative could be found. One leading figure in UFAW, completely misunderstanding the biological issues involved, even argued that rabbits should be preserved at the expense of the snakes![147] For those who did understand the issues, shooting, rather than poisoning, was apparently acceptable, so in 1976 UFAW sent out a marksman, Major J. C. Gouldsbury, who spent six weeks on Round Island; he shot 883 rabbits, significantly reducing numbers, but by no means eliminating them. Being rabbits, they soon bounced back, though some seedling latans and screw-pines were able to establish while numbers were low[148]. After failing to shoot out the rabbits, UFAW admitted their solution did not work. Meanwhile anticoagulant poisons, less problematical, had been found to work well on rabbits in New Zealand, but the Mauritian government, put out by the controversy, withdrew permission for the action. Finally, after Gerald Durrell and the Jersey Wildlife Preservation Trust had negotiated a conservation agreement with the new government in 1984, a team led by Don Merton successfully eradicated the rabbits in 1986[149]. Gouldsbury had reduced goats to just two, which were shot two years later.

While the politicking paralysed the herbivore control work, the vegetation continued to deteriorate. Already by 1973 both Bottle and Hurricane Palms were down to a handful of easily counted individually, and counts over the next 15 years showed how rapidly the end was approaching. Until then no one knew how long these palms lived, but as the surviving trees aged rapidly it became clear that none (including the latans) were long-lived, probably around 150 years[150]. Table 9.3 shows the reduction of old palms during the 21 years 1975–1996; if the rabbits had not been removed in 1986, the Bottle Palms would soon have been lost and latans would have rapidly dwindled to the relict status seen on Gunner's Quoin. By 1986 seeds from the two surviving Hurricane Palms appeared to be unable to germinate on the island (although they still can under controlled conditions). The table also shows the effect of cyclones. Around 90% of mature latans survived over the seven years 1982–89 without severe storms, but dramatically fewer survived when there was a major hurricane (1979: *Claudette*; 1994: *Hollanda*); by *Dina* in 2002 too few old trees remained for comparisons, but younger trees were badly hit. When regeneration is good, cyclones rejuvenate the population naturally, when regeneration is prevented they are clearly disastrous.

Screw-pines grow much faster than palms and appear to live for only 50–60 years; so while several generations must have regenerated in the face of rabbits and goats, they nevertheless inexorably declined, and much of the island had lost screw-pines altogether by 1975. There was a noticeable boost in recruitment after the 1975 cyclone and the near removal of goats in 1976, so there were probably always short windows of opportunity for screw-pines during the 'herbivore period'. Of the original hardwood', only three individuals of two species survived into the 1970s, heavily browsed. One specimen of *Gagnebina pterocarpa*, smothered under an unpalatable creeper in 1982, was cleared and started to revive, but was later ring-barked and killed by rabbits before they were eradicated[152].

No one had any idea of snake numbers, but Günther's Geckos had been estimated at 1,500–1,800 in 1975 before the cyclone, 437–903 after, at risk from loss of latans; Telfair's Skinks *c.* 4,000–5,000, unaffected by the storm. To ensure something survived while the argument on rabbit eradication raged, a team led by Gerald Durrell collected lizards and snakes in 1976 and 1977 for captive breeding at Jersey zoo. The target species were Günther's Gecko, Telfair's Skink and the Keel-scaled Boa; no Burrowing Boas were found [153]. The Telfair's Skinks, more or less inviting themselves to Jersey,

> ... clustered around us, climbing into our laps and partaking of our hard-boiled eggs, tomatoes and passion fruit in their most genteel way, and sipping beer and Coca-Cola out of our glasses with all the decorum of a group of village ladies at a vicarage tea-party. We felt like cads and bounders when at the end of the repast, we simply picked up our well-behaved guests and bundled them into soft cloth bags...[154]

The skinks and geckos bred fairly rapidly; progress was slower with the snake, but sustained breeding was eventually achieved[155]. For Günther's Geckos the initial success proved deceptive. Having bred well for 11 years, they stopped laying and started dying of a syndrome involving excess body fat; the cause remains unclear, but it stopped the breeding programme. Most surviving geckos were repatriated to the open-air enclosures at the Black River aviaries[156]. Many Telfair's Skinks suffered skeletal abnormalities, but this was largely cured using ultraviolet light. Captive breeding continued until the elimination of rabbits made the programme superfluous, most animals being returned to Mauritius[157]. With intensive force-feeding it eventually became possible to rear young Keel-scaled Boas; even the adults are tricky and very eclectic in what they will eat, though two caught in 1977 were still alive in Jersey in 2000. The breeding programme is under review following the de-ratting of Gunner's Quoin and Flat Island, now possible re-introduction sites[158].

Although palms and reptiles attracted the bulk of research and conservation effort, seabirds were not ignored. The enigmatic gadfly-petrels, sitting tamely on their eggs and flying around in vocal displays have

Table 9.3. The decline of mature palms while herbivore removal was delayed[151]; | marks the removal of goats, || the removal of rabbits. Severe cyclones are marked with *; these occurred in Feb 1975 (i.e. before the count), Dec 1979 and Feb 1994. At the time of writing comparative figures from 2003 were not available, though the lone Hurricane Palm survived.

Date	1973	*	1975	*|	1982	||	1989	*	1996	Loss 1973–1996
Latans	?	*	2580	*|	1580	||	(1422)	*	(916)	64%
Bottle Palms	c. 20	*	17	*|	9	||	7	*	2	88%
Hurricane Palms	6	*	5	*|	2	||	2	*	1	80%

BOX 30 WHITE-EYES (ZOSTEROPIDAE)

The white-eyes are successful oceanic island colonists and inhabit numerous islands within the western Indian Ocean; nine endemic species are known. The Mascarenes were colonised twice in their history, the olive white-eyes with two closely related species (Mauritius Olive White-eye *Zosterops chloronothos* and Réunion Olive White-eye *Zosterops olivaceus*), and the Mascarene Grey White-eye *Zosterops borbonicus*, which has subspecies on Mauritius and Réunion. As with flycatchers and cuckoo-shrikes, white-eyes failed to reach Rodrigues. *Z. chloronothos* is now very rare on Mauritius but *Z. olivaceus* is common on Réunion, while *Z. borbonicus* has proved very adaptable and is common everywhere. No *Zosterops* has been identified in the fossil record.

Mascarene Grey White-eye *Zosterops borbonicus* on Réunion. From Barré & Barau, 1982.

Accounts
Buffon (1770–83):
[Mauritius Olive White-eye] *The Viscount de Querhoënt, who has observed this bird in the Isle de France [Mauritius] says that it is not timid, but that nevertheless it rarely approaches inhabited areas; that it lives in groups and feeds on insects.*

[Mascarene Grey White-eye] *They are found everywhere in great numbers in the isle of Bourbon [Réunion] where the Viscount de Querhoënt has observed them. These birds begin to nest in the month of September; one normally finds three eggs in their nest, and it appears that they have several clutches per year. They nest in isolated trees and even in orchards. The nest is composed of dry grass with horsehair on the inside; the eggs are blue. This bird allows a very close approach; it always flies in groups, lives on insects and little soft fruits. When in the countryside it sees a partridge [francolin] running on the ground, a hare, a cat etc., it flutters around it making a particular cry, which serves as an indication to the hunter for finding the game.*

Lesueur (1803, unpublished MSS in Le Havre) in Mauritius in 1803:

[Mascarene Grey White-eye is] *in the genus* Sylvia [which in those days included many disparate genera of warbler-like birds] *we saw the species found* [also] *in Bourbon* [Réunion] *and which bears the name of that island* [i.e. 'Sylvia' borbonicus]. *This species is remarkable for* [the way] *it facilitates catching hares in their hiding places by stopping over the place where one is laid up and fluttering about, and with the calls and noise it makes, together with others of its species scattered through the forest which do not delay in joining it, alerts passers-by, who, going carefully to where the noise is coming from are often able to take the hares that the little birds have brought to their notice. The blacks are skilled at this* [type of] *hunting and even follow the birds in the forest.*

had most attention. Everyone thought their identity had been settled in the 1950s, but in the 1990s it was noticed that they made two distinct types of call, and that some birds sported pale shafts to their primary feathers. This suggested the simultaneous presence of two species, apparently 'Round Island' (Trindade) Petrels and Kermadec Petrels (p. 69)[159]. However preliminary DNA analysis was inconclusive, and a third related form, long suspected, was confirmed in 2007 when a Herald Petrel *Pterodroma heraldica*, ringed off Australia, turned up; the question is still being investigated[160]. Petrel numbers appear to have been slowly increasing, but they are hard to census as breeding is not synchronised, though egg-laying peaks in late winter (July-September). A figure of 400+ pairs published in 2001 is higher than the usually cited '120 pairs', but is probably not unrealistic; regular monitoring confirms that some 150–200 pairs nest in any one year, but ringing studies suggest that over 1000 birds use the island over a period, many not nesting annually[161]. Wedge-tailed Shearwaters appear to have increased from the mid-1970s to 1982, when estimated at 50,000–100,000; it is unclear whether the 2003 figure of 40,000–80,000 really represents stabilisation or simply stable guesstimation. It is by far the largest colony in the Mascarenes[162], and brings huge amounts of nutrients into the ecosystem. Their abundance was mistaken in 2002–03 for banality, and the vegetation restoration project usurped their largest colony, in rare deep soil on the southern slopes, to plant large numbers of native tree seedlings, deliberately collapsing their burrows and netting the whole area to prevent the birds getting back to their nests. This error of judgement pointed up a lack of joined-up thinking in the conservation programmes, since remedied, as has been the shearwater colony; the birds rapidly re-established their burrows once the netting was removed, and plants are now protected with individual tubes[163]. Among seabirds, tropicbirds, especially Red-tailed, have gained most from recent conservation activity, as poachers are now much more likely to be intercepted. The law protecting tropicbirds had always been a dead-letter, and as recently as 1982 a research team caught a raiding party red-handed with 94 adults and 34 young. Since then conservationists have noted only sporadic bird-kills, the last in 1997, though the birds had started to recover their numbers earlier, suggesting that diminishing returns had reduced poaching as early as around 1950. The low point of just a few Red-tailed Tropicbird pairs in 1948 improved to 62+ in 1954, 100 in 1964, 200–300 in 1974–75 and 1982, rising to

500–700 in 1987–89 and 1,500–2,000 in 2003. White-tailed Tropicbirds were not confirmed as nesting on Round Island in the 19th century, though a few were seen in 1860 and 1910. By 1948 they outnumbered the decimated Red-tailed Tropicbirds, though there were still only a handful in 1954. By 1964 they reached 200 pairs, dropping to 40–50 in the mid-1970s, 100 in 1982 and increasing dramatically to 500–1,000 by 2000, nearer the upper figure in 2003[164]. Both these species, like the gadfly-petrels, breed throughout the year, so cannot easily be accurately censused; feeding conditions at sea may affect numbers breeding in any one year. The more stable conditions may also have allowed extra seabirds to colonise. The presumed Kermadec Petrels appeared in the mid-1990s, and Bulwer's Petrels, first seen in 1987, now breed in small numbers. Both are new (as breeders) to the Indian Ocean. The Kermadec Petrels sometimes pair and interbreed with 'Round Island' Petrels[165] – indeed there seems now to be an incipient hybrid swarm, with most birds resembling either Trindade (the majority) or Kermadec Petrels (very few), but there are also some intermediates (see also p. 68–69). Undisturbed, some of these seabirds live a long time. In 2007 there were still petrels ringed in 1973–75 nesting on the island[166].

At present rates of increase Red-tailed Tropicbirds may regain mid-19th century numbers within a decade. We suspect the huge numbers of these aggressive birds occupying all suitable nesting sites restricted petrel numbers, and only when persistent over-harvesting had reduced the tropicbird population could the petrels increase. They both choose very similar rock hollows for nesting, but the tropicbirds are dominant over the smaller petrels when competing for a site. It may therefore require some careful conservation management to maintain a viable petrel presence when the tropicbirds regain their former abundance[167].

Not for the first time in Mauritian ecological history, facts mutated into myth. In the 1960s and 1970s the belief gained currency that Round Island's deep gullies had been caused by erosion following the destruction of the vegetation by rabbits and goats. Although it is perfectly clear from Lloyd's 1844 description that the gullies were already there (then full of soil and sprouting palms), and Jean Vinson had said nothing to suggest otherwise, the prevailing view in Mauritius by the early 1970s was that the gullies had formed since the mid-1800s as a consequence of herbivore overgrazing[168]. This myth was repeated, usually implicitly as a 'given' rather than overtly, throughout the next two decades. In fact, as can easily be seen looking down to the coastline from the summit, the gullies extend below sea-level to the seafloor. This proves that they originated before the post-glacial sea-level rise around 10,000 years ago[169], probably when the cinder-cone was new and the volcanic ash less consolidated. What the rabbits and goats *did* cause was a massive loss of topsoil built up under forest cover over millennia. Another aspect of erosion that has misled many is the assumption that Round Island's landforms are shaped only by water, when in fact many of its features are shaped by wind. The isolated standing rock formerly above the landing place, seen in Figure 9.3 and so many photographs, was a classic 'mushroom rock' sculpted by wind-blown sand. The curious cavities petrels like to nest in, known to geologists as 'honeycomb etching', result from a combination of chemical weathering (slow solution in water) and wind action[170]. On Round Island one can feel the continuous blast of sand grains, and tests in 2003 confirmed that large amounts of fine sand (and salt) are blown across the island when the wind exceeds about force 5. At least 100g, and possibly up to 1kg, is deposited annually per square metre in areas where the wind speed drops, which over time must be responsible for much of the island's sandy soil accumulation. The soil is of the uniform small-grained consistency expected from aeolian deposition, and many deposits are on ridges and spurs, not in valleys where rainwash deposition would occur. In exposed coastal areas (e.g. the 'Big Helipad'), material is stripped off by the wind to be deposited further inland[171]. In short, wind accumulates the soil, rain largely washes it away.

By the 1960s the Burrowing Boa had clearly become desperately rare. None were found between 1953 and 1967, when one was seen in October, and another in August 1975. Since then, despite intense searching and regular monitoring, no more have appeared and the species is considered extinct[172]. The other reptiles survived to see the eradication of rabbits, so their future is more secure, provided rats, house-shrews and house geckos are prevented from stowing away in someone's baggage or stores. However, the results of removing the rabbits were not all desirable. Although there was spectacular regeneration of latans, Bottle Palms and screw-pines, a swathe of introduced plants that had also been held back by rabbits also flourished, threatening to overwhelm the native species. Increased human activity, ironically largely the frequent visits to the island by weeding teams uprooting exotics, accidentally resulted in further species arriving, not only plants but also animals, such as the common house cockroach *Periplaneta americana* first discovered around the camp in 1996[173]. One shrub, *Desmanthus virgatus*, first seen in 1982, could potentially dominate the island if not kept under control, and in recent years the pantropical herbaceous weed *Achyranthes aspera* has covered increasing areas of open ground, as have

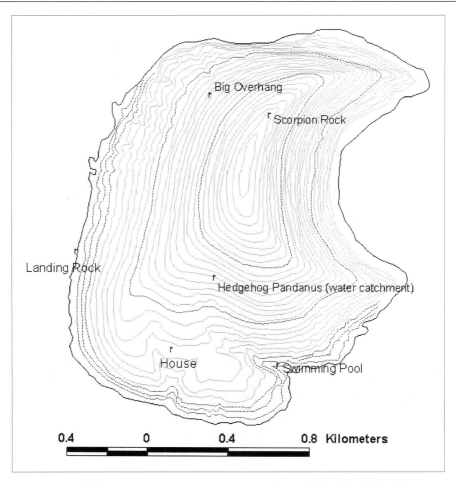

Map 9.3. Round Island in 2003. Digitised contour map first published by Malin Johansson (2003). This is the first topographically accurate map of Round Island, and 'added' 68ha (45%) to the generally accepted island size. We have used the sparsely detailed original here, but note that the west coast is heavily cut with gullies on the lower third of the slope, and the concave contours inland of the 'swimming pool' hide the striking features of the Big Gully. The flat area between the house and the pool is the Big Helipad. Contours at 10m intervals.

various exotic grasses at the expense of native ones. By 2003 *A. aspera* was affected by a rust fungus, and insect damage on other exotics was noted, as was a general increase in insects[174].

The reptiles have responded unevenly to the vegetation changes: a big expansion of skinks, but only moderate increases 1982–1996 in Keel-scaled Boas and Günther's Geckos, with a possible dip in night-geckos (recovered by 2003). By 1996 the density of skinks and boas was levelling out in favourable habitat, but overall numbers were increasing with vegetation cover; boas were estimated at around 1,000 in 2007. In 2003 Telfair's Skink and the snake showed signs of approaching their carrying capacity, though Bojer's was apparently still increasing. Günther's Geckos give visitors the impression of striking increases, but for some reason this has not shown up clearly on formal surveys. Similarly Vinson's Day-geckos became rarer during 1989–96 according to the censuses (though not visibly to ordinary visitors), recovering spectacularly by 2003[175]. Geckos (especially night-geckos) and snakes are much harder to find than skinks in the now much denser vegetation, so they may have been under-counted[176]. Counting reptiles in forest or thick vegetation is fraught with problems. In Guam visual censuses of geckos in exotic popinac forest led to such spectacular undercounting that the visual counts were meaningless. Mourning Geckos were censused visually at 12 individuals per hectare, but when a 'total removal' count was made it turned out there were 1,777 per hectare![177] Visiting Round Island in 1999, the authors of the Guam study attempted to devise a reliable census method:

Round Island is substantially easier to sample than are Guam forests (where our studies were done), but it has a special problem in that as the island vegetation recovers from the rabbits, visibility is likely to decline. Thus visual surveys are likely to experience declining probabilities over time, and these changing detection probabilities are likely to bias estimates of abundance. Visual surveys are not worthless (for some species we have found nothing better), but serious study is needed to quantify the inherent biases, and those biases can be large ... While we were on Round island, Tom Fritts and I spent a lot of time trying to design an accurate and inexpensive way to sample the resident lizards. We did not succeed. Based on what we know, it will cost substantial sums to obtain unbiased and reasonably precise estimates of lizard populations on Round Island. I believe that such costs are justified, but I also realize that MWF and other responsible parties are up against financial limits.[178]

It is likely that the big increase in Telfair's Skinks, largely responsible for a ten-fold surge in reptile biomass from 1982 to 1996, has restricted recruitment of other species. Telfair's will eat young lizards of other species (and probably young snakes), although its principal food is the abundant flightless cockroaches[179]. Both Günther's and Vinson's Day-geckos are closely associated with mature latans with good trunks and an abundant supply of nectariferous flowers. Hence the continued decline of tall palms (Table 9.3), only slowly being reversed by newly maturing trees, has also been an important factor; both geckos also like large mature screw-pines, also in decline until recently[180]. It is not yet possible to predict how the dynamics of the system will develop, so it needs continuous monitoring. Simply removing exotic herbivores does not guarantee a return to the conditions before their introduction, particularly as the tortoises, also grazers, are missing. We know that many other important components of the ecosystem have been lost (e.g. snails, spiders, scorpions), and the lost hardwood forest may have been crucial to the original mix. The reintroduction of tortoises, perhaps Aldabran giants or Radiated Tortoises from Madagascar[181], to replace the extinct native ones would be a step towards restoring a functional ecosystem, though this exposed islet will always be subject to major stochastic effects, notably cyclones.

Figure 9.4. Bottle Palm and latans on Round Island in 1982, with Gunner's Quoin in the distance. Drawing by Brin Edwards in Bullock *et al.* 1983.

Since its degradation by rabbits and goats, Round Island, with no standing water, has been unfriendly habitat for land birds; from 1861 to 1975 Zebra Doves were the only resident species[182]. In 1975 at least one pair of Red-whiskered Bulbuls was nesting, though they have proved sporadic, as have breeding Striated Herons. Spice Finches and House Sparrows, first noted in 1982, now breed regularly. Sparrows remain very few, but Spice Finches and especially Zebra Doves have increased as seeds have become more abundant following the removal of rabbits[183].

The story of the northern islets, and of Round Island in particular, illustrates all too well in microcosm the issues and problems of island conservation – and how much that has survived into the 21st century is there by luck alone. If rats instead of rabbits had got to Round Island there would have been little left for conservationists to argue over or save – we can be thankful for small mercies!

CHAPTER 10

PRACTICAL CONSERVATION ON MAURITIUS AND RODRIGUES

Steps towards the restoration of devastated ecosystems

by Carl G. Jones

The Mascarene Islands share with other oceanic islands a large number of endemic species, high extinction rates, and the habitat degradation and destruction associated with human activity. In a report on conservation in Mauritius, Sir Peter Scott noted that "Mauritius illustrates many of the earth's environmental problems in microcosm"[1]. Looking back at the rich and colourful ecological history of the Mascarenes, one is struck by the loss of so many taxonomically highly distinct species that need not have become extinct. We know that Dodos and giant tortoises survived well in captivity, and it is likely that the flightless rails would have flourished under sympathetic human care. We can but lament the loss of species that could have been saved if only someone had thought to nurture the wild populations or establish captive ones. The clarity of hindsight should, however, be used to help us direct the conservation of the native wildlife on the Mascarenes. Enough in principle is known about the biology of island species and their conservation management to ensure, funding permitting, that there are very few or no future extinctions. However, under the constant pressures of alien species, which are still being introduced, and the impacts of people and their politics, it is likely that for the next few decades at least, we will have to continue to fight to save an increasing number of critically endangered species.

The conservation of species needs to be part of a larger process to preserve wherever possible the plant and animal communities that still exist. While the ultimate aim is the conservation and reconstruction of whole ecosystems, this may be conceptually difficult. It is a sad fact that these systems have become disrupted or totally shattered and we only have an impressionistic image of what these systems were like on the pristine Mascarenes. We are still debating what species, and how many species, were present and how they interacted. An important part of conservation work is to compile ecological histories of the islands, so we can try and understand how these systems functioned, and then to use these data to reconstruct them.

The process of compiling ecological histories offers interesting challenges, deriving information from many disciplines and weaving it together as strands of knowledge to construct explanations. It is perhaps counter-intuitive, but the further we move in time from the pristine Mascarenes, unaffected by humans, the more we are learning about their undisrupted ecology. In the last three decades we have uncovered a great deal from the subfossil record[2], together with more complete archival research into the early written and pictorial accounts; this research has provided a carefully researched history of extinctions and introductions[3]. This information can be interpreted or reinterpreted, based on contemporary knowledge, research and experiences in the Mascarenes and elsewhere. Gaps in our knowledge of what happened in the Mascarenes can be filled by using knowledge from other countries. The well-documented impacts of rats and cats in New Zealand and on other islands can be used to interpret the impact of these exotics in the Mascarenes three or four centuries ago. This method of deriving knowledge using different types of information of different qualities is termed 'cabling'; it is an accepted mode of operation in archaeology[4] and is most appropriate in reconstructing ecological histories.

Applying this knowledge to reconstruct systems

poses further challenges, and the field of restorative ecology is a developing one. The restoration of species and communities should be viewed as part of the whole process of understanding how the system works. The eradication or control of invasive species, and then monitoring how native species respond to their removal, give us some idea of the impact these species were having. For example, in the vegetation restoration work in the native forests and on Ile aux Aigrettes, the control of invasive exotic plants resulted in an increase in the regeneration of native plants, illustrating how the exotics were suppressing native regeneration. Similarly the eradication of Ship Rats on Ile aux Aigrettes resulted in an increase in the germination of ebony seeds. The planting of natives at different densities in Rodrigues gave us information on which species are pioneers and on the competitive relationship between species. Similarly, the differential survival rates of plants on Round Island gives us data on which plants are best-suited to grow on the island, and what their microhabitat needs are. This empirical approach offers insights into how these communities were structured, and allows the formulation of testable hypotheses.

Conservation on Mauritius and Rodrigues

Until the 1970s, conservation on Mauritius and Rodrigues had been influenced by European techniques and was largely passive, with the creation of protected areas and a 'hands off' approach. Indeed, Forestry Service personnel who looked after these reserves expressed great pride in their strict nature reserves that were not touched. However, this system did not halt the increasing invasion of native forest by introduced species and the concomitant decline of the native wildlife. Seeing this led to the increasing realisation that for conservation to be effective, there had to be an active approach to both species and ecosystem management. In recent decades these two approaches have been merging, as it has been necessary to reintroduce and intensively manage species that have declined drastically or become extirpated from areas due to the impact of alien species, and at the same time manage the damaged habitats in which these endangered species occur.

Conservation projects on Mauritius and Rodrigues involving international conservation organisations started in earnest in 1973 with a broad-based report written by Sir Peter Scott, then Chairman of the World Wildlife Fund, and the nearly simultaneous arrival of conservation-based projects from Britain and the USA; Scott's recommendations led to a more comprehensive study by John Procter and Rod Salm in late 1974. Before Scott's visit, the British Ornithologists' Union (BOU) had already organised an expedition to the island, to take place from 1973 to 1975, and Dr Stanley Temple had arrived to study the non-passerines. The results of the BOU's work, together with data subsequently collected on the conservation project, were published together, and this work has provided the baseline data on which much subsequent work has been based[5]. During the same period interest in the ecology of the islands by local naturalists was increasing again – France Staub, Rivaltz Chevreau de Montléhu, Joseph Guého, Claude Michel, Jean-Michel Vinson, Wahab Owadally and others building on the earlier work of Reginald Vaughan, Octave Wiehé, Leo Edgerley and Jean Vinson[6].

The conservation problems in the Mascarenes are very much more urgent than in many continental areas, thus requiring a different approach. While habitat destruction is an obvious cause of species loss, the gradual degradation of ecosystems by the invasion of exotic animals and plants is more insidious and is often difficult to evaluate. Conservation on Mauritius and Rodrigues has been influenced by the work of New Zealand biologists, who since the early 1980s have worked for or with the Mauritian Wildlife Foundation (MWF) most years. The problems that New Zealand biologists have encountered, and their solutions, are more appropriate to solving the ecological problems in the Mascarenes than the experiences of most conservationists in Europe and North America. The New Zealanders have been pioneering work on the restoration of degraded island ecosystems, by the control and eradication of exotic animals, the restoration of the vegetation communities, and the introduction of missing vertebrates[7]. A close informal relationship exists with the New Zealand Department of Conservation, and there has been a regular two-way exchange of staff and ideas with the MWF.

In addition, species management approaches have been developed in Mauritius and Rodrigues. The Peregrine Fund first became involved in Mauritius in 1973. Using techniques first developed on the Peregrine Falcon, they were involved in developing the captive breeding and intensive wild management of Mauritius Kestrels[8]. The success of these intensive management techniques with the kestrel gave the confidence to apply "hands on" intensive management techniques to the Pink Pigeon, Echo Parakeet and Mauritius Fody.

Since 1976, The Jersey Wildlife Preservation Trust (now the Durrell Wildlife Conservation Trust, DWCT) has sponsored the conservation in Mauritius of critically endangered birds, the restoration of Round Island and its reptile populations, and saving the endemic Rodriguan birds and bat[9]. The DWCT initially focused upon the conservation of the critically endangered vertebrates. In addition to this they developed a long-term relationship with the Government

BOX 31 WEAVERS (PLOCEIDAE)

Fodies are small members of the Ploceidae, and the genus is endemic to Indian Ocean islands. One species, the Madagascar Re or Cardinal Fody *Foudia madagascariensis*, has been introduced to the Mascarenes and other Indian Ocean Islands.

Mauritius Fody *Foudia rubra*

The Mauritius Fody *Foudia rubra* is now very rare and extremely restricted in range. 'Sparrows' and 'linnets' of early accounts were presumably this species, when its range extended over much of Mauritius. Again, like many of the small Mascarene passerines, its skeletal remains are rare in fossil deposits.

Rodrigues Fody *Foudia flavicans* by Jean-Michel Vinson, 1974.

Accounts
Hugo in 1673 (translation in Panyandee 2002):

> The last named birds ['sparrows' = fodies] are also a nuisance to the plantation since they are exasperating and irritating as the commander has noticed several times since his arrival. When some grains and beans are sown and have blossomed and become ripe, these birds fall on the plants in such numbers that the people cannot stop them by any means. Shooting is ineffectual as, when the firelock is released, they come and sit on the muzzle. People can still kill dozens of these birds with sticks while they feed on the plants. With time, this vigorous form of protecting the provisions will frighten them away.

Leguat (1707) in 1694:

> There is a species of small bird made like enough to our sparrows except that they have red throats.

Réunion Fody *Foudia delloni*

The Réunion Fody *Foudia delloni*, not previously formally described and of which there are no specimens, is often confused with '*Foudia bruante*' an enigmatic bird in the *Planches Enluminées* which is apparently a colour variant of the Cardinal Fody. Dubois described swarms of 'sparrows' that destroyed entire crops, and Dellon said they were so numerous and unafraid that they flew into kitchens and into open fires. Fodies were not mentioned again after Dubois and it is likely that the species died out very rapidly after rats were introduced. Cardinal Fodies were introduced about a century later and are common everywhere.

Accounts
Dellon (1668):

> While the sparrows are no larger in Mascareigne [= Réunion] than in other countries, their numbers render them inconvenient. They strip sown fields and the houses are full of them, as ours are with flies. One often sees them fall into pots and platters, and burning their wings on fires, which are lit out of doors, the sun providing enough heat for [even] the coolest houses.

Dubois (1674) in 1671–72:

> The sparrows are so thick and in such numbers that they bring great damage to the island, eating a large part of the cereal that is planted, without it being possible to destroy them because there are too many. Several cereal harvests would be made during the year if it were not for these sparrows, because of which one is restricted to a single harvest, in the period they go nesting in the mountains. These sparrows have plumage like those in Europe, except that the males, when breeding, have the throat, head and top of the wings the colour of fire.

Rodrigues Fody *Foudia flavicans*

On Rodrigues, the endemic Rodrigues Fody *Foudia flavicans* was formally described in 1865, although it was mentioned by Tafforet in 1726. It was still found in coastal areas when Edward Newton's visited in 1864, but is rare today and confined to woods on high ground.

Accounts
Tafforet (1726), from A. Newton's 1875 translation:

> There are plenty of goldfinches [i.e. fodies], which have a sweet warbling. Some wagtails [i.e. Rodrigues Warblers] are to be seen, with some other small birds, which have very sweet notes, but they are ever on the look-out for the birds of prey, which are the owls of which I have before spoken.

Pingré (1763) in 1761:

> I saw at Rodrigues only one species of small bird. They have certain traits resembling our mésanges [tits] and are about the same size, perhaps a little smaller. They have a little cry, quite soft, but without modulation, without song. When one calls them they approach from branch to branch to within a hand's grasp, but the slightest gesture makes them retreat.

Specimen collections
Passerine fossil material is rare from the Mascarenes, with fody material being no exception. Specimens are held at the UMZC and UCB.

and helped create the local capacity and infrastructure for effective species management. The rationale was that the conservation work could move from species to species as the expertise grew; hence the progress from Mauritius Kestrel to Pink Pigeon to Echo Parakeet to Mauritius Fody and finally to Mauritius Olive White-eye to restore a community of forest birds. The trust aimed to restore endangered species and to use this work to drive the restoration of habitats and ecosystems. Hence the work on the forest birds has influenced the creation of the National Park and stimulated the restoration of native forest. The initial work on Round Island was on the three keystone reptile species (Telfair's Skink, Günther's Gecko and Keel-scaled Boa) and this helped raise the profile of the reptile community and develop the political momentum and finances to eradicate rabbits and to support the restoration of the island's plant community. On Rodrigues, work on the Rodrigues Fody and Golden Bat acted as a driver for the protection of the most important areas and the habitat restoration work.

The DWCT, and more recently the MWF, have been active in managing, with Government partners, the captive breeding centre for endangered vertebrates. This station, established in 1976, is Government-owned and came under the auspices of the Forestry Service until 1987, and subsequently the newly established National Parks and Conservation Service. In 1997 the centre was named the Gerald Durrell Endemic Wildlife Sanctuary in honour of Gerald Durrell and the long-term contributions of his Trust to Mauritian conservation. This centre has focused on the captive breeding of the Mauritius Kestrel, Pink Pigeon, Echo Parakeet and Mauritius Fody for the restoration programmes, but also holds populations of flying-foxes, reptiles and passerines for captive breeding and research. Parallel to this work, the DWCT has run captive breeding projects for several endemic birds, the Golden Bat and Round Island reptiles at Jersey Zoo[10].

In 1984 the Mauritius Wildlife Appeal Fund (MWAF) was formed to fund-raise and oversee the conservation work being conducted by international conservation organisations, especially the DWCT and the Peregrine Fund. The initial brief was soon expanded and MWAF took on its own conservation projects. The Fund's first project, in 1986, was the long-term lease and conservation management of Ile aux Aigrettes. In recognition of its increasing role in many conservation activities, the Fund became the Mauritian Wildlife Foundation (MWF) in 1994, and has increasingly taken on many projects of its own. The MWF works in partnership with international conservation organisations and has a memorandum of understanding with the Ministry of Agriculture and the Rodrigues Regional Assembly, Government of Mauritius. In Mauritius the MWF works closely with the National Parks and Conservation Service and in Rodrigues with the Forestry Service. These relationships have been very productive but subject to occasional strained relations brought on by the different operating modes of the MWF, a non-government organisation, that is goal-orientated, and government departments that are process-orientated. The MWF (and some other non-government organisations) have greater operational flexibility to run complex projects, while the government has more resources and long-term stability for managing long-term, large and expensive projects such as large-scale habitat restoration projects and National Parks. The MWF is the biggest conservation NGO in the Mascarenes.

It has been MWF policy to work closely with specialist organisations that can bring in state-of-the-art expertise to solve problems. A good example of this is the foundation's relationship with the World Parrot Trust since 1988, to conserve the Echo Parakeet[11]. The foundation not only works on the conservation of critically endangered endemic species and their habitats directly, but also provides logistic support and supervision to Mauritian and international students doing research in Mascarene conservation biology, and helps in training new generations of conservation biologists[12].

Work by the MWF on the endangered flora complements and helps develop work by the Mauritius Herbarium, the Forestry Service and, since 1994, the National Parks and Conservation Service. During the last decade, academics at the University of Mauritius have become increasingly interested in studying the native forest, the impact of invasive plants and forest management[13]. MWF's botanical work has focused on population and distributional studies, nursery propagation and the reintroduction of nursery-produced plants into managed Nature Reserves on Rodrigues, to Conservation Management Areas on Mauritius and to managed offshore islands[14].

Understanding the problem of invasive species

Introduced species are a global problem and are especially problematic on oceanic islands where the plant and animal communities are more susceptible to exotic invasions species than are continental ecosystems[15]. Competition and predation by alien species are probably the main reasons for high extinction rates on islands and the rarity of many endemic species. Understanding the impacts of exotic organisms, and the control and eradication of the worst, are the most important challenges to conservationists in the Mascarenes and other oceanic islands. The problems of invasive species were poorly understood and greatly understated in the Mascarene Islands until the 1970s and 1980s. Even senior Forestry Service personnel felt that invasive weeds had reached equilibrium in the native forest and the highly invasive Strawberry Guava (which produces edible fruits) and the exotic deer (a prized traditional game animal) were an asset to the native forest rather than pests[16]. Regrettably these views still linger. The impact of the introduced Crab-eating Macaque as a nest predator and destroyer of native vegetation was widely accepted, but it has tended to become a scapegoat for damage caused by some other species as well[17]. The impacts of exotic organisms and the dangers of introducing alien species onto the islands are still poorly understood by most of the population, including key decision makers, on Mauritius, Rodrigues and Réunion. In the last two decades the rate of introductions to Mauritius and Rodrigues has not slowed as the dangers became better appreciated by the authorities, but has actually increased (Chapter 8). Invasive exotic trees are still being planted along roadsides for aesthetic purposes, and exotic birds, reptiles and fish continue to be imported for a growing pet trade[18].

The fauna and flora on oceanic islands have in many cases evolved without the presence of weedy plants and terrestrial mammals, and consequently are severely compromised by competitively aggressive introduced species[19]. Alien species introduced to islands are usually free from natural population controls such as predators and diseases. Not all introduced species become invasive, but some are highly so and have spread widely. The Mascarene Islands have many of the most pathogenic tropical plants, vertebrates and invertebrates.

Alien plants and animals often function in synergy to degrade native plant communities. Invasion is exacerbated by exotic herbivores (from snails to deer) that selectively browse native seedlings, and by predators, especially rats and monkeys, which eat native fruits and seeds. Introduced vertebrates often spread exotic plants. Two examples of well-known pairs that are a problem on Mauritius and on other tropical islands are the spread of the thorny weed *Lantana camara* by Common Mynas and the spread of Strawberry Guava by feral pigs[20]. Since many of the same species are involved in the degradation of island ecosystems world-wide, restoration techniques successful on one island can often be successfully applied to other island systems, as has been the case with control measures for many invasive mammals[21].

The impact of exotic species on the native biota is not always easy to evaluate, since full understanding requires data from before, during, and after the species have become established. The issue is further complicated by most oceanic islands having suffered the onslaught of dozens, or hundreds, of introductions, many over the same time period. Many vertebrates were introduced a long time ago and the main phase of colonisation is long past. With only minimal data from before the introductions, one can only speculate on their impacts on the pristine community. Some indication of a species' impact can be gained retrospectively by the timing of extinctions, and currently by the removal or control of exotic species within the ecosystem followed by monitoring of the impacts of these manipulations. The situation with many of the exotic plants is easier to see, since in most cases the process of invasion is still underway, and the stages of penetration can be followed by sampling different areas of native vegetation.

There is no doubt that some species have profound effects upon the ecosystem they are invading and may completely alter trophic levels, resulting in a whole series of extinctions and the subsequent simplification of the system. It is likely that there have been several extinction cascades in the Mascarenes, following the introduction of cats and rats to Mauritius and Rodrigues and probably tenrecs and House Shrews to Mauritius. The impact of the recent shrew invasion in Rodrigues was dramatic and it is not hard to envisage the catastrophic impact it would have had on a more pristine ecosystem. One only has to read of the documented impacts of rats on island endemic species across the world, and the impact of Brown Tree Snakes *Boiga irregularis* in Guam to realise that to describe these as causing catastrophic cascades of extinction is appropriate[22]. In the Mascarenes the introduction of the Wolf Snake *Lycodon aulicum* may have caused the extinction of *Gongylomorphus* skink populations, and also species that feed on skinks (e.g. owls and possibly the Hoopoe Starling[23]) as a result. Several species of Mascarene birds feed on lizards, and they are especially important for feeding young[24].

Most of the invasive species discussed here have been subject to some study, control, exclusion or eradication measures during conservation work. The discussion is mainly limited to the plants, reptiles, birds and mammals that have been the main focus of this work, although it is appreciated that there are invasives in other groups, and there are many exotic invertebrates that have had major impacts on the native wildlife. The problem of invasive disease is poorly documented but it is likely that there are parasitic, viral, fungal and bacterial diseases affecting native wildlife; some pathogens that affect the native birds are being studied.

Invasive plants

Introduced plants have been able to invade island floras to a greater extent than on continents. Mauritius has 731 species of naturalised exotic plants, which make up 53% of the total flora. 56 of these are

 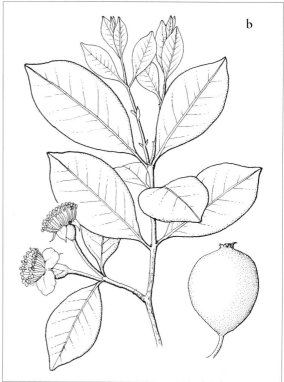

Figure 10.1. Botanical drawings of two of the worst invasive weeds in Mauritius and Réunion, Ceylon Privet (a) and Strawberry Guava (b). From *Flore* 92 (1990) and Cronk & Fuller (1995).

regarded as invasive, and 18 are very invasive. On Rodrigues there are 305 naturalised exotic species of plants that now comprise 78% of the flora. Island ecosystems extensively modified by people are much more susceptible to alien plants than are pristine islands, and tropical islands are probably more susceptible to invasion than islands in temperate regions[25].

Mauritius has a suite of weedy exotic plants common to many tropical oceanic islands[26]; of the 56 species listed as invasive on Mauritius, 26 are also naturalised on Hawaii. The two worst invasive weeds on Mauritius, Strawberry Guava and Ceylon Privet, are vigorous small trees that are not only shade-tolerant, but they will also shade out other vegetation and can form dense impenetrable stands. Since it has the ability to invade undisturbed rainforest; guava dominates large tracts of upland wet evergreen forest, as it does in parts of Hawaii[27]. In some high-rainfall areas Traveller's Palm and Rose-apple combine to form mixed exotic forest; two *Litsea* species are also invasive. In drier lowland areas Brazilian Pepper, Mauritian Hemp and thorny acacias form dense thickets, depressing any native regeneration[28]. On Rodrigues, *Acacia nilotica* introduced in the 1970s (p. 201) has been added to the six highly invasive exotics noted in the 1980s. *A. nilotica* is spreading into all the valleys that have remnants of native vegetation, and is now regarded by some as the most aggressive invasive tree impacting on native plant populations[29]. Low regeneration of some native Rodriguan plants is due to competition from invasive plant species and grazing by domestic livestock[30].

There is a successful history of biological control in Mauritius, and five weed species were controlled in various programmes between 1914 and 1982. Biological control of invasive weeds is one of the best methods, provided an effective control agent exists. It has already been successful with the shrub *Cordia curassavica* on Mauritius, which was one of the worst weeds on the island in the early-mid 20th century. Biological control has also limited the invasive impact of *Lantana* and Prickly Pear *Opuntia vulgaris*[31]. There are, however, risks if the weed has close relatives amongst the endemic flora – Strawberry Guava is, for instance, closely related to the large number of endemics in the genera *Eugenia* and *Syzygium*. Hence control of invasive woody plants in managed reserves has so far been largely limited to mechanical cutting and uprooting. Use of weedkiller on cut stumps has been effective only in drier conditions (Mondrain, Ile aux Aigrettes)[32], though appropriate herbicide use requires careful consideration[33].

Invasive reptiles

Ten species of exotic reptiles have established populations on Mauritius, and two of these, the ubiquitous Cheechak and the Wolf Snake, are thought to have had a profound effect upon the distribution of native reptiles. With one recently discovered exception, nowhere do the endemic *Nactus* night-geckos and the introduced Cheechak occur on the same island. Cheechaks may have displaced *Nactus* geckos by competitive exclusion or by eating their young[34]. On the islands off southeastern Mauritius, Cheechaks are found on Ile de la Passe and Ile aux Fouquets, either side of Ilot Vacoas; there they are absent, whereas the Lesser Night-gecko *N. coindemirensis* survives. Similarly on the six islands north of Mauritius, Cheechaks exist on Ilot Gabriel, which lacks *Nactus*, while the night-geckos occur on Cheechak-free Serpent Island, Round Island, Gunner's Quoin and Pigeon House Rock. No other introduced competitor or predator complements the distribution of night-geckos as well as the Cheechak. Only on Flat Island are the two groups sympatric, and even here the relict Lesser Night-gecko population discovered in 2003 survives only on a tiny area of bare tuff where the surface favours their claws over the Cheechak's foot pads[35]. The main islands of Mauritius, Réunion and Rodrigues have all lost their *Nactus* geckos and all have Cheechaks. However on the main islands it is unlikely that Cheechaks alone were responsible for the extinction of these endemic night-geckos. The Cheechak is largely a lowland species, and in Mauritius subfossil eggs and bones of a large *Nactus* (cf. *serpensinsula*) have been found in upland locations at Bassin Blanc where Cheechaks are not known to occur[36]. Night-geckos on the Mauritian mainland may have been extirpated by rats, Wolf Snakes, House Shrews, or a combination of these exotic predators[37].

Both Macabé Skink *Gongylomorphus fontenayi* and Bojer's Skinks *G. bojeri* once occurred on the Mauritian mainland. The former is now restricted to some upland forested ridges, while a lowland form, the Orange-tailed Skink, which has yet to be formally named, is found on Flat Island; Bojer's is restricted to seven offshore islets. The Réunion Slit-eared Skink *G. borbonicus*, is extinct. The loss of these skinks coincides with the introduction of the Wolf Snake, a specialist lizard (especially skink) predator, as reflected in its jaw morphology[38]. The Wolf Snake is nocturnal and hunts on the ground as well as in trees and among rocks, and presumably searches for sleeping as well as active lizards. Its establishment in or before the 1870s more or less matches the last mainland reports of Bojer's Skink (p. 134). On Réunion the last record of *G. borbonicus* is a specimen collected in 1839, the same date as the first record of the snake![39]. Attempts to control the Wolf Snake on Ile aux Aigrettes by trapping failed when no snakes were caught[40]. On Mauritius Bojer's Skink was apparently limited to the lowlands, and perhaps in pristine Mauritius mainly in the coastal palm forest, while the Macabé Skink still occurs in native forest above 300m in the southwest[41]. Both Wolf Snakes and House Shrews are commonest in dry lowland areas and scarcer in the wetter upland native forest where the Macabé skink survives.

Reptile introductions are still occurring in Mauritius and Rodrigues. The Panther Chameleon *Furcifer pardalis* became re-established in Mauritius during the late 1980s and 1990s. The Great Green Day-gecko *Phelsuma madagascariensis* has been established in large gardens in the upland suburb of Floreal for over a decade. There are also plenty with all age classes in the grounds of the Casela Bird Park near Tamarin in the west, and they have been reported from Grande Baie and Baie du Tombeau in the north, Black River in the southwest and Point d'Esny in the southeast of Mauritius[42]. The Red-eared Slider and the Common Iguana are common pets on Mauritius that have been imported in large numbers in the last decade, and feral populations of the former have been reported. Escaped large lizards, possibly iguanas or monitors, have been seen several times in the last decade; one was caught and identified as a Common Indian Monitor *Varanus bengalensis*. Several other reptiles have occasionally arrived accidentally in shipments of cargo, especially timber, including a large species of *Hemidactylus* gecko and a cobra.

The eradication or control of most invasive reptiles is currently impossible over even quite small areas. An attempt to eradicate Bloodsuckers *Calotes versicolor* on Ile aux Aigrettes failed in 1999, when a concerted effort was made to catch the lizards by hand and to shoot them with catapults. Over 150 were killed, and although there was some short-term control the population had fully recovered a year after the control attempt was halted[43].

Invasive birds and introduced bird diseases[44]

Mauritius has 19 species of naturalised exotic birds, these have broad distributions on oceanic islands, all but one of which (Meller's Duck) have an introduced distribution beyond Mauritius[45]. As with mammals, many of these invasive species are cosmopolitan; 12 are shared with Hawaii. Many of the exotic birds make use of human intervention, and most are found predominantly in urban, agricultural or secondary vegetation, although several do penetrate into the native forest. There they impact upon the native birds, primarily by competing for food and nest sites, though they may also be reservoirs of exotic diseases. Native birds are probably largely restricted to areas of native vegetation because the exotics are competitively superior in secondary vegetation[46], and without the exotics it is likely that many of the native birds would spread into other habitats, as the Mascarene Grey White-eye has. A number of impacts of exotic on native birds have been suggested, including the restriction of the endemic Echo Parakeet to native forest by competition from introduced Ring-necked Parakeets. In forest edge situations, mynas may compete with cuckoo-shrikes for large insects and geckos[47]. The Mauritius Merle may be constrained to native forest by competition with the exotic Red-whiskered Bulbul and Common Myna. Cardinal Fodies may limit the distribution of Mauritius Fodies[48]. Ring-necked Parakeets and Common Mynas occur in the native forest in low numbers, where they compete with the Echo Parakeets for the now-scarce nest holes; when found occupying known or potential Echo Parakeet nest-sites, these birds are removed[49].

The exotic doves (and also the native Malagasy Turtle Dove) are vectors of a flagellate protozoan that causes potentially fatal trichomoniasis in Pink Pigeons[50]. Controlling exotic doves and Cardinal Fodies was tried around the supplementary feeding sites, where they competed with Pink Pigeons for food, and where the doves were suspected of spreading trichomoniasis. Although large numbers were removed, constant re-invasion kept numbers high. It proved more effective to keep them out by designing feeding hoppers to exclude exotic doves by favouring the greater reach of the Pink Pigeon, following which the exotics decreased at the supplementary feeding stations. On Ile aux Aigrettes, attempts to control Cardinal Fodies and Malagasy Turtle Doves proved fruitless because of their high densities and constant re-invasion of the islet from the mainland. Pink Pigeons also carry a protozoan blood parasite: up to 30% were infected with the introduced *Leucocytozoon marchouxi*, which reduced survival in juveniles, though not adults[51].

Several viral diseases have been identified in native bird populations. Some wild Echo Parakeets are affected by the psittacine beak and feather disease (PBFD) and polyoma viruses. The origin of these viral diseases is unclear, although it seems likely that they were brought in with exotic parrots. I caught a Ring-necked Parakeet in the Black River Gorges in 1985 showing the feather lesions typical of PBFD, and I have seen captive Greater Vasa Parrots *Coracopsis vasa* on Mauritius showing similar lesions[52]. Cockatoos in Australia are known reservoirs for the virus[53], and one or two feral, non-breeding cockatoos (*Cacatua galerita* and *C. goffini*) have lived in the Black River Gorges almost continuously for the last 25 years[54]. Released Mauritius Fodies on Ile aux Aigrettes have caught avian pox. Although most soon recovered, some have died[55]; we have never seen avian pox lesions on wild Mauritian Fodies in the uplands. Exotic birds may be a reservoir for avian pox, since lesions have been seen on House Sparrows and Common Mynahs. Native Mascarene Grey White-eyes and Paradise Flycatchers are sometimes infected in the lowlands[56], and the disease has also affected captive and free-living Pink Pigeons and wild Mauritius Kestrels. The control of these viral diseases is problematic, and with PBFD euthanasia of carriers and affected birds is recommended. Two harvested Echo Parakeet chicks, being reared in captivity, that had PBFD had to be destroyed[57].

Introduced fruit eating birds, such as mynas and Red-whiskered Bulbuls, spread the seeds of invasive plants such as *Lantana*, Privet and Strawberry Guava in their droppings[58]. These are highly mobile birds; they move readily between native and secondary vegetation (which the native birds usually do not) and greatly aid the dispersal of these and other invasive species. A recent study of Red-whiskered Bulbuls in native forest has shown the role they play in distributing the seeds of exotic plants. In a sample of 200 droppings, 95% contained seeds from 11 plant species; eight of the nine identified were exotic species. Privet seeds were present in 34% of samples, and the other most frequently found species were *Wikstroemia indica* (17%), *Ossaea marginata* (18%), fig *Ficus reflexa* (25%), Strawberry Guava (10%), *Harungana madagascariensis* (8%) and *Rubus alcaeifolius* (5%) – all invasive exotics except the fig. The passage of privet and *Clidemia hirta* seeds through the gut of the bulbuls facilitates higher and faster germination rates; uneaten, only a few germinate[59]. Red-whiskered Bulbuls also eat insects, and their very high densities in native forest must be affecting invertebrate populations, and hence native insectivorous birds.

Bird introductions are still occurring. Casela Bird Park has free-flying populations of Egyptian Geese *Alopochen aegyptiaca*, White-faced Whistling Duck *Dendrocygna viduata* and Fulvous Whistling Duck *D. bicolor* that were established in the 1990s. All three of these waterfowl now have small but growing feral populations that have been encouraged in the nearby Yemen hunting estate by the construction of ponds where they are fed. This management has also resulted in an increase in the numbers of Mallard *Anas platyrhynchos*, introduced to Mauritius in 1979 (p. 169). These waterfowl are being encouraged, since the hunting estate is being developed into an 'ecotourism' venture where tourists can view the (introduced!) wildlife, including deer and free-living introduced antelopes, from open-backed vehicles[60].

The most successful bird introduction in the last two decades has been the Laughing Dove *Stigmatopelia senegalensis* (p. 168). An "aviary-full" of these birds was released in the Casela Bird Park in about 1989[61],

BOX 32　　　　　　　　　　TORTOISES (TESTUDINIDAE)

Five species of giant tortoise in the endemic genus *Cylindraspis* once inhabited the Mascarenes. On Mauritius there were two species, the smaller domed *C. inepta* and the giant saddle-back species *C. triserrata*. Similarly on Rodrigues, a smaller, domed *C. peltastes* coexisted with the large saddle-backed Carosse Tortoise *C. vosmaeri*, whilst on Réunion just one variable species, *C. indica*, occurred. The genus was characterised by some very distinctive features: the shell often averaged only 2mm in thickness; there were wide open ends to the carapace; an extremely reduced plastron; and a lack of heavy scales on the fore and hind limbs. This evolutionary trend is related to the dismantling of defense mechanisms, since no predators of adult tortoises occurred naturally on the islands.

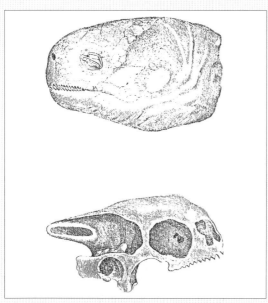

Head and skull of Réunion Tortoise *Cylindraspis indica*. From Petit (1737, in Bour 1978).

Due to their ability to survive on board ship without food and water for months, tortoises were highly prized food animals. They were collected in vast numbers and loaded onto visiting ships. Oil extracted by boiling the tortoises was particularly sought after and 400–500 animals were needed to make a single barrel. They were hunted to extinction by c.1730 on Mauritius and coastal Réunion, whilst on Rodrigues, a few individuals lingered on until the late 1790s. Isolated groups in the Réunion interior and on Round Island survived to the 1840s.

A few carapaces from all three islands have been preserved, with a single stuffed Carosse Tortoise from Rodrigues. Subfossil bones are numerous from all three islands.

Accounts: Mauritius
Jolinck (1601) in 1598:

There are also a heck of a lot of tortoises, big beasts that can carry 2 men.

Manoel d'Almeida in 1617, on Tortoises in Mauritius (from Verin 1983):

One finds there [in Mauritius] large numbers of land tortoises, which are so large that I have seen them easily carry a man on their back for some time. They are ugly and deformed creatures, whose carapace, hard though it is, can nevertheless be shot through by a bullet, as we had occasion to test out.

Mundy (1607–67) in 1638:

Tortoises, wee brought aboard as Many as wee would, good Meat if well Dressed, being very Fatte, having no other tast, butt goode, Some of them have their upper shell off about 3 Foote long, Few exceeding it as I could see; a very untoward, unsightly and unwieldy (although harmelesse) Creature. If they be turned on their backes they cannot helpe themselves butt will soe dye. These are called land tortoises, although they are as Much in the water. They lay their egges in the sand, which are hatched by the Sunne.

Pitot (1905), paraphrasing an account by Hugo of an expedition to the northern islets off Mauritius in 1674:

The largest of these islets [Flat Island] was literally covered in a multitude of tortoises. These did not find there enough food, appeared to be starving and were eating each other. A good number (320) were taken and carried on board, but during this operation the tortoises, which had undoubtedly smelled fresh meat, set about attacking the sailors from behind and biting their calves.

Accounts: Réunion (a selection from numerous accounts)
Carpeau du Saussay (1666):

At daybreak we left this enchanted landscape where we encountered only one inconvenience: it was the large number of tortoises, which came and assaulted us from all sides and often even pushed underneath us. We had a good deal of trouble defending ourselves, and as a result were unable to sleep. The tortoise is a very ugly animal, while being very good to eat; the liver is especially excellent. The oil is also splendid for frying all kinds of things. In addition it has wonderful properties against pain, our surgeons often having happy proof of this. Tortoises are a good bit smaller [than turtles], and stand a foot off the ground on four limbs; they walk over all the mountains.

Melet (1672) in 1671:

The tortoises found there are the size of two serving dishes. They are so thick on the ground here that although we [the crew of General de la Haye's fleet] ate two or three hundred every day, it appeared to have no effect and the numbers did not diminish. One sees them every hour of the day descending the mountains in herds of hundreds. What is best in this animal is the liver and the plastron which are very good, [and] their fat which when rendered into oil, which is as good and very healthy. One sometimes finds in a single [tortoise] three or four hundred eggs which are excellent, once well cooked. This animal has very short fat legs, such that from a distance it is hard to tell whether they walk or creep. They are surprisingly strong, for often on our walks we would put ourselves on their shells to see if they could support us without difficulty, but it was like they had nothing upon them. It is a very good refreshment for ships.

Dubois (1674) in 1671–72:

The whole island is full of tortoises, which are one of the good mannas of this island. They have long necks and their heads are like European tortoises; a thick tail and four legs. They are two to three feet long, a foot and a half wide and a foot or more deep. One of these tortoises can carry a man on its back, but it is all a man can do to lift one. The flesh of the tortoise is like beef, and its tripe has the same taste. The liver of these animals is very large, and is the most delicate morsel one could ever eat. Whoever had these in France would have good cheer on fast-days. There is enough to eat for four people in one of these

> livers. In the flanks of these tortoises are slabs that one extracts for melting down, and from which is extracted oil that never congeals. This oil is as good for all manner of things as the best butter; it is the island's butter. These slabs normally yield 2 pots of oil, more or less if the season allows the finding of fat tortoises, which they are not always. This oil is marvellous for rubbing afflicted limbs; I used it during my paralysis and found it good. Twenty people of good appetite can satisfy themselves in one meal from one of these tortoises.
>
> Luillier in 1703 (in Lougnon 1970):
>
> These tortoises are found at the top of a mountain which is almost completely covered in them. Previously there were even more, but since the island has been settled, many have been destroyed. It is said a tortoise can live up to three hundred years, but since the island has not been long inhabited, this is uncertain. However, one sees them as much as six to seven feet in circumference, and inhabitants say that over several years there is no sign that they have grown. There is a season when they lay their eggs, but it is the sun that incubates them, as it is for sea turtles. Their flesh is better than that of turtles; for nigh on two months we fed on them and did not get tired, such is the pleasure in eating them. One thing worth commenting on is that for months each year they neither eat nor drink, while during the remaining eight months they lay their eggs and take what sustenance they need for the four others . . . Today there is the beginning of a shortage, and according to the inhabitants there are only enough for four [more] years.
>
> *Accounts: Rodrigues*
> Leguat (1707) in 1691–93 (from the original 1708 English translation):
>
> We saw no four-footed creatures, but rats, lizards, and land-turtles, of which there are different sorts. I have seen one that weigh'd one hundred pounds, and had flesh enough about it, to feed a good number of men. This flesh is very wholsom, and tastes something like mutton, but 'tis more delicate: The fat is extremely white, and never congeals nor rises in your stomach, eat as much you will of it. We all unanimously agreed, 'twas better than the best butter in Europe. To anoint one's self with this oil, is an excellent remedy for surfeits, colds, cramps and several other distempers. The liver of this animal is extraordinary delicate, 'tis so delicious that one may say of it, it always carries its own sauce with it, dress it how you will . . .
>
>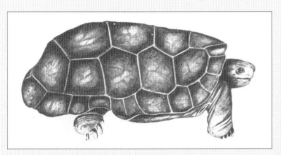
>
> A living Rodrigues Domed Tortoise *Cylindraspis peltastes* depicted by Jossigny, c. 1770.
>
> The bones of these turtles are massy; I mean they have no marrow in them. Every one knows that these animals in general are hatch'd in eggs. The land-turtles lay theirs in the sand, and cover them, that they may be hatch'd. The scale of it, or rather the shell, is soft, and the substance within good to eat. There are such plenty of land-turtles in this Isle, that sometimes you see two or three thousand of them in a flock; so that one may go above a hundred paces on their backs; or, to speak more properly on their carapaces, without setting foot to the ground. They meet together in the evening in shady places, and lie so close, that one wou'd think those places were pav'd with them. There's one thing very odd among them; they always place sentinels at some distance from the troop, at the four corners of the camp, to which the sentinels turn their backs, and look with their eyes, as if they were on watch. This we have always observ'd of them; and this mystery seems the more difficult to be comprehended, for that these creatures are incapable to defend themselves, or to fly.
>
> Pingré (1763), in Rodrigues in 1761:
>
> The tortoise is not a pretty animal, but it was the most useful of those we found at Rodrigues. In the three and a half months that I spent on the island, we ate almost nothing else: tortoise soup, fried tortoise, stewed tortoise, tortoise in godiveau [?], tortoise eggs, tortoise liver – these were pretty much our only savouries. This meat seemed to me as good on the last day as on the first; I did not eat many of the eggs; the liver seemed to me the most delicious part of the animal. After five weeks stay I was attacked by dysentery which I kept secret, because I counted more on myself to heal it than the island's surgeon. Diet and rest put me right in a few days, but it left me with an extraordinary involuntary repugnance for this liver that I had so liked until then. Should I thus regard it as the cause of my indisposition?
>
> Tortoise fat is very abundant and does not congeal; it is what is known as tortoise oil. This oil had no bad taste, it is very healthy, and we seasoned our salads with it, used it in frying and all our sauces. Rodrigues tortoises are a foot and a half long and about a foot across; they were formerly larger, but they are no longer given time to grow. When a bigger one is found, it is called a carrosse. These carrosses cannot harm a waking man, though they have sometimes bitten sleepers hard. The shells of these tortoises served us like baskets to carry oysters and other similar provisions. The flesh of these tortoises is the colour of mutton, and approaches it for taste.
>
> *Specimen collections*
> One unique stuffed Carosse Tortoise *Cylindraspis vosmaeri* can be seen at the MNHN. Carapace and bone material exists at the BMNH, UMZC, MNHN, UCB and MI.

and they are now established in the southwest of Mauritius from the village of Cascavelle down to the coastal villages of Tamarin and Black River, an area of about 40km², supporting a population of several hundred or more birds.

Invasive mammals

Mauritius has 13 species of naturalised exotic mammals and Rodrigues has four. All of these are believed to impact upon native wildlife, and all have established populations on oceanic islands elsewhere. Some, such as the rats, House Shrew, feral cat, Small Indian Mongoose and feral pig, are major problem species. The adverse effect of mammalian predators on the breeding success of native faunas, which have evolved in the absence of such predators, is well-documented[62], and over the last two decades the techniques for their eradication have become increasingly sophisticated. The size of islands that can be cleared of problem mammals has risen steadily[63]. The increasingly sophisticated control of exotic mammals has become a major part of conservation work on

Island	Area (ha)	Mammal	Year of eradication
Round Island	219	Feral Goat *Capra hircus*	1979
Round Island	219	Rabbit *Oryctolagus cuniculus*	1986
Ile aux Aigrettes	25	Ship Rat *Rattus rattus*	1991
Ile aux Aigrettes	25	Feral Cat *Felis domesticus*	1987
Flat Island	253	Ship Rat *Rattus rattus*	1997
Flat Island	253	House Mouse *Mus musculus*	1997
Flat Island	253	Feral Cat *Felis domesticus*	1997
Ilot Gabriel	42	Ship Rat *Rattus rattus*	1995
Gunner's Quoin	65	Norway Rat *Rattus norvegicus*	1995
Gunner's Quoin	65	Black-naped Hare *Lepus nigricollis*	1995
Gunner's Quoin	65	Rabbit *Oryctolagus cuniculus*	1997
Ile de la Passe	2.19	House Shrew *Suncus murinus*	2000
Ile Cocos (Rodrigues)	14	House Mouse *Mus musculus*	1995
Ile aux Sables (Rodrigues)	8	House Mouse *Mus musculus*	1995

Table 10.1. The eradication of exotic mammals from islands around Mauritius and Rodrigues.

Mauritius[64], with a systematic attempt to remove the most problematic species from priority offshore islets[65]. Techniques and protocols developed by the New Zealand Department of Conservation for rat and cat control have been adapted for use on Mauritius, and techniques have been developed for mongoose control[66].

Three species of rodent occur on Mauritius, the Ship Rat, Norway Rat and House Mouse. All are common; the mouse is mainly commensal with man, but both rats occur in the forests, where Ship Rats are by far the most numerous. Ian Atkinson has demonstrated that rats are implicated in the decline and extinction of many species of island birds worldwide[67]. They affect birds primarily by preying upon the eggs and young, and also attack reptiles, large arthropods, land molluscs and seeds and seedlings[68]. On Mauritius rats have been implicated in the extinction of the Didosaurus and the extirpation of Günther's Gecko, Telfair's Skink and the Keel-scaled Boa from the mainland and some of the islets[69].

Eradication of rats on Ile aux Aigrettes started in 1987. Bait was laid regularly over a period of three months and virtually all the rats were killed[70]. However a small population survived; discovered in 1991, all were subsequently poisoned. Some feral cats were also present; three were caught in box traps and others were probably poisoned during the rat eradication; none survive.

Brown or Norway Rats were eradicated from Gunner's Quoin in 1995, again using anticoagulant bait laid on a 25m grid. This poisoning also successfully eradicated the Black-naped Hare. Rabbits were subsequently introduced to the island – these were eradicated by poisoning in 1997[71]. Ship Rats and House Mice have now been eradicated from Flat Island, and mice from Iles Coco and Sable off Rodrigues[72].

The most problematic eradication programme has been the attempt to rid Ile aux Aigrettes of House Shrews. Live traps were placed all over the island, with trapping extended from July through to December, a total of over 100,000 trap nights. However, several animals were missed and the population subsequently recovered. The eradication of shrews from the much smaller and ecologically simpler Ile de la Passe (1.2ha) was successful. This is apparently the first time House Shrews have been eradicated from an island[73].

On Mauritius rats are serious nest predators of native birds, taking eggs and young of several of endangered endemics. There are records of rat predation on the nests of Mauritius Fody, Mauritius Kestrel, Pink Pigeon and Echo Parakeet[74]. Our most complete data on nest predation are for the Pink Pigeon, where up to 48% of unprotected nests have failed due to predators, most often monkeys and Ship Rats[75]. Since 1992 we have been experimentally poisoning rats in some of the managed vegetation plots because of the damage they do to bird populations, and it is assumed that rat control improves fruit production and hence regeneration in these areas. Rats were poisoned weekly during 1992–99 on plots where Pink Pigeon nested (Brise Fer, Mare Longue and Fixon), although this been discontinued as the increased numbers of pigeons can now apparently sustain the levels of predation. Poison and/or snap traps are still laid in a grid around the nests of Echo Parakeets to minimise the risk of rat predation on the eggs and young[76].

In studies of Pink Pigeons, predation was apparently highest during incubation, when eggs were taken by both rats and monkeys, and decreased as the nestlings grew. Large nestlings are not immune from monkey attacks and some have been torn apart. Monkeys also kill free-flying Pink Pigeons, both on the nest and when feeding on the ground; they have been seen grabbing incubating pigeons by the tail, and are probably the cause of the many pigeons observed without tail feathers. Echo Parakeets may also be attacked when on the ground. Monkeys are serious nest predators of Mauritius Fodies, have been recorded taking Mauritius Kestrel eggs, and are undoubtedly major nest predators of other species of native birds[77]. We have trapped problem animals around Pink Pigeon and Echo Parakeet release and breeding sites, and have used various scaring devices, but these studies are still at an early stage. Historically monkeys have been blamed for the decline of many native birds, but such blanket condemnation obscures a more complex problem, with cats, rats, disease and other factors also impacting significantly on bird productivity.

Monkeys also cause considerable damage in both lowland and upland native forest by damaging trees and ruining fruit production. Some trees, such as Colophane *Canarium paniculatum* and *Sideroxylon* spp. may be virtually defoliated by a troop of monkeys passing through an area. The monkeys also rip off flowers and unripe fruit of the endangered endemic Tambalacoque, sampling the fruits before they are ripe and destroying virtually the whole fruit crop. They also strip epiphytic orchids and ferns off trees. In addition to damaging native trees, the monkeys feed on the ripe fruits of the exotic Strawberry Guava, Tamarind and Brazilian Pepper and efficiently disperse their seeds in their droppings[78].

There are two introduced carnivores: feral cat and Small Indian Mongoose. Both are widespread and common in areas of native vegetation and prey upon native vertebrates. The cat was an early introduction to Mauritius, while the mongoose was introduced in 1900. The mongoose is credited with wiping out populations of several introduced gamebirds[79]. On Mauritius, as elsewhere, it has a broad diet feeding on invertebrates, amphibians, reptiles, small mammals, birds and carrion, and has been recorded eating endemic lizards, including day-geckos and the rare Macabé Skink[80].

Cats and mongooses have been identified as the main predators of fledgling kestrels, and are responsible for the deaths of released and wild Pink Pigeons and of Echo Parakeets[81]. Remains of cat-predated pigeons and Echo Parakeets have been found, including chewed feathers and part-eaten carcasses characteristically stashed under bushes or logs[82]. Pink Pigeon remains have been found in cat stomachs, while pigeons and kestrels have been found in mongoose guts; Pink Pigeon feathers were also found in a mongoose scat, and during the same study two Echo Parakeets were also killed by mongooses[83]. Feral cats have even learned to catch Mascarene Swiftlets in the Trois Cavernes cave as they fly past through a low passage[84].

Where not controlled, cats are considered the main threat to Pink Pigeon survival. They learned to prey on birds at sites where high densities gathered, e.g. at supplementary feeding stations. In 1992 11 Pink Pigeons were killed, from a wild total that year of only 19. In one area, 53 Pink Pigeons were killed or disappeared between May–December 1996. This was over half the population at this location; most were thought to have been killed by cats, though some were killed by monkeys. There have also been other episodes of high mortality attributed to cats[85], and episodic mortality is probably a major threat to the population's recovery. A similar pattern is seen in New Zealand Pigeons *Hemiphaga novaeseelandiae*, attributed to predation by introduced Stoats *Mustela erminia*[86]. Cats and mongooses were trapped in areas where we had high concentrations of Pink Pigeons breeding[87]. There have been several studies on rats and mongooses to understand their impact on native bird populations, to gain a better understanding of their feeding ecology and population biology, and to optimise trapping and poisoning regimes[88].

There are two exotic insectivores, the Common Tenrec from Madagascar and the House Shrew from Asia. As both were introduced long ago, their impact on the native biota on the mainland of Mauritius is poorly known. An examination of stomach contents from tenrecs has shown that, among other things, they feed on endemic snails and on the fruits of the highly invasive weed *Ossaea marginata*, which they probably spread in their droppings[89].

The shrew is very prolific and predatory, and its recent accidental introduction to Rodrigues illustrates its invasive nature. The first shrews were recorded in the island capital, Port Mathurin, in 1997, where they remained until early 1998. Their spread over the island was then very rapid. During 1998 they spread to the neighbouring village of Baie aux Huitres. In February 1999 they had reached Mont Lubin, the highest settlement on the island, and by the end of the year they were everywhere – in two years they had spread through the entire 109km². The large centipedes on Rodrigues, *Scolopendra morsitans* and *S. subspinipes*, and a large black field cricket *Gryllus bimaculatus*, all declined precipitously, coincident with the shrew's introduction and spread. *S. subspinipes* used to be very common and was seen daily when working in the field – indeed, it was so abundant that locals used to keep chickens around their homes to help control it. By May 2005 both centipedes, and especially *S. subspinipes*, were exceedingly rare, and few people have seen one since the shrews became established[90].

Bojer's Skink used to occur on Ile aux Fouquets off Mauritius[91]. However when this island was visited in 1987 there were no skinks. It was subsequently discovered that the island had been colonised by shrews, which are assumed to have extirpated the lizard before dying out, possibly in the severe drought of 2002. Bojer's were reintroduced to Fouquets from Ilot Vacoas in January 2007[92].

The irruption of the House Shrew on Rodrigues has many similarities with its eruption on Guam, where it colonised the whole island (541km^2) in just five years. There the shrew has been implicated in the decline or extirpation of several species of lizard and a massive decline in House Mice *Mus musculus*[93]. The shrews are effective predators of large invertebrates, and when they have been controlled (Ile aux Aigrettes) or eradicated (Ile de la Passe) on small islands, the numbers of large terrestrial invertebrates, especially cockroaches, have subsequently increased. Cockroaches are a major food supply for lizards on Round Island, and their presence is thus very desirable for reintroduction projects on other islets to be successful. On Ile aux Aigrettes there is a marked paucity of large terrestrial insects, spiders and centipedes[94]. Attempts to eradicate shrews from Ile aux Aigrettes failed[95], and techniques do not yet exist for controlling or eradicating this species from anything but the very smallest islets.

The Black-naped Hare *Lepus nigricollis* is commonest in lowland open and edge type habitats, but it is found at low densities in native forest at all altitudes, where it prefers rides and open areas; it is presumed to damage seedlings by browsing. Hares are excluded from managed areas of forest by fencing and a population on Gunner's Quoin was removed by poisoning (Table 10.1).

There were formerly rabbits and goats on Round Island. An attempt to shoot them out in 1976 got all but two of the goats (which were shot 1979), but although 883 rabbits were shot, many were left and the population rapidly recovered[96]. In 1986 the rabbits were eradicated with bait treated with anticoagulant; of the estimated 2,500–3,000 animals present before the poisoning began, only 14 are known to have been alive two weeks after the second application of anticoagulant bait, and these were shot[97]. On Round Island rabbits and goats destroyed the vegetation community; several species were extirpated, and natural regeneration was almost entirely prevented. Since their eradication the vegetation has started to recover[98]. On Rodrigues, free-ranging goats have had a big impact upon the last remnants of native vegetation, suppressing regeneration and damaging plants by heavy browsing pressure. There were goats on Grande Montagne in the early 1980s, but these were subsequently removed (Strahm, 1989). Similarly free-ranging pigs are also common, and do considerable damage rooting up young plants; they are major disseminators of guava seeds (and probably other invasives e.g. brambles *Rubus* spp.), spread in their droppings[99]. All herbivores on Rodrigues are now excluded from managed reserves by fencing, although incursions do occur.

Rusa Deer do considerable damage in areas of native vegetation by browsing young seedlings, hence suppressing natural regeneration[100]. The only native herbivores on the pristine Mascarene Islands were the now-extinct giant tortoises *Cylindraspis* spp., so native plants are probably more palatable to deer than exotic species that have evolved in tandem with mammalian herbivores and have developed defensive alkaloids. Wendy Strahm noted that when deer got into an experimental plot they fed selectively not only on native grasses and sedges but also on saplings of several native woody species, and they avoided the invasive introduced grass *Paspalum conjugatum*[101]. All 10 taxa in 4 genera of Campanulaceae in Mauritius may be rare due to their palatability to herbivores; one species, *Nesocodon mauritianus*, is limited to three inaccessible cliffs[102] where it is presumably safe from browsers, including rats.

Restoration of Vegetation Communities

Oceanic island floras are of great importance due to their high levels of endemism. Mauritius has 644 native species, of which 287 (45%) are endemic, and Rodrigues has 132 species of which 44 (33%) are endemic; many more are endemic to the Mascarenes as a whole[103]. About 5% of Mauritius is covered by native forest, but only about 2% is of good quality with an intact or only part-disrupted canopy[104]. The majority of the endemics are threatened, and a species-based approach to their conservation is unrealistic except for the rarest.

Passive protection of vegetation communities is ineffective due to the impacts of alien species. Consequently, since the mid-1980s attempts have been made to manage representative areas of native vegetation both on Mauritius and Rodrigues. This has been the development of work started in the 1930s by Reginald Vaughan and Octave Wiehe, who set up study plots, two of which are still managed[105]. The Forestry Service also established study plots in which they weeded aliens and also tried to manage native trees for commercial forestry[106]. The conservation of representative areas of native vegetation has been expanded into an ecosystem approach; these areas are called Conservation Management Areas (CMAs), where exotic weeds have been largely removed and exotic mammals controlled or excluded. This work was developed during 1982–1998 by Wendy Strahm and Ehsan Dulloo. CMAs are intensively managed and have been weeded and fenced, and in some cases there has been replanting with nursery grown plants.

In addition to the work on CMAs on mainland Mauritius and the restoration work on Rodrigues, there has also been an effort to restore the vegetation communities on satellite islets. The most important island restoration projects have been on Ile aux Aigrettes and Round Island (pp. 244–246).

Conservation Management Areas on Mauritius

The approach to conserving vegetation communities on mainland Mauritius has been to set aside representative areas of native vegetation in good condition and attempting to minimise the impacts of invasive species that were causing these communities to be degraded. The number of native species in these plots is high. In 2000 in the 'Brise Fer (1987) plot', there were 9,025 trees per hectare of 88 species, representing 69 genera[107].

The main priorities in the management of CMAs have been to remove the exotic plants and to fence out deer and pigs. Although some thought fencing was unnecessary, it was subsequently considered important to exclude these animals, since whenever the Forestry Service had planted out nursery-grown native trees in the forest these had invariably been browsed by deer or destroyed by pigs. Even low densities of deer and pigs depress the regeneration of many native plants[108]. The CMAs were fenced with 2m high chain-link fencing, with stones piled along the base of the fence to form a low rock wall, preventing pigs from getting underneath.

About 53% of the Mauritian native flowering plant species are contained within weeded CMAs; if we include other fenced conservation areas and managed offshore islands this figure increases to about 61%. These plots were, and still are, largely experimental since the most efficient ways of managing areas of native forest were unknown. The results show the benefits of fencing out pigs and deer, and the huge effect that weeding has on the health of the vegetation. It was initially considered labour intensive, but after the initial weeding the subsequent removal of exotic seedlings was straightforward, and the amount of labour needed to keep the plots clear decreased substantially.

Mechanical control of invasive plants by uprooting them has been the method of choice in the CMAs, on Rodrigues and on islands. Some exotic trees are left in highly degraded areas, to provide shade and act as nurse trees while native plants establish themselves, and are later removed. Once the established exotics have been removed, the amount of weeding effort per unit area decreases each year as the seed banks of exotics are exhausted[111].

After weeding the upland forest plots on Mauritius, the understory was typically bare, since the exotics had suppressed regeneration and replaced most of the native plants that previously occupied the ground layer. There were gaps in the canopy of large trees, and many were in poor health and dying. The health of the weeded forest improved dramatically over the subsequent three to five years and the trees produced an abundance of flowers and fruit; gaps in the canopy closed[112].

The first plot that was studied in detail was in the upland forest of Macabé in 1937. This is the original site studied by Vaughan and Wiehe, who documented every tree greater than 50cm in height in 1000m^2, a total of 1,785 trees of 69 species[113]. The area was weeded from 1937 to 1952, but thereafter neglected

Table 10.2. Conservation Management Areas on the Mauritian mainland[109]. Data derived from Strahm (1993), Page (1995), Kueffer & Mauremootoo (2002), Mauremootoo et al. (2003).

Name	Date established	Year first weeded	Year first fenced	Area (ha)	Altitude (m)	Vegetation type
Mondrain	1979	1979	1986	5	c. 500–530	Semi-dry Eugenia/Sideroxylon thicket
Macabé[110]	1937	1937	1986	0.4	c. 550	Lower montane rainforest
Brise Fer (old plot)	1987	1987	1987	1.26	c. 570–600	Lower montane rainforest
Mt Cocotte	1987	1987	1987	0.34	c. 770	Superhumid upland cloud forest
Mare Longue	1993	1993	1993	3.46	c. 550	Lower montane rainforest
Bel Ombre, Fixon	1994	1994	1994	4.3	c. 300	Evergreen wet lowland forest
Bel Ombre, Bellouget	1994	Not yet weeded	2002	2.5	c. 300	Evergreen wet lowland forest
Florin	1994	1995	1989	2.53	c. 600	Sideroxylon thicket
Le Petrin	1994	1994	1994	6.2	c. 660	Philippia heath/marsh & Pandanetum
Brise Fer (new plot)	1996	1996	1996	24	c. 570–600	Lower montane rainforest
Morne Seche	1998	Not yet weeded	1998	6	c. 150	Lowland dry forest
Perrier	1969	1969	1969	1.44	c. 550	Sideroxylon thicket

until weeded again in 1978, then finally fenced in 1986; the plot was then thoroughly re-weeded and all the trees measured.

Even with this low level of management the area that had been weeded in 1937 had significantly more native species than the surrounding area. A control plot of 1,000m², by then dominated by Strawberry Guava and privet, in adjacent unweeded upland forest had only 339 native woody plants of 41 species, compared with 1,136 in the weeded area of 56 species[114]. Encouraged by these results, eleven plots in Mauritius have been fenced off, ranging in size from 0.3ha to 24ha and now covering an area of 57.6ha. Comparing the weeded plot with an unweeded control plot over four years, seedling survival was three times better and seedling growth rates were six times better than in the unweeded areas[115]. Ten to twelve years after the initial weeding, the forest had recovered such that the opening up through weeding was undetectable, and it had regained the structure described in 1941[116], although with some changes in species composition.

The 1987 Brise Fer plot shows considerably more regeneration of woody angiosperms and ferns than in non-managed areas, and some species that usually show very poor rates of regeneration such as Colophane and even the endangered Tambalacoque are present as seedlings[117], while fern growth is enhanced[118]. In the unweeded areas the die-off of native trees is higher, suggesting that mortality is accelerated by the presence of the exotic plants[119]. However, 37–53% of the species present are apparently not regenerating, the reasons for which are unclear. It is obvious that regeneration is still depressed in many species, possibly due to the presence of fruit and seed predators such as rats and monkeys (which the fence cannot exclude) and the possible loss of animal pollinators (now-extinct bats, birds, and lizards, not to mention invertebrates).

In some of the CMAs the rates of regeneration were much lower than those seen in the Macabé and Brise Fer 1987 plots. In these cases deer had been accidentally fenced into the plot, or pigs had entered by squeezing under the fence, providing good empirical evidence for the value of effective fencing of managed areas of vegetation.

Impact of forest restoration on animal communities

The weeding of forest plots has profound effects upon animal communities. These are obvious just by walking from a weed-choked area of forest into a managed plot; one is immediately struck by the increase in bird and insect life and the improved health of the native plants. Butterflies are more abundant in CMAs than in adjacent areas of unweeded forest. This response to weeding-out the dense exotic understory is not surprising, since most butterfly species need direct sunlight for feeding and breeding. Butterflies maintain good densities in open areas within the forest and along forest roads and tracks, but numbers are depressed by greater canopy cover, suggesting that to encourage butterfly populations in managed forest some open areas should be maintained[120].

A study on the abundance of canopy dwelling insects in weeded and unweeded areas of upland forest failed to find significant differences in either species richness or evenness[121]. The effect of CMAs on invertebrate assemblages in the litter were more marked, with significantly higher diversity of litter invertebrates at order level, and of beetle families. There was, however, a significantly higher abundance (but less variety) of some of dominant invertebrates such as beetles and hymenopterans in unmanaged plots. For scarab beetles, a lower abundance in the weeded plot may be associated with the exclusion of large vertebrates (and their dung) through fencing[122].

The impact of CMAs on the native snail community is complex[123]. A survey in the Brise Fer 1987 plot in 1995 revealed 36 species of snails and slugs, of which 28 were native (16 endemic). The numbers of *Omphalotropis* species inhabiting the arboreal foliage were about 26 times higher in the unweeded plot compared with the managed area. This was because the snails were found in large numbers on Strawberry Guava leaves in the understory. There are no ecologically similar alien snails, and *Omphalotropis* spp. can evidently adapt to the exotic guava. Removing the guava temporarily removed the habitat of this group of snails, but they will probably increase in the managed plots as the guava understory is replaced with native species.

In contrast to *Omphalotropis*, populations of *Nesopupa* species living on the trunks of native trees were unaffected by weeding, and numbers were similar in weeded and unweeded areas. Similarly, the native streptaxid carnivorous snails were unaffected by weeding. There were, however, fewer large helicarionids in the weeded areas, where they seem to be replaced by ecologically similar alien subulinids. These results are difficult to interpret in isolation, and the weeded plots need to be monitored as they recover their understory to assess whether any of the impacts of weeding are long-term.

Degraded forest is structurally unsuitable for many native birds and is not as productive, resulting in apparent food shortages for many species[124]. Even small managed plots of forest are attractive to native birds, lizards and fruitbats. Based on these results, the CMA concept was expanded from vegetation plots to whole ecosystem management, including controlling introduced predators. In an effort to have areas of sufficient size to support birds, a large fenced CMA of

24ha has been erected in the Brise Fer forest. Here we have released breeding groups of Pink Pigeons[125] and Echo Parakeets. We have seen an increase in the number of Grey-white Eyes, Mauritius Merles and Mauritius Cuckoo-shrikes using and nesting in this area[126]. A recent study showed that managed areas of forest are chosen preferentially by Pink Pigeons[127]; they are attracted to high-quality forest and the CMA's are particularly desirable. Older male Pink Pigeons occupied breeding territories within the CMAs, while the younger males held smaller territories in adjacent areas of unrestored forest.

Point counts in CMAs and in unmanaged forest showed that native bird numbers were related to forest quality, with the highest numbers in areas of full canopy forest. The number of Grey White-eyes declined through progressively more degraded forest. Conversely, exotic birds were fewest in high-quality forest, with higher densities in degraded areas. The most commonly encountered exotic was the Red-whiskered Bulbul. In recently weeded areas of forest Grey White-eye density was lower than in comparable areas of unweeded forest[128]. This was presumably due to the lack of understorey, and it is likely that densities will increase with the regeneration native sub-canopy trees.

Black-spined Flying-foxes use the managed plots heavily at night, feeding on fruits of canopy trees[129], and they apparently use the CMAs more frequently than surrounding unweeded areas. The Upland Forest Day-gecko *Phelsuma rosagularis* is limited to upland climax forest, where it is particularly dependent upon large native trees with cracks, fissures and hollow limbs in which to hide, and cavities in which to lay eggs[130]. This species showed a negative response to weeding, with depressed numbers found in weeded plots compared to unweeded ones[131]. This is likely to be a short-term effect due to the disruption caused by weeding and the removal of large privet trees, which provided a home for the geckos.

The management of native forest needs to be developed further. More managed areas are required and these need to be as large as possible. Studies are required to evaluate the strengths and weakness of different management techniques for CMAs.

Restoration of vegetation on Rodrigues[132]

On Rodrigues there are two fenced and managed reserves, one on Grande Montagne of 33ha, of which about 14ha has been weeded and planted, and the other at Anse Quitor of 30ha, where about 7ha has been weeded and planted. Both these sites, the most important on the island for endemic plants, were declared nature reserves in 1986. Rodrigues does not have naturalised populations of deer, pigs and monkeys, and so the fences do not have to be as elaborate as those used around CMAs on Mauritius, but they do need to exclude domestic goats and cattle. People sometimes breach the Anse Quitor fence to graze their livestock.

There are no intact native forests left on Rodrigues – just a few forest remnants, mainly in Anse Quitor, Grande Montagne, Cascade St Louis and Anse Mourouk, with scattered native trees in valleys and on ridge tops, so it has been necessary to clear areas of exotics and replant with nursery-grown seedlings, using data derived from studies of plant associations in remnant patches of forest[133]. The work on Grande Montagne started in 1986; although badly degraded, this area had the highest diversity of native plants. The work on Grande Montagne was relatively small scale and experimental up until 1996, but intensified when funded by the Global Environment Facility, World Bank Biodiversity Restoration Project (1996–2001). The most effective method has been to thin out the exotics (mainly Rose-apple and *Litsea glutinosa*) and plant natives in the gaps. The exotics are gradually removed over five years when the native saplings have reached 1–3m in height. The use of exotic trees as nurses helps keep the soil moist, provides a windbreak, and gives shade that prevents the young trees from drying out.

Seedlings are grown in MWF's Rodrigues nursery, an effort being made to plant a wide range of species, with easy-to-propagate, fast-growing species being used as pioneers, other species being added later when available[134]. Survival and growth rates for most species have been high, some species growing by as much as a metre a year[135]. A more intensively managed "species recovery area" about 600m^2 is planted with the most endangered plants that have proved problematic to propagate, mixed with some of the commoner species[136]. Within ten years of planting, the tallest trees had reached 5–8m, and several had flowered and fruited[137]. Restored native forest is being used by the endemic Rodrigues Fody and Rodrigues Warbler; the fody only recolonised Grande Montagne in around 1991 after the start of the restoration work, and warbler numbers have increased to some of the highest densities on the island[138].

Anse Quitor is a lower, drier limestone (calcarenite) area, so the plant composition is different, although many of the ecologically more generalised species are found in both areas. This valley was chosen for restoration because it contained good populations of several of Rodrigues's most important endemic plants. It has two of the remaining *Zanthoxylum paniculatum* (the third is nearby), 7 (out of 10) *Gastonia rodriguesiana*, and reasonable numbers of latans *Latania verschaffeltii* and many Bois d'Olive *Cassine orientalis*. Restoration started here in 1997[139]. Planting success was good (90% survival) when there was

BOX 33　　　　　　　　　　　GECKOS (GEKKONIDAE)

The Mascarenes, particularly Mauritius, once harboured one of the most diverse oceanic island lizard faunas anywhere. Unfortunately, little attention was given to these reptiles during the early years of Mascarene history and many had become extinct before being formally described. Evidence as to the huge density of reptiles that once populated mainland Mauritius is exhibited on Round Island, the only place where most of the terrestrial Mauritian reptiles now survive. Round Island is the only large islet that has remained rat-free, and hence it escaped the rats from earlier voyages, most probably the Portuguese in the 1500s, which exterminated all of the larger lizards and snakes on mainland Mauritius prior to the Dutch colonisation. Fossil remains have confirmed the former existence of most Round Island species on the mainland.

Mauritius

The large arboreal Günther's Day-gecko *Phelsuma guentheri*, now confined to Round Island, also occurred on mainland Mauritius and there are four other Mauritian day-gecko species, Vinson's *P. ornata*, Blue-tailed *P. cepediana*, Guimbeau's *P. guimbeaui* and Upland Forest *P. rosagularis*. All described *Phelsuma* species are still extant on Mauritius, which also harbours two terrestrial night-geckos, Mauritius *Nactus serpensinsula* and Lesser *N. coindemirensis* both now confined to islets of the coast.

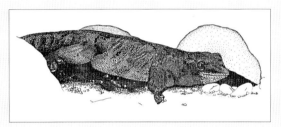

Günther's Day-gecko *Phelsuma guentheri*. Drawing by Brin Edwards in Bullock, North & Grey, 1989.

Accounts

Cossigny (1732–55) describes Vinson's Day-gecko *P. ornata* in a letter to Réaumur in 1736:

I have allowed to dry a very pretty little animal we call gobe-mouches [literally 'fly-gulper'; the term is usually used in France for flycatchers (birds), but in Mauritius at that time evidently served for geckos], but neither in this way nor putting it in spirit have I been able to preserve its colours. This animal, very lively and tame, is variegated along its back to the end of its tail in a very pretty red on a changeable green background. Its head is divided into little stripes, of the same red, of blue, of black and a little white thread edges the whole of its snout, which in all gives it a singular appearance. It is very dextrous in catching ordinary flies. Often several of them stalk and jump on the same prey. Then hops and jumps of jealousy sometimes deliver the fly from the danger it ran.

Milbert (1812), in Mauritius 1801–03, on the Blue-tailed Day-gecko *P. cepediana* in a forest:

Very pretty ultramarine blue lizards, spotted with a fine deep red, showed only their little heads from out of holes in trees, while others ran along the branches hunting insects.

Réunion

Réunion had a much less diverse lizard fauna, and all the species appear to be derived from Mauritian stock. The two *Phelsuma* day-geckos, Manapany *P. inexpectata* and Réunion Forest *P. borbonica*, were originally thought to be subspecies of Mauritian forms, but are now considered full species. An introduced snake may be responsible for their current scarcity. There are no good early travellers' accounts of Réunion lizards, but a 19th-century account pre-dates the formal description of the now endangered *P. inexpectata* by more than 120 years. A small *Nactus* gecko has been discovered from fossil remains but remains undescribed. Nothing is recorded about the species in life but it probably occupied the same niche as the *Nactus* on Mauritius.

Newton's Day-gecko *Phelsuma edwardnewtonii*. From Boulenger, 1885.

Account

Betting (1827) on *P. inexpectata*:

The lézard vert, 4 to 5 inches long, speckled with black and red spots on the back with some horizontal black bands on a grey-green ground. The underparts are of a more tender and uniform green.

Rodrigues

Rodrigues had a large and diverse saurian fauna but again, little mention has been made of them. One of the world's largest geckos, Liénard's Giant Gecko *Phelsuma gigas*, and another large species, Newton's Day-gecko *Phelsuma edwardnewtonii*, were mentioned by a number of observers, and the former was described in great detail. Like Mauritius, Rodrigues has a number of islets within its lagoon and one of these provided a refuge for the larger species long after it had disappeared from the mainland. The naturalist François Liénard described *P. gigas* from five live individuals sent to him on Mauritius in 1841 from Ile Frégate, a rocky islet offshore. He managed to keep one alive for several months but it refused all food except sweetened water from a spoon. This large greyish gecko with a pale yellow underside and pink tongue has not been recorded since. *P. edwardnewtonii* had become very scarce by the 1870s, but survived (possibly on the Rodrigues islets) until the early 20th century. Rodrigues is now known to have also harboured two endemic *Nactus* geckos and two endemic geckos of unknown affinity. Rats were a scourge to the earliest visitors on Rodrigues and probably accounted for the extinction of all of the endemic lizards on mainland Rodrigues. Cats were introduced to control rat numbers and proved to be serious predators of tortoises, and probably the larger lizards as well. Rodrigues has now lost all of its endemic reptiles, the only native survivor being the widespread Mourning Gecko *Lepidodactylus lugubris*.

Réunion Night-gecko *Nactus* sp. From Probst, 1996.

Accounts
Leguat (1707) in Rodrigues 1691–93, on, first, *Phelsuma edwardnewtonii*, with a brief note on *P. gigas*, from the original English translation of 1708:

The palm trees [hurricane palms] and plantanes [latan palms] are always loaden with lizards about a foot long, the beauty of which is quite extraordinary; some of them are blue, some black, some green, some red, some grey, and the colour of each the most lively and bright of any of its kind. Their common food is the fruit of the palm-tree. They are not mischievous, and so tame, that they often come and eat the melons on our tables, and in our presence, and even in our hands; they serve for prey to some birds, especially the bitterns. When we beat 'em down from the trees with a pole, these birds wou'd come and devour them before us, tho' we did our utmost to hinder them; and when we offer'd to oppose them, they came on still after their prey, and still followed us when we endeavour'd to defend them.

[In reference to P. gigas] There's another sort of nocturnal lizard of a grayish colour, and [very] ugly; they are as big and as long as one's arm, their flesh is not bad, they love [being on] plantanes [latan palms].

Pingré (1763), on Rodrigues in 1761, on *P. edwardnewtonii*:

[There are] other lizards nearly a foot long; the brightest colours, blue, green, yellow etc. seem to dispute as to which should clothe these lizards most vividly, though blue and green dominate on the head and back. François Leguat claims to have seen black, blue, green, red and grey, and all most vivid and striking. Here is what I can confirm having seen. I had already met in Rodrigues with several of these lizards, without having had time to examine them [closely]. I saw many on the Islot du Large [now Ile au Fous]; they could not flee far, but hid promptly in the cracks in the rock. I pursued them right into their holes. I had seen a superbly coloured lizard enter a crevice; with a twig I forced out of the same crevice a lizard whose dirty black colour almost horrified me; I had already seen some similarly uncoloured, and did not doubt that they were another species. Finally I watched one carefully; I did not move, for fear of putting it to flight – the brightness of the colours, blue, green, yellow etc. charmed me. A moment later the lizard saw me; black patches erupted in the brightness of its skin; these patches grew rapidly and in a few seconds my lizard was covered in a hideous black all over its body, and took refuge inside a rock. I am confident, as I have said, that I was not mistaken in this observation.

Marragon (1795) on *P. edwardnewtonii*:

With pleasure we see lizards clothed most prettily, 8 to 10 inches long. Amongst them are some whose back is green marked with blue, brown, grey and agreeably variegated with yellow under the throat and the belly. This animal is far from being inconvenient and is most useful: it destroys many flies, which make up its principal food, and other insects. They sometimes change their skin, for I have seen some retaining the sheath of their tail. They are brilliant in the sunshine, their colour iridescent like a pigeon's throat. They do not show up much in winter, but are seen often in houses, on roofs and on coconut palms in spring, summer and autumn.

Liénard (1842) on *P. gigas*:

Toes unequal and enlarged throughout their length; limbs fat and short, body stocky, tail almost as long as the body, an array of pores in front of the anus, upper part of the toes embellished with scales. This Gecko which I owe to the kindness of Captain Descreux, comes from Ile Frégate, situated in the SW of and a mile from Rodrigues; one finds them in large numbers; there are, it is said, some as fat as an arm, 18–20 inches long. They stay hidden between rocks during the day, and come out at dusk to seek their food. It is said that they are both frugivores and carnivores; that they eat pineapple and fruits of the Mauritian Fig (Ficus mauritianus) [actually F. rubra and F. reflexa], [and] that at night they climb trees to take seabirds whose blood they suck, and that they also eat birds' eggs. I have had one free in my room for more than two months, to my knowledge, only sugared water which it drinks from a spoon when offered to it. I have given it successively sugar that skinks are so fond of, grapes, bananas and insects, and it has not touched them. The individual in my possession is 14 inches long; the circumference of its body, taken between the four limbs is 5 inches 9 lines; at the base of the tail it is 3 inches 6 lines. The length of the head is an eighth of that of the [whole] animal; it is flattened, a little hollow between the orbits. There is nearly the same distance from the eye to the snout as there is from that organ to the ear-opening, which is a fairly deep cavity of oval shape. I have noticed that in this animal, as in the Madagascar crocodile, the ear-hole connects from one side of the head to the other. One may easily pass a wire through which meets no obstacle. The corner of the mouth is further back than the perpendicular from the rear side of the eye. It [the mouth] is armed with a single array of small sharp teeth in each jaw; the palate is smooth. The lips are lined each side with large scales, the upper lip 19 and the lower 17. The nasal scale is very large; it fills the distance between the two nostrils. The tongue is very extensible; its tip is opened out like a rose petal, and is the same shape and colour; the base is a delicate pink. The head scales are much larger than those on the body; they get smaller as they get further from the snout. Those that cover the body, both above and below, are excessively small and granular. Those of the tail are large; the form annular rings around that part of the animal. The anus is a transverse slit one third the width of the tail. The toes are of unequal length; the thumb is very short, the first is double its length and breadth; the third is even larger, the fourth the longest of all, double the length of the first toe. The fifth equals the third in length and breadth. All the toes lack claws, and their underside is equipped with 24 transverse lamellae. The body is grey above, marbled and finely spotted with black; the tail is a darker grey without marbling. One notes longitudinal black lines formed by the linking up the spots; all the underside is pale yellow.

[Later, on P. edwardnewtonii] Captain Descreux has brought me from a further trip he has just made to Rodrigues seven Geckos, two in alcohol and five living, of which four are of the species I have just described, and one small one with a double tail six inches long, greenish blue with a yellow throat. The shape, the toes of this little lizard, the scales that are found on its head and along its lips, the number of transverse lamellae under the toes, and the body scales, resemble those of the preceding gecko. It differs only in its colours, its pores that are much larger and whose chevron is more acute; its shape is also less solid than the preceding species. I name it Yellow-throated gecko, because of this colour which is well marked and which contrasts with the greenish blue of the body. Geckos often lose their tails, which break easily. Some naturalists have claimed that this part regrows smooth and without scales. I can confirm the contrary. One of the tails of the Yellow-throated Gecko [and] that of one of the living Gecko Gigas that I have are both entirely enveloped in circular scales. These tails are clearly regrown; and one can easily see the place where they broke. I am happy to find myself in a position to verify a fact that the illustrious Cuvier left in doubt in his Règne Animal vol.2 page 52. I do not know, however, he says, whether the pores in the first subgenus are a mark of sex. In two Geckos of the same species that I've observed, one had pores and the other did not. I assured myself that the pores were not related to [their] sex; the two individuals were females. Their stomachs and their intestines contained only fruits of the Mauritian fig. A few days ago, one of the Geckos for which I am obliged to Captain Descreux, escaped its cage; I asked my maid to recapture it, and the moment she seized it, the tail broke and detached from the body. This end of tail wriggled extremely fast for over twelve minutes. It was only after a quarter of an hour that the convulsions stopped completely. The large wound only issued a single drop of blood from the remaining part of the tail. The animal did not appear weakened by its loss, and continued running at the same speed.

Specimen collections
At least 6 spirit specimens of Newton's Day-gecko have survived and can be seen at the BMNH and MNHN. Fossil material of Newton's Day-gecko and Liénard's Giant Gecko has been recently discovered on Rodrigues and has been deposited at the BMNH.

sufficient rain, but during drought conditions this dropped to 30%. Overall survival is lower than at Grande Montagne because it is a drier part of the island with free draining soil, so more plants die from desiccation[140].

The work at Anse Quitor has been expanded into the adjacent valley of Cavern Tilleul, a huge collapsed cave that has produced a canyon-like valley about 490m long with near vertical walls up to about 20m high. This area has remnants of a native plant community with some endangered species growing out of the canyon sides. The project is a private enterprise, with the MWF providing plants and technical expertise. The aim is to restore the area as near as possible to its pristine state, with a native vegetation community complemented by free-roaming Aldabra and Radiated Tortoises to replace the two extinct tortoise species. The area of about 25ha is being restored with similar vegetation to Anse Quitor, and the tortoises roam over the floor of the collapsed cave. Planting started in July 2004, and the project opened as an educational tourist attraction in 2007 (p. 202) [141].

A remarkable fact about the vegetation restoration in Rodrigues is that a very high percentage of the plants are endangered species, and this work has provided long term security for several species that had seemed doomed.

Small-island restoration

The restoration of representative areas on the mainland of Mauritius is realistic, but the long-term management of these areas is labour intensive due to the constant maintenance required to control the invasion of exotic organisms. A more achievable goal is the restoration of small islands where many invasive species can be completely eradicated and re-invasion rates can be kept low by good quarantine protocols. The aim of island restoration work is to maximise the benefit of restoration for species conservation and to restore plant and animal communities. Where the exact composition of these communities is not known it is assumed to be close to what is known from neighbouring island and coastal communities.

In 1993 all the islets around Mauritius and Rodrigues were surveyed to assess their importance and conservation needs. Most of our work has concentrated on the two islands off Mauritius of greatest conservation importance – Ile aux Aigrettes, important for its coastal forest community[142], and Round Island with its native reptile community and palm-rich forest. In addition, a number of other islands were identified as having conservation potential, cleared of exotic mammals and with their most important species identified. Although only two islets off Mauritius, Pigeon House Rock (0.6ha) and Serpent Island (31ha) remain pristine or near pristine, several have remained free of the most harmful aliens and are refugia for species that have long since become extirpated elsewhere. Islands cleared of exotic vertebrates can be used as reserves on which to "maroon" endangered endemic vertebrates[143]. Pink Pigeons, Mauritius Fodies and Mauritius Olive White-eyes have been released on Ile aux Aigrettes, and the first Round Island reptiles have been moved onto other islands.

The Round Island plant community

Round Island (219ha) is one of our most important islands. It has never had rats and has consequently retained most of its reptile community. As we have seen, introduced goats and rabbits were eradicated in 1976–1979 and 1986. These had modified the ecology of the island considerably; a small hardwood forest mainly on the southern slope near the summit has disappeared, and some species have declined drastically. The island has suffered severe erosion and much of the soil has gone. The Round Island race of the Hurricane Palm is reduced to just one wild individual[144].

The main objectives of the Round Island restoration programme are to re-establish the native plant communities – the palm-rich forest, the hardwood forest and open areas with tussock and other grasses. This involves increasing the floral diversity on the island, since the island has lost many species. In the original restoration plan, Don Merton and his colleagues, after extensive consultation with local botanists, compiled a list of plant species that should be considered for introduction to the island[145]. This list was derived from published records, herbarium specimens, the identification of woody remains still present of hardwood species now vanished, and plants growing in similar plant communities on adjacent islands and in areas of similar altitude on the adjacent mainland. The intention was to plant the species likely to have been found on the island and to establish these, and then to allow nature to select those that survived. Re-establishing plant communities was also expected to prevent further erosion on the island and to rebuild soil deposits.

Eradicating rabbits and goats has had a dramatic impact on the regeneration of native plants[146]. For example, in 1986 the Round Island Bottle palm was reduced to eight adults and 13 young plants. Since the removal of rabbits it has recovered dramatically, and in 1992 the number of young plants had risen to 79 and there were 180 seedlings. Six of the young plants from 1986 had matured by 1996, and seedlings were abundant near adult palms[147].

Vegetation cover monitored on 15 permanent quadrats and transects has shown a marked increase. However, many of the thriving species are exotics which had previously been controlled by the rabbits.

Two peas (*Desmodium incanum*, *Desmanthus virgatus*), a grass (*Cenchrus echinatus*) and a pantropical weed (*Achrynthes aspera*) are proving to be the worst invasive species. *A. aspera* was first seen on the island in 1987 and has since expanded very rapidly, forming a dense gound cover in some places. Attempts to manually control or eradicate the spread of the peas proved impossible, and it is hoped that the importance of these pioneering exotics will decrease as native plants become re-established. When further invasive plants are found that could impact on the restoration efforts (e.g. *Chromolaena odorata*, *Heteropogon contortus* and *Sporobolus capensis*) the plants are manually removed or the patch treated with herbicide.

Experimental soil traps have been set up on steep slopes to help control erosion and to provide areas of soil. These are built by placing crude rock barriers behind which the soil is trapped; build-up can be quite rapid, and on steep slopes 15cm of soil may collect in a year. *Latania*, *Pandanus*, *Hyophorbe* and *Dracaena* have been planted in these soil traps[148]. Pioneering hardwood species are being re-introduced to recreate the hardwood forest.

The best replanting places are areas of good soil with some shade and protection, such as soil blocks and in the gullies, especially at the gully-heads where seeds can easily spread downwards and have a high chance of germination in the moister and deeper soil that accumulates in the gullies. Increasing vegetation helps to bind and trap the soil hence reducing soil erosion further[149].

Initially, apart from the pioneer coastal shrub *Scaevola taccada* which flourished, results of planting were generally poor, due to low rainfall, poor soil and exposure from persistent winds[150], not helped by salt blown in off the sea[151]. Exotic grasses, especially *Digitaria* and *Cenchrus*, smother the young plants. Plants are also dug up by Wedge-tailed Shearwaters when digging their nest burrows. Insufficient hardening off of plants and aftercare also contributed. These disappointing early results showed that to be successful in restoring a plant community on the island more care was necessary in the preparation, planting and management of the plants. A field station was erected on the island in 2002 so that a permanent field presence could be maintained, and long-term conservation management on this island is now a realistic possibility. A small field nursery was created so that plants could be grown for reintroduction, and a rainwater catchment system developed so that seedlings and saplings could be watered[152]. Since 2002 success rates had improved by using larger plants, potted up on the island and re-established before being planted out. In 2004 direct seed sowing was tried again, since this is far easier and less labour intensive than planting out seedlings, and minimises quarantine problems. However, germination rates of broadcast or planted seeds remain patchy[153].

In 2003, trials were carried out to protect the plants from burrowing Wedge-tailed Shearwaters, by laying galvanised wire fencing on the ground, which allowed the plants to grow but prevented the birds from burrowing in. These trials were abandoned because it disrupted the nesting of the shearwaters too drastically, and the fencing was replaced with shaded wire cages surrounding each plant[154], with watering in the dry season[155]. In 2004 a larger water capacity was installed on the island[156]; soil quality and protection from the wind has improved[157]. These enhancements have dramatically improved plant survival, though it is variable between species[158].

Ile aux Aigrettes vegetation restoration

Ile aux Aigrettes is a 25ha low coralline island nature reserve. Ecological restoration started in 1986 with the aim of re-establishing the coastal vegetation community and replacing missing components of the flora and fauna. A secondary aim of the project was to develop a site where rare lowland plants could be grown and carefully managed, secure from the impacts of exotic herbivorous mammals[159]. The island has its own coastal plant community but this has become depauperate through partial clear-felling and invasion by exotic weeds, and browsing by goats from *c*.1953 to 1965; exotic reptiles and mammals were also introduced. Although nominally a reserve since 1965, the island was poorly policed until 1985, and many native trees on the island were cut for firewood; virtually every tree on the island had damage caused by woodcutters cutting off limbs[160]. Since rats were removed in 1991, regeneration of the endangered Ile aux Aigrettes Ebony *Diospyros egrettarum* has been spectacular. The rats ate the seeds and suppressed regeneration, but carpets of seedlings can now be seen beneath the parent plants. The island has the largest and best remnant of Mauritian dry coastal ebony forest, and many of the plants found on the island are endemic to Mauritius, often with most of the surviving population[161]. The forest in the central third of the island is in good condition, about 8m in height, and still growing, with a canopy[162]. A further 30–40% of the island is degraded forest, with the rest badly degraded. The 53 native plant species include four endangered, nine vulnerable and four rare species. Although the vegetation is in better condition than any other coastal remnant it is impoverished in terms of number of species. Early Dutch maps showed the island covered with palm trees, long since gone[163].

From 1986 to the end of 1997 about 60% of the island was cleared of weeds at least once (and much of it several times since the exotic seedbank was large); by 2005 90% of the islet had been weeded. After the first weeding, successive clearing becomes easier as weed biomass decreases. The two worst weeds are Indian Plum (*prune malgache*) *Flacourtia indica* and White Popinac (*acacia*) *Leucaena leucocephala*. These are highly invasive, coppice readily when cut, and suppress the regeneration of natives; popinac had established monotypic stands over 9% of the island. Much work remains before they can be eradicated, although both are under control. Other invasive species have been easier to deal with: Mauritius Hemp (*aloès*) *Furcraea foetida*[164] and she-oak (*filao*) *Casuarina equisetifolia* have been eradicated, and Tecoma *Tabebuia pallida* cut down, but coppicing occurs and seedlings need weeding out[165].

The impact of weeding has been impressive. In areas where repeated weeding has been carried out there has been a major increase in native plant regeneration, which has been enhanced by supplementary planting with nursery grown seedlings. Bois de Rat *Tarenna borbonica* and Bois de Pipe *Ehretia petiolaris* freely regenerate in the open, and where the canopy is fragmented with semi-shade there is regeneration of ebony and Bois Clou *Eugenia lucida*[166]. Survival rate of nursery grown seedlings is high[167] (Ile aux Aigrettes is much wetter than Round Island), though canopy tree saplings (Ile aux Aigrettes Ebony and *Eugenia lucida*) grow slowly until shaded by pioneers[168].

As palms were an important component of the coastal community, three species have been re-introduced from Round Island, latan *Latania loddigesii*, Bottle Palm *Hyophorbe lagenicaulis,* and most importantly, the Hurricane Palm *Dictyosperma album congugatum*. This latter palm has been grown from seeds from the last surviving wild individual on Round Island, and 46 have been planted. Other species typical of coastal palm-rich forest which have been planted are Bois Buis *Fernelia buxifolia*, *Gagnebina pterocarpa*, aloe (*mazanvron*) *Lomatophyllum tomentorii* as understory, and Bois Cabris *Clerodendrum heterophyllum* in coastal regions. A further 21 species not currently occurring on the island but typical of lowland coastal vegetation community have been identified to enhance species diversity, and will be added to plantings across the island, including *Coffea myrtifolia* and *Poupartia pubescens*.

Rare plant propagation and management
Plant endemism is very high in the Mascarenes; of the 885 species of flowering plants native to the three islands, 645 (72%) are endemic and most species are declining and in need of conservation[169]. Many species are critically endangered, and in the short term artificial propagation offers the best hope of securing their future; 27 taxa of threatened Mauritian plants do not grow in any managed nature reserve.

For Mauritius and Rodrigues 120 taxa are known from either less than 20 individuals or just one or two populations, and 28 species have less than ten known individuals in the wild. These are the priority species for artificial propagation. All the critically endangered species need to be brought into cultivation and intensively managed to optimise propagation rates, and to help us learn how to look after reintroduced populations. Meanwhile, where possible the last wild individuals of critically endangered species growing outside of managed reserves are protected from browsing mammals and wood-cutting, and seeds collected. While some can be fenced (e.g. *Ramosmania rodriguesii* on Rodrigues[170]), a shortage of resources and logistics makes this impractical for most species.

Artificial propagation of native plants has a long history on Mauritius. The Forestry Service has experimented sporadically with the growth and planting of native timber trees from the 1880s onwards[171]. In the 1970s and 1980s the Forestry Service grew rare native species for conservation purposes[172], and in 1999 set up a Biodiversity Unit. In the 1980s and 1990s most of the critically endangered plants were successfully propagated in local nurseries, but relatively few of these have resulted in sustained populations. An exception is the lily *Crinum mauritianum* from whose large cultivated population some plants have been transferred into managed reserves – the main wild population was wiped out by a reservoir development[173].

There has been a widely held belief on Mauritius that the cultivation of rare plants in private gardens is a valid conservation strategy. During the late 1980s and 1990s many rare plants derived from wild-collected seed and cuttings were sold to the general public, from Government and private nurseries, to increase interest and participation in plant conservation. Private gardeners have accumulated collections of rare endemic plants in the belief that they are helping to conserve the species. The result is that there are now small numbers of many species in the hands of unknown horticulturists, with little guarantee for the plants' future when the gardeners move or die. Veteran botanist Reginald Vaughan lived in a classic colonial house in a garden full of native plants he had nurtured for 60 years, yet when he died in 1987 it was sold to developers and summarily bulldozed[174]. His executors attempted to rescue of a few of the most important plants before the sale of the house; these were planted around the hunting lodge in Yemen, but few survived in this much drier area. If even Dr Vaughan's place and unique assemblage of plants was not thought worth preserving, how much less shrift will be given to the odd hibiscus or *Trochetia* in an anonymous private garden?

Many attempts have been made to reintroduce

nursery-grown plants. In the 1970s and early 1980s some were planted straight into the wild in Macabé forest, but survival was poor[175]. In the 1980s and 1990s many planted into managed reserves survived better, but survival rates need to be improved further through post-reintroduction care – watering, controlling weeds and excluding pests and disease when appropriate. In the last two decades, standards of rare-plant propagation and record-keeping have improved, but there is no room for complacency as plants are still becoming extinct. In February 1995 the last known *Pandanus pyramidalis* was cut down in the clearing of storm-strewn vegetation following cyclone *Hollanda*, and in November 2001 the last known *Diospyros angulata*, an old female, died[176]. Without urgent action more species are likely to be lost soon.

The propagation technique of choice is to cultivate plants from seed; failing this most woody angiosperms can be grown vegetatively from cuttings or air-layering. These techniques have the advantages of usually being easy to implement, they do not need complex equipment or facilities and they are relatively inexpensive. Tissue culture techniques are useful in the propagation of species that are both rare and difficult to grow by conventional methods, such as some of the very rare orchids that do not set seed. A tree species where tissue culture may be necessary is *Tambourissa tetragona*, which has not been recorded flowering or fruiting for more than two decades; the remaining three very old trees are dying, and very little vegetative material is available. However, even this method has its limits; all attempts to culture the last surviving *Hyophorbe amauricalis* palm, proud but alone in the Curepipe botanic gardens, have failed[177].

The Mauritian Wildlife Foundation has been developing nurseries within the natural range of the species it is propagating. This helps minimise the introduction of exotic disease and maintain the genetic integrity of discrete populations by reducing the possibilities of hybridisation between closely related allopatric species. The close proximity of wild and cultivated populations also favours easy comparison and study.

One of the earliest and most successful *in situ* nurseries is in Rodrigues, where there has been a continuous effort to propagate endangered plants since 1982. The growth of plants for reintroduction increased from 1990–1995 when 8,000 plants of 25 species were produced[178]. Since 1996 there has been a purpose-built nursery with a capacity to produce up to 75,000 plants a year. Propagation has been particularly successful on Rodrigues, where of the 39 surviving endemic angiosperms, 38 have been cultivated at least once. Only the Bois Pasner *Zanthoxylum paniculatum* has yet to be cultivated; only two old adults and a recently discovered young tree survive. The greatest conservation success in numerical terms is *Hibiscus liliiflorus*, which was reduced to just three known individuals in the wild[179]; some 3,000 plants have been reintroduced to Grande Montagne where most still survive, and they are readily flowering and fruiting. Although many species have been propagated artificially and reintroduced successfully, there have also been several that have proved problematic to propagate, notably the Café Marron *Ramosmania rodriguesii*[180], the climbing shrub *Gouania leguatii*[181], and the Bois de Ronde *Carissa xylopicron*[182].

Germination success has been highly variable, but it can be improved by manipulation. One of the hardest to germinate was *Foetida rodriguesiana*; until recently less than 1% of seeds germinated, but by removing the woody epicarp the germination rate has increased to more than 50%. Propagation of critically endangered plant species can yield quick results with some species, but for others it takes many years of careful trial-and-error by experienced personnel to obtain success.

A small nursery established on Ile aux Aigrettes in 1986 was replaced by a larger purpose-built nursery in 1996. Most of the propagation work is concentrated on producing plants for the vegetation restoration, but an increasing effort has been directed on propagating threatened lowland forest, coastal and island species. Several endangered species have been grown and planted on the island, but a few have been more problematic. No seeds have germinated of the critically endangered Bois de Fer *Sideroxylon boutonianum* (one of the impressive Mascarene radiation of Sapotaceae), and of 106 cuttings only three have taken root.

Ile aux Aigrettes has become an important reserve for the experimental planting of some rare species, with variable success. The island has the advantage that there are no browsing mammals or rats, and the young plants are relatively safe on the island. Nevertheless, the rarest species are planted in areas where they can be easily monitored and given supportive care if necessary. In addition to those already planted[183], there are a number of other endangered lowland species that could survive on Ile aux Aigrettes, and plans are being developed to begin propagating two Mascarene endemics virtually extirpated on Mauritius: *Dombeya populnea*, reduced to two known individuals, and *Carissa xylopicron*, which is down to four plants[184]. Several *Dombeya mauritiana* have been planted on the island; this species is extinct in the wild, and this is a good example of how a species can be saved by cultivation and planted into a managed area.

A field nursery has also been constructed at Pigeon Wood, near Montagne Cocotte. This is an area of high rainfall (5m per year) with its own distinctive plant community; several species are limited to this region[185]. In addition, a 1ha fenced plot surrounding the nursery has been weeded and planted with rare native plants that can be carefully managed. The rare species reared in the "field gene bank" will be used as a source of material for propagation and monitored to gain more information on the ecology and life histories of the plants[186].

To summarise, there have been many attempts at propagating rare plants, especially since the 1970s. Most endangered species have been propagated, but relatively few of these attempts have resulted in established long-term viable populations, either in cultivation or in the wild. Over time there has been a tremendous wastage of individuals and species. This has been particularly acute when the plant propagation was not backed up by planting into a gene bank or a managed reserve, where the plants could be carefully looked after – e.g. producing plants for private horticulture, and also planting into nature reserves or CMAs with no post-planting care and little monitoring. In contrast, plants put into the reserves on Rodrigues and Ile aux Aigrettes and, since 2003, on Round Island, have generally had good rates of survival. It remains a high priority to manage critically endangered plants, and current efforts are not yet sufficient to reverse the high rate of plant extinction in Mauritius.

Reptile conservation

There has been continuous interest in Mauritian reptiles since Jean Vinson began working on this group in the 1940s, assisted later by his son Jean-Michel[187]. Their research involved taxonomic and life-history studies on the lizards, ecological studies on the day-geckos, and work on the Round Island reptile community. This provided the base-line for conservation-orientated studies on Round Island reptiles, which have always been the main focus of reptile conservation work because of the relatively intact state of the community. Conservation work started with captive breeding and associated studies of three species by the DWCT at Jersey Zoo: Telfair's Skink *Leiolopisma telfairii*, Günther's Gecko *Phelsuma Güntheri* and Keel-Scaled Boa *Casarea dussumieri*; work on the boa still continues[188]. There have also been captive studies on the smaller day-geckos[189] and *Nactus* night-geckos[190]. The need to be able to safely release captive-bred Round Island reptiles was the major driver of the restoration of Round Island, the first step of which was eradicating the rabbits[191]. This further stimulated work and restoration on other satellite islands. In the 1980s it became apparent that the reptile fauna of the Mascarenes was more diverse than hitherto thought, with several new species and populations being discovered (Chapters 8 and 9), adding considerable complexity to the conservation process.

The conservation strategy of choice is to establish additional populations of the most endangered species on suitable islands that have been cleared of exotic mammals[192]. On some of the larger Mauritian satellite islands it is hoped to sequentially introduce reptiles over a number of years, offering unique opportunities to understand how these species interact and the chance to re-establish near-intact reptile communities. The priority taxa are those that are now restricted to single satellite islands and hence are the most endangered: Telfair's Skink, Günther's Gecko, Keel-Scaled Boa, Orange-tailed Skink and the *durrelli* race of Serpent Island Night-gecko. There are constraints on which islands are suitable. There are historic records for several species on satellite islands, and for most of the larger islands it is probably safe to assume that all or most species of lowland reptiles would once have been found on them before they were modified by the introduction of rats and habitat degradation.

When introducing reptiles onto some of the islands that have rare lizards, a risk assessment must be carried out to evaluate any possible impacts the reintroduced species has on any of the rarer species. For example, introducing Telfair's Skinks to Flat Island would probably result in some predation on Orange-tailed Skinks, although it is unlikely that they would wipe them out, since they shared Flat Island until the mid-19th century before the arrival of rats. Telfair's Skink has been reintroduced to Gunner's Quoin, which has suitable habitat and where the lizard was common in the early 19th century (p. 208). This island is large enough for both Telfair's and Orange-tailed Skinks to be introduced to different parts of the island. By the time the Telfair's Skink population had grown enough to seriously impact on the Orange-tailed Skinks, the latter should be well-established and able to tolerate predation by the larger species. Telfair's Skinks on Gunner's Quoin might also feed on the rare Lesser Night-gecko, but this should be minimised by their different temporal and habitat preferences, and besides they lived together there in the past. Once these species are well-established the island would be suitable for an introduction of the lizard-eating Keel-scaled Boa. Once we know more about how Telfair's and Orange-tailed Skinks interact we should consider introducing Telfair's to Flat Island and neighbouring Ilot Gabriel.

Ile aux Aigrettes may prove an appropriate location for several reptiles now that both rats and cats have been eradicated, and indeed may be closer to the original optimum habitat for Telfair's Skinks than Round Island[193]. Ile aux Aigrettes has suitable food sources for Telfair's Skinks, such as geckos, including common exotic species, cockroaches, snails and other invertebrates and the fruits of the native plants[194], for which it may have been and may become again an important seed-disperser. Thus, combined with its role as a potential pollinator, we are pushing forward not only the reintroduction of the skink itself, but also restoring lost interactions between the skink and many plant species.

Although Ile aux Aigrettes was cleared of rats and cats by 1991, the presence of House Shrews and Wolf Snakes has been considered a disadvantage for the introduction of Round Island reptiles[195]. Attempts to eradicate the shrews and Wolf Snakes have failed[196], and it is unlikely that they can be eradicated by conventional techniques. It was, however, always recognised that a trial introduction should be attempted, even if it proved impossible to get rid of these exotics[197]. While susceptible to rats, Telfair's Skinks may prove able to survive alongside the shrews and snakes, indeed it is possible (even likely) that they may be able to reduce or eradicate these exotics by predation on their young (both) and competitive exclusion (shrews). Competitive exclusion may operate during the dry winter months since shrews and skinks feed on similar foods and the skinks are better adapted to survive periods of drought and seasonal food-shortage. Telfair's Skinks may prove effective predators of young shrews, and are likely to kill Wolf Snakes, since on Round Island they eat young boas and are capable of killing snakes up to the size of small adult Wolf Snakes. On Round Island young Telfair's Skinks are preyed upon by adult skinks and boas, but show a range of predator avoidance behaviours and have different micro-habitat selection to adults; these behaviours are likely to confer the juveniles some protection against the shrews and Wolf Snakes. These interactions could easily be tested in captivity first. The Wolf Snake population on Ile aux Aigrettes is at a very low density despite abundant gecko prey, and current densities of shrews and snakes there are unlikely to have a big impact upon a breeding population of Telfair's Skinks or other species (such as Günther's Geckos) which were susceptible to rats in the past. Telfair's were experimentally released on Ile aux Aigrettes and Gunner's Quoin in late 2006 and early 2007[198].

A reintroduced population of the endangered Günther's Gecko is a high priority since it is limited to Round Island with a population of about 750–1,500 adult animals, and a total population in the region of 3,400–4,500[199]. Of the islands that have been cleared, or partly cleared, of exotic mammals, only Ile aux Aigrettes is suitable for an immediate trial introduction. Using artificially incubated, harvested eggs and head-started young would reduce the impact upon the wild population. Young animals are likely to adapt more easily to the different forest habitat on Ile aux Aigrettes than would wild-caught adults. None of the other islands has enough suitable habitat, though Ilot Gabriel has clusters of latans and screw-pines and could support a small population. Any reintroduction would need to be followed by some habitat restoration, putting back suitable coastal vegetation, including the palms and screw-pines that the species favours on Round Island. Once more is known about the habitat tolerances of Günther's Geckos, Gunner's Quoin and Flat Island should also be considered as introduction sites. Following the introduction of Günther's Geckos to Ile aux Aigrettes, the island should be considered as an introduction site for the Keel-scaled Boa.

The smaller islands in Mahébourg bay are suitable for reptile introductions since most are now free of exotic mammals, and some of the smaller ones are also free of house geckos. The islets are not large enough to support many species but are suitable for Bojer's Skinks, and this skink has been reintroduced to Ilot Fouquets now that the shrews there have died out (p. 238)[200]. There is an urgent need to find more sites for night-geckos, and some of these islands, such as Ile Marianne and Ilot Chat, may be suitable, with the qualification that Ile Marianne has the aggressive and abundant invasive ant *Pheidole megacephala*, which may be a potential predator[201]. The size difference between *N. (s.) durrelli* and *N. coindemirensis* suggests that they should be able to live in sympatry, as they presumably did on the mainland in the past, and this could be tested on Ilot Chat[202]. Enclosure experiments using both species revealed no negative interactions[203].

Bird management

Bird management has been an important component of conservation work on Mauritius and Rodrigues. The approach to bird conservation started with broad-based studies to understand the species and their problems. For rare or failing populations the aim was to improve survival and productivity. For some species, protection and habitat restoration are adequate to increase populations, as with the passerines on Rodrigues (see below). For the most endangered species, such as the Mauritius Kestrel and Echo Parakeet, the approach has been to evaluate the factors impacting upon the population and then to control or mitigate them by management. For example, predators have been controlled where appropriate, pairs have been provided with nest-boxes or nest-sites have been improved, parasitic diseases have been controlled and the birds have received supplementary food. This has been done together with the enhancement of productivity by the manipulation of breeding biology, by breeding birds in captivity and subsequently releasing captive-bred young[204].

BOX 34   SKINKS (SCINCIDAE)

Probably one of the world's largest skinks, *Leiolopisma (Didosaurus) mauritiana* of Mauritius would surely have been noticed by early travellers but it was never mentioned; this suggests that the species may have already been extinct by the time the Dutch arrived in 1598. There was also a smaller but predominantly carnivorous species, Telfair's Skink *L. telfairii*; and two smaller leaf-litter species, Bojer's Skink *Gongylomorphus bojeri* and the Macabé Skink *G. fontenayi;* only the last still survives on the mainland, along with another small species, Bouton's Skink *Cryptoblepharus boutonii*, which is almost restricted to the islets. None of the travellers or early residents described any skinks in detail, but there are a few early, if brief, accounts of *L. telfairii* on Mauritian offshore islets.

On Réunion two skinks, the Réunion Slit-eared Skink *Gongylomorphus borbonicus*, related to Bojer's, and Arnold's Skink *Leiolopisma* sp. (as yet unnamed) close to Telfair's, are now extinct; the last-named is known from the fossil record only. *G. borbonicus* was last collected in 1839 and its extinction can probably be attributed to an introduced snake. Bouton's Skink *Cryptoblepharus boutonii* was recently rediscovered on Réunion. No skink has yet been discovered on Rodrigues.

Account
LaCaille (1763), in Mauritius in 1753, on Telfair's Skink:

On the islet called Coin de Mire [Gunner's Quoin] I saw lizards a foot long and a good inch thick [Telfair's Skink], whereas on the Isle de France [Mauritius mainland] I only saw very small ones [referring to other skinks] running on walls and stones, just as one sees in France.

Specimen collections
Fossil material of *Leiolopisma mauritiana* can be seen at the MNHN, MI and BMNH; specimens and fossil material of *Gongylomorphus borbonicus* and the Réunion *Leiolopisma* is housed at the MNHN.

Bouton's Skinks *Cryptoblepharus boutonii* teasing a centipede (on Aldabra). From Gilham, 2000.

Seabirds

Mauritius currently has 10 known species of breeding seabirds, all but one of which are limited to satellite islands; Rodrigues has seven, of which four are limited to offshore islets. The conundrum of the petrels on Round Island is discussed in Chapters 4 and 9[205]. Seabird conservation on Mauritius and Rodrigues has until recently been a neglected area, but recent studies have demonstrated the international importance of some of these populations. They are also important to the plant and reptile communities on several islands due to the nutrients that they bring to the islands in the form of droppings, regurgitated food, unhatched eggs, dead young and adults.

Poaching of seabirds in Rodrigues and Mauritius has been a persistent problem since the 1870s but has decreased with better protection of breeding islands, more awareness of conservation and an increase in the standard of living. In the 1970s and early 1980s, killing of seabirds, mainly tropicbirds and shearwaters, on Gunner's Quoin and Round Island was common, with many raids on the colonies every year and an annual kill of probably more than a thousand birds. The poachers focused mainly on Red-tailed Tropicbirds or Wedge-tailed Shearwaters, although White-tailed Tropicbirds and 'Round Island' (Trindade) Petrels were taken if they came across them. These raids seriously depressed the population of Red-tailed Tropicbirds, but probably had less impact upon the much commoner Wedge–tailed Shearwaters and the more scattered and difficult to find White-tailed Tropicbirds. On Rodrigues, raids on the tern colonies of Ile Cocos and Ile aux Sables and the accessible tropicbird nests on the mainland were regular in the 1970s and early 1980s. This prevented Sooty Terns from re-colonising, and depressed the populations of tropicbirds and Wedge-tailed shearwaters[206]. Poaching has now stopped on Round Island[207], and but continues on Gunner's Quoin, where several hundred birds are apparently still harvested annually. This may be preventing Red-tailed Tropicbirds from expanding into areas on Gunner's Quoin that are accessible to humans, although is unlikely to be having much impact upon other species. Reduced poaching has led to a spectacular increase in Red-tailed Tropicbirds on Round Island (p. 222) and possibly also Wedge-tailed Shearwaters. Numbers of Wedge-tailed Shearwaters on Round Island appear to have grown as potential nest sites have increased through accumulation of soil and vegetation cover. The 'Round Island' Petrel population is apparently increasing, facilitated by the construction of artificial rock shelters as nest sites, and the enhancement of some of existing sites with rock baffles for shade and protection against inclement weather. Constructing and enhancing nest sites is likely to become more important as the rapidly expanding Red-tailed Tropicbird population increasingly competes with petrels for rock shelters to nest in[208].

On Rodrigues, poaching has decreased substantially with the wardening of the seabird islands of Iles Cocos and Sables with consequent increases in tern populations. Ile aux Sables has been made into a strict reserve with restricted access, and Ile Cocos, an important tourist island, has been zoned, with tourists restricted from visiting the main breeding area.

The main conservation strategies are protecting these seabird colonies by removing predatory exotic mammals. The reintroduction of seabirds onto islands from which they have been extirpated but are now suitable to support them is a possibility. Hence Wedge-tailed Shearwaters and White-tailed Tropicbirds may be reintroduced to Ile aux Aigrettes where their subfossil remains have been found. Other suggested introductions are noddies, Fairy Terns and 'Round Island' Petrels to Ile de la Passe and the other islets on the reef in the Mahebourg Bay[209]. In Rodrigues the failing population of Wedge-tailed Shearwaters may be enhanced by introductions to the Ile Cocos and Ile aux Sables reserves[210].

More ambitious, but more important, reintroductions that should be considered are Abbott's and Red-footed Boobies and Lesser and Great Frigatebirds to the Mascarenes, all lost due to high rates of human predation. These are all species of international significance, especially Abbott's Booby, now down to a single colony; the other three species have all declined dramatically in the Indian Ocean over the last century[211]. Abbott's Booby's last stronghold on Christmas Island, formerly threatened by phosphate mining, is now protected, but the island's ecology was subsequently gravely threatened by introduced Crazy Ants, and in any case it is always safer to have several populations than just one. Re-establishing frigatebirds in Mauritius or Rodrigues would be desirable, but at present only Mauritius, with large tern and tropicbird populations on Round and Serpent Islands, could support their kleptoparasitic lifestyle. In Rodrigues re-establishing Red-footed Boobies (another declining species) would be a pre-requisite to reintroducing frigates. With more interest in seabird conservation and greater protection of the seabird islands, the time is now right to consider bringing these species back. Suitable islands would include the large northern islands in Mauritius and Cocos, Sable or Crabe Islands, Rodrigues[212].

Larger land birds

Our most intensive bird studies have been on the Mauritius Kestrel, Pink Pigeon and Echo Parakeet. All three declined to very small wild populations: the kestrel was down to four wild individuals in 1974, there were only 9–10 Pink Pigeons in 1980 and 8–12 parakeets in 1987[213]. Work started with the Mauritius Kestrel; several conservation management techniques were used in an attempt to increase the productivity of the wild birds[214]. Some pairs were provided with additional food. Nest sites were enhanced to improve their suitability for the kestrels, and in areas where nest sites were few, nest boxes were provided. Predators were controlled around nest and release sites; cats and mongooses were trapped and rats poisoned. Double-clutching was regularly used on established pairs to increase productivity. First clutches were removed soon after laying, hatched in captivity, and the young were then used for captive breeding and reintroduction. Kestrels readily lay second clutches, so these eggs were left for them to incubate and rear any young that hatched. Young kestrels were introduced to the wild by fostering to nesting kestrels or by hacking, a soft-release technique. The reintroduction of captive-bred and captive-reared birds ceased after the 1992–93 season, when 331 birds had been released over a 10-year period. About a third of these were captive-bred and the remainder derived from wild eggs. Since the 1992–93 breeding season 12 rescued and rehabilitated birds have also been released[215]. Today there is little active conservation management applied to this species, although nest boxes are provided; these are used by about 35 pairs. The population in mid-2005 was 800–1,000 individuals[216].

The conservation management of the Pink Pigeon involves predator control, habitat restoration, supplementary feeding of released and wild birds, disease control and the release of captive-bred birds[217]. Ship Rats have been controlled in the main Pink Pigeon nesting areas by poisoning. Cats and mongooses are caught in baited cage traps that are set on a grid around the supplementary feeding stations and nesting areas. There is evidence to suggest that food shortages limit the productivity of wild Pink Pigeons, so feeding stations offering maize and wheat are accessible to all the free-living birds. Both released and wild birds visit the feeding stations, some only infrequently. Since July 1987, 256 birds have been released at four different sites; 248 of these were captive bred. No birds have ever been released into the wild population in Pigeon Wood. The conservation management of this species is still very intensive, and we will have to continue predator control and supplementary feeding together with monitoring of productivity and survival[218]. Diseases have become an

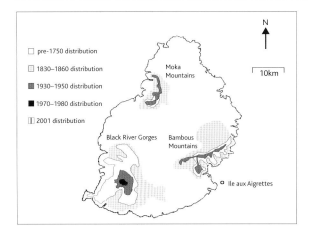

Figure 10.2. Map showing the Mauritius Kestrel's spectacular recovery from near-extinction. From Nicoll et al. (2004).

increasing problem as the population has grown, and these are being studied to find means of controlling them[219]. Feeding ecology and supplementary feeding have been studied with a view to minimising the amount of management while maintaining the current population size[220]. The known population in 2007 was about 380 birds.

Conservation management of the Echo Parakeet has been intensive since 1987[221]. Many different management techniques have been tried. These have included nest site enhancement, provision of nest boxes, clutch and brood manipulations, rescue of eggs and young from failing nests, control of nest parasites, rat control around the nest site, supplementary feeding and habitat restoration. We have experimented with removing the first clutch to encourage the parakeets to lay a second clutch, but their readiness to relay is variable. Fourteen first clutches were taken, but only nine relaid; the clutches removed were artificially incubated and the chicks hand-reared. The success of second clutches is low, and we no longer use this technique to increase productivity. A more successful strategy has been downsizing of broods. Wild parakeets usually lay clutches of 2–4 eggs but generally only rear one, sometimes two, young, as food supply is insufficient for raising whole broods. To avoid this natural wastage of young, wild parakeets are allowed to hatch their eggs and rear the young for the first 5–8 days. In broods of two or more the brood is reduced to one or sometimes two, and the harvested young are either fostered to pairs that have failed to hatch their own chicks or brought into captivity for hand rearing. Since 1997 we have released 139 captive reared birds (84 from harvested eggs or young and 55 captive bred) – these represent 49% of the total recruitment, 143 birds having fledged from free-living birds. The released birds have integrated well into the free-living population and many are breeding. There were 42 breeding (egg-laying) females in 2004–2005, and by 2007 the total population was around 343 birds[222].

Passerine conservation on Mauritius and Rodrigues

There are seven species of native Mauritian passerines; four are endemic species, and two more are Mascarene endemics with Mauritian subspecies. Rodrigues has two native passerines, both endemic. All four Mauritius endemics are threatened; the Mauritius Cuckoo-shrike (500–750 birds) and the Mauritius Merle (500–700) are classed as vulnerable, the Mauritius Olive White-eye (150–300) and the Mauritius Fody (300–400) are critically endangered[223]. The Mauritian race of the Mascarene Paradise Flycatcher (350–500) is endangered[224] and slowly declining. The only native passerine that is common is the Mascarene Grey White-eye with a population of several tens of thousands, though the Mascarene Swallow is stable at an estimated 700–1,500 birds. There have been several studies on the populations, distributions and ecologies of most of these species[225].

The Mauritius Merle and Mauritius Cuckoo-shrike have responded to the restoration of native forest within CMAs, although these areas are far too small to support populations viable in the long term. Merle numbers are believed to be relatively stable, and the cuckoo-shrike population is increasing, apparently recovering from the impacts of organochlorine pesticides that are believed to have caused a decline in the population between the 1950s and 1970s[226]. The cuckoo-shrike died out in the Bamboo Mountains in the east of Mauritius some time before or during the 1950s[227], its disappearance from there mirroring the fate of the local Mauritius Kestrels. There is still suitable habitat for cuckoo-shrikes in this region, and reintroducing birds into this area is a priority. Similarly, translocation of the very patchily distributed paradise flycatcher into areas of suitable habitat where they are now absent is a possible conservation strategy. Paradise flycatchers, which have small territories, should do well in the larger, more established CMAs such as Mare Longue (where they were found in the 1970's), Brise Fer and on Ile aux Aigrettes[228].

The two species that are causing most concern are the Mauritius Fody and Olive White-eye. The fody population is declining through very poor productivity due to heavy nest predation by rats and monkeys. The wild population of Mauritius Fodies was a relict confined to sub-optimal habitat, which misled past researchers into believing it was a specialised insectivorous and nectar feeding species. However, knowing that closely related species from Rodrigues and the Seychelles show considerable ecological and behavioural flexibility, biologists familiar with fodies felt that its adaptability should be tested by releasing some birds on predator-free Ile aux Aigrettes. The translocation of birds onto islands has also begun for the Mauritius Olive White-eye, and is being considered for the cuckoo-shrike, merle and paradise flycatcher[229]. Some of the translocated birds may need management such as supplementary feeding, especially in the short term, although the paradise flycatchers should fare well on Ile aux Aigrettes without any additional feeding. For most of the species young captive-reared birds would be most appropriate for translocation, since these would be more adaptable and amenable to management, and would be more sedentary than adult birds. Wild Mauritius Olive White-eyes, Merles and some Mauritius Fodies are highly mobile during the winter months, making them unsuitable for translocation, since they would be more likely to fly away.

Passerines have been studied in captivity to learn about their biology and to develop incubation and hand rearing techniques. Cardinal Fodies have been used as a model to derive techniques for the critically endangered Mauritius Fody. Work started with the fody in the 2002–03 season, developing hand-rearing techniques[230], and these hand-reared birds were kept for captive breeding studies. In subsequent seasons we harvested egg and young from failing or vulnerable wild nests. In three seasons we have hand-reared 57 fodies, 40 of which have been released on Ile aux Aigrettes[231]. The captive birds started breeding in 2004, and in the 2004–05 season 29 young were reared from three breeding pairs, and the free-living birds have bred well on the islet, reaching a total of 132 by early 2007.

In relation to the conservation of the critically endangered Mauritius Olive White-eye, work is progressing to develop management techniques on the common Mascarene Grey White-eye, a species we have kept in captivity for several years, working on management protocols and suitable diets. In 2004 a Grey White-eye was bred for the first time at the facility in Black River, and a harvested brood of two newly hatched young was successfully hand-reared. The first Mauritius Olive White-eyes, from wild clutches, were hand-reared in 2006 and the first birds released on Ile aux Aigrettes later that year[232].

The Mauritius Fody may well achieve very high densities on seabird islands, as the Seychelles Fody does on islands in the Seychelles[233]. The satellite islands of Flat Island, Gunner's Quoin and Round Island are likely to be suitable for fodies. As the forest recovers, Round Island may prove an important and interesting release site, since it has no Cardinal Fodies or Red-whiskered Bulbuls, and consequently the Mauritius Fodies are likely to behave as they would have done on pristine Mauritius before the introduction of these exotic birds. As forest is restored on Round Island and other islets, suitable habitat will develop to allow us to consider releasing other species such as Mauritius Merles, Cuckoo-shrikes, Paradise Flycatchers and possibly Olive White-eyes.

Cave conservation

The caves on Mauritius and Rodrigues have long been recognised as sites of scientific interest and are a high conservation priority. Mauritian lava tunnels have populations of invertebrates, microbats and swiftlets and are important palaeontological and archaeological sites. Greg Middleton reviewed early accounts of Mauritian caves, and visited 140 and mapped 133 caves between 1992 and 1998, producing a comprehensive report on their conservation and management[234]. Matthew Flinders noted that caves at Trois Cavernes were a refuge for escaped slaves (maroons) in the early 1800s, and Middleton found signs of their former habitation still visible in 1994[235]. In the lower Black River Gorges several small caves show signs of occupation by maroons, including middens with charred tortoise bones, fish bones and *Sideroxylon sessiliflorum* seeds (a tree species that no longer occurs there)[236].

Mauritian caves contain some important palaeontological remains. Several caves, including Trois Cavernes, Plaine des Roches and some overhangs in the Black River Gorges, have remains of Günther's Gecko eggs attached to the walls, mainly around the cave entrances[237]. This large gecko was extirpated on the mainland of Mauritius long ago, probably by rats, and was never seen there alive. Small numbers of tortoise and Dodo bones have recently been found in caves, and subfossil extinct snail shells exist in a few caves. The Rodriguan calcarenite caves are far better known, with spectacular formations and important subfossil deposits[238]. Although these caves were dug over in the 19th century for their subfossils, some untouched deposits remain.

There are important but as yet poorly known invertebrate faunas present in the caves, including endemic species. The first endemic silverfish for Mauritius *Lepidospora mascareniensis* was described in 1996 from Trois Cavernes[239]; the same cave yielded a new species of amphipod *Brevitalitrus strinatii*, the first record of this genus from the Indian Ocean region[240]. Two spiders collected from a Mauritian cave belong to the Telemidae, a family of long-legged cave spiders not previously known from Mauritius[241].

Many Mauritian caves shelter populations of the endemic Mascarene Free-tailed Bat[242]. This bat is widely distributed in caves throughout the island and also roosts in rock fissures on cliff faces in the Black River Gorges[243]. The largest known roost of this species is in Twilight Cavern on Plaine des Roches in northern Mauritius, which in May 2005 held clusters of bats covering an area of 38–57m^2 (400–600 sq. ft) with a total population of between 57,000 and 173,000. On the same day Pont Bondieu cave held c.15,000–30,000 bats and Palma cave c.7,000–14,000[244]. With unknown roosts and those roosting in undetected fissures, the total population must be at least 250,000. Nonetheless, all cave colonies are highly vulnerable to their caves being blocked by rubbish or other forms of human interference.

Middleton's survey recorded Mascarene Swiftlets in 34 caves[245]. In 19 caves, nests had been removed, presumably to make birds'-nest soup, leaving distinct scars on the roof where the nests had been. This species, whose Mauritian population is only some 750 pairs, is considered Near Threatened by Birdlife International[246]. A suggestion in the 1970s that the declining swiftlet population could be helped by placing grills over cave entrances to exclude vandals and poachers was tried in 1992 to the caves at Palma and Petite Rivière/Chebel; but vandals broke them open and they are not now functional[247].

Little has been done to look after the caves of Mauritius and Rodrigues. Vandalism of Mauritian caves was already rife in the 1970s; many have been blocked or part-filled with refuse and even dead livestock; burning old tyres at the entrance of caves is a common practice, and some are used for clandestine rituals[248]. In Rodrigues several caves have been badly vandalised and stalactites and stalagmites have been broken. The entrances of many of the larger ones have been used to provide shelter for livestock with low stone walls built to corral the animals. Cave conservation must now be a very high priority, and raising the profile of these caves by education and with active policies to encourage locals to look after them appears to be a more promising approach. Recently there has been a recent greater interest in the caves on Rodrigues and plans to restore and open a show cave are encouraging.

Invertebrate conservation

Invertebrate conservation on Mauritius and Rodrigues is a neglected area, but it has been addressed indirectly; the assumption has been that by restoring areas of native vegetation on the main and offshore islands the invertebrate populations will respond positively. Although "efforts to save plant communities also conserve the native snails", there is a need to control predators – introduced toads and predatory exotic snails[249], in addition to rats, shrews and tenrecs. We have seen in the examples of leaf litter invertebrates, butterflies and snails discussed above that although this assumption is broadly correct, vegetation restoration can be adapted to better suit the conservation of invertebrate communities that still exist[250].

Islands without mammalian ground predators usually have a wealth of large arthropods and snails, but there is a paucity of large terrestrial invertebrates on Mauritius and Rodrigues. There have been high rates of extinctions for land molluscs[251], and this probably holds in other groups. The records of the invertebrate fauna of pristine Mauritius and Rodrigues are very poor[252], and there are no subfossils except for snails (see below). However, relict populations of large invertebrates have survived on satellite islands free from predatory mammals. The large tenebrionid beetle *Pulsopipes herculeanus*, now confined to Frégate Island in the Seychelles, occurred on Round Island until at least 1869[253]; the endemic centipede *Scolopendra abnormis* is still found on Round Island and Serpent Island in large numbers[254]. A large lizard-eating tarantula *Mascaraneus remotus* survives on Serpent Island[255]. None of these species have been recorded from the Mauritian mainland, but they must originally have occurred there, since the satellite islands were connected to the mainland until only around 10,000 years ago. However, large spiders are mentioned in 18th and 19th century Mascarene literature, although none of these are unequivocally the same as the Serpent Island Tarantula[256]. These records reinforce the view that there was once a rich community of terrestrial invertebrates on these islands. In 1763 Bernardin de St Pierre described seeing centipedes over six inches (150mm) long which cannot be ascribed to any species found today on Mauritius, and may have been *Scolopendra subspinipes* which still occurs in Rodrigues[257].

Any of these large terrestrial invertebrates that survived the onslaught of the 16th century rat invasions would probably have been wiped out from the Mauritian mainland by tenrecs and shrews in the late 18th century. The disappearance of the tenebrionid beetle from Round Island was probably due to the loss of the hardwood forest and the dead wood on which the beetle larvae feed. On Rodrigues the formerly abundant centipede *S. subspinipes* has been virtually wiped out on the main island by shrews introduced in 1997, but it is still found on some of the offshore islets. This shows the huge impact some exotic mammals may have on invertebrate populations, and the value of maintaining small offshore islands as refugia for native species.

The land-snails of Mauritius and Rodrigues show several important radiations and a high degree of endemism. Of the 114 native species recorded from Mauritius, 73 are endemic and 29 already are extinct. Of 26 native species on Rodrigues, 15 are endemic (with seven extinct)[258]. Many of the extinct species are known only from subfossils, suggesting that they may have been wiped out by rats, shrews and tenrecs before the 19th century, when most species were described.

BOX 35   SNAKES (OPHIDIA: BOLYERIDAE)

Quite extraordinarily, an endemic family of snakes (Bolyeridae) comprising two taxa, the Keel-scaled Boa *Casarea dussumieri*, a semi-arboreal species, and a leaf-litter burrowing species, the Burrowing Boa *Bolyeria multocarinata*, is unique to Mauritius; no snakes are known to have inhabited Réunion or Rodrigues. From fossil material found in the Mare aux Songes, *Casarea dussumieri* has been identified along with another extinct endemic species, Carié's Blind-snake, *Typhlops cariei*.

Keel-scaled Boa *Casarea dussumieri*. Drawing by Brin Edwards in Bullock, *et al.* 1989.

No accounts mention the presence of snakes on the Mauritian mainland and the islands were considered 'blessed' due to their absence. However, as with the larger lizard species, snakes certainly once occurred on mainland Mauritius and were probably early victims of rats. Both snake species were found on a number of northern islets until the early 1800s but disappeared as each islet became rat-infested. Both species survived on Round Island until comparatively recently but, due to the degradation of vegetation by rabbits and goats, *Bolyeria multocarinata* was last seen and photographed in 1975 and is now almost certainly extinct. Only one early account described a snake, and it is also the only one to report snakes away from the northern islets.

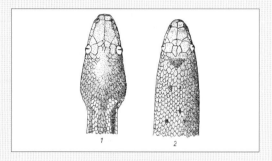

Account
Charpentier de Cossigny (1764), translated in Cheke 1987a:

I should say also that there are snakes, quite long and big, marked with brown and green, on the islets marked on my map. I have seen several shot on the Gunner's Quoin. There were also some on the islet by the Grand Port pass [= Ile de la Passe]. *The workmen working at the battery killed several.*

1. Keel-scaled Boa *Casarea dussumieri*. 2. Burrowing Boa *Bolyeria multocarinata*. Head drawings by Jean Vinson, 1950.

Putting native invertebrates on islands that have been cleared of exotic mammals offers possibilities for conservation and offers the opportunity of at least partly recreating the native terrestrial invertebrate communities. It is likely that these invertebrates would do well on rodent- and shrew-free islands, provided their micro-habitat requirements are met and the prey base for the predatory species exists. The giant tenebrionid beetle could be introduced to Ile aux Aigrettes where there is forest cover but it may have to be provided with dead wood for the larvae to develop in. Ile aux Aigrettes has also been identified as a potential site for re-introducing snails, and it has the advantage that it is relatively well forested, with many different micro-habitats. Subfossil records from the island show that it had a very rich snail fauna, with 31 terrestrial species and 20 species of freshwater, salt marsh and mangrove snails. Only six native species of snail now survive on the island[259]. Two snail re-introduction attempts were made in 1989, with 20 captive bred *Pachystyla bicolor* and 15 juvenile *Tropidophora eugeniae* released in a good area of ebony forest in the middle of the island. They did not survive; they were eaten by the abundant hermit crabs that live in the shells of introduced *Achatina* giant snails[260]; shrews may also have impacted upon them. These re-introductions provided important information, but with the benefit of hindsight they were premature, and future releases should be considered only after the eradication of shrews and *Achatina* snails. Other islands that are now free of rodents, shrews and giant snails should be considered for the re-introduction of invertebrates.

Introducing lost bird and bat species

A number of species that have disappeared from Mauritius and Rodrigues are still found elsewhere. Consideration needs to be given to their reintroduction or to the establishment of managed populations. For Mauritius these include the seabirds mentioned earlier, along with Dimorphic Egret, Reed Cormorant, Réunion Harrier and Golden Bat[261]. Reintroducing Dimorphic Egrets on the eponymous Ile aux Aigrettes is a real possibility. It is likely that all these species could be maintained with little or no management other than perhaps some supplementary feeding. Some species may not be able to sustain wild populations but it may be possible to establish captive or free-ranging populations. These would be useful for educational purposes while helping to drive the restoration of suitable habitats. The coastal mangrove areas surrounding parts of the Mahébourg Bay would support Dimorphic Egrets based on Ile aux Aigrettes; various water bodies could support Reed

Cormorants and wildfowl. Similarly, Malagasy Turtle Doves could be restored to Rodrigues, and several species surviving in Mauritius could be transferred to Réunion where they are extinct. It is important to continually re-examine the list of missing species, with a view to replacing them from elsewhere or with analogues.

The ecological function of tortoises

The Mascarene coastal vegetation evolved with high tortoise densities, and there are probably many plants adapted to tortoise pressure, such as adaptations to intense grazing. The Mascarene endemic tussock-grass *Vetiveria arguta* was the commonest grass on Round Island until the introduced rabbits, which grazed the island heavily, were eradicated in 1986. Subsequently *Vetiveria* declined and was replaced by other grasses[262]. *Vetiveria* probably survives grazing pressure by having coarse, unpalatable leaves that are avoided by herbivores when other grasses are available. Similarly, captive Aldabra Giant Tortoises on Mauritius feed preferentially on softer exotic grasses, ignoring the native grass *Zoysia tenuifolia*. Other species may have adapted to tortoise grazing and browsing pressure: *Aerva congesta*, now found only on open barren areas on Round Island[263], has a small prostrate form that would protect it from grazing. There are several other prostrate native plants that are likely candidates for a tortoise-generated grazing community[264]. Another possible adaptation to deter tortoise grazing is the sharp serrated leaf edges of screw-pines *Pandanus* spp.; the young plants of this genus are also resistant to trampling (as are the young plants of *Dracaena concinna*, *Latania loddigesii* and *Hyophorbe*)[265]. Furthermore, if *Dracaena concinna* is knocked over or damaged in a cyclone, it will readily re-sprout from damaged and flattened stems[266]. Other plants may have had their seeds spread by tortoises, with seed-germination aided by passage through the gut. Between September and November the fallen ripe fruits of the Ile aux Aigrettes Ebony are an important food of the introduced Aldabra Giant Tortoises on Ile aux Aigrettes. They spread the seeds in their droppings and many subsequently germinate[267].

There are other possible examples of coevolutionary plant/tortoise interactions. Many endemic plants show leaves with marked heterophylly, i.e. differing juvenile and adult leaf-shapes. Juvenile leaves, characteristically lanceolate or elongated with marked red venation, typically grow on both young and old plants within about 1.2m of the ground. Feeding experiments offering juvenile leaves to Aldabra Giant Tortoises demonstrated that they would not eat these, but they would readily browse adult leaves of the same species, lending credence to the idea that heterophylly is an adaptation to deter browsing by tortoises[268]. Field observations on the tortoises grazing and browsing on the island have confirmed these findings[269]. On both Rodrigues and Mauritius, which both had an endemic browsing saddle-backed tortoise, heterophylly is common, and reconstructed tortoise models suggest that the largest animals would have been able to browse leaves to a height of about 1.2m.

Since the endemic Mascarene tortoises are extinct, we can never be certain what interactions they had with the native flora. It may, however, be possible to recreate some of these lost interactions by introducing an ecological analogue, and Aldabra or Malagasy tortoises may be suitable substitutes to release on Round Island, where native tortoises survived until 1844 (p. 211)[270]. However, the use of analogues needs to proceed with great care to ensure there are no negative impacts upon native species. Released tortoises are simple to monitor and can, if necessary, be removed, as they are easily found and recaptured. Tortoises could be used on Mauritian and Rodriguan islands to maintain grazed areas, and to provide some browsing pressure, a role that in many restoration projects is fulfilled by domestic livestock – indeed, the experimental release on Ile aux Aigrettes, where they have bred, is already proving very effective[271]. Analogue tortoises would also allow a more accurate evaluation of the ecological role endemic tortoises once had on pristine Mauritius.

Introducing Günther's Gecko and Vinson's Day-gecko to Rodrigues was suggested a decade ago[272] and reconsidered more recently[273]. Rodrigues has lost all of its endemic reptiles and the only surviving native is the small pantropical Mourning Gecko *Lepidodactylus lugubris*[274]. Introducing the large nocturnal Günther's Gecko to one or more of the rat-free islets around Rodrigues would help to establish additional populations of this highly endangered lizard where there is no risk of ecological damage. The large extinct nocturnal endemic Liénard's Giant Gecko *Phelsuma gigas* was closely related to *P. guentheri*, and judging from its morphology may have been ecologically similar[275]. Hence Günther's is likely to be a close ecological analogue for the extinct species; its introduction to the islets would need to be accompanied by habitat enhancement with the planting of palms, screw-pines and various coastal tree species. Similarly, Rodrigues also once had the brightly coloured and partly frugivorous Newton's Day-gecko[276], which became extinct about 1917. Ecologically and taxonomically, Vinson's Day-gecko, a common species from Mauritius, may prove to be an appropriate analogue. Although introducing these two geckos to Rodrigues would be outside their natural range, close relatives were formerly present; all Mascarene day-geckos belong to a unique clade within the genus *Phelsuma*[277]. Through introduction it may prove possible to resurrect species interactions that have been lost. Day-geckos are important pollinators of some endemic Mascarene plants[278] and introducing analogue species to Rodrigues could reactivate lost, and currently unknown, plant–gecko interactions. While these geckos could be introduced to the nature reserves of Iles Cocos and Sables where they could be easily monitored, Ile Frégate would be ecologically superior. It has more endemic trees, rocky outcrops and, significantly, was where Liénard's Giant Gecko was last seen: but there are practical difficulties[279].

There are possibilities for the use of bird analogues to replace extinct species, but many were too unusual and their habits too little known for it to be realistic to believe that accurate ecological replacements could at present be chosen with any confidence[280]. Hence the Dodo and Rodrigues Solitaire must be regarded as quite irreplaceable, and it is stretching the imagination to think that a large cockatoo (a *Calyptorhynchus*?) or a macaw could replace the Raven Parrot *Lophopsittacus mauritianus*. Other parrots should, however, be considered as replacements for those that have been lost. The Echo Parakeet *Psittacula eques* from Mauritius might be used to replace Rodrigues Parakeet *P. exsul* in Rodrigues, and the Alexandrine Parakeet *P. eupatria* could be used as an analogue for the *eupatria*-type parakeets that were found in the Seychelles and the Mascarenes[281]. Both flightless or flying versions of the White-throated Rail could be brought in to replace their extinct relatives on predator-free islets – Ile aux Aigrettes's sharp limestone substrate would be a home from home for the flightless race from Aldabra[282]. There are three species of blue pigeon available to replace the Pigeon Hollandais in Mauritian forests, and restore its function of dispersing fruit of native trees[283]. An accidental analogue, the Egyptian Geese now feral in Mauritius, was mentioned earlier; ducks closely allied to the extinct teal are available, including the endangered Bernier's Teal, much in need of a safe reserve site; Red-knobbed Coots could replace the lost Mascarene Coot[284]. Likewise a new attempt should be made to introduce the Mauritius Fody to Réunion to replace the extinct but similar Reunion Fody, and an extant merle could replace the extinct Rodrigues *Hypsipetes* bulbul[285].

Since it is difficult to predict how analogues are going to interact with the native biota, and whether they are an adequate ecological fit, analogue introductions need to be very carefully monitored. It is probably most appropriate that any analogues are maintained initially in managed populations while their impact is being evaluated[286].

Research and training

A training and research programme has developed to support conservation work on Mauritius and Rodrigues. Volunteer post-graduate biologists have become important in providing a series of workers to carry out much of the fieldwork[287]. The volunteer programme has been developed into vocational training and most volunteers stay for six months, during which they participate in a range of conservation activities. These may include the management and monitoring of the released bird populations, collecting data on survival rates and productivity, managing supplementary feeding stations and controlling exotic predators. Volunteers work on small-island management, including the propagation of native plants, planting and managing vegetation communities. During the last two decades we have taken more than 300 volunteers from 24 different countries[288].

It became clear that to develop a research programme to support the conservation work, we needed the involvement of some academics to answer questions of conservation importance[289]. Staff from the MWF have in the last 15 years supervised or facilitated work on 29 M.Sc. and 22 Ph.D. studies, in addition to supervising or helping more than 70 students doing diploma and B.Sc. honours projects, 40 of which were Mauritian students. All of these studies have supported the conservation work.

Discussion

Conservation management work on Mauritius started in the mid 1970s and initially was mainly species-orientated and focused on the Mauritius Kestrel, Pink Pigeon and Echo Parakeet. A range of techniques have been tried to enhance productivity and survival together with captive-breeding and the release of captive-reared young. Once initial difficulties had been overcome, most of these techniques have been successful, and their application has given good insights into the management requirements of the species. For all three species the released birds were managed carefully to enhance their survival and breeding at liberty.

The intensive management of any endangered species is demanding and requires careful organisation, although conceptually straightforward. The causes of population decline, viewed at their simplest, are reduced breeding success, reduced post-fledging and adult survival, or all of these. In declining or very small populations productivity and/or survival may often be enhanced by management without fully understanding the causes of population failure. It can be argued that empirical evaluation of the factors affecting a declining population can often be the most efficient for understanding the causes of decline or rarity. Indeed, with the kestrel, pigeon and parakeet there was no option other than to approach their conservation problems empirically.

The need for the endangered birds to have viable

habitat to return to stimulated work on the conservation of vegetation communities, and this developed into the establishment of vegetation plots. The initial aim of these plots was to set aside small areas of native vegetation and to "garden" these for the rare plants they contained, but also to use them as locations in which to plant nursery-grown endangered species. In the mid-1980s it was considered unrealistic to restore more than a few hectares of native vegetation. Inspired by the positive results of these restoration efforts, and the development of the techniques to restore areas of native vegetation, the size of these plots has increased, and there has been a closer integration of the vertebrate and vegetation restoration projects. However, there is still a great deal to learn on how to effectively manage areas of native vegetation. The current rates of weeding, three times a year, are not sustainable in the long term. The maintenance of fences has not always been adequate, and incursions by pigs and deer have done considerable damage to regenerating native plants. Concurrent with changes in management, long-term studies are needed to measure the impacts of management, with control studies against which the data can be compared. A challenge for the future is to develop the protocols for efficient weeding and management of CMAs and to be able to increase the size of the plots.

In addition to habitat restoration on the main islands of Mauritius and Rodrigues, effort has been devoted to restoring offshore islands, with similar issues of controlling exotic vertebrates and removing exotic plants. In addition, missing plant species have been reintroduced, since on both Round Island and Ile aux Aigrettes the flora is very impoverished. Work on reintroducing missing vertebrates to islands is still in its infancy although the introduction of Mauritius Kestrels, Pink Pigeons and Mauritius Fodies to Ile aux Aigrettes has been largely successful, though kestrels no longer occur on the island: they moved to the mainland and established themselves there. All of the islands around Mauritius and Rodrigues are of great conservation importance both in terms of restoring the original ecosystems that may have occurred there but also because of their value for conservation introductions of both plants and animals.

The main focus of the work on Mauritius was originally captive breeding, but this broadened to include re-introductions, the management of wild populations and habitat restoration. It is this interface between captive and wild studies that provides such a fertile area for conservation. All of this work has been in partnership with other organisations and the Government of Mauritius, and the role of the government agencies has also increased as the work has grown in importance and success. The high-profile

Figure 10.3. A typical garden scene in Mauritius or Réunion; Mynas, Village Weavers and House Sparrows – not a native to be seen. Drawing done in Réunion by Nicolas Barré, from Barré & Barau (1982).

work with the endangered birds has resulted in the political will to conserve their habitats. For example, the long-standing intention to establish a National Park was turned into reality as a direct result of our work on endemic birds. While debating in Parliament the need for a National Park, the then Prime Minister Sir Anerood Jugnauth, quoting our conservation successes, argued that a National Park was required for the endangered birds to live in. In 1994 the country's first National Park was declared covering 6,574 ha (3.5% of the island). Conservation used to come under the Forestry Service but was in 1987 upgraded to a Conservation Unit and later into the National Parks and Conservation Service. The Forestry Service has in the meantime developed its own Biodiversity Unit, with its own agenda, primarily for plant conservation, which should complement the work of other agencies.

The involvement of universities in the conservation-based research is encouraging and many conservation biologists have been trained in the field. On conservation projects it will always be necessary to balance the need for more knowledge with the need for management. Often the need for conservation action is so urgent that there is no time to study the problem first and the approach has to be inductive, relying upon the experience of others, often on different species. How much knowledge is required to save a species is a question without an easy answer, but the work in Mauritius, both successes and failures, has allowed us to develop a set of guidelines for bird and reptile conservation work that can be widely applied elsewhere[290].

The work on Mauritius and Rodrigues has been very pragmatic, but quite effective. Much of the work has involved captive management and species-oriented approaches. Some ecologists claim that captive-breeding and species-orientated work detracts from

BOX 36   MICROBATS (MICROCHIROPTERA)

Two species of small bat, the Grey Tomb Bat *Taphozous mauritianus* and the Mascarene Free-tailed Bat *Mormopterus acetabulosus*, are comparatively common at present and widespread on Mauritius and Réunion. The tomb bat does not differ morphologically from the same species on the African mainland and Madagascar, but the free-tailed bat appears to be a Mascarene endemic. A DNA analysis would be useful to confirm their affinities. A third species on Réunion, the Pale House Bat *Scotophilus borbonicus*, also found in Madagascar (and probably the same as forms in Africa), has not been seen since 1867 and is presumed to be locally extinct.

Accounts
Bory (1804) in 1801, on the Grey Tomb Bat:

> These trees [enormous tamarinds] *served as refuge to a medium-sized bat, with upper parts all a dark russet-grey; in contrast the underparts are a fine white. It flies quite fast, a bit like certain birds that let themselves fall from canopy height and do not resume flying until the middle of their fall, then rising back up to the place they want to perch.*

Bory's white bat ('*Boryptera alba*') is enigmatic. Whilst describing latan palms near Saint-Joseph, he added:

> *little white bats, of which I was unable to obtain a single individual, find refuge during the day between the petioles of the leaves.*

Later, at Etang Salé, he examined latans but:

> *was unable to see any of the little all-white bats which come seeking asylum against the day's heat in the torn leaves of these trees.*

Grey Tomb Bat *Taphozous mauritianus*. From Peterson et al. (1995).

---

habitat conservation. In Mauritius, on the contrary, it has developed *into* habitat conservation and been the incentive for ecological restoration work. If resources had initially been directed towards habitat conservation we would not have achieved as much and key species might have been lost – in practice, the kestrel, pigeon and parakeet have acted as flagship species from which the broader actions have sprung. The conservation work on these species raised their profile and alerted Mauritians to the problems they face, and hence to the problems of habitat degradation. This has led to the development of the National Park and the restoration of offshore islands, securing the future for many endemic species and their habitats. At the start of this conservation work in the 1970s, the future of the endemic species of Mauritius and Rodrigues seemed bleak, and many thought that few could be saved from extinction. Thirty years later the long-term conservation of all the Mascarene endemics is within our grasp.

Acknowledgements

This work comes under the direction of the Mauritius Wildlife Foundation in partnership with the National Parks and Conservation Service, Mauritius. It is supported by grants from The Global Environment Fund, (World Bank),GEF (UNDP), The Durrell Wildlife Conservation Trust, The Wildlife Preservation Trust Canada, The Peregrine Fund, The World Parrot Trust and other organisations. More complete acknowledgements are given in the respective review papers and by Jones & Hartley (1995).

CHAPTER 11

REFLECTIONS

on island ecology and ecosystem abuse

Except in the case of insects, reptiles and shells, it is improbable that any of the original fauna of the Island exists. When it was first discovered the now extinct Dodo and large fruit eating bats were plentiful, and it is stated by some authors that the Tandrak [tenrec] and Monkeys also belong to the indigenous fauna . . . Many birds have been introduced such as Curlew, Merle, Partridge, Pigeons, Martin, Quail etc. Deer are plentiful and its hunting is the most popular sport. The wild boar is sometimes hunted.

Garbled zoology from the *Mauritius Almanac*, 1922[1]

Breakfast in bed . . . As we eat we are gently besieged by a gaudy army of flycatchers, orange and yellow bulbuls and fussy little crested canaries. The sky is silky and hot and there is a spectacular view of our blue lagoon, Paradise cove . . . There is Casella, a beautifully maintained bird sanctuary, where you can see the rarest bird in the world, the pink pigeon, now almost extinct because (like its ancestor the dodo) it lays its eggs on the ground and they get eaten.

Sun-lounger ornithology from a journalist, 2002[2]

We begin this chapter with two silly quotes for a reason – to illustrate the public perception and (mis)understanding of wildlife, and the need for appropriate education. Scientists habitually ignore this kind of travesty of their efforts and knowledge, but there is much to learn from it, most obviously that, despite their best efforts, the message is either not getting through or is badly distorted in transmission. On islands like the Mascarenes where most native wildlife survives only in remote and little frequented areas, the ordinary citizen simply does not see the biota that really belongs to their island. To the average Mauritian:

> *Every familiar flower, tree, snail, insect, mammal or bird – bar a few butterflies, one bird and a couple of bats and palms – is an exotic, and has been since their great-great-grandparents' lifetimes. What they think of typically Mauritian plants and animals are the everyday tropical species they meet in gardens and countryside, whereas the endemics seen in the museum are as foreign to them as kangaroos or ostriches.*[3]

Furthermore, until recently Mauritian schools used a British syllabus, and Mauritian students took British exams[4]. In Réunion the same applied, the syllabus and exams designed in metropolitan France. Even the language of everyday life, creole, was long regarded officially as a sort of degraded bastard speech, not fit for the educated[5]. Folklore reflected the disparate origins of the population, rather than the islands' realities. Baissac's collection of creole tales from 1880s Mauritius is a mix of European fairytales and local variants of the 'Brer Rabbit' corpus, i.e. African folk stories reworked to suit their new environment. More recent collections reflect the Indian antecedents of most Mauritians[6]. Baissac's tales are peopled with animals, but the only natives are tropicbirds and possibly tortoises; the rest are introductions or domestic (*compère* [= brer] hare, monkey, rat, mouse, cat, giant snail, cow), totally alien (wolf, lion, elephant, crocodile) or blatantly African – the wise grandmother spider[7]. The Indian tales include few animals, but those that are present are again either exotic or domestic – lion, jackal, elephant, monkey, chicken, goat[8].

Much has improved over the past 30 years, with native wildlife regularly featuring in newspaper articles, on television programmes and on postage stamps, and Mauritian schools now take students to the national park and Ile aux Aigrettes to see the vegetation (if not much fauna) at first hand. In Réunion especially there are increasing numbers of educational books featuring local conditions, and particularly in Mauritius, the status of creole has risen as people increasingly appreciate its qualities, no doubt helped by its elevation to an official language in the Seychelles in 1981[9]. There remains, however, a persistent problem of perception. In Mauritius, and even more so in Rodrigues, the remnants of native vegetation are unimpressive to the non-ecologist, and in most places so invaded by exotics that Strawberry Guava and privet overwhelm the endemic experience. Réunion is better served, partly through spectacular scenery, and partly through the upland vegetation being in fairly good condition, full of tree ferns, endemic acacias and attractive flowering trees. Rodrigues is small enough for educational outreach and involvement to work well; everyone at least knows the Golden Bat. In Mauritius hardly anyone has any experience of native fauna, bar the ubiquitous Mascarene Grey White-eye, day-geckos and perhaps the swiftlet; they know the Dodo from tourist tat in the shops, but often consider it mythical. When captive-bred Pink Pigeons were released into Pamplemousses botanic gardens in 1984 their popularity was immediate – but only with small boys, who proceeded to decimate the incipient colony of these easy targets with catapults. They had no clue that they were slaughtering rare and important wildlife![10] In ecologies as transformed as in Mauritius or Rodrigues, it is hard to redress the balance. Some progress has been made in promoting attractive endemic shrubs (e.g. hibiscus and *Trochetia*) as garden plants, but a much wider programme planting native trees, especially those rooted in history (ebonies, *nattes, colophanes, tambalacoques* etc.), is needed to bring them back to general familiarity. For the fauna it is harder, since even if educational plantations were created, endemic birds would still be kept out by monkeys, cats, rats and mongooses. Accessible refuges, such as Ile aux Aigrettes, have an important part to play here, as a great deal can be done on predator-free islets. Unfortunately Aigrettes (25ha) is too small for several bird species, and ideally a much larger islet such as Ile d'Ambre (140ha) or Benitiers (65ha) should be freed of exotics and planted up with natives[11].

Ecological ignorance

Already around 1800 short-term visitors to Mauritius were seeing only exotic vertebrates, and Darwin, on a few days' visit in 1836 saw no native fauna. However for a long time the locals still knew what was in the backwoods, but by the 1920s this had changed, reflecting increasing urbanisation and the decline in natural history expertise in the early 20th century. The first quotation at the chapter head shows how far educated perception had departed from understanding the island's native ecology. It is from the paragraph on 'Flora and Fauna' in the 1922 and subsequent editions of the *Mauritius Almanac*, edited by Albert Walter (earlier editions had no such information). Walter was an astronomer and meteorologist turned statistician, who had edited the almanac since 1912; although born in the UK, he had been on the island since 1897 in charge of the observatory, and must have personally known all the intelligentsia[12], but apparently could find no-one competent to write this piece. Tenrecs and monkeys were of course introduced, while several birds listed as imported are in fact native: the 'curlew' [Whimbrel], the Merle and two surviving pigeons.

The other heading quote illustrates contemporary journalistic writing by a tourist, in which the wildlife information is even more out to lunch than the rest of the article. No 'flycatchers' besiege tourists in hotels – the endemic species is a rare forest dweller – did she see Mascarene Grey White-eyes? There are no 'orange and yellow bubuls' (Village Weavers are the right colour). 'Crested canaries'? This would have been the bulbul (Red-whiskered Bulbuls are the only local birds with a crest[13]). The Pink Pigeon nests up trees like any other pigeon, not on the ground like 'its ancestor' the Dodo (nor in 2002 was it the rarest bird in the world). The 'bird sanctuary' is actually a commercial zoo where elderly Pink Pigeons from the captive breeding facility at Black River are sometimes pensioned off. Some local tourist guide or hotel staffer must have fed this rubbish to the naïve journalist.

Ecological ignorance also leads to pseudo-ecotourism projects, or at least to their success and proliferation. Any visitor expecting a real bird sanctuary at Casela Bird Park is heading for disappointment, and similarly the grandly named Le Val Nature Park is little more than a prawn and fish farm with animal enclosures (deer and monkeys) added. Less obviously spurious to the naïve visitor is the Domaine du Chasseur, though the name is a bit of a give-away: you can not only see the deer but shoot them too! Billed in one tourist guide as 'a nature reserve for monkeys', it is somewhat redeemed by preserving some ebony forest, though the Mauritius Kestrels being fed white mice daily at 3pm is a little tacky[14]. Regrettably these profit-oriented sugar estate offshoots are creaming off tourists who might otherwise visit the real thing – eco-educational places such as Ile aux Aigrettes and the National Park's interpretative centre. We are not suggesting there is no place for commercial projects in eco-tourism – La Vanille 'Crocodile Park' zoo and Owen Griffiths's projected tortoise garden in Rodrigues have a specifically educational and eco-restoration component in addition to their more blatant attractions (such as feeding the crocodiles).

Hopeless natural history in visitor-oriented publications is nothing new in Mauritius. In the early 1900s, Donald d'Emmerez de Charmoy, writing the zoological article in the otherwise authoritative

BOX 37 FRUITBATS (MEGACHIROPTERA)

Flying-foxes were first reported in 1598 and mentioned regularly throughout the occupation of Mauritius and Réunion. Three species are known to have inhabited Mauritius and Réunion but most observers did not distinguish between them. *Pteropus niger*, the Black-spined Flying-fox and *P. subniger*, the Rougette, were sympatric on Mauritius and Réunion, but both had different roosting habits; *P. niger* also reached Rodrigues. A third Mascarene species, the Golden Bat *P. rodricensis*, originally thought to be endemic to Rodrigues, is now known also to have occurred on Mauritius (Hume 2005). Flying-foxes are everywhere considered a delicacy and are caught indiscriminately. This has resulted in a number of extinctions, particularly in the South Pacific, and oceanic island species are everywhere threatened. *P. subniger* roosted in tree cavities and caves and would have been particularly susceptible to hunting. It survived on Mauritius until at least 1864 and on Réunion until *c.* 1860 but is now extinct. *P. niger* survives in reasonable numbers on Mauritius but remains vulnerable; it became extinct on Réunion *c.* 1800. *P. rodricensis* was reduced to perilously low numbers in the 1970s, but the population has now recovered to 3000+.

Golden Bat *Pteropus rodricensis*

Accounts
Pingré (1763) in 1761, on *P. rodricensis*:

> Bats are placed by naturalists among the quadrupeds. Those I saw in Rodrigues were the size of a pigeon but longer. Their head was like enough to that of a fox, the fur is russet, darker on the head and neck than on the rest of the body. The wings are dark grey in colour; extended or deployed they can each be a foot or a foot and a half in span. These bats are otherwise like our European bats. They are very fat. I was told that at the Ile de France [Mauritius], at a certain time of year, each bat supplied a chopine [bottle] of oil, but at that time they are not good to eat, or at least one must de-grease them first. This game did not seem to me bad, but I did not however find that it merited the high reputation it has as a delicacy in some lands in the Indies.

Black-spined Flying-fox *Pteropus niger*
On Réunion, this species was considered common until the mid-17th century but declined drastically shortly after, probably due to persecution and deforestation; the last authentic accounts referable to this species were made before the turn of the century. Later fruit-bat observations are all of the smaller *P. subniger*. On Mauritius, *P. niger* survives in reduced numbers; although roosts are widely spread in forested areas, the animals forage throughout the island.

Accounts
Jolinck in Mauritius in 1598 (in Keuning 1938–51), translated by Henk Beentje:

> I have so far forgot to mention the bat that lives here; there are a lot of bats as large as pied crows, with muzzles like rats, the mouth full of sharp teeth, in front of a colour like a fox, with wings like bats from our country, but half a vadem [fathom] long.

Peter Mundy (1608–67) in Mauritius in 1638:

> Battes wee gotte some off thatt biggnesse thatt they conteyned Foure Foote From the 2 extreamest parts off their wings stretched outt, headed like little Foxes, coulloured like Fitches [polecats], with very sharpe teeth. I never saw any creature Fatter For its biggnesse, For Flaying off the skynne, it remayned yett covered as it were with a coote [coat] off Fatte a good thicknesse, very sweet in the eating, butt in my opinion too lushious or Fulsome, and Dangerous. In the Day tyme they hang on trees by hooke on their wings with their heads Downeward and in the twilighte against Night Fly abroad to feed. They drincke Flying, snatching at the water. If once one on the ground, Not able to rise for want off Feet to spring them into the Ayre to gather wynde."

Leguat (1707) in 1691–92 (from the English translation of 1708), possibly referring to *P. niger*:

> The Batts fly there by day as well as other birds; they are as big as a good hen, and each wing is near two foot long. They never perch, but hang by their feet to the boughs of trees, with their heads downwards, and their wings being supply'd with several hooks, they do not easily fall tho' they are struck. When you see them at a distance, hanging thus wrapt up in their wings, you wou'd take them rather for fruit than birds. The Dutch whom I knew at Maurice Island [Mauritius], made a rare dish with them, and preferr'd it to the most delicate wild-fowl. Every man has his taste; As for us, we found something in these Batts that we did not like, and having a great many things that were much better, at least in our opinion, we never eat any of these filthy creatures. They carry their young about with them; We observ'd they always had two.

Selections from from Lanux's long dissertation (1772) on the Black-spined Flying-fox or Rousette in Réunion, the first published ecological observations of a fruitbat:

> When the animals are resting in a tree, they murmur amongst themselves in a way that is not unpleasing . . . [that fruitbats fly in the day] is true of roussettes but not of rougettes. The former fly in broad daylight – but this means only that one sometimes sees some flying about during the course of a day, but one by one and not in flocks. When doing this they fly very high, such that their size appears less than half [the usual]. They go a great distance and very fast, and I think it very likely that they cross from this island of Bourbon [Réunion] to Ile de France [Mauritius] in little enough time (the distance is less than thirty leagues). They do not soar like birds of prey, like the frigate etc., but at that great height above the ground surface, a hundred or perhaps two hundred toises [fathoms], the movement of their arms is slow; it is brisk when they fly low, and even more so when they are close to the ground . . .

> The mating season for these animals is here around the month of May, that is to say, in general, in the middle of autumn. That for giving birth is about a month after the spring equinox, thus the gestation time is about four and a half to five months. I do not know how long the growth period is, but I know it appears to be complete by about the winter solstice, that is to say at the end of about 8 months since their birth. I know also that after April or May one no longer sees small roussettes, although at this time one can still easily distinguish old from young by the brighter colours of the latter. The old ones, particularly the males, become grizzled, I do not know after how long, and this is when they smell strong, as I have already mentioned, and only blacks will eat them, and the only good thing about them is their fat, of which the species is in general well supplied from the end of spring until the beginning of winter . . .

> It is certainly not any creature's flesh that provides the plumpness of roussettes and rougettes, nor even forms the slightest part of their diet; it is not meat they need. In brief, these animals are in no way carnivores, they are, and are only, frugivores. Bananas, peaches, guavas, a good many sorts of fruit which are provided in succession by our forests, mistletoe berries and others, that is what they eat, and they eat only that. In addition they are very fond of the nectar of certain umbel-type flowers, amongst others our bois puant [*Foetidia mauritiana*] of which the nectarium is very neat. It is these flowers, very abundant in January and February, and generally in the middle of summer, that attract our island's roussettes to the lowlands in large numbers. They make these flowers' numerous stamens fall like rain, and it is very probable that it is to suck nectar from these umbel-flowers that their tongue is as described so

exactly and wisely by Mr Daubenton. I can note that the mango is a fruit with a resinous skin, and our animals do not touch it at all. I know that caged they have been made to eat bread, sugar cane etc, though I have not known that anyone has got them to eat meat, especially raw. But whatever they eat in cages, it is not in a state of slavery that I'm discussing them, it changes the habits and characters of all animals too much. In reality, man has nothing to fear from these [animals], neither personally not for his poultry. It is impossible for them to take, I don't say a hen, but the least little bird. A roussette cannot, like a falcon or a hawk, stoop on a prey. If they aproach too close to the ground, they fall down and cannot take off again without climbing whatever support comes to hand, even a human they may encounter . . .

From the fact that roussettes are sometimes seen skimming the surface of water, much as do swallows, they have been said to eat fish, and made fishers, and this was to be expected when it was assumed that they ate anything and everything. This flesh [i.e. fish] does not suit them any more than any other. Again, they only eat vegetable matter. It is to bathe that they skim the water, and if they are able to sustain flight closer to [the] water [surface] than they can on land, it is that the resistance of the latter affects the beating of the wings, which are free over water. From this [activity] clearly results the natural cleanliness of roussettes. I have seen many, I have killed many, but I have never found on them the least filth; they are as clean as birds generally are.

Head and dentition of Rougette *Pteropus subniger*. From Matschie, 1899.

Rougette *Pteropus subniger*

The thick fur, small ears and ample fat of *P. subniger* suggest that it could tolerate cooler temperatures than *P. niger*, and may have been able to survive at much higher altitudes on Réunion. Furthermore, the unusual ecology of the Rougette may well have evolved due to competitive pressure from the larger species. The Rougette's habit of roosting in caves, rock clefts or hollow trees, and of being totally nocturnal (the proportionally larger orbits indicate that the Rougette was not only behaviourally nocturnal but had evolved larger eyes) may have reduced competition with its larger congener, and its weak, short jaws may have been better adapted to exploit other food resources, e.g. flowers, leaves, soft fruits and perhaps tree sap. The ecological requirements of *P. subniger* might explain why it survived for several decades after the extinction of *P. niger* on Réunion, which died out before 1800; the Rougette was a cryptic species in comparison. On Mauritius, H. Whiteley collected the last known specimen in 1864. Mauritius is comparatively flat compared to Réunion and its mountain ranges are small. This suggests that as *P. subniger* declined in numbers, a lack of montane refugia may have ultimately hastened its extinction, whereas with much larger areas of lowland *P. niger* survived.

Accounts
La Nux (1772), on Réunion:

> I ought to put in here what little I know about rougettes. One never sees them flying by day. They live communally in the large hollows of rotten trees, in numbers sometimes exceeding four hundred. They only leave in the evening as darkness falls and return before dawn. One is assured, and it is taken in this island for granted, that, however many individuals make up one of these associations, there is but a single male. I have not been able to verify this fact. I should only say that these sedentary animals become very fat; that at the beginning of the colony, numerous poorly off and unfastidious people, taught no doubt by the Malacasses [Malagasy slaves], provided themselves plentifully with this fat for preparing their food. I have seen the time when a bat-tree (it is thus that one used to call the retreats of our rougettes) was a real find. It used to be easy, as far as one can judge, to prevent these animals leaving, then to take them out alive one by one, or to suffocate them with smoke, and in one way or another discover the number of males or females of which the association was composed; I do not know any more about this species.

Foucher d'Obsonville (1783) on flying-foxes in Réunion:

> This large variety [of bat] is also found in the islands east of India, in Madagascar, Bourbon [= Réunion] and several other places. Passing through this last island, twenty years ago [c. 1763], I bought a young male, which could still only drink milk; my idea was to take it to Europe, but it died during the crossing. It was of the species of these animals known as rougettes, because the part of the fur from the neck to the waist forms a sort of cravat of a rather brilliant reddish colour. One also sees in that island some, which like those of India, are of a fairly uniform brownish colour: they are called roussettes. These animals have strong canine teeth; however I think they use them only for defence. It is true that if necessary they can live on all sorts of food, and in this way I trained the young one I bought at Bourbon; but it seems that by taste and preferred instinct they are in no way carnivorous. What I can say in this regard is that I neither saw nor heard contradicted in India, where one finds them in some quantity, that they seek and eat anything other than fruits of all species, and particularly of a tree called war [probably the Banyan *Ficus bengalensis*, known by this name in Maharasthra (Cowen 1969); flying foxes are notoriously fond of figs], they also try to drink the liquor that the inhabitants draw from coconut trees and palms. Also the flesh of these animals, when they are fat, is quite good tasting. If nearly all the tribes of India, and even the Mohammedans, have a horror of such a food, it is pure prejudice, based perhaps on their irksome appearance; but the majority of our long-standing inhabitants of the Iles of France and Bourbon [Mauritius, Réunion] eat them very willingly, and even use their fat instead of oil on their salads.

Roch, quoted by Geoffroy (1806):

> The two species of roussettes of the Isle of France [= Mauritius] aggregate together at random on the trees to which they are drawn by their abundance of fruits and flowers; they have however very different habits, for, apart from the time they are busy feeding, the roussettes proper go and attach themselves to large trees in the middle of the forests, while those with the collar, or rougettes, establish themselves in the hollows of old trees or amongst rocks. It is believed that they do not mate together; at least no mules are ever produced.

Clark (1859):

> The species of bat next in size to the flying fox, called Chauve Souris banane, is about two feet across the wings, and the body about the size of a rat. Its aspect differs greatly from that of the flying fox [i.e. Black-spined Flying-fox], its eyes being less lively, and its ears completely hidden in the fur, which is of a more woolly nature than that of other bats, and is very thick. Its head, neck and shoulders are of a light reddish brown, the body of a dark grey, and the rump whitish yellow. Its general character and habits are the same, but it appears less intelligent, and its cerebral development is certainly inferior to that of its larger congeners. I believe it to be less numerous also.

Specimen collections
Skins of *P. subniger* are held in the BMNH, MNHN and the RMNH. Fossil material can be seen at BMNH and MNHN.

Mauritius illustrated, naïvely translated a host of local bird names into English, with disastrous results: "The 'chicken-eater' . . . is a very pretty species of falcon . . . the woodcock and the banana bird are very seldom seen . . . the white bird and the manioc bird are caterpillar eaters . . . the nightingale introduced in 1892 . . . at nesting time the Republican birds assemble in great numbers . . . in the marshes many teal and wild ducks . . . as well as larks"[15]. English visitors might have been forgiven for expecting familiar European birds (Woodcock, Nightingale, lark etc.), and a fearsome eagle ('chicken eater') in addition to the mysterious republican, banana, white and manioc birds. But no, the 'woodcock' (*coq de bois*) is the tiny paradise flycatcher, and the 'chicken eater' (*mangeur de poule*) just the kestrel. Forty years later, d'Emmerez's names were recycled by Bulpin in his historical ramble *Islands in a forgotten sea*, who added a *faux-pas* of his own, using 'musk rat' (mistranslated from the French '*rat musqué*') for the House Shrew (Musk Rats being large North American rodents)[16]. Réunion does not escape either. In 1989 a tourist guide informed us that "the rarest bird on the island now is the white blackbird or crow . . . black blackbirds are common . . . the virgin bird is a pretty little flycatcher . . . martins . . . are common. Among the puffins and petrels . . ."[17]. Actually, in this case, apart from the bad translations of names, the information was reasonably accurate.

Philippe Lenoir's highly illustrated tourist-oriented *Mauritius* used a squirrel photo to represent a mongoose, and a latan picture for '*vacoa*' (screw-pine), a bizarre mistake for a Mauritian. Visiting Australians Coralie and Leslie Rees commented in 1954 on "the bush given by the creoles a name meaning 'private' because the Indians were apt to grow it around their privies". Not exactly: *Privet* in Mauritian French and creole is simply English 'privet'; perhaps someone pulled the Rees's legs[18]. Pasfield Oliver's confused and inaccurate renderings of Réunion fauna from the 1860s are best forgotten altogether, though he later established himself as a respected editor of early Mascarene and Malagasy literature[19]. The list is endless.

Before leaving printed peculiarities, a tribute is necessary to the Mauritian high priest of literary strangeness, the late Malcolm de Chazal. Chazal was a kind of surrealist mystical philosopher and artist who permeated the Franco-Mauritian cultural scene on and off from the 1940s until his death in 1981. His *L'Ile Maurice proto-historique, folklorique et legendaire* includes a visionary view of the creation of Mauritius and the Dodo which can hardly be beaten for exquisite fantasy:

> When the second moon fell to Earth, the bulge of water that had formed at the equator collapsed and in falling back created the Indian Ocean. The Lemurian civilisation must thus have disappeared. The inhabitants of the plains of the Lemurian continent would have drowned, while those of the high plateaus would have asphyxiated. The Lemurian continent [now] swallowed up by the waters, Mauritius became one of its rare remaining peaks . . . It is now that the veil surrounding the Dodo will be lifted . . . The Dodo is a columbiform beast. It can be considered as the dove on an Ark of a Deluge prior to that which entombed Atlantis. This giant dove had a huge sternum, which signifies that it had giant wings. So all is explained . . . The Dodo travelled in the heart of the pre-historic continent, overflying vast expanses. When the cosmic cataclysm that we have spoken of came, a certain number of Dodos found themselves trapped on what is now Mauritius, whose mountains became their dove-cote. The Dodos' wings atrophied. So the animal descended the mountains to the estuaries. The degenerate Dodo ate the hard seed of the Tambalacoque, hence the hard stone found in its stomach and which served to masticate this food. With the Dodo, naïve and innocent animal, one can reconstruct an Earthly Paradise overwhelmed and vanished, of which the Dodo is the last vestige.[20]

Note Chazal's '*is* the last vestige'. Karl Shuker attempted to revive the Dodo in the 1980s, reporting that "cryptozoological explorer Bill Gibbons learned to his astonishment that on this island there is a secluded patch of rainforest called the Plain Champagne, stretching out to the coast where Dodo-like birds have allegedly been seen in recent times at early dawn and dusk, walking upon the silent, half-lit beach". Shuker told readers that Gibbons was returning in 1997 to investigate. He did, but sadly the Dodos eluded him[21].

Uncontrolled introductions and unimportant forests

One consequence of the failure to appreciate the intrinsic value of native ecosystems as opposed to the day-to-day familiar exotic environment is that many people over the years have found Mascarene wildlife lacking, and felt impelled to introduce species to liven it up. We have noted in passing attempts or intentions to introduce songbirds for the quiet forests, gamebirds for under-occupied 'sportsmen', and pretty green geckos for gardens – not to mention all the animals kept as pets that have escaped without control. More insidious have been attempts, official and unofficial, to remedy perceived or real agricultural problems by importing animals for biological control – such species are intrinsically more risky as they are deliberately chosen for their ecological impact. The inspired idea to import mynas in the 1760s to overcome locusts soon had people complaining about depredations to fruit trees, and by 1801 Bory blamed the birds for ruining the entomology of Mauritius[22]. He exaggerated, but was right in principle; well-intentioned introductions are liable to have undesirable knock-on

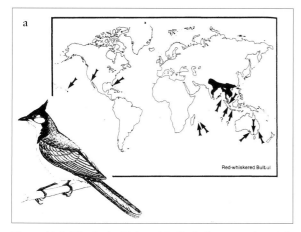 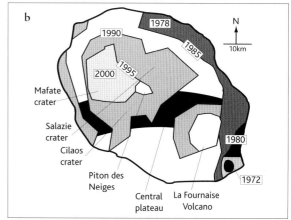

Figure 11.1. The Red-whiskered Bulbul, showing where it has been introduced (a: from Long 1981) and its rapid spread in Réunion (b: from Clergeau & Mandon-Dalger 2001). Note that since Long's map was published, bulbuls from Mauritius have been released on Assumption (near Aldabra, northeast of Madagascar), and from Réunion onto Mayotte, Comoros.

effects. This warning unheeded, introductions continued, usually on the principle that what an animal eats on some distant continent will remain unchanged when it is brought to the Mascarenes. Before the development of ecological science this was perhaps understandable, although even then negative consequences elsewhere were often well-known. We saw in Chapter 7 (p. 136) how the warning signs were deliberately ignored in relation to mongooses in 1900. By the 1930s, when serious attempts were made to introduce Cane Toads to Mauritius, ecological thinking ought to have had an impact, but it clearly did not. Come the 1960s, when Rosy Wolf Snails were released, supposedly to control giant snails, there really was no excuse for government agencies to make so elementary an ecological mistake. However, even in Europe it has taken many decades of unwelcome impacts on natural habitats, and protests from conservationists, for official agricultural agencies to accept the need to look much further than the immediate agricultural domain when introducing new practices, chemicals and species[23]. The widespread adoption of genetic modification in North American agriculture with virtually no environmental impact studies demonstrates that the battle is by no means won, and that potentially very risky technologies can be unleashed by the agricultural industry without proper assessment.

While the quarantine measures to prevent new exotic species reaching the flagship Round Island nature reserve are draconian, and individuals are (officially) forbidden to bring plants or animals into the Mascarenes in their pockets, animal traders or zoos can easily get licences to import interesting pets into Mauritius or Réunion. Once in the country these animals can be freely let loose; Casela Bird Park has irresponsibly released Great Green Day-geckos and several bird species, and certain individuals have admitted releasing various day-geckos in Réunion. Most releases are, however, anonymous; Carl Jones in Mauritius and Jean-Michel Probst in Réunion have assembled lists of escaped birds and reptiles[24]. Neither in Réunion nor Mauritius is any attempt made to eradicate species so released while it is still possible. The authorities are rightly concerned about agricultural pests and diseases, but have yet to grasp the importance of preventing 'non-economic' invaders. Back in 1976 Armand Barau in Réunion took responsibility himself and shot out the newly-released Ring-necked Parakeets[25], and it would have been equally easy in 1998 to have eliminated Rainbow Lizards from their foothold by the new docks at Le Port (p. 184). There is perhaps little incentive to control introductions (e.g. waterfowl) that will have minimal impact on native biota, but more insidious species that invade forests (Red-whiskered Bulbuls, Peking Robins, iguanas, Malagasy day-geckos, ferrets) should be targeted if a release is discovered, or more extinctions are likely to result, although such species are much harder to locate and eliminate than species living in man-made habitats. All this applies with equal force to plants.

Related to the pervasive misapprehensions about fauna is the tendency of the authorities to see 'the forest' as a kind of marginal wasteland to be sacrificed whenever a need for land arises. Significantly, government land under native forest has generally been used despite the existence of large areas of fallow and long-cleared land in private hands. With a little imagination, and some kind of reversionary law for under-used land (as practised in the 18th century), much of the destruction of native forest for geranium and Japanese Red Cedar in Réunion and tea and pine in Mauritius could have been avoided. In Mauritius road-building projects through native forest are still in vogue, although it has long been clear that building roads is one of the best ways to lose forest; the easy access inevitably attracts people with axes and lorries. Furthermore, there are usually perfectly good alternative routes, as was the case for the proposed road through the Bamboo Mountains forest of Ferney, one of the least degraded on the island. After much lobbying by conservation and heritage bodies, the planned road was re-routed in October 2005, with Ferney to become a nature reserve[26].

Education, alienation, perception, legend and official obstinacy are but one aspect of our islands that needs attention. Another is scientific myth and orthodoxy, in particular some aspects of biogeographical theory and also an action-numbing perception of the need for a kind of fundamentalist ecological purity that pervades international conservation bodies. The two are not unrelated, since certain conclusions from ecogeographical theories can powerfully influence conservation decisions, action and funding.

Triage, reintroductions and analogues
Back in the 1970s and 1980s when the numbers of Mauritius Kestrels and Echo Parakeets were in single figures, many assumed that they were past saving. There was controversy in the international conservation community between those wanting to close down 'doomed' projects and others who felt that that species down to the last few birds demanded extra effort – intensive care. Warren King, editor of the ICBP's second bird *Red Data Book*, wrote in 1978 that

> *The active management techniques presently in use that are applicable to endangered species are expensive, short-term and labor-intensive, and must be viewed realistically as only stopgaps. We have all asked ourselves, "Is it worth the effort?" Most of us have answered affirmatively, for we recognise that continued existence of a species is precious, and extinction painfully irreversible. There will be no shortage of critically endangered species on which to work this special and newly-conceived magic, but an administrative mechanism must be found to permit application of realistic budget to the most needy species, and those most likely to respond to treatment, regardless of national boundaries.*[27]

In Norman Myers's triage theory of conservation, he highlighted the kestrel project as being hopeless: "we might abandon the Mauritius Kestrel to its all-but-inevitable fate, and utilise the funds to proffer stronger support for any of the hundreds of threatened bird species that are more likely to survive"[28]. While the ICBP (now BirdLife International) did not necessarily take this on board, Carl Jones has claimed that his original brief from ICBP in 1979 was to wind this 'hopeless' project down[29]. Either way, the project was rescued by Gerald Durrell and Jersey Zoo, who saw the potential for a dramatic success, even if it was a long shot[30]. Although the general principle of conservation as against unrestrained development or other human activities is widely discussed, the detailed decisions on which species or ecosystems to preserve and fund have often been taken by committees behind closed doors. Little of this decision-making is in the public domain, so it is often difficult in retrospect to see why one project was favoured and another not[31]. However, those involved hands-on know that decisions often have to be made fast, and government bureaucracies and international NGOs can be painfully slow and may not fully appreciate the practicalities on the ground. In 1974 one of us (ASC) was in a position to catch surplus Mauritius Fodies displaced from native forest being cleared for pine plantations, and translocate them. The intention was to move them to monkey-free Réunion where the endemic fody was long extinct. This proposal took so long to go through the hoops that when permission was finally granted it was too late – Mauritius was hit by a major cyclone and the BOU expedition had run out of time. Thus only three birds were translocated, too few to establish a population – they were not seen again[32]. In 1976 Gerald Durrell went straight to the Mauritian government for permission to catch Round Island reptiles, Pink Pigeons and Golden Bats, bypassing WWF/ICBP formalities, and thus achieving much quicker results[33].

We are not suggesting that conservation should proceed simply on the basis of untested ideas of fieldworkers and enthusiasts, but that bureaucracy and indecision can be just as deadly to a species as environmental challenges. Related to this is the fact that highly motivated individuals, even with poor resources, can often achieve what highly funded but hierarchically structured organisations cannot. In addition, the field ecologists will be more tuned to conservation priorities than officials thousands of miles away, though the latters' 'experts' are often given more weight[34]. One need only compare the successes in Mauritius and New Zealand with the dismal results achieved in Hawaii, heavily funded but neutered by red tape, rules and inter-departmental rivalry. This is even visible in the way projects are written up: "while one gets the feel for the nail-bitingly close shaves of the Mauritian and New Zealand birds discussed [at a recent conference], the Hawaiian account lacks historical perspective and is curiously laconic, given the on-going disaster steam-rollering these birds into extinction"[35]. However, the long-term management programmes necessary for species and ecological recovery need organisation and structure, thus there is a difficult balance to be struck between dedicated (and sometimes maverick) enthusiasts and the stability provided by institutions. In cases such as Mauritius, New Zealand and Hawaii, where endemic species face multiple intractable problems[36], this is expensive, and there are no quick fixes.

Another reason for abandoning rarities has been the widespread assumption that once species are down to tiny populations, the inevitable inbreeding involved in their recovery would so weaken the species that survival in the wild would be compromised, hence the effort would be wasted[37]. In practice it seems that you cannot tell until you try. The Mauritius Kestrel seems more or less unaffected by its population bottleneck, whereas the Pink Pigeon breeding programme has had to be very carefully managed to avoid congenital defects, even though the founder population was significantly larger[38].

More recently there have been issues around whether to use analogue or replacement species to help reconstruct missing elements of ecosystems. Purists argue that such species are just as much 'alien exotics' as rats, Red-whiskered Bulbuls or Crab-eating Macaques, whereas pragmatists hold that, even if they are not native, they are providing a piece of the ecological jigsaw that can never be retrieved from endemic species now extinct. By filling the gap in the ecosystem left by the lost species, they can perhaps help save a whole web of other endangered animals and plants[39]. Clearly such analogues have to be chosen very carefully, but for the Mascarenes there are many cases of suitable related species surviving in other parts of the Indian Ocean. Detailed suggestions for the more obvious analogues and translocations have been made in Chapter 10 (pp. 255–257), but for more specialised extinct species in endemic genera (Dodos, large parrots, flightless rails, owls, Burrowing Boa, Didosaurus skink) this is trickier, since we know too little of these species' biology to indicate how to replace them with analogues, and there are no congeners to work from. As an understanding of the ecosystem is built up, particularly of plants and their pollination and dispersal requirements, it may become possible to suggest animals to replace the missing species, provided they do not come with a dangerous downside.

On mainland Mauritius feral predators will limit what can successfully be introduced. Where there is a choice, analogues from Madagascar might work better than from the Seychelles or Aldabra, since Malagasy species will be better adapted to predation pressure; lemurs, the Fossa, civets, mongooses and rats should prepare Malagasy pigeons and rails for the monkeys, cats, mongooses and rats in Mauritius. Life is a bit easier in Réunion and Rodrigues (no monkeys or mongooses), which may be good for the birds, but in Réunion Wolf Snakes and rats would make introducing Mauritian analogue lizards (Telfair's and Bojer's skinks, night-geckos) impossible, as it is on the Mauritian mainland. Competitive species, such as rival insectivores (tenrecs and shrews), could also compromise analogue introductions of lizards, even if the predator problem was solved. In any case these alien insectivores will eat young lizards, though Telfair's Skink is probably capable of reciprocating by taking baby shrews. Cryptic competitors, such as the introduced honeybees that rob flower-nectar before birds and day-geckos can get to it in the mornings, may be more common than has yet been discovered[40]; rogue ants are often a problem on islands, but have been little investigated in the Mascarenes[41].

There has been no disagreement in principle about translocating to Réunion birds extinct there but surviving or with close relatives in Mauritius, yet nothing has happened. Re-introductions were first suggested in 1974 and have been supported by many reports since[42]. However desirable this may be, human nature has been cited as an obstacle. Réunionnais are allegedly so addicted to hunting that unfamiliar foreign species (as Mauritian birds will initially be perceived) would be too tempting a target, especially the Echo Parakeets and Mauritius Pink Pigeons[43]. We saw in Chapter 8 (p. 175) that while poaching is still a problem, it is in decline, and the Réunion public is much better attuned to nature and wildlife than it was in the 1970s. The time has come to actually do it; there are now enough wild and/or captive-bred kestrels, parakeets and pigeons for it not to be a major disaster if the experiment failed, and with Mauritius Fodies now breeding freely on Ile aux Aigrettes, they too could be tried again. There is reluctance in Réunion to accept fruitbats, yet fruit-growers have survived mynas and more recently Red-whiskered Bulbuls – the Black-spined Flying-fox would never rival the numbers of these birds. Large-fruited native trees in Réunion are very short of dispersers[44], and re-introducing the bat would remedy this, as established by studies in Mauritius[45]. Two small lizards, Bojer's

Figure 11.2. Grey White-eyes trapped with bird-lime in Réunion. Sketch by Nicolas Barré in Barré & Barau (1982).

BOX 38   DUGONGS (SIRENIA)

Dugongs *Dugong dugon* were once native to the Mascarenes, breeding within the lagoons on Mauritius and Rodrigues, though there were none in Réunion where the lagoon is too small. They were considered a valuable food item and were indiscriminately hunted. Their decline was rapid and they become extremely rare by the 1730s on Mauritius, although they were occasionally seen up to the turn of the century. On Rodrigues, Leguat described the suckling behaviour of nursing Dugongs and vast numbers were found within the lagoon. In 1726, Tafforet noted them as being common but Pingré in 1761 listed them as rare, and the last ones were seen around 1795. Dugongs are now extinct in the Mascarenes.

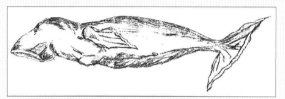

Dugong *Dugong dugon*. Drawn from life by Hubert Hugo c.1673. From Pitot, 1905.

Accounts
Hoffman (1680), in Mauritius 1673–75:

The sea cows [Dugongs], which are also called sea monks and which many people consider actual mermaids, are more numerous around Mauritius than anywhere else. They are of a particular species and differ from those found at the Cape [which were sea lions!]. They are seven feet and more in length, and their body has a cylindrical form; their skin is smooth and without hair. Behind the head they have a sort of hood like a monk. Their eyes are very small. They have no legs, and instead of arms they have two membranous stumps, which I can best compare with the wings of a goose whose feathers have been plucked. Their tail resembles that of fish, with the difference that it is horizontal instead of vertical. They have breasts, with which they suckle their young just like women. Their skin is exactly like Man's, except that it is a little browner. The have a layer of lard a hand's width, lard which is much better than that from pork, and when well prepared, is perhaps preferable to many sought-after foods. Each animal provides 700 to 900 pound and sometimes more. It is certainly a very curious beast, of which I would happily stuff some babies to bring to Europe, were it not that I fear they would putrefy from the grease with which their skin is imbued.

Leguat (1707), in Rodrigues 1691–93, from the 1708 English translation:

The Lamentins, which other Nations call Manati, that is having hands, abound in the sea about this Isle, appearing often in numerous troops. Its head is extremely like that of a hog . . . excepting that its snout was not so sharp. The greatest of them are about twenty foot long, and have no other fins but the tail and two paws. The body is pretty big down to the middle, and a little below it, the tail has this in particular with the whales; that its breadth is horizontal, when the animal lies on its belly. The blood of this creature is hot, its skin is black, very rough and hard, with some hairs, but so few, they are scarce perceivable. Its eyes are small, and it has two holes which it opens and shuts, and for that reason may be call'd [its] gills and its ears. Because it often draws in its tongue, which is not very great, several have assured me it has none. It has hind-teeth, and even tusks like a boar, but no fore-teeth: Its jaws are hard enough to bite grass, its flesh is excellent, and tastes something like the best veal: 'Tis very wholsom meat.

The female has tets like women's; some say its brings forth two young ones at a time, and gives them milk together, carrying them both at its breast with its two things like hands: But since I have never seen it hold but one, I incline to believe it bears no more at once . . .

This fish is very easily taken, it feeds in herds like sheep, about three or four foot under water; and when we came among them did not fly, so that we might take which we wou'd of them, by either shooting them, or falling upon them two or three at a time upon one without arms, and pulling it a-shoar by main force. We sometimes found three or four hundred together feeding on the weeds at the bottom of the water, and they are far from being wild, that they wou'd often let us handle them to feel which was fattest. We put a rope about its tail, and so hale it a-shoar. We never took the greatest of them, because we cou'd not master them so easily, and they might perhaps have master'd us: Besides, their flesh is not so delicate as that of the little ones . . . Their lard is firm and excellent, no body that ever saw and tasted the flesh, took it for any thing but butchers meat. This poor animal dies as soon as it has lost a little blood.

Specimen collections
Fossil material of *Dugong dugon* has recently been found at the first Dutch settlement in southeast Mauritius.

Skink and a night-gecko could be introduced onto rat-free Petite Ile, provided it has been thoroughly checked to ensure no native night-gecko (believed extinct) or introduced house geckos have survived there unseen. Petite Ile is probably too small to accept Telfair's Skinks as well.

In Chapters 8 and 9 we noted that when alien herbivores were removed from Round Island and parts of Rodrigues, it created an opportunity for invasive exotic plants to spread freely. In addition to suppressing native species, the rabbits (Round Island) and livestock (Rodrigues) had also kept the invasives under some control, an effect also noted in the 1970s for deer and the alien giant bramble in Réunion[46]. In Mauritian forest exclosures the problem is less severe and currently controlled by weeding, but where that is impossible (parts of Rodrigues) or unsustainable (Round Island) weeds are a serious menace, the obvious solution to which is introducing a 'mild' herbivore, namely tortoises[47]. A similar paradox has arisen in some places in relation to controlling predators – for instance, eliminating cats on seabird islands can result in rat numbers increasing so much that the supposedly protected seabirds are worse off than when cats were present[48]. In Mauritius cats and rats have generally been tackled simultaneously, whether being eradicated (e.g. Flat Island) or controlled (selected forest areas), but with rat control (but not cat-trapping) being relaxed in some previously targeted forest plots, there may be some risk of 'mesopredator release' as the effect has been called. With the complexity of Mauritian conditions, it would be necessary to look very carefully into knock-on effects on invertebrates, if it were decided to control, say, tenrecs, which currently influence shrew and possibly toad populations in forests. Likewise, as mongooses keep tenrecs down (p. 137), controlling them would increase tenrec numbers which would then impact on soil invertebrates and probably on Macabé Skinks. Rats on Ile aux Aigrettes had so effectively suppressed shrew numbers that researchers did not even know they were there until the rats were eradicated (p. 161); removing rats created ideal conditions for shrews, which have so far proved impossible to eliminate[49]. As yet full food-web studies have not been done, and without such work predicting the effect of specific control measures is extremely hazardous[50].

Island biogeography and exotic introduction theory

Modern island biogeography theory stems largely from a seminal book published in 1967, *The theory of island biogeography*. One of its principal tenets is that the number of species on islands depends on their size and their distance from colonising sources. This is self-evident, but the devil is in the detail. Macarthur and Wilson's original theory reduced everything to island size and distance from mainland source areas, island age being deemed irrelevant because of an associated theory of faunal equilibrium, by which, over time, arrivals were supposed to match extinctions. The variety of habitats and altitudinal range available was subsumed under size – i.e. bigger islands were judged to have more habitats and mountains. Ever since the theory was published, much play has been made of species-area ratios, and how they differ between taxa – i.e. poorly dispersing groups (such as amphibia and snakes) will be fewer on distant islands than good dispersers (such as birds and bats); some taxa are such good dispersers (butterflies, ferns, orchids) that they get everywhere, and have much lower rates of endemism. In the real world the principles work for some taxa some of the time. This is not the place to review the theory, only aspects that relate to our islands; the literature is voluminous[51].

Three studies, published in 1971, 1981 and 1983[52], have applied 'classic' island biogeography theory to the Mascarenes, but all suffered from inadequate data; many bird and reptile species now known from subfossils had not then been discovered, and furthermore the earliest study combined native and introduced species[53]. On a more global scale the Mascarenes have often featured in analyses, but have got lost amongst the world's other islands in attempts to draw general conclusions from the variety of factors that influence faunal and floral diversity of islands[54]. Our graph of pre-human land bird numbers on islands worldwide (Figure 11.3) shows that the Malagasy region shares with southern Australasia (i.e. without New Guinea and satellites) the lowest diversity of all island areas, and that diversity is not only dependent on island size but also on the richness of the source continent – Australia and Madagascar as sources being themselves islands, and relatively poor in species. The lines on the graph are those of maximum species saturation, i.e. the largest number of species to be found per island size for each source region, normally islands on continental shelves; however we have treated the oceanic, but not very distant, Comoros as saturated with respect to Madagascar. The saturation lines are roughly parallel for the different regions, though there is a trend towards steeper lines (a faster decline in species number as islands get smaller) in areas with richer continental avifaunas[55]. If one moves the Malagasy saturation line down parallel to the lower diversity of the Mascarenes it goes neatly through all three islands, suggesting that the immediately pre-human numbers on these islands had reached stability for island size and current distance from sources (much increased since the Ice Age; see p. 63). Aldabra and the Seychelles are between the Mascarene and Madagascar lines – both are better placed than the Mascarenes to receive immigrants, as winds often blow from Madagascar north and northeast[56]. Aldabra, though a low and somewhat uniform island, is relatively close to Madagascar.

The islands of northern Melanesia (the Bismarcks and Solomons, off New Guinea) show much less reduction of bird species with island size than would normally be expected[57], which appears to be because the islands are relatively close together and the species (already island colonisers) are relatively mobile. Stan Temple used a similar pattern within the granitic Seychelles to argue that the islands were over-saturated since the huge Seychelles bank went underwater at the end of the last Ice Age, and thus further extinctions were to be expected[58]. However, it seems more likely that diversity on the smaller islands was maintained (prior to human intervention), as in Melanesia, by the overall presence of many islands close to each other[59]. But plotting the Ice-Age size of the Seychelles bank (c. 43,000km²) suggests that up to 90 species might have been lost to rising sea-levels even against the Mascarene line, so the shallow seabed may conceal many subfossil species. Likewise Rodrigues will have lost about 10 species, and Lord Howe Island, shrunk to 13km² from about 1100km², will have lost about 40 against the southern Australasian line. If the equilibrium position for the Seychelles is similar to that of the Mascarenes (we think it is probably higher) then they might still be 2–3 native species over the stable number, whereas Moheli in the Comoros, now about a quarter of its Ice-Age size, looks seriously over-diverse, with about eight bird species above the Madagascar–Comoros line.

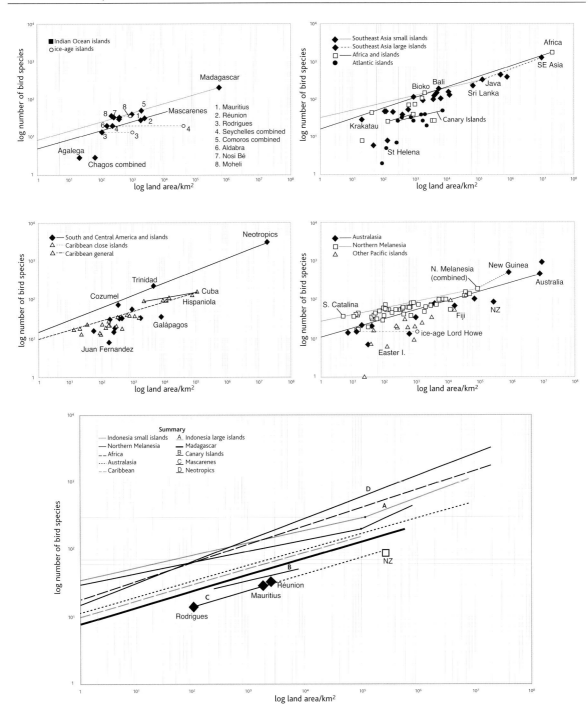

Figures 11.3 and 11.4 (see p. 272). Island native terrestrial bird (11.3) and reptile (11.4) species number against island area, compared by faunal area and against continents or large continental countries (whole continent figures not available for reptiles); log-log plot with saturation lines showing maximal species per km² for each region, usually on land-bridge islands. Some shorter lines link oceanic islands, often by moving the 'main' line down parallel until the lower island diversity is reached, suggesting saturation at the relevant 'distant island' level of diversity. Colonisation rates on very isolated islands may be so low that saturation is never reached within the island's lifetime. Far-oceanic Pacific islands are too disparate in sourcing and history for a meaningful line. Species numbers include Holocene extinctions where known. For 11.3 (birds), data mainly from Case (1996), Mayr & Diamond (2001), Goodman & Benstead (2003), Ricklefs & Lovette (1999), Caribbean and Sundaland handbooks and (western Indian Ocean) this study. **Notes: 1.** See text for discussion of the significance of greater island size in ice-age sea-level low-stands for indicated islands, and for the 'broken-stick' line New Guinea–Northern Melanesia. **2.** Note difference between Cuba–Isle of Pines (on Cuba's shelf) and the main oceanic Caribbean line.

The Hawaiian Islands (when combined equivalent to Jamaica in size and diversity) are exceptional in that, despite their remoteness, they are rich in bird species, due to internal radiations followed by inter-island exchanges[60]. Among Mascarene vertebrates the day-gecko radiation in Mauritius is similar (albeit on a smaller scale) with five species descended from a single original invasion, two of which invaded Réunion and diverged further (p. 67). Several such radiations have made the reptile fauna of Madagascar very rich[61]; in stark contrast to the birds, the Malagasy saturation line for reptiles is above all other island series (Figure 11.4). If, as with birds, we move the line down exactly parallel, we find first that it exactly matches the Comoros, then Mauritius–Rodrigues, whose line extended towards larger islands goes precisely through New Zealand, the largest island in extreme isolation[62]. The Seychelles show up as over-saturated even against the Comores line, and Réunion, Aldabra and Agalega as markedly under-saturated. Compared to birds, reptiles are so poor at over-water dispersal that even three million years is not enough for Réunion to even remotely approach the diversity seen in the older islands of Mauritius and Rodrigues, despite being close to Mauritius and down-current from it. In contrast, the shrinkage of the Seychelles from the vast Pleistocene island has left a reptile fauna over-rich for the current land area. In other words, the effect Temple claimed improbably for birds is very much in evidence for reptiles[63].

Introduced animals in theory and practice

Heavyweight theoreticians have focused on the Mascarenes in recent years to elucidate what favours or negates bird introductions on islands. Attempts using the Mascarenes have been made to support both the 'All or Nothing' theory (some birds always establish, others always fail) and the view that the more exotics already present, the harder it is for a new ones to establish. The Mascarenes' introduction history supports neither theory, despite the authors' contortions to explain what they see as anomalies – in effect, 'exceptions that prove the rule'[64].

Unfortunately the theorists often fail to recognise (or simply dismiss) historical data that throw light on ecological reasons for successes and 'failures', and also take inadequate account of the changing availability of habitats in the islands. Birds that established and thrived for decades or even centuries are deemed 'failed' colonists if they later die out (e.g. Java Sparrow), although there may be good reasons arising from changing agricultural practices or subsequent introductions. For instance, Daniel Simberloff dismissed the ending of cereal cultivation as causing the decline and extinction of Java Sparrows on Mauritius and Réunion because the birds survived some decades after sugar replaced grain. However, he overlooked the arrival of a competitor: House Sparrows, introduced on both islands after the agricultural changes had taken place[65]. It appears that Java Sparrows could survive on largely cereal-free islands provided there was no other bird taking seeds in its favoured size-range, but once the more adaptable House Sparrows arrived they slowly died out.

A curiosity of much of this theoretical literature is that the authors chose to discuss only passerines, as if passerines and non-passerines do not interact or are not being influenced in similar ways by the environment[66]. This bias must make meaningless the calculations by Michael Moulton and colleagues as to whether establishment is affected by numbers of species already present, an error compounded by using spurious early introduction dates for certain Réunion species[67]. By steering clear of non-passerines the authors did not address the rapid 'failure' of several ground-nesting gamebirds in Mauritius when the mongoose was introduced, nor the possible interactions between granivorous ground-feeding doves, Grey-headed Lovebirds and passerines (e.g. Cardinal Fody, House Sparrow) that overlap ecologically. In fact, interspecific competition between exotic species has probably been important in several other species becoming rare or extinct in Mauritius and Réunion: Red Avadavat vs Common Waxbill (both islands), White-rumped Munia vs later seedeaters (Réunion only), Meller's Duck vs Mallard (Mauritius only). In these cases (and others, below) later arrivals established easily and apparently ousted their well-established predecessors, negating the priority theory. The recent establishment in Mauritius of Laughing Doves despite the presence for 200 years or more of two ecologically similar pigeons (Zebra Doves and Spotted Doves) also contradicts this hypothesis. The loss of mixed arable land followed by the addition of a third species appears to have led to an ecological separation of a pair of species that formerly co-existed – Cape Canary versus Yellow-fronted Canary on the arrival of Village Weavers[68]. In the 1860s in Mauritius (and apparently also in Réunion), both canaries occurred together in lowland fields, especially if rich in weeds, but after the establishment of Village Weavers the Cape Canaries retreated to high ground and died out. In Réunion, with higher uplands, the birds survived in clearings and heathlands in the heights, the lowlands occupied by Yellow-fronted Canaries and Village Weavers. Madagascar Partridges have likewise survived in the Réunion uplands, with Grey Francolins confined to lower regions, whereas in Mauritius only the francolin survives[69].

The Grey-headed Lovebird is a particularly interesting case. In Mauritius it arrived early (c. 1735), before any seed-eating passerines. It established

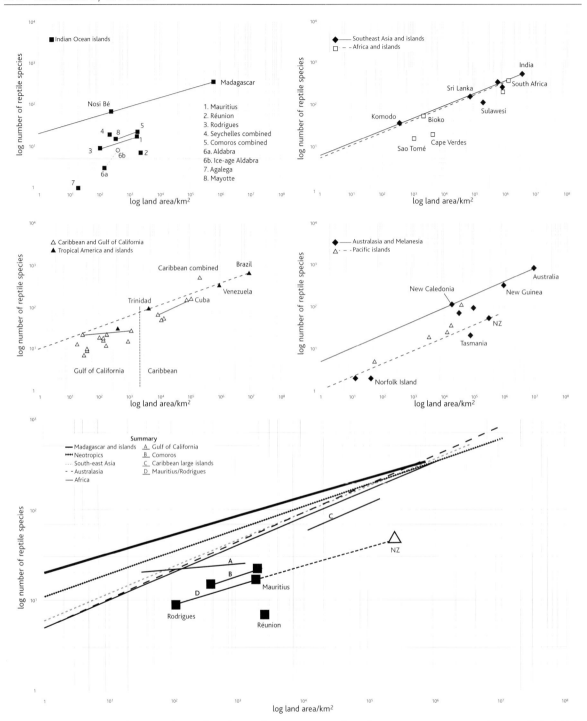

Figure 11.4. See 11.3 (p. 270) for protocol. Data from Case *et al.* (2002), www.earthtrends.wri.org, Mayr & Diamond (2001), Cheke (1984), this study and miscellaneous sources (adequate reptile data is readily available for many fewer islands than for birds). Total reptile numbers on the earthtrends website are generally higher than those from specialist sources for the same islands when available, differences which cannot be accounted for by sea snakes, crocodiles and marine turtles. **Notes 1:** The saturation line for Madagascar is through the offshore island of Nosy Bé (data from Andreone *et al.* 2003). **2:** The Mauritius–Rodrigues line is parallel to the Malagasy line, but Réunion falls far below it. **3:** Seychelles diversity is above Comoros for reptiles, but well below it for birds (see text). **4:** Hawaii is excluded as no reptiles got there. **5:** The combined Caribbean reptile figure is anomalously high due to extensive island radiations and endemism leading to little species overlap between islands (cf. the unexpectedly high bird diversity in Hawaii – see text). **6:** Moving the Australasian 'main' line down parallel to pass through New Zealand also picks up Tasmania and the isolated islands of Norfolk and Lord Howe, suggesting equivalent degrees of isolation (surprising for Tasmania).

rapidly and became a cereal pest (p. 94). In Réunion it was not reported until the 1840s (p. 141), when lots of other exotics were already present, and cereals were on the way out. The lovebird never became common, despite further releases. In Mauritius it declined gradually after cereal-growing ceased (again despite new releases) into the early 20th century, finally disappearing around 1950. The later-arrived Ring-necked Parakeet, supposedly competing for maize seed, has been blamed for its disappearance, but this does not explain the lovebird's demise in Réunion (no parakeets and more maize), and in any case this bird prefers small grass seeds, even wheat or rice being above its optimal item size[70]. We suggest that the Grey-headed Lovebird may have suffered from a lack of nest sites. For a lowland hole-nester, the disappearance on both islands of the last vestiges of lowland forest would have created an acute shortage of suitable over-mature trees – the same problem that forces Ring-necked Parakeets into Mauritian native forest (where they cannot feed) to find nest-holes (p. 169–170). In 1730s Mauritius there were no other hole-nesters in the lovebird's size range, but on both islands first Common Mynas and then House Sparrows were introduced, these probably becoming serious competitors once nest-sites became scarce. Indeed, mynas and sparrows, added to the abandonment of wheat farming, may have prevented the lovebird fully establishing in Réunion. On both islands the arrival of seed-competitors in the form of Village Weavers may have been the last straw. In Rodrigues, with mixed cropping and ample grassland in the cattle walk, the lovebird survived tree loss, mynas and sparrows (though it did not have to contend with Village Weavers) until hunted out as a crop pest in the 1950s (p. 197).

Mascarene history shows that the success or failure of introductions depends on many factors and is not easily predicted, a conclusion also finally reached by some theoreticians[71]. However, one should be cautious of another common argument – that these introduced birds have had little or no impact on the native avifauna. It is true that most damage has been done by introduced mammals, but this is simply because, for mainly cultural reasons, the birds have been in some way anthropophilic, i.e. they inhabit open habitats created by man, or those (e.g. wetlands) where native species had already disappeared. This should not lead to complacency about avian introductions. Red-whiskered Bulbuls penetrate all dry-land habitats, native or anthropogenic, and affect them by predating nests, distributing seeds of exotic plants, and possibly competing for food. Peking Robins, recently established in Réunion (pp. 182–3), are also forest birds likely to impact upon the native fauna and flora. There are plenty of species that would be very dangerous for the endemic forest birds, including several of the numerous parrots that have been recorded as escapes in Mauritius[72], and the same applies to the exotic day-geckos already present. *Any* exotic introduced without prior study carries grave risks for the native biota.

TAILPIECE

LA RAVINE SAINT-GILLES

A poem

The gorge is full of shade, where under the smooth bamboos
The sun at zenith has never lit,
Where the percolation of natural springs
Unite in the flaming silence of midday.

From the hardened lava to the mossy fissures,
Through lichens the waters fall in trickles
Losing themselves, then cutting sudden outflows,
Surge and circle in the depths amid the white gravel.

A pool of blue-black reflections lies there
Gloomy and cold, while along the heavy rocks
A trellised vine hangs its rosy bellflower,
Amongst thick lawns of velvet tussocks.

On the soaring banks where the cactus splashes,
Wandering from *vetiver*[1] to flowering aloes,
The cardinal, clothed in his scarlet plumes,
Bothers the white-eyes in their cottony nests.

The yellow-beaked mynas and the green lovebirds
Watch the sleeping water from the height of pointed bluffs.
And a golden flight is heard turning and humming
In a bright sunbeam around the black hives.

Wheezing their hot breath over the bushes
Hanging over the rough-cut grassy path,
Cattle from Tamatave[2], robust and indolent,
Moisten the air of the ravine that the living water has washed.

And the great butterflies with magnificent wings,
The pink grasshoppers, familiar leapers,
Fearless of the whip, alight in thousands
On their calloused hump and peaceful flanks.

On the rocky slope where the sun-flame penetrates,
The long and supple lizard is drunk with sleep,

Then, suddenly, seized by a frisson of well-being,
Displays its emerald back to the sun.

Under mossy nooks where the well-fed quails
Avoid the burning of the hot savannah,
Prowling cats slink on the velvet of their cautious paws,
Eyes half-closed in food-lust.

And a black slave, sitting on a block of lava,
Keeper of the scattered cattle grazing the bitter herbage,
A red rag around his loins, hums a Sakalava tune[3],
And dreams of the Great Island[4] while gazing at the sea.

Thus, on the two banks of the deep gorge,
All living forms that swarm in this world
Shine, sing and dream in the same moment;
But forms, sounds, colours suddenly cease.

Further down, all is mute and dark in the womb of the gulf –
Ever since the mountain emerged from the waves,
Roaring with jets of granite and sulphur,
And congealed under heaven's vault and knew repose.

Barely does a cloud-gap, sparkling and blue,
Let one glimpse on a patch of pure sky
The flight of tropicbirds towards Rodrigues or Ceylon,
Like a snowflake lost in the azure.

Outside this luminous point that flickers on the waves
The ravine sleeps in motionless night,
And when a rock dislodged from on high falls in,
It wakes not even an echo of its sound.

Charles-René-Marie Leconte de Lisle, 1857[5]

NOTES

1. *Vetiveria* spp. is a genus of fragrant native Mascarene grasses; an introduced species is used for essential oil production.
2. Sakalava is the Malagasy tribe from which the cowherd slave was presumed to come.
3. Great Island – *La Grande Ile* – is a commonly used name for Madagascar.
4. Tamatave is a town in Madagascar.
5. This poem was first published in 1857, and republished in anthologies e.g. Eggli 1943, Damour 1973. Leconte de Lisle, poet and sometime revolutionary, was a major literary figure in France, a member of the Académie Française and the Légion d'Honneur. Born in Réunion in 1818, he was sent by his father to the East Indies to learn about life, and thence to France to complete his education. He returned infrequently to Réunion, the last time in 1843–45. After his political ambitions failed in the turmoil of 1848, poor and jobless in Paris, he wanted to go home. However, Réunion was gripped by the economic consequences of liberating slaves, a policy he had enthusiastically endorsed; refused a job in Saint-Denis in the 1850s, he never returned (*Encyclopaedia Britannica* 11, v16:356, 1911; Eggli 1943, Damour 1973).

La ravine de Saint-Gilles evokes the dry-season waterfall and deep pool in this small west-coast gorge, where day-geckos, Cardinal Fodies, white-eyes, mynas, lovebirds, quails and feral cats could be seen, while tropicbirds flew overhead. We have translated '*perruches*' as 'lovebirds' as it is Mascarene French for *Agapornis cana*, not *Psittacula* spp. which are '*cateau*' (and already extinct in Réunion), and '*colibri*' (= hummingbird in standard French) as white-eye, as it is Olive White-eyes that seek flowers for nectar in Réunion. Leconte de Lisle had clearly been too long in France – '*colibri*' was never used in the Mascarenes, though the Seychelles Sunbird *Nectarinia dussumieri* has been known as *kolibri* since 1787 (Cheke 1982, Skerrett *et al.* 2003). The poem continues for another four verses, omitted here as they leave the *ravine* and wander off into gloomy philosophising.

The Dodo. From Belloc's *A bad child's book of beasts* (1896), featuring Hilaire Belloc's verse with Basil Blackwood's cartoons.

NOTES

In these notes 'pers. comm.' is used for all individual communications (letter, e-mail, conversation), and all are to ASC unless otherwise stated; likewise personal observations are by ASC unless qualified. In Chapter 10 the equivalent references are to or by Carl Jones.

Preface and Acknowledgements

1. Strickland (1848), introduction to *The Dodo and its kindred*.
2. Rees & Rees 1956.
3. Initially through Strickland & Melville's seminal Dodo book (1848), but thereafter popularised in various ways, most notably by Lewis Carroll's *Alice in Wonderland* (1866), with John Tenniel's evocative drawings.
4. This expression completely escaped the linguistic encyclopaedist Eric Partridge: there is no trace of it in either his *Dictionary of Slang and Unconventional English* (1961) or his *Dictionary of Catch-phrases* (1977), where the 'dead as' items include door-nails, mutton, and Julius Caesar! However the *Oxford English Dictionary* has references from 1904 and 1935, and we have found one from 1937 which may originate earlier. 'Elinor Mordaunt' (Evelyn May Clowes) used it in her autobiography *Sinadaba* (Mordaunt 1937) – she had been in Mauritius from 1897–99 (*DMB*:279, P. J. Barnwell). We suspect, like so many expressions, that it had an oral history before it appeared in print, probably arising in mid-Victorian England as a play on 'dead as a doornail'. As suggested by Partridge's failure to record it, it rarely appeared in print before the 1960s (we have found only two further citations: Ley 1941 and Mockford 1950), though Ralfe Whistler, a veteran dodo collector, remembers using it in the late 1930s. 'As extinct as the dodo' is attested earlier, already in metaphorical use by 1870 (Cheke & Turvey, in prep.).
5. 'Dodo' first appeared in print in Thomas Herbert's famous travelogue (1634); in his 1638 edition he claimed it was of Portuguese origin and 'had reference to her [the bird's] simpleness'. This etymology from the Portuguese *doudo* (= stupid) has stuck (see older editions of the *Oxford English Dictionary*), though it is much disputed. The almost simultaneous introduction of *Dodar* (in 1638; see Strickland 1848) from the Dutch *dodaarse* ('rounded/bulbous arse') suggests that the name may have had two independent origins, coincidentally similar. *Dodaarse* dates from the first Dutch visit in 1598 (Grimmaert, in Keuning 1938–51). Herbert was in Mauritius in 1629 but another traveller, Emmanuel Altham, on his way east, wrote home from Mauritius on 18 June 1628 that he was sending to England via a returning ship a live example of "a strange fowle...called by ye portingalls a Do Do" (A. Newton 1874). There is, however, no evidence from Portuguese sources that they used the name (Strickland 1848), or indeed referred to Dodos in any way at all, though North-Coombes (1980) nonetheless thought it likely they did name it first. Gould (1998) and Turvey & Cheke (in prep.) discussed the history of the idea that Dodos were 'fat and stupid' and 'asking to become extinct'.

6. It was commonly argued in the 17th and 18th centuries that even fossils were not really extinct, but they survived undiscovered in an unexplored corner of the planet (Rudwick 1976; Grayson 1984); in a few cases (such as crinoids, Coelacanth and *Metasequoia*) this view turned out to be correct.
7. "M. G. Cuvier and several other naturalists regard the existence of this bird [the Dodo] as very doubtful . . . my compatriot Bory de St.Vincent, during his stay in the Isles of France and Bourbon was unable to procure any hint of these animals"; "M. Cuvier regards the existence of these birds [the Rodrigues Solitaire] as even more doubtful than that of the dodo" – Jean Lamouroux, editorial footnotes to his edition of Buffon's *Histoire naturelle des oiseaux* (Lamouroux & Desmarest 1824–30). Réné Lesson (1828, 1831; Hachisuka 1953), noted French naturalist and visitor to Mauritius, attempted to refute Duncan's careful paper (1828) by claiming the dodo was really a Cassowary, despite Duncan being keeper of the Ashmolean Museum in Oxford, and having an actual Dodo's head in front of him! In fact, decades earlier, before such doubts had arisen, the encyclopaedist Mauduyt (1784) had formally declared the Dodo extinct. Down-to-earth characters, such as the British naval doctor James Prior (1819), on Mauritius in 1801, were also content to accept the extinction as a simple fact: "The Dodo, a large clumsy and singular bird, which like the ostrich could not fly, was formerly found here according to naturalists, but is now extinct." For a discussion of prevailing beliefs on fossils, extinction and the age of the earth in the 18th and early 19th centuries see, *inter alia*, Rachels (1990), Roger (1997), Worthy and Holdaway (2002), and Turvey & Cheke (in prep.). Bishop Ussher's calculated creation date of 4004 BC was widely believed in Protestant Christian circles (Baxter 2003), and included in Bibles from 1701 onwards.
8. Darwin (1839); to his sister Caroline, Darwin wrote from Mauritius: "we are all utterly homesick. All we have seen here is very pleasing. But there is no country which has now any attractions for us, without it is seen right astern, and the more distant and indistinct the better" (Barlow 1945, Burkhardt 1996). See Hays (1973), Jenkins (1978) and Rachels (1990) for further discussion on the origins of evolutionary theory. Alfred Wallace was the inspiration for David Quammen's *Song of the Dodo* (1996).
9. This was two years before land birds were first protected in Newton's native Britain (Barclay-Smith 1964), where seabirds had been protected earlier [1869], spurred by the realisation, explored by Edward's brother Alfred, of an extinction much nearer home – that of the Great Auk *Pinguinus impennis* (Newton & Gadow 1896); Alfred was for many years Professor of Zoology at Cambridge University. Edward had been promoted to Colonial Secretary in Mauritius in 1868, effectively deputy-governor, and thus able to initiate policies for the then colony (Anon. 1897).
10. The 'evil empires' of, respectively, the *Star Wars* and *Star Trek* series. The good guys in *Star Trek* ('The Federation') have strict rules about not interfering with the ecology and culture of newly discovered alien life-forms.
11. Nicholls (2006).
12. See Cheke (1979a, 1987a,b,c,d,e), Cheke & Dahl (1981).

INTRODUCTION

1. Strickland (1848)
2. Interview with University of Mauritius ecologist Vincent Florens in *Weekend*, 9 Jan. 2005.
3. Kingdon (1990) gave an illustrated account of habitat islands in Africa, also touching on islands of the Indian and Atlantic Oceans, though his geological dates for Mascarenes are incorrect.
4. Quammen (1996) has written a very readable account of island biogeography and the effects of human intervention. Menard's *Islands* (1986) is an illustrated introduction to both the formation of islands and their discovery.
5. Bahn & Flenley (1992) showed how careful investigation can reconstruct the past of a completely devastated island, but Hunt's (2007) re-analysis of the ecological history emphasised the importance of *Rattus exulans*, and that the causes of ecological collapse were more complex than often thought.
6. Ziegler (2002) is an up-to-date survey of Hawaii's ecological history. Carlquist's (1970) classic is still worth reading, but more recent subfossil discoveries of extinct birds have dramatically altered the overall picture (Olson & James 1991, James & Olson 1991); Royte (1995) is a popular illustrated account. Mitchell (1989) is a good introduction to wildlife and human history in the Polynesian Pacific as a whole.
7. Stevens et al.(1988) explored the history of Polynesian colonisation of New Zealand; more specifically ecohistorical books are Anderson's (1989) and Worthy & Holdaway's (2002) accounts of moas and their demise. Gill & Martinson (1991) provided an illustrated account of the lost avifauna. King (1984) told the devastating story of introduced predators, in the process giving a general overview of New Zealand's ecological history; Towns et al. (1994) considered reptile extinctions in more detail.
8. In addition to introducing rats *Rattus exulans*, pigs and dogs, the Polynesians cleared almost all the lowland forest on the main Hawaiian islands, eliminating the most diverse forest habitat (Olson & James 1984).
9. Ottino (1976), and more recently Dewar (2003) and Wright & Rakotoarisoa (2003), reviewed the human colonisation of Madagascar; there is some evidence of temporary human visits from c. 100 AD on, with a coastal trading settlement dating from c. 450; the first permanent colonisation dates from c. 800 by people from what is now Indonesia. The island was subject to the Sumatran kingdom of Srivijaya c. 900–1000 AD (Allibert 1988). Preston-Maffham (1991) provided a well-illustrated introduction to Madagascar's unique fauna and flora; Jolly et al. (1984) gave a more detailed taxonomic and ecological overview, while Goodman and Benstead (2003) reviewed the entire biota. A general guide with good wildlife coverage is Bradt (2005).
10. Flacourt's (1658) remarks have been summarised by many authors on Madagascar, e.g. Robyns (1982).
11. The dating and sequence of the break-up of Gondwanaland is still being debated, as is its contribution to the Malagasy fauna (e.g. Fröhlich 1996, Rage 1996, Krause 2003, Vences 2004a), Noonan & Chippindale (2006), Yoder & Nowak (2006).
12. Christmas Island, an elevated coral island of 135km^2 rising to 356m, was first visited by Dampier in 1688, but landing was very difficult and there were very few other visits before 1886. Although the island is only 195 miles (312km) from the nearest land (Java), there are seven endemic land birds, two seabirds and two rats (the latter both extinct). Details and references are in Andrews (1900) and Gray (1981).
13. St Helena, at 122km^2 similar in size to Easter Island and Rodrigues, lies 1,913km off the African coast, and a great deal further from anywhere else except Ascension Island, 1303km to the northwest. It was discovered in 1502 covered in forest and supporting huge seabird colonies, though there were only a few landbirds, all endemic (Ashmole & Ashmole 2000, Rowlands et al. 1998, Olson 1975a). The island was not permanently settled until 1644, but by then pigs and goats (plus rats?) released by the Portuguese in 1513 had seriously damaged the forest, and all but one of the birds were extinct. Grove (1995) discussed the island's eco-political vicissitudes.
14. References to the early literature were listed in Cheke (1987a), supplemented in this book. Ly-Tio-fane (1984) and Grove (1995) covered the personalities and politics involved in the environmental exploration of Indian Ocean islands during the 17th to 19th centuries.

CHAPTER 1. GEOGRAPHY OF THE MASCARENES

1. *A relation of some yeares travaile...* (Herbert 1634).
2. There are numerous general books and guides to the islands; *The Lonely Planet guide to the Mascarenes and the Seychelles*, regularly updated (e.g. Dodd & Philippe 2004), is as good as any, though, like most, a little spare on Rodrigues; Cubitt (1977) and Georges & Vaisse (1998), both highly illustrated with sparse text, are good for general impressions. Island co-ordinates (for roughly the centre of each island) and inter-island distances (measured) are from Admiralty Chart No.4702 *Chagos Archipelago to Madagascar* (1974). Published inter-island distances often differ from each other and from those given here.
3. Further details in Duncan & Richards (1991), Kearey & Vine (1996) and Parson & Evans (2005); other theories of the tectonic and volcanic history of the Mascarene area are discussed by Montaggioni & Nativel (1988), Saddul (1995), and Sheth (2005).
4. Menard (1986), Condie (1997); Courtillot (1999) contains a readable account of hot-spot theory. Although widely accepted, the hotspot explanation for the Deccan Traps and Indian Ocean island series to Réunion has been challenged by Sheth (2005), without disputing Mascarene dating.
5. Miller (1983) gave a general introduction to continental drift; more technical aspects are covered by Condie (1997) and Kearey & Vine (1996). Krause (2003) and Parson & Evans (2005) summarised the timing of events relative to the Malagasy area.
6. Saddul (1995).
7. Montaggioni & Nativel (1988)
8. McDougall et al. (1965), Saddul (1995).
9. Giorgi & Borchellini (1998).
10. Climatic details from Soler (2000), Cadet et al. (2003), with additions from Cadet (1980), Robert (1980), Cheke (1987b). More rain-gauges in recent years have considerably fine-tuned Réunion rainfall maps, which now have contours to 12m, whereas they stopped at 8m in the 1980s; rainfall distribution in Cadet (2003) and Soler (2000) differs significantly from earlier maps, especially in the wet southeast sector of the island.
11. The IGN 1:100,000 tourist map of Réunion is helpful for appreciating the topography, as it highlights the rugged relief with shading; Bénard (2002) has spectacular aerial views. General details are from Robert (1980), Cheke (1987b), Montaggioni & Nativel (1988) and Cadet et al. (2003). The volcano erupts almost annually, often several times, with flows frequently cutting the coastal road, most recently in 2002, 2004, 2005–6 and 2007 (see maps of recent flows in Girard & Notter (2003) and current information from the Observatoire Volcanologique du Piton de la Fournaise at http://volcano.ipgp.jussieu.fr and http://ovpf.univ-reunion.fr/).
12. Hence the highest mountain's name 'Piton des Neiges' (= snowy peak). Snowfalls are very intermittent and decades can pass without any; the most recent was in 2003.
13. The most comprehensive general book on Réunion is Lavaux (1998), but to fully appreciate the landscape, see Bénard's (2002)

aerial views. See Montaggioni & Faure (1980) for an account of the reefs.

14. Basic geographical details are from Saddul (1995, 2002), and various editions of the 1:100,000 map published by Ordnance Survey International, the latest being Y682/DOS529, *Mauritius & Rodrigues* 1994. No book on Mauritius rivals Lavaux (1998) on Réunion, but the better tourist guides (e.g. Heady *et al.* 1997, Richards *et al.* 2006) are a good start; Dormann & Rossi's (1991) aerial photos help appreciate the landscape, but forested areas of ecological interest are somewhat neglected.

15. Saddul (1995), Middleton (1996) and pers. obs.

16. Padya (1989).

17. These islands range from 2.5 miles (4km) to 15 miles (24.5km) offshore.

18. Details from North-Coombes (1971), Montaggioni (1972, 1973), Saddul (1995, 2002). Roberts & Roberts (1998) is an illustrated introduction, and Jauze (1998) an economic geography of Rodriguan life. Brial (1996) is the only adequate survey of the cave system.

19. For fuller discussion of island biogeography see, in order of technical complexity, Quammen (1996), Gorman (1979) or Whittaker (1998).

20. Scarth (1994) discussed types of volcanos and how they work.

21. Condie (1997), Ziegler (2002).

22. Colonisation of remote islands is discussed more fully by Carlquist (1965), Cox *et al.* (1976), Gorman (1979), Menard (1986), Kay (1991) and Whittaker (1998).

23. There is surprisingly little good data on vegetation rafts; Gorman (1979) described one he examined in Fiji.

24. Rafting of living reptiles has only rarely been observed, but there was a dramatic case in the Caribbean after hurricanes in 1995 when several iguanas *Iguana iguana* from Guadeloupe, carried on a huge mat of vegetation, landed on Anguilla some 185 miles (300km) northwest, subsequently breeding; iguanas also reached other islands at the same time (Censky et al. 1998). France Staub saw six unidentified dark geckos land off a log on Albatros I., St. Brandon in December 1973 (Cheke 1975). Day-geckos *Phelsuma* spp., found widely across the Indian Ocean, lay hard-shelled eggs which in some species are glued to the substrate (McKeown 1993). Drew Gardner (1985) found that fresh eggs of *P. sundbergi* from Praslin (Seychelles) could be fully submerged in sea water for 24 hours and still hatch – eggs riding on floating vegetation, perhaps only splashed with salt water, could presumably survive many days or even weeks at sea. Giant tortoises *Dipsochelys dussumieri* float easily, regularly go swimming in the Aldabra lagoon (sometimes being swept out to sea; Coe 1995), have survived several days at sea (Arnold 1979), and can live for months without food (e.g. North-Coombes 1971, Cheke 1987a) – ideal attributes for ocean rafting. In 2000 an Aldabra tortoise was found by Spanish fishermen alive and floating in the middle of the ocean about 300 miles (480km) north of the granitic Seychelles, having drifted some 950 miles (1,500km) northeast from Aldabra (Adrian Skerrett quoting George Rocamora, pers. comm.). Most remarkably, in 2004 one walked out of the sea at Kimbiji (Tanzania), 740km WNW of Aldabra (Gerlach *et al.* 2006); judging by the growth of barnacles on it, this animal had been floating at sea for 6-7 weeks. Back in 1789 Ohier de Grandpré (1803), pondered this question: "Does the land-turtle or tortoise ever swim or undertake long passages by sea? To throw light upon this question it may be useful perhaps to observe that the Sechelles-Islands abound in the species of tortoise. How did they come there? Moreover tortoises taken at the isle of Pralin [Praslin], deposited in the inclosure of that place and marked on the back with a circle made by a cooper's screeving-iron, have been re-taken three leagues off on another island called l'Isle aux Cerfs, near the *barachoas* [landing] at Mahé [now Port Victoria, on Mahé island]. Others, put into the inclosure at Isle aux Cerfs, have been re-taken at Mahé . . . at least the distance of a league." Praslin is in fact 28 miles (45km) across open sea from Port Victoria, and Cerf I. 3 miles (5km) away in sheltered waters (distances from Skerrett *et al.* 2001b).

25. Gray (1981), Atkinson (1985), Dowler *et al.* (2000). It is now thought that all Madagascar's non-flying mammals arrived by sea rather than being stranded when Gondwanaland broke up (articles in Goodman & Benstead 2003, Rabinovitz & Woods 2006), but the island is a huge 'target' and the timescale very long.

26. See Carlquist (1965), Nussbaum (1984a,b), Mielke (1989), Kay (1991). Native stream and lake fish in the Mascarenes are derived from estuarine species dispersed by oceanic currents; they return to salt or brackish water to breed (Staub 1993).

27. Vences *et al.* (2003, 2004).

28. See e.g. Paulay (1994).

29. There were five low-stands >60m of 40–50,000 years duration in the past 500,000 years (Rohling *et al.* 1998, Siddall *et al.* 2003), though periods with water >100m below present were much shorter. If the last eruptions on Rodrigues took place during an early Pleistocene low-stand, there would have been ample land for the biota to survive on.

30. Amongst birds, *Psittacula* parakeets, cave swiftlets *Aerodramus*, cuckoo-shrikes *Coracina*, *Hypsipetes* bulbuls (merles), white-eyes *Zosterops*; these tie in with Asian genera found in the Seychelles and Madagascar (magpie-robins *Copsychus* etc.; see Keith 1980). Asian flying-foxes *Pteropus* spp. and free-tailed bats *Mormopterus* also spread across the Indian Ocean (Chapter 4). 'Flying-fox' is the generic term in English for the large fruitbats in the Indo-Pacific genus *Pteropus*; there are other unrelated fruitbats in Africa and (in addition to flying-foxes) in Madagascar.

31. Saddul (1996), Slingo *et al.* (2005), New *et al.* (2005), Payet (2005) and Admiralty Chart 5126 (1–12).

32. Kieffer *et al.* (1993); Mourer-Chauviré *et al.* (1999) quoted the date range as 230,000–180,000 ya. For Réunion's structural geology see Rivals (1989) or Montaggioni & Nativel (1988).

33. Whittaker (1998).

34. Tirvengadum & Bour (1985), Cheke 1987a, Saddul (1995). Rodrigues had more endemic bird genera than Mauritius (4:3), but Mauritius also has two endemic genera of skinks (which later colonised Réunion) and an endemic family of snakes, the Bolyeridae, related to boas (Guibé 1958, p. 72–74).

35. Strahm (1993): Several plant families and genera have radiated spectacularly in Mauritius but little or not at all in Réunion - examples are *Syzygium* (Myrtaceae), screw-pines *Pandanus*, and ebonies *Diospyros* (Bosser *et al.* 1976–). Rodrigues's current small size and drier climate makes comparisons difficult, but it has nearly as many endemic plant genera as Mauritius and Réunion, including one with no close relatives anywhere – *Mathurina* (Turneraceae). Snail details are in Griffiths (1996) and Griffiths & Florens (2006).

36. Davis & Barnes (1991), Strahm (1993); fern spores and orchid seeds are tiny and wind-dispersed.

37. Reptiles are poorer dispersers than birds, bats and plants, so Réunion acquired only about half the Mauritian lizards and tortoises, and no snakes. See previously cited references for plants, birds, reptiles, snails and butterflies.

38. DNA evidence from reptiles suggests no post-colonisation interchange between Mauritius and Rodrigues (Arnold 2000), and bird lineages likewise look largely separate. Only the flying-foxes *Pteropus niger* and *P. rodricensis* and the Malagasy Turtle Dove *Nesoenas picturata* were shared between Mauritius and Rodrigues, *P. niger* and the dove extending to Réunion.

39. Saddul (1995).

40. DNA studies indicate that ancestors of the tortoise and day-geckos reached Réunion before the pyroclastic events, so must have survived them (Austin & Arnold 2001, Austin *et al.* 2004a).

41. McDougall *et al.* (1965), Upton *et al.* (1967). Their visit to Rodrigues was very short and their samples came from the most

recent lavas; they failed to detect the older rocks described by Giorgi & Borchiellini (1998).

42. Sea level dropped several times during the last million years, sometimes by 140m (Rohling *et al.* 1998), possibly up to as much as 200m (Menard 1986). The last two major low-stands >100m were (with some date variation between studies) roughly 160,000–125,000 ya and 26,000–14,000 ya (Montaggioni 1972, Rohling *et al.* 1998, Siddall *et al.* 2003, Fig 4.1), with a weaker low-stand barely reaching 100m *c.* 80–70,000 ya. A drop of 100m would expand Rodrigues to nearly 400 square miles ($950km^2$, Spencer & Turner 2001, Turner & Klaus 2005; see 50 fathom contour on Admiralty Chart No.715), over half the current size of Mauritius, but only a few metres lowering would leave the current wide lagoon dry. Réunion and Mauritius would only increase by about 1% and 10% respectively in a 100m lowstand.

43. e.g. Dyment *et al.* (2003); Montaggioni (1975) gave an older date, *c.* 15mya, now superceded (see refs in Sheth 2005).

44. Upton *et al.* (1967), Montaggioni (1973).

45. Menard (1986).

46. Giorgi & Borchiellini (1998), who based their dating on a French thesis (Perroud 1982) that has been completely overlooked not only by biological writers (e.g. Strahm 1993, Blanchard 2000, Turner & Klaus 2005), but also by the international geophysical community, who still uncritically use the Upton/McDougall dates as if they applied to the whole island; at least Sheth *et al.* (2003) allowed some leeway in stating that "volcanic activity in Rodrigues ceased *c.* 1.5 million years ago". Back in the 1970s, Cadet (1977) queried the 1.5 my dating because of Rodrigues's ancient endemic plant genera, and in the 19th century it was always assumed to be an old island.

47. See Guého & Staub (1968); waves sometimes wash over the St Brandon atolls and cays in cyclonic weather; they have a depauperate flora and no native land-birds, bats or reptiles (Staub & Gueho 1968, but see note 24).

48. Duncan & Richards (1991).

49. Admiralty chart 4702.

50. Rock samples collected for this purpose by JPH have yet to be dated.

Chapter 2. First contact

1. Anon (1599), commonly referred to as the *True report*. 'Isola de Don Galopes' was actually an early name for Rodrigues (Toussaint 1972), but mariners not infrequently confused the two before longitude could reliably be measured; Mauritius (or in fact Maurits) was the name of the second *Stadtholder* (head of state) of the newly independent Dutch Republic (Moree 1998).

2. Tatton (1625). The English apparently lacked the Portuguese charts used by the Dutch. Before the calendar reform of 1752, the New Year began in England on April 1st, so March was in '1612' not 1613.

3. North-Coombes (1980) summarised what is known of early voyages in the Indian Ocean. Much of the historical argument depends on the interpretation of new islands appearing on dated maps, some reproduced in North-Coombes's book and Visdelou-Guimbeau (1948), but the copies are poor and in monochrome. To appreciate the old maps, better sources are Whitfield (1996, 1998) and Wigal (2000), where many relevant to the 16th-century history of Mascarene discovery are reprinted in colour, though the Malagasy area is often reproduced too small to read the place-names.

4. North-Coombes (1980); he also speculated that Mauritius and Réunion *may* have been seen by the seriously lost accidental 'discoverer' of Madagascar, Diogo Dias in 1500. Other dates are still widely quoted, using older sources now superceded by North-Coombes's research. His book is little known outside the Mas-

carenes, and was overlooked by, *inter alia*, Moree (1998) and Guébourg (1999), and even by Mauritians (e.g. Addison & Hazareesingh 1993).

5. North-Coombes (1980), Cheke (1987a). Describing Réunion for the first time in 1612, the Dutch reported flocks of goats, suggesting a release some time before, though the only known Portuguese landing was over 84 years earlier. Strict rules forbade Portuguese ships from making intermediary stops except at St Helena, which they stocked with cattle, goats, pigs, bananas and other fruit (Ashmole & Ashmole 2000). In Mauritius there were no livestock or fruit trees, nor did the Portuguese discover that valuable commodity, ebony, which was spotted almost instantly by the Dutch. The Portuguese are commonly alleged in the literature to have introduced monkeys, pigs, deer and other animals, but there is no evidence for this (p. 76).

6. The manuscript accounts of the 1598 landing were collated by Keuning (1938–51); Grandidier (1903–20), Verin (1983); Barnwell (1948) reprinted and Pitot (1905) paraphrased several early Dutch accounts. Moree (1998) included references to sources and details of published accounts.

7. Harmenz's fleet reconnoitred Rodrigues in September 1601, but having explored the reefs, decided the landing was too difficult for the fleet as a whole and pushed on to Mauritius. However some men from the *Wachter* landed, catching birds which were shared amongst the ships as fresh meat (North-Coombes 1971, Moree 2001, *précis* in Grandidier 1903–20, Verin 1983).

8. The only recent history of the island is North-Coombes (1971); the *Gelderland*'s artist sketched the island from offshore in 1601 (Moree 2001).

9. Blok's 1612 visit is confirmed by Bruijn *et al.* (1979–1987); his account was published by Brial (2001a). Blok had visited Mauritius with Heemskerk's fleet in 1602 (Bonaparte 1890, Verin 1983, Bruijn *et al.* 1979–87), but contrary winds prevented him reaching Mauritius on the 1611–12 voyage, forcing a landing at port-less Réunion (Brial 2001). Bontekoe (1646, Lougnon 1970) said Blok lost longboats and men in heavy seas in Saint-Paul bay, but Blok himself said nothing of this. Bontekoe (1646) is a classic of Dutch travel literature, the author a hero in the Netherlands to this day; the quote is from the 1929 English translation.

10. Lougnon (1970) published a collection of early visits to Réunion, including Castleton's (Tatton 1625) and Bontekoe's (1646). Grandidier *et al.* (1903–20) assembled and translated early journeys to all islands in the Malagasy area. The published chronicle of Verhoeff's expedition (author unknown) in De Bry's *India orientalis* (see Barnwell 1948), records him travelling (like Blok) with 13 ships, but Moree (1998) and Bruijn *et al.* (1979–87) listed only two ships landing at Mauritius on the return journey, one being Verhoeff's flagship the *Middelburg*. Both Admiral Verhoeff and his successor Simon Hoen died in the East Indies before the fleet started home (Barnwell 1948), so presumably after various battles the returning 'fleet' had shrunk to two.

11. In September 1601 Harmensz's crew found a Frenchman who had been stranded on Mauritius alone since 1599 (Barnwell 1948, Moree 2001). Dutch sailors were wrecked and stranded in 1615 for nearly a year, and some mutineers left there in 1617 were not rescued until 1620; similar punishment exiles occurred in Réunion. Mauritius was occupied by the Dutch from 1638–1658, but, again deserted, it hosted sailors from the wrecked *Arnhem* in 1662 (see Moree 1998). A dozen Dutch sailors were stranded on Rodrigues for three months in 1644 (North-Coombes 1971), and two English pirates were there alone for two years around 1707–9 (Alby 1989).

12. Bizarrely, Hachisuka (1953) attempted to use *Island of Pines* as evidence for the Dodo's survival in 1668, though these fictional dodos were found when the character George Pine arrived on the island in '1559', not when his progeny left five decades later.

13. This classic engraving has been widely reproduced (sometimes mirror-reversed) in both historical and biological publications, e.g.

Strickland (1848), Hachisuka (1953), Verin (1983), van Wissen (1995), Moree (1998).

14. Tortoises are not further described, but Mascarene giant tortoises *Cylindraspis* spp. subsequently became well-known. 'Turtle Dove' seems then to have been used in European languages to describe almost any kind of wild pigeon. The *Gelderland* journal of 1601 contains sketches of one of these, the extinct Pigeon Hollandais *Alectroenas nitidissima*. Mascarene 'parrots' were never clearly described in the 17th century, but the engraving includes an unnamed long-tailed parakeet up a tree, either the extant Echo Parakeet *Psittacula eques* or, more probably, the extinct Thirioux's Grey Parrot *P. bensoni*. The bat is shown absurdly hanging head upwards with wings spread, though the description fits the extant Black-spined Flying-fox *Pteropus niger* well enough. The palms are given peculiar beaded trunks and strange leaves; the 'date-palm' must be *Latania verschaffeltii*, the only Mauritian fan-palm; the one with the edible 'bud' is the Hurricane Palm *Dictyosperma album*, extinct in the wild but cultivated for the edible 'cabbage'.

15. Philips Grimmaert in Keuning (1938–51, vol. 4).

16. See Wissen (1995) and Moree (2001).

17. Hans Bouwer in Moree (2001); however, his report was not written until 1604.

18. Many early accounts (1602 onwards) were not published until 1646–48, so some names could have been inserted by editors; see Strickland's discussion (1848).

19. A. Newton (1874).

20. Macgregor (1983).

21. Wissen (1995) for the *kermis* story. *Dodaars* escapes standard Dutch dictionaries, but is given as the name for Little Grebe in bird books (e.g. Snow & Perrins 1998).

22. The artist's name was unknown until recently (van Wissen 1995), but is now thought to be Joris Laerle (Moree 2001, Hume 2003a). He inveigled himself onto the ship as midshipman, but was sacked for being next to useless; since he had training as an artist, he was re-assigned to keeping a pictorial record of the voyage (Moree 2001). The drawings were rediscovered in 1868 (A. Newton, 1875a), but the journal as a whole remained unpublished until 2001.

23. Several books give world seabird distribution details e.g. Harrison (1985).

24. The French nature artist Nicolas Robert made a series of illustrations for Gaston d'Orleans (Louis XIV's uncle) in the 1650s (Jackson 1993, 1999 Jaussaud & Brygoo 2004), including a Scarlet Macaw *Ara macao* which he labelled *Corbeau d'Inde*. We are indebted to Cécile Mourer-Chauviré for drawing our attention to this picture via the postcard published by the MNHM in Paris.

25. For example, Bontius's *Historiae naturalis et medicae Indiae Orientalis*, a long passage from which was quoted by Ray (1678), and alluded to by Buffon (1770–83) and Oustalet (1897). Ray also discussed and illustrated a 'Horned Indian raven' or 'Rhinocerot [sic] bird', clearly a large *Buceros* hornbill; Edwards (1760) used the same terms.

26. Buffon (1770–1783); Brisson (1760) had perceptively recognised that Bontius was describing a hornbill.

27. Hoffman (1680; see Grandidier *et al*. 1903–20, Cheke 1987a); he was a German employed in 1673–75 by Governor Hugo as a preacher (Moree 1998). In 1668 Jacob Granaet called the birds 'mischievous and unusually large ravens' (Barnwell 1948). The Dutch traveller Nieuhoff (1682), describing Blue-and-yellow Macaws *Ara araruna* seen in Batavia (presumably captive!), called them 'kakataws or Indian ravens'.

28. Oustalet (1897).

29. Verin (1983); Verin is a historian not an ornithologist, and in any case simply borrowed his anthology verbatim from Grandidier *et al*. (1903–20), errors included! Staub (1993) revived Oustalet's views, but suggested African ground hornbills *Bucorvus* spp. rather than Asian *Buceros*. More bizarrely, Bissoondoyal (1968) claimed to prove that Mauritius was discovered by Indians because the first Dutch visitors found '*Indian* crows' there [our italics], and since Indian voyagers were said to carry crows on their ships to help them find land, the island must therefore already have been visited by Indians! This perverse misinterpretation of early visitors' writings has been adopted by later writers keen to present Indians (the ethnic majority in Mauritius) as potential discoverers of the island (e.g. Napal 1979). While the 17th century 'Indian Ravens' were definitely parrots, there *are* now genuine Indian crows in Mauritius: House Crows *Corvus splendens*, which came as stowaways on steamships from India and Sri Lanka. The idea of carrying crows to find land makes sense in the context of House Crow behaviour in fishing villages. In Negombo (Sri Lanka) in June 2005, ASC watched in astonishment as many of these crows, seeking scraps, regularly flew miles out to sea to join fishing catamarans. If they accidentally joined a vessel on a longer voyage, they might spend their enforced journey flying off over the sea in search of land before returning – or not returning if land was discovered. Having seen this behaviour, sailors might then have deliberately carried crows for this purpose, but it has no bearing on the discovery of the Mascarenes, only on the 20th century arrival of House Crows.

30. See, for example, map in Whittaker (1998) on bird penetration into Pacific islands.

31. Herbert (1634), Cauche (1651). Because Herbert was unclear about dates in his narrative, there has been confusion as to when he was in Mauritius. Strickland (1848) said 1627, Barnwell always used 1628 (1948, Barnwell & Toussaint 1949), Alfred Newton (1874) preferred 1629, and others have left it vague. Penrose (1942) established the date as 1629, Herbert travelling on the *Hart*, whose log for Mauritius and Réunion was published by Barnwell (1950–54). The Woodcock is *Scolopax rusticola*, familiar to English squires as a game bird.

32. The story was succinctly told by A.Newton (1868) and Sclater (1915).

33. Moree (1998)

34. Mundy (1608–1667), Keast (1984).

35. Dodos were not reported on the Mauritian mainland; Evertz found some Dodos on an offshore islet in 1662 – see Cheke (2004a).

36. Khan (1927), Cheke (1987a).

37. Hoffman (1680); *toddarschen* is a German version of *dodaarsen*.

38. A.Newton (1868a, Cheke 2006a); Cheke (1982) documented similar name-swaps later in Mascarene history.

39. Cheke (1987a).

40. Wissen (1995).

41. The original illustration for the penguins-as-dodos first appeared in Clusius (1601) on the page facing his Dodo, in the section immediately following his discussion of Dodos; however, he was clearly illustrating and describing a totally different bird, the *Anser Magellanicus*, a penguin. The artist illustrating van West-Zanen's account (Sooteboom 1648) apparently picked the wrong picture in the book he was borrowing from! De Bry's engraving with the cassowaries (and grossly exaggerated tortoise) was clearly concocted from travellers' tales, not drawn from life. Strickland (1848) was scathing about these lapses of accuracy; Hachisuka (1953) and Verin (1983) reproduced the engravings. For further discussion see Hume (2006).

42. Strickland (1848), Hume & Cheke (2004). As Strickland noted, Bontekoe lost everything in a shipwreck and must have written his account from memory many years later.

43. Valentyn's description of Mauritius published in 1726 is a hopeless mishmash of old facts, including a half-hearted attempt to equate Dodos with Leguat's '*geant*'; see Grandidier (1903–20, Verin 1983). Grant (1801) carried this dubious tradition forward into another century!

44. Our view is that the story is probably true, and that most of those writing about it actually witnessed it. Rails reacting to cloths are not confined to the Mascarenes: "I had learned in New Zealand that the ground-dwelling Wekas or Maori Hens [*Gallirallus australis*] could not resist the fluttering of a white handkerchief and could be lured from hiding by brandishing one at ground level. Rowland [a camp-following flightless Aldabra Rail *Dryolimnas (cuvieri) aldabranus*] proved equally susceptible, making off with hankies spread on the bushes to dry. He tried the same tactics with sheets, whose weight defeated him, but he was not put off easily" (Gillham 2000).

45. Leguat's *Voyage et avantures* first appeared in 1707, and has since been reprinted many times in several languages. Full details and references to the '*geant*' saga can be found in Holthuis *et al.* (1971), Cheke (1983, 1987a) and North-Coombes (1983, 1991). The '*Avis indica*' is in the second of Collaert's (1580–1600) two sets of engravings.

46. Leguat (1707).

47. Buffon 1770–83

48. Duquesne (1689); see North Coombes (1991), Racault (1995)

49. Dubois (1674); his Réunion section is reprinted in Lougnon (1970).

50. Dubois's bird list has often been reprinted: e.g., Berlioz (1946), Barré *et al.* (1996).

51. Schlegel (1858) created this chimera, championed subsequently by, *inter alia* Oustalet (1897), Rothschild (1907), Hachisuka (1953) and North-Coombes (1983, 1991). Strickland (1848) followed Buffon (1770–76) in accepting Leguat's birds as flamingos, a view upheld by A. Newton (1907), Mortensen (1934), Carié (1930), Greenway (1967), Holthuis *et al.* (1971) and Cheke (1983, 1987a).

52. North-Coombes (1983) turned the usual argument on its head, arguing that travellers other than Leguat used '*flamant*' as a local name for what were in reality '*géants*' – but was unable to produce a description that could not be easily applied to a flamingo.

53. See Cheke (1987a)

54. Fuller (1987) still gave the *géant* the benefit of the doubt, though by the second edition (2000) he was less sure. Balouet & Alibert (1990) appeared to have no doubts about *Leguatia*. These books reproduced Frowhawk's spectacular plate of the supposed *géant* from Rothschild (1907).

55. Buffon 1770–83

56. Cauche (1651), Strickland (1848)

57. Hachisuka (1953).

58. Paulian (1961:11).

60. The complete list is: [from Schlegel] *Leguatia gigantea* and *Pezocrex herberti* (see text), [from Rothschild] *Necropsittacus francicus* (based on an attribution error, birds were from Réunion not Mauritius: Cheke 1987a), *Bubo leguati* and *Strix sauzieri* (owls from Rodrigues and Mauritius based on single anomalous bones since found to be within normal variation: Cowles 1987, Mourer *et al.* 1999), [from Hachisuka 1937, 1953] *Victoriornis imperialis* (a second dodo/solitaire from Réunion, see text), *Kuina mundyi* (an extra rail from Mauritius, see text), *Testudophaga bicolor* ('Rodrigues Chough', see text). *Foudia bruante* was named in the 18th century after a plate by Martinet in Daubenton's *Planches enluminées* for Buffon's encyclopaedia (1770–76), but although recognised by Rothschild and Hachisuka as the Réunion Fody, it actually appears to depict an aberrant Cardinal Fody *F. madgascariensis* (Berlioz 1946, Cheke 1983b) *contra* Moreau (1960), who thought the island might have supported two fodies. The real Réunion Fody, described by Dubois (1674), was quite different from Daubenton's bird, and the name *bruante* cannot be used for it (p. 42).

61. Schlegel (1858), Hachisuka (1953). Hachisuka was "never one to let the lack of evidence stand in the way of naming new taxa" (Olson *et al.* 2005).

62. The presence of *Dryolimnas* rails in Mauritius (Cowles 1987) might account for a second species. They have straight bills, but the extant White-throated Rail *D. cuvieri* is dull brown and russet rather than 'wheaten' (pale yellow-brown). Mundy's sketch (Mundy 1608–67, Hachisuka 1953, Fuller 2000, Box 11) is equivocal, having a slightly curved bill like the Red Hen, but showing normal-looking primary feathers which the hen lacked (see fig.2.4 and the Prague painting, e.g. Ziswiler 1996 & Fuller 2000), though he clearly described the wings as small and useless for flight; Mundy may have drawn the wing 'conventionally', as for a more normal bird. JPH's discovery (work in progress) that that the Mauritian *Dryolimnas* was a large endemic species, Sauzier's Wood-rail, makes it possible that some travellers *did* see second flightless rail, and as White-throated Rails also respond vigorously to brandished cloths (Gillham 2000), this could account for the similarity in accounts of behaviour. However, no traveller mentioned two species (let alone three), so if the Mauritian *Dryolimnas* survived into human history, then visitors failed to distinguish them from Red Hens, perhaps considering them the young.

63. As noted by Bourne (1968).

64. *Relation du voyage...* (Cauche 1651), edited by Morisot, who added ample marginal notes (Linon-Chapon 2003). The English translation, usually cited as published in 1711, actually first appeared in a partwork in 1710 (copy in Rothschild Library, Tring).

65. Bonaparte (1890), Verin (1983); because the Dutch did not name the French ship, most commentators have assumed it was the *Saint-Alexis* – however, it cannot have been, because it had 14 guns and not 22; the *Saint-Alexis* itself probably never visited Mauritius.

66. The *William* was on at least its third visit to Mauritius (Barnwell 1948); Thomas Herbert sailed in her from Surat to Persia in 1626 before his visit to Mauritius (Penrose 1942).

67. Details are from Bonaparte (1890), Lougnon (1970) and Verin (1983), but the interpretation is ours.

68. Cauche (1651) conflated parrots, weavers and rollers in Madagascar under 'parrots', some described as making weaver-style nests on the leaves of 'palmito trees'. Apart from three Australasian species (one flightless) that nest on the ground, all parrots nest in holes in trees or cliffs (Forshaw 1978). The Malagasy endemic Nelicourvi Weaver *Ploceus nelicourvi* looks like a candidate for the small green 'parrot' – olive back, yellow neck (Sinclair & Langrand 1998) – but Cauche cannot have seen it in Mauritius.

69. The material on Madagascar, published before Flacourt's (1658) *magnum opus*, may be of more value than Cauche's contribution on Mauritius. Several Malagasy names he gave for animals check out correctly (Richardson 1885), and his descriptions of Knob-billed Duck *Sarkidiornis melanotos*, African Openbill Stork *Anastomus lamelligerus*, Glossy Ibis *Plegadis falcinellus*, Greater Vasa Parrot *Coracopsis vasa*, Malagasy Bulbul *Hypsipetes madagascariensis*, Cardinal Fody *Foudia madagascariensis*, and Helmeted Guineafowl *Numida meleagris* are accurate enough, though he confused flamingos and spoonbills, and it is anyone's guess what his 'pheasants' and red-billed fighting 'partridges' might be.

70. .... "they lay but one egg the size of a penny loaf [*pain d'un sol*] by which they place a white stone, as big as a hen's egg, and that on grass they bring together for the purpose, and build their nests in the woods". Cauche (1651, English translation 1710–11) later described pelicans' eggs using the same comparison with a bread roll (see also Strickland 1848).

71. See Lougnon (1970, drawing on Kaeppelin's work from the early 1900s), who doubted that Cauche ever set foot in Réunion. Maurin & Lentge (1979) discussed Cauche at length, pointing out that Étienne de Flacourt (1658) had long ago branded Cauche a comprehensive liar, and reckoned most of his information was gleaned from sailors who really had been to the places mentioned. Flacourt, famous for exploring Madagascar, also stated that Cauche was there for at most three years; he got back to France in 1644, which means he cannot have been in the Indian Ocean before 1640. If

Cauche really visited Rodrigues and Réunion, why did he describe neither island? Linon-Chipon (2003) discussed the veracity and otherwise of 17th century French travellers, with Cauche featuring prominently. However, she did not fully unravel Cauche's comings and goings in the Mascarenes.

72. The *robinsonade* is named after Daniel Defoe's *Robinson Crusoe*, published in 1719, 12 years *after* Leguat's book we are about to discuss. Defoe's direct inspiration was probably English pirate Alexander Selkirk, stranded alone for four years on Juan Fernandez Island off Chile until rescued in 1709 (Souhami 2001, Severin 2002).

73. Leguat (1707); Leguat was a French Protestant or 'Huguenot', banned in France by the edict of Nantes in 1685 and subsequently persecuted; many sought refuge in Switzerland, England and especially Holland, and were prey to resettlement schemes like the one proposed by Henri Duquesne (Racault 1995, Linon-Chapon 2003, North-Coombes 1991).

74. The extraordinary saga of the incompatible views of armchair literary critics and those of naturalists and Indian Ocean historians was forensically examined by North-Combes (1991). For the rest of this discussion we have drawn on his book and also the excellent essays by Jean-Michel Racault that preface the 1984 and 1995 editions of Leguat's book.

75. Günther & Newton (1879), Tafforet (1726)

76. Oliver's heavily annotated English edition (1891) was published after the subfossils and Tafforet's account were discovered, but before Dutch correspondence on the Leguat case had surfaced in the Cape archives.

77. Atkinson (1922). Erwin Stesemann, Dick Hillenius and Otto Helms were prominent zoologists who were persuaded by Atkinson's arguments at the time (see North-Coombes 1991), and more recently David Stoddart (1972) equivocated on Leguat because of Atkinson.

78. Mortensen (1934) was a key contributor, but Dutch accounts of Leguat being landed on Rodrigues and his later imprisonment in Mauritius by governor Deodati on an islet in Mahébourg Bay were published by Leibbrandt (1887, 1896–1907) and Pitot (1905), long before Atkinson developed his thesis. Leibbrandt (1896–1907) contains Deodati's correspondence with the Cape in the volumes *Letters received 1695–1708* and *Letters despatched 1695–1708*, published in 1896, paraphrased by Pitot (1905) and partly reprinted by Barnwell (1948).

79. Both Atkinson and Adams were guilty of selecting evidence to suit their beliefs, especially in ignoring Vivielle's and Deherain's historical detective work in the 1920s (see North-Coombes 1991).

80. In the first edition, 1980, of North-Coombes (1991).

81. Adams (1983), apparently unaware that he was in a hole, carried on digging by declaring that "[in addition to other claimed hoaxers] there is Misson putting together his fireside *Voyage ... de François Leguat* (1707) in which one of his inventions is a bird, the Gelinotte, a composite of three real birds found in previous real travel books, a bird later classified by the *Cambridge Natural History* as *Erythromachus leguati* but dubbed 'extinct', while Newton's *Dictionary of birds* gives this 'singular rail' the title *Miserythrus leguati*". This allegedly fake *gelinotte* is of course Leguat's Rail *E. leguati*, seen and described also by Tafforet (1726), and long since confirmed by subfossil bones (Günther & Newton 1879, Cowles 1987). Even if unaware of North-Coombes, Adams managed to ignore all the 19th century subfossil evidence and the ample historical material to persist in his obtuse insistence on Leguat as fiction. A better-informed critic would have chosen Leguat's '*géan*' as his archetypal fictitious bird – he would then, at least, have got something right.

82. Grove (1995), discussing Leguat and the attempt at founding 'Eden', commented that "there is some debate as to whether the Leguat account is fictional or biographical . . . ultimately the debate is not of vital importance. Even if it is fiction, Misson's Leguat still conveys a valid account of a new Edenic perception". Here a historian of ideas allowed the idea ('Edenism' as a driving force in western exploration and colonisation) to override the facts; see Cheke (1996, 2001b) for further discussion of errors in Grove's book. It should be of academic concern that Atkinson's and Grove's books, albeit 74 years apart, both originated in Ph.D theses, and that neither consulted biologists, let alone those with appropriate expertise, before submitting their work for degrees and publication. Even more recently, Severin (2002), a literary historian discussing Defoe's sources for *Robinson Crusoe*, dismissed without further comment "the writings, probably fictitious, of French traveller François Leguat", because he was only concerned with *true* desert island antecedents!

83. Racault (1995). It seems to have been Misson, not Leguat, who turned events in the narrative into gifts or punishments from on high, and added the classical references.

84. On Rodrigues, in addition to a stray '*géan*', Leguat described eight land birds, five seabirds, a fruitbat, two lizards, tortoises and turtles, all of which match bone finds or extant species (Bourne 1968, Staub 1973b, Cheke 1987a); his dugongs were seen by many later visitors; Balfour (1879b, North-Coombes 1991) was able in 1874 to identify 9 of the 10 plants he mentioned. There were also several birds he did not report, seen by Tafforet in 1726 (Staub 1973b, Cheke 1987a).

85. For example, Adams (1962, 1983). Recent archaeological, historical and linguistic research (Pearson & Godden 2002) has demonstrated that much of Drury's account can be confirmed by outside evidence, and that his fluent southern Malagasy vocabulary could only have been acquired by someone with deep experience of the area – where no Europeans, at that time (1703), had penetrated for more than a few days.

86. '*Necropsar leguati*': Fuller (2000), with the full story (and 'codswallop' quote) in Olson *et al.* (2005); the trembler is *Cinclocerthia gutturalis*. Caution about this specimen was expressed by Greenway (1967) in 1958 and strong doubts by Cheke (1987a); though widely (and wisely) ignored, this mythical starling persisted in some literature until recently (e.g. Day 1981, Cowper 1984, Gupta 2005).

87. The 'grebe' was a migrant Whimbrel *Numenius phaeopus*, the 'darter' a Reed Cormorant *Phalacrocorax africanus* and the 'moorhen' an undescribed large *Dryolimnas* rail; the owls were aberrant examples of the known forms. See Olson (1975b), Cowles (1987), Cheke (1987a); Cowles's reassessment of the rail as a White-throated Rail *D. cuvieri* has been re-assessed again (JPH work in progress). Also, in reconsidering the 'darter', Olson (1975b) compared the bones only with the Afro-Malagasy *P. africanus*, whereas, given the composition of the avifauna in general (Chapter 4), the Little Cormorant *P. niger* from Asia, similar in size, shape and habits, is equally or more likely.

88. Holyoak (1973), Hume (2005).

89. A single ciconiiform bone was assigned by Cowles (1987) to an undetermined stork *Ciconia* sp., but later that year Mourer & Moutou (1987) described, from further bone material, the ibis *Borbonibis latipes*, now known as *Threskiornis solitarius*. ASC recognised immediately that the mysterious Réunion 'Solitaire' had at last been identified, though this connection was not published until 1995 (Mourer *et al.* 1995a,b). Meanwhile Cowles (1994) acknowledged his 'stork' bone was from the ibis, and another puzzle from the past had been resolved.

90. Dubois (1674), often reprinted (Berlioz 1946, Lougnon 1970, Barré *et al.* 1996). Other accounts that would cast doubt on the Réunion 'Solitaire' being a Dodo (Feuilley 1705, Melet 1672) were not discovered until much later (Mourer *et al.* 1995, Hume & Cheke 2004).

91. For example, Buffon (1770–83)

92. Billiard (1822); a couple of Réunion Ibises *were* shipped for France in 1667, but soon "died of melancholy, unwilling to eat or drink" (Abbé Carré in Lougnon 1970, Hume & Cheke 2004).

93. Strickland (1844, 1848), Coquerel (1865a).
94. Coker (1856, illustrated with a poor copy of a white dodo by Pieter Withoos), A. Newton (1869), Newton & Gadow (1896), Rothschild (1907).
95. Renshaw (1938); the Jan Savery picture in Oxford and the Prague skull are both undoubtedly of Mauritius Dodos (Wissen 1995), while the Cambridge skeletons are from a Dodo and a Rodrigues Solitaire (ASC & JPH, pers. obs.).
96. Hachisuka (1937, 1953) blithely ignored the failure of any visitor to see more than one sort of solitaire, highly improbable if two large flightless species were present.
97. Berlioz (1946) rejected Haschisuka's two species, while endorsing the white dodo paintings as representing Réunion birds.
98. First edition (1958) of Greenway (1967); Greenway did not doubt that there had been a dodo in Réunion, and even included it in the same genus as '*Raphus solitarius*'. In addition to rejecting Hachisuka's second Réunion dodo, Greenway also refuted several of Hachisuka's other fantasy species. Halliday (1978) and Fuller (1987) were somewhat circumspect about Réunion dodos, Balouet & Alibert (1990) rather less so. Barré & Barau (1982), reprinting old accounts of Réunion birds, carefully avoided mentioning 'dodo' or '*dronte*' in connection with the *solitaire*.
99. See Cheke (1987a). A few Dutch visitors did visit Réunion in the early-mid 1600s (Hume & Cheke 2004), but there is no evidence they brought home any souvenirs. John Gould (footnote to Coker 1856) thought the picture was of "an albino or white variety of the Dodo", but in suggesting it must have come from Mauritius or *Bourbon* [i.e. Réunion, our italics] he triggered the imaginative frenzy that followed. See Ziswiler (1996) and Hume & Cheke (2004) for fuller accounts of the Holsteyn/Withoos white dodo paintings; the first batch are now attributed to Pieter Holsteyn the younger, not his father as stated in most reference books (e.g. Jackson 1999, Ziswiler 1996).
100. See Chapter 5 for references and details of Dodos brought alive to Europe, and Hume & Cheke (2004) for a fuller account of Savery's 1611 bird. The 1611 painting, appearing without particular comment in Müllenmeister (1988), was independently rediscovered by Valledor (2003), whose conclusion was similar; Hengst (2002) also reproduced it, but did not spot its significance in relation to the Holsteyn/Withoos series.
101. Storer (1970).
102. For example, Monroe & Sibley (1993), Quammen (1996), Gould (1998), Warr (1996: reproducing a white dodo painting by Withoos), Pinto-Correia (2003), Dickinson (2003), Gupta (2005, a book from which Indian children will pick up many misconceptions scattered among the facts – not least a Burrowing Boa *Bolyeria multocarinata* from Round Island up a tree, though all the facts, errors and pictures are lifted directly from Day 1981).
103. Gibbs *et al*. (2001). Fuller (2000) downgraded the White Dodo to the 'hypothetical birds' section of his second edition, but left the door wide open for its resurrection: "Whether seventeenth-century commentators would confuse a dodo with an ibis is largely a matter of opinion. Yet apart from this association of ideas, there is no evidence that the ibis survived into the period of European colonization". Confusion with dodos was never an option, as no French visitor or resident *at the time* connected the two at all, and indeed most had probably never heard of dodos! No 17th-century writer (excepting Bontekoe's probably interpolated account) used any 'dodo' name – they were always *solitaires*, or, rarely, *lourdes* (Lougnon 1970, Cheke 1987a, Hume & Cheke 2004). An important source of confusion is due to Leguat, who never landed on Réunion, borrowing the name *solitaire* from Duquesne for his unrelated Rodrigues bird (which *was* a sort of dodo). Fuller (2002) later accepted that the ibis was contemporary with the first settlers, and that the white dodo paintings had no connection with Réunion, but still entertained a stubborn hope of the Réunion Dodo's return: "Although such finds [ibis bones] may cloud the issue – or even make the possibility of the former existence of a dodo less likely – they certainly do not disprove it. . . it is possible that such a creature once existed."

Chapter 3. The pristine islands

1. Leguat (1707).
2. Herbert (1634).
3. Cadet (1977), Strasberg *et al*. (2005).
4. Strahm (1994).
5. See Brouard (1967). The *Flore des Mascareignes* (Bosser *et al*. 1978–) is a nearly complete modern survey of the Mascarene flora. Dupont *et al*. (1989) illustrated many of Réunion's endangered endemics, likewise Strahm (1989) for Rodrigues; Cadet (1981) covered a selection of both native and exotic Mascarene plants in a popular handbook, while conspicuous Mauritian forest plants have been amply illustrated by Atkinson & Sevathian (2005).
6. Leguat (1707), Tafforet (1726).
7. The best sources on the vegetation are: Réunion: Cadet (1977), Rivals (1952); Mauritius: Strahm (1993), Vaughan & Wiehe (1937), Page & d'Argent (1997); Rodrigues: Wiehe (1949). Blanchard (2000) is a well illustrated survey of vegetation (native and exotic) on all three islands; Guého (1988), more introductory, covers similar ground for Mauritius alone, and Lorence & Vaughan (1992) published an extensive *Annotated bibliography of Mascarene plant life*.
8. Rivals (1952), Cadet (1977) [Réunion]; Vaughan & Wiehé (1937), Strahm (1993) [Mauritius].
9. Atkinson & Sevathian (2005) used the term 'palm-rich forest' rather than 'savanna'.
10. 'Palm savanna' typically indicates a fire-climax grassland, usually artificially maintained, with isolated palms, often *Borassus* spp. (Africa, Asia) or *Chrysalidocarpus* spp. (Madagascar).
11. Rainfall data in Rodrigues has improved recently, and Saddul's map (2002) should be consulted over earlier maps (e.g. Strahm 1989) and published data (North-Coombes 1971, Padya 1984), which do not include records from the newer weather station at the airstrip on the relatively dry Plaine Corail. Feuilley's description (1705) of the forest around the Étang du Gol in Réunion, then still uninhabited and apparently little disturbed, is indicative: "This area is agreeable, and although there is no shortage of trees, they are spaced apart from one another in such a way as to suggest they had been planted like an orchard, with the canopy only of the trees touching to make a fine shade beneath". We have no direct record of a coastal screw-pine in Réunion analogous to those on the other islands, but the widely cultivated *Pandanus utilis* probably originated as a Mascarene endemic (*Flore* 190, 2003, Bosser & Guého) - we suspect it was the 'missing' coastal species, perhaps replaced in the driest areas by *P. sylvestris*.
12. Vaughan & Wiehé (1937) wrote: "the Dutch also record a type of vegetation occurring in coastal plains, particularly in the north and west, where palms were frequent and where they could graze their domestic animals" (p. 290) and later (p. 322) "from the accounts of early explorers it appears that a palm savanna in the narrow zone below the 40 in. (1022mm) isohyet ". We would not disagree with the first statement (though palms were in fact present throughout the lowlands), but it does not imply, as the second statement insists, a 'palm savanna'. Later writers, including even Strahm (1993), have until recently accepted this non-sequitor without question. We have found no Dutch-period accounts that refer to any kind of savanna, and only two mentioned 'grassland', both on the wet south or east coasts, well outside the alleged 'palm savanna' zone. Jolinck in 1598 (Keuning 1938–51, Panyandee 2002) found an open meadow-like area, probably near Flacq, populated by geese, 'cranes' [= flamingos?] and herons. Matelief in 1606 (Barnwell 1948, Verin 1983) referred to areas of coastal grass in the dune zone

along the south coast, which extend from Mahébourg round to Butte aux Sables, the area called 'Savan(n)e' by the French as early as 1725 (Toussaint 1953; *DTIM* 1 :40).

13. The rainfall figure is for nearby Flat Island, where there was a rain-gauge (Padya 1989) before the lighthouse was automated. There is a huge literature relating to Round Island because of the fauna and flora that has survived there.

14. The three editions of Padya's books on Mauritian weather and climate (1989; 1984, revised from *c*.1972) present different amounts of data. Only the 1984 version included a map of all the weather stations and a monthly rainfall summary for each; the full data suggests that the driest zone (under 800mm) is a bit more extensive than the two tiny enclaves shown on Padya's map (1984 and 1989 editions).

15. See Lougnon (1970) for Houssaye's account. Recent editions of the IGN 1:100,000 map of Réunion (3615 IGN, 1992 and 1996) include a rainfall map showing the 750mm isohyet; most published rainfall maps show nothing below the 1000mm isohyet (e.g. Soler 2000, Girard & Notter 2003). The 'benjoin' is the Mascarene endemic *Terminalia bentzoe* (*Flore* 91:10–11, G. E. Wickens).

16. The Pointe des Galets is the stony delta of the Rivière des Galets, and probably has a higher water table than the fast-draining lavas further south.

17. Lougnon (1970); the native plant was presumably seen as the medical equivalent of the closely related Aloe Vera *Aloe barbadensis*, used from ancient times in the Mediterranean area, and was certainly so considered in the 19th century (*Flore* 183: 10, 1978, Rouillard & Guého 1999). *Lomatophyllum purpureum* was originally abundant enough in Mauritius to be considered for processing and export by Lamotius in 1690 (Pitot 1905, Rouillard & Guého 1999). The endemic *Lomatophyllum* species are now rare on all three islands.

18. Tortoises are associated with *Pandanus* seed dispersal on Madagascar and Aldabra (Callmander & Laivao 2003); Leguat (1707) and Pingré (1763) reported tortoises eating *Latania* fruit.

19. Details in note 12.

20. High densities of tortoises on Aldabra maintain a 'tortoise turf' (Merton *et al*. 1976, Hnatiuk *et al*. 1976, Gibson & Hamilton 1983, Coe 1995) analogous to the short turf grazed down by concentrations of Rabbits *Oryctolagus cuniculus* in Europe. The Aldabra Giant Tortoise *Dipsochelys dussumieri*, primarily a grazer, is in a different genus from the Mascarene endemic *Cylindraspis* spp. (Gerlach 2004), whose feeding habits are not known in any detail.

21. Feuilley (1705) described mobile dunes with a few scattered latans at Étang Salé, Réunion, in 1704.

22. Leguat (1707) reported aggregations of 2,000–3,000 tortoises packed together so that he could walk for a hundred paces on their backs without setting foot on the ground. Only one early description of Mauritius (Verhoeff's flotilla 1611: Barnwell 1948) contains similar remarks; although mostly distilled from the 1598 van Warwyck reports (anon. 1601) – it contains some apparently original observations: "all the island is so full of them [tortoises] that while walking in the forest one comes across them assembled and as it were heaped up in great companies, grazing or resting like flocks or herds".

23. The offshore lagoon cays of Iles Sable & Cocos are also composed of dunes; Cocos, lacking tortoises, was covered in low grass (*chiendent*) according to Tafforet (1726).

24. Tafforet (1726); Port Mathurin was submerged to knee height by the tsunami of Boxing Day 2004 (*l'Express* 27 and 29/12/04 from www.lexpress.mu, and *Mauritius Today* 29/12/04 from http://mauritiustoday.com).

25. Cossigny (1732–55), writing to Réaumur in 1732: "[in Mauritius] there are what are called *mangliers*, which grow on the sea shore and 100 fathoms beyond where there are shallows and some silt or fine-grained soil, and these trees make a network of great roots largely covered in oysters which are sometimes very good". The species is *Rhizophora mucronata* (*Flore* 90, 1990, A. J. Scott).

26. There are many similar early accounts of the density of Mauritian wet forest, of which a few follow: "The soil is stony almost everywhere, yet it is fertile in wild trees, which grow so close to each other that it is difficult to pass between them" (anon 1601, as quoted in Barnwell 1948). "And as it [Mauritius] is plentiful in all things, so no one thing exceeds the wood, which is in so great quantitie, that I could hardly procure passage."(Herbert 1634). Leguat (1707) reported that in 1693 there was "in the middle of the island in a great plateau surrounded by mountains, there are woods that are most dangerous to tackle. The branches of the trees are so thick above and so interlaced amongst themselves that it is impossible to see the sun. Thus as one cannot tell where one is going, one gets lost as in a labyrinth, made worse by the fact of being able to find nothing to eat", after which he described how the governor's predecessor (Lamotius) and his men got lost for four days in this way before happening on an exit.

27. Bonaparte (1890), Verin (1983).

27. Thompson (1880).

29. Vaughan & Wiehé (1941), Lorence & Sussman (1988). Trees over 10cm dbh (diameter at breast height): 1,540–1,710 per hectare for Mauritius at around 600m altitude, against 300–800 up to the same altitude elsewhere in the tropics (see Leigh 1999, who assembled a huge table of forest density data from around the world). Likewise the Mauritian figures for bigger trees (20cm dbh, 371 stems/ha) are 1.5–2 times as dense as elsewhere. In world terms the Mauritian forests are 'lowland' forests; higher altitude forest is often denser.

30. Jamaica, Puerto Rico, New Caledonia (Leigh 1999); Lorence & Sussman (1988) discussed of tree density in tropical forests, with a (limited) comparative table. The contrast in the Mascarenes between the resistance of intact native forest in cyclones as against the vulnerability of exotic plantations has been commented on by numerous authors (e.g. Vaughan & Wiehe 1937, Brouard 1967, Cadet 1977, Tassin & Hermet 1994), though isolated or thinned native trees deprived of their natural dense surrounding support are nearly as vulnerable as exotics.

31. "The other woods of this island are easy enough to penetrate" (Leguat 1707, after describing the impenetrable ones in the island's centre).

32. The drawing was first published by Grove (1995), but its true date and locality was established by Cheke (2001b). Poste de Flacq is on the 1600mm isohyet.

33. Reports in Barnwell (1948, 1950–54).

34. This impenetrability protects the fauna and flora, but makes surveying birds very difficult. Curiously, botanists in Réunion have not studied tree-density as in Mauritius, though around 1,500m it must be equivalent to Mauritian values at 600m, or higher. Cadet (1977) mentioned forest density only in stating that the screw-pine thickets are 'extremely dense and more-or-less impenetrable'. The only published figures are for lowland wet forest at 250m (Strasberg 1995, 1996): 1079/ha at 10cm dbh, less dense than on the Mauritian plateau.

35. Such is the demand for palm-cabbage that poachers will go to almost any lengths to harvest them (Doumenge & Renard 1989); steep cliffs and inaccessible plateaux are thus the only places where *Acanthophoenix rubra* can still be seen emerging above the canopy (Cadet 1977, Lavaux 1998; photos in Bénard 2002).

36. Remarkably there is an almost identical acacia, *A. koa*, in Hawaii, with very similar ecological characteristics; see Cadet (1977), Cheke (1987b) and pp. 53–54.

37. Leguat (1707) is the source for most of this paragraph.

38. The palm distribution was confirmed by Tafforet (1726), who added later that the opposite was true of the '*bois de neff*' *Eugenia rodriguesensis*. He did not further describe the appearance of the

forest, but his list of trees, based on his familiarity with Réunion plants, is much more useful than Leguat's. Strahm (1993) concluded that several species described by Leguat, Tafforet and Pingré are unidentifiable, but distinct enough to suggest the travellers had seen plants now extinct and never recorded by botanists.

39. Leguat seems to have exaggerated the size and extent of the Rodriguan fig trees by conflating them with Asian Banyans *Ficus bengalensis* which he may have seen later in Java.

40. The equivalents of old French measurements are discussed in notes 197 and 209.

41. Tafforet (1726); Leguat (1707) remarked on seasonality in respect of the stream by the settlement, but elsewhere stated that the streams "never run dry". The driest months are September–November; Padya (1989) wrongly stated that the driest months are June–August, though the true picture is clear from the tables in his 1984 edition (see also Cheke 1987d, Strahm 1989).

42. Padya (1976) is the most comprehensive book on Mascarene cyclones, though Mayoka (1998) is more graphic with excellent satellite photographs in colour. These storms, technically 'tropical depressions', have acquired different regional names: cyclones (Indian Ocean), typhoons (Far East), and hurricanes (Caribbean).

43. Padya (1976; Mauritius and the region), North-Coombes (1971; Rodrigues), Mayoka (1998, Réunion).

44. Introduced trees from cyclone-free areas (pines, *Casuarina*, fruit-trees etc.) are extremely vulnerable and are badly damaged (Padya 1976, Brouard 1967).

45. Vaughan & Wiehe (1937), Brouard (1967).

46. Strahm (1993)

47. Tortoises, flying-foxes, parrots, Dodos, Rodrigues Solitaires and merles are all known to have had marked fat-cycles, as may the Réunion Ibis. It is unfortunate that this phenomenon can no longer be studied, as its phenology appears confusing. Feuilley (1705; parrots: June–Sept, Réunion), La Motte (1754–7; Mauritius Merles, July–September), Jacob (1861; Réunion Merles: July–September) and Leguat (1707; Rodrigues Solitaires: March–September, Rodrigues) reported birds fat in winter, whereas Lanux (1772) stated that in Réunion Black-Spined Flying-foxes *Pteropus niger* were fattest in quite the opposite season: "throughout the summer and a good part of the autumn" (?November–March). Dubois (1674, Lougnon 1970, Barré *et al*. 1996) reported seasonal fattening in most of the birds, but omitted to say when it occurred; he denied the ibis a fat-cycle, but other observers reported them fat in late summer (March–April; Cheke 1987a). Seasonality is not clear in tortoises or dodos, although Staub (1996, echoing Oudemans 1917) inferred a fat period in winter for the latter. Summer fattening might be an adaptation to food shortage in the drier winter months, while winter fattening could be a precaution against post-cyclone shortages – fallen fruit will be temporarily abundant, then unavailable for many weeks. Echo Parakeets normally live on a very low energy diet, but have an uncanny ability to lay down fat very fast (Jones & Duffy 1993), presumably an adaptation to make the best use of infrequent food surpluses or impending shortages; in captivity they very easily become obese, as evidently did Dodos (e.g. Kitchener 1993).

48. Feeding on the ground was first noted at Jersey Zoo in 1976 with bats caught for captive breeding (Cheke & Dahl 1981), but has recently been observed twice in the wild by Aleks Maljkovic (pers. comm.) and, more consistently, after a cyclone in 2003 (Powell 2004). Maljkovic's animals were surprised in car headlights, and scrambled up trees before flying off; one had been feeding on a '*pomme zaco*', fruit of the introduced *Mimusops bojeri*. No other *Pteropus* has been reported foraging on the ground, but bare-backed fruitbats of the related genus *Dobsonia*, centred in New Guinea, "frequently alight on the ground to feed . . . The adaptations that allow them to do this may have evolved at a time when there were few or no terrestrial mammals with which to compete" (Flannery 1995b). Afficionados of David Attenborough's TV programmes will have seen the New Zealand Lesser Short-tailed Bat *Mystacina tuberculata* foraging on the ground – New Zealand, like Rodrigues, had no native mammalian predators.

49. Cheke & Dahl (1981); Powell (2004) observed bats clustering low down in trees during a severe cyclone.

50. Cheke (1987a); extinction by rats is inferred from the survival of several of the larger reptiles only on rat-free Round Island, and their disappearance from Flat Island and Gunner's Quoin once rats arrived; Towns & Daugherty (1994) documented the analogous situation in New Zealand.

51. No early visitors reported boas or the two large *Leiolopisma* skinks on the mainland of Mauritius; the larger skink or Didosaurus *L. mauritiana* was never reported alive at all, and the large gecko *Phelsuma guentheri* was not discovered until 1869, on Round Island (Vinson & Vinson 1969, Cheke 1987a). However, subfossil bones of Günther's Gecko are known from the mainland (Arnold 1980) and its egg scars are still attached to cliffs in the Black River Gorges (Jones 1988a).

52. Arnold & Jones (1994), Probst (1998).

53. Both Matelief and Herbert referred to more than one type of hawk (Barnwell 1948, Herbert 1634), suggesting they saw harriers as well as kestrels. The absence of Mauritian records of cormorants is very odd, for they survived for more than 50 years in Réunion after settlement, and Mauritius had much more wetland habitat. The Dutch possibly simply saw them as another kind of 'duck', though one would expect European seafarers to be familiar with cormorants.

54. Previously unstudied subfossil bones in the London NHM reveal that *Pteropus rodricensis* was once common in Mauritius (Hume 2005a), where all three species were thus present at some time. A *P. rodricensis* skull was previously known from Round Island (Cheke & Dahl 1981, Bergmans 1988–94, but see Hume 2005a). For *P. niger* in Rodrigues see note 195.

55. Lizards were mentioned only by Herbert (1634) and Mundy (1608–1667) among 17th century visitors, and land crabs only by Mundy. Neither described them, though both said lizards were plentiful, Herbert adding that they were "not a little curious" (meaning 'unusual', not 'inquisitive'), and noted that they liked sap exuding from tapped palm trees.

56. Cheke (1984) reviewed lizard density data. We have no information on land crabs in Mauritius or Réunion to compare with Leguat's observations in Rodrigues but densities, at least in Mauritius, may have been comparable.

57. Respectively *Leiolopisma mauritiana*, *L. telfairi*, *Gongylomorphus bojeri*, *G. fontenayi* and *Cryptoblepharus boutonii*; all are endemic except the last, which is a widespread shoreline species. See Arnold (1980, 2000) for the ancient reptile fauna; Vinson & Vinson (1969) covered the existing and recently extinct lizards. The total size of Didosaurus is estimated from the snout-vent length of 340mm given by Arnold (1980).

58. See Vinson (1975), Bullock (1986), Jones (1993), Freeman (2003) and Pernetta *et al*. (2005) for ecological information on surviving Mauritian reptiles.

59. For example, *Tambourissa* spp., *Syzygium mamillatum*, *Ficus mauritianus*. Staub (1988) discussed some aspects of plant/animal relations in Mauritius, but there is much more to be discovered; see also Blanchard (2000). *S. mamillatum* is currently pollinated by small birds (Mascarene Grey White-eyes, Mauritius Merles and Red-whiskered Bulbuls), but there is no seed dispersal – i.e. the dispersal agent is extinct (Dennis Hansen, pers. comm.).

60. *Nactus serpensinsulae* (two subspecies) and *N. coindemirensis* (Arnold & Jones 1994).

61. *Cylindraspis triserrata* (high-backed) and *C. inepta* (domed) (Arnold 1979, Austin & Arnold 2001, Gerlach 2004).

62. Eskildsen *et al.* (2004); heterophylly in the Mascarenes was first explored by Friedmann & Cadet (1976), but they saw it as an adaptation to reduce transpiration stress on young plants in the dry zone where it is concentrated, without considering the possible influence of browsing animals.

63. Experiments by Eskildsen *et al.* (2004); Ile aux Aigrettes is a 25ha islet in Mahébourg Bay, Mauritius, where the forest habitat is being restored by the MWF (Strahm 1993, Mauremootoo 2003b, p. 245). Hansen *et al.* (2004) showed that juvenile leaves have no additional toxins or chemicals to deter herbivores, so the effect must have been purely visual. Relying perhaps too much on the way Echo Parakeets strip the parenchyma (leaving petiole and midrib), and the lack of heterophylly on other islands with giant tortoises (Aldabra, Galapagos), they preferred Dodos and Solitaires as selective agents to the much more abundant tortoises, while recognising that Aldabran tortoises avoid heterophyllous juvenile leaves.

64. The spiny-trunked species are *Acanthophoenix rubra* and *Tectiphiala ferox* (*Flore* 189; 1984, Moore & Guého); according to Staub (1993) *Hyophorbe vaughani* has spiny leaves when young.

65. *Hyophorbe* palm-heart is toxic to humans, earning the name *palmiste poison* for these species (Staub 1993), but on Round Island this did not deter rabbits and goats from browsing Bottle Palms *H. lagenicaulis*, preventing regeneration until the animals were eliminated (North *et al.* 1994). We are not aware of tests of this genus on tortoises.

66. On Aldabra the principal tortoise nest predators are Coconut Crabs *Birgus latro* and *Cardisoma* land crabs (Swingland & Coe 1979), while these crabs and White-throated Rails *Dryolimnas cuvieri* (plus introduced rats) take hatchlings. *Birgus* is absent from the Mascarenes, but land crabs and the endemic wood-rails (and perhaps, on Mauritius, also Didosaurus and Red Hens) will have been natural predators prior to the arrival of pigs, cats, and rats.

67. No account of Raven Parrots described them as flightless (Herbert 1634, Mundy 1608–67; see also box 18), the closest being Hoffman (1680) who said they flew 'with difficulty'; the van Warwyck engraving of 1598 showed one at the top of a tree. Belief in their inability to fly arose from misinterpretation of subfossil bones; the tiny sternum supposedly associated with the Raven Parrot really belonged to the smaller Thirioux's Grey Parrot (Hume 2005a).

68. Hume (2005a).

69. JPH, work in progress.

70. Servaas (1887), Wissen (1995). The Verhoeff account from 1611 also referred to the risk of injury from the Dodo's 'great thick curved beak' (Wissen 1995, Barnwell 1948)

71. Staub (1996, 2000); Hengst (2003) speculated similarly.

72. This story has become established in the ecological literature as an example of 'obligate mutualism', although it has no basis in fact; for a detailed discussion see Witmer & Cheke (1991), Hershey (2004) and p. 130. Other fanciful theories have arisen because a couple of old pictures by Roelandt Savery showed Dodos eyeing seashells and an eel, but there is no reason to suppose this was anything but artistic juxtaposition; no artist in Europe ever saw Dodos under natural conditions.

73. Hachisuka (1953).

74. Janoo (2000), Shapiro *et al.* (2002).

75. Rodrigues Solitaires also had single gizzard stones.

76. Details in Cheke (1987a).

77. There were no native amphibians in the Mascarenes, as was often commented on by early visitors (Cheke 1987a).

78. Servaas (1887).

79. A reminder that the name 'solitaire' was used for an ibis in Réunion and a dodo-relative (flightless pigeon) in Rodrigues.

80. Feuilley (1705; Cheke 1987a, Barré *et al.* 1996).

81. The Limpkin *Aramus guarauna* is in a monospecific family Aramidae, related to cranes and rails, that feeds 'almost exclusively on large [aquatic] snails' (Palmer 1985).

82. Griffiths (1996), Griffiths & Florens (2006).

83. Jones (2006); we inferred that Red Hens ate snails long before Rungwe Kingdom and Carl Jones noticed the damage in the umbilical area of large subfossil *Tropidophora*, through which they came independently to the same conclusion.

84. Penny & Diamond (1971; Aldabra), Morris & Hawkins (1998; Madagascar). Subfossil bones originally identified as from Common Moorhens *Gallinula chloropus*, were re-assigned to the White-throated Rail *Dryolimnas cuvieri* by Cowles (1987). However, further study (JPH, work in progress) shows that Mauritian birds were probably flightless and markedly larger than typical White-throated Rails, closer to *D. augusti* on Réunion; they have yet to be scientifically named. Sauzier's Wood-rail was not reported by early visitors, but as only Dubois reported the eponymous Wood-rail in Réunion, these birds may have been secretive or inconspicuous, though *D. cuvieri* is quite the opposite on Aldabra (e.g. Gillham 2000). Alternatively, they may have vanished before humans arrived, though this is unlikely since White-throated Rails on Aldabra, also flightless, can survive rats (though not cats).

85. Huxley (1979) said the rails were taking ectoparasites, but none have been detected, so the birds were probably taking flies and mosquitos (Gerlach 2004); Tortoises on Aldabra and the Seychelles respond to any tactile stimulation (e.g. from humans) with the 'tall stretch' posture (Gerlach 2004; pers. obs.)

86. Yamashita (1997).

87. Carlos Yamashita pers. comm. Holyoak (1971) had argued that the Raven Parrot's bill was rather weak and that it probably ate soft fruit, which Yamashita countered by pointing out that the strength is in the horny sheath, which is not preserved in fossils, and that macaws likewise have relatively weak skeletal components.

88. Long John Silver's parrot in *Treasure Island* was based on a well-established tradition; Nieuhoff (1682) saw Blue-and-yellow Macaws *Ara ararauna* in Batavia (= Jakarta, Java) in the 1650s, brought from Dutch Guiana on the other side of the world. Parrots were particularly popular in 16th and 17th century menageries; macaws reached Spain by 1500, and appear in Dutch paintings by 1600 (Jackson 1993).

89. Holyoak (1971), Hume (2007); of early visitors only Reyer Cornelisz in 1602 seems to have noticed the sexual dimorphism, referring to both "great and small Indian Ravens" in addition to "grey and green parrots with long tails" (Strickland 1848, Box 18).

90. Soeteboom (1648), often reproduced (e.g. Hachisuka 1953, Verin 1983). In 1607 Steven van Hagen wrote: "sometimes when we had caught a grey parrot, we made it call out, and at once hundreds more came flying around, and we were able to kill them with sticks" (Barnwell 1948); see Box 18 (p. 172) for West-Zanen's similar comments.

91. Holyoak (1973), Cheke (1987a); from the characteristic sternum, the species has been re-assigned to *Psittacula*, becoming *P. bensoni* (Hume 2007).

92. Of the two hurricane palms *Dictyosperma album* was in the lowlands, while *Acanthophoenix rubra* extended to all elevations. There were three *Hyophorbe* species: *H. lagenicaulis* (Bottle Palm) in dry coastal areas, with *H. vaughani* and *H. amauricaulis* occurring in wetter areas higher up; all are now extinct in the wild or so rare that it is impossible to be sure of their former distribution (*Flore* 189, 1984, Moore & Guého; Maunder *et al.* 2002).

93. Jones & Duffy (1993).

94. *Nesoenas picturata*, see Mourer *et al.* (1999); following DNA analysis by Johnson *et al.* (2001), Cheke (2005b) combined the Malagasy Turtle Dove (formerly shunted between *Streptopelia* and *Columba*) and the Pink Pigeon (sometimes put in *Columba*) in the genus *Nesoenas*.

95. Granivory: ASC's unpublished observations in the Mascarenes, as also reported for the Comoros and Aldabra (Benson 1960, Frith 1977, 1979); Carl Jones (pers. comm.) reports snails. There has

been no full ecological study of *Nesoenas picturata* although it is a Malagasy regional endemic and occurs on many Indian Ocean islands.

96. Until 2007 2007 (bones at *c.* 450m asl, *Mauricien* 30/6/07), Dodo bones were only known from lowland sites (Janoo 2005), and there is only one semi-upland site (Trois Mamelles, 500m) reported for tortoise remains (Florens 2002), but this may be an artefact of collecting or preservation; Telfair's Skink bones have been found at 620m on Corps de Garde mountain (Florens & Turpin 2001). There was also Bouton's Skink, a strictly shoreline species (Jones 1993a).

97. Skink densities on seabird islands in the Seychelles are astonishingly high (Cheke 1984), and there would have been some equivalent areas in primeval Mauritius; see also Rodda & Dean-Bradley (2002).

98. Mourer *et al.* (1994, 1999) reassessed the bones and erected a new genus *Mascarenotus* for Mascarene owls, but argued that, as in two Caribbean species (Puerto Rican Screech Owl *Megascops nudipes* and Bare-legged Owl *Gymnoglaux lawrencii*) and extinct Hawaiian owls (*Grallistrix* spp.), long legs go with catching birds in the absence of small mammals. Lizards were probably more abundant and easier prey than birds in the pristine Mascarenes (as also in the Caribbean), hence the owls' adaptations were for reptile-catching; the Bare-legged Owl is in fact a reptile specialist (Garrido & Kirkonnel 2000). In the absence of native lizards, the Hawaiian owls must have preyed on birds (Olson & James 1991).

99. Cowles (1987), Mourer (2004).

100. Clouet (1978), Bretagnolle et al. (2000c) (morphology), Cheke (1987b) (diet); the closely related Malagasy Marsh Harrier *Circus macroscles* similarly eats birds (44.7% biomass), reptiles (chameleons and snakes, 35.6%) and mammals (rodents and tenrecs, 18.6%); also lots of grasshoppers, though they make up only 1.1% of prey biomass (Rene et al. 2004).

101. Jones (1987), Jones & Owadally (1988), Groombridge *et al.* (2004b).

102. Cheke (1987a), Jones (1987)

103. Vinson (1975), Bullock (1986), Cundall & Irish (1989). Female *Casarea* reach 141.5cm, males only *c.*60cm, rarely weighing over 50g, whereas females weigh 200–500g (Bullock & North 1991, Dulloo *et al.* 1999). Carl Jones (pers. comm.) suggests that in their current restricted habitat on Round Island, males target the small terrestrial *Nactus* geckos, while females also take the much larger and more arboreal *Phelsuma guentheri*.

104. There are very few observations of *Bolyeria* in life (Vinson 1975, Bullock 1977), but its blunt snout and lack of neck are characteristic of burrowing animals; recent specimens have not exceeded 950mm (Bullock 1986), but 19th century museum specimens are quoted as being "up to 1.80m" (Guibé 1958, Vinson 1953). A likely prey on Round Island, Bojer's Skink *Gongylomorphus bojerii*, can burrow extremely fast into soft sand to hide itself (David Bullock & Steve North pers. comm.).

105. In 1732 Cossigny (1732–55) reported large hairy spiders and another sort with bird-catching webs, the latter (*Nephila inaurata*) also noted by Bernardin (1773), Milbert (1812) and Clark (1859). Clark (1859) recorded hairy 'tarentulas' with bodies an inch long that ate house geckos. Dennis Hansen (work in progress, pers. comm.) has recently recorded *Nephila* catching and eating day-geckos *Phelsuma cepediana* and Cheechaks *Hemidactylus frenatus* in upland forests.

106. Vinson (1976b) mapped the four taxa of medium-sized day-geckos; they have largely adapted to man-modified habitats, so current distribution probably reflects the original. Vinson's Day-gecko *Phelsuma ornata* is a largely coastal and lowland species adapted to scrubby vegetation and rocks as well as forest; Guimbeau's Day-gecko *P. guimbeaui* is confined to tall dry forest in the west, while the Blue-tailed Day-gecko *P. cepediana* is a very widespread forest species, scarce only in the eastern dry-forest zone. A fourth form, the Upland Forest Day-gecko, originally described as race *rosagularis* of *P. guimbeaui* (Vinson & Vinson 1969), has been shown by DNA analysis to be more closely related to *P. cepediana* (Austin *et al.* 2004a). It is found in the wettest upland native forest only. All the species are sympatric with at least one other, *cepediana* with all three.

107. On Round Island *Phelsuma guentheri* is remarkably well camouflaged against the trunk and hanging dead inflorescences of latan palms (pers. obs.).

108. Cheke (1987c), Staub (1988), Nyhagen et al. (2001); Olesen et al. (2002) have demonstrated how important broad-spectrum generalist pollinators are in island ecosystems, using *Phelsuma ornata* as an example; see also Staub (1988, with photos) for white-eyes and day-geckos, and Hansen et al. (2002, 2007) and Hansen (2005) for upland forest *Phelsuma*.

109. Staub (1988), Olesen et al. (1998); Day-geckos prefer coloured nectar to clear in artificial flower experiments, and *Trochetia* spp. flowers with red nectar are pollinated by *Phelsuma cepediana* (*T. blackburniana*) and *P. ornata* (*T. boutoniana*) (Hansen et al. 2006a, 2007). Hansen (2005) reported *P. cepediana* pollinating *Roussea simplex*, but has since found that while licking the juicy fruit they also swallow and excrete the tiny seeds, although originally Telfair's Skink may have dispersed them more efficiently.

110. Cheke & Dahl 1981, Nyhagen et al. 2005; the surviving species *P. niger* (Mauritius) and *P. rodricensis* (Rodrigues) readily take fruit of introduced trees.

111. Nyhagen (2004), Nyhagen et al. (2005); *Pteropus niger* groups living in native forest took mainly Natte *Labourdonnaisia glauca* fruit, but also important were *Mimusops petiolaris*, *Diospyros tesselaria*, *Protium obtusifolium* and *Aphloia theiformis* to name just the top five. They also foraged outside the forest on introduced species, such as mango *Mangifera indica*, Indian Almond *Terminalia catappa* and Jackfuit *Artocarpus heterophyllus*.

112. Lanux (1776).

113. Jones (1987), Jones & Owadally (1988).

114. Réunion Merles have been recorded eating palm fruit (Cadet 1977); forest palms are too rare in Mauritius for such observations today.

115. Cossigny (1732–55), Cheke (1987a). There are several species of takamaka (*Calophyllum* spp.) in Mauritius (*Flore* 49, 1980, Robson & Stevens), but the common species is *C. tacamahaca*, formerly an important timber tree, as was the small-leaved natte, *Labourdonnaisia calophylloides* (*Flore* 116, 1981, F. Friedmann). The Seychelles Blue Pigeon *Alectroenas pulcherrima* takes both soft fruit (figs etc.) and fruit with large hard seeds, including another takamaka (*C. inophyllum*; Skerrett et al. 2001) and *Sideroxylon ferrugineum* (allied to *Labourdonnaisia*; Gaymer et al. 1969). Likewise stomachs of *A. sganzini* in the Comoros contained "only fruit pulp and large seeds" (Benson 1960), but there have been no detailed diet studies on *Alectroenas* pigeons (see also Goodwin 1983).

116. Milbert (1812), Cheke (1987a); *Alectroenas* pigeons normally stay strictly in the canopy and do not visit the ground (e.g. Benson 1960, Frith 1977, 1979), despite which Goodwin (1983) accepted this alleged consumption of aquatic molluscs. As Milbert (or at least his informant Mathieu) had collected and possibly dissected the birds, they may well have contained snails, but probably arboreal ones.

117. Hugo, in Panyandee (2002); he referred to fodies as 'sparrows', as was general in the 17th century.

118. Feeding ecology of passerines is from Cheke (1987c), Safford (1991, 1996), Safford & Beaumont (1996) and unpublished obervations by MWF workers, notably on gecko use by merles and cuckoo-shrikes (Carl Jones pers. comm.).

119. Swiftlets and swallows *Aerodramus francicus* and *Phedina borbonica* were studied by Cheke (1987c). There has been no research in Mauritius on the ecology of the microbats *Taphozous*

mauritianus and *Mormopterus acetabulosus*, though known facts were summarised by Cheke & Dahl (1981); Moutou (1982) added data on the same species in Réunion.

120. The swiftlet is endemic to Mauritius and Réunion (Cheke 1987c, Chantler & Driessens 2000); see Chapter 4 for the muddled biogeographical history of the free-tailed bat. Nowadays the bats use house roofs as well as caves for roosting (Cheke & Dahl 1981, Moutou,1982).

121. The Mascarene Swallow *Phedina borbonica* is a well-marked endemic race (and is actually the nominate; another race, *madagascariensis*, is restricted to Madagascar; Cheke 1987c, Morris & Hawkins 1998); The Grey Tomb Bat *Taphozous mauritianus*, although first described from Mauritius (though not until 1813), is a widespread Afrotropical species (Moutou 1989, Peterson et al. 1995).

122. Cowles (1987), Mourer et al. (1999); the birds are respectively *Anas theodori, Alopochen mauritianus, Fulica newtoni, Phalacrocorax africanus* and *Phoenicopterus roseus*. As noted in Chapter 2, subfossil cormorant bones have not been checked against Asian *P. niger*, so the reference to *P. africanus* is provisional.

123. Mourer et al. (1999) found medullary bone, indicative of birds within 10–14 days of laying, in bones of female-sized Greater Flamingos. These flamingos were regularly seen in flocks of up to 500 on Aldabra for years, but were thought to be non-breeding visitors until nests and young were found in 1995 (Skerrett 1995). Breeding on Madagascar is irregular (Morris & Hawkins 1998), as it may be on Aldabra (Skerrett et al. 2001) and we suspect it was in the Mascarenes.

124. Bernier's Teal *Anas bernieri* nests in tree holes in mangroves in Madagascar (Razafindrajao et al. 2001, Young et al. 2001); the Grey Teal *A. gracilis* generally nests in tree-holes when available, as does the Chestnut Teal *A.castanea* in some areas (Frith 1982).

125. Mourer et al. (1999).

126. In 2001 on the north coast of Cuba's Holguin Province, ASC frequently saw both Black-crowned and Yellow-crowned Nightherons (*Nycticorax nycticorax* and *Nyctanassa violacea*) flying into woodland at dusk to catch lizards, in an area of similar rainfall (1,200mm/yr) and forest structure to lowland semi-dry forest in Mauritius and Rodrigues.

127. See Chapter 4.

128. For example, Staub (1976, 1993), Temple (1976), Barré (1984), Barré et al. (1996), Probst (1997) and Showler (2002).

129. For migrants to the Mascarenes see Chapter 4 note 49 (p. 292), for the Seychelles, Skerrett et al. (2001b).

130. "July 7 [1704]. Saw abundance of grey and white birds as big as pintados with forked tails, and also a great many tropic birds. July 8. About noon passed between Hangrock and the Main..." (Log of the *Westmoreland*, Barnwell 1948; the 'grey and white birds' were probably Sooty Terns *Sterna fuscata*; 'Hangrock' = Gunner's Quoin; Barnwell 1955). English ships' logs in the 1600s and 1700s often recorded tropicbirds on the day before reaching Mauritius, and seabird activity in general was a regular matter of interest in the days of sail. Bernardin (1773) noted on 10 July 1768 "I saw a *paillencu* [tropicbird] . . . The sight of this bird denotes the proximity of land" and on the 12th: "We saw tropic-birds" before Mauritius was sighted early on the 13th. Elsewhere in his book Bernardin devoted a whole essay to seabirds and what different species and habits indicate.

131. Jolinck (in Keuning 1938–51); Panyandee's translation (2002) had the birds in 'caves', an error for 'holes'.

132. See Chapter 2.

133. The de Bry brothers (1601) published a Latin version of van Neck's voyage including a paragraph on frigatebirds (Box 5, p. 73); a similar passage is also found in the *Tweede Boek* (anon. 1601, Keuning 1938–51). Staub (1993) wrote that "in 1598 *two* [our italics] species of Frigates . . . were seen nesting in bushes at Rivière des Créoles in Grand Port Bay by the crews of van Warwyck". In fact, without bone material we have no way of telling which species were present: the only description (the legend to the engraving of the Dutch camp) mentioned only white-bellied birds (i.e. females/immatures) with no indication of adult males or their red throat pouches. Leguat (1707), who described the habits of frigatebirds in Rodrigues, made no mention of them in Mauritius, although he was marooned in Grand Port bay for two years; the Dutch had probably wiped them out by then. The birds may have been attracted by or associated with the Abbott's Booby colony (which the Dutch never mentioned).

134. Khan 1927, Cheke 1987a.

135. Leguat (1707) did not describe the terns he called *ferrets*, but commented that they nested tightly packed, laying single eggs on bare sand – typical of Sooty Terns *Sterna fuscata* rather than noddies, which build nests. The dark hole-nesting *plutons* were no doubt Wedge-tailed Shearwaters *Puffinus pacificus*.

136. Barnwell 1955.

137. These seabirds were not formally recorded until Round and Serpent Islands were first surveyed in 1844 (Lloyd 1846).

138. Oddly, as they are so conspicuous and attractive, no early visitor mentioned nesting tropicbirds; apart from sailors on their approach to land, there is no record before 1753 (La Caille 1763). We follow Austin et al. (2004b) in separating the Indo-Pacific Tropical Shearwater *Puffinus bailloni* from Audubon's *P. lherminieri* in the Atlantic. They are strictly nocturnal on land and could easily have been overlooked, being first reported in Mauritius only in 1803 (Lesueur 1803).

139. ". . . and another kind of bird, ash-grey with flat beaks and long necks" – Jolinck (in Keuning 1938–51); see Skerrett et al. (2001) for *Pelecanus rufescens* on St Joseph Atoll, Amirantes (Seychelles).

140. Although these animals were always referred to as sea-cows or manatees in Dutch and English, and later as '*lamentins*' in French, Stoddart (1972) wondered if they might have been mis-described seals. However, he concluded tentatively that the Mascarene animals were *Dugong dugon*, Cheke (1987a) adding further strong evidence in favour of Dugongs. In the Seychelles Stoddart concluded that there really had been unidentified seals, albeit sometimes called 'manatee' and '*vache marine*', though it seems likely to us that both seals *and* Dugongs were present in the Seychelles.

141. Hoffman (1680), Pitot (1905), Stoddart (1972), Verin (1983).

142. Barnwell (1948, 1950–54).

143. Laerle's three sketches include at least two different individuals, one adult, one young (Moree 2001). Published accounts of Loggerheads in the Mascarenes seem very few; Frazier (1984) included Mauritius and Réunion in their distribution in the Indian Ocean, though they only nest in Madagascar (Ratzimbazafy 2003, Gerlach 2004).

144. Although not mentioned by early visitors, the bird-catching orb-web spiders *Nephila inaurata* common today are presumably native in Réunion, as in Mauritius, but it is not clear whether there were tarantulas. It is far from clear how much of Cossigny's report in 1732, which discussed spiders, applied to Réunion as well as Mauritius (Charpentier 1732–55). It is odd that Dubois's (1674) detailed description omitted orb-web spiders, but he barely mentioned lizards and missed the insectivorous bats.

145. Dubois's comments (1674) are the most explicit on vertical migration: "all the birds of this island each have their season at different times, being six months in the flat country, and six months in the mountains, when, returning, they are very fat and good to eat. I except only the waterbirds and the solitaires, the quails and the *oiseaux blues*, which do not change at all." Probst (2001a) found subfossil eggs of *Phelsuma borbonica* in a place way outside the present range, and reported on further relict populations; for *P. inexpectata* see Bour et al. (1995) and Chapter 7. There is a single account of a large, brightly coloured lizard up to 460mm (18") long

allegedly occurring in Réunion, a large day-gecko gecko resembling *P. edwardnewtoni* of Rodrigues. The supposed author of this account, 'Père Brown' (1773), a Jesuit missionary, apparently did not exist (d'Aglosse 1891, Lougnon 1970), and the account itself is cognate with Le Gentil's (1727) from 1717, though not identical (Lougnon 1970); population details in Le Gentil are updated to roughly 1733 by 'Brown' (Cheke 1987a). It seems unlikely that such a creature could have survived unnoticed on Réunion for so long, so if not imaginary it is more likely to represent a hearsay account or captive specimen of Newton's Day-gecko from Rodrigues. Ships visiting the Mascarenes were calling at Rodrigues to collect tortoises from the early 1730s onwards (North-Coombes 1994), which could account for the lizard's appearance in the 'Brown' text but not Le Gentil's – he was in Réunion before French ships had started visiting Rodrigues. In favour of the story being based on a real animal is the report that the ear orifices connected across the head, just as Liénard (1842) later reported for the Rodriguan *P. gigas*.

146. Details from Legendre (1929). Buffon (1777–83) reported that 'seeds [and] berries of *Pseudobuxus* [sic]' had been found in the starling's stomach; '*Pseudobuxus*' was probably the small forest tree *Eugenia buxifolia* (*Flore* 92, 1990, A. J. Scott).

147. Mourer *et al*. (1999, 2006) and Chapter 4.

148. See Cheke 1987a.

149. Dubois (1674, Lougnon 1970, Barré *et al*. 1996); this parrot's plumage suggests a form allied to *Psittacula eupatria* (Hume 2005a) and to the putative full-plumaged male of the Rodrigues Parakeet described by Tafforet (1726), but no bones have yet been found.

150. In Madagascar, Rand (1936) likewise reported that buttonquails "flushed close and their flight was usually short", whereas true quails make longer flights (Roger Safford pers. comm.).

151. See Cheke (1987c,d) for comparative discussions of fody feeding habits and interactions with Cardinal Fodies. *Foudia flavicans* was once again recorded in flocks in 1999 (Impey *et al*. 2002).

152. Dubois (1674) and Dellon (1685 and in Lougnon 1970; see Box 31, p. 228). Seychelles Fodies *F. sechellarum* freely enter the warden's house on Cousin Island, stealing food from plates etc. (pers. obs., 1970s). The Réunion bird, while slightly different in coloration, may have been conspecific with the Mauritius Fody, but in the absence of subfossil or specimen material we think it best to treat it as a full species.

153. Barré (1983), Cheke (1987b); the plants are *Hypericum lanceolatum* and *Sophora denudata* (*Flore* 49, 1980 & 80, 1990). Hering & Hering (2005) published photos of Mascarene Grey White-eyes feeding on the *Sophora*, but called it *S. tetraptera* (a New Zealand species) in error.

154. La Roque in 1709 said the *oiseau bleu* lived 'amongst grasses and aquatic ferns' on the Plaine des Cafres (Lougnon 1970, Barré *et al*. 1996) – but where were the 'aquatic ferns' up there?

155. The fullest description, as usual, comes from Dubois (1674, Barré *et al*. 1996); he was the only observer to describe the *oiseau bleu*'s bill and legs, but his size estimate was probably exaggerated, and we follow Cheke (1987a) in accepting Feuilley's (1704) smaller assessment. All reports called the plumage blue or dark blue (Barré *et al*. 1996, Box 11, p. 128).

156. Swamphens are as secretive in Madagascar (Milon 1951, Morris & Hawkins 1998) as everywhere else in their range.

157. For example, Berlioz (1946), Barau (1980), Barré *et al*. (1996). Milon (1951) preferred an undescribed endemic rail, close to the Purple Swamphen, a view we share. Selys-Longchamps gave it the name *Apterornis coerulescens*, Bonaparte created a new genus '*Cyanornis*', and these combinations were used as if there was no doubt by Rothschild and Hachisuka, who supposed it was similar to the New Zealand Takahe *Porphyrio hochstetteri* (Hachisuka 1953). There was also an unidentified *poule bleue* in the Seychelles (e.g. Lionnet 1984), but the sparse reports before its 18th century extinction suggest it was a fairly ordinary *P. porphyrio*. We have used traditional nomenclature, but DNA analysis (Trewick 2002) indicates that races of '*P. porphyrio*' should be treated as species (e.g. Afro-Malagasy birds become *P. madagascariensis*).

158. Purple Swamphens have long been a popular ornamental and game species; they were certainly imported, but evidence of captive birds being released is indirect (though persuasive)

159. Accounts in Bour (1981), Probst & Brial (2001); notwithstanding the DNA analysis Gerlach (2004) provisionally accepted a second tortoise rather than treat them as a single variable species as accepted by Austin & Arnold (2001), and Bour (in Ribes 2006) noted that hundreds of measured bones fell into four subsets, best accounted for by two sexes each of two species (putatively one highbacked, the other domed). An old carapace in Paris from 'the Indies' has been identified by DNA as a Réunion Tortoise, hence the specific name '*indica*'. This specimen was originally described in 1676 by Charles Perrault, brother of the famous collector of fairy tales (Austin *et al*. 2002, Bour 2004).

160. Bour (1981) maintained that there was no evidence of tortoises on the windward side, but see Tatton (1625) and Melet (1672).

161. Austin & Arnold (2001), Gerlach (2004); Chambers (2004, Galápagos).

162. Merveille in La Roque (1716), Lougnon (1970).

163. McFarland *et al*. (1974), Cheke (1987a:10).

164. Lanux (1772); see Chapter 6 for more of Lanux's contributions.

165. These dates do not tally with modern breeding dates for Mauritius, where mating is recorded in early April, large (nearly independent) young in early December, and no young being carried in February or April (Cheke & Dahl 1981); however, systematic observations are lacking. In captivity, Golden Bats gestate for around 5 months; the young can fly at 11–12 weeks, but suckle until around 6 months old, and remain associated with their mothers for up to a year (West 1986, Young 1987). Nearly independent young *P. niger* in December cannot be the result of matings in April; Golden Bats mate throughout the rearing period of young (West & Redshaw 1987).

166. Roosting in holes was thought to be unique in the genus, but two species from islands off New Guinea, *Pteropus admiralitum* and *P. caniceps*, have recently been found roosting in tree-hollows (Flannery 1995a). Known roosting groups are much smaller (2–30 individuals) than the aggregations Lanux described for *P. subniger*, and there is no evidence of gender bias. In both cases, as in the Mascarenes, there are also branch-roosting flying-foxes sympatric on the same islands.

167. Bory (1804). Brial (2001b) discussed these unique reports and remarked on similar behaviour recorded elsewhere (see also Peterson *et al*. 1995, Schliemann & Goodman 2003, Goodman *et al*. 2007). The Honduran White Bat *Ectophylla alba* rolls *Heliconia* leaves around itself, the Central American genus *Thyroptera* also uses *Heliconia*, while the two small Malagasy endemic *Myzopoda* species are closely associated with *Ravenala*. The latter two bat genera have suckers on their feet to grip the smooth slippery leaves, and both plant genera are in the banana family Musaceae (*sens. lat.*). Peterson *et al*. (1995) noted an additional case, the African *Pipistrellus nanus* roosts in unopened rolled leaves of bananas. There is a wonderful picture of *Ectophylla alba* in Lovett (1997). Brial (2001b) gave Bory's bat a scientific name, *Boryptera alba* (see also Probst & Brial 2002) but it may well have been a *Myzopoda*; Bory was definitely not referring to Grey Tomb Bats *Taphozous mauritianus*, which have white underparts, as he described them quite clearly elsewhere in his text (Bory 1804, v3:227; see Box 36, p. 259).

168. *Scotophilus borbonicus*; Maillard (1863) called it 'upland bat', and his comment that it "lives generally in the forests and is also found in the coastal zone" is the only hint as to its distribution. See Chapter 4 for its confused affinities. It is no longer considered endemic to Réunion though Goodman *et al*.'s suspicions (2005) that the original 18th century specimens were vagrants are unfounded. They overlooked the 19th century evidence given in Cheke & Dahl (1981), despite citing that paper.

169. Probst *et al*. (1995, 1999) described the nesting colonies of *Pterodroma baraui*.

170. Jouanin (1970) elucidated the confused history of *Pseudobulweria aterrima*, then known from only four surviving specimens. Several more (dead and alive) have been found since (Jouanin 1987, Bretagnolle & Attié 1997, Le Corre et al. 2002, Fontaine 2005, p. 187), and Jossigny's 18th century drawing rediscovered (MNHM archives, Paris; pers. obs.). the nesting area is known, but nests have yet to be found, and the breeding season remains unknown.

171. Bory (1804), Jouanin & Gill (1967).

172. Probst (1997), Bretagnolle & Attie (1997) and Chapter 8 give current details of breeding seabirds; there appears to have been little change in the shearwater, noddy and tropicbird distribution over the centuries. Schlegel & Pollen (1868) reported the noddy roost; the birds may also have nested there. Cap Méchant, at Basse Vallée, is called 'Pointe du Brulé du Baril' on the ONF 1:100,000 map.

173. See Cheke (1987a) and Chapter 8 for references on marine mammals and turtles.

174. Leguat (1707) reported seeing 'only very few swallows'; none were recorded by anyone else. It is possible that swiftlets and/or swallows existed in small numbers, or they may have been vagrants from Mauritius.

175. Details of feeding ecology from Leguat (1707) and Tafforet (1726): see Box 17, p. 167. Tafforet said Solitaires ate "seeds and leaves of trees that they pick up off the ground", and Leguat reported the colourful geckos *Phelsuma edwardnewtonii* feeding on latan fruits and cultivated melons.

176. *Phelsuma gigas* (Liénard 1843, Box 33); see Arnold (2004).

177. *Necropsar rodericanus*, known from subfossil bones, was seen only by Tafforet (1726, Box 29, p. 243), who found them only on the seabird colonies on Ile au Mât (now Gombrani). Captive starlings refused seeds, but readily took chopped meat.

178. Leguat (1707), often reprinted. e.g. Strickland (1848) and (rearranged) Hachisuka (1953), Box 17, p. 167.

179. Tafforet gave the maximum weight as 50lbs (23kg), though how he and Leguat estimated their weights is open to conjecture; Leguat estimated the heaviest tortoises at around 100lbs (45kg). Neither Leguat nor Tafforet mentioned the considerable sexual size dimorphism of the Solitaire established from subfossil bones, males standing about 84cm tall, females 15cm shorter (Newton & Gadow 1896).

180. Tafforet (1726) never found a Solitaire's nest, but confirmed seeing only one young at a time.

181. Much of this bird behaviour struck arch-debunker Atkinson (1922) as blatantly unreal – he was so determined to paint Leguat's story as a novel written by Misson that he studiously avoided considering the observations from subfossil remains that confirmed the 'widow's headband', gizzard stones and 'musket-balls' on the wing, though of course it is impossible to confirm what Leguat called the birds' 'marriage' – when they withdrew in pairs after leaving the young in the flock. This behaviour is ornithologically acceptable, though Leguat's anthropomorphic terminology played into Atkinson's hands. If Leguat's description is correct, the aggregations of young birds were not crèches (where many young are guarded by a few adults), but simply flocks of immatures. The weapon on the wing was a "bony knob on the radial side of the metacarpal unlike what is found in any other bird" (Hachisuka 1953). Armstrong (1953) credited Leguat with the first ever description of avian territoriality. There are anomalous whimsical and polemical passages in the account of the Solitaire that Racault (1995) attributed to interpolation by Misson, but most of Leguat's text seems to be based on observation and is in keeping with modern ornithological knowledge.

182. With the exception of flamingoes (Campbell & Lack 1985) pigeons are the only birds to provide 'milk' (Goodwin 1983).

183. *Erythromachus leguati*; Leguat (1707) called them a 'clear grey', while Tafforet (1726) said they were speckled grey and white; no one else saw them.

184. Günther & Newton (1979).

185. Mourer-Chauviré (1999) likened the smaller species (she only had a sternum) to the Pacific genus *Gallicolumba*, but JPH (work in progress) considers it close to *Nesoenas*, suggesting that Rodrigues, like Mauritius, was invaded twice by pigeons of this lineage. Malagasy Turtle Doves are more purplish-brown than slaty-grey, though as some slates are purplish, perhaps Leguat could have described them so, but his birds may have been the other species, or the two looked similar.

186. Cheke (1987a), Hume (2005a); Staub (1973b) suggested the colour-morph hypothesis, supported by Jones (1987).

187. Pingré (1763); the description of the parakeet only occurs in an alternative draft of his chapter on Rodrigues bound into MS 1804 (Cheke 1987a), not in any published version.

188. Only two specimens of *Psittacula exsul* are known, one of each sex, though the male, with no trace of red on the wings, may perhaps not be fully adult; for its affinities see Chapter 4.

189. The fody was the only passerine Leguat (1707) noticed: "somewhat resembling Canaries" but oddly he "never heard them sing", while Tafforet correctly pointed out that it had a "pretty warbling [*ramage*]". In Leguat's day they were so tame as to come and sit on a book he was reading, as Seychelles Fodies still do on Cousin Island. (pers. obs.).

190. Staub (1973), Cheke (1987d).

191. Diamond (1980), Cheke (1987d). Unlike Leguat, Tafforet (1726) noticed the warblers, mentioning "a few *lavandières* ['wagtails'] . . . which have a very pretty song" (full quote in Box 31, p. 228). The warbler often cocks its long tail conspicuously (Staub 1976, Cheke 1987d), hence Tafforet's name for it.

192. Cowles (1987) referred the single sternum of the second passerine provisionally to the babbler subfamily Timaliinae, but as babblers are unknown on any small Indian Ocean islands, this may not be correct, although its affinities remain to be clarified (JPH, unpublished). There are several genera generally assigned to babblers on Madagascar (Morris & Hawkins 1998), but this taxonomy is now disputed (Schulenberg 2003).

193. Hans Brouwer, captain of the *Zeeland*, who did not go ashore himself, reported, all too briefly (in Dutch): "*duijven, papagayen, ganssen, dodoerssen ende ander gevoghelt*" (Moree 2001:157). The translation is in the main text.

194. The Mascarene ducks and geese were normal wetland forms, but in Hawaii ducks evolved into large forest-dwelling browsers (*moa-nalos*) that probably never frequented wetlands (Olson & James 1991).

195. A bat skull found in caves by James Caldwell, presented by him with some Golden Bat skins to the Indian Museum in Calcutta in 1881, was originally listed as *Pteropus rodricensis* (Anderson 1881), but was later re-identified by Knud Andersen (1913, Cheke & Dahl 1981, Hume 2005a) of the British Museum as *P. niger*. There are very few bat remains in Rodrigues caves, and only this one of *P. niger*. Caldwell, temporary magistrate in Rodrigues in 1875, was posted for a few years to Calcutta in 1878 (*DMB*:448–449; 1944, Barnwell & Rae).

196. Pingré (1763, and in Cheke & Dahl 1981).

197. Leguat (1707), Pingré (1763, in Cheke & Dahl 1981). The French measured in feet and inches before the revolutionary government introduced the metric system in 1795; the French *pied*, fixed by Charlemagne around 800AD (Rush & O'Keefe 1962) was equal to 1.065767 Imperial feet (12.7892", *Encyclopaedia Britannica* 11th ed. 1910–11), which should be remembered when using measurements from 18th century French texts (e.g. Buffon 1770–83). The Mascarenes escaped the full force of the French Revolution, and the French foot (mostly in the form of the *toise* [fathom] of 6 *pieds* for measuring land) and an *arpent* of 40,000 square *pieds* (1.043 acres) are still in use (North-Coombes 1971:315 summarised Mauritian units of measurement, the French foot equated to 1.07 Imperial feet).

198. *P. niger* wingspan from La Nux (1772), Pike (1873), Moutou (1982) and recalculated from bone measurements in Bergmans

(1988–97); *P. rodricensis* calculated from photographs in Carroll (1979) using forearm measurements from Cheke & Dahl (1981).

199. The two native species are *Ficus rubra* and *F. reflexa* (Strahm 1989).

200. Strahm (1989); the endemic *Doricera* (formerly *Pyrostria*) *trilocularis*. Balfour (1879) was the first to record the local name and the tree's use by Golden Bats.

201. See e.g. Churchill (1998) for flying-fox breeding; Leguat (1707) and Bertuchi (1923) reported twins, confirmed in the Jersey and Mauritius breeding groups (Carl Jones unpubl. notes.), Chester Zoo (*Zoo Review* 2003: 18), and in the wild in 2003 (Powell 2004).

202. *Phelsuma gigas* was described in detail by Liénard (1842, Vinson & Vinson 1969, Box 33 p. 243), but no specimens have survived. Liénard's type was 14" long (presumably French, so 38cm, SVL *c*.190mm) but he was told of individuals up to 54cm long, with SVL of *c*.270mm, making it the second largest gecko in the world, beating *Rhacodactylus leachianus* from New Caledonia (240mm SVL). The extinct *kawekaweo* of New Zealand, *Hoplodactylus delcourti*, of which a single specimen exists, was larger, reaching at least 62cm in length (370mm SVL) (Bauer & Russell 1986, Russell & Bauer 1986, Towns & Daugherty 1994, illustrated in Flannery & Schouten 2001).

203. Pingré (1763, Vinson & Vinson 1969); all the Mascarene day-geckos have this ability to change colour, turning from bright green-blue to a dull cryptic blackish-brown when threatened (pers. obs.).

204. Day-geckos commonly treat ripe fruits as an alternative nectar source, licking vigorously to get the sweet juice, but no extant species bites off and swallows fruit as implied by Leguat's observations. General information on day-gecko feeding habits can be found in McKeown (1993), Runquist (1995), Cheke (1975, 1984) and Vinson & Vinson (1969).

205. Arnold (2004) reported that *Phelsuma edwardnewtonii*, unlike *P. gigas*, had flat, blade-like teeth consistent with frugivory. *P. gigas* had teeth typical of insectivorous geckos. Marragon (1795) considered Newton's Geckos useful around the house for eating flies, so they were not purely frugivorous, and since Liénard (1843) reported that *P. gigas* ate figs, it clearly was not purely carnivorous.

206. Arnold *et al*. (in prep. a).

207. Nick Arnold, talk at symposium *Evolution and conservation in the Mascarene Islands,* 2nd September 2000. These undescribed lizards are not explicitly mentioned in the circulated summaries of the talks, but are referred to, with very little detail, by Arnold (2000).

208. Nick Arnold, pers. comm.

209. The French pound (*livre* or *poids de marc*) weighed 1.07916 lb avoirdupois (Imperial) or 0.4895kg (*Encyclopaedia Britannica* 11th.ed. 1910–11).

210. Pingré (1763) mentioned carosses; the two species were respectively *Cylindraspis vosmaeri* and *C. peltastes* (Arnold 1979, Burleigh & Arnold 1986, Austin & Arnold 2001). Weights are from North-Coombes (1994); he also sought to resurrect the putative third species, basing his argument on the known reliability of Leguat and Tafforet, and on an anomalous small carapace in the Natural History Museum, London (BMHN76.11.1.356), which Roger Bour said (pers. comm. to North-Coombes) was morphologically distinct from *C. peltastes*. This assessment has not been scientifically published; as the carapace was not included in Austin & Arnold's study (2001), it may not be a Mascarene tortoise at all.

211. The island supported some 200,000 tortoises (p.114).

212. Shearwaters and noddies apparently excluded each other on the smallest islets, though as Tafforet (1726) had shearwaters nesting on several fairly improbable ones, he possibly muddled them when writing up, though his actual species descriptions are accurate enough; unlike Pingré 35 years later, he did not distinguish between the two noddies. Staub (1973) tabulated Tafforet's and Pingré's seabird observations.

213. Cheke (2001a) reviewed Mascarene booby distribution. Tafforet also gave a good description of the '*boeuf*', and a subfossil bone has also been found (Bourne 1976; Cheke 1987a, 2001a).

214. The frigatebird in the NHM (Tring) was collected by C. E. Bewsher and labelled as from Mauritius, but other evidence suggests it was obtained in Rodrigues in 1874 (Cheke 2001a).

215. Suggested by Bourne (1968) and Staub (1973) based on Tafforet reporting two laying seasons and seeing birds feeding up to 30 leagues offshore, Leguat's remarks on bill length (half a foot long), and Pingré's wing-length estimate of 7–8 feet – all implying *Fregata minor* was also present.

216. Gardyne 1846. However, there was no indication of nesting.

217. Leguat apparently only saw White-tailed Tropicbirds and Pingré Red-tailed; see Staub (1973), Showler (2002) and Chapter 8 for recent status of tropicbirds.

218. See Chapter 4.

219. The maximum measured size for a Dugong is 4.06m (Reynolds & Odell 1991), so Leguat and Tafforet may have exaggerated in true fisherman's style – though Leguat's group only harvested smaller animals as they couldn't handle the very big ones. However, before Dugongs were subject to human pressure they may have grown older and larger than they do now. Tafforet's account of both dugongs and turtles seems too close to Leguat's to be entirely independent; Tafforet evidently had access to Leguat's book, though most of his account is clearly based on his own observations.

220. Faure & Montaggioni (1975), Turner & Klaus (2005); the sea grasses present are *Halophila ovalis* and *H. balfouri*.

221. Leguat was right about Dugongs, although the book he disparaged, Corneille's *Dictionnaire des Arts et des Sciences* of 1694 (North-Coombes 1991), was in fact discussing manatees *Trichechus* spp., which do occasionally produce twins (Reynolds & Odell 1991); both were then conflated under the same name '*lamantin*'.

222. The tusks were drawn incorrectly as in the lower jaw; also they are normally only present in males (Reynolds & Odell 1991), whereas Leguat's animal has teats and is holding a calf – the verbal description is more accurate, the illustration made *post hoc*. Atkinson (1922) actually used a source that had borrowed from Leguat to 'prove' that Leguat's *lamantins* (Dugongs) were derivative: the second, 1726, edition of the very encyclopaedia Leguat had criticised improved its text with his new observations, which Atkinson (claiming to use the earlier edition) then used against him for plagiarism (North-Coombes 1991)!

223. Bustard (1972) reported up to six clutches a year, and that large females can lay 200 eggs; the maximum recorded is 226 (Dr Peter Pritchard in North-Coombes 1991). Incubation takes 48–80 days, depending on sand temperature (Bustard 1972), suggesting that Rodriguan sands were pretty hot (32° or more); Tafforet reduced the incubation period to 31 days!

224. One passing reference in his diary (Pingré 1760–2, Cheke 1987a); nothing in his edited account of the island (Pingré 1763).

225. Only to be caught and turned into stuffed souvenirs; most Hawksbills that turn up are too young to be used for tortoiseshell (Cheke 1987a).

226. The Mascarene land crabs are *Cardisoma carnifex* and *C. hirtipes* (Staub 1993a), now much fewer than before.

227. On Christmas Island the principal species is the endemic *Gecarcoidea natalis*, breeding in November-December (Gray 1981). On Aldabra it is *Cardisoma carnifex* (breeding in January), but *C. rotunda* and two *Geograpsus* are also common; altogether there are 12 species, including four shelled hermit crabs (Alexander 1979, Gillham 2000). Echoing Leguat, Alexander reported extensive consumption of seeds and seedlings by *Cardisoma* crabs on Aldabra.

228. Leguat was so caught up with the land crabs that he did not notice the hermit crabs – but Pingré did. The terrestrial species in the Mascarenes and Aldabra are *Coenobita* spp. (Haig 1984, Gillham 2000), still abundant in suitable habitat, (e.g. Ile aux Aigrettes

nature reserve, in Mauritius). The Coconut Crab *Birgus latro*, a giant shell-less hermit crab, was never reported in the Mascarenes – despite its name it does not require coconuts (of which there was a small natural population on Mauritius): it is too big, edible and destructive to have been overlooked. It is or was present on Agalega (Cheke & Lawley 1984), Aldabra (Gillham 2000) and the granitic Seychelles (Haig 1984, High 1976).

229. Pers. obs.; spider's identity from Matjaz Kuntner (per Dennis Hansen).

CHAPTER 4. WHERE DID THE DODO COME FROM?

1. Mundy (1608–1667, slightly modernised by Barnwell 1948). As Evelyn Hutchinson (1962) noted, similar thoughts by Charles Darwin two centuries later led to a revolution in scientific thought! Mundy later (in 1656) reformulated his question more cogently in connection with the flightless Ascension Rail *Atlantisia elpenor*: "I have heretofore asked the question concerning Mauritius henns and dodos, that seeing those could neither fly nor swymme, being cloven footed and without wings on an iland far from any other land, and none to be seen elce where, how they should come thither? . . . the question is, how they shold bee generated, whither [whether] created there from the beginning, or that the earth produces them of its owne accord, as mice, serpents, flies, wormes etts. [etc.] insects, or whither the nature of the earth and climate have alltred the spape [shape] and nature of some other foule into this, I leave it to the learned to dispute of". Philip & Myrtle Ashmole (2000) dedicated *St Helena and Ascension Island* to his memory for having "modestly proposed the idea of evolutionary change".

2. Bory (1804), an insightful geologist and botanist puzzled by the consistent presence of endemic forms on islands, discussed the question in several passages, without arriving at a satisfactory answer. As he could not conceive of non-flying vertebrates and large-seeded plants arriving on their own, while recognising that they were not brought by humans, he invoked (albeit reluctantly) special and continuing creation on islands, which "it cannot be doubted are newer than the continents, and in consequence all that one sees on them is more recent". He clearly thought that once arrived (by whatever means), they evolved and adapted. For an illuminating discussion of biogeographical ideas between the 16th and 18th centuries, see Larson (2001).

3. Whitesell (1964) and Polhill (1990) commented on the similarity of the two island acacias.

4. Altitude and seed details from Whitesell (1964), Scowcroft & Wood (1976), Cadet (1977) and pers. obs.

5. Bretagnolle & Attié (1991) considered the Mottled Petrel *Pterodroma inexpectata* to be possibly closest to *P. baraui* on vocal grounds, with Hawaiian Petrel *P. sandwichensis* a close second, though they only had recordings of the Galápagos Petrel *P. phaeopygia*, from which Hawaiian has recently been split. On other grounds the Hawaiian bird has a better claim. Both Hawaiian and Galápagos Petrels are tropical nocturnal mountain-nesting petrels, whereas the Mottled Petrel breeds in south temperate islands off New Zealand. In his recent monograph, Brooke (2004) took up the suggestion of a relationship between Barau's and Mottled, but ignored Bretagnolle & Attié's almost equal emphasis on *phaeopygia*; Kennedy & Page (2002) placed Barau's closer to *phaeopygia* and Juan Fernandez Petrel *P. externa*, another high nesting tropical species, than Mottled. Imber's (1985) revision put Barau's close to 'Round Island Petrel' *P. arminjoniana* on arbitrary geographical grounds 'in the absence of any other data', despite the latter being a diurnal non-seasonal near-coastal surface nester.

6. Nesting distribution data from Bretagnolle & Attié (1991) and Probst *et al.* (2000) [Réunion], and Berger (1981) and Harrison (1990) [Hawaii]. Around 1800 (Bory 1804, Jouanin & Gill 1967) local inhabitants harvested Barau's Petrel nestlings from cliff-tops of the Remparts and Langevin gorges, in the upland mixed forest-heath transition close to areas of *Acacia heterophylla* (maps of Réunion's original vegetation in Cadet 1977, Girard & Notter 2003).

7. See Whitesell (1964, 1990) for conventional botanical details. Recent work on *Acacia* DNA has unfortunately bypassed both the Réunion and Hawaii species (e.g. Miller *et al.* 2003). The affinity of tetraploid *A. heterophylla* to diploid *A. melanoxylon* has support from protein and cytogenetic studies (Bell & Evans 1978, Coulaud *et al.* 1995), allowing the possibility that *A. heterophylla* and *A. koa* (also tetraploid) arose independently from *A. melanoxylon*.

8. Cadet (1977) recorded seeds from a dried herbarium specimen of *Acacia heterophylla* germinating after 26 years, but suspected a much greater potential longevity.

9. See chapters 8–10 of Hallam's (1994) general overview of world biogeography.

10. Bearman (1989) and Padya (1989) illustrated Indian Ocean wind patterns.

11. "The passage from Diego Reys [Rodrigues] to Mauritius is performed in two days, and from Mauritius to Bourbon in one; but it requires near a month to go from Bourbon either to Mauritius or Diego Reys" (Orme 1763), and "In the South-east monsoon the crossing from Rodrigues to Mauritius is done in two days; fifteen are needed for the return" (Unienville 1838 [as '1830'] in Dupon 1969:33); Dupon commented later (p.38) that taking three weeks to get from Mauritius to Rodrigues was not unusual. In 1761 Abbé Guy Pingré took 19 days for the trip and nearly missed the Transit of Venus in Rodrigues (Pingré 1763, Cheke 2004b).

12. For post-Krakatau pumice movements see Robillard (1885), Blanc (1972) and Thornton (1996). Pumice was still floating around the central Indian Ocean in February 1885 (Meldrum 1886). For details of the currents see Renvoize (1979), Neiman (1970), Duing (1970) and Bearman (1989). Neiman (1970) and Duing (1970) indicated some circulation that would allow floating objects from the east coast of Madagascar to reach the Mascarenes (by a rather long marine detour to the south), in addition to the major movement from Indonesia/Australia, itself subject to seasonal variation and complexity of motion. Leguat (1707) reported viable coconuts washed up in Rodrigues, most probably from the East Indies.

13. Gardyne (1846).

14. New *et al.* (2005).

15. Padya (1989), Soler (2000); Soler presented wind-roses for many stations in Réunion; the westerly component is generally minimal or nil, except in some inland sites with land-generated flow. Similar detail is not easily available for Mauritius, though westerly components are evidently low.

16. Walters (1980), Sibley & Monroe (1990) and Dickinson (2003) gave brief distributions for all the birds in the world, their species limits in some cases differing from ours. The Reed Cormorant *Phalacrocorax africanus* is an Afro-Malagasy species, replaced in India by the very similar Little Cormorant *P. niger*; in reassessing the Mauritian bone material, Olson (1975) did not compare it with *P. niger*, so pending further study, it remains possible that Mascarene birds came from India.

17. Mascarene Moorhens *Gallinula chloropus* are of the Malagasy race *pyrrhorhoa* (Benson 1970–71), originally described from Mauritius (A. Newton 1861b); the Dimorphic Egret *Egretta dimorpha*, confined to Madagascar and adjacent islands, is part of the widespread Little Egret/Reef Heron complex (Morris & Hawkins 1998, Hancock & Elliott 1978, Hancock & Kushlan 1984, Hancock 1999).

18. Charpentier (1799), E. Newton (1858–95), Cheke (1987a).

19. Milbert (1812), on Mauritius in 1801–03, described a *martin-pêcheur* as "verdigris, throat and belly white; lives on the edge of the sea", a passable description of *Butorides striata* that does not fit any other local bird. Sganzin (1840) saw '*crabiers*' in Madagascar

which he said were identical to those "met with in Mauritius, Réunion and part of India". Later 19th century writers apparently denied its presence in Réunion, though they muddled heron species (Cheke 1987a misinterpreted Maillard 1862 and Coquerel 1863); it was not positively confirmed until 1963 (Jouanin 1964).

20. Benson (1984) revised his earlier belief (Benson 1970–71) that Seychelles Moorhens were of African origin, agreeing with Ripley (1977) that African, Seychellois and Asian ones were not reliably separable.

21. White (1951), supported by Benson (1960), came to these rather surprising conclusions about island races of *Butorides striata*; see also Ripley (1969), Bourne (1971) and Carr (2000), who confirmed that Chagos birds *B. s. 'albolimbata'* are paler but otherwise very similar to south Asian *B. s. javanica*; Ripley was, by implication, less certain about the African origin of Seychelles herons *B. s. degens*. That said, Christian Jouanin of the MNHN in Paris (pers. comm.) re-examined the only specimen from Réunion in 2001, and found it intermediate between Malagasy and East Indies birds. He queried the validity of the subspecific distinctions when age and sex variation is fully taken into account.

22. Temple (1976) recorded Cattle Egrets breeding in 1973, but they were not seen again until four showed up in 1981 (Staub 1993); Barré (1984) and Probst (1997) listed occurrences in Réunion. Cattle Egrets breed in the Seychelles (granitics and Aldabra: Skerrett *et al.* 2001b) and are one of the few land birds breeding on the Chagos, where birds introduced from the Seychelles appear to have hybridised with naturally occurring birds from Asia (Bourne 1971, Hutson 1975). Cattle Egrets are widespread in the Old World, crossing the Atlantic to South America around 1887 and thence to the Caribbean and North America in mid-20th century (Crosby 1972, Hancock & Kushlan 1984).

23. *Nycticorax nycticorax*; Skerrett (2001).

24. *Ixobrychus sinensis*; Skerrett *et al.* (2001), Harrison (1999).

25. A 'poule d'eau' (normally a Moorhen) was first reported in 1960 (Bourne 1971); apparently absent in 1971 (Hutson 1975), it reappeared to be positively identified as *Amaurornis phoenicurus* in 1996 (Carr 2000).

26. Le Corre (1999) and Cheke (2001a) discussed Red-footed Booby *Sula sula* colonies and their colour morphs.

27. Tropical Shearwaters *Puffinus bailloni* (as '*P. l'herminieri*') in the Indian Ocean were discussed by Bretagnolle *et al.* (2000a) and Shirihai & Christie (1996), but their conventional taxonomy has been completely overtaken by DNA studies (Austin *et al.* 2004b).

28. Austin *et al.* (2004b); there is a separate lineage (*P. assimilis*) in Australasia.

29. Le Corre (1999: tropicbirds), Steeves *et al.* (2005: boobies).

30. See Table 3.1.

31. *Taphozous mauritianus* and *Scotophilus borbonicus*, the latter often considered conspecific with the *S. leucogaster/viridis* complex on mainland Africa; see Cheke & Dahl (1981), Moutou (1982, 1989), Hill (1980), Peterson *et al.* (1995, but see Goodman *et al.* 2005). Records of '*S. borbonicus*' on Madagascar (e.g. Eger & Mitchell 2003) have recently been shown to be mostly a new species *S. tandrefana*, but there is also an old record of true *borbonicus* from there (Goodman *et al.* 2005), and the related *leucogaster/viridis* group is widespread in Africa; see Hill (1980) & Peterson *et al.* (1995) for the confused nomenclature, the latter also noting its similarity to the southeast Asian *S. kuhlii*. K.Jones *et al.* (2002) allied *S. borbonicus* with *S. viridis* but not *S. leucogaster*, whereas measurement data in Goodman *et al.* (2005) suggest the reverse, and confirm resemblance to *kuhlii*.

32. Moutou (1989).

33. Moutou (1989) cast doubt on the African and Malagasy records of *Mormopterus acetabulosus*, and Eger & Mitchell (1996, 2003) denied the genus's occurrence in Africa and mentioned only *M. jugularis* for Madagascar. No Malagasy specimens could be located, but two African specimens (from Natal and Ethiopia) exist and are indistinguishable from Mascarene examples (Peterson *et al.* 1995) – research is needed on their collection circumstances. Hayman & Hill (1971) noted the South American connections of *M. acetabulosus* and *M. jugularis* (but no other Afro-Malagasy species), while Peterson *et al.* (1995) and K. Jones *et al.* (2002) emphasized similarities with *M. doriae* from Sumatra. Hayman & Hill treated *Mormopterus* as a subgenus of *Tadarida*, but all recent work confirms its distinctiveness (e.g. K. Jones *et al.* 2002), and Peterson *et al.* (1995) suggested that *M. acetabulosus* is the closest among living bats to the ancestral form of the family Molossidae; Eger & Mitchell (1996) mapped the anomalous distribution of *Mormopterus*.

34. Bory (1804), Brial (2001b, naming it *Boryptera alba*), Probst & Brial (2002).

35. *Coleura seychellensis* is closely allied to the North African and Arabian *C. afra* (Cheke & Dahl 1981). The specific status of the Aldabran *Tadarida pusilla* has been questioned (Hill 1971), and it has been allied to both *T. pumila* and *T. leucogaster* (Peterson *et al.* 1995), both found on Madagascar.

36. Vinson & Vinson (1969).

37. Vinson & Vinson (1969), Cheke (1984), Arnold (2000); Andreone & Greer (2002) considered mainland and Nosy Be forms on Madagascar were good species, and suggested that many races of '*Cryptoblepharus boutonii*' were full, if cryptic, species. However Rocha *et al.*'s (2006) DNA study retained all Indian Ocean forms within *C. boutonii*. They argued for a single colonisation event to Madagascar, followed by dispersal, but admitted that "the basal position of the Mauritian haplotypes indicates a need for further analysis". *Cryptoblepharus* clusters with other Australasian skink genera in DNA phylogenies (Austin & Arnold 2006, Carranza *et al.* 2001).

38. See Cuellar & Kluge (1972) and especially Radtkey *et al.* (1995), who traced some *Lepidodactylus lugubris* clones back to Arno Atoll (Marshall Islands, central Pacific). Ineich (1999) was the first to report males on Rodrigues, and has suggested that the sexual forms there and elsewhere should be given separate specific status to the hybrid clones.

39. Jones 1993.

40. Brown & Alcala (1957) found that eggs of three common Asian house geckos (*Hemidactylus frenatus*, *Gehyra mutilata* and *Cosymbotus platyurus*) could survive permanent contact with sea water for up to 11 days (their maximum test period). Gardner (1975) found that day-gecko *Phelsuma* eggs were resistant even to 24 hours' total immersion in sea-water.

41. Vinson & Vinson (1969), Cheke (1984), Barnett & Emms (1998), Arnold (2000). Although Henkel & Schmidt's (2000) assertion that Mourning Geckos now occur in "almost all of the Seychelles" is erroneous (Gerlach 2003), they have reached the Andamans, Maldives and Sri Lanka (Das 2004) in the northern Indian Ocean, but possibly through introduction.

42. The non-anthropophilous populations of *Hemidactylus frenatus* on outlying islands of the Seychelles and Chagos may have arrived by themselves (Cheke 1984), and indeed Peron & Lesueur (MSS in Le Havre) reported a specimen from Coëtivy in 1803 before that island was settled. However DNA studies show that Cheechaks from Madagascar, Comoros, Réunion, Mauritius and Rodrigues matched samples from Sri Lanka and the Andamans (Vences *et al.* 2004b), suggesting introduction from southern India.

43. Racey & Nicoll (1984), *Oryx* 36:105–1–6 (2002; news pages, citing *Sirenews* 36:19, Lousteau-Lalanne 2004); multiple sightings in 2001 and 2003 suggested a possible recolonisation of Aldabra; currents from Madagascar are favourable. Gade (1985) mentioned the 1970 Rodrigues Dugong, but we have not found corroborative confirmation.

44. Dugongs are slow swimmers, averaging 3 km/hr (Reynolds & Odell 1991), so reaching the Mascarenes from Madagascar against a current running at up to 1 knot might be difficult.

45. See Chapter 5 and 6 for Mascarene records, Reynolds & Odell (1991) for a general picture; continued survival in Madagascar looks increasingly tenuous (Cooke et al. 2003, Muir et al. 2004).

46. See e.g. Temple (1976), Staub (1976, 1993), Barré (1984, et al. 1996), Probst (1997).

47. The only wader ringed in Mauritius recovered abroad was a Ruddy Turnstone *Arenaria interpres* ringed by ASC on 14 Feb. 1974, killed at Ratnagiri in India (between Mumbai and Goa) on 18 Aug. 1975 (Safford & Remy 2007). However, Ruddy Turnstones from the Seychelles have been found in Dagestan and Iran; one ringed in November 1982 was in northern Kazakhstan in August 1986, then back in the Seychelles (shot) in November 1988 (Skerrett 2001).

48. On 16 Jan. 1974, 30km west of Rodrigues, ASC saw seven Curlew Sandpipers *Calidris ferruginea* overtake the eastbound ship, and on 21st Jan. 1974, on the return journey to Mauritius, a single Ruddy Turnstone approached the ship 225km west of the island (unpublished notes).

49. Although migratory falcons, Peregrine *Falco peregrinus*, Eleonora's *F. eleonorae* and Sooty *F. concolor*, turn up fairly frequently (Probst 1997; Carl Jones pers. comm.), it is unclear whether they intend to reach the Mascarenes. Two migrant swifts, an unidentified black *Apus* (in 1980) and an African Palm Swift *Cypsiurus parvus* (1997), have also been recorded once each in Réunion. Dozens of migrant passerine species are regularly recorded from the Seychelles (Skerrett *et al.* 2001b), but only two, White Wagtail *Motacilla alba* (Réunion, 1988) and Golden Oriole *Oriolus oriolus* (Mauritius, 1990s) have ever reached the Mascarenes (Probst 1997, Morrad & Baverstock 1998, Carl Jones pers. comm.).

50. Meinertzhagen (1912), discussed in Chapter 9; ASC's observation (unpublished notes) was at Wolmar, Mauritius, 25 May 1974.

51. Swallows from Madagascar irregularly migrate to eastern Africa (Morris & Hawkins 1998, Newman 1998), and also reach Aldabra and the Seychelles (Skerrett *et al.* 2001b).

52. For example, Gibbs *et al.* (2001), though like most books, theirs ignores *Nesoenas picturata*'s presence on Agalega (Cheke & Lawley 1983) and the Chagos (Bourne 1971, Hutson 1975, Carr 2000). There is some doubt as to whether it reached the latter group unaided by man. For its native status in the Mascarenes, see Chapter 3.

53. See e.g. Walters (1980).

54. Mourer *et al.* (1999).

55. Hall & Moreau (1970), Brooke (1972); the swallow genus *Phedina* appears to be a relict of a pre-Pleistocene radiation. The Congo species has very different nesting habits and is sometimes separated as *Phedinopsis* (Brooke 1972). Following Cheke (2005) we are using *Nesoenas* for both Pink Pigeon (*N. mayeri*) and Malagasy Turtle Dove (*N. picturata*); see also note 106.

56. Respectively, *Ixobrychus cinnamomeus*, *Falco amurensis*, *Amaurornis phoenicurus*, *Cuculus poliocephalus*, *Ketupa zeylonensis*, and *Apus pacificus*. We have omitted species that could come from either direction (e.g. Jacobin Cuckoo *Clamator jacobinus*, found in Africa and Asia) and Asian shorebirds (details from Skerrett *et al.* 2001a,b).

57. The significance of land areas exposed during Pleistocene low sea-levels as 'stepping stones' in faunal colonisation of Indian Ocean islands was first noted by Peake (1971).

58. Despite all the research, glacial chronology is still unsettled, particularly the earliest phases (late Pliocene-early Pleistocene), though the overall picture is fairly clear; details drawn from Lowe & Walker (1997), Calkin & Young (1995), Eyles & Young (1994) and Harland *et al.* (1990), plus Whittaker (1998) for the Pliocene sea-level lowstands (evidence from Midway Island, Hawaii).

59. "Suffice it to say that we have stratigraphic, tephrochronologic, radiometric dating and paleomagnetic evidence for at least 14 major glaciations in the Pleistocene and at least two in the Pliocene" – Harland *et al.* (1990), who placed the Pliocene-Pleistocene boundary at the Olduvai-Matuyama magnetic polarity reversal event (1.64mya), the Pliocene having started at 5.2mya.

60. See Rohling *et al.* (1998) for a chronology of sea-level lowstands; around 17–18,000ya sea-level was 110–115m lower than today in Mayotte, Comoros (Camoin *et al.* 2004), and, by inference, similar elsewhere in the region.

61. On Krakatau, devastated by the 1883 eruption, completely sterilised land and fresh tuff had reverted to full canopy forest by 1929, an interval of less than 50 years. Krakatau is very close to Java and Sumatra, so colonisation would be much quicker than on a remote oceanic island, but a few decades' delay is nothing in 50,000 years. Thornton (1996) reviewed the events of the Karakatau explosion and the return of plants and animals to the daughter islands.

62. Cheke & Dahl (1981), Moutou (1988), Mickleburgh *et al.* (1992), O'Brien (2005). The DNA work indicated three independent colonisations, but did not cover the putative fourth, as no DNA could be extracted from the extinct *Pteropus subniger* (John O'Brien pers. comm.). Andersen's (1912) bat relationships based on classic morphological analysis are poorly supported by DNA studies (e.g. Bastian *et al.* 2002, Colgan & Flannery 1995, O'Brien & Hayden 2004, O'Brien 2005), although most of his *rufus* group (including the Mascarene *P. niger*) does hold up, except for *P. voeltzkowi*. This Pemba endemic proves unrelated (*contra* K. Jones *et al.* 2002, but close instead to the Comoros *P. livingstonii*, which shares Moheli and Anjouan with a later arrival, the *rufus*-group species *P. seychellensis*. Moutou (1989) drew attention to reports of the Island Flying-fox *P. hypomelanus* in the southern Maldives, but there is doubt about the sole specimen, and no recent evidence of its presence (Holmes *et al.* 1994).

63. Judging by roosting information given by Mickleburgh *et al.* (1992) no flying-foxes were known to roost in hollows, even *P. subniger* simply "roosted in trees", without mention that it was *inside* them, thus losing a primary clue to its disappearance (see Cheke & Dahl 1981, Moutou 1982) – perhaps they didn't believe it. However they noted that "a roosting bat has been observed only once" in the Philippine *P. pumilus*, suggesting a cryptic roosting site. Tim Flannery (1995), who has probably seen more flying-fox species than anyone else, reported that *P. admiralitatum* and *P. caniceps*, living on islands off New Guinea, regularly roost in hollows. Although these species, like the Rougette, were allied with *P. hypomelanus* by Andersen (1912), DNA work suggests they are not all closely related (Colgan & Flannery 1995, Bastian *et al.* 2002, O'Brien 2005); *P. hypomelanus* itself always roosts on tree branches. The only other species recorded roosting 'anomalously' is *P. alecto* (= *P. gouldi*) in Australia, which, while normally hanging out on trees, sometimes roosts in caves (Churchill 1998).

64. Bastian *et al.* (2002) made the date estimates, but John O'Brien pers. comm.) has "found that the Bastian *hypomelanus* cytochrome-b sequence is aberrant . . . which would significantly skew their dates". Due to what appears to have been rapid evolution in the Pleistocene, the divergence points ('nodes') between *P. rodricensis*, *P. vampyrus* and the *giganteus* lineage (including its '*rufus* group' descendants) have low 'bootstrap' values (O'Brien 2005), i.e. are not necessarily correctly defined on present evidence.

65. Reported by Daniel (1975), but it happened in 1927–29; see also Churchill (1998).

66. Andersen (1912), Hill (1971), Bergmans (1988–97). Bergmans speculated that flying-foxes inhabited Africa in the Miocene and colonised the islands from there, subsequently dying out in Africa. It would require fossil evidence of four *Pteropus* lineages in Africa to consider this theory against the more parsimonious one of island-hopping from south Asia or the East Indies.

67. See Chapter 3 and Hume (2005a); the only other land vertebrate to occur on all three islands was the Malagasy Turtle Dove. Subfossil flying-fox bones have been found in Réunion (Mourer *et al.* 1999), but the few so far measured are from the known species *Pteropus niger* and *P. subniger* (Cécile Mourer pers. comm).

68. Benson (1970–71), Safford (2000); females of the different races of *Coracina tenuirostris* are either Réunion-like (barred underparts) or Mauritius-like (uniform cinnamon-orange) – males are uniformly dark grey, unlike either Mascarene bird. However *C. melanoptera* also has a 'Réunion-like' female and song-call (judging by descriptions in Harrison 1999), though the male looks like the Malagasy species. The genus appears quite malleable in these characters, so it is still premature, pending DNA study, to make inferences about origins of Mascarene cuckoo-shrikes.

69. Cheke & Diamond (1986) discussed differences between Malagasy and Mascarene cuckoo-shrikes; illustrations in Sinclair & Langrand (1998).

70. Chantler & Driessens (2000).

71. Price et al. (2004; supercedes Johnson & Clayton 1998) found the Mossy Nest Swiftlet *Aerodromus salangana* and the Edible-nest Swiftlet *A. fuciphagus* to be close to Mascarene Swiftlet *A. francicus* and Seychelles Swiftlet *A. elaphrus* but did not examine *A. unicolor*'s DNA. However *A. fuciphagus* makes nests of nearly pure saliva, whereas *A. salangana* and Indian Swiftlet *A. unicolor* use mostly twigs, lichen or moss, like the Mascarene and Seychelles species. The Mossy Nest and Edible-nest Swiftlets also share the same species of feather louse *Dennyus carljonesi* as the Mascarene and Seychelles birds (which each have their own subspecies, Clayton et al. 1996), but no lice are yet recorded for the Indian Swiftlet. The widespread Glossy Swiftlet *Collocalia esculenta* would be an ancestral candidate on distribution and nest type, but it does not echolocate and is now put in a different genus (nest and voice details from Chantler & Driessens 2000, Cheke 1987c).

72. Gray (1981).

73. Groombridge (2000), Groombridge et al. (2004a); the Echo Parakeet *Psittacula eques*, considered a race of Ring-necked Parakeet *P. krameri* in the mid-20th century, is now universally recognised as a distinct species (supported by both DNA and its failure to hybridise with introduced *P. krameri* in Mauritius). The DNA work implies that the African form should also upgraded: as the nominate race, the African Ring-necked would remain *P. krameri*, while the split-off Indian Ring-necked (or Rose-ringed Parakeet) becomes *P. manillensis*.

74. Calkin & Young (1995); Groombridge et al.'s (2004a) attempt to relate this to volcanic episodes on Mauritius is less convincing than the timing in relation to temporarily emergent islands.

75. Carié 1916, Cheke 1987a.

76. Forshaw 1989.

77. ASC's pers. obs., and examination of photographs of Ring-necked Parakeets taken in Mauritius in Staub (1993) and Michel (1992); Jones (1996) reported individuals of both Indian races, and also African-type *krameri* birds in Mauritius, suggesting multiple introductions.

78. The Seychelles bird was *Psittacula (eupatria) wardi*; the Mauritius species, formerly *Lophopsittacus bensoni*, thus becomes *P. bensoni* (Hume 2005a).

79. Hume (2007); see Keith (1980), Jones (1987) and Groombridge et al. (2004a) for general comments on Indian Ocean parakeets.

80. Keith (1980) discussed biogeographical affinities of Malagasy-region birds. The endemic Madagascar Starling *Hartlaubius auratus* is often merged generically with the Asian Spot-Winged Starling *Saroglossa spiloptera*, but DNA shows them only distantly related within a clade of African origin (Zuccon 2006; Zuccon et al. 2006 has less complete data). The magpie-robins are *Copsychus seychellensis* and *C. albospecularis*, the owls *Otus insularis* and *Ninox superciliaris*, the cuckoo *Cuculus rochii*, the nightjar *Caprimulgus madagascariensis* (allied to *C. asiaticus*), and the Mayotte Drongo *Dicrurus waldenii* (allied to *D. macrocercus*). *Otus*, *Cuculus*, *Caprimulgus* and *Dicrurus* are widespread genera, but hawk-owls and magpie-robins are otherwise strictly Asian. Rasmussen et al. (2000) suggested that the affinities of *Otus insularis* required re-evaluation.

81. Jones (1987), Boyce & White 1987, Groombridge et al. (2004b); The first two discussed relationships without coming to any convincing conclusion. *Falco punctatus* from Mauritius is larger than the Madagascar and Seychelles Kestrels (*F. newtoni* and *F. araea*), with very different calls (Sinclair & Langrand 1998).

82. Groombridge et al. (2002, 2004b), revising some figures and ideas from his thesis (Groombridge 2000). The dating rules out Common Kestrels from Sri Lanka (Harrison 1999) or northern migrants that straggle south through the Maldives (Phillips 1985) as the source of Indian Ocean kestrels. The three migrant kestrel species that reach the Seychelles (Skerrett et al. 2001) are older lineages, not immediately ancestral to the Indian Ocean clade.

83. Groombridge et al.'s (2002, 2004b) error estimate allows a range of 0.9–2.6mya for the colonisation date; they suggest the arrival coincided with the end of a period of major vulcanism which might have inhibited earlier colonisation (though plenty of older fauna survived). The reappearance of submerged islands as the ice ages began would have made no difference to birds coming from Africa or Madagascar, as there are none between Madagascar and the Mascarenes.

84. Dubois (1674), Mourer et al. (1999).

85. Peterson et al. (1995).

86. This was Keith's view (1980), but Louette & Herremans (1985b, Louette 1988) suggested two waves from Asia both spreading via Madagascar, the second wave swamping the earlier race on the big island.

87. Ben Warren's DNA phylogeny (2003, Warren et al. 2005) suggests the Mauritian species *Hypsipetes olivaceus* is nearest to the Malagasy clade, though he implied (rather unconvincingly) that Madagascar was colonised from Réunion and birds then came back from Madagascar to colonise Mauritius. More plausibly, Réunion was colonised from Mauritius, then subsequently Mauritian birds separately invaded Madagascar. The spread and evolution of the first wave appears to have been so rapid that the DNA phylogeny may not fully resolve the sequence of events – crucially the 'node' where the Réunion and Mauritius/Madagascar lines split is the least well supported in the phylogenetic tree. One of Warren's trees (with less DNA but more samples) shows black bulbuls of the first wave returning to Asia from the Comoros or Seychelles and *then* evolving into the current Asian form *H. (m.) leucocephalus*, a possible but unlikely scenario – showing that this technique, useful though it is, has to be treated with some caution.

88. For fuller discussion of possible sequences of events implied by the DNA see Warren (2003) and Warren et al. (2005).

89. Moreau 1957.

90. See Moreau et al. (1969), Gill (1971), Cheke (1987b,c), Horne (1987) and Safford (1991) for details of the peculiarities of Mascarene white-eyes.

91. Warren (2003).

92. Newton (2003) succinctly summarised the conventional view; see Filardi & Moyle (2005) for the Pacific island birds. Similarly Nicholson et al. (2005) showed that *Anolis* lizards have colonised mainland America from Caribbean islands at least twice.

93. Livesey (1991), Young et al. (1997), Mourer et al. (1999, 2006); Bernier's Teal is *Anas bernieri* and the Sunda Teal *A. gibberifrons*. *A. theodori*'s colour was noted by the crew of the *President* in 1681 (Barnwell 1950–54, Cheke 1987a); although the Madagascar species can look grey at a distance (Morris & Hawkins 1998, Roger Safford pers. comm.) the sailors would have been describing dead specimens seen close-up. Several forms formerly considered races of *A. gibberifrons* are now treated as full species (e.g. Young et al. 1997), so the bones used by Mourer et al. (1999) may have come from the Australian Grey Teal *A. gracilis* rather than *A. gibberifrons* as currently defined.

94. Berlioz (1946) emphasised the similarity of *Circus maillardi*

(Réunion) and *C. (maillardi) mascrosceles* (Madagascar) to Eastern Marsh Harrier *C. spilonotus* (northern Asia), a view supported by Keith (1980) & Cheke (1987b).

95. Simmonds (2000) did not propose a biogeographical route for the Australian birds to reach the western Indian Ocean, though if they evolved first into proto-marsh harriers which then migrated south across the exposed Pleistocene islands it would be possible. Marsh Harriers migrate from the Palaearctic to Sri Lanka and the Maldives (Harrison 1999, Phillips 1985), with stragglers reaching the Seychelles (Skerrett *et al*. 2001).

96. Respectively Booted Eagle *Hieraaetus pennatus* and Black Stork *Ciconia nigra*; see Snow (1978).

97. Madagascar Cuckoo *Cuculus rochii* is derived from Lesser Cuckoo *C. poliocephalus* (Keith 1980), which breeds in the Himalayas, China and Japan, wintering further south in Asia and beyond. Little Cuckoos no longer migrate to Madagascar, although they regularly winter on the east African coast opposite (Johnsgard 1997) – their route there is not known. They also migrate in Asia south to Sri Lanka (Harrison 1999, Johnsgard 1997), with stragglers occasionally reaching the Seychelles (Skerrett *et al*. 2001). Most birds breeding in Madagascar are sedentary, but *C. rochii* still migrate, spending the southern winter in the same part of East Africa as the Asian species spends the northern winter (Johnsgard 1997).

98. Mourer *et al*. 1999, Mourer 2006.

99. Swamphens *Porphyrio* speciate and become flightless very readily (Trewick 2002). Mourer *et al*. 1999 discussed night herons; Ascension Island in the tropical Atlantic also had an endemic night heron (Ashmole & Ashmole 2000).

100. Benson (1971) investigated the affinities of Indian Ocean paradise flycatchers without resolving their relationships; Cheke & Diamond (1986) discussed the similarities of Malagasy and Mascarene Paradise Flycatchers, *Terpsiphone mutata* and *T. bourbonnensis*.

101. See Dowsett-Lemaire (1994) and Leisler *et al*. (1997), who included Seychelles but not Rodrigues Warblers in their studies. Seychelles and Rodrigues Warblers have adapted to forest habitats; the others are all confined to swamp vegetation. Madagascar Swamp-warblers and related African species are sometimes placed in the (sub-) genus *Calamocichla*. The absence of an equivalent warbler in Mauritius and Réunion is at first sight odd; if present, they may have disappeared with the advent of rats. Seychelles Warblers survive only on rat-free islands (Diamond & Feare 1980), and any similarly sensitive warbler would have disappeared on Mauritius before humans arrived, and on Réunion before it was noticed; it is perhaps more surprising that the Rodrigues birds have survived.

102. See Milon (1951), Cheke (1987b), Barré *et al*. (1996), Urquhart (2002). For recent stonechat taxonomy and DNA studies see Wink *et al*. (2002a,b), whose nomenclature we have followed, but is subject to review: e.g. Collar 2005). They split the old '*Saxicola torquata*' into *S. torquata* (Europe), *S. maura* (eastern Eurasia) and *S. axillaris* (Africa and Madagascar), the Réunion bird being derived, but distinct from, *S. axillaris*; the studies did not include the Malagasy *S. (a.) sybilla* or several other African races.

103. The re-assessment of subfossil Mauritian material to the still-undescribed new larger flightless species (JPH work in progress, p. 38) means that the rail shot in Mauritius in 1809 (p. 101) must have been a vagrant, or possibly birds had colonised from Madagascar and survived unnoticed on cat-free offshore islets; it is indistinguishable from *Dryolimnas cuvieri* from Madagascar (Schlegel & Pollen 1868, Ripley 1977).

104. Livesey (1998); Trewick (2002) discussed the tendency of apparently sedentary rails to repeatedly colonise New Zealand.

105. Benson (1970–71), Adler (1994)

106. Johnson *et al*. (2001), supported by the broader study by Shapiro *et al*. (2002). Traditional *Columba* is paraphyletic, the New World species (now *Patagioenas*) being only distantly related. Johnson *et al*. emphasised the similarity in song-calls between Pink Pigeons and Madagascar Turtle Doves, something that had also impressed us in the field, combining with the DNA evidence to unite them in *Nesoenas* (Cheke 2005).

107. Molecular clock dating from Johnson *et al*. (2001), but Shapiro *et al*. (2002) implied *c*.34+mya (*Columba-Streptopelia*) and *c*.27mya (*Nesoenas-Stigmatopelia*). Johnson's text suggests a more recent split (*c*.6–5mya, *Streptopelia-Nesoenas/Stigmatopelia*) than using relative measurements taken from their Figure 6.

108. Mourer *et al*. (1999).

109. JPH (work in progress) considers that '*Columba rodericana*' from Rodrigues, often assigned to *Alectroenas* (e.g. Hachisuka 1953), is related to *Nesoenas*, but may need a genus to itself.

110. Moreau (1960a,b), Warren (2003).

111. Warren (2003).

112. Moreau (1960a), Long (1981), Sinclair & Langrand (1998).

113. The putative Réunion *Alectroenas* differed from the Mauritian Pigeon Hollandais in apparently lacking the latter's distinctive three colours (Barré *et al*. 1996, Cheke 1987a, Mourer *et al*. 1999); three blue pigeons survive, on Madagascar, Seychelles and Comoros/Aldabra (Goodwin 1983, Gibbs *et al*. 2001). Verheyen's view (1957) and Goodwin's hunch that *Alectroenas* is close to *Ptilinopus* is amply confirmed by DNA study (Shapiro *et al*. 2002); the closer New Caledonian relative is *Drepanoptila holosericea*, but only one *Alectroenas* species was sequenced. Our ad hoc dating is based on superposing Johnson *et al*.'s (2001) molecular clock on Shapiro's phylogenetic tree and comparing relevant branch lengths.

114. Arnold (2000), Austin *et al*. (2004a), using DNA analysis.

115. All the other Mascarene day-geckos also show vertical pupils (Vinson & Vinson 1969), especially in juveniles, a morphological character supporting the DNA evidence for a single invasion of the Mascarenes (Austin *et al*. 2004a); additional pupil observations by ASC from photos in McKeown (1993), Rundquist (1995) and Bartlett & Bartlett (2001), using many more species than Mertens (1972), who first noticed this difference between Mascarene and Malagasy day-geckos. However, the DNA phylogenetic trees (Austin *et al*. 2004a) imply that pupils must have changed from vertical to round more than once in *Phelsuma* history, only the Mascarene clade now retaining this character. The Seychelles bronze geckos, formerly all in *Ailuronix seychellensis*, are now split into three species (Gerlach 2002). A southwest African gecko, *Rhoptropella ocellata*, is sometimes considered a *Phelsuma* (Russell 1977), but DNA evidence places it in a sister group that split off before *Phelsuma* appeared.

116. See discussion in Osadnik (1984); additional information from McKeown (1993), Rundquist (1995), Glaw & Vences (1994), Nussbaum *et al*. (2000). There is inconsistency between authors as to which non-Mascarene *Phelsuma* glue their eggs, but only four species, *P. dubia, P. flavigularis, P. modesta* [= *leiogaster*] and *P. malamakibo* seem confirmed as gluers, while *P. barbouri, P. comorensis* and *P. v-nigra* sometimes do so (McKeown 1993, Glaw & Vences 1994, Nussbaum *et al*. 2000, Bartlett & Bartlett 2001); no Malagasy gluers were included in Austin *et al*.'s (2004a) DNA study.

117. The Agalega form is *Phelsuma borbonica agalegae* (Cheke 1975, Cheke & Lawley 1983, McKeown 1993), which must have sea-drifted north from Réunion; Meier (1995) related its origin to a sub-population isolated by lava flows in southern Réunion. Day-gecko eggs are very resistant to seawater (Gardner 1975).

118. Madagascar: Guibé (1958), Raxworthy (2003a); India: Daniel (2002); Australia: all are in *Rhamphotyphlops* (Shine 1991), though subfossil vertebrae are inadequate to distinguish this from *Typhlops*.

119. Indian Ocean currents for each month are given on the series of 'routeing charts' published by the UK Admiralty (series 5126, 1–12, 1970–71); see also brief discussion in Arnold (2000).

120. This is our reinterpretation of the phylogenetic tree in Arnold *et al*. (in prep. a, Jeremy Austin pers. comm.) – they suggested the

first colonisation was to Mauritius and that Rodrigues was colonised from there; we regard this as unlikely due to the prevailing currents; indeed two colonisations of Mauritius from Rodrigues appear more likely than one in the opposite direction. Arnold (2000, contra Arnold & Jones 1994) treated the two races of *Nactus serpensinsula* as separate species (*serpensinsula* and *durrelli*), but this obscures the fact that *serpensinsula* is derived very recently from *durrelli* in the very special conditions prevailing on Serpent Island since its post-glacial isolation (p. 203), whereas the other species evolved over a much longer period on the Mauritian mainland.

121. DNA studies by Arnold *et al.* (in prep. a, Jeremy Austin pers. comm.) supercede earlier interpretations (Bullock *et al.* 1985, Arnold & Jones 1994).

122. Imber (1985) suggested this generic re-alignment, dramatically confirmed in DNA studies by Bretagnolle *et al.* (1998), who also discussed distribution and systematics of *Pseudobulweria*.

123. Information from Jouanin (1970), Olson (1975), Attié *et al.* (1997) and Brooke (2004).

124. Kermadec Petrels *P. neglecta* have different flight-calls but are very similar to Herald-group petrels, and similarly polymorphic in plumage (with light and dark phases), but invariably have white shafts to the primary feathers and appear 'chunkier' in flight, as do birds giving 'Kermadec' calls on Round Island (pers. obs.); see Brooke & Rowe (1996), Brooke *et al.* (2000), Brooke (2004) and Imber (1985) for discussions of this knotty issue. We have included Kermadec Petrels in Table 3.1 in addition to 'Round Island Petrels' (equated, provisionally, with Trindade Petrels), but Mike Brooke caught, measured and blood-sampled only two Kermadec-calling birds, and the DNA evidence was equivocal – so a hybrid swarm with polymorphic vocalisations remains a possibility. Work in progress by Vikash Tatayah and Ruth Brown suggests both species are breeding with some hybridisation. A bird looking like a Kermadec (and display-flying with one) but giving intermediate calls was present in June 2003 (pers. obs.). Kermadec Petrels have recently been recognised on Trindade Island itself and have turned up in the Seychelles (Eikenaar & Skerrett 2006).

125. Carl Jones (1999 & pers. comm.). Although the Herald Petrel's presence was suspected for some time, confirmation came in 2007 with the capture of a bird ringed off Australia. The Black-winged Petrel is one of a sub-group of gadfly petrels (Imber 1985), not recorded as breeding in the tropical Indian Ocean. A small petrel has been found subfossil on Aldabra in the Picard Calcarenites (*P. kurodai*; Harrison & Walker 1978); although not dated by Taylor *et al.* (1979), judging by the stratigraphy given they probably correspond to the sea-level lowstand of 300–250kya given by Rohling *et al.* (1998). Imber (1985) suggested *P. kurodai* might have been a *Bulweria* rather than a gadfly petrel. There was also an undescribed *Pterodroma* on Rodrigues, found subfossil (Cowles 1987).

126. Feare (1978), Nelson (1978), Steadman (2006), Cheke (2001a). Olson & Warheit (1988) considered Abbott's Booby an ancient lineage different enough from other sulids to be given its own genus, *Papasula*.

127. Details are taken from Besse & Courtillot (1988), Duncan & Richards (1991), Courtillot (1999), Saddul (1995), Kearey & Vine (1996), O'Neill *et al.* (2005). There are discrepancies in published volcanic ages for the islands in the hotspot chain, so we have followed the fullest recent source (O'Neill *et al.* 2005), although their map has ODP site 715 too far north against the data in their table 1 and maps in other sources.

128. We have measured inter-island distances from maps because figures quoted in the literature varied enormously. We have taken the edge of the bank on which atolls now sit, at the -200m contour, as the edge of the putative original hotspot island for measurement purposes.

129. The Dodo's age is open to argument, the rails have not been DNA'd and no published molecular date is available for the emergence of the Bolyeridae (but see note 178).

130. Volcanic islands shrink both by erosion and subsidence, the weight of the lump on the crust causing its gradual sinking. The emergent island is often only the tip of a seamount rising off a seabed far below – 4,000m in the case of Réunion and Mauritius. In the tropics corals grow around the margin as they sink, and carry on growing after the bedrock is submerged. Thus an island can survive as an atoll by the continuous addition of coral limestone long after it would have vanished if there were no corals. Coral depth can be more than 1,600m, and it can have been growing for 45 million years or more (Menard 1986). This process, first surmised by Darwin (1839), was not confirmed until deep drilling and magnetic anomaly detection were developed (data from Menard 1986; see also Heatwole 1991, Gray 1993).

131. See the various chapters in Stoddart (1984a), and the putative age of bronze geckos *Ailuronyx* spp. in Austin *et al.* (2004a). However, a more recent origin of the chameleon and, by implication from affinity to Malagasy species, the snakes *Lycognathophis seychellensis* and *Lamprophis geometricus*, have been proposed using DNA analysis (Raxworthy *et al.* 2002, Nagy *et al.* 2003), implying arrival by rafting long after the separation of the islands from India or Madagascar, though the Gondwanan origin of the sooglossid frogs is not disputed (Vences 2004).

132. Livezey (1998), Mourer *et al.* (2000).

133. There are three monotypic extinct flightless New Zealand rails, two subfossil and one extinct but known from skins: *Capellirallus karamu* (mainland), *Diaphorapteryx hawkinsi* (Chatham Islands) and *Cabalus modestus* (Chatham Islands) (Livezey 1998, 2003, reconstructed by Gill & Martinson 1991). Similarity with Mascarene species may be parallel evolution, which Trewick (2002) noted was frequent in flightless rails – e.g. the morphologically distinct *Cabalus modestus* is nested within *Rallus* on DNA analysis, and the two Takahe 'subspecies' (*Porphyrio (m.) mantelli* and *P. (m.) hochstetteri*) prove to have arisen independently from flying ancestors.

134. The major orders of birds had already evolved before Gondwanaland started splitting up in the mid-Cretaceous (e.g. Cooper *et al.* 2001), but there is little to suggest that any recent rail lineages have been isolated that long, so our birds' ancestors probably set off down the chain quite late (nearer 20mya than 60).

135. DNA analysis suggests that, size and shape notwithstanding, dodos and solitaires should be included within the family Columbidae (Shapiro *et al.* 2002; see also Cracraft 1981). If, however, the Columbidae were split into three families as Verheyen (1957) suggested, then dodos would form the subfamily Raphinae of the Caloenidiidae (crested pigeons and Nicobar Pigeon). This arrangement matches the DNA work closely.

136. Strickland & Melville (1848) and Hachisuka (1953) discussed the bizarre taxonomic history of dodos.

137. Lüttschwager (1959); see also Cheke (1985).

138. Osteological cladistics: Janoo (2000), who used two different methods, producing rather inconsistent results; DNA: Shapiro *et al.* (2002).

139. Worthy *et al.* (1999), Worthy (2001).

140. Goodwin (1983).

141. "the gizzard of this bird is very peculiar, being composed of two discs of cartilage as hard as, and of the same texture as bone, slightly convex on the inner surface, between which is a pebble, usually a white quartz a little larger than a fresh pea" – Davison, quoted by Baker (1913).

142. Gibbs *et al.* (2001) mentioned 'grit and stones' (plural), but in essence their description of the Nicobar Pigeon's habits could have been for a Dodo: "It finds all its food on the ground, eating fallen fruits, seeds and occasionally invertebrates. Unlike fruit pigeons, this species is capable of digesting very hard seeds and nuts, owing to its very muscular thick-walled gizzard lined with horny plates and aided by swallowed grit and stones up to 10mm in diameter. It is said to be able to open nuts which a man would need a hammer to crack".

143. Baker (1913), quoting Davison again on feeding habits; Hachisuka (1953) thought Dodos must have used their big bills for digging as well as crushing, but Leguat (1707) made no such observations for Solitaires.

144. Johnson et al. (2001) used the standard cytochrome-b molecular clock rate for birds of 2% change per million years. Although the universality of this has been questioned (e.g. Warren 2003), no one else has come up with a rate as slow as that claimed by Shapiro et al. (2002) for pigeons, i.e. four times slower than expected.

145. Holyoak (1973), Hume (2005a). Berlioz (1940) noted that *Mascarinus* and *Necropsittacus* skulls resembled the Australasian and Indo-Pacific *Eclectus* and *Tanygnathus* parrots, also members of the Psittaculini.

146. *P. eques* and *P. bensoni* (Mauritius, Réunion), *P. exsul* (Rodrigues).

147. The Afro-Malagasy lovebirds *Agapornis* were formerly considered an offshoot of the psittaculines, but their DNA indicates affinities with Australasian lories and platycercines (Kloet & Kloet 2005).

148. See Berlioz (1940), Smith (1975); DNA work (Kloet & Kloet 2005) suggests *Coracopsis* (and the African parrots) shared an ancestor long ago with the South American parrots (Arini), but that its nearest surviving relative is the bizarre monotypic Pesquet's Parrot *Psittrichas fulgidus* of New Guinea.

149. Forshaw (1989), Low (1994); Low was seriously misled in asserting that "photographs of a skin show clearly that '*Mascarinus*' was a *Vasa* parrot (*Coracopsis*) ... there is no evidence the beak was red" – in fact, all original diagnoses of the specimens say the bill was red (see Hachisuka 1953), as did Dubois (1674) in his description from life. Osteologically *Mascarinus* is unlike *Coracopsis* (Mourer et al. 2000), and is clearly of psittaculine stock (Berlioz 1940, Hume 2005a).

150. Mourer et al. (1994, 1999, 2006).

151. See Rasmussen et al. (2000) and references therein.

152. Mourer et al. (1994, 1999, 2006).

153. Günther & Newton (1979), Hachisuka (1953).

154. Morioka (MS), JPH work in progress. Vinous-breasted Starling *Sturnus burmannicus* and immature Black-collared Starlings *S. nigricollis* have plumage similar to the Hoopoe Starling (illustrated in MacKinnon & Phillips 2000); Zuccon's (2006) DNA work confirmed affinity to *Sturnus*.

155. Sibley & Ahlquist (1984) suggested *Sturnus* dates back 4–5 ma, but combining Zuccon's (2006) *Fregilupus* analysis with molecular clock data in Zuccon et al. (2006) gives an earlier date. The Malagasy *Hartlaubius aurata* is a small arboreal species (Morris & Hawkins 1998), often merged with Asian *Saroglossa* (Moreau 1966, Sibley & Monroe 1990, Feare & Craig 1998) but DNA shows the two are not closely related, and both of African origin (Zuccon 2006, Zuccon et al. 2006). African birds are all in one clade except for the Wattled Starling *Creatophora cinerea*, which wanders to Madagascar (Morris & Hawkins 1998) and occasionally reaches the Seychelles (Skerrett et al. 2001).

156. Arnold (1979, 2000), Bour (1980), Le et al. (2006) confirmed the polyphyly of *Geochelone*.

157. Austin & Arnold (2001), Palkovacs et al. (2002), Gerlach (2004); the Radiated Tortoise *Asterochelys radiata* seems closest to the ancestral line.

158. Although Austin & Arnold (2001) argued that *Cylindraspis* first evolved on Mauritius, the mtDNA difference between the various Mascarene species is up to 17%, as it is to the Malagasy clade including *Asterochelys* and *Dipsochelys* (= '*Aldabrachelys*'). They did not attempt a molecular clock dating, perhaps because there is a ten-fold range in the rate change in chelonian mtDNA (0.11 to 1.1%/my; Avise et al. 1992, Near et al. 2005). However, if one assumes both Rodrigues (c.10mya) and Réunion (c.3mya) were colonised when new, then a rate of c.0.75–1.0%/my is indicated, suggesting that *Cylindraspis* split from other Malagasy lineages around 17–23mya; Gerlach (2004), using different analyses, put the split later at 10–15mya. Either way, anything over c.10mya (the earliest possible date for the emergence of Mauritius) suggests initial landfall outside Madagascar on a hotspot island.

159. Austin & Arnold's (2001) mtDNA substitution figures suggest that *Cylindraspis* split into two lineages within a few hundred thousand years of the genus itself differentiating, within the same 17–23mya time frame for the genus itself. As this is well before Mauritius emerged, it appears that proto-*Cylindraspis* colonised different hotspot islands almost simultaneously, though Gerlach (2004) placed the initial split within the genus at 8–9mya on Mauritius.

160. Austin & Arnold (2001), followed (more circumspectly) by Gerlach (2004), proposed that the genus arose in Mauritius (genetic distance notwithstanding), divided there, and one lineage drifted to Rodrigues (against the currents) and Réunion (with them). However there is nothing in their data to prove that the division of the *inepta* line into Rodrigues (*vosmaeri* and *peltastes*) and Mauritius/Réunion (*inepta* and *indica*) clades occurred in Mauritius rather than Rodrigues.

161. Arnold (2000), Austin & Arnold (2006); *Leiolopisma* is in the same group of Lygosomine skinks as *Cryptoblepharus boutonii*, also Australasian in origin (p. 61). Hutchinson et al. (1990) and Patterson & Daugherty (1995) separated Mauritian *L. telfairii* from Australasian skinks formerly included in the old broadly defined *Leiolopisma*.

162. Arnold (1980), Austin & Arnold (2006).

163. Arnold (2000).

164. Greer (1970) Blanc (1972), Cheke (1984) for traditional views, Brandley et al. (2005) for a recent analysis.

165. Notably by Blanc (1972) in his detailed, though now dated, biogeographical analysis of reptiles on Madagascar and other Indian Ocean islands.

166. Carranza et al. (2001), Carranza & Arnold (2003); the DNA work by Jeremy Austin and Nick Arnold concentrating on *Gongylomorphus* has yet to be published.

167. Brandley et al. (2005), Austin & Arnold (2006). Andreone & Greer (2002) had already transferred several species from *Androngo* to *Amphiglossus* (both segregates from the old inclusive *Scelotes*), since when several studies have demonstrated that while various genera under classical taxonomy were polyphyletic, the Malagasy skinks as a whole (+ *Gongylomorphus* [Mascarenes] and *Hakaria* [Socotra]) are, using DNA, a well supported monophyletic clade (Whiting et al. 2004, Schmitz et al. 2005, Brandley et al. 2005).

168. Using data presented in Austin & Arnold (2006).

169. For example, Glaw & Vences (1994).

170. The hinged jaw appears to be an adaptation for catching lizards, allowing the snakes to better grip wriggling cylindrical prey (Cundall & Irish 1989).

171. The Mauritian boas were first elevated to a subfamily by Hofstetter (1946), but the more they are studied the more distinct they appear. Underwood (1976) combined them with four living South American/Caribbean genera in a new family, Tropidophidae, as a sister group of the rest of the boas. The discovery of a new snake *Xenophidion* in Malaya and Borneo has led to a reassessment (Wallach & Günther 1998), and subsequent DNA analysis has yielded a radically different picture. The egg-laying boyerids are shown to be closest to *Xenophidion* and other primitive Asian boa-allies, but not related to the Central American live-bearing group (Lawson et al. 2004, but see Vidal & Hedges 2002, Vidal & David 2004). They appear to be basal to all pythons and boas (Vidal & Hedges 2002, Vidal & David 2004) or near the base of the python lineage (Lawson et al. 2004), thus very distant from Malagasy boas; either way they are an ancient lineage, a relict of a radiation that clearly predates the Mascarenes. A link ultimately with Asia looks more likely than any other.

172. The affinities of the Malagasy boas (subfamily Boinae) have generally been considered South American (Raxworthy 2003), but DNA shows a much closer affinity with the New Guinea-Pacific genus *Candoia* (Vidal & Hedges 2002, Lawson *et al*. 2004), which begs a lot of biogeographical questions (but may connect to the presence of iguanas on Fiji).

173. Nussbaum (1984b).

174. *Lycognathophis seychellensis* is a colubrid in its own endemic genus, related to Malagasy members of its subfamily, Lycodontinae (Nussbaum 1984, though the taxonomy of Malagasy colubrids is in chaos: Glaw & Vences 1994, Cadle 2003). Nonetheless, DNA work suggests that Malagasy colubrids separated from continental relatives in the Eocene or Miocene, and must have colonised across the sea-gap on at least two occasions (Nagy *et al*. 2003), hence *Lycognathophis* must have reached the Seychelles by rafting. The other Seychelles colubrid, *Boaedon* (= *Lamprophis*) *geometricus*, is so similar to some African members of the genus that it may not be a separate species at all, and could have arrived with human help (Nussbaum 1984). For Galápagos, New Hebrides and Fiji see Carlquist (1965), Medway & Marshall (1975) and Mitchell (1989) respectively.

175. Nussbaum 1984.

176. The eco-catastrophe on Guam (Mariana group, Pacific) was documented by Savidge (1987), Quammen (1996) and Fritts & Rodda (1998); in brief, the nocturnal Brown Tree Snake *Boiga irregularis* from New Guinea wiped out most of Guam's endemic birds and lizards in a few years, before anyone worked out why they were disappearing.

177. Lawson *et al*. (2004) invoked Gondwanaland for their 'clade B', pythons and several small allied families. Until recently (e.g. Underwood 1976, Preston-Maffham 1991) Gondwanan 'vicariance' (i.e. fauna stranded as the continent split) was the most popular explanation of Madagascar's unique fauna, but Vences (2004) pointed out that few Malagasy vertebrate taxa originate from lineages known to exist in the Cretaceous, and it now seems that most of the fauna post-dates the last Gondwana split (Yoder & Nowak 2006). The issue is debatable for two freshwater fish lineages, ratites, iguanas, podocnemid terrapins, and boas (Noonan & Chippindale 2006), though Vences *et al*. (2001) suggested that the boas had drifted, like many Mascarene reptiles, from Australasia.

178. Clock estimate by ASC from relative branch lengths of comparable species in Vidal & David (2004), Lawson *et al*. (2004) and Nagy *et al*. (2003) reckoned against time-scale in the last.

179. This island-hopping suggestion was first made by Cundall & Irish (1989), but based on the erroneous belief (Sahni 1984) that the earlier islands were all, like the Seychelles, aseismic (i.e. not of volcanic origin) and existed as continental fragments as Indian moved north. Ridd (1971) suggested that the Indochina/ Malaysia/Borneo/Sunda block was also part of Gondwanaland and travelled up into Asia independently of India – while neatly accounting for the presence of *Xenophidion* in Malaya and Borneo, this theory has had few takers.

180. Although on Round Island it is primarily terrestrial, the boa has occasionally been found up to 4m up palm trees, the only kind of tree now present (Bullock 1986, McAlpine 1982). However, in pristine Mauritius, where there was a wide variety of trees, often growing very tightly together, the snake could well have made more use of its undoubted ability to climb them.

CHAPTER 5. EARLY SETTLEMENT

1. Deodati, reporting to the Dutch East India Company in 1698, exaggerating his predecessor's supposedly wasteful tortoise policy (from Barnwell 1948).

2. Gerrit van Span (from Verin 1983) showed unusual prescience in relation to introducing supposedly beneficial predators. Polecats, Stoats and Weasels, brought in against zoologists' advice to control a plague of rabbits in the 1880s, decimated native wildlife in New Zealand (King 1984); the rabbits were, of course, themselves introduced by humans.

3. See Bulpin (1958), Toussaint (1972) or Grove (1995) for overviews of the general historical scene in the 17th century.

4. North-Coombes 1980.

5. Leibbrandt (1896–1907), Bruijn *et al*. (1979–87), Moree (1998). The company was the *Verenigde Oost-Indische Compagnie*, generally referred to as the VOC.

6. This paragraph was compiled from many sources, principally the histories cited earlier, and Moree (1998), Ashmole & Ashmole (2000), Lougnon (1970). Itineraries of English ships and details of surviving logs were given by Farrington (1999); Barnwell (1948, 1950–54) published several logs pertaining to Mauritius.

7. Although it is difficult to find explicit references to livestock being a standard complement of ships' 'vittles', it appears that carrying live animals to supply fresh meat was routine. To judge by how often livestock was recorded as being released on islands, trading ships must have carried quite a lot of animals – e.g. Matelieff's & l'Hermitte's accounts in Barnwell (1948) and also Herbert (1634). As late as 1695 the standard complement of livestock for large Dutch trading ships was "4 live full-size sows or 12 young pigs" and 40 hens (though no cows or goats) (Bruijn *et al*. 1979–87). Lever (1985) surveyed mammal introductions worldwide, but gave little historical context and is unreliable in detail, particularly for the Mascarenes; Long's more recent survey (2003) is even less reliable!

8. The earliest surviving account of the first Dutch landings in Mauritius is an English translation of a lost Dutch tract entitled *A true report of the gainefull, prosperous and speedy voiage to Iava in the East Indies* . . . (Anon. 1599, reprinted in Keuning 1938–51; see also van Wissen 1995).

9. "The Ile affoords us withall Goats, Hogges, Beeves, and Kine, land tortoyses, (so great that they will creepe with two mens burthens, and serve more for sport, than service or solemne Banquet) Rats and Monkeys, all of which becomes food to such ships as anchor here. They were first brought hither by the Portugall . . . though now for the English and Dutch forces, they dare not rest there, nor owne their firstlings." (Herbert 1634).

10. As a small sample, Mundy (1608–67), La Caille (1763), Grant (1801), Clark (1859, admitting it was hearsay), Pitot (1905, 1914a), Carié (1916, dogmatic on pigs and goats, but allowed that the Dutch may have introduced monkeys), Vinson (1953a, even giving totally unsupported dates of 1528 for monkeys and 1550 for pigs and goats!), Bulpin (1958), Owadally 1981 (repeating Vinson's dates), Lever 1985 (citing no references but giving Owadally as Mauritian contact), Tosi & Coke (2006, citing no pertinent references).

11. The MSS are reproduced in Keuning (1938–51). Heyndrick Jolinck's account (volume 5 of Keuning's compilation, 1947) gave the most details on wildlife, badly garbled in a semi-translation by Panyandee (2002).

12. For example Addison & Hazareesingh (1993), Grove (1995), Moree (1998); Teelock (1998) blamed Portugal only for rats and monkeys, not livestock. Contrary to all the actual evidence, Grove (p.130) asserted "Ecologically, however, the Portuguese must have had some influence and certainly introduced deer, goats, monkeys and pigs" (note that deer, whose introduction by the Dutch in 1639 is well attested have also crept in here!). Sussman & Tattersall (1986), despite arguing that the Crab-eating Macaques came from Java, used a rather convoluted argument (*contra* North-Coombes 1980) to support the Portuguese as the source of monkeys, but were unaware of the full extent and detail of the Dutch reports from 1598 (Keuning 1938–51). Tosi & Coke (2006), alleging that the Portuguese were regular visitors to Mauritius and released monkeys, goats and pigs, ignored the possibility of Dutch input.

13. North-Coombes 1980; see also Cheke (1987a), Ashmole & Ashmole (2000).

14. Keuning (1938–52), Barnwell (1948); the chickens were never reported again.

15. The account of the first voyage to touch at Mauritius, van Warwyck's in 1598, reported that there "are found neither men nor four-footed beasts". On the second visit Harmansz (outward) in 1601 was silent on wildlife, but Willem van West-Zanen, relating a return voyage in 1602, reported that "no quadrupeds occur there except *Katten* [cats], though our countrymen have subsequently introduced goats and swine" (Soeteboom 1648, Strickland 1848, Verin 1983). In 1606 Cornelisz Matelieff de Jonge said that produce would grow well "if not spoilt by rats of which there are multitudes in the island" and later, a little confusingly, that "the only four-footed animals are monkeys. The Dutch left behind them 24 goats, and hogs and sows".

16. We are not aware of cats being introduced to any island through shipwrecks; Abbot (2002) investigated cat arrivals in Australia, concluding that none had established from wrecks. The pattern of extinctions (pp. 80–81) suggests cats were a late 17th-century introduction. Although the probable typographical error had already been pointed out (Cheke 1987a), Roy et al. (1998) still stated that "many cats were noted on Mauritius by the van Warwick expedition in 1601" (error for the '*Katten*' in 1602: note 15 & Soeteboom 1648, Cheke 1987a). The original Dutch publication is in black-letter ('gothic') script, whose upper-case letters 'R' and 'K' are extremely similar, and easily transposed by a typesetter.

17. The monkeys are *Macaca fascicularis*, showing characteristics (pelage colour, hair tufts *etc.*) typical of the populations on Java but not other parts of the species's range (Sussman & Tattersall 1980, 1986). The most detailed DNA studies (Tosi & Coke 2006) link the animals to Sumatra, though Java, favoured by Kondo et al. (1993), is not ruled out. A 10-fold drop in haplotype diversity (Lawler et al. 1995) suggests descent from only "a few individuals" or even "a single female founder". Sumatra was visited by Dutch ships from 1599 (Anon. 1599) onwards. The first homeward-bound ships touching at Mauritius were a group under Hans Schuurman, including Adriaan Blok in July 1602 (Bonaparte 1890, Moree 1998); however its chronicler West-Zanen (Soeteboom 1648) mentioned no monkeys.

18. The best account in English of this shipwreck is in Moree (1998), from the original accounts in Keuning (1938–52).

19. At the time the only rat in Europe, hence the species on European ships, was the Ship or Black Rat *Rattus rattus* (Twigg 1975, Atkinson 1985, Cheke 1987a).

20. The reptiles known only from subfossils on the mainland of Mauritius are: two skinks *Leiolopisma mauritiana* and *L. telfairii*, a large gecko *Phelsuma guentheri*, the two boas *Casarea dussumieri* and *Bolyeria multocarinata*, and the blind-snake *Typhlops cariei* (Arnold 1980, Cheke 1987a); all but the first and last survive (or did until recently) on offshore islets.

21. "Snakes nor toads we saw none nor any venomous thing, neither have any such been heard to be here, although many small lizards, spiders and land-crabs are in the woods" – Peter Mundy in 1638 (Mundy 1608–67, Barnwell 1948).

22. The northern islets, Round, Flat and Gunners Quoin, were a refuge for these reptiles for a long time, and snakes also survived until the mid-18th century on Ile de la Passe in Mahébourg Bay (Cheke 1987a). Rats fortunately never reached Round Island.

23. Subfossil remains assigned to Reed Cormorant *Phalacrocorax africanus* (Cowles 1987) exist, but there is no historical record of cormorants in Mauritius, unlike in Réunion. Others not unequivocally observed were the Golden Bat *Pteropus rodricensis*, the endemic night-heron *Nycticorax mauritianus*, the Harrier *Circus 'alphonsi'* (= *maillardi*), Abbott's Booby *Papasula abbotti*, Sauzier's Wood-rail *Dryolimnas* sp. (see Chapter 3), and, before about 1800, the Malagasy Turtle Dove *Nesoenas picturata* (which may have died out and been reintroduced). The only report of '*coeten*' (= coots, i.e. *Fulica newtoni*) is in 1601 (Bouwer in Moree 2001), but references to 'water-hens' probably refer to them (Moorhens being a later arrival: Cheke 1987a). Herbert's 'bitters' (1634) and Leguat's '*butors*' (1707) may have been night herons (other visitors may have grouped them with 'herons' or 'egrets'; Herbert mentioned herons separately). Matelieff (Barnwell 1948) mentioned 'hawks and falcons' and Herbert (1634) reported 'Goshawkes and Hobbies' so it appears they saw both larger and smaller raptors, i.e. harriers and kestrels. John Marshall's big white birds up a tree, seen in 1666 (Khan 1926), were probably Abbott's Boobies. Sauzier's Wood-rail could have been overlooked or conflated with Red Hens – it is unlikely to have succumbed to rats, as flightless *D. cuvieri* survives them (though not cats) on Aldabra (Penny & Diamond 1971, Gillham 2000).

24. 10–12 pigs were released, (Jacques l'Hermitte, in Barnwell 1948; see also Pitot 1905, Carié 1916), though some sources suggest fewer.

25. Pitot (1905: Verschoor), Barnwell (1948) and Verin (1983: van der Hagen).

26. Cheke (1987a: figure 1). Given ideal conditions, ungulate populations can increase astonishingly quickly – three goats left on Pinta I. (Galápagos) in 1959 multiplied to 20,000 by 1971 (Hamann 1979).

27. Barnwell (1948). Botanists treat coconuts as introduced to the Mascarenes (e.g. *Flore* 189, 1984, H.E. Moore & L.J. Guého), but they are frequently mentioned in early Dutch accounts as growing in small numbers along the south coast and adjacent islets (Barnwell 1948). Rouillard & Guého (1999) considered Dutch reports of native coconuts to refer to latans, but since latans were common and widespread and the 'coconuts' scarce and restricted, we disagree. Leguat (1707) found coconuts washed up and germinating on the beach at Rodrigues, so natural sea-borne colonisation seems probable, though seedlings are vulnerable to crabs (and various mammals; Purseglove 1972). They are generally considered native further north, in the Seychelles (Sauer 1967, Purseglove 1972) and Agalega (Cheke & Lawley 1983, Guého & Staub 1983).

28. Warwijck's men in 1598 called them '*walghvogel*', nauseating birds (Strickland 1848, Barnwell 1948, Hachisuka 1953).

29. Strickland (1948), Verin (1983).

30. Jolinck in 1598 (in Keuning 1938–52, also quoted by Moree 1998).

31. Pitot (1905), Barnwell (1948).

32. Quoted in Barnwell (1948); the grey parrot was probably the bird described by Holyoak (1973) as *Lophopsittacus bensoni* (Cheke 1987a), now re-assigned to *Psittacula* and named in English after the 19th-century fossil hunter Etienne Thirioux (Hume 2007).

33. Warwijck's men in 1598 (Barnwell 1948).

34. Barnwell (1948), Verin (1983).

35. Barnwell (1948), Verin (1983); 5–6 hundredweight is 255–305kg, exactly the average for adult dugongs (250–300kg) given by Reynolds & Odell (1991).

36. As quoted in Barnwell (1948). If the French translation in Verin (1983, *tonneaux*) is correct this means ten 'tuns' or barrels rather than tons weight.

37. An anonymous Dutch sailor talked of Dodos, Red Hens, tortoises and feral livestock in a manuscript travel journal partly published by Servaas (1887), but which is currently untraceable in the Dutch archives (van Wissen 1995).

38. The ships included Both's flagship, our old friend the *Gelderland*; the Dutch salvaged cargo from the wrecked ships during 1615–17, but then avoided the island for several years (Moree 1998). Pieter Both is commemorated by the eponymous mountain crowned by a narrow peak, a conspicuous maritime landmark rising behind Mauritian capital Port Louis.

39. Ships' visits are from Moree (1998) and Bruijn et al. (1979–87).

40. See note 6 for ships' logs.

41. Bonaparte (1890), Grandidier *et al.* (1903–20), Verin (1983).
42. Details from Leibbrandt (1896), Pitot (1905), Barnwell (1948) and Moree (1998).
43. On 8 November 1639, the new commandant or 'governor' Adrian van der Stel arrived from Batavia with 6 pairs each of deer, rabbits, sheep, chickens, geese, ducks and pigeons (Pitot 1905, Carié 1916); the domestic birds were all reported in the Batavia *Dagh-Register* to be thriving in early 1641 (Bonaparte 1890: 54); no mention of sheep or deer, though "the number of head of livestock were continually increasing". The deer are *Cervus timorensis* of the Javan race *rusa* (Owadally & Butzler 1972, Lever 1985).
44. The devastating long-term effect of deer, pigs and other introduced animals on the vegetation is discussed in Chapters 7 and 8. Rats, monkeys and ungulates reached excessively high populations in the mid-late 1600s, the larger animals at intervals hit by epidemics (Cheke 1987a).
45. Mundy's brief visit was in mid-April 1638 (Mundy 1608–67, Barnwell 1948, Keast 1984), Gooyer's settlers arrived on 6 May (Moree 1988); see pp. 28–29 for the full Cauche story.
46. Good faunal lists exist for 1638 (Mundy 1608–1667), 1662 (Evertsz, on mainland; Olearius 1670), 1666 (Granaet; Barnwell 1948), 1668 (Marshall; Khan 1927), 1673–75 (Hoffman 1680), none of which mention Dodos *in situ* (Mundy, having seen captive ones in India, remarked on *not* finding any). Unfortunately there is next to nothing during the crucial period of the first Dutch settlement, 1638–58.
47. Olearius (1670), Cheke (1987a), van Wissen (1995); see Cheke (2004a) for the first English version of this account, from an 18th century MS translation in the Mackenzie Collections in the British Library (*ex* India Office; 1822 Collection, No.13a; see Blagden 1916, who was unaware of the original source). Evertsz also saw on the island, but did not describe, *veldthoenders*, the name used by the Dutch for Red Hens on their early visits, and on the mainland *berghoenders*, possibly another name for the same species.
48. Full details in Cheke (2004a). The myth in Mauritius in the 1990s was that the Dodo's islet was Ile aux Aigrettes off Mahébourg (e.g. Kitchener 1993), and Moree (1998) claimed Evertz and friends 'walked out to one of the small islands in Grand Port Bay'. However all the Grand Port Bay islands are surrounded by deeper water requiring serious swimming to reach, and composed of sharp weathered limestone (*champignon*); anyone running would have their bare feet cut to shreds. Even Ile aux Aigrettes, the largest, is rather small to have supported a viable population of Dodos, tortoises and goats. Ile aux Cerfs, Ile de l'Est and Ile d'Ambre are all larger and basaltic with some flat open areas. The first two have sandy beaches ideal for running (Saddul 1995) but are without streams, whereas Ile d'Ambre has more relief and the 1:25,000 map shows a small stream and path, as mentioned by Desjardins (1829) in his description. Ile d'Ambre was well-known to the Dutch for its high goat population in the 1670s (Pitot 1905:122, Panyandee 2002). Alan Grihault (2005) thinks Ilot Fourneaux near Le Morne (southwest Mauritius) has a better claim, but it much smaller than Ile d'Ambre and the lagoon there is almost dry at low tide, so it would have been accessible to pigs and monkeys. The *Arnhem* survivors evidently landed fairly near Grand Port (Stokram's account, in Panyandee 2002), but Evertsz did not say in which direction he and his friends followed the coastline.
49. Details from Pitot (1905, 1914) and Moree (1998). When Hugo visited Mauritius in November 1662 to rescue shipwrecked sailors from the *Arnhem*, five slaves escaped into the forest, "and were never heard of again" according to Moree. However the slave interviewed by Hugo had been AWOL for 11 years (Pitot 1905, 1914) – i.e. since 1662/63 – surely one of the same men (Cheke 2001b). The man himself, Simon, claimed to have been abducted from the settlement by other fugitives, which Hugo appeared to accept (Pitot 1905), although it was untrue as there was no settlement in 1662/63 (as Hugo should have remembered!). However his story was not all lies, as he said his surviving companion (also captured, but escaped) had killed one of the *Arnhem* survivors, and Pieter Salomonsz had indeed been killed by the escaped slaves from Hugo's ship (Moree 1998). We think Simon fudged the story to avoid reminding his former master who he actually was. Whether he told the truth about Dodos, or whether they were actually Dodos he was reporting, we cannot know. In 1663 Hugo was a privateer in the service of the French; his ship the *Zwarte Arend* picked up 64 *Arnhem* survivors. Evertsz had already left in May on the English ship *Truro* (Moree 1996). Three other survivors published accounts of the *Arnhem* disaster, but their descriptions are partly plagiarised from earlier accounts (van Wissen 1995, Moree 1998); only Evertsz indisputably saw Dodos.
50. Pitot (1905, 1914) only paraphrased the story, so we cannot be sure that all the clues are there. However, Pitot was himself interested in the Dodo's extinction, and aware of the name change, so should have been alert to significant detail.
51. Hume *et al.* (2004); 'other endemic prey' were also recorded (not listed in the paper), but did not include any other putative flightless species (JPH unpublished data).
52. Strickland (1848), Barnwell (1948), Hachisuka (1953), Cheke (1987a).
53. See Box 11 (p. 127) for Marshall's description (from Khan 1927); he obviously didn't have a red handkerchief!
54. This archival discovery was first highlighted by Moree (1998), with further details from Sleigh (2000), Moree (2000) and Hume *et al.* (2004); Moree and Sleigh believed the birds were Dodos, apparently unaware of the name-change issue. Moree was unable in 2000 to substantiate his belief (Cheke 2001b), and further study in the archives in 2004 revealed no descriptions of the birds Lamotius called *dodaarsen* (JPH, pers. obs.). Only Sleigh mentioned 50 birds, Moree (1998) noting only four, and Hume *et al.* mentioning occasions not numbers.
55. Hume *et al.* (2004) vs. Cheke (2006a) – hence in this book we leave the matter open!
56. Pitot (1905), Barnwell (1948).
57. Barnwell (1950–54).
58. Roberts & Solow (2003) used a statistical method intended pinpoint the Dodo's extinction date, but as they included Evertsz's isolated islet population (though not Hugo's slave interview) they got a date of 1690. Running a fuller date series, omitting Evertsz, to get a mainland date, yields an extinction date of 1650 for the Dodo and 1689 for the Red Hen (Cheke 2006a); these studies both excluded the Lamotius 'dodos'. Leguat's *gelinottes* of 1693, although he used term for the related flightless rail on Rodrigues, were also excluded from the Red Hen date series for want of a description. Including the Hugo and Lamotius records in the Dodo series, the algorithm gives an extinction date of 1693 (Hume *et al.* 2004). All these analyses have rather wide 95% confidence limits, 1669–1797 in Roberts & Solow (2003), 1640–1692 (Dodo) and 1675–1743 (Red Hen) in Cheke (2006a), with the smallest range, 1688–1715, in Hume *et al.* (2004). Dates after 1722 are fanciful, as no flightless birds survived into the French period.
59. See Chapter 2 for the doubts around Cauche's account (1651). However, his description of the Dodo's nest and single egg closely matches Leguat's account of the Solitaire's nest in Rodrigues (Leguat 1707).
60. Atkinson (1985, 2000)
61. Pitot (1905), Cheke (1987a).
62. Pig damage to tortoises and turtles in Pitot (1905, Mauritius), Galapagos (Macfarland *et al.* 1974), seabirds (Atkinson 2000, Long 1985), Tuataras *Sphenodon punctatus* (Long 1985). Like the unique bolyerid snakes on Round Island, the Tuatara, sole survivor of the order Rhynchocephalia, survives only on offshore islands free from predators (Towns & Daugherty 1994).

63. Panyandee (2002).
64. Details from many sources, summarised by Noël Brouard (1963). Jacotet was called Ebbenhouts Baay ('Ebony Bay') by the Dutch (Barnwell 1955).
65. Barnwell (1948, 1950–54), Moree (1998).
66. Pitot (1905), Barnwell (1948), Brouard (1963), Moree (1998).
67. The commercial ebony from Mauritius is the endemic species *Diospyros tesselaria*; *bois d'olive* is *Cassine* [= *Eleaodendron*] *orientale* (Cadet 1981).
68. Some idea of the extent of ebony harvesting can be found in Leibbrandt (1896), Barnwell (1948), Brouard (1963) and Moree (1998); no one has collated all the Dutch records in Batavia, the Cape and the Hague to get a fuller picture. Harvesting was apparently greatest during van der Stel's governorship (1639–45) when c.1000 pieces (= trees) per year were exported (Moree 1998).
69. Vaughan & Wiehe (1937) considered that ebony-rich forest (plus their 'palm savanna'; see Chapter 3) originally covered all the island outside the 100" rainfall contour, roughly the 2,400mm isohyet (Padya 1989). By weighing the parts on either side of the 2,400mm isohyet on a tracing of the rainfall map, we estimate that 59% of the island (1100km^2) was under ebony-rich forest. The proportion of good ebony may have been lower in the narrow west coast dry zone under 1,000mm rainfall, though Mundy's account (1608–67, Barnwell 1948) implies good forest within this area, around the mouth of the 'Fresh River' (Grand Rivière Nord-ouest), and the northern plains near Grand'Baie were well wooded (log of the *President*, Barnwell 1950–54).
70. There is no good information on how much arable land was maintained during Dutch rule, but in 1708, just before they left, the settlements were around Port Louis (Noordvester Haven, 16 households), Flacq (Groote Vlakte or Noordwyk Vlakte, 12 households), and Black River (five households) (Barnwell 1948). In 1670 the Flacq farmland covered 65 *morgen* (59ha, Cheke 2001b), but this would have expanded substantially over the next 38 years.
71. "The palmito tree affording meat and drink, the tender top whereof boiled and buttered is as good (if not better) as cabbage, to which end many hundred are already cut down and few remain about the places where our ships use to touch and winter. Besides by cutting the body of the tree there distils a liquor which may be compared to that which comes from the pressed sugar cane, pleasant and wholesome, very good drink" (Mundy 1608–67, Barnwell 1948).
72. Hugo's ban on palm cutting was mentioned by Lamotius in 1683 (Barnwell 1948).
73. Barnwell (1948, 1950–54).
74. After 52 slaves escaped into the forests in 1642 "the Dutch were never able to eradicate this illegal settlement of runaway slaves" (Moree 1998).
75. Although not short of edible animals, Mauritius almost entirely lacked native plants that humans could eat, visitors often complaining about those that made them ill (e.g. Herbert 1634). The palm cabbage was the only plant food universally appreciated.
76. On offshore islets off Rodrigues and on Assumption, lacking large trees, Abbott's Booby *Papasula abbotti* nested in short ones (Tafforet 1726, Stoddart 1981), but on Mauritius apparently chose tall trees as they do on Christmas Island. (Nelson 1978).
77. Rhesus Monkeys *Macaca mulatta*, intended for medical research, released in 1966 on the small former nature reserve (!) of Desecheo Island near Puerto Rico, are considered the primary reason for the local extinction of Red-footed Boobies *Sula sula* and Magnificent Frigatebirds *Fregata magnificens*; the boobies recolonised after most monkeys were removed from 1977 onwards (Gochfeld *et al*. 1994). The Mauritian *M. fascicularis* is a close relative.
78. Abbott's Booby acquired the name *boeuf* in Rodrigues in the early 18th century because its call resembled that of cattle (Tafforet 1726, Cheke 1982a, 2001a). In 1755 the elder Cossigny (Charpentier 1732–55) used that name for a nestling seabird he sent (dried) from Mauritius to Réaumur in France. However, his description is of a White-tailed Tropicbird *Phaethon lepturus*. His use of '*boeuf*' suggested to one of us (Cheke 1987a) that Abbott's Boobies might still have been breeding in Mauritius at least within living memory, but with no other indication of their presence after 1668, we now think this unlikely. Cossigny was in regular contact with the outpost on Rodrigues (trying to get a Solitaire) and may have got the bird (or at least the name) from there. It is odd that he did not recognise a tropicbird, given how well-known *paille-en-queues* were to the locals.
79. This is an assumption; *Lophopsittacus mauritianus* was a poor flyer so may have nested on the ground, though elsewhere only truly flightless parrots, and those living in treeless areas (Australia, New Zealand), do so (Forshaw & Cooper 1989).
80. See Chapter 3.
81. Both adult and young tropicbirds hiss and bite hard when disturbed at the nest, though if their bluff is called they give up quite soon (ASC, pers. obs. on Round Island).
82. "The arsenic sent by the *Soldat* to poison the rats was effectual, but was also destroyed in the fire. It had been before impossible to protect anything from the rats in the stores, and yet though 300 or 400 are killed during the night, it can not be seen that the multitude decreases. To destroy the vermin in the fields and forests is, however, as impossible as to reach the vault of heaven. They are there in thousands of millions, and if they were only about the cultivated lands there would be a chance of eventually rooting them out. But the whole island is covered in them to the top of the highest mountains. We therefore believe it impossible to rear any corn, rice or any other grain here." (despatch from Deodati to the Cape, 12 Dec. 1696: Leibbrandt 1896, *Letters received*:169).
83. The best description of Dutch agriculture is Leguat's (1707) from his visit in 1693–96, but see also La Roque (1716), Leibbrandt (1896), Pitot (1905), Barnwell (1948), Moree (1998).
84. Leguat (1707).
85. The *Berkeley Castle*'s log is well-known because Harry's comments are often considered the last mention of Dodos alive on Mauritius (Strickland 1848, Hachisuka 1953, Fuller 1987), but see earlier discussion (and Cheke 2006a) in relation to the name transfer to Red Hens. Strickland (1848) and Barnwell (1948) printed Harry's account; Barnwell (1950–54) published the *President*'s log in 1951.
86. In 1696 an escaped convict came "to the company's cattle-shed [at Limoen Bosch], where he took three wild ducks" (Barnwell 1948:69; this anonymous account was wrongly attributed to Deodati by Cheke 1987a). A cattle-shed is an odd place to find wild ducks, but since *wild eend* in Dutch can also specifically refer to Mallard *Anas platyrhynchos*, it is perhaps more likely that domestic ducks were stolen. Dutch settlers in Mauritius kept ducks, presumably Mallard derivatives, first brought in by van der Stel in 1639 (Pitot 1905). A few years later, during a dearth of game animals in 1706, the freemen complained that "often . . . no fresh meat is obtained from the forest; they manage however to support themselves by now and then killing a duck or a pair of fowls" (Leibbrandt 1896, *Letters received*: 407). Again the context suggests domestic animals, but is open to interpretation; the original Dutch (an MS in the Cape Archives) may have been less ambiguous. Deodati's remark about geese is in a despatch reprinted by Leibbrandt (1896) and Barnwell (1945).
87. King (1984), Lever (1985), Ashmole *et al*. (1994), Burger & Gochfield (1994), Atkinson (2000), Courchamp *et al*. (2003).
88. Ships' cats could have been offered by visiting sailors to Dutch freemen farmers in exchange for favours (e.g. provision of meat) – there was always tension between the freemen at N-W-Harbour and Black River (Leibbrandt 1896, Barnwell 1948) and the VOC authorities, so the Governor in Fort Hendrick (Grand Port) might be ignorant of any such deal.
89. Sieur de la Merveille's account appeared in La Roque's *Voyage*

de l'Arabie heureuse in 1716, reprinted in Grandidier *et al.* (1903–22), Barnwell (1948, in English), Verin (1983).

90. Granaet (in Barnwell 1948) and Leguat (1707).

91. Pitot (1905); the scarcity at this date was only relative, as tortoises were still easily found until *c*.1690.

92. Piton (1905), Barnwell & Toussaint (1949). Ambergris was an extremely valuable commodity used in medicine and perfumery, second only to ebony in importance as a Mauritian 'product' in the mid-1600s. It is a wax produced in the gut of Sperm Whales *Physeter catodon*, coughed up and washed up on beaches. Sperm Whales were common enough around the Mascarenes, Seychelles and the surrounding Indian Ocean for a major American-based whaling industry to develop in the 19th century (Lionnet 1972); the boats were based in the Seychelles, so whaling is hardly touched on in Mascarene history books. The whalers often visited Rodrigues to get meat and cut wood (North-Coombes 1971).

93. La Roque (1716).

94. See Pitot (1905); it was probably Hugo's old account of wasteful slaughter that Deodati dressed up to denigrate Lamotius.

95. See Pitot (1905) and Barnwell (1949); Flat Island continued to be a source of tortoises after they had become extinct on the mainland (Cheke 1987a, Chapters 6 and 9).

96. In 1681 Benjamin Harry reported "many wild hogs and land turtle which are very good" [to eat] (in Barnwell 1948), but during 1685–88 Lamotius only rarely reported tortoise captures in his diary (JPH pers.obs. in the Cape Archives).

97. Deodati, sent to arrest Lamotius, had political reasons for blaming the colony's parlous state on his predecessor (Pitot 1905, Moree 1998); the tortoise quote is in Barnwell (1948).

98. Courrier de Bourbon (1721).

99. There is no longevity information for Mascarene tortoises, but a Seychelles tortoise brought to Mauritius as an adult in 1769 survived a further 149 years before dying in an accident in 1918; several others are known to have lived for more than 100 years (Lionnet 1972, Bour 1984a,b).

100. Female Aldabran Giant Tortoises take on average 17–23 years to reach maturity, depending on population density (Swingland & Coe 1979).

101. Governor Hugo made strenuous efforts to eliminate feral dogs (Pitot 1905), but they did not thrive in Mauritius. Before the Dutch left in 1710 the VOC planned to release their hunting dogs to destroy the wild livestock and so reduce the island's value to other countries' ships. However Governor Momber pessimistically observed that "these dogs will probably soon become extinct, as dogs have often been left behind in the forests and after six months were not heard of again, the damp woods giving them distemper and killing them" (Barnwell 1948).

102. Pigs are the worst tortoise predators in the Galápagos, rarely missing a nest because tortoise eggs have such a long incubation period (3–8 months); they also take young up to 35–40 cm curved carapace length (*c*.5 years old if similar to Aldabra tortoises, Bourn & Coe 1978). However Galápagos tortoises all nest in beach sand, concentrating eggs and young in a small area (McFarland *et al.* 1974 and literature reviewed in Cheke 1987a:10). Although one report mentioned eggs in "hollow trees exposed to the sun (Hugo, in Pitot 1905), and others nested on beaches, it is not clear from the Dutch accounts where the bulk of Mauritian tortoises laid, but the fact they survived so long suggests to us (*contra* Cheke 1987a) that a significant proportion of nests may have been dispersed in areas less vulnerable to pigs. Aldabran tortoises excavate sites with a permanent temperature of 27–30°C in the middle of the clutch (Swingland & Coe 1978). Any Mauritian tortoises nesting in the forest would have had to tolerate lower egg temperatures. Cats are not considered major tortoise predators on Aldabra (Bourn & Coe 1979; there are no pigs), but they appear to be at low density (Racey & Nicoll 1984).

103. Arnold (1979).

104. The name first appears on a map in 1753; the earliest French map (1721) does not name the islet (Toussaint 1953, Barnwell 1955).

105. Oudemans (1917, 1918) and Hachisuka (1953) made some questionable assumptions when attempting to count Dodos exported alive, including a belief that almost every artist painted from a living subject, and an attempt to identify individual birds from paintings using the number of bill ridges. Their total (the two authors differ a little) came to 15–18, 9–10 in Holland, England 2–3, Belgium 1, Italy 1, India 2 and Batavia/Japan 1; note their excess in Holland compared to our estimates. Stesemann (1958) made a more measured, though incomplete, analysis of living Dodos abroad, and came up with 7 (Prague 1, Holland 1, England 2, India 2 and Batavia 1). Although a complete analysis of cognate and copied Dodo pictures has yet to be made, a good start was made with a diagram by Marijke Besselink in Ziswiler (1996); see also Cheke (1987a) and Hume & Cheke (2004) for the lineage of the Holsteyn/Withoos 'white dodos', and Friedmann (1956) regarding van Kessell's paintings (which are, however, too late to be based on living birds). Oudemans's idea of using bill ridges to identify and estimate age of Dodos could be valid in principle, but there is no consistency in the copy-lineages of the paintings, and we concur with Hengst (2003) that the artists had no concern with such detail, and that the variation is due to artistic licence.

106. From a letter published by Alfred Newton (1874). The outward fleet stopping at Mauritius consisted of four ships (Farrington 1999); Barnwell (1950–54) published three of the logs. Altham says he reached Mauritius on 28 May 1628, corresponding to the timing of the *Star* and the *Hart*; the *Mary* and the *William* (a lone homebound ship) arrived some days later, the *Hopewell*'s arrival is not recorded. Returning to England in 1629 the *Hart* carried Sir Thomas Herbert on his way home via Mauritius and Réunion (Chapter 2). Altham was the first to use the name 'Dodo'. The fate of Altham's Dodo is not known. Newton (1874) noted that Herbert's name appears as a donor to John Tradescant's museum, but that Altham does not. However, as Strickland (1848) ungallantly pointed out, "had the garrulous Sir Thomas actually killed, skinned and brought home a Dodo, he would not have failed to record such an exploit in his Travels" – the more so if he had successfully brought one back alive!

107. Mundy (1608–1667), Barnwell 1948, Keast (1984). The "Suratt howse" (Surat House) was the East India Company's base in Surat, though Mundy did not say where or in what conditions the Dodos were kept; the 'house' owned an extensive garden on the town's outskirts.

108. Jahangir's nature notes are included in his autobiographical *Tuzuk-i-Jahangiri*; see Stresemann (1958), Ali (1968), Das (1973). The consensus is that Mansur often drew from life, and that his Dodo was living when painted. Das noted the exchange of animals for commercial favours. He emphasized that for Mansur to have painted it, the Dodo must have been presented to Jahangir, and been interesting enough for him to commission its image – thus Mundy's birds, seen at the EIC's 'house' in Surat after Jahangir's death, must have been different individuals. Although some writers have assumed Jahangir had a menagerie, no zoo is mentioned in studies of his wildlife interests, nor any report of animals being kept in Mughal gardens (Alvi & Rahman 1968, Verma 1999), so animals he received, apart from those suitable for stables (e.g. his zebra), may have been kept only long enough for his artists to draw them. However, as his predecessor (Akbar, d.1605) and successors maintained menageries in Delhi (Ali 1927), it is probable that these facilities were maintained during Jahangir's reign.

109. "it was kept in a chamber, and was a great fowle somwhat bigger than the largest Turky-cock, and so legged and footed, but stouter and thicker and of a more erect shape, coloured before like a yong cock fesan, and on the back of dunn or deare colour" (Strickland 1848, Hachisuka 1953, Fuller 1987)

110. Mr Gosling's gift was mentioned by Thomas Crosfield in his MS diary (Strickland 1849, Hachisuka 1953, Wissen 1995; Boas 1935 printed the full diary); the source of the specimen is unknown.

111. See Tradescant (1656), Strickland (1848), Gunther (1937), Davies & Hull (1976), Macgregor (1983), Ovenell (1992). Following Strickland it has been generally supposed that the Tradescant Dodo, acquired with the rest of his collection by Elias Ashmole in Oxford, had become rotten through neglect and burnt, some enlightened person snatching the head and foot from the flames. The story was used by museum Keeper Dr Robert T. Gunther in the 1920s to shame the University into more funding for preserving the university's treasures. Ovenell (1992, Wissen 1995) re-examined the evidence for this curatorial disaster, and discovered from archival records that William Huddesford, newly appointed Keeper of the Ashmolean Museum in 1755, following instructions left by Ashmole, made strenuous efforts to rescue what he could of the collections, discarding what was no longer fit for exhibition. He thus deliberately saved the Dodo's head and foot; the body, like *all* the other whole birds in the collection, was deemed beyond preservation - there was no fire! The fate of the two Anatomy School specimens is not known. In 1749 the mounted pair were described as "Dodar-Birds, one of which watches while the other stoops down to drink", but there is no record of them after that (Gunther 1937, Macgregor 1983, Ovenell 1992; A. V. Simcock pers. comm.). No illustration survives, and Hume (2006) speculated that they may not have been Dodos at all.

112. Between 1628 and 1638 numerous British ships called at Mauritius (Barnwell 1948, 1950–54, Farrington 1999) and would have been able, if so inclined, to bring Dodos home, though their logs reveal nothing. Several EIC sea-captains are listed as 'Principal benefactors' in the younger Tradescant's catalogue of his collection, including 'Captain Swanley' (Tradescant 1656); Leith-Ross (1998) suggested this was a Richard Swanley, but conflated two men of the same name in her account. There were in fact three Richard Swanleys captaining ships for the EIC and Courteen's fleet in the 1600s, the younger two of whom visited Mauritius (Barnwell 1948). One visited Mauritius on the *Jonas* in 1630, and then as Mundy's captain on the *Sun* in 1638 (Barnwell 1948, Farrington 1999), and would have opportunities to bring back Dodos, from either Mauritius or Surat. The other, on the *Truro*, rescued sailors from the *Arnhem* in 1662, too late to have produced Tradescant's Dodo. Also listed is Captain John Weddell, who captained the *Charles* (Leith-Ross 1998), visiting the island in 1630 (Farrington 1999). Finally, Thomas Herbert (visit in 1629, Chapter 2), is also listed as a benefactor, though, probably (note 106) did not donate a Dodo. Leith-Ross (1998) argued that as visitors to Tradescant's museum before 1638 (including Mundy in 1634) did not mention the Dodo, it must have been acquired later, and the timing suits Lestrange's bird from 1638 (brought on which ship?). The accounts of the collection from Mundy and others (quoted by Leith-Ross) are brief and very incomplete – would they necessarily have mentioned the Dodo amongst so many other fascinating items? In Leith-Ross's favour, Mundy had just returned from a journey past Mauritius in which he mentioned Dodos (Mundy 1608–1667), but with only negative evidence, the date when one or other Tradescant acquired the Dodo (John the elder died in 1638) must remain unsettled. Nothing else from Mauritius is discernible in the *Musaeum Tradescantianum* (reprinted in Leith-Ross 1998).

113. For example, Ray (1678), George (1985), Staudinger (1990); George noted that in the 17th century most large animals from distant areas were carried alive. Spirit preservation of animals was not discovered until 1663, when demonstrated in London by Robert Boyle; even distillation itself to produce 'spirits of wine' was then a relatively recent invention (George 1985, Asma 2001). 17th century taxidermy was rudimentary; preservation techniques were not perfected until the middle 1700s (Schultze-Hagen *et al.* 2003).

114. van Wissen (1995) suggested that the British Museum's foot may have been the one described by Clusius in 1605, but the latter's history after 1617 is unknown; see also Hachisuka (1953, who confused the Oxford and BM feet), Strickland (1848) and Hume (2006). The BM specimen can no longer be found in the NHM collections (Fuller 2002, Kallio 2004, Hume *et al.* 2006), and has apparently been missing for decades.

115. The most detailed account of Savery's life and work is Müllenmeister (1988, 1991), listing most of his paintings and reproducing many in colour or black and white; Ziswiler (1996) and especially Hengst (2003) published more complete sets of Savery's Dodo paintings. The 1611 picture, in Berlin, is the second of 23 Savery did between 1610 and 1628 depicting the Greek myth of Orpheus charming the animals, three of which include Dodos. Savery was prolific and formulaic, often recycling the same image in different settings (Vignau-Wilberg 1990, Hume & Cheke 2004). Dodos are but one of many animals re-used; a classic example is an incongruous flying condor, appearing in an identical pose in three of his Dodo pictures (see Ziswiler 1996) and commented on by Jackson (1993).

116. The Savery sketch, in the E. B. Crocker Art Gallery, Sacramento, California was researched by Friedmann (1956); van Wissen (1995) and Ziswiler (1996) gave its date as 1626, but others date it to Savery's time in Prague (Hengst 2003, Pinto-Correia 2003). However the expert on Savery's drawings, Joaneath Spicer, while initially attributing the drawing (and other similar ones) to the Prague period, later revised her view to "approximately in the decade 1616 to 1626" (Spicer 1984). Ziswiler published a diagram of Dodo-picture lineages; see also Hume & Cheke (2004) for details of pre- and post-1626 Saverys and fuller references. The Hondecoeter Dodo, in a painting entitled *Perseus and Andromeda*, is still in Sion House, London (Arturo Valledor de Lozoya pers. comm.) *contra* Ziswiler (1996), who said it had been moved to Alnwick Castle (UK). Until fairly recently (Broderip 1853, Oudemans 1917, Hachisuka 1953, Friedmann 1956) it was thought to be by Jan Goeimare, who does not feature in reference works by Jackson (1993, 1999) or Turner (2000). Paul Huvenne (in Geeraerts 1982) pointed out that is signed with Hondecoeter's initials and dated 1627 by the artist (not 1626 as cited by Besselink 1995).

117. English translation from Besselink (1995). Although the caption gives the impression that Venne's drawing was made from the live bird (as is generally assumed), it doesn't specifically say so, and we think it much too similar to the 'stock' Savery image to be independent of it; it most closely resembles a Hondecoeter version of Savery's standard, also from 1626 (Hengst 2003: 58–59). The Leiden Dodo was reported by Johannes Walther (translated in Brial 1998b; his *Ornithographie* of 1657 not seen by us), who claimed his odd-looking Dodo illustration (e.g. Hachisuka 1953:119) derived (albeit indirectly) from the live bird – though it appears to us to be embellished from Johnstonus's c.1650 version of Clusius's sketch (illus. in Hume 2006).

118. Hengst (2003) thought the Saftleven Dodo derived from Hondecoeter's *Perseus* painting, but this seems to us rather far-fetched, particularly as the Saftleven is much more lifelike, though there are points of similarity in the details of the head. Saftleven feathered the forehead in front of the eyes – was this an error in copying Hondecoeter or a real feature of an immature Dodo?

119. This painting is reproduced in Jackson (1993) and Ziswiler (1996). The copy-lineage diagram in Ziswiler's book suggests a link with a Dodo picture of a preserved specimen in Vienna, part of Emperor Rudolf II's *Bestiaire*. Although Jackson (1993, 1999) gave no indication that Jan Brueghel visited Prague, Müllenmeister (1988) implied that he did, and may have met Savery there. However the birds in *The Element of Air* are not cognate with Rudolf's *Bestiaire*, only the Dodo and the Crowned Crane having any resemblance, though Breughel may have seen the stuffed specimens from which it was drawn.

120. Millies (1868) suggested the bird(s) painted by Savery and van

der Venne arrived in February 1626 on one of the ships that spent three months at Mauritius in late 1625, the *Maeght van Dordrecht, Wesop* (= *Weesp*) and *Leewinne* (= *Leeuwin*; spellings in brackets are those of Bruijn *et al*. 1979–87 and Moree 1998). Since no Dutch ships visited Mauritius between 1617 and 1625 (Moree 1998), it seems that Johannes Walther (note 117) is likely, 30 years later, to have been mistaken about the date of his Leiden bird, and that it was in fact 1626. Saftleven's bird could have reached Holland on the *Petten*, whose crew were cutting ebony in Mauritius from late 1634 to early 1636 (Moree 1998).

121. For example, by Renshaw (1933), Hachisuka (1953), Staub (1993). This picture, known as 'Edwards's Dodo', belonged in the mid-18th century to George Edwards, author of *Gleanings of natural history* (1760), who used it to compose his famous if curious depiction of a Dodo standing by a Guinea Pig *Cavia porcellus*. This Savery has been much copied, featuring as the frontispiece of Hachisuka (1953), the cover of Ziswiler (1996) etc. The original, acquired by Sir Hans Sloane and presented with his collection to the British Museum, is now in the NHM in South Kensington. The 19th century copies (Hachisuka's frontispiece by Keulemans, and another by Louisa Günther in the History of Science Museum, Oxford) tend to enhance the *Aphanapteryx*-like appearance of what we think are Eurasian Bitterns *Botaurus stellaris*. The well-known oversize Dodo painting by Jan (= Hans) Savery dating from *c*.1655, now in the Oxford University Zoology Museum (reproduced in van Wissen 1995), is clearly copied from the same source – the wing elements and colouring are identical, although it looks even more obese!

122. An identical-looking bird also appears in Jan Brueghel's *Noah's Ark*, in the bottom left-hand corner next to a heron (Jackson 1993: 34) – again apparently a poorly rendered Eurasian Bittern. Renshaw (1936) claimed another similar bird in a Savery painting was a Red Hen. This bird with full normal wings, a posture more heron- than rail-like, and a stoutish straight bill, appears to be another bittern. *The animals entering the Ark* by Jacopo da Ponte (known as Bassano) has been claimed to include a Red Hen (e.g. Ripley 1977; see Cheke 1987a). Were the identification correct, the dating to *c*.1570 (Bettagno *et al*. 1996, Berdini 1997) would be suspect on historical grounds (could a Red Hen have reached Italy in 1570?). However, the bird is not identifiable; only a dark brown (not russet) head and neck are visible, the bill curved, but barely longer than the head.

123. Clusius (1605).

124. Emperor Rudolf II had a world-class menagerie, some animals from which (once dead and stuffed) were painted in oil on parchment, and preserved in two bound volumes now in the national library in Vienna (Haupt *et al*. 1990, van Wissen 1995, Besselink 1995, Jackson 1999). These pictures have been variously attributed to Joris (= Georg) Hoefnagel (Besselink 1995: 90, *contra* van Wissen in the same book, p.64), his son Jakob (Wissen 1995, Jackson 1999, Hume 2006) and Dirck de Quade van Ravesteyn (Hengst 2003) or his workshop (Ziswiler 1996). Although Jakob Hoefnagel was paid a good sum in 1610, often said to be for the illustrations (e.g. Jackson 1999), further study indicates that this money was for other items, and that the best candidate for the bulk of the pictures is van Ravesteyn (Vignau-Wilberg 1990), as stated by curator Daniel Fröschl in his catalogue (Bauer & Haupt 1976). However the varying styles and quality of the paintings show that several artists, possibly including Savery, were involved (Vignau-Wilberg 1990, pers. obs.), probably under the direction of Fröschl as curator/artist, who may himself have contributed (Irblich 1990). The *opus* is reproduced complete in colour in *Le bestiaire de Rodolphe II* (Haupt *et al*. 1990). Fröschl's manuscript inventory of Rudolf's *kunstkammer* (collection of curiosities) was published by Bauer & Haupt (1976); see Impey & Macgregor (1985) and Mauriès (2002) for overviews of these remarkable early museums, and Marshall (2006) for the intellectual hothouse that was Rudolf's

Prague. The Prague painting of *Aphanapteryx* (Haupt *et al*. 1990, Ziswiler 1996) was a crucial link associating the accounts of red-handkerchief-chasing 'hens' with the bones of a large rail found in the Mare aux Songes (see Milne-Edwards 1868, Sclater 1915), achieving fame in the title and associated vignette of Nicholas Pike's (1873) book on Mauritius. The *Bestiaire* was rediscovered by Frauenfeld (1868), but his reproductions were not fully accurate. As with the Dodo, the painting is from a stuffed or dried specimen, the legs thin and shrivelled compared to the stout limbs portrayed in the *Gelderland* drawing of a freshly dead bird (Moree 2001 and Figure 2.4). Staudinger (1990) neither identified the flying-fox nor discussed its origin, but its pelage clearly establishes it as a *Pteropus niger* from Mauritius (Cheke 2007b). The specimen presumably arrived in Prague via Holland with the Dodo(s) and the Red Hen.

125. Bauer & Haupt (1976), van Wissen (1995); see Chapter 2 and Hume & Cheke (2004) on the question of white dodos.

126. In all Savery's other paintings, Dodos are portrayed as mid to dark grey, paler underneath, though never black like Mansur's Indian miniature. This suggests that the Prague specimen Savery painted was albinistic, and that he adjusted the colour on returning to Holland (1617 onwards) where more information and perhaps further specimens were available. However he retained the same appearance and general pose of his Dodos until he saw live bird(s) in 1626 – the early Dodos (six paintings) are bulbously fat, neck held back and facing left, though one (in Reims) leans forward and downwards (reproductions in Müllenmeister 1988, 1991, Ziswiler 1996, Hengst 2003).

127. The Narodny Museum in Prague still retains a number of Dodo bones (not subfossils), including an upper mandible of an adult (illustrated with some leg bones by Kallio 2004), which may have come from the bird pictured in the *Bestiaire* or the 'whitish' stuffed specimen (Staudinger 1990, van Wissen 1995); a lower mandible formerly present (van Wissen 1995) is apparently now missing (Valledor 2003, Kallio 2004). Staudinger, noting the colour difference between the description and the painting, but arguing that the pictured bird was alive, suggested the juvenile may have been dark and the adult pale. However we are convinced, like most observers, that the illustration is from a dead specimen; the twisted position of the wing, the shrivelled skin of the bill (compare the fresh bill in the Gelderland Dodo; Moree 2001 and Figure 2.2), and the unstable front-heavy body position all point to a museum mount.

128. Staudinger (1990); de Bry (1601, Strickland 1848, Keuning 1938–40, vol. 2) reported that Dutch sailors brought a Dodo back to Holland.

129. Millies (1868), Newton (1868).

130. Nieuhoff (1682); his Dodo sketch is another clone of a fat, left-facing Savery-type bird. He described the Dodo's iris as black, as did his source, Piso's edition of Bontius's encyclopaedia (Piso 1658; see Strickland 1848, van Wissen 1995), whereas it is white in most of Savery's paintings, and also, independently, in Mansur's (reproduced in Ziswiler 1996). Nieuhoff was in Batavia in 1753 and 1758.

131. Moree (1998).

132. One in England, one in Holland, two in Surat (India) and the one sent from Batavia to Japan.

133. Tatton (1625).

134. "All [the wild birds] fell to our hands without effort, so no one found it necessary to go and catch the goats that we saw several times in bigger and smaller flocks in the mountains . . ." (Adriaan Blok in Brial 2001a). "We also saw some goats, but they were so wild that they could not be approached. We were able to catch only one; so old was it that its horns were eaten by worms and we found it impossible to eat" (Bontekoe 1646 and in Grandidier 1903–20 and Lougnon 1970). Blok and Bontekoe landed at Saint-Paul, Tatton at Saint-Denis; in 1613 the goats may only have been in the west.

135. There was a Portuguese landing prior to 1528, but no others are known for certain (North-Coombes 1980).

136. Herbert was told there was no livestock by a sailor on the *Hart* who had been ashore with Castleton on the *Pearl* in 1613 (Herbert 1634, Barnwell 1950–54).

137. Flacourt (1661, Lougnon 1970); this ambiguous remark might imply that harmless snakes were present, but later travellers made it clear there were none.

138. "I gave it the name of Bourbon, not finding any name which could more aptly define its goodness and fertility, nor suit it better than that one" (Flacourt 1661, Lougnon 1970). Since the French company of which he was a director had just acquired a royally approved monopoly, he no doubt felt such brown-nosing was a necessary *quid pro quo*. Mauritius was similarly named after Prince Mauritz of the Netherlands.

139. Details from Lougnon (1970) and Hébert (1999): extracts from Flacourt, Ruelle and Souchu de Rennefort.

140. Historical details from Melet (1672), Defos Du Rau (1960), Lougnon (1970), Toussaint (1972) and Barassin (1989). The ratio of Europeans to black slaves was still 54% to 46% in 1713.

141. Feuilley (1705); his first name is unknown.

142. Leibbrandt (1896).

143. Quoted from Boucher (1710); see also Guet (1888). Boucher arrived in Réunion only in 1702 (Alby 1989), so some of this harking back to greater past abundance may have been hearsay. The post of storekeeper (= treasurer) was second only to the Governor in importance (Alby 1989).

144. Lougnon (1956), Toussaint (1970), Vaxelaire (1999).

145. Boucher's most famous piece of writing is his other *Mémoire* (Alby 1989), a rant about the shortcomings of all the island's inhabitants, demolished one by one in short biographies. Nonetheless it gives a remarkable snapshot of the lives of French colonists in the early 1700s, without which our understanding of Réunion's human history would be much poorer. Barassin (1989) gave a more general account of early 18th century Réunion.

146. Barassin (1989), citing a manuscript of governor Parat.

147. "There are neither serpents, nor scorpions, nor any other sort of dangerous reptile or insect: the goodness of the air kills them! The French have experience of this with respect to rats." (Dellon, 1668, in Lougnon 1970). "Snakes, vipers, scorpions, rats and other similar creatures cannot survive there, so inimical is the air to everything that could be injurious" (Père Vachet, 1671, in Lougnon 1970).

148. Dubois (1674) provided the fullest account.

149. Boureau-Deslandes in Lougnon (1970).

150. Melet (1672). 'Thrushes' would be Merles, 'partridges' buttonquails and 'buzzards' harriers. The pigeons/doves and parrots are not identifiable further, there being several of each to choose from. We have not been able to trace the word '*pepeux*', but *pepier* is to peep or chirp, so *pepeux* may have meant 'small birds', passerines.

151. Dubois (1674), who left the island in 1672, reported an absence of rats and mice, and Boureau-Deslandes (in Lougnon 1670) writing of his visit in 1674, gave no indication of any agricultural problems; he did not mention rats, present or absent. The boat-wreck was reported by François Martin (Martineau 1931–34, Lougnon 1970), who had been on the island when it was rat-free (in 1665, 1667), but had heard about their arrival (but gave no actual date) before writing up his notes in India c.1684–85 (Kaeppelin 1908). Martin is regarded as the founder of the French trading colony of Pondicherry. Absence of rats was referred to belatedly by Duquesne (1689) and Leguat (1707), but Duquesne's description was lifted from Dubois (1674), and Leguat, who saw the island from offshore but could not land, based his on Dellon (visiting in 1668; Lougnon 1970) and Duquesne.

152. A similar rapid population explosion occurred when Ship Rats invaded Bermuda in 1613, reaching plague proportions within a year (Lever 1985); likewise on the Big South Cape Islands (New Zealand; Bell 1978). In most places poor documentation precludes confirming whether explosive expansion normally follows reaching a new island, though this is likely. In Mauritius and Rodrigues, where rats had arrived before anyone wrote accounts, no such observations were possible.

153. From Guet (1888) and Barassin (1953), also paraphrased in Farchi (1937).

154. Dellon described the inconvenient habits of the *moineaux* in 1668, while Dubois's description confirmed the birds as fodies - an endemic species of which no specimens were collected or bones discovered (*Foudia delloni*; see p. 42). Kitchen-visiting was not confined to Réunion Fodies – in the 1970s Seychelles Fodies *Foudia sechellarum* on Cousin Island nature reserve used to come into the warden's house and perch on crockery after scraps, ignoring any humans present (pers. obs.).

155. Cheke (1987c) summarised research on rats and fodies; Safford (1997) further emphasised the devastating predation of Ship Rats on Mauritius Fody nests.

156. There may have been some domestic cats earlier, though Dubois's (1674) remark that "the cats have nothing to do; there are no rats or mice for them" may have been metaphorical, meaning that they *would* have nothing to do (had there been any).

157. Borghesi's report (in Lougnon 1970) is confusing because he called the pigeons 'de tour' (tower pigeons, i.e. Rock Dove/Feral Pigeon). There is no other contemporary indication of domestic doves gone wild, and his description of their former abundance and absurd tameness fits the native bird, described as very numerous by several earlier writers, better than feral *Columba livia*. In Combaluzier's translation into French (from Italian) the birds were supposedly abundant "dans les jours que nous passames dans l'île" [our italics], but also had already disappeared! We suggest (following Cheke 1987a) that the passage should read '*avant*' [before] rather than '*dans*' [in, during]. Borghesi's letter was written soon after his visit, so cannot reflect the same kind of before-and-after perspective that François Martin had for rats.

158. Bernardin (1687); Merveille saw ducks and geese on the Étang de Saint-Paul in 1709 (La Roque 1716, Lougnon 1970), but the context suggests these were domesticated.

159. Hébert (1708), Compagnie des Indes (1711).

160. Reports quoted by Lougnon (1970), also Melet (1672), Feuilley (1705); see Cheke (1987a), Mourer *et al*. (1995b). Although they landed at Saint-Denis, Tatton's men in 1613 also visited Saint-Paul – it is not clear where he saw his large white flightless birds, presumably ibises; Blok did not report any such bird at Saint-Paul in 1612. From Melet's account *Threskiornis solitaria* was apparently still common around Saint-Denis in 1671.

161. Hébert (1708), Cheke (1987b), Barré *et al*. (1996).

162. Sauzier (1891), claiming to cite Lesson, who allegedly heard the story from Freycinet. However he gave no reference, and we cannot find the story in Lesson's published works. Furthermore, as the bird was always called a *solitaire* in Réunion, asking about '*drontes*' (i.e. dodos) might have elicited an unreliable response – the old man might equally have been talking about *oiseaux bleus*. He supposedly told Freycinet that he had "heard much talk of the *dronte* in his childhood, and that they still survived in this area [Saint-Joseph] during his father's first years of life". Assuming the exchange took place at all, we can only say that the man was told in his youth of a large bird now extinct that had survived into his father's childhood. Freycinet apparently asked around generally, and this was the only positive account he got. Freycinet was governor 1821–26 (Reydellet 1978); Lesson visited Mauritius (but not Réunion) in 1824–25 (Lesson 1830).

163. Mourer *et al*. (1999).

164. Cat predation: Jones *et al*. (1999). Both Pink Pigeons (1607–1709) and Pigeons Hollandais (1638–1755) also had long gaps without records, suggesting scarcity, probably predator-induced (see Appendix 2).

165. Penny & Diamond (1971), Gillham (2000).
166. While beaches were likely to be their preferred nest-site, it seems unlikely that tortoises that lived far inland (even the Plaine des Cafres, *fide* de Villers, in La Roque 1716, Lougnon 1970) would have come down to the coast to breed. This may explain how tortoises survived in Cilaos into the 1830s (Bour 1981, Probst & Brial 2001, pp. 141–2), having disappeared from the lowlands by 1740 (Cheke 1987a).
167. Historical information from Melet (1672) and Lougnon (1970); Durot in 1705 cited the number of tortoises per canoe, also noting that tortoises on his ship survived for three months at sea unfed.
168. Since Boucher was de Villers's deputy and hatchet-man for nine years (Villers 1701–1710, Vaxelaire 1999), his complaints against the Company's failure to control hunting and waste during that period look more like an attempt to shift blame for a food-supply disaster than a serious desire to conserve!
169. Villers (1701–10).
170. Probst & Brial (2001).
171. Guet (1888).
172. Governor Drouillard made these rules (Guet 1888, Maurin & Lentge 1979); Guet characterised the colonists' philosophy as "no hunting, no settlers".
173. Defos du Rau (1960) gave a succinct account of the hard but independent and self-sufficient life of the more mutinous colonists, supplemented by retired pirates and other fugitives from authority (see also Barassin 1989). The COI's governors had no back-up to enforce regulations, so anything the locals disagreed with was effectively a dead letter. For details of individual governors see Guet (1888), Barassin (1953), Reydellet (1978) and the *DBR*.
174. Hébert (1708, Bour 1981, Probst & Brial 2001).
175. Tortoises were reported as absent from the east in 1694 (Bour 1981); Bour thought they had never existed in the east, but Tatton (1625) reported them common near his initial landing place in 1613 (Saint-Denis), and Melet (1672) wrote that there were still plenty there during his visit. An absence of sandy beaches in the east may have constrained numbers through shortage of nest-sites.
176. Villers (1701–10); the tortoise trips suffered worse problems than distance – the slaves were getting restless, and inclined to evict the white supervisor and escape with the canoe, each of which had to have first two (1704) then three (1705) Frenchmen to keep control. Villers sympathized with those losing canoes because "that prevents you from enjoying the same privilege that the others have in going for tortoise ['*aller à la tortue*'], which is such a major part of your food supply". On game hunting in the forest, Villers complained as much about killing by uncontrolled dogs as by the inhabitants themselves.
177. From Feuilley (1705). Feuilley used '*ces*' for both modern French '*ces*' (these) and '*ses*' (his, their) throughout his MS; only by translating '*ces bestiaux*' as 'their [own] animals' and not 'these animals' does the piece make sense. Rendering the original into modern French, Probst & Brial (2001) missed this usage, making the text self-contradictory.
178. François Martin in Martineau (1931–34) and Lougnon (1970); repeated in Bour 1981, Probst & Brial 2001.
179. Hébert (1708), Hermann (1898), Lougnon (1956), Bour (1981), Cheke (1987a), Probst & Brial (2001).
180. Lougnon (1956).
181. Peletyer *et al.* (1701).
182. Kaeppelin (1908)
183. Thomas Herbert (1634) was somehow aware of tortoises and 'dodos' (i.e. Solitaires) on Rodrigues. He did not land, and the 1601 and 1611 Dutch accounts were still unpublished, and in any case mentioned no tortoises and 'birds' only rather vaguely (Verin 1983, North-Coombes 1994); Brouwer's account of '*dodoersen*' in 1601 did not appear in print until 400 years later (Moree 2001). Herbert must have tapped into an effective mariners' bush-telegraph. French ships visited in 1638 or 1640 (Cauche's account) and 1641 (the *Concorde*). A Dutch crew marooned there for three months in 1644 reported only that the island contained "neither ebony nor anything else that could be of use to the Company". The next recorded landing was Leguat in 1691 (North-Coombes 1971, 1994). Brouwer saw "doves, parrots, geese, dodos, and an abundance of other birds and fish". The 'geese' were probably boobies (large, white, with webbed feet) rather than anatids – there is no wetland habitat on Rodrigues and no other reports of geese, either alive or subfossil.
184. See North-Coombes (1971) for a general history, and Leguat (1707) for details of his stay on the island.
185. The two pirates escaped to Mauritius in a home-made dugout canoe, stealing a boat there to reach Réunion, where they featured in Boucher's *Mémoire* of 1710 (Alby 1989), leaving no account of their adventures (see also North-Coombes 1994).
186. North-Coombes (1971).
187. Tafforet's ship the *Ressource* brought settlers, partly to stop the English getting a foothold, partly to rid Réunion of troublesome characters. There were men and women; it was a serious attempt to install a viable colony. However, the ship missed the best anchorage, and while Tafforet and four others were ashore in the longboat, the *Ressource* was blown from its moorings, and unable to reverse against the prevailing wind had to return to Réunion for assistance. It took until June 1726 to mount a rescue party to pick up the stranded men (Lougnon 1956, North-Coombes 1971).
188. Nonetheless, Leguat (1707) reported that the pigeons only nested on rat-free islets, and Tafforet (1726) found starlings, parrots and pigeons largely confined to the southern islets.
189. Sea level in the Mascarenes rose steadily until about 7,000 years ago, thereafter slowly until stabilizing at current levels $c.3,000$ years ago. The final shrinking of Rodrigues from the size of the lagoon ($c.400km^2$) to the present size (104km8 pt) at $c.7,000$ years ago would have been very rapid as the sea flooded the shallow shelf developed during an earlier Pleistocene sea-level high-stand (Rees *et al.* 2005, Camoin *et al.* 1997).

CHAPTER 6. UNITED UNDER FRANCE

1. Bory (1804), impressed by the wild nature of central Mauritius, seen from the Piton du Milieu in 1801.
2. Astronomer Le Gentil (1779–81), perceptively diagnosing a serious problem while cooling his heels in Mauritius between the Transits of Venus in 1761 and 1769.
3. Details in this paragraph from Guet (1888), Lougnon (1956), Barnwell & Toussaint (1948), Toussaint (1972).
4. 'Peter Pepper' (or Piper) of the English tongue-twister: "Peter Pepper picked a peck of pickled pepper".
5. Historical summary from Barnwell & Toussaint (1949) and Toussaint (1972).
6. Barnwell (1948), Selvon (2001).
7. Lougnon (1956) – there was a particularly large increase in slaves going AWOL in 1730 when the authorities banned male and female slaves from living together.
8. Brouard (1963), Lagesse (1972).
9. Lougnon (1944–46), Toussaint (1956), Lagesse (1972), Pineo (1993). "It's a real loss to us, game becoming rarer than ever, and it not being possible to catch more than one or two animals [per day] when hunting on land. Nonetheless hunting and fishing are the only resources we have to live on" (letter from the administrators of Mauritius to those in Réunion, February 1731; Lougnon 1944–46), after a hunter and his crew were lost in a boat wreck in 1731.
11. Editorial note in the 1937 printing of Mahé de La Bourdonnais (1740).
12. Barnwell & Toussaint (1949), Gandon (1732); Lagesse (1972)

documented a similar attack in Flacq in 1731, apparently not the same.

13. Toussaint (1956).
14. Toussaint (1972); Pineo (1993) published a map of the concessions granted by 1731, the agricultural lands being almost all in Grand Port, Grande Rivière Sud-Est and Moka, with a few in Flacq and on the plains north of Port Louis. However, when the main port and administrative centre moved from Grand Port to Port Louis in 1731, some settlers abandoned the southeast and started afresh around Pamplemousses (Pineo 1993).
15. *Courrier de Bourbon* 1722; three other ships collected tortoises from Flat Island in 1721–22 (the *Triton*, HMS *Lyon*, the *Diane*; Garnier 1722, Barnwell 1950–54), and the northern islets (rather than Rodrigues) were still designated by the Company as the proper source for Mauritius in 1732 (Lougnon 1933–49, Cheke 1987a); see Chapter 9.
16. This islet was where the Dutch originally found coconuts in 1607 (Barnwell 1948, van der Hagen's account), and was always named after them. In the 1700s it was 'Isle des Cocos', being reduced to 'Deux Cocos' (two coconuts) by 1808 (Barnwell 1955).
17. Quoted by Milne-Edwards (1875) and North-Coombes (1971), though both thought the 'islands' referred to Rodrigues not the northern islets of Mauritius.
18. Gandon (1732, and in Barnwell 1948).
19. Ducros (1725), from the English translation in Barnwell (1948).
20. Barnwell & Toussaint (1948).
21. Baron Grant in 1741 (in Grant 1801).
22. Quoted in Grant (1801).
23. Barnwell & Toussaint (1948): the stock of game and cattle had been much reduced through recent famines; North-Coombes (1994) discussed the tortoise trade.
24. Baron Grant in 1741 (in Grant 1801).
25. Baron Grant (in Grant 1801). Monty Python could hardly have put it more eloquently; Grant did manage to harvest some bananas, 'monkies' notwithstanding.
26. Pitot (1899), citing a '*lettre à M Malartic*' by Laurent de Reine, written in 1792, when he must have been an old man, addressed to the then governor of the islands, General Malartic (Trouette 1898). De Reine, an infantry captain, had arrived in 1729; his letter is cited as published in local periodical *Mauritiana* in 1909 (Rouillard & Guého 1999), which we have not seen; the complete letter might include further natural history interest. His undated references to rats and locusts evidently refer to his early years on the island. Ducros had called Mauritius "the island of rats" in 1725 (Barnwell 1948) but appears to have been referring to the same kind then present in Réunion (*Rattus rattus*) to which he compared them.
27. See Twigg (1975), Atkinson (1985).
28. Cossigny (1732–55), Grant (1801).
29. In 1729 Jonchée de la Goléterie wrote to the Company in Paris pointing out that Mauritius suffered no insect damage, unlike Réunion (Lagesse 1972: 47); Gandon (1732) and Gennes (1735) only mentioned rats and monkeys as pests. As late as 1740 a letter from the island council giving agricultural results to the Company (printed in Pineo 1993) mentioned no pests at all – clearly selective reporting! Poivre believed the locusts had arrived in La Bourdonnais's time (i.e. 1735–46; Malleret 1974), we think between 1735 and 1740. Rouillard & Guého (1999) quoted de Reine "[who] arrived in Mauritius in 1729" as reporting clouds of locusts, but the date he saw them is not given.
30. Cossigny (1732–55).
31. Cheke (1987a): *Lepus nigricollis* was not positively identified in Mauritius until de Querhoënt's description was published by Buffon (1776), but it was no doubt long established.
32. Carié 1916; If the other *lièvres* were rabbits, their decline after initial establishment coincided with the arrival of Norway Rats. Rabbits only thrive in the tropics if rainfall is under 1,000mm/yr (Atkinson & Atkinson 2000), so the Mascarenes are generally too wet, and in practice they have only survived on offshore islets in the absence of cats and usually also of rats. If rabbits did establish on the Mauritian mainland, the Norway Rat invasion would have tipped the balance against them.

33. See Oustalet (1897) and Cheke (1987a) – the first and last were collected and illustrated by Sonnerat (1782), who contrasted the Chinese Francolin with another 'ordinary' kind, presumably the Grey. Baron Grant, writing in 1849, said the gazelles "resemble the roebuck in size and figure; their ears are large, and lined with a very black hair . . . their horns are black, with fluted rings to half their length and resemble the antique lyre . . . as to their colour, the greater part are fallow on the upper parts of their body, and white under the belly, with a brown stripe which separates the two colours on the lower part of the flanks" (Grant 1801); the lyre-shaped horns favour Heglin's (Red-fronted) *Gazella* (*Eudorcas*) *rufifrons* over the sympatric Dorcas Gazelle *G. dorcas* (Hallenorth & Diller 1980). Barthélemy David, previously in Senegal, was Governor 1746–52 (Grant 1801, *DMB* 330–1; 1943, A. Toussaint).
34. La Caille (1758), as translated by Grant (1801). Contemporary writers did not always distinguish between land 'cleared' and that under cultivation. Thus Grant's 'not an eighth cleared' in 1740 is not incompatible with La Caille's a tenth 'cleared and under cultivation' in 1753. La Caille as a competent surveyor was no doubt fairly accurate, although the map by Bellin in 1763 derived from his surveys (e.g. Toussaint 1953 or Staub 1976, endpapers) does not fully indicate boundaries of cultivated land and forest (map 6.1 is a version of this).
35. Poivre and Fusée-Aublet were bitter rivals – for their feud, and also their respective achievements in horticulture and acclimatisation see Malleret (1974) and Ly-Tio-Fane (1958, 1980). Ly-Tio-Fane (1996) described development of tropical botanic gardens and the pivotal role of Pamplemousses.
36. Céré's correspondence, as published (Pope-Hennessy 1889, Ly-Tio-Fane 1958, 1970, *Rev.Hist.Litt.Ile Maurice*, vol.4, 1890), contains very little about wildlife, possibly due to editorial bias towards cultivated plants.
37. Rochon (1791).
38. See Bernardin (1773), Fusée-Aublet (1775), Brouard (1963).
39. Flinders (1814). Boullanger (1803) also commented on an abundance of wayside raspberries; see also Brouard (1963), *Flore* 81 (1997, F. Friedmann). The 'wild tobacco' was probably the shrubby African weed still called *tabac marron* – *Solanum mauritianum* (Cadet 1981 *Flore* 128, 2000, A. J. Scott).
40. Bernardin (1773) – these cagebirds were apparently the only animals on board not seasick in very rough weather! The chickens nearly all died and the dogs were very unhappy.
41. Although Sonnerat's drawings kept with Commerson's MSS in Paris confirm that *Serinus mozambicus*, called '*oiseau du Cap*', was established by *c*.1770 (Oustalet 1897, Cheke 1987a), Le Gentil (1779–81) actually said the birds were 'yellow and grey' (not 'green' as misquoted by Cheke 1987a), which better fits the other southern African canary *S. canicollis*, not formally reported until 1801 (Cheke 1987a). Possibly both were released around the same time but not initially distinguished by locals. By the 1850s the Yellow-fronted Canary had become '*serin du pays*', the Cape Canary being '*serin du Cap*' (Cheke 1982a).
42. Le Gentil (1781). The '*moineaux de Chine*' were identified from a description in Commerson's MSS in the MNHN archives (pers. obs.).
43. Toussaint (1956: 24), Epinay (1890), Mamet (1993: 114). In 1770 Poivre amended this to 10 birds and 20 rats (no monkeys), with a fine of 5 *sols* per head of bird or rat tail not supplied. There were 20 *sols* to the *livre* (or franc) (acccording to the *National Encyclopaedia*, *c*.1867); as a guide to value, the government price to farmers for wheat in 1767 was 20 livres/quintal (= hundredweight, i.e. 112 French pounds = 51kg) (Ly-Tio-Fane 1968). In 1772 beef was 1

livre/lb (0.4kg) and "an indifferent fowl or duck" cost 5 *livres* (John Colpoys in Barnwell 1948). The 1792 rate was at 1 *sou*/head, with a fine of 5 *sous*/head of birds not supplied (Unienville 1982; *sou* = *sol*).

44. *Règlement* 184 of 7 May 1770 (Delaleu 1826); the annual prize was omitted when the regulations were reissued by Poivre's successors in 1775 (*Ordonnance* 223 of 14 Oct.1775, Delaleu 1826).

45. "The Colonial Assembly, Considering how important it is to employ all possible methods to revive in this colony the cultivation of cereal grains, from which [activity] many inhabitants have been discouraged solely by the prodigious number of birds that devour the harvests; [and] Considering that apart from a few species of bird that destroy only insects, the others are of no use to the inhabitants, even to those who devote themselves to export crops, and thus all inhabitants are equally interested in removing a scourge the effect of which is to deprive the colony of subsistence that its soil could offer; After having deliberated urgently decrees..." (Unienville 1982).

46. Fusée-Aublet (1775), presumably discussing the situation when he was on the island (1753–61; Ly-Tio-Fane 1978); by the time his book was published the problem had been solved by introducing mynas.

47. d'Epinay (1890: 169).

48. Malleret claimed (1974) that the first mynas were brought in by Poivre in 1751, without supporting evidence. Poivre was travelling in China and the Philippines during 1750–53, as Malleret himself documented, and we believe this date was an off-the-cuff error. Brouard (1963), also with no supporting source, attributed the importation of mynas to Governor Barthélemy David (in charge 1746–53), coinciding with Malleret's 1751. Brouard could have misread an ambiguous paragraph in Bulpin (1958: 118) to imply David's involvement, though Bulpin's book is not in his bibliography. Contemporary authors implied a first release date of *c.*1759, with a later release in *c.*1762 (see note 50), so we are treating the 1751 date with caution.

49. Foucher d'Obsonville (1783).

50. The history of myna introductions was discussed by Oustalet (1897) and Cheke (1987a); we differ a little from their conclusions. Both Buffon (1770–83, writing in 1775), and Foucher d'Obsonville (1783) mentioned a first release, destruction by the inhabitants, and then a second release, implying that it was on both islands. However, Buffon conflated the events on the two islands into a single story; he alone had a complete account from Réunion (from Lanux), but muddled it with details Sonnerat sent him about Mauritius. All the published first-hand evidence comes from Mauritius alone. As far as we can tell the release followed by a massacre occurred only on Réunion, as none of the Mauritian contemporary reports refer to it. Maudave (or Modave) was a Mauritian resident, soldier, politician and entrepreneur (Toussaint 1972) who dabbled in trade with Madagascar and later (1773) settled in India where he died in 1777 (Ly-Tio-Fane 1978; see also *DMB*:462). He was credited by Buffon (as 'M de Morane') as responsible for the second release, Céré (in Pope-Hennessy 1889) also saying it was he who brought the birds to Mauritius. Foucher clearly stated that Maudave made the arrangements at the Indian end, but that he (Foucher) and a M. Beylier collected and sent the birds; to which island he does not say, but we think it was Mauritius. The younger Cossigny wrote to Céré in 1784 claiming that he, not Maudave, sent the birds from India, on a ship captained by a M. Martin who released them on Maudave's estate in 1763 – not 1762 (Rouillard & Guého 1999:646). Desforges probably commissioned mynas from various sources, so Cossigny's birds were additional to Foucher's. Oustalet misread Maudave as 'Mandane' in Desjardins's notes. Buffon's claim that the second release was eight years after the first does not fit other facts, nor does his claim that it all started "hardly more than 20 years ago" (i.e. *c.*1755, a date then used by others, e.g. Milne-Edwards & Oustalet 1893, Decary 1962). All are agreed (including Buffon) that the original move came from governor Desforges-Boucher, but he took office in 1759 (Toussaint 1972 and all other sources). Foucher, based in India and only sporadically visiting the islands, dated the first release to 1759 (not stating which island), but for the second, said only that it was "towards the end of the last war" (i.e. The Seven Years War, 1756–63). Rouillard & Guého assumed that the bird was called *martin* after Cossigny's ship's captain, but the Portuguese *martinho*, already current in India for at least a century, is a more likely origin (Cheke 1982a).

51. Delaleu (1826); 500 *livres* was equal to 25 sacks of grain or the price of 100 chickens.

52. De Querhoënt (1773), Céré (in 1777, Pope-Hennessy 1889), Buffon (1770–83) and Sonnerat (1782) were certain the absence of locusts was due to mynas, though Cossigny (Charpentier 1799) carefully noted the lack of locusts was 'attributed' to the birds, not committing himself. The 1770 locust upsurge is only reported in contemporary correspondence between Commerson and Cossigny (Cap 1861, Mamet 1996); the definitive cyclone was reported by d'Epinay (1890). Cossigny later (1799) said that no more locusts were seen after 1770, but de Querhoënt (1773), writing closer to the time, was less sweeping: the mynas had "almost entirely destroyed the locusts that ravaged the harvests before its arrival". Buffon, writing in 1775, said the locusts were "entirely destroyed", but his account is confused. Poivre, in his preamble to the 1770 law about bird-pests, said "the scourge of locusts, the most destructive of all, has noticeably diminished as a result of the intensive drives made against these insects" (Delaleu 1826) – no mention of mynas, but perhaps he wanted to congratulate farmers on their hard work; later in these regulations he specifically re-iterated the protection for mynas and the fines consequent on killing them.

53. Bernardin (1773); Flinders (1803–14) described a myna roost near Henrietta in 1805 estimated at between 2,000 and 10,000 birds, noting that they accumulated from miles around as such large numbers were never seen in the fields and woods.

54. Ly-Tio-Fane (1968).

55. This locust resurgence, only in eastern Réunion, provoked new control regulations (Wanquet 1974).

56. Ly-Tio-Fane (1968); Poivre's plan was to carry on with the mass killing of hoppers until the population was reduced enough for the mynas to keep it under control thereafter (letter from Poivre to his Minister in Paris, February 1768, cited by Malleret 1974).

57. Mamet (1993: 112); Greathead (1971) discussed the history of biological control, citing the myna in Mauritius as the first example.

58. Mamet (1993) identified the 18th century locusts using the only contemporary description (Pingré 1760–62), the type of spine on the hind leg being diagnostic. The Migratory Locust *Locusta migratoria* also occurs in the Mascarenes, although the general consensus is that the Red Locust was responsible for the 18th century ravages (Williams 1963, Hemming 1964). The Red Locust is a grass specialist (Hemming 1964), whereas in Réunion at least, the locusts ravaged nearly all crops: "nothing escapes, except coffee, from these insects. They kill peach-trees, orange-trees, lemon-trees and bananas; at Saint-Paul one now only rarely sees vegetables, pineapples, cabbages etc. . . . which were previously common" (letter from the *Conseil Supérieur* at Réunion to the Company in Paris, 20 Oct.1731 in Lougnon 1934–49, vol.1: 142). The Red Locust has damaged sugar-cane (a large grass) in Mauritius more recently, in 1933 and 1962–63 (Williams 1963, Mamet 1993). Hemming (1964) attributed the 1960s resurgence to decimation of mynas by the "most severe cyclone for at least 800 years", *Carol* in 1960, a view supported by the occurrence (unknown to Hemming) of a severe cyclone in 1931 (Padya 1976), two years before the 1933 outbreak. Most severe cyclones however do not result in locust outbreaks, although mynas can be badly hit (e.g. 50–70% mortality after *Gervaise* in 1975, Cheke 1975b). Both locust species are native to Madagascar, and occur widely in Africa.

59. Buffon (1770–83), offering rare caution in 1775 on the dangers of ill-thought-out introductions.

60. Bory (1804: v1, 225).

61. Bernardin (1773).

62. Philibert Commerson to Joseph Charpentier de Cossigny, 29 Sept. 1770 (in Cap 1861).

63. Charpentier (1799) – the animals were probably *Rhysomys pruinosus* (Cheke 1987a).

64. Charpentier (1803); he was upset that feral cats were routinely killed for raiding poultry when they could be out there, he imagined, killing rats. He failed to appreciate that cats go for the easiest catch, not what humans think they ought to control!

65. Sonnerat (1782).

66. Bernardin's *ami du jardinier* was brown and "the size of a large sparrow", whereas the 1774 birds, brought from the Cape, were "black with grey and white patches", a little smaller than a myna, hopped like a magpie and bobbed their tail a lot (Oustalet 1897 and, more fully, in *Rev. Retrospective Ile Maurice* 4:339, 1953); nine were released at Pamplemousses. Clearly two species were involved, neither identifiable; the most likely candidate for the 1774 birds is the Common Fiscal *Lanius collaris*, which is common around Cape Town, but has white underparts like other South African shrikes (Newman 1998). When the law requiring heads of seed-eaters was renewed in 1775, to the ban on killing mynas was added "the shrikes recently imported for the destruction of beetles and the eggs of little birds" (Delaleu 1826).

67. "It has been tried, but without success, to import frogs, which eat the eggs that mosquitos lay on stagnant waters" (Bernardin 1773); the species is not known. The brothers Genève are generally thought to have introduced *Pychadena mascareniensis* in 1792 from Madagascar (Unienville 1838, Clark 1859), though Péron (*c*.1803; MS 15031) claimed Lislet Geoffroy was responsible (no date given), and Milbert (1812) asserted that the source, about 20 years previously (again *c*.1792), was the Seychelles, echoing Betting's (1827) report for Réunion. The frogs were already abundant in 1801 (Milbert), and have been ever since. They appear to have been introduced to Mauritius and Réunion independently, at around the same time. DNA studies show that the haplotypes on the two islands differ (Mauritius sharing one with the Seychelles), confirming they came from different source populations (Vences *et al.* 2004a). Lislet was in southern Madagascar (Baie Saint-Luce) in 1787 (*DMB*:167–8; 1942, R.O.Bechet) where the Réunion haplotype is prevalent, so perhaps he released them there not Mauritius; he also went to the Seychelles in 1793. However the Genève brothers hired a ship, the *Julien*, which traded in Madgacasar in 1792 (Toussaint 1967: 425). The Mauritius/Seychelles haplotype has not yet been found in Madagascar.

68. Common Tenrecs fatten up and hibernate underground during the southern winter, when they are hunted and dug up for food, especially in Réunion (e.g. Telfair 1831, Schlegel & Pollen 1868, Carié 1916, Staub 1993). See Bory (1804), Lesueur (*c*.1803) and especially Milbert (1812) for the 1801–03 observations; in his day any tenrec, fat or thin, was fair game for the "*noirs*" (slaves). Clark (1859) credited Mr Mayeur, and Cossigny's interest (Charpentier 1803) was also noted by Carié (1916). Nicolas Mayeur, more of an interpreter and explorer than a slaver, was in 1777 one of the first Europeans to penetrate central Madagascar (Oliver 1904, Bulpin 1958, *DMB* 215 [1942, Pelte & Toussaint]). He left voluminous manuscripts on his Malagasy explorations (some given by Barthélemy Huet de Froberville to Matthew Flinders in 1806: Flinders 1803–14, Pineo 1988: 138, now preserved in the British Library, Farquhar papers); it has not been possible to scour these for the date he introduced tenrecs, though there is nothing in material printed in the *Bull. Acad. Malgache* 10: 52–156 in 1913. Mayeur apparently left Madagascar for the last time in 1787 (Oliver 1904: 628), and died at Trois Islots in 1809. Trois Islots, now a village near Bel Air (Flacq), was in 1800 a large administrative area in the centre-west of the island without an eponymous settlement (map in Grant 1801). Having served in Paris as ambassador for Mauritius during the revolution, Cossigny returned in 1800 to run the Powder Mills (munitions factory), but resigned and returned to France when suspected of wanting to liberate the slaves (Toussaint 1972). Two other species of tenrec may have been feral for 30–40 years. Milbert (1812) added "*Erinaceus setosus*" as also present, and both Bojer and Telfair sent specimens of "*Centenes setosus*" to Europe from Mauritius in the 1830s, Bojer also including "*C. semispinosus*". In principle these are the Greater Hedgehog Tenrec *Setifer spinosus* and the Spiny Tenrec *Hemicentetes semispinosus*, but tenrec taxonomy was still developing at the time, and some doubt must remain, especially as the Zoological Society's catalogue (Waterhouse 1838) also includes (from Telfair) "Spiny tenrec *Centenes* [sic] *ecaudatus*?", *ecaudatus* being the Latin name of the common species, not spiny except when young. Futhermore, a pair of Lesser Hedgehog Tenrecs escaped from Telfair's house in Mauritius in 1833, leaving a new-born young which he sent to London (Telfair 1833), later named after him as *Echinops telfairi*. Telfair's specimens to the Zoological Society in London (see Wheeler 1997 for the sorry tale) are lost, but Bojer's sent to Vienna (mammal accessions log, courtesy of Dr H.Schifter in 1978, to ASC) may still exist. Although based in Mauritius, both Telfair and Bojer (but not Milbert) collected in and received specimens from Madagascar. A similar uncertainty surrounds chameleons collected by Telfair.

69. Céré (1781); The early introduction was *A. fulica*; *A.immaculata* was a later arrival (Germain 1921).

70. Buffon (1776) cited de Querhoënt on a new kind of rat with "the strongest odour of musk" that had established, of which it was said "that when it passes through a place where there is wine it sours it"; this reputation belongs in the Mascarenes to the Indian House Shrew (e.g. Milbert 1812, Carié 1916). It soon acquired the name *rat musqué* (Céré 1781), still current today (e.g. Baker & Hookoomsing 1987, Probst 1997).

71. In a misguided attempt to encourage agriculture, Réné Magon withdrew all restrictions on woodcutting and forest clearance, resulting in widespread abuse; a few years later he had to re-impose constraint, but much damage had been done. After resigning in 1759, he pursued a short career abroad, before retiring back to Mauritius, where, despite his foolhardy forest policy, he was respected by Poivre and others for having promoted agricultural development (Epinay 1890, Barnwell & Toussaint 1949, Toussaint 1972).

72. The arid hillsides around Port Louis apparently eroded rapidly, filling the harbour with silt, such that major dredging works by Bernard de Tromelin were necessary in the 1770s before it could take large ships again (Rochon 1791, Barnwell & Toussaint 1949). A small sandy seabird island halfway between Mauritius and northern Madagascar is named after this engineer and former sea-captain.

73. Preambles to *Ordonnances* 175 (24 Oct.1767) and 177 (18 Jan.1768); Delaleu (1826), Ly-Tio-Fane (1968).

74. Poivre and his friend Bernardin de Saint-Pierre were against slavery – a step too far for a plantation economy in the mid-1700s, especially as the military governor (Dumas's successor Desroches) was militantly pro-slavery, and busy trying to create exclusive black and white zones in Port Louis (Bénot 1983).

75. Charpentier (1764).

76. Ly-Tio-Fane (1968); Poivre's figure was based on detailed data compiled in 1766 by Desforges-Boucher (Pinco 1998), which gave an overall figure of 8% *en valeur* (i.e. under cultivation), subject, no doubt, to errors of estimation. Overall 45% had been conceded, but much was still uncultivated, although partly cut-over for timber; Desforges estimated, with 100,000 *arpents* reserved for forests, and 32,000 unusable on mountains and rivers, that 300,000 *arpents* could be brought into cultivation, 69% of the total 432,000. The concessions, broken down by administrative area, showed great

Notes on Chapter 6 311

disparity in different districts. Pineo's text throughout confuses '% conceded' with '% *en valeur*', the figures for the former in fact applying to the latter.

77. Grove (1995); the Navy ministry was responsible for overseeing colonies.
78. Grove (1995); the 30% figure is from Desforges-Boucher's reckoning in 1766, adopted by Poivre (Pineo 1998; see also note 76).
79. *Réglement Économique* of 16 Nov. 1769 (Delaleu 1826; see also Grove 1995).
80. "The proprietor shall be held . . . to making a sowing of trees of all species, both foreign and native to the country, and shall prefer for these sowings native seeds, those of *bois de natte* [*Mimusops*, *Labourdonnaisia* and *Sideroxlon* spp.], *bois puant* [*Foetidia mauritiana*], *tacamahaca* [*Calophyllum tacamahaca*], *benjoin* [*Terminalia bentzoe*], *pomme* [*Syzygium glomeratum* and other spp.] and *cannelle* [*Ocotea* spp.]" Article 5 of the *Réglements Économiques* of 16 Nov.1769 (Delaleu 1826).
81. Essentially the same list as recommended for replantings, with two species of *natte* listed (à grande et à petite feuilles) and the island's largest tree *colophane* [*Canarium mauritianum*], then much used for making *pirogues*, dug-out canoes.
82. Lougnon (1956) gave a history of what were later called the *pas géometriques*, still in force today, at least in theory (see Brouard 1963).
83. Brouard (1963), Toussaint (1972); as Brouard pointed out, relicts of this survive today in the ancient plantations at l'Asile and Powder Mills near Pamplemousses.
84. Ly-Tio-Fane (1968).
85. La Motte (1754–57), Cossigny (1764), Pitot (1899).
86. This was undoubtedly an exaggeration, as others (e.g. Bernardin 1773) reported them at least locally common.
87. *Ordonnance* 175 of 24 Oct.1767 (Delaleu 1826), infractions attracting a 1000 *livre* fine for a first offence, or a whipping for a second ('blacks' got the whipping first time). Individual deer doing damage to cultivation could be shot, but the carcass had to be delivered to the hospital.
88. *Ordonnance* 176 of 24 Oct.1767 (Delaleu 1826).
89. Article 5 of *Règlement* 184 of 7 May 1770 (Delaleu 1826). The '*mesanges*' are not identifiable: in modern French this term refers to tits *Parus* spp., absent from Mauritius. It was probably used by Poivre for birds with roughly similar ecology, e.g. the Mascarene Grey White-eye. However the term was used rather loosely in the 18th century, as Cossigny (1799) used it ("une espèce de *mesange*") for an unnamed seed-eater, and Bernardin (1773) described "a pretty *mesange* with wings spotted with white", evidently the Red Avadavat.
90. Grove (1995) claimed that this 1792 legislation indicated a budding understanding that endemic birds needed protection: "the categories '*messanges*' and '*merles*' were probably meant to cover all endemic birds". There is no basis for this interpretation (Cheke 1996); the preamble (note 45) was crudely utilitarian, though the law itself was loosely based on Poivre's more imaginative original; d'Unienville (1982) printed it in full. The doves would have been Zebra Doves, Spotted Doves and possibly also Malagasy Turtle Doves. Grove claimed (citing a communication from Raymond d'Unienville to P. J. Barnwell in 1986) "that '*tourterelles*', an endemic species, were also pests and voracious seed eaters". He should have checked with an ornithologist! The first two are introduced, only the third native (but not endemic); they are nowhere recorded as grain pests.
91. Bernardin (1773).
92. Céré (1781): *nattes*: *Mimusops* and *Labourdonnaisia* spp., *canelle*: *Octoea*.
93. Toussaint (1956: 25), Lamusse (1990), Rouillard & Guého (1999).
94. Epinay 1890, Brouard (1963); see also Céré (1781).
95. Brouard (1963).
96. Flinders (1814).
97. Prentout (1901), Brouard (1963), Unienville (1989).
98. Brouard (1963). Barnwell & Toussaint (1949) noted that stonemasons were freemen needing payment, whereas wood was cut by slaves whose labour cost nothing.
99. Prentout (1901), Brouard (1963), Flinders (1814), Rouillard & Gerého (1999); François Peron (1807) made the most penetrating analysis of the forest cover and water flow issue (p. 102).
100. Brouard (1963: 13) attributed to Poivre (without reference) the importation of "several game-birds from China (in 1749–50)". Malleret (1974), in a footnote, reported an invoice in French archives which recorded "peacocks, waterhens . . ." (full list not given) purchased by Poivre in China destined for Mauritius, without saying whether these arrived. Poivre had enough trouble keeping his plants alive on his long tortuous sea-journeys (Malleret 1974, Ly-Tio-Fane 1958); could he also have looked after live birds? The waterhens ('*poules d'eau*') may have been Purple Swamphens, as transporting Common Moorhens seems unlikely. Although Moorhens were first recorded in Mauritius soon afterwards, they were not imports from China as they are of the Malagasy race *pyrrhorhoa* (Rountree et al. 1952), and are generally thought to have arrived on their own (p. 59).
101. This expedition was described by Malleret (1968); the quoted parts are from his paper, not Poivre's original. We cannot identify all the birds required – 'green turtle doves' may have been small fruit-pigeons (*Treron* or *Ptilinopus* spp.), and the locust-eating *merles* were presumably mynas. *Huamei* or *hwamei* is a general term for babblers, the name being used more specifically for species of laughing-thrush *Garrulax*, particularly *G. canorus* (Cheng 1963, Mackinnon & Phillips 2000) esteemed in China as a cagebird for mimicry and song, and beneficial to agriculture through eating insects. Poivre had earlier given Réaumur a '*hoamy*' collected c.1750 in Canton, described by Brisson (1760) and identified as *G. canorus* (Stresemann 1952).
102. Malleret (1968), Laissus (1973).
103. Buffon (1770–83); Milbert (1812) also mentioned captive breeding of swamphens.
104. Charpentier (1803, Epinay 1890). Oustalet (1897, Cheke 1987a) gave the date, from Desjardins's notes, as 1781, but Cossigny himself said it was 1767. Mauritian *Stigmatopelia chinensis* are of the race *tigrina* that extends from Bengal to Indonesia (Lever 2005, Hawkins & Safford in prep.), confirming Cossigny's claim. They are not *suratensis* from peninsular India as usually cited (e.g. Rountree et al. 1952), which is much more boldly marked on the back and wing-coverts (pers. obs. in Sri Lanka).
105. Bernardin (1773); one may have been the Malagasy Turtle Dove which we now know was native, though it is likely the indigenous population was supplemented or replenished by birds brought from Madagascar. The other was almost certainly the Zebra Dove.
106. Sonnerat's *Voyage aux Indes Orientales et à la Chine* (1782) contains descriptions of many new bird species, including the first published illustration of the Mauritian Pigeon Hollandais. See Ly-Tio-Fane (1978) for details of Sonnerat's life, and Oustalet (1897) and Cheke (1987a) for further comments on his Mascarene birds.
107. Charpentier (1803), Milbert (1812); Milbert clearly read Cossigny's book rather superficially. We cannot tell what species of duck was involved, though the African Black Duck *Anas sparsa* answers to being a 'mountain duck' (Newman 1998). Cossigny added that the Cape tortoises were all "stolen by the blacks, without leaving any descendants".
108. Bernardin (1773), Flinders (1803–14,1814), Prior (1819). Flinders was a British naval explorer, famous for charting the coasts of Australia. Arriving in Mauritius with a crippled ship and what he thought was a valid safe-conduct, he was imprisoned by governor Decaen as a spy, then held under virtual house arrest at La Marie (near Vacoas), then the Mauritian 'outback' (e.g. Pike 1873, Pineo 1988). There he was able to explore forests, and make obser-

vations on wildlife, forests, agriculture etc., but he could have missed ducks on lowland meres. The site of his internment is a minor pilgrimage place for Australians (e.g. Rees & Rees 1956, whose *Westward from Cocos* includes an amusingly irreverent look at Franco-Mauritian life in the 1950s).

109. ". . . passengers come from Europe have brought out of curiosity *serins jaunes* [domesticated *Serinus canaria*], *bovreuils* [*Pyrrhula pyrrula*] and *chardonnerets* [*Carduelis carduelis*]; some have escaped into the woods, where they will increase in due course the number and variety of forest songsters" (Milbert 1812).

110. La Motte (1754–7); the Ortolan Bunting *Emberiza hortulana* is a European species much prized by gourmets and "so celebrated for the delicate flavour as to become proverbial" (A. Newton & Gadow 1896).

111. Cossigny (1764).

112. The Réunion *oiseau bleu* is another candidate, but its date of extinction is unknown. Details of four other species that *were* sent to Europe failed to get noticed in the literature: a Réunion Cuckooshrike *Coracina newtoni* skin (see below under Réunion), and good drawings by Jossigny of the Mauritius Lizard-owl, Rodrigues Parakeet (Oustalet 1897), and the Réunion Black Petrel (pers. obs. in the MNHN archives, Paris).

113. Bernardin (1773).

114. Cossigny (1799); he may, however, have been remembering earlier days.

115. Cossigny (1732–55), Grant (1801).

116. D'Heguerty (1754), Cossigny (1732–55).

117. On 24 March 1755 Cossigny wrote to Réaumur, sending, amongst other things, unidentified seeds of two species he'd found in the gizzard of a Pink Pigeon, and commenting on the bird's toxicity (Charpentier 1732–55).

118. Cheke (1987a) and Jones (1987) discussed the evidence; the last recorded toxic episode was in 1943.

119. La Motte (1754–57).

120. Bernardin (1773). Clearly *Latania loddigesii* still survived in the wild at that date; there seems to be only one subsequent record of the species on the mainland (Milbert 1812) in 1801–03; see Maunder et al. 2002 for a general survey of the fate of Mascarene palms). The kestrel was not formally described until 1823 (Rountree et al. 1952).

121. Milbert (1812) covered much of the island on various excursions, while both George Clark (1859) and Nicholas Pike (1873) gave good accounts of circumambulating the island. Engineer Jacques-Henri Bernardin de Saint-Pierre sailed to Mauritius with the Comte de Maudave to help him establish a settlement in Madagascar (Wilson 2002); after quarrelling with his boss, he decided to stay in Mauritius, working as a mason. He was befriended by Poivre, who became his mentor in natural history. The same Maudave had helped bring mynas to the island in 1762 (p. 94).

122. "It is claimed that there used to be many flamingos; it is a big and beautiful sea bird pink in colour. It is said that only three remain; I have not seen any" (Bernardin 1773).

123. Brown (1776), Fox (1827); the specimen disappeared later in the 19th century. The dismal fate of this large and important 18th century collection was discussed by Jessop (1999).

124. Tuijn (1969). Arnout Vosmaer is the Dutch zoologist after whom the Carosse Tortoise *Cylindraspis vosmaeri* is named. Tuijn reproduced two pictures by G. Haasbroek showing the long neck feathers erected in a frill around the face, as also noted by Temminck and illustrated by Pauline de Courcelles (later Knip) (Temminck & Knip 1811, Temminck 1813–15, Milne-Edwards & Oustalet 1893) – perhaps Temminck also saw Vosmaer's bird.

125. Adapted from Cheke (1987a). We do not know when Cossigny (Charpentier 1764) saw these snakes; he was in Mauritius on and off between 1732–1759 (*DMB* : 10–11, A.Toussaint, 1941).

126. Charpentier (1732–55); the letters are preserved in the *Académie des Sciences* in Paris (Torlais 1962). Jean-François Charpentier de Cossigny was a combative military engineer working for the French East India Company, with a reputation for ill-tempered relations with other senior officials, but he comes over quite differently in letters to Réaumur, revealing a deep love of nature and its workings (Lacroix 1936, *DMB*:10 [1941, A.Toussaint]); his son Joseph-François is also important in our story.

127. Rochon (1791), Laissus (1974).

128. Oustalet (1897). Laissus (1974) gave a rough taxonomic breakdown of the Commerson drawings: 'worms' 3, crustaceans 23, arachnids and myriapods 22, insects 169, reptiles 22, birds 205 and mammals 32 (not all from the Mascarenes); they include a Cheechak from Mauritius (pers. obs.), the first identified house gecko from the Mascarenes.

129. Sonnerat's life, work and controversies were discussed by Ly-Tio-Fane (1978), where references to the non-Mascarene inaccuracies can be found (see also ASC's 1981 review in *Ibis* 123:255). In relation to the Mascarenes, Sonnerat (1782) made the Chinese Francolin *Francolinus pintadeanus* originate in Madagascar rather than China (creating confusion for decades), and reported the Madagascar Partridge *Margaroperdix madagascariensis* and Spotted Dove only from Madagascar and China respectively, although both were already present in Mauritius, where he may have collected them. Jossigny drew a Madagascar Partridge in Mauritius before 1772 (Oustalet 1897; Jossigny left Mauritius for Réunion in 1772, Ly-Tio-Fane 1978); the dove was introduced by the younger Cossigny in 1767.

130. Buffon (1770–83, 1776); A manuscript also survives of de Querhoënt's bird notes that Buffon intended for his projected bird supplement (Querhoënt 1773). De Querhoënt does not feature in Mauritian histories or the *DMB*, so his reason for being there remains unknown.

131. Buffon's anomalous 'Mascarene' birds were discussed by Cheke (1983b). Buffon worked largely from specimen labels and published literature; it seems that many birds in the *Cabinet du Roi* were wrongly labelled.

132. Edwards (1760); Savery's painting, now in the UK Natural History Museum in London, is perhaps the best known of all Dodo paintings, widely copied and reproduced (e.g. the frontispiece to Hachisuka 1953, the cover of Ziswiler 1996 etc.).

133. Buffon (1770–83), vol 2.

134. Morel (1778); The French had been in Réunion for more than 110 years, and the Rodrigues Solitaire had been seen in 1755 and possibly 1761. Morel usefully sank the mythical 'Isle de Nazare', equating it with the tiny Isle de Sable (now Tromelin) a few hundred miles north of Mauritius, "on which an infinity of seabirds come to lay their eggs". Strickland (1848) was nonetheless still in a muddle about the 'Isle de Nazare' 70 years later.

135. Mauduyt (1784), *Histoire naturelle des oiseaux*. Morel's article (unacknowledged) was probably the source of this 'new' information.

136. Milbert (1812).

137. Cuvier, originally a leading doubter, recanted in 1830 when sent Solitaire bones by Julien Desjardins (Strickland 1848, North-Coombes 1991).

138. "The scientific contingent travelling with Baudin on the *Géographe* and the *Naturaliste* was an especially large one consisting of five zoologists, five gardeners, four artists, four astronomers and hydrographers, three botanists and two mineralogists. However thanks to Baudin's incompetence and lack of sympathy with the scientists, many of his team, including three of the artists, left the expedition at Mauritius" (Jacobs 1995). Ingleton (1986: 302), who favoured Baudin's view that the problems arose from, the scientists' inexperience and indiscipline, listed nine men "discharged at the Ile de France", Pineo (1988) said 10 left, together with 49 deserting sailors. Cornell (1974) translated Baudin's version adding a complete list of the ships' crews; Bory (1804) and Brown (2000) also

detailed the scientists' difficulties with Baudin; see also Milbert (1812), Peron & Freycinet (1807–16), Ly-Tio-Fane (2003) and entries for Milbert, Delisse, Dumont, Bory, Peron and Baudin in the *DMB*. Baudin himself "was not very sorry about the officers and scientists who had abandoned the expedition" (Cornell 1974). Nicolas Baudin became ill during the voyage, dying in Mauritius in September 1803.

139. Milius's diary (1800–1804), not published until 1987. On the return journey Milius assumed command of the expedition after Baudin's death, and later (1818–21) became Governor of Réunion (Reydellet 1978, Bonnemains & Hauguel 1987, Ly-Tio-Fane 2003).

140. See Ly-Tio-Fane (1965) on the societies. Dumont's specimens are mentioned by Lesson (1828) and Oustalet (1896); he emigrated to Réunion after the British took Mauritius, worked as a doctor, dying there in 1822 (*DMB*:424, 1944, J.Vinson). Chapotin published a book (1812).

141. Dumeril & Bibron (1834–54). The endemic day-gecko *Phelsuma cepediana* was dedicated to Lacépède by Peron in his MS 78122 and Lesueur in his MS 15037 (c.1803, preserved at Le Havre), Milbert borrowing that name in his book, though Merrem in 1820 is usually credited (Vinson & Vinson 1969, Pasteur & Bour 1992). Milbert also recorded as '*Gecko francicus*' another similar day-gecko – this appears in the MSS as an illegible Latin diagnosis by Peron, identifiable from the proportions of Lesueur's accompanying pencil sketch as *P. guimbeaui*, 160 years before it was formally described! Peron & Lesueur also brought back the type of the Burrowing Boa, the first house gecko specimen from the island (*Gehyra mutilata*; '*Gecko loncurus*' in their MSS; '*G. lonhurus*' in Milbert's book) and probably the first specimen of Telfair's Skink, on which Dumeril & Bibron founded the genus *Leiolopisma*. Following Baudin's death in Mauritius in 1803 (Ly-Tio-Fane 2003), Peron got the job of writing up the expedition (Peron & Freycinet 1807–1816), but when he died in 1810, Lesueur (who had joined the expedition as a gunner!), bitter at not being asked to continue the work, emigrated to America. There he pursued a career in natural history, geology and art, before returning to France in 1837 (Bonnemains 2003). His voluminous manuscripts (with some of Peron's), which are kept in the natural history museum at Le Havre (Bonnemains & Ly-Tio-Fane 2003), indicate that they explored little beyond the settled parts of northern Mauritius, and that specimens from the northen islets were given to them, not collected *in situ*.

142. A military surgeon named Roch brought a Black-spined Flying-fox *Pteropus niger* alive to Europe in 1803, intending it for the Paris museum's menagerie. However his ship was intercepted by the British near Gibraltar; stranded at Cadiz, he tried to send the animal to Paris from there, but it died *en route* (Geoffroy 1806). Another soldier, Col. Mathieu, brought home a Grey Tomb Bat that became the type of *Taphozous mauritianus*, described by Geoffroy in 1813 (Moutou 1982).

143. Louis Dufresne, chief taxidermist at the Paris natural history museum, sold his collection in 1819 to the new university museum in Edinburgh (Sweet 1970); not all his specimens survive, but his catalogue does, so we still have a complete list of Mathieu's birds from Mauritius in Dufresne's collection (Bob McGowan pers. comm.). For details of other specimens see Lesson (1828: rail, *Drylimnas cuvieri*), Moutou (1982: Grey Tomb Bat) and Dumeril & Dumeril (1851, Bojer's Skink *Gongylomorphus bojerii*; Vinson & Vinson 1969 wrongly assigned it to Réunion). Milbert (1812) noted that Mathieu supplied Dufresne with Mauritian birds, and an MS note by Dufresne in the Paris Museum records his recent return from Mauritius in July 1811, after '8–10 years' there, with a collection of carefully prepared natural history specimens, some of which he gave to the museum (Milne-Edwards & Oustalet 1893:47, footnote). Mathieu's first name is not recorded; we have found no reference to him in standard French biographical reference books. Ly-Tio-Fane (1965, 2003) gave details of the *Société Littéraire*. Mathieu, not a member of the *Société des Sciences et Arts de l'Isle de France* founded in 1801 (Ly-Tio-Fane 2003), probably arrived in 1803 with the new governor Decaen, then leaving with him in 1810. He will have encountered Milbert, and no doubt also Peron and Lesueur, in Mauritius in 1803. Milbert's engravings were made in part from Lesueur's drawings (Ly-Tio-Fane 2003), and he acknowledged Peron & Lesueur as source for his natural history notes, so Milbert and Mathieu could have explored together the forests that Peron and Lesueur had no time to visit, although Milbert mentioned no travelling companions by name. The Malagasy Turtle Dove *Nesoenas picturata*, given a rather muddled description by Milbert (1812), was formally described by Temminck (1813–15), presumably from Mathieu's specimen in Dufresne's collection – elsewhere in the book Temminck frequently mentioned Dufresne's specimens, and the species does not appear in his earlier book (Temminck & Knip 1811). Mathieu's specimen of Bojer's Skink was collected more than 25 years before the species was described by Desjardins (1831).

144. Lesson (1828), Hartlaub (1878); With the endemic *Dryolimnas* being long extinct, the Malagasy form may have attempted to colonise. In the past it has reached the atolls of the Aldabra group from Madagascar (Stoddart 1971, Taylor & Van Perlo 1998), and also the Mascarenes long before, forming the now-extinct endemic Réunion and Mauritian species. Carl Jones (1987) believes the Dufresne parakeet came from Réunion, as it has a more complete pink neck-ring than male Mauritian Echo Parakeets. There is only one indisputably Réunion bird in the Dufresne collection, an Olive White-eye *Zosterops olivaceus* (pers. obs.), but if there is one there could have been others. Mathieu may have visited Réunion, but there is no other evidence of Echo Parakeets surviving there into the early 1800s (i.e. nothing in Bory 1804, Renoyal 1811–38, Betting 1827).

145. Common Moorhens were first mentioned by Cossigny (Charpentier 1799, Cheke 1987a) and also seen soon after by Flinders (1814). In 1863 Newton (1858–95) wrote that "[James] Caldwell told me that an old man of 82, one Paul Baron, told him that he had always known *poules d'eau* to exist in the marsh at the foot of Mont Orgeuil [now Mondrain reserve, Vacoas Ridges] since he was a child, and before the days of de Chazal, Mrs Moon's father." Toussaint de Chazal bought the estate in 1796 (*DMB*:1314–15, 1987 [G. Rouillard & R. d'Unienville]; Malcy de Chazal, later Mrs Moon, was born on the estate in 1803 (Guého & Staub 1979). This is where Flinders, billeted next door, would have seen them. It would appear that they were already well established in the mid-1790s. See Chapter 4 for the biogeography and colonisation potential of *Gallinula chloropus*.

146. Milbert (1812). This appears to be the first intimation in Mauritius of an ecological sensitivity extending from the general (forests and mountains) to the particular ('I will not cut this tree'). For all his exploration of the growth of environmentalism, especially in Mauritius, Grove (1995) missed this illuminating example.

147. Probably *Phelsuma cepediana*; see Box 33, p. 242.

148. '*Cardinal*' could then have been either Cardinal or Mauritius Fody, or both. *Coq des bois* is the Mascarene Paradise Flycatcher (Cheke 1982a), the two parakeets the Grey-headed Lovebird and the Echo Parakeet, and the doves Zebra Doves, whose persistent calling is still a characteristic sound of an otherwise rather silent forest.

149. The large spiders with huge webs are *Nephila inaurata*, formerly abundant, but becoming scarce in the early 1900s (p. 136). '*Bengali*' in Milbert's time was the Red Avadavat *Amandava amandava*, extinct in Mauritius since; the name subsequently transferred to the Common Waxbill (Cheke 1982a).

150. Milbert excelled in artwork and observational writing, but his attempt at a systematic list of animals is a hodge-podge compiled uncritically from several sources, muddling those he saw himself with real and imaginary ones from books; it contains some useful information, but is unreliable in detail.

151. Peron (1807); total rainfall on oceanic islands is now considered more dependent on topography and oceanic weather patterns than on tree cover (as Peron surmised), although where there is extensive persistent montane cloud, trees (if present) filter off water droplets and capture 'horizontal precipitation' unrecorded by normal rain-guages (Padya 1989, Whittaker 1998, Ashmole & Ashmole 2000). This is less important on high-rainfall islands such as Mauritius or Réunion (rainfall range 500–10,000+ mm/yr), than dry ones, e.g., Tenerife (rainfall range 100–800mm/yr), where much of the island's water supply is captured from clouds by montane pine forests (Ashmole & Ashmole 1989). Mauritius, much lower than Réunion or Tenerife, has a small area under persistent cloud, and even that is intermittent; on higher islands cloud-banks form invariably every day (Soler 2000, Ashmole & Ashmole 1989, pers. obs.).

152. See Ly-Tio-Fane (2003) and Prentout (1901) for details of the expedition's return visit to Mauritius, and Peron's dealings with Decaen.

153. Lesueur (c.1803; MSS 15035, 15037); only Lesueur mentions the name *coupe-vent*, though it is clearly cognate with *taille-vent* used for Barau's Petrel *Pterodroma baraui* in Réunion, then (Bory 1804) and now (Cheke 1982a), and for Wedge-tailed Shearwaters *Puffinus pacificus* in Mauritius in the mid-19th century (Clark 1859). '*Coupeur-d'eau*' was cited for an unidentified Mauritian seabird blown inland by cyclones by (Abercromby 1888).

154. Lesueur (c.1803, MS 15037); *Coturnix sinensis* is the only quail small enough to be compared to a sparrow. The first confirmed records of larger quails are from 1850s when Jungle Bush Quail *Perdicula asiatica* was present (pp. 124, 133). The Malagasy race of Common Quail *C. coturnix* is, in 1801, perhaps more likely than the Indian species; Milbert (1812), using Peron and Lesueur's notes, mentioned no quails. Lesueur's expedition drawings, some reproduced in monochrome by Bonnemains & Chapuis (1985), include species omitted from his notes. The endemic paradise flycatcher *Terpsiphone bourbonnensis* drawing has an identifying caption, but *Serinus canicollis* and '*Lonchura striata*' do not. These authors assigned the canary to the Cape (although common in Mauritius at the time), and the supposed White-rumped Munia to Mauritius, where it has never been recorded. In fact, seen in colour (kindly sent to ASC by Gabrielle Baglione at Le Havre), the excellent munia painting is of two species, *L. quinticolor* and *L. pallida* (identified from Restall 1996), both from the Lesser Sunda Islands in the East Indies.

155. Hawkins & Safford (in prep.).

156. Ly-Tio-Fane (1970, 1978).

157. Lesueur (1803). Given the location then deep in native forest, the snake was perhaps the endemic *Typhlops cariei*, known only as a subfossil (Hofstetter 1946c), but *T. cariei* was 'notably bigger' than *Rhamphotyphlops braminus* (Hoffstetter had only vertebrae, so did not estimate its length). The Flowerpot Snake only reaches 170mm (Daniel 2002), so the endemic species was probably at least 200+ mm in size. The 1803 specimen may not have reached Paris; it was not catalogued by Dumeril & Bibron (1834–54) or Dumeril & Dumeril (1851).

158. North-Coombes 1993.

159. Toussaint (1972).

160. For details of events in Haiti see e.g. Boulenger (1935).

161. Grant (1801), Barnwell & Toussaint (1949), Unienville (1982, 1989); early on much of this sugar was distilled into 'arack' (i.e. rum) (North-Coombes 1993).

162. Prentout (1901), Barnwell & Toussaint (1949), North-Coombes (1993).

163. Toussaint (1972), discussing Réunion; but Mauritius was also struck in December 1806 by a "hurricane doing great harm to crops", followed by two more in February 1807 (Padya 1976).

164. Barnwell & Toussaint (1949).

165. Grant (1801).

166. North-Coombes (1993: 18% cultivated), Brouard (1967: 25% cultivated, 50–75% "under forest cover, mostly dense native forest). Gleadow (1904) estimated as much as 71% forest cover remaining in 1804, and 65% as late as 1836.

167. Toussaint (1972).

168. Hermann (1898).

169. Lougnon (1956) and Barassin (1989, poorly) reproduced Champion's beautiful map.

170. Lougnon (1944–46).

171. Settlement information from Defos du Rau (1960); see also Toussaint (1972), Scherer (1980).

172. Lougnon (1970).

173. *Athalanthe* (1722), Lougnon (1956, 1970).

174. Lougnon (1956).

175. Hermann (1898) explained the frequent reversals of tortoise protection policy by suggesting there were two rival factions, one for conserving game and against colonising the southwest ('*parti du tortue*'), the other wanting to expand settlements and eat the wildlife. Desforges, probably still unpopular with older inhabitants for his robust assessment of their shortcomings in 1710 (Alby 1989), supported the new settlers; he became governor in 1723, but died in 1725. His successor Dumas tried playing to both sides, but was eventually pressured, too late, into a conservationist position.

176. Hermann (1898).

177. Hermann (1898), Bour (1981), Probst & Brial (2001).

178. Lougnon (1944–46).

179. Gandon (1732), Cossigny (1732–55), Caulier (1764).

180. Petit (1741, Bour 1978, 1981).

181. Hermann (1902, Bour 1981). Bour's paper is a detailed history of the Réunion Tortoise *Cylindraspis indica*, summarised also by Probst & Brial (2001).

182. Lanux (1753–6); his *poux* may have come from imported Rodrigues tortoises.

183. Bory (1804, Bour 1981); the tortoise was found (alive?) at the Mare d'Azule (southern slopes of the Volcan above Saint-Philippe).

184. For example, Le Gentil in 1717, Gaubil in 1721 (Lougnon 1970), Cossigny (1732–55) in 1732, d'Heguerty (1754) in the 1730s.

185. Gaubil (in Lougnon 1970), Charpentier (1732–55); d'Heguerty (1754) also mentioned "parrots of several species", without further detail; Cossigny discussed both islands together, so it is not certain that both parrots were still present on Réunion.

186. Foucherolle *et al.* (1714).

187. Lougnon (1956), paraphrasing the original list which he did not reproduce.

188. Le Gentil (1727 and in Lougnon 1970), Charpentier (1732–55).

189. Lougnon (1956).

190. Hermann (1898).

191. Heguerty (1754); *perdrix* were also mentioned (but not described) by Lanux (1753–6) in 1754.

192. Whistler (1949).

193. Cossigny (1732–55), Lougnon (1934–49, 1944–46).

194. Lecoq (1740) and Heguerty (1754); see Grant (1801:227) and Hermann (1898) for the 1744 locust plague.

195. Charpentier (1732–55). He added that "this is much repented of, for these extremely large animals, which really do obscure the sky, devour everything, even the bark of certain trees. Happily Providence has not yet given them a taste for coffee trees, for if they had it, the country would be ruined".

196. Lecoq (1740); also reported in official Company despatches (Lougnon 1956).

197. Lougnon (1934–49).

198. Letter from the *Conseil Supérieur of Réunion* to the company in Paris, 25 Mar.1741 (Lougnon 1934–49, vol.3: 181).

199. Heguerty (1754); he was military deputy to the Governor in Réunion in the 1730s. Stories of monkeys smashing mollusc shells with rocks are told in Mauritius: "They are also very expert at catching shrimps, and the freshwater cray fish which abound in many of our rivers, as well as the fluviatile nerites [freshwater snails], the shells of which I have been told they crack with a stone. I cannot assert this from personal observation; and from having met with heaps of shells on rocks where there were plenty of stones at a short distance, but none close to the shells, I am inclined to believe they crack them in their teeth, but it is not unlikely that they use both methods" (Clark 1859). Mauritian monkeys do indeed appear to use stones for this purpose (Jones 1993b); D'Heguerty may have borrowed a monkey story from Mauritius and applied it to Réunion rats!
200. Mahé (1740), Vaxelaire (1999).
201. "The are very fond of fowls which they slaughter and drag into their holes, for they bury themselves and dig like moles . . . Brother Barnard told me he had found a good sackful of grain in a hole of these rats" (Lecoq 1740).
202. Twigg (1975).
203. Lanux (1753–6).
204. Lecoq (1740): "in addition [to rats] there is a species of bird called *bengalis* which have multiplied greatly all over the island, but even more in the Sainte-Suzanne area, which eat rice and wheat when in ear; to the extent that it requires five or six negresses for each piece [of land], and even then they have difficulty in saving any [of the crop]. It happens that nothing is harvested despite all the effort. They go in flocks. They are hard to kill with rifles because they are so small. Some are taken with bird-lime, but they are so wily that they get round everything". Lougnon (1944–46) quoted a letter of 10 July 1741 from the Réunion administration to Mauritius asking for gunpowder so these birds could be shot. 'Bengali' was a catch-all name for weaver-finches in the 18th century (see the muddled discussion in Buffon 1770–83), though by the 1760s in the Mascarenes it became attached to the Red Avadavat *Amandava amandava*, before transferring, in Mauritius, to the Common Waxbill *Estrilda astrild* – from Africa, not Bengal (Cheke 1982a).
205. Brisson (1760), Buffon (1770–83).
206. "This island is however much infested by caterpillars, locusts and other insects; by rats and small birds, which make prodigious havoc among the crops" (anon. 1763). Grant (1801), citing as author an 'officer of the British navy', altered the passage, replacing 'caterpillars' by 'snails' and 'locusts' by 'grasshoppers'. The 'officer', possibly brigadier-general Richard Smith (Hill 1916), was imprisoned on the island during the Seven Years War.
207. Pingré (1763, Cheke 1987a) recorded that each landowner had to deliver annually to the authorities, per slave owned, 50lbs of grasshoppers, 100 rat tails and 100 heads of "small birds which damage grain crops especially wheat"
208. *Ordonnance* 201 of 28 April 1769 (Delaleu 1826) required planters to supply four rat tails per slave per month (48 per year), and four heads of *oiseaux de Malgache* per slave per year; the requirement for birds' heads was raised to 12 per slave per year in 1770 (Maillard 1863; rats still at 48), but reduced to one bird and four rats in 1774 (Wanquet 1974). Did this reduction reflect success in controlling the pests or a difficulty in achieving the larger numbers? Luckily Lanux (1753–56) referred in 1754 to "little *jacobins* called here *oyseaux de Malgache*" or we would have assumed they were Cardinal Fodies, the only Mascarene passerine granivore genuinely from Madagascar.
209. See discussion in Stresemann (1952), Long (1981) and Cheke (1983b, 1987a); by historical accident, Réunion is officially the 'type locality' for *Lonchura striata*. Unknown to the above authors, beyond the specimen seen by Brisson, the *jacobin* was mentioned in 1769 legislation, and Lanux in 1754 sent Réaumur a nest with four eggs (Lanux 1753–56, Torlais 1936).
210. Berlioz (1946), Cheke (1983b, 1987a); this bird, a uniform dull red ('*mordoré*' = 'bronze', but it is not so dark) without any black, has been considered a Réunion native fody (e.g. Hachisuka 1953, Moreau 1960), although it does not fit Dubois's (1674) description; Berlioz commented that colour sports of this kind are not rare in Cardinal Fodies; Forbes-Watson (1969) saw one in the Comoros, and ASC has seen one in Mauritius.
211. Buffon (1770–83) gave no source for his assertion, but had correspondents in both islands.
212. Delabarre (1844), Decary (1962): "the birds whose destruction is ordered are principally *calfats*, *cardinals*, *tarins*, *bengalis* and in general all those whose beak shape shows they eat grain" – i.e., respectively, Java Sparrows, Cardinal Fodies, Yellow-fronted Canaries and Red Avadavats (Coquerel 1864, Vinson 1868). Mynas, '*tourterelles*' (presumably Zebra Doves) and '*teque-teques*' (Réunion Stonechats) were specifically excluded from the order's remit. Renoyal (1811–38) is heavy going; Probst (2000a) summarised the animals recorded in his diaries. Mees (2006) described a new subspecies of Spice Finch *Lonchura punctulata insulicola* from 19th century Réunion specimens (and assumed Mauritian ones were the same), but could not pinpoint where in Asia they originated. Having examined Mauritian specimens, we feel it is more probable that the Mascarene populations are of multiple origin, and that the 'subspecies' is a result of introgression of different races of escaped cagebirds.
213. '*Tourterelles*'. (Delabarre 1844); Renoyal (1811–38, Probst 2000a) reported more specifically '*tourterelle de Sumatra*' in 1829, and '*tourterelles*' (in addition to '*pigeons ramiers*', i.e. *Nesoenas picturata*) were listed in the 1839 game law (Delabarre 1844).
214. Pingré (1763), Anon (1763), Ly-Tio-Fane (1968). Vinson (1867, 1868) thought the birds had been introduced to Réunion by Poivre, and cited the (Mauritian) law of 24 October 1767 on their protection for Réunion.
215. Wanquet (1974).
216. For legislation see Delaleu (1826), summarised by Maillard (1863); some wildlife-related legislation is also in Wanquet (1974).
217. Le Gentil (1717, Lougnon 1970), repeated in 'Brown' (1773).
218. Anon. (1763), somewhat amended in Grant (1801).
219. For example, Strickland (1948), Hachisuka (1953).
220. Brisson (1760), Buffon (1770–83); Brisson said "I do not know in which country it is found, I saw it alive in Paris". Buffon's specimen was illustrated by Martinet in the *Planches enluminées*.
221. Mauduyt (1784), Cheke (1987a).
222. Bory 1804; for Paris specimens see Newton & Newton (1876), Milne-Edwards & Oustalet (1893), Hachisuka (1953), Barré *et al.* (1996). The three specimens Levaillant saw were owned by Mauduyt, Aubry and the natural history museum – only the last survives.
223. Sir Ashton Lever had a menagerie and large private museum at Alkington Hall near Manchester, in England (Largen 1987), where John Latham (1781–85) saw a stuffed Mascarin Parrot in the early 1780s, describing it in his *General Synopsis of birds*. There is no record of when or how the bird was acquired; it may originally have been kept alive in the aviary. After Lever's death in 1788, the museum was catalogued (King & Lochée 1806), then sold by auction in 1806, this Mascarin (cat.No.5828: 'Parrot, a curious variety, America') being among some 400 specimens bought by Leopold van Fichtel for the Imperial Collection in Vienna (Pelzeln 1873, Largen 1987). Catalogue Nos.1050 and 5730 were parrots labelled '*Psittacus obscurus*', a name also used at the time for *P. mascarinus* – e.g. Buffon (1770–83). These may also have been Mascarins, but the synonymy of *obscurus* remains unresolved. As Latham had noted, the Vienna specimen of *P. mascarinus* is a partial albino (e.g. Sassi 1940, Forshaw & Cooper 1989). Fuller (2000) mistakenly thought the bird was *collected* in 1806, claiming its history was unknown!
224. Hahn (1834), details repeated in e.g. Hachisuka (1953),

Greenway (1967), Cheke (1987a), Forshaw & Cooper (1989), Barré *et al*. (1996), Fuller (2000); only the Newton brothers showed signs of caution: "Hahn's figure, published in 1834, was taken, *he says*, from a living bird then in the menagerie of the king of Bavaria . . ." (Newton & Newton 1876, our italics).

225. Anon (1826), kindly sent to JPH by Dr Hans Puchta of the Bayerisches Hauptstaatsarchiv, Munich (see also Hume 2007, Hume & Prys-Jones 2005).

226. Brisson (1760), Buffon (1770–85); what was probably the same specimen was also illustrated in a superior plate by Levaillant (1805; see Hume 2007). Brisson's Echo Parakeet had left Réunion before 1760, and there are no specific *in situ* reports after Feuilley in 1704, though Gaubil's "several species of parrot" in 1721 (Lougnon 1970) and d'Heguerty's (1754) similar remarks from *c*.1740 no doubt included them, as well as the grey parrot (note 227).

227. Cossigny (1732–55), writing in 1732 from Réunion on the fauna of both islands, mentioned both green and grey parakeets, but, for parrots, failed to indicate which island (or both) he was discussing.

228. Buffon (1776), Lanux (1772); see Lacroix (1936) and Couteau *et al*. (2000) for brief biographies of Lanux. The genus, *Nuxia*, of the Pink Pigeon's favourite food tree is named after him.

229. Lacroix (1936), Torlais (1936), Ly-Tio-Fane (1974), Couteau *et al*. (2000).

230. After Réaumur's death in 1757 his museum, left to the *Académie des Sciences*, was diverted by royal order to the *Cabinet du Roi*, under Buffon's direction. Buffon disliked Brisson for having usurped his (self-appointed) monopoly on encyclopaedic monographs, and denied him access to the royal collection – Brisson then abandoned ornithology for a successful career in physics (Torlais 1936, Farber 1982).

231. Ly-Tio-Fane (1974), Couteau *et al*. (2000).

232. The local names in Lanux's letters (Lanux 1753–56) must also have been on specimen labels used by Brisson (1760). Several (*tectec*, *tuituit*, *oiseau vert*, *merle*) have survived to this day (Cheke 1982a, Barré *et al*. 1996), whereas others (*petit simon*, *oiseau rouge*, *jacobin*) have vanished. *Petit simon* was the Mascarene Grey White-eye and *oiseau rouge* the Paradise Flycatcher. This latter sounds like a name for the Cardinal Fody, but Lanux described it: ". . . *oyseaux rouges*. The male is distinguished by the head, of which the feathers are a dark blue".

233. The *tuituit* specimen (*Coracina newtoni*) evidently did not last long, as Brisson (1760) did not describe it; the species did not surface again until the 1860s (p. 140).

234. For example, Dubois (1674).

235. De Querhoënt, although clearly based in Mauritius, also visited Réunion where he watched passerines and reported briefly on their habits to Buffon, amongst them the *oeil-blanc* (Réunion Olive White-eye, which Lanux called *oiseau vert*, still current) and the *petit-simon* (Mascarene Grey White-eye). Dubois (1674) had mentioned the Réunion Fody in the 17th century.

236. Bory (1801), Brial (2001).

237. Lanux (1753–56); these letters, with Réaumur's papers in the archives of the Paris *Académie des Sciences*, now the *Institute de France* (Torlais 1936, 1962; E. Couteau pers. comm.), have been transcribed by Evelyne Couteau, who kindly transmitted them to ASC. Le Gentil (1779–81, Grant 1801) published Lanux's letters on astronomical matters, and similar ones to Pingré, equally without wildlife interest, are in the Bibliothèque Sainte-Géneviève in Paris (Lacroix 1936; E. Couteau pers. comm.).

238. The bat account is Lanux (1772).

239. All the last section is from La Nux (1772).

240. Lougnon (1933–49), Cheke & Dahl (1981).

241. Bory (1804, Brial 2001); Tombe (1810) reported that "there is a considerable quantity of large bats which are eaten en civet [stewed] or roast with as much pleasure as [eating] hare, especially when they are fat" – he had seen the other species in Mauritius "as large as a big hen".

242. Cheke & Dahl (1981). In Mauritius *Pteropus subniger* died out and *P. niger* survived. As Mauritius lacks high-altitude forest, *P. subniger* may have been more vulnerable there (see Hume 2005 and p. 134).

243. Lanux (1753–56) was unable to culture silk-worms in the open as they were eaten by "ants, spiders, mice, rats, cats, dogs, lizards, wasps, hens, partridges, guinea-fowl, [and] other birds".

244. Defos du Rau (1960), Toussaint (1972).

245. Le Gentil (1779–81), Pingré (1763); bird-lime in the islands is made from the sticky sap of sapotaceous trees; Pingré was referring to both *Labourdonnaisia revoluta* and *Mimusops maxima*.

246. Historical data from Bory (1804), Defos du Rau (1960), Toussaint (1972), Wanquet (1974) and Sellhausen's maps (Guébourg 1991, Cadet *et al*. 2003, Map 6.2)

247. Defos du Rau (1960), Reydellet (1978).

248. Cornu (1974), Ly-Tio-Fane (1974). In 1768 Crémont and Bellecombe were the twin sub-governors of Réunion, under Poivre and Dumas in Mauritius. Jean-Baptiste Lislet-Geoffroy, son of a French engineer and his domestic slave, was a rare example in those times of an child of mixed blood being educated and accepted by the white father; he became a scientist, surveyor and chart-maker in Mauritius, training under Tromelin (Cornu 1974). We meet him again in Chapter 9.

249. Bory (1804), Jouanin & Gill (1967). Jossigny's drawings for Commerson include a Réunion Black Petrel, 60 years before it was formally described, though there is no indication of where on the island it was found.

250. According to an account in Grant (1801), Admiral Kempenfelt saw the deer in 1758, but P. J. Barnwell (*DMB*: 919, 1969) doubted he was in Réunion in 1758, pointing out that Grant's 'Kempenfelt' text was largely derived from Noble (1756), although with additions. Military governor Desroches noted in 1770 that "Desforges had also introduced deer which he had got sent over from the Ile de France [Mauritius], and bred them in his grounds at Le Gol, whence they have dispersed into the mountains and forests" (Malleret 1974). Desforges "developed and embellished the family property" during his spell as interim governor in Réunion 1757–59, creating "magnificent gardens" and a "menagerie with rare animals, deer and large areas of water full of fish" (Pineo 1999), though some of this may date from his retirement a decade later.

251. Caulier (1764), Ly-Tio-Fane (1970), Malleret (1974), Reydellet (1978), Pineo (1999).

252. *Ordonnances* 160 of 26 December 1767 and 220 of 11 March 1786 (Delaleu 1826).

253. Bory (1804), Brial (2001).

254. Vinson (1868), Bouton (1869).

255. Betting (1827).

256. The native buttonquail may have had a resurgence, and larger birds such as the Madagascar Partridge were regularly called 'quails' in the 18th-19th centuries, so we cannot put a name to the 1786 birds.

257. Germain (1921) retold a story current in the 1820s that *Achatina fulica* had been imported from Ile Sainte-Marie (Madagascar) to treat "une affection de poitrine" suffered by "Madame Mothey" the *intendant*'s wife. Augustin Motais de Narbonne was *ordonnateur* and acting commissioner-general in Réunion 1780–89 (*DMB*:29, 1941, A.Toussaint). Betting (1827) wrote of the frogs that "it is barely 40 years since they were brought from the Seychelles". DNA analysis however suggests they came directly from Madagascar (note 67).

258. Pingré (1760–62), Tombe (1810).

259. Bory (1804), Brial (2001).

260. See Ly-Tio-Fane (1970) and Grondin (1992) for Hubert and his

activities, and the clove trade; we have been unable to consult Emil Trouette's compilations of Hubert's notes and memoirs.

261. Neither Rose-apple nor guava are mentioned in Trouette's small monograph (1898) on plants introduced into Réunion! For guava and bats see Caulier (1764).

262. Milne-Edwards & Oustalet (1893) and Legendre (1929), citing Levaillant.

263. Probst (2000a) summarised faunal references in Renoyal's voluminous diaries, (Renoyal 1811–38, published 1990); by 1827, according to Renoyal, tenrecs had "much multiplied and become indigenous [*sic*]".

264. All the paragraph is from Bory (1804); Bory probably saw both Mascarene Swallows *Phedina borbonica* and Mascarene Swiftlets *Aerodramus francicus*. The noddies were at Le Brûlé du Baril in the south, where they still roost and possibly nest today (Barré *et al.* 1996, Probst 1997). For petrel harvesting at the Caverne de Cotte, see also Jouanin & Gill (1967), who confirmed Bory's story by finding Barau's Petrel bones in the same cave; the birds no longer breed in this relatively accessible area (Jouanin & Gill 1967, Bretagnolle & Attié 1991).

265. Dr Macé first arrived in the area in 1791 (Ly-Tio-Fane 1982), he explored Mauritius with Petit-Radel (1801) in 1793, but we do not know when he visited Réunion. The Colonial Assembly later considered him a spy for the revolutionary Directorate in Paris, deporting him from Mauritius in 1798 (Epinay 1890, Unienville 1989). His (undated) bat specimens presumably came back with him to Paris when he left the islands; *Scotophilus borbonicus* was described by Geoffroy Saint-Hilaire in 1803, but may be conspecific with the African *S. leucogaster/viridis* complex (Hill 1980). The only physical evidence of this bat on Réunion is one of Macé's cotypes, in the Leiden museum, though it apparently survived until the 1860s (p. 142).

266. Prentout (1901).

267. Chanvalon (1804).

268. '*Avalasse*' is a word in Réunion creole that has no exact equivalent in English. Albany (1974) defined it as follows: "During cyclones the smallest watercourses swell and become disproportionate torrents that overflow in veritable avalanches known as *avalasses*. The *avalasse* of 1807 lasted nearly 22 days – and ruined the country". In essence the entire surface of any sloping ground becomes a fast-flowing sheet of water, sweeping everything mobile before it (pers.obs. by ASC during cyclone *Gervaise* in Mauritius in 1975), its depth and force depending on the intensity of the rain. The intensity of the 1806–07 *avalasse* in Réunion appears to have been exceeded only by that recorded by the Dutch in Mauritius in 1695 (Leibbrandt 1896, Barnwell 1948), where a water depth of 7ft (2.2m) was recorded inland "fully two hours from the shore" and "cattle and stags were swept away into the sea"; all crops were lost, even sugar-cane.

269. A 'yellow' sea is now a regular result of cyclones, as nothing prevents topsoil being washed away in sugar-cane fields (pers.obs. by ASC of Mauritius from the air, a few days after cyclone *Gervaise* in February 1975).

270. The quotation from Pajot's *Simple renseignements sur l'Ile Bourbon* (published 1887, not seen by us) was reproduced in Prentout (1901).

271. Trees, particularly those (like coffee) originating outside hurricane zones, can be so shaken by cyclones that their damaged roots never recover and they slowly die (e.g. Brouard 1960).

272. Prentout (1901), Defos du Rau (1960); for cloves, Ly-Tio-Fane (1970).

273. Not only was the island full of tortoises, but Desforges thought that coffee would do well there (it does, there is a fine fruiting tree at Solitude (pers. obs.). The story of this expedition was first fully elucidated by Lougnon (1956), who identified Tafforet as the author of the previously anonymous *Relation de l'île Rodrigue*; see also North-Coombes (1971).

274. Tafforet (1726), Staub (1973).

275. "We have never sent to Rodrigues to take tortoises, even when our need was most pressing . . . we have here neither fishing nor hunting and we keep the cattle and fowls for the ships. It would appear reasonable for the company not to deprive us in an time of need of a saving resource when foreigners can procure abundantly and without moderation any time they feel inclined to go there" – Superior Council of Réunion to the Company, 12 Oct.1733 (Lougnon 1933–49, 2:131; also paraphrased in North-Coombes 1994). In fact the authorities in Mauritius had, in 1731, ordered La Fontaine and the now experienced Tafforet to divert via Rodrigues and collect "as many tortoises as they can to populate the islets adjacent to our island to ensure [the availability of] victuals for the Company's ships to supplement hunting which has become so sterile . . ." (Lougnon 1944–46, p. 205).

276. Gandon (1732), North-Coombes (1994).

277. "The Company feels the need to warn you that it has this year forbidden [the inhabitants of] Mauritius, under any pretext whatsoever, from sending to Rodrigues for tortoises, and to keep for the island's needs to those which are easily found on Round Island or Flat Island, although they are less good and smaller than those one goes to fetch at Rodrigues Island, which the Company has judged should be preserved for vessels returning from the Indies who have orders to put in there. This ban, which applies to you as much as to Mauritius, will apply until the Company can procure a stopover for its ships on this side of the Cape of Good Hope"– The Company to the Superior Council of Réunion, 17 September 1732 (Lougnon 1933–49, 2: 131; see also North-Coombes 1994).

278. Aldabra was unknown to them, and the Seychelles unexplored – both were well off the then current routes to and from India (see e.g. Pineo 1993).

279. The Company's instructions were given in December 1734; see Labourdonnais (1740; footnote in Lougnon's edition of 1937, p. 120), North-Coombes (1994). Reading between the lines suggests that Labourdonnais took the job on condition that he could take Rodrigues tortoises.

280. North-Coombes (1994), whose research replaces the previous belief that the outpost was set up in 1736.

281. These big ones, needless to say, 'got away'!

282. Gennes (1735); the two ships were the *Philibert* and the *Duchesse*.

283. The rare occasions when the planet Venus passes across the sun's disc, known to astronomers as the 'transit of Venus', were of great significance in 18th century astronomy, as they enabled the earth's distance from the sun to be computed. The 1761 transit was the first occasion that this exciting opportunity had arisen since Edmund Halley had worked out the maths. The calculation depends on several observers timing the transit from different places on the earth's surface (Woolf 1959, Sellers 2001). Rival astronomers Pingré and Le Gentil came to the Indian Ocean to record the 1761 event. Both failed due to bad weather and other circumstances, Le Gentil staying on until 1769 to record the next transit (North-Coombes 1971, Serviable & Alby 1993, Cheke 2004b). They both left useful accounts on Mascarene natural history (Pingré 1760–62, 1763, Le Gentil 1779–81). The next transit, in 1874, led to a British expedition to Rodrigues, the first to properly survey the island's geology, botany and zoology (pp. 151–2).

284. Noble (1756): "In the harbour are two rocks or small islands on which are built two stone windmills; there is also a small basin or saltwater pond formed by nature, wherein they keep the sea tortoise they receive from the island of Rodrigue or Diego Raijs, about 90 leagues to the Eastward of Mauritius, for the use of the Governor's table, the hospital etc. An iron rail goes across the mouth of this little creek or basin, whereby the water is not hindered from ebbing and flowing, and the turtle hindered from making their escape. This Fishery is thought so useful at Mauritius, that they have always a Serjeant's party, on that little island of Rodrigue, who collect all the

fish they can, for the boats that are sent to bring them, at certain times, and the ships, that generally touch there, in their way to Mauritius. There is also a particular spot of ground, inclosed here, for keeping and breeding [sic] the Land Tortoise, for the same purposes."

285. The route took them east in the roaring forties to 60–70° (even to 80–90°) E, whereupon they would turn northwest and sail back on themselves in the Southeast Trades to reach Rodrigues, and thence the other Mascarenes (d'Heguerty 1754, Dupon 1969, Pineo 1993).

286. Heguerty (1754), writing of his time in Réunion 1735–46 (Labourdonnais 1740 and Lougnon 1944–46, *DBR* 3:99–100, 1998). D'Heguerty probably saw the captive bird(s) in Réunion, though he no doubt visited Mauritius and possibly saw them there; he does not seem to have been to Rodrigues himself.

287. Billiard (1827) gave no indication where he got the story from, and there is no mention of solitaires of any kind in Labourdonnais's writings (Labourdonnais 1740, 1827), biographies (Herpin 1905, Crépin 1922b), or in the Company's correspondence of his time (Lougnon 1933–49).

288. Jonchée (1729).

289. Charpentier (1732–55, writing on 24 March 1755).

290. Pingré (1763). Pingré's fascinating MSS in the Bibiothèque Sainte-Geneviève remain only partly published, the full version of MS 1804 (an edited and re-arranged account of his travels) being first published only in 2004. MS 1803, still unfortunately unpublished, is an amusing day-to-day diary of his voyage (pers. obs.).

291. Pingré gave an extended list of native trees, identified by Strahm (1993); it included three kinds of *palmiste* recognised by the locals. Only two are known today, the *palmiste poison Hyophorbe verschaffeltii* and the Hurricane Palm *Dictyosperma alba* – was there originally a third species (an *Acanthophoenix?*) that became extinct before the island's flora was first studied in 1874? Pingré did not suggest a reason, but poor regeneration of palms was probably due to rat predation on seeds, as noted for latans in Rodrigues in the 19th century (Chapter 7) and on Gunners Quoin (Mauritius) in the 20th (Chapter 9).

292. Pingré recorded mango, banana, papaya, custard-apple and citrus (oranges and lemons) – only the latter self-propagate in forests in the Mascarenes, normally not pathologically, though citrus became invasive in Rodrigues in the 19th century.

293. Cheke (1987a).

294. Tafforet (1726).

295. Buffon (1770–83), Woolf (1959), Cheke (2004b).

296. The Black-spined Flying-fox *Pteropus niger*, probably the species described by Leguat, was not reported by Tafforet or Pingré, but is unlikely to have been distinguished from the Golden Bat *P. rodricensis* which Pingré described in some detail. There is no information on when it disappeared, though it is unlikely to have been affected by cats. For the enigmatic red-winged parrots reported by Tafforet see Chapter 3 and Box 20 (p. 181).

297. As noted earlier, rats may have driven doves, parrots and the starling to nest on offshore islets, but the first two still frequented the mainland to feed, where cats could have caught them.

298. Marragon (1795).

299. Rodrigues has no inaccessible terrain and or impenetrable forests that appear to have allowed some tortoises in Réunion to escape predators.

300. Pingré (1763); the fire had been set by a slave 'either in malice or by negligence', prefiguring similar activity during Marragon's time (p. 115).

301. Details for this paragraph from Pingré (1760–62, 1763), North-Coombes (1971, 1993) and Cheke (2004b). The British warships were scouting out the island as an assembly point prior to an intended invasion of Mauritius. The islanders were totally outgunned, and rapidly surrendered. To ensure no news of their visit leaked out, the British even smashed the tiny local fishing boats, though destroying the rice on the burnt ship may have been accidental. A fortnight later, two more British ships stole 3,000 tortoises assembled in the pound awaiting shipment to Mauritius (Puvigné in Dupon 1969, North-Coombes 1971; his 1994 account is confused). Pingré escaped when a ship arrived from Mauritius on 6 September, leaving on the 8th; just in time, as four more British ships arrived on the 14th, building up to 13 in early November (North-Coombes 1971).

302. Nichelson (1780).

303. Pingré (1763), Serviable & Alby (1993).

304. An average of around 3,500 tortoises per shipment were taken off the island during 1735–65, often two or three times per year (North-Coombes 1993).

305. Bourn & Coe (1978), Coe et al. (1979). The higher average rainfall on Rodrigues (1,200mm/yr) than on Aldabra (946mm/yr, Coe et al. 1979), might allow a higher tortoise density, but on both islands it fluctuates enormously, and the limit will be set by drought years rather than the average. The presence of two species on Rodrigues, a large browser (*Cylindraspis vosmaeri*) and a small grazer (*C. peltastes*), might also have increased the overall carrying capacity.

306. Even the full 280,000 animals (equivalent to 26 per ha) is within the possible numbers given the higher rainfall and presence of two ecologically separated species.

307. Quoted in North-Coombes (1993).

308. North-Coombes (1971, 1994); several specimens of Rodrigues tortoises reached France (Austin & Arnold 2001), including two hatchlings (Bour 2005). From 1773 to *c*.1810 regular cargoes of tortoises from the new Seychelles settlement replaced those from Rodrigues, at least 10,000 accounted for in official records (Chambers 2004); Toussaint (1967) listed incoming cargoes from the Seychelles between 1776–1809; tortoises no longer feature in the manifests after 1801. After the Seychelles supply dwindled, Aldabra became the sole source (Mondini 1990, Chambers 2004).

309. Après (1775), Dupon (1969).

310. Morel (1778). The tortoises were *supposed* to go to the hospital, but as Pingré wryly pointed out: "The tortoise is an excellent remedy against sickness acquired at sea, but by a quite opposite effect, the appetite for this meat sometimes makes ill those whose health would seem susceptible of no change. On the arrival of a corvette laden with tortoises, the influential of Mauritius are suddenly struck with sea-sickness; they remove three-quarters of the cargo, [only] the remainder going to the hospital" (Pingré 1763).

311. North-Coombes (1971, 1994); see note 313.

312. The drawings were reproduced in Oustalet (1897, parakeet) and, for the tortoise, Vaillant (1898) and North-Coombes (1993). Jossigny's itineraries are in an MS document published in Ly-Tio-Fane (1978); Commerson's intention to visit Rodrigues was expressed in a letter to his brother in 1768 (La Lande 1775).

313. A description of exploring two caves (and finding a tortoise) in Valgny's time (1786–91), transcribed by Etienne Rochetaing (in Rodrigues 1802–05), is attributed to one Captain de La Bistour (or 'Labistour'), visiting the island in 1786 (North-Coombes 1991, 1994; *contra* Strickland 1848, North-Coombes 1971 and Middleton 1995: '1789'). Although the account mentions neither date nor bones, the captain's son-in-law, M Roquefeuil, told Julien Desjardins (to whom he gave the bones in 1830) that they had been found in 1786 (Desjardins 1831b and in *Proc. Comm. Sci. Corresp. Zool. Soc. Lond.* 1:111, 1831, Strickland 1848).

314. Marragon (1795). The younger Cossigny (Charpentier 1799) considered exploitation more responsible than predators, but there was little harvesting after the mid-1770s. However, the population did not recover, which suggests that, while exploitation had slashed the numbers, predators prevented any recovery.

315. Marragon (1802, Berthelot 2002; mis-cited as '1803' in Cheke

1987a); in letters to Mauritius until 1809 Marragon referred occasionally to *tortues* (Berthelot 2002), but the context always suggests marine turtles.

316. Although there is no evidence of tortoise survival beyond 1802 (Stoddart & Peake 1979, Cheke 1987a), early 19th century publications often recycled mid-18th century observations of tortoises and dugongs (Dupon 1969, Cheke 1987a), and as late as the 1840s tortoises were still thought by Mauritians to exist, though rare: d'Unienville (1830) paraphrased Marragon's 1795 report apparently without having checked the current situation, and Froberville (1848) was equivocal but implied some tortoises still survived.

317. Letters from Marragon in Berthelot (2002).

318. This factor was specifically commented on later in the 19th century (North-Coombes 1971); such accumulations of dead latan leaves can be seen today on Round Island, Mauritius (pers. obs.).

319. The public position was that evacuation was to avoid helping the British, but this was strategic nonsense; the real reason seems to have been that Decaen wanted to banish all Mauritian lepers to a safe distance; in the end they were sent even further, to Diego Garcia! (North-Coombes 1971).

320. Cheke (1987a).

321. North-Coombes (1971).

322. *Scolopendra morsitans* and *S. subspinipes* (Lewis 2002); oddly, the latter does not occur in Mauritius – it is not known whether it did so in the past and has been wiped out by House Shrews (as recently in Rodrigues, pp. 200, 237–238), or whether it never occurred there. There are also (mostly smaller) endemic species on the different islands.

323. Telfair (1809).

CHAPTER 7. A CENTURY OF SUGAR

1. Billiard (1822) on arrival in March 1817.
2. Cited by Gleadow (1904).
3. Historical details from Toussaint (1972).
4. Michel (1935), Ly-Tio-Fane (1972).
5. Maure (1840), referring to the unconceded part of the reserved forests in the (then) Pamplemousses district, marked 'Grandes Réserves' on Lislet Geoffroy's 1807 map (Toussaint 1953). In 1745 the reserves stretched north and west from Pamplemousses village, but about half were conceded subsequently (*DTIM*: 57, 1998) before Poivre called a halt around 1770 (Toussaint & Adolphe 1956; map in Lenoir 1979: 74, dating from 1772); the other half remained uncut in 1807. The British invading force in 1810 passed through "a wood, impenetrable on both sides" for three or four miles on the road from Grand Baie to Pamplemousses (John Blakiston, in Barnwell 1948).
6. King (1827).
7. This paragraph largely from Brouard (1967), Toussaint (1972), North-Coombes (1993), Storey (1997).
8. Stirling (1833).
9. No formal land surveys were done, but the boundary between cleared land and forest on Fraser's map of 1835 closely follows the 200m contour on a modern map (Cheke 1987a). Fraser's excellent map (Grove 1995: 207, much reduced; Map 7.1, p. 123) was used by Gleadow (1904) and Vaughan & Wiehe (1937) in calculating the progressive decline of forest cover.
10. Brouard (1963), North-Coombes (1993). Port Louis burnt down in a great fire in 1816; reconstruction was supposed to be in stone, but a great deal of wood must also have been used (Brouard 1963, Toussaint 1966).
11. Unienville (1838), Brouard (1963).
12. Storey (1997).
13. French scientific expeditions continued to visit – Jean-Marie-Constant Quoy and Paul Gaimard collected important specimens and made observations on seabirds when in Mauritius in 1818 on the *Uranie*, but the expedition reports say little on the state of the island (Quoy & Gaimard 1824). The expedition was led by Louis Freycinet (1824–44), a veteran of the Baudin expedition of 1801–4, and co-editor of its reports (Chapter 6). In 1824 René-Primevère Lesson and Prosper Garnot collected for Duperrey's expedition on the *Coquille* (Lesson 1830). Although Lesson claimed to have "traversed the country in detail", they seem only to have explored around Port Louis (trips to the Pouce and the Petite Rivière caves are recounted). The only native animals encountered were Moorhens and day-geckos; their lists and collections being a catalogue of common introduced species (mammals, birds, the frog). They were the first to record the Common Waxbill (as '*bengali rayé ou marteau*').

14. Desjardins (1837), Oustalet (1897), Ly-Tio-Fane (1972); the first 5 manuscript reports were abstracted by the Zoological Society of London in their *Proceedings: Proc. Comm. Sci. Corresp. Zool. Soc.* 1: 45–46 (1831), 2: 111–112 (1832), *Proc. Zool. Soc* 1: 117–118 (1933), 3: 204–208 (1835). ASC tried unsuccessfully to trace Desjardins's manuscripts in the 1970s; they had been acquired by Prof. Alphonse Milne-Edwards in Paris, but were apparently dispersed with his library and papers after he died in 1901 (Cheke & Dahl 1981). Desjardins's memoir on the island's mammals, read to the natural history society in 1829 or 1830 (Ly-Tio-Fane 1972), might add to our meagre knowledge of the Rougette.

15. The owl specimen subsequently vanished. According to Edward Newton (1861–82), writing in the 1870s, it was destroyed in 1870 when a cyclone waterlogged the museum through damaging the roof (Hume & Prys-Jones 2005). However, there was no cyclone in 1870, though a very severe storm in 1868 caused much structural damage in Port Louis (Pike 1873, Walter 1914, Padya 1976); however, there is no mention of any damage to the museum in the RSAS proceedings around that date or in the annual report for that year (*TRSAS* NS 3 (2):102 ff, Bouton 1869b). It is odd, if the specimen really had survived until 1868/1870, that Newton neither saw it, nor mentioned it in discourses on vanishing Mauritian birds (E. Newton 1875, 1878a, 1888). Equally Bouton (1869b) said nothing about it when discussing the owl's extinction. Possibly an earlier cyclone destroyed the specimen, and Newton subsequently decided to quietly ignore the inaccurate hearsay information. There were no very severe storms between 1836 (when the owl was collected) and 1861, so perhaps the 1861 storm was responsible, before Newton (arrived 1859) had explored the museum. However, we have found no reports confirming cyclone damage to the Royal College while the museum was there (1842–84; Cheke 2003), although the building was later destroyed by the notorious storm of 1892 (Macmillan 1914). We suspect there was no cyclone, the specimen's disappearance simply ascribed to the standard local explanation for any disaster!

16. Desjardins (1837), Milne-Edwards & Oustalet (1893), Oustalet (1897), Cheke (1987a). Desjardins (1837) was given four live birds by a Mr Sauzier from Réunion, two escaping after a few months (Milne-Edwards & Oustalet 1893, Legendre 1929). The bird shot later is usually cited as collected in 1837, but it was presented to the society already stuffed on 5 January (Milne-Edwards & Oustalet 1893, Legendre 1929), so had no doubt been killed in 1836. Desjardins presented two specimens in alcohol to the Paris museum (Milne-Edwards & Oustalet 1893), probably the two that did not escape; the shot bird is presumably the still-surviving specimen in the Mauritius Institute (Cheke & Jones 1987, Cheke 2003a). One alcohol specimen was sent in exchange to the American Museum of Natural History, New York in 1933 (Berlioz 1933), and the other later to Harvard (Hachisuka 1953, Berger 1957). The Desjardins alcohol specimens are generally cited as having been collected in 1839, but that was the year he went to Paris (*DMB*:203, 1942, S. Pelte), so is probably the accession date to the Paris collections. Hoopoe Starlings were popular cagebirds in Réunion and commonly taken home by visiting Mauritians. Bouton (1878) recorded

two carried by a passenger in 1817, and Renoyal (1811–38, Probst 2000a) reported four sold for sending to Mauritius in 1822. Whether any escaped or were released before Desjardins's birds is not known, though Gustave Autard de Bragard, who shot the '1837' specimen, claimed to have seen 'considerable flocks' in Mauritius (Desjardins MS, in Milne-Edwards & Oustalet 1893, Legendre 1929). Milne-Edwards & Oustalet thought this represented a successful introduction, but Legendre pointed out that a pair could not generate such numbers in less than two years. Perhaps earlier imports escaped and bred, but as there were no wild *huppes* in Mauritius in 1826 (Desjardins in Milne-Edwards & Oustalet 1893, Legendre 1929), any feral population lasted less than a decade.

17. Desjardins (1831a) reported Bojer's Skinks *Gongylomorphus bojerii* as "in very great quantity at Flacq and in the vicinity of the town of Port Louis". Strangely, assiduous collectors such as Peron & Lesueur (in 1801 and 1803), Quoy & Gaimard (in 1818) and Lesson (1830; visit in 1824), failed to find it (see Dumeril & Bibron 1834–54, Dumeril & Dumeril 1851) for specimens; also Lesueur 1803). See Desjardins (1830, 1834), for mammals and quail respectively.

18. Laplace (1841–54, vol. 2).

19. Julien Desjardins, son of a corsair (government-sponsored pirate), was born in Mauritius in 1799, and trained in natural history and science at the Paris natural history museum and botanic gardens with his contemporary Louis Bouton. These two, with Bohemian émigré botanist Wenceslas Bojer, formed the active core of the *Société d'Histoire Naturelle de l'île Maurice* in its early years. Bouton succeeded Desjardins as secretary and held the post until his death in 1879 (Staub 1993b).

20. The official, but disingenuous, reason for cancelling the science post was lack of funds (Vaughan 1958). Desjardins (1832) acidly commented that "the Natural History class which had been established at the Royal College, and which I regret to say has been so suddenly withdrawn and for such weakly based reasons, has given Mr Charles Telfair the opportunity to offer certain encouragements to those young people of our island who attended the lessons of our two colleagues W. Bojer and H. Faraguet" (by publicly promoting their private tuition). Telfair himself, in a letter to the Zoological Society in London on 8 Nov. 1832, wrote "our young men have been disheartened in their pursuits of science by the presence of misery. I told you that a Cath. Bishop had upset our chair of Natural History, and having done all the harm he could, stole away furtively from the island and has not since been heard of. We have now no professorship of science in our college, neither Natural Philosophy [= physics], nat. History and Botany nor Chemistry – all of which I had introduced into the courses of Education in the college of Port Louis of which I am Vice-President". He goes on to ask his unidentified correspondent in London to use influence to get the decision reversed (MS in the Zoological Society's archives). The 'Cath. Bishop' in question must have been Edward Bede Slater who was based in Port Louis. He apparently made himself unpopular with all sections of society, including his own congregation, and was sacked by the Vatican in December 1831; he left in June 1832, only to die at sea three days later (*DMB*: 784–5, 1964, J. Mamet). For biographies of Bojer, Telfair and Desjardins see Michel (1935), Sornay & Koenig (1941), Vaughan (1958), Staub (1993) and *DMB* 741–42, 532–33, 203–04.

21. Telfair (1831): "In Mauritius they [*Tenrec ecaudatus*] sleep through the greater part of the winter, from April to November, and are only to be found when the summer heat is felt". Sganzin (1840), in Réunion 1831–32, also reported hibernation in the Mascarenes soon afterwards. Only four of Telfair's letters survived a regrettable archival cull in the 1950s, and only one from Desjardins, who also regularly wrote to the society (ASC pers.obs. in the Zoological Society's library, January 2005; the librarian said 'unimportant' letters had been destroyed); see Wheeler (1997) on the fate of the Society's collections.

22. Strickland (1831–34), Sganzin (1840).

23. Darwin (1845); "The scenery cannot boast the charms of Tahiti and still less of the grand luxuriance of Brazil; but yet it is a complete and very beautiful picture. But there is no country which now has any attractions for us, without it is seen right astern, and the more distant and indistinct the better. We are all utterly home sick" (letter to his sister Caroline from Port Louis, 29 April 1836, reprinted in Burkhardt 1996). See also Barlow (1945) and Barnwell (1948), who wrongly said Darwin's visit was in 1835.

24. Holman in Barnwell (1948). Java Sparrows were probably declining by this time (and were rare by 1859, Clark 1859), but lovebirds were still common, so it is odd that it was worthwhile importing them.

25. Freeman (1851). He was a Church of England missionary, seeking out Malagasies because many had been converted to Anglicanism in Madagascar. However, these early inhabitants of Curepipe are unlikely themselves to have come from Madagascar, but were probably descendants of Malagasy slaves who had retained some ancestral culture.

26. 34,525 Indians were imported in 1843 alone, more than the total in the ten previous years (North-Coombes 1993).

27. The now relatively up-market settlement of Phoenix started life as an ex-slave village, as did the coastal town of Grande Gaube in the north (Barnwell & Toussaint 1949).

28. North-Coombes (1993).

29. Brouard (1963), Toussaint (1966). The road network was expanded over the next 15 years by Surveyor-general John Augustus Lloyd, explorer of Round Island, best known for a cameo appearance in the *Voyage of the Beagle* (Darwin 1845) as Darwin's host in 1836, impressing the traveller by owning the island's only elephant (see also *DMB*:143–44, 1942, P. J.Barnwell).

30. Barnwell & Toussaint (1949), Cheke (1987a). Data presented by North-Coombes (1993) suggests a peak in sugar production was reached in 1855, an increase of 49% over 1854, suggesting a short burst of clearance at a rate not seen since 1843–46. There is no forest data to confirm this, and some of it could result from fallow land being re-used. The standstill in total sugar acreage during the rest of the century did not preclude forest clearance for cane; some lands were abandoned, and others, in higher areas, opened up (Barnwell & Toussaint 1949).

31. Figure 7.1 and its notes are unchanged from Cheke (1987a); if the exact date of the main forest destruction was a few years later, it does not affect the overall pattern, furthermore 'forest' was defined in different ways – for some it was any land with trees on it, for others only primeval forest was included. The best reference on forest decline is Gleadow (1904), principal source for Vaughan & Wiehe's maps (1937); see also Thompson (1880), Walter (1908), Koenig (1914a) and Brouard (1963).

32. Brouard (1963).

33. *TRSAS* NS 2 (1):178–196, *séance* of 25 Nov. 1857; Guinea Grass, locally *fataque*, is *Panicum maximum*, an ubiquitous introduced tall forage grass (Baker 1877).

34. Mann (1860), Brouard (1963); Grove (1994) reviewed the 'deforestation-desiccation discourse' worldwide. Increasing demand and the need for reliable supplies eventually led to the damming of the outlet streams of the Mare aux Vacoas, creating a reservoir opened in the late 1880s and enlarged in 1895 (Pitot 1914a, *MA* 1895:303–04).

35. Thompson (1880), Vaughan & Wiehe (1937, 1941); Thompson wanted to preserve the best parts of Concession Dayot in the Kanaka Block, but this area was cut over in the early 1900s (Koenig 1914). Storey (1997) discussed on rival cultural viewpoints in 19th century Mauritius in relation to the understanding and tackling of problems in sugar agronomy; similar issues applied to forest conservation. In the 1970s analogous cultural differences strongly influenced perceptions of the role of deer in native forest (p. 161).

36. Bouton (1860b), Brouard (1963).

37. See Ly-Tio-Fane (1979) for the history of the Baker's *Flora* (1877). Paintings of native flowers by Malcy Moon were important to Baker's descriptions, as he acknowledged; he never visited Mauritius, compiling the work entirely at Kew.
38. Bouton (1846).
39. Anon (1963); Wise was told that the trees had been killed by the last cyclone, but that was the stock local explanation for dieback in the 19th century. Wise's account is cited from a brief published abstract of his visit; the original MS, formerly in the library of the Royal Commonwealth Society, was found to be missing when the collection was catalogued in 2003 after being moved to Cambridge in 1993–94 (Rachel Rowe pers. comm.), though there is a typed transcript in the Carnegie Library in Curepipe, Mauritius (Guy Rouillard pers. comm.).
40. In the mid-1860s Charles Boyle (1867) described Bois Sec in similar terms: "A wide flat surface of many acres thickly dotted with the tall gaunt ghastly utterly and entirely denuded stems of hundreds and thousands of dead forest trees."
41. Gordon (1894).
42. Bishop Ryan (1864) crossed the island in June 1855 only a few days after his arrival: "Fine specimens of the fern-tree abounded on the elevated plain in the centre of the island. And here one feature of the scenery . . . struck us very forcibly, viz. the large number of bare trees, which looked as if blasted by lightning or stripped and shattered by whirlwinds, but which are really eaten by white ants. The large nests of the ants near the tops of some of them had a very strange appearance." The termites will in fact have taken over the trees after they had died.
43. Clark (1859), Pike (1873), E. Newton (1858–95, MS journals). 'Doody' was, according to Newton's notes, near Fressanges (now Bananes). No such place is marked on maps, but Pike also visited the '*hangar*' (shooting lodge) at 'Dhoodie' and described it as 45 minutes' walk from the Fressanges *hangar*, and "on a little bend of the Rivière du Bois". This stream runs under the north slopes of the western end of the Lagrave range, joining the Grande Rivière SE just below the Diamamou(ve) Falls. There is also to the northwest a short river Doudy marked on the 1:25,000 map feeding into the Grande Rivière SE at La Pipe; it would appear that much of the area now known as Midlands was called Doody or Dhoodie in the 1860s.
44. Brouard (1967).
45. Padya (1976); the cyclone of February 1818 may have been as bad as *Carol*, but was too early to have been connected with the death of trees in the 1840s. A fairly strong storm in 1848, after Bouton's observations, could have exacerbated the situation by killing already weakened trees.
46. Paul Koenig (1932), Conservator of Forests in the early 1900s, supported the boggy ground/stagnant water theory. He added that Frank Gleadow (1904), forest consultant from India, had recommended the exploitation of these patches of dead trees. They were then, as Koenig noted, over 60 years old, showing that even when dead, Mauritian trees withstand all the weather can throw at them. Many dead trunks resulting from cyclone *Carol* in 1960 are likewise still standing over 40 years later (pers. obs.).
47. *List of severe droughts*, MA 1913: C58.
48. Koenig (1895).
49. Gleadow (1904).
50. Chris Lyal (Natural History Museum), pers. comm.
51. For example, Owadally (1980).
52. Vinson only listed six scolytid beetles for Mauritius in 1935, two of which were obvious introductions, but the total had reached 18 by 1967 (Vinson 1935, 1967). In Réunion, Gomy (1973) noted that while only two species were known in 1967, he had since found 18! Emphasising that these were "a potential threat" to the forest.
53. Gleadow's picture clearly shows a curculionid, but was it the culprit or simply another inhabitant of the dying tree?
54. Wilkinson (1978), Harde (1994), Chinery (1993); beetles rarely kill directly, but introduce a fungus that kills the tree.
55. Thompson (1880), Koenig (1895), d'Emmerez (1901); 'dry rot' is nowadays used to describe a fungus (*Merulius lachrymans*) that attacks dead wood, but Thompson apparently used the term generically for fungi attacking living trees. D'Emmerez never followed up the beetle Gleadow left him to identify, no doubt because he favoured the wind theory of ultimate causation, and had dismissed insects as a cause of the epidemic (d'Emmerez 1901).
56. It is possible that in the early 20th century a predator or parasite was (accidentally?) introduced, which happened to control the bark beetle. Greathead (1971) included no likely candidates in his review of biological control, but a parasite could easily arrive unseen. Carl Jones (pers. comm.) has suggested the Red-whiskered Bulbul, introduced in 1892, might have controlled the beetle, as it affected other arthropods. Although it does eat beetles (Carié 1916), it is primarily frugivorous and does not feed by searching bark (pers. obs.).
57. Wilkinson (1978) told of a storm in Colorado, USA, that blew down huge numbers of spruce (*Picea*) trees that, once dead, are host to a beetle that lives under their bark. Normally in North America, woodpeckers keep bark beetle numbers under control, but the woodpeckers ignored the fallen trees, so beetle numbers exploded to the extent they attacked healthy trees and infected them with a fungus that killed millions over six years. Gleadow (1904) recommended introducing woodpeckers from India to Mauritius to control bark beetles.
58. Troup (1940: 279).
59. Clark (1859), Boyle (1867), Pike (1873).
60. Bouton (1860), Brouard (1963).
61. Bouton (1871).
62. Ellis (1859); this area hosted the first industrial plantations of *Furcraea* in the 1820s; the plant was introduced in the mid-18th century (Rouillard & Guého 1999).
63. Baker (1877).
64. Mann (1860) estimated that 3,200 acres (1,295ha, *c*.0.7% of island area) of forest were consumed annually for domestic fuel alone by the then population of Mauritius. Much of this was from fallow land under regenerating exotic scrub-forest, but also, as Brouard (1963) put it: "during the 1860s the indiscriminate exploitation of private [native] forests continued without any appreciable increase in the areas under cane and other agricultural crops". Edward Newton (1858–95) often noted timber felling by landowners around Grand Bassin and Tamarin Falls in the early-mid 1860s.
65. Barnwell & Toussaint (1949), Toussaint (1972). There had been earlier serious epidemics – smallpox in 1792, rabies in 1813 (most dogs were killed to control it), cholera in 1819 and four times from 1850 to 1861, but none had major ecological effects on the mainland, though cholera precautions included building a quarantine station which devastated Flat Island's biota (Chapter 9). Pike (1873) wrote a vivid eye-witness account of the 1865–68 malaria epidemic. Before the migration to high ground, the only house in Curepipe was a wayside inn for travellers to and from the south (Mouat 1851, Clark 1859).
66. *DMB*:628 (1947, P. J. Barnwell). While 'Midlands' currently refers to a huge area north of Montagne Lagrave (Currie's *chasse* lands, then 'Doody'), the area converted to sugar was around the (then) railway stations of Midlands and Fressanges (now Bananes), west and south of the Lagrave range (information from maps of various dates).
67. Walter (1914b).
68. The history of the Desjardins Museum, absorbed in 1885 into the Mauritius Institute, was covered *inter alia* by Koenig (1939), Tirvengadum (1980) and Cheke (2003).
69. E. Newton (1861–82), letter to [E.] Blyth.
70. Vinson (1968).

71. Clark (1959).
72. Probably *Acacia farnesiana* (*Flore* 80, 1990).
73. *Jamblong* is *Syzygium cumini*, a rather inferior fruit-tree, naturalised in drier forests of Mauritius and Réunion (*Flore* 80, 1990).
74. The 'goldfish' is the introduced oriental carp *Carassius auratus*, known locally as *dame Céré* after the the 18th century botanist's wife. It has disappeared from many waters since the tilapia *Tilapia niloticus* was introduced around 1950 (Staub 1993a).
75. Clark (1859); he mistakenly thought the owl was probably introduced, and also that it would have fed largely on rats. Bernard Dalais lived before 1860 by Montagne Créole at Cent Gaulettes, which is probably where he shot the owls; he was 73 in 1859, dying in 1869 (*DMB*:1680–1682; R. d'Unienville, 1999).
76. Apparently then unaware of the extinct endemic owl (and Desjardins's unique specimen), Newton (1858–95) noted in September 1861 that "the Governor (Stevenson) told me a day or two ago that some night-bird he thought an owl settled most nights on a tree to the NE of Réduit and close to his bedroom window and uttered a very peculiar note – I wonder what he can be thinking of?" At that time Tropical Shearwaters still nested in ravines surrounding the Réduit, and they certainly make a 'most peculiar' call; but if the Governor's bird was really in a tree, then it cannot have been the shearwater . . . a flying-fox?
77. *Alectroenas madagascariensis* takes palm fruits *inter alia* (e.g. Hawkins & Goodman 2003). Benson (1960) suggested the *Alectroenas* habit of perching conspicuously might have contributed to *A. nitidissima*'s extinction in Mauritius.
78. Pink Pigeons, when distinguished, were called *gros ramier*, but also sometimes '*pigeon hollandais*', a name transferred from the extinct *Alectroenas nitidissima* (Cheke 1982a); the name also transferred to vagrant (but colourful) Broad-billed Rollers *Eurystomus glaucurus* (Cheke 1987a). Edward Newton's writings in the 1860s reveal the confusion then reigning about local pigeons.
79. Anon (1858) specifically reported Echo Parakeets from the Camisard peak in the Bamboo Mountains, the only record from the southeastern range.
80. Clark said sparrows had been introduced "within two or three years" of his date of writing (1859), but Edward Newton (1861–82), writing to his brother Alfred in 1862, said it was "some ten years ago by a sergeant of the 5th Reg." i.e. *c.*1852 – the 5th Northumberland Fusiliers were in Mauritius 1847–57 and 1858–63 (Barnwell 1948), unhelpfully spanning both dates. Sparrows were often taken to foreign shores by British colonists in the 19th century as a reminder of home – Newton himself released birds from Port Louis at his house at Pailles in April 1862: "it is jolly to hear their familiar chirp" (E. Newton 1861–82).
81. In 1859 feral pigeons were common in Port Louis and on cliffs elsewhere, e.g. Corps de Garde mountain; Bernardin (1773) and Milbert (1812) had mentioned domestic pigeons, and indeed they were first imported in 1639 by the Dutch (Bonaparte 1890).
82. The 'partridges' were the Grey and Chinese Francolins, the 'quails' included the Madagascar Partridge, the Painted Quail (first reliably noted by Peron & Lesueur 1803, 60 years earlier) and what Clark believed was the Common Quail *Coturnix coturnix*; he did not describe this bird "commonly brought from India", which was more probably a *Perdicula* bush quail.
83. Germain (1921), Griffiths & Florens (2006). Koenig (1932) attributed the introduction ('1850') of *Achatina pantherina* (now *A. immaculata*) from Madagascar to a Mr Dunford, about whom we have discovered nothing, nor why he imported it.
84. Desjardins (1837), Dumeril & Bibron (1834–54), Walond (1851: 149, for the 1840s sighting), *TRSAS* NS 17:17 (1885) and 18: 28 (1886; mentioned *passim* in proceedings of meetings); Telfair sent a number of chamelons from Mauritius to the Zoological Society of London in the 1830s; the collecting locality was not recorded but some, possibly all, were collected in Madagascar, in common with other specimens of his (Cheke 1987a); Telfair confirmed that provenance for one in a letter dated 26 Feb. 1833 in the Zoological Society's library. Lesson and Quoy could have been given in Mauritius animals originating in Madagascar, but the 1880s records appear to confirm some were feral.
85. Clark's discursive list of species includes 'shearwater' under the local name *taillevent*, the only recorded use of this term in Mauritius (it is used for Barau's Petrel in Réunion; Cheke 1982a). However the birds described are clearly Wedge-tailed Shearwaters; living in Mahébourg, he would have been within ear-shot of the (then) colonies on Ile de la Passe and Ile aux Fouquets, Leguat's *plutons* from the 1690s. Only near a colony would shearwaters' calls be audible onshore at night as Clark reported. We doubt that he would have used 'sea fowl' for tropicbirds, which he discussed separately, listing breeding sites which did not include the Grand Rivière SE.
86. Benson (1970–71); birds collected in 1860 (one) and 1865 (two) are in Cambridge, the latter two from Coromandel in the lower Grande Rivière NO gorge.
87. Anderson (1918).
88. Anderson (1918), Ricaud (1979). Charles Meldrum, weatherman and polymath, first proposed draining of the marshes in 1881; work did not start until 1901, after Daruty & d'Emmerez published a paper on mosquitos in Mauritius. The malariologist Ronald Ross (1923) visited in 1908, his report stimulating an acceleration of drainage works.
89. Ross (1923).
90. Overhunting and mongooses are usually blamed for the extinctions (Carié 1916, Staub 1973a; see Cheke 1987a), but the role of wholesale habitat loss through drainage has been overlooked.
91. Meinertzhagen (1912), pictures in Macmillan (1914); Mamet's (1979) comprehensive account of malaria control measures included where and when marshes and ponds were drained.
92. Anon (1897), anon (1908).
93. E. Newton (1858–95, 1861–82); This is no doubt why Newton got on with the island's Franco-Mauritian elite, who largely shared his prejudices, though he didn't like the French either! (Barnwell & Toussaint 1949).
94. "The Colonial Secretary [Newton] suffers from frequent attacks of fever [malaria] but is, on the whole, in better preservation than his colleagues. He has however, within the last year, lost both his wife and only child." Gordon (1894; letter dated June 1871); see also anon (1908).
95. E. Newton (1861–82). His forecast took longer to materialise than he expected, but the scenario envisaged is eerily close to the actual situation in the early 1970s (Chapter 8). From other letters we know he was aware that *Passer domesticus* was a recent arrival, but he wrote this passage before the gamebirds, gallinules and ducks being released around this time had become established.
96. Newton (1858–95) called the tree a 'citron', clearly a *Citrus*; although some are naturalised in the forest, this nest was apparently close to a hunting lodge, so may have been in a cultivated tree.
97. Anon (1858).
98. Boyle had breakfast in the Gorges at "the point where the valleys meet" (presumably where the Grandes Gorges meet the Mare aux Joncs valley): "there was a dish of young parrots just shot, and a very good one too."
99. E.Newton (1858–95, 1888), Roch & Newton (1862), Pike (1873), Carié (1904).
100. Newton did most of his birding on a hill he called 'Mont Orgeuil'. This appears on no map, and the name was unknown also to France Staub (pers. comm.) who knew the area all his life. However, Ida Pfeiffer (1861), visiting the Moons and "their equally obliging relative [*sic*] Mr Caldwell" in April 1857, was taken up 'Mont Orgeuil' where she was able to see from the one spot Black River Peak, The Morne, Tamarin mountain, Rempart, Trois Mam-

melles, Corps de Garde, the Moka Range and Nouvelle Decouverte. This confirms our inference from Newton's notes that he was working on what were known to Vaughan & Wiehe (1937) as the 'Vacoas Ridges'. Only the small topmost ridge survives under native forest, now the Mondrain nature reserve (Guého & Staub 1979); the native birds are long gone (Cheke 1987c, Strahm 1988).

101. Guého & Staub (1979). Pfeiffer (1861) said Malcy Moon was "a very accomplished lady" whose "talent for drawing is quite remarkable". In June 1867 Bishop Ryan visited the house: "At Mr Moon's we looked over the exquisite drawings of island flowers made by Mrs Moon, which are the admiration of all who see them." Rouillard & Guého (1999) reproduced many of her paintings.

102. E. Newton (1858–95: 600); the long sentences and ampersands are his. 'Old Moon' had by then died, hence the 'poor'. The bird was red, *white*, and blue – no green!

103. *DMB*: 1275–6 (1986, G. Rouillard), Guého & Staub (1979); it was Malcy de Chazal's second marriage, having divorced her first husband. William Moon died in 1862, and thereafter she associated with James Caldwell, travelling with him to Rodrigues, Australia, New Caledonia and India, where she died at Calcutta in 1880. Caldwell had a long and varied career in Mauritius, though his biography (*DMB*: 448–9, 1944, P. J. Barnwell & W.C. Rae) omits mention of his long 'companionship' with the flower artist, admitted in hers, written at a less censorious time.

104. E. Newton (1858–95).

105. Barnwell (1948) dated Simpson's departure as 1838, though *DMB*: 91 (1941, P. J. Barnwell) has 1837.

106. Desjardins (1832), Cheke (1987a).

107. "Colonel Pike's science is very superficial, but he has a large smattering of knowledge on many subjects. [John] Horne's science is real, but limited strictly to botany" – Arthur Gordon (1894) commenting on fellow passengers, *en route* to the Seychelles in 1871.

108. Pike (1873). Before his brief diplomatic career in mid-life (Portugal and Mauritius), Pike ran a wallpaper business in New York, but had a life-long interest in natural history (Gudger 1929).

109. "Col. Pike has willingly consented to put at the society's disposal his work 'Natural history of birds of Mauritius' to be published in the next *Transactions* of the society" (minute of the RSAS session of 21 October 1871, in *TRSAS* NS 6: 4, 1872).

110. "I have sought diligently to find this manuscript, running down every possible clue, but I can find no trace of it, nor indeed any person who has even heard of it." (Gudger 1929). Specimens collected by Pike in Mauritius are in the US National Museum (Smithsonian), Harvard and the AMNH; the latter also has his fish drawings and associated notes.

111. Newton (1861–82).

112. Boyle (1867); Barnwell (1948: 265) and *DMB*: 876 [1967, P. J. Barnwell] gave brief biographical notes.

113. Pike (1973).

114. Brouard (1963).

115. Regnaud was evidently an all-purpose medico-biological expert that the government called on at will; Bouton was the botanist with long-standing concerns about forests; Meller, a former member of Livingstone's East African expeditions, was Director of the Pamplemousses Botanic Garden (*DMB*: 366, 1943, Béchet & Barnwell) and Caldwell "a nearsighted Irish freemason with a somewhat checkered colonial history as a teacher, translator and diplomat" (Storey 1997) and Malcy Moon's boyfriend, had succeeded Edward Newton as Assistant Colonial Secretary. He regularly accompanied Newton on birdwatching excursions at Mondrain/Tamarin Falls (E. Newton 1858–95); there is still a road named after him in Henrietta. Sclater (1864) named *Anas melleri*, the Malagasy duck introduced to Mauritius, after Meller.

116. ". . . of late years much of the soil that was formerly very productive in its sugar yield has become to the extent of many thousands of acres waste and unemployed. This applies chiefly to the low lying lands on the northern and western districts of the Island . . . To replace the land thus rendered unproductive, hundreds of acres of the primeval forest have been recklessly cleared away and by many it is believed that from that cause the climate of the Island has been injuriously affected, the temperature increased and the humidity of the soil therefore correspondingly lessened." – acting Governor E.Selby Smyth (1872), writing in late 1870, presumably on the advice of his staff (he had only arrived May and assumed office in June).

117. Meteorologist and statistician Albert Walter (1908) compiled all local rainfall data to settle the forest and rain issue once and for all, demonstrating conclusively that, in Mauritius at least, forest removal had no effect on overall rainfall. Walter was prominent in early 20th century Mauritius, writing numerous reports and editing the later issues of the *Mauritius Almanach;* he left Mauritius in 1926, dying at 95 in 1972 (*DMB*: 1121–22, 1981, P. J. Barnwell); his books and papers, deposited in Oxford's Rhodes House library, have provided ASC with much key literature used in this book.

118. Smyth (1872), Brouard (1963). Brouard apparently thought there were *two* water and forest commissions, in 1867 and again in 1870, but these were the dates of its appointment and completion. Although the findings were published in 1871 (not 1870), sections of the report are actually dated 1867 (Bouton 1871) and 1868 (Regnaud 1871); we have not seen the full government-published version. There *was* a second commission, again under Regnaud, convened in 1872, reporting in 1873 and 1874 (Toussaint & Adolphe 1956). Toussaint & Adolphe refer to these reports under the title 'Water Pollution Commission', but the parts published in the RSAS *Transactions* are headed 'Water *Supply* Commission', which makes more sense.

119. Map 'Mauritius Forests in 1872' in Gleadow (1904), and map in Walter (1908; Map 7.2); see also Cheke (1987a).

120. "Some measure of protection of the few yet remaining forests, and for the replanting of trees in some of the absolutely denuded districts of the island is also a matter of the first necessity" (Gordon 1894, diary entry for 23 May 1871).

121. Curiously, historians (Pitot 1914a, Barnwell & Toussaint 1949, Bulpin 1958, Toussaint 1972) have completely ignored the long-standing local current in favour of protecting forest cover, and ascribed all 19th century forest legislation to decisions of British governors imposing their well-meaning preservationist views on an unwilling local plantocracy (who stood to lose rights and some land). They feature British expert Thompson of 1880, but never the vocal Mauritian ones. Mauritian forester Noel Brouard (1963) redressed the balance, but largely ignored the political context. More recent general histories (Addison & Hazareesingh 1984, Selvon 2001, Teelock 2001) have, worryingly, ignored the environment almost entirely. The 1875 laws were Ordinances No.13 and 14 of 1875 (*MA* 1876:105–115, 116–119). Gordon hated Mauritius (an unwanted promotion after ruling Trinidad), but worked conscientiously to improve forest and labour legislation. He enjoyed nature, but did not get on with Newton, his naturalist colonial secretary, who "is a gentleman, cultivated and agreeable; he is an authority on natural history, but not, I should say, on political science. His 'policy' is the narrow one of 'stamping out' the French language, French laws, French manners, French religion, thereby making everything English as unpopular as in Wales or Ireland. My policy is just the reverse . . ."; the antipathy was mutual (E. Newton 1858–95). Gordon's glowing accounts of trips to the Seychelles (one of them with Nicolas Pike and John Horne) contrast with his morose descriptions of Mauritius: "It was a dull and dismal wet day, as detestable as everything else in this detestable place" (Gordon 1894).

122. Rouillard & Guého (1999) published a table of ebony exports 1859–1900. We have not located where the huge quantities of ebony cut in 1877–78 came from; there are hints, but no details, of recent

clearances in Thompson's report (1880). A candidate area would be the $c.20\text{km}^2$ between Quartier Militaire and Montagne Blanche that was forested in 1872 (Gleadow 1904) but completely unforested in 1903 (map 9 in Toussaint 1953), but illegal logging of live trees in the mid-1870s by contractors supposedly removing dead ones may also have generated a lot of ebony.

123. Gleadow (1904), implying that crown forests had been extensively, though surreptitiously, logged of valuable timber.

124. Regnaud (1871); the commission also commented, as had Rochon (1791) a century earlier, on the silting up of Port Louis harbour and elsewhere from eroded land, particularly mentioning the 'Mer Rouge'. This became a tidal mudflat favoured by migrant waders and terns, and thus birders, in the 1970s (Temple 1976), but subsequently reclaimed for industrial use (Guébourg 1999).

125. Bouton (1871), Brouard (1963).

126. Thompson (1880). Napier-Broome succeeded Newton as Colonial Secretary, and was *de facto* governor for five years, the titular governor, Bowen, being absent for most of his term of office (Barnwell & Toussaint 1949 and list of British governors in North-Coombes 1971).

127. Proclamation No.37 of 1900 protected tenrecs on Crown Lands (see *MA* 1901).

128. The collection and exhibition of prize antlers as trophies was, and is, practiced as seriously in Mauritius as ever in Scotland (e.g. Antelme 1914, 1932). The ritual of *la chasse* features in every travel book from Milbert (1804) onwards, with varying degrees of enthusiasm; Rees & Rees (1956) provided an entertaining Aussie slant. Willox (1989) summed it up succinctly: "during the hunting season (June to September) the Franco-Mauritian gentry take to their estates and blast the hides off 2,500 Java deer each year". This passage has vanished from Lonely Planet's recent editions!

129. Brouard (1963).

130. Koenig (1895); in Koenig's defence, there may have genuinely been a decrease in visible damage for a while, as the deer were decimated by a rinderpest epidemic in the 1880s (Antelme 1914, Unienville 1991).

131. List of woods growing in Mauritius (*MA* 1877: 330–334), probably compiled by John Horne. The *pirogue* was originally a dugout canoe fitted with a sail and used for fishing, but the term is now used for sailing dinghies made with planks or fibreglass. In earlier centuries such dugouts were the standard inshore fishing boats.

132. Vaughan & Wiehe (1941).

133. Fuller details of this ecological equivalent of the urban myth are given in Cheke (1987a), Witmer & Cheke (1991) and Hershey (2004), with complete references; the key ones are Hill (1941), King (1946), Iverson (1987), Temple (1977, 1983), Jackson *et al.* (1988), Vaughan (1984), Strahm (1993), Baider & Florens (2006). Strahm reported rare 'young trees' in a study plot, implying some regeneration, without details of size or age; the smallest tree seen by François Friedmann in the 1970s (*Flore* 116: 10, 1981) was 10cm dbh (6–7m tall), estimated at 30–50 years old; see Baider & Florens (2006) for the 2001–03 survey. The fungus rotting the seeds is unidentified (Jackson *et al.* 1988), but very likely exotic, and not within the tree's adaptations – though Vincent Florens reports (pers. comm.; Baider & Florens 2006) that it only invades overmature seeds; healthy seeds are not susceptible. John Iverson, who suggested that tortoises might be more important than Dodos in Tambalacoque seed biology, offered his perfectly well-argued paper to *Science*, where Temple had published, but Temple, as referee, rubbished a draft, and because "Temple is a well respected ecologist at a big university and I was just a young turk at a small college" *Science* rejected it. He even had difficulty publishing elsewhere (pers. comm.). Pioneering ecologist of Mauritian vegetation Reginald Vaughan died in 1987 at the age of 91 (Lorence & Vaughan 1992; *DMB*: 1784–86, 2000, G. Rouillard).

134. Gould (1980) incorporated Temple's story in an essay on mutualism, then, apparently tipped off that it was perhaps not entirely sound, included a postscript reprinting Owadally's refutation (1979), with further discussion on the problems of disproof. Proving Gould's point, the canard is still turning up in serious literature – e.g. in a review paper by Walter Bock (2003). There is no hint that this author, a respected 'elder statesman' of ornithology, had read beyond Temple's original paper. Baider & Florens (2006) cite more 21st century sightings in ecological texts.

135. Vincent Florens, pers. comm.; although he and Claudia Baider reported their discoveries to a British Ecological Society symposium in July 2003, publication was delayed (Baider & Florens 2006). They have also succeeded in germinating large numbers of seeds, with and without the epicarp, with and without abrasion and with no sign of the serious fungus damage reported in the 1980s. Most seeds used in past germination tests were apparently dead prior to sowing, probably because the fruits were broken off and dropped by monkeys before they were ripe. During the 2001–03 study, increased human activity around the weeded plots apparently discouraged monkey activity, allowing the trees to complete their cycle. A lone tree in monkey-free urban Curepipe (source of Temple's seeds in 1974) always fruits well. Native wildlife was no doubt important in dispersal, but had no effect on germination.

136. Ordinances No.1 and 10 and Proclamations No. 23 and 24 of 1881 (see *MA* 1882:175–76). Since Mauritius became a republic in 1992 Crown Lands have become 'State Lands'.

137. Napier-Broome (1884); the proposed extension of mountain and river reserves led to "explosive protest that culminated in a Public Meeting. This Meeting, the most stormy Mauritius had ever seen, registered a 'Declaration of colonial property rights' – and the government backed down" (Magny 1882).

138. Bruce (1910). Hawley was in charge because Pope-Hennessy (who favoured forest conservation) had been recalled to Britain to answer charges of maladministration, later dismissed (Barnwell & Toussaint 1949); he and Hawley were at daggers drawn (Pope-Hennessy 1964; *DMB*: 126, 1942, P. J. Barnwell), so the suspension of the forest policy probably had no more rational basis than pique.

139. Passage quoted by Pope-Hennessy (1964; original source not given).

140. Koenig (1895); Gleadow (1904) was unexpectedly sanguine about the 1877 contract and suspected the charges against the contractors were somewhat exaggerated. However, these irregularities do coincide with the sudden upsurge in ebony exports in 1877–78, and may be related.

141. Gleadow (1904).

142. Bruce (1910).

143. Koenig, the first trained forester to take the post, was appointed acting Director of Forests in 1903 on the death of his predecessor, then full Director in 1907 (*DMB*: 1209–10,1984, G. Rouillard). Ex-forest guard Joseph Vanckeirsbilck, who ran the department 1893–1903 (*DMB*: 1154; 1982, G. Rouillard), had a reputation for weak leadership, explaining the failure to control large landowners; he was apparently more interested in Pamplemousses Gardens (also then under the Department) than the forests.

144. Gleadow's report (1904) was "so severely criticised for its 'undeserved and unjustifiable attacks against the Mauritian Community' (*Annual Report* for 1905) that it had to be amended by the deletion of 'injurious passages' in 1906" (Brouard 1963; see also Bruce 1910). Brouard left the accuracy of the accusations open, but Pitot (1914a) singled out Gleadow amongst other visitors who habitually "abuse the Mauritians at large in papers and magazines": "Mr Gleadow sent in a lengthy report on reafforestation, wherein he indulged in some most offensive and unjustifiable comments on the Mauritians, which caused such indignations that a part of this peevish gentleman's tirade was suppressed, it is said, by Government order". Amongst those criticised for bending the borders of their mountain reserves were Sir Virgile Naz, jurist and former President of the RSAS, and planter-agronomist Dr Edmond

Icery, long-term stalwart of the Chamber of Agriculture (though by 1904 dead 20 years: *DMB*: 1042–3, 1975, R. d'Unienville).
145. See Sale (1935a), Vaughan & Wiehe (1937), King (1946), Brouard (1963) and Owadally (1980). The privet, generally called *Ligustrum walkeri* in the Mauritian literature, is now considered a variety of *L. robustum* (*Flore* 199: 6, 1981, A. J. Scott; Rouillard & Guého 1999); DNA work shows that Mascarene plants originated in Sri Lanka (Milne & Abbott 2004). Gabriel d'Argent, a veteran forester, told Strahm (1993) the story of Koenig on the train. According to Guy Rouillard (*DMB*: 1209, 1984), Koenig also expected the privet to suppress the invasive bramble *Rubus alceifolius*. The privet was already used as hedging in the Curepipe Botanic Gardens in 1895 (Rouillard & Guého 1990).
146. In 1891 Daruty was simultaneously curator of the museum, secretary of the RSAS, president of the *Société d'Acclimatation* (*MA* 1891) and founder-manager of the *Revue Agricole de l'Ile Maurice* (*DMB*: 14, 1941, A. d'Emmerez de Charmoy).
147. "quantities of ducks, quail (*Margaroperdix striata*), *Porphyrio* and other birds are brought by almost every ship from Tamatave" (E.Newton 1861–82, writing in 1862), and "large quantities" of whistling ducks, White-faced *Dendrocygna viduata* and Wandering *D. arcuata* were "brought over alive in nearly every bullock ship" (E. Newton 1863) – though only the former had gone feral. Although long-established in the island, "vast numbers" of lovebirds "are brought over yearly to Mauritius" (Roch & Newton 1862–63), presumably for the cagebird trade.
148. The Painted Quail had been around several decades (p. 102).
149. Clark (1959), E. Newton (1858–95), Daruty (1878), Carié (1916).
150. Newton (1858–95) initially identified the ducks as Red-billed Teal *Anas erythrorhyncha*, although the bill was not red. An egg of his labelled *A. erythrorhyncha* is still in Cambridge, but is identical to another dated 1877 labelled *A. melleri*, the first Mauritian specimen of which dates from 1866 (also in Cambridge). Meller's Duck was not described until 1864 (Sclater 1864), explaining Newton's identification problem in 1863. Prior to finding these ducks at Mare Longue, Newton (1863) thought that White-faced Whistling Duck was the only wild duck. James Currie, who released the Malagasy ducks, was a British planter, who in 1865 owned two sugar estates (*MA* 1865), becoming by 1874 chairman of the Chamber of Agriculture (Robert 1914; *DMB*: 628, 1947, P. J. Barnwell); he leased the Doudy *chasse* with the dying forest.
151. An egg in Cambridge dated 1866 and labelled '*Coturnix communis*' (= *C. coturnix*) was assumed to be from wild Common Quail by Cheke (1987a), but Newton's notebooks (1858–95) reveal it was from a captive California Quail *Callipepla californica* belonging to the Governor's wife (Lady Barkly). Carié (1916) claimed Common Quails were released frequently during the 18th century, but gave no source for this assertion, which we cannot corroborate; Rountree *et al.* (1952) repeated this claim, identifying Mauritian birds as the African race *C. c. africana*. Meinertzhagen (1912) did not include this species in his Mauritian bird list
152. Sganzin (1840), Meinertzhagen (1912); Carié (1916) said nothing about the buttonquails' origins, and failed to mention them in an earlier paper (Carié 1904). Newton (1861–82) reported killing a 'button quail' in 1863, but we think this was an error for '*bush* quail' (*Perdicula*).
153. Carié (1916). Although Meinertzhagen (1910–11) claimed Purple Swamphens "have been here for over a century", the fact that Clark (1859) did not mention them, and that large shipments came in soon afterwards, suggests they probably did not become feral until the 1860s. They had bred in captivity in Mauritius in the 1770s (Buffon 1770–83, Chapter 7), and only tame birds were present in 1801–03 (Milbert 1804). Although supposed to be fully protected by the 1880 regulations, hunters continued treating swamphens as game, and by 1910 they had disappeared from the list of protected species (Meinertzhagen 1912).

154. The nearest equivalent publication of such wide scope that has appeared since is Staub (1993a).
155. Bouton (1847, 1848), *PRSAS* 1849–51: lviii (published 1851), meeting of 7 March 1850. Augustus Lloyd, Government Surveyor, suggested to his deputy Thomas Corby, who lived near Grand Port, that he excavate the ruins of the Dutch settlement; if he did so, no report was published.
156. Exhibited to the RSAS at their meeting on 8 November 1860 (*TRSAS* NS 2 (1):157, 1861), and noted briefly in *Proc. Zool. Soc. Loud.* 1860: 443; he found the bone in "one of the caverns of the Rivière Noire [district]"; exact location not specified.
157. Clark (1865, 1866); Vinson (1968) found the Society's laconic response to Clark's discovery in the RSAS archives; that particular session was not printed in their *Transactions*. Once the excitement in Europe filtered back to Mauritius, Clark was belatedly given honorary membership of the Royal Society on 26 June 1866 (*TRSAS* NS 3 (1): 12, 1869); Vinson (1968) said this move originated with Sir Henry Barkly at the AGM on 13 April 1866, but Edward Newton (1861–82) had been encouraging Clark's activities since October 1865. Our text is from Clark's account, but there is another version, from railway engineer Harry R. Higginson, who claims it was he in October 1865 who first spotted 'coolies' making piles of bones when cutting peat from the Mare aux Songes, and took samples to Clark to identify (Grihault 2005, Hume & Prys-Jones 2005). Higginson certainly collected Dodo remains, as he later presented skeletons to three UK museums, but Clark had already got bones in September (Newton 1861–82). The subfossil sites in the Mascarenes are described and discussed by Hume (2005b).
158. It is odd that the Dodo should have become so iconic when attention was also focused on the Great Auk *Pinguinus impennis*, vanishing nearer home in the North Atlantic. However, the Dodo was definitely no more, whereas it was many years before the Great Auk could be confirmed as extinct (Turvey & Cheke, in prep.), although the last birds were killed in 1844 (Newton & Gadow 1896, Fuller 2000). The fullest bibliography of 19th century Dodo publications is in Hachisuka (1953); Owen (1866) was the first to study Clark's bones. Livezey's bibliography (1993) is very extensive and updates the osteological references.
159. Thirioux never published, but his activities were briefly described by d'Emmerez (1903) and Pitot (1914b); the University Museum of Zoology, Cambridge, preserves correspondence with Alfred Newton about his finds. For Carié's life see Crépin (1931) and *DMB*: 69 (1941, L. Halais).
160. E. Newton & Gadow (1893) summarised the Sauzier material and referred to earlier work on Clark's collections; Réné Hofstetter used Carié's finds in several papers on fossil reptiles in the 1940s, reviewed and extended by Arnold (1980), where the references can be found.
161. Hume (2005a).
162. Arnold (1980), Cheke (1987a); Bonnemains & Bour (1997) did not indicate where Lesueur saw the *Asterochelys radiata*, but given their itinerary, Mauritius is the most likely place. In his MSS Lesueur only mentioned tortoises imported from the Seychelles, for which he urged captive breeding to ensure future supplies for the hospital (Lesueur 1803, MS 15037).
163. Desjardins (1837), Clark (1859).
164. *Bufo melanostictus*, Günther (1874), Boulenger (1882); see Daniel (2002) for a recent survey of Indian reptiles and amphibians. Bouton (1875) alleged that "toads are now and then found in digging or clearing ditches", but there is no corroboration for a toad population then, and Bouton, by then 76 years old (Ly-Tio-Fane 1972), may have recalled the 1830s rather than described the position in the 1870s. The specimen he sent to Günther, although labelled 'Bouton' (Boulenger 1882), perhaps as the immediate source, was almost certainly one of the two brought to Desjardins. There is another Mauritian *B. melanostictus* specimen in Vienna,

from Bojer's collection, together with, more bizarrely, a European *B. bufo*, dated 1833, but with no record of the collector; Desjardins (1837) himself referred to the animals as '*Bufo bufo*', but was this a positive identification, or an assumption through lack of reference material? Later references to *B. melanostictus* as present in Mauritius (e.g. Mertens 1934) all relate to the 1830s collections and their identification by Günther (1874).

165. d'Unienville (1838), Desjardins (1830).

166. Günther (1869), *TRSAS* NS 6: 12 (1872), Daruty (1883a,b): *Rhamphotyphlops braminus*, *Lycodon aulicus*. The blind-snakes found in 1869 and 1871 were originally identified by Günther as *Typhlops flavoterminatus* (a larger Indian species), and the Wolf Snake was called '*Proepeditus lineatus*' by Daruty, a name Bour & Tirvengadum (1985) were unable to trace.

167. Flowerpot Snakes are parthenogenic, and so widespread in the tropics that their original home is unknown.

168. *Pelusios subniger*: Edwards (1872, already "in big enough numbers"), *TRSAS* NS 13: 8 (for 1878); at the time Daruty apparently did not recognise the Diego Garcia animals as the same, but only this species of *Pelusios* is known from there (see Bour 1984a,b).

169. Dupont (1860); apparently the only reason he thought crows could control rats was that Bernardin believed it some 100 years earlier!

170. Marie-Dominique Wandhammer, curator of the Musée Zoologique in Strasbourg, pers. comm. Bergmans (1988–97) speculated that the 1876 date might indicate the Rougette's survival well beyond the mid-1860s. It was collected (or presented?) by one Schneider (Aellen 1957, Bergmans 1988–97), about whom we have discovered nothing. However, two *Phelsuma guentheri* from 'Hr. G. Schneider' dated 1882 are in the St.Petersburg museum (Strauch 1887; accession or collection date not stated), and an 'M. Schneider' collected three small bats *Mormopterus acetabulosus* from Mauritius accessioned to the Leiden Rijksmuseum in 1888 (Jentinck 1888). Whether either was the same as for Strasbourg is not known, but all are later than the 1860s, so it is not impossible that the Rougette was collected in the 1870s. For Clark's account see Box 37 (p. 263); the name '*chauve souris banane*' for the Rougette transferred at an unknown date to the Grey Tomb Bat *Taphozous mauritanus*.

171. For a fuller discussion of *Pteropus subniger*'s demise see Cheke & Dahl (1981). Henry Whiteley travelled to Japan in 1864 (anon. 1893; pre-Suez Canal, so presumably via Mauritius), but in addition a 'Whiteley' is mentioned by E. Newton (1858–95) as being in Madagascar in 1862; the specimens were presented to the BM(NH) in 1866. The Rougette was listed as if extant by d'Emmerez (1914), Vinson (1953) and Michel (1972); Temple (1974) was the first to label it extinct. Carl Jones (pers. comm.) reported that elderly Mauritian hunters interviewed since 1979 recalled rare smaller fruitbats being shot in the 1920s, but without specimens it is difficult to be sure these were not young free-flying *Pteropus niger*.

172. E. Newton (1861–82); according to Peters (1877), Möbius collected *Gongylomorphus bojerii* on Ile aux Fouquets (Mahébourg Bay) and 'Black River', but according to the Berlin museum accessions register both specimens deposited were from 'Fouquet'; only one remains (Rainer Günther, pers. comm.). Möbius himself (1880) mentioned visiting Black River, but it did not feature in his discussion of lizard distribution and habitat; see Chapter 9 for the problems with his Round Island specimens.

173. Pike (1870) contains his clearest discussion of lizards, though the Tamarin sighting is in *Subtropical rambles* (1873: 297) – the lizard was darting in and out of sea-water lapping a jetty, behaviour typical of Bouton's Skink (Jones 1993a) but alien to other Mauritian species. In his book (p.161) Pike confused the names, and a sentence about day-geckos is transposed into the skink discussion.

174. D'Emmerez (1914).

175. On bolyerid and *Lycodon*'s adaptations see Cundall & Irish (1987, 1989), Jackson & Fritts (2004). Carl Jones (1993a) suggested that *Lycodon* probably caused the skink's extinction; Auguste Vinson (1868) had long before blamed the decline in Réunion day-geckos on snakes. In its native India *L. aulicus* "prefers lizards of the gecko family but takes any small animal it can overcome; mice and skinks are also eaten" (Daniel 2002); in the 1980s François Moutou found both house geckos and Bloodsuckers in stomachs of Réunion specimens (accession data; MNHN, Paris). Its recent introduction to Christmas Island (1980s, Fritts 1993) will be an unintentional experiment on its effects on snake-naïve lizards. Jones (pers. comm.) suggests that bolyerids are better adapted to catching geckos than skinks, whereas the Wolf Snake has longer teeth that can grip and pierce both slippery hard-bodied skinks and non-slippery soft-bodied geckos.

176. Although Vinson & Vinson (1969) wrote as if they had seen the Ile aux Fouquets lizards themselves, Jones (1993a) was told by Jean-Michel Vinson that neither author had visited Fouquets prior to their publication. Jones thus assumed there was a mix-up on islet names and that Bojer's Skinks had never occurred there, but Möbius (1880) who camped on Ile aux Fouquets in 1874 studying the reefs and collected Bojer's Skinks there (note 172), has both Fouquets and 'I. Vacquoas' [*sic*] correctly located on his map. Jones (pers. comm.) has since confirmed that there is a Fouquets specimen dated 1972, and that the skink must have subsequently (before 1987: Jones 1988b) succumbed to the shrews first formally recorded in 1993 (Bell *et al.* 1993); see also Chapter 10.

177. E. Newton (1878a,b); the endemic birds Newton proposed to protect were the Mauritius Kestrel, Merle, and Cuckoo-shrike, plus the Echo Parakeet and Pink Pigeon.

178. Newton (1878b).

179. Newton (1878b) reported each of his listed species being shot, often in the breeding season, but apart from the token reference to protecting the native flora (see the text quotation) nowhere did he mention threats to their habitat after his 1863 letter quoted earlier. The diaries are E. Newton (1858–95). Newton did sign one of the 1874 forest ordinances into law (Thompson 1880, Appendix M), but it was promoted by governor Gordon, not himself.

180. The list of protected birds under Ordinance No. 40 of 1880, as printed in the *Mauritius Almanac* for 1881 (p.111–12), not only includes Mauritian species, but also, intermixed and not identified as such, native birds of the Seychelles (then a Mauritian dependency) and also, too late, the Rodrigues Parakeet. Both Mauritian white-eyes appear twice under different Latin names, and the Mauritius and Seychelles Kestrels are conflated as "*Pennunculus* [*sic*] *Punctatus Gracilis*"! There are numerous typographical errors in scientific names – it is a complete mess. The source(s) of this bodged list is not clear, but appears to have been the RSAS and the *Société d'Acclimatation*, as part of their lobbying to extend Newton's list (*TRSAS* NS 14: 22, *séance* of 28 July 1880), though Newton's notoriously illegible handwriting may also have contributed. Ordinance 19 of 1881 revised all the game legislation, but let stand the proclamations under the previous (1869) Ordinance. As cereal-growing died out under the advance of sugar, the regulations requiring land-owners to kill granivores fell into disuse (Rouillard 1866–69), and by 1880 the ex-pests were back to being prized cage-birds: Cardinal Fodies, Java Sparrows, Spice Finches, Common Waxbills and both canaries were all on the 1880 protected list (*MA* 1881: 111–112).

181. See Chapter 9 for details of the slaughter of tropicbirds on Round Island.

182. Proclamation No. 37 of 1900 (Lane 1946, *MA* 1901). Tenrecs remained technically protected in state lands until the Wildlife Act of 1983 which listed them as 'Unprotected Wildlife' (Act No.33 of 1983, Second Schedule; anon.1983b).

183. *Mauritius Almanac*, *passim*; Meinertzhagen (1912), Koenig (1932), Vinson (1956a).

184. Barclay-Smith (1964).

185. Carié (1916) is the main source of introduction details, but for

the bulbul see also Carié (1910), d'Emmerez (1941), Cheke (1987a) and Lever (2005); Meinertzhagen (1912) gave a different account of the arrival of the weaver, and a later date [1892]. A century later *Nephila inaurata* began to show signs of recovery, re-appearing in forest areas (e.g. Brise Fer) where they had not been noted for decades, even catching day-geckos in their webs (Dennis Hansen pers. comm. 2005).
186. Pers. obs.
187. Edward Newton's diaries and letters (1858–95, 1861–82) make it clear that in the 1860s all the endemic passerines were common in native forest between Mondrain/Tamarin Falls and Grand Bassin, although sometimes the forest appeared silent and empty (e.g. Bassin Blanc in November 1865). In 1868 at Bois Sec Merles "abounded, and pity the poor maladroit hunter who missed his deer – often he would hear, above his head, the whistle of a mocking merle" (Charles Peyrebère in d'Unienville 1991).
188. *Corvus splendens*; Meinertzhagen (1912), Guerin (1940– 53), Cheke (1987a), Feare & Mungroo (1990). Meinertzhagen said the crows arrived in 1910, but Carié (1904) had already reported unidentified crows breeding around Port Louis harbour. House Crows ride freely on boats.
189. Carié (1904, 1916), Meinertzhagen (1912), Cheke (1987a). The 1892 cyclone was later demonised for causing the disappearance of these three species, and also (more anachronistically) Grey-headed Lovebirds (e.g. Rountree et al. 1952), but no contemporary accounts mentioned the cyclone.
190. Meinertzhagen (1912) drew attention to the connection between Java Sparrows and cereal crops, and ASC (in Jones 1996, Lever 2005) to the possible impact of House Sparrows.
191. James Holman reported cagebirds imports in 1829–30 (in Barwell 1948); Carié (1904) noted Red Avadavat prices; see also Cheke (1987a).
192. Newton (1861–82) would undoubtedly have introduced *Centropus toulou* had his contact in Madagascar (one Mr Plant) not proved totally unreliable. Newton's flimsy justification was that "it is always climbing about the leaves and smaller boughs of trees, and I fancy it would do the same in the sugar canes and eat many a moth before she had laid her eggs" (letter to Plant, Feb.1864), yet the next year he excoriated an unnamed Franco-Mauritian for similarly fanciful proposals: "People in these parts have the most extraordinary ideas concerning birds which are believed to be useful to agriculturalists. A Frenchman once introduced a Nightingale, several Blackcaps, a Whitethroat and some Sky Larks for the purpose of destroying the borer which lives in the interior of the cane itself" (letter to Alfred Newton, 15 Sept. 1865; the birds are all European: *Luscinia megarhynchos, Sylvia atricapilla, S. communis* and *Alauda arvensis*). See d'Emmerez (1914), Vinson & Vinson (1969), Greathead (1971) for the lizard *Calotes versicolor*, known in Mauritius and Réunion as *caméléon*; d'Emmerez never said exactly when he released the lizards; Koenig (1932) put it at "a few years" after it arrived in Réunion (c.1865), but d'Emmerez was hardly in a position to do so before 1891 when he joined the museum aged 18 as assistant to Albert Daruty (*DMB*: 44–45, 1941, L. Halais & A. d'Emmerez de Charmoy). The borer, one of several related cane pests present, is *Sesamia calamistis*, since controlled by a parasitic braconid wasp *Apanteles sesamiae* (Greathead 1971), and other parasitoids (Rouillard & Guého 1999).
193. Bruce (1910); Charles Bruce left Mauritius in 1904 (Barnwell & Toussaint 1949), perhaps before the full consequences of the mongoose disaster were evident. The mongoose was initially identified as the Indian Grey Mongoose *Herpestes edwardsii* (e.g. Carié 1916); it is much greyer than the *H. auropunctatus* ASC had seen in north India, leading him (Cheke 1987a *contra* Lever 1985) to perpetuate the error, soon rectified by Carl Jones (1988b). Mauritian animals are often referred to as *H. javanicus*, but Veron et al. (2007) have shown that *auropunctatus* and *javanicus* are different species.

194. Lever (1985).
195. Clark (1859), Cheke (1987a).
196. Robert (1914).
197. Details of the mongoose story from Bruce (1910), Carié (1916) and Greathead (1971); their diet was still very similar 100 years later, the same three mammals predominating (Roy et al. 2002), though toads (not present in 1910) are also now important (Carl Jones, pers. comm.). Carié (1916) blamed "Mr Nash, then manager of the Albion Dock" for persuading the government to introduce the animals, Barnwell & Toussaint (1949) claiming he made the mistake over their sex. Frederick Nash was president of the Chamber of Agriculture (Toussaint 1972), which "by a large majority deprecated the experiment" and strongly advised in 1899 *against* releasing mongooses (Robert 1914). Nash had been overruled by the Chamber before (in 1896) – could he have been overruled again and imported mongooses with both sexes to spite them? Nash was completely written out of Robert's (1914) history of the Chamber, but was briefly covered in *DMB*: 642 (1947, P. J. Barnwell); he died in 1902. There is no evidence of snakes being introduced, but "14 owls from Madagascar" (species not stated) were released in January 1902 (*MA* 1914: A61), though according to Bruce (1910) the originally intended source was South Africa; they were not heard of again.
198. Antelme (1914); Henrietta is near where Edward Newton found this species common in the 1860s. George Robinson owned the Henrietta estate in the late 19th century until his death in 1902 (*DMB*: 528, 1945, P. J. Barnwell).
199. Meinertzhagen (1912), Carié (1916), Varnham et al. (2002). Tenrec densities are much higher in the Seychelles (up to 1.5 adults/ha, no mongooses) than in Madagascar (native predators including mongooses) or contemporary Mauritius (Nicoll 2003).
200. Meller's Duck was always confined to lakes and streams in the central uplands (Carié 1916), numbers apparently peaking in the late 19th century when bird recorders were few. Slater's comments were in Hartlaub (1877), and 'P. H. G.' (anon 1941), reminisced on the former abundance of 'mallard' (= Meller's Duck) and 'teal' (= White-faced Whistling Duck); however, neither Meinertzhagen (1912) nor Carié (1916), writing soon after the mongoose introduction, suggested Meller's Duck had formerly been more abundant. We have been unable to identify 'P. H. G.'. Carié (1916) reported Purple Swamphens having already declined (when many marshes still survived), but at that stage whistling ducks appeared unaffected.
201. Huron (1923).
202. Meinertzhagen was a soldier with the British garrison; his published account of birds (1912) is more comprehensive than his diaries (1910–11), but less specific, and contains material from local naturalists (especially the ornithologically unreliable d'Emmerez), without it being clear what he saw himself and what they told him (see also Chapter 9). He had a long and bizarre later career (e.g. *DMB*: 1493–4, 1993, P. J. Barnwell), dying in 1967 aged 89.
203. The 1814 Treaty of Paris returned most of France's colonies following the Napoleonic Wars, the British retaining, in the Indian Ocean, Mauritius and its dependencies. General historical and agricultural details in this paragraph and the next are from Billiard (1822), Defos du Rau (1960) and Toussaint (1972).
204. Billiard (1822) was scathing about this atavistic constitution, and considered English rule less oppressive!
205. Defos du Rau (1960), Fuma (2002).
206. From Billiard (1822), Betting de Lancastel (1827), Defos du Rau (1960) and Fuma (2002). The figures from different sources are not entirely consistent, and can differ even within one work; we have used Defos du Rau's more probable figure for maize area in 1851 on p.162, rather than the 29,000 ha cited on p.150.
207. Billiard (1822), Betting (1827); Thomas's (1828) substantial compilation on the island covered trade and agriculture in detail, but it is hard to extract significant ecological information.
208. Billiard (1822), 117–120.

209. Renoyal (1811–1838; Probst 2000a), Betting (1827), Sganzin (1840). Betting de Lancastel was Réunion's first 'Directeur de l'Intérieur', or deputy governor in charge of infrastructure (Toussaint 1972, Reydellet 1978); apart from initiating a round-the-island road, little seems recorded of him, even his first name! Lever (2005) cited Simberloff (1992, Moulton et al. 1996) who, despite Cheke's (1987a) review, claimed an introduction date of c.1700 for Red Avadavats, Common Waxbills and Spice Finches, based on a total misreading of Barré & Barau (1982), who never suggested anything of the kind.

210. Delabarre (1844), Decary (1962): "the introduction into the colony of all species of food-grain destroying birds is banned from today on. It is equally forbidden to have them in cages, the general interest not supporting dependence on an individual watchfulness. In consequence all birds that destroy food-grains shall be killed, both free and in cages". This applied from 1 January 1821, with a 10 franc fine for keeping a caged *granivore*.

211. Two of the Hoopoe Starlings were caught in the Bras des Chevrettes near Saint-André (Renoyal 1811–38).

212. Betting (1827) said tenrecs had been naturalised for only about a dozen years, alleging previous attempts (including Renoyal's in 1801/02?) had failed. He also said a second species was present, with less spiny hairs "that has received from naturalists the name of *setosus*" (i.e. the Lesser Hedgehog Tenrec *Setifer setosus*); this may, as in Mauritius, have been naturalised for a while.

213. Betting's (1827) description of the *lézard vert* (Box 33, p. 242) was the first of *Phelsuma inexpectata*. The House Shrew, though reported in the 1730s, was not recorded again in Réunion until 1864, contra Cheke (1987a) who assumed continuity. If present in any numbers, local naturalists are unlikely to have missed it (e.g. Coquerel 1859, Maillard 1863), yet Vinson (1867), writing on introductions, was unaware of it, and it escaped mention in local literature until 1916 (Carié 1916).

214. Sganzin (1840) muddled kites with harriers (because '*papang(ue)*' is used for kites in Madagascar and harriers in Réunion), Madagascar Partridges with Chinese Francolins (a prevalent error at the time), white-eyes and '*thzeiri*' [= *tsery*?] (confusing a Buffon white-eye from Réunion with a garden bird in Ile Sainte Marie, Madagascar), and his descriptions of 'doves' appear to include unidentifiable cuckoos! Although *tsery* is now applied to sunbird-asities *Neomixis*, Sganzin's *thzeiri* was surely different, as *Neomixis* spp. bear no resemblance to Mascarene Grey White-eyes. His name lives on in the Comoro Blue Pigeon *Alectroenas sganzini*.

215. "The *rousette* of Madagascar is, I think, the same as that met with in Réunion and Mauritius" (Sganzin 1840).

216. Vinson (1888).

216. Sganzin described the *Turnix* adequately from Malagasy examples, including its lack of a hind toe, but it is less clear how well he examined Mascarene quails

218. Common Moorhens may have invaded very recently, for around the same time Desjardins noted their *absence* from Réunion, in contrast to their abundance in Mauritius (Oustalet 1897); there is a specimen from Réunion in the Pisa museum (Carlo Violani, pers. comm.), collected at the same time (*c*.1840) as the 'Italian' batch of Hoopoe Starlings.

219. Vinson & Vinson (1969); Eydoux was surgeon on the *Favorite* (Desjardins 1837), which, before Mauritius, had visited Réunion (22 April–3 March 1830; Laplace 1833–35). He returned briefly on the *Bonite* in July 1837, but due to an urgency to refit and leave for France, only the captain (Vaillant) and botanist Gaudichaud were allowed beyond Saint-Denis (La Salle 1845–52). Bour et al. (1995) cast doubt on the Réunion origin of Eydoux's Manapany Daygeckos, suspecting they were Mauritian *Phelsuma ornata*; two certainly appear to be so (pers. obs.), though the others (apart from a *P. borbonica*!) are *inexpectata*. This also makes the source of his skinks suspect; a promised fuller analysis of these specimens has yet to appear.

220. Dumeril & Bibron (1834–54, vol.7), Vinson & Vinson (1969), Cheke (1987a).

221. Maillard (1863); when Maillard first published his book (in 1862) he had been Réunion's chief engineer for 26 years (Simonin 1862), so some of his status assessments for animals may have been from memory of earlier times rather than current at the time of publication. Deso & Probst (2007) reported *Lycodon* in *Cryptoblepharus* habitat in Réunion; it eats *C. egeriae* on Christmas Island (Fritts 1993); Cole (2005) reported Cheechaks eating the skink in Mauritius.

222. Cheke (1987a); Duméril & Bibron (1834–54) were confused about which collectors had been in Réunion and which in Mauritius for *Chamaeleo* (= *Furcifer*) *pardalis* and sometimes for other species, although later Dumeril & Dumeril (1851) cited Nivoy's chameleon as from Réunion. In the case of the frog *Ptychadena mascareniensis* they just lumped both islands (plus the Seychelles) together! Cossigny in Mauritius had received a Panther Chameleon from Madagascar in 1755 (Charpentier 1732–55, Cheke 1987a). Betting (1827) said of the frog that "it is barely forty years that it was imported from the Seychelles" (cf. Milbert 1812 for Mauritius); see Chapter 6 for a fuller discussion of likely origins. Nothing seems to be known about the collector Nivoy. Raxworthy et al. (2002) suggested, based on differences in dorsal crest and hemipenis morphology (summarised, somewhat confusingly, in Bourgat 1970, 1972), that Panther Chameleons may be native in Réunion, not introduced. However, Bourgat considered them introduced, and as Malagasy populations are proving very variable (Glaw & Vences 1994, Chris Raxworthy pers. comm.), introduction remains the most likely option, all seven Réunion reptiles known to be native having arrived with the currents from Mauritius, not against them from Madagascar.

223. Delabarre (1844).

224. Trouette (1898).

225. 43,000 contract labourers arrived between 1848 and 1859, with 4,000–5,000 per year thereafter until about 1865 (Defos du Rau 1960).

226. Leal (1878), discussing his first visit to Salazie in 1846, described it as "almost virgin forest that we frequently wandered through".

227. Details in this paragraph from Billiard (1822), Defos du Rau (1960) and Lavergne (1998).

228. Vinson (1888), also paraphrased by Lavaux (1998: 391). Auguste Vinson was a medical doctor, entomologist and folklorist who died in 1903 (*DBR* 2: 97–8). In his 1868 retelling of a local folktale 'Le Bassin du Taureau' in Roussin's *Album* (1860–67, vol.5: 117–120), Vinson evoked local hunters who told this tale and recounted their exploits, speaking, "of grey *fouquets* [Barau's Petrels] taken from nests in the vertiginous ramparts, young tropicbirds which resembled pompoms of white down, and finally the quantities of *collets-rouges* [Rougettes] collected, that bat as big as a hen which provides so delicate a viand"; he also mentioned the legendary bull-hunters as hearing "the monotonous bleating of Merles, the song of *huppes* [Hoopoe Starlings] and the murmurings of a few birds scattered in the foliage". The interest here is the only field reference to Barau's Petrels between Bory's account in 1801 (Bory 1804) and the bird's rediscovery in the 1960s (Jouanin 1967), and the only indication that Hoopoe Starlings had a song.

229. Vinson (1888).

230. For example, the detailed suggestions by Thomas (1828).

231. Fuma (1998).

232. Defos du Rau (1960: 429).

233. Details from Defos du Rau (1960). Defos's map shows settlement contours for this period (1848–80) lower downslope than details in the text; we have assumed the text information is the more accurate. The settlement history map in Girard & Notter (2003) also differs in detail.

234. Gonthier (1860), Azéma (1926); the Science and Arts society was inaugurated by Eduard Manès, *Directeur de l'Intérieur* (effectively deputy governor), brother (see *DBR* 3: 81, 1998) of Gustave Manès, mayor of Saint-Denis and founder of the museum (*DBR* 2: 143–44, 1995). The acclimatisation society, absent from history books, is of more interest to our theme.

235. Details in Maillard (1863), Morel (1926); the Science and Arts society redesigned the building, and Louis Morel, a lawyer, took over as director after both Manès and his successor, botanist Bernier, died unexpectedly. Morel had a good reputation as an icthyologist (e.g. Simonin 1862), but his writings on birds and geckos in Roussin's *Album* are dire. Manès started the collection with specimens bought in Paris out of his own pocket (*DBR* 2:143–44, 1995).

236. Maillard (1863; revised from the 1862 original), Coquerel (1863, 1864), Vinson (1868), Roussin (1862–8). Maillard established his list largely from specimens in Paris, with brief locally-informed tags on abundance, such as TR (= *très rare*), PA (= *peu abondant*), etc.

237. Pollen (1865, 1866), Schlegel & Pollen (1868).

238. Coquerel (1859); few observers distinguished between rat species – Coquerel was the first since d'Heguerty (1754) in the 1730s.

239. Coquerel (1863), Gervais & Coquerel (1866).

240. Pollen (1868), Schlegel & Pollen (1868).

241. Henri (1865); Schlegel & Pollen (1868) blamed Henri for the release, but Henri himself, supported by Coquerel (1865b), said he merely fed and encouraged birds already escaped from elsewhere.

242. Vinson (1868).

243. Schlegel & Pollen (1868). Coquerel (1864) also noted swamphens as found occasionally at the Étang de Saint-Paul, but there is no clear evidence of any feral population as in Mauritius. Vinson's remarks 20 years later (1887), perhaps suggest a spread of Malagasy Turtle Doves, as he included lowland orchards in its distribution in addition to the interior. The lowland birds were said to eat castor-oil (*Ricinus communis*) seeds, "giving the flesh an oily taste and a particular smell".

244. E. Newton (1858–95), Hartlaub (1877), Milon (1951:168); the egret specimen is still in Cambridge (Mike Brooke pers. comm.). The local literature (Coquerel 1864, Maillard 1963) referred to "*Aigrette Herodias calceolata* or *Ardea calceolata*" as very rare, Vinson (1861, 1868) contrasting visiting '*aigrette blanche*' (Cattle Egret) with a "grey species of the same genus that lives and reproduces in our rivers" (Dimorphic Egret, probably conflated with Striated Heron); see also Chapter 3.

245. Schlegel & Pollen (1868).

246. Schlegel & Pollen (1868) used the name '*Procellaria aterrima*' (i.e., in principle the Réunion Black Petrel) for shearwaters commonly caught, according to his informants, using flares, but which must, from their abundance, in fact have been Tropical Shearwaters. They used '*Procellaria obscura*' (= *Puffinus obscurus*, properly a name for the Little/Audubon's/ Tropical Shearwater complex) for a rarer species, probably Wedge-tailed Shearwaters, and finally '*Procellaria cinerea*' for specimens in the Saint-Denis museum that were presumably Barau's Petrels. They took the last name from the existing museum labels, but neither saw the birds live nor brought specimens home. It was probably these same specimens that puzzled Milon (1951) in 1948, not properly identified until 1963 (Jouanin 1964a). Coquerel (1864), illustrating the armchair nature of the locals' activities, conflated noddies and shearwaters under his "Puffin brun (*Puffinus oequinoctialis*, Gmelin) Hirondelle de mer" said to be "Common, nests on rocks by the sea, the young are covered in thick down and their flesh is esteemed as a delicacy".

247. Vinson (1861, 1868); Legras (1863) reported a single pair of quails introduced at Saint-Pierre "nearly 20 years ago", followed by a good bag only two years later. Vinson reported repeated later introductions, but complained that the birds disappeared. The 19th century authors all referred to them as Rain Quails '*C. textilis*' from Southern India (now *C. coromandelicus*), but Carié (1916) identified Réunion specimens as Common Quails *C. coturnix* (probably from Madagascar), as have all subsequent authors (e.g. Barré *et al.* 1996). Vinson did not reveal who was busy introducing gamebirds in the 1840s.

248. Lantz (1887); Crépin (1887) confirmed that the birds were well-established at the forest edge starting between 400 and 700m asl, and also in heathland higher up.

249. Vinson (1876); Lantz (1887) did not report them. Carié (1916) did not mention Chinese Francolins for Réunion, nor did he collect any in the early 1900s (no specimens in Paris, ASC pers. obs.); Hermann (1909) mentioned just one *perdrix*. Maillard (1863) and Coquerel (1864) included it, but probably due to the then-prevalent confusion with Madagascar Partridges (Cheke 1987a); Vinson (1868) omitted it. Berlioz (1946) accepted it (without supporting evidence), but considered the presence of Grey Francolins doubtful – Carié's specimens of the latter did not reach the Paris museum until 1964 (MS accessions book, seen by ASC). The first definitive record of Chinese Francolins is from Milon (1948), but its true date of introduction remains obscure.

250. Coquerel (1864), Eggli (1943), Damour (1973). Leconte's last visit to his home island was in 1842–45.

251. Schlegel & Pollen (1868), Vinson (1861), Coquerel (1864). The ducks identified by Pollen as '*Anas erythrorhyncha*' were more likely to have been, as in Mauritius, Meller's Duck (though other Malagasy *Anas* are possible). The Egyptian Goose is *Alopochen aegyptiacus*; the crows may have been Pied Crows *Corvus albus* from Madagascar.

252. Coquerel (1865b), Vinson (1863, 1868, 1887); see also Cheke (1987a). Charles Coquerel was a Dutch naval doctor, born in 1822, who first visited Réunion in 1846. He was later based in Paris, but travelled widely, visiting Réunion again 1851–54, and finally taking up a surgical post there in 1862, dying at Salazie in 1867 (Grandidier 1867).

253. Maillard (1863); Maillard's '*Sciurus tristriatus*' is now called *Funambulus palmarum* (Prater 1965). The Saint-Denis museum acquired a specimen in 1863 (accession book seen by ASC); Vinson (1868), writing on introductions, mentioned no squirrels.

254. Coquerel (1859), Maillard (1863), Vinson (1868), Cheke (1987a); the introduction of lemurs must pre-date 1851 when Coquerel had last visited Réunion (Grandidier 1867). He (followed by Maillard 1863) tentatively used the name *Lemur* (= *Eulemur*) *mongoz* (Mongoose Lemur) for the Réunion lemurs, but included Mayotte (Comoros) in their distribution, which means he intended *L.* (*E.*) *fulvus* (Brown Lemur; Louette 1999), which fits his pelage description of animals from Ile Saint-Marie and Mayotte. Coquerel and Maillard may not have seen the Réunion lemurs themselves, as Vinson (1868), who clearly had, gave a good description of the black-and-white Ruffed Lemur *L.* (=*Varecia*) *variegatus* and named them *maki vari*, as earlier had Lantz in a list of wildlife sent to Maillard in 1864 (MS in Saint-Denis museum; Cheke 1987a). Newton (1861–82) was told in 1863 of black-and-white lemurs in Réunion. The last mention is from Regnaud (1878) – he was a Mauritian visitor, and followed Maillard's nomenclature so may have been relying on old hearsay (in the original MS the bit about lemurs is an afterthought in the margin). Possibly more than one species was present at various times.

255. As reported by Leal (1878; Cheke 1987a); the intention to introduce birds is retained, but the 1990 edition of Leal omitted the list of species.

256. Vinson (1870, Cheke 1987a) reported the arrival of *Calotes versicolor* off the *Saint Charles* "about five years ago", with details of the lizard's ecology and breeding in Réunion. Trevise said they were introduced 'since 1861' (note 257).

257. Vinson (1868) reported that the snake "had been imported from Madagascar having slipped into sacks of bones, which were

being transported to our colony to make bone black. Harmless to humans, this snake has much reduced the number of mice, and has almost entirely destroyed our gecko with pretty green, blue and red colours that Lamarck had dedicated to Lacépède (*Platydactylus cepedianus*). This changeable gecko is now found but rarely in the eastern part of the island, and will finish by disappearing completely under the voracity of the Madagascar snake." It was in fact *Phelsuma cepediana* from Mauritius that was dedicated to Lacépède (Chapter 6). Given that it was found in the east, Vinson must have been referring to *P. borbonica*, not *P. inexpectata*. The only Malagasy snakes reported from Réunion are *Acranthophis dumerili* and *Liophidium vaillantii* (e.g. Bour & Moutou 1982, Tirvengadum & Bour 1985), there are single specimens of each in Paris with inadequate data (pers. obs.). The former feeds on small mammals (Glaw & Vences 1994, Raxworthy 2003), but *Liophidium* spp. are lizard specialists (Cadle 2003) and may have been temporarily established and attacked day-geckos (Bour & Moutou 1982 – though they also cited an unreferenced quote from 1875 blaming *Lycodon*). *P. inexpectata* only occurs in the dry west; the Marquis de Trevise, whose 1860s sketches are thought to have been done near Saint-Pierre, remarked in 1891 that "at present I am told that not a single one is left, the *caméléons* [Bloodsuckers] introduced to Réunion since 1861 have destroyed them all" (MS quoted in Bour *et al.* 1995). Mahy (1891: 45), another Saint-Pierrois, also blamed the disappearance of day-geckos on the Bloodsucker, which he was told also ate baby chicks, birds and their eggs.

258. Deso & Probst (2007) reported *Lycodon* hunting *P. inexpectata*; stomachs of two killed in the 1980s contained a Bloodsucker and a Cheechak (label note in MNHN, Paris). See Harris (2000) and Cole (2005) for Cheechaks competing with day-geckos. There is little information on when house geckos reached Réunion, though at least one species was apparently present by 1801 (Bory 1804) and both *Hemidactylus frenatus* and *H. brookii* by the 1860s (Paris museum records, pers. obs.); Maillard (1863) included *Gehyra mutilata* based on information from Duméril in Paris, but there are no records or specimens of that date or earlier. From their different haplotypes, *H. brookii* in Réunion appear not to come from the same source population as those in Mauritius and Rodrigues (Rocha *et al.* 2005).

259. The 1864 shrew record is in the Saint-Denis museum accession book; the Strasbourg specimen was sent from the Réunion museum in 1865 (M-D. Wandhammer, pers. comm.), the Leiden one is listed by Jentick (1888). Carié (1916) noted the shrew's presence in Réunion; see also Moutou (1986, unaware of the 19th century specimens), and Cheke (1987a, overlooked the long gap without sightings after the 1730s).

260. Vinson (1868), Hermann (1898), Bour (1981), Probst & Brial (2001). Bour conflated extracts from Hermann (1898) and Mac-Auliffe (1902) into a single quotation (and cited a different book of Hermann's in his bibliography!), leading Probst & Brial to attribute it all to Mac-Auliffe. Bour (without giving a source) added that Rochefeuille, born in coastal St-Louis in 1819, was a well-known character in Cilaos; but we do not know when he moved there and have been unable to trace anything about him.

261. Vinson (1868), Bour (1981); both Radiated and Aldabra Tortoises were imported – Bour gave some figures from contemporary newspaper advertisements, e.g. 6,000 from Madagascar in January 1838, for sale wholesale and retail! Importation from Madagascar and Aldabra was well-established by the mid-1820s (Betting 1827); the animals were often kept alive – Betting even listed forage plants suitable for backyard tortoises. The trade from Madagascar continued into the 1950s (Bour 1981).

262. Hermann (1902); see also Bour (1981).

263. Maillard (1863); Pollen (in Schlegel & Pollen 1868) merely repeated Maillard's remarks. Pasfield Oliver (1881), a British officer visiting from Mauritius in 1864, claimed to have seen *both* flying-foxes in the Rivière Saint-Denis, but his frequent mentions of wildlife appear to have been added *post hoc* to enhance his narrative, with Latin names dutifully copied from Maillard. Seeing 'puffins' and 'cormorants' in Mauritius after a cyclone was equally dodgy! We have excluded all his claimed observations from our tables. Despite this unpromising start, he later did a respectable job of editing English editions of Leguat's and Dubois's voyages (Oliver 1891, 1897), and writing lives of Benyowsky and Commerson (Oliver 1904, 1909).

264. Fuller details in Cheke & Dahl (1981). The two putative 19th century Réunion specimens both date from the 1830s, but could have come from Mauritius. Jacob's letter was to George Mason, and is kept at the London Natural History Museum with a specimen from Desjardins dated 1839, allegedly from Réunion. The other skin is from Bojer's collection, in Vienna; but since the Vienna collection has a Keel-scaled Boa (NMW 21648; Franz Tiedemann, pers. comm.) labelled 'Réunion' (an impossibility), the bat's provenance must be in doubt. The silence in the Saint-Denis museum accessions book is indicative of absence, not proof, but there is also nothing in Roussin's *Album* (1860–68), nor in Lantz (1887). The fact is that there are no post-1831 eye-witness accounts of the Rougette in Réunion.

265. The black parrot was also said to be very rare and confined to remote forests in the 1860s (Vinson 1861, Maillard 1863, Coquerel 1864), but as for the tortoise and bat, there was only hearsay, no eye-witness accounts; Pollen (Schlegel & Pollen 1868) simply repeated Coquerel's comments.

266. Cheke & Dahl (1981), Moutou (1982b); the last record of *Scotophilus borbonicus* is a specimen (since lost) accessioned by the Saint-Denis museum in 1867. The only surviving Réunion specimen is the lectotype in Leiden, see Chapter 4 (note 31) for a discussion of its unresolved relationships. It is possible that a nocturnal animal, if sufficiently rare, could survive unseen, though attempts in the 1970s and early 1980s (Cheke & Dahl 1981, Moutou 1982b) drew a blank, as did a brief survey with a bat detector in 2003 (Jude Smith, pers. comm.).

267. Legras (1861, Milne-Edwards & Oustalet 1893, Cheke 1987a). Vinson (1861, 1868), Coquerel (1863, 1864) and Maillard (1863) had little useful to say, apart from noting the starling's rarity.

268. Schlegel & Pollen (1868); Pollen was in the island on and off during 1864–65.

269. Lantz's letter was paraphrased by Hartlaub (1877), without giving its date – interestingly, Milne-Edwards failed to mention it in his own definitive study of the subject (Milne-Edwards & Oustalet 1893). Vinson's short note on extinctions (1877) adds little to our knowledge of *Fregilupus varius*. There are no *huppes* noted in the Saint-Denis museum accession book starting in 1855, and none in the collection in the 1860s (Legras 1861), though three were later acquired (Milne-Edwards & Oustalet 1893). In one sense Lantz was right, he *did* procure specimens, but not from the wild! Meinertzhagen (1910–11) saw two specimens in Saint-Denis in 1910, but by 1948 there was only one (Milon 1951), still preserved (Barré *et al.* 1996, seen by JPH in 2000).

270. Vinson's account at Salazie in 1831 (Vinson 1888); four live birds sent to Desjardins in 1835, specimens collected (Legendre 1929) in 1832 (Verreaux; skeleton in Cambridge), 1833 (Nivoy, specimen in Paris), c.1833 (Riocour collection, in NHM Tring), c.1835 (Chaumet, four specimens in Troyes).

271. Salvadori (1876), Legendre (1929); we have been unable to locate further information about Lombardi, but he was probably a member of the *spiritains* (= Congrégation du Saint-Esprit) who were the sole order providing priests in Réunion at the time (G. Bernard: *Histoire religieuse de l'île de la Reunion*. This can be seen at http/:col-r.verges.ac-reunion.fr/Dossiers/ Reunion/cdrReligions/Catholiques2.htm).

272. Manders (1911, Brasil 1912); the letter (published in English, but presumably translated from French) goes on to discuss goes on to discuss the birds' diet and the ease of keeping them in captivity.

273. Details of Jacob's life are from Lincoln (undated), summarised in Cheke (1987a); in his flora, Jacob (1895) wrote, rather vaguely, of first botanising in the island "more than 30 years ago".

274. *Huppes* do not appear in 18th century game legislation (Delaleu 1826), but are listed as *gibier* in the 1839 game law (Delabarre 1844). However, no one after Feuilley (1705) referred to eating them: "it ate berries, seeds and insects; the *créoles*, disgusted by the last fact, considered it impure game" (Vinson 1877).

275. Brasil (1912), Cheke (1987a).

276. See Chapter 3 for feeding habits of Mauritian birds; Carl Jones (pers. comm.) suggests a shortage of lizard food for the young may have added to the starling's woes, and competition with mynas may have restricted use of other sources (large insects). See Box 29 (p. 216) and note 274 for Jacob's and Vinson's remarks on its diet; Desjardins (MSS, quoted in Milne-Edwards & Oustalet 1893) said his captive ones in 1835 "ate everything".

277. This was the Spotted Stem Borer *Chilo sacchariphagus*, possibly spread from Mauritius, having appeared there in 1850 (Rouillard & Guého 1999; see Greathead 1971 for details of the confusing nomenclature).

278. Vaxelaire (1999), Leguen (1979). Unlike in Mauritius, malaria in Réunion only gets a footnote in the history books, although it would appear to have been a severe problem until the 1950s. Wealthier residents of Saint-Denis moved uphill to La Montagne and Le Brulé, but in general there does not appear to have been the major population movement (with all the knock-on effects) seen in Mauritius.

279. The leaf rust *Hemileia vastatrix* first appeared in Mauritius in 1880 (Rouillard & Guého 1999), reaching Réunion in 1882 (Jacob 1925).

280. This paragraph is drawn from Defos du Rau (1960), Leguen (1979) Vaxelaire (1999) and Sudel (2002).

281. The history of geranium cultivation is largely from Defos de Rau (1960).

282. Kerourio (1900); see also Rivals (1952).

283. The species is the Black Wattle *Acacia mearnsii*, with some Silver Wattle *A. dealbata* also found (*Flore* 80: 50, 1990).

284. Kerourio (1900), Defos du Rau (1960).

285. Leal (1878) was rather shocked by people in Cilaos being unable to repair their dilapidated houses, while a Mauritian friend (unnamed) of the chief engineer was able to get a holiday home built very rapidly! In 1881 in Mafatte houses were reportedly dismantled to make coffins when foresters refused to allow wood to be cut (Defos du Rau 1960).

286. Mahy (1891).

287. Kerourio (1900); railway sleepers were made from Mascarene endemics *bois maigre Nuxia vertillata* and *petite natte Labourdonnaisia calophylloides*.

288. Kerourio (1900).

289. Oliver (1896).

290. Miguet (1957).

291. Kerourio (1900), Garsault (1900); Garsault's publisher's excuse was that "several very interesting and well documented studies" were too long to include, and others had missed the publication deadline.

292. Jacob (1925); the Étang Salé dunes were planted up to prevent them encroaching on agricultural land, in addition to providing firewood and timber.

293. Respectively *Grevillea robusta*, *Albizia lebbek*, *Acacia* and *Robinia* spp., *Eucapytus* spp. and *Melia azederach*.

294. Miguet (1957, 1973, 1980).

295. Although considered the father of Réunion forestry, Goizet's first name remains a mystery, though his initial was 'G' (Trouette 1898)!

296. From Miguet (1957, 1973) – we have not seen Goizet's original report, which may only exist in manuscript in the Réunion archives; Miguet (1955, 1957, 1973, 1980) never published a formal reference.

297. Oliver (1881); the abundance of palms is despite a massive harvest from further down the Plaine when it was settled: one early settler (date not mentioned) claimed to have exported 80,000 palm hearts in a few years (Lavaux 1998). Apart from a few cultivated ones, there are now no palms on the Plaine des Palmistes (Lavaux 1998) – only a few on steep inaccessible cliffs of the nearby Ilet à Patience and the Takamaka ravine (Cadet 1977, photos in Bénard 2002 and Lavaux 1998: 215). The upland species is *Acanthophoenix rubra* (*Flore* 189, 1984).

298. No palms are visible in photos in Garsault (1900), Mac-Auliffe (1902) or Jacob (1904); Oliver (1896) reported palms on the track from Le Brulé to the Plaine des Chicots – there are none now (Cheke 1987b).

299. La Salle (1845–52); this was no doubt *Rubus rosifolius*.

300. Miguet (1957), Jacob (1895), Lavergne (1978), *Flore* 81 (1997).

301. Cheke (1987b), Lavergne (1978).

302. Lavergne (1978) summarised the distribution (with map) and characteristics of the worst exotic weed species in Réunion, with introduction dates where known; most are shared with Mauritius. **Dry lowlands**: *Prosopis juliflora* (introduced around 1913), *Litsea glutinosa*, *Lantana camara*, *Furcraea foetida*, *Hiptage benghalensis*, *Schinus terebinthifolius*, *Leucena leucocephala* – most of which were well-established in the late 19th century; Mauritius Hemp *Furcraea* can be seen in contemporary photographs (e.g. Garsault 1900:64). **Wet lowlands**: *Casuarina equisetifolia* (invasive on fresh lava, restricting establishment of natives), already so established by the 1890s (Jacob 1895). **Mixed wet forest** (all altitudes, additional to those mentioned in text): *Solanum mauritianum* (pioneer weed, but not persistent under canopy), *Tibouchina viminea* (may prevent regeneration in cleared areas). **High altitude**: Gorse, *Ulex europea*.

303. Oliver (1896); Gorse was present in Réunion before 1825 (Lavergne 1978), but it is unclear when it became invasive in the heath zone.

304. Lavergne (1978), Cadet (1981).

305. *Fuchsia magellanica* and *Fuchsia x exoniensis* are similar shrubby small-flowered species (see Chapter 8), and *F. boliviana* a taller large-flowered form. An unidentified species was cultivated in Mauritius in the late 19th century (Rouillard & Guého 1999).

306. Leal (1878) gave no date for his trip, but Mario Serviable revealed it introducing the 1990 reprint. Maillard (1863) had already noted that some pigs (and also cattle) had recently gone feral at the Plaine des Cafres.

307. Vinson (1887), Daruty (1889). According to *DBR* 2: 97–98 (1995), Vinson was preparing a work on Réunion fauna when he died aged 84 in 1903; if rediscovered, this manuscript might prove interesting.

308. Lantz (1887) was his only paper. The Réunion natural history museum thrived in the late 1800s under Lantz's dynamic curation; he collected in Madagascar and the Seychelles, and exchanged with museums worldwide (Jouanin 1970a). Leal (1878) and Keller (1901; in Réunion c.1887) praised the museum in Lantz's time. Although Lantz published little, the accessions book during his tenure holds much useful information (pers. obs.).

309. In a model piece of archival investigation, Christian Jouanin (1970a) explored the interactions between Newton, Lantz and de Villèle, though he never discovered *where* de Villèle had caught the petrels. For the 1970 specimen, and another in 1973, see Jouanin (1987). There is a brief inadequate biography of de Villèle in *DBR* 2: 93–94 (1995). His grandson, also Auguste and also a sugar engineer, has been a pillar of the local nature conservation society, the SREPEN, since the 1970s; Dubois's Wood-rail *Dryolimnas augusti* is named after him.

310. Carié (1904).

311. The report is from nearly a century later (Albany 1974), confirmed by Armand Barau, successor in the 1970s to Bellier's sugar business (pers. comm.), who also supplied the approximate date of the short-lived *débarcadère*. Another version (Auguste de Villèle, pers. comm.) has it that Bellier, on a boat wrecked on the coast near his property, deliberately released the birds before the ship foundered.

312. Oliver (1896) visited Réunion in 1895; the shipwreck was 'some years earlier'; he noted that "behind the Museum there is a menagerie on a small scale".

313. Hermann (1909), Carié (1916).

314. The introduced Red Junglefowl *Gallus gallus*, of wild Asian stock, escaped ornithological attention until 1948 (Milon 1951), its true identity not confirmed until 1964 (Jouanin 1964b). Jouanin quoted Armand Barau as saying that 'sénateur Crespin' had imported Red Junglefowl from Indochina in the early 1900s. Felix Crépin was Procurator-General in Tonkin (now northern Vietnam) in the late 1890s before retiring back to Réunion (*DBR* 3:57–58, 1998; no birds mentioned). However, Red Junglefowl were present earlier – one was accessioned in the museum in 1892 (Cheke 1987a), though it was possibly of aviary origin.

315. The nearest Carié got to publishing on Réunion natural history was a geographical paper (Carié 1920; contains only fragmentary wildlife information), and occasional comments in his paper on introductions to Mauritius (Carié 1916).

316. Details from Hoart (1825) and North-Coombes (1971, 1994). The trees mentioned are *Foetidia rodriguesiana* and *Terminalia bentzoe*, and the shearwaters *Puffinus pacificus*. 'Squine' (or *esquine*) was used in Rodrigues for *Chloris barbata* (Balfour 1879), but Hoart may have simply meant 'rough grass'.

317. Telfair (1832); John S. Duncan and his brother Philip B. Duncan were successively (1823–26, 1826–54) Keepers of the Ashmolean Museum (Davies & Hull 1976, *CDNB*: 859–860) – it was John who wrote an important Dodo paper (Duncan 1828), and was probably the brother who contacted Telfair, although by then no longer Keeper.

318. Telfair (1832), Telfair *et al.* (1833), Strickland (1848); Strickland said Telfair started enquiring in 1831, which led North-Coombes (1971) to suppose that Eudes had collected the bones in 1831. However Dawkins's trip on the *Talbot* was in 1832 (Telfair 1832, North-Coombes 1991), and Eude's searches arose from assisting Dawkins during his visit (Eudes 1832). Given the Mauritian administration's lack of interest in Rodrigues in the 1830s, with official visits so rare, it is odd that the *Talbot*'s visit does not feature in North-Coombes's history except in relation to caves and bones. Dawkins gets a brief biography in *DMB*:1066–7 (1977, P. J. Barnwell).

319. Eudes (1832); Marragon (1795, 1802) never referred to the *boeuf*, presumably subsuming it under *fous* (boobies).

320. Before the discovery of Eudes's letter, and with the London specimen identified as a Red-footed Booby (editor's note to Telfair *et al.* 1833, Cheke 1987a, 2001a), it was unclear whether Abbott's Booby still survived in 1832 or whether the name *boeuf* had transferred to its smaller and commoner congener (Cheke 1987a, 2001a).

321. Liénard (1842; Box 33, p. 243); see *DMB*:670–1 (1948, L. N. Regnard) for his interesting life; we been unable to find out anything about Descreux or his purpose on Ile Frégate.

322. Bertuchi (1923): "There are a great number of rabbits on this island. Rats are plentiful, some of great size" – very large rats can be assumed to be *Rattus norvegicus*.

323. North-Coombes (1971), Berthelot (2002). Charles Anderson's report (1838) on apprentices in the Mauritian dependencies, referred to livestock and soil erosion but said little on wildlife, and a post-liberation report from magistrate H. M. Self (1841) was similarly reticent.

324. Corby (1845); see also North-Coombes (1971). Corby's mission was to investigate Anderson's recommendation (1838, Dupon 1969) that "the attempt to cultivate Rodrigues should be entirely relinquished, and the whole of the island excepting its shores ought to be devoted to the rearing and pasturage of cattle" to render Mauritius independent of supplies from Madagascar.

325. Cheke (1987a: 51); the fody (passed on to Col. Lloyd) remained unstudied in Cambridge until Edward Newton, visiting Rodrigues in 1864, realised that it was a new species, described by his brother as *Foudia flavicans* (E. Newton 1865a, A. Newton 1865a); the batch of birds from 1845 that ended up in Cambridge included two Red-footed Boobies, a Great Crested Tern and a Sooty Tern (Cheke 1987a: 51).

326. Bouton (1848), Strickland (1848); while on leave in England, Cunninghame's ear was bent by Strickland's naturalist friend Sir Walter Trevelyan (*CDNB*: 3013). Some Solitaire bones were later found and sent to Strickland by Cunninghame in 1849 (*PRSAS* 1849–51:lviii–lix, session of 7 March 1850).

327. Saddul (2002).

328. Corby (1845) estimated 3,000 wild pigs and 30 feral cattle (plus 64 tame ones); oddly he did not mention goats, reported as numerous by Anderson (1838) and Self (1841). Rights to the wild cattle were the subject of rivalry and litigation in the 1840s (North-Coombes 1971: 80).

329. North-Coombes (1971).

330. Gardyne (1846), Higgin (1849).

331. Strickland & Melville (1848); the only earlier illustrations were Leguat's (1707), done from memory years later, and a view of the island from off Port Mathurin, too small to show significant detail, drawn up by Nichelson's squadron in 1761 (reproduced in Pineo 1998).

332. Higgin and Gardyne neither saw nor heard of day-geckos or bats, both recorded by Corby (1845) the previous year.

333. Foreword to *Phil. Trans. Roy. Soc.* 168 (extra volume), 1879; North-Coombes (1991), Cheke (2004b).

334. Duncan (1857), North-Coombes (1971): the latan "is a splendid palm, the seed which it yields in great abundance is excellent food for pigs, so that the inhabitants can keep as many of these animals as they please without having to provide any other food for them" – producing, in due course, a legacy of near total regeneration failure. Duncan's report printed the previously manuscript reports of Hoart (1825), Self (1841) and Captain Kelly's annexe to Corby's report in 1845.

335. North-Coombes (1971).

336. E. Newton (1865a); Edward Newton sent his rough notes to his brother Alfred who published them *verbatim* in *Ibis*, prompting Edward to complain to Alfred in August 1865 that "I did not write my Rodrigues notes for publication, had I done so I think I should have made a better paper of it" (E.Newton 1861–82, letter 3 Aug. 1865) – and no doubt also put species names to the shearwaters and boobies. After Alfred described the fody and warbler in the Zoological Society's *Proceedings* (A. Newton 1865a), Edward whinged further: "I wish Sclater [PZS editor] would not put your papers in the Proc: next to indecent articles; I can not show the Gov[r]. the plate of the Rodrigues birds for fear Lady B [Barkly] should ask what the 'os penis' is"! (letter 3 Sept. 1865).

337. A. Newton & E. Newton (1870), North-Coombes (1971, 1991); the money was voted by the British Association for the Advancement of Science in 1865 (E. Newton 1861–82).

338. Newton was also told of a '*serin*', but given local usage (Cheke 1982a), notwithstanding Newton's own guess and assumptions by Vinson (1964b) and Staub (1976), it is more likely to have been the Fody (*serin zaune* to this day) than *Serinus mozambicus*; the latter was not recorded until 1963 (Cheke 1987a). Lever (2005) followed Staub (1973b, 1976, *contra* Cheke 1987a) who alleged that Zebra Doves and Common Waxbills arrived in 1764, but Staub was misquoting Vinson (1964b) who simply said they had arrived before

1864 (i.e. Newton's visit); Lever cited Vinson, without, we suspect, having read his paper.

339. Colin, in Kennedy (1893) – 36 francolin and 12 quails (*Perdicula* sp.?) were brought from Tranquebar (near Nagapatnam south of Madras); the quails were reportedly killed off by feral cats.

340. Bertuchi (1923); we have not traced any independent record of either '*Teemayma*' or '*Gemima*'.

341. Colin (in Kennedy 1893) claimed that the first pair of deer (1862) came from Borneo (on the *Gazelle*, under Captain Worth) and only the second (1863) from Mauritius, the obvious source (on the schooner *l'Espoir*); the imports (according to Berthelot 2002) were organised by Jenner. The *Espoir* belonged to a Rodriguan trader shipping produce to Mauritius; it had brought Jenner to the island in 1862 (Berthelot 2002), and was lost in December 1863 (North-Coombes 1971). Barthélemy Colin administered the island a number of times during 1890–1900, alternating with other magistrates (North-Coombes 1971); his information seems more reliable than Bertuchi's.

342. Kennedy (1893), Cheke (1987a), Berthelot (2002); The *Rodrigues Game Regulations 1883* (see Cheke 1974, 1987a) were among a number of laws brought in by magistrate O'Halloran, after being given authority to legislate independently for Rodrigues (North-Coombes 1971).

343. North-Coombes (1971: 126) reported up to 200 deer being killed in a day! See also Kennedy (1893, 1900), Broome (1904), Fremantle (1914).

344. Broome (1904).

345. North-Coombes (1971), Bertuchi (1923). Bower (1903) reported that "there was formerly good deer hunting, but the deer have been greatly reduced in number by the inhabitants". Deer are a menace to crops, so it is possible that 'the inhabitants' took matters into their own hands after Jerningham's remarks in 1893. By 1914 the deer had "considerably increased" from the low point a decade earlier (Bertuchi 1923).

346. *Illustrated London News* 28 Nov. 1874 (mostly reprinted in Dupon 1969).

347. Slater *et al*. (1874), Slater (1875, 1879a,b), Balfour (1879a,c); Balfour, still only 21 in 1874, went on to be successively Professor of Botany at Glasgow, Oxford and Edinburgh universities, and was knighted in 1920, dying in 1922 (*CDNB* 1: 128).

348. Balfour (1879b).

349. Balfour (1979b); *Doricera* (= *Pyrostria*) *trilocularis* was abundant, *Foetidia rodriguesiana*, *Fernelia buxifolia* and *Dictyosperma alba* were very common, and *Hyophorbe verschaffeltii* still widespread, though other species were already rare (e.g. *Gastonia rodriguesiana* and *Zanthophyllum paniculatum*) – see Strahm (1989).

350. *Furcraea foetida*, *Agave angustifolia*; *Citrus medica*, *C. aurantium* (and possibly *C. aurantifolia*; *Flore* 65, 1979, M. J. E. Coode).

351. Gulliver (1874–75), Slater (1879b); Slater's 'ravine' is clearly Caverne Tilleul (Brial 1996), where a new educational and tourist-oriented tortoise and native plant reserve has been set up by Owen Griffiths (p. 202).

352. North-Coombes (1971); numbers were 1,108 in 1871, 1,431 in 1881.

353. Slater *et al*. (1874); the Seychelles, a granitic micro-continent, has an ancient lineage of endemic frogs dating back to the break-up of Gondwanaland (e.g. Nussbaum 1984a, Biju & Bossuyt 2003).

354. A. Newton (1872, 1875c); the first paper is dated 10 Nov. 1871.

355. Slater *et al*. (1874).

356. Caldwell (1875), Newton & Newton (1876), Vinson & Vinson (1979). Newton's friend and colleague Caldwell had gone to Rodrigues in May 1875 to investigate Jenner's successor Henry Bell for corruption, and stayed on briefly as magistrate until Desmarais took over in July (North-Coombes 1971). '*Coulevec*' appears to be a distortion of the French *couleuvre* (= harmless snake); this name was reported only the once. The 'long lizard' of that name may in fact have been *Phelsuma gigas*, 'as long as an arm' (Leguat 1707), rather than *P. edwardnewtonii*.

357. Boulenger (1884), and editorial note attached. Vinson & Vinson (1969) suggested O'Halloran had found the day-geckos on an offshore islet. There is no evidence for this; the editor simply said O'Halloran had "sent for the [Zoological] Society a specimen of a large lizard said to be found only in that Island [Rodrigues] and to be very rare there" (the second specimen was sold to the BM(NH) by O'Halloran's wife in 1887, Vinson & Vinson 1969).

358. Mourning Geckos were first collected in 1944 (Vinson & Vinson 1969); they are too abundant in almost all habitats to have been missed by the Transit collectors (who only got Stump-toed Geckos, Günther 1879), although Gulliver's letters (1874–75) refer to dark and light varieties, the former, putatively Mourning Geckos, being 'very beautiful on trees' – suggesting he saw (without realising it) two species, but collected only one. He was ill for most of his time there. A passage in Arnold (2000) suggested subfossils had been found, but this requires confirmation.

359. Slater (1875), Balfour (1879b).

360. Newton (1865a) saw only one frigatebird, so breeding had probably ceased well before Slater's comment.

361. Slater (1875).

362. Slater (1875).

363. Dobson (1879), Balfour (1979b); Balfour was referring to *Doricera trilocularis*, but *Fernelia buxifolia* also acquired the same name (e.g. Koenig 1914b).

364. Günther (1879), *Gehyra mutilata*.

365. Bertuchi (1923), Dobson (1979).

366. Newton & Newton (1976); Vandorous was the half Native American island pilot, unrivalled seaman and general jack of all trades, noted for regularly sailing to Mauritius in small boats to draw attention to emergencies on the island (North-Coombes 1971).

367. North-Coombes (1971).

368. Le Juge (1876), A. Newton (1878), Slater (1879a), Newton & Gadow (1896), Anderson (1913), Cheke & Dahl (1981).

369. Jenner was in charge from 1862 to 1871; details from North-Coombes (1971).

370. North-Coombes (1971); *Casuarina* was absent from Balfour's plant list compiled in 1874.

371. Berthelot (2002), Koenig (1914b); Ordinance No.3 of 1881 established the principle and Regulation No.6 of 1882 set up the reserves.

372. North-Coombes (1971: 137); Colin failed to maintain existing water-works and abandoned "all idea of providing a better quality of water to the inhabitants of Port Mathurin" – by implication his feud with the difficult doctor got the better of his civic duties.

373. W. Morrison in 1863 (in Dupon 1969) noted English and American whalers calling 'frequently', as many as 30 at once in 1864–65 (Beane 1905, per Storrs Olson); Jenner's restrictions are from North-Coombes (1971), and the ban from a rare article on Rodrigues in *MA* 1902: 132.

374. Koenig (1914b), Bertuchi (1923), North-Coombes (1971); Balfour did not mention Grande Montagne, though it is hardly credible that he did not explore the area. It is possible that its vegetation recovered somewhat after mountain reserves were established – it remains the best area for native vegetation today (Strahm 1989, Mauremootoo & Payandee 2002). '*Gandine*' is *Mathurina penduliflora*, a monotypic endemic genus in the Turneraceae (*Flore* 98, 1990, M. M. Arbo).

375. Bertuchi (1923), photos opposite p.59 and 52.

376. North-Coombes (1971) tentatively referred the insect to *Icerya seychellarum*; see also Strahm (1989). *Bois puant* is *Foetidia rodriguesiana*, Lecythidaceae.

377. Details from Dupon (1969: 99), North-Coombes (1971); the situation was further regularised by new rules in 1904 and 1914,

when, for the purpose of the cattle-walk, 'cattle' were defined to include bovines, equines, goats, sheep and pigs.

378. Koenig (1914b).
379. Snell & Tams (1919), Bertuchi (1923).
380. Bertuchi (1923), photo opposite p.53; Snell & Tams (1919) noted "many fine old indigenous trees" in "large rifts . . . often 30 feet or more deep" on Plaine Corail, but commented on the old burnt stumps on the flat surface of the calcarenite outside the collapsed caves. Snell & Tams's paper needs to be treated cautiously in parts, as matter from the literature is indiscriminately interpolated with Snell's actual observations – e.g. under geckos (p.287) there are observations on abundance and egg-sites (probably for *Lepidodactylus*, *Hemidactylus* or *Gehyra*), followed by a misleading unreferenced quote on the species involved taken from the Transit of Venus papers, suggesting *Phelsuma* 'cepediana' (= *P. edwardnewtonii*) was merely rare, when it was by the time of Snell's visit (1918) almost certainly extinct.
381. Bertuchi (1923) has numerous errors of detail, largely because he was a fairly naïve observer, and clearly took no expert advice on historical minutiae, plant and animal names etc. Hence it reads more like a 17th or 18th century traveller's account than 20th century writing (the islanders are all 'natives'!), but also has a certain freshness and lack of prior bias that inspires confidence in the direct observations. A telegraph cable from Australia to South Africa via Rodrigues and Mauritius had been laid in 1901 (North-Coombes 1971); Bertuchi worked at the cable station during the 1914–18 war. Curiously North-Coombes (1971) mentioned both man and book only in his introduction, where Bertuchi's contribution is given 'B-': "cannot be considered satisfactory from the historical standpoint, though his description of what he saw for himself is excellent".
382. It was the synergy between the bulbul *Pycnonotus jocosus* and *Cordia curassavica* that caused the latter's explosive spread in Mauritius (Rouillard & Guého 1999).
383. North-Coombes (1971: 123), Cheke (1987a).
384. Snell & Tams (1919); North-Coombes (1971) generously allowed Rodriguans five years of cropping before abandoning a plot; it was probably variable, depending on site and farmer. By 1914 much of this shifting cultivation must have been on land under invasive Rose apple or popinac – squares of such clearance are visible in Bertuchi's photos of Mt.Limon and St.Gabriel (1923).
385. Koenig (1914) North-Coombes (1971). Attitudes to state lands were still the same in the mid-1970s (Cheke 1974); the 1975 *Rodrigues Island Land Suitability Map* (FAO/UNDP, 1:20,000) echoed Koenig in designating La Ferme as the best area for mixed cropping.
386. Bertuchi (1923) mentioned former Sooty Tern colonies on Coco, Sable and Gombrani islets. Snell & Tams (1919) mentioned "a few white terns and boobies" nesting on Iles Coco and Sable, but 'boobies' must be an error for noddies, possibly taken from the by then-outdated Transit reports.
387. Slater (1875), Sharpe (1979); Slater gave no evidence of Great Crested Terns breeding, and his specimens were in non-breeding plumage, but the existence of the local name, the specimen from 1845 and the subsequent virtual absence of the species suggests a breeding population that suffered the same fate as the Sooty Terns. See Cheke (1982a) for creole bird names.
388. North-Coombes (1971: 175).
389. Bertuchi (1923) said *Casuarina* was first planted on Iles Cocos and Sable in 1884, which fits the size of tree in which a nestling Fairy Tern is shown (p.82).

CHAPTER 8. THE LIMITS TO GROWTH

1. Moutou (1983b).
2. Clergeau & Mandon-Dalger (2001) discussing the release of Red-whiskered Bulbuls on Réunion.
3. Crepin (1931), *DMB* 69–70 (1941, L. Halais), *MA* 1912: D29, E48.
4. "The role of the Royal Society of Arts and Sciences was thus necessarily eliminated, and its President Albert Daruty de Grandpré and its members succumbed, one, in his great modesty, to the inevitable influence of progress in the principal industries of the country, while the others were absorbed by the new elements of transformation in the rural economy of the colony" (Koenig 1932; see also notes on the RSAS in *MA* 1914: A69 and Ly-Tio-Fane 1979).
5. As an entomologist d'Emmerez had a good reputation, though Carié's almost hagiographical praise (1919) rather over-egged the pudding; beyond insects his standing was poor. The self-taught but meticulous palaeontologist Etienne Thirioux, complaining in 1906 of d'Emmerez's intention to publish on Mauritian reptiles, wrote that "he's not up to it, that is to say that he's not done any specialist study for such a task, and in any case he's occupied with other things; he's an *arriviste*, no more" (MS correspondence between Thirioux and Alfred Newton, Cambridge University Zoology Museum). Thirioux's opinion is confirmed by the muddled references to lizards in *Mauritius Illustrated* (d'Emmerez 1914). Colonial official Alfred Harding referred to d'Emmerez in 1912 as "the ex-Curator of a slovenly and useless Museum" (Storey 1997; see *DMB*: 1561–63, 1995, P. J. Barnwell for details on Harding); he was in fact still curator in 1912 (Tirvengadum 1980, *DMB*: 44–45, 1941; L. Halais and A. d'Emmerez de Charmoy).
6. Carié (1904, 1916); there is a photo of a captive Pink Pigeon in the 1916 paper; Meinertzhagen (1912) reported that "several pairs of this [Pink] Pigeon are kept in private aviaries, but nobody has yet been successful in breeding them in captivity". Georges Antelme corresponded with James Greenway, probably in the late 1930s, providing the only recent(ish) information from Mauritius for his book on extinct and endangered species (Greenway 1958). For deer see Antelme (1932). Pelte (1938), Mamet (1996b) and *DMB*: 193 (1942, L. Halais) gave biographical details; Antelme died in 1938. Georges's brother Henri's article on hunting (Antelme 1914) has notes on monkeys, hares, gamebirds *etc*,. as well as deer (see *DMB*: 2, 1941, J. L. Urruty).
7. Huron (1920, 1923). On Mascarene Grey White-eyes he wrote (1923) that these birds, "formerly very numerous, had almost disappeared, but for a while now we have, with great pleasure, been seeing them again more or less everywhere" and that "they must have found a way of taking cover from the execrable bulbuls". Paradise flycatchers were evidently still widespread, though Huron did not mention most endemic forest species, suggesting he spent little time in native forests.
8. See Cheke (1987c), Jones (1987), Cheke & Dahl (1981). There is some published information for birds in Guérin (1940–53), but it is patchy and not always reliable.
9. A small area of dry forest (7.8ha) at Yemen is now a private nature reserve under RSAS auspices (Guého 1996); there is some good lowland forest on state land in the lower parts of the Black River Gorges National Park (Morne Sèche, p. 239).
10. Land around Grand Bassin (including Bois Sec) was compulsorily purchased by the state following Thompson's recommendations, and the Bois Cherie tramway, built for sugar-cane extraction, was extended to Kanaka in 1905 (*MA* early 1900s, *passim*, Koenig 1914, Brouard 1963, *DMB*:1209: 1984, G. Rouillard).
11. Edgerley (1963); a few isolated *nattes* (*Labourdonnaisia glauca*) from this experiment were still standing in the 1970s (pers. obs.).
12. Lamusse (1990).
13. Total Crown forests were given as 64,000 *arpents* (66,816 acres) in 1935 (Sale 1935b) and 67,151 acres in 1950 (Allan & Edgerley 1950). In 1928 alone 129 tons of *natte Labourdonnaisia glauca* and 13 of Makak *Mimusops petiolaris* were taken from the Midlands reservoir site, but this had dwindled to insignificance by 1931 (*Annual Report of the Forest Department* for 1928–31).

14. Governor John Pope-Hennessy is credited with introducing Chinese Pine from Hong Kong in 1884, Loblolly Pine *P. taeda* was first planted in 1910 and Slash Pine *P. elliotti* in 1929–30 (Sale 1935a, Edgerley 1963, Brouard 1963).

15. Most of this paragraph is from Brouard (1963), with details also from Koenig (1926), Sale (1935a,b), King (1946), Edgerley (1963) and *Annual Report of the Forest Department* for 1921–23, 1928–31. In Mauritian literature the Chinese Pine is often called *Pinus sinensis* and the Slash Pine *P. caribea*, but see *Flore* 28:1–2 (1997, W. Marais).

16. Information from Gleadow (1904), Sale (1935a,b), *Annual Reports of the Forestry Department* for 1921–47, Koenig (1926), Vaughan & Wiehe (1937), King (1947), Edgerley (1963), Brouard (1963, 1966). From about 61ha in 1904, by 1935 there were 1,683ha of conifer plantations, as against 43,500ha of 'broadleaved' forest (much of it native, at least in part) (Sale 1935b). In the 1939–45 war "indigenous forests were drained of 168,000 cu. ft log volume. This came mainly from poor relic forest covers without shelterwood, in areas where removal was due under the pine plantation working plan" (King 1947). The clearances around Nicolière included the construction of a dam and large reservoir in 1929.

17. Jones (1987: 280); the picture for the Mauritius Fody is similar (Cheke 1987c: 199), but with an additional outlier on the (then) still-native Montagne Nicolière.

18. Although Vaughan & Wiehe's map (1937) showed no native forest in the Candos–Vacoas–Trois-Mamelles triangle, Frank Rountree told ASC in 1976 (pers. comm) that degraded native forest persisted along streams and around Candos hill, sheltering native fodies and paradise flycatchers (Cheke 1987c).

19. Carver, *Annual Reports of the Forestry Department* for 1946, 1947. Edgerley (1963) mentioned *Cryptomeria* and other non-pine conifer plots in the Grand Bassin area dating from *c*.1918–20, but existing plots are of younger trees (pers. obs.), apparently from the 1946–7 plantings.

20. Details of economics and areas under cane from North-Coombes (1993), Rouillard & Guého (1999); see also graphs in Cheke (1987a) and Rouillard (1990) and 1854 and 1905 maps in Toussaint (1953) compared with Vaughan & Wiehe (1937). Some land around Albion in the west and Cap Malheureux in the north had apparently already reverted by the mid-1930s to *Furcrea* or *Cordia* scrub (Vaughan & Wiehe 1937).

22. d'Emmerez (1914), Greathead (1971).

23. Only Regnard's biography in *DMB*: 30–31 (1941, A. d'Emmerez de Charmoy) mentions that he released this toad; all other sources are mute about how the toad arrived. Despite his ecological gaffes, Regnard was no out-of-control maverick, but a respectable businessman and agriculturalist specialising in introducing and improving fruit-trees, also sometime Vice President of the RSAS.

24. For toad introductions to Mauritius see Jepson (1936), Moutia & Mamet (1948), Vinson (1953) and Greathead (1971), and note 23. The first mention of *Bufo gutturalis* in print was a laconic "toads have been introduced recently" in 1927 (faunal summary, 1927–28 *Mauritius Almanac*). Easteal (1981) and Lever (2001, 2003) reviewed Cane Toad releases worldwide, but had incomplete data for Mauritius. Its effects on native wildlife are still under-researched, but the species is omnivorous and takes vertebrate and invertebrate prey up to the size of a mouse, and has poison glands which can kill cats, dogs and snakes (Tyler 1994, Sherley 2000, Lever 2001); its invasion of Australia (see Tyler 1994, Lever 2001, 2003) is notorious. The Cane Toad's failure to establish in Mauritius implies a tadpole predator in the largish ponds it uses for breeding, as there is nothing that could tackle the adults. Successfully introduced *B. gutturalis* breed successfully in tiny temporary pools free from aquatic predators (pers. obs.). Cane Toads, introduced in the 1980s via the American base, have invaded Diego Garcia (Lever 2001, 2003). The atoll was formerly administered from Mauritius, which still claims it; should sovereignty revert, great care would be necessary to ensure no toads were brought to Mauritius. Bour's claim (1984b) that Mauritius already harboured Cane Toads is in error.

25. Griffiths (1996), Griffiths & Florens (2006).

26. See Arnold & Jones (1994), Cole et al. (2005); the exception is on Flat Island in very particular circumstances (Chapter 9). On Pacific islands Cheechaks can displace even widely successful species like Mourning Geckos (Hanley et al.. 1998).

27. Arnold (2000), Harris (2000).

28. Moutia & Mamet (1948); Greathead (1971) wrongly gave the *Echinops telfairi* introduction as 1933, the date the details were published. Rouillard (1990) and Rouillard & Guého (1999) discussed *Tenrec ecaudatus* as beetle predator.

29. Bour (1984b), Lever (2003); The Wattle-necked Softshell Turtle was first reported by Vinson (in 1953: "a small [!] soft-bodied marsh species", not identified); they escaped from a Dr Athouam (about whom we can find nothing). Das (1991) and Daniel (2002) mentioned waterbirds killed by *Aspideretes* (=*Palea/Trionyx*) *gangeticus*.

30. Bour (1984b) found no 20th century records.

31. Amending Ordinance No.10 of 1905 prohibited the introduction of "noxious animals, birds or reptiles"(*MA* 1927–28: D25); this was revised by Proclamation No.30 of 1923 to "restrict the importation of mammals, birds, reptiles and fishes" (*MA* 1924/25: D96). Regnard's biographer André d'Emmerez recorded in 1941 that "for a time it [the toad] was subject to an extermination campaign" (*DMB*: 31).

32. There are numerous examples of official agencies with very specific targets persistently introducing ecologically disastrous species without regard for the wider consequences – examples are the Small Indian Mongoose *Herpestes auropunctatus*, Rosy Wolf Snail *Euglandina rosea* and Western Mosquitofish *Gambusia affinis*, which despite their well-known effects continue to be widely introduced in misguided attempts to control rats, giant snails and mosquitos respectively (Lowe 2001).

33. Details from Wiehe (1946a,b), Greathead (1971) and Rouillard & Guého (1999); the controlling insects were a beetle *Schematiza cordiae* and a chalchid wasp *Eurytoma howardi*.

34. Vaughan & Wiehe (1937, 1941); Vaughan died aged 91 in 1987 (see Antoine 1988, Lorence 1988 and *DMB*: 1784–86, 2000, G. Rouillard). Wiehe went on to be the first director of the Mauritius Sugar Industry Research Institute (MSIRI) and in 1968 became the first Vice Chancellor of the new University of Mauritius, dying in office in August 1975 (Rouillard 1979, Lorence & Vaughan 1992:16; *DMB*:1794–96, 2000, P. Harel & G. Rouillard).

35. King (1946).

36. This legislation was part of a raft of measures promoted by Governor Sir Donald Mackenzie-Kennedy to improve Mauritian cultural coherence in the wake of the 1939–45 war, which had brought food shortages, riots, inter-communal tension and an Indo-nationalist movement based on experience in India (Barnwell & Toussaint 1949). See Tirvengadum (1980) for a history of the Mauritius Institute; *DMB*: 1784–86 (2000, G. Rouillard) covered Vaughan's involvement in conservation legislation.

37. Vaughan's 1951 remarks (*PRSAS* 1: 95) came in discussion following a talk by Frank Rountree to the RSAS on 20 September 1950.

38. Echo Parakeets and Mauritius Cuckoo-shrikes favoured (and still favour) the Macabé-Brise Fer area, but the other species have progressively declined there, as have all species (apart from fruit-bats and cuckoo-shrikes) in Bel Ombre (Cheke 1987c, Safford 1997c), though reintroduced Pink Pigeons and Mauritius Kestrels do not avoid these areas (Chapter 10). Vaughan is reported to have tried to prevent then Conservator Leo Edgerley from driving roads through native forest in and near 'his' reserves, but Edgerley got his way (Carl Jones, pers. comm.) this must have been in the 1950s, as

Edgerley retired in 1962, though Sale had started a network of motorable roads in the 1930s (Brouard 1963).

39. King (1946).

40. In reverse, when visiting northern Somerset in the 1980s, ASC recognised immediately that coastal woods west of Porlock showed virtually no regeneration and symptoms of heavy deer browsing – Red Deer *Cervus elephas* were indeed over-abundant, crammed into the steep narrow forest strip where the Exmoor hunt was unable to follow them.

41. Brouard (1966).

42. Owadally & Butzler (1972), Douglas (1982), Procter & Salm (1975). Owadally & Butzler commented on de-barking of one endemic tree *Pittosporum senacia*, but saw no other evil.

43. In the 1970s deer density was estimated at 10–40 animals per km^2 in native forest, their least favoured habitat; in forested hunting estates it reached 70–200/km^2 (Noel 1974, Douglas 1982, Cheke 1987a). In Europe only <2.5/km^2 is acceptable in forestry plantations, though in New Zealand, forest regeneration is considered acceptable at 5–7/km^2, though the composition is modified towards the least susceptible tree species (Cheke 1987a).

44. Cheke (1978a, 1987a).

45. Owadally (1980), Jones & Owadally (1982), Lorence & Sussmann (1986, 1988).

46. For a fuller discussion of deer density see Cheke (1987a: 12–13); current policy is stated in Case Study 2.14 in Wittenberg & Cock (2001).

47. Proposals ASC put forward in 1985 (Cheke 1985b) for a forest dynamics study fell foul of covert resistance from Wahab Owadally, then Conservator of Forests, and Wendy Strahm, the botanist promoting (non-experimental) weeded exclosures – apparently they saw the project as treading on their toes, although the intent was complementary research to discover how best (and with least effort) to promote forest regeneration. In her thesis, however, Strahm (1993) stressed the negative impact of introduced animals on the forest; see also Strahm (1994, 1996a) and equivalent work in Hawaii (Cabin *et al*. 2000, 2002). Owadally once told Carl Jones that, in relation to conservation and forest policy "what I can't control I'll destroy".

48. Strahm (1993, 1996a), Dulloo *et al*. (1996), Mauremootoo *et al*. (2002), Mauremootoo & Towner-Mauremootoo (2002). Vaughan himself established the first fenced and weeded forest patch in 1969 – the tiny 1.44ha Perrier reserve, by 1960 surrounded by pine, eucalyptus and invasive weeds (Vaughan 1969, Owadally 1971 and subsequent Forestry Service reports), Guého & Thiolley 1990 (with erroneous dates), Strahm 1993). This exclosure demonstrated that native forest could be induced to regenerate, and thus inspired all the later work (Strahm 1993).

49. Carl Jones, pers. comm., partly based on David Hall's unpublished rat studies. Shrew density on Ile aux Aigrettes is higher than on the mainland to a spectacular degree (pers. obs.). Rats were almost eliminated in 1987, but a few survived and were poisoned in 1991 (Strahm 1993); subsequent attempts to eliminate shrews failed (Varnham *et al*. 2002). Probst (1995c) attempted to census shrews in 1990 following the first rat eradication attempt, estimating 400–600 for the 25ha island; the trapping programme in 1999 caught 759 shrews in three months (Varnham *et al*. 2002). A similar increase in shrews has also occurred in upland forest where rats have been poisoned for predator control (Carl Jones, pers. comm.).

50. Antelme collected birds and other animals which he later presented to the Mauritius Institute. Guérin's *Faune ornithologique* (1940–53) is largely an uncritical compilation from the literature with little input from his own observations. D'Emmerez's contribution (1928) lacked substance.

51. Cheke (1987c), Jones (1987).

52. Through a misinterpretation by Oustalet (1897), compounded by Meinertzhagen (1912) mistaking Ile Plate (Seychelles) for Flat Island (Mauritius), and total confusion by Guérin (1940–53), the Red-footed Booby *Sula sula* became established in the literature as breeding on Mauritian offshore islets (e.g. Rountree *et al*. 1952, Feare 1978, Le Corre 1999). In fact it never bred in or around Mauritius at all (Cheke 1987a, 2001a); the nearest colonies were Rodrigues and St.Brandon.

53. Frank Rountree's birdwatching notebooks (copies in ASC's possession). Rountree was first posted from the UK to Mauritius as a policeman in the 1920s, but only started making systematic bird notes in 1949. He retired to Sark in the Channel Islands, where ASC discussed Mauritian birds with him in 1976; he died in 1987 (*Ibis* 130: 478, 1988).

54. Rountree *et al*. (1952).

55. Rountree (1951).

56. Vinson (1956) wrote in March 1953 that "indigenous plant communities cover 66 sq.miles [170.9km^2] or 9 per cent of the island (though much of this area is invaded by exotics), but with the post-war shortage of timber this area is being rapidly reduced in favour of quick-growing exotics, mainly conifers, and it is to be feared that in the near future it will only consist of some 5,000 acres [20.2km^2] of forests proclaimed National Reserves by law and intended to be preserved in their original state". In the same paper he proposed new and more comprehensive bird protection legislation. The ICBP branch's original members were René Guérin, Stanislas Pelte, Jean Vinson and Robert Newton (*PRSAS* 1(4): 409–10, annual meeting of 16 Mar. 1955).

57. Newton (1958b), including the full text of the new law; Robert Newton (1958a) was on the island 1954–57.

58. Newton (1958a, 1959), Gill (1971b).

59. Cheke (1987a); until 1977 bird protection depended on proclamations under the Game Ordinances (last revised 1939) that had to be regularly renewed, but a new law that year made the protection permanent. The active involvement of internationally sponsored wildlife research was the trigger for improved laws from 1973 onwards, as it was for extending nature reserves (see below). The Wildlife Act was No.33 of 1983.

60. Jackson (2001); Barnwell & Toussaint's account (1949) is rather brief.

61. Roy (1969) summarised the history of tea in Mauritius to 1968; see also North-Coombes (1948).

62. Allan & Edgerley (1950); see also Brouard (1966), who argued, like Allan & Edgerley, in favour of retaining large areas of prime catchment under forest, while recognising the job-creation potential of tea.

63. Figures from Roy (1969), Meade (1960) and Anon. (1968).

64. Figure from the government's 4-year plan (Anon. 1971a). The next plan (Anon. 1976) cited the same acreage, but by 1983 the area had shrunk slightly to 3,900ha (Anon. 1985). Guébourg (1999) quoted a figure of 6,230ha for 1968, but as this matches the *projected* figure cited in 1966 (i.e. 6,273 ha), we think this area hypothetical, though presumably supplied by the Ministry of Agriculture; Guébourg's figure for 1985, 3,770ha is clearly about right. Devaux (1983) quoted 5,800 ha under tea (4,150ha on government land); it seems that the 4-Year Plan figure (4,221ha) was only for government plantations, and excluded the large private estates of Bois Chéri and Corson (rough count of km squares on OSI 1:100,000 map 1994 and Devaux's map). If so, it does not affect our discussion, as the expansion of the Bois Chéri plantation 1967 onwards was into derelict land, not forest (comparing DOS map Y682 of 1971, using vegetation data from 1967, with the 1994 map, also Y682).

65. This is most clearly seen on OS International (formerly Directorate of Overseas Surveys) 1:100,000 map Y682/DOS 529, edition 5, published 1994 but clearly using 1980s land-use data.

66. Safford (1997b) gave a figure of 4,100ha of native forest within the 6,570ha National Park.

67. Area figures and economic information from Guébourg (1999) and *Food and agricultural statisics of Mauritius for the '90s* at http://ncb.intnet.mu/moa; other economic activities, mostly industrial, had enormously improved Mauritian prosperity in the decade 1985–95 (Bowman 1991, Dommen & Dommen 1999, Guébourg 1999, pers. obs.) resulting in higher urban wages militating against arduous low-income rural jobs like tea-picking. ASC saw the bulldozing in 1996.

68. Parc-aux-Cerfs is shown as 'tea' on the forest map in Anon. (1968), but was not taken up.

69. 2.98km^2 of the projected 4.38km^2 was operational in 2004, all but 28ha of which was 'forest/scrub land' – the rest being a squatter village and tea (Jogoo 2003, Anon 2004, Sahai 2004); when complete it will be the largest reservoir in Mauritius.

70. Bowman (1991) described *Travail Pour Tous* as "little more than a government-sponsored make-work scheme".

71. Anon. (1971a); no details were given in the Plan, but the land was first turned over to firewood-cutters and charcoal-burners (there was little commercial timber); when all usable wood had gone, it was then manually cleared before being planted (pers. obs.).

72. The 4-Year Plan (anon. 1971a) referred to the International Development Agency for tea, but only to 'external finance' for the *Travail pour Tous* programme. Procter & Salm (1975) noted that 'funds and technical direction for this forestry programme have been drawn from overseas'. However to those active at the time (ASC, John Procter, Stanley Temple etc.) it was well-known to be funded by the World Bank, and was so attributed by Cheke (1987a). Temple petitioned Robert Macnamara, then World Bank president, through ICBP president S. Dillon Ripley, to halt the habitat destruction at Les Mares (ASC's personal records). Mannick (1979) mentioned World Bank funding for rural development and agriculture, but like other recent politico-economic books (e.g. Ramgoolam & Mulloo 1982, Dommen & Dommen 1999), and the World Bank's own papers (e.g. Devaux 1983), made no mention of forests. By contrast Meade's famous report (1960) covered forestry in some detail.

73. Absorbed by the Ministry of Agriculture and renamed 'Forestry Service' in 1968 (Brouard 1970).

74. The Forestry Service *Annual Reports* give scant details of the DWC's plantation work, but the areas cleared for new plantations were roughly mapped by Cheke (1987c), revised by Safford (1997b).The DWC's forestry section was disbanded in 1984, all work reverting to the Forestry Service (Owadally 1985). "In 2003 a major decision was taken to set aside 50% of the planted forests found on catchment areas as protection forests for soil and water conversion [*sic* = conservation]; these areas will not be logged except for hygienic fellings and for salvaging timber after cyclones" (*National report to the 5th* session of the UN Forum on Forests for Republic of Mauritius, October 2004 on www.un.org/esa/forests).

75. Anon.(1971a, 1976, 1983a, 1985), Scott (1973), Procter & Salm (1975); for Gerald Durrell's involvement see Durrell (1977b) and many subsequent articles in *On the Edge* and *Dodo*.

76. Cheke 1987e.

77. The BOU Mascarene Islands Expedition (led by ASC) and Stanley Temple's project sponsored by ICBP, WWF and the New York Zoological Society (Diamond & Cheke 1987). After an intended expedition to Cuba had to be abandoned, a timely visit from France Staub attuned the BOU to the Mascarenes. Temple's project arose from an American initiative in November 1969: "while we were there [in Mauritius], Mr Edgerley received a letter from some American enthusiasts suggesting that three pairs [of Mauritius Kestrel] be caught and bred in captivity, in an attempt to prevent this species from becoming extinct. His reply was that "nobody can say whether even one pair exists today!" (Fabian 1970). Leo Edgerley was a retired Conservator of Forests, and the approach was presumably from Tom Cade, later Temple's sponsor. In 1970 France Staub chaired the Mauritian ICBP branch, and for many years was almost the only Mauritian ornithologist; he died aged 85 in 2005 (Cheke 2006b).

78. Cheke (1987c,e).

79. Safford (1997c); cuckoo-shrikes extended their range into lower altitude forest (Black River Gorges, Bel Ombre & Combo), some of it partly exotic, between 1975 and 1992. The Mauritius Merle was always more catholic in its choice of habitat, and very mobile (Safford 1997c).

80. Jones (1987); Jones suggested parakeet numbers in 1973–74 may have been lower than generally accepted, and had certainly dropped to perhaps only 14 by mid-1975. Of the six kestrels in 1974–75, two were in captivity and four wild, only one pair breeding.

81. The name *Pigeon des Mares* (first reported 1904) appears to have been coined only after the bird became restricted to this part of the island (Cheke 1982a).

82. Cheke (1987c), Jones (1987), Safford (1997a,d), Jones *et al.* (1999). In addition to nest predation, adult pigeons feeding on the ground are attacked by cats and mongooses; neither would have found untouched Les Mares to their taste.

83. Safford (1997d); screw-pines *Pandanus* spp. are generally found to be good nesting sites (refs in Safford 1997d).

84. In the 1970s ASC never saw cats in the upland forest, despite often finishing fieldwork at nightfall and driving along the deserted forest roads at night. In 1996 and 1999, cats were often seen, not only in broad daylight at tourist spots (raiding litter bins) but also at night on roads throughout the area. Since neither Norway Rats *Rattus norvegicus* nor cats were studied in the 1970s there are no hard figures, but, in the absence of other major changes in the forest, an apparent increase in Norway Rats (Lance Woolaver, pers. comm.), providing regular year-round ground-living prey, would account for the increase in cats. However, Roger Safford (pers. comm.) says Norway Rats were still scarce in the early 1990s, and cats rarely seen, so the increases may have more to do with recent tourism and consequent litter than the DWC's clearances. The Forestry Service (Anon. 2002c) has highlighted an additional source of cats: "Domesticated cats (*Felis* spp.) that have been released by visitors have started to proliferate in the forests and are causing some harm to wildlife. This practice must be curbed and action will have to be taken to get rid of the ones found in the forest". Abandoning pets in the wild also happens in Rodrigues. Although only 9% of feral cats' diet is birds (46% rats, 23% shrews and tenrecs), Pink Pigeons are particularly vulnerable to them (Roy *et al.* 1998).

85. Roy *et al.* (2002).

86. Vaughan & Wiehe (1937) described Les Mares as it was; for feeding ecology of Echo Parakeets, Pink Pigeons, Mauritius Fodies and Mauritius Olive White-eyes, see Cheke (1987c), Jones (1987) and Safford (1991).

87. Jones (1987), Jones *et al.* (1998).

88. Jones *et al.* (1998); Mauritius Merles and immature Mauritius Fodies also wander outside the breeding season, but are not as tied to specific foods as Mauritius Olive White-eyes (Cheke 1987c, Safford 1996, 1997g).

89. *Mangeur de poule* is a common French dialect name for any bird of prey, and the Kestrel was probably so called for that reason, not through any predilection for chicks (Cheke 1982a). In fact it prefers arboreal geckos, though will take other lizards, insects and small birds (Jones 1987, Jones & Owadally 1988, Jones *et al.* 1994b).

90. "The hawk *Falco punctatus* has, almost certainly, been reduced to below the point at which any revival is possible. I do not know the reason" (Newton 1958b), and "this falcon seems to be reduced to a handful, and, for some unexplained reason, to be on the verge of extinction" (Newton 1960). Newton (1958a) added: "Presumably [it] was too closely adapted to some special feature of the old Mauritian environment to be capable of meeting change, and although I

see no reason why it should not still flourish in the south-western forests and mountains, I have only seen it four or five times since I arrived in 1953". In the mid-1970s the Mauritius Kestrel was still thought to require 'mature evergreen forest' for survival (e.g. Temple 1975), and in 1978 Temple (1978b) blamed the crash on a supposed unadaptive shift in nesting from cliff cavities to tree-holes, in fact entirely mythical (Jones 1987). As late as 1980, Jones (1980a) commented that "the low density of the kestrels within the habitat now available is difficult to explain, but may be natural".

91. This paragraph largely follows Safford & Jones's review (1997), which developed ASC's suggestion (in Pasquier 1980) that DDT might have caused the Kestrel's decline. For details of past Mauritius Kestrel numbers see Jones (1987) and Jones et al. (1994). For the history of malaria control see Mamet (1979; also Ricaud 1975 and Cheke 1987a); DDT spraying was reduced after 1965 when the principal vector *Anopheles funestus* was controlled, continuing at lower intensity against *A. gambiae* until 1973. The only endemic birds still found in the Bamboo Mountains are Mauritius Merles and Mascarene Grey White-eyes (Malcom Nicoll, pers. comm.).

92. Nichols *et al.* (2002).

93. *MWF Newsletter* 7(1): 3, April 2007, see also Chapter 10. Like other island fodies, Mauritian birds, freed from nest predators, have responded spectacularly, with some females laying five clutches per year! (Carl Jones, pers. comm.).

94. Safford (1991), Nichols (1999), Nichols *et al.* (2004, 2005b). Olive White-eye nests have rarely been found, and nearly all have failed (Nichols *et al.* 2005a); by early 2006 only two successful nests had been seen (in 2000 and 2005: Nichols & Woolaver 2003, Carl Jones, pers. comm. to Roger Safford). In the 2005–06 season, four young were reared in captivity from nests at risk of predation (*On the Edge* 100: 10, 2006); releases were reported on www.mauritianwildlife.org (accessed 4/2007). In the absence of other candidates, the ubiquitous bulbul looks the most likely candidate for depressing breeding success, possibly across all native passerines, which managed quite well with rat predation alone before the bulbul arrived but seem unable to cope with both together (Carl Jones, pers. comm.). If correct, we can expect declines in Réunion once bulbuls become thoroughly established throughout the habitat of native species. The area of suitable habitat in Mauritius should support much larger populations than currently exist of all passerines (except perhaps cuckoo-shrikes), judging by the densities of relatives in Rodrigues (fodies), Réunion (white-eyes, merles, paradise flycatchers), the Seychelles (fodies, merles) and elsewhere.

95. Iben Hansen & Cristian Marcet, work in progress (pers. comm.). Merles also raided white-eye nests, though they are so much rarer than bulbuls that they cannot be considered a major source of loss.

96. Hansen *et al.* (2002).

97. Cheke (1987c) gave the most detailed account.

98. No monkeys were seen during 1973–75 (Cheke 1987c), though they reportedly arrived subsequently (Cheke 1987e), and Safford (1994) implied some presence in "monkeys are certainly rare there". Damholdt & Linnebjerg (2003) said they were 'abundant' in 2002, though Owen Griffiths (pers. comm.) differs: "there are not a lot there, and trapping is pretty fruitless" – there were still plenty of paradise flycatchers. Monkeys were allegedly brought to Bras d'Eau by flamboyant politician Gaëtan Duval, who had a *campement* (beach cottage) nearby, and reputedly enjoyed eating *zaco* (monkey) (ASC's notes 1985, source not recorded).

99. For paradise flycatchers in Mauritius see Staub (1971), Cheke (1987c), and Safford (1997c,e). In 1996 ASC revisited sites with birds in 1973–75 but not re-surveyed by Safford. No birds were found in plantations under Le Pouce (habitat unchanged), near Piton du Milieu (native forest patch recently bulldozed), at Pamplemousses Gardens or in the plantations at Powder Mills.

100. Nichols (1999), Damholdt & Linnebjerg (2003). The dense cluster of paradise flycatchers around Bassin Blanc in 1973–75, which persisted to 1993 (Cheke 1987c, Safford 1994, 1997c), had gone by 1996 (pers. obs.); none were seen there in 2002, though one was found dead on the adjacent road (Damholdt & Linnebjerg 2003: 22).

101. Safford (1994: 52) suggested the temporary crash at Pigeon Wood was caused by paradise flycatchers eating poisoned flies after a campaign against cats and mongooses. The apparent increase at Combo is harder to explain. Damholdt & Linnebjerg (2003) suggested that the end of deer-hunting (and removal of deer) following Combo's addition to the National Park might have improved understorey conditions, but if correct, why do they not thrive elsewhere under similar conditions? The increase may be illusory, just reflecting increased recent fieldwork there (Carl Jones, pers. comm.); certainly neither ASC (Cheke 1987c) nor Roger Safford (1994, 1997c) spent much time in Combo, then considered peripheral. The relatively healthy population is paralleled by Mauritius Olive White-eyes; Nichols (1999) noted that both species (for very different reasons) were at their highest density in the Rose-apple thickets that occupy much of lower Combo.

102. Damholdt & Linnebjerg (2003), Owen Griffiths (pers. comm.); the Baie du Cap (Mont sur Mont) *chasse* where Damholdt & Linnebjerg found flycatchers in 2003 appears from their map to be exactly where ASC saw them in 1973 (Cheke 1987c: 181).

103. See Cheke (1987c) for general information on these species.

104. Cheke (1987e). Total numbers are not known precisely, but unlike in Réunion there are no longer any very large swiftlet colonies in Mauritius. In June 1996 ASC visited most caves local caver Jörg Hauchler knew to contain swifts: the largest colony had 310+ nests (Surinam, cave 64), with only two others having more than 50 (Quinze Cantons 'N'= cave 63 [70] and Palma, cave 18, [170, with *c*.10,000 bats]). Hauchler's comprehensive cave surveys (unpubl. data 1996) indicated a population of *c*.730–800 pairs. Safford (2001) reported the Pont Bondieu colony had "100–1000 pairs" (37–50 according to Hauchler in 1996); in 2002 and 2005 it held 100+ nests (Roger Safford, pers. comm; Carl Jones, pers. comm). In 2003 Hauchler reported little overall change since 1996 (pers. comm.); e.g. the Palma cave held 185 nests in mid-2005 (Carl Jones, pers. comm). It is clear from Réunion that given the chance, swiftlets can multiply very quickly.

105. Cheke (1987c and pers. obs.), Safford (1993b), Middleton & Hauchler (1998).

106. Mascarene Swiftlet nests are made of vegetation glued with saliva (Cheke 1987c), whereas the best Asian 'white nests' are of pure saliva; nonetheless Mauritian nests were reported in 1996 to fetch Rs 500 (£20/US$35) for three (Jörg Hauchler, pers. comm.).

107. In 1978 ASC had recommended protective grilles on caves (Cheke 1978a, 1987c), but in August 1996 a security fence round the Palma cave had a broken padlock (although no sign of vandalism inside), while a heavy grille installed in 1992 (Safford 1993b, Jones & Hartley 1995) at the well-known Petite Rivière cave was lying on the ground, wrenched off. The latter cave was full of such acrid smoke that penetration was impossible. This cave, harbouring thousands of nests 40 years earlier, is a notorious black magic site, and was broken into twice within a year of the grille being installed (Safford 1993b); the owners' (Medine Sugar Estate) promise to maintain the gate had clearly ceased to be effective well before 1996.

108. Pers. obs.: 1973–75, maximum 26 nests (Dec.1974, 'Cave N': Cheke 1987c), 70 nests in August 1996.

109. When Khemraj Sooknah (1987) investigated 17 caves with his schools project, swift numbers averaged around four birds per cave, with similar low numbers still being present in the 1990s (Safford 2001). ASC and Jörg Hauchler visited three Roche Noires caves in August 1996; two held an estimated 8,000 and 18,000 roosting bats, but only one swift nest, though 30+ birds were roosting in one cave. The larger cave had no bats in 2003 (Jörg Hauchler, pers. comm.) but *c*.100,000+ in mid-2005 (Carl Jones, pers. comm.).

110. Bat data from Jörg Hauchler (pers. comm.) and ASC's observations in 1996 and 1999. The largest colony (*c*.50,000) of freetailed bats was then at Pont Bondieu (Hauchler, cave 26; cited as 'large numbers' by Safford 2001); two other caves held more than 10,000, both at Roches Noires: French/Twilight (cave 4) & Bat (cave 22). Carl Jones (Chapter 10 and pers. comm.) says that earlier estimates were too low, underestimating the density of roosting bats on cave roofs, which can cover dozens of square metres at *c*.2,500 bats per m^2. His careful estimates in 2005 of roosting density and roof area covered indicated much larger bat numbers than before, probably not representing an increase in population. His figures for Palma (7–14,000) and Pont Bondieu (15–30,000) in 2005 were, however, broadly similar to 1996, but the Twilight cave figure had increased ten-fold (note 108). Jones has found bat roosts in cliff clefts in the Black River Gorges.

111. Middleton & Hauchler (1998, never released by government and not seen by us) noted that most caves were damaged by vandals, swift-nest robbers and rubbish dumping, and called for cave protection legislation and the creation of a special caves national park at Plaine des Roches (details from *Weekend* 22 Apr. 2001). See also Anon. (2000) and Chapter 4 for affinities of *Mormopterus acetabulosus*.

112. See Cheke (1987c); there was no sign of nests at colonies under bridges at Jacotet and Rivière des Galets in June 1996 (pers. obs.), but the visit was out of season, and swallows have nested there again since (Carl Jones, pers. comm.).

113. The MWF campaigned unsuccessfully against the flying-fox cull, arguing that proper research was needed first. Carl Jones (pers. comm.) has seen tomb bats inland at Vacoas, Phoenix and Palma; he estimates the total at 'the low thousands at least', perhaps optimistically given the paucity of data.

114. Vinson & Vinson (1969).

115. '*Phelsuma cepediana*' turned out to include two distinct additional forms, Vinson's Day-gecko *P. ornata* and Guimbeau's Daygecko *P. guimbeaui*; *P. cepediana* is now used for the Blue-tailed Day-gecko, the most widespread species (Mertens 1963, 1966, Vinson & Vinson 1969). The fourth species is *P. rosagularis*, originally considered a race of *P. guimbeaui* (Vinson & Vinson 1969), but now recognised from DNA studies as closer to *P. cepediana* (Arnold 2000, Austin *et al*. 2004a). The Macabé Skink *Gongylomorphus fontenayi*, originally considered a race of *G. bojerii* (Vinson 1973), has been upgraded to full species on anatomical and DNA evidence (Arnold 2000, Arnold *et al*. in press b).

116. Vinson (1976b), Jones (1993a), Freeman (2003) and pers. obs.).

117. d'Emmerez (1914) reported Bouton's skink as common, but by the 1960s it barely maintained a toehold on the mainland (Vinson & Vinson 1969). Data for turtles is fragmentary: Cheke (1987a: 54) and Thompson (1981) referred to Green Turtles nesting in 1977. News briefs in the *Marine Turtle Newsletter* 54: 31 (1991), 95: 21 (2002) and Nallee (2004) are imprecise: "despite years of persecution, marine turtles still nest on Mauritian beaches" (2002) or "occasional turtle nesting reported around Mauritius and Rodrigues" (2004). The last nesting known to the Albion Fisheries Research Centre in Mauritius was at La Cambuse (south coast) in 1987 (V. Mangar, pers. comm.); Mangar & Chapman (1996) gave additional general data.

118. David McKelvey, then running the Kestrel and Pink Pigeon breeding project, was also employed by Medine sugar estate to design Casela (Keeley 2000). Robert Keeley called his book "'factual fiction', that is, a mostly true story that has been embellished by passing it through the memory function of the human mind" – like most books. Some names have been changed – the then Conservator of Forests, Wahab Owadally, appears as 'Mr Dollawally'.

119. By ASC in 1978, though not published until later (Cheke 1987a:77).

120. See Chapter 10; there are also free-flying Fulvous Whistling Ducks *Denrocycgna bicolor*. The dates these species were released are not all known, though Laughing Doves *Stigmatopelia senegalensis* were released in 1989 (Chapter 10) and first recorded wild in 1995 (Jones 1996); they are of the Asian race *cambayensis* (Lever 2005, Hawkins & Safford in prep.).

121. Cheke (1987a: 76); Diamond Doves *Geopelia cuneata* are from Australia.

122. Information from Carl Jones and Debbie de Chazal, 2003. Vikash Tatayah (pers. comm.) says that imported reptile pets are increasingly popular, and that MWF regularly receives reports of feral day-geckos (including *Phelsuma standingi*) and iguanas (*Iguana iguana*).

123. ASC (pers. obs.), Staub (1993), Safford (1995); by 2003 *Anas platyrhynchos* could also be seen in the lowlands (e.g. at Yemen turn on the coast road, pers. obs.).

124. The extinct *Alopochen mauritianus* was an endemic derivative of the Egyptian Goose *A. aegyptiacus* (Mourer *et al*. 1999).

125. Bour (1984b) found terrapins *Trachemys scripta*, *Amyda sinensis* and *Cuora amboinensis* on sale in Port Louis market, plus the Indian tortoise *Geochelone elegans*. Nik Cole (pers. comm.) has seen Red-eared Sliders *T. scripta* in several hotel ponds, and knows of many pets released unwanted once they grew too large. The 'new' blind-snake *Typhlops porrectus* is widespread in southern Mauritius, and has even reached Ile de le Passe in Mahébourg Bay.

126. *Weekend* 9 Jan. 2005.

127. Greathead's (1971) account was fuller than Griffiths *et al*. (1993) or Griffiths (1996), though he prematurely claimed that *Euglandina* was "credited with causing a marked reduction in the abundance of *A. fulica*"; *Euglandina* is native to North America, hence not a 'natural enemy' of African giant snails. Indian glow-worms *Lamogera tenebrosus* brought in 1956–58 to control *Achatina* failed to establish (Greathead 1971).

128. Griffiths *et al*. (1993), Cowie (2001); in some circumstances *Euglandina* can reduce *Achatina* numbers, mostly in anthropogenic habitats with few other snails as prey, and where giant snails are diseased ('leucodermic lesions'; Gerlach 2001).

129. Griffiths *et al*. (1993), Griffiths (1996), Gerlach (2001), Griffiths & Florens (2006); *Gonaxis quadrilateralis* successfully established in man-made habitats but is less of a threat, being scarcer than *Euglandina* and lacking contact with native species.

130. Griffiths & Florens (2006).

131. Guérin (1940–53), Newton (1958a), Cheke (1987a). A birding tourist in 2002 claimed to have seen two Madagascar Buttonquails on the "Macchabée-Bel Ombre forest reserve track" (the main road through Les Mares?) and a single male Madagascar Partridge "between the information center and lookout/Chamarel crossing" (near Black River Gorges viewpoint?) (Hottola 2002). MWF personnel have never seen such birds in 25 years fieldwork (Carl Jones pers. comm,. *contra* errors in Sinclair & Langrand 1998: see Jones 1999); however, both are common in Réunion and might attempt to colonise.

132. Cheke (1987a), Staub (1993a), Jones (1999); Carl Jones (pers. comm.) considers *Francolinus pondicerianus* has increased since the mid-1980s.

133. Rountree *et al*. (1952); Newton (1958a) recorded neither, and Staub (1973, 1976) regarded both duck and swamphen as extinct; a swamphen seen in 1976 (Michel 1981, 1992) may have been a vagrant.

134. Darby (1978), Safford (1995), Morris & Hawkins (1998); *Anas melleri* was considered until recently 'near threatened' in Madagascar (Stattersfield & Capper 2000, Lever 2005), but its status has since worsened (www.birdlife.org, species factsheet 2005). Lever (2005) was unaware of its probable extinction in Mauritius.

135. Cheke (1987a).

136. See discussion in Cheke (1987a); in Réunion *Serinus canicollis* is found only above *c*.600m, hence perhaps the last Mauritian birds

lingering at one of the highest points. In the mid-19th century they were common at low elevations on both islands (e.g. E. Newton 1861b, Schlegel & Pollen 1868) – their subsequent restriction to higher altitudes may be connected on both islands with the arrival of Village Weavers *Ploceus cucullatus* (Chapter 11).

137. Cheke (1987a).

138. North-Coombes (1948); from 125 *arpents* (52ha) in 1932, rice cultivation expanded during the wartime shortages, fading out after 1948 with a resumption of imports until it was revived in 1967 (Wiehe 1969, Rouillard & Guého 1999). Promoted by the government into the mid-1980s (Anon. 1985), it was abandoned again as uncompetitive with imports (Rouillard & Guého 1999): none has been grown since 1990 (*agricultural statistics on www.gov.mu/portal/site/cso*). In the early-mid 1970s the paddyfields at Post de Flacq supported a tiny temporary colony of Cattle Egrets *Bubulcus ibis* nesting in nearby mangroves (Temple 1976, ASC pers. obs.). A group of c.30 arrived at a pond near Grand Baie in 1967, disappearing after some months, some having been shot (unnamed officer on the *Mauritius*, pers. comm.) – the survivors may have moved on to settle in Flacq.

139. Dennis Hansen, pers. comm.

140. Cheke (1987a). Village Weavers are still lowland breeders (<300m, Lahti 2003), but ASC saw a feeding group at c.500m by Tamarin Falls in 1996. There has been too little work on Red-whiskered Bulbul, but Linnebjerg (2006) has looked at abundance, diet and seed dispersal.

141. Information on *Psittacula krameri* is from Jones (1987: 292–94); replacement of maize from ASC's pers. obs. The Ring-necked Parakeet is one of several competitors for Echo nest-sites: mynas, White-tailed tropicbirds, bees and termites all use the same cavities (Woolaver 1999). Total maize acreage in Mauritius fell from 1817ha in 1986 to < 60 since 1997.

142. Cheke (1987a, citing Claude Courtois and Frank Rountree), Feare & Mungroo (1989, 1990), Mauremootoo *et al.* (2003), ASC's pers. obs.; the 2002 estimate is from *JR* 19 Aug. 2002 (on www.clicanoo.com; throughout these notes any *JR* references after 1996 are from this website). In 1950 a pair arrived on the *Ikauna* from Colombo (Rountree *et al.* 1952), where House Crows habitually ride on fishing boats. Ryall (2002) has reviewed the ongoing spread of this species. Hicks & Knight (c.1999) commented unfavourably on the MSPCA's crow traps, adding that "we have not seen a crow problem in Mauritius, just an efficient way of killing them" – but they were animal welfare specialists, not ornithologists or health experts.

143. Roy *et al.* (2002).

144. Tirvengadum & Bour (1985), Staub (1993); see Rodda *et al.* (2002) for attempts to trap Wolf Snakes on the islet.

145. Sussman & Tattersall (1980, 1986).

146. 130/km^2 in secondary scrub, 33/km^2 in native forest, within the expected range in their native lands (Sussman & Tattersall 1986); they chose Bel Ombre for their native forest study, a poor choice as it is surrounded by degraded/secondary forest and cane-fields, giving monkeys ample scope for not spending too much time there. Macabé and many parts of the National Park are too far from alternative foraging areas, so monkeys have to rely on native forest, and thus are likely to be a greater danger to endemic wildlife.

147. Cheke (1987c: 205) reported monkeys frequently birdwatching (mostly from tree-tops near Pigeon Wood in 1974: ASC, unpublished), but most work has relied on signs left on destroyed nests and (more recently) hidden cameras (Carter & Bright 2002).

148. McKelvey (1976), Jones *et al.* (1992, 1999), Safford (1997a), Carter & Bright (2002). McKelvey not only observed nest-robbing by monkeys, but actually saw them grab at incubating Pink Pigeons, sometimes retaining tail-feathers from the escaping birds; Pink Pigeons are notoriously tame and 'naïve' to potential predators.

149. Bertram & Ginsberg (1994) thought monkeys had not increased, merely that Sussman & Tattersall (1986) had underestimated them. Owen Griffiths (pers. comm.) reckoned c.60,000 animals on the basis of 4,000 wild-caught annually having "no visible impact on the population".

150. Bertram & Ginsberg (1994), Stanley (2003). Crab-eating Macaques are called 'cynomolgus monkeys' by medical researchers. While free of bacterial and viral diseases, Mauritian monkeys harbour a tapeworm *Bertiella studeri*, recently found to infect humans (Bhagwant 2004).

151. Paupiah (2001), Stanley (2003), Anon (2004); 5,410–7,870 monkeys were exported annually from 1995–2003 (Anon. 2004), each generating US$50 for the National Parks and Conservation Fund (Paupiah 2001, Mauremootoo & Towner-Mauremootoo 2002), thus $284,000–387,000 per year.

152. Owen Griffiths was answering questions following the conference paper presented by his colleague Mary Ann Stanley (2003).

153. Following attacks on UK staff by animal rights terrorists, British Airways announced on 28 May 2005 that they would no longer transport monkeys intended for medical experimentation; Air Mauritius claimed they had not flown monkeys into the UK for some years after representations from the RSPCA and BUAV (British Union Against Vivisection). In fact, Air France and Air Mauritius flew the monkeys to Paris and trucked them to Britain via Dover; the animal welfare pressure thus caused longer and more stressful journeys for the animals. Following revelations about the continued imports, a new animal rights group called 'Gateway to Hell' picketed travel agents and called for a tourist boycott of Mauritius. This panicked Air Mauritius into telling the exporters that it was "no longer prepared to transport their macaques off the island" as neither the airline nor the island's economy could afford a tourist boycott (*The Guardian*, 28 May 2005). As with the rabbit control issue on Round Island (Chapter 9), some animal rights activists cannot see beyond their narrow perspective – the broader question of 'ecosystem rights', and the risk of losing whole species, is apparently quite outside their purview, and any linkage, even if acknowledged, is treated as irrelevant.

154. Staub (1976), Temple (1976a), Safford & Basque (2007); in early 1974 ASC ringed a Ruddy Turnstone *Arenaria interpres* that was killed in 1975 near Bombay, and saw a ringed Common Tern *Sterna hirundo*, presumably from Europe.

155. Background: Toussaint (1973). Mer Rouge was infilled with silt dredged to deepen the harbour; 20ha had been reclaimed by 1971 (Anon. 1971a), but the job was not finished until 1991 (Anon. 1993). A 1965 photo in Toussaint (1973: 49) shows the mudflat before work started, Heady *et al.* (1997: 210) illustrated the container park it became.

156. Staub (1993), Abhaya & Probst (1995) and Safford & Basque (2007) gave details of birds seen at Terre Rouge, slightly fewer species than at Mer Rouge in the 1970s. Proclamation No.13 of 1999 established the bird sanctuary. For Ramsar Convention site details see www.ramsar.org/profiles; Terre Rouge is the sole Mauritian site, Ref.1MU001. In 1995–96 the Terre Rouge stream feeding the sanctuary was badly polluted with industrial and agricultural effluent (Anon. 1996).

157. Temple (1976a), Staub (1993).

158. Anon.(1971a), Scott (1973), Procter & Salm (1975); see Jones & Hartley (1995), Safford (2001). The Macabé-Bel Ombre nature reserve, declared a UNESCO Biosphere Reserve in 1977, formed the core of the park (Safford 2001).

159. Anon. (2000).

160. Cheke (1987c:160–161) and ASC's unpublished observations.

161. Paupiah (2001), Anon. (2004); total households rose from 236,000 in 1990 to 288,000 in 2000.

162. Ruhomaun (2003) highlighted the threats; the huge expansion of coastal tourist hotels since the early 1970s has monopolised

many beaches (Cazes-Duvat & Paskoff 2004), not only usurping the tranquil space of the *pas géometriques* but depriving the Mauritian public of previously open-access beaches (pers. obs.). For the Ferney forest's narrow escape from a new road, see p. 266.

163. Much important material was lost or damaged during this tragedy for the museum's collections and reputation. The failure to appoint a fit director arose from a lack of government interest, and the way the director is chosen by the civil service without due appreciation of the qualities required for running museums (Cheke 2003; this paper was editorially censored – the situation was worse than described!). The *Mauritius Institute Bulletin*, first published in 1937, ceased publication in 1988 (*contra* Cheke 2003; '1984', though only one further issue appeared) but was revived in 2003 under new management (who were also unaware of the 1988 issue! The preface to volume 11 (1) mentioned a 19-year break since 1984).

164. This boom was short-lived, but once beet sugar was back on stream the French authorities ring-fenced price and quotas for colonial cane-sugar which prevented a renewed slump in Réunion (Vaxelaire 1999).

165. This surge is not explained in the literature; the principal use of geranium oil is in perfumery and formerly as an ubiquitous ingredient in fragrance formulations for soaps and toiletries, now largely replaced by synthetics (Robbins 1985, Vaxelaire 1999).

166. Defos du Rau (1960: 174).

167. Jacob (1925). Paris-based bank Crédit Foncier acquired land in Réunion from bad debts and bankruptcies in the lean years after 1865, becoming a major landowner by default. The Réunion branch moved out of banking and in 1920 renamed itself the *Société de Sucreries Coloniales* (later *Sucreries de Bourbon*) (Toussaint 1972, Fuma 2002). Jacob's section on forests was largely based on Kerourio (1900), including comments on the Crédit Foncier's good husbandry, so it is unclear how far their enlightened policy survived into the 1920s – however, their lands were mostly in the east, where at the Plaines des Chicots and des Fougères forests below the *ligne domaniale* survived into the 1950s at much lower altitudes than those in the west (Rivals 1952).

168. Based on Rivals (1952), there having been little change in the 1930s and 1940s. The *enclos* is the cliff-bounded area between the Volcan and the sea into which most lava flows erupt.

169. Emile Hugot told ASC in 1974 about his father's early 20th century Merle hunting parties; the 'gizzard man' story came from M Picard, warden of the Volcan *gîte*, also in 1974 (Cheke 1987b). Emile Hugot (1904–1993), a pioneer of hydroelectricity in Réunion, war hero and internationally renowned sugar engineer (*DMR* 3: 105–106), had a lifelong interest in natural history.

170. From Rigotard (1934), by then retired from running both Agriculture and Forestry in Réunion; he added that the *Conseil Générale* (the island council) were discussing the reserves suggestion, with a proposal for a commission to locate suitable sites.

171. Antoine (1973), Vaxelaire (1999: 506).

172. The late Armand Barau told Harry Gruchet in 1979 that toads were introduced in 1927 (Cheke 1987a), but de Villèle's grandson (also Auguste) was less precise (between 1926 and 1930, pers. comm.). The only published reference to de Villèle's involvement (Gérard 1974) gave the date as 'around 1920' – too early, as toads were not released in Mauritius until 1922. Most faunal information in Gérard's book is "a load of muddled nonsense" (ASC's comment in notebook).

173. Decary (1948, 1962, Cheke 1987a), Harry Gruchet pers. comm. Decary (1948) gave the toad's introduction from Mauritius as "a score of years ago" (i.e. *c*.1928). He did not mention that his information dated from 1937; we have not seen his book describing that visit.

174. Lavaux (1998), quoting an old inhabitant of Grand Étang. The lake periodically dries up and is rather sterile (pers. obs.); Lavaux's informant blamed the birds' disappearance on food shortage.

175. Toussaint (1972), Vaxelaire (1999).

176. Miguet (1973).

177. Bertile (1987).

178. Miguet (1957, 1980: 'natural regeneration' of *Acacia heterophylla*), Miguet (1973; felling in 1949), Doumenge & Renard (1989; revocation of concessions).

179. Isnard (1951).

180. Rivals (1952), pers. obs. by ASC at the Plaine des Fougères in 1974; see Vaxelaire (1999:506) for geranium exports 1889–1997.

181. Mauritian deer herds are heavily keepered against poachers (pers. obs.), and tenrecs are still hunted by country people (Tatayah & Driver 2000).

182. Amedée (2000).

183. The cry was, in creole (Chaudenson's orthography, 1974), "kabèb, kabèb, zwazó d mómô" (oye!, oye!, mother's birds). 'Kabèb' (not explained by Chaudenson) appears to be an attention cry without specific meaning; it could be related to 'bébèt', animal.

184. Isnard (1951); *JR* 13 Apr. 1998.

185. Schlegel & Pollen (1868) discussed tenrec hunting and scaling cliffs after seabirds, and many 19th century authors discussed the joys of hunting and poaching Merles (e.g. Renoyal 1811–38, Jacob ['J. C.'] 1861); the elite used guns to hunt Merles legally, once they had got wise to being hit with sticks.

186. Vaxelaire (1999: 587).

187. The *gobe* had many versions: Albany (1974) described a deadfall with a stone suspended on a trigger system of sticks, whereas Chaudenson (1974) listed several variants all using sieves or baskets, designed to trap birds alive rather than kill them.

188. Milon (1951).

189. Cheke (1976, 1987b), Barré & Barau (1980). Chaudenson (1974) also reported a drop in birds and professional poachers: "[bird] numbers are, it is said, declining; previously there were professionals who were relatively well-off", including a once notorious 'Bruny la colle' in the Cirque of Salazie. The disappearance of cuckoo-shrikes from the lower eastern slopes of the Plaine des Chicots in the mid-1970s and again in the 1990s was blamed (Cheke 1976, 1987b, Ghestemme 2002) on poachers from La Bretagne – former woodcutters from Salazie, who had settled there in two waves (early 1900s, 1930s) to grow geranium (Lavaux 1998).

190. Cherel *et al*. (1989), Barré *et al*. (1996).

191. Barré & Barau (1982).

192. Amedée (2000).

193. The local birdwatchers' group SEOR devoted a special number of their journal *Taille-vent* to poaching in 2000: see Amedée (2000), Ghestemme & Rochet (2000).

194. Ghestemme & Rochet (2000: harriers, seabirds), Marc Salamolard (pers. comm.: deer). Maunder *et al*. (2002) seriously underestimated the intensity and effects of palm-heart harvesting in their survey of endangered Mascarene palms, mentioning it as a threat to *Acanthophoenix rubra*, but not addressing the issue when discussing conservation measures.

195. Milon (1951); Berlioz's bird list (1946) was compiled from museum specimens.

196. Gill (1972, 1973a) simply confirmed that cuckoo-shrikes were limited to the Plaine des Chicots area.

197. Cheke (1976, 1987b); the research was part of the BOU Mascarene Islands Expedition.

198. Milon (1951); Berlioz (1946) made essentially armchair assessments of status for Réunion birds, apparently based on conversations with Paul Carié before he died in 1930 – he considered cuckoo-shrikes on both islands near extinction, and even merles (then considered conspecific) were allegedly "probably maintaining itself in a few isolated spots in the two islands where forest habitat persists", when the Réunion species was actually widespread and common. In general Berlioz failed to always distinguish clearly between populations on the two islands.

199. Despite ample 19th century evidence, Berlioz (1946) did not admit *Francolinus ponticerianus* to the Réunion list, so Milon's observations and discussions with locals reinstated it.

200. See Barré & Barau (1982) and Probst (1997) for bird status in the 1970s and 1990s respectively.

201. The harrier and cuckoo-shrike are discussed separately.

202. Heim de Balsac (1956), later surveyed and mapped by François Moutou (1986, Moutou & Vesmanis 1986), who found shrews everywhere up to the tree-line. Moutou (1980) found mice rather scarce (perhaps through competition with shrews), but rats common and widespread, though Norway Rats were absent from the dry west-coast fringe except in towns. For reptiles see Bour & Moutou (1982); Moutou (1983c) mapped native *Phelsuma* species with scant data, fuller mapping having to wait until the 1990s (Bour et al. 1995, Turpin & Probst 1998).

203. Moutou (1979, 1981), Moutou (1987).

204. Mertens (1966, 1970), Vinson & Vinson (1969), Bour & Moutou (1982). Mertens described the geckos as races of their Mauritian counterparts – both since elevated to full species (Cheke 1982b, Bour et al. 1995, Austin et al. 2004a). The Forest Day-gecko was later found to have two distinct populations: nominate *borbonica* in most of the range, and a predominantly yellow form *mater* (Meier 1995, Turpin & Probst 1998), first noted in 1974 (Cheke 1975a), in a restricted area in the extreme south (separated from *borbonica* by lava flows). Bour et al. (1995) described the spread of the Manapany day-gecko. Spray-drift from nearby canefields, possibly helped by lizard collectors, apparently eliminated a thriving sub-population near Saint-Joseph in 1996, after a shrubby barrier between the field and the lizards' screw-pine trees was removed (Probst & Turpin 1997).

205. Moutou (1984).

206. Bourgat (1970), Nougier (1971), Anon. (1973), Probst (1997, 2002).

207. Rebeyrol (1966); the decision to integrate Réunion politically with France was made in 1946, but only took effect in 1948 (Toussaint 1972, Vaxelaire 1999).

208. Moulin & Miguet (1968).

209. Doumenge & Renard (1989) rightly called these patches 'token reserves'. Miguet (1973) claimed that the tiny Mare Longue reserve at Saint-Philippe was set up "solely on the foresters' initative", but according to Cadet the idea came from the ORSTOM botanist Jean Bosser (Cadet & SREPN 1977), who later returned to propose a series of much larger reserves.

210. *Société Réunionnaise pour l'Etude et la Protection de la Nature*, becoming in the 1990s, SREPEN, replacing "*la Nature*" with "*l'Environnement*".

211. Cadet (1971); Thérésien Cadet was making a comprehensive study of the island's vegetation, and was appalled at the way it was scheduled to disappear, though his criticism was characteristically oblique; his thesis (Cadet 1977) remains the standard work on Réunion native vegetation.

212. Moulin & Miguet (1968) wrote that 25% of zoned land was planted, rising to 33% of target for Japanese Cedar *Cryptomeria japonica* – we think these figures were overstated, as two years earlier only 250ha out of 2,000 intended (12.5%) had been converted to 'natural regeneration' of *nattes* etc. at Saint-Phillepe, and 800ha (16–27%) of 3,000–5,000 projected for *Cryptomeria* (Rebeyrol 1966). Curiously the same Moulin-Miguet map was used 20 years later by Bertile (1987) to illustrate forestry intentions long after the policy had been substantially scaled down, as his text confirmed (see also map in Doumenge & Renard 1989: 56).

213. Miguet (1955, 1957, 1980). Rebeyrol (1966) published a map showing the distribution of forest supposedly '*figée*'. Ironically this concept would have aptly described the condition of Mauritian forests, but completely misrepresented those in Réunion.

214. Rigotard (1934, 1935), Rebeyrol (1966), Miguet (1973, 1980).

215. Miguet (1955, 1957).

216. Rivals (1952), Cadet (1977); the Koa *Acacia koa* in Hawaii has a parallel biology (Scowcroft & Wood 1976).

217. In an interview in 1977, Cadet explicitly stated that "the forest may appear to be congealed [*figée*] in time, but this is normal because it is a climax, i.e. it is in equilibrium with its environment" (Cadet & SREPN 1977). Thébaud & Strasberg (1997) looked at how differently aged lava flows were colonised from patches of native forest; the larger the fruit, the slower the colonisation – only the smallest of the original frugivores (the Réunion Merle) survives.

218. Miguet (1955, 1957), Moulin & Miguet (1968). In 1955 Miguet even envisaged *exporting* native hardwoods from Réunion.

219. Gruchet (1973).

220. According to an ONF document (Anon. 1974b), 1,600ha of the (then) proposed 3,500ha of Japanese Cedar plantations were on land previously deforested (presumably areas on the Plaines des Palmistes and des Cafres, and in Mafatte: map in Bertile 1987: 84).

221. Rebeyrol (1966), Miguet (1973, 1980).

222. Doumenge & Renard (1969), especially p.74–75 on cabbage-palms and tree-ferns. The commercial use of tree-fern trunks (*fanjans*) to support orchids etc. in horticulture is a relatively recent development, but these plants are now under severe pressure. In the 1970s "the opening of the [hydroelectric] work site at the Rivière de l'Est has permitted a massacre of the cabbage palms in the heights of Sainte-Rose, all in the space of 2–3 years" (Cadet & SREPN 1977). It is likely that regeneration of these palms is compromised by rats, so that harvesting by humans is the *coup de grace* (see Hunt 2007 for rat predation on Pacific island palms).

223. Cheke (1976).

224. The confrontation with the ONF provoked a bitter rift in the SREPN, with seven members of the council resigning (Gruchet & Gomy 1973), including Thérésien Cadet (despite his environmental views) and Armand Barau, birdwatching sugar baron and Miguet's friend since they were at college together in France in the 1940s (*DBR* 3: 17–18, 1998). The June 1973 AGM report (Gruchet & Gomy 1973) said only that the SREPN council "felt obliged to withdraw from the journal the three articles by staff of the ONF", but at the time (November 1973) Harry Gruchet told ASC that they were under immense pressure from the ONF, including threats of legal action. Not all copies were bowdlerized – many complete copies were surreptitiously distributed (including to ASC) once the fuss had died down (H. Gruchet, pers. comm.).

225. Anon (1974; see also Cheke 1976), a short dossier produced by the ONF in Réunion in response to a letter from Claude Benard to President Giscard d'Estaing complaining about what was happening to Réunion forests. She was not then prominent in the SREPN, but was on their council a few years later.

226. Cheke (1976, 1987b).

227. Miguet (1973); assigning a 'vocation' or 'calling' to a tree was pretty extraordinary, but his argument (that the Japanese Cedar was like a missing bit of the island's natural jigsaw) is further confused by the comparison with sugar cane and potato, arable crops that even Miguet cannot have seen as enhancing biodiversity! The man was a clearly more of a missionary than a scientist, and indeed a long-time associate described him in an eulogy as "intimately convinced of the major utility of his mission" and cited his "passion for work, elevated almost to a priestly calling" (Baillif 2000).

228. Moutou (1984).

229. Miguet (1980).

230. Paulian (1980); Miguet (1980: 13) quoted (without reference) a piece Paulian had written 25 years earlier (in 1955) suggesting that Miguet-type enrichment silviculture should be more widely known and employed. It is not clear from the short quote whether Paulian advocated replacing indigenous ecosystems with this technique or using it to re-afforest previously cleared areas in a more ecologically acceptable way – judging by his 1980 article, the latter seems more

likely. Paulian had seen Miguet's early work in Réunion in 1958, and re-appraised it in 1979 – he was chairing an important environmental meeting on the island (notice in *Info-Nature, Ile Réunion* 17: 9–10, 1979, on the 6th SEPANRIT conference). Paulian died aged 90 in 2003.

231. Bosser (1982); we have not seen the joint ONF/SREPN 1978 report (details are from Doumenge & Renard 1989), though the SREPN was still critical of the ONF's policy thrust in 1981 (Anon. 1981b). The SREPN claimed that the initiative to "solicit the visit of a botanist" was theirs, which is probably correct as it was stated at their AGM in June 1980 (Anon. 1981a), nearly two years before Bosser's visit. Bosser was research director of ORSTOM, and an editor of the *Flore des Mascareignes*.

232. Obituary by Alain Dupuis in *JR* 1 Nov. 2000 and article on the *Conservatoire* 26 Mar. 1999. Miguet, born in France, came to Réunion in 1949, and became the ONF's regional director in 1973 (though he had long been the power behind the throne).

233. See Doumenge & Renard (1989); a 'ZNIEFF' is a *Zone Naturelle d'Intérêt Écologique, Faunistique et Floristique*.

234. According to Michon (1998) the new *Directive Locale d'Amenagement* of February 1993 effectively accepted the conservationists' case; we have not seen the actual document. This official policy change was prefigured by reports from Sigala in 1988 and 1991, and for the first time involved a consultative committee consisting of 'independent scientific personalities' (Sigala & Soulères 1991). For the activities of the *Direction Régional de l'Environnement* (DIREN), Réunion, see www.reunion.environnement.gouv.fr.; also a brief mention in Barré *et al.* (1996).

235. Anon (1981b); following Cheke (1976, 1978), the SREPN wanted a 16km² reserve at the Plain des Chicots for the cuckoo-shrike; this was echoed by Barré (1988) and Doumenge & Renard (1989), Barré emphasizing the disparity by superposing Bosser's proposal on a map of the cuckoo-shrike's distribution.

236. Barré (1988), Doumenge & Renard (1989). In 1985 ASC's discussions and correspondence with new ONF director Pierre de Montaignac led to him accepting that Bosser's proposed cuckoo-shrike reserve was too small. Doumenge & Renard's proposals followed the SREPN's ideas from 1981 (Anon. 1981b), with additional reserves from Bosser (1982); Barré suggested reserving much larger areas of native forest.

237. Sigala & Soulères (1991), Doumenge & Renard (1989). The big reserves were the Hauts de Saint-Philippe (4,073ha, all habitats) on the southern slopes of the volcano, and the inaccessible high plateau of Mazerin (1,869ha, wet *Pandanus* forest and heath) – the former 500ha larger than Bosser had proposed! A 935ha heathland reserve (Les Mares) had also been created on the northern slopes of the Volcan.

238. Sigala (1998), Probst (2000b).

239. Ghestemme (2003), *JR* 23 Feb. 2002, 29 Jan. 2003 and 12 Jul. 2003. The SEOR was founded in 1997 by Mathieu Le Corre, then of the Saint-Denis natural history museum (Tassin & Le Corre 1997). Not all Réunion reserves have the same conservation status (definitions in Doumenge & Renard 1989, Le Corre & Safford 2001). The only fully inalienable sites ('Nature Reserves') are Plaine des Chicots/Roche Écrite and Mare Longue; the others are mostly 'State Biological Reserves', integral or managed, the latter subject to some forestry activity, and both of fixed terms, subject to review and renewal.

240. Jacques Trouvilliez, pers. comm.

241. LeCorre & Safford (2001).

242. *JR* 7 & 8 March 2007, Anon (2003), Girard & Notter (2003); see also www.parc-national-reunion.prd.fr. Girard & Notter's atlas is an excellent summary of Réunion's ecological geography, natural and human. The park proposals were summarised in Robert & Hoarau (2000).

243. Doumenge & Renard (1989). According to *JR* 14 Nov. 2003, the administration intended to acquire a further 69ha of primary forest near the Mare Longue reserve.

244. Pers. obs.

245. Figures from Defos du Rau (1960), Rebeyrol (1966), Anon. (1971), Bertile (1987), Doumenge & Renard (1989), Michon (1998), Anon. (2002), and Girard & Notter (2003) – updated by Bruno Navez (pers. comm., with tables from an unpublished report to FAO. The ONF website in 2004 referred to 25km² of *tamarin* plantations, contradicting their more formal documentation (Anon. 2002). A 1989 figure of 17km² planted (Doumenge & Renard 1989) had shrunk to a quoted 15.4km² by 1998 (Michon 1998), but has risen again since (we think the changes are in figures issued, not area planted). In 1989 expanding *tamarin* plantations to 40km² was still planned, but this was scaled down in the 1993 re-appraisal (Michon 1998), though Michon's figures generally appear too low. Navez's figure for camphor (89ha) does not match the larger area shown on the forestry map in Girard & Notter (2003).

246. 1,880ha is from Anon. (2002) and Bruno Navez (pers. comm.; in fact 1,745ha in *Cryptomeria* and 134ha in pine), compared with 2,500ha on the ONF's website in November 2004. But the site is self-contradictory; in one place it mentions 6,300ha of plantation forest (incuding 2,500 of *tamarin* and 2,500 of Japanese Cedar), while in another says 'productive forest' totals 3,660ha! The more formal ONF paper (Anon. 2002) cited 4,030ha (4,119 according to Navez) of timber plantations, of which 3,430ha (3,705) was converted from primary forest (the rest on old agriculture or coastal dunes). However, native forest converted but not in production accounts for at least another 445ha, mid-altitude (*c*.500m) wet forest converted to camphor a further *c*.285ha, and plantations where we have been unable to determine the previous vegetation could add another *c*.900ha. There is also *c*.20km² of exotic recreational and protection forest on land cleared of native forest long ago.

247. See Anon. (2002) and articles in *JR* 26 Jan. and 18 Feb. 2002, 1 Nov. 2003 and 19 Jan. 2004. The areas under productive forestry in 2003, with the principal species indicated, are mapped in Girard & Notter (2003:41).

248. Anon. (2002, Réunion), Bruno Navez (pers. comm.) vs. Paupiah (2001, Mauritius).

249. "The extensive development of *vigne marronne* [*Rubus alceifolius*], favoursied by this 'brutal' silviculture, has imposed much more frequent and intensive maintenance than was initially anticipated (two cut-backs per year for three years, then seven intensive cleansings, biannually at first then every three years)" (Bordères 1991).

250. Baret *et al.* (2004), Baret *et al.* (2005).

251. In 1975 there were *c*.500–1,500 deer *Cervus timorensis* on *c*.1,200ha of the Plaine des Chicots (Cheke 1976), i.e. 40– 120/km², a very high density (Cheke 1987a).

252. Cheke (1975f, 1976 and unpublished notes).

253. Pers. obs. Current deer numbers at the Plaine des Chicots are unknown, though all are agreed that there are fewer than in the 1970s. A newspaper article in 2003 suggested 500 free-range deer in the whole island (*JR*, 8 Sept. 2003), as against *c*.1500 in 1999, and a rumoured "5,000" in the 1970s (*JR* 18 July 1999). These figures seem excessive compared to published estimates from the 1970s (note 251).

254. Cheke (1987a, citing letter from Pierre Rivals) for the 1940s, Cazal (1974). Cazal's article is somewhat vague and muddled, but as the only history from a deer-hunter, we have used his dates for introductions (1954, 1964), rather than those in Probst (1997) and Lavaux (1998) (1955, 1966).

255. Editorial note to Cazal (1974), Cheke (1975f, 1976).

256. Moutou (1979, 1981), Probst (1997), Anon. (2002b).

257. *JR* 3 Jun. 2004.

258. The New Caledonian Kauri is *Agathis moreli* (Le Bel *et al.* 2001); Attié (1994) referred to the ONF deer report (not seen by us).

259. Cazal (1974).
260. Anon. (1971b).
261. Cheke (1987a); according to Probst (1997) goats seen on remote cliffs etc. are now all owned animals.
262. *JR* 8 Sep. 2003.
263. Milon (1951).
264. *Anas bernieri* is in the same species group as the extinct Mascarene *A. sauzieri* (see Chapter 4), whereas *A. hottentota* (= Milon's '*A. punctata*') is not closely related.
265. Cheke (1987a); parakeets and bulbuls were designated as pests (to be shot at any time) in 1974 (Servat 1974). Importing bulbuls was banned in 1972 (Probst 1996), already too late.
266. Probst (1996) gave the fullest account of the early records, the first public report dating from December 1972 (*JR* 13 Dec. 1972, in Probst 1996). Staub (1973a, 1976) cited the same article reprinted in the *Cernéen* (a Mauritian newspaper) of 27 Dec. 1972. On 25 Sep. 2003 the *JR* claimed (on what basis?) that the tourists were an (unnamed) English couple who, charmed by the birds, brought them back by boat – surely a folkloric attempt to blame foreigners rather than admit that Réunionnais brought this disaster on themselves! Although most commentators (Staub 1973a, 1976, Probst 1996, Mandon-Dalger *et al.* 1999) refer to "tourists (originating) from Mauritius" (*touristes en provenance de l'île Maurice*), the original 1972 report is vaguer and indicates multiple imports: "numerous bird-fanciers have brought them back from Mauritius, to enhance their living space" – clearly implying the offenders were Réunionnais. In Réunion the Mauritian name (*zwazo*) *condé*, was rapidly supplanted by *merle de Maurice* (French) or *merl moris* (creole), giving the species a kind of spurious Mascarene validity.
267. Cherel (1989), Mandon-Dalger *et al.* (1999), Besnard (2000), Clergeau & Mandon-Dalger (2001). On recent visits in the southern winter (1996, 1999, 2003), ASC noted fewer bulbuls in the uplands than expected from the literature, supporting Probst's suggestion (1996) that birds may retreat seasonally from the cold.
268. Amiot *et al.* (2007) found bill-size and other changes in different parts of the island. ASC (pers. obs.) noted *Nephila inaurata* was scarcer in 1996 and 1999 than in 1973–75.
269. Mandon-Dalger *et al.* (2004); for most fruits tested, bulbuls improved germination similarly to cleaning the seeds artificially, but for Brazilian Pepper *Schinus terebinthifolius* the bird's gut was 20% more effective than cleaning. Linnbjerg (2006) reported similar results in Mauritius for *Clidemia* and privet.
270. Macdonald *et al.* (1991) suggested ASC was wrong to call *Fuchsia magellanica* 'a non-pathogenic invader' in 1974 (Cheke 1987b: 309), but in fact its status appears since to have changed (pers. obs.); Cadet (1977) evidently did not think it invasive enough to mention in his seminal vegetation study. The plants formerly considered '*F. magellanica*' are in fact two similar species, a hybrid *F.* x *exoniensis* that rarely fruits, and *F. magellanica* proper, which fruits freely (*Flore* 97: 11, 1990, A. J. Scott). It appears that since the 1970s the more invasive *F. magellanica* has progressively displaced *F.* x *exoniensis*, and that the former's fruits are eaten by Mascarene Grey White-eyes, Réunion Merles and mynas (Christophe Lavergne, pers. comm.), and possibly other species; Red-whiskered Bulbuls have not (yet) been seen eating them, and may not have contributed to the spread.
271. Cherel (1989).
272. Articles in *JR*: 5 Oct. 1999, 7 Nov. 2001, 25 Sep. 2003.
273. Clergeau & Mandon-Dalger (2001).
274. *Leiothrix lutea* is a smallish babbler from China, a popular cagebird in Asia (Mackinnon & Phillips 2000), and all too successfully introduced to Hawaii (Long 1981, Le Corre 2000, Lever 2005).
275. Probst (1997), Le Corre (2000); before ornithologists caught on, the bird had already acquired the name *rossignol pays* ('native nightingale'), implying a right to be there . . .

276. Tassin & Rivière (2001) fed captive *Leiothrix* fruits of *Ligustrum robustum*, *Hedychium gardnerianum* and *Psidium cattleianum*.
277. Probst (1995, 1997), Forlacroix (2000) and Manglou (2003) for the whydah *Vidua macroura*, Riethmuller (2001) and *Chakouat* 14: 13 & 17: 14 for *Quelea quelea*, and ASC's pers. obs. in July 2003 for the dove *Streptopelia risoria*, with verbal information from Jean-Michel Probst; since then a few reports have appeared in *SEOR Lettre d'information* and its successor *Chakouat*. Also in July 2003, ASC saw from the car, on wires near Le Port, a pair of doves that may have been *Stigmatopelia senegalensis* – another cagebird likely to establish, as it has in Mauritius (above). SEOR observers regularly report other escaped cagebirds, and crows (*Chakouat*, *passim*).
278. Barré & Barau (1982), Probst (1997), Couzi & Salamolard (2002).
279. Salamolard (2004).
280/281. Turpin & Probst (1997) listed 8 further Malagasy day-gecko species kept by reptile fanciers in Réunion. Details from Moutou (1995), Probst (1997a and pers.comm.), Mozzi *et al.* (2005) with more on *P. laticauda* from Moutou (1995), Turpin & Probst (1997), Auguste de Villèle (pers.comm.), and ASC's pers. obs. 1996–2003.
282. Deso (2001), who also noted day-geckos warming up on sun-heated car bonnets, and reported a Great Green Day-gecko riding some 5km from the heights of Sainte-Suzanne to Sainte-Marie.
283. Vinson & Vinson (1969) were the first to report *Hemidactylus brookii* (as '*H. mercatorius*', see Vences *et al.* 2004b) and *Hemiphyllodactylus typus*, but ASC has since found that *brookii* specimens (as '*H. maculatus*') reached the Paris museum in 1863.
284. Guillermet *et al.* (1998).
285. Jean-Michel Probst, pers. comm.
286. Turpin *et al.* (2001).
287. Courbet (1998) and Jean-Michel Probst, pers. comm. Ferrets *Mustela 'furo'* are generally considered a domesticated variety of Polecat *Mustela putorius* and/or *M. eversmanni*. The Réunion mongoose was not identified, but *Herpestes auropunctatus* is the likely species.
288. Probst (1997), Lever (2003). Champagne *et al.* (1997) and Boulay & Probst (1998) listed chelonians kept captive on Réunion, including seven other species of terrapin; an Alligator Turtle *Macroclemys temminckii* had also been caught in the wild (probably only a stray individual). Importing exotic pets has increased in Réunion since the 1990s – many get released or escape (*JR* 2 Feb. 2001, 14 Apr. 2001).
289. Probst (1997).
290. Greathead (1971), Griffiths (1996).
291. Meyer & Picot (2001); '*Achatina pantherina*' (now *A. immaculata*) was the culprit on plots studied. Dupont *et al.* (1989) had already noted that snails devoured *Lomatophyllum*, and that seeds would only germinate if the fruit pulp was removed.
292. Griffiths & Florence (2006); the *Achatina* effect may explain why the Mauritian endemic *Lomatophyllum tormentorii* is confined to snail-free northern islets, although *L. purpureum* survives in restricted areas on the mainland (*Flore* 183, 1978, W. Marais & M. J. E. Coode).
293. Macdonald *et al.* (1991), Lavergne *et al.* (1999); privet did not feature in Lavergne's (1978) survey of invasive plants, nor in Cadet's (1977) vegetation studies, though he collected a herbarium specimen in 1969. It was noted without comment as naturalised at Cilaos in *Flore* (119: 6, A. J. Scott, 1981).
294. Lavergne & Barré (1997), Lavergne *et al.* (1999). Abhaya (2001) included the following invasive species among fruits eaten by the endemic merle in Réunion: *Ardisia crenata*, *Flacourtia indica*, *Fuschia boliviana* (but not *F. magellanica*), *Lantana camara*, *Ligustrum robustum* (as *L. walkeri*), *Litsea glutinosa*, *Psidium*

cattleianum, *Solanum mauritianum* (as *S. auriculatum*), *S. torvum* and *Syzygium jambos*.

295. Sigala (1998), Lavergne et al. (1999); Sigala (2001) was more specific on areas treated: 250ha in 1995–96, 'a bit less' in 1997–98 with an additional 80ha in 'proper forest'. Weeding is combined with herbicides: glyphosphate ('Roundup') on cut stumps.

296. Lavergne (1999), Le Bourgeois et al. (2003), Kueffer & Lavergne (2004). Without diseases or insect predators, privet in Réunion produces six times as many fruits as in its native Sri Lanka, and the density of plantlets beneath the trees is 14,000 times higher! (Le Bourgeois et al. 2003).

297. Four species in all: *Flore* 171 (R. M. Smith, 1983); Lavergne (1978) said the commonest and most widespread species was the Yellow Ginger *Hedychium flavescens*, though Macdonald et al. (1991) considered *H. gardnerianum* a worse weed.

298. Cheke (1987a).

299. Macdonald et al. (1991); current understanding of invasive plant ecology was summarised by Tassin et al. (2006), and the degree of invasion in native habitats assessed by Strasberg et al. (2005); see also Kueffer & Lavergne (2004).

300. The added species are *Solanum mauritianum* (= *S. auriculatum*), *Boehmeria macrophylla*, *B. penduliflora*, *Erigeron karwinskianus*, *Lantana camara*.

301. Tassin & Rivière (1999). The plants are *Hiptage benghalensis*, *Acacia mearnsii*, *Schinus terebenthifolius*, *Ulex europaea* and *Zantedeschia aethiopica*. These and more were listed by Macdonald et al. (1991), but at lower priority; see also Kueffer & Lavergne (2004).

302. Baret & Strasberg (2005).

303. Sigala (2001), Wittenberg & Cock (2001), Kueffer & Mauremootoo (2004), Kueffer & Lavergne (2004). When ASC proposed biological control for guava in Mauritius in the 1970s it was rejected, as the plant was too important economically (Cheke 1987a: 16), but rural communities currently depend less on this source of food, fruit wine and wood (pers. obs.). Nonetheless, guava is still a "highly prized fruit in Mauritius, and any biocontrol programme for it is likely to be strongly opposed by the general public", while controlling bramble would pose no problems (Mauremootoo & Towner-Mauremootoo 2002). According to Sigala (2001), biological control is being considered for Helicopter Vine, Gorse and gingers; he did not mention guava.

304. Jouanin (1964a), Jouanin & Gill (1967).

305. Jouanin (1987), Brooke (1978).

306. Bretagnolle & Attié (1991), Stahl & Bartle (1991), Probst et al. (1995); mountain guide Pascal Colas found the first nests.

307. Le Corre & Safford (2001); J-M. Probst, quoted in *JR* (19 Oct. 2001), reported censusing 19 colonies with more than 6,000 burrows since 1995; Probst (1992) re-iterated the figure of 6,000, mentioning 20 colonies. The 'old' figure of 3,000 pairs was re-stated in the local press in April 2004 (*JR* 7 Apr. 2004). By 2003 Probst (pers. comm.) was claiming 30 colonies, discovering more every time he abseiled into appropriate areas. Minatchy (2004), calculating with very tenuous figures, arrived at a population of 10,000–20,000 pairs.

308. See discussion in Cheke (1987a: 30–31) vs. Stahl & Bartle (1991).

309. Jean-Michel Probst, pers. comm.

310. Seabirds nesting well inland need a way of recognising water to aim for, and moonlight on the sea may be an important clue; also, many petrels fish for bioluminescent squid, so they may mistake artificial light for food (Le Corre et al. 2002). Barau's Petrels are attracted more to mercury vapour lights (bluish-white) than other sources, and are apparently unaffected by sodium vapour (yelloworange) (Minatchy 2004).

311. House-building and population rose in parallel until 1974, since when housing has rapidly outstripped population growth (Bertile 1987, Pavageau 2000, Berthier 2004); for roads see Guébourg (1999), or various dates of the IGN 1:100,000 map. Le Corre et al. (2002) noted the rise in electricity production since the 1940s, the graph taking off exponentially in the 1970s; floodlit sports grounds increased from 46 in 1970 to an estimated 400 in 2000.

312. Le Corre et al. (2002), Minatchy (2004), Salamolard (2007); the largest numbers were collected in Cilaos, under the colonies, but significant numbers were also grounded where major gorges reach the sea at Saint-Denis, Pointe des Galets, and Saint-Louis/Saint-Pierre. Apart from the 1,643 Barau's Petrels recovered or found dead in 1996–99, there were 674 Tropical Shearwaters (28.7%), 28 Wedge-tailed Shearwaters (1.2%), and 3 Réunion Black Petrels; by season's end in 2005 more than 7,000 birds had been recovered (*JR* 6 Jun 2005). Jouanin (2006) reported the April 2006 kill.

313. Jean-Michel Probst (pers. comm.).

314. J-M.Probst, quoted in *JR*, 19 Oct. 2001; see also Probst (1996), Le Corre & Safford (2001) and, for cats in another context, Salamolard & Ghestemme (2004). Earlier, before studying fledging success, Probst considered the risk was mainly from rats (Probst et al. 2000 – despite Probst 1996, where the table of cat data was inadvertently replaced by a duplication of rat data).

315. Probst et al. (2000).

316. Bretagnolle & Attié (1993, 1995, 1997), Barré et al. (1996); Thiollay (1996), who had recently seen 4,000 birds together offshore on one afternoon, argued that Bretagnolle & Attié exaggerated the drop in population due to shooting.

317. Le Corre & Safford (2001); the cliff reserves are 'Protected Biotopes', the second strictest form of reserve.

318. Jouanin (1970a, 1987); the species is *Pseudobulweria aterrima*.

319. Attié et al. (1997), Le Corre et al. (2003b), *JR* 7 Apr. 2004, 15 Dec. 2004, Fontaine (2005).

320. Stattersfield & Capper (2000), Le Corre et al. (2003b), *JR* 1 Mar. 2003; details of the recently discovered cliff-top nesting sites have yet to be formally published for security reasons, due to the Réunion Black Petrel's extreme rarity (Thomas Ghestemme, pers. comm.).

321. *JR* 23 Nov. 2000, Probst (2001); Probst showed us the (then) captive animal in June 1999.

322. Jean-Michel Probst and Roger Bour pers. comm.

323. Vinson & Vinson (1969), Maillard (1982).

324. Landing tracks (but no eggs) were found in 1986 (Bertrand et al. 1986), and laying recorded three times during 1994–2000 (Sauvignet et al. 2000). For 2004: *JR* 21 Aug. and 20 Dec. 2004; the paper wrongly claimed (citing no source) that the last clutch had been c.60 years previously. Beaches are made more turtle-friendly by planting shrubby shelter belts on the upper beach.

325. Details from Lebeau & Lebrun (1974), Sauvignet et al. (2000), www.tortuemarine-reunion.org and *JR* 3 June, 21 Aug. and 20 Oct. 2004. During its commercial period the farm was notorious for polluting the adjacent reef (Auguste de Villèle, pers. comm.); Bour & Moutou (1982) expressed a (then) rather pious hope that the farm "would play a real role in conserving the species" – it was François Moutou who alerted ASC in 2004 to the turtles' return, hope having become reality!

326. Recent information on *Asterochelys radiata* from *JR* 22 Nov. 2003, historical data from Vinson (1868), Bour (1981) and Cheke (1987a). Serious importing started around 1830, easing off in the 1880s but continuing until the 1950s; although animals escaped (bones found in cave deposits), no feral population established. No doubt they were too easily found by poachers, any hatchlings falling to cats and rats. Selling tortoises was banned in January 2006, and transfers/gifts need to be registered with the local authority (*JR* 14 Jan. 2006). Réunion *A. radiata* have been used for reintroduction to Madagascar (Boullay 1995).

327. Probst & Coujou (1998), *JR* 10 May 2003, Le Corre et al.

(2003), *Chakouat* 11: 14 (2004). Maillard (1862) mentioned, without details, *Phaethon rubricauda* amongst birds occasionally reported after cyclones, and there is a specimen in the Saint-Denis museum, presumed to be of local origin (Probst 1997).

328. Milon (1951), Bretagnolle & Attié (1997), Ghestemme & Rochet (2001), *JR* 17 Aug. and 8 Sept. 2000; the poacher in 2000 claimed to be catching his food on his own land using only traditional local methods: a stuffed lure waved around on a pole, plus a gun (indeed, exactly as described by Milon!). However, it was not his first offence, so he could not claim ignorance of the law. White-tailed Tropicbird numbers from Probst (2002, 200–500 pairs; 250–300 according to Abhaya & Probst 1997). A pair nested on a building in Saint-Denis in early 2006 (*JR* 11 Mar. 2006).

329. Renman (1994), Bretagnolle & Attié (1996b), Louisin *et al*. (1997). *Bébet* in Réunion Creole can mean either 'small animal' (often an insect) or 'ghost, spirit' (Chaudenson 1974), the folkloric *bébet t'out* being the latter.

330. Arquetout (1972), Louisin & Probst (1997). Oddly the *bébet t'out* legend escaped the folklore compilers (Albany 1974, Chaudenson 1974), but it is assimilated into the better-documented belief in *gran'mer kal* (Arquetout 1972, Louisin & Probst 1997), a bogy-woman whose frightening nocturnal cries are associated with *fuké* (shearwaters) (Chaudenson 1974) or quails (Albany 1974) – indeed Albany reported a species of quail called *kay merkal* or *kay gran'merkal*, so evidently someone knew the biology of the call.

331. J-M.Probst, pers. comm.; Skerrett *et al*. (2001).

332. Lavergne *et al*. (2004, 2005) and pers. comm. The *Latania commersoni* copse is at Cap de l'Abri, near Manapany.

333. Milon (1951), Gill (1964), Gruchet (1975), Cheke (1987a), Barré (1988). There appears to be no record of what triggered the harrier's upgrade, but the date suggests that Christian Jouanin's visit in 1964 may have stimulated Armand Barau to lean on the authorities.

334. Clouet (1978), Cheke (1987b); the upper figure of 200 pairs, increasing, was generally accepted (e.g. Barré & Barau 1982), and later duly revised upwards to 300 pairs (Barré *et al*. 1996, citing a pers. comm.), a progression also noted by ASC on brief visits in 1978, 1985 and 1996/99.

335. Bretagnolle *et al*. (2000c, from a conference in 1998); the map shows 77 recorded pairs (74 in text, 36–39 positive, *c*.35 probable, with many 2x2km squares not covered), to which they added *c*.40 pairs (20 missed in covered squares, 20 from unsurveyed ones), thus minimum *c*.59 and maximum 117, shrunk in their text to 50–100. Their later fuller report (Bretagnolle *et al*. 2000b) was vaguer on numbers in the text and allowed 400–600 individuals, including non-breeders. A revised map actually shows 105 territories, with much likely habitat unsurveyed – adding the estimated 20 pairs from unsurveyed squares takes the total to 125. Failure to survey a square is not, under Réunion conditions, a criticism – in fact they surveyed many very difficult areas. The alternative analysis, by the SEOR, was in Ghestemme (1998, cited in Ghestemme 2000); the 1999 surveys were reported by Ghestemme (2000). Bretagnolle & Attié (1995) had earlier been more alarmist, claiming only 50 pairs, and possibly felt unable later to publicly increase their figure too much.

336. Stattersfield & Capper (2000). As Bretagnolle *et al*. (2000b,c) pointed out, the Réunion Harrier is now rarer than the Mauritius Kestrel, once the world's most endangered bird – due, of course, to the kestrel's spectacular recovery through intensive management.

337. Ghestemme & Rochet (2001).

338. Bretagnolle *et al*. (2000b), Ghestemme & Rochet (2001).

339. The large spreading canopies of *Acacia heterophylla* appear suited to the cuckoo-shrike's very deliberate food-search strategy, whereas, although richer in insects, the upland mixed evergreens are too dense – this also suggests that the species may be best adapted to the large crowns (and emergent palms?) of lowland mixed forest, like the Mauritius Cuckoo-shrike (Thiollay & Probst

1999). However, birds are currently at highest density where mixed forest grades into *tamarins* (Ghestemme & Salamolard 2007), suggesting a sort of 'better suboptimal' zone above the pure upland mixed forest.

340. In the 1970s, entomologist Yves Gomy failed to locate any endemic *Oryctes* beetles (Gomy 1973a, Cheke 1987b), but they have since been rediscovered: *O. borbonicus* lives in *tamarin* forest and more commonly in *Philippia* heath further up, the larvae eating its roots, the adults fallen berries; the related *Marronus borbonicus* is common and *O. tarandus* much scarcer, also in heathland (Gomy 2000, Christian Guillermet pers. comm.). Guillermet also doubts that the Cuckoo-shrike's bill is big enough to crush the very hard elytra of *Oryctes* spp., so did Pollen mistake the genus?

341. Thomas Ghestemme, pers. comm.; planting *Acanthophoenix* palms in the Plaine des Chicots forests is being investigated (Ghestemme & Salamolard, circular e-mail on management proposals, 2005), but this may be irrelevant (note 340).

342. Ghestemme (2002), Ghestemme & Salamolard (2007). *Quotidien de la Réunion* 22 Jan. 2006, 21–22. Payet (2006), *World Birdwatch* 28(3):8 (2006), *JR* 9/12/06:19. Information on fires from Jean-Michel Probst (verbally 1999, damage seen by us), *Temoignages* 9/11/2006 (www.temoignages.re) and *JR* 9/12/06.

343. Cats are now common (Salamolard & Ghestemme 2004), whereas they were present but scarce in the 1970s (pers. obs., Théophane Bègue, pers. comm.). A curious anomaly in French law means that all cats, even feral ones causing problems, are technically 'domestic animals', and controlling them (unlike 'vermin') is subject to serious constraints (Ghestemme & Salamolard, circular e-mail on management proposals, 2005). Réunion scenery is so spectacular that helicopter flights for tourists are now a thriving business.

344. Population estimates from Thiollay & Probst (1999), Ghestemme (2005) and *World Birdwatch* 28(3):8 (2006). Not only are Cuckoo-shrikes hard to watch for any length of time, they are very hard to catch, so it is difficult to establish a marked population to investigate longevity and accurately map territories and movements (Ghestemme 2002, Salamolard & Ghestemme 2004), though ringing nestlings (*JR* 9/12/06) may remedy this.

345. Gill (1973b) on *Zosterops borbonica* variation, Gill (1971a) on numbers; later estimates were: Cheke (1987a, for 1973–75: *c*.400,000/*c*.150,000), Barré (1983: 465,000/156,000), Probst (2002: 450,000+/ 150,000).

346. Cheke (1987b), Barré (1983).

347. Cheke (1987b) estimated Réunion Merles at *c*.25,000 pairs (i.e. *c*.50,000 birds), whereas Barré's figure (1983) was only 20,000 birds; the difference was probably more due to sampling differences than a decline from 1974 to 1979–81, though estimates for other species were less disparate.

348. Cheke (1987b) for the 1960s and 70s; Tassin & Rivière (1998) found the only native passerines at Étang Salé were Grey White-eyes; Jacques Tassin (pers. comm.) never saw paradise flycatchers there (1993-) and suggested that harvesting mature she-oak stands might also be a factor. Probst (1997) claimed that the merles were, in addition to suffering from poachers, being "pushed back by the establishment of the Red-whiskered Bulbul" and "appear to be steadily declining throughout the island", but offered no supporting evidence.

349. A misprint in Barré's paper (1983) gave the swallow population as '3,500' instead of 300–500 (Nicolas Barré, pers. comm.).

350. Jadin & Billiet (1979), Cheke (1987b), Grimaud & Probst (1997), Probst (2002), Jean-Michel Probst & Marc Salamolard (pers. comm.). The Cilaos colony increased from *c*.10,000 nests in 1997 to *c*.15,000 in 2002. Relative swiftlet abundance on the two islands had reversed by 1985 (Cheke 1987e), as they declined in Mauritius and increased in Réunion. A low figure of 350 *individuals* was still quoted for swallows by Probst (1997a), upped to 500–700 five years later (Probst 2002), but casual observation suggests an fivefold increase or more since the 1970s (ASC, contrasting

1973–75 with 1996–2003), indicating a population of 2,000+ individuals/1000+ pairs.

351. Rompillon (2003). It would be fascinating to document the way in which protected species acquire desirable characteristics once they become legally unavailable. A parallel case concerns *Ramosmannia rodriguesii*, a nearly extinct Rodriguan plant (p. 196). The idea in Mauritius that soup from cooked guts of the introduced soft-shelled pond-turtle *Palea steindachneri* has aphrodisiac properties (Bour 1984b) is equally mysterious, though this animal is not protected. It seems that many people cannot grasp the idea that a species can be protected for its own intrinsic value, so assume it must have a special cachet that 'they' (the government, landowners etc.) want to stop ordinary people using. An 'urban myth' then arises of supposed beneficial qualities, causing intensified poaching and/or harvesting of animals or plants previously of little cultural interest. Of course, the Chinese community has always sought swiftlet nests for soup, but only on a fairly limited scale in the Mascarenes due to their inferior quality.

352. Cheke & Dahl (1981), Moutou (1982). As lava tunnels are few and small in Réunion, most *Mormopterus* roost in relatively small numbers (by Mauritian standards) on cliffs in rock clefts and in buildings. In the 1970s there was a colony of several hundred in Armand Barau's roof at Bois Rouge; only one known cave, at Trois Bassins, held greater numbers – 3,000+ were found dead there after the 1980 cyclone. Probst (1997) noted the continued scarcity of *Taphozous* in summer, and major *Mormopterus* roosts were mapped in Girard & Notter (2003), probably from Probst's data.

353. Jouanin (1987), Probst (1995b, 2002), Bretagnolle *et al.* (2000). Prior to the 1970s neither breeding sites nor numbers were known.

354. Wedge-tailed Shearwaters were not mentioned by 19th century visitors or locals, although undated specimens from Réunion in Leiden and Paris were noted by Hartlaub (1877). The lack of mentions *in situ* was partly due to confusion with Réunion Black Petrels, as noted by Jouanin (1970). For recent records see Jouanin (1987) and Probst (1995b, 1997, 2002); confusingly Probst reduced the number of colonies from 13 (1995b, 1997) to 6 (2002), apparently through combining adjacent breeding sites into larger, more dispersed 'colonies'.

355. Milon (1951), Jouanin (1987), Le Corre (2001).

356. Bretagnolle & Attié (1997) reported a decline in poaching of Tropical Shearwaters for food; they made up 28% of birds recovered after lightfalls (674 during 1996–99, Le Corre *et al.* 2002); only 1% (28 birds) were Wedge-tailed Shearwaters.

357. Details from Bertile (1987), Guébourg (1999) and *JR* 15 Jan. 1999, 9 Nov. 2002, 28 Jun. 2003, 15 Jan., 22 Apr., 1 July and 20 July 2004; a massive landslide on 24 Mar. 2006 killed two people and ignited a debate on the road's future (*JR* 24 Mar. 2006).

358. Probst (1997, 2003), *JR* 13 Apr. 1998.

359. *JR* 13 Apr. 1998 and Probst (1998b), who noted a nesting attempt by Lesser Noddies in 1997, and a prospecting Fairy Tern in 1995, another potential colonist. Petite Ile is a 'Protected Biotope' (see Le Corre & Safford 2001), the second strictest form of reserve.

360. Populations of the other introduced birds were estimated as: Common Myna 59,000, canaries 21,000 (Yellow-fronted) and 3,000 (Cape), Spice Finch 17,000, Village Weaver 8,000 (Barré 1983). Barré did not attempt to enumerate gamebirds or buttonquails.

361. The decline of several gamebirds is blamed on reduced keepering and increased poaching in areas where they were once preserved and shot, together with extending urbanisation in the dry west (Barré *et al.* 1996, Probst 1997). However, the impact of increasingly industrial methods in cane-fields may also be significant, as seen with Grey Partridges *Perdix perdix* in Britain (e.g. Potts 1986), where reduced cover and food shortage due to pesticides produced a dramatic decline, despite landowners being the people keenest to preserve this traditional quarry.

362. Couzi & Salamolard (2002).

363. Casual observation (ASC's unpublished notes 1973–75 vs. 1996–2003) and Couzi & Salamolard's map and comments (2002) confirm that *Nesoenas picturata* has increased since Barré's (1983) 'rare everywhere' (1,300 estimated population). Although evident to long-term residents (Auguste de Villèle, pers. comm.), this increase is not acknowledged in the local literature (e.g. Probst 1997, 2002 or Couzi & Salamolard 2002). The main increase may thus have happened in the 1980s, before these observers came to Réunion, implying it happened before formal protection – indeed an incipient expansion was already noted around 1980 (Barré & Barau 1982).

364. Servat (1974), Barré (1988).

365. Le Corre (2001), Probst (2002).

366. Ghestemme & Rochet (2001), Dalleau-Coudert (2005).

367. Correspondence of Émile Hugot with Jacques Berlioz kept by Alain Vauthier in Réunion (copies in ASC's library). Roger Bour has been unable to locate the tortoise bones in the MNHN (pers. comm.).

368. Cowles (1987, 1994), Bour (1979), Kervazo (1979), Arnold (1980), Brial (1998), Mourer *et al.* (1999, 2006). Auguste de Villèle showed ASC the half-built bypass in 2003. The subfossil bat bones mentioned by Mourer *et al.* (1999) have not been fully studied, but both the known flying-foxes are present (Cécile Mourer, pers. comm.).

369. Girard & Notter (2003). There are anomalies – the map of the historical progress of settlement differs in some areas materially from Defos du Rau while the new map looks more precise, it includes inaccuracies, such as showing the southeast coastal fringe as colonised in the 19th century, when in fact it remains unsettled due to frequent lava flows. There is also an apparently meaningless map of insect groups. Another recent atlas, Cadet (2003), is largely socio-geographic, with little on the environment.

370. Durand (2001).

371. See Guébourg (1999), Pavageau (2000), Cadet *et al.* (2003) and Berthier (2004). There is substantial urban development at higher altitude around Le Tampon, but forest here (lower Plaine des Cafres) was devastated in the 19th century (Chapter 7). The 2005 census figure is from *JR* 2 Feb. 2006.

372. In the 1970s there was only the (then) SREPN active in nature conservation; there are now numerous NGOs and also the government's environment arm, the DIREN.

373. Pidgeon (1932), North-Coombes (1971); Hampton (1922) stated that there were "no wheel [*sic*] vehicles, no roads for wheel vehicles" [*sic*], but it is unclear when he was there. Lone sailor Harry Pidgeon saw no "plow or animal-drawn vehicle on the island" in October 1923; motor vehicles first arrived in 1954 (North-Coombes 1971).

374. Bertuchi (1923), Hampton (1922); Latans are conspicuously absent from most of Bertuchi's landscapes, so they must by then have been husbanded in certain properties only.

375. Hampton (1922).

376. Most details from Hampton (1922) and Bertuchi (1923), population figure from North-Coombes (1971).

377. Thirioux was visiting or staying with his son who was working on Rodrigues (*DMB*:222–223, 1942, L. Halais); correspondence from him about Rodrigues in the RSAS archives (Toussaint & Adolphe 1956) is now apparently missing (Rosemay Ng, pers. comm.). The *Phelsuma edwardnewtonii* specimens sent to Thirioux's sponsor Paul Carié were accessioned by the Paris museum in 1917 (Vinson & Vinson 1969); there is nothing to corroborate the common supposition that they were collected on a lagoon islet. The six known specimens of Newton's Day-gecko were unexpectedly supplemented in 1995 by another, acquired by German herpetologist Herbert Rösler, in circumstances concealed by a confidentiality agreement, though he has given a detailed description (Rösler 1995). While the specimen appears to be old (19th century), he

knows neither its collection date nor who collected it (pers. comm.).

378. North-Coombes (1971: 190); enthusiasm and good planning were lamentably rare qualities in the administration of Rodrigues.

379. North-Coombes (1971).

380. We have been unable to discover when the state acquired Solitude, but plantations were begun in the 1940s (Cheke 1987c, citing discussion with Leo Edgerley).

381. *Rodrigues protection of produce regulations 1923, Rodrigues birds protection regulations 1923* (Cheke 1974, 1987a). The bird law also failed to include breeding terns, but included the Lesser Flamingo, a vagrant seen in 1922 (Guérin 1942–53; '1923' in error, Showler & Cheke 2002).

382. *Lantana camara* is nowhere mentioned in Wiehe's (1949) vegetation paper, but he had found three plants c.10 years old which he ordered to be destroyed (Wiehe 1938, Strahm 1989:34), but it was presumably too late. Balfour (1879) was shown a sample allegedly from Rodrigues, but saw none himself and doubted he would have overlooked it. The 1949 quote is from Strahm (1989); Cadet (1971b) found *Lantana* dominant on calcarenite and common throughout the western cattle-walk.

383. Summarised from North-Coombes (1971). The effects of rats as seed predators has been underestimated in the Mascarenes – if anything like Pacific islands (e.g. Hunt 2007), it is likely that many shorter-lived plants (i.e. less than 400 year life-span) with edible seeds may have been exterminated by their action alone.

384. Wiehe (1949) mentioned the planting programme and noted abundant wild latans.

385. Wiehe (1938, 1941, 1949), North-Coombes (1971).

386. Wiehe's map does not show Cascade Saint-Louis as wooded, so he must have missed it. In 1964 it had '200 acres' of indigenous forest (*Annual Report on Rodrigues for the year 1964*: 12).

387. In 1918 the water-catchment dam in Cascade Pigeon had silted up from erosion on the slopes (North-Coombes 1971: 198), so vegetative cover may still have been inadequate in 1938.

388. The photos are in Wiehe's unpublished reports (Wiehe 1938, 1941); details of birds seen from discussion with ASC in 1974 (Cheke 1987d). Jean Vinson and Frank Rountree, in 1930 and 1941–42 respectively, also saw warblers and fodies throughout the Rose-apple thicket then blanketing the uplands around Monts Malartic, Lubin and Limon and down to St.Gabriel (Vinson 1964b; F. Rountree, pers comm. in Cheke 1987d)

389. *Psidium cattleianum, Litsea glutinosa*.

390. North-Coombes (1971).

391. North-Coombes (1971) on the tax, Wiehe (1949) on charcoal.

392. Vaughan *et al*. (1948). It was a high powered commission: Vaughan was Director of the Mauritius Institute, the other members included senior forester Leo Edgerley, agriculturalist Alfred North-Coombes and the Mauritius Director of Agriculture William (Bill) Allan. North-Coombes's amusing personal account (1996) of the commission's visit contains no natural history observations. The island's historian died aged 91 in 1998 (*DMB*:1809–12, 2002, P. Harel).

393. North-Coombes (1996) – adding "O my poor heavenly isle! What a tragedy!".

394. North-Coombes (1971); Hotchin's departure with his team, at the time of Mauritian independence in 1968 (Selvon 1983), was presumably not coincidental.

395. Leo Edgerley (1961), conservator of forests in Mauritius, wrote laconically that "under the Rodrigues Development Plan large areas of land previously under the control of the Forest Department have now passed over to agricultural use, and it is estimated that approximately 750 acres [304ha] now remains under the control of the Forest Department" – probably just Cascade Pigeon, Solitude and Grande Montagne. Although Edgerly claimed forest and agricultural work were amicably integrated, foresters L. Remy and G. Elysée, there at the time, told ASC in 1974 of distress at the tree-cutting and resentment at being overruled. Cadet (1975, Strahm 1989) mentioned massive *déboisement* in '1945–50', but we think he was misled on the dates of Hotchin's clearances. There is no corroboration for deforestation in 1945–50, and Cadet, if the '1945–50' events were additional, would surely have noted Hotchin's more recent activities.

396. Cascade Victoire was not listed as an area with indigenous forest in the *Annual Report on Rodrigues for the year 1964* (Colony of Mauritius, No.22 of 1965), and certainly had no forest in 1974 (Cheke 1974).

397. North-Coombes (1971: 311); Cadet (1975) noted that "the *Leucena* communities established on outward slopes on skeletal soils and on the steep flanks of valleys have practically disappeared, to be replaced by savanna in the face of attempts at cultivation, livestock rearing, the necessity for producing fuel-wood for an ever increasing population and finally the teeth of goats".

398. Staub (1968).

399. ASC, pers. obs., 1974; in 1981 a Dutch expert, N. H. Vink, reporting for the European Development Fund, wrote that "today there are no unit farms or dairy farms and many terraces are neglected. In 1968 all technicians left and Rodrigues was largely forgotten. Also farmers expected Government to upkeep the work as they had been paid to execute the work" (Anon 1981c; see also Selvon 1983 [who identified Vink as the author], Gade 1985). In 1998, after episodic terracing and erosion-control work over the previous 40 years, including the EDF work in the 1980s, "poor management verging on abandonment of anti-erosion works" still ruled (Jauze 1998). A new anti-erosion programme funded by the EDF for 1999–2003 was lambasted for disorganised and inefficient planning and implementation by a government audit in c. 2000 (Ministry for Rodrigues, audit statement on Anti-Erosion Programme in Rodrigues on ncb.intnet.mu/audit/rep00/revrod.htm).

400. Gade (1985).

401. Anon (1971a).

402. Gade (1985).

403. See e.g. Jauze (1998:50) on pervasive alcoholism in Rodrigues. Adams & Carwardine (1990) noted "not much goes on in Rodrigues other than home entertainments" i.e. sex and booze – a little harsh on the Rodriguais, but there is certainly little alternative leisure activity on the island.

404. Strahm (1989), Adams & Carwardine (1990), Verdcourt (1996), Stearns & Stearns (1999); the surviving plant is *Ramosmannia rodriguesii*, the original *café marron* now split into two species; the other, *R. heterophylla*, is extinct.

405. The supposedly last wild *Hibiscus liliiflorus* was killed by being repeatedly chopped back, including by foresters trying to squeeze it into too small a fence! Once fenced it became, like *Ramosmannia*, the centre of attention – as a 'magic tree', bits were cut off, candles burnt on it and money placed to bribe the spirits (Strahm 1989, Stearns & Stearns 1999). However a cutting in the local priest's garden survived, and two more old plants found in Cascade Mourouk, from which cuttings have been successfully propagated (Strahm 1989).

406. Cheke (1974).

407. Cheke (1987d); *tecoma* is *Tabebuia pallida* (*Flore* 133, J.Bosser & H. Heine, 2000). The 1964 *Annual Report on Rodrigues*, whose upbeat agricultural section was presumably drafted by Hotchin, disingenuously stated that "filao [= she-oak] has proved an excellent tree for giving a quick cover on poor eroded windswept ridges".

408. Vinson (1964).

409. Vinson (1964b), Gill (1967), Cheke (1987d).

410. Bourne (1963), Watson *et al*. (1963), Greenway (1967); the first bird *Red data book* was Vincent (1966). Bourne's review stung the Smithsonian into producing a supplement later in 1963 which urged restraint in collecting endangered island species, without

really discouraging it. A list of endangered birds was included, omitting the Rodrigues endemics, though there was a catch-all rider that "generally all native land birds on Christmas, Rodriguez, Mauritius, Réunion and the Seychelles are endangered". There was no mention of local laws that might be breached by collecting. Whether or not Gill had permission to collect in Rodrigues, the fact remains that the 1923 regulations then in force did not protect the endemic birds!

411. Gomy (1973b), Staub (1973); for Rodrigues cyclones to 1987 see North-Coombes (1971), Cheke (1987d) and Strahm (1989).

412. Local informants generally attributed the decline of guineafowl and lovebirds to cyclones (Gill 1967, Cheke 1987a), but more careful enquiries revealed agricultural pest control as the main factor (Staub 1976, Cheke 1987a), though lovebirds will also have lost nest-sites (tree-holes). Lovebirds were abundant enough before 1956 to be exported to Mauritius as cage-birds (Claude Courtois, cited by Cheke 1987a). Both species had survived numerous cyclones over the previous century; the lovebird's sudden crash in 1956 coincided with Hotchin's arrival in 1955, but not with any significant cyclone.

413. The deer were probably shot, though Courtois (pers. comm., Cheke 1987a) said they 'died out'. Courtois preceded Hotchin in charge of agriculture (North-Coombes 1971), and may have been reluctant to blame his high-profile successor for crimes against deer, sacred to many Franco Mauritians. In fact eliminating introduced species was probably the best thing Hotchin did!

414. Vinson (1964b); Gill (1967) also saw canaries in 1964. Not mentioned by Bertuchi (1923), they were presumably still absent in 1914–17.

415. Cheke (1974, 1979a, 1987d), Cheke & Dahl (1981). In 1974 ASC allowed for up to 85 bats, but hindsight (with better knowledge of the animals' habits) suggests few would have escaped the actual counts, 53+ in Feb/Mar, and 69 in July, when some previously dependent young had become free-flying (Cheke & Dahl 1981). In 1974 the endemic birds' closest relatives, *Foudia sechellarum* and *Acrocephalus sechellensis*, were also largely confined to a tiny area – Cousin Island in the Seychelles (e.g. Collar & Stuart 1985) – but even that was larger at 28ha than the area occupied by *F. flavicans* in Rodrigues.

416. Respectively *Mangifera indica*, *Terminalia catappa* and *Tamarindus indicus*.

417. Cheke & Dahl (1981), Cheke (1987d) and recollections of foresters G.Elysée and F. Rémy, 1955–73 (ASC's unpubl. notes).

418. Cheke (1987d), citing Jean Montocchio (pers. comm.) who, with France Staub and Joseph Guého, did the searching. Cheke (1979a) discussed the effects of cyclones on bird numbers.

419. Cheke & Dahl (1981), citing information from René-Paul Alès. Reviewing Dr Alès's comments (in ASC's 1974 notebooks), we find that Cheke & Dahl's statement that he "did not see the Cascade Pigeon flock" is incorrect, suggesting there was a flock to see! Alès, who scoured the forests looking for warblers, simply stated that he had never seen bats in Cascade Pigeon, or indeed anywhere but the Baie aux Huitres area.

420. Vinson (1964b), Cheke & Dahl (1981).

421. Anon (1971a, 1976); Dr Z. Arlidge's land-use report, apparently intended to accompany the *Rodrigues land use suitability map* of 1975 (FAO & MSIRI, Mauritius), seems never to have been issued, but his leaked recommendations on reserves were quoted by Staub (1973b) and discussed by Cheke (1974).

422. Cheke (1974); the botanists were Jean Bosser and François Friedmann.

423. Jocelyn Forget asked the *parquet* (civil service) in April 1974 to protect the bat under the 1923 bird law (Tony Gardner, then Assistant Conservator of Forests, pers. comm.). The abortive *Rodrigues Wildlife Protection Regulations 1974* (draft copy in ASC's library) were intended to replace the 1923 law.

424. *An Act to amend and consolidate the law relating to game, camarons and shrimps, and to make better provision for the protection of wildlife* (Act No.33 of 1983, 12pp.); this also repealed the *Rodrigues Bird Protection Regulations 1923* (see Cheke 1987a).

425. Heseltine, styled Resident Commissioner, was formally island governor (Berthelot 2002), while Hotchin was technically junior to the magistrate – in practice, with the colonial government's support, he was *de facto* island ruler (North-Coombes 1971; pers. comms from numerous persons resident in the 1960s).

426. About 30 bats had been shot in July 1977, and c.15 in September 1978, but there were nonetheless 151+ in October (Cheke & Dahl 1981).

427. Durrell (1977a,b) – the second, *Golden Bats and Pink Pigeons*, is an entertaining popular account of the first Mauritian expeditions.

428. Cheke (1979a).

429. There were c. 650 bats in 33 zoos worldwide in 2005 (O'Brien *et al*. 2007). West (1986) summarised the first decade's results in Jersey. The atolls of the Chagos group, with overgrown coconut plantations mixed with fruit trees, were considered most suitable for the proposed reserve population, but the plans were essentially sunk by political problems around the American airbase on Diego Garcia (e.g. Durrell 1990), and, notwithstanding the recommendation in Mickleburgh *et al*. (1992), the increasing unwillingness of conservation bodies to introduce species outside their natural range – even where, as in the Chagos, there was no endemic (and hardly any native) fauna to be affected.

430. It was nearly five years before young were reared in Jersey, and another before any survived to breeding age (Darby *et al*. 1985). The birds lingered on, breeding fitfully, until a lone surviving female was exported in 1994 (*Dodo* 31: 161, 1995), probably to London Zoo where ASC saw a pair in c.1995.

431. Mungroo 1979, Cheke & Dahl (1981), Cheke (1987d). Fody numbers were down to 42 pairs (+some extra males), warblers to 8 pairs found, and bats to 66 individuals, with a few possibly missed.

432. Cheke (1974).

433. Cheke (1979a); in fact drought conditions persisted, and indeed October 1978 was one of the driest months on record (only 9mm of rain). Starvation during "*la grande sécheresse*" (the great drought) cut cattle numbers by half (5,911 to 2,337) by mid-1976, after which conditions gradually improved, though exports, usually several hundred, dropped to 10 in 1978. Pig numbers dropped 20% by mid-1976. Goats and sheep numbers crashed initially (to early 1975) by c.30%, but then recovered by mid-1976 to more than in 1974 (statistics in *Rodrigues Almanach 1982*). The drought was worst in 1974 and 1975, easing somewhat in 1976–78, but rainfall (38% of average in 1974, 56% in 1975) only returned to normal in 1979 (Strahm 1989). Illegal woodcutting was still a major problem in 1981, though 100ha of fuelwood plantations had been established (Carroll 1982).

434. Strahm (1989); So many plants were endangered that Strahm filled a 241-page book.

435. Cadet (1975), from observations in 1970; there were few mangroves left in 1974 (pers. obs.).

436. Barclay *et al*. (2001).

437. Jauze (1998); by the end of the century accelerated erosion from inadequate controls and overgrazing etc. was seriously silting the over-fished lagoon and further affecting fish stocks (Jauze 1998, Turner & Klaus 2005).

438. Gade (1985).

439. Berthelot (2002: 195) for paraffin imports in 1978; Jauze (1998) gave the 1990 figure, and also estimated that in 1985 the c.5000 households using wood consumed some 11,700 tonnes per year. Removing tax on bottled gas in c.1985 was probably the most significant single factor in the change.

440. Pers. obs. June 1999; Mary-Jane Raboude, cited in Impey *et al*. (2002:300), also made the point about cooking on gas.

441. North-Coombes (1971), *Rodrigues Almanach 1982*: 27, Strahm (1989), Jauze (1998:).
442. Arlidge (in Staub 1973b), Cheke (1974, 1978b), Lesouef (1975), Strahm (1983). Suggestions as late as 1981 (e.g. Carroll 1982), following Wiehe, to preserve Cascade Victoire, reflected inadequate fieldwork – the valley had long been devastated, Carroll elsewhere citing the report (Cheke 1974) that made this clear.
443. Thingsgaard (1986), Strahm (1989).
444. Stearns & Stearns (1999: 224); the reserve was Grande Montagne. This huge agricultural project was in fact funded by the European Development Fund (Jauze 1998, hence the ECUs), not the FAO. Fencing off forested watersheds from livestock was an important component of the work.
445. Pers. obs., 1999.
446. Mauremootoo & Payendee (2002), Payendee (2003), and articles in Mauritian newspaper *Weekend* 9 Sept. and 9 Dec. 2001.
447. Mauremootoo & Payandee (2002), John Mauremootoo pers. comm. (plus other correspondence and an unpublished MWF impact statement report in ASC's possession).
448. Articles in *Weekend* 9 Sept. and 9 Dec. 2001; Mary-Jane Raboude, with an Australian university degree, is unusual – most well-educated Rodriguans leave for greener pastures elsewhere. She was appointed educational outreach worker for conservation in 1998 with support from Philadelphia Zoo (USA), much involved in Golden Bat conservation projects (Jamieson 1999); Trewhella *et al.* (2005) reviewed successful conservation outreach in relation to bats.
449. Carroll (1982), Safford (1991), Impey *et al.* (2002), Showler *et al.* (2002).
450. Where possible, immature Rodrigues Fodies settle within 100–150m of their natal territory (Cheke 1987d), though as numbers and density increased some must have been forced into longer movements.
451. The 1999 census was in April/May (Showler 2002), outside the breeding season, no doubt why Dave Showler found so many single birds, and made no estimate of breeding pairs.
452. See graph in Showler *et al.* (2002: 213).
453. Safford (1992), Impey *et al.* (2002); Impey's distribution map lacks the precision of previous ones, making detailed comparisons difficult. In 1981 and 1982, despite cyclone *Damia* intervening, fody numbers were stable at 72–76 pairs, but unaccountably dropped to 60 pairs in 1983. Since then censuses have been too infrequent to pick up similar fluctuations. JPH noticed a marked increase on Grande Montagne between 1999 and 2005.
454. Powell & Wehnelt (2003), O'Brien *et al.* (2007), also Strahm (1989:37) for a 1984 figure. Vicki Powell and her MWF team had one count of 5,076, the rest varying from 2,497 to 3,842 during March 2001–February 2002. Although totals were supposed to be from simultaneous counts at different sites, we suspect some inadvertent duplication in January 2002, the date of the extra-high score. Some fluctuation over a year is due to first 'fledging', then mortality, of newly independent young.
455. Hume (2003c) and various MWF staff verbally to ASC.
456. Cheke & Dahl (1981), Powell & Wehnelt (2003).
457. Hume (2003c).
458. Hume (2003c) and JPH pers. obs.
459. Carl Jones pers.comm. and JPH pers. obs.
460. Probst (1997a, Abhaya & Probst 1997b) said Bloodsuckers were introduced in 1990, but Alecks Majlkovic (then MWF botanist on Rodrigues) told ASC in 1999 that it was "about 13 years ago". Blue-tailed Day-geckos first appeared at Anse aux Anglais, apparently originating from the bottling plant at Rose-Hill in Mauritius (Carl Jones, pers. comm.).
461. Jauze (1998: 72), confirmed by Alecks Majlkovic (pers. comm.). European readers of Jauze's book should note that 'caméléons' in Mascarene usage are Bloodsuckers *Calotes versicolor*, not chameleons *Chamaeleo* or *Furcifer*.

462. Distribution from Carl Jones, pers. comm. No surviving Mascarene day-gecko (apart from the crepuscular *Phelsuma guentheri*) is comparable in size to *P. edwardnewtonii*, so the nearest brightly coloured diurnal equivalent might be *P. madagascariensis* (Madagascar) or *P. sundbergi* (Seychelles). Also, *P. cepediana* is not the best adapted to the dryish Rodriguan climate – *P. ornata* would be a more appropriate small Mauritian species. Jones (1993, 2002) proposed putting *P. ornata* and *P. guentheri* (as an analogue to *P. gigas*) on Iles Coco and Sable, and (1993) additionally suggested using the islets as an extra refuge for Bojer's Skinks and Keel-Scaled Boas.
463. Gomy (1973b), Lewis (2002).
464. Varnham *et al.* (2002); in 1995 there was no sign of predation on centipedes *Scolopendra subspinipes* and *S. morsitans* (Lewis 2002), but in mid-1999 we only saw two centipedes in a week on Rodrigues, both on Grande Montagne, and one of those was seriously injured. There is no direct evidence of how the shrews arrived, but since a quay was built for ships to come alongside in 1980 (Jauze 1998, Berthelot 2002), the risk of animal self-introduction clearly increased. Cheke (1979b) warned about this in relation to Ship Rats rather than shrews, though it has since turned out that the brown-coloured rats already present included the brown variety of *Rattus rattus* and were not all *R. norvegicus* as previously thought (Carl Jones, pers. comm.).
465. *Hemiphyllodactylus typus* (Vinson & Vinson 1969), *Hemidactylus brookii* (specimen collected by Carl Jones, pers. comm.). The 1874 expedition recorded only Stump-toed Geckos *Gehyra mutilata* (Günther 1879), movement between Rodrigues and Mauritius increased in the last quarter of the 19th century century, and Cheechaks are unlikely to have escaped introduction then.
466. ASC recorded four species in June 1999 (*Hemiphyllodactylus typus*, *Hemidactylus brookii*, *Gehyra mutilata*, and the native *Lepidodactylus lugubris*); Carl Jones (pers. comm.) has also seen Cheechaks *H. frenatus* with the others.
467. *Ramphotyphlops braminus* was first reported in print in the 1960s (Vinson 1964b), but possibly on the basis of 1930s collections, as with other anthropophilous reptiles.
468. Alecks Majlkovic, pers. comm.
469. Neither Bertuchi (1923) nor the more invertebrate-oriented Snell & Tams (1919) mentioned snails at all, any more than did Koenig (1914) or Stockdale (1914). North-Coombes (1971) mentioned Rodriguans reduced to eating the 'big snails' in droughts, but was vague as to when this happened.
470. Greathead (1971).
471. North-Coombes (1996: 13) has a photo of "'Piquant loulou' – a young plant introduced by Nigel Heseltine".
472. *L'Express* [Mauritius] 18 Dec. 1982 reported a French aid programme that proposed planting "screw-pine, latan, hurricane palm, eucalyptus, popinac and *eburnea* [= *Acacia nilotica*]" at Graviers in Rodrigues. Strahm (1989) reported it "widely planted and naturalized", and by 1999 Turnbull (1999) commented on its adverse effects on native species (see also p. 231). It was apparently absent before 1970 (Cadet 1971b, 1975) and in 1974 (pers. obs.), so may well have been introduced by Heseltine in the mid-1970s.
473. Rountree (1943); Wiehe (1949: 286) saw Fairy Terns nesting on *Pisonia grandis* scrub in 1938 or 1941 on "Ile aux Fous and Ile aux Chats", described as "no more than elevated sandbanks" – i.e. he meant Iles Coco and Sable.
474. d'Unienville (1954), Branegan (in Bourne 1959), Gill (1967); Alix d'Unienville gave no dates, but travelled to Rodrigues on the *Floreal* (entered service 1949, Berthelot 2002), together with the Anglican Bishop of Mauritius, at the end of a cyclone season. The bishop's only late summer visit during 1949–54 was in April 1952 (North-Coombes 1971), which pinpoints d'Unienville's trip. Branegan claimed he was on 'Sand Island', but assuming the distribution of Fairy Terns was similar to 1964 (Coco: 400, Sable: 100; Gill 1967), he was probably in fact on Coco; he also conflated the two

noddy species, combining them (1,000 birds) as Brown Noddies. His shipmate D. M. Neale (Bourne 1959:20), who saw only 250 noddies and 50 Fairy Terns, was presumably, as he claimed, on Sable.

475. See discussion in Safford (2001); the same caveat applies to tropicbird numbers on Round Island (Chapter 9).

476. Cheke (1974); Staub, visiting in 1967, 1968 and 1970, noticed Fairy Terns had declined relative to noddies, but did no counts (Staub 1973b).

477. Cheke (1974); the Fairy Tern massacre was previously unpublished. Nigel Heseltine (pers. comm.) had spoken to the witness; various local people told versions of the story to ASC in 1974.

478. Mungroo (1979), April–May 1979; few birds had been breeding six months earlier (Oct. 1978), so no count was made (Cheke 1979).

479. Strahm (1989: 38), visit in October 1985. Breeding appears aseasonal – plenty of nests in October 1985, virtually none in the same month seven years earlier (Cheke 1979). Mungroo's figures (1979) in April–May were very similar (800 individuals *Anous tenuirostris* and 200 *A. stolidus*) to the October 1985 figures, albeit not on the same islet. Staub (1973b) reported hearsay evidence of Sooty Terns trying to nest on Gombrani and Pierrot (= Chat) islets, but that "labourers collect their eggs and scare them away".

480. Owadally (1984), Cheke (1987a). Reservation was first requested by the Ancient Monuments and Reserves Board in 1972 (Claude Michel, pers. comm.) and supported in 1974 by Heseltine who said "I will get these two sandy islands declared a 'nature reserve', in fact do it myself as a local UDI" (pers. comm.; 'UDI' = 'unilateral declaration of independence', a term then current in relation to the Smith regime in Rhodesia, now Zimbabwe). Any local regulation he may have made was not recognised in Mauritius.

481. Vinson (1964b), Cheke (1974, 1979); shearwaters are (or were) used as fish bait in the Seychelles (Tony Diamond, pers. comm.).

482. Unlike the other islets, Frigate is privately owned (Bell *et al.* 1993), which complicates the position.

483. Bell *et al.* (1993), Dulloo (1996), Bell & Bell (1996), Bell (2002); in 1993 there was a cat on Coco, but it had gone by 1995. Noddy numbers in 1993 were similar to counts by an Irish university expedition in 1991: 4,620 nests of *Anous tenuirostris*, 288 of *A. stolidus* (Jones 1993, Showler 2002), though in January 1996 there were apparently fewer Lesser (600+ pairs) and more Brown (800+ pairs) (Abhaya & Probst 1997). The two noddies may not be synchronised for peak breeding. Carl Jones (unpubl.) saw the big Fairy Tern flock in 2001.

484. Abhaya & Probst (1997), Showler (2002), Hume (2003); noddy numbers in 2003 were comparable to 1993, but some had recolonised Sable; Carl Jones's estimates for 2005 are from Chapter 9 (Table 9.2).

485. Unlike the mainland forest areas, the seabird islets get a high profile in tourist guidebooks (e.g. Moreau 1993, Roberts 1998) and internet travel sites.

486. Abhhaya & Probst (1997b).

487. Carl Jones, pers. comm.

488. Alecks Majlkovic (pers. comm.) tried to kill the cats with poisoned fish in 1999; JPH could find no shearwaters in October 2003, nor could French seabird researcher Vincent Bretagnolle also in 2003 (Carl Jones, pers. comm.). Bretagnolle confirmed the continued presence of small numbers of *Puffinus pacificus* on the mainland, perhaps 50–70 pairs.

489. Carl Jones, pers. comm.. Over 20 years, Ile aux Aigrettes in Mauritius has similarly been unwanted host to two dogs and at least one cat.

490. Cheke (1987a); there was no sign of rabbits in 1993 (Bell *et al.* 1993), but plenty of *Rattus rattus*. However, no rats were detectable in 1999 (Alecks Majlkovic, pers. comm.). When did Ship Rats replace Norway Rats on Frégate?

491. Neither Vinson (1964b) nor Gill (1967) saw tropicbirds in 1963–64; Staub (1973) rediscovered them in 1967. For later records see Cheke (1974), Strahm (1989), Abhaya & Probst (1997), Showler (2002), Chapter 9 (Table 9.2). Both species were covered by the 1923 bird protection law, so Red-tailed Tropicbirds may have nested sporadically in the interval.

492. Pers. obs. and letter from ASC to N. Heseltine 30 Dec. 1974.

493. Pers. obs., *Marine turtle Newsletter* 95: 21 (2002), diving information on www.mysterra.org/webmag/rodrigues-island.html.

494. Cheke (1974, 1987a and unpubl. notes), Jouanin (1987), Showler & Cheke (2002); the specimen (head only) was deposited in the Mauritius Institute.

495. Shoals: e.g. Burnett *et al.* (2001); Prince William: e.g. *Guardian*, travel section 22 Sept. 2000; Prize: www.whitley-award.org (Whitley Award to John Mauremootoo in 2002).

496. *Population and vital statistics: Republic of Mauritius – year 2004* on http://statsmauritius.gov.mu. The birth-rate declined sharply after 1982 following successful implementation of birth-control measures, despite the population being overwhelmingly Roman Catholic (Jauze 1998).

497. Two desalination plants were projected in 2002 (*Business Mag-online*, 9 Jan. 2002: www.businessmag.mu); see Cazes-Duvat & Paskoff (2004) for details of hotel development and impact.

498. The Giant Tortoise Reserve and Cave Project has been developed by tortoise breeder Owen Griffiths, using Mauritian-bred tortoises (Middleton 2007).

Chapter 9. A miraculous survival

1. Barnwell (1950–55), who unfortunately left gaps in the published transcript. 'Cabish' = cabbage, i.e. hurricane palms.

2. Noble (1756). The earliest known account of a landing on Round Island (although tortoises were harvested there much earlier). Curiously we have found no independent record of the *Sumatra* – the account implies it was British, but there was no East Indiaman of that name (Farrington 1999), she was not a Royal Navy ship or licensed privateer (Phillips 2005), nor have we found any Dutch or French vessels of that period so named.

3. Saddul (1995) summarised the geology of the northern islets; rising sea-levels would have cut off Round and Serpent Islands. (-63m) around 12,000ya and Flat Island and Gunner's Quoin (-54m) about 11,000ya (Siddall *et al.* 2003, Camoin *et al.* 2004).

4. Island sizes and heights from Bell *et al.* (1993) and Safford (2001); Round Island was re-mapped in 2003, proving to be 45% larger than previously thought (219ha vs 151ha; *MWF Plant News* April to September 2003, Johansson 2003); likewise Serpent Island is now reckoned to be 31ha not 19. Distances from the mainland measured on the 1:100,000 map (Ordnance Survey International Y682 = DOS 529, 1994).

5. Padya (1989 and earlier eds., especially 1984 for Flat Island).

6. Foucherolle *et al.* (1714); *Latania loddigesii*.

7. The screw-pine is *Pandanus vandermeeschii*; although named for a 19th century lessee of Round Island, A. Vandermeersch (Pike 1873), Balfour (in Baker 1877) misspelt the name without the second 'r'. As scientific names (even if misspelt) cannot later be altered, we are stuck with *vandermeeschi*, although the 'corrected' version *vandermeerschii* has been widely used in the literature (see *Flore* 190: 26, 2003, Bosser & Guého).

8. Barnwell (1955).

9. Pitot (1905:164–65).

10. Calculated from the average Aldabra forest density of 20/ha (see p. 114). Tortoises should not have been starving at the end of the wet season, but perhaps the population was peaking before a crash. The Aldabra tortoise population has cycles of abundance, overgrazing and mass mortality (triggered by years with low rainfall),

after which vegetation recovers and tortoises multiply again (Bourn *et al*. 1999).

11. Barnwell (1950–55, *passim*; 1955).
12. Barnwell (1950–55): *Rising Sun*, *Scarborough* and *Lyon*; Barnwell (1948): *Westmoreland*.
13. Foucherolle *et al*. (1714), Garnier (1722).
14. *Courrier de Bourbon* (1723), Le Housec (1721–22).
15. Lougnon (1933–49, 1944–46); see also Chapter 6. A boat crew harvesting latan leaves was trapped on Gunner's Quoin by a cyclone in 1728. The intended 1731 tortoise run to Rodrigues apparently never happened.
16. La Caille (1763).
17. Cossigny (1764), de Querhoënt (in Buffon 1770–83).
18. The specimen, now in London's Natural History Museum, is the type of *Testudo schweigeri* (Arnold 2004), shown by DNA to be the same species as the Mauritian subfossil *Cylindraspis triserrata* (Austin & Arnold 2001).
19. Barnwell (1948); a Rodrigues tortoise would have been less surprising, given the British naval presence there in 1761 (Chapter 6).
20. See Mauries (2002).
21. The 'reed' was presumably the bullrush *Typha dominguensis* (*Flore* 191, 1984, W. Marais), still "abundant in the swampy hollows" in 1885 (Horne 1887, as *T. angustifolia*), but apparently since vanished (not recorded by Bell *et al*. 1993). 'Veloutier' refers to the common beachfront shrubs *Tournefortia argentea* and *Scaevola taccada*, still present (Bell *et al*. 1993).
22. Lislet-Geoffroy (1790); a 28–30' (71–76cm) snake could be either endemic boa.
23. Tombe (1810), Barnwell & Toussaint (1949).
24. Michel (1935).
25. Prior (1819); Telfair's Skink's inquisitive behaviour is familiar to all visitors to Round Island. The 'tufted' grass was probably *Vetiveria arguta*, and Prior's 'cotton-shrub' was presumably cultivated *Gossypium* sp.
26. Chapotin (1812). A reference to Cossigny (Charpentier 1799) reporting lizards on Flat Island (Cheke 1987a) was an error.
27. Pitot (1910).
28. Jones (1988) reported egg traces found by botanist Joseph Guého (date not recorded).
29. Milbert (1812).
30. Milbert (1812); Cheke (1987a) assumed that Prior's "small . . . hares" were rabbits, but we now consider this identification doubtful.
31. See Chapter 6.
32. Quoy (1833); Charles Telfair, who owned a sugar estate at Mapou opposite Gunner's Quoin (Desjardins 1831), provided Quoy and Gaimard with a boat.
33. Cheke (1987a) reported the rabbit skull, though subsequent re-examination by Carl Jones (pers. comm.) suggests that it is indeterminate. See below for hares and their decline; Dulloo (1994) reported the 1934 release.
34. Ayres (1860), Horne (1887), Vinson (1953), Staub (1993), Bell *et al*. (1993), Souchon (1994).
35. La Caille (1763, Grant 1801) listed only Gunner's Quoin, Flat Island ('Ile Longue') and Round Island as having snakes. For his map see Toussaint (1953, map 4); Barnwell (1955) only mentioned 'Ile Parasol' as used by La Caille for Serpent Island. The earliest French map to show all the islets, from 1735 (Lenoir 1979:19), names only 'Ile Ronde', leaving Serpent Island unnamed. The Mauritius half of a Mascarene pair dated 1763 in Toussaint (1972), attributed to Rigobert Bonne in *DTIM* 1: 1 (not included in Toussaint & Adolphe 1956), still shows Serpent Island as 'I. Paras', whereas Bonne's 1782 map uses 'I. au Serpent' (note 36). The two sets are stylistically similar (pers. obs.), but as Bonne was based in Paris (*DTIM* 1: 1) he probably just copied from La Caille, as he did later from d'Après.

36. d'Après (1775) published a small map, echoed by Bonne's (reproduced in Lenoir 1979) in a world atlas published in 1782 (Toussaint & Adolphe 1956: 779); Grant (1801) included a large fold-out map.
37. Bory (1804), Tombe (1810).
38. Quoy & Gaimard (1824).
39. Dumeril & Bibron (1834–54).
40. Lesueur collection, Le Havre, MS 15037. In another manuscript (MS.78092) Peron commented that "it is of note that all the snakes brought from Round Island or Serpent Island to Mauritius itself cannot survive there", so they did become aware of another island with snakes.
41. Reported in the Mauritian press in 1816 (Ly-Tio-Fane 1972); Lislet-Geoffroy, a leading member of the Society (Ly-Tio-Fane 1965), may have promoted this more rational approach.
42. Lesson (1830), Quoy (1833), Magon (1839).
43. Barkly (1870), Vinson (1950). Corby was with Lloyd's team on Round Island in 1844 (Lloyd 1846), and mapped Flat Island in 1856 (Toussaint & Adolphe 1956); we have no record when he visited Gunner's Quoin, although he had been there before 1844, as he compared skinks there with those on Serpent Island. Corby presented a *Round Island* snake to the Desjardins Museum in 1851 (Bouton 1851). Barkly could have been mistaken the origin, but it is very likely that Corby also collected on Gunner's Quoin.
44. Dumeril & Bibron (1834–54); Dumeril & Dumeril (1851) narrowed the alleged origin to 'Port Jackson'.
45. Lesueur MSS, Le Havre, No.15037. Had he and Peron actually been to the islets he would surely have been clear where the snakes occurred!
46. Bélanger (1834), who was there c.1825–6, reported meeting Dussumier in Mauritius, and a Mauritius Merle in the Paris museum collected by Dussumier was accessioned in 1826 (ASC pers. obs.); see also Guibé (1958). However, Dussumier crisscrossed the Indian Ocean many times on his voyages, frequently stopping at Réunion or Mauritius (Laissus 1973b); as the specimen was not described until 1844, it could have been collected later, though it pre-dates the surviving register begun in 1839 (pers. obs.).
47. Barkly (1870); Vinson (1950) bemoaned the loss of this vital specimen.
48. Dumeril & Bibron (1839–54) mentioned three specimens from Manilla [!] which they equated with Desjardins's *Scincus telfairii*, erecting for the species the new genus *Leiolopisma*; they stated neither collector nor date for the 'Manilla' specimens; Dumeril & Dumeril (1851) listed Desjardins's type and, unhelpfully, "other specimens from Mauritius and Manilla". Of the five 19th century specimens still present in 2006, only the type has a specific locality, the others simply labelled 'Mauritius'; they are all much larger (144–194mm) than the maximum SVL of 110mm cited by Vinson & Vinson (1969) for Round Island, but Dulloo *et al*. (1999) found lizards up to 168mm there in 1996.
49. Desjardins (1831); the paper was read in 1830 – he named the lizards after his colleagues in the *Société d'Histoire Naturelle de l'île Maurice*. Quoy saw Telfair's skink specimens from Gunner's Quoin in Desjardins's collection in 1828 (Quoy 1833), though the type specimen came from Flat Island (Dumeril & Dumeril 1851).
50. Details from Ayres (1860) and Pike (1873); Bouton's remarks were in a footnote to the 1872 printing of Lislet (1790). Order No.3 of 1857 (*MA* 1858:176–88) formally established the quarantine station, but it had been set up provisionally in 1856 because of a cholera crisis, and the island so used (without facilities) in 1854 (Anderson 1918). Telfair's Skinks were first stated as confined to Round Island by Pike (1873), who had presumably seen none on Flat Island in 1869.
51. Newton (1858–95, 1861), Pike (1873).
52. Horne (1885).
53. Gleadow (1904).

54. Horne (1885). No other native palms survived; '*Eugenia cordifolia*' (now *E. lucida*), the only native broadleaved tree he had found previously (in 1871), had vanished, browsed to death, he thought, by goats.

55. Respectively *Furcraea foetida*, *Opuntia* sp. and *Leucena leucocephala*.

56. Besant's (1902) 'vivid recollection' seriously misjudged the size of the rock (much too small) and its distance from Flat Island (much too close). Besant (later Sir Walter, novelist, reformer and brother-in-law of theosophist Annie Besant; *CDNB*: 229) did not date his visit, but was in Mauritius 1861–67 (*DMB*:324–5, 1943, P. J. Barnwell) and Ayres died in 1863 (Pike 1873, *DMB*: 229–30, 1943, A. Toussaint).

57. Edward Newton (1858–95), *en route* to Flat Island in December 1862, saw on the lee side (i.e. under the big cliff) "over 100 red-tailed paille-en-queues flying and screaming around; they were evidently breeding"; fellow passenger Longbourne had collected eggs the previous year. Corby had told him they bred only on Round Island, which is odd as Corby was familiar with Gunner's Quoin and was aware of Red-tailed Tropicbirds there in the mid-1840s (Lloyd 1846); W.H. Power gave Newton a shearwater skin and an egg collected on Gunner's Quoin in 1864. Daruty (1883) mentioned restriction of snakes (cited in error as '1879' in Cheke 1987a).

58. The Gunner's Quoin rats were identified as *Rattus rattus* by Bullock *et al.* (1982), but their measurements matched small *R. norvegicus* (c.g. Corbct & Southcrn 1977), since confirmed (Bell *et al.* 1993, Dulloo *et al.* 1996).

59. Boobies: Quoy & Gaimard (1824), Cheke (2001a); Sooty Terns: Barnwell (1948: 80) – "abundance of grey and white birds as big as pintados, with forked tails", seen the day before the ship "passed between Hangrock [Gunner's Quoin] and the Main[-land]".

60. In Mauritius *boeuf* had transferred to the Masked Booby from Abbott's, *cordonnier* (noddies), and *goelette* (Sooty Terns) remained the same (Cheke 1982a).

61. Vinson (1950), Jones (1993a).

62. Lloyd (1846) called the large lizards "*Gecko* sp.", but is more likely to have seen the conspicuous Telfair's Skinks than the similar-sized but cryptic and as yet undiscovered Günther's Day-gecko.

63. E. Newton (1861a), Rountree *et al.* (1952), Brooke (1976), Probst (1996, Réunion).

64. Barkly (1870) claimed that Corby, one of the 1844 party, had released the rabbits, but they were already abundant at that date. Vinson (1965) thought Corby had released them on an earlier visit, but we think it more likely that they were released in the 1790s when rabbits and/or hares were put on the other islands. Corby was in Mauritius from 1840 till he died of cholera in 1865 (Vinson 1965), so there was hardly time for him to release rabbits and for them to reach such abundance by late 1844. Furthermore, there is no hint in Lloyd's paper that the rabbits had been introduced by his co-explorer.

65. Barkly (1870); Kerr was Mauritius auditor-general in 1844 (*DMB*:143, 1942, P. J. Barnwell)

66. Layard (1863), Brooke (1976, 1978, 1981). Thousands of Tropical Shearwaters breed in nearby Réunion.

67. At this point Newton sketched (badly!) a Bottle Palm *Hyophorbe lagenicaulis*.

68. E. Newton (1859–1878).

69. E. Newton (1861a).

70. Barkly (1870). Since Corby had leased the island, he probably introduced the goats for 'sport', i.e. recreational shooting.

71. Newton (1861–82) mentioned lizards and snakes on his 1869 visit, which may be when he collected two Bojer's Skinks now in the NHM (Boulenger 1885–87).

72. Immortalised (misspelt) in the screw-pine *Pandanus vandermeeschi*.

73. Pike's printed account (1870a) was published after the second expedition had taken place, so Barkly must have sent Hooker a draft; see also Barkly (1870).

74. Pike (1873); Newton (1861–82) also described Pike's stormy night.

75. Barkly (1870), Strahm (1993). Barkly wanted goats and rabbits eliminated: "I trust, therefore, that Mr Vandermeersch will engage the sportsmen of his acquaintance to extirpate both these kinds of animals" – his trust was, unfortunately, misplaced.

76. Boulenger (1893); young specimens of the Keel-scaled Boa differ in colour from adults (Dulloo *et al.* 1999, Korsós & Trócsány 2002), no doubt contributing to the confusion.

77. "The *S. bojerii* unlike the *Telfairii* I found mostly in open places on smooth rocks, also great numbers of its eggs deposited in rows on the underside of the branches of Vacoa [screw-pine]" (Pike 1870b). Skinks lay eggs in soil, while Mascarene day-geckos glue eggs to branches. In his book, Pike (1873) nearly recovered from this, but a misplaced sentence transposed his identification of day-geckos as source of these eggs to making them "identical" with Bojer's Skink! On the distribution of Bojer's and Bouton's Skinks, Pike (1870) appears right and Barkly (1870) wrong.

78. Vinson (1950, 1964a, 1976).

79. Pike (1870a, 1873). Trindade Petrels lay larger, rounder eggs than Wedge-tailed Shearwaters (Newton 1956, Vinson 1976a), very like a domestic hen's. Wedge-tailed Shearwaters nest out of sight *under* rocks (also in burrows in soil). Pike's 'great gorge' can only be Big Gully, the gorge north of the Big Helipad. Newton's set of the *TRSAS* is in Oxford's Radcliffe Science Library.

80. Vinson (1976) gave the fullest account of nest, eggs and habits of petrels and shearwaters on Round Island.

81. E. Newton (1861–82), letters to his brother Alfred 4 Dec. 1864, 29 Jan. 1865 and 1 Oct. 1869; neither Pike nor Newton's friend Barkly mentioned his presence! Several other specimens from Lloyd (collected by Corby *et al.*) are in Cambridge (Cheke 1987a), but no Round Island Petrel. In 1844 Lloyd (1846) himself had described the *fouquet*'s beak as black (like the gadfly-petrel), but the shearwater's bill is dark grey, so this is inconclusive. It seems when Newton 'got' his petrel he meant 'seen' rather than 'collected'.

82. The two quotes come from Pike (1873 and 1870b).

83. Pike (1872).

84. Vinson (1964a), Gerlach *et al.* (1997); the beetle is *Pulsopipes herculeanus*. Pike (1870b) made clear that some items in the case were, for comparative purposes, from mainland Mauritius, but there is no indication that it included any Seychelles specimens. Newton (1861–82) wrote in 1869 that on Round Island he "couldn't find a beetle", suggesting there was a special beetle known to be there; he never mentioned insects elsewhere in his letters.

85. Barkly (1970).

86. Gordon (1893, vol.2: 680); Le Clezio's novel *La quarantaine* (1995) is an account of Flat Island in 1891, purporting to derive from notes made in quarantine there by the author's grandfather; vegetation and birds are mentioned, but the source is too unreliable to be used.

87. Vinson (1950), Daruty (1883), Strahm (1993). Daruty's brief report on his 1879 visit, is mainly on geology, mentioning vegetation only in passing, and fauna not at all. Möbius (1880), or rather someone on his behalf, collected on Round Island (specimens of *Leiolopisma telfairii*); he mentions no such trip in his itinerary, and his data is apparently compromised. He claimed to have found a pickled Keel-scaled Boa (as '*Leptoboa dussumieri*') in a jar in a chemist's in Mahé (Seychelles), and to have collected '*Hemidactylus maculatus*' (a large Indian species) on Round Island. Dr Rainer Günther from the Berlin Museum reports that the gecko resembles *H. mercatorius* (pers. comm.), thus, probably in fact *H. brookii*, a species never since found on Round Island or any Mauritian offshore islet. The snake, bizarrely, is actually the extinct Burrowing Boa, only the seventh specimen known. The alleged finding

location is clearly suspect, Bauer & Günther (2004) suggesting the pharmacy was in Mahébourg (near where Möbius was based in Mauritius), rather than Mahé Island in the Seychelles. Strauch (1887) listed *Phelsuma guentheri* in the St. Petersburg museum collected in 1882 and 1885 by G. Schneider and E. Riebeck.

88. Johnston (1984), Strahm (1993). Johnston may have been referring to the severe droughts in 1880–81 and 1885–86 (*MA* 1913:C58); he was Surgeon General to the British troops 1887–90, writing several papers on Mauritian plants; he died aged 83 in 1939 (*DMB* 246, 1943, P. J. Barnwell).

89. Andersen (1907), Mason (1907), Vinson (1964a), Cheke & Dahl (1981), Cheke (1987a); Hume (2005) re-examined the bat skull, which appears to be from a young animal and cannot be identified to species with certainty. Mason also reported rabbit and goat bones in the deposit.

90. E. Newton (1878); see also Chapter 7.

91. The *Graphic* of 3 Mar. 1883 published a short, anonymous article on Red-tailed Tropicbirds on Round Is., illustrated with a sketch by 'Lieutenant-Colonel H. Robley, Sutherland & Argyll Highlanders'. Horatio G. Robley was in Mauritius with this regiment around 1870 (www.aucklandartgallery.govt.nz) and from 1879–1881 (Barnwell 1948); his party plucked about 2,000 tail feathers (i.e. from *c*.1000 sitting birds) in a day. The laws were Proclamation No.40 of 1880, No.6 of 1886 and No.39 of 1897, see *MA* 1881: 111–112 and 1887, 1901: 187; the texts of the 1886 and 1897 orders were not published in the almanac, but the latter was summarised as making "the shooting, taking or destroying of the Red-tailed 'Paille-en-queue' (*Phoeton* [sic] *Phoenicurus*) of Round Island punishable by a fine of not less than Rs.50 and not more than Rs.100" with imprisonment for repeat offences (*MA* 1902: 127).

92. Rasmussen & Prys-Jones (2003). Meinertzhagen later removed and relabelled specimens from the British Museum and other collections to make it appear that he had collected them when visiting Mauritius, the Seychelles and numerous other places, but there is no indication, when in Mauritius 1910–11, that he either collected anything or deliberately falsified any data.

93. There are inconsistencies, however. In his paper, Meinertzhagen (1912) said he was on Flat Island in July, August, October and November, whereas the diary (Meinertzhagen 1910–11) records visits only in September and November. Both paper and diary muddle the scientific names of two shearwater species; the paper records shearwaters nesting on Round Island, while the diary says he found no evidence of breeding. He completely confused the details of alleged breeding records of Red-footed Boobies (Cheke 2001a). The information on land birds seems, by contrast, to be sound.

94. Lyttleton (1883), Oliver (1891; vol.2: 376, footnote), Chambers (2004); Horne (1887) did not mention tortoises. Chambers (2004) concentrated on Galápagos and Aldabra tortoises, and was unfortunately inadequate, inaccurate and out-of-date on Mascarene tortoise history, having missed important recent historical references (e.g. Bour 1980, Cheke 1987a). However in pursuing Günther's efforts to save the Aldabra Tortoise he uncovered the reason for its introduction to Flat Island, not evident from Pasfield Oliver's laconic footnote cited by Cheke (1987a). Günther was the man after whom Boulenger (1885–87) named *Phelsuma guentheri*. See Bour (1984b) for breeding performance in Mauritian tortoise enclosures; in the right circumstances they breed readily, as Owen Griffiths has amply shown at his facility at La Vanille (Griffiths 2003).

95. The editor queried the Whimbrel nest: "some confirmation is here needed. The Whimbrel breeds in the North in May" (footnote to Meinertzhagen 1912). ASC saw a Whimbrel in territorial display at Wolmar, Mauritius in 1974, but in May, not November – other Whimbrels ignored the displaying bird.

96. Meinertzhagen (1910–11).

97. Meinertzhagen (1912).

98. Vaughan & Wiehé (1937) did not date Wiehé's visit; '1935' comes from Strahm (1993), probably from dates on herbarium specimens.

99. *Lantana camara* and *Albizzia lebbek*.

100. Dulloo (1994); Gunner's Quoin was leased to a sugar estate as a hunting reserve in the 1930s; Père Henri Souchon remembers hunting parties there after deer in the early 1940s (pers. comm. to Vikash Tatayah).

101. Richard Barnwell showed his father's honeymoon photos to ASC and David Bullock in December 2001; P. J. Barnwell died in 1997. The Mauritius Institute has a tortoise specimen from Flat Island, collected in 1943 (Bour 1984b). Souchon (1994 and pers. comm. to Vikash Tatayah) reported seeing tortoises including a juvenile on Flat Island *c*.1940, also claiming (*contra* Tatayah, note 102) they were later destroyed in a fire.

102. Information given to Vikash Tatayah (pers. comm. to ASC, 2003) from former lighthouse keeper Mr Cangy, confirmed by Tristan Bréville.

103. In 1909 a Royal Commission on Mauritian finances under Sir Frank Swettenham recommended closing the Flat Island quarantine station (Bruce 1910), but we have not been able to find out when this actually happened. According to Macmillan (1914: 273) both Flat Island and Cannoniers' Point quarantine stations were still active, but the text appears to have been written in 1912.

104. Newton (1956).

105. Meinertzhagen (1912), Alec Forbes-Watson pers. comm. Newton (1956, 1958a) named only waxbills among otherwise unspecified landbirds.

106. Bell *et al*. (1993), who included vegetation maps of Gunner's Quoin and Flat and Gabriel Islands.

107. "The eastern fringe of the lagoon is fronted by *Scaevola* and *Argusia* [=*Tournefortia*] with *Suriana* immediately behind this" (Bell *et al*. 1993), all typical beach-front shrubs in the tropical Indian Ocean.

108. Nichols & Freeman (2004), Arnold (2000); the distinct Flat Island variety is known as the 'Red-tailed Skink'.

109. Arnold & Jones (1994) and Arnold (2000) suggested *Nactus* extinctions were caused by *Hemidactylus frenatus*; The Flat Island sympatry data is in Cole (2005b), Cole *et al*. (2005).

110. Nogales *et al*. 2004, Chapter 10 (*contra* Safford 2001); Nogales claimed that "probably recent releases by campers" had put cats on Flat Island, whereas they had in fact been present since the 1860s at least.

111. Newton (1956), Rowlands (1982).

112. Bell & Bell (1996), Chapter 10. Lavergne (2007) suspected rats had recolonised from the numerous unquarantined tourist boats.

113. Alec Forbes-Watson (unpubl. data): one White-tailed and six Red-tailed Tropicbirds.

114. Newton (1956), Staub (1973, 1976), Rowlands (1982); Newton (1958) later said the tropicbirds on Pigeon Rock were White-tailed. The birds Rowlands claimed to see in October 1970 are exactly those found on Serpent Island, yet he did *not* report any tropicbirds (either species) on Pigeon Rock. However France Staub (pers. comm.) confirmed seeing boobies, noddies and Sooty Terns settled on the stack on several occasions, admitting his presumption of breeding could not be confirmed. Temple (1976) stated cautiously that "small numbers [of Sooty Terns] roost and may nest occasionally on Pigeon House Rock".

115. Tezier (1995); the video shows the birds and skink, giving no indication of numbers, but the impression is of only a few dozen pairs each of the tropicbirds. The gecko (not on the video) was recorded by Probst (1998a).

116. Vinson (1950), where he omitted the skink, which features only later in his lizard monograph (Vinson & Vinson 1969).

117. Arnold (2000) and Chapter 3.

118. ASC, unpublished notes. Seabird counts in Table 9.1 are

mostly from offshore; Jean-Michel Vinson landed in 1969 and 1971 judging by plants in the Mauritius Herbarium (Safford 1993).

119. Newton (1956, 1958a, 1960).

120. Safford (1993a), Bell *et al.* (1993), Arnold & Jones (1994), Jones (1993a). The tarantula is not only a new species but a new endemic genus: *Mascaraneus remotus* (Gallon 2005). Captive Serpent Island Tarantulas are adept at catching and killing house geckos, tactics presumably used also on night-geckos (Arnold & Jones 1994). Tarantulas probably originally occurred on the mainland – Bernardin (1773) described a large hairy spider that ate scorpions and centipedes. Clark (1859) mentioned "a species of Tarentula of a dull brown colour and covered with a sparse pubescence is very common in houses. It is about an inch long and its legs cover a circumference of about three inches." . . . "these spiders are very powerful. They often catch large cockroaches and I once saw one holding in its jaws a gecko full three inches in length which it killed and sucked dry"; 'sparse pubescence' is reminiscent of *M. remotus*. Presumably some predator(s) introduced since 1859 eliminated them on the mainland. Despite their ability as adults to kill house geckos, vigorous species like Cheechaks might take immatures; Bloodsuckers are diurnal and live exclusively out of doors, but the Wolf Snake (nocturnal and anthropophilous) is a strong candidate. Carl Jones (pers. comm.) reports another unrelated unidentified large gecko-eating spider found on Ile aux Aigrettes.

121. Lewis (2002); the endemic centipede is *Scolopendra abnormis*; Corby saw these centipedes in 1844, but did not collect any.

122. Arnold & Jones (1994) treated the night-geckos as subspecifically distinct, but Arnold (2000) later treated them as full species (*Nactus serpensinsula* and *N. durrelli*), as did Cole (2005b); the race of Bojer's Skink on Serpent Island has yet to be formally described.

123. References are: 1. Corby (1846); 2. Layard (1863), Brooke (1976, 1978); 3. Vinson (1950); 4. Newton (1956); 5. Newton (1958, 1960); 6. Rowlands (1982); 7. ASC notes and Feare (1984); 8. Temple (1976); 9. Safford (1993a); 10. Bell *et al.* (1993); 11. Tatayah (*Report of expedition to Serpent Island 25–26 November 2003* [unpubl.]). The inconsistencies between the disparate sets of figures deriving from ASC's boat trip round Serpent Island in 1973 are unresolved, as the reasoning behind the increased figures given to Chris Feare is now lost. The original (lower) figures were more or less plucked from the air as clouds of birds rose off the island in response to the boat's horn, while the later (larger) numbers appear to have been a guesstimate based on the island's size and the lack of unused space (as were Vinson's figures for 1948). The reversed abundance of the two noddies remains unexplained, though Bell *et al.* (1993) also recorded much larger numbers of Lesser than Brown Noddies. Carl Jones (verbally to ASC, 2005) says the most recent estimate for Sooty Terns is from a mosaic of photographs.

124. Vinson (1950) reported similar numbers in June 1949 to those the previous November, many with eggs. This implies a rolling population of breeders, the total being higher than numbers visible at any one time. If they are more synchronised in some years than others, it might account for some of the apparently random variability in numbers estimated.

125. ASC's notes from a visit on 6 Oct. 1973. For Bojer's Skink see Vinson & Vinson (1969), who also noted a Bouton's Skink collected in 1964.

126. ASC's notes, Staub (1973, 1976), Temple (1976).

127. Cheke (1979, 1987a), Bloxam (1983), Bullock (1983), Bullock *et al.* (1983); due to finding a rabbit skull, ASC assumed the dominant lagomorphs in 1978 were rabbits (none were seen by the landing party), but hares were the only species found alive in 1982. Another 1978 skull was identified as a Norway Rat (Cheke 1987a), confirmed in 1993 as the only rodent present (Bell *et al.* 1993).

128. Bullock *et al.* (1985), Arnold & Jones (1994).

129. Bullock *et al.* (1983); rats are notorious palm-seed predators (e.g. Hunt 2007).

130. Only Dulloo (1994) and Souchon (note 100) refer to the former presence of deer on Gunner's Quoin. After Souchon (early 1940s) no visitor or account mentioned them. Although Jean Vinson never wrote specifically on Gunner's Quoin it is clear from remarks in his various papers that he went there several times.

131. Bell *et al.* (1993), Dulloo (1994), Bell & Bell (1996), Dulloo *et al.* (1999); Safford (2001) suggested hares had survived, but was probably misled by reports of the briefly reintroduced rabbits (Bell 2002, Smith *et al.* 2004). The 2003 NCPS report (Smith *et al.* 2004) included good vegetation maps but advocated some bizarre approaches to ecosystem restoration, and appeared designed to delay reptile reintroductions indefinitely – though NCPS subsequently supported (Carl Jones, pers. comm.) MWF's release of Telfair's Skinks there, and on Ile aux Aigrettes in January 2007 (*MWF Newsletter* 7(1): 3, April 2007).

132. Bell (2002).

133. Bullock *et al.* (1983), Smith *et al.* (2004); ASC (unpublished notes) estimated 30–50 Red-tailed Tropicbird nests on cliffs in October 1973, counted from offshore (Temple 1976 cited 30 pairs), and found 11 on the island itself (inaccessible cliffs not included) in October 1978, a decline from 95+ in 1955 (Newton 1956). Poaching was not mentioned in the 1982 and 1996 reports (Bullock *et al.* 1983, Dulloo *et al.* 1999), possibly an oversight. Lizards were noted but numbers not assessed in 2003 (Smith *et al.* 2004).

134. Bird lists in Bullock *et al.* (1983), Bell *et al.* (1993), Smith *et al.* (2004); also ASC's notes in 1973 and 1978. Dulloo (1994) reported the mobile bulbuls. In addition Spice Finches were seen in 1982.

135. Vinson (1950).

136. Vinson (1953b), Murphy & Pennoyer (1952); the Herald Petrel is often considered conspecific.

137. Hoffstetter (1946a), Anthony & Guibé (1951), Vinson (1953b).

138. Newton (1958a,b); Vaughan (1968) dated the nature reserve declaration. On Round Island tropicbirds (and petrels) in fact breed all the year round, with only shearwaters being strictly seasonal (Gill *et al.* 1970, Vinson 1976a, Ashcroft & Cheke 2007).

139. Vinson (1964a). Darwin's great-grandson Quentin Keynes (1956) published the first colour photos of Round Island; he did not date his visit, but as he talked to the RSAS on 30 Oct. 1952 (*PRSAS* 1:282, 1953), and Jean Vinson accompanied him to the islet, it must have been in November 1952 (see Vinson 1964a).

140. Newman & Bannister (1965).

141. *Rapport du conseil d'administration sur les activités de la Société* for 1964 and 1965 read at the sessions on 29 Jan. 1965 and 26 Jan. 1966 (see *PRSAS* 3: 135–138 and 145–148, 1968).

142. For Vinson's life see *PRSAS* 3: 149–150 and 155 and *DMB*: 1361–62 (1988, G. Rouillard); Vinson, an entomologist by training, had been Director of the Mauritius Institute since 1954. Staub (1973a) described a hunting expedition, combined with bird-ringing activities – as goat and rabbit carcasses were piled up for loading on the boat, Telfair's Skinks, ever opportunistic, rushed out and lapped up their leaking blood.

143. *PRSAS* 4: 70 and 73 (1973).

144. Vinson (1974), who mentioned only shooting (not cyclones) as a potential reason for regeneration. Some of these *Latania* had reached full crown size by 1973–75 (ASC's notes, 1973; photos in Vinson 1974, Durrell 1977b [US edition only, pictures by ASC in 1973–75, unacknowledged], Bullock & North 1984), gaining trunks by 1982. North & Bullock (1986, *contra* Bullock & North 1984), wrongly assigned this recruitment to the reduced rabbit population in 1975–76. Their analysis in the expedition report (Bullock *et al.* 1983) is much clearer than in the vegetation paper published later.

145. Procter & Salm (1975); Temple (1973, 1974b) reported on the 1973 field test.

146. i.e. Mauritius Society for the Prevention of Cruelty to Animals and Universities Federation for Animal Welfare (a UK charity). In 1974 the MSPCA devoted an issue of their newsletter entirely to

Round Island, including an article by Peter Scott referring to "a programme for eradicating rabbits and goats" – but Dr Jalal-ud-Din Shuja, its Secretary and organiser, who had the ear of government, soon became an implacable opponent of the poisoning programme. This was common knowledge at the time, but the only reference in print appears to be a slightly oblique account in Durrell (1990). The MSPCA's Council in 1973–74 included several government ministers (MSPCA *Annual Report* for 1973); a decade later, there were none (*Ann. Rep.* for 1984). In 1999 the Government invited International Animal Rescue to investigate Shuja and the MSPCA following complaints. John Hicks and Alan Knight (undated but c.1999) produced a damning report on the MSPCA's cruel dog-catching activities, also accusing them of running a protection racket (dogs would be killed unless owners paid up). The MSPCA was subsequently disaffiliated from the international network of the RSPCA (UK's venerable animal welfare organisation), and a rival government-sponsored animal welfare group (PAWS) was set up in Mauritius (see www.iar.org.uk).

147. This controversy largely took place behind closed doors at the WWF/IUCN headquarters in Switzerland, but a column by John Hillaby in *New Scientist* of 10 Oct. 1974 supporting UFAW against the 'feudal gents' of the WWF led to a vigorous response from Bill Bourne (1975) on the general problem of introduced mammals on islands, the strange letter from J. R. Bareham for UFAW (4 Sept. 1975) and a reply from ASC (25 Sept. 1975); there is a more cryptic reference in Durrell (1977a) and a fuller (if pseudonymous) one in Durrell (1990), where John Hillaby features as 'Dr Glenfiddis Balmoral'.

148. Rabbit numbers were down to 650–1,500 in winter 1975 after cyclone *Gervaise*, but despite Gouldsbury's efforts, were back up to 2,450–2,900 in winter 1982 (North & Bullock 1986), maintaining that level in 1986 (Merton 1988). After the lower numbers in 1975–76 some *Latania* and *Pandanus* regeneration was noted in 1978 (Cheke 1979), resulting in a cohort of trees with full-size crowns (still without trunks) in 1996 (Dulloo *et al*. 1999).

149. Hartley (1984), Merton (1988), Durrell (1990). The poison used was rabbit pellets laced with the 'new generation' anticoagulant brodifacoum, particularly effective against rabbits. Despite precautions, about 100 Telfair's Skinks also died, apparently of heat stress caused by the poison (skinks had shown no interest in the bait in extensive trials); this was only a tiny fraction of the skink population, estimated at over 100,000 in 1996 (Dulloo *et al*. 1999). The eradication operation was facilitated by an Australian warship which, under the guise of an exercise for the crew, lent its helicopter to deliver stores to the island (Durrell 1990). Rabbit control on Round Island and rodent poisoning elsewhere (e.g. Ile aux Aigrettes, Chapters 8 and 10) was done before the toxicity of anticoagulants to snails was documented (Gerlach & Florens 2000, Gerlach 2006); brodifacoum does not affect arthropods, but is lethal to all snails at very low doses, and is apparently persistent for some years – a serious complication for de-ratting in areas with endemic snails. The new conservation agreement in 1984 was helped by the collapse of the old post-independence Mauritian government in the 1982 elections, followed by political realignments and the emergence of a more pragmatic administration after further elections in 1983 (Addison & Hazareesingh 1993). Nonetheless a subsequent White Paper on conservation strategy (Ministry of Agriculture 1985), retained a very cautious flavour – one objective was "to alleviate Round Island from heavy grazing pressure exerted by rabbits"; no methods were mentioned.

150. Studies on the growth rates of mature trees (21–66mm/yr, average 34), and the time it takes to form a trunk from the seedling stage, suggest the majority of old latans (trunk >5m) were 120–130 years old in 1996 (Dulloo *et al*. 1999, Bullock & North 2006), and must have become established after the introduction of goats and rabbits. After cyclones reduced herbivore numbers (and after shooting in 1976) there was significant regeneration of latans and screwpines, but much less of the other two palms. However the growth rate of *Hyophorbe* seems comparable to *Latania*, so probably the few that survived into the 1970s and 1980s had also established after the rabbits and goats, but the low numbers show how rarely this happened: Bottle Palm seedlings were much more palatable to rabbits than latans or screw-pines (Bullock 1977, 1982, Bullock & North 1991, North *et al*. 1994).

151. Data are largely from Bullock *et al*. (1983), Bullock & North (1991, 2006), Dulloo *et al*. (1999), Bullock *et al*. (2002). 1973 details are from ASC's unpublished notes, Vinson (1974) and Temple (1974b); Temple recorded five Hurricane Palms, ASC counted six. ASC and Temple counted fewer Bottle Palms (14 and 16) than Vinson, albeit on the same visit in November! The numbers of mature *Latania* in 1989 and 1996 are estimates based on sample counts and average rates of decline: the full numbers were not counted in those years. Cyclone *Gervaise* in February 1975, severe on the mainland (Padya 1976, 1989), seems to have left Round Island's vegetation (though not the rabbits) largely unaffected. Cyclone *Claudette* was on 22–23 Dec. 1979 (not '22–23 Jan. 1980' as in Cheke 1987c); another very wet but less severe storm, *Hyacinthe*, struck in late January 1980 (Padya 1989). See Jones (1994) and Mayoka (1998) for *Hollanda* in 1994. Without rabbit eradication Bottle Palms would have lingered longer than the table suggests, because a small group that germinated in 1975 actually survived (the first for decades), possibly due to low rabbit and goat numbers in 1975–76 (Bullock *et al*. 1983, Bullock & North 1991, 2005). At the time of writing the full data for 2003 were not available.

152. Bullock & North (1991), Strahm (1993).

153. Durrell (1977a,b); lizard numbers from Vinson (1975) before *Gervaise*, Bullock (1977, 1986) after. Bullock's earlier estimate of 200–300 Günther's post-cyclone was subsequently revised upwards, but his 1982 estimate, 181–553, showed further decline.

154. Durrell (1990).

155. References in Jones & Hartley (1996).

156. Cooper *et al*. (1999).

157. Cooper & West (1988), Tonge & Barlow (1996), Tatayah (2000).

158. Tatayah 2000 and Chapter 10.

159. Brooke & Imber (2000); by 2005 there were 10–15 pairs of Kermadec Petrels breeding annually (Table 9.2).

160. Ruth Brown from Queen Mary College in London started work in 2005 on petrel genetics and DNA; preliminary results do not point to a clear resolution of the identity problem. The Herald Petrel was ringed on Raine Island (Gt. Barrier Reef) 22 years earlier (Carl Jones, pers. comm.).

161. The spread of breeding throughout the year, but varying in intensity, was documented by Vinson (1976a). His estimate of 105–120 pairs breeding in October 1974 (excluding birds from other seasons) has been widely recycled, without any very visible justification – though the 1989 study examined 244 different individuals in August (Bullock and North 1991). Gill *et al*. (1970) had estimated 75 pairs active in October 1964; Safford (2001) gave the 400+ figure. The breeding cycle lasts 147–160 days (5 months; incubation 55–60 days, fledging 92–99, Vikash Tatayah pers. comm.; see also Gardner *et al*. 1985). The figure for ringed birds is from Carl Jones (unpubl.)

162. Merton *et al*. (1989), Safford (2001). Temple (1976) and Vinson (1976) estimated 2,000–3,000 and 3,500 pairs respectively, though ASC (in Duck & Gardner 1979) thought the 1973 number nearer 10,000 pairs. Bullock's (1977) fuller study in 1975 suggested 18,000 pairs; Duck & Gardner (1979) reckoned 15,000–20,000 in 1978, and Bullock, North & Grieg (1983) and Bullock & North (1984) 50,000–100,000 in 1982, increased since 1975. It is unclear how far these figures represent real trends or just reflect evolving estimation methods.

163. Although this lapse in MWF's project management is in many ways best forgotten, we have included it as an example of how, with

the best intentions, conservationists can make serious mistakes. The plant restoration project over-interpreted the aim in the island management plan to "protect planted areas against shearwater burrowing as necessary" (anon. 2003b) as *carte blanche* permission to displace 3–4 hectares of the densest shearwater colony, several thousand pairs, without having considered the ramifications, and without fully consulting colleagues. The netting was removed after ASC raised the issue with MWF management and Carl Jones (who was absent in the UK), but the incident shows how sectional interests can sometimes override the broader issues, even in non-bureaucratic organisations (see also p. 245). The management plan included seabird numbers, but the quoted shearwater totals, "4,000–8,000" pairs, were low by an order of magnitude.

164. Tropicbird numbers can be traced through E. Newton (1861a), Meinertzhagen (1912), Vinson (1950), Gill *et al.* (1970), Bullock (1977), Bullock *et al.* (1983), Merton *et al.* (1989), Safford (2001), Ashcroft & Cheke (2007); they were not estimated on the 1989 and 1996 expeditions. The figures are for birds on the island at census time, but given that breeding is neither seasonal nor synchronised and a breeding cycle lasts *c*.130 days, it is possible that over a year the breeding total is 2–2½ times the census figure (i.e. 3,000–5,000 pairs of Red-tailed Tropicbirds on 2003 figures, Ashcroft & Cheke 2007).

165. See Brooke *et al.* (2000), Merton & Bell (2003). Both Kermadec and Bulwer's Petrel breeding numbers are small; in June 2003 a bird making a call intermediate between Kermadec and 'Round Island' Petrels paired up in display flights with a normal Kermadec Petrel (pers. obs.), and Carl Jones has seen apparent hybrids (pers. comm). The two species are very similar in appearance, but their calls are usually completely different; the bird with intermediate calls looked like a pale-phase Kermadec – as are all Kermadecs seen on Round Island (Carl Jones and Vikash Tatayah, pers. comm.). Other potential colonists are Flesh-footed Shearwater *Puffinus carneipes* (adult with brood patch; Carl Jones, pers. comm. to Roger Safford) and Black-winged Petrel *Pterodroma nigripennis* (on ground as if intending to nest; Carl Jones, pers. comm.).

166. Carl Jones (pers. comm.); these old birds are survivors of only 19 ringed by ASC in November 1973 (12 chicks, 4 adults) and January 1975 (1 chick, 2 adults); in all over 2,000 petrels have been ringed.

167. Ashcroft & Cheke (2007); given the similarity in nest-site, it seems no coincidence that petrels were rarely recorded, even by competent birdwatchers, until after Red-tailed Tropicbird numbers had been drastically reduced. A similar case occurred in Bermuda, where in the suboptimal habitat where the last Cahows *Pterodroma cahow* survived, White-tailed Tropicbirds invaded their burrows and killed their chicks (before conservation intervention; Wingate 1978). Burger & Gochfield (1991) compared nest-sites of 'Round Island' Petrels and White-tailed Tropicbirds, but, oddly, did not consider Red-tailed; White-tailed Tropicbirds do occasionally displace petrels on Round Island (Vikash Tatayah, pers. comm.), but generally occupy less open sites .

168. The idea first appeared in print in Temple (1974b), followed by Schauenberg (1975).

169. Both ASC and Steve North (pers. comm.), noticed this independently during the 1996 expedition, acknowledged in a short sentence in the report (Dulloo *et al.* 1999). Strahm (1993) argued that the presence of deep gullies, and hence advanced erosion, in 1844 (Lloyd 1846) suggested that goats might have been on the island far longer than previously thought – in the light of the pre-Holocene date of the gullies this is clearly not tenable. The northern islets formed during the eruption of the 'Recent Series' lavas (25,000–700,000ya), but are not precisely dated (Saddul 1995).

170. Of all Round Island's scientific visitors, only Don Merton (1988) fully recognised in print the significant synergy of wind and water: "Physical weathering from wind and water is very active and the overlapping ash beds [tuff] have been sculpted into numerous steps, pedestals, cavernous overhangs and other weird shapes". Despite his comments (Merton *et al.* 1989: 24) that much soil "accumulated . . . as a result of wind-blowing", soil restoration efforts have concentrated on rainwash. Dulloo *et al.* (1999), following ASC's observations, inserted a footnote suggesting wind-blown sand should be incorporated in soil conservation efforts. An illustrated account of these processes can be found in Mather (1964) and Murphy & Doherty (2000). The 'mushroom rock' had fallen by mid-2003 (pers. obs.), apparently a year or two earlier.

171. Cheke (2007a); $100g/m^2/yr$ would yield a depth of about 4cm over 1,000 years, while $1kg/m^2/yr$ would give 40cm, using a standard mineral density of $2.4 g/m^2$ for the sand. Island soils currently have low organic content, though before the devegetation and erosion caused by rabbits, there were no doubt soils with more humus (Johnston 1993).

172. Vinson (1975), Bullock (1977), Bullock & North (1991), Korsós & Trócsányi (2001b, 2002); *Bolyeria* is not even mentioned in the 1996 expedition report (Dulloo *et al.* 1999).

173. *Periplaneta americana*, see Dulloo *et al.* (1999); in 1999 ASC found another 'new' winged cockroach as well (smaller, not identified) – this does not appear in the report. Other cockroaches, wingless and fossorial, have been abundant in recent times on the islet, important in the diet of lizards and probably also young snakes (Vinson 1976, Bullock 1986); although often cited as *Blatta orientalis*, they are probably in fact endemic (probably two species, pers. obs.); they are referred to as *Blattodea* sp. by Pernetta *et al.* (2005); see also note 179.

174. Strahm (1993), North *et al.* (1994), Dulloo *et al.* (1999), Bullock *et al.* (2002), Bullock & North (2005). *Achyranthes*, like the invasive grasses *Cenchrus* and *Heteropogon*, has hooked seeds that attach easily to mammalian fur and, significantly on Round Island, researchers' clothing. Johansson (2003) produced the first island's vegetation map, noting that plant cover had increased from only 10–20% in 1975 to over 50% in 2003.

175. The first surveys after rabbit removal, in 1989 (Tonge 1990, North *et al.* 1994), suggested an increase in all reptiles, but subsequent monitoring, and fuller surveys in 1996, could not confirm any increase in *Casarea* or *Phelsuma guentheri*, though daytime area counts showed a consistent rise for both (Bullock *et al.* 2002, Fig. 4), not mirrored in transects or night counts. Both ASC (2003 vs 1999/1996) and Dennis Hansen (2005 vs 2000; pers. comm.) had the impression *P. guentheri* was increasing well, but the systematic counts in 2003 (Bullock & North 2007) only partially reflected this; a sample census in 1999 (Korsós & Trócsányi 2000, 2001a) estimated Günther's Gecko numbers at 4,430 as against 3,426 estimated in 1996, though their calculation seemed somewhat arbitrary. Bullock & North (2007) did not estimate population for 2003. Korsós & Trócsányi (2002) published excellent colour photos of the reptiles.

176. Dulloo *et al.* (1999); the results of different census techniques, both by day and night, were to some extent contradictory, particularly for snakes and Günther's Day-geckos, though the authors believed they did not under-count, and did not mention this potential problem in their later paper (Bullock & North 2002). Korsós & Trocsányi (2000) reckoned that 50+% of adult and still more juvenile *P. guentheri* were missed even with intense searching – they found 3.26 geckos per hour compared with 0–1 by Dulloo *et al.* (1999) in 1996, and pointed out that the lizards see humans coming and slip behind trunks, leaves etc. (as ASC also noticed; see also Bullock 1986). Bullock (in Bullock & North 2007) commented that *Nactus* had become harder to locate as leaf litter increased.

177. Rodda & Campbell (2002). 'Total removal' involves screening off a small plot (5x5m on Guam) and literally taking it apart, thus finding every lizard present. Clearly this is not only expensive but ecologically destructive, though as Round Island is recovering well, it might be worth sacrificing a few latan clumps to get accurate reptile numbers. The Guam lizards were several nocturnal arboreal

geckos and two diurnal terrestrial skinks, counts being done at night (some in popinac, some in native forest) – the encounter rates for one skink and one gecko (*Gehyra mutilata*) were a roughly predictable, though small, fraction of the real numbers – the skink was 13 times more abundant/ha than numbers seen/hr, the gecko 6,105 times! (Rodda *et al*. 2005a). The other geckos (the ubiquitous *Hemidactylus frenatus* and *Lepidodactylus lugubris*) were too unpredictable – ratios varied widely from census to census and plot to plot. Studies using sticky traps (glueboards) produced similar results, only skinks giving reliable results (Rodda *et al*., 2005b), the day-gecko equivalent (*Anolis cristatellus*) almost entirely avoiding capture. Conditions on Round Island are clearly different but the same principles apply, and one would expect night-geckos to be censused least accurately using visual counts. Korsós & Trocsányi (2000) did a pilot total removal (without fully foolproof boundaries) on a 0.12ha plot for Telfair's Skinks, catching 46 animals, estimated as 75% of those initially present. Their resulting (rather rough) estimate for the island population was c.33,300, 67% *less* than the c.100,800 estimated in 1996 (Dulloo *et al.* 1999) – which confirms the problem! Korsós & Trocsányi considered the earlier work's capture-recapture technique had overestimated numbers by not controlling for skink mobility in and out of the study plots.

178. Gordon Rodda pers. comm.
179. Bullock *et al*. (2002), Pernetta *et al*. (2005); Estimated reptile biomass increased from c.4kg/ha in 1982 to c.40kg/ha in 1996, 82% due to *Leiolopisma telfairii*. Invertebrate monitoring methods (pitfall traps and sweep nets) have failed to sample the flightless cockroach population, but from casual observation it has continued at the same high level seen in 1973 (ASC unpubl.) and 1975 (Bullock 1977). Telfair's skinks may eat young snakes, but adult snakes certainly eat young Telfair's (Korsós & Trocsányi 2001b).
180. Bullock *et al*. (2002) and ASC (pers. obs.) for associations, Vinson (1975) and Dennis Hansen (work in progress; pers. comm.) for nectar-feeding habits of Round Island *Phelsuma*. Bullock & North (2007) documented a new cohort of trunked latans.
181. *Asterochelys radiata*; although now rare in Madagascar, thousands are kept and bred in Réunion, with smaller numbers in Mauritius (Bour & Moutou 1982, Cheke 1987a). A dry-country animal, it can probably survive without access to water better than the Aldabra species. Round Island lacks permanent water; rainwater pools are short-lived in the persistent winds, and absent during the dry season (pers. obs.). Aldabra Tortoises released on Ile Aux Aigrettes nature reserve since 2000 (Mauremootoo & Towner Mauremootoo 2002), initially in large pens (seen by ASC in 2003) and more recently completely free, have successfully bred (Carl Jones and Vikash Tatayah, pers. comms.). Their influence on the vegetation is being monitored. Releasing tortoises on Round Island was advocated by Jones (2002) and Johansson (2003); the first Aldabra Giant Tortoises were taken to an enclosure there in May 2007 (Dennis Hansen, pers. comm.).
182. E. Newton (1861a), Vinson (1964a), Bullock & North (1976).
183. Bullock, North & Grieg (1983), Merton *et al*. (1989), Dulloo *et al*. (1999), ASC's pers. obs. 1973–2003, Mario Allet (pers. comm.; *Butorides striata*). Spice Finches were reported to have disappeared between 2003 and 2007 (Carl Jones pers. comm.)

CHAPTER 10. PRACTICAL CONSERVATION ON MAURITIUS AND RODRIGUES

1. Scott (1973)
2. For example, Cowles (1987, 1994), Mourer *et al*. (1999), Hume (2005)
3. Cheke (1987a), Chapters 5–8.
4. Wylie (1989).
5. Scott (1973), Procter & Salm (1975); the BOU expedition report was not published until 12 years later (Diamond 1987); Temple initiated the captive breeding project, though it was many years before it thrived (see Jones 1987).
6. See Chapter 8 for further details.
7. Merton (1978), Atkinson (1990), Craig & Veitch (1990), Towns *et al*. (1990).
8. Cade & Jones (1994), Jones *et al*. (1995).
9. See Durrell (1977a,b), Durrell & Durrell (1980), Jones & Hartley (1995) and various papers on Mascarene species published in the journal *Dodo* 1976–2001 for summaries and reviews on this work.
10. Jones & Hartley (1995) and papers in *Dodo*.
11. The Echo Parakeet work is in partnership with the World Parrot Trust with the support of the North of England Zoological Society (Chester Zoo). We have had technical staff from Paradise Park (Hayle, England), Chester Zoo, Adelaide Zoo (Australia) and the Department of Conservation (New Zealand). These have worked on all aspects of the field work and captive breeding, including artificial incubation, hand-rearing, release and veterinary care. Vets from the International Zoo Veterinary Group help us with all of our animal health problems, but have been intimately involved with the parakeet work.
12. Jones & Hartley (1995).
13. The University of Mauritius's involvement in plant conservation was started by John Mauremootoo, and has been developed by Vincent Florens (Mauremootoo 1997, Florens 2003). The work of the MWF on plant conservation in Mauritius and Rodrigues, was driven by Wendy Strahm (1982–1993), and later by Ehsan Dulloo and John Mauremootoo. Gabriel d'Argent, MWF botanical advisor, has for three decades been a fount of knowledge on the distribution and biology of Mauritius' plants and has inspired and nurtured a generation of botanists, many of whom are now active in conservation. He is also the driving influence behind the management of the Mondrain Nature Reserve. On Rodrigues the plant work has been successful under the management of Richard Payendee.
14. Strahm, (1989, 1993), Dulloo *et al*. (1996). This work has been supported by the World Wide Fund for Nature, the DWCT and the Global Environment Facility with some support and direction from the Royal Botanical Gardens, Kew, the World Conservation Union (IUCN), Botanic Gardens Conservation International and Chester Zoo.
15. Vitousek (1990), Williamson (1996), Myers & Bazely (2003), Cox (2004).
16. Cheke (1987a)
17. The Crab-eating Macaque *Macaca fascicularis* has been blamed for all ornithological ills for 250 years.
18. CGJ pers. obs. On Rodrigues there is a policy of planting exotic invasive Traveller's Palms *Ravenala madagascariensis* along roads when it would be more appropriate to plant endemic palms.
19. Cronk & Fuller (1995) and references therein.
20. Elton (1958), Cronk & Fuller (1995).
21. Towns *et al*. (1990), Veitch & Clout (2002)
22. Atkinson (1985), Fritz & Rodda (1998)
23. Cheke (1987a). See also Chapter 7 for further details and other theories on the extinction of the Hoopoe Starling *Fregilupus varius*; there is no record of it eating lizards (ASC, pers. comm.), but it likely to have fed large insects and small vertebrates to its young.
24. Mauritius Kestrel (Jones, 1987), Cuckoo-shrike (Safford & Beaumont, 1996), Merle (CGJ, pers. obs.).
25. Strahm (1993)
26. Rouillard & Guého (1999)
27. Lorence & Sussman (1986, 1988): Mauritius; Smith (1989): Hawaii.
28. *Ravenala madagascariensis* (ravinal), *Syzygium jambos* (jamrose), *Schinus terebinthifolius* (poivrier marron), *Furcraea foetida*

(aloès). Local names are given in italics in brackets when the species is first mentioned. Strahm (1996) and Page (1995) gave lists of the most invasive weeds with good agreement between authors; for Rodrigues see Strahm (1989).
29. CGJ pers. obs., R. Payendee pers. comm., Turnbull (1999).
30. Félicité (2000) looked at the status and regeneration of *Hyophorbe verschaffeltii* (palmiste poison), *Fernelia buxifolia* (bois boureille) and *Carissa xylpicron* (bois de ronde).
31. Wiehe (1946, 1949): *Cordia*; Fowler *et al.* (2000): *Lantana*, *Opuntia* and a review of biological control on Mauritius; also Kueffer & Mauremootoo (2004).
32. McIntyre (1997), Mungroo (1997), Mungroo & Tezoo (2000), Newfield *et al.* (2003).
33. Kueffer & Mauremootoo (2004)
34. Arnold & Jones (1994), Arnold (work in progress). This work stimulated the Ph.D study by Cole (2005).
35. Cole *et al.* (2005).
36. CGJ, pers obs.; the *Nactus* eggs and bones were collected by Owen Griffiths and are now in the NHM (London).
37. Cheke (1987a), Jones (1993), Arnold & Jones (1994).
38. Cundall & Irish (1988), Jackson & Fritts (2004).
39. See Vinson & Vinson (1969), Cheke (1987a), and Chapter 7.
40. Rodda *et al.* (2002).
41. Vinson & Vinson (1969), Jones (1993a), Freeman (2003). There is also an unconfirmed sighting of a small skink in the forest of Le Pouce in the Moka Range of Mountains in the North of Mauritius (Fleur Maseyk, pers. comm.).
42. CGJ, pers. obs.; see also p. 169.
43. Ashok Kedun, pers. comm. See Brooks (2002) for a study on *Calotes versicolor* on Ile aux Aigrettes.
44. Jones (1996).
45. Long (1981), Lever (2006). The duck is *Anas melleri*.
46. Jones (1996).
47. Common Mynas regularly feed themselves and their young on large insects and geckos, both house geckos and day-geckos *Phelsuma* spp. (CGJ, pers. obs.).
48. Jones (1996).
49. Mallam (2005).
50. For trichomoniasis in Pink Pigeons see Swinnerton *et al.* (2005a).
51. Swinnerton *et al.* (2005b), Bunbury *et al.* (2007).
52. Jones & Duffy (1993) and Jones (1996) warned of the potential danger to the endemic Echo Parakeet of exotic parrot diseases brought in through the trade in parrots, and called for stricter controls.
53. McOrist *et al.* (1984), McOrist (1989).
54. Jones (1996). A Goffin's Cockatoo *Cacatua goffini* has also been seen using the same nest cavities that Echo Parakeets nest in.
55. CGJ, pers. obs.
56. CGJ pers. obs. for sparrows and mynas; for native birds see Cheke (1987c) who also commented on the uneven distribution and unknown pathogenicity in Mauritius and Réunion of introduced avian malaria, *Leucocytozoon fringillarum* and the apparently endemic *L. zosteropis* reported by Peirce *et al.* (1977). Blood parasites also occur in Rodrigues birds (Peirce 1979).
57. Since 2004–2005 there have been outbreaks of a feather condition that affected free-living Echo Parakeets, causing some feather loss, or discolouration from green to yellow in some feathers, about 25–30 birds (10%) of the population are affected and PBFD has been confirmed as the cause (pers. obs.; *MWF Newsletter* 7(1): 4, April 2007).
58. Cronk & Fuller (1995), Strahm (1996), Linnebjerg (2006).
59. Linnebjerg (2006); see Mandon-Dalger *et al.* (2004) for Réunion.
60. See Chapter 11 for comment on this kind of pseudo-ecotourism, and also the failure of the authorities to prevent Casela Zoo from releasing new species into the Mauritian environment. It is not yet clear if the wildfowl would be self-sustaining without management.
61. Caroline de Ravel, pers. comm.
62. King 1984, Atkinson 1985, Johnson & Stattersfield (1990).
63. Veitch & Bell (1990), Moors *et al.* (1992), Towns & Ballantine (1993), Thomas & Taylor (2002).
64. Roy *et al.* (1997).
65. Bell *et al.* (1993) assessed the conservation value of the islets around Mauritius and Rodrigues and suggested strategies or their restoration.
66. New Zealand conservation biologists worked with MWF and NPCS; some helped to set up the predator control grids. Studies on the ecology and control of invasive mammalian species have been supported by the Mammal Research Group of Bristol University under the direction of Stephen Harris. They supported Sugoto Roy's Ph.D study on mongooses and to a lesser extent on feral cats, and David Hall's Ph.D on the ecology and control of rats in the native forest. This research group also attempted to eradicate shrews from Ile aux Aigrettes.
67. Dingwall *et al.* (1978), Atkinson (1985, 1996).
68. Whitaker (1973), Ramsay (1978), Campbell (1978), Atkinson (1985).
69. Arnold (1980), Cheke (1987a).
70. Poisoned bait (Talon 20P) was placed on a grid of about 50m.
71. Brodifacoum bait was laid on a 25m grid. The new rabbits, of a domestic breed, were apparently released by people hoping to hunt them there (Bell 2002), possibly ignorant of the conservation initiatives in progress! See also Chapter 9.
72. As mice were also involved, a 10m grid using two poisons (Talon 20P and Ridrat, which has an active ingredient of 0.05mg bromadiolone) was used on Flat Island (Bell 2002); see also Bell & Bell (1996) for Rodrigues.
73. Varnham *et al.* (2002), Seymour *et al.* (2005); both Longworth Traps and Trip Traps were used to catch shrews; a 12.5m grid was used on Ile aux Aigrettes.
74. Cheke (1987c), Jones (1987), Jones *et al.* (1989), Jones & Duffy (1993), Jones *et al.* (1995), Safford (1997a).
75. The actual causes of nest failure can be difficult to evaluate with problems of interpreting field signs after a predation event. Further to this, there are proximate and ultimate considerations; for example, Pink Pigeons may abandon a nest of infertile eggs after 14 days of incubation, and the unguarded eggs are then taken by rats. The real cause of failure would have been infertility, but the field signs would have suggested rats. See McKelvey (1976, 1977a), Jones (1987), Jones *et al.* (1989, 1992) and Swinnerton (2001a) for discussions on the roles of rats and monkeys as nest predators.
76. The rat poisons used have been Klerat and Talon (active ingredient 0.5% Brodifacoum); in the pigeon plots; poisoned baits were set out on a 50m grid; see Hall (2003).
77. McKelvey (1976) and Swinnerton (2001a) discussed the impact of monkeys on Pink Pigeons.
78. The damage to trees is based on personal observations and Strahm (1996a)
79. For example, Cheke (1987a), Jones (1996); see also Chapter 7.
80. Jones (1988b, 1996), Roy (2001); see also Chapter 7.
81. Kestrels: Jones (1987), Jones & Owadally (1988), Jones *et al.* (1991, 1994); pigeons: Jones *et al.* (1992).
82. CGJ pers. obs. and Roy (2002). Veitch (1985) gave details of cat-predated carcasses.
83. Roy (2001, 2002).
84. CGJ, pers. obs.
85. Swinnerton (2001a).

86. Clout *et al.* (1995).
87. The traps, baited with salt fish, were laid on a grid of about 100–200m, along access points such as paths and dry stream-beds and in a perimeter ring around the breeding area; about 15–20 traps were set at any one time.
88. Carter (1998), Roy (2001), Hall (2003).
89. CGJ, unpublished.
90. CGJ, pers. obs.
91. Vinson & Vinson (1969)
92. CGJ, unpublished data and *MWF Newsletter* 7(1):3, April 2007.
93. Fritts & Rodda (1998).
94. CGJ, pers. obs.
95. Varnham *et al.* (2002)
96. Bullock (1977), Chapter 9.
97. Merton (1987), Bell (2002). Bait was treated with 'Talon 20P' (brodifacoum) at 0.02mg per kg., and was applied on a 10m grid pattern in two 'pulses' at 14-day intervals at a rate of 4–6kg/ha.
98. For Round Island see North & Bullock (1986), North *et al.* (1994) and Chapter 9.
99. Strahm (1999).
100. For a full discussion of *Cervus timorensis* damage, see Cheke (1987a).
101. Strahm (1988) recorded selective deer damage in native forest; she also listed native plants recorded as having been eaten by deer. The grasses eaten were *Isachne mauritianus* and possibly *Oplismenus* sp., the sedge *Cyperus longifolius*, and the rush *Juncus effusus*. They also fed on young woody plants: *Aphloia theiformis* (Flacourtiaceae), *Psathura* sp. *Chassalia coriacea* (Rubiaceae), and *Faujasiopsis flexuosa* (Compositae).
102. Blanchard (2000), CGJ pers. obs.
103. Strahm (1993); a much higher proportion is endemic to the Mascarenes as a whole.
104. Strahm (1993), Page & d'Argent (1997), Safford (1997b).
105. Vaughan & Wiehe (1941).
106. See Chapters 7 and 8.
107. Ah-King (2000) provided a good study of upland forest CMAs, looking at species composition and the rates of regeneration inside and outside the managed areas.
108. Forest-compatible deer densities were discussed by Cheke (1987a).
109. Vegetation types are based on Strahm (1993); see also Kueffer & Mauremotoo (2004). The as-yet (2007) unweeded Morne Seche plot is not heavily invaded by exotics. The Brise Fer conservation management area was extended from 1.26ha to 24ha in 1996–97; the original 1987 plot is included within the larger fenced area. All these sites are managed by the National Parks & Conservation Service (NPCS), excepting Perrier (managed by the Forestry Service) and Mondrain (managed by the RSAS and MWF). There are additional unfenced managed vegetation plots (Kueffer & Mauremootoo 2004, Mauremootoo *et al.*, in press) that are not included in the table.
110. The original pre-1941 Macabé plot was only 0.1ha. This was weeded until 1952. The Forestry Service restarted work on this plot in 1978, and it was fenced and extended to 0.4ha. in 1986.
111. The physical control methods we have developed in Mauritius were quoted as a case study in Cronk & Fuller (1995). See Mauremootoo & Towner-Mauremootoo (2002) and Kueffer & Mauremootoo (2004) for further details on weed control.
112. CGJ, pers. obs.
113. Vaughan & Wiehe (1941).
114. Strahm (1993), Dulloo *et al.* (1996).
115. Strahm (1993), also quoted in Cronk & Fuller (1995).
116. Vaughan & Wiehe (1941)
117. Colophane *Canarium paniculatum*, Tambalacoque *Sideroxylon grandiflorum*; see Baider & Florens (2006) for the discovery of natural Tambalacoque seedlings. For the supposed relationship of Dodos to Tambalacoque regeneration, see p. 130.
118. Seegoolam (1999).
119. John Mauremootoo, pers. comm.
120. Mauremootoo *et al.* (in sub.).
121. Budulla (2002). Differences would be expected if the canopy of the weeded plot had not fully recovered from the effects of weeding all of the exotic understory. Canopy health does improve post-weeding, and the result might be significant if repeated once the forest canopy had had more time to recover. Only long-term studies will truly reveal how weeding affects invertebrate communities.
122. Sharp (2004)
123. Information summarised from Florens (1996), Florens *et al.* (1998)
124. Jones & Owadally (1989).
125. Jones *et al.* (1992)
126. CGJ, pers. obs.
127. Edmunds (2005).
128. Boyla (2000).
129. Bats feed on *Labourdonnaisia glauca* and *Mimusops petiolaris* (Sapotaceae) and *Diospyros tesselaria* (Ebenaceae); see Nyhagen (2004), Nyhagen *et al.* (2005).
130. Vinson & Vinson (1969), Vinson (1976b).
131. Padayatchy (1998); the differences between managed and unmanaged areas of forest were not great. In a 144m^2 quadrat the mean number of geckos found in the weeded and unweeded plots was 3.6 and 4.8 in the austral winter, rising to 4.2 and 5.2 in summer when geckos are more active.
132. This work is done jointly with the Rodrigues Forestry Service and the MWF, much of it driven and directed by Richard Payendee and his conservation team.
133. Strahm (1989), Kell (1996).
134. On Grande Montagne, 69,500 plants of 36 native species were planted between 1997 and 2002.
135. Rodrigues data is updated from Mauremootoo & Payendee (2002) and Kueffer & Mauremootoo (2004). The main plants used (fuller details in Strahm 1989) are listed with the number of adults known to exist in the wild in parenthesis (in May 2005). Bois Balais *Turraea laciniata* (17), Bois Carotte *Pittosporum balfourii* (30), Bois Malay *Olea lancea* (<1,000), Bois de Ronde *Carrisa xylopicron* (65), *Psiadia rodriguesiana* (<200); these have all been particularly successful with good survival. Ground cover is provided by *Senecio boutoni* (>1,000).
136. The species planted in this plot are listed, followed in parenthesis by the number of known wild adults on Rodrigues. Bois Mangue *Scyphoclamys revoluta* (17), Bois Papaye *Badula balfouriana* (8); the most successful canopy tree species have been *Gastonia rodriguesiana* (10), *Pleurostylia putamen* (<200), *Terminalia bentzoe rodriguesensis* (<50), *Diospyros diversifolia* (<250), *Cassine orientalis* (<1,000), Palmiste Marron *Hyophorbe verschaffeltii* (33), Rodrigues Latan *Latania verschaeffeltii* (>250), screw-pine (vacoas) *Pandanus heterocarpus* (>1,000), Bois Chauve-souris *Doricera trilocularis* (Rubiaceae) (<250), and Bois Gandine *Mathurina penduliaflora* (<500). In the understorey Bois de Fer *Sideroxylon galeatum* (34), Mandrinette *Hisbiscus liliiflorus* (3), *Foetidia rodriguesiana* (<50), Figue Marron *Obetia ficifolia* (<50), *Eugenia rodriguesensis* (<30), Bois Lubine *Poupartia castanea* (6), Bois Troisfeuille *Vepris lanceolata* (<20).
137. *Pittosporum, Mathurina, Sideroxylon, Premna obtusifolia, Draecena reflexa, Poupartia castanea* and *Hibiscus* flowered and fruited rapidly. *Latania, Sideroxylon, Mathurina, Pandanus, Pittosporum, Doricera* and *Turrea* are the principal canopy species.
138. This data derived from discussions with Richard Payendee and the Rodrigues team of MWF, and personal observations.

139. The main species planted are listed followed in parenthesis by the number of wild adults known on Rodrigues where not previously mentioned. *Gastonia*, *Latania*, *Pandanus*, *Cassine*, *Terminalia*, Hurricane Palm *Dictyosperma album* var. *aureum* (2), *Myoporum mauritianum* (<100) and *Dodonea viscosa* (3). Smaller numbers of *Clerodendrum laciniatum* (<250) and *Sarcanthemum coronopus* (38).
140. CGJ pers. obs. and Richard Payendee, pers. comm.
141. Owen Griffiths has created a new entity, Francois Leguat Ltd, to run this project (pers. comm); see also references in Chapter 8.
142. Bell *et al*. (1993), Bell & Bell (1996). An earlier attempt at conservation on Ile aux Aigrettes had been made between 1953–1964 when Bernard Guimbeau "leased the island from the government with the object of restoring the indigenous flora and fauna" (Niven 1964; the lessee misspelt as 'Grimbeau'). In practice, according to Guimbeau (pers. comm.) and those who recalled this exercise in 1974 (France Staub pers. comm.), he had a series of aviaries with exotic birds, many of which he liberated on the island, including Budgerigar *Melopsittacus undulatus* and various Mauritian exotics including Spice Finch *Lonchura punctulata*, Common Waxbill *Estrilda astrild* and Yellow-fronted Canary *Crithagra mozambica*, none of which now occur on the island. Little was done by way of botanical conservation, although he kept goats to control popinac (Dulloo *et al*. 1997). Guimbeau was the (re-) discoverer in 1963 of the daygecko *Phelsuma guimbeaui* that Mertens (1966) named after him.
143. Williams (1977).
144. The race *conjugatum* of the palm *Dictyosperma alba*; for historical details on Round Island see Chapter 9.
145. Merton *et al*. (1997).
146. As the amount of vegetation cover has increased, so have some of the lizard populations (North *et al*. 1994). The vegetation community has been studied and monitored since 1975 (North & Bullock 1986, North *et al*. 1994, Dulloo *et al*. 1999).
147. North *et al*. (1994), Bullock *et al*. (2002); for pre-eradication data see North & Bullock (1986) & Chapter 9.
148. Soil traps, small walls of rocks, were constructed in sloping barren rock areas in order to reduce soil loss from the island. The majority of soil traps were constructed in 1993 by the Raleigh International Expedition (Daszak 1994), with a few added up to January of 2004. Of the traps constructed in 1993, most have collected enough soil to plant in. Several methods of planting were tried in these soil traps in 2004, including direct seeding and planting. Both planting of nursery-grown seedlings and direct seeding into the traps has been tried, but the former method, if watered at planting time, was much more successful (nearly 100% survivorship in 2004), as less than 10% of seeds sown into traps germinated.
149. Experimental use of weed mats to suppress the growth of weedy grasses has shown considerable promise and together with appropriate aftercare of translocated plants should greatly improve survival (CGJ, pers. obs.).
150. Other pioneers that have been (re-)introduced are Mauritius Dragon Tree (bois chandelle) *Dracaena concinna*, *Tarenna borbonica* and *Dodonea angustifolia*; some of the first have thrived, but most nursery-grown seedlings and young plants of the others have died. Some critically endangered plants were also reintroduced, including five *Dictyosperma album congugatum*, but these also died. Clearly rare species should not be included in the restoration until numbers in cultivation have greatly increased, and the problems of post-planting survival solved – even then they will need extra care.
151. Cheke (2007a).
152. A water-catchment system was built in 2002 to trap and store rain water in ten 900 litre tanks. All plants were watered weekly for the first month after planting; to conserve water, only those plants actually wilting were watered subsequently.
153. Germination rates of some species have been reasonable, such as *Hyophorbe lagenicaulis* (2–8%) and *Latania lodigesii* (9–20%), while for others it is low: *Lomatophylum tormentorii* (1–2%), *Scaevola taccada* (1%). No seeds germinated of *Dodonea viscosa*, *Vetiveria arguta*, *Dracaena concinna* or *Pandanus vandermeeschii* despite several hundred seeds of each species being sown.
154. All hardwood species were planted in wire mesh cages to reduce mortality from shearwater burrowing. The cage is cylindrical with an open top and bottom, 30cm diameter and 50cm high. The wire mesh sides are covered with shade cloth (70% light exclusion) to provide some early protection against the wind and sun. Cages were dug into the soil down to rock level (ranging from 20–40 cm in depth) to prevent loss of soil by shearwater burrowing adjacent to the plants.
155. Zuel *et al*. 2005, Zuel & Smith, 2005. After field trials established that no watering or watering monthly caused high mortality, twice-monthly watering was substituted.
156. Five more 900 litre water tanks were added to the catchment system, increasing the storage capacity to 13,500 litres.
157. Zuel & Smith, 2005.
158. Survival to one year for the annual cohorts was 0% for 2000, 49% for 2002, 37% for 2003 and 87% for 2004. Survival to two years was 30% for the 2002 plantings and 9% for those in 2003; this latter poor survival rate reflects the experimental minimal post-planting management of this cohort. Survival to three years for the 2002 plantings was 28%, which is encouraging since it shows that once the young plants are established survival is excellent – i.e. of plants that survived two years, 92% were still alive at the end of year three.
159. Dulloo *et al*. 1997.
160. Parnell *et al*. (1989).
161. The island has the most important populations of the following endemic plants: *Gastonia mauritiana*, *Diospyros egrettarum*, *Phyllanthus revaughanii*, *Pandanus vandermeeschii*, and *Dracaena concinna*. Ile aux Aigrettes also has the most important Mauritian population of the Mauritius and Reunion endemic *Tarenna borbonica*.
162. 8m may not seem very tall, but the island is subject to the full force of the southeast trade winds that cut back trees as at any windy coastal site.
163. For example, De Bry (1601).
164. Although called 'Mauritius Hemp', the species is native to Central America, but gained its name through being cultivated in Mauritius as a fibre crop (Rouillard & Guého 1999).
165. A detailed weed management strategy for the island (Newfield *et al*. 2003) provides details of weeding done to 2002, and plans for the long term management of weeds.
166. Early in the restoration effort (1986–1990), some plants were grown in a makeshift nursery, but many seedlings were transplanted from high regeneration areas. In areas left with little shade through weed-tree removal, fast growing (possibly native) *Terminalia catappa* were planted to act as nurse trees. However, good survival of transplanted seedlings, together with natural regeneration from the seed-bank, made the use of nurse trees unnecessary.
167. Early plantings were at about one per m^2, but it was later considered that the plants could be planted at higher densities. Trials were conducted with plants at 4, 8, and 12 per m^2. Survival was high, and a maximum density of 4/m^2 found most suitable (Newfield *et al*. 2003).
168. These were the principal species found in this coastal community (fuller details in Atkinson & Sevathian 2005), and also Bois de Renette *Dodonea viscosa*, Bois d'olive *Cassine orientalis*, *Tarenna borbonica*, Var *Hibiscus tiliaceus*, Bois Sureau *Premna serratifolia*, Sainte Marie *Thespesia populnea*, Mauritius Latan *Latania loddigesii*, *Ehretia petiolaris*. An excess of the pioneer dragon trees *Dracaena* planted in 2001 was subsequently balanced by planting a larger proportion of other species.

169. Strahm (1993).
170. Strahm (1989) described the discovery and fencing of this last individual in 1980, after the species had been pronounced extinct. After much cutting back by locals for its alleged medicinal properties, a present triple fence, together with improved local understanding seems to be working. In May 2005 the plant looked thriving and was in full flower.
171. Thomson (1880), King (1946); see Chapter 8
172. Wyse-Jackson *et al*. (1988).
173. The long-delayed Midlands dam, finally completed in 2003 on land clear-felled in the 1920s (Jogoo 2002, Chapter 8). A small additional population of this lily was found along the Perrier River in 2000 (CGJ, unpublished).
174. CGJ, pers. obs..
175. For example, Owadally (1981) for the planting, CGJ pers. obs. for the survival.
176. CGJ, pers. obs.; see also *Flore* 190 (Bosser & Guého, 2003) on *Pandanus*, of which seven Mauritian species are now thought to be extinct.
177. Maunder *et al*. (2002).
178. Jones & Hartley (1995).
179. See Strahm (1989) for full details of the the plight of Rodrigues endemic plants.
180. Verdcourt (1996); five plants have been grown from cuttings on Rodrigues, and 11 were recently repatriated from the Royal Botanic Gardens, Kew; one was planted out on Grande Montagne in 2002, where it is growing well and flowering (CGJ, pers. obs.).
181. Strahm (1989); *Gouania* has been propagated from cuttings, but all subsequently died; recent attempts to propagate it from the sole surviving individual have failed.
182. Richard Payendee, pers. comm. Seeds *Carissa* of were collected in 1877 but not seen again until 2001, when one tree produced three fruit; of 43 seeds obtained, 41 germinated, producing seedlings that were planted out in Grande Montagne and Anse Quitor, where most still survive.
183. *Inter alia Myoporum mauritianum* (vanished from Mauritius, plants come from endangered Rodrigues population), *Zanthoxylum heterophyllum*, Baume de l'île Plate *Psiadia arguta*, and *Aerva congesta*. This last is possibly a grazing climax species that would have grown in areas of high tortoise density. Propagated from seeds, *Aerva* has been planted out in clumps to test habitat requirements, with the idea of testing grazing by Aldabra Giant Tortoises.
184. A list of 17 priority species has been drawn up for propagation on Ile aux Aigrettes; Strahm (1996b) listed 30 of the most threatened Mauritian endemic plants.
185. Vaughan & Wiehe (1937).
186. Critically endangered plants targeted for this project include *Tambourissa tetragona* (three known), *T. cocottensis* (one known), and *Monimia ovalifolia* (under 10), all endemic to Montagne Cocotte, and the Blue Bell-Flower *Nesocodon mauritianus*, which is only found in three small populations on cliffs. Other important species (with numbers of plants in the wild) include *Claoxylon linostachys* var *linostachys* (<50), *Tectiphiala ferox* (40), *Badula reticulata* (<20), and *Tetrataxis salicifolia* (<40). A high priority species is Bois dentelle *Elaeocarpus bojeri* (two known); successful grafts onto Ceylon Olive *E. serratus* root-stock are to be planted into the field gene bank (this grafting was done by staff at the Plant Genetic Resources Unit of the Ministry of Agriculture). A number of other endangered species have been planted in this gene bank: *Hubertia ambavilla* var *ambavilla*, *Hibiscus boryanus*, *Pandanus microcarpus*, *Claoxylon linostachys* var *pedicellare*, *Albizia vaughani*, *Hyophorbe vaughani*, *Chionanthus broomeana* var *broomeana* and *Pilea cocottei*.
187. Vinson & Vinson (1969), Vinson (1973, 1975, 1976b).
188. Bloxam (1976), Bloxam & Vokins (1978), Tonge & Barlow (1986), Bloxam & Tonge (1989).

189. McKeown, (1989, 1993)
190. Arnold & Jones (1994).
191. Merton (1988).
192. Towns & Ferreira (2000).
193. The coastal forest with scrubby areas and open patches is probably the habitat in which Telfair's Skinks evolved, and is the likely optimum habitat for the species. Ile aux Aigrettes also has grassy areas that are the favoured microhabitat of juvenile skinks; there are small open rocky areas and beaches that will provide some habitat variability for the skinks. Skink releases began in autumn 2006.
194. Including *Diospyros egrettarum*, *Eugenia lucida*, *Gastonia mauritiana*, *Dracaena concinna*, *Ehretia petiolaris*, *Scevola taccada*, *Lomatophyllum tormentorii*, *Turraea casimiriana*, *Pandanus vandermeeschi* and *Ficus reflexa*.
195. Jones (1993a).
196. Rodda *et al*. (2002), Varnham *et al*. (2002), Seymour *et al*. (2005).
197. Jones (1993a); given the very large numbers of Telfair's Skink on Round Island, there is no conservation risk if the experiment fails.
198. On the mainland both Telfair's Skinks and Günther's Geckos were wiped out by rats long before shrews or Wolf Snakes were introduced (Chapter 3), so they have never encountered these particular aliens.
199. The population size of Günther's Gecko has been estimated at 3,400 (Dulloo *et al*. 1999) and 4,500 (Korsós & Trócsányi 2001a), although the effective population size of breeding adults may only be about 750–1,500 animals.
200. The most appropriate animals for this translocation are those from the neighbouring islet of Ilot Vacoas. If these (from such a small population) are considered too inbred, an experimental mix of animals from several islands could be released to see if there is selection for particular traits and gene combinations.
201. Cole (2005).
202. Nick Cole (Cole *et al*. 2005) looked at the ecology of the night-geckos *Nactus* spp. and Cheechaks *Hemidactylus frenatus*, providing information on the requirements of the night-geckos and on how they are excluded by Cheechaks.
203. Cole (2005).
204. See Jones (2004) for a discussion on these techniques.
205. 'Round Island' Petrels are being studied by Vikash Tatayah from the University of Mauritius and Ruth Brown of Queen Mary College, London. These studies, looking at the population and evolutionary genetics of the petrels with detailed studies looking at the morphological and plumage variation, will hopefully help clarify their taxonomic status. The breeding biology and success of these petrels is being studied to quantify the effects of nest-site location and structure on fledging success. Some 2,000 petrels have been ringed on Round Island; many individuals are recaptured regularly, with some now (2007) at least 34 years old.
206. The once-huge Rodrigues colonies of Sooty Terns were exploited to extinction in the late 19th century (Cheke 1987a, Chapter 7).
207. The last recorded poaching on Round Island was of several hundred Wedge-tailed Shearwaters, two 'Round Island' Petrels and a few White-tailed Tropicbirds in early 1997. Since a permanent presence on Round Island was established in 2002, there have been few uninvited visitors and no poaching.
208. This work forms part of Vikash Tatayah's Ph.D research; see also Ashcroft & Cheke (2007).
209. Noddies and shearwaters bred on these islets in the 17th century (Leguat 1707, Chapter 5), but there is no record of Fairy Terns nesting in Mauritius.
210. The most suitable birds for release would be young birds hand-reared and released on the island, on which it is hoped they would

breed. Some preliminary work has already been done by CGJ with artificially incubated eggs and hand-reared chicks of tropicbirds (both species) and Sooty Terns.

211. For boobies see Feare (1978), Cheke (2001a); Cheke (2004c) briefly reviewed western Indian Ocean frigatebird history and distribution. For the devastation caused by Crazy Ants *Anoplolepis gracilipes* on Christmas Island, and their subsequent partial control see Slip (2002) and Abbott (2006).

212. The knowledge and infrastructure exist on Mauritius to implement these reintroductions, perhaps after some experience reintroducing some of the commoner seabirds. On Rodrigues there is still need to develop local seabird management knowledge, and the restoration of Ile Crabe, the largest lagoon islet, is a high conservation priority. This island would be an excellent prospective site for seabird introductions, but only once local staff had gained experience in managing Iles Cocos and Sable. The MWF have put in a proposal to manage Ile Cocos and Ile aux Sable. This would be based on the established experiences of those managing seabird islands in the Seychelles. Management of these islets would include training of local wardens, zoning the islands for ecotourism and conservation, and long term population monitoring.

213. During the mid-1980's (e.g. Cheke 1987e) only five parakeets were known, including two females. It is now clear that several birds were being overlooked, and the 1987 figure is a minimum since additional birds were subsequently found in Bel Ombre Forest.

214. See Jones *et al.* (1991, 1995), Cade & Jones (1994) for a review of this work. The Mauritius Kestrel work was in partnership with the Peregrine Fund (Boise, Idaho); they provided a skilled technician for ten seasons between 1982 and 1993, who helped with captive breeding, artificial incubation and hand-rearing of the birds.

215. Details in Cade & Jones (1994), Jones *et al.* (1995) and Jones & Swinnerton (1997).

216. See note 289 for details of ongoing work on kestrels.

217. Pink Pigeon research was reviewed by Jones *et al.* (1992). Jones (1995) reviewed the work on captive birds and Swinnerton (2001a) reviewed much of the work on wild and reintroduced birds.

218. See note 289 for ongoing pigeon research.

219. Swinnerton *et al.* (2005a,b).

220. Swinnerton (2001a), Jones (2004).

221. Jones & Duffy (1993), Thorsen & Jones (1998), Jones *et al.* (1998).

222. Malham (2005), *MWF Newsletter* 7(1): 4, April 2007.

223. Stattersfield & Capper (2000).

224. Jones (1996).

225. For example, Cheke (1987c), Safford (1994, 1996, 1997a,c,d,e, 2000), Safford & Beaumont (1996). Summaries of the conservation work on the passerines are given in Stattersfield & Capper (2000); Safford & Jones (1998) considered in depth the main conservation strategies for the native land birds. More recently, Damholdt & Linnebjerg (2003) studied the Paradise Flycatcher on Mauritius, and Nichols *et al.* (2002, 2004, 2005a,b) the Mauritius Fody and Mauritius Olive White-eye.

226. Safford & Jones (1997).

227. Cheke (1987c).

228. Research so far has not explained the disappearance of paradise flycatchers from many areas of native forest, and some exotic forest, since the 1970s. The nests are very visible, and may be more vulnerable to predation by monkeys, rats, Red-whiskered Bulbuls and Common Mynas than other species, but that begs the question of why they only recently became rare. Competition with and egg predation by Red-whiskered Bulbuls seem to be problems that need careful study; see also p. 165.

229. Safford & Jones (1997), Stattersfield & Capper (2000).

230. Owen (2004).

231. The Mauritius Fody introduction to Ile aux Aigrettes has been supported by staff from the DWCT (Jersey Zoo); Andrew Owen developed the hand-rearing protocols (Owen 2004) and Markus Handschu conducted the first release. We have had support from Chester Zoo, and the RSPB facilitated the visit of Trace Williams, who worked on hand-rearing, see *MWF Newsletter* 7(1): 3 (April 2007) for numbers.

232. *On the Edge* 100: 10 (2006).

233. For example, Bathe & Bathe (1982); see also Cheke (1987d): 393.

234. Middleton (1995), Middleton & Haüchler (1998).

235. Flinders (1814), Middleton (1995).

236. CGJ, pers. obs.

237. Jones (1987) and CGJ unpublished observations; Günther's Gecko glues its eggs to the substrate, like all Mascarene day-geckos (Vinson & Vinson 1969).

238. Cowles (1987), Saddul (1995)

239. Mendes (1996), also discussed in Middleton & Haüchler (1998).

240. Stock (1997), also discussed in Middleton & Haüchler (1998).

241. Middleton & Hauchler (1998); the telemid spiders were found in L'Esperance Window Cave; telemids are known from Tanzania, Cameroon, South Africa and the Seychelles. Middleton and others have found further cave invertebrates in Rodrigues and Mauritius, as yet unidentified.

242. This species does not have a vernacular name in Mauritius (beyond *susuri* = 'bat'), though we have called it the Mascarene Free-tailed Bat (see Chapters 3 and 4). Nowak (1994) referred to the genus *Mormopterus* as Little Goblin Bats.

243. While visiting Mauritius Kestrel nest sites (Petite Gorge and Site 3) in the Black River Gorges, CGJ found fissures in the cliff faces that held unknown numbers of bats. The characteristic bat smell was very strong and there was staining around the fissure entrances. This bat is common in areas of native forest at night.

244. CGJ, pers. obs. In June 1994 a roost in a cave on Plaine des Roches covered about $45m^2$ of the cave roof and contained an estimated 60,000–120,000 bats; in June 2001 it covered a roughly circular area $c.6m$ across, estimated to contain 30,000–50,000 bats; bats apparently move their roosts from cave to cave periodically.

245. Middleton (1998); in 16 of the 19 caves 75–100% of nests had been removed.

246. Stattersfield & Capper (2001), Chapter 8.

247. Cheke (1978a), Safford (1992); Chapter 8.

248. Cheke (1987c), Middleton (1996), CGJ and ASC pers. obs.

249. Griffiths (1996).

250. A study being conducted by Saoud Matola and financed by the Darwin Initiative is looking at invertebrate diversity and reviewing what is known about the native insect fauna. An objective of this study is to provide a basis for making conservation recommendations for the remaining native arthropods.

251. Griffiths (1996), Griffiths & Florens (2006).

252. Mamet (1993, 1996a) surveyed early entomological reports in the Mascarenes, unfortunately excluding spiders and centipedes.

253. Gerlach *et al.* (1997); the former presence of this beetle on Round Island has been much disputed (Chapter 9).

254. Lewis (2002).

255. Safford (1993a), Gallon (2005), CGJ pers. obs.; the spider is a member of the Theraphosidae.

256. Bernardin (1773), Charpentier (1732–1755), Clark (1859); Bernardin confused large hairy spiders (= tarantulas) with others that made webs strong enough to catch small birds (*Nephila inaurata*, these are not hairy). Clark specifically mentioned that his large hairy spiders ate geckos.

257. Bernardin (1773); centipedes up to 6" (150mm) long were also reported by Clark (1859), but they were scarcer than smaller species. Recent specimens from the islands appear not to exceed 85–90mm (Lewis 2002).

258. Griffiths (1996), Griffiths & Florens (2006).
259. Griffiths (1996).
260. Griffiths (1996).
261. Some of these reintroductions were suggested by Jones (1996), and Ghestemme & Rochet (2001) suggested reintroducing Réunion Harriers to Mauritius as part of a conservation strategy. Various Indian Ocean islands (Agalega, Diego Garcia) were proposed to host additional populations of Golden Bats (e.g. Durrell 1990), but as we now know (Hume 2005) that they formerly occurred in Mauritius, that is the obvious place to release them. However, as in Réunion, there is anxiety about adding further frugivorous species to an island rich in cultivated fruit trees.
262. Bullock *et al.* (2002). *Vetiveria arguta* survives is low numbers on Mauritius and Rodrigues, but only where there is intensive grazing by domestic livestock (CGJ, pers. obs.). Wiehe (1949) attributed the loss of native grasses in Rodrigues to their replacement by exotic species.
263. *Aerva congesta* is endemic to Mauritius and Rodrigues, although last seen in Rodrigues in 1874 (Strahm, 1989). On Mauritius it has been recorded from two mainland locations in recent decades, although only the Round Island population is known to survive. The species has been (re-)introduced to Ile aux Aigrettes and to new sites on Round Island.
264. There are many plants in the coastal region on Mauritius and Rodrigues with prostrate and compact forms. In some cases these are xerophytic characters to help cope with drought conditions, but in others it may also be to deter grazing tortoises. The two grasses *Stenotaphrum dimidiatum* and *S. micranthum* may grow 45cm high, but when grazed by tortoises they form dense prostrate mats (CGJ, pers. obs.); both are widespread on Indian Ocean islands and are probably native (Hubbard & Vaughan 1940).
265. This has been obvious in our restoration trials; when these species are accidentally trampled they usual show little or no damage. More convincingly, Aldabra Giant Tortoises often trample these seedlings without permanent damage. See also Griffiths (2005).
266. Griffiths (2005) and CGJ pers. obs.
267. Maseyk (2001), Griffiths (2005). Old decomposed piles of droppings often have clusters of ebony seedlings growing out of them (CGJ, pers. obs.).
268. Eskildsen *et al.* (2004); see also Chapter 3.
269. Maseyk (2001).
270. There are large captive populations of Radiated Tortoise *Asterochelys radiata* in the Mascarenes (especially on Réunion, Cheke 1987a), or another option would be the Ploughshare Tortoise *A. yniphora*; both are from Madagascar and endangered. Ploughshares are Critically Endangered with a very limited distribution, and one could argue that a conservation introduction to Round Island would be appropriate, since there it would also fulfil an ecological role. This Round Island initiative would not be the first use of analogue tortoises on Indian Ocean islands, as Aldabra Giant Tortoises were released on Flat Island, Mauritius in 1883 (though not deliberately as analogues; Chambers 2004, Chapter 9), to islands in the granitic Seychelles (Frégate, Curieuse, Cousin), and also to Assumption Island (CGJ, pers. obs.). These introduced tortoises have or had well-established breeding populations on these islands (Hambler 1994). However, in the Seychelles the Aldabra Giant Tortoises should perhaps be considered reintroductions, as recent DNA studies suggest the Aldabra and granitic Seychelles forms were probably conspecific (Austin *et al.* 2003).
271. In late 2005 there were 45 baby tortoises from recovered clutches on Ile aux Aigrettes, and at least four 1–2 year olds from undiscovered nests (Vikash Tatayah, pers. comm.).
272. Jones (1993a): *Phelsuma guentheri*, *P. ornata*.
273. Jones (2002).
274. See Chapter 4 for the curious history of *Lepidodactylus lugubris*.
275. See Austin *et al.* (2004) and Chapter 4 for the relationships of Mascarene day-geckos.
276. *P. edwardnewtoni*.
277. Austin *et al.* (2004a).
278. Nyhagen *et al.* (2001), Dennis Hansen, work in progress.
279. See Chapter 7; Frégate would have to be purchased from its private owner and freed of rodents and cats (see Chapter 8). Ile Crabe, which has forest remnants, would also be suitable for these lizards, as well as analogue tortoises and reintroduced seabirds once cleared of rats; unlike the other islets, Crabe is currently used for livestock grazing.
280. This does not preclude the most closely related or ecologically similar species being used for educational purposes to help illustrate what has been lost. Hence a case can be made for a local zoo to display captive Nicobar Pigeons *Caloenas nicobarica* as the Dodo's closest living relative, one of the large black cockatoos as an approximate ecological equivalent of the Raven Parrot, or the large *Corucia* skink from the Solomon Islands to illustrate the lost giant skink Didosaurus of Mauritius.
281. Respectively: *Lophopsittacus mauritianus*, *Psittacula eques*, *P. exsul* & *P. eupatria*; the *P. eupatria*-like Thirioux's Grey Parrot *P. bensoni* was a rather atypical *Psittacula*, and *P. eupatria* itself might not in fact be a close enough analogue for use in the Mascarenes, but it would probably be suitable to replace *P. wardi* from the Seychelles..
282. *Dryolimnas (cuvieri) aldabranus* (Aldabra, flightless) and *D. c. cuvieri* (Madagascar, volant).
283. Jones (1996); *Alectroenas sganzini* (Comoros and Aldabra), *A. pulcherrima* (Seychelles) and *A. madagascariensis* (Madagascar) (Sinclair & Langrand 1998).
284. Respectively: *Anas bernieri*/*A. theodori*, *Alopochen aegyptiacus*/*A. mauritianus.*; no ecological information on the endemic wildfowl was recorded before they became extinct. Bernier's Teal breed freely in captivity, and there is a captive population at the DWCT in Jersey (Jeggo *et al.* 2002). For affinities, see Mourer *et al.* (1999) and Chapter 4.
285. Mauritius Merles were successfully translocated to Diego Garcia, probably in the 1940s (Hutson 1975, Cheke 1987c), but they died out after mynas were released in 1954–55; the Seychelles species *Hypsipetes crassirostris*, abundant despite mynas, might prove a better choice for Rodrigues.
286. See Atkinson (1990) for a useful discussion on the possible introduction of the Rock Wren *Xenicus gilviventris* to suitable islands as a replacement for Stead's Bush Wren *X. longipes*, which became extinct on Big South Cape Island (New Zealand) after Ship Rats invaded in 1963–64.
287. Jones & Hartley, 1995.
288. Volunteering on conservation projects has become an important stepping-stone in the careers of a large percentage of MWF ex-volunteers within the conservation and biological sciences. They occupy jobs as zoo directors and curators, veterinarians, biology lecturers, researchers, trainers in conservation practice, administrators of conservation projects, wildlife photographers, wildlife artists, wildlife writers, eco-tourism operators, and there are several who now run their own conservation programmes.
289. Work on the ecology and management of the native forest is being conducted by the University of Mauritius under the supervision of Vincent Florens. Links were established with the Institute of Zoology, London and Queen Mary College, London, to help with population genetics and the management of small populations (Professors Mike Bruford and Richard Nichols). The long-term population studies on the Mauritius Kestrel have been developed with Professor Ken Norris and Dr Malcolm Nicoll (University of Reading). Professor Lucas Keller (University of Zurich) has been supervising Steven Ewing in his studies on inbreeding and survival in the Mauritius Kestrel. Work on the biology and control of exotic

mammals has been supported by the Mammal Research Group, University of Bristol (Professor Stephen Harris). Professor Jens Mogens Olesen (University of Aarhus) and Dr Christine Müller (University of Zurich) work on maintaining the health of plant communities, focusing on plant-animal interactions. The University of East Anglia annually provide MSc students to work in Mauritius or Rodrigues. A research programme with the UEA is looking at Pink Pigeon population biology, with Nancy Bunbury studying the epidemiology of diseases that affect this species, under the supervision of Dr Diana Bell and Professor Paul Hunter.

290. Jones (2002, 2004).

CHAPTER 11. REFLECTIONS

1. From the paragraph on 'Flora and Fauna' in the 1922 edition of the *Mauritius Almanac*.
2. Calkin (2002).
3. Cheke (2003).
4. "Secondary education is oriented principally around five sets of examinations, all of which are written in English and prepared in the UK, but have not yet undergone the sort of modernisation to which examinations elsewhere have already been submitted" (anon. 1971). By 1985 the aim was to "adapt the schools to the evolving socio-economic and cultural system of the country" and, as "less than a third of all children who are in secondary schools actually pass the School Certificate Examination [i.e. UK 'O level'], Government is, therefore, considering the introduction of a more relevant alternative syllabus" (anon. 1985). By 2001 the examination system had been "'Mauritianised' at the primary and secondary [levels]", and the pass rate for School Certificate (local version) had risen to 77% (Teelock 2001).
5. Mauritian Creole or minor variants is spoken in Mauritius, Rodrigues and the Seychelles; Réunion Creole, spoken only there (and on Mayotte), is not mutually comprehensible with Mauritian (e.g. Baker & Corne 1982, Cheke 1982).
6. Ramsurrun (1982), Baissac (1888); Ramsurrun's collection lacks Baissac's authenticity, and is in English, i.e. translated from creole and Bhojpuri – inaccurately, to judge by his re-rendering of Baissac's tales.
7. Some native trees (*colophane*, *tambalacoque*) appear, though in no way distinguished from exotics (*bois noir*, *badamier*) – respectively *Canarium paniculatum*, *Sideroxylon grandiflorum*, *Albizzia lebbek* and *Terminalia catappa*.
8. Ramsurrun (1982).
9. Lee (1999), Guébourg (1999:158–160) on Mauritian Creole. While the recent flush of locally-generated history books is welcome (e.g. Addison & Hazareesingh 1993, Teelock 2001, Selvon 2001 for Mauritius, Vaxelaire 1999, Leguen 1979 for Réunion), they include (Vaxelaire excepted) much less environmental information than earlier books from a more colonial perspective, concentrating on socio-political issues. We suspect this is a reflection of the alienation, particularly of Mauritians, from the natural environment.
10. Jones & Owadally (1988), Durrell (1990); despite the exotic parkland habitat the birds thrived and even bred – but they were no match for the catapults.
11. Bell *et al*. (1993) proposed a kind of forest botanic garden on Ile d'Ambre, but we think this should be taken further to provide adequate habitat for birds needing larger territories, which would then be easy for schools or any Mauritian to visit. Ile d'Ambre, the largest lagoon islet, is government land, so would be an ideal candidate for an educational endemic restoration site. Benitiers is privately owned and subject to hotel building proposals, but, unlike Iles d'Ambre and Aigrettes is in the island's lee, thus more sheltered; its sandy substrate also offers different options for restoration to d'Ambre (basalt) and Aigrettes (limestone). A good survey of the islets (including area figures) can be found on www.intnet.mu/iels.
12. Toussaint & Adolphe (1956) gave details of the *Mauritius Almanac*; *DMB*: 1121–22 (1981, P. J. Barnwell) has a biography of Walter.
13. Calkin (2002). Britain is not immune from equivalent nonsense – in May 2006 the *Guardian* newspaper distributed a poster of 'garden birds' – several of which required visiting gardens in northern Scandinavia.
14. See various tourist websites and guidebooks. Bindloss *et al*. (2001) referred to the 'nature reserve for monkeys' without apparently being tongue-in-cheek; see Heady *et al*. (1997), which boasted a real ornithological consultant in Adrian Skerrett, for a balanced account of the Domaine du Chasseur; Richards *et al*. (2006), while not entirely error-free, is the best tourist guide for wildlife information. A tourist website in Hong Kong (www.travel.com/mauritu.htm) offered in 2005 the following insight into Mauritian wildlife: "a diverse variety of animal life – wild boar, Java deer, Javanese stags, white-tailed tropical birds and even the extinct Mauritian Kestrel", also including the 'Dodo bird' amongst fauna to be seen on the island.
15. Macmillan (1914); Eurasian Woodcock and Nightingales are *Scolopax rusticola* and *Luscinia* spp. Additional glossary (Cheke 1982): 'banana bird' (*oiseau banane*) = Mauritius Fody, 'republican bird' (French not recorded!) = Village Weaver, 'white/manioc birds' (*oiseau blanc/manioc*) = Grey/Olive White-eye, 'nightingale' (*bulbul*, ex Persian) = Red-whiskered Bulbul, 'lark' (*alouette* [*de mer*]) = small waders, 'teal/wild duck' (*sarcelle*, *canard*) = whistling and Meller's Duck. D'Emmerez was curator of the Mauritius Institute museum!
16. Bulpin (1958); Musk rat is *Ondatra zibethicus*.
17. Willox (1989); the 'translations' are from *merle blanc* (cuckoo-shrike), *merle* [*noir*] (i.e. *black* blackbird), *oiseau de la Vierge* (flycatcher); *martin* is local French for myna, *puffin* is French for shearwater, not *Fratercula arctica* as in English. Where he got 'crow' from we cannot guess.
18. Lenoir (1974); the errors had vanished by the second edition (Lenoir 1979). Rees & Rees (1956: 92) were referring to the invasive *Ligustrum robustum*; Baker & Hookoomsing (1987) are clear on the derivation of *privet*, but *privé* (without the final 't', pronounced in Mauritius but not in France) is used in creole for 'privy'.
19. Oliver (1881). Le Clezio (1997), an expatriate Mauritian novelist recounting his visit to Rodrigues in the 1990s, claimed to see, in addition to terns and Striated Herons, 'magnificent seagulls' and 'black frigates carrying their red goitres' – there were never any gulls (though *goilon* is used for Roseate Terns) and he would need to travel a century back through time to see frigates . . .
20. Chazal (1973), see *DMB*: 1899–1903 (2003, R. d'Unienville) for a biography. Chazal's use of the Dodo and Tambalacoque myth shows its prevalence 'in the air' locally, influencing Stanley Temple's notorious theory of obligate mutualism, published four years later (Temple 1977b; see Chapters 3 and 7). The 'second moon' forms part of an occult cosmological theory attributed by Chazal to Professor Hoerbiger (see www.seachild.net). Chazal was brought up in the Swedenborgian church, which no doubt helped prepare him for his visions; he was generally more celebrated in Paris than in Mauritius. The editor of *The pictorial tour of the world* (anon. 1879), no doubt in a time-warp since Chazal's Atlantean times, told his armchair readers that "some of the mountains of Mauritius are between two and three thousand feet in height, and are covered in snow during a great part of the year". At least he got the altitude right.
21. Shuker (1997: 104), chapter entitled *How dead is the Dodo*. The birds at half-light on the beach were no doubt the greyish Striated Herons, which commonly stalk the tideline at undisturbed times. Plaine Champagne is on the high plateau and does not connect with the coast. Bill Gibbons is a creationist, apparently seeking to disprove extinctions (see www.cryptozoologicalrealms.com).
22. Bory (1804); in fact mynas mostly avoid native forests, thus sparing endemic insects.

23. For example, Manchester & Bullock (2000) for the UK.

24. Jones (1996), Probst in *Bulletin Phaethon*; see Chapter 8.

25. Cheke (1987a); Probst (1997) saw parakeets, including a pair, in 1996, since when *Chakouat* has carried more reports, without any indication of breeding... yet.

26. Cheke (1987e) reported on the downside of the road built past Bassin Blanc; see the Mauritian press *passim* on the proposed road through Ferney (e.g. *L'Express* 9 July and 2 Aug. 2005, and earlier articles in this paper and *Weekend*). As Henri Marimootoo put it (*Week-End* 31 July 2005) "Let us remember that, for the sake of a few million rupees indemnity to the Chinese construction firm Beijing Chang Construction, the former government was ready to sacrifice the environment, and were it not for the election result, the cause would have been irretrievably lost". The new Government announced on 15 Oct. 2005 that an alternative, less damaging route would be built, thus saving both the forest and the cancellation indemnity to the contractors (*Weekend* 16 Oct. 05). See Chapter 8 for the similar case of the Rodrigues airstrip.

27. King (1978); his *Red Data Book* was King (1978–79).

28. *The sinking ark* (Myers 1979, Quammen 1996) was intended to focus conservationists and their funding on 'do-able' projects.

29. "In 1978 I went to a lecture by Professor Tom Cade, who was the world expert on falcons. This changed my life. He said something about the rarest falcon in the world, the Mauritius Kestrel, which might be saved by captive breeding. I thought 'Christ, this is meant for me. I'm good at captive breeding and I can save a whole species!'. I couldn't sleep for weeks. So I applied to the ICBP... which was running the Mauritius Kestrel project. They said how pleased they would be to send me to Mauritius because they wanted someone to close the project down. They didn't want to throw good money after bad. There was one pair in captivity which weren't breeding, and two pairs in the wild. They wanted to shut their book on it" – Carl Jones (Whiteley 1992: 161; also, in similar terms, in Adams & Carwardine 1990: 177). Elsewhere (Adams & Carwardine 1990: 176–78) Jones called the by-then very successful project 'his biggest failure', so completely had he failed to close it down! A good story, but a little enhanced: in fact, Warren King (ICBP) told Jones he had two years' funds, maximum, whether he succeeded or failed (Carl Jones pers. comm.), though in practice they extended the deadline.

30. The Jersey Wildlife Preservation Trust (as was) had adopted the iconically extinct Dodo as its logo, and in consequence had decided in 1976 to get involved with conservation in Mauritius (Durrell 1977a, Botting 1999). ICBP, having already run down the funding leaving the project perilously close to closure (Stearns & Stearns 1999), allowed a couple of further years due to specific funding by Ian Pollard before finally pulling out in March 1985 (Jones & Hartley 1995).

31. In the case of the ICBP's withdrawal from the Mauritius conservation project, there is published material, as Carl Jones and Wendy Strahm subsequently discussed the issue with the interviewers cited (note 28), and Quammen (1996: 550–59); see also more oblique references in Stearns & Stearns (1999) and Young (1999). There was, however, no discussion on priorities in ICBP's flagship book *Save the birds* (Diamond & Schreiber 1989) nor in Mountfort's (1988) *Rare birds of the world*, and the implied criterion in 20th century Red Data Books (e.g. Collar *et al.* 1994) was that degree of endangerment is the prime basis for action, when in fact many other factors are involved. Nonetheless, this simple criterion is in 2006 explicit on BirdLife International's website, www.birdlife.org (but see Mace & Collar 2002 for a more nuanced BirdLife view). Even the more recent additional (and welcome) prioritisation by habitat blocks ('Important Bird Areas') is done on a pre-determined formulaic basis – non-specialist readers of IBA publications (e.g. Fishpool & Evans 2001 for Africa, incuding Indian Ocean islands) might easily infer that all IBAs were of equal conservation significance. A controversial case, that of the California Condor *Gymnogyps californianus*, has been aired in some detail by Alagona (2004), though even this account skips over the early failures of the captive breeding programme. In fairness to ICBP, the first years of the Mauritius project were plagued with failures due to both difficult birds and personnel who were not always ideal – but by the time they pulled out, Carl Jones had turned the project around and was successfully breeding Mauritius Kestrels and Pink Pigeons. However, Jones's reluctance to produce regular reports (Nigel Collar & Paul Goriup pers. comm.) appeared to count against him more than his avicultural achievements counted in his favour. In public, ICBP praised "the very dedicated and skilful work of Carl Jones" (*ICBP Newsletter* 6, 1984). As Jones himself admitted later (Whiteley 1992), "if I didn't answer the telephone, which didn't work, and didn't answer any letters, I could just get on with it" – not the way to endear yourself to distant funding organisations. As ICBP (then small and understaffed) had no money of its own and had to rely on soliciting outside funding for each project, difficult or apparently failing programmes were always vulnerable, whereas Jersey's finances allowed them to allocate funds without specific reference to donors (Roger Safford, pers. comm.; see note 33). In contrast, ICBP's contemporary Cousin Island project was extremely successful (e.g. Mountfort 1988, Bell & Merton 2002) – but did not involve captive breeding.

32. Cheke (1980, 1987e), Barré & Barau (1982). Small numbers *can* suffice for a translocation. Only five Seychelles Fodies were moved to D'Arros island in the Amirantes in 1965, but they founded a still-thriving population (e.g. Rocamora & Skerrett 2001).

33. Cheke (1980), and Gerry Durrell (pers. comm.); it can be read between the lines in Durrell (1977a,b). As Carl Jones later noted "Gerry... was outside all committees and in-fighting that ties up so much of the conservation movement. He wasn't like some of the bigger conservation organisations who deal only in high-profile animals, put in a load of money then pull out. He put local people in on the ground and stuck with the project hands-on *in situ* through thick and thin" (Botting 1999: 494).

34. As Carl Jones said to David Quammen: "One organization, which shall be nameless, told us that they would not work with us [the Mauritius project] unless they had complete control of the worker they supported. *They* would tell him what to do. *They* would define the mission. Of course we said 'sod that, mate, we're not having any of it'". On Quammen asking 'which nameless organization?', Jones answered "The ICBP" (Quammen 1996: 559).

35. Cheke (2001c); see also Walters (2006) for the sorry saga of the Hawaiian Crow or 'Alala *Corvus hawaiiensis*.

36. Bell & Merton (2002), Jones (2004).

37. See, for example, chapters 1 and 6 in Norris & Pain (2002).

38. Groombridge *et al.* (2001), O'Brien *et al.* 2007, Swinnerton *et al.* (2004).

39. Jones (2002) discussed analogues more fully in respect of reptiles, Atkinson (2000) more generally (with particular reference to New Zealand birds); Hutton *et al.* (2007) likewise in relation to restoring Lord Howe Island off Australia; Burney (2003) suggested establishing an analogue-based 'Pleistocene Park' in Madagascar. Hansen *et al.* (2006b) discussed analogues in the context of failing reproduction in Mauritian forests.

40. Hansen *et al.* (2002), Chapter 8; Traveset & Richardson (2006) briefly reviewed how invasives upset native plant reproduction worldwide.

41. There are several invasive ant species, endemics have become rare (Ward 1990) and one exotic is accused of nectar-robbing orchid flowers (Roberts & McGlynn 2004).

42. Cheke (1974b, 1978a, 1985), Temple (1981), Barré & Barau (1982), Barré (1988), Safford (1997). There are 8 vertebrate species (or sibling species) surviving in Mauritius but extinct on Réunion. Birds: *Falco punctatus, Nesoenas mayeri, Psittacula eques, Foudia rubra*; Bats: *Pteropus niger*; Lizards: *Leiolopisma telfairii, Gongylomorphus bojerii, Nactus* spp.

43. For example, Safford (1997), and the attached editor's note suggesting 25–30 years of educational programmes might, perhaps, suffice to prepare the Réunion public.
44. Thébaud & Strasberg (1997).
45. Nyhagen *et al.* (2005).
46. See Chapters 8 and 9, and Cheke (1975f) for *Rubus alceifolius*.
47. As already recommended by Jones (2002).
48. Courchamp *et al.* (1999, 2003).
49. Varnham *et al.* (2002), Seymour *et al.* (2005). Controlling rats in upland forest has similarly resulted in locally increased shrew numbers (Carl Jones, pers. comm.).
50. Courchamp (2003) discussed many of the pitfalls in mammal control.
51. We consider Whittaker (1998), Case *et al.* (2002), and Mayr & Diamond (2003) particularly pertinent to issues we discuss here; Gorman (1979) is a readable introduction, and Quammen (1996) covered the ground well for non-specialists. See also Hubbell (2001) and chapters 6 and 7 of Newton (2003). Above all we recommend close study of Steadman's (2006, chapters 18–20) recent critique of classical island biogeography theory, published too recently to be fully incorporated here, but which explores more deeply many of the issues we have raised.
52. Peake (1971), Temple (1981), Moutou (1983a); a fourth study by Adler (1994) looked at Indian Ocean bird distribution in a rather different way, more in terms of isolation than island size or age, but still lacked some data available now. See Diamond (1985) for a fuller critique of Temple's paper.
53. We have used birds, as they are the best-studied and data is widely available for both living and subfossil species – biogeographically it probably makes more sense to combine birds and bats, but on a world scale bat data is not always available, though some islands have so many they may influence the number of potential bird colonists. In the Malagasy area and Australia bats are 9–18% of bird numbers, but in Borneo it is 24%, in northern Melanesia 30%, and up to 50% in the Lesser Antilles (Ricklefs & Lovette 1999). Over moderate distances bats can be more mobile than birds: 14% in Madagascar rises to 17–18% in Mauritius/Réunion, and 7% in New Guinea rises dramatically to 30% in Melanesia.
54. See Newton (2003), Whittaker (1998).
55. Probably because richer continental avifaunas contain a higher proportion of sedentary genera and groups that do not cross water.
56. Padya (1989), Jury (2003).
57. Mayr & Diamond (2001); the same low slope is evident in the Caribbean. Ricklefs and Bermingham (2004) and especially Steadman (2006) address this question in detail.
58. Temple (1981).
59. In contiguous areas (i.e. land masses, not islands) Hubbell (2001) predicts an S-shaped curve, with the log-log straight-line relationship holding only at intermediate areas (roughly 1–1,000,000 km² for birds), but not over larger or smaller areas. In the intermediate areas this curve is posited as relatively flat over six orders of magnitude. It appears to us that in close-knit archipelagos (e.g. northern Melanesia, granitic Seychelles, Hawaiian main islands, Caribbean islands) there is a flattening of this kind, but at different area ranges depending on circumstances, probably for the reasons mentioned in the main text (see also Steadman 2006). For reptiles the effect is noticeable at much smaller scales (Case *et al.* 2002).
60. The spectacular radiations of honeycreepers and ducks/geese in Hawaii were on islands never reached by reptiles; without reptiles, birds may have had more scope to diversify than would normally be the case.

61. There are 343 known reptile species in Madagascar, a hundred more than land birds (211) and bats (32) together; no other large land mass, apart from Australia, has more reptiles than birds – the ratio normally reverses only on islands of a few hectares! Given the low avian diversity in Madagascar and Australia, it is likely the reverse of the Hawaiian scenario occurred; reptiles diversifying in the relative absence of birds. A similar reptile explosion on relatively bird-poor islands occurred in the Greater Antilles, particularly in *Anolis*, a genus of small iguanids that are ecological parallels to Malagasy day-geckos *Phelsuma* (e.g. Losos 2004). A temperate climate less favourable to reptile development may have prevented the same happening in New Zealand.
62. Unlike for birds, the Comoros are nowhere near the saturation line for reptiles, emphasizing reptiles' much poorer ability to disperse over water. The line extension from the Mascarenes through New Zealand for reptiles' also applies to birds (Figure 11.1), so it appears that the same factors of New Zealand's isolation and climate apply to both sets of taxa.
63. Temple (1981).
64. Simberloff (1992), Moulton *et al.* (1996).
65. As already pointed out by ASC (in Jones 1996). Similarly in Rodrigues, Java Sparrows were established in the 1870s but had disappeared by 1914 – House Sparrows being introduced in the interim (Chapter 7). Revisiting the issue, Simberloff & Gibbons (2005) were still puzzled by vanishing Java Sparrows, but, despite citing Jones (1996), did not consider the role of House Sparrows.
66. Simberloff (1992), Moulton *et al.* (1996), Moulton & Sanderson (1997, 1999), Duncan & Young (1999).
67. Moulton *et al.* (1996) took their introduction dates from Simberloff (1992) who erroneously assumed "species described by Barré & Barau (1982) as 18th century introductions" to have been introduced in 1700, leading them to believe that three estrildid finches had been introduced to Réunion before the 'failed' White-rumped Munia; In fact Barré and Barau made no claims for 18th century arrivals of passerines, remaining wisely vague where they had no information. At the time there were no good dates for Common Waxbills, Red Avadavats or Spice Finches, though there was nothing to suggest they had arrived before the 1750s. The '1750' assumed date for the two canaries was equally arbitrary, but less arguably inaccurate, but the '1806' date for Cardinal Fodies was plain wrong. They were known (Cheke 1987a) to be present before 1770; see Chapter 7 for revised dates for introduced seed-eaters. Lever (2005) was inclined to accept most of Simberloff's spurious dates.
68. See Chapters 7 and 8 for details; Jones (1996) made the connection between canaries and Village Weavers, which Simberloff & Gibbons (2005) dismissed on the grounds of sympatric survival on Réunion – but they may not have been aware of the absence of high-altitude heath habitat on Mauritius.
69. Although mongooses supplied the *coup de grace* in Mauritius, Madagascar Partridges might have died out anyway, as they had done poorly once several other gamebirds were introduced in the 19th century (Chapter 7). Although several gamebirds survive in Réunion, some lowland species are not thriving (Couzi & Salamolard 2002), possibly due to modern agricultural practices (p. 191).
70. Cheke (1987a), Jones (1996), Lever (2005); Grey-headed Lovebirds feed on small grass seeds taken on the ground (Benson 1960, Morris & Hawkins 1998).
71. Case (1996), Duncan *et al.* (2003), Sol *et al.* (2005). Duncan had apparently missed Jones's important paper (1996).
72. Jones (1996) listed 29 species seen as escaped or attempted introductions between 1979–1995, including 15 different parrots! Two of his 29, Mallard and Laughing Dove, have become established.

Appendix 1. Land vertebrates introduced to the Mascarenes

Notes on this table are on page 390
Selected invertebrates are also included. Only species for which there is good evidence supporting establishment on at least one island are included.

- ● = present and established in 2006
- i = only established on offshore islets
- X● = died out, but successfully reintroduced later
- ⓘ = present but may not be established
- (X) = extinct, but never fully established
- Xⓘ = died out, reintroduced but not clear if established
- X = extinct
- [X] = extinct, not certain if ever truly established
- n = native on specified island
- X? = probably extinct

	Species	Réunion	Mauritius	Rodrigues
Mammals	Ruffed Lemur *Varecia variegata*	X		
	Crab-eating Macacque *Macaca fascicularis*		●	
	Common Tenrec *Tenrec ecaudatus*	●	●	
	Greater Hedgehog Tenrec *Setifer setosus*	[X]	[X]	
	House Shrew *Suncus murinus*	●	●	●
	Rabbit *Oryctolagus cuniculus*		Xi	X
	Black-naped Hare *Lepus nigricollis*	●	●	
	Ship Rat *Rattus rattus*[1]	●	●	●
	Norway Rat *Rattus norvegicus*[1]	●	●	●
	House Mouse *Mus musculus*	●	●	●
	Palm Squirrel *Funambulus palmarum*	X		
	Small Indian Mongoose *Herpestes auropunctatus*		●	
	Cat *Felis domesticus*	●	●	X?
	Horse *Equus caballus*	X	X	
	Pig *Sus scrofa*	X	●	X
	Rusa Deer *Cervus timorensis*	X●	●	X
	Heuglin's Gazelle (Red-fronted Gazelle) *Gazella rufifrons*		[X]	
	Cattle *Bos taurus*	X	X	X
	Goat *Capra hircus*	X?	X	X
Birds	Meller's Duck *Anas melleri*		X?	
	Mallard *Anas platyrhynchos*		●	
	White-faced Whistling Duck *Dendrocygna viduata*		X●	
	Egyptian Goose *Alopochen aegyptiacus*		ⓘ	
	Grey Francolin *Francolinus pondicerianus*	●	●	●
	Chinese Francolin *Francolinus pintadeanus*	X	X	
	Malagasy Partridge *Margaroperdix madagascariensis*	●	X	
	Common Quail *Coturnix coturnix*	●	X	
	Painted Quail *Coturnix chinensis*	●	X	
	Jungle Bush Quail *Perdicula asiatica*	●	X	
	Helmeted Guineafowl *Numida meleagris*	X	●	X
	feral chicken *Gallus gallus*	ⓘ	X	X
	Red Junglefowl *Gallus gallus*	●		
	Malagasy Buttonquail *Turnix nigricollis*	n	X	
	Purple Swamphen *Porphyrio porphyrio*		X	
	Feral Pigeon *Columba livia*	●	●	●
	Spotted Dove *Stigmatopelia chinensis*		●	
	Laughing Dove *Stigmatopelia senegalensis*		●	
	Barbary Dove *Streptopelia risoria*	●		
	Zebra Dove *Geopelia striata*	●	●	●
	black parrot *Coracopsis* sp.	X		
	Ring-necked Parakeet *Psittacula krameri*	(X)	●	
	Grey-headed Lovebird *Agapornis canus*	X?	X	X
	Red-whiskered Bulbul *Pycnonotus jocosus*	●	●	
	Peking Robin (Red-billed Leiothrix) *Leiothrix lutea*	●		

Appendix 1 (cont.)

	Species	Réunion	Mauritius	Rodrigues
Birds	Yellow-fronted Canary *Serinus mozambicus*	●	●	●
	Cape Canary *Serinus canicollis*	●	X	
	Common Waxbill *Estrilda astrild*	●	●	●
	Red Avadavat *Amandava amandava*	●	X	
	Spice Finch (Scaly-breasted Munia) *Lonchura punctulata*	●	●	
	White-rumped Munia *Lonchura striata*	X?		
	Java Sparrow *Lonchura oryzvora*	X	X	X
	Pin-tailed Whydah *Vidua macroura*	●		
	Village Weaver *Ploceus cucullatus*	●	●	
	Cardinal Fody (Madagascar Red Fody) *Foudia madagascariensis*	●	●	●
	House Sparrow *Passer domesticus*	●	●	●
	Common Myna *Acridotheres tristis*	●	●	●
	(Indian) House Crow *Corvus splendens*	?	●	
Reptiles	Aldabra Giant Tortoise *Dipsochelys dussumieri*[2]		Xi●i	
	East African Box Terrapin *Pelusios subniger*		X?	
	Wattle-necked Softshell Turtle *Palea steindachneri*		●	
	Red-eared Slider *Trachemys scripta*	?	●	
	Green Iguana *Iguana iguana*	?		
	Rainbow Lizard *Agama agama*	●		
	Bloodsucker *Calotes versicolor*	●	●	●
	Warty Chameleon *Furcifer verrucosus*		X	
	Panther Chameleon *Furcifer pardalis*	●	●	
	Gold-dust Day-gecko *Phelsuma laticauda*	●		
	Great Green Day-gecko *Phelsuma madagascariensis*	●	●	
	Lined Day-gecko *Phelsuma lineata*	●		
	Seychelles Small Day-gecko *Phelsuma astriata*	●		
	Blue-tailed Day-gecko *Phelsuma cepediana*	●	n	●
	Stump-toed Gecko *Gehyra mutilata*[3]	●	●	●
	Cheechak *Hemidactylus frenatus*	●	●	●
	Indian House Gecko *Hemidactylus brookii*	●	●	●
	Common Worm Gecko *Hemiphyllodactylus typus*	●	●	●
	Square-toed Gecko *Ebenavia inunguis*[3]		●	
	Wolf Snake *Lycodon aulicus*	●	●	
	Flowerpot Snake *Rhamphotyphlops braminus*	●	●	●
	Slender Worm Snake *Typhlops porrectus*		●	
Amphibians and invertebrates	Malagasy Grass Frog *Ptychadena mascareniensis*	●	●	
	Guttural Toad *Bufo gutturalis*[4]	●	●	
	giant snail *Achatina fulica*	●	●	●
	giant snail *Achatina immaculata*	●	●	●
	Rosy Wolf Snail *Euglandina rosea*	●	●	●
	locust *Nomadacris septemfasciata*	●	●	

Appendices 2–8

Key to the timeline tables (Appendices 2–8) for native and introduced fauna

These tables show all observations of species over time. They represent the reports of observers, not an interpretation by us. The symbols indicate reports (or lack of them) without any specific comment on abundance or other pertinent information. Records are given at five-year intervals until 1900, after which the interval is increased to 12.5 years. The first data for each island appears in the column to the right of the grey-filled area to the left of the matrix, which represents the last years of 'prehistory'. We have included scientific names where no English specific name exists. Animals in the native fauna tables are split by biome into marine and terrestrial guilds; these two environments were subject to different kinds of pressures following human colonisation, and different observational biases; for example, most seabirds were recorded much less frequently than landbirds. Some timelines are generic (e.g. 'rats'), when species are not distinguished, though most are species specific; once the species are identified the data from the general line may transfer to the specific (shown by ⌐, with the transfer also shown on the specific line by ⌐). In some cases the general line also continues, covering those observers that still mentioned just the generic group, such as 'rats' or 'partridges'.

The following symbols apply to the matrices of observations in appendices 2–8.

●	abundant
•	common/unexceptional
·	rare
?	uncertain record
X	extinction/absence noted
#	species observed, but no evidence of breeding
—	likely but unconfirmed presence: used where there is a sizeable interval between records and an animal could have died out and recolonised or been reintroduced, and where the first record showed a species already well-established so it must have been present for some time.
✳	several species observed but not separated
○	no record, but present as observed subsequently
§	not reported by visitors
▶	recorded release
▷	approximate date of introduction
c	captive only

Notes on Appendices 2–9 are on page 390

Appendix 2. Native fauna in Mauritius (mainland; for islets see Appendix 8)
Timelines of observations from the first account to the present.

Appendix 2 (*cont.*)

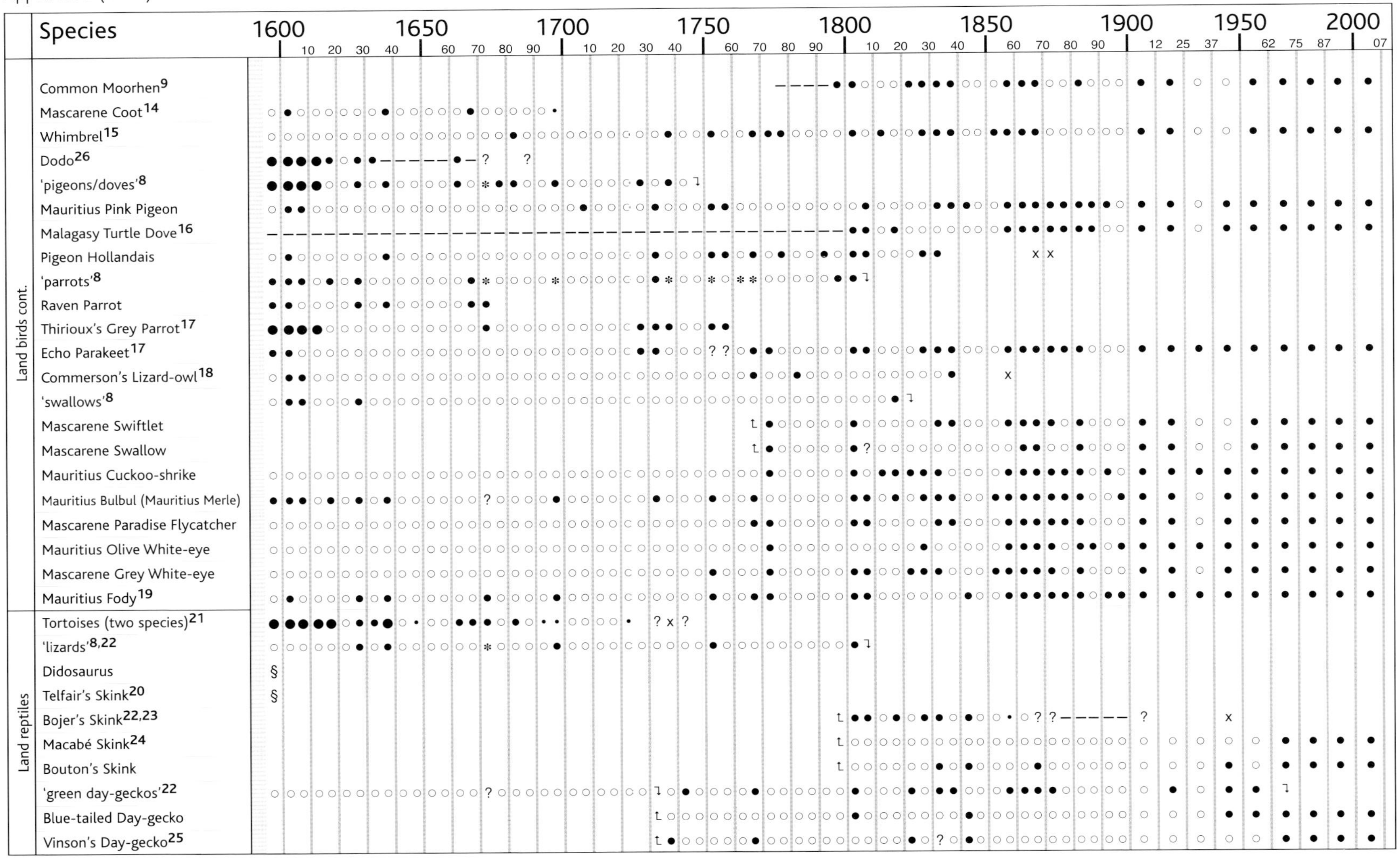

Appendix 2 (cont.)

Species		1600	1650	1700	1750	1800	1850	1900	1950	2000
Land reptiles cont.	Guimbeau's Day-gecko[25]	§			t ○○○○○○○○○○○○●●○○○○○○○○○○○○○○○○○○○○○				○○○○	●●●●
	Upland Forest Day-gecko[25]	§			t ○○○○○○○○○○○○●●○○○○○○○○○○○○○○○○○○○○○				○○○○	●●●●
	Günther's Gecko[20]	§								
	night-geckos (two species)[20]	§								
	Carié's Blind-snake	§								
	Mauritius boas (two species)[20]	§								

Appendix 3. Native fauna in Réunion

Timelines of observations from the first account to the present.

Appendix 3 (*cont.*)

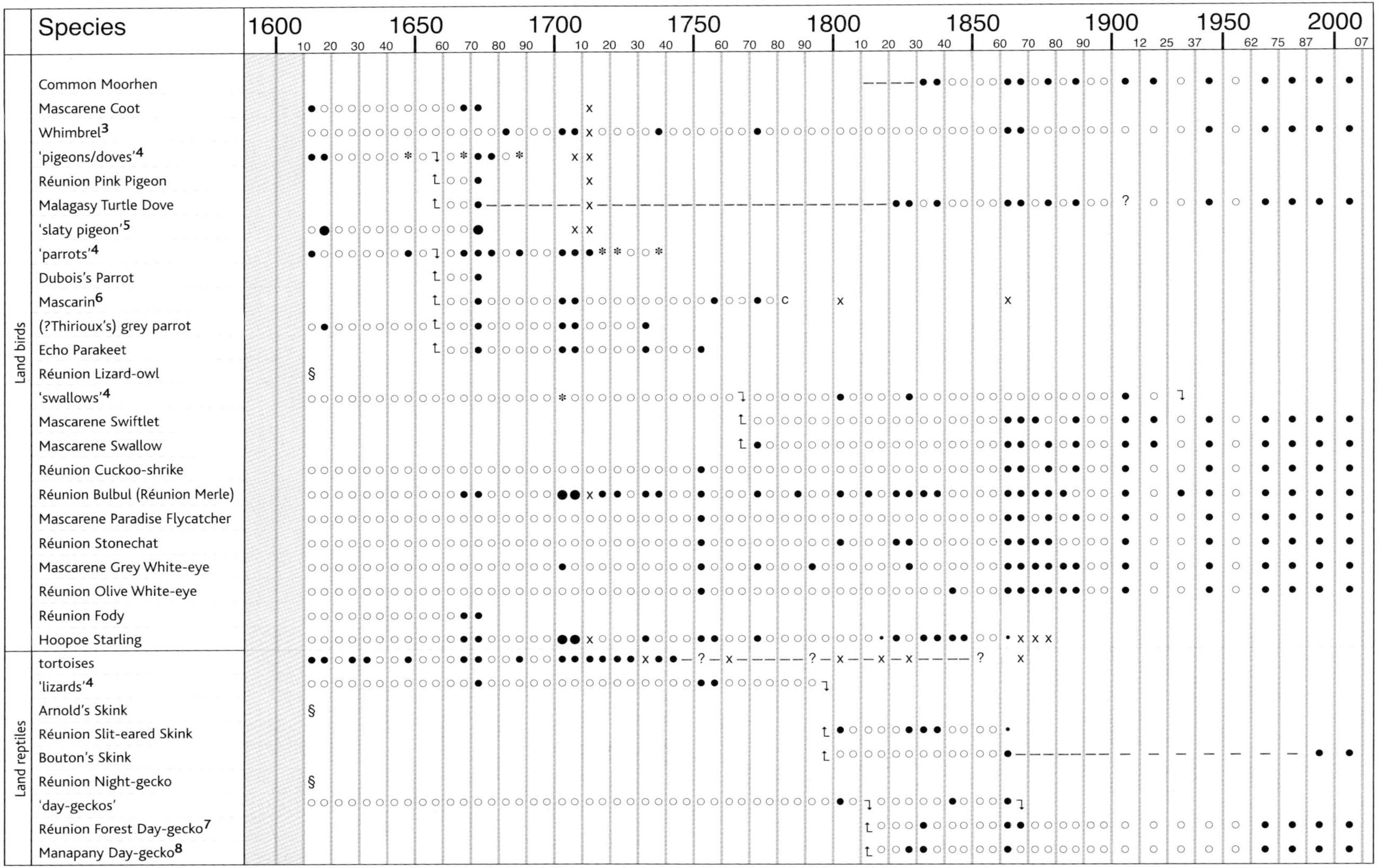

Appendix 4. Native fauna in Rodrigues

Timelines of observations from the first account to the present.

Appendix 4 (*cont.*)

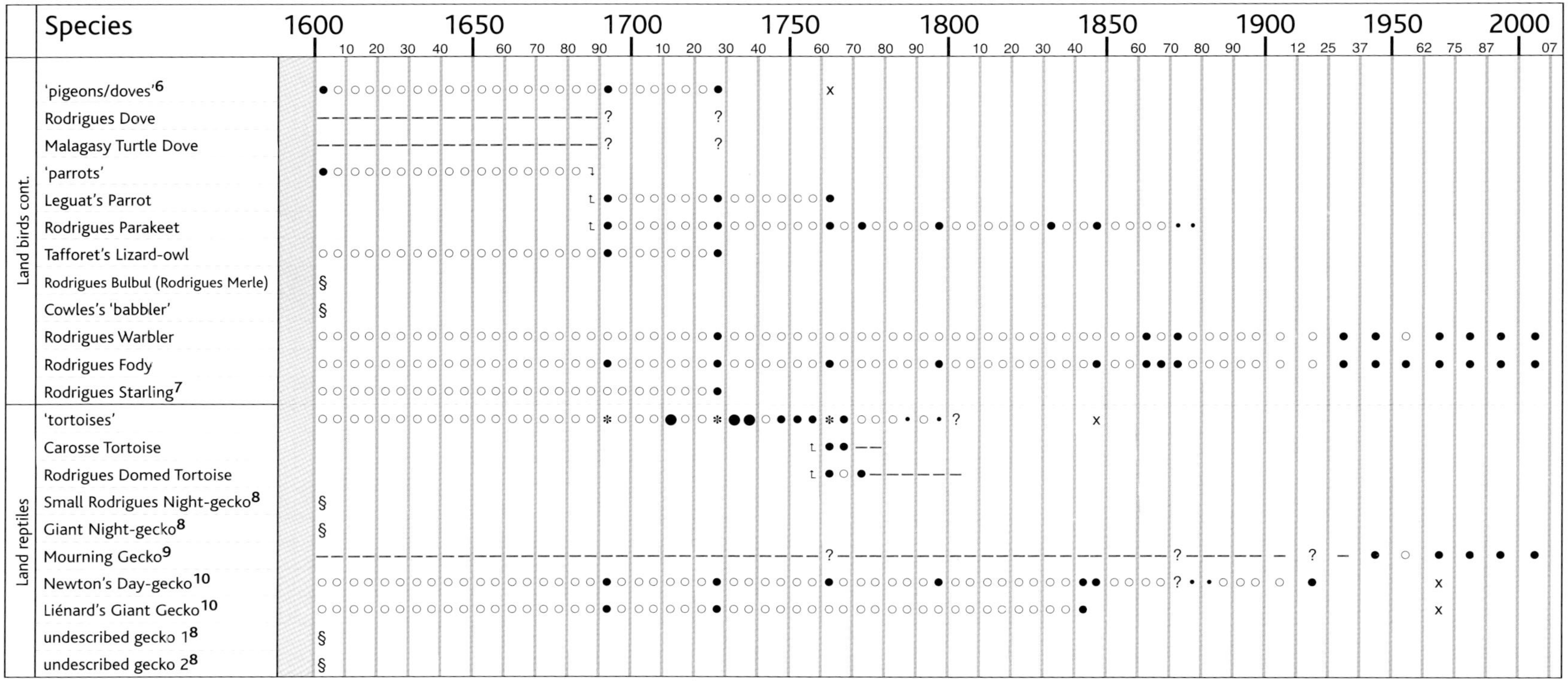

Appendix 5. Introduced fauna in Mauritius

Timelines of observations from the first account to the present.

Appendix 5 (*cont.*)

Species	1600	1650	1700	1750	1800	1850	1900	1950	2000
Birds cont.									
Feral Pigeon									
Spotted Dove									
Laughing Dove									
Zebra Dove									
Ring-necked Parakeet									
Grey-headed Lovebird									
Red-whiskered Bulbul									
Yellow-fronted Canary									
Cape Canary									
Common Waxbill									
Red Avadavat									
Spice Finch									
Java Sparrow									
Village Weaver									
Cardinal Fody									
House Sparrow									
Common Myna									
(Indian) House Crow									
Reptiles[10]									
East African Box Terrapin									
Wattle-necked Softshell Turtle									
Red-eared Slider									
Bloodsucker									
Warty Chameleon[7]									
Panther Chameleon[7]									
Great Green Day-gecko									
'house geckos' (5 spp.)									
Stump-toed Gecko									
Cheechak[8]									
Indian House Gecko[9]									
Common Worm Gecko									
Square-toed Gecko									

Appendix 5 (*cont.*)

	Species	1600				1650			1700			1750			1800			1850			1900			1950			2000	
		10	20	30	40	60	80	90	00	20	30	40	60	70	80	90	00	10	20	30	40	60	70	80	90	12	25	37	62	75	87	07
Reptiles cont.	'snakes'	x		x					x		x		x				x		x			?		x x ⌐								
	Wolf Snake[11]																				○	○	○	○	○	●	●	○	●	●	○	●
	Flowerpot Snake[12]																				○	○	● ●	●	○	●	●	○	●	●	—	●
	Slender Worm Snake[13]													— —	—	—	—	—	—	—	—	—	— —	—	—	—	—	—	—	—	—	—
Amphibians	Malagasy Grass Frog							x	x								▲	○	○	○	●	○	● ●	●	○	●	●	○	●	●	●	●
	'toads'																					x	x x ⌐									
	Guttural Toad[14]																	○	○	○	⌐	○	⌐									
Invertebrates	giant snails *Achatina*															●	○	○	○	○	○	○	●	●	○	●		○	●	●	●	●
	giant snail *Achatina fulica*[15]														●	—	—	○	○	○	○	○	○	○	●	●	○	○	●	●	●	●
	giant snail *Achatina immaculata*[15]														—	—	—	○	○	○	○	○	▲	○	○	●	○	○	○	○	●	●
	Rosy Wolf Snail												● ●	● ●	x	x																
	locusts[16]																									▲			▲			

Appendix 6. Introduced fauna in Réunion

Timelines of observations from the first account to the present.

Appendix 6 (*cont.*)

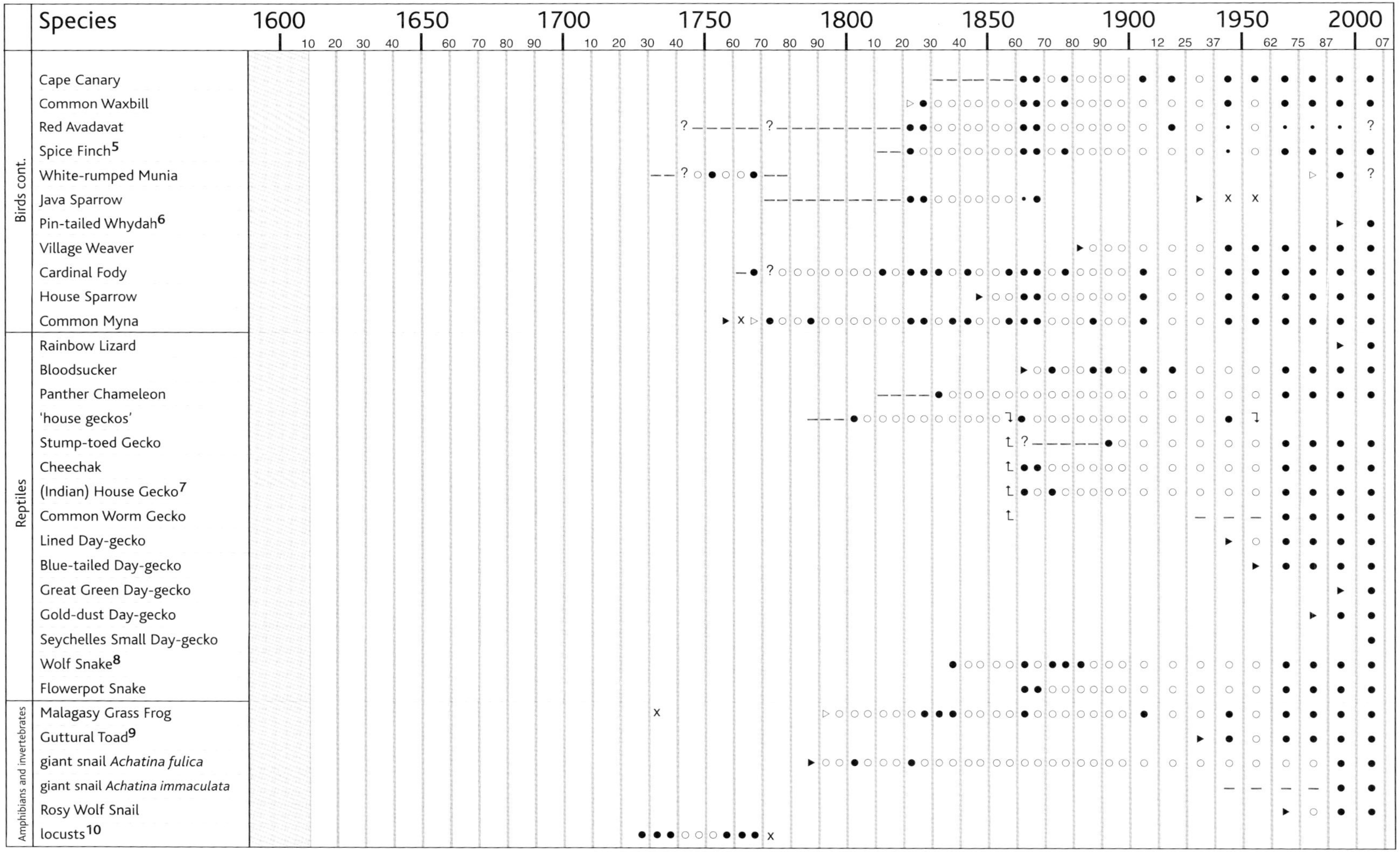

Appendix 7. Introduced fauna in Rodrigues

Timelines of observations from the first account to the present.

Appendix 8. Faunal observations on the northern islets of Mauritius

Timelines of observations from the first account to the present. Both native and introduced species are included, with introduced species italicised. References for all records in this table are in Chapter 9.

Appendix 8 (cont.)

Flat Island

Appendix 8 (cont.)

Appendix 8 (cont.)

Species	1600				1650				1700				1750				1800				1850				1900				1950				2000	
	10	20	30	40	50	60	70	80	90	10	20	30	40	50	60	70	80	90	10	20	30	40	50	60	70	80	90	12	25	37	62	75	87	07
Gunner's Quoin (cont.)																																		
Zebra Dove																															●	●	●	○
Red-whiskered Bulbul																															●	●	●	○
Common Waxbill																															●	—	●	○
Spice Finch																															●	●	●	○
Common Myna																													—	—	●	●	○	●
Telfair's Skink	○	○	○	○	○	○	○	○	○	○	○	○	○	●	—	—	—	—	—	—	—	—	—	×				—	—	—				●
Bojer's Skink	○	○	○	○	○	○	○	○	○	○	○	○	○	○	○	○	○	○	○	○	○	●	○	○	○	○	○	○			●	●	●	●
Bouton's Skink	○	○	○	○	○	○	○	○	○	○	○	○	○	○	○	○	○	○	○	○	○	○	○	○	○	○	○	○	—	—	—		●	●
Vinson's Day-gecko	○	○	○	○	○	○	○	○	○	○	○	○	○	○	○	○	○	○	○	○	○	○	○	○	○	○	○	○	○	○	●	●	●	●
Lesser Night-gecko	○	○	○	○	○	○	○	○	○	○	○	○	○	?	○	○	○	○	?	○	○	●	—	—	—	—	○	○	○	○	●	●	●	●
bolyerid snakes	○	○	○	○	○	○	○	○	○	○	○	○	○	○	○	○	○	○	○	○	○	○	○	○	○	×					○			

Appendix 9. Sources for appendices 2–8.

This table lists the principal references for records in the timeline tables (Appendices 2-8). Only the major sources are listed for those 5-year intervals in which there are several accounts. Simple references (author + date, no brackets) direct the reader to entries in the bibliography; numbers refer to the notes below this table. A name in square brackets with an associated number indicates an account in an anthology or secondary source; a number on its own in square brackets indicates a minor source, or a passing reference in a secondary source. Names followed by '/' and a museum acronym indicate a collector whose specimens are in the specified museum, which we have personally seen or had supplied by respective curators. Names in italics are names of ships, the reference being to the ship's log. 'L/' followed by a reference signifies 'legislation in [ref.]'.

Year	Mauritius	Réunion	Rodrigues
1595-00	[(van) Warwijck & crew 1, 2, 3, 61]		
1600-05	[West-Zanen;1, 3, 4]; [*Gelderland* 44]		[Bouwer: 44]
1605-10	[Matelief, van der Hagen: 1, 2, 3]		
1610-15	['Verhuff': 1, 2, 5]	[Blok: 8]; Tatton 1625	['Verhuff': 5]
1615-20	[van den Broecke, Almeida: 1, 5]	Bontekoe 1648 [6]	
1620-25			
1625-30	Herbert 1634; [*Mary*; 7]	Herbert 1634 [*Hart*: 7]	
1630-35	[*Discovery*: 7]; [anon: Servaas 1887]		
1635-40	Mundy 1608–67 & 2; Cauche 1651 & 1	Cauche 1651	45
1640-45	[van der Stel: 9]		45
1645-50		[exiles to Flacourt: 6]	
1650-55	[3]	[Thoreau: 6]	
1655-60			
1660-65	[*Arnhem* survivors: 3,5,10]		
1665-70	[Granaet: 2]; [Marshall: 11]; [2]	[Martin, Carpeau, Ruelle *et al.*: 6]	
1670-75	Hoffmann 1680; [3, 5]	Dubois 1674 [Bellanger *et al.*: 6]	
1675-80		[Boureau-Deslandes: 6]	
1680-85	[Harry: 1, 2]; [*President*: 7]		
1685-90		Bernardin 1687; [Houssaye: 6]	
1690-95	Leguat 1707; [Deodati: 2, 12]		Leguat 1707
1695-00	[Deodati: 2, 12]		
1700-05		Feuilley 1705; [Borghesi: 6]	
1705-10	[Merveille: 2, 5]; [Momber: 2, 12]	Feuilley 1705; [Merveille: 5, 6]	45
1710-15		Boucher 1710; [Lougnon 1956]	[British survey: 46]
1715-20		Le Gentil 1727 & [6]; [Parat: 76]	
1720-25	*Courrier de Bourbon* 1721; [13, 41]	*Athalanthe* 1722; [Gaubil: 6]	
1725-30	[Ducros: 2, 5]; [Jonchée: 13, 39]; [25]	L/Hermann 1898	Tafforet 1726
1730-35	Gandon 1732; Cossigny 1732–55; [25]	Cossigny 1732–55; Gandon 1732	Gennes 1735
1735-40	d'Heguerty 1754; Gennes 1735; 42	d'Heguerty 1754	d'Heguerty 1754
1740-45	Grant 1741–44 (in Grant 1801)	Lecoq 1740/20; [25, 31]	
1745-50	Grant 1749 (in Grant 1801)		
1750-55	La Motte 1754–7; La Caille 1763	Fréri 1751; [La Nux: 21]	
1755-60	Cossigny 1732–55, 1764; [Flotte: 49]	[La Nux: 21]	Cossigny 1732–55
1760-65	Noble 1756; Pingré 1763; [25]; 50	Pingré 1763; Caulier 1764; 22; 50	Pingré 1763; Nichelson 1780
1765-70	Bernardin 1773; Le Gentil 1781; [Commerson 14, 15, 47]	Crémont 1768; L/Delaleu 1826	
1770-75	[Querhoënt: 14, 40]	[Querhoënt, La Nux: 14] [Commerson: 47]	[Commerson/Jossigny: 15, 47]
1775-80	Sonnerat 1782; [Buncle: 2]		
1780-85	Céré 1781		
1785-90		L/Delaleu 1826	[Labistour: 51]
1790-95	[E.Newton 1858–95]; Petit-Radel 1801; 74	Petit-Radel 1801	
1795-00	Cossigny 1799; Guignes 1808		Marragon 1795
1800-05	Tombe 1810; Chapotin 1812; Milbert 1812; 19; 48	Bory 1804; Renoyal 1811–38; 66	Marragon 1802
1805-10	Flinders 1803–14, 1814; [Mathieu: 71]	Tombe 1810	
1810-15	Prior 1819	Renoyal 1811–38	

Appendix 9 (*cont.*)

Year	Mauritius	Réunion	Rodrigues
1815-20	Quoy & Gaimard 1824/MNHN/15	Billiard 1822 [Quoy & Gaimard: 15]	
1820-25	Lesson 1830/MNHN/King 1827	Renoyal 1811–38 [L/ 54]	
1825-30	[Holman: 2]; [Desjardins: 17]; 28; 70	Renoyal 1811–38; Betting 1827	Hoart 1825
1830-35	Desjardins/17; Sganzin 1840; 33	Renoyal 1811–38 " Sganzin 1840; 15; 17; 29; 64	[Dawkins: 1, 32]
1835-40	Desjardins: 17; Desnoyers 1837; 36; 62; 79	Rousseau/MNHN & 29; [L/54]	Anderson 1838; Backhouse:79
1840-45	Marchal/83	[Leconte de Lisle: 57]	Self 1841
1845-50	Walond 1851; [Kelsey: 78]	Coquerel 1848	Corby 1845: 34; Higgin 1849; Gardyne 1846
1850-55	Mouat 1852; Flemyng 1853; Ellis 1859		
1855-60	Clark 1859; E. Newton/UMZC: 38: [Layard: 84]	Beaton 1860; Coquerel 1859; 73	Duncan 1857
1860-65	E. Newton 1861b: 38; Boyle 1867; BMNH: 37; 29	Maillard 1863; Coquerel: 29, 68; 43; 63; 73	E. Newton 1865a/BMNH
1865-70	E. Newton 1865b/UMZC: 38; Pike 1873; 37	[Pollen: 23]; Vinson 1868; 65; 73	[Jenner 38]
1870-75	E. Newton 1875; Möbius 1880: 35; 18; 38	Vinson 1870; Möbius 1880; RMNH: 55; 73	Slater 1875 & [Sharpe 1879]
1875-80	E. Newton 1878b, 1878; Audebert/RMNH 80	Leal 1878; RMNH; Slater: 75; 72; 73	Broome 1904
1880-85	Broome 1904; 60	LaHuppe 1882; 77; Dyboski 81	
1885-90	Abercromby 1888; Schneider/RMNH 59; 82	Lantz 1887; Vinson 1887; 72; 73	Fremantle 1914
1890-95	Carié 1916, 1921; Kennedy 1893, 1900	Alluaud & Belly/MNHN	Kennedy 1893, 1900
1895-00	Carié 1916; Walter 1968		
1900-12	Carié 1904; Meinertzhagen 1912; 58	Manders 1911; Naidoo/AMNH; 58; 67; 69	[Bower: 34]
1912-25	Emmerez 1914; Carié 1916; Huron 1923	Carié 1916, 1920/MNHN	Bertuchi 1923; Snell & Tams 1919
1925-37	BMNH	Decary 1948	Vinson 1964
1937-50	Rountree: 85; Guérin 1940–53	Milon 1951; Mockford 1950	Rountree 1943; [Wiehé: 52]
1950-62	Rountree 1952; Newton 1958a	Decary 1962; Bourgat 1967	Vinson 1964; [Courtois: 52]
1962-75	Cheke: 26; Staub 1973; Gill/USNM 29	Gill 1963 & USNM; Cheke: 27, 29	Gill 1967 & USNM; Cheke 1987d; 30
1975-87	Cheke 1987c,d; Staub 1976	Barré, Bour, Moutou: 24	Cheke 1979, 1987d
1987-00	Staub 1993; Jones 1993, 1996	Probst 1997; Barré *et al.* 1996	Showler 2001; pers. obs.
2000-07	[pers. comms]	[pers. comms]	[pers. comms]

Notes for Appendices 2–9

Notes for Appendix 1

1. English names for rats follow standard New Zealand usage (e.g. Atkinson & Atkinson 2000, Veitch & Clout 2002), as New Zealanders have pioneered rat control methods and spread their expertise very widely. The names 'Black Rat' (for *Rattus rattus*) and 'Brown Rat' (for *R. norvegicus*) are widely used in the UK and USA, but given that nearly all rats are brown, including multiple races of *R. rattus*, these are unsuitable names in a wider context. The attempt to replace 'Brown Rat' with 'Common Rat', pioneered by Corbett & Southern (1977), and followed by Cheke (1987a), although appropriate, has not found wide favour, and we have dropped it in favour of the more familiar, if biogeographically inaccurate, 'Norway Rat'.

2. We follow current Seychelles usage (e.g. Palcovacs *et al.* 2003, Gerlach 2004) in calling the Aldabra Giant Tortoise *Dipsochelys dussumieri* not *Aldabrachelys gigantea*. Bour (1984) established that Schweigger's *Testudo gigantea* was based on a Mascarene specimen now lost (a Réunion Tortoise *Cylindraspis indica*) not an Aldabra or Seychelles one, and that the first valid name for the Aldabra species was Gray's *dussumieri*. Bour also pointed out that as *Aldabrachelys* was founded on Schweigger's animal (or at least his name), then *Aldabrachelys* fell into junior synonymy of *Cylindraspis*. Hence he created the new genus *Dipsochelys* for the Aldabran species. Despite this Bour himself has, without explanation, been co-author on more recent papers using *Aldabrachelys* (e.g. Arnold *et al.* 2002).

3. *Ebenavia* is assumed to be introduced to Mauritius (e.g. Vinson & Vinson 1969), but it is not an anthropophilous species (Glaw & Vences 1994, Cheke 1987a *contra* Vinson & Vinson 1969) so may have arrived naturally. Although we have retained the familiar name for Mascarene *Gehyra*, Ivan Ineich (in Glaw & Vences 1994: 277) considers that Indonesian and nearby continental forms (and introductions therefrom) are specifically different from those in the type locality (Phillipines); if confirmed, Mascarene lizards would become *G. peronii*, based on specimens from Mauritius (Dumeril & Bibron 1834–54).

4. Although the toad introduced to the Mascarenes was identified some time ago as the segregate *gutturalis* from the *Bufo regularis* complex (e.g. Bour & Moutou 1982), Henkel & Schmidt (2000) and Lever (2003) nonetheless still used the old aggregate name.

Notes for Appendix 2

1. Ship Rats were already present when the Dutch arrived in 1598 (see Chapter 3). The introduction dates of many species likely to have had an impact on the native vertebrates are not known, amongst them the mosquito *Culex (pipiens) quinquefasciatus* (which carries avian malaria), and the Cheechak *Hemidactylus frenatus*.

2. Early reports appear to refer to Green Turtles or did not distinguish the species. Hoffman (1680) described combs and other objects made from carapaces, so Hawksbills are implied, though the first explicit reference is in Clark (1859). Loggerheads were illustrated in the *Gelderland* journal (Moree 2001), but never appear in the literature. They still breed in southern Madagascar (Ratzimbazafy 2003).

3. Jolinck (Keuning 1938–51, Panyandee 2002) reported edible grey birds that lived in holes (not caves, as Panyandee wrongly translated) that screamed or cried all night like an unhappy rabble. These appear to have been Wedge-tailed Shearwaters, the only record from the mainland. Tropical Shearwaters are conspicuously black and white, and have a different, less human-like call, and generally nest on cliffs or steep slopes; however, Tropical Shearwater calls may also have been heard, if only Wedge-tailed Shearwaters were actually found.

4. Following Austin *et al.* (2004b), we are using the name Tropical Shearwater for the Indo-Pacific *Puffinus bailloni* which, on DNA evidence, has now been split from the Atlantic *P. lherminieri*.

5. Jolinck (Keuning 1938–51) reported tropicbirds, frigatebirds and "white herons almost as large as white storks" and "another kind of bird, ash-grey with flat beaks and long necks" off Mauritius the day before the ships anchored in Grand Port Bay. Herons are not going to be seen out to sea, so the 'herons' are likely to have been boobies (large, white with black wingtips like storks). Both Masked and Abbott's Boobies fit the description, so either or both may have been seen. The ash-grey birds are likely to have been pelicans; the Pink-backed Pelican *Pelecanus rufescens* is greyish. They have long necks and flat beaks, though the only known Indian Ocean colony was on St. Joseph atoll in the Amirantes group. This population was extinct by the 1930s (Skerrett *et al.* 2001).

6. References to large bats (flying-foxes) are assumed to be to the large, common Black-spined Flying-fox, unless the smaller Rougette is specifically described or mentioned. There were no reports that can now be ascribed to the Golden Bat, but it was probably present undetected until the mid-18th century at least.

7. Microbat species were not distinguished before the late 18th century, so are lumped together in the table until 1770. After that date the two species are listed separately, but an overlap allows for general accounts that did not distinguish them.

8. Many early reports refer to 'herons', 'pigeons/doves', 'parrots', 'swallows' or 'lizards' without further detail. These are all included on the unspecified family lines.

9. Herbert's 'bitters' [= bitterns] and Leguat's 'butors' are taken to refer to the endemic night heron. The Striated Heron is apparently a more recent immigrant, as is the Common Moorhen; see Chapters 4 and 6.

10. Although the literature asserts that both Greater Flamingos *Phoenicopterus rubeus* and Lesser Flamingos *P. minor* may have been present on pristine Mauritius (Oustalet 1896, Rountree *et al.* 1952), re-examination of the bones by JPH reveals only Greater Flamingos (very variable in size), the same applying in Réunion (Mourer *et al.* 1999). However groups of Lesser Flamingos appeared in the Mascarenes in 1870 (Oustalet 1896) and 1872–3 (Bouton 1875) – these are not included in the table. The 'cranes' reported by Jolinck in 1598 (Keuning 1938–51) were probably flamingos – they shared habitat with geese and herons.

11. Herbert's (1634) 'goshawkes' were probably harriers, while his 'hobbies' would have been kestrels. He also wrote of 'kites' which he listed separately from the hawks (between 'swallows' and 'blackbirds'), and elsewhere noted they came to drink sap oozing from palm trees, together with parrots and lizards. These 'kites' remain unidentified, but do not appear to have been raptors. Herbert used an idiosyncratic nomenclature – he also saw 'powts', but these can be identified from old usage as pigeons. Likewise van West-Zanen (Soeteboom 1648) reported 'sparrowhawks' (sparwer) and 'falcons' (falke).

12. The 1680–5 doubtful record of the Red Hen is Benjamin Harry's inedible 'dodo'. Jolinck's 'Indise riviers houdt-snippen' the size of a hen in 1598 are taken to be this species, though the *Dryolimnas* wood-rail cannot be ruled out.

13. The record of a White-throated Rail in 1809 may have been a vagrant from Madagascar, as there is no other report of living birds, though Cowles (1987) suggested Herbert's picture of a straight-billed 'hen' may have been *Dryolimnas*. Herbert's text however, though giving no plumage description, suggests he was discussing Red Hens *Aphanapteryx* (Chapter 2). The subfossil *Dryolimnas* bones (Cowles 1987, Mourer *et al.* 1999) are now known to be a flightless form close to *D. augusti* of Réunion, which we are calling Sauzier's Wood-rail. In early accounts the wood-rails may not have been distinguished from Red Hens or may have been considered to be their young.

14. Granaet's 'waterhens', Mundy's 'moorhens' and Leguat's 'poules d'eau' are taken to refer to Mascarene Coot (see Chapter 3 for explanation).

15. The whimbrel (the only migrant regularly reported by early visitors) is treated as a resident for comparative purposes (presence indicated by ● rather than #).

16. Malagasy Turtle Doves, although found in subfossil deposits, were not explicitly recorded by any 17th- or 18th-century visitor. Milbert's (1804) descriptions of pigeons are equivocal, but he mentioned that his '*Columba triangularis*' (probably the Malagasy Turtle Dove), together with the Pigeon Hollandais (and an unidentified third form) had been collected by his friend M. Mathieu (who was in Mauritius at the same time) and passed to Dufresne. There is a specimen in Edinburgh from the Dufresne collection (the type), so, like the Pigeon Hollandais there (Cheke 1987: 76), it is probably the bird Milbert mentioned, thus presumably collected between 1801 and 1810 (Milbert's and Mathieu's dates in Mauritius).

17. 17th- and 18th-century reports refer clearly to 'grey parrots'; we have included Hoffman's 'mottled [*'bigarrés'*] parrots' here also. These appear to be the same as subfossil bones described as *Lophospittacus bensoni* by Holyoak (1973, Cheke 1987a) and now referred to *Psittacula* (Hume 2007). 17th- and 18th-century accounts also refer to green parrots without any indication of size or other details, but the first green parrot to be described well enough to be identified is Jossigny's drawing of an Echo Parakeet around 1770 (Oustalet 1897); as there was only one green species all are assumed to have been referrable to this species.

18. We have included both West-Zanen and Matelief's owl records (1602 and 1606) respectively, but their bird lists are virtually identical and thus certainly cognate – since neither account was published until the 1640s it is impossible to say which one was original.

19. West-Zanen's and Hugo's 'sparrows' (Strickland 1848, Panyandee 2002), Herbert's 'robbins' (Herbert 1634), Mundy's 'little bird like a linnet' (Mundy 1608–1667) and Leguat's 'little birds made rather like our sparrows except they have red throats' (Leguat 1707) are all taken to be Mauritius Fodies, as they pre-date the introduction of other small finches and weavers.

20. No snakes, Telfair's Skinks, Round Island Geckos or *Nactus* nightgeckos were ever recorded on the mainland. Records all refer to offshore islands (see Chapter 9 and Appendix 8).

21. No visitors distinguished between the different species of tortoise.

22. Lizard records apply to the mainland only; for the offshore islets see Chapter 9 and Appendix 8. The 'day-gecko' line includes only the green species, not Günther's Day-gecko *Phelsuma guentheri*. There is doubt about the identity of 'common' skinks seen in coastal areas in the 1860s (see Chapter 7). Post-1750 gecko reports that clearly apply to house geckos are omitted (see Appendix 5).

23. Bojer's Skinks are extinct on the mainland, but survive on offshore islets. It is very unclear whether Pike (1873) or Möbius (1880, Peters 1877) found the species on the mainland (Chapter 7). Pike was very muddled about lizard identity, and Peters reported a specimen allegedly collected at Black River, but it is now lost and the museum jar and catalogue entries refer only to 'Fouquet' (i.e. Ile aux Fouquets, Mahébourg Bay) (Rainer Günther pers. comm.).

24. Including the 'orange-tailed skink' from Flat Island.

25. Mathieu's *c*. 1805 specimens of these two forms were not recognised until the 1960s.

25. Cossigny (1732–55) gave a detailed description of a Vinson's Daygecko in 1738, nearly 100 years before the first specimen was collected. Similarly Peron and Lesueur's manuscripts in Le Havre contain a Latin diagnosis and drawing of *P. guimbeaui* that pre-dates its formal description by 160 years. The upland form *rosagularis* (Vinson and Vinson 1969), originally described as a race of *P. guimbeaui*, is now considered a full species (Austin and Arnold 2004), the Upland Forest Day-gecko; Matthieu had already collected on in *c*. 1805 (in MNHN, Paris).

26. See p. 78 for the isolated record on the inshore islet of Ile d'Ambre in 1662.

Notes for Appendix 3

1. Goats were already well-established in 1612 at the time of the first visit for which we have an account.

2. References to large bats (flying-foxes) are assumed to relate to the large common Black-spined Flying-fox unless the smaller Rougette is specifically described or mentioned.

3. The Whimbrel is included as the only example of a migrant that was regularly reported by early visitors (due to being favoured as a game bird). Dubois (1674) referred to 'bécasses', normally meaning 'Woodcock', but here probably meaning, generically, 'waders'.

4. Many early reports referred to 'pigeons' and/or 'doves', 'parrots', 'swallows' (= swifts and true swallows) or 'lizards' without further detail – these are all included on the unspecified family lines.

5. The slaty pigeon is generally assumed to be a 'blue pigeon' *Alectroenas* but no confirmatory bones have been found (see text).

6. The alleged 1834 captive record is excluded, see Hume (2007) and text (Chapter 6).

7. The 1860s record is an MNHN specimen of *P. borbonica* that was marked as a *P. cepediana* from from Peron & Lesueur, but apparently assigned to them later in error, and bears a reticketing date of 1864 (Pasteur & Bour 1992); it is likely to have been collected by or for Maillard, *c*. 1860.

8. The early records are specimens from *c*. 1830 in the Paris museum (MNHN, Vinson & Vinson 1969) and a drawing dating from 1861 (Bour *et al*. 1995).

Notes for Appendix 4

1. Ship Rats were already enormously abundant in 1691 when Leguat arrived (Leguat 1707), but prior reports are so poor that we have no idea when they arrived.

2. Jossigny's drawings for Commerson in the MNHN archives, Paris include a Red-tailed Tropicbird and Red-footed Booby, both without data apart from a note by Commerson to include them in the 'natural history of Bourbon [= Réunion]'; the birds never occurred in Réunion but were common in Rodrigues, from which Jossigny also drew better annotated drawings. Hence we have taken these two seabirds to have come from Rodrigues. Backhouse (1844) saw frigatebirds, tropicbirds and 'sooty petrels' (Wedge-tailed Shearwaters) off Rodrigues in 1838, but did not land, so these records do not confirm breeding.

3. Tafforet's 'fouquets de montagne' were presumably one or both of the two *Pterodroma/Pseudobulweria* petrels, otherwise known only from subfossil bones.

4. There may have been two species of frigatebird originally, but we only have documentary evidence of *F. ariel*.

5. There is no positive evidence that Great Crested Terns bred, but see Chapter 7 (note 387).

6. We do not know which doves Leguat (1707) and Tafforet (1726) saw, but Leguat said they were all slaty coloured, which is a possible but unlikely description of Malagasy Turtle Doves. Either only one kind still survived then, or he did not distinguish between the two species (as with the parrots). Tafforet gave no description of his 'tourtourelles'.

7. A second alleged starling, described as *Necropsar leguati* from a unique skin in Liverpool Museum acquired in 1850 and labelled 'Madagascar', was long supposed (with no evidence whatsoever) to have come from Rodrigues or even Mauritius (e.g. Hachisuka 1953, Cowper 1984). Cheke (1987a) expressed strong reservations, and it has since been shown to be an albino Martinique or Grey Trembler *Cinclocerthia gutturalis* (Olson *et al*. 2005 & Chapter 2).

8. Referred to by Arnold (2000), see Chapter 3; full details remain to be published.

9. Pingré (1763) reported small lizards similar to those in France, which may have been of this species.

10. The last definite mainland record for *Phelsuma edwardnewtonii* was in 1876. Vinson & Vinson (1969) presumed the 1884 and 1917 specimens were from Ile Frégate, which was also the location for the 1842 specimen of *P. gigas*, but there is no evidence for this supposition (Chapters 7, 8). The latter was probably extinct on the mainland before Pingré's visit in 1761.

Notes for Appendix 5

1. There is some doubt as to the species (see Chapter 6); the Spiny or Yellow-streaked Tenrec *Hemicentetes semispinosus* may also have been present during part of the same period.

2. Ship Rats, 'House' Mice (*Mus (musculus) gentilulus*) and House Shrews in Madagascar came not from Europe and India, but Yemen and Africa via Arab traders before the European penetration of the Indian Ocean (rat, mouse), or possibly more recently (shrew) (Hutterer & Tranier 1990, Duplantier & Duchemin 2003). These authors correctly imply the first date they were aware of (1858) for the shrew in Madagascar was due to under-recording; in fact Sganzin (1840) reported them in 1831–32, already evidently widespread and common, and probably already present for many decades if not centuries (although unrecorded in archaeological sites). It is thus possible any or all of these were introduced to the Mascarenes from Madagascar, not on ships from Europe (rat, mice) or direct from India (shrew) as is normally assumed (Cheke 1987a). Contact with Madagascar was frequent from the early Dutch and French settlements (Moree 1998, Toussaint 1972). No studies have been done on the DNA of Mascarene rats, mice or shrews, but it should be possible to determine their origin in this way.

3. While rabbits were kept domestically from the beginning of Dutch settlement, there is no clear evidence that they were feral anywhere except on inshore and offshore islets. The offshore islets are considered separately (Chapter 9); definite feral rabbit records in the table all refer to Ile d'Ambre, within the lagoon on the northeast coast. Likewise there is no evidence that the domestic chickens or ducks brought in by the first Dutch settlers ever went wild.

4. Dogs are omitted from the table as there is no convincing evidence that self-sustaining populations were ever established, though hunting dogs often escaped into the forest during Dutch times (Pitot 1905). Wild dogs were reported around Mahébourg in the 1850s (Clark 1859), though Boyle (1867) expressly stated that, contrary to local hearsay, there were no feral dog populations.

5. The bush quail in Mauritius has generally been considered to be *Perdicula argoondah* (e.g. Newton 1888, Ogilvie-Grant 1896, Carié 1916, Rountree *et al.* 1952, Cheke 1987a), but specimens collected by I. Legentil and S. Roch in Mauritius for Edward Newton in 1860 and 1863, preserved in Cambridge, are in fact *P. asiatica* (pers. obs. in 1975, *contra* Cheke 1987a) Rountree *et al.* noted two specimens of *P. asiatica* in the Paris museum collected by Carié, but nonetheless accepted that the bird feral in the 19th century was *P. argoondah*. Carié himself (1916) clearly thought the birds in Mauritius were *P. argoondah*, even the extra ones he imported and released. The photo of a stuffed male specimen in his paper is too poor to allow firm identification; it appears to have no stripe below the eye, which is characteristic of *P. argoondah*. In contrast the photo of a Port Louis museum specimen in d'Emmerez (1914) appears to be a *P. asiatica*; this is the species still surviving in Réunion (Barré *et al.* 1996), and is the commoner and more widespread species in its native India (Ali & Ripley 1968–74, Grimmett *et al.* 1999). Both species may have been imported and released at different times without being distinguished locally.

6. The Madagascar Buttonquail was cited as occurring in Mauritius in 1831 by Sganzin (1940), but as there are no other records until 70 years later, this may have been in error for another type of quail. Edward Newton (1858–95) mentioned 'button quails' twice in his journals, but appears to have been referring to Painted Quails or bush quails, as nowhere in his writings or collections is there any sign of *Turnix*.

7. After a long absence feral chameleons reappeared in the 1990s, being frequently brought alive to the Crocodile Park at Senneville (Rivière des Anguilles) (Owen Griffiths, pers. comm.) All those in the zoo are *C. pardalis* (pers. obs.). It is not impossible that a small number survived from the 19th century and have recently increased (as has the population of *C. pardalis* in Réunion, Probst 1997), though it is more likely that they have been reintroduced, perhaps from Réunion. We have taken the date of Quoy & Gaimard's Warty Chameleon to be from their second visit in 1828, but it may date from 1818.

8. The 1830–35 record for the Cheechak *Hemidactylus frenatus* is a specimen from Desjardins in Paris (Dumeril & Dumeril 1851); the earliest (1770s) record is an excellent drawing in Commerson's MSS (MNHN, Paris, pers. obs).

9. The house gecko identified as the Malagasy '*Hemidactylus mercatorius*' by Vinson & Vinson (1969, Cheke 1987b) turns out in fact to be the Indian House Gecko *H. brookii* (Lever 2003, Vences *et al.* 2004b). Specimens of '*H. maculatus*' allegedly collected on Round Island by the Möbius expedition in 1874 (Peters 1877) resemble *H. mercatorius* (R. Guenther, pers. comm.) so are probably in fact *H. brookii*; the specimens from Quoy & Gaimard in 1818 referred to '*maculatus*' by Dumeril & Bibron (1834–54) are indeed *brookii* (pers. obs)

10. This table does not include Aldabra Giant Tortoises which are feral only on Ile aux Aigrettes, and formerly on Flat Island.

11. It appears from Desjardins's report (1830) that the snake '*Coluber rufus*' he reported on was a one-off rather than part of a feral population; it is not possible to identify the species involved.

12. See Chapter 6 for the blind-snake found in 1803, which being c. 13cm (5") long we have assumed to be *Ramphotyphlops braminus*. Found in recent years on the mainland (Nik Cole, pers. comm.), *Typhlops porrectus* (much longer but just as thin) was also discovered in 2004 on Ile de la Passe in Mahebourg Bay (Nik Cole, pers. comm.).

13. *B. gutturalis* is a segregate from the *B. regularis* species group; the Mauritian toad has often been cited under the latter name. The oriental toads *Bufo melanostictus* found in 1837 may have maintained a feral existence briefly, but were certainly extinct by 1859 (Clark 1859, Cheke 1987a).

14. It is generally agreed (e.g. Germain 1921) that the early introductions of giant snails on both Mauritius and Réunion were of *Achatina fulica* and that *A. immaculata* arrived later. Koenig (1932), using contemporary local information, states that the latter was released in Mauritius in 1850, while Germain (relying on derivative sources) gave a first record of 1847.

15. Locusts were not actually extinct after 1800, but are treated so in the table as they only very rarely become troublesome (e.g. in 1962, Rouillard & Guého 1999, possibly as a result of reduced numbers of mynas after the 1960 and 1962 cyclones).

Notes for Appendix 6

1. The feral lemur was originally reported as the Mongoose Lemur *Lemur (Prosimia) mongoz* (Coquerel 1859, Maillard 1862), but this was soon corrected to the Ruffed Lemur (Coquerel 1863, Vinson 1868); see Chapter 7.

2. European Hares *Lepus europaeus* and Red Deer *Cervus elaphas* introduced in, respectively, the late 1950s and 1966, appear not to have persisted (Moutou 1981, Probst 1997), though Probst reported a Red doe shot in 1995, 15 years after the last stag had supposedly been killed.

3. Goats were already well-established at the time of the first account in 1612 (Brial 2001a), having presumably been introduced by the Portuguese who left no accounts (Chapter 5).

4. Both Coquerel (1964) and Vinson (1968) listed not *C. coturnix* but '*C. textilis*' (= *C. coromandelicus*, Rain Quail), though Carié (1916), confirmed by Berlioz (1946), identified Réunion specimens as *C. coturnix* of the African race. It is possible that, as both Coquerel and Vinson claimed the birds came from India, that the Rain Quail, a resident bird in south India (Grimmett *et al.* 1999), was introduced and was temporarily successful, but was replaced later in the 19th century by the Common Quail. At the time Pondicherry in south India was still French territory.

5. An editorial footnote in Renoyal (1811-38:1392) glosses his 'damier' (a stuffed specimen recorded in 1822), as 'pétrel tacheté' (i.e. Cape Pigeon *Daption capensis*), but the term has also long been used for *Lonchura* spp., and in the Réunion context it is much more likely to refer to the Spice Finch ('capucin damier' of Barré *et al.*, 1996).

6. There is a single 19th-century record of a Paradise Whydah *V. paradisaea* caught wild in Réunion (Vinson 1861, Coquerel 1864; date of capture not given). It was thought at the time to be a vagrant, it is more likely to have been an escape or deliberately released.

7. A gecko specimen from Réunion in Berlin, collected by the Möbius expedition in 1874, was identified by Peters (1877) as *Hemidactylus maculatus*. Dr Rainer Günther (pers. comm.) says it in fact resembles *H. mercatorius*, so it is probably in fact *H. brookii*, the only *Hemidactylus* other than *H. frenatus* now present (Vences *et al.* 2004b); 19th-century Paris specimens of '*maculatus*' from the Mascarenes are all *H. brookii* (pers. obs.), and we have treated the Berlin specimen as this species.

8. 19th-century snake records are presumed to refer to *Lycodon*, but other names were used and some were thought to come from Madagascar, so there may have been other species involved (see Chapter 7).

9. Gerard's claim (1974) that the toad *Bufo gutturalis* was introduced from Mauritius 'around 1920' is in error because the animal did not reach Mauritius until 1922 (Vinson 1953, Cheke 1987a). The date given by Armand Barau to Harry Gruchet in 1979 ('1927', Cheke 1987a), confirmed by Auguste de Villèle ('1926-30', verbally to ASC in 1985), is probably correct.

10. Locusts are treated as extinct after they ceased to be pests, but at least one species can sometimes be seen in some numbers in the dry west coast savannas (pers. obs., 1996).

Notes for Appendix 7

1. Rabbit bones were found in caves on Plaine Corail in 1874 (Dobson 1879), but from at least 1892 the species was confined to offshore islets, until dying out between 1978 and 1983 (Cheke 1987a, Chapter 8). North-Coombes (1971) believed they were introduced by Marragon who certainly intended to do so (Marragon 1802). A rabbit seen at Cascade Victoire in 1999 may have been an escape (Showler 2002).

2. The 'uncertain' records for rats and mice in 1874 are bones found in caves (Dobson 1879), so do not represent 'live' records at that date, but they do indicate the prior presence of these introduced species, and constitute the first record for Norway Rat.

3. Feral cats have not been reported on the mainland since 1918 (Snell & Tams 1919), but are certainly present on some offshore islets (Bell & others 1993), and may simply have been overlooked on the mainland.

4. Although cattle and goats have not been feral for many years, until recently they were free-range with a vengeance, though they are now excluded from several forest areas (Chapter 8).

5. Lever (2005) cited Rountree et al. (1952) as saying that Cattle Egrets *Bubulcus ibis* had "been successfully introduced to the island of Rodrigues" – there is, in fact, no such claim in Rountree's paper, and indeed Cattle Egrets have never been recorded there (Showler & Cheke 2002).

6. Apparently feral guineafowl with young were recorded in 1999, possibly re-establishing themselves from domestic stock (Showler 2002).

7. Both *Achatina fulica* and *A. immaculata* are present (Griffiths 1994, Griffiths & Florens 2006), but there are no details of when either species arrived.

Notes for Appendix 8

1. The tarantula discovered in 1992 was finally described in 2005 as *Mascaraneus remotus* in a new monotypic endemic genus (Gallon 2005).

2. The identity of the subfossil fruitbat has been questioned (Hume 2005) as the skull appears anomalous; however, either Golden Bats or Black-spined Flying-foxes are likely to have occurred.

3. In addition to the northern islets, Bouton's Skink occurs on the Mahébourg Bay islets of Iles de la Passe, Vacoas, Fouquets, Rocher des Oiseaux and Fous, as well as on Ile d'Ambre and (rarely) the mainland. Vinson's Day-gecko is also found on the east coast islets (d'Ambre and Cerfs groups), Benitiers and Fourneaux, as well as Ile aux Aigrettes and the mainland (Bell et al. 1993, Carl Jones unpubl. data 2005).

4. Apart from the northern islets, the only other place that endemic boas were found in historical times was on Ile de la Passe in the 18th century (Chapter 6).

5. Sea turtles almost certainly nested on Flat Island and Gabriel Islet in the past, but there seem to be no unequivocal records.

6. Includes a herd of feral donkeys *Equus asinus* present from 1912 to c. 1940 (Meinertzhagen 1910–11, Souchon 1994).

7. The smaller *Gongylomorphus* skink on Flat Island, referred to informally as the 'orange-tailed skink', is different enough from the mainland population of Macabé Skinks to be treated as a separate subspecies (Freeman 2003), though it has not been formally described.

8. The night-gecko discovered on Flat Island in 2003 is very similar to Lesser Night-geckos on neighbouring islands (and Ilot Vacoas), but smaller with some physical differences (Nik Cole, pers. comm.), so it is possibly distinct subspecifically.

9. See Chapter 9; there is doubt as to whether rabbits were present in addition to hares, though it appears they were there in the early 19th century. A re-introduction after 1996 was rapidly dealt with.

Notes for Appendix 9

1. Strickland 1848
2. Barnwell 1948
3. Pitot 1905
4. Soeteboom 1648 and Cornelisz (in Strickland 1848)
5. Grandidier et al. 1903–20, for Mauritius also reprinted in Verin (1983)
6. Lougnon 1970
7. Barnwell 1950–54
8. Brial 2001
9. Bonaparte 1890
10. Stokram 1663, Olearius 1670, Panayandee 2002, Cheke 2004a
11. Khan 1927
12. Leibbrandt 1896–1906
13. Pineo 1993 (and, in part, Brouard 1967)
14. Buffon 1770–83
15. Oustalet 1897, Milne-Edwards & Oustalet 1893; also MNHN specimens and Jouanin 1970.
16. Ly-tio-Fane 1972
17. Desjardins 1831, 1834, 1835, 1837 and 15, 16
18. MCML and AMNH collections (H. H. Slater & A. C. Smith); also Slater in Hartlaub (1877) and Cowper (1984).
19. Also for birds: Bory (1804), Oustalet (1897; specimens in MNHN from Dumont), and Mathieu/Dufresne specimens in RSME (Bob McGowan pers. comm). Lesueur (1803) was added or confirmed birds, mammals, reptiles and amphibians (see also 48).
20. Père Lecoq was formerly thought to be Père Lebel, and the printed account in 1938 edited by Albert Lougnon appeared under this name; see Barassin (1983) for the re-assignment to Lecoq.
21. Specimens sent by Lanux to Réaumur and described by Brisson (1756, 1760) or mentioned by Torlais (1936: 330). Lanux's correspondence with Réaumur spanned 1753–56 (Tourlais 1962), but the birds were all sent in 1754 (Lanux 1753–56). Animals in Brisson not mentioned in Lanux's letters (parrots, flying-foxes) are assigned to 1750–55.
22. "Officer of the British Navy" (Grant 1801; see text, Chapter 6).
23. Schlegel & Pollen 1868, Pollen 1866, RMNH
24. Barré & Barau 1982, Barré 1983, Bour & Moutou 1982, Moutou 1981, 1982a,b, 1984 and Cheke 1987b
25. Lougnon 1956
26. Cheke & Dahl 1981, Cheke 1987c,e
27. Cheke & Dahl 1981, Cheke 1975f, 1987b
28. Kennedy 1827
29. Vinson & Vinson (1969). Lizards and frogs collected by Nivoy, Eydoux, Rousseau, Maillard and Morel & Bernier. Nivoy's Hoopoe Starling was presented in 1832 (Legendre 1929); we assume the same for his lizards. Eydoux was the surgeon on the *Favorite* (Laplace 1833–5, Desjardins 1837) in Réunion and Mauritius in 1830; Rousseau was in Réunion in 1839 (Cheke 1987a, Jaussaud & Brygoo 2004), Maillard and Morel & Bernier deposited specimens in 1862 and 1863. Unfortunately other important 19th-century reptile collections (Leschenault, Bosc, Hugot, Vangrignouse) are undated; Leschenault was with Baudin's expedition at Mauritius in 1801, then later (1816–22) travelled widely in India and Ceylon (Brown 2000). He could therefore have passed Mauritius and/or Réunion and collected on several occasions.
30. Vinson 1964, Vinson & Vinson 1969
31. Lougnon 1933–7
32. Telfair (1833b) and MS letters in Zoological Society of London archives.
33. UMZC – Algernon Strickland's specimens (see also Strickland 1831–34).
34. UMZC – specimens labelled 'Col. M.B.S. Lloyd' but probably collected by Thomas Corby (see Cheke 1987a: 51).
35. Peters (1877) wrote up Möbius's reptile collections.
36. The only vertebrate that Charles Darwin collected in Mauritius was a frog, *Rana mascareniensis*, now in the NHM, London (Boulenger 1882, Keynes 2000).
37. Reptiles collected by Edward Newton and Sir Humphrey Barkly are in the NHM, London (Boulenger 1882, 1885–87); there are others from the Mascarenes, but mostly without dates.
38. Edward Newton's bird specimens from Mauritius, not all noted in his published work, are in Cambridge; the collection's bird catalogue can be

browsed online at www.zoo.cam.ac.uk. Also his field journals and letters (E. Newton 1858–95, 1861–82) and specimen lists in Newton 1860–61. He also recorded fodies sent from Rodrigues by magistrate Jenner in 1865 in his letters.

39. Brouard 1963
40. Querhoënt 1773
41. Lagesse 1972
42. Charpentier 1732–55; his detailed description of a Vinson's Day-gecko *Phelsuma ornata* antedates the first specimen collection by nearly 100 years.
43. Bour *et al.* 1995
44. Moree 2001; Gelderland drawings, plus Bouwer's account in the *Zeeland*.
45. Landings in Rodrigues in 1638 or 1640 (Cauche 1651), 1641 and 1644 (North-Coombes 1971), *c.* 1706–8 (Alby 1989) yielded no account of wildlife.
46. North-Coombes 1971.
47. Cap (1861) and Commerson/Jossigny MSS in the MNHN Bibliothèque Centrale archives. Jossigny's drawings of Red-footed Booby and Red-tailed Tropicbird, although glossed '*hist. nat. de Bourbon*', are assumed to have come from Rodrigues, as these birds do not occur in Réunion.
48. Mammals: Geoffroy 1806; reptiles: Peron & LeSueur in Dumeril & Bibron (1834–54)/MNHN
49. Epinay (1890: 146) quoted a passage on flying-foxes from La Flotte's *Essais historiques sur l'Inde* (1769).
50. Foucher-d'Obsonville 1783
51. North-Coombes 1993
52. pers. comm.
53. Probst 2000
54. Pest and game legislation for 1820 and 1839 given in Delabarre (1844), also partly (1820 only) cited in Decary (1962).
55. List of birds, a snake and invertebrates collected in Réunion by Bewsher in 1870–71, and presented to the Desjardins Museum (TRSAS NS 6: 19–20, séance du 24 juin 1871); at least one collected at the same time, a Réunion Harrier, is now in the NHM (Tring).
56. Gordon 1894
57. Leconte de Lisle mentions several birds, feral cats and day-geckos in his poem *Ravine Saint-Gilles*, published in the 1857 but no doubt written in (or recording images from) his last visit to the island in 1842–45 (Eggli 1943, Damour 1973). His reference to 'vertes perruches' is the first record of Grey-headed Lovebirds for Réunion. Images of Réunion, including wildlife, recur in Leconte's poems from the 1880s and 90s (Eggli 1943), but are more ethereal and we have exluded them.
58. Swettenham 1912; Meinertzhagen's diaries (1910–11) add flying-foxes to the bird list in his 1912 paper on Mauritius.
59. Williams 1903
60. Proclamation no. 40 of 1880 – protecting birds. We have only used this first version of the law in the table, as subsequent renewals were based on it, and not on any re-checking of the status of the birds' status.
61. Keuning 1938–51
62. Anon 1838, Thomas A. Wise (Anon 1963).
63. Roussin 1860–67; articles by various authors.
64. Vinson 1888; also, abridged, in Lavaux 1998.
65. Jeantet in Roussin (1860–68).
66. Chanvalon 1804
67. Mac-Auliffe 1902; this book includes animals referred to in the text as if familiar to the author, and others in a list supplied by Edmond Bordage, curator of the St Denis museum. Although the latter is less reliable, it appears to have been tailored to Mac-Auliffe's locality (Cilaos), and so the records are also included – fortunately duplicated by other sources.
68. Coquerel 1863, 1864; also Pollen 1868.
69. Hermann 1909.
70. Dussumier specimens in Paris (Lesson 1828, Dumeril & Bibron 1834–54, Oustalet 1897).
71. Mathieu collected birds in Mauritius for Louis Dufresne; surviving specimens are now in NMS, Edinburgh, supplemented by Dufresne's catalogue (Bob McGowan, pers. comm.). There are also Mathieu specimens in Paris; birds (Lesson 1928), tomb-bat (Moutou 1982), and a skink supposedly from Réunion (Vinson & Vinson 1969) but actually labelled 'Isle de France' (= Mauritius).
72. François de Mahy, politician and even a French government minister, had a fondness for Common Moorhens but little other natural history – apart from being amused by seeing someone put a cigarette in a Bloodsucker's mouth and have it smoke itself to death! (Mahy 1891).
73. Accession and exchange logs, Muséum d'Histoire Naturelle, Saint-Denis; accession register MNHN Paris.
74. Tuijn 1969; Vosmaer commented on a living *Alectrenas nitidissima* received in Holland in 1790.
75. Henry Slater's Réunion specimens collected in Dec/Jan 1874–5 are in AMNH (J. Farrand, pers. comm.) and MCML (Cowper 1984); also 1875 specimens from de l'Isle and Lantz received in Paris (recorded in museum accessions book).
76. Some details from Parat's 1716 MS *Productions de l'isle Bourbon* in the Paris Archives Nationales are quoted by Barassin (1989).
77. *PRSAS* NS 16: 5–6 (1885, session of 21.09.1882) recorded a *Lycodon* sent from Réunion; Lantz sent a harrier from Réunion to the Paris Museum (recorded in accession book).
78. Two paintings by H. J. Kelsey made in 1845 (reproduced in Pitot & Lenoir *c.* 1980: 25–26) show wildlife, (White-tailed) tropicbirds in one, (Black-spined) flying-foxes in the other.
79. Backhouse 1844.
80. Lt. Haig in anon. (1879).
81. Collection of birds from Dyboski, accessioned to the Paris Museum in 1886 (details from accession book).
82. Bird specimens given by Théodore Sauzier to the Paris Museum in 1896, but presumably collected during his visit in 1888–89 (*DMB*: 89–90, d'Emmerez, 1941).
83. Marchal collected the type specimen of the Mauritius Pink Pigeon (Temmick *et al.* 1838–1843, in vol. 2, 1843), and nests and specimens of both Mauritius Fody and Cardinal Fody (Lafresnaye 1850) and a number of lizards; his collection dates are not known but must have been *c.* 1840. His collection was mostly sold and dispersed after he died (Lafresnaye 1850), though his widow presented his lizard specimens to the MNHN in 1846.
84. Brooke 1976, 1978.
85. Frank Rountree's field notebooks, copy in ASC's files.

Appendix 10. Contributions of travellers, collectors and fossils to knowledge of the Mauritius land vertebrate fauna

This table should be read in conjunction with Table 3.1. Note the failure of travellers to record microbats and aerial feeding birds, also missing from the subfossil record. Seabirds are particularly under-represented as fossils, as are lizards and small passerines in travellers' accounts.
Key to the categories:

1 = total seen by travellers before 1750	4 = species that have colonised since 1750	7 = known only from travellers and subfossils
2 = total found as subfossils	5 = known only from travellers reports	8 = species without skins/pickled specimens
3 = total as skins or pickled[e]	6 = known only from subfossils	(i.e. the sum of 5–7)

Group	Total known	1	2	3[e]	4	5	6	7	8
seabirds	13–14	6	3+	11 (3[d])	2–3[d]	1[h]		1[k]	2
wildfowl, cormorants	3	2	3				1	2	3
herons, ibises, flamingos	4	3	2+	1[b]	1[b]	1[h]		2	3
rails, buttonquails	4	2+	3	1[c]	1[c]		1	2	3[f]
raptors, owls	3	3	3	1				2	2[f]
pigeons	4	3	4	4					
parrots	3	3	3	1				2	2
flying-foxes	3	nd (1)[i]	3	2			1		1
microbats, swifts, swallows	4			4					
large passerines	2	1		2					
small passerines	4	1	1+	4					
skinks	5	1[a]	3	4 (1)			1		1
geckos	7	1	4+	7 (5)					
tortoises	2	nd (1)[i]	2	1			1		1
snakes	3	1+[a]	2	2					
total resident land fauna	64+	29+	36+	45	4–5	2	5	11	18
migrant waders	c. 25	1	1	(sight)[g]					

a. Only seen on offshore islets.
b. Striated Heron (see Chapter 3).
c. Common Moorhen (see Chapter 3).
d. Kermadec, Bulwer's and 'Round Island'/Trindade Petrels were recognised post-1945; the former two arrived recently (1990s, Chapter 9) and are represented by photographs rather than skins. The 'Round Island' Petrel may have arrived after 1750.
e. Numbers in brackets are those species recognised post-1945; note the geckos in particular.
f. Good illustrations (but no skin specimens) exist of the Red Hen and Commerson's Lizard-owl.
g. Migrant shorebirds have been included to show how under-represented they are in both subfossils and travellers' accounts – only Whimbrels were seen and preserved as bones. Most identifications are recent sight records, not specimens.
h. Frigatebirds and Dimorphic Egrets are the only species known solely from travellers' accounts.
i. 'nd' means 'not distinguished'; travellers did not distinguish species of flying-fox or tortoise.
j. Reed Cormorant, *Dryolimnas* wood-rail, Golden Bat, Didosaurus and the 'second' tortoise.
k. Abbott's Booby.

Appendix 11. Contributions of travellers, collectors and fossils to knowledge of the Réunion land vertebrate fauna

This table should be read in conjunction with Table 3.1. Note the relatively large number of species known only from traveller's accounts compared with Mauritius or Rodrigues, although, as usual, lizards and small passerines are under-represented by travellers.
Key to the categories:

1 = total seen by travellers before 1750	4 = species that have colonised since 1750	7 = known only from travellers and subfossils
2 = total found as subfossils	5 = known only from travellers reports	8 = species without skins/pickled specimens
3 = total as skins or pickled[f]	6 = known only from subfossils	(i.e. the sum of 5–7)

Group	Total known	1	2	3[f]	4	5[k]	6[h]	7	8
seabirds	6	3	4	6					
wildfowl, cormorants	4	3	3			1	1	2	4
herons, ibises, flamingos	5	4	3	2[b]	1[b]			3	3
rails, buttonquails	5	4	2	2[c]	1[c]	1[j]		2	3
raptors, owls	3	2	3	1			1	1	2
pigeons	3	2	2	1		1		1	2
parrots	4	4	1	2[a]		2			2
flying-foxes	2	nd (1)[d]	2	2					
microbats, swifts, swallows	6	nd (3)[d]		5		1[g]			1
large passerines	3	2	1	3					
small passerines	5	1		4		1[i]			1
skinks	3		1	2			1		1
geckos	3	nd (1)[d]	1	2 (1)			1		1
tortoises	1	1	1	1					
snakes	0								
total resident land fauna	53	31+	24	33	2	7	3	10	20
migrant waders	14	1	1	(sight)[e]					

a. At least one skin of Echo Parakeet from Réunion formerly existed, from which the species was described in detail by various writers (Brisson 1760, Buffon 1770–83, Levaillant 1805).
b. Striated Heron (see Chapter 3).
c. Common Moorhen (see Chapter 3).
d. 'nd' means 'not distinguished'; travellers did not distinguish species of bats or tortoises, apart from Bory (1804) who distinguished 2 microbats and reported swallows (included despite being post-1750 because of his important bat record: see note 'g').
e. Migrant shorebirds have been included to show how under-represented they are in both subfossils and travellers' accounts – only Whimbrels were seen and preserved as bones. Most identifications are recent sight records, not specimens.
f. Numbers in brackets are those species recognised post-1945.
g. Bory's White Bat – a late traveller's account (in 1801), but no specimens exist.
h. Pochard, Lizard-owl, Arnold's Skink, *Nactus* night-gecko.
i. Réunion Fody.
j. 'oiseau bleu' (unidentified swamphen, see Chapter 3).
k. Apart from those listed above, no cormorant, slaty pigeon, grey parrot or Dubois's Parrot bones have been found.

Appendix 12. Contributions of travellers, collectors and fossils to knowledge of the Rodrigues land vertebrate fauna

This table should be read in conjunction with Table 3.1. Note particularly that no animals are known only from travellers' accounts, the large proportion of lizards known only from subfossils, and the high proportion of seabirds recorded by travellers compared to Mauritius and Réunion.

Key to the categories:

1 = total seen by travellers before 1750	**4** = species that have colonised since 1750	**7** = known only from travellers and subfossils
2 = total found as subfossils	**5** = known only from travellers reports	**8** = species without skins/pickled specimens
3 = total as skins or pickled specimens	**6** = known only from subfossils	(i.e. the sum of 5–7)

Group	Total known	1	2	3	4	5	6	7	8
seabirds	13-15[i]	11	4+	11[j]			1	2	3[h]
wildfowl, cormorants	0								
herons, ibises, flamingos	2	2	1	1				1	1
rails, buttonquails	1	1	1					1	1
raptors, owls	1	1	1					1	1
pigeons	3	nd (2)[a]	3				1	2	3[h]
parrots	2	'3'[b]	2	1				1	1
flying-foxes	2	nd (1)[a]	2	1				1[k]	1
microbats, swifts, swallows	0								
large passerines	3	1	3				2[f]	1[l]	3
small passerines	2	2	2	2					
skinks	0								
geckos	7	2	6	2 (1[c])	1[c]		4[g]	1[e]	5
tortoises	2	'3'[b]	2	2					
snakes	0								
total resident land fauna	**38+**	**27 ('29')**	**27**	**20**	**1**		**8**	**11**	**19**
migrant waders	13	1		(sight)[d]					

a. 'nd' means 'not distinguished'; travellers did not distinguish species of flying-foxes or pigeons; the second 'pigeon' seen was the Solitaire.

b. Leguat claimed to see three species of tortoise and Tafforet three species of parrot, neither supported by subfossils; see Chapter 3 for the interpretation of Tafforet's parrots.

c. The Mourning Gecko *Lepidodactylus lugubris* was not collected by the thorough 1874 expedition (Chapter 7), so may have arrived recently.

d. Migrant shorebirds have been included to show how under-represented they are in both subfossils and travellers' accounts – only Whimbrels were seen and none were recorded preserved as bones. Most identifications are recent sight records, not specimens.

e. A very full description exists of a living Liénard's Giant Gecko *Phelsuma gigas*, but no specimen (apart from subfossils).

f. A bulbul *Hypsipetes* sp. and an undescribed 'babbler' (Cowles 1987).

g. All undescribed (Arnold 2000); two *Nactus* species and two possibly new genera.

h. Travellers mentioned only one sort of pigeon (Leguat 1707, Tafforet 1726) and one 'mountain petrel' (Tafforet 1726), when two of each are known as subfossils; the other seabird without a skin collection is Abbott's Booby.

i. '13-15' allows for one or two species of frigatebird, and the possible presence of breeding Great Crested Terns.

j. Includes Great Crested Tern, and Red-tailed Tropicbird from post-1967 sight records (no specimens).

k. Treating Leguat's flying-fox description as Black-spined, *Pteropus niger*.

l. Rodrigues Starling *Necropsar rodericanus*.

Appendix 13. Excavations in 2005 and 2006

This section brings our story up to date, with a brief review of some stunning palaeontological discoveries on the islands in the field seasons of 2005 and 2006.

Mare aux Songes

When George Clark excavated the Mare aux Songes (see Chapter 7), labourers were sent in to the deepest part and felt around with their hands and feet for large bones. Few details were obtained about how the fossils got there, the presence of complete individuals, or how the sediments in which the fossils were found. A second excavation was organised in 1889 under Théodore Sauzier, where smaller fossil remains were recovered and six new but now-extinct Mauritian species were described. A huge number of introduced animals were found as well. Apart from a collection made at the turn of the century by amateur naturalist Paul Carié, whose family owned the estate, the marsh was then left to fall into neglect for the next 100 years.

On 28 October 2005, a Dutch research team was undertaking archaeological studies on Mauritius, looking at human-induced changes to the environment. This involved taking samples of sediment with hollow pipes (cores). By chance, they were given permission to study cores already taken from the Mare aux Songes by Japanese scientists, and within them were fragments of bone. Permission was then given by the Mon Tresor, Mon Desert sugar estate to dig a test-pit. Heavy machinery was necessary to break through a thick layer of hard core put in place by the British to prevent malaria. From the first scoop of the digging machine's bucket a large number of bones were recovered, including Dodos. In June the following year an international team of Dutch, British (including JPH) and Mauritian experts were put together to excavate the marsh on a larger scale. Geologists, geophysicists, palaeobotanists, palaeontologists and DNA experts were assembled to excavate three pits. A sieving machine was constructed with three mesh sizes, for collecting the largest and smallest fossils. The work began with the geophysicist determining the underground structure of the marsh. Cores were then taken to locate the best areas for digging.

It was very quickly realised that the marsh still harboured huge numbers of fossil bones, but what differed from any previous work was that no introduced animals were found. This was a natural accumulation. Each scoop was carefully placed on the sieving machine and all fossil material removed by the scientists. Thousands of bones, egg shell fragments, seeds, branches, leaves, insects, snails and even fungi were recovered. Each specimen was packed and labelled, ready for

A view inside Caverne Bambara. Photo courtesy of Lorna Steel.

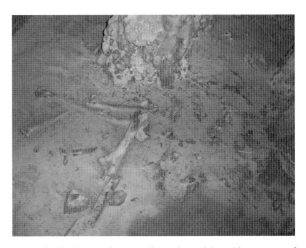

An articulated Rodrigues Solitaire Pezophaps solitaria *skeleton. Photo courtesy of Lorna Steel.*

A bone-packed scoop of marsh sediment from the Mare aux Songes. Photo courtesy of Ranjith Jayasena.

study. After just one month, 4,000 bones and over 5,000 plant remains were recovered. Because the fossils were uncontaminated, a detailed reconstruction of the pre-human habitat of the Dodo can be determined. About 4,000 years ago, the marsh was a small freshwater lake, which had formed inside a vast collapsed lava tunnel. Being set in a dry lowland setting it probably acted as an oasis, supporting a large and diverse community of plants and animals. Tall canopy trees, palms and vacoas grew on its shores, while giant tortoises wallowed in the mud and birds came down to drink and feed. Over millennia, their remains slowly accumulated and the lake became a marsh, finally providing science with an insight into the lost world of the Dodo.

Rodrigues caves

Long before fossil remains of the Dodo were discovered in the 1860s, lime-encrusted Solitaire bones were collected from the caves on the Plaine Corail, Rodrigues in 1786 (see Chapter 6), which were finally illustrated and described in Strickland and Melville's 1848 monograph. However, it was not until Edward Newton showed interest in

The main body of Dodo Fred lies at the base of the cave. Courtesy of Greg Middleton.

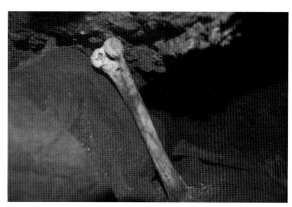

The articulated leg bones of Dodo Fred. Courtesy of Greg Middleton.

the caves that serious excavation took place, and George Jenner, a Rodrigues magistrate, was provided with four men to accomplish the task. This resulted in a haul of over 2,000 bones that Newton and his brother Alfred presented to the Royal Society of London in 1868. Henry Slater, who had accompanied the 1874 Transit of Venus expedition (see Chapter 7), excavated the caves with nine men, and so efficient was Slater's work that almost nothing was left behind. Few caves escaped intact, and as with the Mare aux Songes, little contextual data was obtained.

After the caves were so thoroughly excavated during the 19th century, further work seemed superfluous. It was not until 1974 that another expedition was organised, this time by the British Ornithological Union. Graham Cowles re-examined the caves and found more fossils including new but as-yet undescribed birds. Recently, an excavation explored three pristine caves to obtain contextual data. Entrance to these caves was extremely difficult, hence their survival unscathed by humans. This resulted in the first stratigraphic data and photographic images of complete, articulated fossil species. From the excavation, it was apparent that the build-up of fossil material occurred in two phases. The Plaine Corail has been dated at 80,000 years; as time went by, cracks and fissures appeared due to the action of acidic water eroding the limestone. This allowed the first fossil material to enter the caves, which included snails and individual bones. These were found in the deepest sediments. Over time and probably comparatively recently, the cave entrances grew large enough to become natural pit-fall traps for the Rodrigues fauna. Solitaires, giant tortoises and other animals fell in and their often perfectly preserved remains accumulated on the surface. Further analysis of this material, which includes plants, pollen and invertebrate remains, will provide an opportunity to determine the full extent of human impact on the pristine Rodrigues ecology. Results of this excavation and the Mare aux Songes are expected in 2007/08.

Discovery of a cave Dodo

Two American cave biologists, Fred Stone and Debbie Ward, were looking for Mauritian cave cockroaches in late 2006. They were shown a lava tunnel near Bois Cheri, southeast Mauritius, by Australian cave expert Greg Middleton. Debbie discovered some large bones and photographed them, suspecting they were Dodo bones. These images were sent to JPH, who visited the cave in June 2007 and confirmed the identification. Not only did the bones turn out to be Dodo, they also belonged to a single individual in its position of death, a unique discovery. The Dodo skeleton – affectionately called 'Dodo Fred' – was carefully removed, but many elements had already crumbled. However, these fragments are potentially suitable for DNA studies (unlike the material recovered from the Mare aux Songes), making Fred the most scientifically important Dodo in the world.

Trapped in a cave and too weak to move, Dodo Fred died and his body collapsed into a small crevice, leaving part of the bill and one foot on the surface.

Appendix 14. Species lists for artwork in the colour section

◀ **a) Etang de Saint Paul.** Greater Flamingo *Phoenicopterus roseus* (**1**); Reed Cormorant *Phalacocorax africanus* (**2**); Mascarene Coot *Fulica newtoni* (**3**); Réunion Night Heron *Nycticorax duboisi* (**4**); Pochard *Aythya* cf. *innotata* (**5**); Mascarene Teal *Anas theodori* (**6**); Mascarene Swallow *Phedina borbonica* (**7**); Mascarene Swiftlet *Aerodramus francicus* (**8**); and Dimorphic Egret *Egretta garzetta dimorpha* (**9**).

◀ **b) Ile Frégate, Rodrigues.** Dugong *Dugong dugon* (**1**); Abbott's Booby *Papasula abbotti* (**2**); Red-footed Booby *Sula sula* (**3**); Brown Noddy *Anous stolidus* (**4**); Lesser Frigatebird *Fregata ariel* (**5**).

◀ **c) Mauritian lowland dry forest.** Dodo *Raphus cucullatus* (**1**) beside a critically endangered Tambalacoque *Sideroxylon grandiflorum* (**2**); Mauritius High-backed Tortoise *Cylindraspis triserrata* (**3**) and Mauritius Domed Tortoise *Cylindraspis inepta* (**4**); Burrowing Boa *Bolyeria multocarinata* (**5**); Blue-tailed Daygecko *Phelsuma cepediana* (**6**) on Screw Pine *Pandanus* sp.; Didosaurus *Leiolopisma mauritiana* (**7**) on rock; Telfair's Skink *Leiolopisma telfairi* (**8**); Mascarene Paradise Flycatcher *Terpsiphone bourbonnensis* (**9**); Black Ebony *Diospyros tessellaria* (**10**) and the palm *Hyophorbe amauricalis* (**11**) are endangered; the palm is reduced to just one individual.

◀ **d) The Mare aux Songes.** Réunion Harrier *Circus maillardi* (**1**); Pigeon Hollandais *Alectroenas nitidissima* (**2**); Black-spined Flying-fox *Pteropus niger* (**3**); Mauritius Pink Pigeon *Nesoenas mayeri* (**4**); Commerson's Lizard-owl *Mascarenotus sauzieri* (**5**); Greater Flamingo *Phoenicopterus roseus* (**6**); Mascarene Teal *Anas theodori* (**7**); Mauritius Sheldgoose *Alopechen mauritianus* (**8**); Mauritius Kestrel *Falco punctatus* (**9**); Mauritian High-backed and Domed Tortoises *Cylindraspis triserrata* and *Cylindraspis inepta* (**10**); Reed Cormorant *Phalacrocorax africanus* (**11**); Thirioux's Grey Parrot *Psittacula bensoni* (**12**); Dodo *Raphus cucullatus* (**13**); Mauritius Night Heron *Nycticorax mauritiana* with Burrowing Boa *Bolyeria multocarinata* (**14**); Raven Parrot *Lophopsittacus mauritianus* (**15**); Echo Parakeet *Psittacula echo* (**16**); Mascarene Coot *Fulica newtoni* (**17**); Mauritius Fody *Foudia rubra* (**18**); Didosaurus *Leiolopisma mauritiana* (**19**); Red Hen *Aphanapteryx bonasia* (**20**).

▼ **e) Round Island.** Keel-scaled Boa *Casarea dussumieri* (**1**); Burrowing Boa *Bolyeria multocarinata* (**2**); Günther's Gecko *Phelsuma guentheri* (**3**); Telfair's Skink *Leiolopisma telfairi* (**4**); Bojer's Skink *Gongylomorphus bojerii* (**5**); Mauritian Domed Tortoises *Cylindraspis inepta* (**6**) feed alongside nesting Red-tailed Tropicbirds *Phaethon rubricauda* (**7**).

▼ **f) Seabirds.** Réunion Black Petrel *Pseudobulweria aterrima* (**1**); Barau's Petrel *Pterodroma baraui* (**2**); and Wedge-tailed Shearwater *Puffinus pacificus* (**3**). A Mascarene Free-tailed Bat *Tadarida acetabulosus* (**4**) flies below.

Appendix 15. Local names of Mascarene land vertebrates and turtles

Names cited here are those in use by ordinary non-ornithologists; the local French names are often not those in use in standard French bird books – e.g. the Réunion Black Petrel is simply another *fouquet* locally, rather than the ornithological *Pétrel Noir de Bourbon*. Only those animals with genuine current local names are included (see Cheke 1982 for obsolete names, derivations and historical development). Many local names of native birds in Réunion can be traced back to the 1670s (non-passerines, Hoopoe Starling) and 1750s (passerines). Names in continuous use since 1800 or earlier are asterisked *, with a • used to signify whether Mascarene French usage is unique or unusual. Introduced species appear in italics. Only surviving species (or those still known in captivity) are included. Créole orthography follows Baker & Hookoomsing (1987), except that the nasaliser 'ñ' is used instead of 'n'.

English name	Mauritius French	Mauritius Créole	Réunion French	Réunion Créole	Rodrigues Créole
Common Tenrec	tangue *•	tañg [1]	tangue *•	tañg [1]	
House Shrew	rat musqué *•	lera miske	rat musqué *•	rat miske	lera miske
Flying-foxes	chauve-souris *	susuri [2]	chauve-souris *	sovsuri	susuri [2]
Microbats	chauve-souris banane	susuri banan	chauve-souris banane	sovsuri banan	
Crab-eating Macaque	jacot *•	zako [3]			
Black-naped Hare	lièvre *	yev	lièvre	lyev	
feral pig	cochon marron	kosoñ maroñ [4]			
Rusa Deer	cerf *	serf	cerf	serf	
Small Indian Mongoose	mangouste	mañgus			
feral cat	chat marron	sat maroñ [4]	chat haret	sat maroñ [4]	
rats	rat *	lera	rat *	ra	lera
House Mouse	souris *	suri	souris *	suri	suri
Wedge-tailed Shearwater	*fouquet* *•	*fuke* [5]	*fouquet* *•	*fuke* [5]	*fuke* [5]
Tropical Shearwater			fouquet noir	fuke (nwar) [5]	
'Round Island' Petrel	fouquet *•	fuke			
Barau's Petrel			taille-vent *•	tayvoñ [6]	
Réunion Black Petrel			fouquet	timize [7]	
Wilson's Petrel				tipolka *•	
tropicbirds	paille-en-queue *	payañke	paille-en-queue *	payañke	payañke
Red-footed Booby	fou *	fu [8]			
Masked Booby	boeuf *•	bef [8]			
frigatebirds	frégate *	fregat	frégate *		
Striated Heron	gasse •	gas	butor	bitor	begas •[9]
Meller's Duck/Mallard	canard sauvage	kanar sovaz			
White-faced Whistling Duck	sarcelle	sarsel			
Réunion Harrier			papangue *•	papang, pye zon [10]	
Mauritius Kestrel	mangeur de poule *•	manzer (d') pul [11]			
Grey Francolin	perdrix *	perdri	perdrix *	perdri	perdri
Malagasy Partridge			caille malgache	frañkoleñ, kay malgas [12]	
Common Quail			caille patate	kay bwad'patat [12]	
Painted Quail			caille de Chine	kay boyo ruz [12]	
Jungle Bush Quail			caille de l'Inde	kay zabo ruz [12]	
Helmeted Guineafowl	pintade *	peñtad	pintade *	peñtad	peñtad
Red Jungle Fowl			coq de bruyère	kok d'bruyer	
southern skua sp.	poule mauve *•	pulmov [13]		asasin	
Roseate Tern					goloñ •[14]
Sooty Tern	goëlette *•[14]	golet, yeye [14]			golet [14]
Fairy Tern					golet (blañ) *[14]
Brown Noddy	macoua *•	makwa	macoua *•	makwa	malan [15]
Lesser Noddy	marianne	maryan, kordonye *•			malan [16]
Whimbrel	corbijeau *•	korbizo [17]	courlis	kurli	korbizo [17]
small waders	alouette [18]	zalwet			zalwet, kordonye
Common Moorhen	poule d'eau *	puldo	poule d'eau *	puldo	
Madagascar Buttonquail			caille pays	kay tañbur, kay kravat	
Feral Pigeon	pigeon *	pizoñ	pigeon *	pizoñ	pizoñ
Mauritius Pink Pigeon	pigeon des Mares	pizoñ demar			
Malagasy Turtle Dove	(pigeon) ramier *	(pizoñ) ramye	touterelle malgache	turtrel malgas	
Spotted Dove	(grosse) tourterelle	(gro) tutrel			
Zebra Dove	(petite) tourterelle *	(ti) tutrel	tourterelle (pays)	tutrel (pei)	tutrel
Echo Parakeet	grosse cateau *•	kato, katover			

Appendix 15 (Cont.):

	English name	Mauritius French	Mauritius Créole	Réunion French	Réunion Créole	Rodrigues Créole
Birds cont.	Ring-necked Parakeet	petite cateau	kato, katover [19]			
	Grey-headed Lovebird	perruche *	peris	perruche *	peris	peris
	Mascarene Swiftlet	petite hirondelle *	(ti) zirondel	petite hirondelle *	(ti) zirondel	
	Mascarene Swallow	grosse hirondelle *	(gro) zirondel	grosse hirondelle *	(gro) zirondel	
	Réunion Cuckoo-shrike			tuit-tuit *•, merle blanc	titwit [20]	
	Mauritius Cuckoo-shrike	merle cuisinier	merl kwizinye			
	Réunion Bulbul (Merle)			merle (noir) *	merl (pei)	
	Mauritius Bulbul (Merle)	merle (noir) *	merl (pei)			
	Red-whiskered Bulbul	oiseau condé	(zwazo) konde	merle de Maurice	merl moris	
	Peking Robin				rosinyol pei	
	Mascarene Paradise Flycatcher	coq de bois [21]	kokdebwa	oiseau de la vierge [21]	sakwat, tyakwat	
	Réunion Stonechat			tectec *•	tektek [20]	
	Rodrigues Warbler					(zwazo) loñbek
	Mascarene Grey White-eye	oiseau manioc	(zwazo) pikpik	oiseau blanc	zwazo blañ	
	Réunion Olive White-eye			oiseau vert *	zwazo ver	
	Mauritius Olive White-eye	oiseau lunettes				
	Yellow-fronted Canary	serin du pays	sereñ dipei	serin	tisereñ	sereñ
	Cape Canary			moutardier	mutardye	
	Common Waxbill	bengali	beñgali	bec rose	bekroz	beñgali
	Red Avadavat	bengali * moucheté	beñgali muste	petit coq	tikok	
	Spice Finch	pingo	pingo [22]	coutil	tulit, titulit	
	Java Sparrow	calfat *•	kalfat, galfat [23]	calfat	kalfat, kalfa [23]	
	Pin-tailed Whydah			veuve		
	Village Weaver	serin du Cap	sereñ dikap [24]	oiseau Bellier	(zwazo) belye	
	Mauritius Fody	oiseau banane	zwazo banan			
	Rodrigues Fody					sereñ zon
	Cardinal Fody	cardinal *•	kardinal	cardinal *•	kardinal, mal ruz	sereñ (ruz)
	House Sparrow	moineau	mwano	moineau	mwano	mwano
	Common Myna	martin *	marteñ [23]	martin *	marteñ [23]	marteñ [23]
	(Indian) House Crow	corbeau	korbo	corbeau		
Reptiles and amphibians	Tortoises and turtles (all)	tortue *	torti	tortue *	torti	torti
	Bloodsucker	caméléon	kamaleoñ	caméléon	kamaleoñ	kamaleoñ
	Panther Chameleon			endormi	leñdormi, krokodil	
	Lizards (general)	lézard *	lezar [24]	lézard *	lezar [24]	lezar [24]
	Day-geckos (general)	lézard vert	lezar ver	lézard vert	lezar ver	
	House-geckos (general)	lézard, margouillat	lezar (gri)	lézard, margouillat	lezar (gri)	lezar (gri)
	Snakes (general)	serpent *	serpañ	serpent *	serpañ	
	Wolf Snake	couleuvre	kulev	couleuvre	kulev	
	Malagasy Grass Frog	grenouille	grenwi	grenouille	grenwi	
	Guttural Toad	crapaud	krapo	crapaud	krapo	

Mostly from Cheke (1982a), supplemented by Barré & Barau (1982) and Probst (1997), and checked with Baker & Hookoomsing (1987) and Chaudenson (1974). Mascarene French bird names, and their Créole derivatives, often differ from those in use in France, usually reflecting usage by mariners and colonists in the 17th and 18th centuries now long obsolete in metropolitan France. The more interesting are commented on below.

1. *Tangue* is a contraction of Malagasy *tandrak*; English 'tenrec' is likewise from the Malagasy.

2. The 18th century flying-fox names *rousette* (*Pteropus niger*) and *rougette* (*P. subniger*) have long vanished, though the former was adopted into standard French for fruitbats generally, and we have revived the latter in this book for *P. subniger*.

3. *Jacot* possibly derives from Yoruba *jako*, imported by early West African slaves (Baker & Hookoomsing 1987).

4. *Marron* originally meant an escaped slave, but came to mean 'wild' or 'feral' in both local French and Créole.

5. *Fouquet* (little *fou*, see below) is not used today in standard French, but it was widely used by travellers in the 18th century.

6. *Taille-vent* means 'wind-cutter', akin to the English term 'shearwater'.

7. The etymology of *timize* is not known (Thomas Ghestemme, pers. comm.).

8. Abbott's Booby (now locally extinct) was named *boeuf* for its cattle-like call, the name transferring later to the similar Masked Booby (which does not have the appropriate call). Boobies generally are known as *fou* (= idiot-bird).

9. *Butor* is bittern *Botaurus* spp. in standard French; juvenile Striated Herons have similar streaky brown plumage. *Gasse* appears to be a contraction of *bégasse* (*begas* in Créole), itself a variant of *bécasse*, French for Woodcock *Scolopax rusticola* and (as *bécasse de mer*) Oystercatcher *Haematopus ostralegus* – wading birds with long beaks, like the heron.

10. *Papangue* is borrowed from the Malagasy *papango* for the Black Kite *Milvus migrans*, so presumably given originally to the harrier by Malagasy slaves. Harriers in France are called *busard*.

11. *Mangeur de poule* ('chicken-eater') is, or was, a common generic French folk-name for birds of prey, and does not particularly imply that the Mauritius Kestrel was a major predator of domestic fowl, though it acquired the reputation by dint of the name!

12. Quails and partridges have numerous local names in Réunion; only the most common are quoted here.

13. *Mauve* is a gull-name, long obsolete in France, cognate with *mew* in English and *meeuwe* in Dutch/German. The Réunion name, from the French *assassin*, means 'murderer'.

14. *Goëland* is French for gull, used originally in the Mascarenes for Great Crested Terns, but shifting to Roseate Terns when the larger species ceased to breed. *Goëlette*, its diminutive, survives in metropolitan French only as a name for a schooner, a type of sailing boat. Although *golet* is the usual Créole term in Mauritius itself, Mauritian fishermen stationed at St Brandon call Sooty Terns the onomatopoeic *yeye*.

15. *Macoua* is the name of a tribe from Mozambique, whose skin colour the noddy was thought to resemble.

16. With no origin known, Cheke (1982) proposed that, as many islets are named after birds, someone found these noddies on Isle Marianne (there is one each in Mauritius and Rodrigues) and assumed the birds were '*mariannes*'. This name, of Mascarene origin (as *noddi marianne*) has been adopted as the standard metropolitan French for the species. *Kordonye* (from French *cordonnier*, 'bootmaker') is probably a reference to the birds' long, narrow, pointed bill.

17. *Corbijeau* is old dialect French for curlew, obsolete in France.

18. In the absence of larks, the French *alouette de mer* has contracted to just '*alouette*'.

19. In the 18th century *cateau* or *catau* was dialect French for parrots, cognate with the current *kakatoës* (i.e. without the first syllable) and English 'cockatoo'.

20. These names are both onomatopoeic. The name *tectec*, cited in Brisson (1760), was adopted by Gmelin as the specific name (mispelt as *tectes*) for the Réunion Stonechat.

21. Both French names allude to the flycatcher being the first bird heard at dawn.

22. *Pingo* is unusual in Créole as an animal name of Indian origin, with a Sanskrit root cognate with the English word 'finch'.

23. *Calfat* and *martin* are now standard French names, but they originated in the Mascarenes and were popularised by Buffon (1770-83). *Calfat* means 'caulker', and is onomatopoeic of the tapping of a caulker's hammer in a shipyard; *martin* is probably from the Portuguese *martinho*.

24. Adrien Bellier was the man said to have introduced the Village Weaver to Réunion.

25. While several kinds of native lizards have Créole names in the Seychelles (Cheke 1984), names are more limited in the Mascarenes, possibly because there are no longer any mainland skinks. Bloodsucker males can change colour quite spectacularly, hence perhaps the use of chameleon names for it, while the real chameleon in Réunion is called *endormi*, 'asleep', i.e. 'dozy'.

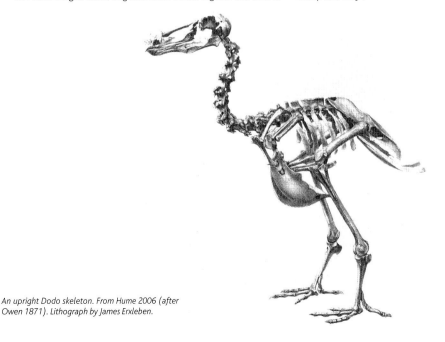

An upright Dodo skeleton. From Hume 2006 (after Owen 1871). Lithograph by James Erxleben.

BIBLIOGRAPHY

This bibliography follows standard practice in most respects, but a few special conventions should be noted:

1. Where a book or paper has more than three authors, the first author only is given, followed by *et al.*

2. French names of the form 'de X' & 'd'X' appear under 'X', and 'de la X' under 'La X'. Dutch names of the form 'van W', 'van der W' or 'den W' appear under 'W'. Titles (e.g. *Père*, *Abbé*, *Lord*) are given in italics to distinguish them from first-names (which are sometimes not known).

3. Special abbreviations 1: The Réunion journal *Receuil trimestriel de documents et travaux inédits pour servir à l'histoire des Mascareignes françaises* is abbreviated to *Rec. Trim.*, and the *Transactions/Proceedings of the Royal Society of Arts and Sciences of Mauritius* to *TRSAS* and *PRSAS* (as also in endnotes). The full title of the local Réunion bird newsletter shortened below to *Taille-vent* is *Taille-vent. Bulletin Trimestriel de la Société d'Études Ornithlogiques de la Réunion*, and the forestry journal cited as *Bois For. Trop.* is *Bois et Forêts des Tropiques. Revue Trimestriel du Centre Technique Forestier Tropicale*. The regrettably discontinued Jersey Zoo/Durrell Wildlife Conservation Trust journal has had various extended titles over the years; here all are listed simply as *Dodo*. The towns of Port Louis and Saint-Denis cited in the references are respectively the capitals of Mauritius and Réunion; other Mascarene towns of publication have their island added.

4. Special abbreviations 2: The *Dictionary of Mauritian biography* (Toussaint *et al.* 1941-) is cited as *DMB* + page number(s) in endnote references, followed by date & authors of the individual biographical article (in brackets where appropriate). Pagination is continuous through the issues, as it is with the *Dictionnaire Toponymique de l'Ile Maurice* (Goldsmith *et al.* 1997-), cited as *DTIM*, followed by page number and date issued. Similarly endnote references to information in proceedings of meetings or lists in the *TRSAS* or *PRSAS* and the *Mauritius Almanac* is given as *TRSAS/PRSAS* and *MA* respectively, followed by volume (*TRSAS/PRSAS* only), date and pages, but proper articles are fully referenced in the bibliography. The society itself is abbreviated in notes to 'RSAS'. Rather than cite dozens of separate parts (family monographs) of the *Flore des Mascareignes* (Bosser *et al.* 1978-), citations are usually given as *Flore* + issue and page number, followed by date and author in brackets. References to the *Dictionnaire biographique de la Réunion* (Verguin & Serviable 1993-8) are given as *DBR* followed by volume number and pages, and the *Journal de la Réunion* is cited as *JR*, followed by publication date. Finally the British *Concise dictionary of National Biography* (anon. 1992) is abbreviated in endnotes to *CDNB* followed by volume and page numbers.

5. Standard acronyms used in the bibliography and endnotes are expanded as follows:
BMNH – see NHM
BOU – British Ornithologists' Union.
CABI – Commonwealth Agricultural Bureaux International [UK]
CIPO – Conseil International pour la Protection des Oiseaux [=ICBP].
CSIRO – Commonwealth Science & Industry Research Organisation [Australia]
DDASS – Direction Départementale de l'Action Sanitaire et Sociale.
DIREN – Direction Régionale de l'Environnement.
FAO – Food & Agriculture Organisation (of the United Nations).
FFPS – Fauna & Flora Preservation Society.
GISP – Global Invasive Species Programme.
HMSO – Her Majesty's Stationery Office.
ICIPE – International Centre of Insect Physiology and Ecology [Nairobi].
ICBP – International Council for Bird Preservation/Protection [now Birdlife International].
INRA – Institut National de la Recherche Agronomique.
INSEE – Institut National de la Statistique et des Études Économiques.
IUCN – International Union for the Conservation of Nature.
MWF/MWAF – Mauritian Wildlife Foundation/Appeal Fund.
MNHN – Muséum Nationale d'Histoire Naturelle, Paris
NPCS – National Parks & Conservation Service [Mauritius].
NHM – Natural History Museum, London & Tring, UK [formerly British Museum (Natural History)].
ONF – Office National des Forêts.
ORSTOM – Office de la Recherche Scientifique et Technique Outre-Mer.
SEOR – Société d'Etudes Ornithologiques de la Réunion.
SREPN/SREPEN – Société Réunionnaise pour l'Étude et la Protection de la Nature/l'Environnement.
UNEP – United Nations Environment Programme.
WWGBP – World Working Group on Birds of Prey.

6. We have where possible given standard printed references, but some material is only available on-line, in which case we have given the URL (internet address), usually of the site's home page. As internet material is less stable than printed matter, it is possible that some such references cited may not be available when sought later.

7. All literature cited has been seen by the authors, except for a

few marked '*', which are cited on the authority of others and considered important enough to retain in bibliography.

8. Although extensive, this is far from being a complete bibliography of Mascarene vertebrate biology. Those seeking further references should consult Hachisuka (1953), Diamond (1987) and Jones & Hartley (1995).

Abercromby, Ralph. 1888. *Seas and skies in many latitudes – or – wanderings in search of weather*. London: Edward Stanford. 447pp.

Abbot, Ian. 2002. Origin & spread of the cat *Felis catus* on mainland Australia, with a discussion of the magnitude of its early impact on native fauna. *CSIRO Wildl. Res.* 29: 51–74.

Abbott, Kirsti L. 2006. Spatial dynamics of supercolonies of the invasive yellow crazy ant, *Anoplolepis gracilipes*, on Christmas Island, Indian Ocean. *Diversity Distrib.* 12: 101–110.

Abhaya, Keshava & Probst, Jean-Michel. 1995. Plaidoyer pour l'estuaire de Terre Rouge, un lieu d'observation unique de limicoles et des oiseaux d'eaux dans les Mascareignes (Ile Maurice). *Bull. Phaethon* 2: 107.

— & — 1997a. La falaise du littoral ou "Route en corniche", un site de nidification de première importance pour le paille-en-queue à brins blancs. *Bull. Phaethon* 6: 107.

— & — 1997b. Lists commentée des oiseaux et de quelques vertébrés observés à l'Ile Rodrigues du 12 au 24 janvier 1996. *Bull. Phaethon* 5: 25–34.

— 2001. Note sur le régime alimentaire du Merle pays *Hypsipetes borbonica*, oiseau endémique de l'île de la Réunion. *Bull. Phaethon* 13: 53.

Adams, Douglas & Carwardine, Mark. 1990. *Last chance to see...* London: Heinemann. 208pp.

Adams, Percy G. 1962. *Travelers and travel liars*. Berkeley: Univ. California Press. 292pp.

— 1983. *Travel literature and the evolution of the novel*. Lexington: University Press of Kentucky. 368pp.

Addison, John & Hazareesingh, Kissonsingh, 1993. *A new history of Mauritius*. Rev.ed. Rose-Hill, Mauritius: Editions de l'Océan Indien. 122pp.

Adler, Gregory H. 1994. Avifaunal diversity and endemism on tropical Indian Ocean islands. *J. Biogeog.* 21: 85–95.

Aellen, V. 1957. Les chauves-souris africains du Musée Zoologique de Strasbourg. *Rev. Suisse Zool.* 64(6): 189–214.

Aglosse, P.d'.[pseud.] 1891. Le prétendu relation du P.Brown. *Rev. Hist. Litt. Ile Maurice* 5: 337–340.

Ah-King, J. 2000. Monitoring regeneration of native vegetation in two managed areas in upland Mauritian forest. B.Sc. (Hons.) thesis. University of Mauritius.

Alagona, Peter S. 2004. Biography of a "feathered pig": the California Condor conservation controversy. *J. Hist. Biol.* 37: 557–583

Albany, Jean. 1974. *P'tit glossaire. Le piment des mots créoles*. Paris [author]. 116pp. [reprinted 2001 – Saint-Denis: Hi-Land].

Alby, Jean (ed.). 1989. *Antoine Boucher. Mémoire pour servir à la connaissance particulière de chacun des habitants de l'isle de Bourbon, suivi des notes du Père Barassin*. Sainte-Clotilhde, Réunion: Ars Terres Créoles. 335pp.

— & Serviable, Mario (eds.) 1993. *Abbé Alexandre-Gui Pingré: Courser Venus, voyage scientifique à Rodrigue en 1761*. Ste. Clothilde, Réunion: Ars Terres Créoles & Rose-Hill, Mauritius: Editions Ocean Indien. 123pp. [extracts from Pingré 1763, q.v., with a historical introduction]

Alexander, J.B. Duck, C.D. & Gardner, Andrew S. 1978. *Edinburgh University Expedition to Round Island 1978*. Edinburgh: [authors]. 9pp + maps; photocopied.

Alexander, H.G.L. 1979. A preliminary assessment of the role the terrestrial decapod crustaceans in the Aldabra ecosystem. *Phil. Trans. Roy. Soc. Lond.* B 286: 241–246. [*in* Stoddart 1979, q.v.]

Ali, Salim. 1927. The Moghul emperors of India as naturalists and sportsmen. Part 1. *J. Bombay Nat. Hist. Soc.* 31: 833-61.

— 1968. Dodo. Pp.15–17 *in* Alvi & Rahman (1968), q.v.

— & Ripley S.Dillon. 1968–74. *Handbook of the birds of India and Pakistan*. Bombay: Oxford University Press. 10 vols. [also 2nd.ed., 1978-, Delhi: OUP, in progress].

Allan, W. & Edgerley, Leo F. 1950. *White paper on Crown Forest Land (land utilization) and Forestry*. Port Louis: Govt.Printer. 37pp.

Allibert, Claude. 1988. Les contacts entre l'Arabie, le Golfe Persique, l'Afrique orientale et Madagascar: confrontation des documents écrits, des traditions orales et des données archéologiques récentes. Pp.111–126 *in* J.F.Salles (ed.) *L'Arabie et ses mers bordières: itinéraires et voisinages*. Lyon: Maison de l'Orient. 199pp.

Alvarez, Yolands *et al*. 1999. Molecular phylogeny and morphological homoplasy in fruitbats. *Mol. Biol. Evol.* 16: 1061–1067.

Alvi, M.A. & Rahman, A. (eds.) 1968 *Jahangir the naturalist*. New Delhi: National Institute of Science of India.

Amédée, Christy. 2001. Définition du braconnage; Aspects sociologiques du braconnage. Pp. 6–7, 9 *in* Le Corre (2001), q.v.

Amin, Mohamed, Willetts, Duncan & Skerrett, Adrian. 1995. *Aldabra – World Heritage Site*. Nairobi: Camerapix Publishers International (for Seychelles Islands Foundation). 189pp.

Amiot, Christophe *et al*. 2007. Rapid morphological divergence of introduced Red-whiskered Bulbuls *Pycnonotus jocosus* in contrasting environments. *Ibis* 149: 482–489.

Andersen, Knud. 1907. Some remarks on *Pteropus mascarinus*. *Ann. Mag. Nat. Hist.* (7)20: 351–355.

— 1912. *Catalogue of the Chiroptera in the collection of the British Museum. Vol.1. Megachiroptera*. 2nd.ed. London: Trustees of the British Museum. 854pp.

— 1913. A subfossil bat's skull from Rodrigues I. *Rec. Ind. Mus.* 9: 337.

Anderson, Atholl. 1990. *Prodigious birds – Moas and Moa-hunting in New Zealand*. Cambridge: Cambridge University Press. 256pp.

Anderson, Charles. 1838. Report on apprentices in the islands dependent on Mauritius. MS in UK Public Record Office (CO167 vol.204, No.105) & [copy] in Mauritius Archives (Y/3/2).

Anderson, Daniel R. 1918. *The epidemics of Mauritius*. London: H.K.Lewis & Co. 312pp.

Anderson, John. 1881. *Catalogue of the Mammalia in the Indian Museum, Calcutta*. Calcutta: Trustees of the Indian Museum. 2vols.

Andreone, Franco & Greer, Allen E. 2002. Malagasy scincid lizards: descriptions of nine new species with notes on the morphology, reproduction and taxonomy of some previously described species (Reptilia: Scincidae). *J. Zool., Lond.* 258:

139–181.

— *et al.* 2003. The amphibians and reptiles of Nosy Be (N.W.Madagascar) and nearby islands: a case study of diversity and conservation of an insular fauna. *J. Nat. Hist.* 37: 2119–2149.

*Andrews, C.W. 1900. *A monograph of Christmas Island (Indian Ocean)*. London: British Museum (Natural History). 321pp.

anon. 1599. *A true report of the gainefull, prosperous and speedy voiage to Iava in the East Indies performed by a fleet of eight ships from Amsterdam*. London: W.Apsley [reprinted in Keuning 1938–51, q.v., vol.2]

— 1601. *Het Tweede Boeck. Journael oft Dagh-register inhoudende een warachtig verhael ende historische vertellinghe van de Reyse geaen door de acht schepen van Amstelredame...* Amsterdam: Cornelis Claesz. [reprinted 1942 in vol.3. of Keuning (1938–51), q.v.]

— 1763. An account of the island of Bourbon in 1763. [pub. 1794 in] *Oriental Repertory* 2(2): 201–212. [ex MS in Orme Collection, British Library (ex India Office); also *in* Grant (1801, q.v., as by an "officer of the British Navy") who 'improved' the grammar & slightly altered the text content. Hill (1916, q.v.) suggested it was written by Brigadier-General Richard Smith].

— 1826. Verzeichnis ausländischer Thiere welche in dem Königlichen Garten zu Nymphenburg. Pamphlet, 4pp. [copy in the Bavarian State Archives, Munich]

— ['Bengal Civilian' ?=E.O.Irwin] 1838. *Journal of five months residence in the Mauritius*. Calcutta: Samuel Smith.

— [? P.B.Ayres] 1858. Sketches in Mauritius. *Mauritius Almanac & Official Directory* 1858: cxxx-cxliii.

— n.d. [1879]. *The pictorial tour of the world, containing pen and pencil sketches of travel, incident, adventure and scenery in all parts of the globe*. London: James Sangster & Co. 507pp. [Mauritius pp.228–232].

— 1893. [obituary] H.Whiteley. *Ibis* (6)5: 287–288.

— [? Alfred Newton] 1897. Obituary. Sir Edward Newton. *Ibis* (7)3: 475–479.

— 1908. [biographies of the] Original Members [of the British Ornithologists' Union] Sir Alfred Newton, Sir Edward Newton. *Ibis, Jubilee Supp.*: 107–116, 117–120.

— 1963. Mauritius in 1838. *Roy. Commonwealth. Soc. Libr. Notes* NS 82: 1–3 [for traveller's identity as Thomas Wise see *DMB*: 986 (1971, P.J.Barnwell)]

— ('P.H.G.') 1941. After wild deer in Mauritius. *Field* 177: 676–677.

— 1971a. *4-year Plan for social and economic development*. Port Louis: Government Printer. 2 vols.

— 1971b. La Réunion. *Marchés Trop. & Mediterr.* 27: 3635–3656.

— 1973. Caméléon de l'Ile Bourbon. *Info-Nature, Ile Réunion* 10: 66–72.

— 1974. Fiche sur la restauration de la forêt réunionnaise. Saint-Denis: Office National des Forêts. Cyclostyled, 4pp.

— 1976. *Mauritius 1975–1980. 5-year Plan for social and economic development*. Port Louis: Govt.Printer. 200pp.

— 1981a. Compte-rendu de l'assemblée générale de la SREPEN le 13 juin 1980 au Lycée Agricole de Saint-Joseph. *Info-Nature, Ile Réunion* 18: 3–7.

— 1981b. Reflexions pour la création des reserves. Document SREPEN. *Info-Nature, Ile Réunion* 18: 83–87.

— 1981c [=N.H.Vink] Agricultural development in Rodrigues. Cyclostyled Report for the European Development Fund [no place of publication given]. 94pp.

— 1983a. *Government of Mauritius Wildlife Research and Conservation Programme. Phase I 1984–1985. Initiation and integration*. Cambridge: ICBP. 45pp.

— 1983b. *The Wildlife Act 1983. Act. No.33 of 1983*. Port Louis: Govt. Printer. 12pp.

— 1985. *1984–1986 Development Plan*. Port Louis: Ministry of Economic Planning & Development. 252pp.

— 1992. *The concise dictionary of national biography. From earliest times to 1985*. Oxford: Oxford University Press. 3334pp, continuously paginated through 3 vols.

— 1993. Republic of Mauritius, Port of Port Louis: Port extension and free port development project. Consulting services for environmental impact statement. Washington, DC: World Bank. 9pp.

— n.d. [c.1994]. *Ile aux Aigrettes*. Port Louis: Mauritian Wildlife Appeal Fund. 12pp.

— 1996. *Enquêtes sur les pollutions et degradations de la zone cotière de Maurice*. Quatre Bornes, Mauritius: Commission de l'Océan Indien. 45pp. [despite title, this report is in English]

— 2000. *Republic of Mauritius. First national report to the Convention on Biological Diversity*. Réduit, Mauritius: National Parks & Conservation Service. 57pp.

— 2002a. *Ile de la Réunion. Gestion durable des milieux naturels forestiers et developpement stratégique des entreprises locales du bois*. Saint-Denis: ONF.

— 2002b. Rapport de presentation. Aménagement de Bélouve (3418ha). Révision d'aménagement 2002–2016. Saint-Denis. Réunion: ONF. 5pp.

— 2002c. *Annual Report of the Forestry Service for the year 1998*. Port Louis: Govt. Printer (No.11 of 2002).

— 2003a. *Les Hauts de la Réunion. Projet Parc National. Principes pour un parc national de nouvelle génération. 1. Démarche. 2. Propositions* [3.= Girard & Notter (2003), q.v.] *+ Résumé*. Saint-Denis: Mission de Création du Parc National. 4 vols.

— 2003b. Round Island management plan 2003–2007. Mauritius: MWF & NPCS. 19pp + appendices.

— 2004. *Government of Mauritius. Annual digest of statistics 2003*. [on line at www.statsmauritius.gov.mu].

Antelme, Henri. 1914. Hunting. Pp.169–173 *in* Macmillan (1914), q.v.

*Anthony, J. & Guibé, J. 1951. Les affinités anatomiques de *Bolyeria* et *Casarea* (Ophidiens). *Mem. Inst. Sci. Madagascar* A 7(2): 189–201.

Antoine, E. 1973. La forêt tropical à la Réunion. *Info-Nature, Ile Réunion* No. Spécial hors série 'La Forêt' : 39–47.

Antoine, Robert. 1988. Reginald E. Vaughan, OBE DSc FRIC (1895–1987). *PRSAS* 5(1,2): 5–6.

Après de Mannevillette, J.B.N.D.d'. 1775. *Le Neptune Orientale*. Paris: Demonville. 194 cols.(2/page) + 61 charts. [Rodrigues col.19–20]

Armstrong, Edward A. 1953. Territory and birds. A concept which originated from study of an extinct species. *Discovery* July 1953: 223–224.

Arnold, E. Nicholas. 1979. Indian Ocean giant tortoises: their systematics and island adaptations. *Phil. Trans. Roy. Soc. Lond.* B 286: 127–145. [& *in* Stoddart 1979, q.v.]

— 1980. Recently extinct reptile populations from Mauritius and

Réunion, Indian Ocean. *J. Zool., Lond.* 191: 33–47.

— & Jones Carl.G. 1994. The night geckos of the genus *Nactus* in the Mascarene Islands, with a description of the distinctive population on Round Island. *Dodo* 30: 119–131.

— 2000. Using fossils and phylogenies to understand evolution of reptile communities on islands. *Bonn. Zool. Monogr.* 46: 309–323.

— 2004. Are islands a gateway to extinctions? [video of talk given at the NHM, London on 29/3/2004, on line at www.nhm.ac.uk under 'Darwin Centre Live']

*Arnold, E. Nicholas & Bour, Roger. [in prep.] Recent fossil lizards from Réunion, Indian Ocean, with description of two new species.

—, Austin, Jeremy & Jones, Carl G. [in prep.] a. Using mtDNA to explore a devastated island radiation: the *Nactus* geckos (Reptilia: Gekkonidae) of the Mascarene Islands. ?? [early draft only seen by ASC]

— & — [in prep.] b. Interrelationships and history of the slit-eared skinks (Scincidae: *Gongylomorphus*) of the Mascarene Islands, Indian Ocean, based on mtDNA.

Arquetout, F. 1972. "Bebête Tou't". *Info-Nature, Ile Réunion* 6: 25.

Ashcroft, Ruth [E]. & Cheke, Anthony [S]. 2007. Red-tailed Tropic-bird (*Phaethon rubricauda*) numbers on Round Island in 2003. *In* Bullock & North (2007), q.v.

Ashmole, [N.] Philip & Ashmole, Myrtle. 1989. *Natural history excursions in Tenerife. A guide to the countryside, plants and animals*. Peebles, Scotland: Kidston Mill Press. 252pp.

—, — & Simmons Ken[neth] E.L. 1994. Seabird conservation and feral cats on Ascension Island, South Atlantic. Pp.94–121 *in* Nettleship *et al*. (1994), q.v.

— & — 2000. *St. Helena and Ascension Island: a natural history*. Oswestry (UK): Anthony Nelson. 475pp.

Asma, Stephen T. 2001. *Stuffed animals and pickled heads. The culture and evolution of natural history museums*. New York: Oxford University Press. 302pp.

Atkinson, Geoffrey. 1922. *The extraordinary voyage in French literature. Vol.2. From 1700 to 1720*. Paris: Champion [reprinted 1969, New York: Burt Franklin]

Atkinson, Ian A.E. 1985. The spread of commensal species of *Rattus* to oceanic islands and their effects on island avifaunas. Pp.35–81 *in* Moors (1985), q.v.

— 1990. Ecological restoration on islands: Prerequisites for success. Pp.73–90. in Towns *et al*. (1990), q.v.

— 1996. Introductions of wildlife as a cause of species extinctions. *Wildl. Biol.* 2, 135–141.

— 2000. Introduced mammals and models for restoration. *Biol. Conserv.* 99: 81–96.

— & Atkinson, Toni J. 2000. Land vertebrates as invasive species on the islands of the South Pacific. Pp.19–84 *in* Sherley (2000), q.v.

Atkinson, Rachel & Sevathian, Jean-Claude. 2005. *A guide to the plants in Mauritius*. Vacoas, Mauritius: MWF. 188pp.

*Attié, M. 1994. Impact du cerf de Java *Cervus timorensis russa* à la Plaine des Chicots, et proposition de restauration du milieu. Le Tampon, Réunion: ONF.

Attié, Carole, Stahl, J.C. & Bretagnolle, Vincent. 1997. New data on the endangered Mascarene Petrel *Pseudobulweria aterrima*: a third twentieth century specimen and distribution. *Colonial waterbirds* 20: 406–412.

Austin, Jeremy J. & Arnold, E. Nicholas. 2001. Ancient mitochondrial DNA and morphology elucidate an extinct island radiation of Indian Ocean giant tortoises (*Cylindraspis*). *Proc. Roy. Soc. Lond.* B 268: 2515–2523.

— . & Arnold, E.Nicholas. 2006. Using ancient and recent DNA to explore relationships of extinct and endangered *Leiolopisma* skinks (Reptilia: Scincidae) in the Mascarene Islands. *Molec. Phylogen. Evol.* 39: 503–511.

—, Arnold, E. Nicholas & Bour, Roger. 2002. The provenance of type specimens of extinct Mascarene Island giant tortoises (*Cylindraspis*) revealed by ancient mitochondrial DNA sequences. *J. Herpetol.* 36: 280-285

—, Arnold, E. Nicholas & Bour, Roger. 2003. Was there a second radiation of giant tortoises in the Indian Ocean? Using mitochondrial DNA to investigate speciation and biogeography of *Aldabrachelys* (Reptilia: Testudinidae). *Molec. Ecol.* 12: 1415–1424.

—, Arnold, E. Nicholas & Jones, Carl G. 2004a. Reconstructing an island radiation using ancient and recent DNA: the extinct and living day-geckos (*Phelsuma*) of the Mascarene Islands. *Molec. Phylogen. & Evol.* 31: 109–122.

—, Bretagnolle, Vincent & Pasquet, Eric. 2004b. A global molecular phylogeny of the small *Puffinus* shearwaters and implications for systematics of the Little-Audubon's Shearwater complex. *Auk* 121: 847–874.

Avise, James C. *et al*.. 1992. Mitochondrial DNA evolution at a turtle's pace: evidence for low genetic variability and reduced microevolutionary rate in the Testudines. *Molec. Biol. Evol.* 9: 457–473.

Ayres, Philip B. 1860. On the geology of Flat and Gabriel Islands. *TRSAS* NS 1: 220–232.

Azéma, Henri. 1926. *Histoire de la ville de Saint-Denis de 1815 à 1870*. Paris: Aristide Quillet [reprinted (n.d., c.1998), with 4 other short texts by Azéma, Saint-Denis: Grand Océan, 313pp.]

Backhouse, James. 1844. *A narrative of a visit to the Mauritius and South Africa*. London: Hamilton Adams & Co. lvi+648pp.

Baider, Claudia & Florens, Vincent. 2006. Current decline of the 'Dodo-tree': a case of broken down interactions with extinct species, or the result of new interactions with alien invaders. Chapter 11 *in* W.Laurance & C.Peres (eds.) *Emerging threats to tropical forests: new and poorly understood pressures on tropical ecosystems and their biota*. Chicago & London: Chicago University Press. 520 pp.

Bahn, Paul & Flenley, John. 1992. *Easter Island, earth island*. London: Thames & Hudson. 240pp.

Baillif, Thérèse. 2000. Au revoir Monsieur l'Ingénieur. *JR* 7/11/2000 [on-line at www.clicanoo.com].

Baissac, Charles. 1888. *Le folk-lore de l'Ile Maurice (texte créole et traduction française)*. Paris: Maisonneuve & Leclerc viii+xix+468pp. [reprinted 1967 – Paris: Maisonneuve & Larose].

Baker, E.C.Stuart. 1913. *Indian pigeons and doves*. London: Witherby & Co. 260pp.

Baker, J.G. 1877. *Flora of Mauritius and the Seychelles. A description of the flowering plants and ferns of those islands*. London: Reeve. 558pp. [reprinted 1999 – New Delhi: Asian Educational Services]

Baker, Philip & Corne, Chris. 1982. *Isle de France créole. Affinities and origins*. Ann Arbor, Michigan: Karoma Publications. viii+299pp.

— & Hookoomsing, Vinesh Y. 1987. *Diksyoner kreol morisyen/*

Dictionary of Mauritian Creole. Paris: Editions L'Harmattan. 365pp.
Balfour, I.Bayley. 1879a. The physical features of Rodriguez. *Phil. Trans. Roy. Soc. Lond.* 168: 289–292.
— 1879b. [The collections from Rodriguez] Botany. *Phil. Trans. Roy. Soc., Lond.* 168: 302–387.
— 1879c. Reports of Proceedings of the Naturalists. 1. Report of Dr Is.B.Balfour. *Phil. Trans. Roy. Soc., Lond.* 168: 293.
Balouet, Jean-Christophe & Alibert, Eric. 1990. *Extinct species of the world*. New York: Barrons Educational Series & Auckland (NZ): David Bateman. 192pp.
Bannermann, David A. 1953. *The birds of west and equatorial Africa*. Edinburgh & London: Oliver & Boyd. 2 vols.
Barassin, Jean. 1953. *Bourbon, des origines jusqu'en 1714*. Saint-Denis: Librarie Cazal.
—, 1983. *Histoire des établissements religieux de Bourbon au temps de la Compagnie des Indes, 1664–1767*. Saint-Denis: Fondation pour la Recherche et le Developpement dans l'Ocean Indien. 218pp.
— 1989. *La vie quotidienne des colons de l'Ile Bourbon à la fin du regne de Louis XIV, 1700–1715*. Saint-Denis: Académie de la Réunion. 274pp.
Barau, C.Armand. 1980. L'histoire des oiseaux de la Réunion du dodo à nos jours. *Bull. Acad. Réunion* 24: 41–72.
Barclay, Jamie, Raphael, Marie L. & Lynch, Tara. 2001. Mangrove impact assessment, Rodrigues island. P.20 *in* Burnett, Kavanagh & Spencer (2001), q.v.
Barclay-Smith, Phyllis. Protection. 1964. Pp.666–667 in A.Landsborough-Thomson (ed.) *A new dictionary of birds*. London: Thomas Nelson & Sons. 928pp.
Baret, Stéphane *et al.* 2004. Altitudinal variation in fertility and vegetative growth in the invasive plant *Rubus alceifolius* (Rosaceae) on Réunion island. *Plant Ecol.* 172: 265–273.
— & Strasberg, Dominique. 2005. The effects of opening trails on exotic plant invasion in protected areas on La Réunion island (Mascarene Archipelago, Indian Ocean). *Rev. Ecol. (Terre Vie)* 60: 325–332.
—, Le Bourgeois, Thomas & Strasberg, Dominique. 2005. Comment *Rubus alceifolius*, une espèce exotique envahissante, pourrait-elle progressivement coloniser la totalité d'une forêt tropicale humide ? *Can. J. Bot.* 83: 219–226.
Barkly, Henry. 1870. Notes on the fauna and flora of Round Island. *TRSAS* NS 4: 109–130 & *Mauritius Almanac & Colonial Register* 1870: 92–101.
*Barlow, Nora. 1945. *Darwin and the voyage of the Beagle. Unpublished letters and notes*. London: Pilot Press. 279pp.
Barnett, L.K. & Emms, C. 1998. An annotated checklist of the Chagos Archipelago terrestrial fauna (omitting birds) recorded during the 1996 'Friends of the Chagos' expedition. *Phelsuma* 6: 41–52.
Barnwell, Patrick J. 1948. *Visits and despatches, 1598–1948*. Port Louis: Standard Printing Establishment. 306pp.
— & Toussaint, Auguste. 1949. *A short history of Mauritius*. London: Longmans, Green. vii+268pp.
— (ed.) 1950–54. Extracts from ships' logs visiting Mauritius. *Rev. Retrospective Ile Maurice* 1: 195–8, 259–62, 324–6; 2: 17–24, 79–84, 139–44, 201–8, 265–70, 317–20; 3: 17–22, 77–82, 139–46, 199–210, 267–274, 327–32,; 4: 1–6, 131–6, 197–200; 5: 85–94.
— 1955. Early place-names of Mauritius. *Rev. Retrospective Ile Maurice* 6: 167–180.
Barquissau, Raphaël, Foucque, Hippolyte & Jacob de Cordemoy, Hubert. 1925. *L'Ile de la Réunion (ancienne île Bourbon)*. Paris: Emile Larose. 282pp.
Barré, Nicolas & Barau, [C.] Armand. 1982. *Oiseaux de la Réunion*. Saint-Denis: [authors].
— 1983. Distribution et abondance des oiseaux terrestres de l'île de la Réunion (Océan Indien). *Rev. Ecol. (Terre Vie)* 37: 37–85.
— 1984 ('1983'). Oiseaux migrateurs observés à la Réunion. *Oiseau Rev. Fr. Orn.* 53: 323–333.
— 1988. Une avifaune menacée: les oiseaux de la Réunion. Pp.167–196 in J-C.Thibault & I.Guyot (eds.) *Livre Rouge des oiseaux menacés des régions françaises d'outre-mer*. Paris: CIPO/ICBP. 258pp.
—, Barau, [C.] Armand & Jouanin, Christian. 1996. *Oiseaux de la Réunion*. 2nd.ed., revised. Paris: Éditions du Pacifique. 208pp.
Bartlett, Richard D. & Bartlett Patricia. 2001. *Day geckos*. Hauppauge, NY: Barrons Educational Series. 46pp.
Bastian, S.T. *et al.* 2002. Evolutionary relationships of Flying Foxes (genus *Pteropus*) in the Philippines inferred from DNA sequences of Cytochrome b gene. *Biochem. Genet.* 40(3/4): 101–116.
Bathe, Helen [V.] & Bathe, Graham [M.] 1982. Feeding studies of three endemic landbirds, *Acrocephalus sechellensis, Foudia seychellarum* and *Nectarinia dussumieri* on Cousin Island, Seychelles. *Cousin Is. Res. Station Tech. Rep.* 26. Cambridge, UK: ICBP.
Bauer, Aaron M. & Russell, Anthony P. 1986. *Hoplodactylus delcourti* n.sp. (Reptilia: Gekkonidae), the largest known gecko. *NZ. J. Zool.* 13: 141–148.
— & Günther, Rainer. 2004. On a newly identified specimen of the extinct Bolyerid snake *Bolyeria multocarinata* (Boie, 1927). *Herpetozoa* 17: 170–181.
Bauer, Rotraud & Haupt, Herbert. 1976. Das Kunstkammerinventar Kaiser Rudolf II, 1607–1611. *Jahrb. Kunsthistorsichen Samml. Wien* 72: xlv+191pp.
Baxter, Stephen. 2003. *Revolutions in the earth. James Hutton and the true age of the world*. London: Orion/Weidenfeld & Nicholson. 245pp.
*Beane, J.F. 1905. *From forecastle to cabin. The story of a cruise in many seas. . .* New York: Editor Publishing Co. 341pp.
Bearman, Gerry (ed.) 1989. *Ocean circulation*. Milton Keynes & Oxford: Open University & Pergamon Press. 238pp.
Beaton, Patrick. 1860. *Six months in Réunion. A clergyman's holiday and how he passed it*. London: Hurst & Blackett. 2 vols.
Bélanger, Charles *et al.* 1834. *Voyage aux Indes Orientales. Zoologie*. Paris: Arthus Bertrand. 535pp.
Bell, Ben D. & Merton, Don V. 2002. Critically endangered bird species and their management. Pp.105–138 *in* Norris & Paine (2002), q.v.
Bell, Brian D. 1978. The Big South Cape Islands rat irruption. Pp.33–45 *in* Dingwall *et al.* (1978), q.v.
— , Dulloo, Ehsan, & Bell, Michael. 1993. *Mauritius offshore islands survey report and management plan*. Wellington, NZ: Wildlife Management International Ltd. 176pp.
— & Bell, Elizabeth A. 1996. *Mauritius offshore islands project, phase II, Implementation of management recommendations*. Wellington, NZ: Wildlife Management International Ltd.

56pp.

— 2002. The eradication of alien mammals from five offshore islands, Mauritius, Indian Ocean. Pp.40–45 *in* Veitch & Clout (2002), q.v.

Bell, E.Arthur & Evans, Christine S. 1978. Biochemical evidence of a former link between Australia and the Mascarene Islands. *Nature* 273: 295–6.

Belloc, Hilaire. 1896. *A bad child's book of beasts*. Oxford: Alden & Co. & London: Duckworth. 48pp.

Bénard, Roland. 2002. *Scènes de la Réunion, vue d'en haut*. Réunion: Editions Orphie. 127pp.

Bénot, Yves. 1983. Introduction. Pp.7–22 *in* the 1983 reprint of Bernardin (1773), q.v.

Benson, Con[stantine] W. 1960. The birds of the Comoro Islands: results of the British Ornithologists' Union Centenary Expedition 1958. *Ibis* 103b: 5–106.

— 1970–71. The Cambridge collection from the Malagasy region. *Bull. Brit. Orn. Club* 90: 168–172 & 91: 1–7.

— 1971. Notes on *Terpsiphone* and *Coracina* spp. in the Malagasy region. *Bull. Brit. Orn. Club* 91: 56–64.

Berdini, Paolo. 1997. *The religious art of Jacopo Bassano. Painting as visual exegesis*. Cambridge: Cambridge University Press. 208pp.

Berger, A.J. 1957. On the anatomy and relationships of *Fregilupus varius*, an extinct starling from the Mascarene Islands. *Bull. Am. Mus. Nat. Hist.* 113: 227–272.

— 1981. *Hawaiian bird life*. 2nd ed. Honolulu: Univ. Hawaii Press. xv+260pp.

Bergmans, Wim. 1988–97. Taxonomy and biogeography of African fruit bats (Mammalia, Megachiroptera). *Beaufortia* 38(5), 39(4), 40(7) [includes *Pteropus*], 44(4) & 47(2) [discussion/conclusions].

Berlioz, Jacques. 1933. Au sujet de *Fregilupus varius*. *Oiseau Rev. Fr. Orn.* NS 3: 41.

— 1940. Note critique sur une espèce éteinte de Psittacidé de l'île Maurice: *Lophopsittacus mauritianus* (Owen). *Bull. Mus. Nat. Hist., Paris* (2)12: 143–8.

— 1946. *Oiseaux de la Réunion*. Faune de l'Empire Français 4. Paris: Larose. 81pp.

Bernardin, *Père*. 1687. Mémoire du R.P.Bernardin sur l'Ile de Bourbon (1687). [ed.A.Lougnon] *Rec. Trim.* 4: 57–70.

Bernardin de St.Pierre, Jacques-Henri. 1773. *Voyage à l'Isle de France, à l'Isle de Bourbon, au Cap de Bonne Espérance; &c. par un officier du Roi*. Neuchâtel: Société Typographique [2 parts with separate pagination, bound in one; reprinted 1983 – Paris: Editions la Découverte. 262pp.; English tr., 1800, London: Vernon & Hood; reprinted 1999, New Delhi: Asian Educational Services. Also new tr. & intro. by Jason Wilson as *Journey to Mauritius*, Oxford: Signal Books, 2002, 290pp.]

Berthelot, Lilian. 2002. *La petite Mascareigne. Aspects de l'histoire de Rodrigues*. Port Louis: Centre Culturel Nelson Mandela pour la Culture Africaine. 255pp.

Berthier, Colette (ed.) 2004. *Tableau Économique de la Réunion. Édition 2004–5*. Saint-Denis: INSEE. 219pp.

Bertile, Wilfrid. 1987. *La Réunion. Atlas thématique et régionale*. Saint-Denis: Arts Graphiques Modernes. 162pp.

Bertrand, J., Bonnet, B. & Lebrun, G. 1986. Nesting attempts of *Chelonia mydas* at Réunion Island (S.W.Indian Ocean). *Marine Turtle Newsl.* 39: 3–4.

Bertuchi, A.J. 1923. *The island of Rodrigues*. London: John Murray. 117pp.

Besant, Walter. 1902. *Autobiography of Sir Walter Besant*. London: Hutchinson & Co. 292pp.

Besnard, Nicolas. 2000. Répartition du Bulbul orphée sur l'île de la Réunion. *Taille-vent* 5–6: 2–3.

Besse, Jean & Courtillot, Vincent. 1988. Paleogeographic maps of the continents bordering the Indian Ocean since the early Jurassic. *J. Geophys. Sci.* 93 (B10): 11791–11808.

Besselink, Marijke. 1995. Images of the dodo/ het beeld van de dodo. Pp.84–96 *in* Wissen (1995), q.v.

Bettagno, Alessandro *et al.* 1996. *The Prado Museum*. Madrid: Fundacion amigos de Museo del Prado. 669pp.

Betting de Lancastel. 1827. *Statistique de l'île Bourbon. Présenté en exécution de l'article 104s28 de l'Ordonnance royale de 21 août 1825*. Saint-Denis: LaHuppe, Imprimerie du Gouvernement. 199pp.

Bhagwant, S. 2004. Human *Bertiella studieri* (Fanily Anoplocephalidae) infection of probable Southeast Asian origin in Mauritian children and an adult. *Am. J. Trop. Med. Hyg.* 70(2): 225–228.

Billiard, Auguste. 1822. *Voyage aux colonies orientales ou lettres écrites des îles de France et de Bourbon pendant les années 1818, 1819 et 1820*. Paris: Libraire Ladvocat. 490pp. [also 1829, Paris: J.L.J.Brière & 1990, Ste.Clotilde, Réunion: Ars Terres Créoles, 254pp.]

Biju, D.D. & Bossuyt, Franky. 2003. New frog family from India reveals an ancient biogeographical link with the Seychelles. *Nature* 425: 711–714.

Bindloss, Joseph *et al.* 2001. *Mauritius, Réunion & Seychelles*. 4th ed. Melbourne: Lonely Planet Publications. 336pp.

Bissoondoyal, Basdeo. 1968. *The truth about Mauritius*. Bombay: Bharatiya Vidya Bhavan. 246pp.

Blagden, C.O. 1916. *Catalogue of manuscripts in European languages belonging to the library of the India Office. Vol.1 The Mackenzie Collections. Part 1. The 1822 Collection & the Private Collection*. London: Oxford University Press. 302pp.

Blanc, Charles P. 1972. Les reptiles de Madagascar et des îles voisines. Pp.501–611 *in* R.Battistini & G.Richard-Vindard (eds.) *Biogeography and ecology of Madagascar*. The Hague: W.Junk. 765pp.

Blanchard, Frédéric. 2000. *Guide des milieux naturels: La Réunion, Maurice, Rodrigues*. Paris: Éditions Eugen Ulmer. 384pp.

Bloxam, Quentin M.C. 1977 ('1976'). Maintenance and breeding of the Round Island skink *Leiolopisma telfairii* (Desjardins). *Jersey Wildl. Preserv. Trust. Ann. Rep.* 13: 53–56.

— & Vokins, A.Mark A. 1979 ('1978'). Breeding and maintenance of *Phelsuma guentheri* (Boulenger 1885) at the Jersey Zoological Park. *Dodo* 15: 82–91.

— 1983 ('1982'). Feasibility of reintroduction of captive-bred Round Island skink to Gunners' Quoin. *Dodo* 19: 37–42.

Boas, F.Samuel. 1935. *The diary of Thomas Crossfield*. London: Oxford University Press. xxix+169pp.

Bock, Walter J. 2003. Ecological aspects of the evolutionary processes. *Zool. Sci.* 20: 279–289.

Bonaparte, Roland. 1890. *Le premier établissement des Néerlandais à Maurice*. Paris: [author]. 60pp.

Bondini, Elizabeth. 1990. Les tortues géantes terrestres et le choc de l'histoire. *Cahiers d'Outre Mer* 43(172): 555–559.

Bonnemains, Jacqueline & Chappuis, Claude. 1985. Les oiseaux de la collection C.A.Lesueur du Muséum d'Histoire Naturelle du Havre (dessins et manuscrits). *Bull. Trim. Soc. Geol. Normandie Amis Mus. Havre* 72: 25–78.

— & Hauguel, Pascale (eds). 1987. *Récit du voyage aux terres australes par Pierre Bernard Milius, second sur le "Naturaliste" dans l'expédition Baudin (1800–1804)*. Le Havre: Société havraise d'Études Diverses & Muséum d'Histoire Naturelle de Havre. vi+82pp.

— & Bour, Roger. 1997. Les chéloniens de la collection Lesueur du Muséum d'histoire naturelle du Havre. *Bull. Trim Soc. Geol. Normandie Amis Mus. Havre* 83: 5–45.

— 2003. Biographie resumée de Charles Alexandre Lesueur (1778–1846). Annexe d, pp.xxv-xxvi *in* Ly-Tio-Fane (2003), q.v.

— & Ly-Tio-Fane, Madeleine. 2003. Le voyage de découvertes aux terres australes. La collection Lesueur du Muséum d'Histoire Naturelle du Havre. Dossier 15. Catalogue. Pp.i-xxv *in* Ly-Tio-Fane (2003), q.v.

Bontekoe, Willem Y. 1646. *Iournal ofte Gedenckwaerdige beschrijvinghe vande Oost-Indische Reyse van Willem Ijbrantz Bontekoe van Hoorn*. Hoorn, Netherlands: Jan Jansz Deutel. [English tr. = *Memorable description of the East Indian voyage 1618–25*. London: Routledge, 1929; reprinted 1992, New Delhi: Asian Educational Services. 168pp.]

Boonsong Lekagul & Cronin, Edward W. 1974. *Bird guide of Thailand*. 2nd. ed. Bangkok: Association for the Conservation of Wildlife. 316pp. + index.

Bordères, Michel. 1991. Histoire d'une silviculture – le Tamarin des Hauts. *Bois For. Trop.* 229 (special issue on Réunion): 35–42.

Bory de Saint-Vincent, Jean-Baptiste G.M. 1804. *Voyage dans les quatre principales îles des mers d'Afrique, fait par order du gouvernement pendant les années neuf et dix de la République (1801 et 1802)*. Paris: F.Buisson. 2 vols. + 'atlas' [plates]. [Reprinted 1980 – Marseille: Lafitte Reprints]

Bosser, Jean, Cadet, Théresien, Guého, Joseph & Marais, W. (eds.) 1978-[ongoing]. *Flore des Mascareignes: La Réunion, Maurice, Rodrigues*. Réduit, Mauritius: Sugar Industry Research Institute, Paris: ORSTOM, Kew, UK: Royal Botanic Gardens.

— 1982. Projet de constitution de réserves biologiques dans le domaine forestier à la Réunion. Rapport de mission de J.Bosser. Paris: ORSTOM & Saint-Denis: ONF. 36pp + maps [unpublished photocopied document]

Botting, Douglas. 1999. *Gerald Durell – the authorized biography*. London: HarperCollins. 644pp.

Boucher, Antoine. 1710. Mémoire d'Antoine Boucher sur l'Ile Bourbon en 1710 [ed.A.Lougnon] *Rec. Trim.* 5: 279–355 (1941).

Boulanger, Charles-Pierre. 1803. Excursion au Grand Bassin. [MS 07007 of the Collection Lesueur, Le Havre Natural History Museum, published pp.xxxi-xxxii *in* Ly-Tio-Fane (2003), q.v.]

Boulay, Stéphanie & Probst, Jean-Michel. 1998. Note sur la capture d'une Tortue alligator introduite *Macroclemys temminckii* pêchée dans la Rivière de l'Est (île de la Réunion). *Bull. Phaethon* 7: 52.

Boulenger, George A. 1882. *Catalogue of the Batrachia Salientia s.Ecaudata in the collection of the British Museum*. London: British Museum (Natural History). 495pp. + plates.

— 1884. [note on *Phelsuma newtoni*]. *Proc. Zool. Soc. Lond.* 1884: 1–2.

— 1885–87. *Catalogue of the lizards in the British Museum (Natural History)*. London: British Museum (Natural History). 3 vols.

— 1893. *Catalogue of the snakes in the British Museum (Natural History)*. London: British Museum (Natural History). 3 vols.

Boulenger, Jacques (ed.). 1935. *Voyage d'un naturaliste en Haïti 1799–1803 par M E.Descourlitz*. Paris: Librairie Plon. 232pp.

*Boullay, Stéphane. 1995. Repatriation of radiated tortoises *Geochelone radiata* from Réunion Island to Madagascar. *Chelonian Conserv. Biol.* 1: 319–320.

Bour, Roger. 1979. Première découverte de restes osseux de la Tortue terrestre de la Réunion, *Cylindraspis botbonica*. *C.R. Acad. Sci., Paris* 288D: 1223–1226 [also in *Info-Nature, Ile Réunion* 17: 53–59 (1979), with an additonal illustration].

— 1981. Histoire de la tortue terrestre de Bourbon. *Bull. Acad. Ile Réunion* 25: 98–147.

— & Moutou, François. 1982. Reptiles et amphibiens de l'île de la Réunion. *Info-Nature, Ile Réunion* 19: 119–156.

— 1984a. Taxonomy, history and geography of Seychelles land tortoises and freshwater turtles. Pp.281–307 *in* Stoddart (1984a), q.v.

— 1984b. Données sur la repartition géographique des tortues terrestres et d'eau douce aux îles Maurice et Rodrigues. *Info-Nature, Ile Réunion* 21: 7–38 & *Mauritius Inst. Bull.* 10: 75–102.

—, Probst, Jean-Michel & Ribes, Sonia. 1995. *Phelsuma inexpectata* Mertens 1966, le lézard vert de Manapany-les-Bains (La Réunion): données chorologiques et écologiques (Reptilia, Gekkonidae). *Dumerilia* 2: 99–124.

—, 2004. Holotype of "La grande tortue des Indes" in Perrault, 1676 – *Cylindraspis indica* (Schneider, 1783). *Emys* 11: 33–35.

—, 2005. Type specimens of *Testudo rotunda* Latreille, 1801. *Emys* 12(6): 23–27.

Bourgat, Robert M. 1970. Recherches écologiques et biologiques sur le *Chameleo pardalis* Cuvier 1929 de l'île de la Réunion et de Madagascar. *Bull. Soc. Zool. Fr.* 95: 259–269.

— 1972. Biogeographical interest of *Chamaeleo pardalis* Cuvier, 1829 (Reptilia, Squamata, Chamaeleonidae) on Reunion Island. *Herpetologia* 28: 22–24.

Bourn, David [M.] & Coe, Malcolm J. 1978. The size, structure and distribution of the giant tortoise population of Aldabra. *Phil. Trans. Roy. Soc. Lond.* B 282: 139–175.

— & Coe, Malcolm J. 1979. Features of tortoise mortality and decomposition on Aldabra. *Phil. Trans. Roy. Soc. Lond.* B 286: 189–193 [& *in* Stoddart (1979), q.v.].

— *et al.* 1999. The rise and fall of the Aldabra giant tortoise population. *Proc. Roy. Soc. Lond.* B 266: 1091–1100.

Bourne, William (Bill) R.P. 1959. Notes on reports received: 1958–1959. *Sea Swallow* 12: 6–17, 20–21.

— 1963. [review of Watson Zusi & Storer (1963), q.v.] *Ibis* 105: 411–12.

— 1968. The birds of Rodriguez, Indian Ocean. *Ibis* 110: 338–344.

— 1971. The birds of the Chagos group, Indian Ocean. *Atoll Res. Bull.* 149: 175–207.

— 1975. Mammals on islands. *New Scientist* 21 Aug,1975: 422–425.

— 1976. On subfossil bones of Abbott's Booby from the Mascarene Islands, with a note on the proportions and distribution of the Sulidae. *Ibis* 118: 119–123.

Bouton, Louis [S.] 1846. [note sur l'état actuel des forêts] *P.V. Soc. Hist. Nat. Ile Maurice* 6 Oct.1842–28 Aug.1845: 175–6.

— 1848. Annual address of the secretary of the Royal Society of Arts and Sciences of Mauritius for 1846–7 read at the annual general meeting August 24, 1847. Port Louis: RSAS. 17pp.

— 1851. *Rapport Annuel des travaux de la Société Royale des Arts et des Sciences de Maurice* 1851, 16pp.

— 1860a. Sur le mode de repartition de certaines plantes à Maurice. *TRSAS* NS 2: 113–4.

— 1860b. Annual Report of the proceedings of the Royal Society of Arts and Sciences 1859–60. *TRSAS* NS 2: 130–146.

— 1869a. Séance du vendredi 19 juin 1868. *TRSAS*, NS 3: 109–111.

— 1869b. Annual Report. Royal Society of Arts and Sciences [for 1868]. *TRSAS* NS 3(2): i–xviii.

— 1871. Water Supply commmission. Answers to questions for 4th committee. *TRSAS* NS 5: 170–178.

— 1875. Report of the Secretary to the Royal Society of Arts and Sciences, 3 May 1873 to 10 October 1874. *TRSAS*, NS 8: i–xxiii+.

— 1878. [letter to Edward Newton on the *huppe de Bourbon* and the *pigeon hollandais*, dated 21 February 1878] MS in bound volume entitled 'Indian Ocean 3. Madagascar – Mascarene Islands MSS.', Newton Library, Cambridge University Zoology Department.

Bower, G. 1903. Mauritius. Report for 1901. *Colonial Rep.-Ann.* 379. London: HMSO.

Bowman, Larry. 1991. *Mauritius: democracy and development in the Indian Ocean.* Boulder, Colorado: Westview Press. 208pp.

Boyce, Douglas A. & White, Clayton W. 1987. Evolutionary aspects of kestrel systematics: a scenario. Pp.1–21 *in* D.M.Bird & R.Bowman (eds.) *The ancestral kestrel.* Quebec: Raptor Research Foundation.

Boyla, K. 2000. The impact of habitat management on native and exotic birds in a native forest remnant on Mauritius. M.Sc. Thesis. University of East Anglia, U.K.

Boyle, Charles J. 1867. *Far away. Sketches of scenery and society in Mauritius.* London: Chapman & Hall. 368pp.

Bradshaw, T. 1832. *Views in the Mauritius or Isle of France, drawn from nature by T.Bradshaw.* London: J.Carpenter. 14pp + 40 plates.

Bradt, Hilary. 2005. *Madagascar. The Bradt travel guide.* 8th ed., 2004, amended 2005. Chalfont St.Peter, UK: Bradt Publications. 496pp.

Brandley, Matthew C., Schmitz, Andreas & Reeder, Tod W. 2005. Partitioned Bayesian analyses, partition choice and phylogenetic relationships of scincid lizards. *Systemat. Biol.* 54(3): 373–390.

Bretagnolle, Vincent & Attié, Carole. 1991. Status of Barau's Petrel (*Pterodroma baraui*): colony sites, breeding population and taxonomic affinities. *Colonial Waterbirds* 14: 25–33.

— & — 1993. Massacre d'une espèce protégée sur le terrtoire français: le pétrel de Barau. *Courrier Nat.* 138: 40.

— & — 1995. Réunion. Une deuxième vague d'extinction en perspective ? *Courrier Nat.* 154: 14.

— & — 1996a. Coloration and biometrics of fledgling Audubon's Shearwaters *Puffinus lherminieri* from Réunion Island, Indian Ocean. *Bull. Brit. Orn. Club.* 116: 194–197.

— & — 1996b. Comments on a possible new species *Otus* sp. on Réunion. *Bull. Afr. Bird Club* 3: 36.

— & — 1997. Statut et conservation des oiseaux marins dans l'ouest de l'Océan Indien. *Courrier Nat.* 163: 24–29.

—, Attié, Carole & Pasquet, Eric. 1998. Cytochrome-B evidence for validity and phylogenetic relationships of *Pseudobulweria* and *Bulweria* (Procellariidae). *Auk* 115: 188–195.

—, Attié, Carole & Mougeot, François. 2000a. Audubon's Shearwaters *Puffinus lherminieri* on Réunion Island, Indian Ocean: behaviour, distribution, biometrics and breeding biology. *Ibis* 142: 399–412.

— *et al.* 2000b. Distribution, population size and habitat use of the Réunion Marsh Harrier *Circus m.maillardi*. *J. Raptor Res.* 34: 8–17.

—, Thiollay Jean-Marc & Attié, Carole. 2000c. Status of Réunion Marsh Harrier *Circus maillardi* on Réunion island. Pp.669–676 *in* R.D.Chancellor & B-U.Meyburg (eds.) *Raptors at risk. Proc. V World Conf. Birds of Prey & Owls, Johannesburg 4–11 August 1998.* Berlin: WWGBP & Surrey, BC: Hancock House.

Brial, Pierre. 1997 ('1996'). Atlas des cavernes de l'Ile Rodrigue. Réunion: [author]. Unpaginated.

— 1998a. La Caverne de la Tortue. *Info-Nature, Ile Réunion* 24: 116–125.

— 1998b. Le Dodo blanc, et autres faits inexpliqués concernant le Solitaire de la Réunion. *Taille-vent* 4:2–7.

— 2001a. La relation d'Adriaen Martensz Block. *Bull. Soc. Geog. Réunion* 1: 3–4.

— 2001b. Une espèce inconnue de microchiroptère observée par Bory de Saint-Vincent à l'île de la Réunion en 1801. *Bull. Soc. Geog. Réunion* 1: 5.

— 2001c. Observations de Bory de Saint-Vincent sur la faune de l'île de la Réunion dans le *Voyage dans les quatres principales îles des mers d'Afrique 1801–1802. Bull. Soc. Geog. Réunion* 1: 14–19.

Brisson, M.J. 1760. *Ornithologie.* 6 vols. Paris: J-B.Bauche.

Broderip, William J. 1853. Notice of an original painting, including a figure of the Dodo, in the collection of his Grace the Duke of Northumberland. *Proc. Zool. Soc. Lond.* 1853: 54–57.

Brooke, Michael de L. 1978. Inland observations of Barau's Petrel *Pterodroma baraui* on Réunion. *Bull. Brit. Orn. Club* 98: 90–95.

— 2004. *Albatrosses and petrels across the world.* Oxford: Oxford University Press. 499pp.

—, Imber, M.J. & Rowe, G. 2000. Occurrence of two surface-breeding species of *Pterodroma* on Round Island, Indian Ocean. *Ibis* 142: 139–158.

Brooke, Richard K. 1972. Generic limits in old world Apodidae and Hirundinidae. *Bull. Brit. Orn. Club* 92: 53–57.

— 1976. Layard's extralimital records in his Birds of South Africa and in the South African Museum. *Bull. Br. Orn. Club* 96: 75–80.

— 1978. XIX century Indian Ocean seabirds eggs in the South African Museum. *Bull. Br. Orn. Club* 98: 75–80.

— 1981. Layard's bird hunting visit to Tromelin or sandy Island in December 1856. *Atoll Res. Bull.* 255: 73-82.

Brooks, S.E. 2002. The ecology and behaviour of the introduced Agamid lizard *Calotes versicolor* on Ile aux Aigrettes, Mauritius. M.Sc. thesis. University of East Anglia.

Broome [Barker], Mary A. 1904. *Colonial memories.* London: Smith, Elder & Co. xxii+301pp.

Brouard, Noel R. 1960. A brief account of the 1960 cyclones and

their effect upon exotic plantations in Mauritius. *Emp. For. Rev.* 39: 411–416.
— 1963. *A history of woods and forests in Mauritius*. Port Louis: Government Printer. 86pp.
— 1966. Les problèmes forestiers de Maurice. *Rev. Agric. Sucrière Ile Maurice* 45: 220–230.
— 1967. *Damage by tropical cyclones to forest plantations, with particular reference to Mauritius*. Port Louis: Government Printer. 7pp.
— 1970. *Annual report of the Forestry Service for the year 1968*. Port Louis: Govt.Printer (Mauritius, No.13 of 1970).
Brown, Anthony L. 2000 *Ill-starred captains. Flinders and Baudin*. Hindmarsh, S.Australia: Crawford House & Mechanicsburg, Penn.: Stackpole Books. 512pp.
Brown, 'Père'. n.d. Lettre du révérend *père* Brown, missionaire de la Compagnie de Jesus, à Madame la Marquise de Benamont. *Lettres Édifiantes, Mémoires des Indes* 30: 321–335 (1773) [also 13: 302–319 (1781 ed.) & 7: 450–460 (1819 ed.); this text is cognate with Le Gentil (1717), q.v.]
Brown, Peter. 1776. *New illustrations of zoology*. London: B.White. 136pp.
Bruce, Charles. 1910. *The broad stone of empire. Problems of colonial administration with records of personal experience*. London: Macmillan & Co. 2 vols.
Bruijn, J.R., Gaastra, F.S. & Schoffer, I. 1979–87. *Dutch Asiatic Shipping in the 17th and 18th centuries*. The Hague: Martinus Nijhoff. 3 vols.
Bry, Theodore de & Bry, Israel de. 1601. *Pars Quarta Indiae Orientalis* & *Quinta Pars Indiae Orientalis*. Frankfurt: M.Becker. [first Dutch voyage to Mauritius in the second section of part 4 (the *Vriesland*'s return journey in 1599) and in part 5 (abridged translation of the *Tweede Boeck* (Anon. 1601), q.v.)]
Budulla, Z. 2002. A preliminary study on effects of management and season on canopy-dwelling insects in Brise Fer. B.Sc. (Hons) thesis. University of Mauritius.
Buffon, G.L.LeClerc, Comte de. 1770–1783. *Histoire naturelle des oiseaux*. Paris Imprimerie Royale. [3 original editions with variable numbers of volumes, published over more or less the same dates; one, in 10 vols., includes the *Planches Enluminées* by Martinet. Numerous later reprints; see E.Genet-Varcin & J.Roger 1954 *Bibliographie de Buffon*, Paris: Presses Universitaires de France, and N.Mayaud *Alauda* 9: 18–32 (1939)]
— 1776. *Histoire naturelle générale et particulière, servant de suite à l'histoire des animaux quadrupèdes*. Supplément, vol.3. Paris: Imprimerie Royale.
Bullock, David [J.] & North, Steven. 1977. Report of the Edinburgh University expedition to Round Island, Mauritius, July & August 1975. Edinburgh: [authors]. Photocopied, 115pp.
— 1977. Round Island – a tale of destruction. *Oryx* 14: 51–58.
— (ed.) 1982. Round Island Expedition 1982. Preliminary Report. St.Andrews, Scotland: [author/expedition]. 15pp.
—, North, Steven & Grieg, Susan. 1983. Round Island Expedition 1982. Final Report. St.Andrews, Scotland: [authors/expedition]. 100pp + appendices.
— & — 1984. Round Island in 1982. *Oryx* 18: 36–41.
—, Arnold, E.Nicholas & Bloxam, Quentin. 1985. A new endemic gecko from Mauritius. *J. Zool., Lond.* (A) 206: 591–599.
— 1986. The ecology and conservation of reptiles on Round Island and Gunner's Quoin, Mauritius. *Biol. Conserv.* 37: 135–156.
— & North, Steven G. (eds.) 1991. Round Island Expedition 1989. Final Report. Leicester, UK: [editors/expedition]. 76pp. + appendices.
— et al. 2002. The impact of rabbit and goat eradication on the ecology of Round Island, Mauritius. Pp.53–63 *in* Veitch & Clout (2002), q.v.
— & North, Steven G. (eds.) 2007. Report of the expedition to Round Island, Flat Island and Gabriel Island, Mauritius, June/July 2003. Issued on CD only.
Bulpin, T.V. 1958. *Islands in a forgotten sea*. Cape Town: Books of Africa. 346pp. [the '2nd edition' of 1969 appears to differ only by the addition of a series of colour photographs]
Bunbury, Nancy et al. 2007. Avian blood parasites in an endangered columbid: *Leucocytozoon marchouxi* in the Mauritian Pink Pigeon. *Parasitology* 134 [online: doi:10.1017/S0031182006002149].
Burckhardt, Frederick (ed.). 1996. *Charles Darwin's letters. A selection*. Cambridge: Cambridge University Press. 249pp.
Burger, Joanna & Gochfield, Michael. 1991. Nest-site selection by the Herald Petrel and White-tailed Tropicbird on Round Island, Indian Ocean. *Wilson Bull.* 103: 126–130.
— & — 1994. Predation and effects of humans on island-nesting seabirds. Pp.39–67 *in* Nettleship et al. (1994), q.v.
Burnett, Juliet, Kavanagh, Jessica & Spencer, Tom. 2001. *Shoals of Capricorn. Field Report 1998–2001*. London: Royal Geographical Society. 108pp.
Burney, D.A. 2003. Madagascar's prehistoric ecosystems. Pp.47–51 in Goodman & Benstead (2003), q.v.
Bustard, H.Robert. 1972. *Sea turtles. their natural history and conservation*. London: Collins. 220pp.
"J.C." – see Jacob de Cordemoy.
Cabin, Robert J. et al. 2000. Effects of long-term ungulate exclusion and recent alien species control on the preservation and restoration of a Hawaiian tropical dry forest. *Conserv. Biol.* 14: 439–453.
— & — 2002. Effects of microsite, water, weeding and direct seeding on the regeneration of native and alien species within a Hawaiian dry forest preserve. *Biol. Conserv.* 104: 181–190.
Cade, Tom J. & Jones, Carl G. 1994. Progress in restoration of the Mauritius Kestrel. *Conserv. Biol.* 7: 169–175.
Cadet, Frédéric et al. (eds.) 2003. *Atlas de la Réunion*. Saint-Denis: Université de la Réunion & INSEE. 143pp.
Cadet, [L.J.] Thérésien. 1971a. Flore et végétation. *Info-Nature, Ile Réunion* 3:31–32.
— 1971b. Flore de l'Ile Rodrigue: espèces spontanées introduites depuis Balfour (1874). *Mauritius Inst. Bull.* 7: 1–12.
— 1975. Contribution à l'étude de la végétation de l'Ile Rodrigue. *Cah. Centre Univ. Réunion* 6: 5–29.
— 1977. *La végétation de l'Ile de la Réunion: étude phytoécologique et phytosociologique*. Thesis, University of Marseille [published 1980: St-Denis, Réunion: Imprimerie Cazal. 312pp.]
— & SREPN. 1977. Questions er réponses concernant la thèse de Monsieur Thérésien Cadet. *Info-Nature, Ile Réunion* 15: 43–57.
— 1981. *Fleurs et plantes de la Réunion et de l'Ile Maurice*. Papeete, Tahiti: Éditions du Pacifique. 131pp.
Cadle, J.E. 2003. Colubridae, Snakes. Pp.997–1004 *in* Goodman

& Benstead (2003), q.v.

Caldwell, [W.] James. 1875. Notes on the zoology of Rodrigues. *Proc. Zool. Soc. Lond*. 1875: 644–647.

Callmander, M.W. & Laivao, M.O. 2003. Pandanaceae, *Pandanus*. Pp.460–467 in Goodman & Benstead (2003), q.v.

Calkin, Jessamy. 2002. The perfect beach. *Telegraph Magazine* 5 January 2002. 3pp. [article on Mauritius]

Calkin, P.E. & Young, G.M. 1995. Global glacial chronologies and causes of glaciation. Pp.9–75 in J.Menzies (ed.) *Modern glacial environments*. Oxford: Butterworth-Heinemann. 621pp.

Camoin, G.F. *et al*. 1997. Holocene sea level changes and reef development in the southwestern Indian Ocean. *Coral Reefs* 16:247–259.

—, Montaggioni, Lucien F. & Braithwaite, C.J.R. 2004. Late glacial to post glacial sea-levels in the western Indian Ocean.. *Marine geol*. 206: 119–146.

Campbell, Bruce & Lack, Elizabeth (eds.). 1985. *A dictionary of birds*. Calton, Staffordshire, UK: T. & A.D. Poyser. 670pp.

Campbell, D.J. 1978. The effects of rats on vegetation. Pp.99–120 in Dingwall *et al*. (1978), q.v.

Cap, P.A. 1861. *Philibert Commerson, naturaliste voyageur. Étude biographique suivi d'un appendice*. Paris: Victor Masson et fils. 197pp.

Carié, Paul. 1904. Observations sur quelques oiseaux de l'Ile Maurice. *Ornis* 12: 121–128.

— 1910. Note sur l'acclimatation du Bulbul (*Otocompsa jocosa*) à l'Ile Maurice. *Bull. Soc. Natl. Acclim., Paris* 57: 462–464.

— 1916. L'acclimatation à l'île Maurice. *Bull. Soc. Natl. Acclim., Paris* 63: 10–18, 37–46, 72–79, 107–10, 152–59, 191–98, 245–50, 355–63 & 401–4 [also reprinted by the society with new pagination, 62pp.]

— 1919. L'oeuvre de la Direction de l'Agriculture à l'Ile Maurice. *Bull. Soc. Natl. Acclim. Paris* 1919: 317–326.

— 1920. Les îles soeurs de l'Océan Indien (La Réunion – Maurice). *La Géographie* 23: 385–404.

— 1921. Le Merle Cuisinier de l'Ile Maurice (*Lalage rufiventer* Sw.). *Rev. Hist. Nat. Appliq., Oiseau* 2: 2–5.

— 1930. Le *Leguatia gigantea* Schlegel (Rallidae) a-t-il existé ? *Bull. Mus. Natl. Hist. Nat. Paris* (2)2: 204–213.

Carlquist, Sherwin. 1965. *Island life*. New York: Natural History Press. 451pp.

— 1970. *Hawaii, a natural history*. New York: Natural History Press . 463pp.

Carr, Peter. 2000. Expedition report Diego Survey 97. Part two – the landbirds. *Sea Swallow* 49: 30–35.

Carranza, S. *et al*. 2001. Parallel gigantism and complex colonization patterns in the Cape Verde scincid lizards *Mabuya* and *Macroscincus* (Reptilia: Scincidae) revealed by mitochondrial DNA sequences. *Proc. Roy. Soc. Lond*. B 268: 1595–1603.

— & Arnold E.Nicholas. 2003. Investigating the origin of transoceanic distributions: mtDNA shows *Mabuya* lizards (Reptilia: Scincidae) crossed the Atlantic twice. *Systematics & Biodiversity* 1: 275–282.

Carroll, J.Bryan. 1979. The general behavioural repertoire of the Rodrigues Fruit Bat in captivity. *Dodo* 16: 51–59.

— 1982 ('1981'). The wild status and behaviour of the Rodrigues Fruit Bat *Pteropus rodricensis*. A report of the 1981 field study. *Dodo* 18: 20–29.

Carroll, Lewis [pseud. = Charles Dodgson]. 1866. *Alice's adventures in Wonderland*. London: Macmillan & Co. [+ many subsequent reprints]

Carter, Steven P. 1998. Mammalian predation on endangered Mauritian birds: predator identification, impacts and refuges in exotic vegetation. BSc. Hons. Thesis, Royal Holloway, University of London.

— & Bright, Paul W. 2002. Habitat refuges as alternatives to predator control for the conservation of endangered Mauritian birds. Pp.71–78 in Veitch & Clout (2002), q.v.

Case, Ted. J. 1996. Global patterns in the establishment and distribution of exotic birds. *Biol. Conserv*. 78: 69–96.

—, Cody, Martin L. & Excurra, Exequiel. 2002. *A new island biogeography of the Sea of Cortés*. New York: Oxford University Press. 669pp.

Catchpole, Clive & Komdeur, Jan. 1993. The song of the Seychelles Warbler *Acrocephalus sechellensis*, an island endemic. *Ibis* 135: 190–195.

Cauche, François. 1651. *Relation du voyage que François Cauche de Rouen à fait à Madagascar, îles adjacentes et coste d'Afrique. Receuilly par le sieur Morisot, avec notes en marge*. Paris: Augustin Courbe. x+194pp. [An English translation appeared in vol.2. of John Stevens (ed.) 1711, *A new collection of voyages and travels...*, and the entire French text was reprinted in Grandidier *et al*., 1903–20, vol.7., q.v.]

Caulier, Père. 1764. Fragments sur l'île Bourbon par le R.P.Caulier en 1764 [ed.A.Lougnon]. *Rec. Trim*. 3: 149–169 (1938).

Cazal, Charles. 1974. Une réserve de chasse. *Info-Nature, Ile Réunion* 11: 56–58.

*Cazes-Duvat, Virginie & Paskoff, Roland. 2004. *Les littoraux des Mascareignes entre nature et aménagement*. Paris: l'Harmattan. 186pp.

Censky, Ellen J., Hodge, Karim & Dudley, Judy. 1998. Over-water dispersal of lizards due to hurricanes. *Nature* 395: 556.

Céré, Nicolas. c. 1781. [letter to P.Poivre, undated]. Published in *Rev. Hist. Litt. Ile Maurice* 4 (Archives coloniales): 281–320+.

Chambers, Paul. 2004. *A sheltered life – the unexpected history of the giant tortoise*. London: Hodder-Headline (John Murray). 306pp.

Champagne, Antoine, Turpin, Agnès & Probst, Jean-Michel. 1997. Inventaire préliminaire des tortues marines, d'eau douce et de terre de la Réunion et des îles de l'Océan Indien. *Bull. Phaethon* 6: 65–67.

Chantler, Phil & Driessens, Gerald. 2000. *Swifts – a guide to the swifts and treeswifts of the world*. 2nd.ed. Robertsbridge, Sussex, UK: Pica Press. 272pp.

Chanval[l]on, J.B.T. de. 1804. [Mémoire de Chanvalon sur la Réunion en 1804] [MS, published 1942, ed.A.Lougnon] *Rec. Trim*. 6: 285–332.

Chapotin, Charles. 1812. *Topographie médicale de l'île de France*. Paris: Didot jeune. 104pp.

Charpentier de Cossigny, Jean F. 1732–1755. [Treize lettres de Cossigny à Réaumur, ed. A.LaCroix]. *Rec. Trim*. 4: 168–96, 205–82, 305–16 (1939–40).

— 1764. Mémoire sur L'Isle de France. MS published pp.51–91 *in* Crepin (1922a), q.v.

Charpentier [de] Cossigny, Joseph F. 1799 [An VII]. *Voyage à Canton, capitale de la province de ce nom, à la Chine: par Gorée, le Cap de Bonne espérance et les îles de France et de la*

Réunion. Paris: André. viii+608pp.
Charpentier-Cossigny, Joseph F. 1803 [An XI]. *Moyens d'amélioration et de restauration proposées au gouvernement et aux habitans des colonies...* Paris: Marchant. 3 vols.
Chaudenson, Robert. 1974. *Le lexique du parler créole de la Réunion.* Paris: Honoré Champion. 2 vols.
Chazal, Malcolm [E.] de. 1973. *L'Ile Maurice, protohistorique, folklorique et légendaire.* Port Louis: G. de Spéville. 45pp.
Cheke, Anthony S. 1974a. Report on Rodrigues. London: BOU & Curepipe, Mauritius: Forestry Service. 24pp. (cyclostyled)
— 1974b. Proposition pour introduire à la Réunion des oiseaux rares de Maurice. MS privately circulated, published 1975 in *Info-Nature, Ile Réunion* 12: 25–29.
— 1975a. An undescribed gecko from Agalega: *Phelsuma agalegae* sp.nov. *Mauritius Inst. Bull.* 8: 33–48.
— 1975b. Cyclone Gervaise – an eye-witness comments. *Birds Int.* 1: 13–14.
— 1975c. [untitled letter to the editor about deer at the Plaine des Chicots]. *Info-Nature, Ile Réunion* 13: 128–129.
— 1975d. Un lézard malgache introduit à la Réunion. *Info-Nature, Ile Réunion.* 12: 94–96.
— 1976. Rapport sur la distribution et la conservation du Tuit-tuit, oiseau rarissime de la Réunion. London: BOU (Mascarene Islands Exped. Conserv. Memo 2, 16pp, cyclostyled), reprinted 1977 in *Info-Nature, Ile Réunion* 15: 21–42.
— 1978a. A summary of A.S.Cheke's recommendations for the conservation of Macsarene vertebrates. Conservation Memorandum 3 arising out of the British Ornithologists' Union Mascarene Islands Expedition. Oxford: [author], photocopied, 13pp. [published in French translation in *Info-Nature Ile Réunion* 16: 69–83, 1978]
— 1978b. Habitat management for conservation in Rodrigues. Conservation Memorandum 4 arising out of the BOU Mascarene Islands Expedition. Oxford: [author] 10pp. (photocopied).
— 1979a. The Rodrigues Fody *Foudia flavicans*. A brief history of its decline, and a report on the 1978 expedition. *Dodo* 15: 12–19.
— 1979b. The threat to the endemic birds of Rodrigues (Indian Ocean) from the possible introduction of Ship Rats *Rattus rattus* from vessels coming alongside the proposed new wharf at Port Mathurin. Conservation Memorandum 5 arising out of the BOU Mascarene Islands Expedition. Oxford: [author]. 4pp. (photocopied).
— 1980. Urgency and inertia in the conservation of endangered island species, illustrated by Rodrigues. *Proc. 4th. Pan-Afr. Orn. Congr.*: 355–359.
— & Dahl, Jeremy D. 1981. The status of bats on western Indian Ocean islands, with special reference to *Pteropus*. *Mammalia* 45: 205–238.
— 1982a. *Les noms créoles des oiseaux dans les îles francophones de l'Océan Indien.* Paris: Institut International d'Ethnosciences (Collection l'Homme et son milieu). 64pp.
— 1982b. A note on *Phelsuma* Gray 1825 of the Agalega Islands, Indian Ocean. *Senckenbergiana Biol..* 62: 1–3.
— 1983a. A note on the *Album of a hundred birds* by Kono Bairei, a nineteenth century Japanese artist, with new light on the 'Avis indica' of Collaert. *Arch. Nat. Hist.* 11(2): 291–297.
— 1983b. The identity of Buffon's *Grand Traquet* and other nomenclatural problems in eighteenth century descriptions of endemic Mascarene birds. *Bull. Brit. Orn. Club* 103: 95–100.
— 1984. Lizards of the Seychelles. Pp.331–360 *in* Stoddart 1984a, q.v.
— & Lawley, Jonathan C. 1984 ('1983'). Biological history of Agalega, with special reference to birds and other land vertebrates. *Atoll Res. Bull.* 273: 65–107.
— 1985a. Dodo. Pp.152 *in* Campbell & Lack (eds.) 1985, q.v.
— 1985b. Conservation in Mauritius – the forest ecosystem. A proposal for a long-term study of the ecological dynamics of the native forest. Unpublished report to World Wildlife Fund. 9pp. [photocopied].
— & Diamond, Anthony W. 1986. Birds on Mohéli and Grande Comore (Comoro Islands) in February 1975. *Bull. Brit. Orn. Club* 106: 138–148.
— 1987a. An ecological history of the Mascarene Islands, with particular reference to extinctions and introductions of land vertebrates. Pp. 5–89 *in* Diamond (1987), q.v.
— 1987b. The ecology of the surviving native landbirds of Réunion. Pp.301–358 *in* Diamond (1987), q.v.
— 1987c. The ecology of the smaller land-birds of Mauritius. Pp.151–207 *in* Diamond (1987), q.v.
— 1987d. Observations on the surviving endemic birds of Rodrigues. Pp.364–402 *in* Diamond (1987), q.v.
— 1987e. The legacy of the dodo – conservation in Mauritius. *Oryx* 21: 29–36.
— & Jones, Carl G. 1987. Measurements and weights of the surviving endemic birds of the Mascarenes and their eggs. Pp.403–422 *in* Diamond (1987), q.v.
— 1996. The ecological history of Mauritius and its contribution to the history of environmental conservation. A review of *Green Imperialism* by Richard H.Grove. *PRSAS* 6: 213–218.
— 2001a. Booby *Sula* colonies in the Mascarene area (Indian Ocean): extinctions, myths and colour morphs. *Bull. Brit. Orn. Club.* 121: 71–80.
— 2001b. Is the bird a Dodo ? The wildlife of a mid-seventeenth century drawing of Dutch Mauritius. *Arch. Nat. Hist.* 29: 347–351.
— 2001c. [review of] Roth, T.L., Swanson, W.F. & Brattman, L.K. (eds.) Seventh World Conference on breeding endangered species. *Ibis* 143: 512–3.
— 2003. Treasure Island. The rise and decline of a small tropical museum, the Mauritius Institute. Pp.197–206 *in* Collar, Fisher & Feare (2003), q.v.
— 2004a. The Dodo's last island. *PRSAS* 7: 7–22.
— 2004b. The natural history and human drama of the 18th and 19th century transit expeditions [to Mauritius & Rodrigues]. *Emmanuel Coll. Mag.* 86: 141–147.
— 2004c. Seabirds on Agalega (Indian Ocean) – survival of boobies and frigate-birds into the 1870s, with comments on other species. *PRSAS* 7: 1–5.
— 2005. Naming segregates from the *Columba-Streptopelia* group of pigeons following DNA studies on phylogeny. *Bull. Brit. Orn. Club* 125: 293–295.
— 2006a. Establishing extinction dates – the curious case of the Dodo and the Red Hen. *Ibis* 148: 155–158.
— 2006b. [obituary] France Staub (1920–2005). *Ibis* 148: 610–611.
— 2007a. Wind-blown sand and the accumulation of soil on Round Island. *In* Bullock & North (2007), q.v.
— 2007b. The identity of a flying-fox in Emperor Rudolf II's Bes-

tiaire. *J. Natl. Mus. Praha, Nat. Hist. Ser*. 176.
Cheng Tso-Hsin (ed.) 1963. [*China's economic fauna: birds*] Peking: Science Publishing Society. vi+694pp. [English tr.: Washington, DC: U.S.Dept. of Commerce, Joint Publications Research Service, 1964]
Cherel, J.F. *et al*. 1989. Le point sur *Coracina newtoni* (Tuit-tuit). *Info-Nature, Ile Réunion* 23: 23–45.
Chinery, Michael. 1993. *Insects of Britain & northern Europe*. 3rd.ed. London: HarperCollins. 320pp.
Churchill, Sue. 1998. *Australian bats*. Sydney: Reed New Holland. 230pp.
Clark, George. 1859. A ramble round Mauritius with some excursions into the interior of that island; to which is added a familiar description of its fauna and some subjects of its flora. Pp.i-cxxxii *in* Palmer & Bradshaw, compilers, *The Mauritius Register: Historical, official & commercial, corrected to the 30th June 1959*. Port Louis: L.Channell.
— 1865. The Dodo. Commercial Gazette, Port Louis, 15 Nov. 1865 & [revised] *Mauritius Almanac & Colonial Register* 1869: 37–40. [written in the third person, but almost certainly by Clark; the reprint is stated as revised by him].
— 1866. Account of the late discovery of Dodo's remains in the island of Mauritius. *Ibis* (2)2: 141–146.
Clayton, Dale H., Price, Robert D. & Page, Roderic D.M. 1996. Revision of *Dennyus* (*Collodennyus*) lice (Phiraptera: Menoponidae) from swiftlets, with descriptions of new taxa and a comparison of host-parasite relationships. *Syst. Entomol.* 21: 179–204.
Clergeau, Philippe & Mandon-Dalger, Isabelle. 2001. Fast colonization of an introduced bird: the case of *Pycnonotus jocosus* on the Mascarene Islands. *Biotropica* 33(3): 542–546.
Clouet, Michel. 1978. Le Busard de Maillard (*Circus aeruginosus maillardi*) de l'île de la Réunion. *Oiseau Rev. Fr. Orn.* 48: 95–106.
Clout, Mick N., Karl, B.J., Pierce, R.J. & Robertson, H.A. 1995. Breeding and survival of New Zealand pigeons *Hemiphaga novaeseelandiae*. *Ibis* 137: 264–271.
Clusius, Carolus [Charles LeCluse]. 1605. *Carolus Clusii atrebatis ... Exoticum libri decem: ....* Leiden: Frans Raphelengius. 2 vols.
Coe, Malcom J., Bourn, David & Swingland, Ian R. 1979. The biomass production and carrying capacity of giant tortoises on Aldabra. *Phil. Trans. Roy. Soc., Lond*. B 286: 163–176.
— 1995. Kingdom of the giant tortoise. Pp.111–128 *in* Amin *et al*. (1995), q.v.
Coker, William W. 1856. Rare birds [with engraving of "rare goose and white dodo"]. *Illus. Lond. News* 29 (821): 303.
Cole, Nik [C.]. 2005a. The impact of invasive species on reptile communities. Ph.D Thesis, University of Bristol.
— 2005b. The new noisy neighbours: impacts of alien house geckos on endemics in Mauritius. *Aliens* 22: 8–10.
—, Jones, Carl G. & Harris, Stephen. 2005. The need for enemy-free space: the impact of an invasive gecko on island endemics. *Biol. Conserv.* 125: 467–474.
Colgan, D.J. & Flannery, Tim F. 1995. A phylogeny of Indo-west Pacific Megachiroptera based on ribosomal DNA. *Syst. Biol.* 44: 209–230.
Collaert, Adriaan. 1580–1600. *Avium vivae icones* & *Avium iconum editio secunda*. Antwerp: [publisher unknown] [reprinted in 1 vol., 1967, with intro. by J.Balis, Brussels: Culture et Civilisation. Unpag.]
Collar, Nigel J. & Stuart, S.N. 1985. *Threatened birds of Africa and related islands. The ICBP/IUCN Red Data Book Part 1*. Cambridge: ICBP & IUCN. 761pp.
—, Crosby N.J. & Stattersfield, Alison J. 1994. *Birds to watch 2. The World List of threatened birds*. Cambridge: Birdlife International [Birdlife Conservation Series 4]. 407pp.
—, Fisher, Clem[entine] & Feare, Chris[topher J.] (eds.) 2003. *Why museums matter. Avian archives in an age of extinctions. Bull. Brit. Orn. Club* 123A, Supplement. 360pp.
— 2005. Family Turdidae, Thrushes. *In* vol.10 of J.del Hoyo, A.Elliott & D.Christie (eds.). 1992-[ongoing]. *Handbook of the birds of the world*. Barcelona: Lynx Editions.
Compagnie des Indes. 1711. Mémoire sur l'Isle Bourbon adressé par la Compagnie des Indes au Gouverneur Parat le 17 fèvrier 1711. [MS, published 1940, ed. A.Lougnon] *Rec. Trim.* 5: 183–276.
Condie, Kent C. 1997. *Plate tectonics and crustal evolution*. 4th ed. Oxford: Butterworth-Heinemann. 282pp.
Cooke, A., Lutjeharms, J.R.E. & Vasseur, P. 2003. Marine and coastal ecosystems [around Madagascar]. Pp.179–209 *in* Goodman & Benstead (2003), q.v.
Cooper, Alan *et al*. 2001. Complete mitochondrial genome sequences of two extinct moas clarify ratite evolution. *Nature* 409: 704–707.
— & Penny, David. 1997. Mass survival of birds across the Cretaceous-Tertiary boundary: molecular evidence. *Science* 275: 1109–1113.
Cooper, John E. & West, C.D. 1988. Radiological studies on endangered Mascarene fauna. *Oryx* 22: 18–24.
—, Bloxam, Quentin M.C. & Tonge, Simon J. 1999 ('1998'). Pathology of Round Island geckos *Phelsuma guentheri*: some unexpected findings. *Dodo* 34: 153–158.
Coquerel, Charles. 1848. Note sur les habitudes des Tanrecs et de l'Ericule. *Rev. Zool., Paris* 11: 33–35.
— 1859. Notes de mammalogie. *Rev. Mag. Zool.* 1859: 457–468.
— 1863a. Discours de M le Docteur Coquerel [on Réunion wildlife] *C.R. 1er Séance Générale Ann., Soc. Acclim. Hist. Nat. Réunion* 1863: 9–20.
— 1863b. *Des animaux perdus qui habitaient les îles Mascareignes*. Saint-Denis: [author]. [reprinted *in* Roussin (1860–68), q.v., 3: 74–86, 1865]
— 1864. Catalogue des animaux qui se rencontrent à la Réunion: Oiseaux. *Bull. Soc. Acclim. Hist. Nat. Ile Réunion* 2(1): 7–27.
— 1865b. Les fringilliens. Pp.169–173 *in* Roussin (1860–68), q.v., vol.3.
Corbett, Gerald B. & Southern, H.N. 'Mick'. 1977. *The handbook of British mammals*. 2nd.ed. Oxford: Blackwell Scientific Publications. 520pp.
Corby, Thomas. 1845. [letter to J.A.Lloyd, H.M.Surveyor General in Mauritius, dated 30/10/1845, on the subject of Rodrigues]. MS in UK Public Record Office: C.O.167, vol.262 [the MS is unattributed, but the ship's log, HMS *Conway*, confirms the author's identity; see also North-Coombes 1971].
— 1846. Report on Serpent Island. Appendix, pp.161–2 to Lloyd (1846), q.v.
Cornell, Christine (tr.). 1974. *The journal of post-captain Nicola Baudin, commander-in-chief of the corvettes* Géographe *and* Naturaliste. Adelaide: Libraries Board of South Australia.

609pp.

Cornu, Henri. 1974. L'Homme et la Fournaise. *Cah. Centre. Univ. Réunion, No. Spécial Colloque Commerson*: 124–130.

Cossigny: see – Charpentier de Cossigny / Charpentier-Cossigny.

Coulaud, Joelle, Brown, Spencer G. & Siljak-Yakovlev, Sonja. 1995. First cytogenetic investigation in populations of *Acacia heterophylla*, endemic from La Réunion island, with reference to *A.melanoxylon*. *Ann. Bot.* 75: 95–100.

Courbet, Gérard. 1998. Capture d'un Furet *Mustela furo* à Saint-André ! (île de la Réunion). *Bull. Phaethon* 7: 48.

Courchamp, Franck, Langlais, Michel & Sugihara, George. 1999. Cats protecting birds: modelling the mesopredator effect. *J. Anim. Ecol.* 68: 282–292.

—, Chapuis, Jean-Louis & Pascal, Michel. 2003. Mammal invaders on islands: impact, control and control impact. *Biol. Rev.* 78: 347–383.

Courrier de Bourbon, log of the. 1720–22. [extracts from] Journal du voyage des Indes sur le vaisseau *Le Courrier de Bourbon*, commandé par M Gillet [1720–1722]. *Rev. Retrospective Ile Maurice* 3: 149–166 (1952).

Courtillot, Vincent. 1999. *Evolutionary catastrophes – the science of mass extinction*. Cambridge: Cambridge University Press. 188pp.

Courts, Sian E. 1997. Insectivory in captive Livingstone's and Rodrigues fruit bats *Pteropus livingstonii* and *P. rodricensis* (Chiroptera: Pteropodidae): a behavioural adaptation for obtaining protein. *J. Zool. Lond.* 242: 404–410.

Couteau, Evelyne, Lanux, Patrick D.de & Murat, Georges. 2000. D'argent au noyer de sinople. Jean-Baptiste-François de Lanux. *Hommes et Plantes* 35: 27–32.

Couzi, François & Salamolard, Marc. 2002. *Étude des Phasianidés, Turnicidés et Colombidés à la Réunion. Bilan de l'enquête 2001–2002*. Saint-André, Réunion: SEOR & ONF. 34pp.

Cowen, D.V. 1969. *Flowering trees and shrubs in India*. 5th rev.ed. Bombay: Thacker & Co. 159pp.

Cowie, Robert H. 2001. Can snails ever be effective and safe biocontrol agents ? *Int. J. Pest Manag.* 47(1): 23–40.

Cowles, Graham, S. 1987. The fossil record. Pp.90–100 in Diamond (1987), q.v.

— 1994. A new genus, three new species and two new records of extinct Holocene birds from Réunion Island, Indian Ocean. *Geobios* 27(1): 87–93.

Cowper, S.G. 1984. Birds from the Mascarene islands in the collections of the Merseyside County Museums. *Mauritius Inst. Bull.* 10: 7–14.

Cox, C.Barry, Healey, Ian N. & Moore Peter D. 1976. *Biogeography. An ecological and evolutionary approach*. Oxford: Blackwell Scientific Publications. 194pp.

Cox, G.W. 2004. *Alien Species and Evolution*. Washington, DC: Island Press. 392pp.

Cracraft, Joel. 1981. Towards a phylogenetic classification of the recent birds of the World. *Auk* 98: 681–714.

Craig, J.L. & Veitch, C.R ('Dick'). 1990. Transfer of organisms to islands. Pp.255–260 *in* Towns *et al.* (1990). q.v.

Crémont, H.de. 1768. Rélation du voyage au volcan de l'Isle de Bourbon les 27 et 28 octobre 1768 par Monsieur de Crémont, commissaire ordonnateur en la dite isle. [ed A.Lougnon] *Rec. Trim.* 3: 15–18+ (1937).

Crépin, [?]. 1887. Sur la Caille de Madagascar. *Bull. Soc. Natl. Acclim., Paris* (4)4: 240–241.

Crépin, Paul. 1922a. *Charpentier de Cossigny, fonctionnaire colonial, d'après ses écrits et ceux de quelques-uns de ses contemporains*. Thesis, Université de Paris, Faculté des Lettres. 107pp.

— 1922b. *Mahé de Labourdonnais*. Paris: Leroux. 485pp.

— 1931. Paul Carié 1876–1930. *Bull. Soc. Natl. Acclim. France* 78: 302–394.

Cronk, Quentin C.B. & Fuller, Janice L. 1995. *Plant invaders. The threat to natural ecosystems*. London: Chapman & Hall. 241pp. [reprinted 2001, London: Earthscan Publications].

Crosby, Gilbert T. 1972. Spread of the Cattle Egret in the Western Hemisphere. *Bird-Banding* 43: 205–212.

Cubitt, Gerald. 1977. *Islands of the Indian Ocean*. London: Cassell. 176pp.

Cuellar O. & Kluge, A.G. 1972. Natural parthenogenesis in the gekkonid lizard *Lepidodactylus lugubris*. *J. Genet.* 61: 14–26.

Cundall, David & Irish, Frances J. 1987 ('1986'). Aspects of locomotor and feeding behaviour in the Round Island Boa *Casarea dussumieri*. *Dodo* 23: 108–111.

— & Irish Frances J. 1989. The function of the intramaxillary joint in the Round Island Boa *Casarea dussumieri*. *J. Zool., Lond.* 217: 569–598.

Dalleau-Coudert, S. 2005. Braconnage de Merle pays. *Chakouat* 13: 10–11.

Damholdt, Marianne & Linnebjerg, Jannie F. 2003. Distribution, foraging behaviour and habitat selection of the Mauritius Paradise Flycatcher *Terpsiphone bourbonnensis desolata*. Unpubl. report to MWF, 44pp., photocopied.

Damour, Christian (ed.). 1973. *Leconte de Lisle (anthologie)*. Rodez, France: Éditions Subervie. 102pp.

Daniel, J.C. 2002. *The book of Indian reptiles and amphibians*. Bombay: Bombay Natural History Society & Oxford University Press. 238pp.

Daniel, M.J. 1975. First record of an Australian fruit bat (Megachiroptera: Pteropodidae) reaching New Zealand. *NZ J. Zool.* 2: 227–231.

Darby, P.W.H. 1979 ('1978'). The breeding of Meller's Duck *Anas melleri* (Sclater) at the Jersey Zoological Park. *Dodo* 15: 29–32.

— , Jeggo David F. & Redshaw, M.E. 1985 ('1984'). Breeding and management of the Rodrigues Fody *Foudia flavicans*. *Dodo* 21: 109–126.

Daruty de Grandpré, [J.M.R.] Albert. 1878. [on *Margaroperdix striata*]. *TRSAS* NS 10: : 55.

— 1883a. Rapport annuel de secrétaire, 5 février 1879. *TRSAS* NS 12: 134–137.

— 1883b. Rapport annuel du Secrétaire. 3 février 1881. *TRSAS* NS 13: 72–86.

— 1889. *Compte-rendu de la séance annuelle du 28 janvier 1889*. Port Louis: Royal Society of Arts & Sciences. 25pp.

Darwin, Charles R. 1839. *Journal of researches into the natural history and geology of the countries visited during the voyage round the world of H.M.S. Beagle*. London: Henry Colburn [revised editions in 1845 & 1860; often reprinted as *The Voyage of the Beagle*]

Das, Ashok K. 1973. The Dodo and the Mughals. *History Today* 23: 60–73.

Das, Indraneil. 1991. *Colour guide to the turtles and tortoises of the Indian subcontinent*. Portishead, UK: R&A Publishing.

133pp.

— 2004. *Lizards of Borneo*. Kota Kinabalu: Natural History Publications. 83pp.

Daszak, P. 1994. The 1993 Raleigh International Round Island Expedition, including a survey of intestinal parasites collected from animals of Round Island, Mauritius, and observations on Round Island reptiles. Unpubl report; copy in library of Kingston University, UK.

Davies, K.C. & Hull, James. 1976. *The zoological collections of the Oxford University Museum*. Oxford: Oxford University Museum. 136pp.

Davis, P.M.H. & Barnes, Michael J.C. 1991. The butterflies of Mauritius. *J. Res. Lepidopt.* 30: 145–161.

Day, David. 1981. *The Doomsday book of animals – a unique natural history of vanished species*. London: Ebury Press. 288pp.

Decary, Raymond. 1948. L'île de la Réunion. Pp.87–144 in H.Deschamps (ed.) *L'Union Française: Côte des Somalis, Réunion, Inde*. Paris: Berger-Levrault. 209pp.

— 1962. Sur des introductions imprudentes d'animaux aux Mascareignes et à Madagascar. *Bull. Mus. Hist. Nat., Paris* (2)34(5): 404–407.

Defos du Rau, Jean. 1960. *L'Ile de la Réunion*. Bordeaux: Institut de Géographie. 716 pp.

Delabarre de Nanteuil. 1844. *Legislation de l'île Bourbon. Repertoire raisonné des lois ... en vigeur dans cette colonie*. Paris: J-B.Gross. 3 vols.

Delaleu, J.B.Étienne. 1826. *Code des îles de France et de Bourbon*. 2nd.ed. Port Louis: Tristan Mallac. 2 vols.

Dellon, Gabriel. 1685. *Relation d'un voyage des Indes orientales*. Paris: Claude Barbin. 284pp.

Desjardins, Julien. 1829. Description physique de l'Ile d'Ambre. MS published in *TRSAS* NS 6: 35–41 (1872) as 'L'Ile d'Ambre il y a 42 ans'.

— 1830. Premier Rapp. Ann. Trav. Soc. Hist. Nat. Ile Maurice [MS published in Ly-Tio-Fane (1972), q.v.]

— 1831a. Sur trois espèces de lézard du genre scinque qui habitent l'Ile Maurice (Ile de France). *Ann. Sci. Nat.* 22: 292–299.

— 1831b. Deuxième Rapp. Ann. Trav. Soc. Hist. Nat. Ile Maurice. [MS, published pp.25–51 in Ly-Tio-Fane (1972), q.v.]

— 1832. Troisième Rapp. Ann. Trav. Soc. Hist. Nat. Ile Maurice. [MS published pp.53–112 in Ly-Tio-Fane (1972), q.v.]

— 1834. Cinquième Rapp. Ann. Trav. Soc. Hist. Nat. Ile Maurice [MS published pp.139–166 in Ly-Tio-Fane (1972), q.v.]

— 1835. 6e *Rapp. Ann. Trav. Soc. Hist. Nat. Ile Maurice*.

— 1837. 8e *Rapp. Ann. Trav. Soc. Hist. Nat. Ile Maurice*.

— 1840. 9e *Rapp. Ann. Trav. Soc. Hist. Nat. Ile Maurice* [read 24 August 1838]

Desnoyers, J.L. 1837. Aperçu d'une topographie médicale de l'Ile Maurice, *Rapp. Ann. Trav. Soc. Hist. Nat. Ile Maurice* 8: 21–27.

Deso, Grégory. 2001. Note sur le transport insolite des geckos verts: le cas de *Phelsuma inexpectata*. *Bull. Phaethon* 13: 56.

— & Probst, Jean-Michel. 2007. *Lycodon aulicus* Linnaeus 1758 et son impact sur l'herpétofaune insulaire à la Réunion (Ophidia: Colubridae: Lycodontidae). *Bull. Phaethon* 25: 29–36.

Devaux, Michel J.C. 1983. *Mauritius. Economic memorandum: recent developments and prospects*. Washington: The World Bank. 122pp.

Dewar, R.F. 2003. Relationship between human ecological pressure and the vertebrate extinctions [in Madagascar]. Pp.119–122 in Goodman & Benstead (2003), q.v.

Diamond, Anthony W. 1980. Seasonality, population structure and breeding ecology of the Seychelles Brush Warbler *Acrocephalus sechellensis*. *Proc. 4th. Pan-African Orn. Congr.*: 253–266.

— & Feare, Christopher J. 1980. Past and present biogeography of central Seychelles birds. *Proc. 4th. Pan-African Orn. Congr.*: 89–98.

— 1985. The conservation of land birds on islands in the tropical Indian Ocean. Pp.85–100 in Moors (1985), q.v.

— 1987. (ed.) *Studies of Mascarene Island birds*. Cambridge: Cambridge University Press. 458pp.

— & Cheke Anthony S. 1987. Introduction. Pp.1–4 in Diamond (1987), q.v.

— & Schreiber, Rudolf L. (eds.). 1987. *Save the birds*. Cambridge, UK: Cambridge University Press & [Rev.ed. 1989] Boston, USA: Houghton Mifflin. 384pp.

Dickinson, Edward C. (ed.) 2003. *The Howard & Moore complete checklist of the birds of the world*. 3rd.ed, rev. & enlarged. London: A&C Black/Christopher Helm.

Dictionary of Mauritian Biography. 1941-. [part-work compilation under a series of editors & numerous contributors]. Port Louis: Société d'Histoire de l'Ile Maurice. [separate parts rebound in 2 vols to 1997 & again to 2003 + ongoing parts].

Dingwall, P.R., Atkinson, Ian A.E. & Hay, C. (eds.) 1978. *The ecology and control of rats in New Zealand nature reserves*. Wellington: Department of Lands & Surveys, Information Series No.4. Pp.237.

Dobson, G.E. 1879. [the collections from Rodriguez] Mammalia. *Phil. Trans. Roy. Soc., Lond.* 168: 457–458.

Dodd, Jan & Philippe, Madeleine. 2004. *Mauritius, Réunion and the Seychelles*. 5th. ed. Hawthorn, Australia: Lonely Planet Publications. 304pp.

Dommen, Edward & Dommen Bridget. 1999. *Mauritius. An island of success*. Wellington, NZ: Pacific Press & Oxford: James Currey. 100pp.

Dormann, Geneviève & Rossi, Guido. 1991. *Mauritius from the air*. Paris: Nouvelles Editions du Pacifique. 135pp.

Douglas, Malcolm J.W. 1982. Biology and management of rusa deer on Mauritius. *FAO/UN Tigerpaper* 9(3): 1–10.

Doumenge, C. & Renard, Y. 1984. *La conservation des écosystèmes forestiers de l'île de la Réunion*. Gland, Switzerland & Cambridge, UK: IUCN -&- Saint-Denis: SREPEN. 95pp.

Dowler, Robert C.; Carroll, Darius & Edwards, Cody W. 2000. Rediscovery of rodents (genus *Nesoryzomys*) considered extinct in the Galapagos Islands. *Oryx* 34: 109–117.

Dowsett-Lemaire, Françoise. 1994. The song of the Seychelles Warbler *Acrocephalus sechellensis* and its African relatives. *Ibis* 136: 489–491.

Dubois. 1674. *Les voyages fait par le sieur D.B. aux isles Dauphine ou Madagascar et Bourbon ou Mascarenne es années 1669,70,71 et 72*. Paris: Claude Barbin. 234pp. [for Eng.tr. see Oliver (1897); the section on Réunion was reprinted in full in Lougnon (1970), q.v., and his list of birds by, *inter alia* Berlioz (1946) & Barré, Barau & Jouanin (1996), qq.v.]

Duck, Callan D. & Gardner, Andrew S. 1979. The Trindade Petrel *Pterodroma arminjoniana* and other birds on Round Island, Mauritius. Unpubl. draft of paper [of which only the section

on the petrel was published: see Gardner *et al.* 1985]
Ducros, Père. 1725. Lettre du *Père* Ducros, missionaire de la Compagnie des Indes, à M l'abbé Raguet, directeur de la Compagnie des Indes. *Letres Édifiantes, Mémoires des Indes* 18: 1–26 (1728 ed.), 13: 320–340 (1781 ed.) & 7: 461–473 (1819 ed.) [parts also in Barnwell (1948) & Lougnon (1970), qq.v.]
Duing, Walter. 1970. *The monsoon regime of the currents in the Indian Ocean*. Honolulu: East-West Center Press. 68pp.
Dulloo, M.Ehsan. 1994. Botanical survey of Gunners Quoin, Mauritius. *Rev. Agric. Sucrière Ile Maurice* 73: 27–36.
— 1996. The flora and conservation potential of offshore islets around Rodrigues, Indian Ocean. *PRSAS* 6: 13–36.
— *et al.* 1996. Ecological restoration of native plant and animal communities in Mauritius, Indian Ocean. Pp.83–91 *in* D.L.Pearson & C.V.Klimas (eds.) *The role of restoration in ecosystem management*. Madison, Wisconsin: Society for Ecological Restoration.
— *et al.* 1997. *Ile aux Aigrettes management plan 1997–2000*. Port Louis: MWAF (Tech. Ser. 1/97).
—, Bullock, David J. & North, Steve. n.d.[1999]. *Report on the expedition to Round Island and Gunner's Quoin, Mauritius. July/August 1996*. Port Louis: Mauritian Wildlife Foundation. 62pp.
Duméril, A.M.C. & Bibron, G. 1834–54. *Erpetologie générale ou histoire naturelle complète des reptiles*. Paris: Librairie Roret. 9 vols + 'atlas'.
— & Duméril [A.H.] Auguste. 1851. *Catalogue méthodique de la collection des reptiles du Muséum d'Histoire Naturelle de Paris*. Paris: Gide & Baudry. 224pp.
Dumont-d'Urville, Jules S.C. 1830–33. Histoire de voyage. 5 vols. [part of *Voyage de découvertes de l'Astrolabe, exécuté par ordre du Roi, pendant les années 1826–1827– 1828–1829 sous le commandement de M J.Dumont-d'Urville*] Paris: J.Tastu.
Duncan, James. 1857. Observations respecting the island of Rodrigues. Pp.11–14 *in* J.R.Mann (ed.) *Report on Rodrigues as a quarantine station*. Port Louis: Government Printer.
Duncan, John S. 1828. A summary review of the authorities on which naturalists are justified in believing that the Dodo, *Didus ineptus* Linn., was a bird existing in the Isle of France, or the neighbouring islands until a recent period. *Zool. J.* 3: 554–567.
Duncan, R.A. & Richards, M.A. 1991. Hotspots, mantle plumes, flood basalts and true polar wander. *Revs. Geophysics* 29(1): 31–50.
Duncan, Richard P. & Young, Jim R. 1999. The fate of passeriform introductions on oceanic islands. *Conserv. Biol.* 13: 934–936.
—, Blackburn, Tim M. & Sol, Daniel. 2003. The ecology of bird introductions. *Ann. Rev. Ecol. Evol. Systematics* 34: 71–98.
Duplantier, J-M. & Duchemin, J-B. 2003. Introduced small mammals and their ectoparasites: a description of their colonization and its consequences. Pp. 1191–1198 *in* Goodman & Benstead (2003), q.v.
Dupon, J.F. 1969. *Receuil de documents pour servir à l'histoire de Rodrigues*. Port Louis: Mauritius Archives, Publ.10. 121pp.
Dupont, Evenor. 1860, Moineaux et corbeaux. *TRSAS* NS 1: 326–328.
Dupont, Joël, Girard, Jean-Claude & Guinet, Marcel. 1989. *Flore en détresse. Le livre rouge des plantes indigènes menacées à la Réunion*. Saint-Denis: SREPEN & Région Réunion. 133pp.
Duquesne, Henri. 1689. *Receuil de quelques mémoires servant d'instruction pour l'établissement de l'île d'Eden*. Amsterdam: H.Debordes. [reprinted 1984 & 1995 as an appendix to editions of Leguat (1707), q.v.]
Durand, Hélène. 2001. Apport d'un SIG nomade pour cartographier la végétation naturelle de l'île de la Réunion. *Rev. XYZ Assoc. Fr. Topog.* 89: 43–46.
Durrell, Gerald M. 1977a ('1976'). The Mauritian expedition. *Jersey Wildl. Preserv.Trust Ann.Rep.* 13: 7–11.
Durrell, Gerald [M]. 1977b. *Golden Bats and Pink Pigeons*. London: Collins [also New York: Simon & Schuster, with photos (none in UK ed.)]
— & Durrell, Lee M. 1980. Breeding Mascarene wildlife in captivity. *Int. Zoo Yearb.* 20: 112–119.
— 1990. *The ark's anniversary*. London: HarperCollins & (1991) New York: Little, Brown/Arcade. 179pp.
Dyment, Jerôme *et al.* 2003. Interaction between the Deccan-Réunion hotspot and central Indian Ocean ridge. *[proceedings of] InterRidge Symposium & Workshop: Ridge-Hotspot interaction, 8–10 September 2003*: 16–17.
Easteal, S.O. 1981. The history of introductions of *Bufo marinus*, a natural experiment in evolution. *Biol. J. Linn. Soc.* 16: 93–113.
Edgerley, Leo F. 1961 ('1962'). *Annual Report of the Forest Department for the year 1960*. Port Louis: Government Printer (Colony of Mauritius No.24 of 1961).
— 1962. The conifers on Mauritius. *PRSAS* 2: 190–202.
Edmunds, K. (2005). Resource utilisation: territory distribution and supplementary food use by the endangered Pink Pigeon *Columba mayeri*. M.Sc. Thesis. University of East Anglia.
Edwards, A. 1872. [Quelques renseignements au sujet d'une tortue...]. *TRSAS* NS 6: 12–13.
Edwards, George. 1760. *Gleanings of Natural History*. London: George Sidney. 3 vols.
Edwards, Scott V. *et al.* 2002. A genomic schism in birds revealed by phylogenetic analysis of DNA strings. *Syst. Biol.* 51: 599-613.
Eeckhout, Paul (ed.) 1954. *Roelandt Savery 1576–1639*. Gand/Ghent: Muséee des Beaux-Arts. Pp.50 + plates.
Eger, Judith L. & Mitchell, Lorelie. 1996. Biogeography of the bats of Madagascar. Pp.321–328 *in* Lourenço (1996), q.v.
— & — 2003. Chiroptera, Bats. Pp.1287–1298 *in* Goodman & Benstead (2003), q.v.
Eggli, Edmond (ed.). 1943. *Leconte de Lisle. Poèmes choisis*. Manchester: Manchester University Press. 132pp.
Eikenaar, Cas & Skerrett, Adrian. 2006. First record of Kermadec Petrel *Pterodroma neglecta* for Seychelles. *Bull. Afr. Bird Club* 13: 88–90.
Ellis, William. 1859. *Three visits to Madagascar during the years 1853, 1854, 1856 ...* London: John Murray. 470pp.
Elton, Charles. 1958. *The ecology of invasions by animals and plants*. London: Methuen. 181pp.
Emmerez de Charmoy, André d'. 1941. Le Boul-boul (*Otocompsa jocosa peguensis* W.L.Sclater). *Rev. Agricole Ile Maurice* 20: 154–158.
Emmerez de Charmoy, [P.] Donald d'. 1901. Du rôle des insectes dans la mortalité des arbres forestiers à Maurice. *Rev. Agricole Ile Maurice* NS 3(1): 1–8.
— 1903. *Rapport sur la faune ornithologqiue éteinte de l'Ile Maurice*. Port Louis: Mauritius Institute. 8pp. [reprinted 1931,

Rev.Agricole Ile Maurice 58: 134–140]
— 1914. Fauna. Pp.94–101 in Macmillan (1914), q.v.
— 1928. Mauritius: the surviving fauna. *J. Soc. Preserv. Fauna Empire* NS 8: 108–109.
Epinay, Adrien d'. 1890. *Renseignements pour servir à l'histoire de l'Isle de France jusqu'à l'année 1810 inclusivement*. Port Louis: Nouvelle Imprimerie Dupuy. vii+577+xxiii pp.
Eskildsen, Louise I., Olesen, Jens M. & Jones, Carl G. 2004. Feeding response of the Aldabra giant tortoise (*Geochelone gigantea*) to island plants showing heterophylly. *J. Biogeog.* 31; 1785–1790.
Eudes, [Honoré]. 1832. [letter to Charles Telfair from Rodrigues, dated 20/4/1832, kept in the library of the Zoological Society of London].
Evers, Sandra J. & Hookoomsing, Vinesh Y. (eds.) 2000. *Globalisation and the south-west Indian Ocean*. Leiden: International Institute for Asian Studies & Réduit, Mauritius: University of Mauritius. 235pp.
Eyles, N. & Young. G.M. 1994. Geodynamic controls on glaciation in Earth history. Pp.5–22 in M.Deynoux et al. (eds.) *Earth's glacial record*. Cambridge, UK: Cambridge Univ.Press.
Fabian, D.T. 1970. The birds of Mauritius. *Bokmakierie* 22: 16–17, 21.
Farber, Paul L. 1982. *Discovering birds. The emergence of ornithology as a scientific discipline, 1760–1850*. New York: D.Reidel [reprinted 1997 with new preface – Baltimore: Johns Hopkins Univ.Press, xxiii+191pp.]
Farchi, Jean. 1937. *Petite histoire de l'Ile Bourbon*. Paris: Presses Universitaires de France.
Farrington, Anthony. 1999. *A catalogue of the East India ships' journals and logs, 1660–1834*. London: British Library. 800pp.
Faure, Gerard & Montaggioni, Lucien. 1975. Le récif corallien de l'Ile Rodrigue: géomorphologie et repartition des peuplements. *Cah.Centre Univ. Réunion* 6: 63–96.
Feare, Christopher J. 1978. The decline of booby (Sulidae) populations in the western Indian Ocean. *Biol.Conserv.* 14: 295–305.
— 1984a. Seabird status and conservation in the tropical Indian Ocean. Pp. 457–471 in J. J. Croxall, P. G. H. Evans & R. W. Screibers (eds.) *Status and conservation of the world's seabirds*. Cambridge: ICBP [Tech.Pub.2].
— 1984b. Seabirds as a resource: use and management. Pp.593–606 in Stoddart (1984), q.v.
— & Mungroo, Yousoof. 1989. Notes on the House Crow *Corvus splendens* in Mauritius. *Bull. Brit. Orn. Club.* 109: 199–201.
— & Mungroo, Yousoof. 1990. The status and management of the House Crow *Corvus splendens* (Vieillot) in Mauritius. *Biol. Conserv.* 51: 63–70.
— & Craig, Adrian. 1998. *Starlings and mynahs*. London: A&C Black (Christopher Helm). 285pp.
— , Jacquement, Sébastien & Le Corre, Mathieu. 2007. An inventory of Sooty Terns *Sterna fuscata* in the western Indian Ocean, with special reference to threats and trends. *Ostrich* 78: 423–434.
Feuilley. 1705. Mission à l'île Bourbon du Sieur Feuilley en 1704 [ed. A.Lougnon]. *Rec.Trim.* 4: 3–56, 101–167 (1939) [the part listing Réunion birds is reprinted in Barré et al. 1996, q.v.]
Filardi, Christopher E. & Moyle, Robert G. 2005. Single origin of a pan-Pacific bird group and upstream colonization of Australasia. *Nature* 438:216–219.
Fishpool, Lincoln D.C. & Evans, Michael I. 2001. *Important birds areas in Africa and associated islands. Priority sites for conservation*. Cambridge: Birdlife International & Newbury, UK: Pisces Publications (Birdlife Conservation Series No.11).
*Flacourt, Etienne de. 1658. *Histoire de la grande isle Madagascar*. Paris: J.Henault [& expanded ed. 1661, Troyes, France: Nicolas Oudot. 471pp.; many times reprinted, e.g. Saint-Denis: Ars Terres Créoles, 1991 (1658 ed.)].
Flannery, Tim. 1995a. *Mammals of the South-west Pacific & Moluccan Islands*. Chatsworth, NSW: Reed Books & Ithaca, NY: Cornell University Press. 464pp.
— 1995b. *Mammals of New Guinea*. Rev.ed. Chatsworth, NSW: Reed Books & Ithaca, NY: Cornell University Press. 568pp.
— & Schouten, Peter. 2001. *A gap in nature. Discovering the world's extinct animals*. London: Random House (Heinemann). 184pp.
Flinders, Matthew. 1803–14. *Private journal of Matthew Flinders 1803–1814. The Mitchell Library manuscript*. Guildford, UK: Genesis Publications & Alphington, Victoria: Hedley Australia. 366pp. [facsimile, ed. G.C.Ingleton] [also on-line at www.sl.nsl.gov.au/flinders/manuscripts]
— 1814. *A voyage to Terra Australis, undertaken for the prupose of completing the discovery of that vast country and prosecuted in the years 1801, 1802 and 1803...* London: G. & W.Nicol. 3vols.
Florens, F. B.Vincent. 1996. A study of the effects of weeding invasive exotic plants and rat control on the diversity and abundance of land gastropods in the Mauritian wet upland forest. B.Sc. (Hons.) thesis. University of Mauritius.
— 2003. Research undertaken at the University of Mauritius of relevance to the management of alien invasive species for the restoration of biodiversity. Pp. 76-80 in Mauremootoo (2003a), q.v.
—, Daby, D. & Jones, R. 1998. The impact of controlling alien plants and animals on the snail fauna of forests on Mauritius. *J.Conchol., Special Publ.* 2: 87–88.
— & Turpin, Agnès. 2001. Découverte de materiel ostéologique de Scinque de Telfair *Leiolopisma telfairi* au Corps de Garde (île Maurice). *Bull. Phaethon* 14: 67–70.
— 2002. osteological finds on Trois Mamelles mountain extends the known ecological range of the extinct endemic Mauritian tortoise *Cylindraspis* sp. *Phelsuma* 10: 56–58.
Fontaine, W. 2005. Découverte d'un Pétrel Noir. *Chakouat* 13: 6.
Forbes-Watson, A.D. 1969. Notes on birds observed in the Comores on behalf of the Smithsonian Institution. *Atoll Res.Bull.* 128: 1–23.
Forshaw, Joseph M. & Cooper, William T. 1989. *Parrots of the World*. 3rd.ed. London: Cassell & Melbourne: Lansdowne Press. 672pp.
[Foucher-d'Obsonville]. 1783. *Essais philosophiques sur les moeurs de divers animaux étrangers, avec des observations relatives aux principes & usages de plusieurs peuples, ou extraits des voyages de M*** en Asie*. Paris: Couturier fils & veuve Tilliard et fils. 430pp.
Foucherolle, Soullet & Tardif. 1714. Mémoire pour faire porter à l'isle de Bourbon des plants de café de Moka. Instruction pour la prise de possession de Cirné ou Maurice. MS in Mauritius

Archives (X14/1/1) publ. in *Rev. Retrospective Ile Maurice* 5: 131–33 (1954) & Document III in Pineo (1993), q.v.

Fowler, S.V. *et al..* 2000. Biological control of weeds in Mauritius: past successes and present challenges. *Proc. X Int. Symp. Biol. Control of Weeds, Bozeman, Montana 1999*: 43–50.

Fox, George T. 1827. *Synopsis of the Newcastle Museum, late the Allan, formerly the Tunstall or Wycliffe Museum*. Newcastle: T.& J.Hodgson. 312pp.

Frauenfeld, G.von. 1868. *Neu aufgefundene Abbildung des Dronte und eines zweiten kurzflügeligen Vogels*. Vienna: C.Uebererreuter'sche Buchdrückerei. 12pp. [abridged versions also in *Ibis* (2)4: 480–482 & *J. Orn.* 16: 138–140, both 1868]

Frazier, Jack. 1984. Marine turtles in the Seychelles and adjacent territories. Pp.417–468 *in* Stoddart (1984a), q.v.

Freeman, Joseph J. 1851. *A tour in South Africa with notices of Natal, Mauritius, Madagascar, Ceylon, Egypt and Palestine*. London: J.Snow. xii+492pp.

Freeman, Karen J.M. 2003. *The ecology and conservation genetics of the* Gongylomorphus *skinks of Mauritius*. Ph.D. Thesis, Queen Mary College, London University.

Fremantle, E.R. 1914. *The navy as I have known it, 1849–1899*. London: Cassell & Co. 472pp.

Fréri [?]. 1751. [a visit to the *volcan* on Bourbon, ed. A.Lougnon] *Rec. Trim*. 3:8–12, 1937.

Freycinet, Louis [C.D. Saulces] de. 1816: [see Peron 1807]

— (ed.). 1824–44. *Voyage autour du monde exécuté sur les corvettes de S.M. l'Uranie et la Physicienne pendant les années 1817, 1818, 1819, et 1820*. Paris: Pillet ainé. 9 vols + 4 vol. atlas [illustrations].

Friedmann, François & Cadet, Thérésien. 1976. Observations sur l'hétérophyllie dans les îles Mascareignes. *Adansonia* (2)15: 423–440.

Friedmann, Herbert. 1956. New light on the Dodo and its illustrators. *Smithsonian Rep*. for 1955: 475–481.

Frith, Clifford B. 1977. Life history notes on some Aldabran land birds. *Atoll Res. Bull*. 210: 1–17.

— 1979. Feeding ecology of land birds on West Island, Aldabra Atoll, Indian Ocean: a preliminary study. *Phil. Trans. Roy. Soc. Lond*. B 286: 195–210.

Frith, Harry J. 1982. *Waterfowl in Australia*. Rev. ed. London & Sydney: Angus & Robertson. 332pp.

Fritts, Thomas H. 1993. The common wolf snake, *Lycodon aulicus capucinus*, a recent colonist of Christmas Island in the Indian Ocean. *Wildl. Res*. 20: 261–265.

— & Rodda, Gordon H. 1998. The role of introduced species in the degradation of island ecosytems: a case history of Guam. *Ann. Rev. Ecol. Syst*. 29: 113–140.

Froberville, Eugène de. 1848. Rodrigues, Galéga, Les Séchelles, Les Almirantes etc. Pp.64–114 *in* Part 3 of M.A.P. d'Avezac (ed.) *Iles de l'Afrique*. Paris: Firmin Didot. 3 vols.

Fröhlich, François. 1996. La position de Madagascar dans le cadre de l'évolution géodynamique et de l'environnement de l'Ocean Indien. Pp.19–26 *in* Lourenço (1996), q.v.

Fuller, Errol. 1987. *Extinct birds*. London: Viking Rainbird/ Penguin Books Ltd. 256pp.

— 2000. *Extinct birds*. Oxford: Oxford University Press. 398pp. [text extensively rewritten from Fuller (1987), q.v.]

— 2002. *Dodo – from extinction to icon*. London: HarperCollons. 180pp.

Fuma, Sudel. 1998. *L'abolition de l'esclavage à la Réunion*. St.André, Réunion: GRAHTER & Océan Editions. 178pp.

— 2002. *Histoire d'une passion. Le sucre de canne à la Réunion*. Réunion: Muséum Stella Matutina & Océan Editions. 312pp.

Fusée-Aublet, Jean-Baptiste C. 1775. Notices pour servir à l'histoire naturelle de l'Ile de France. Mémoire 10, pp.129–160 *in* vol.2 of same author's *Histoire des plantes de la Guiane Française, rangées suivant la methode sexuelle, avec plusieurs mémoires*. Paris: P.F.Didot. 4 vols.

Gade, Daniel W. 1985. Man & nature in Rodrigues: tragedy of an island common. *Env. Conserv*. 12: 207–216.

Gallon, Richard C. 2005. On a new genus and species of therapsoid spider from Serpent Island, Mauritius (Araneae, Theraphosidae, Eumenophorinae). *Bull. Br. Arachnol. Soc*. 13: 175–178.

Gandon, Abbé. 1732. Les Mascareignes vues par l'*Abbé* Gandon en 1732 [ed. A.Lougnon]. *Rec.Trim*. 5: 74–83 [1940] [Eng.tr. of the Mauritius part only in Barnwell (1948), q.v.]

Gardner, Andrew [S]. c1985. Viability of the eggs of the day-gecko *Phelsuma sundbergi* exposed to sea water. MS typescript, 5pp. [copy in ASC's library]

— , Duck,, Callan D. & Grieg, Susan. 1985. Breeding of the Trindade Petrel *Pterodroma arminjoniana* on Round Island, Mauritius. *Ibis* 127: 517–522.

Gardyne, Alexander. 1846. *The shipwreck of the bark* Trio *at Rodrigues, 6 March 1846*. [MS ed. & pub. by J.Brouard, Mauritius, n.d. (c.1999), 66pp]

Garnier du Fougeray. 1722. Lettre du chevalier Garnier du Fougeray aux Directeurs de la Compagnie des Indes. Groix, le 22 mars 1722. *Rev. Retrospective Ile Maurice* 6: 67–80 [1955].

Garrido, Orlando H. & Kirkonnel, Arturo. 2000. *Birds of Cuba*. London: A&C Black (Christopher Helm) & Ithaca, NY: Cornell University Press. 253pp.

Garsault, A.G. (ed.). 1900. *Notice sur la Réunion*. Paris: J.André/Librairie Africaine et Coloniale. 308pp.

Gaymer, Roger, Blackman, R.A.A., Dawson, P.G., Penny, Malcolm & Penny, C.M. 1969. The endemic birds of the Seychelles. *Ibis* 111: 157–176.

Geeraerts, Jan (ed.). 1982. *Het Aards Paradijs. Dierenvoorstellingen in de Nederlanden van de 16de en 17de eeuw*. Antwerp: Koninklijke Maatschappij voor Dierkunde van Antwerpen. 175pp.

Gennes de la Chancelière, 1735. Observations sur les îles de Rodrigue et de France en mars 1735 [ed.A.Lougnon]. *Rec. Trim*. 1: 210–238 (1933).

Geoffroy de Sainte-Hilaire, Étienne. 1806. Note sur quelques habitudes de la grande chauve-souris de l'Ile de France, connu sous le nom de *roussette*. *Ann. Mus. Hist. Nat. Paris* 7: 227–230.

George, Wilma. 1985. Alive or dead. Zoological collections in the seventeenth century. Pp.245–255 *in* Impey & MacGregor (1985), q.v.

Georges, Eliane & Vaisse, Christian. 1998. *The Indian Ocean*. Köln: Benedikt Taschen. 157pp. [tr. from French ed. of 1997].

Gerard, Gabriel. 1974. *Petit Album de la Réunion*. Saint-Denis: [author]. 300pp. + index.

Gerlach, Justin, Daszak, P. & Jones, Carl G. 1997. On *Polposipus herculeanus* Solier 1848 (Tenebrionidae: Coleoptera) on Round Island, Mauritius. *Phelsuma* 5: 43–48.

— & Florens, F. B. Vincent. 2000. Toxicity of 'specific' rodenticides and the risk to non-target taxa. *Phelsuma* 8: 75–76.

— 2001. Predator, prey and pathogen interactions in introduced snail populations. *Anim. Conserv.* 4: 203–209.

— 2002. The enigmatic Giant Bronze Gecko *Ailuronyx trachygaster*. Part 1. Identity. *Gekko* 3(1): 29–38.

— 2003. The biodiversity of the granitic islands of the Seychelles. *Phelsuma* 11 (Supplement B). 47pp.

— '2004'. *Giant tortoises of the Indian Ocean. The genus Dipsochelys inhabiting the Seychelles Islands and the extinct giants of Madagascar and the Mascarenes*. Frankfurt am Main: Edition Chimaira. 207pp.

— 2006 ('2005'). The impact of rodent eradication on the larger invertebrates of Fregate island, Seychelles. *Phelsuma* 13: 43–54.

— , Muir, Catherine & Richmond, Matthew D. 2006. The first substantiated case of trans-oceanic tortoise dispersal. *J. Nat. Hist.* 40: 2403-2408.

Germain, Louis. 1921. *Faune malacologique terrestre et fluviatile des Iles Mascareignes*. Angers: Gaulthier & Th'bert. iv+495pp.

Gervais, Paul & Coquerel, Charles. 1866. Sur le Dronte, à propos d'os de cet oiseau récemment découvertes à l'île Maurice. *C.R. Séances Acad. Sci. Paris* 62: 924–928 [amended version reprinted in Roussin (1860–67), q.v., 4: 92–96, 1867, attributed solely to Coquerel]

Ghestemme, Thomas. 2000. Contribution à l'étude de la distribution et de la reproduction des Papangues: compte-rendu des sorties concertées de la SEOR les 19 juillet, 24 juillet et 2 octobre 1999. *Taille-vent* 5–6: 4–10.

— & Rochet, Myriam. 2001. La braconnage aujourd'hui: 3 cas. Pp.10–18 *in* Le Corre (2001), q.v.

— 2002. *Contribution au plan de gestion de la Reserve Naturelle de la Roche Écrite. Evaluation patrimoniale de la faune et des habitats et propositions de définitions des objectifs de gestion*. Dissertation pour 'Stage de DESS', University of Réunion. 53pp + maps & charts.

— 2003. La réserve naturelle de la Roche Écrite: un nouveau chantier de travail. *SEOR lettre d'information* 9: 3–4.

— 2005. Saison de reproduction 2004 du Tuit tuit: mauvaises nouvelles pour cette espèce menacée. *Chakouat* 14: 4-5.

— & Salamolard, Marc. 2007. L'échenilleur de La Réunion, *Coracina newtoni*, espèce endémique en danger. *Ostrich* 78: 255-258.

Gibbs, David, Barnes, Eustace, & Cox, John. 2001. *Pigeons and doves. A guide to the pigeons and doves of the world*. Robertsbridge, Sussex (UK): Pica Press. 615pp.

Gibson, Charles W.D. & Hamilton, Julie. 1983. Feeding ecology and seasonal movements of Giant Tortoises on Aldabra Atoll. *Oecologia, Berlin* 56: 84–92.

Gill, Brian & Martinson, Paul. 1991. *New Zealand's extinct birds*. Auckland, NZ: Random-Century. 109pp.

Gill, Frank B. 1964. [notes on Réunion birds] Typescript MS [copy in Alexander Library, Edward Grey Institute, Oxford University Zoology Dept.].

— 1967. Birds of Rodriguez Island (Indian Ocean). *Ibis* 109: 383–390.

— , Jouanin, Christian & Storer, Robert.W. 1970. Notes on the seabirds of Round Island, Mauritius. *Auk* 87: 514–521.

— 1971a. Ecology and evolution of the sympatric Mascarene white-eyes *Zosterops borbonica* and *Z. olivacea*. *Auk* 88: 35–60.

— 1971b. Endemic landbirds of Mauritius Island, Indian Ocean.

Typescript MS [copy in Alexander Library, Edward Grey Institute, Oxford University Zoology Department & ASC's papers].

— 1972. Causerie de Franck B. Gill [transcript of untitled talk given in Réunion in 1967] *Info-Nature, Ile Réunion* 6: 8–10.

— 1973a. Passé, present et avenir des oiseaux des forêts de l'Ile de la Réunion. *Info-Nature, Ile Réunion* No. spécial hors série 'La Forêt': 62–64.

— 1973b. *Intra-island variation in the Mascarene white-eye* Zosterops borbonica. *Orn. Mongr.* 12. 66pp.

Gillham, Mary E. 2000. *Islands of the trade winds. An Indian Ocean odyssey*. London: Minerva Press. 462pp.

Giorgi, Loïc & Borchellini, Serge. 1998. *Carte géologique de l'Ile Rodrigues au 1: 25000. Le schema hydrogéologique. La notice explicative*. Paris: Ministère Délégué à la Cooperation et de la Francophonie & Géolab. 28pp.+ maps.

Girard, Magali & Notter, Jean-Cyrille (eds.) 2003. *Les hauts de la Réunion – Projet Parc National. Principes pour un parc national de nouveau génération. 3. Premier état des lieux*. Saint-Denis: Mission Parc. 46pp. [large format atlas]

Glaw, Frank & Vences, Miguel. 1994. *A fieldguide to the amphibians and reptiles of Madagascar. Second edition including mammals and freshwater fish*. Köln: [authors]. 480pp.

Gleadow, F. 1904. *Report on the forests of Mauritius with a preliminary working plan*. Port Louis: Govt.Printer 230pp. [there was a 'revised' (actually censored) ed. in 1906]

Gochfield, Michael *et al.*. 1994. Successful approaches to seabird protection in the West Indies. Pp.186–209 *in* Nettleship *et al.* (1994), q.v.

Gomy, Yves. 1973a. L'insecte et la forêt. *Info-Nature, Ile Réunion*, No.spécial hors série 'La Forêt': 65–76.

— 1973b. Voyage en île d'amertume. *Info-Nature, Ile Réunion*. 9: 72–90.

— 2000. *Nouvelle liste chorologique des Coléoptères de l'archipel des Mascareignes*. Saint-Denis: Société des Amis du Muséum. 140pp.

Gonthier, Marius. 1860. Aperçu historique. Pp.i-xxxiii in vol.2 of Roussin (1860–68), q.v.

Goodman, Steven M. & Benstead, Jonathan P. (eds.) 2003. *The natural history of Madagascar*. Chicago: University of Chicago Press. 1709pp.

— , Jenkins, Richard K.B. & Ratrimomanarivo, Fanja H. 2005. A review of the genus *Scotophilus* (Mammalia, Chiroptera, Vespertilionidae) on Madagascar, with the description of a new species. *Zoosystema* 27: 867-882.

— ., Rakotondraparany, Félix & Kofoky, Amyot. 2007. The description of a new species of *Myzopoda* (Myzopodidae: Chiroptera) from western Madagascar. *Mamm. Biol.* 72(2): 65–81.

Goodwin, Derek. 1983. *Pigeons and doves of the world*. 3rd.ed. New York: Cornell University Press & London: British Museum (Natural History). 363pp.

Gordon, Arthur H. 1894. *Mauritius. Records of private and public life, 1871–1874*. Edinburgh: R.& R.Clark. 2 vols.

Gorman, Martyn. 1979. *Island ecology*. London: Chapman & Hall. 79pp.

Gosse, P. 1938. *St.Helena 1502–1938*. London: Cassell [reprinted 1990, Oswestry, UK: Anthony Nelson].

Gould, Stephen J. 1980. *The panda's thumb. More reflections in natural history*. NY & London: W.W.Norton & Co. 343pp.

[also Harmondsworth: Penguin Books, 1983].
— 1998. *Leonardo's mountain of clams and the diet of worms. Essays on natural history.* London: Random House/ Jonathan Cape. 422pp.
Grandidier, Alfred. 1867. Charles Coquerel [obituary]. Pp.201–205 *in* vol.4 of Roussin (1860–68), q.v.
— et al. 1903–20. *Collection des ouvrages anciens concernant Madagascar.* Paris: Comité de Madagascar. 9 vols.
Grant, Charles. 1801. *The history of Mauritius or the Isle of France and the neighbouring islands from their first discovery to the present time....* London: W.Bulmer & Co. 571pp + maps. [reprinted 1995 – New Delhi: Asian Educational Services]
Gray, Howard S. 1981. *Christmas Island, naturally. The natural history of an isolated oceanic island, the Australian territory of Christmas Island, Indian Ocean.* Perth (Australia): [author]. 133pp.
Gray, William. 1993. *Coral reefs & islands. The natural history of a threatened paradise.* Newton Abbot (UK): David & Charles. 192pp.
Grayson, Donald A. 1986. Nineteenth century explanations of Pleistocene extinctions: a review and analysis. Pp.5–39 *in* P.S.Martin & R.G.Klein (eds.) *Quaternary extinctions. A prehistoric revolution.* Tucson: University of Arizona Press. 892pp.
Greathead, D.J. 1971. A review of biological control in the Ethiopian Region. *Tech.Comm.Commonwealth Inst.Biol. Control* 5. 162pp.
Greenway, J.C. 1958. *Extinct and vanishing birds of the world.* New York: American Commission for International Wildlife Protection., Special Pub.13. [also 2nd.ed., 1967, New York: Dover. 520pp.]
Greenwood, Andrew [G.] 1997. Echo parakeet season 1996 – progress! *Psittascene* 9(1): 10
Greer, A.E. 1970. The systematics and evolution of the sub-saharan Africa, Seychelles and Mauritius scincine Scincid lizards. *Bull. Mus. Comp. Zool., Harvard* 140: 1–24.
Griffiths, C. 2005. Plant-tortoise interactions on Ile aux Aigrettes and the potential use of Aldabra giant tortoises (*Geochelone gigantea*) as analogues for the extinct Mauritian tortoises. Unpublished report to the Mauritian Wildlife Foundation.
Griffiths, Owen [L.], Cook, A. & Wells, Susan M. 1993. The diet of the introduced carnivorous snail *Euglandina rosea* in Mauritius and its implications for threatened island gastropod faunas. *J. Zool. Lond.* 229: 79–89.
— 1994. A review of the land snails of Rodrigues Island (Indian Ocean) with notes on their status. *J. Conchol., Lond.* 35: 157–166
— 1996. Summary of the land snails of the Mascarene Islands, with notes on their status. *PRSAS* 6: 37–48.
— 2003. Breeding of giant tortoises and Nile crocodiles in Mauritius 1985–2003. Presentation to 'Symposium agriculture', session 6 [conference in Mauritius], published on-line (illustrations only) at www.prosi.net.mu/session_6/ Ogriffiths.pdf.
— & Florens, F.B.Vincent. 2006. *A field guide to the non-marine molluscs of the Mascarene islands (Mauritius, Rodrigues and Réunion) and the northern dependencies of Mauritius.* Mauritius: Bioculture Press. 185pp.
Grihault, Alan. 2005. *Dodo – the bird behind the myth.* Cassis, Mauritius: IPC Ltd. 171pp.
Grimaud, Luc & Probst, Jean-Michel. 1997. Un oiseau à tendance cavernicole, la Salangane des Mascareignes *Collocalia francica* (Gmelin, 1788). *Bull. Phaethon* 6: 11.
Grimmett, Richard, Inskipp, Carol & Inskipp, Tim. 1999. *A guide to the birds of India, Pakistan, Nepal, Bangladesh, Bhutan, Sri Lanka and the Maldives.* London: A&C Black. 888pp.
Grondin, Dominique C. 1992. *Joseph Hubert, naturaliste.* Saint-Denis: Éditions CNH (Cahiers de notre histoire No.36). 26pp.
Groombridge, James J. 2000. *Conservation genetics of the Mauritius Kestrel, Pink Pigeon and Echo Parakeet.* Ph.D. Thesis, Univ. of London (Queen Mary & Westfield Colleges). 173pp.
— et al. 2000. 'Ghost' alleles of the Mauritius kestrel. *Nature* 403: 616.
— et al. 2001. Evaluating the severity of the population bottleneck in the Mauritius Kestrel *Falco punctatus* from ringing records using MCMC estimation. *J. Anim. Ecol.* 70: 401–409.
— et al. 2002. A molecular phylogeny of African kestrels with reference to divergence across the Indian Ocean. *Molec. Phylogenet. Evol.* 25: 267–277.
— et al. 2004a. Molecular phylogeny and morphological change in the *Psittacula* parakeets. *Molecular Phylogeny & Evol.* 31: 96–108.
— et al. 2004b. Associations of evolutionary & ecological distinctiveness amongst Indian Ocean kestrels. Pp.679–692 *in* R.D.Chancellor & B-U.Meyburg (eds.) *Raptors worldwide, Proc. VI World Conf. on Birds of Prey & Owls, Budapest 18–23 May 2003.* Budapest: WWGBP & MME [=Birdlife, Hungary]. 867pp.
Grove, Richard H. 1994. A historical review of early institutional and conservationist responses to fears of artificially induced global climate change: the deforestation-desiccation discourse 1500–1860. *Chemosphere* 29: 1001–1013.
— 1995. *Green imperialism. Colonial expansion, tropical island Edens and the origins of environmentalism.* Cambridge, UK: Cambridge University Press. 540pp.
Grubb, Peter. 1971. The growth, ecology and population structure of giant tortoises on Aldabra. *Phil. Trans. Roy. Soc. Lond.* B 260: 327–372.
Gruchet, Harry. 1973. Note sur la conservation de la nature intéressant plus spécialement la Réunion. *Info-Nature, Ile Réunion* 10: 29–49.
— & Gomy, Yves. 1973. Compte-rendu de l'assemblée générale de la SREPN tenue le lundi 25 juin 1973. *Info-Nature, Ile Réunion* 10: 4–7.
— 1975. Oiseaux nuisibles à la Réunion. *Info-Nature, Ile Réunion* 14: 91–93.
Gudger, E.W. 1929. Nicolas Pike and his unpublished paintings of the fishes of Mauritius, western Indian Ocean, with an index to the fishes. *Bull. Am. Mus. Nat. Hist.* 58: 489–530.
Guébourg, Jean-Louis. 1991. La representation cartographique de l'île de la Réunion du XVIIe au XIXe siècle. *Mappemonde* 1991(2): 107.
— 1999. *Petites îles et archipels de l'Océan Indien.* Paris: Editions Karthala. 570pp.
Guého, Joseph & Staub [J.J.] France. 1979. Le Mondrain nature reserve. *Cent-cinquantaire de la Société Royale des Arts et des Sciences de l'Ile Maurice*: 75–102.
— & Staub [J.J.] France. 1983. Observations botaniques et ornithologiques à l'Atoll d'Agaléga. *PRSAS* 4(4): 15–110.
— 1988. *La végétation de l'Ile Maurice.* Rose-Hill, Mauritius:

Éditions Océan Indien. 57pp + appendices.

— & Thiolley, Bernard. 1990. Un micro-réserve à l'Ile Maurice: la réserve de Perrier. *Cah. Outre Mer* 43(172): 551–554.

— 1996. The Sir Emile Sériès Reserve. *PRSAS* 6: 207–212.

Guérin, Réné. 1940–53. *Faune ornithologique ancienne et actuelle des îles Mascareignes, Seychelles, Comores et des îles avoisinantes*. Port Louis: General Printing & Stationery Co. 3 vols.

Guet, Isidore. 1888. *Les origines de l'Ile Bourbon et de la colonisation française de Madagascar. D'après des documents inédits tirés des Archives Coloniales du Ministère de la Marine et des Colonies, etc.*. 2nd.ed. Paris: Charles Bayle. 303pp.

Guibé, J. 1958. Les serpents de Madagascar. *Mem. Inst. Sci. Madagascar*, ser.A, 12: 189–260.

Guignes, C.L.J. de. 1808. *Voyages à Peking, Manille et l'Ile de France faits dans l'intervalle des années 1784 à 1801*. Paris: Imprimerie Impériale. 3 vols + atlas.

Gulliver, George. 1874–5. [letters to T.W.Rolleston, kept in the Historical Correspondence files, Oxford University Museum of Zoology].

Günther, Albert. 1869 (1870). [letter to L.Bouton, dated 3/11/1869, on various zoological specimens] *TRSAS* NS 4: 139, 1870.

— 1874 (1875). [letter to L.Bouton, dated 5/3/1874, on various zoological specimens] *TRSAS* NS 8: 87, 1875.

— 1879. Reptiles [of Rodrigues]. *Phil. Trans. Roy. Soc. Lond.* 168: 470.

— & Newton, Edward. 1879. The extinct birds of Rodrigues. *Phil. Trans. Roy. Soc., Lond.* 168: 423–437.

Gunther, Robert T. 1937. The Oxford Dodos. *Bird Notes & News* 17(6): 141–142.

Gupta, Kiran. 2005. *Know about vanished species*. New Delhi: Young Learner Publications. 48pp.

Hachisuka, Masauji. 1937. Revisional note on the didine birds of Réunion. *Proc. Biol. Soc. Washington* 50: 69–72.

— 1953. *The Dodo and kindred birds, or the extinct birds of the Mascarene Islands*. London: H.F. & G.Witherby. 250pp.

Hahn, C.W. 1834–41. *Ornitologische Atlas oder naturgetreue Abbildung und Beschreibung der aussereuropäischen Vögel. Erste Abteilung: Papageien*. Nürnberg: C.H.Zeh'sche Buchhandlung. 100pp.

Haig, Janet. 1984. Land and freshwater crabs of the Seychelles and neighbouring islands. Pp.123–139 *in* Stoddart (1984a), q.v.

Hall, B.P. & Moreau, Reginald E. 1970. *An atlas of speciation in African Passerine birds*. London: British Museum (Natural History). 423pp.

Hall, David G. 2003. The ecology of the black rats *Rattus rattus* on Mauritius and how management affects native birds. Unpublished PhD. Thesis, University of Bristol.

Hallam, Anthony. 1994. *An outline of phanerozoic biogeography*. Oxford: Oxford University Press. 246pp.

Halliday, Timothy R. 1978. *Vanishing birds*. London: Sidgwick & Jackson. 296pp.

Haltenorth, Theodor & Diller, Helmut. 1980. *Mammals of Africa including Madagascar*. London: Collins. 400pp.

Hamann, Ole. 1979. Regeneration of vegetation on Santa Fé and Pinta Islands, Galápagos, after the eradication of goats. *Biol. Conserv.* 15: 215–236.

Hambler, Clive. 1994. Giant tortoise *Geochelone gigantea* translocation to Curieuse Island (Seychelles) – success or failure ? *Biol. Conserv.* 69: 293–299.

Hampton, George G. 1922. *Rodrigues, Indian Ocean. A synopsis (illustrated)*. Port Mathurin, Rodrigues: [staff at the cable station]. 32pp.

Hancock, James, & Elliott, Hugh F.I. 1978. *The herons of the world*. London: London Editions. 304pp.

— & Kushlan, J.[A.] 1984. *The herons handbook*. London: Croom Helm. 288pp.

— 1999. *Herons and egrets of the world: a photographic journey*. London & San Diego: Academic Press. 208pp.

Hanley, Kathryn A, Petren, Kenneth & Case, Ted J. 1998. An experimental investigation of the competitive displacement of a native gecko by an invading gecko: no role for parasites. *Oecologia* 115: 196–205.

Hansen, Dennis M., Olesen, Jens & Jones, Carl G. 2002. Trees, birds and bees in Mauritius: exploitative competition between introduced honey bees and endemic nectarivorous birds. *J. Biogeog.* 29: 721–734.

— 2005. Pollination of the enigmatic Mauritian endemic *Roussea simplex* (Rousseaceae): birds or geckos. *Ecotropica* 11: 69–72.

—, Beer, Karin and Müller, Christine B. 2006a. Mauritian coloured nectar no longer a mystery: a visual signal for lizard pollinators. *Biology Letters* 2: 165–168.

—, Kaiser, Christopher N. & Müller, Christine B. 2006b. Endangered endemic plants on tropical oceanic islands: seed dispersal, seedling establishment, and ecological analogues. Pp. 35–70 in C. N. Kaiser, *Functional integrity of plant-pollinator communities in restored habitats in Mauritius*. Doctoral thesis, University of Zürich. 251pp. [& submitted to *J. Ecol.*]

— *et al*. 2007. Positive indirect interactions between neighboring plant species via a lizard pollinator. *Am. Nat.* 169: 534–542.

Hansen, Ina, Brimer, L. & Mølgaard, P. 2004. Herbivore deterring secondary compounds in heterophyllous woody species of the Mascarene Islands. *Perspectives Plant Ecol. Evol. Systemat*. 6(3): 187–203.

Harde, K.W. 1994. *A field guide in colour to beetles*. London: Octopus Books & [1998] Leicester, UK: Blitz Editions. 334pp.

Harland, W.Brian *et al*. 1990. *A geologic time scale 1989*. Cambridge: Cambridge Univ. Press. 263pp.

Harmon, Luke J. 2005. *Competition and community structure in Day Geckos (Phelsuma) in the Indian Ocean*. PhD Thesis. University of Washington, USA.

Harris, Donna. 2000. *A comparative study of the distribution, abundance and habitat use of two exotic and an endemic reptile species on Mauritius and Ile aux Aigrettes*. MSc. dissertation, University of East Anglia, UK.

Harrison, Colin J.O. & Walker, Cyril A. 1978. Pleistocene bird remains from Aldabra Atoll, Indian Ocean. *J. Nat. Hist.* 12: 7–14.

Harrison, Craig S. 1990. *Seabirds of Hawaii. Natural history and conservation*. Ithaca, NY: Cornell Univ. Press. 249pp.

Harrison, John. 1999. *A field guide to the birds of Sri Lanka*. Oxford: Oxford University Press. 219pp.

Harrison, Peter. 1985. *Seabirds, an identification guide*. Rev.ed. London & Sydney: Croom Helm. 448pp.

Hartlaub, G. 1877. *Die Vögel Madagascars und der benachbarten Inselgruppen*. Halle: H.W.Schmidt. 435pp.

Hartley, John R.M. 1984. Our mission in Mauritius. *On the edge*

47: 4–6.
Haupt, Herbert *et al.* 1990. *Le bestiaire de Rodolphe II. Cod.min.129 et 130 de la Bibliothèque nationale d'Autriche.* Paris: Editions Citadelles. 495pp.
Hayman, R.W. & Hill, John E. 1971. Part 2: Order Chiroptera. *in* J.Meester & H.W.Setzer (eds.), *The mammals of Africa: an identification manual.* Washington: Smithsonian Institution.
Hawkins, A. Frank A. & Goodman, Steven M. 2003. Introduction to the birds. Pp.1019–1044 in Goodman & Benstead (2003), q.v.
— & Safford, Roger J. in prep. *Birds of the Malagasy Region.* London: A&C Black.
Hays, H.R. 1973. *Birds, beasts and men. A humanist history of zoology.* London: J.M.Dent & sons. 383pp.
Heady, Sue *et al.* 1997. *Spectrum guide to Mauritius.* Nairobi: Camerapix Publishers International. 342pp.
Heatwole, Harold. 1993. Coral islands. Pp.141–153 *in* F.H.Talbot & R.E.Stevenson (eds.) *Oceans and islands.* New York: Smithmark. 240pp.
Hébert, G. 1708. Rapport de G.Hébert sur l'Ile Bourbon en 1708 avec les apostilles de la Compagnie des Indes. [ed.A.Lougnon]. *Rec. Trim.* 5: 34–73 [1940].
Hébert, Jean-Claude. 1999. La relation du voyage à Madagascar de Ruelle (1665–1668). *Etudes Ocean Indien* 25/26: 9–94.
Heguerty, Pierre-André d'. 1754. Discours prononcé devant le roi de Pologne Stanislas le 26 mars 1751 sur l'île Bourbon. *Mem. Soc. Roy. Sci. Belles Lettres Nancy* 1: 73–91.
Heim de Balsac, Henri & Heim de Balsac, M-H. 1956. *Suncus murinus* (L.) à la Réunion et en Nouvelle Guinée. Considérations sue le commensalisme et la vie domiciliaire des Soricidés. *Nat. Malgache* 8: 143–147.
Hemming, C.F. 1964. I. Red locusts in Mauritius. II. The Anti-Locust Research Centre and international locust control. *Mauritius Sugar Industry Res.Inst. Tech.Circ.* 22. 40pp.
Hengst, Jan den. 2003. *The Dodo – the bird that drew the short straw.* Marum, Netherlands: Art Revisited. 119pp.
Henri, [?]. 1865. Note sur l'acclimatation du moineau à l'île de la Réunion. *Mem. Soc. Imp. Sci. Nat. Cherbourg* 11: 252–256.
Herbert, Thomas. 1634. *A relation of some yeares travaile, begunne Anno 1626. Into Afrique and greater Asia, especially the territories of the Persian Monarchie: and some parts of the Orientall Indies and Iles adiacent.* London: William Stansby. 225pp. [also amplified & rewritten editions in 1638 & 1677, and translations into Dutch (1658) & French (1663)]
Hering, Jens & Heidi. 2005. Mascarene Grey White-eye with yellow forehead. *Bull. Afr. Bird Club* 12: 55 [with correction, 13: 8, 2006].
Hermann, Jules. 1885. *La fondation du quartier Saint-Pierre – La Bourdonnais.* Saint-Denis: Gaston LaHuppe. 89pp. [reprinted 1990 in Reverzy (1990), q.v.]
— 1902. Prise de Cilaos par Mussard (1743–53). Pp.60–84 *in* Mac-Auliffe (1902), q.v.
Hermann, Paul. 1909. *Histoire et géographie de l'Ile de la Réunion.* Paris: Ch.Delagrave.
Herpin, Eugène. 1905. *Mahé de la Bourdonnais et la Compagnie des Indes.* Saint-Brieuc: Réné Prudhomme. 256pp. [reprinted 1997 – Rennes: La Découvrance]
Hershey, David R. 2004. The widespread misconception that the Tambalacoque or Calvaria Tree absolutely required the Dodo bird for its seeds to germinate. *Plant Sci. Bull.* 50(4): 105–109.
Hicks, John & Knight, Alan. c.1999. Report on the problems of controlling the stray dog population of Mauritius. Uckfield, UK: International Animal Rescue. 7pp. [on-line at www.iar.org.uk]
Higgin, Edward. 1849. Remarks on the country, products and appearance of the island of Rodriguez, with opinions as to its future colonization. *J. Roy. Geog. Soc.* 19: 17–20.
High, Jeremy. 1976. *Natural history of the Seychelles.* Port Victoria, Seychelles: [?Government Printer]. 63pp.
Hill, A.W. 1941. The genus *Calvaria* with an account of the stony endocarp and germination of the seed, and description of a new species. *Ann. Bot.* NS 5: 587–606.
Hill, John E. 1971. The bats of Aldabra Atoll, western Indian Ocean. *Phil. Trans. Roy. Soc. Lond.* b 260: 573–576.
— 1980. The status of *Vespertilio borbonicus* E.Geoffroy 1803 (Chiroptera: Vespertilionidae). *Uitg. Rijksmus. Nat. Hist., Leiden* 55: 287–295.
Hill, S.C. 1916. *Catalogue of the manuscripts in European languages belonging to the library of the India Office. Vol.2, part 1. The Orme Collection.* London: Oxford University Press. 421pp.
Hnatiuk, Roger, Woodell, Stanley & Bourn, David M. 1976. Giant tortoise and vegetation interactions on Aldabra Atoll. Part.2: coastal. *Biol. Conserv.* 9: 305–316.
Hoarau, Alain. 1993. *Les îles éparses. Histoire et découverte.* Saint-Denis: Azalées Éditions. 237pp.
Hoart, C.T. 1825. Observations générales sur l'île Rodrigue. MS, Mauritius Archives LH45 [published, pp.22–27 *in* J.R.Mann (ed.) 1857. *Report on Rodrigues as a quarantine station.* Port Louis: Government Printer]
Hoffman, Johann C. 1680. *Oost-Indianische Voyage...* Cassel: [author] [Reprinted 1931, The Hague: Martinus Nijhoff; section on Mauritius reproduced (in French) in Grandidier *et al.* (1903–20) 3: 368–380, q.v.]
Hoffstetter, R. 1946a. Remarques sur la classification des ophidiens et particulièrement des Boidae des Mascareignes (Bolyerinae subfam. nov.). *Bull. Mus. Natl. Hist. Nat., Paris* (2)18: 132–135.
— 1946b. Les Typhlopidae fossiles. *Bull. Mus. Natl. Hist. Nat., Paris* (2)18: 309–315.
Holmes, M., Hutson Anthony M. & Morris, J. 1994. *The Maldives Archipelago, Indian Ocean – a report on an investigation of fruit bats and birds, November 1993.* London: Bat Conservation Trust. 32pp.
Holthuis, L.B., Muller, H.E. & Smeenk, Chris. 1971. Vogels op Nederlandse 17de eeuwse tegels naar gravures van Adriaen Collaert en iets over *Leguatia gigantea. Bull. Mus. Boymans van Beuningen* 21(1): 3–19.
Holyoak, David T. 1971. Comments on the extinct parrot *Lophopsittacus mauritianus. Ardea* 59: 50–51.
— 1973. An undescribed extinct parrot from Mauritius. *Ibis* 115: 417–418.
Horne, Jennifer F.M. 1987. Vocalisations of Mascarene Island birds. Pp.101–150 *in* Diamond (1987), q.v.
Horne, John. 1887. Notes on Flat Island -&- Notes on the flora of Flat Island. *TRSAS* NS 19: 101–112 & 115–151.
Hottola, Petri. 2002. Bring your own scope! An alternative approach to birding on Mauritius and Réunion. Trip report

(14pp.) on www.camacdonald.com/birding.
Hubbard, Charles E. & Vaughan, Reginald E. 1940. *The grasses of Mauritius and Rodriguez*. London: Crown Agents.
Hubbel, Stephen P. & Foster, R.B. 1983. Diversity of canopy trees in a neotropical forest and implications for conservation. Pp.25–41 in S.L.Sutton, T.C.Whitmore & A.C.Chadwick (eds.) T*ropical Rain Forest: ecology and management*. Oxford: Blackwell Scientific Publications. 498pp.
— 2001. *The unified neutral theory of biodiversity and biogeography*. Princeton: Princeton University Press. 375pp.
Hume, Julian P. 2003a. The journal of the flagship *Gelderland*: Dodo and other birds on Mauritius, 1601. *Arch. Nat. Hist*. 30: 13–27.
— 2003b. Survey of the seabird colonies on Ile aux Cocos and Ile Sable. Unpubl. report to the Rodrigues administration. 2pp.
— 2003c. Report on the Rodrigues Fruit Bat *Pteropus rodericensis* [sic] population October 2003. Unpubl. report to MWF. 1pp.
— & Cheke, Anthony S. 2004. White Dodos – unravelling a scientific and historical myth. *Arch. Nat. Hist*. 31: 57–79.
— , Martill, David M. & Dewdney, Christopher. 2004. Dutch diaries and the demise of the dodo. *Nature* [10/6/2004 on-line only: www.nature.com/nature]
— 2005a. *The vertebrate palaeontology of the western Indian Ocean*. PhD thesis, Portsmouth University & NHM, Tring.
— 2005b. Contrasting taphofacies in ocean island settings: the fossil record of Mascarene veertebrates. Pp.129–144 in J. A. Alcover & P. Bover (eds.) Proceedings of the International Symposium "Insular vertebrate evolution: the palaeontological approach". *Monogr. Soc. Hist. Nat. Balears* 12: 129–144.
— & Prys-Jones, Robert P. 2005. New discoveries from old sources, with reference to the original bird and mammal fauna of the Mascarene Islands. *Zool. Med., Leiden* 79–3(8): 85–95.
— 2006. The history of the Dodo *Raphus cucullatus* and the penguin of Mauritius. *Historical Biol*. 18: 65–89.
— , Datta, Anna & Martill, David M. 2006. Unpublished drawings of the Dodo *Raphus cucullatus* and notes on Dodo skin relics. *Bull. Brit. Orn. Club* 126A: 49-54.
— 2007. Reappraisal of the parrots (Aves: Psittacidae) from the Mascarene Islands, with comments on their ecology, morphology and affinities. *Zootaxa* 1513: 1–76.
Hunt, Terry L. 2007. Rethinking Easter Island's ecological collapse. *J. Arch. Sci*. 34: 485-502.
Huron, Joseph. 1920. *Chasses et pêches de l'Ile Maurice: Le chasse se meurt. Vive la chasse*. Port Louis: General Printing & Stationery Co. 94pp.
— 1923. *Croquis rustiques*. Port Louis: [author]. 127pp.
Hutchinson, G.Evelyn. 1962. *The enchanted voyage and other studies*. New Haven, USA: Yale University Press. [article on Dodo & Solitaire, pp.77–89]
Hutchinson , Mark N. *et al*. 1990. Immunological relationships and generic revision of the Australian lizards assigned to the genus *Leiolopisma* (Scincidae: Lygosominae). *Australian J. Zool*. 38: 535–554.
Hutson, Anthony M. 1975. Observations on the birds of Diego Garcia, Chagos Archipelago, with notes on other vertebrates. *Atoll Res. Bull*. 175. 25pp.
Hutton, I., Parkes, J.P. & Sinclair, A.R.E. 2007. Reassembling island ecosystems: the case of Lord Howe Island. *Anim. Conserv*. 10: 22-29.
Hutterer, Rainer & Tranier, M. 1990. The immigration of the Asian house shrew (*Suncus murinus*) into Africa and Madagascar. Pp.309–319 in Peters & Hutterer (1990), q.v.
Huxley, Christopher R. 1979. The tortoise and the rail. *Phil. Trans. Roy. Soc. Lond*. B 286: 225–230.
Imber, Michael J. 1985. Origins, phylogeny and taxonomy of the gadfly petrels *Pterodroma* spp. *Ibis* 127: 197–229.
Impey, Andrew J., Côté, Isabelle & Jones, Carl G. 2002. Population recovery of the threatened endemic Rodrigues fody (*Foudia flavicans*) (Aves, Ploceidae) following reforestation. *Biol. Conserv*. 107: 299–305.
Impey, Oliver & MacGregor, Arthur (eds.). 1985. *The origins of museums. The cabinet of curiosities in sixteenth and seventeenth-century Europe*. Oxford: Clarendon Press. 431pp. [reprinted 2001, Thirsk (UK): House of Stratus]
Ineich, Ivan. 1999. Spatio-temporal analysis of the unisexual-bisexual *Lepidodactylus lugubris* complex (Reptilia, Gekkonidae). Pp. 199–228 in H. Ota (ed.) *Tropical island herpetofauna: origin, current diversity and conservation. Proceedings of the International Symposium Diversity of Reptiles, Amphibians, and Other Terrestrial Animals on Tropical Islands: Origin, Current Status, and Conservation*, Okinawa, Japan, 6–7 June 1998. Amsterdam: Elsevier. 368pp.
Ingleton, Geoffrey C. 1986. *Matthew Flinders, navigator and chartmaker*. Guildford, UK: Genesis Publications & Alphington, Victoria: Hedley Australia. 467pp.
Irblich, Eva. 1990. Étude codilogique et historique du "Muséum" de Rodophe II. p.65–89 *in* Haupt *et al*. (1990), q.v.
Isnard, Hildebert. 1951 ('1952'). L'Ile de la Réunion. Pp.243–276 in R.Decary *et al*. *La France de l'Océan Indien*. Paris: Société d'Éditions Géographiques, Maritimes et Coloniales. 314pp.
Iverson, John B. 1987. Tortoises, not dodos, and the tambalacoque tree. *J. Herpetol*. 21: 229–230.
Jackson, Ashley. 2001. *War and empire in Mauritius and the Indian Ocean*. Basingstoke, UK: Palgrave. 241pp.
Jackson, Christine. 1993. *Great bird paintings of the world. Vol.1. The Old Masters*. Woodbridge, UK: Antique Collectors Club. 143pp.
— 1999. *Dictionary of bird artists of the world*. Woodbridge, Suffolk (UK): Antique Collectors' Club. 550pp.
Jackson, Kate & Fritts, Thomas H. 2004. Dentitional specialisations for durophagy in the Common Wolf Snake *Lycodon aulicus capucinus*. *Amphibia-Reptilia* 25: 247–254.
Jackson, Peter W., Cronk, Quentin C.B. & Parnell, John A.W. 1988. Notes on the regeneration of two rare Mauritian endemic trees. *Trop. Ecol*. 29: 98–106.
[Jacob de Cordemoy, (?Eugène), as 'J.C.'] 1861. Le merle de Bourbon. Pp.78–80 in vol.2 of Roussin (1860–67), q.v.
Jacob de Cordemoy, Eugène, 1895. *Flore de l'Ile de la Réunion*. Paris: P Klincksieck. 574pp. [reprinted 1972, Lehre, Germany: J.Cramer].
Jacob de Cordemoy, Hubert. 1904. *Étude sur l'île de la Réunion. Géographie physique, richesses naturelles, cultures et industries*. Marseille: Institut Colonial. 71pp.
— 1925. La vie et l'évolution économique à la Réunion. Pp.144–243 *in* Barquissau *et al*. (1925), q.v.
Jacobs, Michael. 1995. *The painted voyage. Art, travel & exploration 1564–1875*. London: British Museum Press. 160pp.
James, Helen F. & Olson, Storrs L. 1991. Description of thirty

two new species of birds from the Hawaiian Islands. Part II: Passeriformes. *Orn. Monog.* 46: 1–88.

Jamieson, Heidi. c1999. Conservation on the island of Rodrigues, Mauritius, Indian Ocean. [article, ? from a printed original, on www.batconservation.org]

Janoo, Anwar. 2000. Rooting the Dodo *Raphus cucullatus* Linnaeus 1758 and the Solitaire *Pezophaps solitaria* Gmelin 1789 within the Ornithurae: a cladistic reappraisal. *Ostrich* 71 (*Proc. 9th. Pan-Afr. Orn. Congr.*): 323–329.

— 2005. Discovery of isolated dodo bones [*Raphus cucullatus* (L.), Aves, Columbiformes] from Mauritius cave shelters highlights human predation, with a comment on the status of the family Raphidae Wetmore, 1930. *Ann. Paléontol.* 91: 167–180.

Jaussaud, Philippe & Brygoo, Edouard-Raoul. 2004. *Du jardin au museum en 516 biographies*. Paris: MNHN. 630pp.

Jauze, Jean-Michel. 1998. *Rodrigues. La troisième île des Mascareignes*. Paris: L'Harmattan & St-Denis, Réunion: Université de la Réunion. 269pp.

Jeggo, David, Young, H.Glyn & Darwent, Mark 2002 ('2001'). The design and construction of the Madagascar teal aviary at Jersey Zoo. *Dodo* 37: 50–69.

Jenkins, Alan. 1978. *The naturalists. Pioneers of natural history*. London: Hamish Hamilton. 199pp.

Jentinck, F.A. 1888. *Muséum d'Histoire Naturelle des Pays-Bas, tome xii. Catalogue systematique des Mammifères*. Leiden: E.J.Brill. 280pp.

Jepson, Walter F. 1936. *A summary of the results of the Phytalus investigation 1933 to 1936, with recommendations as to further lines of work*. Port Louis: Govt. Printer. 19pp.

Jessop, L. 1999. The fate of Marmaduke Tunstall's collections. *Arch. Nat. Hist.* 26: 33–49.

Jogoo, Vasant K. 2003. EIA of the proposed Midlands dam project: Mauritius. Pp.261–267 in M.McCabe & B.Sadler (eds.) *UNEP Case Studies of EIA practice in developing countries*. Geneva: UNEP. 310pp.

Johansson, Malin C. 2003. Vegetation monitoring and change on Round Island, Mauritius. Msc. Thesis, University of Birmingham, UK. 73pp. + appendices.

Johnsgard, Paul A. 1997. *The avian brood parasites – deception at the nest*. New York: Oxford University Press. 409pp.

Johnson, T.H. & Stattersfield, Alison J. 1990. A global review of island endemic birds. *Ibis* 132, 167–180.

Johnson, Genevieve. 2003. Bringing the Giant tortoises back to the Mascerenes [*sic*]. Interview with Owen Griffiths, published on-line on www.pbs.org/odyssey.

Johnson, Kevin P. & Clayton, Dale H. 1999. Swiftlets on islands: genetics and phylogeny of the Seychelles and Mascarene Swiftlets. *Phelsuma* 7: 9–13.

— *et al.* 2001. A molecular phylogeny of the dove genera *Streptopelia* and *Columba*. *Auk* 118: 874–887.

Johnston, H.H. 1894. Report on the flora of Round Island, Mauritius. *Trans. Bot. Soc. Edinburgh* 20: 237–264.

Johnston, Stuart. 1993. Round Island soil survey 1993. Soil fertility and erosion risk assessment. Unpubl. report to MWF. 23pp.

Jolly, Alison; Oberle, P. & Albignac, R. 1984. *Key environments: Madagascar*. Oxford: Pergamon Press. 239pp.

Jonchée de la Goleterie. 1729. [letter to the *Compagnie des Indes* on Mauritius and its potential] MS in Mauritius Archives (X14/4/5b/1) published as Document VI in Pineo (1993), q.v. [also extracts in Brouard (1963), q.v.]

Jones, Carl G. 1980a. The Mauritius Kestrel. *Hawk Trust Ann. Rep.* 10: 18–29.

— & Owadally, A.Wahab 1982a. Conservation priorities in Mauritius. Unpublished photocopied report submitted to ICBP. 32pp.

— 1987. The larger landbirds of Mauritius. Pp.208–300 in Diamond 1987, q.v.

— 1988a. Remains of Gunther's Gecko eggs found in the Black River Gorges. *PRSAS* 5(1,2): 135–136.

— 1988b. A note in the Macchabée Skink with a record of predation by the Lesser Indian Mongoose. *PRSAS* 5(1,2): 131–134.

— & Owadally, A.Wahab. 1988. The life histories and conservation of the Mauritius Kestrel *Falco punctatus* (Temminck 1823), Pink Pigeon *Columba mayeri* (Prevost 1843) and Echo Parakeet *Psittacula eques* (Boddaert 1783). *PRSAS* 5(1): 79–130.

—, Todd, David M. & Mungroo, Yousouf. 1989. Mortality, morbidity and breeding success of the pink pigeon *Columba (Nesoenas) mayeri*. *Proceedings of the Symposium on Disease and Management of Threatened Bird Populations. ICBP Technical Publication No. 10.*, Ontario: ICBP.

— *et al.* 1991. A summary of the conservation management of the Mauritius Kestrel *Falco punctatus* 1973–1991. *Dodo* 27: 81–99.

— *et al.* 1992. The release of captive-bred Pink Pigeons *Columba mayeri* in native forest on Mauritius. A progress report July 1987-June 1992. *Dodo* 28: 92–125

— 1993a. The ecology and conservation of Mauritian skinks. *PRSAS* 5(3): 71–95.

— 1993b. Do Long-tailed Macaques use rocks to smash open snails ? *PRSAS* 5(3): 139–140.

— & Duffy, Kevin. 1993. Conservation management of the Echo Parakeet. *Dodo* 29: 126–148.

— 1994. Birds survive cyclone Hollanda. *On the edge* 70: 6.

— *et al.* 1994. The restoration of the Mauritius Kestrel *Falco punctatus* population. *Ibis* 137, Supplement: 173–180.

— 1995. Studies on the biology of the Pink Pigeon *Columba mayeri*. Unpublished PhD. thesis, University College of Swansea in the University of Wales.

— & Hartley, John. 1995. A conservation project on Mauritius and Rodrigues: an overview and bibliography. *Dodo* 31: 40–65.

— *et al.* 1995. The restoration of the Mauritius Kestrel *Falco punctatus* population. *Ibis* 137: 173–190.

— 1996. Bird introductions to Mauritius: status and relationships with native birds. Pp.113–123 in J.S.Holmes & J.R.Simons (eds.) *The introduction and naturalisation of birds*. London: HMSO. viii+136pp.

— & Swinnerton, Kirsty. J. 1997. Conservation status and research for the Mauritius Kestrel, Pink Pigeon and Echo Parakeet. *Dodo* 33: 72–75.

— *et al.* 1998. The biology and conservation of the Echo Parakeet *Psittacula eques* of Mauritius. Pp.110–121 *in* W.Kiessling (ed.) *IV International Parrot Convention, 17–20 September 1998*. Tenerife: Loro Parque. 230pp.

— 1999. [review of Morris & Hawkins (1998) and Sinclair & Langrand (1998), qq.v] *Ibis* 141: 515–517.

— 1999. The restoration of the free-living populations of the Mauritius Kestrel (*Falco punctatus*), Pink Pigeon (*Columba mayeri*) and Echo Parakeet (*Psittacula eques*). Pp.77–86 in

Roth, T.L., Swanson, W.F. & Brattman, L.K. (eds.) *Seventh World /Conference on breeding endangered species: linking zoo and field research to advance conservation*. Cincinnati, USA: Cincinnati Zoo & Botanical Gardens. 345pp.

— 2002. Reptiles and amphibians. Chapter 18, pp.355–375 *in* vol.1 of M.R.Perrow & A.J.Davy (eds.) *Handbook of ecological restoration*. Cambridge, UK: Cambridge University Press. 2 vols.

— 2004. Conservation management of endangered birds. Chapter 12, pp.269–301 *in* W.J.Sutherland, I.Newton & R.E.Green (eds.) *Bird ecology & conservation*. Oxford: Oxford University Press. 386pp.

— 2006. Preface to Griffiths & Florens (2006), q.v.

Jones, Kate E. *et al*. 2002. A phylogenetic supertree of the bats (Mammalia: Chiroptera). *Biol. Rev.* 77: 2323–259.

Jouanin, Christian. 1964a ('1963'). Un petrel nouveau de la Réunion, *Bulweria baraui*. *Bull. Mus. Natl. Hist. Nat., Paris* (2)35: 593–597.

— 1964b. Notes sur l'avifaune de la Réunion. *Oiseau Rev. Fr. Orn.* 34: 83–84.

— & Gill, Frank B. 1967. Recherche de Petrel de Barau, *Pterodroma baraui*. *Oiseau, Rev. Fr. Orn.* 37: 1–19.

— 1970. Le petrel noir de Bourbon, *Pterodroma aterrima* Bonaparte. *Oiseau Rev. Fr. Orn.* 40: 48–68.

— 1987. Notes on the nesting of Procellariiformes in Réunion. Pp.359–363 *in* Diamond (1987), q.v.

— 2006. A Cilaos les pétrels se ramassent à la pelle. *Courrier Nat.* 228: 13.

Jury, M.R. 2003. The climate of Madagascar. Pp.75–87 *in* Goodman & Benstead (2003), q.v.

Kaeppelin, Paul. 1908. *La Compagnie des Indes Orientales et François Martin*. Paris: Augustin Challamel. xv+673pp.

Kallio, Harri. 2004. *The Dodo and Mauritius Island – imaginary encounters*. Stockport, UK: Dewi Lewis Publishing. 120pp.

Kay, E.Alison. 1991. The Pacific islands. Pp.159–165 *in* F.H.Talbot & R.E.Stevenson (eds.) *Oceans and islands*. New York: Smithwark Publishers Inc. 240pp.

Kearey, Philip & Vine, Frederick J. 1996. *Global tectonics*. Oxford: Blackwell. 333pp.

Keast, John (ed.). 1984. *The travels of Peter Mundy, 1597–1667*. Redruth, UK: Dyllasnsow Truran. 120pp. [these selections from Mundy's journals are much abridged without this being clearly indicated]

Keeley, Robert V. 2000. *The great Phelsuma caper (A diplomatic memoir)*. Washington, DC: Five & Ten Press. 147pp.

Keith, Stuart. 1980. Origins of the avifauna of the Malagasy region. *Proc. 4th. Pan-African Orn. Congr.*: 99–108.

Kell, S.P. 1996. *A vegetation survey of a potential genetic reserve of the Indian Ocean island of Rodrigues*. MSc thesis, University of Birmingham, UK.

Keller, Conrad. 1901. *Madagascar, Mauritius and the other East African islands*. London: Swan Sonnenscheim. xiii+242pp. [tr. from German ed of 1898; reprinted 1969, New York: Negro Universities Press]

Kennedy, John. 1827. [letter to William Leiper in Glasgow, from Mahébourg, Mauritius, 18.11.1827]. MS in Scottish United Services Museum, Edinburgh [photocopy in Rhodes House, Oxford]

Kennedy, Martyn & Page, Roderic D.M. 2002. Seabird supertrees: combining partial estimates of procellariiform phylogeny. *Auk* 119: 88–108.

Kennedy, William R. 1893. Notes on a visit to the islands of Rodriguez, Mauritius and Réunion. *J. Bombay Nat. Hist. Soc.* 7: 440–446.

— 1900. *Hurrah for the life of a sailor!* London: Wm. Blackwood & Sons. 356pp.

Kerourio, G. 1900. Les forêts de la colonie. Pp.33–62 *in* Garsault (1900), q.v.

Kervazo, Bertrand. 1979. Fouilles de la Grotte dite des "Premier Français". *Info-Nature, Ile Réunion* 17: 47–52.

Keuning, J. (ed.) 1938–51. *De tweede schipvaart der Nederlanders naar Oost-Indië onder Jacob Cornelisz. van Neck en Wybrant Warwijk 1598–1600. Journalen, Documenten andere beschieden*. s'Gravenhage: Martinus Nijhoff. 5 vols (in 6) + 2 of maps.

Keynes, Quentin. 1956. Mauritius, island of the Dodo. *Natl. Geog. Mag.* 109: 77–104.

Keynes, Randall. 2000. *Charles Darwin's zoology notes and specimen lists from HMS Beagle*. Cambridge: Cambridge University Press. 430pp.

Khan, S.A. (ed.) 1927. *John Marshall in India. Notes and observations in Bengal*. Oxford: Oxford University Press. 471pp.

King & Lochée (auctioneers). 1806. *Catalogue of the Leverian Museum*. London: King & Lochée. Pp.296+17 + 69pp. handwritten appendix. [reprinted 1979, London: Johnson & Hewitt, bound with *A companion to the Leverian Museum, 1790*].

King, Carolyn. 1984. *Immigrant killers. Introduced predators and the conservation of birds in New Zealand*. Auckland (NZ): Oxford University Press. 224pp.

King, H.C. 1946. *Interim report on indigenous species in Mauritius*. Port Louis: Govt. Printer [for Colony of Mauritius Forest Department] 21pp.

— 1947. *Colony of Mauritius. Empire forests and the war*. Statement to British Empire Forests Conference 1947. Port Louis: Government Printer.

King, Phillip P. 1827. *Narrative of a survey of the intertropical and western coasts of Australia performed between the years 1818 and 1822*. London: John Murray. 2 vols.

King, Warren B. 1978. Endangered birds of the world and current efforts towards managing them. Pp.9–17 *in* Temple (1978a), q.v.

— 1978–9. *Endangered birds of the world. The ICBP bird Red Data Book*. Washington DC: Smithsonian Press & ICBP. 2 parts, reissued as single vol. in 1981. Unpaginated.

Kingdon, Jonathan. 1990. *Island Africa – the evolution of Africa's rare animals and plants*. London: Collins. 587pp.

Kirsch, John A.W. *et al*. 1995. Phylogeny of the Pteropodidae (Mammalia: Chiroptera) based on DNA hybridisation, with evidence for bat monophyly. *Aust. J. Zool.* 43: 395–428.

Kitchener, Andrew C. 1993. On the external appearance of the dodo, *Raphus cucullatus* (L., 1758). *Arch. Nat. Hist.* 20: 279–301.

Kloet, Rolf S. & Kloet, Siwo R. 2005. The evolution of the spindlin gene in birds: sequences analysis of an intron of the spindlin W and Z gene reveals four major divisions of the Psitttaciformes. *Molec. Phylogen. Evol.* 36: 706–721.

Kluge, A.G. 1983. Cladistic relationships among gekkonid lizards. *Copeia* (1983): 465–475.

Koenig, Paul. 1895. *Suggestions for the management of the*

'Grand Bassin' forests. Mauritius [no place or publisher, but ? Port Louis: Govt.Printer]. 24pp.
— 1914a. Economic flora. Pp.102–109 in Macmillan (1914), q.v.
— 1914b. Report on Forestry at Rodrigues. MS, Mauritius Archives SD 166.
— 1926. Reafforestation in Mauritius. *Empire Forestry J.* 5: 218–222.
— 1932. Actes et comptes-rendus de la Société Royale des Arts et des Sciences de l'Ile Maurice. Pp.39–97 in *Centenaire de la Société Royale des Arts et des Sciences de l'Ile Maurice 1829–1929*. Port Louis: RSAS. 171pp.
— 1939. Le Muséum Desjardins. *TRSAS* C 8: 39–51.
Kondo, M. *et al.* 1993. Population genetics of crab-eating macaques (*Macaca fascicularis*) on the island of Mauritius. *Am. J. Primatol.* 29: 167–182.
Korsós, Zoltán & Trócsányi, Balász. 2000. Report on the Hungarian herpetological expedition to Round Island, Mauritius, November 1999. Unpubl. report. Budapest: [authors]. 12pp.
— & — 2001a. Population assessment of Gunther's Gecko in it's natural habitat, Round Island, Mauritius. Abstract only; *15th Ann. Meet. Soc. Conserv. Biol.*, Hawaii 27/7–1/82001. (on www.conbio.org/SCB/Activities/Meetings/2001).
— & — 2001b. Report on the *Bolyeria* search expedition on Round Island, Mauritius, April 2001. Unpubl. report. Budapest: [authors]. 16pp.
— & — 2002. Herpetofauna of Round Island, Mauritius. *Biota, Rače* 3: 77–84.
Krause, David W. 2003. Late Cretaceous vertebrates of Madagascar: A window into Gondwanan biogeography at the end of the age of dinosaurs. Pp.40–47 in Goodman & Benstead (2003), q.v.
Kuchling, G. Lippai, C. & Behra, O. 2003. Crocodylidae: *Crocodilus niloticus*, Nile Crocodile, *Voay, Mamba*. Pp.1005–1008 in Goodman & Benstead (2003), q.v.
Kueffer, Christoph & Lavergne, Christophe. 2004. *Case Studies on the status of invasive woody plants species in the western Indian Ocean. 4. Réunion.* Rome: FAO (Forestry Department Working Paper FBS/4–4E). (iv)+37pp.
— & Mauremootoo, John. 2004. *Case Studies on the status of invasive woody plants species in the western Indian Ocean. 3. Mauritius (islands of Mauritius and Rodrigues)*. Rome: FAO (Forestry Department Working Paper FBS/4– 3E). (iv)+35pp.
Labourdonnais – see Mahé de Labourdonnais.
La Caille, Nicolas L., *Abbé* de. 1763. *Journal historique du voyage fait au Cap de Bonne Espérance par feu M. l'Abbé de la Caille.* Paris: Guillyn. 380pp.
Lacroix, Alfred. 1936. Notice historique sur les membres et correspondents de l'Académie des Sciences ayant travaillé dans les colonies françaises des Mascareignes et de Madagascar au XVIIIe siècle et au début du XIXe. *Mem. Acad. Sci. Inst. France* (2)62: 1–119. [also published in book form – Paris: Gauthier Villars, usually cited as '1934', the date of Lacroix's lecture]
Lagesse, Marcelle. 1972. *L'Ile de France avant La Bourdonnais (1721–1735)*. Port Louis: Mauritius Archives; Publ. 12. 110pp. [reprinted 1999, Port Louis: Editions IPC].
Lahti, David C. 2003. A case study of species assessment in invasion biology: the Village Weaverbird *Ploceus cucullatus*. *Anim. Biodivers. & Conserv.* 26(1): 45–55.
LaHuppe, Thomy. 1882. Cinq jours aux Salazes. *Bull. Soc. Sci. Arts Ile Réunion* 1881: 92–133.
Laissus, Yves. 1973a. Note sur les manuscrits de Pierre Poivre (1719–1786) conservés à la bibliothèque central du Muséum nationale d'histoire naturelle, Paris. *PRSAS* 4(2): 31–56.
— 1973b. Note sur les voyages de Jean-Jacques Dussumier (1792–1883). *Ann. Soc. Sci. Nat. Charente-Maritime* 5:387–406.
— 1974. Catalogue des manuscrits de Philibert Commerson conservés à la bibiothèque centrale du Muséum nationale d'Histoire Naturelle (Paris). *Cah. Centre Univ. Réunion, No. Spécial Colloque Commerson*: 76–101 [reprinted 1978 in *Rev. Hist. Sci., Paris*: 31: 131–162]
La Lande, de. 1775. Éloge de M Commerson. *Obs. Phys. Hist. Nat. & Arts* 5: 89–120.
La Motte, D. de. 1754–77. Voyages de Sieur DDLM contenant dix-neuf lettres écrites sur les lieux à un de ses amis en Europe dans les années 1754 à 1757. MS [location unstated; may be E.126 of Toussaint & Adolphe (1956), q.v.] of which the parts referring aux Mascarenes were published, ed. P.d'Aglosse [pseud.], in *Rev. Hist. Litt. Ile Maurice* 5: 205–209, 220–223, 231–235, 243–246, 257–260, 270–272, 284–288, 297–300, 303–307, 319–323 (1891).
Lamouroux, J.V.F. & Desmarest, A.G. (eds.) 1824–1830. *Oeuvres complètes de Buffon avec les descriptions anatomiques de Daubenton son collaborateur. Nouvelle Édition. Oiseaux.* Paris: Ladrange & Verdière. 11 vols.
Lamusse, Axel. 1990. L'histoire exemplaire des ébéniers de l'Ile Maurice. *Cah. Outre Mer* 43(172): 561–564.
Lantz, Auguste. 1887. Sur les mammifères et les oiseaux de l'île de la Réunion. *Bull. Soc. Nat. Acclim. Paris* (4)4: 657–659.
Lanux, Jean-Baptiste F.de. 1753–56. [correspondances de Jean-Baptiste de Lanux à ses correspondants à l'Institut de France] MSS in Archives de L'Institut de France, Paris [photocopies in ASC's possession, Oxford; offered to Oxford University Libraries System, but refused...]
— 1772. [letter on *roussettes* and *rougettes*] Published pp.253–262 in Buffon 1776, q.v. [reprinted 1979 in *Info-Nature, Ile Réunion* 17: 35–41 &, edited, in Probst & Brial (2002)]
Laplace, Cyrille P.T. 1833–5. *Voyage autour du monde par les mers de l'Inde et de Chine, executé sur la corvette de l'état 'La Favorite' pendant les années 1830, 1831 et 1832*. Paris: Imprimerie Royale. 4 vols.
— 1841–54. *Campagne de circumnavigation de la frégate 'l'Artemise' pendant les années 1837, 1838, 1839 et 1840*. Paris: Bertrand. 6 vols.
Largen, M.J. 1989. Bird specimens purchased by Lord Stanley at the sale of the Leverian Museum in 1806, including those still extant in the collections of the Liverpool Museum. *Arch. Nat. Hist.* 14: 265–288.
La Roque, Jean de. 1716. *Voyage de l'Arabie Heureuse*. Paris: Cailleau. xvi+416pp. [includes Gollet de La Merveille's acount of Mauritius (pp.171–185) & Réunion (pp.185–211), extracted in Barnwell (1948) & Lougnon (1970)].
Larson, Edward J. 2001. *Evolution's workshop. God and science on the Galapagos Islands.* New York: Perseus/Basic Books. 320pp.
La Salle, A.de. 1845–1852. *Relation de voyage.* Vols.1–3 of *Voyage autour du monde executé pendant les années 1836 et 1837 sur la corvette La Bonite.* Paris: Arthus Bertrand. 9 vols.
Lavaux, Catherine. 1998. *La Réunion – du battant des lames au*

sommet des montagnes. Réunion & Paris: Éditions Cormorans. 455pp.

Lavergne, Christophe & Barré, Sébastien. 1997. Les problèmes de méthologie dans l'étude de la dissémination des graines du Troène *Ligustrum robustum walkeri* par les oiseaux. *Taille-vent* 3: 11–13.

—, Raneau, Jean-Claude & Figier, Jacques. 1999. The invasive woody weed *Ligustrum robustum* subsp. *walkeri* threatens native forests on La Réunion. *Biol. Invas.* 1: 377–392.

—, Duret, Clément & Gigord, Luc. 2004. The last wild Red Latan population in the Mascarene archipelago. *Plant Talk* 36: 32–33.

—, Duret, Clément & Gigord, Luc. 2005. La plus importante population sauvage de Lataniers Rouges dans l'Archipel des Mascareignes. *Mag. Palmeraie Union Latania* 13: 22-27.

— 2007. Les palmiers menacées de Maurice et de ses îlots. *Rapport du mission 22 mars au 2 avril 2006*. Réunion: Association Palmerai-Union. 45pp.

Lavergne, Roger. 1978. Les pestes végétales de l'Ile de la Réunion. *Info-Nature, Ile Réunion* 16: 9–60.

Lawler, Susan H., Sussman, Robert W. & Taylor, Linda L. 1995. Mitochondrial DNA of the Mauritian macaques (*Macaca fascicularis*): an example of the founder effect. *Am. J. Phys. Anthropol.* 96(2): 133–141.

Lawson, Robin, Slowinski, Joseph B. & Burbrink, Frank T. 2004. A molecular approach to discovering the phylogenetic placement of the enigmatic snake *Xenophidion schaeferi* among the Alethinophidia. *J. Zool. Lond.* 263: 285–294.

Layard, Edgar L. 1863. Ornithological notes from the antipodes. *Ibis* 5: 241–250.

Le, Minh *et al.* 2006. A molecular phylogeny of tortoises (Testudines: Testudinidae) based on mitochondrial and nuclear genes. *Molec. Phylogenet. Evol.* 40: 517–531.

Leal, Charles H.[D.] 1878. *Un voyage à la Réunion. Récits, souvenirs et anecdotes*. Port Louis: General Steam Printing Co. 284pp. [reprinted 1990, Saint-Denis: Ars Terres Créoles, pp.328, incl. introduction by M.Serviable]

Le Bel, Sébastien *et al.* 2001. Présence du cerf rusa dans le massif de l'Aoupinié en Nouvelle Calédonie et impact sur le reboisement en kaoris. *Bois For. Trop.* 269: 5–18.

Lebeau, A. & Lebrun, G. 1974. Élevage experimental de la tortue marine *Chelonia mydas*. *Info-Nature, Ile Réunion* 11: 7–27.

Le Bourgeois, Thomas *et al.* 2003. Actions de recherche et de réglementation sur les espèces exotiques envhaissantes à la Réunion. Pp.43–55 in Mauremootoo (2003), q.v.

Le Clezio, J.M.G. 1986. *Voyage à Rodrigues*. Paris: Gallimard. 146pp.

— 1995. *La Quarantaine*. Paris: Gallimard. 540pp.

Lecoq, Étienne (attrib.) 1740. Lettre d'un frère de Saint-Lazare sur les paroisses de Bourbon en 1740. *Rec.Trim.* 3: 236–265+ (1938). [MS, ed.A.Lougnon, who thought it was by L.Lebel; for re-attribution see Barassin (1984), q.v.]

Le Corre, Mathieu. 1996. The breeding seabirds of Tromelin island (western Indian Ocean): population sizes, trends and breeding phenology. *Ostrich* 67: 155–159.

— 1999. Plumage polymorphism of red-footed boobies (*Sula sula*) in the western Indian Ocean: an indicator of biogeographical isolation. *J. Zool., Lond.* 249: 411–415.

— & Jouventin, Pierre. 1999. Geographical variation in the White-tailed Tropicbird *Phaethon lepturus*, with description of a new subspecies endemic to Europa Island, southern Mozambique Channel. *Ibis* 141: 233–239.

— 2000. Le rossignol du Japon *Leiothrix lutea* (Sylvidées, Timaliinés), nouvelle espèce introduite à la Réunion (Océan Indien). *Alauda* 68: 68–71.

— (ed.) 2001. Numéro Special braconnage. *Taille-vent* (spécial issue), 28pp.

— & Safford, Roger. 2001. La Réunion et les Iles Éparses. Pp.693–702 in Fishpool & Evans (2001), q.v.

— *et al.* 2002. Light-induced mortality of petrels: a 4-year study from Réunion Island (Indian Ocean). *Biol. Conserv.* 105: 93–102.

—, Salamolard, Marc & Portier, Marie C. 2003. Transoceanic dispersion of the Red-tailed Tropicbird in the Indian Ocean. *Emu* 103: 183–184.

Le Corre, Mathieu *et al.* 2003b. Rescue of the Mascarene Petrel, a critically endangered seabird of Réunion Island, Indian Ocean. *Condor* 105: 391–395.

Lee, Jacques K. 1999. *Mauritius: its creole language*. London: Nautilus Publishing Co. 149pp.

Legendre, Marcel. 1929. La Huppe de la Réunion (*Fregilupus varius* Boddaert). *Rev. Hist. Nat. Appl., Oiseau* 10: 645–654, 729–743.

Le Gentil de la Barbinais, Guy. 1727. *Nouveau voyage autour du monde*. Paris: Flahaut. 3 Vols. [+ Paris: Briasson, 1729]

Le Gentil de la Galaisière, Guillaume J.H.J-B. 1779–81. *Voyage dans les mers de l'Inde fait par ordre du roi à l'occasion du passage de Venus sur le disque du soleil, le 6 juin 1761 et le 3 du même mois 1769*. Paris: Imprimerie Royale. 2 vols.

Legras, A. 1861. Le Martin. Pp.81–84 in vol.2 of Roussin (1860–67), q.v.

Legras [?C.] Dr. 1863. De la Grande Caille de Madagascar et son acclimatation à l'île de la Réunion et en France. *Bull Soc. Acclim. Hist. Nat., Réunion* 1: 75–76.

Leguat de la Fougère, François. 1707 ('1708'). *Voyage et avantures de François Leguat & de ses compagnons en deux îles désertes des Indes Orientales*. Amsterdam: J.J.de Lorme. 2 vols. [see also Oliver (1891); reprinted 1984 with a long introduction by J-M.Racault (Paris: Editions la Découverte, 244pp.) and again in 1995, with expanded introduction (Paris: Editions de Paris, 269pp.)]

Leguen, Marcel. 1979. *Histoire de l'Ile de la Réunion*. Paris: L'Harmattan. 263pp.

Le Housec, G. 1721–22. Extraits de journal du vaisseau la *Diane*. *Rev. Retrospective Ile Maurice* 4: 65–84. (1953).

Leibbrandt, Hendrik C.V. 1887. *Rambles through the archives of the colony of the Cape of Good Hope*. Cape Town: J.C.Juta & Co. xvi + 205 pp.

— 1896–1906. *Precis of the archives of the Cape of Good Hope*. Cape-Town: W.A.Richards & sons. 17 vols. [refs. in this book are all from the volumes entitled *Letters received 1695–1708* and *Letters despatched 1695–1708*, both published in 1896]

Leigh, Egbert G. 1999. *Tropical forest ecology. A view from Barro Colorado Island*. New York: Oxford University Press. 245pp.

Leisler, B. *et al.* 1997. Taxonomy and phylogeny of *Acrocephalus* warblers: an integrated analysis. *J. Orn.* 138: 469–496.

Leith-Ross, Prudence. 1998. *The John Tradescants. Gardeners to the Rose and Lily Queen*. 2nd.ed. London: Peter Owen. 333pp.

Le Juge, Léonce. 1876. Rapport ... sur l'exposition Inter-Coloniale ... au Collège Royale les 9 et 10 septembre 1875 ... sous les auspices de la Société Royale des Arts et des Sciences. *TRSAS* NS 9: [Caldwell's solitaires p.162]

Lenoir, Philippe. 1974. *Mauritius. Isle de France en mer indienne*. Port Louis: Éditions Isle de France. 147pp.

— 1979. *Mauritius, former Ile de France*. Port Louis: Éditions du Cygne. 237pp.

Lesouëf, Jean-Yves. 1975. Proposal by monsieur Lesouëf for a reserve in Rodrigues: restoration of the vegetation of Grande Montagne at Rodrigues. Unpubl. typescript report to IUCN Threatened Plant Committee [+ detailed comment by ASC in letter to M.J.E.Coode, 14/1/1976].

Lesson, Réné-Primevère. 1828. [review of Duncan (1828, q.v.)] *Bull. Sci. Nat. Geol.* 15: 304–5.

— (ed.). 1830. *Zoologie*. 2 vols.+ atlas. Part of L.I.Duperrey (ed.) *Voyage autour du monde exécuté par ordre du roi sur la corvette de sa Majesté La Coquille pendant les années 1822, 1823, 1824 et 1825*. Paris: Bertrand. 5 vols.

— 1831. *Traité d'Ornithologie*. Paris: Levrault. 2 vols.

Lesueur, Charles-Alexandre. 1803. [untitled MSS, 15035 & 15037, in the Lesueur collection in the Le Havre *Muséum d'Histoire Naturelle*; see Bonnemains & Ly-Tio-Fane 2003; copies in ASC's possession, Oxford]

Levaillant, François. 1805. *Histoire naturelle des perroquets*. Paris: Levrault, Schoell & Cie. 2 vols.

Lever, Christopher. 1985. *Naturalised mammals of the world*. London: Longman. 487pp.

— 2001. *The Cane Toad. The history and ecology of a successful colonist*. Otley, UK: Westbury Academic & Scientific Publishing. xvii+230pp.

— 2003. *Naturalized reptiles and amphibians of the world*. Oxford: Oxford University Press. 318pp.

— 2005. *Naturalised birds of the world*. 2nd ed, revised & expanded. London: T & AD Poyser. 352pp.

Lewis, J.G.E. 2002. The scolopendromorph centipedes of Mauritius and Rodrigues and their adjacent islets (Chilopoda: Scolopendromorpha). *J. Nat. Hist.* 36: 79–106.

Ley, Willy. 1941. *The lungfish & the unicorn*. New York: Modern Age Books. 305pp. [also London: Hutchinson (1948) &, revised ed., as *The lungfish, the dodo & the unicorn*, NY: Viking Books (1948)]

Liénard, François. 1843 ('1842'). Zoologie: Reptiles. *Rapp. Ann. Soc. Hist. Nat., Ile Maurice* 13 [for 1841–2]: 55–57 [reprinted in Vinson & Vinson (1969:312–14), q.v.].

Lincoln, R. n.d. Philippe Eugène Jacob de Cordemoy (1835–1911). MS biographical note in the *Archives de la Réunion*. [Copy in ASC's papers, Oxford]

Linnebjerg, Jannie. 2006. The ecological impacts of the invasive Red-whiskered Bulbul *Pycnonotus jocosus* in Mauritius. MSc thesis. University of Aarhus. 86pp.

Linon-Chipon, Sophie. 2003. *Gallia orientalis. Voyages aux Indes Orientales 1529–1722. Poétique et imaginaire d'un genre littéraire en formation*. Paris: Presses de l'Université de Paris-Sorbonne. 691pp.

Lionnet, J.F.Guy. 1984. Extinct birds of the Seychelles. Pp.505–511 in Stoddart (1984a), q.v.

Lislet-Geoffroy, Jean-Baptiste. 1790. Ile Plate [extract from Lislet's journal] *TRSAS* NS 6: 9 (1872).

Livezey, Bradley C. 1991. A phylogenetic analysis and classification of recent dabbling ducks (tribe Anatini) based on comparative morphology. *Auk* 108: 471–499.

— 1993. An ecomorphological review of the dodo (*Raphus cucullatus*) and solitaire (*Pezophaps solitaria*), flightless columbiformes of the Mascarene Islands. *J. Zool., Lond.* 230: 247–292.

— 1998. A phylogenetic analysis of the Gruiformes (Aves) based on morphological characters, with an emphasis on the rails (Rallidae). *Phil. Trans. Roy. Soc. Lond.* B 353: 2077–2151.

— 2003. Evolution of flightlessness in rails (Gruiformes: Ralidae): phylogenetic, ecomorphological and ontogenetic perspectives. *Orn. Mongr.* 53. x+654pp.

Lloyd, John A. 1846. [letter read to the Society on 2.10.1845 on the subject of Round & Serpents Islands.] *P.V. Soc. Hist. Nat. Ile Maurice*, 6 Oct. 1842–28 Aug. 1845: 154–162.

Lockwood, William B. 1984. *The Oxford book of British bird names*. Oxford: Oxford University Press. 174p.

Long, John L. 1981. *Introduced birds of the world*. Sydney: A.H. & A.W.Reed. & Newton Abbot, UK: David & Charles. 528pp.

— 2003. *Introduced mammals of the world*. Collingwood, Victoria: CSIRO & Wantage, UK: CABI Publishing. xxi+589pp.

Lorence, David H. & Sussman, Robert W 1986. Exotic species invasion into Mauritius forest remnants. *J. Tropical Ecol.* 2: 147–162.

— 1988. [Deaths] Thérésien Cadet (1937–1987); Reginald Vaughan (1895–1987). *Taxon* 37: 229–230.

— & Sussman, Robert W. 1988. Diversity, density and invasion in a Mauritius wet forest. *Monogr. Syst. Bot. Missouri Bot. Gard.* 25: 187–204.

— & Vaughan Reginald E. 1992. *Annotated bibliography of Mascarene plant life*. Lawai, Hawaii: National Tropical Botanic Garden. 274pp.

Losos, Jonathan B. 2004. Adaptation and speciation in Greater Antillean anoles. Pp.335-343 *in* U. Dieckmann *et al.* (eds) *Adaptive speciation*. Cambridge: Cambridge University Press. 476pp.

Louette, Michel & Herremans, Marc. 1985. Taxonomy and evolution in the bulbuls (*Hypsipetes*) on the Comoro Islands. *Proc. Int. Symp. Afr. Vert., Bonn*: 407–423.

— 1988. Les Oiseaux des Comores. *Ann. Mus. Roy. Afr. Centrale, Sci. Zool.* 255. 192pp.

— 1999. La faune terrestre de Mayotte. *Ann. Mus. Roy. Afr. Centr. Sci. Zool.* 284. 247pp.

Lougnon, Albert (ed.). 1934–49. *Correspondance du conseil supérieur de Bourbon et de la Compagnie des Indes. 1724–1750*. Saint-Denis: [author]. 5 vols. [NB only vols. 1–3 were available to us for consultation]

— (ed.) 1944–46. Correspondance des administrateurs de Bourbon et ceux de l'Ile de France. Première série 1727–1735. *Rec.Trim.* 7: 32–80, 114–169, 247–312 (1944); Deuxième série 1735–1746. *Rec.Trim.* 8: -38- (1946)

— 1956a. *L'Ile Bourbon pendant la Régence. Desforges-Boucher et les débuts du café*. Paris: Université de Paris, Fac. des Lettres [printed thesis] & [1957] Paris: Larose. 371pp.

— 1956b. *Classement et inventaire du Fonds de la Compagnie des Indes (Serie C°) 1665–1767*. Nérac: G.Coudert/ Archives Départementales de la Réunion. 392pp.

—, 1970. *Sous la signe de la tortue. Voyages anciens à l'Ile Bourbon (1611–1725)*. 3rd.ed. Saint-Denis: [author]. 284pp. [reprinted 1992, with new illus. added by A.Vaxelaire, as '4th

ed.'; Saint-Denis: Azalées Éditions]
Lousteau-Lalanne, M. 2004. Seychelles. Pp.45–47 *in* Muir *et al.* (2004), q.v.
Louisin, Jean-Marie & Probst, Jean-Michel. 1997. A la recherche de l'identité de "Bêbête tou't". *Bull. Phaethon* 6: 102–103.
— *et al.* 1997. A propos de la présence d'un Hibou à l'île de La Réunion. *Bull. Phaethon* 6: 99–101.
Lourenço, Wilson R. (ed.). 1996. *Biogéographie de Madagascar. Actes du Colloque Internationale Biogéographique de Madagascar, Paris, 26–28 septembre 1995*. Paris: ORSTOM. 588pp,.
Lovett, Sarah. 1997. *Extremely weird animal disguises*. Santa Fé, NM: John Muir Publications. 32pp.
Low, Tim. 2002. *Feral future. The untold story of Australia's exotic invaders*. 2nd.ed. Chicago: Chicago University Press. 420pp.
Lowe, J.J. & Walker, M.J.C. 1977. *Reconstructing quaternary environments*. London: Longman. 446pp.
Lowe, Sarah, n.d.[c2001]. *100 of the world's worst invasive species*. Auckland, NZ: Invasive Species Specialist Group. 11pp.
Lüttschwager, J. 1959. Zur systematischen Stellung der ausgestorbenen Dronte-vögel *Raphus* und *Pezophaps*. *Zool. Anz.* 162: 127–148.
Ly-Tio-Fane, Madeleine. 1958. *Mauritius and the spice trade. The odyssey of Pierre Poivre*. Mauritius Archives Publ.4. Port Louis: Esclapon. 148pp.
— 1965. Joseph Hubert and the *Société des Sciences et Arts de l'Isle de France (1801–1802)*. *PRSAS* 2: 221–246.
— 1968. Problèmes d'approvisionnement de l'Ile de France au temps de l'Intendant Poivre. *PRSAS* 3: 101–115.
— 1970. *Mauritius and the spice trade 2. The triumph of Jean Nicolas Céré and his Isle Bourbon collaborators*. Paris: Mouton & Co. 302pp.
— (ed.). 1972. *Société d'Histoire Naturelle de l'Ile Maurice. Rapports annuels I-IV, 1830–1834*. Port Louis: Royal Society of Arts & Sciences of Mauritius. 200pp.
— 1974. Le séjour de Commerson à l'Isle de Bourbon, 1770–1771. *Cah. Centre Univ.Réunion, No.Spécial Colloque Commerson*: 111–116.
— 1978 ('1976'). *Pierre Sonnerat 1748–1814. An account of his life and work*. Mauritius: (author). xvi+157pp.
— 1979. Notice historique. *Cent-cinquantaire de la Société Royale des Arts et des Sciences de l'Ile Maurice*: 3–27.
— 1982. Contacts between Schönbrunn and the Jardin du Roi at Isle de France (Mauritius) in the 18th century. An episode in the career of Nicolas Thomas Baudin. *Mitt. Östereich. Staatsarchivs* 35: 85–109.
— 1984. Indian Ocean islands – naturally. Some account of their natural history as depicted in the literature. Pre-print of paper for 2nd.Int.Conf. on Indian Ocean Studies, Perth (Australia) 5–12 Dec.1984. [an abridged version also in *Indian Ocean Newsl.* 5(2): 1–5, 1984]
— 1996. Botanic gardens: connecting links in plant transfer in the Indo-Pacific and Caribbean regions. *Harvard Papers in Botany* 8: 7–14.
— 2003. *Le Géographe et Le Naturaliste à l'île de France 1801, 1803. Ultime escale du capitaine Baudin*. Port-Louis, Mauritius: [author]. 169pp.
Lyttleton, W. 1883. Memorandum on Aldabra Island Tortoises. *Nature* 28: 398.
McAlpine, Donald F. 1982 ('1981'). Activity patterns of the Keel-scaled Boa. *Dodo* 18: 74–78.
Mac-Auliffe, J.M. 1902. *Cilaos pittoresque et thermal. Guide médicale des eaux thermales*. Saint-Denis: Albert Dubourg. 291pp.+ appendices. [reprinted 1996, with added biography of author by A.Vauthier; Saint-Denis: Azalées Editions]
Macdonald, Ian A.W., Thébaud, Christophe & Strahm, Wendy A. 1991. Effects of alien plant invasions on native vegetation remnants on La Réunion (Mascarene Islands, Indian Ocean). *Env. Conserv.* 18: 51–61.
McDougall, I., Upton, B.G.I. & Wadsworth, W.J. 1965. A geological reconnaissance of Rodrigues Island. *Nature* 206: 26–27.
Mace, Georgina M. & Collar, Nigel J. 2002. Priority setting in species conservation. Pp.61–73 *in* Norris & Pain (2002), q.v.
McFarland, C.G., Villa, J. & Toro, B. 1974. The Galapagos tortoises (*Geochelone elephantopus*). Part 1. Status of the surviving populations. Part 2. Conservation Methods. *Biol. Conserv.* 6: 118–133, 198–212.
McGregor, Arthur (ed.). 1983. *Tradescant's rarities. Essays on the foundation of the Ashmolean Museum 1683, with a catalogue of the surviving collections*. Oxford: Clarendon Press. xiii+382pp.
McIntyre, L.F. 1997. Control of unwanted plants. Pp.47–50 *in* Mungroo, Maureemootoo & Bachraz (1997), q.v.
McKelvey, David S. 1976. A preliminary study of the Pink Pigeon. *Mauritius Inst. Bull.* 8: 145–175.
— 1977. Observations on the Mauritian Pink Pigeon. *Birds. Int.* 2: 36–38.
McKeown, Sean. 1988. Breeding and maintenance of the Mauritius Lowland Forest Day Gecko (*Phelsuma g. guimbeaui*) at the Fresno Zoo. *Int. Zoo Yearb.* 28: 116–121.
— 1993. *The general care and maintenance of day geckos*. Lakeside, California: Advanced Vivarium Systems. 143pp.
MacKinnon, John R. & Phillipps, Karen. 2000. *A field guide to the birds of China*. Oxford: Oxford University Press. 586pp.
Macmillan, Allister (ed.). 1914. *Mauritius illustrated*. London: W.H.& L. Collingridge. 456pp. [reprinted 2000, New Delhi: Asian Educational Services].
McOrist, S., Black, D.G., Pass, D.A., Scott, P.C. and J. Marshall. 1984. Beak and feather dystrophy in wild sulphur-crested cockatoos (*Cacatua galerita*). *J. Wildl. Dis.* 20, 120–124.
— 1989. Some diseases of free-living Australian birds. Diseases and Threatened Birds. *ICBP Technical Publication* 10: 63–68.
Magny, Henri. 1882. *Maurice à vol d'oiseau*. Port Louis: Merchants & Planters Gazette. 263pp.
Magon de Saint-Elier, Ferdinand. 1839. *Tableaux historiques, politiques et pittoresques de l'Ile de France, aujourd'hui Maurice, depuis sa découverte jusqu'à nos jours*. Port Louis: [author]. 246+vi pp.
Mahé de La Bourdonnais, Bertrand F. 1740. *Mémoire des isles de France et de Bourbon*. [publ. 1937, ed.A.Lougnon] Saint-Denis: [editor] & Paris: Gaston Daudé. 203pp.
Mahé de Labourdonnais, L.C. (ed.). 1827. *Mémoires historiques de B.F.Mahé de Labourdonnais, gouverneur des îles de France et de Bourbon, receuillis et publiés par son petit-fils*. Paris: Felicier & Chatet. vi+367pp.
Mahy, François de. 1891. *Autour de l'île Bourbon et de Madagascar*. Paris: Alphonse Lemerre. 287pp. [reprinted 1993, with a new preface, pp.v-xii, by J.Hedo; Saint-Denis: Grand Océan]
Maillard, Louis. 1863. *Notes sur l'île de la Réunion*. 2nd.ed.,

revised & expanded. Paris: Dentu. 2 vols.
Mallam, J. 2005. *Echo Parakeet Management Report 2005.* Unpublished report to the Mauritian Wildlife Foundation.
Mamet, J.Raymond. 1979. Chronology of events in the control and eradication of malaria in Mauritius. *Rev. Agric. Sucrière Ile Maurice* 58: 107–146.
— 1993. L'entomologie aux îles Mascareignes. L'époque pre-Linnéenne (1619–1771). *PRSAS* 5(3): 97–132.
— 1996a. L'entomologie aux îles Mascareignes. L'époque pre-Linnéenne II (1619–1771). *PRSAS* 6: 123–157.
— 1996b. L'entomologie aux îles Mascareignes III. Notes bio-bibliographiques sur les entomologistes et autres naturalistes ayant recolté ou étudié les insectes à l'Ile Maurice ... pendant les années 1772 à 1910 jusqu'à leur décès. *PRSAS* 6: 159–195.
Manchester, S.J. & Bullock J.M. 2000. The impacts of non-native species on UK biodiversity and the effectiveness of control. *J. Anim. Ecol.* 37: 845–864.
Manders, Neville. 1911. An investigation into the validity of Müllerian and other forms of mimicry, with special reference to the islands of Bourbon, Mauritius and Ceylon. *Proc. Zool. Soc., Lond.* 1911: 696–749.
Mandon-Dalger, Isabelle *et al.* 1999. Modalités de la colonisation de l'Ile de la Réunion par le Bulbul orphée (*Pycnonotus jocosus*). *Rev. Écol. (Terre Vie)* 54: 282–295.
Mangar, V. & Chapman, R. 1996. The status of sea turtle conservation in Mauritius. Pp.121–124 *in* S.L.Humphrey & R.V.Salm (eds.) *Status of Sea Turtle conservation in the Western Indian Ocean.* Nairobi: IUCN/UNEP (UNEP Regional Seas Reports No. 165). 162pp.
Manglou, Jimmy. 2003. La Veuve dominicaine. *SEOR Lettre Info.* 8:5.
Mann, James R. 1860. Observations on the water supply of Mauritius. *TRSAS* NS 2: 63–81.
Marragon, Philibert. 1795. Mémoire sur l'Isle de Rodrigue. MS TB5/2 in the Mauritius Archives, 17pp. [abridged version printed in Dupon (1969) & Berthelot (2002), qq.v.]
— 1802. [letter addressed to the 'Citoyens Adminstrateurs Généraux des Établissements Français à l'Est du Cap de Bonne Esperance à l'Isle de France' dated '19 Thermidor an 10' (= 6 August 1802)]. MS TB5/1 in the Mauritius Archives, 3pp. [printed Pp.53–55 *in* Berthelot (2002), q.v.]
Marshall, Peter. 2006. *The theatre of the world. Alchemy, astrology and magic in renaissance Prague.* London: Random House (Harvill Secker). 276pp.
Martineau, A. (ed.). 1931–34. *Mémoires de François Martin, fondateur de Pondichéry (1668–1696).* Paris: Société des Éditions Géographiques, Maritimes et Coloniales. 3 vols.
Maseyk, F.J.F. 2001. Giant Aldabra tortoise trial release on Ile aux Aigrettes; tortoise-vegetation interactions- preliminary analysis and comment October 2000 – July 2001. Unpublished report to the Mauritian Wildlife Foundation.
Mason, George E. 1907. On an extinct undescribed fruit-bat of the genus *Pteropus* from the Mascarenes. *Ann. Mag. Nat. Hist.* (7)20: 220–222.
Matschie, P. 1899. *Die Megachiroptera des Berliner Museums für Naturkunde.* Berlin: G.Reimer. 105pp.
Mather, Kirtley F. 1964. *The earth beneath us. The fascinating story of geology.* London: Thomas Nelson & Sons / New York: Random House. 320pp.
Mauduyt, P.J.E. 1784. *Histoire naturelle des oiseaux.* Pp.321–391 of vol.1 & all vol.2 (544pp.) of the *Encyclopédie Methodique.* Paris & Liège: Panckoucke & Plomteux.
Maunder, Mike *et al.* 2002. The decline and conservation management of the threatened endemic palms of the Mascarene Islands. *Oryx* 36: 56–65.
[Maure, André] 1840. *Souvenirs d'un vieux colon de l'Ile Maurice.* La Rochelle, France: Frédéric Boulet. 536pp.
Mauremootoo John R. 1997. University of Mauritius undergraduate projects of relevance to the conservation of native ecosystems. Pp. 30-46 *in* Mungroo, Mauremootoo, & Bachraz (1997), q.v.
— & Payandee, Richard. 2002. Restoring the endemic flora of Rodrigues. *Plant Talk* 28: 26–28.
— & Towner-Mauremootoo, Clare V. 2002. Restoring paradise: alien species management for the restoration of terrestrial ecosystems in Mauritius & Rodrigues. Pp.55–70 *in* I.A.W.Macdonald *et al.* (eds.) *Prevention and management of invasive alien species. Proceedings of a workshop on forging cooperation throughout southern Africa, 10–12 June 2002, Lusaka, Zambia.* Cape Town: GISP.
— *et al.* 2002. The effectiveness of weeded and fenced 'conservation management areas' as a means of maintaining the threatened biodiversity of mainland Mauritius [abstract only]. Pp.408–9 *in* Veitch & Clout (2002), q.v.
— (ed.) 2003a. *Proceedings of the regional workshop on invasive alien species and terrestrial ecosystem rehabilitation in western Indian Ocean island states, Seychelles 13–17 October 2003.* Cape Town: IUCN Indian Ocean Plant Specialist Group/GISP (& on www.gisp.org). xlvi+160pp.
— 2003b. Conservation work undertaken by the Mauritian Wildlife foundation: our history, the secrets of our success and where do we go from here. Pp.81–94 *in* Mauremootoo (2003a), q.v.
— *et al.* 2003. Mauritius. Pp.12–37 *in* I.A.W.Macdonald *et al. Invasive alien species in southern Africa. National reports and directory of resources.* Cape Town: GISP. 125pp.
Mauriès, Patrick. 2002. *Cabinets of curiosities.* London: Thames & Hudson. 256pp.
Maurin, Henri & Lentge, Jacques. 1979. *Le memorial de la Réunion.* St-Denis, Réunion: Australe Éditions. 6 vols.
Mayoka, Mireille. 1998. *Cyclones à la Réunion.* Réunion: Météo-France. 48pp.
Mayr, Ernst & Diamond, Jared. 2001. *The birds of northern Melanesia. Speciation, ecology and biogeography.* New York: Oxford University Press. 492pp.
Mazin, Rosine, Coulon, Gérard & Lagesse, Marcelle. 1986. *Mauritius from the air.* Singapore: Times Editions/ Éditions du Pacifique. 128pp.
Meade, J.E. *et al.* 1960. *The economic and social structure of Mauritius.* Port Louis: Govt.Printer & [1961] London: Methuen & Co. 246pp.
Meads, M.J., Walker, K.J. & Elliott, G.P. 1984. Status, conservation and management of the land snails of the genus Powelliphanta (Mollusca: Pulmonata). *NZ J. Zool.* 11:277–306.
Medway, [Gathorne], Lord & Marshall, Adrian. 1975. Terrestrial vertebrates of the New Hebrides: origin and distribution. *Phil. Trans. Roy. Soc. Lond.* B 272: 423–465.
Mees, Gerlof F. 2006. The avifauna of Flores (Lesser Sunda Islands). *Zool. Med., Leiden* 80(3): 1-261.
Meier, Harald. 1995. Neue nachweise von *Phelsuma borbonica*

auf Réunion, Maskarenen, mit dem versuch einer taxonomischen Einordnung. *Salamandra* 31: 33–40.

Meinertzhagen, Richard. 1910–11. Diary [the 1910 & 1911 volumes cover his stay in Mauritius & brief visit to Réunion]. Multi-volume typescript MS, kept in Rhodes House (Bodleian Library), Oxford.

— 1912. On the birds of Mauritius. *Ibis* (9)6: 82–108.

Meldrum, Charles. 1886. Table showing when and where pumice or volcanic dust was observed in the Indian Ocean in 1883–84. *TRSAS*. NS 18: 165–177.

Melet, [?Jean-Jacques] de. c1672. *Relation de mon voyage aux Indes Orientales par mer...* [Publ. 1999, ed.A.Sauvaget (with intro., pp.95–102). *Études Ocean Indien* 25/26: 103–289].

Menard, H.W. 1986. *Islands*. New York: Scientific American Books. 230pp.

Mendes, Luis F. 1996. Further data on the Nicolettidae (Zygentoma), with description of a new species from Mauritius. *Revue Suisse de Zoologie* 103: 749-756.

Mertens, Robert. 1934. Die Insel Reptilien ihre Asbreitung, variation und Artbildung. *Zoologica, Stuttgart* 32(84): 1–209.

— 1963. The geckos of the genus *Phelsuma* on Mauritius and adjacent islands. *Mauritius Inst. Bull.* 5: 299–305.

— 1966. Die nichtmadagassischen Arten und Unterarten der Geckonengattung *Phelsuma*. *Senckenbergiana Biol.* 47: 85–110.

— 1970. Neues über einige Taxa der Geckonengattung *Phelsuma*. *Senck. Biol.* 51: 1013.

— 1972. Senkrecht-ovale Pupillen bei Taggeckos und Skinken. *Salamandra* 8: 45–47.

Merton, Donald V. 1978. Controlling introduced predators and competitors on islands. Pp.121–128 *in* Temple (1978a), q.v.

— 1985. Round Island nature reserve, Mauritius: a reserve for native wildlife or rabbits ? A rabbit eradication proposal. Unpubl. report to the Mauritius government, ICBP & JWPT. 37pp.+ appendices.

— 1988 ('1987'). Eradication of rabbits from Round Island, Mauritius: a conservation success. *Dodo* 24: 19–43.

— et al. 1989. *A management plan for the restoration of Round Island, Mauritius*. Jersey, CI: Jersey Wildlife Preservation Trust. 46pp.

— & Bell, Mike. 2003. New seabird records from Round Island, Mauritius. *Bull. Br. Orn. Club* 123: 212–215.

Merton, L.F.H., Bourn, David M. & Hnatiuk, R.J. 1976. Giant tortoise and vegetation interactions on Aldabra. – Part 1: inland. *Biol. Conserv.* 9: 293–304.

Meyer, Jean-Yves & Picot, Frédéric. 2001. Achatines attack! The impact of Giant African land snails on rare endemic plants in La Réunion island (Mascarene Is., Indian Ocean). *Aliens Newsl.* 14: 13–14.

Michel, Claude. 1972. [Annexe II] Zoologie. Pp.174–182 *in* Ly-Tio-Fane (1972), q.v.

— 1981. *Notes on the birds of Mauritius*. Réduit, Mauritius: Mauritius Institute of Education. 54pp.

— 1992. *Birds of Mauritius*. 3rd.ed. Rose-Hill, Mauritius: Éditions de l'Ocean Indien. 46pp.

Michel, Lélio. 1935. Conférence sur Charles Telfair. *TRSAS* C 3: 19–48.

Mickleburgh, Simon P., Hutson, Anthony M. & Racey, Paul R. 1992. *Old world fruit bats. An action plan for their conservation*. Gland, Switzerland: I.U.C.N. 252pp.

Middleton, Greg. 1995. Early accounts of caves in Mauritius. *Helictite* 33:(1): 5–18 & *PRSAS* 6: 49–87 (1996) [the 1995 version, not seen by us, may be abridged].

*— & Hauchler, Jörg. 1998. *Conservation and management of the caves of Mauritius (including Rodrigues)*. [report to Mauritius government, much quoted but not officially published; not seen by us].

— 2007. How to spend six weeks underground – opening a new show cave in Rodrigues. *Speleo Spiel, Newsl. Southern Tasmanian Caverneers Inc.* 361: 16–17.

Mielke, Howard W. 1989. *Patterns of life. Biogeography in a changing world*. Boston, Mass.: Unwin Hyman. 370pp.

Miguet, Jean-Marc. 1955. Boisement et régénération de forêts reliques en zone tropicale humide. Les forêts de Saint-Philippe à la Réunion. *Rev. For. Fr.* 7: 1876–200.

— 1957. Mise en valeur et régénération de la forêt de tamarins des hauts en zone d'altitude, à la Forêt de Bélouve à la Réunion. *Rev. For. Fr.* 8: 285–310.

— 1973. Forêt et equilibre biologique. *Info-Nature, Ile Réunion* No. spécial hors série 'La Forêt': 48–55 [officially withdrawn by author (Chapter 8), but circulated nonetheless]

— 1980. Régénération et reconstitution de forêts naturelles à l'île de la Réunion. *Rev. Ecol. (Terre Vie)* 34: 3–22.

Milbert, J.G. 1812. *Voyage pittoresque à l'Ile de France, au Cap de Bonne Esperance et à l'Ile de Ténériffe*. Paris: A.Nepveu. 2 vols.+ atlas.

Milder, Sharon L. & Schreiber, Ralph W. 1989. The vocalizations of the *Acrocephalus aequinoctialis*, an island endemic. *Ibis* 131: 99–111.

Milius, Pierre-Bernard. 1800–1804. [untitled journal kept on board the *Naturaliste* and the *Géographe*, 1800–1804]. p.1–66 *in* Bonnemains & Hauguel (1987), q.v.

Miller, Joseph T., Andrew, Rose & Bayer, Randall J. 2003. Molecular phylogenetics of the Australian acacias of subg. Phyllodineae (Fabaceae: Mimosoideae) based on the trnK intron. *Austral. J. Bot.* 51: 167–177.

Miller, Russell. 1983. *Continents in collision*. Amsterdam: Time-Life Books. 176pp.

Millies, H.C. 1868. Over eene nieuw ontdekte afbeelding van den Dodo (*Didus ineptus*). *Natuurk. Verh. K. Akad. Wet., Amsterdam* 11: 1–20.

Milne, R.I. & Abbott, R.J. 2004. Geographic origin and taxonomic status of the invasive privet *Ligustrum robustum* (Oleaceae) in the Mascarene Islands, determined by chloroplast DNA and RAPDs. *Heredity* 92(2): 78–87.

Milne-Edwards, Alphonse. 1868 (1869). Researches into the zoological affinities of the bird recently described by Herr von Frauenfeld under the name of *Aphanapteryx imperialis*. *Ibis* NS 5: 256–275 [tr. from French original in *Ann. Sci. Nat. (Zool.)* (5)10: 325–346.

— 1875. Nouveaux documents sur l'époque de la disparition de la faune ancienne de l'île Rodrigue. *Ann. Sci. Nat. Zool.* (6)2, art.4.

— & Oustalet, Emile. 1893. Notice sur quelques espèces d'oiseaux actuellement éteintes qui se trouvent représentées dans les collections du Muséum d'Histoire Naturelle. Pp.190–252 *in Centenaire de la fondation du Muséum d'Histoire Naturelle*. Paris: Muséum d'Histoire Naturelle.

Milner, Angela C., Milner Andrew R. & Evans Susan E. 2000.

Amphibians, reptiles and birds: a biogeographical review. Pp.316–332 in S.J.Culver & P.F.Rawson (eds.) *Biotic response to global change. The last 145 million years.* Cambridge: Cambridge Univ. Press. 512pp.

Milon, Philippe. 1951. Notes sur l'avifaune actuelle de l'ile de la Réunion. *Terre Vie* 98: 129–178.

Minatchy, Nelly. 2004. *Stratégie de réduction de la mortalité des pétrels induite par les éclairages artificiels.* Saint-Denis: SEOR & Université de la Réunion. 142pp.

Ministry of Agriculture, Fisheries and Natural Resources, 1985. *White paper on objectives and policies for a national conservation strategy.* Port Louis: M.A.F. & N.R. 73pp.

Mitchell, Andrew. 1989. *A fragile paradise. Nature and man in the Pacific.* London: Collins. 256pp.

Möbius, Karl A. 1880. Eine Reise nach der Insel Mauritius im Jahre 1874–75. Pp.1–61 in Möbius, K., Richters, F. & Martens, E von (eds.) *Beiträge zur Meeresfauna der Insel Mauritius und der Seychellen.* Berlin: Verlag der Gutmann'schen Buchhandlung. 352pp.

Mockford, Julian. 1950. *Pursuit of an island.* London & New York: Staples Press. 155pp.

Molnar, Peter & Stock, Joann. 1987. Relative motions of hotspots in the Pacific, Atlantic and Indian Oceans since late Cretaceous time. *Nature* 327: 587–591.

Monnier, Jeannine et al. 1993. *Philibert Commerson, le découvreur du bougainvillier.* Châtillon-sur-Chalaronne: Association Saint-Guignefort. 191pp.

Monroe, Burt L. & Sibley, Charles G. 1993. *A world checklist of birds.* New Haven, Conn. & London: Yale Univ. Press. 393pp.

Montaggioni, Lucien 1972. Essai de chronologie relative des stationnements marins quaternaires à l'Ile Maurice (Archipel des Mascareignes, Océan Indien). *C.R. Acad. Sci. Paris* 274 (D): 2936–2939.

— 1973. Histoire géologique de l'ile Rodrigue. *Info-Nature, Ile Réunion* 9: 52–59.

*— & Faure, Gerard. 1980. *Les récifs coralliens des Mascareignes (Océan Indien).* Coll. Trav. Centre Universitaire, Réunion. 151pp.

— & Nativel, Pierre. 1988. *La Réunion, Ile Maurice. Géologie et aperçus biologiques.* Paris: Masson. 192pp.

Moors, Philip J. (ed.) 1985. *Conservation of island birds.* Cambridge, UK: ICBP (Tech. Pub. 3). 271pp.

—, Atkinson, Ian A.E. & Sherley, Greg H. 1992. Reducing the rat threat to island birds. *Bird Conserv. Int.* 2:93–114..

Morad, Joan & Baverstock, Bernard. 1998. Birds on Mauritius. *Avicult. Mag.* 104: 157–160.

Mordaunt, Elinor [pseud. = Evelyn May Clowes] 1937. *Sinadaba.* London: Michael Joseph. 352pp.

Moreau, Chantal, 1993. *A la découverte de Rodrigues. Guide touristique.* Rose-Hill, Mauritius: Éditions Océan Indien. 119pp.

Moreau, Reginald E. 1957. Variation in western Zosteropidae. *Bull. Br. Mus. (Nat. Hist.), Zool.* 4(7): 311–433.

— 1960a. The Ploceine weavers of the Indian Ocean islands. *J.Orn.* 101: 29–49.

— 1960b. Conspectus and classification of the Ploceine weaverbirds. *Ibis* 102: 298–471.

— 1966. *The bird faunas of Africa and its islands.* London & New York: Academic Press. 424pp.

—, Perrins, Mary & Hughes, J.H. 1969. Tongues of the Zosteropidae (white-eyes). *Ardea* 57: 29–47.

Moree, Perry J. 1998. *A concise history of Dutch Mauritius, 1598–1710.* London: Kegan Paul International, & Leiden: International Institute of Asian Studies. 127pp.

— 2000. Discovering the undiscovered country. Pp.3–6 in Evers & Hookoomsing (2000), q.v.

— 2001. *Dodo's en galjoenen. Die reis van het schip* Gelderland *naar Oost-Indië, 1601–1603.* Zutphen, Netherlands: Walburg Pers. 348pp.

Morel, 1778. Sur les oiseaux monstrueux nommés *Dronte, Dodo, Cygne capuchoné, Solitaire* & *Oiseau de Nazare* & sur la petite isle de sable à 50 lieues environ de Madagascar. *Obs. Phys. Hist. Nat.& Arts, Paris* 12: 154–157.

Morel, Louis [as 'M...'] 1860. Musée d'Histoire Naturelle. Pp.158–160 in vol.1. of Roussin (1960–68), q.v.

Morioka, H. in prep. The skull of the Réunion Starling *Fregilupus varius* and the Sickle-billed Vanga *Falculea palliata*, with notes on their relationships and feeding habits.

Morris, Pete & Hawkins, Frank. 1998. *Birds of Madagascar – a photographic guide.* Robertsbridge, Sussex, UK: Pica Press. 316pp.

Mortensen, T. 1934. On François Leguat and his "Voyage et Avantures", with remarks on the dugong of Rodriguez and on *Leguatia gigantea. Ardea* 23: 67–77.

Mouat, Frederic J. 1852. *Rough notes of a trip to Réunion, Mauritius and Ceylon.* Calcutta: Thacker, Spink & Co. vi+140pp. [reprinted 1984 & 1999, New Delhi: Asian Educational Services].

Moulin, Pierre & Miguet, Jean-Marc. 1968. L'Office National des Forêts au service de l'homme dans un département d'outre-mer. Unpaginated reprint from *La Revue Française*, supplément au No.213.

Moulton, Michael P., Sanderson, James G. & Simberloff, Daniel. 1996. Passeriform introductions to the Mascarenes (Indian Ocean): an assessment of the role of competition. *Écologie* 27:143–152.

— & Sanderson, James G. 1997. Predicting the fates of Passeriform introductions on oceanic islands. *Conserv. Biol.* 11: 552–558.

— & — 1999. Fate of passeriform introductions: reply to Duncan and Young. *Conserv. Biol.* 13: 937–938.

Mountfort, Guy. 1988. *Rare birds of the world.* London: Collins. 256pp.

Mourer-Chauviré, Cécile & Moutou, François. 1987. Découverte d'une forme récemment éteinte d'ibis endémique insulaire de la Réunion: *Borbonibis latipes* n.gen. n.sp. *C.R. Acad. Sci. Paris* 305(II): 419–423.

— et al. 1994. *Mascarenotus* nov.gen. (Aves, Strigiformes), genre endémique éteint des Mascareignes et *M.grucheti* n.sp. espèce éteinte de la Réunion. *C.R. Acad. Sci. Paris* 318, ser.2: 1699–1706 [reprinted 1998, *Info-Nature, Ile Réunion* 24: 80–93].

—, Bour, Roger & Ribes, Sonia. 1995a. Was the Solitaire of Réunion an ibis ? *Nature* 373–568.

— & — 1995b. Position systématique de Solitaire de la Réunion: nouvelle interprétation basée sur les restes fossiles et les récits des anciens voyageurs. *C.R. Acad. Sci. Paris* 320(IIa): 1125–1131 [reprinted 1998, *Info-Nature, Ile Réunion* 24: 94–106].

—, Bour, Roger, Ribes, Sonia & Moutou, François. 1999. The

avifauna of Réunion Island (Mascarene Islands) at the time of the arrival of the first Europeans. *Smithsonian Contrib. Paleobiol.* 89: 1–38.

—, Bour, Roger & Ribes, Sonia. 2004. The taxonomic identity of *Circus alphonsi* (Newton & Gadow 1893), the extinct harrier from Mauritius. *Ibis* 146: 168–172.

—, Bour, Roger & Ribes, Sonia. 2006. Recent avian extinctions on Réunion (Mascarene Islands) from paleonto-logical and historical sources. *Bull. Brit. Orn. Club* 126A: 40-48.

Moutia, André & Mamet, J.Raymond. 1948. A review of twenty-five years of economic entomology in the island of Mauritius. *Bull. Entomol. Res.* 36: 439–472.

Moutou, François. 1979. Les mammifères sauvages de l'île de la Réunion. *Info-Nature, Ile Réunion* 17: 25–34.

— 1980. *Enquête sur la faune murine dans le département de la Réunion.* Réunion: DDASS. 131pp.

— 1981. Les mammifères sauvages de l'île de la Réunion, notes complimentaires. *Info-Nature, Ile Réunion.* 18: 29–42.

— 1982. Note sur les chiroptères de l'île de la Réunion (Océan Indien). *Mammalia* 46: 35–51.

— 1983a. Les peuplements des vertébrés des îles Mascareignes. *Rev. Ecol. (Terre Vie)* 37: 21–35.

— 1983b. Introductions dans les îles: l'éxemple de l'île de la Réunion. *C.R. Soc. Biogeog.* 59(2): 201–211 & *Info-Nature, Ile Réunion* 20: 39–48.

— 1983c. Identification des reptiles réunionnais + carte de repartition de 3 espèces de reptiles réunionnais. *Info-Nature, Ile Réunion* 20: 53–62, 63–64.

— 1984. Wildlife on Réunion. *Oryx* 18: 160–162.

— 1986. *Suncus murinus* à la Réunion. *Info-Nature, Ile Réunion* 22: 17–23.

— & Vesmanis, Indulis E. 1986. Zur Kenntnis der Moschusspitzmaus *Suncus murinus* (Linnaeus 1766) von Réunion (Indischer Ozean) (Mammalia, Insectivora, Soricidae). *Zool. Abh. Staat. Mus. Tierk. Dresden* 42: 41–52.

— 1987. Les carnivores des îles françaises d'Outre-Mer. Part 20, 10pp., of M.Artois & P.Delattre (eds) 1986-2002 *Encyclopédie des carnivores de France*. Paris: Société Française pour l'Étude et la Protection des Mammifères.

— 1988. Biogéographie des chauves-souris de l'Océan Indien occidental. *C.R. Soc. Biogeog.* 64(3): 89–97 [reprinted in *Info-Nature, Ile Réunion* 23: 73–88, 1989].

— 1995. *Phelsuma laticauda*, nouvelle espèce de lézard récemment introduite à la Réunion. *Bull. Phaethon* 1: 33–34.

Mozzi, Robert, Deso, Grégory & Probst, Jean-Michel. 2005. Un nouveau gecko vert introduit à la Réunion: Le *Phelsuma astriata semicarinata* (Cheke, 1982). *Bull. Phaethon* 21: 104.

Muir, Catherine, Ngusaru Amani & Mwakanema, Lydia (eds.) 2004. The status of dugongs in the western Indian Ocean and priority conservation actions. Dar-es-Salaam: WWF.

Müllenmeister, Kurt J. 1988. *Roelant Savery. Die Gemälde mit kritische Oeuvrekatalog.* Frerens, Germany: Luca Verlag. 427pp.

— 1991. *Roelant Savery. Neues und Ergänzungen zur Oeuvreverzeichnis, Freren 1988.* Bremen: Carl Ed. Schünemann KG. 87pp.

Mundy, Peter. 1608–1667. *The travels of Peter Mundy in Europe and Asia* [MS publ. in 5 vols., 1905–1936, ed. R.C.Temple. London: Hakluyt Society]

Mungroo, Yousoof. 1979. Report of post-Celine II survey of the endemic Passeriformes and bats and sea birds of Rodrigues. Curepipe, Mauritius: Forestry Service. Cyclostyled, 5pp.

— 1997. Restoration of highly degraded and threatened native forests of Mauritius. *In* Mungroo, Maureemootoo & Bachraz (1997), q.v.

—, Maureemootoo, John & Bachraz, V. (eds.) 1997. *Proceedings of the workshop on restoration of highly degraded and threatened native forests in Mauritius, University of Mauritius, Reduit 8-12 September 1997.* Réduit, Mauritius: National Parks & Conservation Service.

— & Tezoo, Vishnu. 2000. Control of alien invasive species in Mauritius. Pp.18–24 in E.E.Lyons & S.E.Miller (eds.) *Invasive species in Eastern Africa: Proceedings of a workshop held at ICIPE, July 5–6, 1999.* Nairobi: ICIPE. 108pp.

Murphy, Pat & Doherty, Paul. 2000. *Traces of time. The beauty of change in nature.* San Francisco: Chronicle Books. 120pp.

Murphy, R.Cushman & Pennoyer, Jessie M. 1952. Larger petrels of the genus *Pterodroma*. *Am. Mus. Novitates* 1580. 43pp.

Myers, J.H. & Bazely, D.R. 2003. *Ecology and control of introduced plants.* Cambridge, UK: Cambridge University Press. 328pp.

Myers, Norman. 1979. *The sinking ark. A new look at the problem of disappearing species.* Oxford, Pergamon Press. xiii+307pp.

Nagy, Zoltán et al. 2003. Multiple colonization of Madagascar and Socotra by colubrid snakes: evidence from nuclear and mitochondrial gene phylogenies. *Proc. Roy. Soc. Lond.* B 270: 2613–2621.

Nallee, M. 2004. Status of marine turtles in Mauritius. [minimalist 'power-point' file on www.ioseaturtles.org/ eleclib]

Napal, Dayachand. 1979. *Dutch Mauritius and Ile de France (1638–1810).* Port Louis: [author]. 145pp.

Napier-Broome, Frederick. 1884. Speech of his excellency the Lieutenant-Governor on opening the session of the Council of government for 1883. *Mauritius Almanac & Colonial Register* 1884: 128–133.

Near, Thomas J., Meylan, Peter A. & Shaffer, H.Bradley. 2005. Assessing concordance of fossil calibration points in molecular clock studies: an example using turtles. *Am. Nat.* 165(2): 137–146.

Neiman, B.G. 1970. *Nobye karty tyechenii indiiskogo okeana [New maps of Indian Ocean currents].* Dokladi Akad. Nauk SSSR NS 195: 948–951.

Nelson, J. Bryan. 1978. *The Sulidae. Gannets and Boobies.* Oxford: Oxford Univ. Press. 1012pp.

— 2006. *Pelicans, Cormorants and their relatives. Pelicanidae, Sulidae, Phalacrocoracidae, Ahingidae, Fregatidae, Phaethontidae.* Oxford: Oxford University Press. 680pp.

Nettleship, David N., Burger Joanna & Gochfield, Michael (Eds.). 1994. *Seabirds on islands, Threats, case studies and action plans.* Cambridge, UK: Birdlife International. 318pp.

New, A.L. et al. 2005. Physical and biochemical aspects of the flow across the Mascarene plateau in the Indian Ocean. *Phil. Trans. Roy. Soc. Lond.* A 363: 151–168.

Newfield, M. et al. 2003. *Ile aux Aigrettes weed management strategy, 2003–2008.* Port Louis: Mauritian Wildlife Foundation (Tech. Ser. 5/03).

Newman, Kenneth B. & Bannister, A.B. 1965. Did the Dodo die in vain ? *Animals* 7: 199–205 [on Round Island]

— 1998. *Newman's birds of South Africa.* 7th.ed. Halfway

House, S.Africa: Southern Book Publishers/Struik. 510pp.
Newton, Alfred. 1860–61. [editorial notes on specimens sent from Mauritius by Edward Newton] *Ibis* 2: 201–2, 3:115–6.
— 1861. Description of a new species of water-hen (*Gallinula*) from the island of Mauritius. *Proc. Zool. Soc. Lond.* 1861: 18–19.
— 1865. On two new birds from the island of Rodriguez. *Proc. Zool. Soc. Lond.* 1865: 46–48.
[Newton, Alfred]. 1868. Recent ornithological publications. 2.Dutch; 3.German. [publications reviewed & abstracted by the editor, A.Newton] *Ibis* NS 4: 476–482.
Newton, Alfred. 1869. On a picture supposed to represent the Didine Bird of the Island of Bourbon (Réunion). *Trans. Zool. Soc. Lond.* 6: 373–376.
— & Newton, Edward. 1870 ('1869'). On the osteology of the Solitaire or Didine bird of the island of Rodrigues, *Pezophaps solitaria* (Gmel.). *Phil. Trans. Roy. Soc. Lond.* 159: 327–362 [advance abstract in *Proc. Roy. Soc. Lond.* 16: 428–433, 1868].
— 1872. On an undescribed bird from the island of Rodriguez. *Ibis* (3)2: 30–34.
— 1874. On a living Dodo shipped for England in 1628. *Proc. Zool. Soc. Lond.* 1874: 447–449 & *TRSAS* NS 9: 34–38 (1876).
— 1875a. [on some unpublished sketches of the Dodo and other extinct birds of Mauritius] *Proc. Zool.Soc.Lond.* 1875: 349–350 [reprinted in *TRSAS* N.S.11: 107–109, 1883]
— 1875c. A note on *Palaeornis exsul*. *Ibis* (3)5: 343–344.
— & Newton, Edward. 1876. On the Psittaci of the Mascarene Islands. *Ibis* (3)6: 281–289.
— 1878. [on a gizzard stone sent by Mr Caldwell]. *Proc. Zool. Soc. Lond.* 1878: 291–2.
—, & Gadow, Hans. 1896. *A dictionary of birds*. London: A.& C.Black. 1088pp.
— 1907. Leguat's giant bird. *Ornis* 14 (*Proc. 4th Int. Orn. Congr.*): 70–71.
Newton, Edward. 1858–95. [egg collecting & birdwatching journals] MSS kept with Alfred Newton's papers at Cambridge University Library.
— 1861a. Ornithological notes from Mauritius. No.I. A visit to Round Island. *Ibis* 3: 180–182.
— 1861b. Ornithological notes from Mauritius. No.II. A ten days' sojourn at Savanne. *Ibis* 3: 270–277.
— 1861–82. [letter books, 6 vols., missing Feb.1866-Mar.1869] MSS kept with Alfred Newton's papers at Cambridge University Library.
— 1863. Notes of a second visit to Madagascar. *Ibis* 5: 333–50, 452–61.
— 1865a. Notes of a visit to the island of Rodriguez. *Ibis* NS 1: 146–153.
— 1865b. [address to the Society], abstracted *in* L.Bouton, Annual Report[of the Secretary]. *TRSAS* NS 2: 225.
— 1875. [address to the Society]. Pp.xxii-xxiii *in* Bouton (1875), q.v.
— 1878a (1883). Discours de l'Hon.Edward Newton [delivered to the RSAS, 6 February 1878]. *TRSAS* NS 11: 137–141.
— 1878b (1883). [letter to the Hon.V[irgile] Naz, dated 26 February 1878, published 1883 in:] *TRSAS* NS 11: 70–73.
— 1888. Address by the President, Sir Edward Newton KCMG FLS CMZS, to the members of the Norfolk and Norwich Naturalists' Society. *Trans. Norfolk & Norwich Nat. Soc.* 4: 537–554.
— & Gadow, Hans. 1893. On additional bones of the Dodo and other extinct birds of Mauritius obtained by Mr Théodore Sauzier. *Trans. Roy. Soc. Lond.* 13: 281–302.
Newton, Ian. 2003. *The speciation and biogeography of birds*. London & San Diego: Elsevier (Academic Press). 668pp.
Newton, Robert. 1956. Bird islands of Mauritius. *Ibis* 98: 296–302.
— 1958a. Ornithological notes on Mauritius and the Cargados Carajos Archipelago. *PRSAS* 2: 39–71.
— 1958b. Bird preservation in Mauritius. *Bull. Int. Counc. Bird Preserv.* 7: 182–185.
— 1959. Notes on the two species of *Foudia* in Mauritius. *Ibis* 101: 240–243.
— 1960. Birds in Mauritius. *Devon Birds* 13: 36–40.
Nichelson, W. 1780. Mr Nichelson's account of a passage from Madras to and from the island Diego Rayes, or Rodrigues, in the south-west monsoon. CCXVII. A description of the island Rodrigues. Pp.272–276 *in* S.Dunn (ed.) *A new directory for the East Indies*. 5th.ed. London: Henry Gregory.
Nicholls, Henry. 2006. Digging for Dodo. *Nature* 443: 138-140.
Nichols, Richard A. & Freeman, Karen. 2004. Using molecular markers with high nutation rates to obtain estimates of relative population size and to distinguish the effects of gene flow and mutation: a demonstration using data from endemic Mauritian skinks. *Molec. Ecol.* 13: 775–787.
Nichols, Rina [K.] 1999. Endangered passerines on Mauritius. Management report 1998/1999. Port Louis: Mauritian Wildlife Foundation. Photocopied, 27pp.
— *et al.* 2002. Status of the critically endangered Mauritius Fody in 2001. *Bull. Afr. Bird Club* 9: 95–100.
—, & Woolaver, Lance. 2003. First observation of a successful nest for the endangered Mauritius Olive White-eye *Zosterops chloronothos*. *Bull. Afr. Bird Club* 10:101–106.
—, Woolaver, Lance & Jones, Carl [G.] 2004. Continued decline and conservation needs of the endangered Mauritius Olive White-eye *Zosterops chloronothos*. *Oryx* 38: 291–296.
— , Woolaver, Lance & Jones, Carl G. 2005a. Breeding biology of the endangered Mauritius Olive White-eye *Zosterops chloronothos*. *Ostrich* 76: 1–7.
—, Woolaver, Lance & Jones, Carl G. 2005b. Low productivity in the critically endangered Mauritius Olive White-eye *Zosterops chloronothos*. *Bird Conserv.Int.* 15: 297–302.
Nicholson, Kirsten E. *et al.* 2005. Mainland colonization by island lizards. *J. Biogeog.* 32: 929-938.
Nicoll, Malcolm A. C. , Jones, Carl G. & Norris, Ken. 2003. Declining survival rates in a reintroduced population of the Mauritius Kestrel: evidence for non-linear dependence and environmental stochasticity. *J. Anim. Ecol.* 72: 917-926.
Nicoll, Martin E. 2003. *Tenrec ecaudatus*, Tenrec, *tandraka*, *trandraka*. Pp.1283–1287 *in* Goodman & Benstead (2003), q.v.
Nieuhoff, John. 1682. [as *John Nieuhoff's voyages*], pp.1–305 *in* Churchill's *voyages and travels*, 2nd.ed.1752, vol.2., London: Thomas Osborne. [reprinted 1988, Singapore: Oxford University Press. 326pp., & 2001, as *Voyages and travels into Brasil and the East Indies*, New Delhi: Asian Educational Services, viii+370pp.]
Niven, Cicely. 1964. A field day on the island of Mauritius. *Bokmakierie* 16: 6–7.
Noble, Charles F. 1756. Some remarks made at the French islands

of Mauritius and Bourbon, 1755. [MS pub. 1794 *in Oriental Repertory* 2(1): 99–133; also reprinted (in French) in *Rec. Trim*.1: 299–322, 359–370, 398–421, & 458–472 (1934), but there attrib. to Alexander Dalrymple, who *edited* the *Oriental Repertory* !]

Nogales, Manuel *et al.* 2004. A review of feral cat eradication on islands. *Conserv. Biol.* 18: 310–319.

Norris, Ken & Pain, Deborah (eds.) 2002. *Conserving bird biodiversity*. Cambridge, UK: Cambridge University Press. 337pp.

North, Steven G., Bullock, David J. & Dulloo, M.Ehsan. 1994. Changes in the vegetation and reptile populations on Round Island, Mauritius, following eradication of rabbits. *Biol. Conserv.* 67: 21–28.

North-Coombes, G. Alfred. 1948. *Agricultural production other than sugar cane*. Unpubl. submission to the Mauritius Economic Commission 1947–8. 51pp.+ appendices. [copy in Rhodes House Library, Oxford]

— 1971. *The island of Rodrigues*. Port Louis: [author]. 338pp. [reprinted 2002, with new intro. by M. & P.North-Coombes; Mauritius: [editors]]

— 1980. *La découverte des Mascareignes par les Arabes et les Portuguais. Retrospective et mise à point*. Port Louis: Service Bureau. 175pp. [reprinted 1994]

— 1983. François Leguat, *le géant* and the flamingo in the Mascarene Islands. *PRSAS* 4(3): 1–30.

— 1991. *The vindication of François Leguat*. 2nd.ed., revised. Rose-Hill, Mauritius: Éditions de l'Océan Indien. 306pp. [1st ed was in 1980; a 3rd ed., 1995, has only minor corrections].

— 1993. *A history of sugar production in Mauritius*. [reissue with new chapter of *The evolution of sugar-cane culture in Mauritius*, 1937] Floreal, Mauritius: [author]. 173pp.

— 1994. *Histoire des tortues de terre de Rodrigues*. 2nd.ed. Port Louis: [author] x+100pp.

— 1996. *Two visits to Rodrigues: 1948 and 1949*. Quatre Bornes, Mauritius: Éditions Capucines. 40pp.

Nougier, Paul. 1971. Nos derniers caméléons sont-ils sauvés ? *Info-Nature, Ile Réunion* 4: 34–35.

Nowak, Ronald M. 1999. *Walker's Bats of the world*. Baltiimore: Johns Hopkins University Press. 287pp.

Nussbaum, Ronald A. 1984a. Amphibians of the Seychelles. Pp.379–415 *in* Stoddart (1984a), q.v.

— 1984b. Snakes of the Seychelles. Pp.361–377 *in* Stoddart (1984a), q.v.

— *et al.* 2000. New species of day-gecko, *Phelsuma* Gray (Reptilia, Squamata: Gekkonidae), from the Reserve Naturelle Intégrale d'Androhahela, southern Madagascar. *Copeia* 2000(3): 763–770.

Nyhagen, Dorte F. *et al.* 2001. Insular interactions between lizards and flowers: flower visitation by an endemic Mauritian gecko. *J. Trop. Ecol.* 17: 755–761.

— 2004. A study of the bat-fruit syndrome on Mauritius, Indian Ocean. *Phelsuma* 12: 118–125.

— *et al.* 2005. An investigation into the role of the Mauritian flying fox, *Pteropus niger*, in forest regeneration. *Biol. Conserv*, 122:491-497.

O'Brien, John & Hayden, Thomas J. 2004. The application of genetic research to the conservation of fruit bats in the western Indian Ocean. *Phelsuma* 12: 141–146.

— 2005. Phylogeography and conservation genetics of *Pteropus* fruit bats (Megachiroptera: Pteropodidae) in the western Indian Ocean. PhD Thesis. University College Dublin, Ireland.

— *et al.* 2007. Rodrigues Fruit Bats (*Pteropus rodricensis*, Megachiroptera, Pteropodidae) retain genetic diversity despite population declines and founder events. *Conserv. Genet.* 8: 1073–1082.

O'Neill, Craig, Müller, Dietmar & Steinberger, Bernhard. 2005. On the uncertainties in hot spot reconstructions and the significance of moving hot spot reference frames. *Geochem. Geophys. Geosyst.* 6(4) [on-line only at www.agu.com/pubs, ref. Q04003 doi:10.1029/2004 GC000784, 2005]

Ogilvie-Grant, W.R. 1893. *Catalogue of birds in the British Museum. Vol.22. Game Birds (Pterocles, Gallinae, Opisthocomi, Hemipodii)*. London: British Museum (Natural History). xvi+585pp.

Ohier de Grandpré, Louis M.J. 1803. *A voyage in the Indian Ocean and to Bengal*. London: G.& J.Robinson [reprinted 1995; Delhi: Asian Educational Services. Tr. from orig. French ed. of 1801, not seen]

Olearius, A. (ed.) 1670. *De Beschryving der Reizen van Volkert Evertsz naar Oostindien...* Pp.91–130 *in* J.H.Glazemaker (tr.& ed.) *Verhaal van drie voorname Reizen naar Oostindien...* Amsterdam: Jan Rieuwertsz & Pieter Arentsz [originally appeared in German in 1669]

Olesen, Jens M. 1998. Mauritian red nectar remains a mystery. *Nature* 393: 529.

— , Eskilden, Louise I. & Venkatasamy, Shadila. 2002. Invasion of pollination networks on oceanic islands: importance of invader complexes and endemic super generalists. *Diversity & Distrib.* 8: 181–192.

Oliver, S.Pasfield. 1881. *On and off duty. Being leaves from an officer's notebook*. London: W.H.Allen & Co. 386pp.

— (ed.) 1891. *The voyage of François Leguat of Bresse to Rodriguez, Mauritius, Java, and the Cape of Good Hope, transcribed from the first English edition, edited and annotated by Capt.Pasfield Oliver*. London: Hakluyt Society. 2 vols. [reprinted c1974, New York: Burt Franklin & 2003, Chestnut Hill, Maryland: Adamant Media Corp.]

— (tr. & ed.) 1897. *The voyages made by the Sieur D.B. to the islands Dauphine or Madagascar and Bourbon or Mascarenne in the years 1669,70,71 and 72*. London: David Nutt. xxxv+160pp.

— (ed.) 1904. *Memoirs & travels of Mauritius Augustus Count de Benyowsky*. London: Kegan Paul Trench Trübner & Co. xxxvi+636pp. [apparently revised from the 1893 ed., London: T.Fisher Unwin]

— 1909. *The life of Philibert Commerson, D.M., Naturaliste de Roi. An old-world story of French travel and science in the days of Linnaeus*. [ed. by G.F.Scott Elliott] London: John Murray. xvii+242pp.

Oliver, William D. 1896. *Crags and craters. Rambles in the island of Réunion*. London: Longmans, Green & Co. 213pp.

Olson, Storrs L. 1975a. Paleornithology of St.Helena Island, south Atlantic Ocean. *Smithsonian Contrib. Paleontol.* 23. 49pp.

— 1975b. An evaluation of the supposed *Anhinga* of Mauritius. *Auk* 92: 374–376.

— & James, Helen F. 1984. The role of Polynesians in the extinction of the avifauna of the Hawaiian Islands. Pp.768–780 *in* P.S.Martin & R.G.Klein (eds.) *Quaternary extinctions: a pre-*

historic revolution. Tucson: Univ. of Arizona Press.
— & Warheit, K.I. 1988. A new genus for *Sula abbotti*. *Bull. Brit. Orn. Club* 108: 9–12
— & James Helen F. 1991. Description of thirty-two new species of birds from the Hawaiian Islands. Part I: Non-Passeriformes *Orn. Monogr*. 4: 1–88.
— et al. 2005. Expunging the 'Mascarene Starling' *Necropsar leguati*: archives, morphology and molecules topple a myth. *Bull. Brit. Orn. Club*. 125: 31–42.
Orme, Robert. 1763. *A history of the military transactions of the British nation in Indostan from the year MDCCXLV*. London: John Nourse. 415pp.
Osadnik, Gerhard. 1984. An investigation of egg-laying in *Phelsuma* (Reptilia: Sauria: Gekkonidae). *Amphibia-Reptilia* 5: 125–134.
Ottino, P. 1976 ('1974'). Le moyen age de l'Océan Indien et le peuplement de Madagascar. *Annuaire Pays Océan Indien* 1: 197–221.
Oudemans, A.C. 1917. Dodo-Studien, naar aanleiding van de vondst van een gevelsteen met Dodo-beeld van 1561 to vere. *Verh. K. Akad. Wet. Amsterdam* (2)19(4). Pp.vii+140+plates
— 1918. [studies on the dodo]. *Ibis* (10)6: 316–321. [untitled summary in English of Oudemans (1917), q.v.]
Oustalet, Emile. 1897 ('1896'). Notice sur la faune ornithologique ancienne et moderne des Iles Mascareignes, et en particulier de l'Ile Maurice. *Ann. Sci. Nat. Zool.* (8)3: 1–128.
Ovenell, R. F. 1992. The Tradescant dodo. *Arch. Nat. Hist.* 19: 145–152.
Owadally, A.Wahab. 1971. *Annual Report of the Forestry Service for the year 1969*. Port Louis: Government Printer. 15pp.
— & Butzler, W. 1972. *The deer in Mauritius*. Port Louis: [authors]. 32pp.
— 1979. The dodo and the Tambalacoque tree. *Science* 203: 1363–1364 [with a reply from S.A.Temple, p.1364.; both reprinted in Gould (1980), q.v.]
— 1980. Some forest pests and diseases in Mauritius. *Rev. Agricole Sucrière Ile Maurice* 59: 76–94.
— 1984. *Annual Report of the Forestry Service ... for the year 1981*. Port Louis: Government Printer [No.3 of 1984]. 28pp.
— 1985. *Progress report 1980–84 by the Forestry Service of the Ministry of Agriculture, fisheries and Natural Resources, prepared for the Twelth Commonwealth Forestry Conference, 1985*. Port Louis: Government Printer. 14pp.
Owen, Andrew. 2004. Hand-rearing the Mauritius Fody *Foudia rubra*. *Avicult. Mag.* 110(3): 119–129.
Owen, Richard. 1866. On the osteology of the Dodo. *Trans. Zool. Soc. Lond*. 6: 49–86.
Padya, B.M. 1976. *Cyclones of the Mauritius region*. Port Louis: Mauritius Printing Co. & Government Printer. 151pp.
— 1984. *The climate of Mauritius*. 2nd ed. Vacoas, Mauritius: Meteorological Office. 217pp. [revised & extended from original undated ed. (c1972)].
— 1989. *Weather and climate of Mauritius*. Moka, Mauritius: Mahatma Gandhi Institute. 283pp.
Padayatchy, N. (1998). The short-term effects of weeding on populations of endemic day geckos in a Mauritian upland forest Conservation Management Area. B.Sc. (Hons). thesis. University of Mauritius.
Page, Wayne. 1995. A preliminary survey of the indigenous forest vegetation of southwest Mauritius to locate priority areas for conservation management. Unpublished Report for the Mauritius Wildlife Appeal Fund and the National Parks and Conservation Services, Government of Mauritius.
— & d'Argent, Gabriel. 1997. *A vegetation survey of Mauritius to identify priority rainforest areas for conservation management*. Port Louis: Mauritian Wildlife Foundation. Unpaginated.
*Palkovacs, Eric P., Gerlach, Justin & Caccone, Adalgisa. 2002. The evolutionary origin of Indian Ocean tortoises (*Dipsochelys*). *J. Molec. Phylogenet. Evol.* 24: 216–227.
— et al. 2003. Are the native tortoises from the Seychelles really extinct ? A genetic perspective based on mtDNA and microsatellite data. *Molec. Ecol.* 12: 1403–1413.
Palmer, Ralph S. 1985. Limpkin. Pp.328–9 *in* Campbell & Lack 1985, q.v.
Panyandee, Sitradeven. 2002. *The Dutch odyssey. Encounter with Mauritius*. Moka, Mauritius: Mahatma Gandhi Institute. 91pp.
Parnell, John *et al.* 1989. A study of the ecological history, vegetation and conservation management of Ile aux Aigrettes, Mauritius. *J. Trop. Ecol.* 5: 355–374.
Parson, Lindsay M. & Evans, Alan J. 2005. Seafloor topography and tectonic elements of the western Indian Ocean. *Phil. Trans. Roy. Soc. Lond*. A 363: 15–21.
Pasquier, Roger. 1980. Report and management plan on ICBP's project for the conservation of forest birds of Mauritius. Washington: ICBP. Photocopied report, 11pp.
Pasteur, Georges & Bour, Roger 1992. Priorité de *Phelsuma cepediana* (Milbert 1812) sur *Phelsuma cepediana* (Merrem 1820) dans la désignation de l'espèce type du genre *Phelsuma* Gray (Sauria, Gekkonidae). *Bull. Soc. Herp. Fr.* 63: 1–6.
Patterson, G.B. & Daugherty, C.H. 1995. Reinstatement of the genus *Oligosoma* (Reptilia: Lacertilia: Scincidae). *J. Roy. Soc. NZ*. 25: 327–331.
Paulay, Gustav. 1994. Biodiversity on oceanic islands: its origin and extinction. *Am. Zool.* 134–144.
Paulian, Renaud. 1961. *La zoogéographie de Madagascar et des îles voisines*. Faune de Madagascar 13. Tananarive: Institut de Recherche Scientifique. 481pp.
— 1980. Une originale experience de sylviculture restauratrice en milieu montagnard tropical à la Réunion, et quelques reflexions qu'elle suggère. *Rev. Ecol. (Terre Vie)* 34: 23–30.
Paupiah, Hans. 2001. *Forestry outlook studies in Africa (FOSA). Mauritius*. Rome: FAO. 29pp. [on-line at www.fao.org].
Pavageau, Colette. 2000. *Tableau économique de la Réunion. Édition 2000–2001*. Saint-Denis: INSEE. 210pp.
Payendee, Richard. 2003. Restoration projects in Rodrigues carried out by the Mauritian Wildlfe Foundation. Pp.95–98 *in* Mauremootoo (2003), q.v.
Payet, Rolph. 2005. Research, assessment and management on the Mascarene Plateau: a large marine ecosystem perspective. *Phil. Trans. Roy. Soc. Lond*. A 363: 295–307.
Payet, Sébastien (ed.). 2006. Bilan des activités de la SEOR en 2005, présentée à l'Assemblée Générale du 17 juin 2006. *Chakouat* 18: 1–15.
Peake, J.F. 1971. The evolution of terrestrial faunas in the western Indian Ocean. *Phil. Trans. Roy. Soc. Lond*. B 260: 581–610.
Pearson, Mike P. & Godden, Karen. 2002. *In search of the red slave. Shipwreck and captivity in Madagascar*. Stroud, UK:

Sutton Publishing. 217pp.

Peirce, Michael A., Cheke, Anthony S. & Cheke, Robert A. 1977. A survey of blood parasites of birds in the Mascarene Islands, Indian Ocean, with descriptions of two new species and taxonomic discussion. *Ibis* 119: 451-461.

— 1979. Some additional observations on haematozoa of birds in the Mascarene Islands. *Bull. Brit. Orn. Club* 99: 68–71.

—, Greenwood, Andrew G. & Swinnerton, Kirsty [J.] 1997. Pathogenicity of *Leucocytozoon marchouxi* in the Pink Pigeon (*Columba mayeri*) in Mauritius. *Vet. Rec.* 140: 155–156.

Peletyer *et al.* 1701. Ordre et instructions que Messieurs le Directeurs généraux de la Compagnie des Indes Orientales desirent être executez en l'isle de Bourbon par le Sieur de Villers... MS in New York Public Library, published 1909 in *Bull. New York Public Lib.* 13(1): 7–12.

Pelte, Stanislas. 1938. Éloge funèbre de M Georges Antelme. *TRSAS* C 8: 2–7.

Penny, Malcolm J. & Diamond, Anthony W. 1971. The Whitethroated Rail *Dryolimnas cuvieri* on Aldabra. *Phil. Trans. Roy. Soc. Lond.* B 260: 529–548.

Penrose, Boies. 1942. *Urbane travellers 1591–1635*. Philadelphia: Pennsylvania University Press. 251pp.

Pernetta, Angelo P., Bell, Diana J. & Jones, Carl G. 2005. Macro- and microhabitat use of Telfair's Skink (*Leiolopisma telfairii*) on Round Island, Mauritius: implications for their translocation. *Acta Oecol.* 28: 306–312.

Peron, François. 1803. Untitled MSS 15022, 15031 & 15033 in the Lesueur Collection in the Muséum d'Histoire Naturelle, Le Havre [copies in ASC's papers, Oxford; see Bonnemains & Ly-Tio-Fane, 2003]

— (1807) & Freycinet, Louis (1816). *Voyage de découvertes aux Terres Australes executée par ordre de sa majesté l'Empereur et roi sur les corvettes* Le Géographe, Le Naturaliste, *et la goelette* Le Casuarina *pendant les années 1800, 1801, 1802, 1803, et 1804*. Paris: Imprimerie Imperiale. 2 vols. [vol.1 by Peron, vol.2 by Freycinet] + Atlas [book of plates].

*Perroud, B. 1982. *Étude volcano structural des îles Maurice et Rodrigue (Océan indien occidental), origine du volcanisme*. Thesis, University of Grenoble, France. 210pp.

Peters, Gustav & Hutterer, Rainer (eds.) 1990. *Vertebrates in the tropics. Proceedings of the International Symposium on Vertebrate Biogeography and Systematics in the Tropics, Bonn, June 5–8, 1989*. Bonn: Alexander Koenig Zoological Research Institute & Mueum. 424pp.

Peters, W. [C.H.] 1877. Über die Herrn Prof. Dr. K.Möbius 1874 auf den Maskarenen und Seychellen sowie über die von Herrn Dr. Sachs im vorigen jahr in Venezuela gesammelten Säugethiere und Amphibien. *Monatsber. K. Preuss. Akad. Wiss., Berlin* 1877: 455–460.

Peterson, R.L., Eger, J.L. & Mitchell, L. 1995. *Faune de Madagascar 84: Chiroptères*. Paris: Museum Nationale d'Histoire Naturelle. 204pp.

Petit, F. P. du. 1741. Description anatomique des yeux de la grenouille et de la tortue. *Mém. Acad. Roy. Sci., Paris* '1737': 142–169.

Petit-Radel, P. 1801. *De amoribus Pancharitis et Zoroae*. Paris: Didot. [the part of the introduction on the author's visit to Réunion, tr. into French by F.Cazamian, was published under the title 'Un voyage à la Réunion en 1794' *in* Roussin (1860–68, q.v.) 1:1–16, and the Mauritian part, tr. by E.Avirignet, in *TRSAS* 19: 52–56, 1887].

Pfeiffer, Ida. 1861. *The last travels of Ida Pfeiffer, inclusive of a visit to Madagascar*. London: Routledge, Warne & Routledge. 338pp.

Phillips, Michael. [2005] Ships of the Old Navy 2. A history of the ships of the 18th century Royal Navy. [online on www.cronab.demon.co.uk, a work in progress, checked 1/11/05]

Phillips, W.W.A. & Sims, R.W. 1985. Some observations on the fauna of the Maldive Islands. Part III. birds. *J. Bombay Nat. Hist. Soc.* 55: 195–217.

Pike, Nicholas, 1870a. A visit to Round Island. *TRSAS* NS 4: 11–22.

Pike, Nicholas. 1870b. Notes on the fauna of Round Island, with special reference to the prepared case sent to his excellency Sir Henry Barkly, K.C.B. *TRSAS* NS 4: 131–135 & *Mauritius Almanac & Colonial Register* 1870: 102–104.

— 1872. A visit to the Seychelles Islands. *TRSAS* NS 6: 83–142 & *Mauritius Almanac & Colonial Register* 1873: 74–98.

— 1873. *Subtropical rambles in the land of Aphanapteryx. Personal experiences, adventures and wanderings in and around the island of Mauritius*. London: Sampson Low, Marston, Low & Searle; New York: Harper & Brothers. 510pp.

Pineo, Huguette L.-T.-F. 1988. *In the grips of the eagle. Matthew Flinders at Ile de France 1803–1810*. Moka, Mauritius: Mahatma Gandhi Institute. 223pp.

— 1993. *Ile de France 1715–1746. Tome 1. L'Émergence de Port Louis.*. Moka, Mauritius: Mahatma Gandhi Institute. 357pp.

— 1999. *Ile de France 1747–1767. Port Louis, base navale*. Moka, Mauritius: Mahatma Gandhi Institute. 365pp.

Pingré, [A-] Gui. 1760–62. *Relation de mon voyage de Paris à l'île Rodrigue*. MS 1803, Bibliothèque Ste-Génévieve, Paris [day-to-day diary].

— 1763. *Voyage à l'isle Rodrigue*. MS 1804, Bibliothèque Ste.Geneviève, Paris [edited & re-arranged version of his diary, MS 1803 (1760–62). Published 2004 with intro. by S.Hoarau, M-P. Janiçon & J-M.Racault, Paris: Le Publieur, 373pp.; also, abridged, *in* Alby & Serviable 1993]

Pinto-Correia, Clara. 2003. *Return of the crazy bird. The sad strange tale of the Dodo*. New York: Springer-Verlag (Copernicus Books). 216pp.

*Piso, Gulielmus. 1658. *Medici Amstelaedamensis de Indiae utruisque Re Naturali et Medica Libri*. [incl. an ed. of & commentary on: Bontius, J. *Historia naturalis et medicae libri sex*].

Pitot, Albert. 1899. *L'Ile de France. Esquisses historiques (1715–1810)*. Port Louis: Imprimerie Pezzani. 450pp.

— 1905. *T'Eylandt Mauritius. Esquisses historiques (1598–1710). Précédés d'une notice sur la découverte des Mascareignes et suivies d'une monographie du Dodo, des Solitaires de Rodrigue et de Bourbon et de l'Oiseau Bleu*. Port Louis: Coignet Frères & Cie. xv+372pp.

— 1910–14. L'Ile Maurice. *L'Ile Maurice. Esquisses historiques*. Port Louis: Coignet Frères. 3 vols.

— 1914a. History. Pp.11–70 *in* Macmillan (1914), q.v.

— 1914b. Extinct birds of the Mascarene Islands. Pp.82–93 *in* Macmillan (1914), q.v.

Pitot, R.M. & Lenoir, Philippe. c. 1980. *Mauritius seen by the artists, 1800–1980*. Port Louis: Éditions de la Table Ovale. 62pp.

Polhill, M. 1990. Legumineuses. Fasc.80 *in* Bosser *et al.* (1978-), q.v.

Pollen, François P.L. 1865. Note sur l'*Oxynotus ferrugineus*. *Bull. Soc. Imp. Acclim. Hist. Nat., Réunion* 3: 7–14 [reprinted as 'Le Tui-tuit' in Roussin (1860–68), q.v., 3: 193–4].

— 1866. On the genus *Oxynotus* of Mauritius and Réunion. *Ibis* NS 2: 275–280.

— 1868. Relation de voyage. Vol.1. of *Recherches sur la faune de Madagascar et de ses dépendences, d'après les découvertes de François P.L.Pollen et D.C.van Dam*. Leiden: J.K.Steenhoff.

Pope-Hennessy, James. 1964. *Verandah. Some episodes in the crown colonies 1867–1889*. London: George Allen & Unwin. 313pp. [reprinted 1984 – London: Century Publishing]

Pope-Hennessy, John. 1889. Speech of His Excellency the Governor delivered at the annual meeting of the Royal Society of Arts and Sciences of Mauritius held at Réduit on the 28th January 1889. *C.R. Séanc. Ann.* [28.1.1889] *Roy. Soc. Arts Sci. Mauritius*: i-viii.

Potts, G.Richard ['Dick']. 1986. *The Partridge. Pesticides, predation and conservation*. London: Collins. 274pp.

Powell, Vicki J. & Wehnelt, Stephanie C. 2003. A new estimate of the population size of the critically endangered Rodrigues fruit bat *Pteropus rodricensis*. *Oryx* 37: 353–357.

— 2004. The ecology and conservation of the Rodrigues fruit bat *Pteropus rodricensis*. Ph.D. Thesis, Manchester Metropolitan University. 170pp.

Prater, S.A. 1980. *A book of Indian mammals*. 3rd rev.ed. Bombay: Bombay Natural History Society. xxii+a-g+324pp.

Prentout, Henri. 1901. *L'Ile de France sous Decaen (1803–1810). Essai sur la politique coloniale du premier empire*. [doctoral thesis, University of Paris] Paris: Hachette & Cie. 688pp.

Preston-Mafham, Ken. 1991. *Madagascar – a natural history*. Oxford & New York: Facts on File Ltd. 224pp.

Price, J.Jordan, Johnson, Kevin P. & Clayton, Dale H. 2004. The evolution of echolocation in swiftlets. *J. Avian Biol.* 35: 135–143.

Prior, James. 1819. *Voyage in the Indian Seas in the* Nisus *frigate, to the Cape of Good Hope, Isles of Bourbon, France and Scychelles* [sic]; *to Madras; and the isles of Java, St.Paul and Amsterdam, during the years 1810 and 1811*. London: Richard Phillips. iv+112pp. [reissued 1820 as vol.1. of *New Voyages & Travels*, same publisher].

Probst, Jean-Michel. 1995a. Un nicheur nouveau pour la Réunion: la Veuve dominicaine *Vidua macroura*. *Bull. Phaethon* 1:49.

— 1995b. Note sur plus de 40 colonies de nidification nouvelles de deux espèces de Procellariformes indigènes de la Réunion: *Puffinus pacificus* et *Puffinus lherminieri*. *Bull. Phaethon* 2: 50–57.

— 1995c. Note sur la présence de la musaraigne des maisons *Suncus murinus* à l'île aux Aigrettes (île Maurice). *Bull. Phaethon* 2: 82–84.

—, Colas, Pascal & Douris, Hervé. 1995. Premières photos d'un site de nidification du pétrel de Barau à l'île de la Réunion. *Courrier Nat.* 150: 16.

— 1996. Sur la colonisation du Bulbul orphée *Pycnonotus jocosus emeria* à l'île de la Réunion et carte de repartition de l'espèce en 1991. *Bull. Phaethon* 3: 5–11.

— 1997a. *Animaux de la Réunion. Guide des oiseaux, mammifères, reptiles et amphibiens*. Sainte-Marie, Réunion: Azalées Éditions. 167pp.

— 1997b. Le rossignol du Japon *Leiothrix lutea* (Scopoli) – une espèce introduite nouvelle pour l'île de la Réunion. *Bull. Phaethon*. 6: 86–88.

— & Turpin, Agnès. 1997. Disparition d'une population de Gecko de Manapany dans le secteur littoral de Saint-Joseph. *Bull. Phaethon* 6:104.

— 1998a. Découverte d'une population de *Nactus coindemirensis* sur le Rocher aux Pigeons. *Bull. Phaethon* 8: 102.

— 1998b. Petite Ile, une réserve et un conservatoire précieux pour les espèces littorales. *Bull. Phaethon* 8: 111.

— 1999. Redécouverte d'un reptile considéré comme disparu depuis plus de 130 ans à la Réunion, le Scinque de Bouton *Cryptoblepharus boutonii*. *Bull. Phaethon* 9: 1–3.

— 2000a. Temoignages sur la faune des vertbrés terrestres rapportés par Jean-Baptiste Renoyal de Lescouble (l'île de la Réunion, entre 1812 et 1838). *Bull. Phaethon* 12: 62–71.

— 2000b. Une nouvelle réserve naturelle à l'île de la Réunion: le massif de la Roche Écrite. *Courrier Nat.* 185: 14–15.

—, Le Corre, Mathieu & Thébaud, Christophe. 2000. Breeding habitat and conservation priorities in *Pterodroma baraui*, an endangered gadfly petrel of the Mascarene archipelago. *Biol. Conserv.* 93: 135–138.

— 2001a. Découverte d'oeufs sub-fossiles de *Phelsuma borbonica* dans l'ouest de la Réunion, et observation littorale de l'espèce. *Bull. Soc. Géog. Réunion* 1: 10–11.

— 2001b. Le Scinque de Bouton *Cryptoblepharus boutonii*. *Bull. Phaethon* 14: 104–105.

— & Brial, Pierre. 2001. Temoignages anciens sur la tortue terrestre de Bourbon, disparu il y plus de 100 ans à la Réunion. *Bull. Soc. Géog. Réunion* 1: 20–29.

— 2002. *Faune indigène protegée de l'île de la Réunion*. Le Port, Réunion: Association Nature et Patrimoine. 111pp.

— & Brial, Pierre. 2002. *Récits anciens de naturalistes à l'île Bourbon. Le premier guide des espèces disparus de la Réunion (reptiles, oiseaux, mammifères)*. Le Port, Réunion: Association Nature et Patrimoine. 112pp.

Procter, John & Salm, Rod. 1975. *Conservation in Mauritius*. Morges, Switzerland: IUCN. Cyclostyled report, 133pp.

Purseglove, J.W. 1972. *Tropical crops: monocotyledons*. London: Longman. 607pp.

Quammen, David. 1996. *The Song of the Dodo – island biogeography in an age of extinctions*. London: Random House (Hutchinson). 702pp.

Querhoënt, [Vicomte] de. 1773. [MS 369, Bibliothèque Centrale, Muséum Nationale d'Histoire Naturelle, Paris. Copies, in Buffon's hand, of notes made in 1773 on birds of Mauritius, intended for a supplementary volume on birds of the *Histoire Naturelle*, never published; photocopy in ASC's papers]

Quoy [Jean-Réné-Constant] & Gaimard, [J-] Paul. 1824. *Zoologie*. Special vol., 710pp + atlas [illustrations] of L.C.D.Freycinet (1824–44), q.v. [articles extracted from this volume also appeared in *Ann. Sci. Nat.* 5: 123–155 & 476–491 (1825)]

Quoy, [Jean-Réné-Constant]. 1833. Notes/Extrait du journal de M Quoy. Pp.643–651 *in* vol.5 of Dumont (1830–33), q.v.

Rabinowitz, Philip D. & Woods, Stephen. 2006. The Africa-Madagascar connection and mammalian migrations. *J. Afr. Earth Sci.* 44: 270–276.

Racault, Jean-Michel. 1995. Introduction. Pp.5–40 *in* the Éditions de Paris edition of Leguat (1707), q.v. [revised & expanded from the 1984 version published by Éditions la

Découverte].

Racey, Paul A. & Nicoll, Martin E. 1984. Mammals of the Seychelles. Pp.6–7–626 *in* Stoddart (1984a), q.v.

Rachels, James. 1990. *Created from animals. The moral implications of Darwinism*. New York: Oxford University Press. 245pp.

Radtkey, Ray R. *et al.* 1995. When species collide: the origin and spread of an asexual species of gecko. *Proc. Roy. Soc.Lond.* B 259: 145–152.

— 1996. Adaptive radiation of day-geckos (*Phelsuma*) in the Seychelles archipelago: a phylogenetic analysis. *Evolution* 50: 604–623.

Rage, Jean-Claude. 1996. Le peuplement animal de Madagascar: une composante venue de Laurasie est-elle envisageable ? Pp.27–35 *in* Lourenço (1996), q.v.

Ramdoyal, Ramesh. 1981. *More Tales from Mauritius*. London: Macmillan Press.

Ramgoolam, Seewoosagur & Mulloo, Anand. 1982. *Our struggle. 20th century Mauritius*. London: East-West Publications & New Delhi: Vision Books. 208pp.

Ramsay, G.W. 1978. A review of the effects of rodents on the New Zealand invertebrate fauna. Pp.89–97 *in* Dingwall *et al.* (1978), q.v.

Ramsurrun, Pahlad. 1982. *Folk tales from Mauritius*. New Delhi: Sterling Publishers. 120pp.

Rasmussen, Pamela C. *et al.* 2000. Geographic variation in the Malagasy Scops-Owl (*Otus rutilus* auct.): the existence of an unrecognised species on Madagascar and the taxonomy of other Indian Ocean taxa. *Bull. Brit. Orn. Club* 120: 75–102.

— & Prys-Jones, Robert P. 2003. History vs. mystery: the reliability of museum specimen data. Pp.66–94 *in* Collar *et al.* (2003), q.v.

Ratsimbazafy, R. 2003. Sea turtles. Pp.210–213 *in* Goodman & Benstead (2003), q.v.

Raxworthy, Christopher J., Forstner, M.R.J. & Ron A. Nussbaum. 2002. Chameleon radiation by oceanic dispersal. *Nature* 415:784–787.

— 2003a. Introduction to the reptiles. Pp.934–949 *in* Benstead & Goodman (2003), q.v.

— 2003b. Boidae, Boas. Pp.993–997 *in* Benstead & Goodman (2003), q.v.

Ray, John. 1678. *The ornithology of Francis Willughby ... wherein all the birds hitherto known ... are accurately described*. London: John Martin. 3 vols in 1. [Tr. & enlarged from Latin ed. of 1676; reprinted 1972, Newport Pagnell, UK: P.B.Minet]

Razafindrajao, Felix *et al.* 2001. Discovery of a new breeding population of Madagascar Teal *Anas bernieri* in north-west Madagascar. *Dodo* 37: 60–69.

Rebeyrol, Y. 1966. Les problèmes forestiers à la Réunion. *Rev. Bois Appl.* 21: 51–53.

Rees, Coralie & Rees, Leslie. 1956. *Westward from Cocos. Indian Ocean travels*. London: George Harrap & Co. & Sydney: Australasian Publishing Co. 268pp.

Rees, Siwan, A. *et al.* 2005. Coral reef sedimentation on Rodrigues and the western Indian Ocean and its impact on the carbon cycle. *Phil. Trans. Roy. Soc. Lond.* A 363: 101–120.

Regnaud, Charles. 1871. Water supply commission. Report of 3rd Committee. *TRSAS* NS 5: 158–170.

— 1878. Tec-tec [letter to Sir Edward Newton dated 14 February 1878, kept in the Newton Library, Zoology Department, Cambridge; published 1984: *Info-Nature, Ile Réunion* 21: 79–81]

Rene de Roland, Lily-Arison, Rabearivony, Jeanneney & Randriamanga, Ignace. 2004. Nesting biology of the Madagascar Harrier (*Circus macrosceles*) in Ambohitantely special reserve, Madagascar. *J. Raptor Res.* 38: 256–262.

Renman, Eric. 1995. A possible new species of Scops Owl *Otus* sp. on Réunion. *Bull. Afr. Bird Club* 2: 54.

Renoyal de Lescouble, Jean-Baptiste. 1811–38. *Journal d'un colon de l'Ile Bourbon*. [published 1990, ed. N.Dodille] Paris: L'Harmattan. 3 vols.

Renshaw, Graham. 1933. Poulet rouge and Corbeau Indien. *J. Soc. Preserv. Fauna Emp.* NS 19: 16–20.

— 1936. A third specimen of the Poulet Rouge. *J. Soc. Preserv. Fauna Emp.* NS 29: 60–61.

— 1938. Some extinct birds. 4. The White Dodo. *Bird Notes & News* 18(4): 85–86.

Restall, Robin. *Munias and mannikins*. Robertsbridge, UK: Pica Press. 264pp.

Reverzy, Jean-François (ed.) 1990. *La fondation du quartier Saint-Pierre et autres textes. Oeuvres de Jules Hermann, tome 1*. Saint-Denis: Editions de Tramail. 318pp.

Reydellet, Dureau. 1978. *Bourbon et ses gouverneurs*. Saint-Denis: Imprimerie Cazal. 96pp.

Reynolds, John E. & Odell, Daniel K. 1991. *Manatees and dugongs*. New York: Facts on File Inc. 192pp.

Ribes (-Beaudemoulin), Sonia (ed.). 2006. *Les animaux disparus*. Saint-Denis: Muséum d'Histoire Naturelle. 32pp.

Ricaud, Claude. 1975. Pesticide use and hazards in Mauritius agriculture. *Rev. Agricole Sucrière Ile Maurice* 54: 143–148.

Richards, Alexandra, Ellis, Royston & Schuurman, Derek. 2006. *Mauritius * Rodrigues * Réunion*. 6th ed. Chalfont St.Peter, UK: Bradt Travel Guides. 342pp.

Richardson, J. 1885. *A new Malagasy-English dictionary*. Antananarivo [=Tananarive]: London Missionary Society. 832pp. [reprinted 1967; Farnborough, UK: Gregg International Publishers]

Ricklefs, Robert E. & Lovette, Irby J. 1999. The roles of island area per se and habitat diversity in the species-area relationships of four Lesser Antillean faunal groups. *J. Anim. Ecol.* 68: 1142-1160.

— & Bermingham, Eldredge. 2004. History and the species-area relationship in Lesser Antillean birds. *Am. Nat.* 163: 227–239.

Ridd, M.F. 1971. South-east Asia as part of Gondwanaland. *Nature* 234: 531–533.

Rietmuller, Martin. 2001. Observations ornithologiques réalisés à La Réunion entre août 2000 et août 2001. *SEOR Lettre d'Info*. 1:6.

Rigotard, Marcel. 1934. Les conditions forestières de l'île de la Réunion. *Rev. Int. Bois* Jan.1934: 67–73.

— 1935. Une essence forestière de l'île de la Réunion. Le tamarin des hauts (*Acacia heterophylla* W.). *Rev. Int. Bois* Jun-Jly 1935: 541–551.

Ripley, S.Dillon. 1969. Comment on the Little Green Heron of the Chagos Archipelago. *Ibis* 111: 101–102.

— 1977. *Rails of the World. A monograph*. Boston, Mass.: David R.Godine. 406pp.

Rivals, Pierre. 1952. *Etudes sur la végétation naturelle de l'île de la Réunion*. Toulouse: Douladoure, & *Trav. Lab. For. Toulouse* 5(3) vol.1 art.2. 214pp.

— 1989. *Histoire géologique de l'île de la Réunion*. Saint-Denis:

Azalées Éditions. 400pp.

Robbins, S.R.J. 1985. Geranium oil: market trends and prospects. *Trop. Sci.* 189–196.

Robert, Henri. 1914. The Chamber of Agriculture. Pp.191–207 *in* Macmillan (1914), q.v.

Robert, René. n.d. 1980. *Géographie physique de l'île de la Réunion.*. St-Denis, Réunion: [author]. 78pp.

Robert, Réné & Hoarau, M. 2000. A propos de la création d'un parc naturel dans les hauts de l'île de la Réunion (Ocean Indien). *Rev. For. Fr.* 52: 159–168.

Roberts, David L. & Solow, Andrew R. 2003. When did the dodo become extinct ? *Nature* 426: 245.

— & McGlynn, T.P. 2004. *Tetramorium insolens* Smith (Hymenoptera: Formicidae): a new record for Mauritius, Indian Ocean. *Afr. Ent.* 12: 265–267.

Roberts, Katerina & Roberts, Eric. 1998. *Rodrigues – île escale.* tr. N.Louis. Vacoas, Mauritius: Editions le Printemps. 87pp.

Robillard, V. de. 1885. Note sur les pierres ponces trouvés sur divers points des côtes de Maurice. *TRSAS* 17: 49–52.

Robyns de Schneidauer, Thierry. 1982. *Guide-nature de l'Océan Indien.* Brussels: Institut Royale des Sciences Naturelles de Belgique. 264pp.

Rocamora, Gérard & Skerrett, Adrian. 2001. Seychelles. Pp.751–768 *in* Fishpool & Evans (2001), q.v.

Roch, S. & Newton, Edward. 1862–3. Notes on birds observed in Madagascar. *Ibis* 4: 265–275, 5: 165–177.

Rocha, Sara *et al.* 2006. Deciphering patterns of transoceanic dispersal: the evolutionary origin and biogeography of coastal lizards (*Cryptoblepharus*) in the Western Indian Ocean. *J.Biogeog.* 33: 13–22.

Rochon, Alexis M. de. 1791. *Voyage à Madagascar et aux Indes Orientales.* Paris: Prault. 322pp. [English tr., London: E.Jeffrey, 1793, 406pp.]

Rodda, Gordon H. *et al.* (eds.) 1999. *Problem snake management. The Habu and the Brown Treesnake.* Ithaca, NY: Cornell University Press. 534pp.

— & Campbell, Earl W. 2002. Distance sampling of forest snakes and lizards. *Herpetol. Rev.* 33: 271–274.

— & Dean-Bradley, Kathryn. 2002. Excess density compensation of island herpetofaunal assemblages. *J. Biogeog.* 29: 623–632.

— *et al.* 2002. Practical concerns in the eradication of island snakes. Pp.260–265 *in* Veitch & Clout (2002), q.v.

— & — 2005a. The predictive power of visual searching. *Herpetol. Rev.* 36: 259–264.

—, Fritts, Thomas, & Dean-Bradley, Kathryn. 2005b. Glueboards for estimating lizard abundance. *Herpetol. Rev.* 36: 252–259.

Rohling, Eelco J. et al.. 1998. Magnitudes of sea-level lowstands of the past 500,000 years. *Nature* 394: 162–165.

Rolfin, Adolfo (ed.) 1990. *Mauritius in your pocket.* Milan: System Bank. 512pp.

Rompillon, Sylvie. 2003. Porquoi tant de haine ? [braconnage de salanganes]. *SEOR Lettre d'Info.* 8: 12.

Rösler, Herbert. 1995. Description of a female *Phelsuma edwardnewtonii* Vinson & Vinson 1969 (Sauria: Gekkonidae). *Dactylus* 2(2): 71–75.

Ross, Ronald. 1923. *Memoirs, with a full account of the great malaria problem and its solution.* London: John Murray. 547pp.

Rothschild, Walter. 1907. *Extinct birds.* London: Hutchinson. xxix + 244pp.

Rouillard, Guy. 1979. Obelisque Liénard. Hommage rendu à la mémoire de James Duncan, John Horne, Jean Vinson et Octave Wiehe. *Cent-cinquantaire de la Société Royale des Arts et des Sciences de l'Ile Maurice 1829–1979*: 103–109.

— 1990. *Historique de la canne à sucre à l'Ile Maurice, 1639–1989.* Mauritius: Chamber of Agriculture. 50pp.

— & Guého, Joseph. 1990. *Le jardin botanique de Curepipe.* Curepipe, Mauritius: Municipalité de Curepipe. 49pp.

— & Guého, Joseph. c. 1999. *Les plantes et leur histoire à l'Ile Maurice.* Mauritius: [authors]. 752pp.

Rouillard, John. 1866–69. *A collection of laws of Mauritius and its dependencies.* Port Louis: L.Channell. 9 vols.

Rountree, Frank R.G. 1943. Seabirds and waders at Rodriguez. Typescript MS [photocopies in Alexander Library, Edward Grey Institute, Oxford & ASC's papers].

— 1951. Some aspects of bird life in Mauritius. *PRSAS* 1: 83–96.

— *et al.* 1952. Catalogue of the birds of Mauritius. *Bull. Mauritius Inst.* 3: 155–217.

Roussin, Albert (ed.) 1860–68. *Album de la Réunion.* St-Denis, Réunion: A.Roussin. 5 vols. [reprinted 1879–93, by same publisher & Paris: L.Vanier with the articles re-arranged amongst the volumes, but otherwise unaltered; 1st ed reprinted 1975, Marseille: Editions Jeanne Lafitte in 2 vols + *atlas*].

Rowlands, Beau W. 1982. Tropic birds and other seabirds in Mauritius. *Bokmakierie* 34: 9–12.

— *et al.* 1998. *The birds of St.Helena.* B.O.U.Checklist No.16. Tring, UK: British Ornithologists' Union. 295pp.

Roy, B. D. 1969. The development of the tea industry in Mauritius. *Rev. Agric. Sucrière Ile Maurice* 48: 156–164.

Roy, Sugoto [S.], Jones, Carl G. & Swinnerton, Kirsty [J.] 1998. The feral cat in Mauritius. *Cat News* 28: 18–20.

— 2001. *The ecology and management of the Lesser Indian Mongoose* Herpestes javanicus *in Mauritius.* Ph.D Thesis, University of Bristol, UK.

—, Jones, Carl G. & Harris, S. 2002. An ecological basis for control of the mongoose *Herpestes javanicus* in Mauritius: is eradication possible ? Pp.266–273 *in* Veitch & Clout (2002), q.v.

Royte, Elizabeth. 1995. On the brink – Hawaii's vanishing species. *Natl. Geog.* 188 (3): 2–37.

Rudwick, Martin S.J. 1976. *The meaning of fossils. Episodes in the history of palaeontology.* 2nd.ed. New York: Science History Publications. 287pp.

Ruhomaun, Kevin. 2003. *State of forest and tree genetic resources in Mauritius.* Rome: FAO (Forestry Department Working Paper FGR/58E). 8pp+appendices.

Rundquist, Eric M. 1995. *Day geckos.* Neptune City, N.J.(USA): TFH Publications. 63pp.

Rush, Philip & O'Keefe, John. 1962. *Weights and measures.* London: Methuen & Co.Ltd. 96pp.

Russell, Anthony P. 1977. The genera *Rhoptropus* and *Phelsuma* (Reptilia: Gekkonidae) in southern Africa: a case of convergence and a reconsideration of the biogeography of *Phelsuma*. *Zool. Afr.* 12: 393–408.

— & Bauer, Aaron M. 1986. Le gecko géant *Hoplodactylus delcourti* et ses relations avec le gigantisme insulaire chez les Gekkonidae. *Mesogée* 46: 25–28.

Ryall, Colin. 2002. Further records of range expansion in the House Crow *Corvus splendens*. *Bull. Brit. Orn. Club* 122:

231–240.

Ryan, Vincent W. 1864. *Mauritius and Madagascar: journals of an eight years' residence in the diocese of Mauritius, and of a visit to Madagascar*. London: Seely, Jackson & Halliday. xi+340pp.

Saddul, Prem. 1995. *Mauritius – a geomorphological analysis*. Moka, Mauritius: Mahatma Gandhi Institute. 340pp.

— (ed.) 2002. *Atlas of Mauritius*. 3rd.rev.ed. Rose Hill, Mauritius: Éditions de l'Océan Indien & London: George Philip. 48pp.

Safford, Roger J. 1991. Status and ecology of the Mauritius Fody *Foudia rubra* and Mauritius Olive White-eye *Zosterops chloronothos*: two Mauritian passerines in danger. *Dodo* 27: 113–138.

— 1992. Status in 1991 of the smaller native landbirds of Mauritius and Rodrigues. *Work. Group Birds Madagascar Region Newsl.* 2(1): 1–4.

— 1993a. Serpent Island in 1992. *PRSAS* 5(3): 33–39.

— 1993b. Conservation of the Mascarene Swiftlet *Collocalia francica* on Mauritius by protection of a nesting cave. Unpubl. report to FFPS. 7pp. [photocopy in ASC's papers]

— 1994. *Conservation of the forest-living native birds of Mauritius*. PhD thesis, University of Kent, Canterbury, UK. 166pp + appendices.

— 1995. Meller's Duck in Mauritius. *Threatened Waterfowl Res. Group Newsl.* 7: 17.

— 1996. Notes on the biology of the Mauritius Black Bulbul *Hypsipetes olivaceus*. *Ostrich* 67: 151–154.

— & Beaumont, J. 1996. Observations on the biology of the Mauritius Cuckoo-shrike *Coracina typica*. *Ostrich* 67: 15–22.

— 1997a. Nesting success of the Mauritius fody in relation to its use of exotic trees as nest sites. *Ibis* 139: 555–559.

— 1997b. A survey of the occurrence of native forest vegetation remnants on Mauritius in 1993. *Biol. Conserv.* 80: 181–188.

— 1997c. Distribution studies of forest living native passerines of Mauritius. *Biol. Conserv.* 80: 189–198.

— 1997d. The destruction of source and sink habitats in the decline of the Mauritius Fody *Foudia rubra*, an island endemic bird. *Biodiversity & Conservation* 6: 513–527.

— 1997e. Mascarene Paradise Flycatcher *Terpsiphone bourbonnensis*. *Bull. Afr. Bird Club* 4: 130–131.

— 1997f. Introduction à la Réunion d'espèces menacées de l'Ile Maurice. *Taille-vent* 3: 7–9 [translated & adapted from section 9.5 of Safford (1994), q.v.]

— 1997g. The annual cycle and breeding behaviour of the Mauritius fody *Foudia rubra*. *Ostrich* 68: 58–67.

— & Jones, Carl G. 1997. Did organochlorine pesticide use cause declines in Mauritian forest birds ? *Biodiversity & Conservation* 6: 1445–1451

— & Jones, Carl G. 1998. Strategies for land-bird conservation on Mauritius. *Conserv. Biol.* 12: 169–176.

— 2000. The Mauritius Cuckoo-shrike *Coracina typica* from egg to adult. *Bull. Afr. Bird Club* 7: 29–30.

— 2001. Mauritius. pp.583–596 in Fishpool & Evans (2001), q.v.

— & Basque, Rémy. 2007. Records of migrants and amendments to the status of exotics on Mauritius in 1989-1993. *Bull. Afr. Bird Club* 14: 26-35.

Sahai, I.M. 2004. Power plans in Mauritius. *Int. Water Power & Dam Construct.* 56 (9), 10-13.

Sahni, Ashok. 1984. Cretaceous-Paleocene terrestrial faunas of India: lack of endemism during drifting of the Indian Plate. *Science* 226: 441–443.

Salamolard, Marc. 2004. Le cas du Corbeau familier, *Corvus splendens*. *Chakouat* 11:5.

— & Ghestemme, Thomas. 2005 ('2004') *Plan de conservation de l'Echenilleur de la Réunion (Coracina newtoni)*. Saint-André, Réunion: SEOR. 46pp. [the final version, although dated 2004 like the draft, was issued in August 2005]

— 2007. *Plan de conservation du Pétrel de Barau Pterodroma baraui*. Document de travail. St André, Réunion: SEOR. 58pp.

Sale, G.N. 1935a. *Exotics in Mauritius* [paper given at the British Empire Forestry Conference, South Africa, 1935]. Port Louis: Government Printer. 12pp.

— 1935b. *British Empire Forestry Conference, South Africa 1935. Quinquennial statement on forestry in Mauritius*. Colony of Mauritius: Government Printer. 16pp.

Salvadori, Tommaso. 1876. Nota intorno al *Fregilupus varius* (Bodd.). *Atti Real. Accad. Sci. Torino* 11: 482–489.

Sandlund, Odd T., Schei, Peter J. & Viken, Aslang. 1996. *Proceedings of the Norway/UN Conference on Alien Species, Trondheim 1–5 July 1996*. Trondheim, Norway: Directorate for Nature Management. 233pp. [also published as: *Invasive species and biodiversity management*. Dordrecht: Kluwer Academic Publishers, 1999]

Sassi, M. 1940. Die wertvollsten Stücke der Wiener Vögelsammlung. *Ann. Naturhist. Mus. Wien* 50: 395–409.

Sauer, J.D. 1967. *Plants and man on the Seychelles coast: a study in historical biogeography*. Madison: Univ. of Wisconsin Press. 132pp.

Sauvignet, Hendrik et al. 2000. Premiers resultats des campagnes de dénombrement aérien des tortues marines sur les côtes Ouest et sud de la Réunion. *Bull. Phaethon* 11: 8–18.

Sauzier, Théodore. 1891. Conference de M.Sauzier sur les animaux disparus des îles Mascareignes [edited extracts of his talk]. *Bull. Soc. Sci. Arts Ile Réunion* 1889/90: 100–108.

Savidge, J.A. 1987. Extinction of an island forest avifauna by an introduced snake. *Ecology* 68: 660–668.

Scarth, Alwyn. 1994. *Volcanoes*. London: UCL Press, & College Station, Texas: A&M Univ. Press. 273pp.

Schauenberg, Paul. 1975. L'île Ronde agonise. *Musées de Genève* NS 16(159): 11–18.

Schlegel, Hermann. 1858 (1866). On some extinct gigantic birds of the Mascarene Islands. *Ibis* N.S.2: 146–168 [translation from the Dutch original in *Versl. Med. K. Ned. Akad. Wet., Afd. Natuurkunde* 7: 116–137, 1858]

— & Pollen, François P.L. 1868. *Mammifères et oiseaux*. Vol.2 of *Recherches sur la faune de Madagascar et de ses dépendances, d'après les découvertes de François P.L.Pollen et D.C. van Dam*. Leiden: J.K.Steenhoff.

Schliemann, H. & Goodman, Steven M. 2003. *Myzopoda aurita*, Old World Sucker-footed Bat. Pp.1303–1306 in Goodman & Benstead (2003), q.v.

Schmitz, A. et al. 2004. Opening the black box: phylogenetics and morphological evolution of the Malagasy fossorial lizards of the subfamily "Scincinae". *Molec. Phylogenet. & Ecol.* 34: 118–133.

Schulenberg, T.S. 2003. The radiations of Passerine birds in Madagascar. Pp.1130–1134 in Goodman & Benstead (2003), q.v.

Schultz-Hagen, Karl et al. 2003. Avian taxidermy in Europe from

the Middle Ages to the Renaissance. *J. Orn.* 144: 459–478.
Sclater, Philip L. 1864. Description of a new species of duck from Madagascar. *Proc. Zool. Soc. Lond.* 1864: 487–88.
Sclater, William L. 1915. The "Mauritius Hen" of Peter Mundy. *Ibis* (10)3: 316–319.
Scott, Peter [M.]. 1973. *Conservation in Mauritius*. Unpubl. report to the Prime Minister of Mauritius. Slimbrdge, UK: World Wildlife Fund. 22pp.
Scowcroft, P.G. & Wood, H.B. 1976. Reproduction of *Acacia koa* after fire. *Pacific Sci.* 30: 177–186.
Seegoolam, V. 1999. Effect of conservation management on native fern regeneration in an upland Mauritian forest. B.Sc. (Hons.) thesis. University of Mauritius.
Self, H.M. 1841. Report on the state of Rodrigues, 1841. MS in Public Record Office, London: C.O.167, v.231, art.78.
Sellers, David. 2001. *Transit of Venus. The quest to find the true distance of the sun*. Leeds: Magavelda Press. 222pp.
Selvon, Sidney. c. 1983. Agriculture. 1983 a bien commencé. *Rodrigues Almanach 1984*: 22–24.
— 2001. *A comprehensive history of Mauritius from the beginning to 2001*. Port Louis: Superdist Ltd. 437pp.
Servaas van Rooijen. 1887. Hongersnood in Suratta a° 1631. *De Navorscher, Amsterdam* 37(1): 4–8.
Servat, J. [signatory]. 1974. Règlement permanent sur la police de la chasse dans le département de la Réunion. Paris: Secretaire de l'État à l'Environnement; 8 April 1974. Cyclostyled, 8pp.
Severin, Tim. 2002. *In search of Robinson Crusoe*. New York: Basic Books. 333pp.
Seymour, Adrian *et al.* 2005. Mechanisms underlying the failure of an attempt to eradicate the invasive alien musk shrews *Suncus murinus* from an island nature reserve. *Biol. Conserv.* 125: 23–35.
Sganzin, Victor. 1840. Notes sur les mammifères et sur l'ornithologie de l'Ile de Madagascar (1831 et 1832). *Mem. Soc. Mus. Hist. Nat. Strasbourg* 3(1): 1–49.
Shapiro, Beth *et al.* 2002. Flight of the Dodo. *Science* 295: 1683. [on Dodo & solitaire DNA & phylogeny]
Sharp, M. 2004. Conservation of Mauritian upland wet forest: conservation Management Areas – restoring native vegetation structure or restoring ecosystem function and dynamics? A study using leaf-litter invertebrates as indicators of ecosystem health. B.Sc. (Hons.) thesis. University of Plymouth.
Sharpe, R.Bowdler. 1879. Birds [of Rodrigues]. *Phil. Trans. Roy. Soc. Lond.* 168: 459–469.
Sherley, Greg [H.] (ed.). 2000. *Invasive species in the Pacific: a technical review and draft regional strategy*. Samoa: South Pacific Regional Environment Programme. 190pp.
Sheth, Hetu C., Mahoney, J.J. & Baxter, A.N. 2003. Geochemistry of lavas from Mauritius, Indian Ocean: mantle sources and petrogenesis. *Int. Geol. Rev.* 45: 780–797.
— 2005. From Deccan to Réunion: no trace of a mantle plume. Ch.29 in G.R.Foulger *et al.* (eds.) *Plates, plumes & paradigms*. Geological Society of America, Special Paper 388.
Shigesada, Nanako & Kawasaki, Kohkichi. 1997. *Biological invasions: theory and practice*. Oxford: Oxford University Press. xiii+205pp.
Shine, Richard. 1991. *Australian snakes – a natural history*. Chatswood, NSW: Reed Books Australia & Ithaca, NY: Cornell University Press. 223pp.
Shirihai, Hadoram, Sinclair, Ian & Colston, Peter [J.]. 1995. A new species of *Puffinus* shearwater from the western Indian Ocean. *Bull. Brit. Orn. Club*. 115: 75–87.
— & Christie, David A. 1996. A new taxon of small shearwater from the Indian Ocean. *Bull. Brit. Orn. Club* 116: 180–181.
— 2001. The Mascarene shearwater. *Birding World* 14(2): 78–85.
Showler, Dave A. 2002. Bird observations on the Indian Ocean island of Rodrigues, March-June 1999. *Bull. Afr. Bird Club* 9: 17–24.
— & Cheke, Anthony [S.] 2002. Checklist of the birds of Rodrigues. Pp.22–24 *in* Showler (2002), q.v.
—, Côté, Isabelle M. & Jones, Carl J. 2002. Population census and habitat use of Rodrigues Warbler *Acrocephalus rodericanus*. *Bird Conserv. Int.* 12: 211–230.
Shuker, Karl P.N. 1997. *From flying toads to snakes with wings*. St.Paul, Minnesota: Llewellyn Publications. 222pp.
Sibley, Charles G. & Ahlquist, John E. 1984. The relationships of the starlings (Sturnidae: Sturnini) and the mockingbirds (Sturnidae: Mimini). *Auk* 101: 230–243.
— & Monroe, Burt L. 1990, 1993. *Distribution and taxonomy of birds of the world.* + *Supplement to...* New Haven, Conn. & London: Yale University Press.
Siddal, Mark *et al.* 2003. Sea-level fluctuations during the last glacial cycle. *Nature* 423: 853–858.
Sigala, Pierre & Soulères, Olivier. 1991. La politique de protection des milieux naturels. *Bois For. Trop.* 229 [special issue on Réunion]: 67–72.
— 1998. Le problème des espèces exotiques envahissantes en milieu insulaire fragile. Un exemple: La Réunion. *Courr. Env. INRA* 34: 6pp [on-line at www.inra.fr].
— 2001. La lutte contre les pestes végétales sur le domaine forestier à la Réunion. *Rev. For. Fr.* 53: 156–162.
Simberloff, Daniel. 1992. Extinction, survival and effects of birds introduced to the Mascarenes. *Acta Oecol.* 13: 663–678.
Simberloff, Daniel & Gibbons, Leah. 2005. Now you see them, now you don't! Population crashes of established introduced species. *Biol. Invasions* 6: 161–162.
Simmons, Robert. 2000. *Harriers of the world*. Oxford: Oxford Univ. Press. 352pp.
Simonin, L. 1862. Voyage à l'île de la Réunion (île Bourbon). *Tour du Monde, Nouv. J. Voy.* 1862(2): 145–176.
Sinclair, Ian & Langrand, Olivier. 1998. *Birds of the Indian Ocean Islands*. Cape Town: Struik. 184pp.
Skerrett, Adrian. 1995. "Birds of almost all description". Pp.129–150 *in* Amin *et al.* (eds.) (1995), q.v.
— 2001. Ringing recoveries from Seychelles. *Birdwatch (Seychelles)* 37: 19–23.
— *et al.* 2001a. The second report of the Seychelles bird records committee. *Bull. Afr. Bird Club* 8: 23–29.
—, Bullock, Ian & Disley, Tony. 2001b. *Birds of Seychelles*. London: A.& C. Black (Croom Helm). 320pp.
—, Matyot, Pat & Rocamora, Gérard. 2003. *Zwazo Sesel. The names of Seychelles birds and their meanings*. Victoria, Seychelles: Island Conservation society. 94pp.
Slater, Henry H., Gulliver, George & Balfour I.Bayley. 1874. Letters received from the naturalists attached to the Transit-of-Venus Expedition at Rodrigues. *Proc. Roy. Soc. Lond.* 23: 132–136.
— n.d. (c. 1875). Notes on the birds of Rodrigues. Unpaginated 16pp. MS in bound vol. entitled 'Indian Ocean 3. Madagascar – Mascarene Islands (MSS)' in Newton Library, Cambridge

University Zoology Department.

— 1979a. Reports of proceedings of the naturalists. 2. Report of Henry H. Slater, Esq., B.A. *Phil. Trans. Roy. Soc. Lond.* 168: 294–295.

— 1879b. Observations on the bone caves of Rodrigues. *Phil. Trans. Roy. Soc. Lond.* 168: 420–422.

Sleigh, Daniel. 2000. The economy of Dutch Mauritius during the second Dutch occupation (1664–1710). Pp.51–56 in Evers & Hookoomsing (2000), q.v.

Slingo, Julia *et al*. 2005. The meteorology of the western Indian Ocean, and the influence of the East African highlands. *Phil. Trans. Roy. Soc. Lond.* A 363: 25–42.

Slip, D.J. 2002. Control of the invasive exotic yellow crazy ant (*Anoplolepis gracilipes*) on Christmas Island, Indian Ocean. P.413 in Veitch & Clout (2002), q.v.

Slocum, Joshua. 1900. *Sailing alone around the world*. New York: Century Co. 294 pp. [often reprinted, e.g. 2006 London: Adlard Coles].

Smith, C.W. 1989. Non-native plants. In C.P. & D.B. Stone (eds.), *Conservation Biology in Hawaii*. University of Hawaii Press, Honolulu.

Smith, G.A. 1975. Systematics of parrots. *Ibis* 117: 18–68.

Smith, Michael A.K. *et al*. 2004. *Management plan for Gunner's Quoin*. Mauritius: Ministry of Agriculture, Food & Natural Resources. 44pp.

Smyth, E.Selby. 1872. Position of the colony in 1870. *Mauritius Almanac & Colonial Register* 1872: ?–99.

Snell, H.J. & Tams, W.H.T. 1919. The natural history of the island of Rodrigues. *Proc. Cambridge Philos. Soc.* 19: 283–292.

Snow, David W. (ed.) 1978. *An atlas of speciation in African non-passerine birds*. London: British Museum (Natural History). 390pp.

— & Perrins, Christopher M. 1998. *The birds of the Western Palearctic. Concise Edition*. Oxford: Oxford university Press. 2 vols.

Soeteboom, H. (ed.) 1648. *Schipper Willem van West-Zanen's Reys na de Oost-Indien, A°1602 &c*. [this is the running head – for the long & obscure full title see Wissen (1995) or Strickland (1848)] Amsterdam: H.Soeteboom. viii+60pp.

Sol, Daniel *et al*. 2005. The ecology and impact of non-indigenous birds. Pp.13-31 in Josep de Hoyo, Andrew Elliott & David Christie (eds) *Handbook of the birds of the world*. vol.10. Barcelona: Lynx Editions. 895pp.

Soler, Olivier. 2000. *Atlas climatique de la Réunion*. Sainte-Clotilde, Réunion: Météo-France. 79pp.

Sonnerat, Pierre. 1782. *Voyage aux Indes Orientales et à la Chine...* Paris: Froulé. 2 vols.

Sooknah, Khemraj. n.d.[1987]. The distribution and abundance of cavedwelling swifts (*Collocalia francica*) in lowland exotic forest of Bras D'Eau and Roches Noires. Unpubl. typescript report to ICBP (Cambridge, UK). 23pp.

Sornay, Pierre de & Koenig, Paul. 1941. Centenaire de la mort de Julien Desjardins. *TRSAS* C 10: 38–45.

Souchon, H. 1994. L'Ile au tresor. Pp.27–30 in PB.Pyamootoo & R.Poonoosamy (eds.) *Maurice – le tour de l'île en quatre-vingts lieux*. Port Louis: Publication Immedia.

Souhami, Diana. 2001. *Selkirk's island*. London: Orion Publishing (Weidenfeld & Nicholson). 256pp.

Spencer, Tom & Turner, John. 2001. An introduction to the Mascarene Plateau and its region. Pp.xxxii-xxxv *in* Burnett *et al*. (2001), q.v.

Spicer, Joaneath A. 1984. [review of] Drawings from the Holy Roman Empire, 1540–1680, a selection from North American collections. Exhibition catalogue by Thomas Dacosta Kaufmann 1982–83, 237pp., 120 illus. Princeton University Art Museum [etc.]. *Master Drawings* 22: 320–328.

Stahl, J.C. & Bartle, J.A. [Sandy]. 1991. Distribution, abundance and aspects of the pelagic ecology of Barau's Petrel (*Pterodroma baraui*) in the south-west Indian Ocean. *Notornis* 38: 211–225.

Stanley, Mary Ann. 2003. The breeding of naturally occurring B Virus-free Cynomolgus monkeys (*Macaca fascicularis* on the island of Mauritius. Pp.46–8 & [comments] p.59 in S.Vaupel (ed.) *International perspectives: the future of non-human primate resources. Proceedings of the workshop held April 17–19, 2002*. Washington, DC: National Academies Press. 248pp.

Starmühlner, F. 1979. Results of the Austrian Hydrobiological Mission, 1974, to the Seychelles-, Comores- and Mascarene Archipelagoes. Part 1. Preliminary report. *Ann. Naturhist. Mus. Wien* 82: 621–742.

Stattersfield, Alison J. & Capper David R. (eds.) 2000. *Threatened birds of the world. The official source for the birds on the IUCN Red List*. Cambridge: Birdlife International & Barcelona: Lynx Editions. 852pp.

Staub, [J.J.] France. 1968. [observations receuillies récemment à Rodrigues]. *PRSAS* 3: 158–161.

— & Guého, Joseph. 1968. The Cargados Carajos Shoals or St.Brandon: resources, avifauna and vegetation. *PRSAS* 3: 7–46.

— 1971. Actual situation of the Mauritian endemic birds. *Bull. ICBP* 11: 226–227.

— 1973a. *Oiseaux de l'Ile Maurice et de Rodrigue*. Port Louis: Mauritius Printing Co. 69pp.

— 1973b. Birds of Rodrigues Island. *PRSAS* 4: 17–59.

— 1976. *Birds of the Mascarenes and Saint Brandon*. Port Louis: Organisation Normale des Entreprises. 110pp.

— 1988. Evolutionary trends in some Mauritian phanerogams in relation to their pollinators. *PRSAS* 5(1): 7–78.

— 1993a. *Fauna of Mauritius and associated flora*. Mauritius: [author]. 97pp + unpag.plates.

— 1993b. Julien François Desjardins 1799–1840. Conference pour commémorer le 150ème anniversaire de sa mort. *PRSAS* 5(3): 1–10.

— 1996. Dodos and solitaires, myths and reality. *PRSAS* 6: 89–122.

— 2000. New hypothesis on the Dodo's true morphology from an ecological consideration of its available diet. Pp.67–76 *in* Evers & Hookoomsing (2000), q.v.

Staudinger, Manfred. 1990. Études descriptives de zoologie historique. Pp.91–486 *in* Haupt *et al*. (1990), q.v.

Steadman, David. W. 2006. *Extinction and biogeography of tropical Pacific birds*. Chicago: Chicago University Press. 594pp.

Stearns, Beverley P. & Stearns, Stephen C. 1999. *Watching, from the edge of extinction*. New Haven, Conn.: Yale University Press. 269pp.

Steeves, Tammy E., Anderson, David J. & Friesen, Vicki L. 2005. A role for nonphysical barriers to gene flow in the diversification of a highly vagile seabird, the masked booby

(*Sula dactylatra*). *Molec. Ecol.* 14: 3877-3887.
Stephens, James F. 1819–26. Birds. Vol.9–14 of G.Shaw & J.F.Stephens (1800–26) *General zoology or systematic natural history*. London: G.Kearsely & Longman & Co. 14 vols.
Stevens, G., McGlone, M & McCulloch, B. 1988. *Prehistoric New Zealand*. Auckland, NZ: Heinemann Reed. 128pp.
Stirling, E. 1833. *Cursory notes on the Isle of France made in 1827, with a map of the island*. Calcutta: Baptist Mission Press. iv + 50pp.
Stock, Jan H. 1997. A new species of *Brevitalitrus* (Crustacea, Amphipoda, Talitridae) from Mauritius – first record of the genus from the Indian Ocean. *Rev. Suisse Zool.* 104 (1): 3–11.
Stockdale, Frank A. 1914. Report on Rodrigues. MS in Mauritius Archives, SD 166.
Stoddart, David R. 1971. White-throated Rail *Dryolimnas cuvieri* on Astove Atoll. *Bull. Brit. Orn. Club* 91: 145–147.
— 1972. Pinnipeds or sirenians at western Indian Ocean islands ? *J. Zool. Lond.* 167: 207–217.
— (ed.) 1979. *The terrestrial ecology of Aldabra*. London: Royal Society. 263pp.+index.
— & Peake, J.F. 1979. Historical records of Indian Ocean giant tortoise populations. *Phil. Trans. Roy. Soc. Lond.* B 286: 147–161 [& *in* Stoddart 1979, q.v.].
Stoddart, David R. 1981. Abbott's Booby on Assumption. *Atoll Res.Bull.* 255:27–32 [originally circulated as a 1977 pre-print]
— (ed.) 1984a. *Biogeography and ecology of the Seychelles Islands*. The Hague: W.Junk (*Monographiae Biologicae* vol.55). 691pp.
— 1984b. Breeding seabirds of the Seychelles and adjacent islands. Pp.575–592 *in* Stoddart (1984a), q.v.
— 1984c. Scientific studies in the Seychelles. Pp.2–15 *in* Stoddart (1984a), q.v.
Stokram, Andries. 1663. *Korte beschryvinge van de ongeluckige weer-om-reys van het schip* Aernhem... Amsterdam: Jan van Duisberg. 19pp. [facsimile reprint 1966, Pretoria: State Library] & Amsterdam: Gillis J. Saegman [illus. ed.].
Storer, Robert W. 1970. Independent evolution of the Dodo and the Solitaire. *Auk* 87: 369–370.
Storey, William K. 1997. *Science and power in colonial Mauritius*. Rochester, NY: University of Rochester Press. 238pp.
Strahm, Wendy. 1988. Mondrain nature reserve and its conservation management. *PRSAS* 5(1,2): 137–177.
— 1989. *Plant red data book for Rodrigues*. Konigstein, Germany: Koetlz Scientific Books (for IUCN). 241pp.
— 1993. *The conservation and restoration of the flora of Mauritius and Rodrigues*. PhD thesis, University of Reading, UK. 239pp + appendices.
— 1994. Regional overview: Indian Ocean islands -&- Mascarene Islands: Mauritius (including Rodrigues) and Réunion. Pp.265–270 & 282–287 *in* vol.1. of S.D.Davis, V.H.Heywood & A.C.Hamilton (eds.) *Centres of plant diversity. A guide and strategy for their conservation*. Gland, Switzerland: WWF & IUCN. 3 vols.
— 1996a. Invasive species in Mauritius: examining the past and charting the future. Pp.167–174 *in* Sandlund *et al.* (1996), q.v.
— 1996b. Conservation of the flora of the Mascarene Islands. *Curtis's Bot. Mag.* 13(4): 228–237.
Strasberg, Dominique. 1995. Processus d'invasion par les plantes introduites à á Réunion, et dynamique de la végétation sur les coulées volcaniques. *Ecologie* 26: 169–180.
— 1996. Diversity, size composition and spatial aggregation among trees on a 1-ha rain forest plot at La Réunion. *Biodiversity Conserv.* 5: 825–840.
— *et al.* 2005. An assessment of habitat diversity and transformation on La Réunion island (Mascarene islands, Indian Ocean) as a basis for identifying broad-scale conservation priorities. *Biodiversity Conserv.* 14: 3015–3032.
Strauch, Alexander. 1887. Bemerkungen über die Geckonidensammlung im zoologischen Museum der Kaiserlich Akademie der Wissenschaften zu St.Petersburg. *Mem. Acad. Imp. Sci. St.Petersbourg* (7)35(2). 72pp.
Stresemann, Erwin. 1952. On the birds collected by Pierre Poivre in Canton, Manila, India and Madagascar (1751– 1756). *Ibis* 94: 499–523.
— 1958. Wie har die Dronte (*Raphus cucullatus* L.) ausgesehen. *J. Orn.* 99: 441–459.
Strickland, Algernon. 1831–34. [Letters to his father from Cape Town, Mauritius and other ports, while serving on HMS *Undaunted*]. MS in Bodleian Library, Oxford [Misc.Eng.Letters d.104; contemporary fair copies, not originals]
Strickland, Hugh E. 1844. On the evidence of the former existence of struthious birds distinct from the Dodo in the islands near Mauritius. *Proc. Zool. Soc. Lond.* (1844/12): 77–79.
— 1848. History and external characters of the Dodo, Solitaire and other extinct brevipennate birds of Mauritius, Rodriguez and Bourbon. Part 1, pp.3–65 of Strickland & Melville (1848), q.v.
— & Melville, A.G. 1848. *The Dodo and its kindred*. London: Reeve, Benham & Reeve. 138pp.
— 1949. Supplementary notices regarding the Dodo and its kindred. Nos.1,2,3. *Ann. Mag. Nat. Hist.* (2)3: 136–139.
Sussman, Robert W. & Tattersall, Ian. 1980. A preliminary study of the Crab-Eating Macacque (*Macaca fascicularis*) in Mauritius. *Mauritius Inst.Bull.* 9: 31–52. [a revised version appeared in *Primates* 22: 192–205, 1981]
— & — 1986. Distribution, abundance and putative ecological strategy of *Macaca fascicularis* on the island of Mauritius, southwestern Indian Ocean. *Folia Primatol.* 46: 28–43.
Sweet, Jessie M. 1970. The collection of Louis Dufresne (1752–1832). *Ann. Sci.* 26: 32–71.
Swettenham, Frank. 1912. *Also and perhaps*. London & New York: John Lane/Bodley Head. 304pp.
Swingland, Ian R. & Coe Malcolm J. 1978. The natural regulation of giant tortoise populations on Aldabra Atoll: reproduction. *J. Zool., Lond.* 186: 285–309.
— & — 1979. The natural regulation of giant tortoise populations on Aldabra atoll: recruitment. *Phil. Trans. Roy. Soc. Lond.* B 286: 177–188.
Swinnerton, Kirsty J. 2001a. *The ecology and conservation of the Pink Pigeon* Columba mayeri *in Mauritius*. Ph.D. thesis. University of Kent.
— 2001b. Ecology and conservation of the pink pigeon *Columba mayeri* on Mauritius. *Dodo* 37: 99.
— *et al.* 2004. Inbreeding depression and founder diversity among captive and free-living populations of the endangered pink pigeon *Columba mayeri*. *Anim. Conserv.* 7: 353–364.
— *et al.* 2005a The incidence of the parasitic disease Trichomoniasis and its treatment in reintroduced and wild Pink Pigeons

Columba mayeri. Ibis 147: 772–782.
— et al. 2005b. Prevalence of *Leucocytozoon marchouxii* in the endangered Pink Pigeon *Columba mayeri. Ibis* 147: 725–737.
Tafforet, Julien [attrib.]. ca.1726. *Relation de lisle Rodrigue*. MS in the Archives Nationales in Paris. [published in full, in original orthography, in *PRSAS* 4: 1–16].
Tassin, Jacques & Hermet, Michel. 1994. Les dégâts du cyclone Hollanda à la Réunion. *Boi For. Trop.* 240: 29–34.
— & Le Corre, Mathieu. 1997. En guise d'introduction. *Taillevent* 1–2: 1.
— & Rivière, Jean-Noël. 1998. Evaluation de l'impact des plantations forestières sur l'avifaune. Application du littoral réunionnais. *Bois For. Trop.* 258: 37–47.
— & — 2001. Le rôle potentiel du Leiothrix Jaune *Leiothrix lutea* dans la germination de plantes envahissantes à la Réunion (Océan Indien). *Alauda* 69: 34–41.
— et al. 2006. Bilan des connaissances sur les consequences écologiques des invasions de plantes à l'île de la Réunion (Archipel des Mascareignes, Océan Indien). *Rev. Ecol. (Terre Vie)* 61: 35–52.
Tatayah, R. Vikash. 2000. *Husbandry protocols for the Round Island boa* Casarea dussumieri *at Jersey Zoo, and an assessment for a captive breeding programme on Mauritius*. Diploma dissertation, University of Kent & Durrell Wildlife conservation Trust. 61pp.
— & Driver, B.M.F. 2000. An evaluation of the carcass quality of male tenrec (*Tenrec ecaudatus*), a non-conventional source of meat protein in Mauritius. *Univ. Mauritius Sci. Technol. Res. J.* 6: 69–82.
Tatton, J. 1625. A Journall of a voyage made by the Pearle to the East India wherein went as Captain Master Samuel Castleton of London and Captain George Bathurst as Lieutenant: written by John Tatton, Master. Pp.343–354 in vol.3 of S.Purchas (ed.) *Hakluytus Posthumus, or Purchas his Pilgrimes, containing a history of the world in sea voyages and lande travells by Englishmen and others.* [reprinted 1905–7, Glasgow: James MacLehose & Sons, 20 vols.]
Taylor, Barry & Perlo, Ber van. 1998. *Rails. A guide to the rails, crakes, gallinules and coots of the world*. Robertsbridge, Sussex, UK: Pica Press. 600pp.
Taylor, J.D. et al. 1979. Terrestrial faunas and habitats of Aldabra during the Pleistocene. *Phil. Trans. Roy. Soc. Lond.* B 286: 47–66 [& *in* Stoddart 1979, q.v.]
Teelock, Vijayalakshmi. 1998. *T'Eylandt Mauritius: a history of the Dutch in Mauritius*. Rose Hill, Mauritius: [author; Children's history Series]. 141pp.
— 2001. *Mauritian history. From its beginnings to modern times*. Moka, Mauritius: Mahatma Gandhi Institute. 434pp.
Telfair, Charles. 1831. [letter from Mauritius on tenrecs & fish]. *Proc. Comm. Sci. Corresp. Zool. Soc. Lond.* 1: 89–90.
— 1832. [letter on Rodrigues dated 8/11/1832 to the Zoological Society of London, preserved in the society's archives].
— 1833a. [letter from Mauritius to the Zoological Society of London on tortoises, chamaeleons and a tenrec called 'sokinah', dated 26/2/1833, preserved in the society's archives; the part on tenrecs abstracted in *Proc. Zool. Soc. Lond.* 1833: 81, 1833]
—, Dawkins, [Francis H.] & Eudes [Honoré]. 1833 [edited extracts from letters dated 1832 to the Zoological Society of London, about Rodrigues] *Proc. Zool. Soc. Lond.* 1833: 31–32.

Telfair, William. 1809. [Rodrigues in 1809]. MS in the Public Record Office, London, Ref. WOI 721, p.701, printed in *Rev. Retrospective Ile Maurice* 6: 311–2 (1955).
Temminck, Coenraad J. & Knip, Pauline. 1811. *Les pigeons par Madame Knip née Pauline de Courcelles, premier peintre d'histoire naturelle de S.M. l'Impératrice Reine Marie Louis, le texte par Coenraad C.J. Themminck* [sic], *directeur de l'Academie des Sciences et des Arts de Harlem* [sic] *etc*. Paris: Knip & Garnery.
— 1813–15. *Histoire naturelle des Pigeons et des Gallinacées*. Amsterdam: J.C.Sepp & fils & Paris: G.Dufour. 3 vols.
—, Prevost, Florent & Knip, Pauline. 1838–43. *Les Pigeons par Madam Knip...* 2nd rev.ed. 2 vols. [vol.1. revised from 1811 ed, vol.2 entirely new with text by Prevost].
Temple, Stanley A. 1973. A report on the preliminary poisoning of rabbits and goats on Round Island. Unpubl. cyclostyled report, 2pp.
— 1974a. Wildlife in Mauritius today. *Oryx* 12: 584–590.
— 1974b. Last chance to save Round Island. *Wildlife* 16: 370–374.
— 1975. The native fauna of Mauritius: 1. The land birds. Appendix 6 *in* Procter & Salm (1975), q.v.
— 1976. Observations of seabirds and shorebirds on Mauritius. *Ostrich* 47: 117–125.
— 1977. Plant-animal mutualism: co-evolution with Dodo leads to near extinction of plant. *Science* 197: 885–886.
— (ed.) 1978a. *Endangered birds*. Madison: University of Wisconsin Press & London: Croom Helm. 466pp.
— 1978b. Manipulating behaviour patterns of endangered birds. Pp.435–443 *in* Temple (1978a), q.v.
— 1981. Applied island biogeography and the conservation of endangered island birds in the Indian Ocean. *Biol. Conserv.* 20: 147–161.
— 1983. The Dodo haunts a forest. *Animal Kingdom* 86(1): 20–25.
Tezier, Remy. 1995. *Le Rocher. A l'île Maurice: ascension du Rocher aux Pigeons et plongée parmi les requins* [video]. St-Denis, Réunion: RFO Réunion. 36mins.
Thébaud, Christophe & Strasberg, Dominique. 1997. Plant dispersal in fragmented landscapes: A field study of woody colonization in rainforest remnants of the Mascarene archipelago. Chapter 21, pp.321–332 *in* W.F.Laurance & R.O.Bierregaard (eds.) *Tropical forest remnants. Ecology, management and conservation of fragmented communities*. Chicago & London: Chicago University Press. xv+616pp.
Thingsgaard, Karen. 1986. Rodrigues. *Threatened Plants Newsl.* 17: 10–11.
Thiollay, Jean-Marc. 1996. Oiseaux de la Réunion: vers une nécessaire union des efforts de tous les ornithologues. *Courrier Nat.* 158: 4.
— & Probst, Jean-Michel 1999. Ecology and conservation of a small insular bird population, the Réunion cuckoo-shrike *Coracina newtoni*. *Biol. Conserv.* 87: 191-200.
Thomas, P.L.V. 1828. *Essai de statistique de l'Ile Bourbon, considéré dans sa topographie, sa population, son agriculture, son commerce etc... suivi d'un projet de colonisation de l'intérieure de cette ile*. Paris: Bachelier. 2 vols.
Thomas, B.W. & Taylor, Rowley H. 2002. A history of ground-based rodent eradication techniques developed in New Zealand, 1959–1993. Pp.301–310 in Veitch & Clout (2002), q.v.
*Thompson, K. 1981. Nesting of the green turtle *Chelonia mydas*

(Linnaeus) 1758, in Mauritius. *Rev. Agric. Sucr. Ile Maurice* 60: 125-130.

Thompson, Richard [H.E.]. 1880. *Report on the forests of Mauritius, their present condition and future management*. Port Louis: Mercantile Record Co. 124pp.

Thornton, Ian. 1996. *Krakatau. The destruction and reassembly of an island ecosystem*. Harvard: Harvard University Press. 346pp.

Thorsen, Mike & Jones, Carl G. 1998. The conservation status of Echo Parakeet *Psittacula eques* of Mauritius. *Bull. African Bird Club* 5(2):122–126.

Tilbrook, E.M. 1968. *Annual report of the Forest Department for the year 1966*. Port Louis: Govt.Printer. 31pp.

Tirvengadum, Deva D. 1980. Le Mauritius Institute – une institution centenaire au service de la communauté. *Mauritius Inst. Bull*. 9(1): i-xviii.

— & Bour, Roger. 1985. Checklist of the herpetofauna of the Mascarene Islands. *Atoll. Res. Bull*. 292: 49–60.

Tombe, Charles F. 1810. *Voyage aux Indes Orientales pendant les années 1802, 1803, 1804, 105 et 1806...* Paris: Arthus Bertrand. 2 vols. + atlas.

Tonge, Simon J. & Barlow, S.C. 1986 ('1985'). Aspects of the biology of the Round Island skink *Leiolopisma telfairii*. *Dodo* 22: 97–109.

— 1990 ('1989'). Changes in reptile populations on Round Island following the eradication of rabbits. *Dodo* 26: 8–17.

Torlais, Jean. 1936. *Réaumur. Un esprit encyclopédique en dehors de "l'Encyclopédie", d'après des documents inédits*. Paris: Desclée de Brouwer & Cie. 448pp.

— 1962. Inventaire de la correspondence et des papiers de Réaumur conservés aux archives de l'Académie des Sciences à Paris. Pp.13–24 in P-P.Grassé (ed.) *La vie et l'oeuvre de Réaumur (1683–1757)*. Paris: Presses Universitaires de France. 188pp.

Tosi, Anthony J. & Coke, Cathryn S. 2006. Comparative phylogenetics offer new insights into the biogeographic history of *Macaca fascicularis* and the origin of the Mauritian macaques. *Molec. Phylogen. Evol*. 42: 498-504.

Toussaint, Auguste *et al*. (eds.) 1941-[continuing]. *Dictionary of Mauritian biography*. Port Louis: Société de l'Histoire de l'Ile Maurice.

— (ed.) 1953. *Atlas souvenir de l'Abbé de la Caille*. Port Louis: Mauritius Archives Publications. Unpaginated.

— 1956. *Repertoire des archives de l'ile de France pendant la régie de la compagnie des Indes (1715–1768)*. Port Louis: Mauritius Archives (Publ. No.3). 46pp. &, as 'Inventaire du fonds de la Compagnie des Indes des Archives de l'Ile de France', *in* Lougnon (1956b: 347–392), q.v.

— & Adolphe, Harold. 1956. *Bibliography of Mauritius (1502–1954), covering the printed record, manuscripts, archivalia and cartographic material*. Port Louis: Esclapon. 884pp.

— 1967. *La routes des îles: contribution à l'histoire maritime des Mascareignes*. Paris: SEVPEN. 540pp.

— 1972. *Histoire des Iles Mascareignes*. Paris: Berger-Levrault. 351pp.

— 1973. *Port Louis, a tropical city*. London: George Allen & Unwin. 144pp. [tr. by W.E.F.Ward from the French edition of 1966, with new epilogue].

Towns, D.R., Daugherty, Charles H. & Atkinson, Ian A.E. (eds.) 1990. *Ecological restoration of New Zealand islands*. Wellington, NZ: Department of Conservation (Conservation Sciences Publ.2). 370pp.

— & Ballantine, W.J. 1993. Conservation and restoration of New Zealand island ecosystems. *Trends. Ecol. Evol*. 8: 452–457.

— & Daugherty, Charles H. 1994. Patterns of range contraction and extinctions in the New Zealand herpetofauna following human colonisation. *NZ J. Zool*. 21: 325–339.

— , and Ferreira, S.M. 2000. Conservation of New Zealand lizards (Lacertilia: Scincidae) by translocation of small populations. *Biol. Conserv*. 98: 211–222.

Tradescant, John. 1656. *Musaeum Tradescantianum, or a collection of rarities preserved at South Lambeth near London*. London: Nathaniel Brooke. 179pp. [reprinted as Appendix 2 *in* Leith-Ross (1998), q.v.]

Traveset, Anna & Richardson, David M. 2006. Biological invasions as disruptors of plant reproductive mechanisms. *Trends Ecol. Evol*. 21: 208–216.

Trewhella, W.J. *et al*. 2005. Environmental education as a component of multidisciplinary conservation programs: lessons from conservation initiatives for critically endangered fruit bats in the western Indian Ocean. *Conserv. Biol*. 19: 75–85

Trewick, R.A. 2002. Flightlessness and phylogeny amongst endemic rails (Aves: Rallidae) of the New Zealand region. *Phil Trans. Roy. Soc. Lond*. B 352: 429–446.

Trouette, Emile. 1898. *Introductions de végétaux à l'Ile de la Réunion. Notes historiques*. [orig. publisher unknown; reprinted 1983, Saint-Denis: SREPN. 65pp.]

Troup, Robert S. 1940. *Colonial forest administration*. Oxford: Oxford University Press. 476pp.

Tuck, Gerald S. & Heinzel, Hermann. 1978. *A field guide to the seabirds of Britain and the world*. London: Collins. 292pp.

Tuijn, P. 1969. Notes on the extinct pigeon from Mauritius *Alectroenas nitidissima* (Scopoli, 1786). *Beaufortia* 16(218): 163–170.

Turnbull, Stephen. 1999. Invasion of *Acacia nilotica* into an area of semi-natural vegetation on Rodrigues, Indian Ocean. Unpubl. student thesis, Aberdeen University.

Turner, Jane (ed.). 2000. *From Rembrandt to Vermeer. 17th-century Dutch artists*. London: Macmillan (Grove Art Series). 422pp.

Turner, John & Klaus, Rebecca. 2005. Coral reefs of the Mascarenes, western Indian Ocean. *Phil. Trans. Roy. Soc. Lond*. A 363: 229–250.

Turpin, Agnès & Probst, Jean-Michel. 1997. Nouvelle répartition du Gecko vert malgache *Phelsuma laticauda* (Boettger, 1880) dans l'ouest de la Réunion. *Bull. Phaethon* 5: 3–4.

— & — 1998. Essai d'une carte de répartition des deux taxons endémiques du Gecko vert des forêts *Phelsuma b.borbonica* et *P.b.mater*. *Bull. Phaethon* 8:109–110.

—, Barroil, Pierre & Vatel, Pierre. 2001. Au sujet des dernières captures et sur la determination des sous-espèces d'iguanes verts introduites à l'île de La Réunion. *Bull. Phaethon* 13: 26–28.

Turvey, Samuel T. & Cheke, Anthony S. (submitted). Dead as a dodo: the fortuitous rise of an extinction icon.

Twigg, G. 1975. *The brown rat*. Newton Abbot, Devon, UK: David & Charles. 150pp.

Tyler, Michael J. 1994. *Australian frogs. A natural history*. Rev.ed. Ithaca, NY: Cornell University Press. 192pp.

Underwood, Garth. 1976. A systematic analysis of boid snakes.

Pp.151–175 *in* A.de A.Bellairs & C.B.Cox (eds.) *Morphology and biology of reptiles*. London: Linnean Soc. (*Linn.Soc.Symp. Ser*.3) & Academic Press. 290pp.

Unienville, Alix d'. 1954. *Les Mascareignes. Vieille France en mer indienne*. Paris: Albin Michel. 286pp.

Unienville, M. C. A. M. d'. 1838. *Statistique de l'Ile Maurice et ses dépendances, suivie d'une notice historique sur cette colonie et d'un essai sur l'Ile de Madagascar*. Paris: Gustave Barba. 3 vols.

Unienville, Raymond V. d'. 1982. *Histoire politique de l'Isle de France (1791–1794)*. Port Louis: Mauritius Archives (Publ.No.14). 214pp.

— 1989. *Histoire politique de l'Isle de France (1795–1803)*. Port Louis: Mauritius Archives (Publ.No.15). 343pp.

— (ed.) 1991. *Le cerf et sa chasse à l'île Maurice, 1639–1989*. Port Louis: Société des Chasseurs de l'Ile Maurice.

Upton, B.G.J., Wadsworth, W.J. & Newman, T.C. 1967. The petrology of Rodriguez Island, Indian Ocean *Geol. Soc. Am. Bull*. 78: 1495–1506.

Urquhart, Ewan. 2002. *Stonechats. A guide to the genus* Saxicola. London: Christopher Helm. 320pp.

Vaillant, L. 1898. Dessins inédits de chéloniens tirés des manuscrits de Commerson. *Bull. Mus. Hist. Nat. Paris* 4: 133–139.

Valledor de Lozoya, Arturo. 2003. An unnoticed painting of a white dodo. *J. Hist. Coll*. 15: 201–210.

Varnham, K.J. *et al*. 2002. Eradicating musk shrews (*Suncus murinus*) from offshore islands. Pp.342–349 *in* Veitch & Clout (2002), q.v.

Vaughan, Reginald E. & Wiehe, P. Octave. 1937. Studies on the vegetation of Mauritius. 1. A preliminary survey of the plant communities. *J. Ecol*. 25: 289–343.

— & Wiehé, P. Octave. 1941. Studies on the vegetation of Mauritius. 3. The structure and development of the upland climax forest. *J. Ecol*. 29: 127–160.

— *et al*. 1948. *Report on Rodrigues by the Natural Resources Board*. Port Louis, Mauritius. Cyclostyled.

— 1958. Wenceslaus Bojer 1795–1856. *PRSAS* 2: 73–98.

— 1968. Mauritius and Rodrigues. *Acta Phytogeog. Suec*. 54: 265–272 [special vol. on *Conservation of vegetation in Africa south of the Sahara*, ed. I. & O. Hedberg]

— 1970. The Mauritius Herbarium. *Rep. Mauritius Sugar Ind. Res. Inst*. [for] 1969: 157–165.

— 1984. [untitled letter on Dodos and Tambalacoques]. *Animal Kingdom* 87(1): 6–7.

Vaxelaire, Daniel. 1999. *L'histoire de la Réunion*. Paris: Éditions Orphée. 2 vols.

Veitch, C.R. 1985. Methods of eradicating cats from offshore islands in New Zealand. Pp.125–141 *in* Moors (1985), q.v.

— ['Dick'] & Bell, Brian D. 1990. Eradication of introduced mammals from the islands of New Zealand. Pp.137–146 *in* Towns *et al*. (1990), q.v.

— ['Dick'] & Clout, Mick (eds.) 2002. *Turning the tide: the eradication of invasive species. Proceedings of the International Conference on Eradication of Island Invasives*. Gland, Switzerland & Cambridge, UK: IUCN & Auckland, NZ: Invasive Species Specialist Group. 424pp.

Vences, Miguel *et al*. 2001. Phylogeny of South American and Malagasy Boine snakes: molecular evidence for the validity of *Sanzinia* and *Acranthophis* and biogeographical implications. *Copeia* 2001: 1151–1154.

— & — 2003. Multiple dispersal overseas in amphibians. *Proc. Roy. Soc. Lond*. B 270: 2435–2442.

— 2004. Origin of Madagacar's extant fauna: a perspective from amphibians, reptiles and other non-flying vertebrates. *Ital. J. Zool*., Suppl.2.: 217–228.

— *et al*. 2004a. Phylogeography of *Ptychadena mascareniensis* suggests transoceanic dispersal in a widespread African-Malagasy frog lineage. *J. Biogeog*. 31: 593–601.

— *et al*. 2004b. Natural colonization or introduction? Phylogeographical relationships and morphological differentiation of house geckos (*Hemidactylus*) from Madgascar. *Biol. J. Linn. Soc*. 83: 115–130.

Verdcourt, Bernard. 1996. *Ramosmannia rodriguesii*. *Curtis's Bot. Mag*. 13: 2–4–209.

Verguin, Michel & Serviable, Mario (eds.) 1993–98. *Le dictionnaire biographique de la Réunion*. Saint-Denis: Edition CLIP & Ars Terres Créoles. 3 vols.

Verheyen, Réné. 1957. Analyse du potentiel morphologique et project de classification des Columbiformes (Wetmore 1934). *Mem. Inst. Roy. Sci. Nat. Belg*. 33(3): 1–42.

Verin, Pierre. 1983. *Maurice avant l'Isle de France*. Paris: Fernand Nathan. 128pp. [this is a subset relating to Mauritius of accounts taken verbatim, errors included, from Grandidier *et al*. 1903–20, q.v.]

Verma, Som P. (ed.) 1999. *Fauna and flora in Mughal India*. Bombay: Marg Publications. 164pp.

Veron, Geraldine *et al*. 2007. Systematic status and biogeography of the Javan and small Indian mongooses (Herpestidae, Carnivora). *Zool. Scripta* 36: 1-10.

Vidal, Nicolas & Hedges, S. Blair. 2002. Higher level relationships of snakes inferred from four nuclear and mitochondrial genes. *C.R. Acad. Sci. Paris, Biol*. 325: 977–985.

— & David, Patrick. 2004. New insights into the early history of snakes inferred from two nuclear genes. *Molec. Phylogen. Evol*. 31: 783–787.

Vignau-Wilberg, Thea. 1990. Le "Muséum de l'empereur Rodolphe II" et le Cabinet des arts et curiosités. Pp.31–64 *in* Haupt *et al*. (1990), q.v.

Villers, Jean-Baptiste de. 1701–10. *Journal de l'île de Bourbon*. MS in the New York Public Library [published (in part) in *Bull. NY Public Lib*. 13(1): 12–63, 1909]

Vincent, J. 1966. *The Red Data book. 2 (Aves)*. Morges, Switzerland: IUCN.

Vinson, Auguste. 1861. De l'apparition d'oiseaux étrangers aux îles de la Réunion et Maurice. [apparently published in Réunion in 1861, *fide* Coquerel (1864, q.v.), but no copy appears to have survived. An MS text submitted to the Académie des Sciences in Paris was noticed in 1862 (*C.R. Hebd. Séanc. Acad. Sci., Paris* 54: 275) & is held in the Académie's archives; photocopy in the Alexander Library, Edward Grey Institute, Oxford]

— 1867. Le Martin (*Acridotheres tristis* Vieillot), son utilité pour les pays exposés à l'invasion des sauterelles. *Bull. Soc. Imp. Zool. Acclim., Paris* (2)4: 181–189.

— 1868. De l'acclimatation à l'île de la Réunion. *Bull. Soc. Imp. Zool. Acclim., Paris* (2)5: 579–590, 625–638 [also in *Bull. Soc. Sci. Arts Réunion* 1868: 35–65]

— 1870 [Letter dated 1870 to L.Bouton on *Agama versicolor*] *TRSAS*, NS 5: 31–34. (1871)

— 1876. [letter to the RSAS dated 27 July 1876, on *perdrix pin-

tadées] *TRSAS* NS 10: 41 (1878).
— 1877. Faune détruite: les Aepiornidés et les Huppes de l'île Bourbon. *Bull. Hebdo. Assoc. Sci. Fr.* 20: 327–331.
— 1887. Étude sur les colombes des Mascareignes et les espèces importées. *Bull. Soc. Nat. Acclim., Paris* (4)4: 640–651.
— 1888. *Salazie, ou le Piton d'Anchaine.* Paris: Ch.Delagrave.
Vinson, Jean. 1935. Contribution à l'étude des coleoptères des îles Mascareignes. *TRSAS* C 3: 153–216.
— 1950. L'Ile Ronde et l'Ile aux Serpents. *PRSAS* 1: 32–52.
— 1953a. The fauna of Mauritius. Part 1. The vertebrates. *Mauritius Police Gazette* 1: 37–41.
— 1953b. Some recent data on the fauna of Round and Serpent Islands. *PRSAS* 1: 253–257.
— 1956a. The problem of bird protection in the island of Mauritius. *PRSAS* 1: 387–392.
— 1963. The extinction of endemic birds in the island of Mauritius, with a possible way of saving some of the remaining species. *Bull. Int. Counc. Bird Preserv.* 9: 99–101.
— 1964a. Sur la disparition progressive de la flore et de la faune de l'Ile Ronde. *PRSAS* 2: 247–261.
— 1964b. Quelques remarques sur l'Ile Rodrigue et sur sa faune terrestre. *PRSAS* 2: 263–277.
— 1968. Le centenaire de la découverte à l'île Maurice des ossements du Dronte ou Dodo, *Raphus cucullatus* Linné. *PRSAS* 3(1): 1–5.
— & Vinson Jean-Michel. 1969. The saurian fauna of the Mascarene Islands. *Mauritius Inst. Bull.* 6: 203–320.
Vinson, Jean-Michel. 1973. A new skink of the genus *Gongylomorphus* from Macabé Forest (Mauritius). *Rev. Agric. Sucrière Ile Maurice* 52: 39–40.
— 1974. Round Island: conservation problems. *Mauritius Soc. Prevent. Cruelty Animals Newsl.* 3(1): 2–3.
— 1975. Notes on the reptiles of Round Island. *Mauritius Inst. Bull.* 8: 49–67.
— 1976a. Notes sur les procellariens de l'Ile Ronde. *Oiseau Rev. Fr. Orn.* 46: 1–24.
— 1976b. The saurian fauna of the Mascarene Islands. II. The distribution of *Phelsuma* species in Mauritius. *Mauritius Inst. Bull.* 8: 177–195.
Visdelou-Guimbeau, Georges de. 1948. *La découverte des îles Mascareignes.* Port Louis: General Printing & Stationery Co. 65pp.
Vitousek, P.M. 1990. Biological invasions and ecosystem processes; towards an integration of population biology and ecosystem studies. *Oikos* 57: 7–13.
Wallach, Van & Günther, Rainer. 1998. Visceral anatomy of the Malaysian snake *Xenophidion*, including a cladistic analysis and allocation to a new family (Serpentes: Xenophiidae). *Amphibia-Reptilia* 19: 385–404.
Walond, R.F. [attrib.] 1851. *A transport voyage to the Mauritius and back, touching at the Cape of Good Hope and St.Helena.* London: John Murray [this book was formerly attributed to A.Blenkinsop (e.g. Toussaint & Adolphe 1956, Cheke 1987a, qq.v.); for new attribution see *DMB*: 1082 (1988, Barnwell)]
Walter, Albert. 1908. *On the influence of forests on rainfall and the probable effect of "déboisement" on agriculture in Mauritius.* Mauritius: Government Printer. 51pp. + appendix.
— 1914. The sugar industry. Pp.208–231 in Macmillan (1914), q.v.
— c. 1968. Echoes of a vanishing empire, being the memoirs of a meteorologist & civil servant in the colonial empire 1897–1947. Bound typescript MS, 2 vols, in Rhodes House library, Oxford.
Walters, Michael. 1980. *The complete birds of the world.* Newton Abbot, Devon (UK): David & Charles. 340pp.
Walters, M.J. 2006. *Seeking the Sacred Raven. Politics and Extinction on a Hawaiian Island.* Washington, DC: Island Press/Shearwater Books. 293pp.
Wanquet, Claude. 1974. Bourbon dans les débuts de l'époque royale. *Cah. Centre Univ. Réunion, No. Spécial Colloque Commersion*: 14–75.
Ward, Philip S. 1990. The endangered ants of Mauritius: doomed like the Dodo ? *Notes from Underground* 4: 3–5.
Warham, John. 1996. *The behaviour, population biology and physiology of the petrels.* London: Academic Press. 613pp.
Warr, Frances E. 1996. Manuscripts and drawings in the ornithology and Rothschild libraries of the Natural History Museum at Tring. *Brit. Orn. Club Occ. Pub.* 2. 100pp.
Warren, Ben [H.]. 2003. *Phylogeography and evolution of species-rich bird lineages of the western Indian Ocean islands.* PhD thesis, University of East Anglia & Natural History Museum, Tring. 257pp.
— et al. 2005. Tracking island colonization history and phenotypic shifts in Indian Ocean bulbuls (*Hypsipetes*: Pycnonotidae). *Biol. J. Linn. Soc.* 85: 271–287.
— et al. 2006. Immigration, species radiation and extinction in a highly diverse songbird lineage: white-eyes on Indian Ocean islands. *Molec. Ecol.* 15: 3769-3786.
Watson, George E., Zusi, Richard L. & Storer, Robert W. 1963. *Preliminary field guide to the birds of the Indian Ocean. For use during the International Indian Ocean Expedition.* Washington, DC: Smithsonian Institution. 214pp.
Wayne, R.K. *et al.* 1994. Molecular genetics of endangered species. Pp.92–117 in P.J.S.Olney, G.M.Mace & A.T.C.Feistner (eds.) *Creative conservation: interactive management of wild and captive animals.* London: Chapman & Hall. xiv+517pp.
West, C.C. 1986. Reproductive biology of *Pteropus rodricensis*. *Myotis* 23/24: 137–141.
— & Redshaw, M.E. 1987. Maternal behaviour in the Rodrigues Fruit Bat. *Dodo* 24: 68–81.
Wheeler, Alwyne. 1997. Zoological collections in the early British Museum: the Zoological Society's Museum. *Arch. Nat. Hist.* 24(1): 89–126.
Whistler, Hugh. 1949. *Popular handbook of Indian birds.* 4th.ed. [rev. by N.B.Kinnear] Edinburgh: Oliver & Boyd. 560pp. [reprinted 1963]
Whitaker, A.H. 1973. Lizard populations on islands with and without Polynesian rats *Rattus exulans* (Peale). *Proc. NZ. Ecol. Soc.* 20: 121–130.
White, C.M.N. 1951. Systematic notes on African birds. (2). The affinities of the races of *Butorides striatus* in the eastern Indian Ocean. *Ibis* 93: 460–461.
Whiteley, Edward. 1992. *Gerald Durrell's army.* London: John Murray (& London: Pan-Macmillan Publishers, 1993). 241pp.
Whitesell, Craig D. 1964. Silvical characteristics of Koa (*Acacia koa* Gray). *US Forest Res. Pap.* PSW16. 12pp.
— 1990. Koa. *Acacia koa* A Gray. Pp.17–28 in R.M.Burns & B.H.Honkala (eds.) *Silvics of North America.* Washington, DC: US Deptartment of Agriculture. 2 vols.
Whiting, Alison S., Sites, Jack W., & Bauer, Aaron M. 2004. Mol-

ecular phylogenetics of Malagasy skinks (Squamata: Scincidae). *Afr. J. Herpetol.* 53 (2):135-146.
Whitfield, Peter. 1996. *Charting the oceans.* London: British Library & Rohnet Park, Ca.: Pomegranate Artbooks. 136pp.
— 1998. *New Found Lands. Maps in the history of exploration.* London: British Library & New York: Routledge. 200pp.
Whitmore, Timothy C. 1975. *Tropical rain forests of the Far East.* Oxford: Clarendon Press. 282pp.
Whittaker, Robert J. 1998. *Island biogeography.* Oxford: Oxford University Press. 285pp.
Wiehe, Alain. 1969. Projet de la culture du riz à Maurice. *Rev. Agric. Sucrière Ile Maurice* 48: 173–180.
Wiehe, P. Octave. 1938. Report on a visit to Rodrigues. Typescript MS; copy in the Mauritius Herbarium, Réduit, Mauritius.
— 1941. Report on plant pathologists's visit to Rodrigues. MS in Mauritius Herbarium, Réduit, Mauritius.
— 1946a. L'herbe condé et la lutte contre les mauvais herbes. *Rev. Agric., Ile Maurice* 25: 51–61.
— 1946b. *Report on a visit to Trinidad, Louisiana and other countries.* Port Louis: Government Printer. 119pp.
Wiehé, P. Octave. 1949. Vegetation of Rodrigues Island. *Mauritius Inst.Bull.* 2: 279–305.
Wigal, Donald. 2000. *Historic maritime maps. 1290–1699.* New York: Parkstone Press. 264pp.
Wilkinson, Gerald. 1978. *Epitaph for the elm.* London: Hutchinson. 160pp.
Williams, G. R. (1977): Marooning – a technique for saving threatened species from extinction. *International Zoo Yearbook* 17: 102–106.
Williams, Frédéric C. 1903. *From journalist to judge. An autobiography.* Edinburgh: George A.Morton. 319pp.
Williams, John R. 1960. The control of black sage (*Cordia curassavica*) in Mauritius: the introduction, biology and bionomics of a species of *Eurytoma* (Hymenoptera, Chalcidoidea). *Bull. Entomol. Res.* 51: 123–133.
— 1963. Locusts as pests of sugar cane in Mauritius. *Mauritius Sugar Industry Res.Inst.Tech.Circ.* 21. 20pp.
Williamson, Mark H. 1996. *Biological invasions.* London: Chapman & Hall. xii+244pp.
Willox, Robert. 1989. *Mauritius, Réunion and the Seychelles.* 1st ed. Hawthorn, Australia: Lonely Planet Publications. 274pp.
Wilson, Jason. 2002. Introduction. Pp.1–51 of his translation of Bernardin (1773), q.v.
Wingate, David B. 1978. Excluding competitors from Bermuda Petrel nesting burrows. Pp.93–102 *in* Temple (1978a), q.v.
Wink, Michael *et al.* 2002a. A molecular phylogeny of stonechats and related birds. Pp.22–30 *in* Urquhart (2002).
—, Sauer-Gürth, Hedi & Gwinner, Eberhard. 2002b. Evolutionary relationships of stonechats and related species inferred from mitochondrial DNA sequences and genomic fingerprinting. *British Birds* 95: 345–348.
Wissen, Ben van (ed.). 1995. Dodo *Raphus cucullatus.* Amsterdam: Zoöligisch Museum, Amsterdam University. 102pp.
Witmer, Mark C. & Cheke, Anthony S. 1991. The dodo and the tambalacoque tree: an obligate mutualism reconsidered. *Oikos* 61: 133–137.
Wittenberg, Rüdiger & Cock, Matthew J.W. 2001. *Invasive alien species. How to address one of the greatest threats to biodiversity. A toolkit of best prevention and management practices.* Wallingford, UK: Global Invasive Species Programme & CABI. 240pp.
Woolaver, Lance. 1999. Wild echoes: Echo Parakeet field season 1998/99. *Psittascene* 11(1): 6–7.
Woolf, Harry 1959. *The transits of Venus: a study of eighteenth century science.* Princeton: Princeton University Press. 258pp.
Worthy, Trevor H., Anderson, Atholl J. & Molnar, Ralph E. 1999. Megafaunal expression in a land without mammals – the first fossil faunas from terrestrial deposits in Fiji (Vertebrata: Amphibia, Reptilia, Aves). *Senckenbergiana Biol.* 79: 237–242.
— 2001. A giant flightless pigeon gen. et sp.nov. and a new species of *Ducula* (Aves: Columbidae) from Quaternary deposits in Fiji. *J. Roy. Soc. NZ* 31: 763–794.
— & Holdaway, Richard N. 2002. *The lost world of the Moa. Prehistoric life in New Zealand.* Bloomington: Indiana Univ. Press. 718pp.
Wright, H.T. & Rakotoarisoa, J.A. 2003. The rise of Malagasy societies: new developments in the archaeology of Madagascar. Pp.112–119 *in* Goodman & Benstead (2003), q.v.
Wylie, A. 1989. Archaeological cables and tacking: the implications of practice for Bernstein's options beyond objectivism and relativism. *Philosophy of Science* 19: 1–18.
Wyse-Jackson, Peter S., Cronk, Quentin C.B. & Parnell, John A.N. 1988. Notes on the regeneration of two rare Mauritian endemic trees. *Trop. Ecol.* 29: 98–106.
Yamashita, Carlos. 1997. *Anodorhynchus* macaws as followers of extinct megafauna: an hypothesis. *Ararajuba* 5(2): 176–182.
Yoder, Anne & Nowak, Michael D. 2006. Has vicariance or dispersal been the predominant biogeographic force in Madagascar? Only time will tell. *Ann. Rev. Ecol. Syst.* 37: 405–431.
Young, H.Glyn, Sorensen, M.D. & Johnson, K.P. 1997. A description of the Madagascar teal *Anas bernieri* and an examination of its relationships with the grey teal *A.gracilis*. *Wildfowl* 48: 174–180.
— Lewis, Richard E. & Razafindrajao, Felix. 2001. A description of the nest and eggs of the Madagascar Teal *Anas bernieri*. *Bull. Brit. Orn. Club* 121: 64–67.
Young, J.A. 1987. A note on the hand-rearing and reintegration of an infant Rodrigues Fruit Bat. *Dodo* 24: 82–86.
Young, Louise B. 1999. *Islands. Portraits of miniature worlds.* New York: W.H.Freeman. 297pp.
*Ziegler, Alan C. 2002. *Hawaiian natural history, ecology and evolution.* Honolulu: University of Hawaii Press. xviii+477pp.
Ziswiler, Vincent. 1996. *Der Dodo. Fantasien und fakten zu einem vershwunderen Vogel.* Zürich: Zoologisches Museum der Universtät Zürich. 94pp.
Zuccon, Dario. n.d. (=2006). A molecular phylogeny of starlings (Aves: Sturnini): evolution, biogeography and diversification in a passerine family. Thesis, Università degli Studi di Torino. 110pp.
Zuccon, Dario *et al.* 2006. Nuclear and mitochondrial sequence data reveal the major lineages of starlings, mynahs and related taxa. *Molec. Phylogenet. Evol.* 41: 333–344.
Zuel, Nicolas & Smith, K. (2005). Round Island Summary of Plant Restoration 2002–2004. Unpublished report to the Mauritian Wildlife Foundation.
—, Smith, K., & Stepnisky, D. (2005). Round Island 2004 Flora Report. Unpublished report to the Mauritian Wildlife Foundation.

INDEX

Page numbers in **bold** refer to main subjects/species descriptions in box features; in *italic* refer to illustration captions; and in ***bold italic*** refer to map or maritime chart captions.

abbreviations used in box features **12**
Acacia heterophylla 35, 53–4, 146, 174, 176
  koa 53
  melanoxylon 54
  nilotica 231
  simplicifolia 54
Accipitridae **120**
Achatina sp. 225
  fulica 95, 169, 184
  *immaculata*169, 184
Achyranthes aspera 223, 224, 245
Acridotheres tristis 94, 97, 106, 215, 216, 218, 230, 233, 258, 273
Acrocephalus newtoni 66
  rodericanus 46, 50, 56, 66, 150, 152, 154, 196, 197, 199, **208**, *208*, 241
  sechellensis 66
Aerodramus elaphrus 64
  francicus 40, 42, 44, 50, 56, 64, 166, **189**, *189*, 190, 237, 254
  unicolor 64
Aerva congesta 256
Agalega Island 52, 68, 271
Agama agama 184, 191, 265
Agapornis canus 94, 118, 132, 140, 141, 150, 151, 152, 169, 175, 184, 197, 271, 273
Agathis moreli 182
agricultural crops 77, 138
  clove 109, 110, 137, 138
  coffee 92, 104, 109, 110, 138, 144
  geranium 144, 147, 173, 174
  sugar-cane 103, 116, 117, 125, 137, 138, 147, 158, 173
  tea 162–3
  vanilla 144, 147
agricultural pests 94, 105–6, 124, 182
Aigrettes, Ile aux 23, 37, 81, 91, 161, 165, 170, 171, 227, 233, 236, 238, 244, 245–8, 249, 255, 256, 267
Ailuronyx sp. 68
albino specimens
  Dodos 30–1
  passerines 30
  see also white specimens
Albizzia lebbeck 109, 122, 137, 146, 153, 215, 218
Aldabra 21, 34, 38, 48, 52, 59, 60, 61, 65, 66, 87, 269, 271
Alectroenas madagascariensis 150
  nitidissima 40, 50, 56, 99, 100, 101, 117, 124, 126, 134, **143**, *143*, 150, 257
aloes 184, 218, 246
Alopochen sp. 56, 66
  aegyptiaca 141, 168, 169, 233, 257

  kervazoi 42, 49, **112**
  mauritianus 49, **112**
Altham, Emmanuel 82
altitudinal zonation in Réunion birds *180*
Amandava amandava 106, 136, 138, 140, 169, 175, 191, 271
Amaurornis phoenicurus 59, 63
ambergris 75, 81
Ambre, Ile d' 78, 79, 170, 207, 261
Amphiglossus sp. 73
Anas gibberifrons 66, 112
  hottentota 182
  platyrhynchos 169, 233, 271
  theodori 40, 42, 49, 56, 66, 80, **112**, *112*
Anatidae **112**
Andersen, Knut 63
Androngo sp. 73
Anodorhynchus sp. 38
Anous stolidus 41, 44, 50, 55, 219
  tenuirostris 41, 50, 55, 201, 217, 219
Antelme, Georges 156, 161
Antelme, Henri 137, 156
ants 249, 251, 267
Aphanapteryx bonasia 25–6, *26*, 28, 29, 37, 38, 39, 43, 50, 57, 70, 78, 80, 82, 83, **127**, *127*, 133
Apodidae **189**
Après de Mannevillette 114, 207
Apus apus 193
  pacificus 63
Araucaria 165
Ardeidae **86**
Arnold, Nick 72
Asio otus 124
associations, animal 38
Assumption 66, 265
Asterochelys radiata 133, 142, 187–8, 225, 244
Atkinson, Ian 236
Australia 54, 64, 68, 160, 188
Avadavat, Red 106, 136, 138, 140, 169, 175, 191, 271
Ayres, Philip 133, 208
Aythya sp. (subfossil pochard) 88, **112**
  ?*innotata* 49, 55, 61, 112

'Babbler', Rodrigues 37, 50
Baker 119
Balfour, Bayley 151, 152
bamboos 35
Barau, Armand 182, 186, 187, 265
Barkly, Sir Henry 208, 211, 212, 213, 220
Barnwell, Patrick 215, 217
Barré, Nicolas 190, 191
Bat, Aldabran Free-tailed 61

  Bory's White 44, 49, 60–1, 109, **259**
  Golden 35–6, 46, 49, 56, 64, 77, 152, 154, 196, 197, 198, 199–200, 214, 229, 255, 261, **262**, 266
  Grey Tomb 40, 42, 44, 49, 55, 60, 101, 109, 168, 190, **259**, *259*
  Mascarene Free-tailed 40, 42, 44, 49, 56, 60, 65, 166, 190, 253–4, **259**
  Pale House 44, 49, 55, 60, 109, 142, **259**
Baudin, Nicolas 101
bees, introduction of 85, 267
beetles 189
  bark 121
  scarabeid 159, 240
  tenebrionid 213, 254, 255
Bègue, Théophane 181
Bell, Sir Hesketh 192
Bellanger 85
Bellecombe 108
Bellier, Adrien 147
Belloc, Hilaire 275
Benitiers, Ile aux 170, 261
Benson, Con 64
Berger, Andrew 71
Bernache, Ile 170
Bernardin de St. Pierre 93, 94, 95, 97, 98, 99, 254
Bertuchi, A.J. 150, 153, 154, 192, 202
Besant, Walter 209
Betting de Lancastel 137, 138
Billiard, Auguste 112, 116, 137
biogeography 54, 269–71
biological controls 95, 136–7, 141, 159–60, 169, 186, 231
bird-lime 108, 174, 181, 182, 267
bird management 249–52
bird species and island area 270
Birdlife International 179, 254
bishop-birds 67
Bismarck Islands 269
Bissy, Gaston de 133
Bittern, Cinnamon 63
  Yellow 59, 63
Blackwood, Tasmanian 54
Blind-snake 169
  Carié's 39, 51, 56, 68, 255
Blok, Commander Adriaen Martensz 22, 83
Bloodsucker 134, 136, 141, 147, 160, 170, 184, 200, 232
Boa, Burrowing 39, 51, 74, 77, 207, 208, 219, 223, 255
  Keel-scaled 39, 51, 74, 208, 220, 224, 229, 236, 248, 249, 255, *255*
Boiga irregularis 74, 230

bois buis 246
cabris 246
clou 218, 246
de campêche 121
de fer 247
de pipe 246
de rat 246
de ronde 247
d'oiseaux 121, 147, 194, 241
d'olive 241
noir 109, 122, 137, 146, 153, 215, 218
pasner 241, 247
Bojer, Wenceslas 117–18
Bolyeria sp. 57
  multocarinata 39, 51, 74, 77, 207, 208, 219, 223, 255
Bolyeridae 73, **255**
Bonne 207
Bontekoe, Willem 30, 83, 85
Booby, Abbott's 37, 39, 41, **47**, 47, 48, 49, 56, 69, 80, 113, 148, 251
  Masked 41, 47, 49, 55, 60, 210, 211, 217, 219
  Red-footed 47, 48, 49, 55, 60, 80, 148, 152, 251
Borer, Pink Stem 136
Borghesi 85, 87
Bory de St Vincent, Jean–Baptiste 44, 53, 61, 90, 95, 101, 107, 108, 109, 110, 207, 264
Boryptera alba 44, 49, 60–1, 109, **259**
Bosser, Jean 179
botanic gardens *see* Pamplemousses, Mauritius
Bottle Palm *see* Palm, Bottle
BOU *see* British Ornithologists' Union
Boucher, Antoine 84, 85, 87, 88, 90; *see also* Desforges–Boucher, Antoine–Marie
Bour, Roger 142, 192
Bourne, Bill 196
Bouton, Louis 119, 121, 128, 131, 132, 133, 208
Bouvet de Lozier 108
Bouwer 112
box features, comments on **12**
Boyle, Charles 126, 128
Brasil, Louis 144
Brebner, William 192
Brevitalitrus strinatii 253
Brisson 107
British Ornithologists' Union (BOU) 227
Bromme, Mary 151
Brouard, Noel 129
Brown, Peter 99
Bruce, Sir Charles 131, 136
Brueghel, Jan (the Elder) 82
Bubulcus ibis 59
Buffon, Comte de 44, 100, 106, 107, 113, 138
Bufo gutturalis 159, 160
  marinus 159, 265
  melanostictus 133
Bulbul 56, 63, 65, **200**
  Mauritius 39, 40, 50, 99, 124, 125, 136, 144, 164, **200**, 233, 241, 252, 253
  Red–whiskered 135, 136, 154, 156, 160, 165, 169, 182, 189, 191 194, 215, 218, 225, 233, 241, 253, 261, 265, 265, 267, 273
  Réunion 42, 50, 138, 173, 175, 186 190, 191, **200**, *200*
  Rodrigues 37, 50, 257
Bullock, David 218
Bulweria bulwerii 49, 69, 219, 223
Butorides striata 40, 48, 49, 55, 59, 86, 115, 124, 138, 141, 215, 219, 225
butterflies 191, 240
Buttonquail, Madagascar 42, 50, 55, 61, 105, 138, 188, 191

Cabbage Palm *see* Palm, Cabbage
'cabling' 226
Cacatua galerita 233
  goffini 233
Cadet, Thérésien 178, 179, 198
café marron 196, 247
Caldwell, James 129, 132, 152
Calipepla californica 141
Calla 186
Caloenas nicobarica 70, 71, *71*
Calotes versicolor 134, 136, 141, 147, 160, 170, 184, 200, 232
Calvaria major see *Sideroxylon grandiflorum*
Campephagidae **196**
Canarium mauritianum/paniculatum 130, 158, 161, 237, 240, 261
Canary, Cape 94, 136, 169, 271
  Yellow-crowned 215
  Yellow-fronted 94, 106, 138, 197, 271
Capra hircus 236
captive breeding programmes 165, 168, 169, 198, 221, 229, 253
Caretta caretta 41, 51, 55
Carié, Paul 126, 133, 136, 147, 155, 161
Carissa xylopicron 247
Carlsberg Ridge *see* Mid–Indian Ocean Ridge
Carré 85
Carver, J.E. 158
Casarea sp. 57
  dussumieri 39, 51, 74, 208, 220, 224, 229, 236, 248, 249, 255, 255
Casela Bird Park 168, 232, 233, 260, 261, 265
Cassine orientalis 241
Castleton, Samuel 22, 83
Casuarina equisetifolia 146, 149, 153, 154, 174, 190, 192, 196, 209, 215, 246
Cat, Feral 189, 202, 236
  eradication attempts 85
  introduction of 76, 80, 85
  predation on native fauna 81, 87, 112, 113, 114, 115, 237
cattle
  introduction of 78, 83
Cauche, François 28–9, 78, 79, 80
Caulier, Père 104
caves
  conservation 166, 253–4
  degradation 166
Cazal, Charles 182
Cedar, Japanese Red 155, 158, 164, 165, 175, 176, 178, 179, 180
Cedrela odorata 131
Cenchrus echinatus 245

census methods 224–5
centipedes 115, 149, 200, 202, 210, 214, 217, 220, 237, 238, 254
Céré, Nicolas 93, 94, 95, 97, 103, 109
Cerfs, Ile aux 78, 170
Cervus elaphus 181
  timorensis 77, 151, 180, 181, 238
Ceutorhynchus 121
Chagos 59, 61, 63, 69
Chagos Bank 70
Chameleon
  Panther 124, 139, 170, 175, 232
  Warty 124
Chapotin, Charles 101, 206, 208
Charpentier de Cossigny (the elder), Jean–François 91, 95, 97, 99, 100, 104, 105, 112, 206
Charpentier de Cossigny (the younger), Joseph–François 95, 97, 98, 101, 103
Chat, Ilot 154, 249
chats **205**
Chazal, Malcolm de 264
Chazal, Malcy de *see* Moon, Malcy
Chazal, Toussaint-Antoine de 103, 126
Cheechak 61, 160, 170, 200, 215, 217, 232
Cheke, Anthony 178, 192, 266
Chelonia mydas 41, 44, 48, 51, 55, 115, 168, 187
Chevreau de Montléhu, Rivaltz 227
China 66, 163
cholera epidemics 149, 208
Christmas Island 12, 19, 47, 48, 52, 60, 64, 67, 69, 80, 251
Chromolaena odorata 245
Cicadabird, Black-headed 64
  Black-tipped 64
  Common 64
  Indochinese 64
  Madagascar 64
Circus aeruginosus 66
  approximans 66
  maillardi 39, 42, 50, 52, 56, 66, **120**, 188, 255
  melanoleucos 66
  spilonotus 66
CITES regulations 187, 188
citrus canker 193
Clark, George 119, 122, 123, 124, 132, 133, 134
Clergau, Philippe 155
Clerodendrum heterophyllum 246
Clidemia hirta 182, 233
climate, effects of deforestion on 129
CMAs *see* Conservation Management Areas
cobras 232
cockroaches 223, 225, 238, 249
Cocos, Ile 48, 91, 148, 152, 154, 201, 219, 236, 250, 251, 257
Coëtivy 61
Coffea myrtifolia 246
Colas, Paul 187
Coleura seychellensis 61
Colin, Barthélemy 153
Collocalia esculenta 64
Colophane 130, 158, 161, 237, 240, 261
Columbia livia 124
Columbidae **143**, **150**, **156**, **162**, **167**–**8**

Commerson, Philibert 94, 95, 100, 107, 108, 114
Comoros 15, 59, 60, 64, 65, 67, 71, 269, 271
conservation
   and bureaucratic indecision 266
   concept of 6
   theories of 266
conservation biologists, training of 258
Conservation Management Areas (CMAs) 239–40, 252, 258
Coot, Common 135
   Mascarene 40, 42, 50, 56, 66, 135, *135*, 257
   Red-knobbed 135, 257
Coquerel, Charles 140
Coracina sp. 56, 63, 64, 164
   cinerea 64
   melanoptera 64
   newtoni 42, 50, 107, 140, 175, 179, 181, 189, **196**
   polioptera 64
   schisticeps 64
   tenuirostris 64
   typica 40, 50, 100, 125, 144, 165, **196**, *196*, 241, 252, 253
Coracopsis 71, 108, 109, 118, 138, 139, 142
   vasa 233
Corby, Thomas 148–9, 153, 208, 210, 211, 212
Cordia curassavica 160, 231
   macrostachya 218
Cormorant, Little 51
   Reed 37, 40, 42, 49, 55, 59, **60**, *60*, 255
Corvus splendens 136, 170, 184
Cossigny *see* Charpentier de Cossigny
Coturnix chinensis 102, 108, 117, 137, 138, 169, 191
   coromandelica 141
   coturnix 133, 137, 141, 169, 191
Courchant, Beauvollier de 90
Cowles, Graham 191, 192
Crab, Coconut 52
   Hermit 52, 255
Crabe, Ile 48, 149, 152, 153, 201, 251
crabs
   land 37, 48, 52, 79, 88, 113, 115
   red 52
Crémont 108
creole, status of 260–1
crickets 237
Crinum mauritianum 246
crops *see* agricultural crops
Crow, Indian House 136, 170, 184
Cryptoblepharus sp. 68
   boutonii 37, 42, 51, 55, 61, 134, 187, 191, 208, 213, 215, 218, 250, *250*
Cryptomeria japonica 155, 158, 164, 165, 175, 176, 178, 179, 180
Cuban Bast 151
Cuckoo, Little 63, 66
Cuckoo-roller 141
Cuckoo-shrike 63, 64, 164
   Mauritius 40, 50, 100, 125, 144, 165, **196**, *196*, 241, 252, 253
   Réunion 42, 50, 107, 140, 175, 179, 181, 189, **196**
cuckoo-shrike surveys 178–9
Cunninghame, George 149

Curlew, Common 40
currents *see* ocean currents
Currie, James 122, 133
Cuvier, Georges 114
cyclones 168, 193, 196; (1731) 91; (1734) 104; (1771) 94; (1806) 103, 109, 137; (1825) 139; (1829) 139; (1844) 210; (1861) 126; (1875–76) 152–3; (1886) 153; (1892) 135; (1896) 121; (1945) 157, 170; (1948) 175; (1960) 220; (1960: *Carol*) 119, 164; (1962) 164; (1968) 195; (1969: *Monica*) 197; (1972) 195; (1972: *Fabienne*) 197, 199; (1975: *Gervaise*) 164, 266; (1979) 198; (1979: *Celine II*) 199; (1979: *Claudette*) 221; (1980: *Hyacinthe*) 190; (1982: *Damia*) 199; (1991: *Bella*) 199; (1994: *Hollanda*) 221, 247; (2002) 180; (2002: *Dina*) 221; (2003: *Kalunde*) 199
cyclones, adaptations to 34, 35–7
Cylindraspis sp. 57, 72, 238
   indica 43–4, 45, 51, 72, **234**
   inepta 51, 72, **234**
   peltastes 51, 72, 89, 114, **234**, *235*
   triserrata 45, 51, 72, **234**
   vosmaeri 47, 51, 72, 114, **234**

Dam, D.C. van 140
Daruty de Grandpré, Albert 132, 134, 147, 156, 213
Darwin, Charles 6, 118, 261
David, Governor 97
Dawkins, Francis 148
Day-gecko 39, 42, 46, 56, 67
   Blue-tailed 51, 175, 184, 200, **242**
   Gold-dust 184
   Great Green 168, 169, 170 184, 232, 265
   Guimbeau's 51, **242**
   Günther's 39, 51, 67–8, 206, 213, 220, 221, 224, 225, 229, 236, **242**, *242*, 249, 253, 256
   Liénard's *see* Gecko, Liénard's Giant 44–5, 51, 148, 152, **242**, *243*, 256, 257
   Malagasy 184, 265
   Manapany 41, 51 138, 139, 141, 175, 184, **242**
   Newton's 46, 51, 86, 113, 115, 148, 152, 192, **242**, *242*, **243**, 256
   Réunion Forest 41, 51, 141, 175, **242**
   Seychelles Small 184
   Upland Forest 51, 241, **242**
   Vinson's 51, 141, 160, 213, 215, 218 220, 224, 225, **242**, *242*, 256
DDT 164–5
Decaen, General 98, 102, 103, 109, 115, 116, 137
deer 108, 129–30, 161, 218
   and control of invasive plants 180–1
Deer, Red 181
   Rusa 77, 151, 180, 181, 238
deforestation 11, 34, 84, 92, 97, 98, 102, 103, 104, 116, 117, *123*, 125, 129, *132*, 135, 139, 140, 144–7, 149, 195–7
Defos du Rau 144
Delisse, Jacques 101
Dendrocygna bicolor 233
   viduata 124, 125, 132, 137, 141, 160, 168, 169, 233

Deodati, Roelof 75, 80, 81
Desbassyns, Charles 137
Descreux, Captain 46, 148
Desforges-Boucher, Antoine–Marie 94, 104, 108, 110; *see also* Boucher, Antoine
Desjardins, Julien 117, 118, 122, 124, 126, 133, 206, 207, 208
Desmanthus virgatus 223, 245
Desmarais 153
Desmodium incanum 245
d'Heguerty 105, 106, 111, 112
Dicanthium nodosum 215
Dictyosperma album congugatum 32, 34, 35, 80, 113, 124, 146, 192, 210, 221, 244, 246
Didosaurus 37, 51, 72, 76, 133, 236
Didunculus strigirostris 70
Digitaria sp. 245
Diospyros angulata 247
   egrettarum 245, 246, 256
Dipsochelys dussumieri 114, 170, 187, 202, 225, 244, 256
DIREN 179, 192
diseases
   bird 233, 251–2
   human 11
DNA analysis
   conflict with conventional taxonomy 63
Dodo 12, 25, 50, 57, 109, **162**, *162*, 261
   ancestors 70–1
   call 28, 29
   diet 37–8
   distribution 39
   eggs 28, 29, 79
   extinction 6, 10
      date 79
      realisation of 100
   last eye-witness account 78
   nest 28, 29, 79
   paintings/sketches of 82–3, *162*
   poems about 275
   population decline 79–80
   preening 9
   as seed dispersal agents 130
   slaughter by early settlers 77
   subfossil discoveries 133, 140
   transported from Mauritius 81–3
   'white' 30–1, *31*
   for early descriptions see walckvogel
dodo, 'Réunion'
   literature on 30–1
The Dodo (poem) 275
dogs, hunting
   introduction of 76, 81
Dombeya mauritiana 247
   populnea 247
Dove, Barbary 183
   Diamond 169
   Laughing 67, 168, 233, 271
   Malagasy Turtle 38, 39, 41, 45, 50, 52, 55, 59, 61, 67, 89, 101, 124, 126, 134, 138, 140, **143**, *143*, **150**, 156, 191, 233, 256
   Rodrigues 45, 50, 56, 67, 156
   Spotted 67, 98, 124, 218, 271
   Zebra 98, 99, 106, 124, 138, 150, 152, 191, 212, 215, 218, 225, 271
doves **143**, **150**, **156**

Dracaena sp. 245
  concinna 218, 256
Dragon Tree 218, 256
Drepanoptila holosericea 67
drift *see* ocean currents
Dromas ardeola 171
drongos 65
droughts 91, 104, 109, 121, 193, 194, 198, 213, 238, 244
Drury, Robert 30
Dryolimnus sp. 37, 56, 66–7, **128**
  augusti 50, 87, 128
  cuvieri 38, 67, 87, 101, 128, 257
    c. abbotti 128
    c. aldabranus 128
  sp. 38, 42, 50, 77, 87
Dubois 85, 86, 87, 107, 144
Duck, Fulvous Whistling 233
  Meller's 125, 133, 137, 169, 232, 271
  White-faced Whistling 124, 125, 132, 137, 141, 160, 168, 169, 233
ducks **112**
Dufresne, Louis 101
Dugong 41, 44, 48, 49, 55, 59, 61, 77, 88, 99, 111, 114, 115, 148, 205, **268**, *268*
Dulloo, Ehsan 239
Dumas, Daniel 95, 96
Dumont, Jean-Baptiste 101
Dupont, Evenor 134, 213
Durrell, Gerald 163, 198, 220, 221, 266
Durrell Wildlife Conservation Trust (DWCT) 227, 229, 248
Dussumier 208
Dutch
  arrival in Mascarene Islands 21–1
  life on Mauritius (1598) *24*

Eagle, Booted 66
ear-closing by skinks 72–3
East India Company 75, 82, 90, 95, 107, 108, 205
Easter Island 10–11, 16
Ebenavia inunguis 170
ebony 101, 210, 212, 261
  harvesting 75, 78, 79–80, 98, 117, 129
Ebony, Ile aux Aigrettes 245, 246, 256
Echinops telfairi 160
ecological histories, reconstruction of 226–7
ecological restoration projects 202
ecological study 160–2, 164–6, 168
ecosystems
  degradation 10
  reconstruction 267
Edgerley, Leo 227
Edwards, Brin 225
Edwards, George 100
egg-glueing by geckos 68
Egret, Cattle 59
  Dimorphic 40, 42, 49, 55, 59, 81, 85, **86**, 91, 140, 182, 255
Egretta dimorpha 40, 42, 49, 55, 59, 81, 85, **86**, 91, 140, 182, 255
Ehretia petiolaris 246
  serrata 121
elephant birds 11
Emballonura sp. 61
Emmerez de Charmoy, Donald d' 121, 134, 136, 155, 160, 261
endemic species 10
  close relatives elsewhere 62–9
  conservation 134–6
  obscure or distant affinities 69–74
  plant propagation and management 246–8
endemism 238
Eretmochelys imbricata 41, 44, 48, 51, 55, 115, 168, 202
Erythromachus leguati 44, 45, 50, 57, 70, **127–8**
Est, Ile de l' 78, 170
Estrilda astrild 106, 136, 138, 140, 150, 183, 191, 215, 218, 271
Eudes, Honoré 148
Eugenia sp. 231
  lucida 218, 246
Euglandina rosea 169, 184, 201, 265
Eulemur fulvus 141
Euplectes sp. 67
Europa Island 60
Evertsz, Volkert 78
evolution
  rates 15
  theories of 6
extinction
  cascades 230
  concept of 6
  see also under names of extinct species
Eydoux, Joseph-Fortuné 139

Falco sp. 56
  amurensis 63
  duboisi 42, 50, 65, 87, **120**
  punctatus 39, 42, 50, 65, 87, 99, **120**, *120*, 125, 137, 164, 165, 227, 229, 233, 236, 237, 249, 251, 252, 261, 266, 267
  rupicoloides 65
  tinnunculus 65
Falcon, Amur 63
falcons **120**
famines 92, 96, 97, 109–10, 174
Farquhar, Sir Robert 116
fattening, seasonal
  and cyclone survival 35
Felis domesticus 76, 80, 81, 85, 87, 112, 113, 114, 115, 189, 202, 236, 237
fencing of nature reserves 199, 239, 240
Fernelia 213
  buxifolia 212, 246
ferrets 265
Feuilley, Sieur 84, 85, 86, 88, 107
Ficus reflexa 233
figs, strangling 35
Fiji 54, 74
Finch, Bengalese *see* Munia, White-rumped
  Spice 94, 138, 175, 183, 225
fire-climax species 35, 53, 146
fires 96, 113, 115, 149, 151, 166, 189, 206
fish stocks 85
Flacourt 11, 83
Flacourtia indica 218, 219, 246
Flamingo, Greater 40, 49, 55, **96**, *96*
flamingos 27, 99
Flat Island 14, 41, 61, 76, 81, 91, 149, 170, 203, 205, 206, 207, 208–10, *209*, *210*, 213, 214, 215, 218, 219, 221, 232, 236, 248
flightlessness 19
  evolution of 37
Flinders, Matthew 94, 97, 99, *118*, 126, 137, 253
Florens, Vincent 10
Flycatcher, Malagasy 66
  Mascarene Paradise 40, 42, 50, 56, 66, 164, 165, 189, **212**, *212*, 233, 252, 253
  Seychelles 66
Flying-fox, Black-spined 39, 41, 42, 44, 46, 49, 56, 64, 65, 83, 107, 124, 134, 153, 168, 241, **262–3**, 267
  Island 63
  Little Red 64
  Seychelles 65
flying-foxes
  colonisation from Asia 63–4
Fody, Aldabra 85
  Cardinal 42, 67, 106, 124, 138, 154, 189, 191, 193, 215, 233, 253
  Madagascar Red *see* Fody, Cardinal
  Mauritius 39, 40, 50, 67, 85, 99, 125, 126, 136, 158, 164, 165, 166, 198, 206, 227, **228**, 229, 233, 236, 237, 244, 252, 253, 257, 266, 267
  Red *see* Fody, Cardinal
  Réunion 42–3, 50, **228**, 257
  Rodrigues 42, 46, 51, 149, 150, 152, 154, 196, 197, 199, **228**, *228*, 229, 241, 253
  Seychelles 85
Foetida rodriguesiana 247
food-web studies 269
Forbes, H.O. 30
Forbes-Watson, Alec 215
forest clearance *see* deforestation
forest conservation 96–7, 119, 131, 153
forest cover
  and watershed protection 102
forest die-back 119, 121–2
forest diseases 121, 193
forest pests 121, 129–30, 153, 161, 238
forest protection legislation 102, 116, 128–9, 131, 144, 174
forest regeneration 130–1, 161, 245
forest restoration 153, 154, 200, 229, 240, 241, 244
forest types 239
  cloud forest 129
  dwarf (giant heather) forest 35, 43, *43*
  lowland dry forest 32, 34
  lowland wet forest *43*, 176, 180
  mangrove forest 34
  screw-pine forest *43*
  tamarin forest 35, *43*, 53, 178, 180
  upland 38
  upland wet forest 34, *43*
forestry policies 156–8
  conflict with forest conservation 175–9
forests, government attitudes to 265
Foucher d'Obsonville 94
Foudia sp. 56
  aldabrana 85
  delloni 42–3, 50, **228**, 257
  flavicans 42, 46, 51, 149, 150, 152, 154, 196, 197, 199, **228**, *228*, 229, 241, 253
  madagascariensis 42, 67, 106, 124, 138,

154, 189, 191, 193, 215, 233, 253
  rubra 39, 40, 50, 67, 85, 99, 125, 126, 136, 158, 164, 165, 166, 198, 206, 227, **228**, 229, 233, 236, 237, 244, 252, 253, 257, 266, 267
  sechellarum 85
Fouquets, Ile aux 41, 170, 219, 232, 238, 249
Fourneau, Ilot 170
Fragaria vesca 109
Francolin, Chinese 93, 137, 139, 141, 169, 175
  Grey 93, 105, 137, 141, 150, 169, 175, 191, 197, 215, 218, 271
Francolinus pintadeanus 93, 137, 139, 141, 169, 175
  pondicerianus 93, 105, 137, 141, 150, 169, 175, 191, 197, 215, 218, 271
Fregata ariel 48, 49, 55, 73, 251
  minor 48, 49, 73, 251
Frégate, Ile 46, 47, 48, 148, 152, 242, 257
Fregatidae 73
Fregilupus sp. 57
  varius 41–2, 51, 71–2, 109, 117, 138, 139, 140, 141, 142, 144, **216**, *216*, 230
Freycinet, Louis–Henri de 87, 207
Frigate Island, Seychelles 213
Frigatebird, Great 48, 49, 73, 251
  Lesser 48, 49, 55, 73, 251
frigatebirds 39, 48, **73**, 73, 80, 152, 210; *see also rabos forcados*
Fritts, Tom 225
Frog, Malagasy Grass 95, 108, 138, 139
frosts 35
fruitbats 36, **262–3**
fuchsias 147, 186
Fulica atra 135
  cristata 135, 257
  newtoni 40, 42, 50, 56, 66, 135, *135*, 257
Funambulus palmarum 141
Furcifer pardalis 124, 139, 170, 175, 232
  verrucosus 124
Furcraea foetida 94, 121, 151, 209, 231, 246
Fusée Aublet 93

Gabriel Islet 170, 203, 205, 206, 207, 209, 214, 215, 217, 219, 248, 249
Gade, Daniel 195
Gagnebina pterocarpa 221, 246
Gaimard, Paul 207, 208
Galápagos Islands 19, 74, 79, 81
Gallinago macrodactyla 182
Gallinula chloropus 40, 50, 52, 55, 59, 101, 124–5, 135, 138, 140
Gallirallus philippensis 67
Galloys, Abbé 98
Gallus gallus 147, 191
Gandon, Abbé 92, 104, 111
Gardyne, Alexander 149, 152
Gastonia rodriguesiana 241
Gaulbil, Père 104
gazelles 93
Gecko, Bronze 68
  Common Worm 170, 200
  Gulliver's Stump-toed 152, 170, 215
  Indian House 170, 200
  Liénard's Giant 44–5, 51, 148, 152, 242, **243**, 256, 257

Mourning 51, 55, 61, 152, 224, 242, 256
  Rodrigues 57
  Square-toed 170
Gekkonidae **242–3**
geese **112**
Gehyra mutilada 152, 170, 215
Gennes de la Chancelière 111
Gentil, Guy le 90, 105, 108
Geopelia cuneata 169
  striata 98, 99, 106, 124, 138, 150, 152, 191, 212, 215, 218, 225, 271
Germain de Fleurimond, Governor 85
'giant' bird, Leguat's account of 27–8
Gibbons, Bill 264
Gill, Frank 175, 188, 189, 196, 197
Ginger 186
gizzard stones 45, 71, 111, 167, 168
glaciation *see* sea–level changes
Gleadow, Frank 121, 129, 131, 156
Global Environment Facility 199, 241
Goat, Feral 236
  eradication 219, 220, 238, 244
  introduction of 78
Goizet 146, 178
Gombrani, Ile 113, 149, 154
Gonaxis quadrilateralis 169
Gondwanaland 13, 70, 74
Gongylomorphus sp. 57, 72–3, 230
  bojeri 37, 39, 51, 101, 117, 124, 134, 208, 210, 213, 215, 217, 220, 224, 232, 238, 249, 250, 267
  borbonicus 42, 51, 88, 138, 139, 144, 232, 250
  fontenayi 37, 51, 124, 168, 215, 232, 237, 250, 269
Goodwin, Derek 71
Goose, Egyptian 141, 168, 169, 233, 257
Gooyer, Cornelius 34
Gordon, Arthur 119, 129
Gorry 148
gorse 147, 186
Gouania leguatii 247
Gould, Stephen Jay 130
Gouldsbury, J.C. 220
Grand Port, Mauritius 23, 40, 79
Grant, Baron 92, 99, 129, 207
Grant, Charles 103
grassland 34
Great Chagos Bank 69
Greater Sundas 75
Greenway, James 31, 196
Grevillea robusta 146
Griffiths, Owen 171, 261
ground-living animals
  diet 37–9
  see also flightlessness
Gryllus bimaculatus 237
Guam 224, 230
guano 154, 210, 213
Guava, Chinese *see* Guava, Strawberry
  Strawberry 94, 109, 121, 131, 146, 147, 151, 153, 186, 191, 194, 230, 231, *231*, 233, 237, 240
Gudger, E.W. 126
Guého, Joseph 227
Guérin, Réné 161
guineafowl 92, 105, 115, 148, 149, 152
Guineafowl, Helmeted 169

gullies 245
  causes of 223
Gulliver, George 151, 152
Gunner's Quoin 14, 37, 41, 91, 99, 170, 203, 205, *206–7*, 207, 208–10, *210*, 214–15, 218–19, 221, 232, 236, 238, 248, 250
Günther, Albert 133, 214
Gygis alba 48, 50, 55, 113, 154, 201, 219, 251

Habitat conservation
  resulting from species conservation 259
Hachisuka, M. 70
Haematoxylon campechianum 121
Hagen, Steven van der 77
Hahn 107
Hanning, Wallace 193, 198
Hare, Black-naped 93, 108, 138, 181, 207, 236, 238
hares 92, 207, 215
Harmenszoon, Wolfert 112
Harrier, Australian Swamp 66
  Eastern Marsh 66
  Malagasy Marsh 66
  Pied 66
  Réunion 39, 42, 50, 52, 56, 66, **120**, 188, 255
  Western Marsh 66
Harry, Benjamin 78, 79, 80
Hart, William Edward 156
Harungana madagascariensis 233
Hawaii 11, 53, 232, 266
Hawaiian Islands 15–16, 271
hawk owls 65
hawks **120**
Hawley, William 131
heathers 35
Hébert 87, 88
Hedychium gardnerianum 186
Hemidactylus brooki 170, 200
  frenatus 61, 160, 170, 200, 215, 217, 232
Hemiphaga novaeseelandiae 237
Hemiphyllodactylus typus 170, 200
Hemp, Mauritius 94, 121, 151, 209, 231, 246
Hen, Red 25–6, 26, 28, 29, 37, 38, 39, 43, 50, 57, 70, 78, 80, 82, 83, **127**, *127*, 133
Herbert, Thomas 13, *27*, 32, 76, 83
Hermann, Jules 141
Heron, Striated 40, 48, 49, 52, 55, 59, **86**, 115, 124, 138, 141, 215, 219, 225
Herpestes auropunctatus 136, 137, 235, 237
Heseltine, Nigel 197, 201
heterophylly 37, 53, 256
Heteropogon contortus 209, 215, 245
Hibiscus sp. 196, 261
  lilliflorus 247
  tiliaceus 151
Higgin, Edward *148*, 149
Hill, A.W. 130
hinged jaws in boas 220
Hirundinidae **193**
Hirundo rustica 193
historical wildlife descriptions, interpretation of 22–31
Hoart, C.T. 148
Holman, James 80, 118

Hondecoeter, Gillis de 82
honeycomb etching 223
Hooker, Sir Joseph 119, 212
Hoopoe 182, 216
Horne, John 207, 209, 212, 215
Hotchin, Philip 195, 197
Houssaye, Guillaume 33, 34
Hubert, Joseph 108, 109, 137
Hubert-Deslisle, Henri 140
Hugo, Governor 78, 79, 80, 205
Hugot, émile 173, 191
hunting
  bans 91, 97, 104, 105
  la chasse 108, 129, 147, 151, 156, 174, 181, 260, 267
  regulations 191
  see also poaching and trapping
Huron, Joseph 137, 156
Hurricane Palm *see* Palm, Hurricane
Hyophorbe sp. 245
  amauricalis 247
  lagenicaulis 210, 212, 221, 223, 225, 244, 246
Hypericum lanceolatum 35, 43
Hypsipetes sp. 56, 63, 65
  borbonicus 42, 50, 138, 173, 175, 186, 190, 191, **200**, *200*
  olivaceus 39, 40, 50, 99, 124, 125, 136, 144, 164, **200**, 233, 241, 252, 253
  sp. (Rodrigues Bulbul) 37, 50, 257

Ibis, Réunion 41, 45, 49, 56, 66, **103**, *103*, 192
  Sacred 103
  Straw-necked 103
ibises **103**
ICBP 161, 164, 266; *see also* Birdlife International
Iguana, Common 232
  Green 184
iguanas 265
inbreeding 267
India 82
Indian Ocean, western 16
  volcanic origin of islands 17
Indian Raven 22, 23, *24*, 25
Indonesia 67
insecticides 252; *see also* DDT
introduced species 10, 264–5
  amphibians 133, 200
  birds 98–9, 115, 124, 132–3, 134, 141, 168–9, 182–3
  mammals 141, 184, 200
  molluscs 184, 200–1
  plants 76, 146–7
  reptiles 124, 133–4, 141, 168–9, 170, 182, 200
  see also agricultural pests; biological controls; Cat, Feral; deer; dogs; invasive species; livestock *and* rats
invasive species
  birds 232–3
  mammals 235–8
  plants 93–4, 131, 160, 184, 186, 229–31
    arrival of 94, 151
    spread 121–2, 180, 186, 201
  reptiles 232

invertebrate conservation 254–5
island area
  and bird species 270
  and reptile species 270
islands, remote tropical
  colonisation 15–20
  ecosystems 10
  endemism 238
islands, small
  ecological restoration 244–5
Isnard, Hildebert 174
IUCN 220
Ixobrychus cinnamomeus 63
  sinensis 59, 63

Jack-fruit 154
Jacob de Cordemoy, Eugène 140, 142, 144, 146, 178
Jahangir, Emperor of India 82
jamblong 121
Janetaescincus sp. 73
Janoo, Anwar 70
Jenner, George 150, 152, 153
Jepson, W.F. 160
Jerningham, Sir Hubert 151
Jersey Wildlife Preservation Trust 164, 220, 221, 227; *see also* Durrell Wildlife Conservation Trust (DWCT)
Jersey Zoo 169, 171, 198, 229, 248, 266
Johansson, Malin *224*
Johnston, Henry H. 213
Jonchée de Goleterie 112
Jones, Carl 158, 226, 265, 266
Jossigny, Paul 100, 108, 114
Jouanin, Christian 175, 186, 187
Jugnauth, Sir Aneerood 258
Julienne, Yves 114
Junglefowl, Red 147, 191

Kauri 182
Kelly, W. 149
Kerourio, G. 145, 146
Kerr, William 211
Kervazo, Bertrand 191
Kestrel, Common 65
  Dubois's 42, 50, 65, 87, **120**
  Greater 65
  Mauritius 39, 42, 50, 65, 87, 99, **120**, *120*, 125, 137, 164, 165, 227, 229, 233, 236, 237, 249, 251, 252, 261, 266, 267
Ketupa zeylonensis 63
King, H. C. 158, 161
King, Warren 266
Koa 53
Koenig, Paul 121, 131, 132, 153, 154, 155, 156, 157, 158, 192, 193, 195
Krakatau volcano 54

La Caille, Nicolas de 93, 99, 206
La Fontaine 205
La Giroday, Vincent de 184
La Haye, Jacob de 87
La Merveille 81
La Motte 97, 99
La Passe, Ile de 170, 219, 232, 236
Lakshadweep Islands 69
Lamotius, Isaac 75, 78, 79, 80, 81
Lantana 182, 186, 193, 215, 217, 218, 231, 233
Lantz, Auguste 140, 141, 142, 147, 187
Lanux, Jean-Baptiste de 44, 106, 107, 108
Laplace, Admiral Cyrille 117, 118, 139
Latan 34, 35, 38, 44, 45, 48, 80, 101, 113, 137, 149, 151, 192, 196, 203, 205, 206, 207, 208, 209, 210, 218, 220, 221, 223, 225, 241, 245, 249
  Red 188
  Blue 246, 256
Latania sp. 34, 35, 38, 44, 45, 48, 80, 101, 113, 137, 149, 151, 192, 196, 203, 205, 206, 207, 208, 209, 210, 218, 220, 221, 223, 225, 245, 249
  loddigesii 246, 256
  lontaroides 188
  verschaffeltii 241
Layard, Edgar 211
Le Pouce Mountain, Mauritius 38, *98*
Leal, Charles 147
Leconte de Lisle, Charles-René-Marie 141, 276
Lecoq, Frère 105
legal protection
  bats 168
  birds 97, 135
  forests 102, 116, 128–9, 131, 144, 174
  reptiles 162
Legras, A. 142
Leguat, François 27, 32, 35, 41, 45, 46, 47, 48, 80, 81, 86, 88, 89, *89*, 113
  exile on Rodrigues 22, 23, 29–30
Leiolopisma sp. 57, 68, 72
  mauritiana 37, 51, 72, 76, 133, 236
  sp. (Arnold's Skink) 51, 72, 88, 250
  telfairii 37, 39, 42, 51, 72, 124, 206, 208, 210, 217, 218, 220, 221, 224, 225, 229, 236, 248, 249, **250**, 267
Leiothrix lutea 182–3, 186, 189, 191, 265, 273
Leiothrix, Red-billed *see* Robin, Peking
Lemur, Brown 141
  Ruffed 141
lemurs, giant 11
Lenoir, Philippe 264
Lenoir, Pierre 91, 92, 205
Lepidodactylus lugubris 51, 55, 61, 152, 224, 242, 256
Lepidospora mascareniensis 253
Leptosomus discolor 141
Lepus nigricollis 93, 108, 138, 181, 207, 236, 238
Lesser Sunda Islands 54
Lesson, René-Primevère 100
Lestrange, Sir Hamon 82
Lesueur, Charles-Alexandre 101, 102, 103, 207, 208
Leucena leucocephala 94, 121, 151, 153, 195, 209, 215, 246
Leucocytozoon marchouxi 233
Levaillant 107
Lever, Sir Ashton 107, 206
lichens 35
Liénard, François 148, 152, 242
light pollution 187, 190
Ligustrum robustum 132, 186, 231, *231*, 233, 240
lime-sticks *see* bird–lime

Limpkin 38
Lislet-Geoffroy 101, 108, 206, 207, 208
Litsea glutinosa 121, 147, 194, 241
   monopetala 131
livestock 198
   release of by passing ships 76
   see also cattle; Goat, Feral; pigs
Lizard, Rainbow 184, 191, 265
Lizard-owl 57, 71, 123–4
   Commerson's 50, 88, **185**, *185*
   Réunion 37, 42, 50, 88, **185**
   Tafforet's 46, 50, **185**
lizards 37
   widespread species 61
Lloyd, Col. John Augustus 149, 210, 213, 223
Locust, Red 95
locusts 94–5, 96, 105
logos, Bourbon Beer *31*
Lomatophyllum macrum 184
   tomentorii 246
Lonchura oryzivora 94, 106, 118, 124, 136, 138, 140, 151, 152, 169, 271
   punctulata 94, 138, 175, 183, 225
   striata 106, 184, 271
Lophopsittacus mauritianus 25, *25*, 30, 37, 38, 39, 50, 57, 71, 80, 81, **172**, *172*, 257
Lord Howe Island 269
Lovebird, Grey-headed 94, 118, 132, 140, 141, 150, 151, 152, 169, 175, 184, 197, 271, 273
Lumley, Captain 206
Lycodon aulicus 74, 133–4, 139, 141, 170, 171, 184, 230, 232, 249, 267
Macaque, Crab-eating 76, 171, 230, 267
macaws 38
Macé, Jean 109
Machaerina iridifolia 35
Madagascar 11, 15, 19, 21, 22, 38, 53
   species shared with Mascarene Islands only 55, 61–2
   species with affinities or close relatives elsewhere 65–8
Magon de Saint-Elier 208
magpie-robins 65
Mahé de Labourdonnais, Bertrand 90, 91, 92, 93, 96, 111, 112, 205
Maillard, Louis 139, 140, 142, 187
Majastre 147
Makak *38*
malaria 128, 144, 153, 163, 173
malaria control measures *see* mosquito control
Maldives 21, 59, 63, 64, 69
Maliku 69
Maljkovic, Alecks 198
Mallard 169, 233, 271
manatees *see* Dugong
Manders, Neville 142
Mandon-Dalger, Isabelle 155
Manès, Gustave 140
Mangénie *see* Est, Ile de l'
mangroves 198, 255
Mann, James 119, 129
Mare aux Songes, Mauritius 7, 27, 38, 133, 155
Margaroperdix madagascariensis 93, 132, 141, 169, 175, 191, 271
Marianne, Ile 249
Marragon, Philibert 46, 114, 115, 148
Martin, François 85, 88
Martin, Sand 193
Martinet 106
Mascaraneus remotus 254
Mascarene Islands 12
   biogeography 269–71
   colonisation 19–20
   discovery and early description 21–31
   flora and fauna
      endemics with no obvious affinities 57, 69–74
      inter-island movement and colonisation 54, *55*, 59–61
      species shared with Madagascar only *55*, 61–2
      species with congeners elsewhere 56–7, 62–9
   nearest land 53
   volcanic origins 13
   see also Mauritius; Réunion *and* Rodrigues
Mascarenotus sp. 57, 71
   grucheti 37, 42, 50, 88, **185**
   murivorus 46, 50, **185**
   sauzieri 50, 88, **185**, *185*
Mascarinus mascarinus 42, 50, 57, 71, 107, **177**, *177*
Matelieff 77
Mathieu 101
Maudave, Comte de 94
Mauduyt 100, 107
Mauremootoo, John 202
Mauritian Wildlife Foundation (MWF) 218, 227, 229, 241, 244, 247, 257
Mauritius
   agriculture 93
   conservation 227, 229, 239–40
   forests
      clearance *123*, *132*
      distribution 93
      original forests 32, 33
      remnant native forest (1997) 159
   geography 14
   history
      (1598–1715) 76–83
      (1715–1810) 91–104
      (1810–1914) 116–37
      (1914–) 155–73
   islets off 204
   rainfall 33
   towns, roads and nature reserves 157
   volcanic origins 13, 19
Mauritius Institute 161, 173
Mayotte 19, 265
Megachiroptera **262–3**
Meinertzhagen, Richard 125, 133, 136, 137, 161, 214, 215
Meldrum, Charles 116
Melet 85, 87
Melia azedarach 146
Meller, Charles 129
Mer Rouge, Mauritius 171
Mertens, Robert 175
Merton, Don 221, 244
Merle 56, 63, 65, **200**

Mauritius 39, 40, 50, 99, 124, 125, 136, 144, 164, **200**, 233, 241, 252, 253
   Réunion 42, 50, 138, 173, 175, 186 190, 191, **200**, *200*
   Rodrigues 37, 50, 257
Messiter, Edward 149, 153
Michel, Claude 227
microbats **259**
Microchiroptera **259**
Middleton, Greg 253, 254
Mid-Indian Ocean (Carlsberg) Ridge 69
migratory birds 61
Miguet, Jean-Marc 178, 179, 180
Milbert, Jacques 99, 100, 101, 102, *102*, 109, 207, 208
Milius, Pierre Bernard 101
Milne-Edwards, Alphonse 142
Milon, Philippe 174, 175, 182, 190
Mimusops maxima 38
Minicoy *see* Maliku
moas 11
Möbius, Karl 134, 213
Moheli 269
Monarchidae **212**
monarchs **212**
Mongoose, Small Indian 136, 137, 235, 237
mongooses 164, 170–1, 265
Monitor, Common Indian 232
monkeys, 92, 147, 161
   forest damage 237
   introduction of 76, 78
   predation on native fauna 79, 237
   studies of 171
Moon, Malcy 126
Moorhen, Common 40, 50, 52, 55, 59, 101, 124–5, 135, 138, 140
Moreau, Reg 66
Morel 114
Morne du Bras Panon, Réunion *110*
mosquito control 95, 122, 137, 164–5
Moulton, Michael 271
Mourer, Cécile 67, 192
Mouse, House 236, 238
Moutou, François 155, 175, 179
Mundy, Peter 53, 77, 80, 82
Mungroo, Yousoof 171
Munia, Scaly-breasted *see* Finch, Spice
   White-rumped 106, 184, 271
Murphy, Cushman 220
Mus musculus 236, 238
museum exhibits 30
museums
   Ashmolean Museum 148
   British Museum 82, 133, 212, 213
   Lever's 206
   Mauritius 173
   Natural History Museum, Leiden 140
   Natural History Museum, London 134
   Natural History Museum, Paris 100, 107, 139, 191
   Newcastle (Hancock) 99
   Réunion 140, 142
   South African Museum 211
   Vienna 107
Mustela erminia 237
mutualism 130
MWF *see* Mauritian Wildlife Foundation
Myers, Norman 266

Myna, Common 94, 97, 106, 215, 216, 218, 230, 233, *258*, 273
mynas 94–5, 148, 149, 152

Nactus sp. 37, 39, 47, 56, 89
  coindemirensis 51, 68, 215, 217, 218, 232, 242, 248, 249
  serpensinsula 51, 68, 232, 242
    serpensinsula 217
    s. durrelli 248, 249
Napier–Broome, Frederick 129, 131, 135
Nastus borbonicus 35
national parks 161, 163–4, 171, 179, 192, 229, 258
Natte 39, 176, 261
nature reserves 161, 179–80, 187, 191, 198, 199, 201, 219, 244, 245, 266
Nazareth Bank 19, 27, 70, 71, 72
Neck, Admiral Cornelisz van 22
Necropsar sp. 57
  rodericanus 46, 51, 110, **216**
Necropsittacus rodricanus 46, 50, 57, 71, **181**
Neem 146
Nephila ardentipes 52
  inaurata 136
Nesocodon mauritianus 238
Nesoenas duboisi 42, 50, 67, 87, **150**
  mayeri 39–40, 50, 62, 67, 87, 99, 124, 125, 126, 137, **143**, 156, 164, 171, 227, 229, 233, 236, 237, 241, 244, 251, 261, 266, 267
  picturata 38, 39, 41, 45, 50, 52, 55, 59, 61, 67, 89, 101, 124, 126, 134, 138, 140, **143**, *143*, **150**, 156, 191, 233, 256
  rodericana 56, 67, 156
Nesopupa sp. 240
Neville, Henry 22
New Caledonia 67, 182
New Guinea 67, 70, 71, 269
New Zealand 15, 64,69, 70, 79, 135, 161, 220, 227, 266, 271
Newton, Alfred 7, 30, 125, 150, 213
Newton, Edward 101, *118*, 119, 122, 125–6, 132, 133, 134, 135, 136, 137, 140, 147, 150, 152, 208, 211, 212, 213, 214
Newton, Robert 161, 214, 217, 218, 220
Nichelsen 114
Nieuhoff, John 83
Night-gecko, Giant 46, 51
  Lesser 51, 68, 215, 217, 218, 232, 242, 248, 249
  Mauritius 51, 68, 232, 242
  Réunion 51
  Serpent Island 217, 248, 249
  Small Rodrigues 51
night-geckos 37, 39, 47, 56, 89
Night Heron, Black-crowned 59, 66
  Mauritius 37, 40, 49, **86**, 91
  Réunion 42, 49, **86**
  Rodrigues 45, 49, **86**, *86*
nightjars 65
Nivoy, de 138
Noble, Charles 203
Noddy, Brown 41, 44, 50, 55, 109, 141, 191, 201, 217, 219
  Lesser 41, 50, 55, 201, 217, 219
Nomadacris semifasciata 95

North-Coombes, Alfred 21, 29, 76, 114, 154, 163, 195
nose-closing by skinks 72–3
Nussbaum, Ron 74
Nycticorax sp. 56
  duboisi 42, 49, **86**
  mauritianus 37, 40, 49, **86**, 91
  megacephalus 45, 49, **86**, *86*
  nycticorax 59, 66
Nymphoea stellata 122

Oak, Silver 146
ocean currents 54, 59, 68, 70
O'Halloran, Joseph 153
oiseau bleu 43, 50, 66, 87, 106–7, 108, **128**
Oliver, Pasfield 146, 264
Oliver, William 145, 146, 147
Omphalotropis sp. 240
Ophidia **255**
Opuntia vulgaris 193, 209, 231
Orgeret, Henri d' 87
ORSTOM 179
Oryctes sp. 189
Orytolagus cuniculus 236
Ossaea marginata 233, 237
Otus insularis 71
  magicus 71
Oustalet, Emile 117
overgrazing 194
Owadally, Wahab 198, 227
Owl, Brown Fish 63
  Long-eared 124
  see also Lizard-owl
owls, fossil 39

Pachystyla bicolor 255
Padua Bank 69
Palaeocorynchus 121
Palea steindachneri 160, 170
Palm
  Bottle 210, 212, 221, 223, 225, 244, 246
  Cabbage 101, 109, 137, 146, 178, 189
  Hurricane 32, 34, 35, 80, 113, 124, 146, 192, 210, 221, 244, 246
  Traveller's 44, 94, 121, 126, 151, 192, 231
'palm savanna' 32
Pamelaescincus sp. 73
Pamplemousses, Mauritius 93, 109, 209, 215, 261
Pandanus sp. 32, 35, 113, 203, 209, 211, 213, 214, 218, 221, 223, 245, 256
  pyramidalis 247
Papasula abbottii 37, 39, 41, **47**, *47*, 48, 49, 56, 69, 80, 113, 148, 251
Parakeet 63
  Alexandrine 65, 257
  Echo 38, 39, 42, 50, 56, 64, 71, 99, 101, 107, 124, 126, 158, 164, 169–70, **172**, *172*, **177**, 227, 229, 233, 236, 237, 241, 249, 251, 252, 257, 267
  Mauritius Ring-Necked *see* Parakeet, Echo
  Ring-necked 64, 135, 164, 169, 182, 184, 233, 265, 273
  Rodrigues 46, 50, 56, 65, 71, 152, **181**, *181*, 257
Parat, Governor 85
Parrot, Broad-billed *see* Parrot, Raven

Dubois's 42, 50
Greater Vasa 233
Leguat's 46, 50, 57, 71, **181**
Mascarin 42, 50, 57, 71, 107, **177**, *177*
Raven 25, 25, 30, 37, 38, 39, 50, 57, 71, 80, **81**, **172**, *172*, 257
Rodrigues *see* Parrot, Leguat's
Thirioux's Grey 37, 42, 50, 56, 71, 77, 99, **172**, *172*, **177**
parrots 12
  black 71, 108, 109, 118, 138, 139, 142
  grey 38, 39
Partridge, Madagascar 93, 132, 141, 169, 175, 191, 271
Paspalum conjugatum 238
Passer domesticus 124, 134, 136, 140, 151, 154, 215, 225, 233, 258, 271, 273
passerine conservation 252
Paulian, Renaud 179
Payendee, Richard 199
Pelican, Pink-backed 41
Pelusios subniger 134, 160, 170
Pemba 64
Pepper, Brazilian 182, 186, 231, 237
Perdicula asiatica 133, 141, 169, 191
Peregrine Fund 227, 229
Periplaneta americana 223
Peron, François 101, 102, 207, 208
pest controls 106, 148, 149, 150, 160; *see also* biological controls
Petite Ile 44, 174, 187, 190, 191, 268
Petrel, Barau's **18**, 44, 49, 53, 56, 69, 102, 109, 186–7, 202
  Black-capped 18
  Black-winged 69
  Bourne's 49
  Bulwer's 49, 69, 219, 223
  Hawaiian 53
  Herald 69, 222
  Kermadec 49, 56, 69, 219, 222, 223
  Réunion Black 18, 44, 49, 56, 68, 147, 173, 187
  'Round Island' 18, 41, 49, 56, 68, 69, 213, 219, 220, 222, 223, 250, 251
  St Helena 68
  Tahiti 68
  Trindade *see* Petrel, 'Round Island'
Pezophaps solitarius 3, 29–30, 55, 67, 100, 111–13, 114, 133, 148, **167–8**, *167*
Phaethon lepturus 41, **42**, *42*, 44, 55, 60, 80, 141, 154, 164, 188, 191, 202, 212, 217, 218, 219, 223, 250, 251
  rubricauda 41, **42**, 55, 135, 152, 188, 202, 206, 208, 210, 211, *211*, 214, 215, 217, 218, 219, 220, 222–3, 250, 251
Phaethontidae **42**
Phalacrocoracidae **60**
Phalacrocorax africanus 37, 40, 42, 49, 55, 59, **60**, *60*, 255
  niger 51
Phasianus colchicus 184
Pheasant, Common 184
Phedina borbonica 40, 42, 44, 50, 55, 61, 125, 166, 168, *189*, 190, **193**, 252
Pheidole megacephala 249
Phelsuma sp. 39, 42, 46, 56, 67
  astriata 184
  borbonica 41, 51, 141, 175, 242

cepediana 51, 175, 184, 200, **242**
edwardnewtoni 46, 51, 86, 113, 115, 148, 152, 192, **242**, *242*, **243**, 256
gigas 44–5, 51, 148, 152, 242, **243**, 256, 257
guentheri 39, 51, 67–8, 206, 213, 220, 221, 224, 225, 229, 236, 242, *242*, 249, 253, 256
guimbeaui 51, 242
inexpectata 41, 51 138, 139, 141, 175, 184, **242**
laticauda 184
lineata 184, 265
madagascariensis 168, 169, 170 184, 232, 265
ornata 51, 141, 160, 213, 215, 218, 220, 224, 225, **242**, 256
rosagularis 51, 241, 242
Philippia montana 35
Phoenicopteridae **96**
Phoenicopterus roseus 40, 49, 55, **96**, *96*
Phyllophaga smithi 159
Pigeon, Cloven-feathered 67
   Feral 124
   Hollandais 40, 50, 56, 99, 100, 101, 117, 124, 126, 134, **143**, *143*, 150, 257
   Malagasy Blue 150
   Mauritius Pink 39–40, 50, 62, 67, 87, 99, 124, 125, 126, 137, **143**, 156, 164, 171, 227, 229, 233, 236, 237, 241, 244, 251, 261, 266, 267
   New Zealand 237
   Nicobar 70, 71, *71*
   Réunion Pink 42, 50, 67, 87, **150**
   Réunion Slaty 45, 50, **150**
   Rodrigues see Dove, Rodrigues
   Tooth-billed 70
Pigeon House Rock 170, 203, 206, 209, 217, 218, 219, 232, 244
Pigeon Wood, Mauritius 164, 165, 247–8, 251
pigeons 77, **143**, **150**, **156**
   blue 42, 67
   crowned 70, 71
pigs
   introduction of 78, 83
   predation on native fauna 79, 81, 83, 84, 88, 104
Pike, Nicholas 119, 126, 128, 134, *134*, 209, 212, 213, 214, 220
Pine, Chinese 156
   Slash 158
Pingré, Alexandre-Guy 46, 47, 86, 108, 111, 113, 114, 152, 154, 210
Pinus elliotti 158
   tabuliformis 156
piquant lulu 201
Pisonia sp. 152, 154
Pitot, H. 158
Ploceidae **228**
Ploceus cucullatus 135, 136, 147, 169, 175, 258, 261, 271, 273
Plover, Crab 171
Plum, Indian 218, 219, 246
plumage colour phases 60
poaching and trapping 174, 175, 187, 188, 189, 191, 250
Pochard, Madagascar 49, 55, 61, 112

sp. (subfossil) 88, **112**
Poivre, Pierre 93, 95, 96, 97, 98, 100, 103, 106, 114
Pollen, François 140, 141, 142, 189, 191
pollination by animals 39, 43, 249
Pope-Henessy, Sir John 131, 214
population, human 92, 108, 122, 125, 139, 149, 162, 163, 192, 194, 198, 202
Porphyrio sp. 56, 66
   porphyrio 43, 98, 124, 125, 132, 133, 137, 138, 160, 169
Port Louis, Mauritius 97, 98
Port Mathurin, Rodrigues 35
Portuguese discovery of Mascarene Islands 21
Portulaca oleracea 206
Post de Flacq, Mauritius 34
potato trees 186
Poupartia pubescens 246
predation on native fauna
   by cats 81, 87, 112, 113, 114, 115, 237
   by pigs 79, 81, 83, 84, 88, 104
   by rats 76–7, 79, 85, 87, 236
   native vertebrates 39
Prickly Pear 193, 209, 231
Prior, James 99, 206, 207, 208
Privet, Ceylon 132, 186, 231, *231*, 233, 240
Probst, Jean-Michel 186, 187, 188, 201, 217, 265
Procellariidae **18**
Procter, John 161, 164, 227
Pronis, Jacques 83
Pseudobulweria aterrima 18, 44, 49, 56, 68, 147, 173, 187
   rostrata 68
   rupinarum 68
Psiadia 215
Psidium cattleianum 94, 121, 131, 146, 147, 151, 153, 186, 191, 194, 230, 231, 231, 233, 237, 240
Psittacidae **172**, **177**, **181**
Psittacula sp. 63
   bensoni 37, 42, 50, 56, 71, 77, 99, **172**, *172*, **177**
   eques 38, 39, 42, 50, 56, 64, 71, 99, 101, 107, 124, 126, 158, 164, 169–70, **172**, *172*, 227, 229, 233, 236, 237, 241, 249, 251, 252, 257, 267
   eques 177
   eupatria 65, 257
   exsul 46, 50, 56, 65, 71, 152, **181**, *181*, 257
   krameri 64, 135, 164, 169, 182, 184, 233, 265, 273
Pterodroma 'arminjoniana' 18, 41, 49, 56, 68, 69, 213, 219, 220, 222, 223, 250, 251
   baraui **18**, 44, 49, 53, 56, 69, 102, 109, 186–7, 202
   hasitata 18
   heraldica 69, 222
   neglecta 49, 56, 69, 219, 222, 223
   nigripennis 69
   sp. 49
Pteropus hypomelanus 63
   niger 39, 41, 42, 44, 46, 49, 56, 64, 65, 83, 107, 124, 134, 153, 168, 241, **262–3**, 267
   rodricensis 35–6, 46, 49, 56, 64, 77, 152,

154, 196, 197, 198, 199–200, 214, 229, 255, 261, **262**, 266
   scapulatus 64
   subniger 39, 42, 44, 49, 56, 63, 64, 107, 108, 117, 124, 138, 139, 140, 142, 158, **263**, *263*
Ptilinopus sp. 67
Ptychadena mascarensis 95, 108, 138, 139
public ignorance of ecology and wildlife 260–1, 263
Puffinus bailloni **18**, 41, 44, 49, 55, 60, 102, 124, 187, 190, 211
   chlororhynchus 213
   lherminieri 102
   pacificus **18**, 40, 41, 44, 48, 49, 55, 152, 190, 191, 201, 210, 211, 212, 213, 214, 217, 218, 219, 222, 245, 250, 251
Pulsopipes herculeanus 213, 254
Pycnonotidae **200**
Pycnonotus jocosus 135, 136, 154, 156, 160, 165, 169, 182, 189, 191 194, 215, 218, 225, 233, 241, 253, 261, 265, 265, 267, 273

Quail, California 141
   Common 133, 137, 141, 169, 191
   Jungle Bush 133, 141, 169, 191
   Painted 102, 108, 117, 137, 138, 169, 191
   Rain 141
quarantine stations 149, 208, 209, 213, 215
Quelea, Red-billed 184
Quelea quelea 184
Querhoënt, Vicomte de 98, 100, 107
Quoy, Jean-Réné 207, 208

Rabbit 135, 152, 202, 211, 218, 236
   eradication 219, 220, 221, 223, 229, 238, 244
rabos forcados 22, 23, *24*, 40–1
Raboude, Mary Jane 199
Rail, Aldabra 128
   Assumption Island 128
   Buff-banded 67
   Dubois's 50, 87, 128
   Great Chatham Island Rail 70
   Hawkins see Rail, Great Chatham Island
   Leguat's 44, 45, 50, 57, 70, **127–8**
   Red see Hen, Red
   White-throated 38, 67, 87, 101, 128, 257
rainfall 32–3, 35, 203
Raleigh International 202
Rallidae **127–8**, **135**
Ramosmania rodriguesii 196, 246, 247
Ramsar sites 171
Raphus cucullatus 6, 10, *12*, 25, 28, 29, 30–1, *31*, 37–8, 39, 50, 57, 70–1, 77, 78, 79–80, 81–3, 100, 109, 130, 133, 140, **162**, *162*, 261
Rat, Black see Rat, Ship
   Brown see Rat, Norway
   Norway 92, 105–6, 140, 154, 164, 210, 236
   Ship 106, 164, 189, 202, 210, 215, 216, 217, 227, 236
rats 92
   arrival of 75, 76, 85, 88
   control measures 150
   eradication 218, 227, 236

predation on native fauna 76–7, 79, 85, 87, 236
Rattus norvegicus 92, 105–6, 140, 154, 164, 210, 236
rattus 106, 161, 164, 189, 202, 210, 215, 216, 217, 227, 236
Ravenala madagascariensis 44, 94, 121, 126, 151, 192, 231
La Ravine Saint-Gilles 275
Réaumur, Réné-Antoine 100, 107, 112
Rees, Coralie and Leslie 6, 264
Regnard, Gabriel 136, 159
Regnaud, Charles 128, 129, 131
Regnault, étienne 84
Reinhardt, Professor 70
reintroduction projects 267
　birds
　　land birds 251–2, 253, 257
　　seabirds 251
　historical ecology 271, 273
　invertebrates 238, 255
　reptiles 238, 248–9, 256
Rempart Mountains, Mauritius *134*
Renman, Eric 188
Renoyal de Lescouble, Jean-Baptiste 106, 109, 137, 138
reptiles
　and island area 272
research programmes 257–9
reservoirs 163, 169
Réunion
　agricultural crops 138
　forests *105*
　　clearance *145*
　　original forest 33–4, 36, *43*
　　remnant native forest 183
　geography 13–14
　history
　　(1598–1715) 21, 75, 83–5, 87–8
　　(1715–1810) 104–10
　　(1810–1914) 137–47
　　(1914–) 173–92
　lava flow zone *14*, 85
　rainfall 33
　towns, roads and nature reserves 176
　volcanic origins 13, 19
Rhamphotyphlops braminus 103, 133–4, 170, 200
Ribes, Sonia 192
Riparia riparia (Sand Martin) 193
Rivals, Pierre 174, 178, 179
Rivière du Poste, Mauritius *130*
road construction 98, 178, 190–1, 192, 265
Robin, Peking 182–3, 186, 189, 191, 265, 273
Robinson Crusoe 22
Roch, S. 213
Rochefeuille, Ambroise 142
Rochon, Abbé Alexis 93, 100
Rodrigues
　conservation 227, 229
　ecological ruin 110
　forests 32, 35
　　restoration 241, 244
　geography 14–15
　history
　　(1598–1715) 88–9
　　(1715–1810) 110–15

(1810–1914) 147–54
(1914–) 192–202
rainfall 32
and sea-level changes 89
settlements, road sand protected forest 194
volcanic origins 13
Rogotard, Marcel 173
Rose–apple 94, 109, 121, 131, 137, 146, 147, 149, 153, 154, 165, 186, 192, 193, 194, 197, 199, 231, 241
Ross, Ronald 125
Rothschild, Walter 30
Rougette 39, 42, 44, 49, 56, 63, 64, 107, 108, 117, 124, 138, 139, 140, 142, 158, **263**, *263*
Round Island 14, 32, 33, 37, 39, 41, 42, 74, 77, 149, 170, 171, 219, 203, 205, 206, 207, 208, 210–11, *211*, 213, 214, 215, 219–25, 227, 232, 238, 244–5, 248, 250
Rountree, Frank 161
Rousseau, Emmanuel 139
Roussel, L.E. 153
Roussin, Albert 140
Rowlands, Beau 217
Royal Society of Arts and Sciences (RSAS) 141, 156, 161
RSAS *see* Royal Society of Arts and Sciences
Rubus alceifolius 121, 146, 147, 233
　rosifolius 94, 109, 121
Rudolph II, Emperor 82
Ryan, Vincent 119

Sables, Ile aux 148, 201, 219, 236, 250, 251, 257
Safford, Roger 217
Saftleven, Cornelius 82
Saint-Denis, Réunion 84, 144
Sale, Gilbert 158
Salm, Rod 171, 227
Samoa 54
Sandalwood 218
Santalum album 218
Sauzier, Theodore 133
savanna habitat *43*
Savery, Roelant 82, *83*, 100
Saxicola axillaris 66
　tectes 42, 50, 56, 66, 189, **205**, *205*
Saya de Malha Bank 69–70, 71
Scaevola taccada 245
scale insects 153
Scelotes sp. 73
Scincidae **250**
Scolopendra abnormis 254
　morsitans 237
　subspinipes 237, 238, 254
Scops Owl 65
　Indo-Pacific 71
　Seychelles 71
　see also Lizard-owl
scorpions 220
Scotophilus borbonicus 44, 49, 55, 60, 109, 142, 259
Scott, Sir Peter 164, 171, 220, 226, 227
Screw-pine 32, 35, 113, 203, 209, 211, 213, 214, 218, 221, 223, 256
sea-level changes

islands created 203
islands exposed 19, 62, 63, 64
islands shrunk 89, 269
species lost 269
sedges 35
seed dispersal
　by birds 54, 130, 233
　by mammals 39, 238, 267
　by reptiles 249, 257
SEOR 188
Serinus mozambicus 94, 106, 138, 197, 271
Serpent Island 14, 37, 39, 41, 47, 170, 219, 203, 205, 207–8, 210–11, 217–18, 232, 244
Sesamia calamistis 136
Seychelles 13, 15, 19, 37, 40, 59, 60, 61, 63, 64, 65, 66, 67, 68, 70, 71, 74, 114, 149, 269, 271
Seychelles Bank 269
Sganzin, Victor 118, 133, 138
Shearwater, Audubon's 102
　Tropical *18*, 41, 44, 49, 55, 60, 102, 124, 187, 190, 211
　Wedge-tailed *18*, 40, 41, 44, 48, 49, 55, 152, 190, 191, 201, 210, 211, 212, 213, 214, 217, 218, 219, 222, 245, 250, 251
shearwaters *18*, 39
Sheldgoose 66
　Mauritius 49, **112**
　Réunion 42, 49, **112**
she-oak 146, 149, 153, 154, 174, 190, 192, 196, 209, 215, 246
shipwrecks 11, 76, 78, 84, 105, 147, 149, 205, 206
Shoals Rodrigues lagoon conservation programme 202
Showler, Dave 202
Shrew, House 95, 106, 115, 124, 134, 138, 141, 149, 175, 200, 230, 232, 235, 236, 237, 249
Shuker, Karl 264
Sideroxylon sp. 237
　boutonianum 247
　grandiflorum (formerly *Calvaria major*) 37, 130, 158, 237, 240, 261, 264
　sessiliflorum 130, 253
Sigala, Pierre 179
silverfish 253
Simberloff, Daniel 271
Singe, Ile aux 170
Sirenia **268**
Skink, Arnold's 51, 72, 88, 250
　Bojer's 37, 39, 51, 101, 117, 124, 134, 208, 210, 213, 215, 217, 220, 224, 232, 238, 249, 250, 267
　Bouton's 37, 42, 51, 55, 61, 134, 187, 191, 208, 213, 215, 218, 250, *250*
　Macabé 37, 51, 124, 168, 215, 232, 237, 250, 269
　Orange-tailed 52, 232, 248
　Réunion Slit-eared 42, 51, 88, 138, 139, 144, 232, 250
　Telfair's 37, 39, 42, 51, 72, 124, 206, 208, 210, 217, 218, 220, 221, 224, 225, 229, 236, 248, 249, **250**, 267
Skua, Southern 61
Slater, Henry 137, 151, 152, 154
slavery, abolition of 118–19, 139, 148

slaves, escaped (*marrons*) 91, 96, 253
Slider, Red-eared 169, 170, 184, 232
slugs 240
Snail, Rosy Wolf 169, 184, 201, 265
snails, land 38, 160, 220, 240, 254, 255
 giant 95, 124, 169, 184, 200, 255
Snake, Blind *see* Blind-snake
 Brown Tree 74, 230
  Flowerpot 103, 133–4, 170, 200
  Slender Worm 170
  Wolf 74, 133–4, 139, 141, 170, 171, 184, 230, 232, 249, 267
snakes 39, 73–4, 99, 205, 208, **255**
 absence of 76, 206, 207
Snell, H. J. 154
Snipe, Madagascar 182
soil erosion 195, 244
Solanum auriculatum 146
Solitaire, Réunion *see* Ibis, Réunion
 Rodrigues 3, 29–30, 50, 57, 100, 111–13, 114, 133, 148, **167–8**, *167*
solitaires
 ancestors 70–1
 extinction 6, 10
 population declines 85, 87
 subfossil discoveries 152, 192
Solomon Islands 269
Sonnerat, Pierre 95, 99, 100, 102
Sophora denudata 35
Soudan Bank 20, 70
Span, Gerrit van 75, 80
Sparrow, House 124, 134, 136, 140, 151, 154, 215, 225, 233, 258, 271, 273
 Java 94, 106, 118, 124, 136, 138, 140, 151, 152, 169, 271
species saturation 269
specimen collections **12**
Sphenodon sp. 79
spiders 39, 52, 101, 136, 253
 orb-web 136, 169, 182
Sporobolus capensis 245
Squirrel, Palm 141
SREPN 176, 178, 179, 192
Sri Lanka 64
St Brandon 54, 60, 68, 70, 71, 72
St Brandon Bank 19, 20
St Helena 12, 75
St John's Wort 35, 43
Starling, Hoopoe 41–2, 51, 71–2, 109, 117, 138, 139, 140, 141, 142, 144, **216**, *216*, 230
 Rodrigues 46, 51, 110, **216**
starlings, endemic pied 45
Staub, France 37, 195, 196, 197, 202, 227
Stel, Van de 79
Stenotaphrum dimidiatum 215
Sterna bengalensis 171
 bergii 50, 152, 154, 171
 dougallii 50, 55
 fuscata 41, 48, 50, 52, 55, 113, 115, 149, 154, 201, 210, 211, 217, 219, 250
 hirundo 61, 171
Stigmatopelia sp. 67
 senegalensis 67, 168, 233, 271
Stoat 237
Stockdale, Frank 154, 193
Stonechat, African 66
 Réunion 42, 50, 56, 66, 189, **205**, *205*

stones, gizzard 45, 71, 111, 167, 168
Stork, Black 66
Storm-petrel, Wilson's 61
Strahm, Wendy 198, 238, 239
Streptopelia chinensis 67, 98, 124, 218, 271
 risoria 183
Strickland, Algernon 118
Strickland, Hugh 6, 10, 22, 70, 106, 118, 133
Strigidae **185**
Sturnidae **216**
Sturnus sp. 72
Suez Canal, impact of 144
Sula dactylatra 41, 47, 49, 55, 60, 210, 211, 217, 219
 sula 47, 48, 49, 55, 60, 80, 148, 152, 251
Sulidae **47**
Suncus murinus 95, 106, 115, 124, 134, 138, 141, 149, 175, 200, 230, 232, 235, 236, 237, 249
Swallow, Barn 193
 Mascarene 40, 42, 44, 50, 55, 61, 125, 166, 168, *189*, 190, **193**, 252
Swamphen, Purple 43, 98, 124, 125, 132, 133, 137, 138, 160, 169
Swift, Common 193
 Pacific 63
Swiftlet, Glossy 64
 Indian 64
 Mascarene 40, 42, 44, 50, 56, 64, 166, **189**, *189*, 190, 237, 254
 Seychelles 64
swiftlets, cave-nesting 63
Sylviidae **208**
Syzigium cumini 121
 jambos 94, 109, 121, 131, 137, 146, 147, 149, 153, 154, 165, 186, 192, 193, 194, 197, 199, 231, 241

Tabebuia pallida 246
Tadarida 'pusilla' 61
 acetabulosus 40, 42, 44, 49, 56, 60, 65, 166, 190, 253–4, 259
Tafforet, Julien 45, 47, 48, 89, 110, 113, 154, 205
Tamarin Falls, Mauritius *118*, 126
tamarin des hauts 35, 53–4, 146, 174, 176
Tamarind 197, 237
Tambalacoque 37, 130, 158, 237, 240, 261, 264
Tambourissa tetragona 247
Taphozous mauritianus 40, 42, 44, 49, 55, 60, 101, 109, 168, 190, **259**, 259
tarantulas 217, 254
Tarenna borbonica 246
Tatton, J. 21
Teal, Bernier's 40, 66, 182, 257
 Hottentot 182
 Mascarene 40, 42, 49, 56, 66, 80, **112**, *112*
 Sunda 66, 112
Tecoma 246
Telfair, Charles 116, 117, 118, 148, 206
Tempany, Harold 158, 163, 193
Temple, Stanley 130, 220, 227, 269
Tenrec, Common 95, 118, 124, 135, 138, 160, 237
 Lesser Hedgehog 160
Tenrec ecaudatus 95, 118, 124, 135, 138,

160, 237
Tern, Common 61, 171
 Fairy 48, 50, 55, 113, 154, 201, 219, 251
 Great Crested 50, 152, 154, 171
 Lesser Crested 171
 Roseate 50, 55
 Sooty 41, 48, 50, 52, 55, 113, 115, 149, 154, 201, 210, 211, 217, 219, 250
Terpsiphone sp. 63
 bourbonnensis 40, 42, 50, 56, 66, 164, 165, 189, **212**, *212*, 233, 252, 253
 corvina 66
 mutata 66
Terrapin, East African Box 134, 160, 170
Testudinidae **234–5**
theories of conservation 266
Thirioux, Étienne 133, 192
Thompson, Richard 119, 121, 129, 130, 131, 156
Threskiornis solitarius 41, 45, 49, 56, 66, **103**, *103*, 192
 solitarius 103
 spinicollis 103
Threskiornithidae **103**
thrushes **205**
Thuillier 113
timber
 demand for 155
 for firewood 171, 173
 restrictions on use of 97, 98, 144
Toad, Black–spined 133
 Cane 159, 265
 Guttural 159, 160
toads 173, 200
Tombe, Charles 107, 207
Toon 131
Tortoise, Aldabra Giant 114, 170, 187, 202, 225, 244, 256
 Carosse 47, 51, 72, 114, **234**
 Galápagos 44
 Mauritius Domed 51, 72, **234**
 Mauritius High-backed 45, 51, 72, **234**
 Radiated 133, 142, 187–8, 225, 244
 Réunion 43–4, 45, 51, 72, **234**
 Rodrigues Domed 51, 72, 89, 114, **234**, *235*
tortoise parks 187
tortoises 12, 34, 35, 47–8, 39, 206, 211, 215, **234–5**
 associations 38
 browsing 37
 conservation 87, 88
 ecological function 256–7
 evolution 72
 extinction 104, 115, 141–2
 hunting of 104, 110–11, 205
 predation
  by cats 81, 87, 113, 114, 115
  by pigs 81, 83, 84, 87, 88
 re-introduction 225
 slaughter by early settlers 75, 77, 81, 84, 87, 88
 subfossil discoveries 191, 192
 trading in 110–11, 114
tourism 155, 173, 182, 187, 189, 201, 202, 233, 251, 252
Tournefourtia argentea 214
Trachemys scripta elegans 169, 170, 184,

232
Tradescant, John 82
trading routes
  strategic island landfalls 75
training, conservation 257–9
Transit of Venus expeditions
  (1761–3) 86, 111
  (1874) 149, 151–2
Travail pour tous 163–4
Travellers' Palm *see* Palm, Travellers'
tree-ferns 34, 35
tree-nettles 186
trees
  adaptation to tortoise browsing 37
Trembler, Martinique 30
trichomoniasis 233
Trochetia sp. 261
Trois Mamelles, Mauritius 79, *134*
Tromelin 27, 60
Tropicbird, Red-tailed 41, **42**, 55, 135, 152, 188, 202, 206, 208, 210, 211, *211*, 214, 215, 217, 218, 219, 220, 222–3, 250,251
  White-tailed 41, **42**, *42*, 44, 55, 60, 80, 141, 154, 164, 188, 191, 202, 212, 217, 218, 219, 223, 250, 251
Tropidophora eugeniae 255
Trouvillliez, Jacques 179
tuataras 79
Tunstall, Marmaduke 99
Turdidae **205**
Turnix nigricollis 42, 50, 55, 61, 105, 138, 188, 191
Turtle, Green 41, 44, 48, 51, 55, 115, 168, 187
  Hawksbill 41, 44, 48, 51, 55, 115, 168, 202
  Loggerhead 41, 51, 55
  Wattle-necked Softshell 160, 170
turtles
  hunting 88, 202
  trade in 187
Typhlops cariei 39, 51, 56, 68, 255
  porrectus 170

UFAW 220
Ulcoq, L. 193
unemployment 163
upland exploitation 139–40
Upupa epops 182, 216
urbanisation 155
Usnea sp. 35

Vacoas, Ilot 41, 170, 218, 219, 232, 238

Vandermeersch 212
Vandorous, William 152
Vanuatu 74
Varanus bengalensis 232
Varecia variegata 141
Vaughan, Reginald 119, 130, 160, 161, 178, 195, 214, 227, 238, 239, 246
Vaxelaire, Daniel 174
vegetation communities
  restoration 238–9
  see also forest types
Venne, Adriaen van der 82
Verheyen, René 70
Vetiveria sp. 275
  arguta 256
Vidua macroura 183
Villèle, Auguste de 147, 173, 192
Villers, Governor de 85, 87, 88
Vine, Helicopter 186
Vinson, Auguste 138, 139, 140, 141, 142, 144, 147
Vinson, Jean 168, 175, 196, 197, 202, 213, 217, 219, 220, 223, 227, 248
Vinson, Jean-Michel 227, 248
Vosmaer 99
Voyage et avantures 29–30, 89

wading birds, migratory 61, 156, 171
walckvogel 22–3, *24*, 82
Wallace, Alfred Russel 6
Walter, Albert 261
Warbler, Madagascar Swamp 66
  Rodrigues 46, 50, 56, 66, 150, 152, 154, 196, 197, 199, **208**, *208*, 241
  Seychelles 66
Warren, Ben 66
wars
  (1756–62) 94
  (1914–18) 157, 173
  (1939–45) 158, 161, 162, 173–5
Warwyck, Admiral Wybrant van 21, 22, 23, 77
water supplies 98
Waterhen, White-breasted 59, 63
watershed protection 102, 116, 119, 128, 129, 131, 144, 199
Wattle, Black 186
Waxbill, Common 106, 136, 138, 140, 150, 183, 191, 215, 218, 271
Weaver, Village 135, 136, 147, 169, 175, 258, 261, 271, 273
weavers **228**
weeding of nature reserves 199, 239, 240,

245–6, 258
West-Zanen, Willem van 77, 86
wetland drainage 125, 137
Whimbrel 40, 61, 113, 115, 117, 124, 206, 208, 214, 261
White Popinac 94, 121, 151, 153, 195, 209, 215, 246
White-eye 63, 65–6
  Chestnut-flanked 66
  Mascarene Grey 39, 40, 42, 43, 50, 56, 66, 124, 156, 165, 175, 189, 190, **222**, 222, 233, 241, 252, 253, 261, 267
  Mauritius Olive 39, 50, 56, 124, 125 164, 165, 166, 174, **222**, 229, 244, 252, 253
  Réunion Olive 35, 42, 43, 50, 56, 190, 222
white specimens
  bats 44, 60–1, 109
  see also albino specimens
Whiteley, Henry 134
Whydah, Pin-tailed 183
widespread species
  inter-island movement and Mascarene colonisation 54, 55, 59–61
Wiehe, Octave 119, 130, 160, 178, 193, 195, 214, 227, 238, 239
Wikstroemia indica 121, 233
Wildlife Act 1983 164, 197
winds, prevailing 54, 59
Wise, Thomas 119
Wood-rail 37, 56, 66–7, **128**
  Sauzier's 38, 42, 50, 77, 87
World Bank 164, 241
World Parrot Trust 229

Xenophidion sp. 73, 74

yatis 131

Zanthoxylum paniculatum 241, 247
ZNIEFF 179
Zosteropidae **222**
Zosterops sp. 63, 65–6
  borbonicus 39, 40, 42, 43, 50, 56, 66, 124, 156, 165, 175, 189, 190, **222**, 222, 233, 241, 252, 253, 261, 267
  chloronothos 39, 50, 56, 124, 125 164, 165, 166, 174, **222**, 229, 244, 252, 253
  maderaspatana 65
  olivaceus 35, 42, 43, 50, 56, 190, 222
  semiflava 66
Zoysia tenuifolia 256